The Genetics of Cattle, 2nd Edition

The Genetics of Cattle, 2nd Edition

Edited by

Dorian J. Garrick

Iowa State University, USA

and

Anatoly Ruvinsky

University of New England, Australia

www.cabi.org

CABI is a trading name of CAB International

CABI	CABI
Nosworthy Way	38 Chauncy Street
Wallingford	Suite 1002
Oxfordshire OX10 8DE	Boston, MA 02111
UK	USA
Tel: +44 (0)1491 832111	Tel: +1 800 552 3083 (toll free)
Fax: +44 (0)1491 833508	E-mail: cabi-nao@cabi.org
E-mail: info@cabi.org	
Website: www.cabi.org	

A catalogue record for this book is available from the British Library, London, UK.

Library of Congress Cataloging-in-Publication Data

The genetics of cattle/edited by Dorian J. Garrick, Iowa State University, Ames, IA 50011, USA, and Anatoly Ruvinsky, University of New England, Armidale, NSW 2351, Australia. -- 2nd edition.
  pages ; cm
  Preceded by: Genetics of cattle / edited by R. Fries and A. Ruvinsky. c1999.
  Includes bibliographical references and index.
  ISBN 978-1-78064-221-5 (hbk : alk. paper)
1. Cattle--Genetics. 2. Cattle--Breeding. I. Garrick, Dorian J. (Dorian John), 1960- editor. II. Ruvinsky, Anatoly, editor. III. C.A.B. International, issuing body.
[DNLM: 1. Cattle--genetics. SF 201]

  SF201.G45 2014
  636.2'0821--dc23
                                    2014011562

ISBN-13: 978 1 78064 221 5

Commissioning editor: Sarah Hulbert
Editorial assistant: Emma McCann
Production editors: Simon Hill and Tracy Head

Typeset by SPi, Pondicherry, India.
Printed and bound by CPI Group (UK) Ltd, Croydon, CR0 4YY.

Contents

Contributors

D.L. **Adelson**, The University of Adelaide, Adelaide, South Australia, Australia
Y. **Aida**, RIKEN Brain Science Institute, Saitama, Japan
C.L. **Baldwin**, University of Massachusetts, Amherst, Massachusetts, USA
R.G. **Banks**, University of New England, Armidale, New South Wales, Australia
D.C. **Beitz**, Iowa State University, Ames, Iowa, USA
D.P. **Berry**, Teagasc, Moorepark, Fermoy, Ireland
D.S. **Buchanan**, North Dakota State University, Fargo, North Dakota, USA
H.M. **Burrow**, CSIRO Animal, Food and Health Sciences, Armidale, New South Wales, Australia
J.J. **Crowley**, University of Alberta, Edmonton, Alberta, Canada
V. **Ducrocq**, UMR1313 Génétique Animale et Biologie Intégrative (GABI) INRA, Jouy-en-Josas, France
M. **Farré**, Royal Veterinary College, London, UK
M. **Felius**, Rotterdam, The Netherlands
R. **Fernando**, Iowa State University, Ames, Iowa, USA
D.J. **Garrick**, Iowa State University, Ames, Iowa, USA
A. **Hassanin**, Université Pierre et Marie Curie and Muséum national d'Histoire naturelle, Paris, France
Z.-L. **Hu**, Iowa State University, Ames, Iowa, USA
N. **Hunter**, University of Edinburgh, Easter Bush, Midlothian, UK
A.K. **Kaushik**, University of Guelph, Guelph, Ontario, Canada
E. **Kennedy**, Teagasc, Moorepark, Fermoy, Ireland
B.P. **Kinghorn**, University of New England, Armidale, New South Wales, Australia
B.W. **Kirkpatrick**, University of Wisconsin-Madison, Madison, Wisconsin, USA
D.M. **Larkin**, Royal Veterinary College, London, UK
J.A. **Lenstra**, Utrecht University, Utrecht, The Netherlands
S.L. **Lim**, The University of Adelaide, Adelaide, South Australia, Australia
F. **McCarthy**, University of Arizona, Tucson, Arizona, USA
R.G. **Mateescu**, University of Florida, Gainesville, Florida, USA
R.A. **Nafikov**, Iowa State University, Ames, Iowa, USA
F.W. **Nicholas**, University of Sydney, Sydney, New South Wales, Australia
K. **Nicholas**, Monash University, Clayton, Victoria, Australia
C.A. **Park**, Iowa State University, Ames, Iowa, USA
Z. **Qu**, The University of Adelaide, Adelaide, South Australia, Australia
J.M. **Raison**, The University of Adelaide, Adelaide, South Australia, Australia
J.M. **Reecy**, Iowa State University, Ames, Iowa, USA
C. **Robert**, Université Laval, Québec, Canada.

A. Ruvinsky, University of New England, Armidale, New South Wales, Australia
J.A. Sharp, Deakin University, Geelong, Victoria, Australia
H.A. Shojaei Saadi, Université Laval, Québec, Canada
G. Simm, Scottish Agricultural College, Edinburgh, UK
H. Soyeurt, University of Liège, Gembloux, Belgium
S.-n. Takeshima, RIKEN Brain Science Institute, Saitama, Japan
A. Teale, Retired, UK
M.L. Thonney, Cornell University, Ithaca, New York, USA
P. Wiener, University of Edinburgh, Easter Bush, Midlothian, UK
G. Wiggans, United States Department of Agriculture, Beltsville, Maryland, USA

Preface

Fifteen years have passed since publication of the first edition of *The Genetics of Cattle*. During this time a deep transformation has occurred in biological sciences. Just two decades ago the chromosomal location of only a few genes was known in cattle. By 2009 the bovine genome was sequenced and annotated, and all this information became easily accessible. The consequences of such an incredible scientific and technological explosion will follow; some of them are still unknown and others are discussed in this book. All this provides a strong case for the publication of the second edition of *The Genetics of Cattle*.

Since domestication more than 10,000 years ago, cattle have played an increasingly important role in development of human civilizations around the world. It is not easy to find a country that does not have a more or less significant population of cattle. The ability to effectively digest rough plant mass allows cattle to occupy a special ecological position in the global environment. Cattle have always provided essential human needs like food, clothing, draught, soil improvement and more, including meeting cultural and religious necessities. The current number of cattle worldwide exceeds 1300 million and continues to grow.

Traction power was probably the initial reason for bovine domestication, which marked a turning point in the development of agriculture. Over time, cow's milk steadily became a staple source of food in many geographical areas. This process is continuing, and milk, as well as numerous milk products, is spreading into countries that were not traditional dairy consumers. The total world production of cow's milk was 600 billion kilogrammes in 2010. Another valuable product is beef. The worldwide production of beef and veal exceeds 65 billion kilogrammes per year.

Progress in cattle breeding and selection over the past century was impressive. Breeding programmes developed to exploit principles of quantitative genetics, artificial insemination and embryo transfer. Scientifically designed breeding schemes along with increasing computerization of the industry were the main causes of the tremendous increase in milk production per cow. Previously separated, quantitative and molecular genetics have now become a unified approach in identification of loci underlying important cattle traits (quantitative trait loci). However, lengthy and convoluted pathways from genes to complex traits affected by numerous factors create significant impediments in both theoretical understanding and practical applications.

The purpose of this book is to present in one location a complete, comprehensive and fully updated description of cattle genetics. It is our intention to combine essential knowledge from various fields of genetics and biology of cattle in this reference book. The 24 chapters of the book can be partitioned into five connected sections. The first five chapters cover systematics, phylogeny, domestication, breeds and factorial genetics of cattle. The next two chapters provide crucial information about the structure of bovine chromosomes and the genome, as well as gene mapping in cattle. Chapters 8–10 cover the foundations of immune response and disease resistance. The following section, Chapters 11–14, discusses genetics of behaviour, reproduction and development. Chapters 15–23 are devoted to genetics applied to cattle improvement. Standard genetic nomenclature for cattle is presented in the final chapter.

This book is the result of truly international cooperation. Scientists from Australia, Belgium, Canada, France, Japan, Ireland, Netherlands, the UK and the USA made valuable contributions to this book. The editors are very grateful to all of them. The authors have made every attempt to highlight the latest and most important publications in the area of cattle genetics. However, we realize that omissions and errors are unavoidable and apologize for possible mistakes.

The book is addressed to a broad audience, which includes researchers, lecturers, students, farmers and specialists working in the industry. The 2nd edition of *The Genetics of Cattle* is the latest book in a series of monographs on mammalian genetics recently published by CAB International. Two other recent books, *The Genetics of the Pig* (2011) and *The Genetics of the Dog* (2012) are based on similar ideas and have comparable structure.

It is our hope that this book will be useful to many people throughout the world interested in cattle genetics. Perhaps it will support consolidation and further progress in this field of science and its implementation in order to advance practical agriculture.

Dorian J. Garrick
Anatoly Ruvinsky

1 Systematics and Phylogeny of Cattle

Alexandre Hassanin

*Université Pierre et Marie Curie and Muséum
national d'Histoire naturelle, Paris, France*

Introduction

Cattle belong to the subtribe Bovina, a taxonomic group represented by a single genus, *Bos*, which contains six wild species that were widely distributed in the Palearctic, Nearctic and Indomalayan regions during the Late Pleistocene and Holocene epochs (Plate 1): (i) *Bos primigenus* – aurochs (now only represented by domestic forms, humpless cattle and humped cattle or zebu); (ii) *Bos bison* – American bison (in the western half of North America) and European bison (reintroduced in Poland and a few adjacent countries); (iii) *Bos gaurus* – gaur (in India and mainland Southeast Asia); (iv) *Bos javanicus* – banteng (in Southeast Asia); (v) *Bos mutus* – yak (throughout the Tibetan Plateau); and (vi) *Bos sauveli* – kouprey (formerly in northern and eastern Cambodia and adjacent countries). The present classification differs from that of the IUCN (2012) by the fact that American and European bison are not considered as two separate species of the genus *Bison*, but are treated as two subspecies of *Bos bison* (see below for details).

The most fascinating aspect in the evolution of *Bos* is that four species have been domesticated since the Neolithic period: *B. primigenius*, *B. mutus*, *B. javanicus* and *B. gaurus*. Today, there are more than 1.5 billion cattle and zebu around the world (FAO, 2011), 14 million domestic yaks in the Tibetan Plateau and adjacent Asian highlands (Leslie and Schaller, 2009), 2.6 million Bali cattle, a domesticated banteng of Indonesia (Martojo, 2012) and more than 100,000 mithun (or gayal), a domesticated form of gaur found in the hill regions of Bangladesh, Bhutan, northeast India, Myanmar and China (Simoons and Simoons, 1968; Mondal et al., 2010). Domestic cattle provide an important part of the food supply for many of the world's people, either as livestock for meat or as dairy animals for milk. In many rural areas, they are still used as draught animals. Before the Neolithic period, populations of wild cattle were very successful and widely distributed across Europe, North Africa, Asia and North America (Plate 1). Today, domestic forms are present on all arable land on Earth, whereas wild species are restricted to small and isolated populations in a few countries. The aurochs, *Bos primigenius*, which was the ancestor of most breeds of domestic cattle, became extinct in 1627 (Van Vuure, 2005). Field scientists have not reported a living specimen of kouprey (*B. sauveli*) in Southeast Asia since the 1980s, suggesting that it is also extinct (IUCN, 2012).

Despite the obvious importance of cattle in the emergence and development of human civilizations, several aspects of their evolutionary history still remain poorly understood. In this chapter, I review the systematic position of the genus *Bos*, give a brief description of wild species and discuss ancient and recent phylogenetic hypotheses of interspecies relationships. I also propose a biogeographic scenario explaining their past and current geographic distributions.

Systematic Position of the Genus *Bos*

Cattle belong to the order Cetartiodactyla (Fig. 1.1), which is the second most diversified order of large mammals after Primates (IUCN, 2012). Members of this taxonomic group were originally divided into two different orders: Artiodactyla and Cetacea (e.g. Wilson and Reeder, 2005). Artiodactyls are even-toed ungulates including ruminants, pigs, hippos and camels. They were originally present on all continents, except Antarctica and Australasia, and most domestic livestock come from this group, including cattle, sheep, goats, pigs and camels. They

are characterized by two main limb features: a paraxonic foot, which means that the axis of the limb support passes between the third and fourth digits; and in the ankle, the astragalus is 'double-pulleyed', i.e. with a trochlea for the tibia and an opposing trochlea for the navicular, which enhances hind limb flexion and extension and allows very limited lateral rotation of the foot. Cetaceans include whales, dolphins and porpoises. All are marine animals except a few species of freshwater dolphins. The common ancestor of Cetacea acquired many adaptations for an aquatic life, such as a fusiform body, forelimbs modified into flippers, no hindlimbs or rudiments and a tail fin (fluke) used for propulsion (Muizon, 2009; Uhen, 2010).

Molecular studies have recovered a sister-group relationship between cetaceans and hippos, indicating that Artiodactyla is paraphyletic (Irwin and Arnason, 1994; Gatesy et al., 1996; Montgelard et al., 1997). In addition, they have shown that Ruminantia, Cetacea and Hippopotamidae form a clade named the Cetruminantia by Waddell et al. (1999) (e.g. Shimamura et al., 1997; Gatesy et al., 1999; Hassanin et al., 2012). To render the classification compatible with the molecular phylogeny, Montgelard et al. (1997) proposed to place all species of Artiodactyla and Cetacea into the same order, called Cetartiodactyla. After several years of controversy between molecular biologists and morphologists, a paraxonic foot and a double-pulley astragalus were found in Eocene whales (Gingerich et al., 2001), confirming that cetaceans evolved from terrestrial cetartiodactyls.

Cattle belong to the suborder Ruminantia (Fig. 1.1), which is the most diversified group of Cetartiodactyla, with 214 species related to goats, sheep, deer, pronghorn, giraffes and chevrotains. Ruminants are herbivores, which are primarily defined by rumination, i.e. the digestion is done through a process of regurgitation, rechewing and reswallowing of foregut digesta (Mackie, 2002). Since this process greatly facilitates the digestion of plant fibres, it is clear that rumination largely explains the evolutionary success of ruminants. All ruminants are able to digest cellulose through the enzymes produced by various microorganisms (bacteria and eukaryotes, such as ciliates and fungi) that are contained in the rumen, the most

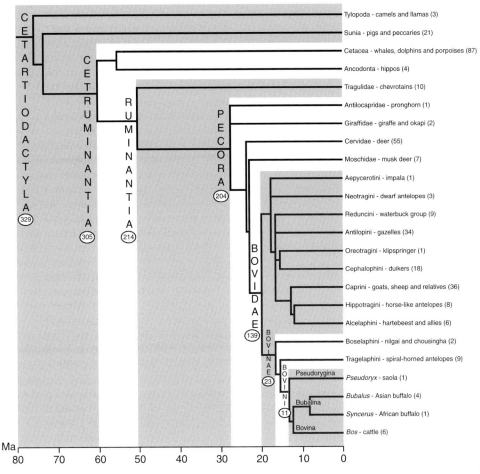

Fig. 1.1. Systematic position of the genus *Bos*. The tree is a chronogram, which means that branch lengths are proportional to divergence times. A time scale is provided at the bottom. Divergence times were deduced from the molecular studies of Meredith *et al.* (2011) and Hassanin *et al.* (2012). The number of species is indicated on the nodes for higher taxa, and between round brackets for terminal taxa (data extracted from IUCN, 2012).

developed compartment in their stomach. The ruminant stomach is composed of three other compartments: the reticulum, omasum and abomasum. The reticulum retains particles larger than 1 mm in the rumen (Zharova *et al.*, 2011). The main function of the omasum remains little known, but it filters particles depending on their size, serves as a suction pump controlling the flow of digesta (liquid and particles) between the reticulum and abomasum, facilitates the reabsorption of water, volatile fatty acids, ammonia, sodium, potassium and carbon dioxide, and participates in fibre digestion (Hackmann and Spain, 2010). The last

compartment, the abomasum, corresponds to the glandular stomach: secreting acid, its function is very similar to that of the stomach of monogastric mammals such as humans.

Ruminants have well-adapted teeth characterized by the presence of an incisiform lower canine, which is adjacent to the lower incisors, and a horny pad that replaces the upper incisors. All modern and fossil representatives of the group are diagnosed by an osteological autapomorphy, which corresponds to the fusion of the cuboid and navicular bones in the tarsus. Molecular studies have confirmed the monophyly of Ruminantia, as well as the major division between Tragulina and Pecora (Hassanin and Douzery, 2003; Matthee et al., 2007; Hassanin et al., 2012). The infra-order Tragulina is represented by only a few species of chevrotains, which are found in tropical forests of the Old World. The infra-order Pecora contains all other families, i.e. Bovidae (see below), Cervidae (deer), Moschidae (musk deer), Giraffidae (giraffes and okapi) and Antilocapridae (pronghorn). According to the most recent estimations (Hassanin et al., 2012), Pecoran families diverged rapidly from each other at the Oligocene/Miocene boundary, between 27.6 ± 3.8 Ma and 22.4 ± 2.4 Ma. During the Early Miocene, most habitats in Western Europe and Africa were forested; whereas those of Eastern Europe, Asia and North America were more open (Prothero and Foss, 2007). The spread of open habitat grasses at the Oligocene/Miocene was promoted by a global environmental change, when the warming Late Oligocene was interrupted by a brief but deep glacial maximum at the Oligocene/Miocene boundary (Stromberg, 2005). All these paleoecological data suggest that the emergence of modern pecoran families took place in the northern hemisphere, most probably in Eastern Eurasia. Then, the families dispersed and diversified rapidly: Antilocapridae and a group of extinct Moschidae (Blastomerycinae) entered into North America; Cervidae and other Moschidae stayed in Eurasia; Bovidae and Giraffidae appeared suddenly throughout the Old World (Prothero and Foss, 2007).

Cattle belong to the family Bovidae (Fig. 1.1), which is the most successful family of the suborder Ruminantia (139 species). Bovids are characterized by the structure of the horns. Present in all males and sometimes females, these consist of a permanent bone core covered by a non-branched and non-deciduous sheath of keratin. Wild representatives of the family are found on all continents except Australia, Antarctica and South America. Most older classifications recognize between five and eight subfamilies within the Bovidae (e.g. Simpson, 1945; McKenna and Bell, 1997; Wilson and Reeder, 2005), but molecular studies have concluded there exists a major division within the family Bovidae, separating the subfamily Bovinae from all other species of Bovidae (Hassanin and Douzery, 1999a, b; Matthee and Robinson, 1999; Matthee and Davis, 2001;

Ropiquet et al., 2009; Hassanin et al., 2012). This result is consistent with the taxonomic view of Kingdon (1982, 1997), who defined only two bovid subfamilies on the basis of morphology and behaviour: on the one hand, the subfamily Bovinae includes the three tribes Bovini (cattle, buffaloes and saola), Boselaphini (nilgai and four-horned antelope) and Tragelaphini (bongo, eland, bushbuck, kudu, nyala and sitatunga); on the other hand, the subfamily Antilopinae includes all other bovid tribes: Antilopini (gazelles), Aepycerotini (impala), Alcelaphini (hartebeest and allies), Caprini (goats, sheep and relatives), Cephalophini (duikers), Hippotragini (horse-like antelopes), Neotragini (dwarf antelopes), Oreotragini (klipspringer) and Reduncini (reedbucks, kob and rhebok) (Fig. 1.1). Molecular estimations of divergence times have suggested that the tribal diversification of both Bovinae and Antilopinae occurred simultaneously, between 16 and 14 Ma (Hassanin and Douzery, 1999b; Hassanin et al., 2012). This evolutionary event coincides with the Middle Miocene Climatic Optimum (MMCO), between 17 and 15 Ma, which was the last of a series of global warming events that have punctuated the Cenozoic Era (Zachos et al., 2008). This interval was the warmest time since 35 Ma in Earth's history. The warmer and more humid conditions prevailing during the Middle Miocene favoured the expansion of evergreen forests (Woodruff, 2003; Utescher et al., 2007; Patnaik and Chauhan, 2009; Senut et al., 2009). However, the MMCO was directly followed by a drastic global cooling, from 15 to 13 Ma, marking a transition from a greenhouse world to an 'icehouse-world' (Zachos et al., 2008). These rapid climatic changes created a mosaic of ecosystems, which may have promoted the emergence of most bovid tribes. Biogeographic inferences have suggested that the two bovid subfamilies first diversified in two different continents, Bovinae in Eurasia and Antilopinae in Africa (Hassanin and Ropiquet, 2004; Ropiquet and Hassanin, 2005).

Cattle belong to the tribe Bovini (Fig. 1.1). Originally, this group was defined on the basis of morphological similarities between cattle and buffaloes (Gentry, 1992; Geraads, 1992).

There are currently five wild buffalo species (IUCN, 2012): Syncerus caffer (African buffalo in sub-Saharan Africa), Bubalus arnee

(Asian water buffalo; only a few small populations in India and mainland Southeast Asia; but domesticated forms occur everywhere in Asia), *Bubalus depressicornis* (lowland anoa) and *Bubalus quarlesi* (mountain anoa) (both anoa species are only found on Sulawesi and Buton Islands), and *Bubalus mindorensis* (tamaraw; endemic to the Philippine island of Mindoro).

In 1992, a new bovid species, *Pseudoryx nghetinhensis*, was discovered in the Annamite Range, the mountainous jungle that separates Vietnam and Laos (Dung *et al.*, 1993). The species is listed as critically endangered by the IUCN (2012) because it is facing an extremely high risk of extinction in the wild. The saola, as called by local hunters, is characterized by an unusual morphology, including very long horn cores (ca. 40–50 cm) and a tricoloured pelage. Schaller and Rabinowitz (1995) suggested that the saola shares morphological affinities with members of the tribe Bovini: the frontal sinus of the skull extends well into the base of the horn cores; it has equally sized incisors; and its body shape, hooves and horns are similar to those of anoas. Most phylogenetic studies based on DNA sequences have confirmed that the saola belongs to the tribe Bovini (Hassanin and Douzery, 1999a; Gatesy and Arctander, 2000; Hassanin and Ropiquet, 2004; Hassanin *et al.*, 2012; Hassanin *et al.*, 2013). The association of the saola with Bovini is robust and reliable, as it is supported by both mitochondrial and nuclear genomes. Molecular estimations have revealed that the most recent common ancestor of crown Bovini experienced rapid diversification into three divergent lineages corresponding to the subtribes Bovina (cattle; species of the genus *Bos*), Bubalina (African and Asian buffaloes) and Pseudorygina (saola) during the Late Middle Miocene, between 13.5 and 13.0 ± 2.0 Ma (Hassanin *et al.*, 2012; Hassanin *et al.*, 2013). The identification of the sister-group of the saola was problematic, but a recent study based on 18 autosomal markers suggested that it is the clade uniting Bovina and Bubalina, a node supported by 12 molecular signatures detected in several independent markers and including three diagnostic deletions (Hassanin *et al.*, 2013). South Asia is likely to constitute the centre of origin of Bovini, as supported by the fossil record and biogeographical analyses (Hassanin and Ropiquet, 2004; Bibi, 2007).

Description of *Bos* Species

Wild species of *Bos* are heavily built mammals (400–1000 kg) with head carried low, possessing a large muzzle, short legs and a long tail with a terminal tuft of hair. There are no facial, pedal or inguinal glands. Two pairs of teats are present. Both sexes have typically smooth horns, which are located near the top of the skull; those of males being larger and more complex. The cross-section of horns is less angular than in other bovids. There is a strong sexual dimorphism characterized by differences in body weight (males are about 30% larger than females), horn shape and size (cow horns tend to be thinner and more upright), and pelage (adult males tend to be darker than adult females and young) (Lydekker, 1913; Huffman, 2013). Sexual dimorphism is generally explained by polygyny, a mating system in which dominant males can reproduce with multiple females, the reproductive success of males being directly correlated with strength and horn size (McPherson and Chenoweth, 2012).

Bos primigenius Bojanus, 1827

Today, the aurochs (*Bos primigenius*) is extinct in the wild, but in the Late Pleistocene, more than 11,700 years ago, it occurred in a wide geographic area covering Western Europe to East Asia and India, through North Africa, the Middle East and Central Asia (Plate 1). In the Middle Ages, the aurochs' range was already restricted to central Europe. The last aurochs died in 1627 at Jaktorów in Central Poland. The extinction of the aurochs was caused by man because of increasing hunting pressure and the development of agriculture, in particular, the competition with domestic livestock for food resources (Van Vuure, 2005).

The aurochs was first domesticated in the Tigris-Euphrates Valley from the nominate subspecies *Bos primigenius primigenius* at around 8500 BC. An independent episode of domestication occurred at around 6500 BC in the

Indus Valley (Pakistan) from the smallest Indian subspecies *Bos primigenius namadicus* (Chen *et al.*, 2010; Ajmone-Marsan *et al.*, 2010; Vigne, 2011). Today, cattle breeds are distributed worldwide. The two domestic forms were listed as *Bos taurus* (humpless cattle) and *Bos indicus* (humped cattle or zebu) in the 10th edition of *Systema Naturae* (Linnaeus, 1758), whereas the aurochs was described later as *Bos primigenius* Bojanus, 1827. According to the zoological code of nomenclature, the names based on wild forms should be therefore synonymized with those proposed for domestic derivatives. However, the International Commission on Zoological Nomenclature (2003) stated that two different species names could be used for wild and domestic forms. Alternatively, the commission considered that subspecies names could be also used for wild and domestic forms (Gentry *et al.*, 2004). In this chapter, I consider domestic forms as subspecies of *B. primigenius*, *B. p. taurus* and *B. p. indicus*. Indeed, using different species names for humpless and humped cattle is not appropriate because hybrids between these two domestic forms are fully fertile.

The morphology of aurochs is well described in Van Vuure (2005). Its general appearance was similar to modern cattle breeds, but with considerably longer and more slender legs, and with larger and more elongated horns. In addition, sexual dimorphism was expressed more strongly, including body size, coat colour and horn size and shape. The bulls were significantly bigger and more muscular than the cows. The shoulder height of European bulls varied from 160 to 180 cm, whereas that of European cows was around 150 cm. Pleistocene aurochs were apparently 10 cm higher than those of the Holocene, and the size varied by region, Indian populations being smaller than European populations. The coat colour was reddish-brown for calves and cows, and changed to a dark brown or black in bulls, with apparently a white dorsal stripe running down the spine. Both sexes had a pale muzzle. Another feature often attributed to the aurochs is blond forehead hairs. The horns of the aurochs were different from those of domestic breeds: they were longer and thicker, and typically forward-pointing and inward-curving.

The horns of bulls were larger, with the curvature more strongly expressed than in cows.

Bos bison Linnaeus, 1758

Bison were widespread in North America and probably in most parts of northern Eurasia during late prehistoric times (Plate 1). By the end of the 19th century, the species was close to extinction in both North America and Europe, mainly owing to overhunting. In 1903, the number of American bison fell to 1644. Today, there are approximately 500,000 bison, but most of them are captive commercial populations; there are 15,000 free-ranging individuals occurring as geographically isolated populations in prairies and woodlands of North America (Meagher, 1986; IUCN, 2012). In Europe, the last wild population survived in the Caucasus until 1927, and captive populations were subsequently reintroduced in several areas, including mixed deciduous forests in Białowieża (Poland and Belarus) and Western Caucasus. Today, there are around 1800 free-living bison in Europe and a few hundred captive animals (Pucek *et al.*, 2004; IUCN, 2012). Bison are gregarious, forming herds of females and their young. Males are either solitary or found in small groups.

The brown coat of bison is well characterized by long and woolly hair on the head, neck, hump and forequarters. The tail is short by comparison with other species of Bovina, and tufted only near the tip. Sexes are similar in appearance, although males develop larger body size, larger hump, and longer and more conical horns. Body weight is around 700 kg for males and 450 kg for females. Both sexes have short black horns that curve upward and inward (Meagher, 1986; Pucek *et al.*, 2004).

Some authorities place the bison in a different genus, *Bison*, and recognize two distinct species, *B. bison* (American bison) and *B. bonasus* (European bison) (e.g. McKenna and Bell, 1997; Wilson and Reeder, 2005; IUCN, 2012). In addition, two subspecies are often considered in North America, *Bison bison bison* (plains bison) and *B. b. athabascae* (wood bison). Two subspecies are also distinguished in Europe, *Bison bonasus bonasus* (lowland bison) and *B. b. caucasicus* (Caucasian bison). In agreement with some molecular and morphological data (see below), all bison are here included into a single species *Bos bison*. American and European bison are treated as two distinct subspecies, *Bos bison bison* and *B. b. bonasus*. American bison have a much darker coat colour, being blackish brown, passing into black on the long hair of the head and forequarters. In addition, they are stockier in appearance but smaller than European bison.

Bos gaurus C.H. Smith, 1827

The gaur is the largest bovid species in the world. The biggest males can reach 2.2 m high at the shoulder, and can weigh more than 1500 kg. They are found in forested areas of India, Indochina and the Malay Peninsula. Gaur formerly occurred in Sri Lanka. The number of wild gaur is estimated at between 13,000 and 30,000 animals, but the populations are heavily fragmented and most of them are in serious decline, in particular in Southeast Asian countries (IUCN, 2012).

The general coloration is dark brown, with white 'stockings' on their lower legs (from knees or hocks down to the hoofs) and pale muzzle. The hair is short, fine and glossy. On the forehead, there is a convex ridge connecting the horns, which is enhanced by grey or blonde hair. Both sexes have white or yellow horns that turn black at the tips. At the base, they are flattened and go outward; then, they curve inward and tend to point at each other. Sexual dimorphism is important, as males are larger and heavier than females, and exhibit a high muscular ridge on back, as well as a dewlap under the neck.

Two subspecies are currently recognized: *Bos gaurus gaurus* in India and Nepal and *Bos gaurus laosiensis* in Southeast Asia. Indian specimens are smaller, with relatively longer nasal bones, a wider horn span and a larger occiput (Groves and Grubb, 2011).

There are several forms of semi-wild gaur, known as mithun, gayal or dulong, in the hill regions of Bangladesh, Bhutan, northeast India, Myanmar and China (Mondal *et al.*, 2010). These domestic forms were described before wild representatives as *Bos frontalis* Lambert, 1804. In this chapter, I will treat them as a subspecies of *B. gaurus*, *B. g. frontalis*. The population of mithun was estimated to be between 100,000 and 150,000 individuals (Simoons and Simoons, 1968). Morphologically, mithun are similar to wild gaur, but they have a smaller body size (400–500 kg *versus* 600–1000 kg) and their horns show different and variable shape and size. They are reared under free-living conditions in dense forests at altitudes of between 1000 and 3000 m, and are primarily used for work and meat production. Variable hybrid forms between semi-wild gaur and domestic cattle also occur in these regions.

Bos javanicus d'Alton, 1823

The banteng occupies a variety of forest types in Java, Borneo, Thailand, Cambodia, Laos, Vietnam and Myanmar, where it generally occurs as small isolated populations (< 500 individuals) (IUCN, 2012). Recent field surveys in the Eastern Plains Landscape of Cambodia have however suggested that a large population, including more than 2000 individuals, may still survive within protected areas of this region (Gray *et al.*, 2012). The world population of banteng is estimated to be less than 8000 individuals (IUCN, 2012).

Banteng exhibit a characteristic white rump patch that contrasts sharply with the colour of the body, which is rufous-brown in females and young, and which generally turns dark brown or dark chocolate in adult males. The pelage of banteng shares similarities with that of gaur: the hair is short, but not glossy; there are white 'stockings' on the legs; and the mouth is surrounded by white hair. Banteng are however smaller and more lightly built than gaur. In addition, banteng males have a less developed dorsal ridge, which does not form a distinct hump, and their dewlap is smaller. The horns of both sexes are clearly distinct from those of other species of *Bos*. The horns of males emerge laterally, and then curve upward and inward, and are typically connected by a horny patch of thick skin on the forehead. The short horns of females emerge more upright and point inward at the tips.

Two subspecies of banteng can be distinguished, mainly on the basis of the coat colour of adult males. On Java and Borneo, adult bulls of *B. javanicus javanicus* are blackish brown or black. On the Asian mainland, the colour observed in adult bulls of *B. j. birmanicus* ranges from dark fawn, orange, chestnut, to chocolate, with different colours between face and body (either lighter or darker).

Banteng have been domesticated in Indonesia, probably at around 3500 years BC. Domestic banteng, which are known as Bali cattle, are kept on several islands, including Bali, East Java, Sumatra and Sulawesi. Bali cattle are similar in appearance to wild banteng, but they are smaller (adult females: 200–300 *versus* 500–650 kg; adult males: 350–400 *versus* 600–800 kg). They should be named differently, e.g. *Bos javanicus domesticus*, in order to avoid confusion between wild and domestic forms. There are around 2.6 million domestic banteng, which are used as working animals and for their meat (Martojo, 2012). Mating between Bali cattle and zebu occurs freely in Indonesia, resulting in high levels of genetic introgression into Indonesian cattle breeds (Mohamad *et al.*, 2009).

Bos mutus (Przewalski, 1883)

The wild yak occurs on the Tibetan Plateau at an altitude of 3000–6000 m, which covers three different countries, i.e. China, India and Nepal. Therefore, this species developed specific physiological adaptations for living under extreme conditions, including low temperatures, low oxygen availability, high solar radiation and aridity. Now, there are probably no more than 15,000 wild yaks, which are restricted to remote high-elevation areas of the Tibetan Plateau (Leslie and Schaller, 2009; IUCN, 2012).

The coat colour is black with rust-brown hues. The pelage is dense and woolly, with long draping hair on chest, flanks, rump and tail. The withers are elevated forming a conspicuous hump. Limbs are short and stout and have broad hooves and large dewclaws, as an adaptation to cold snow-covered environments.

The wide-spaced horns of yak are smooth and nearly circular in section: firstly, they are orientated transversely; then, they curve forward, and finally, they point upward and inward, frequently with a more or less marked backward inclination at the tips. Those of females are much more slender. There is an important sexual dimorphism in body size, as females may be only one-third of the weight of large males (350 *versus* 1000 kg; Olsen, 1990; Leslie and Schaller, 2009). Yaks live in herds of 10–300 individuals, most of which are females and their young.

Yaks have been domesticated across most of their range. The domestic forms were described before wild representatives as *Bos grunniens* Linnaeus, 1766. The 14 million yaks currently herded in the Tibetan Plateau and adjacent Asian highlands (North India, Pakistan, Kyrgyzstan, Mongolia and Russia) have originated from one or two domestication events during the Neolithic period (Guo *et al.*, 2006; Wang *et al.*, 2010). Domestic yaks are still used by nomadic pastoralists for transportation, and for providing milk, meat, wool, leather and even dried dung, which is used as fuel.

Bos sauveli Urbain, 1937

The kouprey was described as a new species, *Bos sauveli*, on the basis of a calf captured in the Preah Vihear province of Cambodia and kept alive at the Vincennes Zoo near Paris until 1940 (Urbain, 1937). In the middle of the 20th century, its range was already limited to open deciduous dipterocarp forests found in northern provinces of Cambodia, and slightly beyond the borders with Thailand, Laos and Vietnam (Sauvel, 1949). Populations have declined dramatically during the past six decades due to multiple possible factors including overhunting, deforestation and competition with domestic livestock. No living specimen has been observed during the past four decades, suggesting that the species is definitively extinct.

The kouprey was a graceful animal when compared to other wild cattle found in the Indochinese region, i.e. banteng and gaur. Adult females had a characteristic grey colour, which gave the animal its other local name (grey ox), and readily distinguished them from the reddish-brown females of banteng and the dark, blackish-brown females of gaur. The male lost its grey flanks with age, becoming increasingly black. Both sexes had notched nostrils, and a long tail. Adult males had a very large dewlap, which may nearly touch the ground. They had wide-spaced horns, which raised laterally, dropped below the base, and then curved upward and backward. Typically, the tips tended to split in old bulls. The horns of females were lyre-shaped.

The analyses of a taxidermy mount preserved in the collections of the Natural History Museum of Bourges (N° 1871-576) have suggested that the kouprey may have been domesticated in Cambodia (Hassanin et al., 2006). This stuffed specimen shares morphological similarities with the kouprey, but differs in several aspects, including its smaller size, its coat colour, as well as the shape of its horns. Such differences are generally observed between domestic and wild forms of Bos species. A preliminary molecular study has indicated that the enigmatic specimen of Bourges was a male possessing a mitochondrial genome of kouprey. It was therefore interpreted as being either a domestic kouprey, or alternatively, a hybrid between kouprey and domestic cattle (Hassanin et al., 2006). My recent analyses of nuclear sequences have revealed that the specimen of Bourges was an F1 hybrid resulting from a mating between a domestic bull of B. primigenius and a female kouprey (B. sauveli) (unpublished data). Indeed, two non-coding fragments of the Y chromosome (SRY-5' and SRY-3' fragments; Hassanin and Ropiquet, 2007) were found to be identical to those of B. primigenius, but different from those of all other species of Bos. In addition, the intronic sequences of two autosomal genes (FGB and TG; Hassanin and Ropiquet, 2007) were found to contain several heterozygous nucleotide sites, as expected for an F1 hybrid between B. primigenius and B. sauveli (for more details on the theory, see Hassanin and Ropiquet, 2007). These new molecular data show therefore that kouprey have been affected by occasional hybridization with domestic cattle, which may have accelerated its extinction in the wild.

Phylogenetic Relationships Among Species of Bovina

Classifications of the tribe Bovina based on morphology

In the popular classification of Simpson (1945), the species of wild cattle were included in three distinct genera: Bos (aurochs, domestic cattle and yak), Bibos (gaur and banteng) and Bison (European bison and American bison) (Fig. 1.2A). Bohlken (1958) largely followed this view, but treated the European bison as a subspecies of Bison bison. Simpson (1945) did not mention the kouprey in his classification, whereas Bohlken (1958) interpreted the kouprey as a hybrid between banteng and domestic cattle. Three years later, Bohlken (1961) recognized however that the kouprey belongs to a distinct species closely related to the banteng, but he reduced Bibos to a subgeneric rank under Bos.

In the morphological study of Groves (1981), the kouprey was grouped with the aurochs, and the yak with the bison (Fig. 1.2B). As a consequence, the genera Bibos and Bison were synonymized with Bos. The phylogenetic results of Geraads (1992) were very similar, but both yak and bison were found to be the sister-group of the aurochs, banteng, gaur and kouprey. Geraads (1992) retained apparently only two genera within the subtribe Bovina: Bison for bison and yak, and Bos for the four other species.

In more recent classifications, such as those of McKenna and Bell (1997) (Fig. 1.2C), Wilson and Reeder (2005) and IUCN (2012) (Fig. 1.2D), Bos and Bison were also distinguished, but the yak was ranged into the genus Bos. In addition, McKenna and Bell (1997) recognized four distinct subgenera: Bos was restricted to the aurochs and domestic forms, Bibos for banteng and gaur, Bison for American and European bison, and Poephagus for the yak.

Molecular phylogenies

The question of interspecific relationships among species of Bovina remains highly debated among molecular biologists, as recent different datasets have produced conflicting phylogenetic results.

I discuss below that there are only two cases of robust incongruence between mitochondrial and nuclear phylogenies of Bovina: one concerning the monophyly of banteng, and the other concerning the monophyly of bison. Moreover, I show that most conflicts between nuclear studies can be explained by a lack of robust phylogenetic signal.

The relationships among species of Bovina were first studied with DNA sequences of the mitochondrial genome, such as the subunit II of the cytochrome c oxidase gene, the cytochrome b gene, the small and large subunits ribosomal RNA genes (12S and 16S rRNAs), the control region (also named D-loop) or the complete mtDNA genome (Miyamoto et al., 1989; Janecek et al., 1996; Hassanin and Douzery, 1999a; Hassanin and Ropiquet, 2004; Verkaar et al., 2004; Hassanin et al., 2012). The phylogenetic relationships supported by

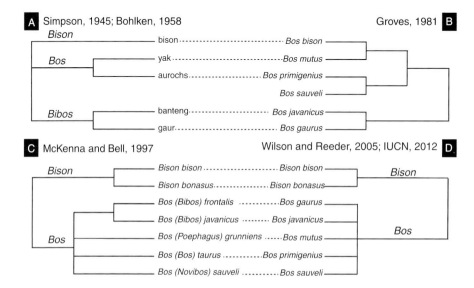

Fig. 1.2. Interspecies relationships within the subtribe Bovina. (A) In the classifications of Simpson (1945) and Bohlken (1958), the species of Bovina were ranged into three different genera: aurochs and yak into *Bos*, banteng and gaur into *Bibos* and bison into *Bison*. (B) By analysing 30 skull characters with the cladistic method, Groves (1981) confirmed that banteng and gaur are closely related, but suggested that the yak is allied to bison, whereas the aurochs is associated to kouprey. Groves (1981) considered that all species should be included in the genus *Bos*. (C) In their classification, McKenna and Bell (1997) recognized *Bison* as a separate genus, and split the genus *Bos* into four subgenera: *Bos* (aurochs, cattle and zebu), *Bibos* (banteng and gaur), *Novibos* (kouprey) and *Poephagus* (yak). (D) In the classifications of Wilson and Reeder (2005) and IUCN (2012), *Bison* was also treated as a distinct genus, whereas all other species were included in the genus *Bos*. (E) Bayesian tree reconstructed using the Y chromosomal sequences published in Nijman et al. (2008) (see Appendix F in Hassanin et al., 2013 for more details). Dashed branches indicate nodes that were not highly supported by the data. (F) Bayesian tree reconstructed using the nuclear data published in MacEachern et al. (2009) (see Appendix H in Hassanin et al., 2013 for more details). Dashed branches indicate nodes that were not highly supported by the data. The terminal branch with a danger sign indicates that the specimen was possibly concerned by genetic introgression from domestic cattle (see text for details). (G) Tree summarizing the phylogeny of Bovina as reconstructed from a DNA alignment of complete mitochondrial genomes (Hassanin et al., 2012). A terminal branch with a danger sign highlights a taxon concerned by an ancient mitochondrial introgression (see text for details). Note that the complete mitochondrial genome was not available for the kouprey (*Bos sauveli*). The phylogenetic position of *B. sauveli* was however inferred using three mitochondrial markers, corresponding to the cytochrome b gene (cytb), subunit II of the cytochrome c oxidase (CO2) and D-loop (DL) (Hassanin and Ropiquet, 2007). (H) Tree summarizing the analyses of 18 autosomal genes (Hassanin et al., 2013). Only the nodes considered to be reliable are shown in the figure.

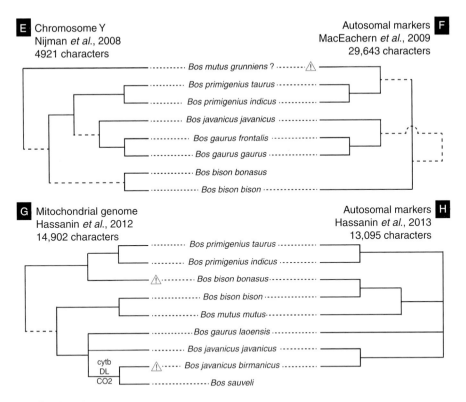

Fig. 1.2. Continued.

the mitochondrial data are summarized in Fig. 1.2G. Surprisingly, the genus *Bison* and the species *Bos javanicus* were not found to be monophyletic with mtDNA data. Since these results are in strong disagreement with the analyses of morphological characters and nuclear markers, it is now obvious that the phylogenetic signal of the mtDNA genome can be misleading for inferring relationships among species of Bovina (Hassanin and Ropiquet, 2004; Verkaar *et al.*, 2004; Hassanin and Ropiquet, 2007; Nijman *et al.*, 2008).

The monophyly of *Bison* has been confirmed by analysing three nuclear markers from the Y chromosome (Verkaar *et al.*, 2004; Nijman *et al.*, 2008; Fig. 1.2E) and 18 autosomal genes (Hassanin *et al.*, 2013; Fig. 1.2H). Although discordant with mtDNA data, these results are in good agreement with the morphology, and the fact that hybrids between American and European bison are fertile in both sexes (Van Gelder, 1977). Verkaar *et al.*

(2004) proposed two hypotheses for explaining the anomalous divergence of the mtDNAs from the two bison species: the first hypothesis is lineage sorting, which implies that two distinct mitochondrial lineages coexisted until the recent divergence of American and European bison; alternatively, the second hypothesis is that the European bison has emerged by species hybridization initiated by introgression of bison bulls in another ancestral species. A recent study has shown that American and European bison have very similar autosomal sequences (Hassanin *et al.*, 2013). Their nucleotide variation is similar to that found for intraspecific variation in *Bos gaurus*, *Bos javanicus* and *Bos primigenius*. There is no trace of hybrid origin in the nuclear markers of the European bison. Therefore, the data suggest that the mitochondrial genome of European bison was acquired by introgression after one or several past events of interspecific hybridization between a male of European bison and a female of an extinct species, which was related to *Bos primigenius*. The mtDNA introgression probably arose somewhere in Europe after the divergence between American and European bison, i.e. during the Middle Pleistocene according to the fossil record (Scott, 2010). From the taxonomic point of view, these molecular results suggest, first, synonymy of the genus *Bison* with *Bos*, and second, treatment of the American and European bison as subspecies of *Bos bison*.

The sister-group relationship between Cambodian and Javan banteng, i.e. the monophyly of *Bos javanicus*, has been confirmed by analysing DNA sequences from the SRY gene and different autosomal genes (Hassanin and Ropiquet, 2007; Hassanin *et al.*, 2013). Although slightly different, the Cambodian and Javan banteng share many morphological characteristics that are not observed in other species of Bovina, including a large white patch on the hindquarters, the typical reddish-brown colour of females and young, the shape and size of the horns in females and the horny shield that connects the bases of the horns on the forehead of adult males. As in the case of bison, mtDNA introgression has been invoked to explain mtDNA sequence similarity between Cambodian banteng (subspecies *Bos javanicus birmanicus*) and kouprey (*Bos sauveli*) (Hassanin and Ropiquet, 2007). According to this hypothesis, the mitochondrial genome of kouprey was transferred into the ancestor of Cambodian banteng by natural hybridization. Molecular dating estimates have suggested that the hybridization occurred during the Pleistocene epoch, at 1.34 ± 0.45 Ma (Hassanin and Ropiquet, 2007). The mitochondrial introgression hypothesis assumes that at least one kouprey female, which was probably young in order to overcome interspecific ethological barriers, was adopted into a herd of banteng. The event may have happened in open, dry, deciduous forests of Northern Cambodia, where several field biologists have reported the existence of temporary mixed herds between banteng and kouprey (Edmond-Blanc, 1947; Wharton, 1957; Pfeffer, 1969).

These two detected cases of ancient mtDNA introgression suggest that interspecific hybridization may have been a relatively common process during the evolutionary history of Bovina. Several arguments can be advanced to support that idea. First of all, several hybrids have been described between wild and domestic species of Bovina, such as the yakow, which is a hybrid between yak and domestic cow, the selembu, which is a hybrid between gaur and zebu, and the beefalo, which is a hybrid between domestic bull and bison (Mamat-Hamidi and Hilmi, 2009). Such hybrids can be easily obtained because most species of Bovina share very similar karyotypes with the same diploid number of chromosomes (2n = 60): *Bos primigenius*, *Bos javanicus*, *Bos mutus*, *Bison bison* and *Bison bonasus*. Only *Bos gaurus* has slightly different karyotypes (2n = 58 or 56), which can be, however, easily explained by one or two Robertsonian fusions, (1;29) and (2;28) (Ropiquet *et al.*, 2008). Similar Robertsonian fusions (1;29) and (2;28) have also been described in the subspecies *Bos javanicus birmanicus*, as well as in several individuals of *Bos primigenius taurus* and *Bos primigenius indicus* (Ropiquet *et al.*, 2008). All these cytogenetic data explain why viable hybrids have been produced in captivity among various species of Bovina (Van Gelder, 1977). In general, the males of the first hybrid generation are sterile, whereas the females are not (e.g. Steklenev and Elistratova, 1992; Steklenev, 1995; Qi *et al.*, 2010). The sterility of hybrids of the heterogametic sex (males XY) is commonly encountered in mammals (Haldane's rule; Coyne and Orr, 2004), and this characteristic explains why introgression occurred more often on the maternally inherited mtDNA rather than on the paternally inherited markers of the Y chromosome (Ballard and Whitlock, 2004).

By analysing 18 autosomal markers, Hassanin *et al.* (2013) have concluded that bison and yak are sister-groups. This hypothesis is robust, as the node was recovered in the separate analyses of seven independent nuclear markers. In addition, it confirms the morphological studies of Groves (1981) and Geraads (1992), as well as the paleogeographic scenario of Tibetan mammals proposed by Deng *et al.* (2011). The analyses of amplified fragment length polymorphism (AFLP) fingerprinting published in Buntjer *et al.* (2002) have also provided a signal in favour of the clade uniting bison and yak. By contrast, other molecular studies have concluded different relationships, including a basal position of *Bison* (microsatellites; Ritz *et al.*, 2000), a basal position of *Bos mutus*

(Y chromosomal genes; Nijman et al., 2008) or an association of Bos mutus with Bos primigenius (autosomal genes; MacEachern et al., 2009). However, the reanalyses of Hassanin et al. (2013) have shown that the nuclear datasets used in Nijman et al. (2008) and MacEachern et al. (2009) do not contain a strong signal for the position of both bison and yak (BP$_{ML}$ < 80). For instance, the early divergence of Bos mutus is only supported by the fact that all other species of Bovina share a G nucleotide in position 3186 of the Y chromosomal alignment. Another problem is that MacEachern et al. (2009) provided no information on the origin of the yak sequenced in their study, while this point seems crucial for interpreting their results. Indeed, nomad pastoralists of the Tibetan Plateau have traditionally used hybridization of domestic yak with domestic cattle for over 3000 years, because yak–cattle F1 hybrids are preferred to both parental types for meat and milk. Although F1 hybrid males are sterile, F1 hybrid females remain fertile, which has promoted cattle introgression into populations of domestic yak, as demonstrated in Qi et al. (2010). Therefore, the association of yak with domestic cattle in the tree of MacEachern et al. (2009) can be easily explained if they sequenced a domestic yak, which has been introgressed from cattle several generations earlier. I consider therefore that the hypothesis of a sister-group relationship between yak and bison cannot be rejected by the nuclear studies of Nijman et al. (2008) and MacEachern et al. (2009).

To conclude, only a few nodes can be considered as being robust and reliable within the subtribe Bovina. They include the association of American and European bison, their sister-group relationship with the yak and the monophyly of the species Bos javanicus, Bos gaurus and Bos primigenius. All other relationships are unstable and need further testing with additional molecular data. However, it has been suggested that the species Bos javanicus and Bos gaurus share close phylogenetic affinities using three independent datasets: mtDNA genome (Hassanin et al., 2012); Y chromosomal genes (Nijman et al., 2008); and autosomal genes (MacEachern et al., 2009). Interestingly, these two species share with Bos sauveli a few morphological characteristics that are not observed in other species of Bovina, including a dewlap in adult males and the occurrence of white socks in both sexes (Pfeffer and Kim-San, 1967).

Fossil Record and Biogeography

Rapid radiation and dispersal of Bovina during the Middle Pliocene

Several extinct genera, which are possibly related to Bovina, first appeared during the Pliocene: Simatherium and Pelorovis in Africa, Leptobos in Africa and Europe. During the Early Pleistocene, Pelorovis spread in the Middle East, Leptobos was present in northeastern India and China, and Epileptobos was found on Java (Duvernois, 1992; Vislobokova, 2005; Martínez-Navarro et al., 2007; Dong, 2008). The phylogenetic relationships between extinct and extant species of Bovina remain problematic (Geraads, 1992; Bibi, 2009), but the morphometric analyses of Martínez-Navarro et al. (2007) have evidenced two major groups: Pelorovis + Bos primigenius, and Leptobos + Bison. Unfortunately, most extant species of Bovina were not included in the analyses (i.e. B. gaurus, B. javanicus, B. mutus and B. sauveli). However, the authors proposed to synonymize the genus Pelorovis with Bos, and suggested that Bos originated in the Late Pliocene of eastern Africa, and dispersed into the Middle East at around 1.4 Ma, and finally into Europe at around 0.6–0.5 Ma. My interpretation is that this biogeographic scenario could hold only for the lineage leading to Bos primigenius. Molecular and cytogenetic studies have shown that the genus Bison should be treated as a synonym of Bos. If the extinct genera Epileptobos, Leptobos and Pelorovis are confirmed to be related to living species of Bos (and Bison), then this would imply their synonymy with Bos as well. This would result in the genus Bos being already represented by different morphological and biogeographic lineages by the Pliocene of Africa, Europe and northeastern South Asia. Molecular dating analyses support this early age, suggesting that the last common ancestor of extant Bovina underwent a rapid radiation during the Early to

Middle Pliocene (3.3–4.8 Ma in Hassanin and Douzery, 1999b; 4.7 ± 0.8 Ma in Hassanin and Ropiquet, 2004; 3.3 + 0.9 Ma in Hassanin *et al.*, 2012; 3.7 ± 1.1 Ma in Hassanin *et al.*, 2013). At that time, most of the modern arid and semi-arid climate zones in Africa and Arabia were covered with temperate and tropical xerophytic shrublands and grasslands (Salzmann *et al.*, 2011). Simulations have also indicated that there was a northward expansion of temperate forests and grasslands in Eurasia, with vast tracts of grassland in Siberia (Haywood *et al.* 2009). I suggest that these palaeoenvironmental conditions triggered the radiation and dispersal of Bovina in a huge area covering North and eastern Africa, southern Europe, western Asia, central Asia and Siberia.

Evolution of the yak–bison lineage in the northern hemisphere

As discussed previously, nuclear data strongly support the association of American and European bison, and their sister-group relationship with the yak. Interestingly, yak and bison share at least two osteological synapomorphies, which may be useful for identifying their ancestry in the fossil record: unlike other species of Bovina, the premaxillae doesn't touch the nasal (Olsen, 1990); and they have 14 thoracic vertebrae and 5 lumbars, whereas other Bovina have 13 thoracic vertebrae and 6 lumbars (Groves, 1981). Unfortunately, these characteristics are only rarely preserved in the fossil remains. To date, fossils related to the yak have been found dating to the Late Pleistocene in Siberia, Tibet and Nepal (Leslie and Schaller, 2009),

whereas fossils ascribed to *Bison* are much older, with *Bison sivalensis* from the Late Pliocene of the Upper Siwaliks (dated between 3.3 and 2.6 Ma; Khan *et al.*, 2010). Bison appeared in Eastern Europe at 1.77 Ma, and were present in the Middle East at around 1.6–1.2 Ma, and thereafter in Western Europe at around 1.5 Ma (Martínez-Navarro *et al.*, 2011). Bison entered North America at around 240–220 ka, and then they rapidly spread out across the continent, where they diversified into different (sub)species (Scott, 2010). The paleontological data suggest therefore that the common ancestor of yak and bison emerged somewhere in Asia, and possibly during the Pliocene. Based on the discovery of a Himalayan woolly rhino in the Pliocene (dated to 3.7 Ma), Deng *et al.* (2010) have proposed that some Ice Age megaherbivores, such as the woolly rhino and yak, first evolved in Tibet before the beginning of the Ice Age. Such a scenario is compatible with molecular estimates, since the yak/bison lineage separated from other Bovina between 3.7 ± 1.1 Ma and 2 ± 0.5 Ma (Hassanin *et al.*, 2013). Subsequently, yak remained in East Asia, while bison spread in the Siwaliks, Middle East and Europe, and crossed into North America through Beringia in the Middle Pleistocene.

Acknowledgements

I am grateful to Daniel Foidl for the illustration of aurochs and Nicolas Puillandre for the picture of American bison. I also acknowledge Anatoly Ruvinsky and Dorian J. Garrick for their useful comments.

References

Ajmone-Marsan, P., Garcia, J.F., Lenstra J.A. and the GLOBALDIV Consortium (2010) On the origin of cattle: how aurochs became cattle and colonized the world. *Evolutionary Anthropology* 19, 148–157.

Ballard, J.W. and Whitlock, M.C. (2004) The incomplete natural history of mitochondria. *Molecular Ecology* 13, 729–744.

Bibi, F. (2007) Origin, paleoecology, and paleobiogeography of early Bovini. *Palaeogeography Palaeoclimatology Palaeoecology* 248, 60–72.

Bibi, F. (2009) Evolution, systematics, and paleoecology of Bovinae (Mammalia: Artiodactyla) from the Late Miocene to the Recent. PhD Thesis, Yale University, New Haven, Connecticut.

Bohlken, H. (1958) Vergleichende untersuchungen an wildrindern (Tribus Bovini Simpson, 1945). *Zoologische Jahrbücher* 68, 113–202.

Bohlken, H. (1961) Der Kouprey, *Bos* (*Bibos*) *sauveli* Urbain 1937. *Zeitschrift für Saügetierkunde* 26, 193–254.

Buntjer, J.B., Otsen, M., Nijman, I.J., Kuiper, M.T. and Lenstra, J.A. (2002) Phylogeny of bovine species based on AFLP fingerprinting. *Heredity* 88, 46–51.

Chen, S., Lin, B.Z., Baig, M., Mitra, B., Lopes, R.J., Santos, A.M., Magee, D.A., Azevedo, M., Tarroso, P., Sasazaki, S., Ostrowski, S., Mahgoub, O., Chaudhuri, T.K., Zhang, Y.P., Costa, V., Royo, L.J., Goyache, F., Luikart, G., Boivin, N., Fuller, D.Q., Mannen, H., Bradley, D.G. and Beja-Pereira, A. (2010) Zebu cattle are an exclusive legacy of the South Asia neolithic. *Molecular Biology and Evolution* 27, 1–6.

Coolidge, H.J. (1940) The Indo-Chinese forest ox or kouprey. *Memoir of the Museum of Comparative Zoology, Harvard* 54, 421–531.

Coyne, J.A. and Orr, H.A. (2004). *Speciation*. Sinauer Associates, Sunderland, UK.

Deng, T., Wang, X., Fortelius, M., Li, Q., Wang, Y., Tseng, Z.J., Takeuchi, G.T., Saylor, J.E., Säilä, L.K. and Xie, G. (2011) Out of Tibet: Pliocene woolly rhino suggests high-plateau origin of Ice Age megaherbivores. *Science* 333, 1285–1288.

Dong, W. (2008) Nouveau matériel de *Leptobos* (*Smertiobos*) *crassus* (Artiodactyla, Mammalia) du Pléistocène inférieur à Renzidong (Chine de l'Est). *Geobios* 41, 355–364.

Dung, V.V., Giao, P.M., Chinh, N.N., Tuoc, D. and MacKinnon, J. (1993) A new species of living bovid from Vietnam. *Nature* 363, 443–445.

Duvernois, M.-P. (1992) Mise au point sur le genre *Leptobos* (Mammalia, Artiodactyla, Bovidae); Implications biostratigraphiques et phylogénétiques. *Geobios* 25, 155–166.

Edmond-Blanc, F. (1947) A contribution to the knowledge of the Cambodian wild ox or kouproh. *Journal of Mammalogy* 28, 245–248.

FAO (2011) Food and Agriculture Organization of the United Nations. FAOSTAT. http://faostat3.fao.org/home/index.html.

Gatesy, J. and Arctander, P. (2000) Hidden morphological support for the phylogenetic placement of *Pseudoryx nghetinhensis* with bovine bovids: a combined analysis of gross anatomical evidence and DNA sequences from five genes. *Systematic Biology* 49, 515–538.

Gatesy, J., Hayashi, C., Cronin, M.A. and Arctander, P. (1996) Evidence from milk casein genes that cetaceans are close relatives of hippopotamid artiodactyls. *Molecular Biology and Evolution* 13, 954–963.

Gatesy, J., O'Grady, P. and Baker, R.H. (1999) Corroboration among data sets in simultaneous analysis: hidden support for phylogenetic relationships among higher level artiodactyl taxa. *Cladistics* 15, 271–313.

Gentry, A., Clutton-Brock, J. and Groves, C.P. (2004) The naming of wild animal species and their domestic derivatives. *Journal of Archaeological Science* 31, 645–651.

Gentry, A.W. (1992) The subfamilies and tribes of the family Bovidae. *Mammal Review* 22, 1–32.

Geraads, D. (1992) Phylogenetic analysis of the tribe Bovini (Mammalia, Artiodactyla). *Zoological Journal of the Linnean Society* 104, 193–207.

Gingerich, P.D., ul Haq, M., Zalmout, I.S., Khan, I.H. and Malkani, M.S. (2001) Origin of whales from early artiodactyls: hands and feet of Eocene Protocetidae from Pakistan. *Science* 293, 2239–2242.

Gray, T.N.E., Prum, S., Pin, C. and Phan, C. (2012) Distance sampling reveals Cambodia's Eastern Plains Landscape supports the largest global population of the endangered banteng *Bos javanicus. Oryx* 46, 563–566.

Groves, C. and Grubb, P. (2011) *Ungulate Taxonomy*. Johns Hopkins University Press, Baltimore, Maryland.

Groves, C.P. (1981) Systematic relationships in the Bovini (Artiodactyla, Bovidae). *Zeitschrift für Zoologische Systematik und Evolutionsforschung* 19, 264–278.

Guo, S., Savolainen, P., Su, J., Zhang, Q., Qi, D., Zhou, J., Zhong, Y., Zhao, X. and Liu, J. (2006) Origin of mitochondrial DNA diversity of domestic yaks. *BMC Evolutionary Biology* 6, 73.

Hackmann, T.J. and Spain, J.N. (2010) Invited review: ruminant ecology and evolution: perspectives useful to ruminant livestock research and production. *Journal of Dairy Science* 93, 1320–1334.

Hassanin, A. and Douzery, E.J. (1999a) Evolutionary affinities of the enigmatic saola (*Pseudoryx nghetinhensis*) in the context of the molecular phylogeny of Bovidae. *Proceedings of the Royal Society B: Biological Sciences* 266, 893–900.

Hassanin, A. and Douzery, E.J. (1999b) The tribal radiation of the family Bovidae (Artiodactyla) and the evolution of the mitochondrial cytochrome b gene. *Molecular Phylogenetics and Evolution* 13, 227–243.

Hassanin, A. and Douzery, E.J. (2003) Molecular and morphological phylogenies of Ruminantia and the alternative position of the Moschidae. *Systematic Biology* 52, 206–228.

Hassanin, A. and Ropiquet, A. (2004) Molecular phylogeny of the tribe Bovini (Bovidae, Bovinae) and the taxonomic status of the Kouprey, *Bos sauveli* Urbain 1937. *Molecular Phylogenetics and Evolution* 33, 896–907.

Hassanin, A. and Ropiquet, A. (2007) Resolving a zoological mystery: the kouprey is a real species. *Proceedings of the Royal Society B: Biological Sciences* 274, 2849–2855.

Hassanin, A., Ropiquet, A., Cornette, R., Tranier, M., Pfeffer, P., Candegabe, P. and Lemaire, M. (2006) Has the kouprey (*Bos sauveli* Urbain, 1937) been domesticated in Cambodia? *Comptes Rendus Biologies* 329, 124–135.

Hassanin, A., Delsuc, F., Ropiquet, A., Hammer, C., Jansen van Vuuren, B., Matthee, C., Ruiz-Garcia, M., Catzeflis, F., Areskoug, V., Nguyen, T.T. and Couloux, A. (2012) Pattern and timing of diversification of Cetartiodactyla (Mammalia, Laurasiatheria), as revealed by a comprehensive analysis of mitochondrial genomes. *Comptes Rendus Biologies* 335, 32–50.

Hassanin, A., An, J., Ropiquet, A., Nguyen T.T. and Couloux, A. (2013) Combining multiple autosomal introns for studying shallow phylogeny and taxonomy of Laurasiatherian mammals: application to the tribe Bovini (Cetartiodactyla, Bovidae). *Molecular Phylogenetics and Evolution* 66, 766–775.

Haywood, A.M., Chandler, M.A., Valdes, P.J., Salzmann, U., Lunt, D.J. and Dowsett, H.J. (2009) Comparison of mid-Pliocene climate predictions produced by the HadAM3 and GCMAM3 General Circulation Models. *Global and Planetary Change* 66, 208–224.

Huffman, B. (2013) Your guide to the world's hoofed mammals. Available at: http://www.ultimateungulate.com (accessed 8 April 2013).

International Commission on Zoological Nomenclature (2003) Opinion 2027 (Case 3010). Usage of 17 specific names based on wild species which are predated by or contemporary with those based on domestic animals (Lepidoptera, Osteichthyes, Mammalia): conserved. *Bulletin of Zoological Nomenclature* 60, 81–84.

Irwin, D. and Arnason, U. (1994) Cytochrome b gene of marine mammals: phylogeny and evolution. *Journal of Mammalian Evolution* 2, 37–55.

IUCN (2012) IUCN Red List of Threatened Species. Version 2012.2. www.iucnredlist.org (accessed 8 April 2013).

Janecek, L.L., Honeycutt, R.L., Adkins, R.M. and Davis, S.K. (1996) Mitochondrial gene sequences and the molecular systematics of the artiodactyl subfamily bovinae. *Molecular Phylogenetics and Evolution* 6, 107–119.

Kerley, G.I.H., Kowalczyk, R. and Cromsigt, J.P.G.M. (2012) Conservation implications of the refugee species concept and the European bison: king of the forest or refugee in a marginal habitat? *Ecography* 35, 519–529.

Khan, M.A., Kostopoulos, D.S., Akhtar, M. and Nazir, M. (2010) Bison remains from the Upper Siwaliks of Pakistan. *Neues Jahrbuch für Geologie und Palaontologie-Abhandlungen* 258, 121–128.

Kingdon, J. (1982) *East African Mammals: An Atlas of Evolution in Africa*, Volume IIIC. Academic Press, London.

Kingdon, J. (1997) *The Kingdon Field Guide to African Mammals*. Natural World Academic Press, London.

Leslie, D.M. Jr and Schaller, G.B. (2009) *Bos grunniens* and *Bos mutus* (Artiodactyla: Bovidae). *Mammalian Species* 836, 1–17.

Linnaeus, C. (1758) Tomus I. Systema naturae per regna tria naturae, secundum classes, ordines, genera, species, cum characteribus, differentiis, synonymis, locis. Editio decima, reformata, Holmiae, Impensis Direct, Laurentii Salvii.

Louys, J., Curnoe, D. and Tong, H. (2007) Characteristics of Pleistocene megafauna extinctions in Southeast Asia. *Palaeogeography Palaeoclimatology Palaeoecology* 243, 152–173.

Lydekker, R. (1913) *Artiodactyla, Family Bovidae, Subfamilies Bovinae to Ovibovinae (Cattle, Sheep, Goats, Chamois, Serows, Takin, Musk-Oxen, etc.). Vol. 1 of Catalogue of the Ungulate Mammals in the British Museum*. British Museum (Natural History), London.

MacEachern, S., McEwan, J. and Goddard, M. (2009) Phylogenetic reconstruction and the identification of ancient polymorphism in the Bovini tribe (Bovidae, Bovinae). *BMC Genomics* 10, 177.

Mackie, R.I. (2002) Mutualistic fermentative digestion in the gastrointestinal tract: diversity and evolution. *Integrative and Comparative Biology* 42, 319–326.

Mamat-Hamidi, I.I. and Hilmi, M. (2009) Karyotype of Malayan Gaur (*Bos gaurus hubbacki*), Sahiwal-Friesian cattle and Gaur × cattle hybrid backcrosses. *Pakistan Journal of Biological Sciences* 12, 896–901.

Martínez-Navarro, B., Perez-Claros, J.A., Palombo, M.R., Lorenzo, R. and Palmqvist, P. (2007) The olduvai buffalo *Pelorovis* and the origin of *Bos*. *Quaternary Research* 68, 220–226.

Martínez-Navarro, B., Ros-Montoya, S., Patrocinio, E.M. and Palmqvist, P. (2011) Presence of the Asian origin Bovini, *Hemibos* sp. aff. *Hemibos gracilis* and *Bison* sp., at the early Pleistocene site of Venta Micena (Orce, Spain). *Quaternary International* 243, 54–60.

Martojo, H. (2012) Indigenous Bali cattle is most suitable for sustainable small farming in Indonesia. *Reproduction in Domestic Animals* 47 (Suppl. 1), 10–14.

Matthee, C.A. and Davis, S.K. (2001) Molecular insights into the evolution of the family Bovidae: a nuclear DNA perspective. *Molecular Biology and Evolution* 18, 1220–1230.

Matthee, C.A. and Robinson, T.J. (1999) Cytochrome b phylogeny of the family bovidae: resolution within the alcelaphini, antilopini, neotragini, and tragelaphini. *Molecular Phylogenetics and Evolution* 12, 31–46.

Matthee, C.A., Eick, G., Willows-Munro, S., Montgelard, C., Pardini, A.T. and Robinson, T.J. (2007) Indel evolution of mammalian introns and the utility of non-coding nuclear markers in eutherian phylogenetics. *Molecular Phylogenetics and Evolution* 42, 827–837.

McKenna, M.C. and Bell, S.K. (1997) *Classification of Mammals Above the Species Level.* Columbia University Press, New York.

McPherson, F.J. and Chenoweth, P.J. (2012) Mammalian sexual dimorphism. *Animal Reproduction Science* 131, 109–122.

Meagher, M. (1986) *Bison bison. Mammalian Species* 266, 1–8.

Meredith, R.W., Janečka, J.E., Gatesy, J., Ryder, O.A., Fisher, C.A., Teeling, E.C., Goodbla, A., Eizirik, E., Simão, T.L., Stadler, T. *et al.* (2011) Impacts of the Cretaceous Terrestrial Revolution and KPg extinction on mammal diversification. *Science* 334, 521–524.

Miyamoto, M.M., Tanhauser, S.M. and Laipis, P.J. (1989) Systematic relationships in the artiodactyl tribe Bovini (family Bovidae), as determined from mitochondrial-DNA sequences. *Systematic Zoology* 38, 342–349.

Mohamad, K., Olsson, M., van Tol, H.T., Mikko, S., Vlamings, B.H., Andersson, G., Rodríguez-Martínez, H., Purwantara, B., Paling, R.W., Colenbrander, B. and Lenstra J.A. (2009) On the origin of Indonesian cattle. *PLoS One* 4, e5490.

Mondal, M., Karunakaran, M., Lee, K.B. and Rajkhowa, C. (2010) Characterization of Mithun (*Bos frontalis*) ejaculates and fertility of cryopreserved sperm. *Animal Reproduction Science* 118, 210–216.

Montgelard, C., Catzeflis, F.M. and Douzery, E. (1997) Phylogenetic relationships of artiodactyls and cetaceans as deduced from the comparison of cytochrome b and 12S rRNA mitochondrial sequences. *Molecular Biology and Evolution* 14, 550–559.

Muizon de, C. (2009) Origin and evolutionary history of cetaceans. *Comptes Rendus Palevol* 8, 295–309.

Nijman, I.J., van Boxtel, D.C.J., van Cann, L.M., Marnoch, Y., Cuppen, E. and Lenstra, J.A. (2008) Phylogeny of Y chromosomes from bovine species. *Cladistics* 24, 723–726.

Olsen, S.J. (1990) Fossil ancestry of the yak, its cultural significance, and domestication in Tibet. *Proceedings of the Academy of Natural Sciences of Philadelphia* 142, 73–100.

Patnaik, R. and Chauhan, P. (2009) India at the cross-roads of human evolution. *Journal of Biosciences* 34, 729–747.

Pfeffer, P. (1969) Considérations sur l'écologie des forêts claires du Cambodge oriental. *Terre et Vie* 1, 3–24.

Pfeffer, P. and Kim-San, O. (1967) Le kouprey, *Bos* (*Bibos*) *sauveli* Urbain, 1937; Discussion systématique et statut actuel. Hypothèse sur l'origine du zébu (*Bos indicus*). *Mammalia* 31, 521–536.

Prothero, D.R. and Foss, S.E. (2007) *The Evolution of Artiodactyls.* Johns Hopkins University Press, Baltimore, Maryland.

Pucek, Z., Belousove, I.P., Krasinska, M., Krasinska, Z.A. and Olech, W. (2004) European bison. Status survey and conservation action plan. IUCN/SSC Bison Specialist Group. IUCN, Gland, Switzerland.

Qi, X.B., Jianlin, H., Wang, G., Rege, J.E. and Hanotte, O. (2010) Assessment of cattle genetic introgression into domestic yak populations using mitochondrial and microsatellite DNA markers. *Animal Genetics* 41, 242–252.

Ritz, L.R., Glowatzki-Mullis, M.L., MacHugh, D.E. and Gaillard, C. (2000) Phylogenetic analysis of the tribe Bovini using microsatellites. *Animal Genetics* 31, 178–185.

Ropiquet, A. and Hassanin, A. (2005) Molecular phylogeny of caprines (Bovidae, Antilopinae): the question of their origin and diversification during the Miocene. *Journal of Zoological Systematics and Evolutionary Research* 43, 49–60.

Ropiquet, A., Gerbault-Seureau, M., Deuve, J.L., Gilbert, C., Pagacova, E., Chai, N., Rubes, J. and Hassanin, A. (2008) Chromosome evolution in the subtribe Bovina (Mammalia, Bovidae): the karyotype of the Cambodian banteng (*Bos javanicus birmanicus*) suggests that Robertsonian translocations are related to interspecific hybridization. *Chromosome Research* 16, 1107–1118.

Ropiquet, A., Li B. and Hassanin, A. (2009) SuperTRI: a new approach based on branch support analyses of multiple independent data sets for assessing reliability of phylogenetic inferences. *Comptes Rendus Biologies* 332, 832–847.

Salzmann, U., Williams, M., Haywood, A.M., Johnson, A.L.A., Kender, S. and Zalasiewicz, J. (2011) Climate and environment of a Pliocene warm world. *Palaeogeography Palaeoclimatology Palaeoecology* 309, 1–8.

Sanderson, E.W., Redford, K.H., Weber, B., Aune, K., Baldes, D., Berger, J., Carter, D., Curtin, C., Derr, J., Dobrott, S. *et al.* (2008) The ecological future of the North American bison: conceiving long-term, large-scale conservation of wildlife. *Conservation Biology* 22, 252–266.

Sauvel, R. (1949) Le Kou-Prey ou bœuf gris du Cambodge. *Terre et Vie* 96, 89–109.

Schaller, G.B. and Rabinowitz, A. (1995) The saola or spindle-horn bovid *Pseudoryx nghetinhensis* in Laos. *Oryx* 29, 107–114.

Scott, E. (2010) Extinctions, scenarios, and assumptions: changes in latest Pleistocene large herbivore abundance and distribution in western North America. *Quaternary International* 217, 225–239.

Senut, B., Pickford, M. and Segalen, L. (2009) Neogene desertification of Africa. *Comptes Rendus Geoscience* 341, 591–602.

Shimamura, M., Yasue, H., Ohshima, K., Abe, H., Kato, H., Kishiro, T., Goto, M., Munechika, I. and Okada, N. (1997) Molecular evidence from retroposons that whales form a clade within even-toed ungulates. *Nature* 388, 666–670.

Simoons, F.J. and Simoons, E.S. (1968) *A Ceremonial Ox of India: The Mithan in Nature, Culture, and History, with Notes on the Domestication of Common Cattle.* University of Wisconsin Press, Madison, Wisconsin.

Simpson, G.G. (1945) The principles of classification and a classification of mammals. *Bulletin of the American Museum of Natural History* 85, 1–350.

Sipko T.P. (2009) European bison in Russia – past, present and future. *European Bison Conservation Newsletter* 2, 148–159.

Steklenev, E.P. (1995) The characteristics of the reproductive capacity of hybrids of the bison (*Bison bison* L.) with the domestic cow (*Bos* (*Bos*) *primigenius taurus*). 1. The reproductive capacity of hybrid males. *T͡Sitologii͡a i Genetika* 29, 66–76.

Steklenev, E.P. and Elistratova, T.M. (1992) The characteristics of the reproductive capacity of hybrids of banteng (*Bos* (*Bibos*) *javanicus* d'Alton) with the domestic cow (*Bos* (*Bos*) *primigenius taurus*). *T͡Sitologii͡a i Genetika* 26, 45–57, 75.

Stromberg, C.A.E. (2005) Decoupled taxonomic radiation and ecological expansion of open-habitat grasses in the Cenozoic of North America. *Proceedings of the National Academy of Sciences* 102, 11980–11984.

Uhen, M.D. (2010) The origin(s) of whales. *Annual Review of Earth and Planetary Sciences* 38, 189–219.

Urbain, A. (1937) Le Kou Prey ou bœuf gris cambodgien. *Bulletin de la Société Zoologique de France* 62, 305–307.

Utescher T., Erdei, B., François, L. and Mosbrugger, V. (2007) Tree diversity in the Miocene forests of Western Eurasia. *Palaeogeography Palaeoclimatology Palaeoecology* 253, 226–250.

Van Gelder, R.G. (1977) Mammalian hybrids and generic limits. *American Museum Novitates* 2635, 1–25.

van Vuure, C. (2005) *Retracing the Aurochs – History, Morphology and Ecology of an Extinct Wild Ox.* Pensoft Publishers, Sofia-Moscow.

Verkaar, E.L., Nijman, I.J., Beeke, M., Hanekamp, E. and Lenstra, J.A. (2004) Maternal and paternal lineages in cross-breeding bovine species. Has wisent a hybrid origin? *Molecular Biology and Evolution* 21, 1165–1170.

Vigne J.D. (2011) The origins of animal domestication and husbandry: a major change in the history of humanity and the biosphere. *Comptes Rendus Biologies* 334, 171–181.

Vislobokova, I. (2005) On Pliocene faunas with Proboscideans in the territory of the former Soviet Union. *Quaternary International* 126–128, 93–105.

Waddell, P.J., Okada, N. and Hasegawa, M. (1999) Towards resolving the interordinal relationships of placental mammals. *Systematic Biology* 48, 1–5.

Wang, Z.F., Shen, X., Liu, B., Su, J.P., Yonezawa, T., Yu, Y., Guo, S.C., Ho, S.Y.W., Vila, C., Hasegawa, M. and Liu, J.Q. (2010) Phylogeographical analyses of domestic and wild yaks based on mitochondrial DNA: new data and reappraisal. *Journal of Biogeography* 37, 2332–2344.

Wharton, C.H. (1957) An ecological study of the kouprey, *Novibos sauveli* (Urbain). *Monographs of the Institute of Science and Technology, Manila, Philippines* 5, 1–107.

Wilson, D.E. and Reeder, D.M. (2005) *Mammal Species of the World: a Taxonomic and Geographic Reference* (3rd edn). Johns Hopkins University Press, Baltimore, Maryland.

Woodruff, D.S. (2003) Neogene marine transgressions, palaeogeography and biogeographic transitions on the Thai-Malay Peninsula. *Journal of Biogeography* 30, 551–567.

Zachos, J.C., Dickens, G.R. and Zeebe, R.E. (2008) An early Cenozoic perspective on greenhouse warming and carbon-cycle dynamics. *Nature* 451, 279–283.

Zharova, G.K., Naumova, E.I., Chistova, T.U. and Danilkin, A.A. (2011) Digestion of cellulose fibers in the digestive tract of wild ruminants. *Doklady Biological Sciences* 441, 370–372.

2 Genetic Aspects of Domestication

J.A. Lenstra[1] and M. Felius[2]

[1]Utrecht University, Utrecht, The Netherlands;
[2]Rotterdam, The Netherlands

Introduction

The appearance, characteristics and genetic makeup of cattle have been heavily influenced by domestication and by their dynamic history right up to the present day. Paleontology and molecular-genetic analysis have revealed the approximate place and date of domestication (Zeder et al., 2006). However, documentation of the subsequent history of cattle, although closely connected with human history, is scarce until 250 years ago. In this chapter we summarize the available data on the domestication and history of cattle, including the appearance of phenotypes that today are the subject of genetic research.

The First Domestic Cattle

Domestic cattle appeared in northern and western parts of the Fertile Crescent in Southwest Asia not long after domestic sheep and goats and at about the same time as domestic pigs (Hongo et al., 2009). Together with the introduction of crops, livestock fundamentally changed human demography and eventually led to our present complex society. It is plausible that cattle husbandry, requiring more labour and organization than the keeping of smaller sized sheep, goats and pigs, contributed to the earliest stratification of society (Ajmone-Marsan et al., 2010).

The oldest evidence of taurine domestic cattle was found on both sides of the Turkish–Syrian border northeast of Aleppo and dates from 10,300–10,800 BP (Ho and Shapiro, 2011; Vigne, 2011; Bollongino et al., 2012). Modelling of cattle autosomal DNA sequence variation suggested a predomestic population bottleneck, which was possibly induced by a glaciation period (Murray et al., 2010; Teasdale and Bradley, 2012). Coalescence analysis of mtDNA sequences from Iranian Neolithic and Iron Age cattle led to an estimate of c.80 female aurochs (*Bos primigenius*) being the maternal ancestors of almost all present day taurine cattle (Bollongino *et al.*, 2012). The present taurine (T) mtDNA is clearly less variable than the mtDNA of yak or bison and diverged less than 15,000 BP, indicating taurine population expansion after domestication (Ho and Shapiro, 2011).

Zebu (*B. indicus*) (Plate 4) emerged in the Indus valley by a separate domestication of a different aurochs subspecies (*B.p. namadicus*, Chen *et al.*, 2010). Archeological evidence dates the domestication of zebu 2000 years after the taurine domestication (Bradley, 2006; Fuller, 2006; Jarrige, 2006). This is in line with Bayesian estimates of mtDNA variants (Ho and Shapiro, 2011) and with the diversity pattern and broad geographic distribution of the mtDNA haplogroup I1 (Chen *et al.*, 2010; Teasdale and Bradley, 2012). However, the absence in East Asia of the second haplogroup I2, modelling of autosomal gene variation and a more complex I2 diversity cline suggests that there may have been an additional zebu domestication, possibly including introgression of wild females into domestic herds (Murray *et al.*, 2010; Teasdale and Bradley, 2012).

Remarkably, the characteristic hump, which is caused by an overdevelopment of the thoracic part of the rhomboid major muscle relative to the cervical part, does not appear on rock paintings of *B.p. namadicus*. Fossil remains from Mehrgarh in Baluchistan have been attributed to zebu and were dated at 8000 BP (Jarrige, 2006), but the earliest convincing clay figurines of humped cattle are dated at 5000 to 6000 BP, suggesting that the hump emerged after domestication. The earliest clear depictions of humped cattle are from a seal from 4450–4200 BP found in Harappa in the Indus valley and in pictures from south Indian Neolithic sites (Allchin and Allchin, 1974).

Since there are no reproductive barriers between zebu and taurine cattle, they should zoologically be considered as subspecies with designations *Bos primigenius indicus* and *B.p. taurus*, respectively. However, *B. indicus* and *B. taurus* are the more common designations.

As in other domestic species, adaptation to the habitat of early human settlements was accompanied by profound genetic changes in morphology, physiology and behaviour (Hall, 2004). This included decrease in size, reduction of the outspoken sexual dimorphism of the aurochs and increase in intramuscular fat content. Taming implies an attenuation of behaviour, but feral populations such as the Chillingham and Heck cattle regain the typical behaviour of wild herd species with male dominance. The selection signatures are likely to be different in indicine and taurine cattle because of their separate domestications, but selection may have targeted some of the same genes.

Taurine Cattle Spread over the Old World

Early farms and dairying

The expansions of the first, well-populated agricultural communities from Southwest Asia to the rest of Eurasia and to Africa have so far been reconstructed on the basis of sporadic pictorial representations and by dating of the earliest farms and paleontological remains of livestock. Domestic taurine cattle probably reached central Anatolia between 10,000 (Vigne, 2011) and 8500 BP (Arbuckle and Makarewicz, 2009). This was possibly preceded by intensified management of wild cattle populations, which may have interacted with the smaller imported domesticates.

The subsequent colonization of Europe proceeded via the Mediterranean coast and along the Danube River. Traces of dairy products in remains of pottery and nitrogen isotope ratios as signs of early weaning of calves showed that dairying followed soon after the arrival of domestic cattle (Payne and Hodges, 1997; Price, 2000; Tresset, 2003): in the 9th millennium BP in Southwest Asia; in the 7th millennium in Africa (Dunne *et al.*, 2012); in the 8th millennium in southeastern (Evershed *et al.*, 2008) and northern (Salque *et al.*, 2013) Europe; and in the late 7th millennium in the UK (Copley *et al.*, 2003) and France (Balasse and Tresset, 2002). This was accompanied by a gene flow from the Southwest Asian agricultural societies into the European communities of hunter-gatherers (Pinhasi *et al.*, 2012; Rasteiro and Chikhi, 2013). The emergence of lactase persistence in adult humans in European and African Neolithic populations may be regarded as an example of human–animal coevolution (Beja-Pereira *et al.*, 2003; Gerbault *et al.*, 2011).

Maternal lineages

Archaeological observations do not rule out secondary domestications of taurine cattle outside the Fertile Crescent. Separate domestications have been postulated for African cattle on the basis of fossil remains (Brass, 2012) and East Asian cattle on the basis of the high frequency of the T4 mtDNA haplotype (Fig. 2.1; Mannen *et al.*, 2004). However, the Mesopotamian origin of almost all taurine cattle is supported by a phylogeny of

the common taurine mtDNA haplotypes without deep splits and by their geographic distribution (Bradley *et al.*, 1998; Ajmone-Marsan *et al.*, 2010). Southwest Asia has a high haplotype diversity with haplogroups T, T1, T2 and T3 (Fig. 2.1). In contrast T1 is almost fixed in Africa, whereas T3 is dominant in Europe and north-central Asia (Troy *et al.*, 2001; Beja-Pereira *et al.*, 2006; Achilli *et al.*, 2009; Kantanen *et al.*, 2009; Bonfiglio *et al.*, 2010; Jia *et al.*, 2010; Ginja *et al.*, 2010; Stock and Gifford-Gonzalez, 2013).

The shift from ~29% T1 in Southwest Asia to almost 100% in Africa indicates strong maternal founder effects during migrations from Southwest Asia to North Africa and then to West and Central Africa (Fig. 2.1, Bonfiglio

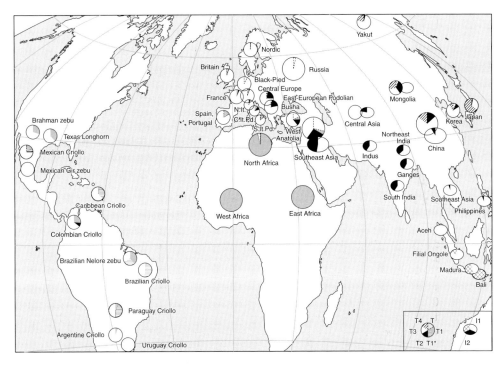

Fig. 2.1. Continental distribution of taurine (circles) and zebu/banteng (ovals) mtDNA haplogroups (Lenstra *et al.*, 2014). Data are from Cymbron *et al.*, 1999; Troy *et al.*, 2001; Magee *et al.*, 2002; Miretti *et al.*, 2002, 2004; Carvajal-Carmona *et al.*, 2003; Kim *et al.*, 2003; Mannen *et al.*, 2004; Komatsu *et al.*, 2004; Lai *et al.*, 2006; Lei *et al.*, 2006; Cortes *et al.*, 2008; Kantanen *et al.*, 2009; Mohamad *et al.*, 2009; Bonfiglio *et al.*, 2010; Chen *et al.*, 2010; Ginja *et al.*, 2010; Armstrong *et al.*, 2013; Horsburgh *et al.*, 2013; and Ludwig *et al.*, 2013. N.It., North Italy; C.It.Pod, central Italian Podolian; S.It.Pod., South Italian Podolian; T to T4, taurine haplogroups; T1*, T1c1a1 subgroup expanded in American cattle; I1, I2, zebu haplogroups; J, banteng mtDNA sequence. Large circles and ovals represent cattle from a continental region. With the exception of haplotypes of Texas Longhorn and Brahman, no mtDNA sequences are available for USA, Canadian, or Australian cattle. Indonesian cattle have not been typed for I1/I2 differentiation.

et al., 2012). Likewise, the T4 is a subvariant of the closely related T3 (Achilli et al., 2009) and probably spread over East Asia by a founder effect during the eastward migration of cattle.

The frequency of the T3 haplogroup increases from ~40% in Southwest Asia to 100% in northwest Europe (Beja-Pereira et al., 2006; Bonfiglio et al., 2010) with a concomitant decrease of T2 (Fig. 2.1). This suggests a large influence of T3 carrying founders, although a predomestic origin of the high T3 frequency in Europe cannot be ruled out (Beja-Pereira et al., 2006; Mona et al., 2010; Lari et al., 2011). Ancient DNA confirmed that most Neolithic European cattle already carried T3 haplotypes (Bollongino et al., 2006). This agrees with Bayesian analysis of the coalescence of taurine mtDNA variants showing population expansion during the last 10,000 years (Finlay et al., 2007).

A few examples show that haplotypes in European cattle other than T3 and T2 may provide additional information on the history of cattle:

- A low frequency (c.1/1000) of the P haplotype from European aurochs in European domestic cattle and the sporadic finding of R haplotypes suggests a rare recruitment of cows from the European aurochs population (Stock et al., 2009; Bonfiglio et al., 2010).
- A frequency of 13% T1 in Iberian breeds reflects prehistoric or later gene flow across the Strait of Gibraltar (Cymbron et al., 1999; Beja-Pereira et al., 2006; Ajmone-Marsan et al., 2010; Ginja et al., 2010), which recently has been confirmed by SNP genotyping (Decker et al., 2014).
- The high frequency of T1 in Sicilian and south Italian Podolian breeds may also indicate African influence.
- Podolian breeds in central Italy have appreciable frequencies of both T1 and T2 (Bonfiglio et al., 2010), which for T2 is also observed in east European cattle (Fig. 2.1). Since there are no records of intensive demographic contacts between Africa and central Italy, secondary gene flow from Anatolia (Pellecchia et al., 2007) or Greece (Kron, 2004, see below) well after the introduction of cattle may explain the high mtDNA diversity in central Italy.

Paternal lineages

In contrast to mtDNA, which shows the maternal origin and therefore stays with the herds, Y chromosomal haplotypes are markers of paternal origin and male introgression. So far two major well diverged Y chromosomal haplogroups have been identified in taurine bulls, Y1 and Y2. Y1 is predominant in northern European and in north Spanish breeds, has a low frequency in Southwest Asian bulls and is carried by male offspring of recent European imports (Edwards et al., 2011). Y2 is dominant in central European, Mediterranean, Asian and African taurine bulls.

Remains of European aurochs bulls for which their wild origin was validated via their mtDNA all carried Y2 haplotypes (Bollongino et al., 2008). Since these cannot yet be differentiated from European or Southwest Asian Y2 haplotypes, this neither proves nor disproves wild male introgression. Wild–domestic crossbreeding was suggested by intermediate-sized Neolithic bones found in what is now the Czech Republic (Kyselý and Hájek, 2012). The Y1 distribution pattern is interpreted as reflecting later expansions of dairy breeds (Edwards et al., 2011, see below; Bollongino et al., 2008; Svensson and Gotherstrom, 2008).

The finding of African-specific Y2 haplotypes provides evidence for introgression of African aurochs in domestic herds (Perez-Pardal et al., 2010a,b; Stock and Gifford-Gonzalez, 2013). An African origin of taurine cattle, in spite of a Southwest Asian maternal origin, has been confirmed by SNP analysis (Decker et al., 2014).

Adaptation

Fossil remains reveal that skeletal morphology of cattle kept changing after domestication. In Europe the size of taurine cattle continued to decrease in the Stone Age, Bronze Age and Iron Age (Jewell, 1962; Zeuner, 1963; Bökönyi, 1974; Barker, 1985). A selective disadvantage of large cattle may have been imposed by: (i) slaughtering of the largest animals just before the winter; (ii) food shortage during winter in the temperate zones; and (iii) castration of the strongest bulls for use as work animals (Barker, 1985; Clutton-Brock, 1989).

The first domestic cattle were long-horned. This phenotype still persists in several British, French, Mediterranean, Podolian and zebu breeds (see Plates 10, 12, 16, 21, 22, 24 and 26), but did not suit the domestic habitat. About 2000 years after the domestication of long-horned cattle, short-horned cattle appeared in Mesopotamia and reached Africa 6000–5000 BP (Payne and Hodges, 1997), southern and central Europe 4500–5000 BP and Britain 3000–4000 BP. Long-horned cattle persisted in the Bronze Age in eastern Europe when the majority of Swiss and Austrian cattle were already short-horned (Bökönyi, 1974). From 3600 BP short-horns were predominant in Africa and from 3000 BP in Europe (Epstein and Mason, 1984).

Horns became dispensable in captivity because domestic cattle are protected against predation and the bulls do not fight for dominance. Hornless skulls found in Switzerland, Poland and Germany date as early as the 6th millennium BP (Bökönyi, 1974), while English hornless skulls date from 2700 BP (Jewell, 1962). Polled cattle were depicted in the 2nd millennium BC in Egypt, although it is not clear if these animals were also born hornless (Strouhal, 1992; Bard and Shubert, 1999). Herodotus mentioned hornless cattle kept by Scythians (Rawlinson, 1985), while in the north of the Netherlands most hornless skulls date from the Roman era (Lauwerier, 2011).

Molecular analysis localized the horned/polled causative mutation in an intergenic region on BTA1 (Medugorac et al., 2012; Allais-Bonnet et al., 2013; Wiedemar et al., 2014). The congenital absence of horns correlated in a broad panel of breeds from Switzerland, France, the UK and Scandinavia with the presence of one particular indel, whereas a different indel in the same region was associated with polledness in the dairy breeds from northwestern-continental Europe. Remarkably, in spite of the old origin of the trait, its autosomal transmission and an obvious advantage to the farmer, the absence of horns has rarely been fixed within breeds, possibly because of association with unfavourable traits (Allais-Bonnet et al., 2013). However, polledness was favoured in the ancestors of Scottish Angus (Plate 2) and Galloway and of several Nordic breeds, possibly because of the necessity to house cattle during long winter periods (Felius, 1995; Medugorac et al., 2012).

Domestication also favoured a diversity in coat colours, ever the most visible trait (see Plates 2–29; Chapter 4).

Zebus around the Equator

Zebu spread after domestication to eastern and southwestern Asia, in the latter region becoming introgressed into pre-existing taurine populations (Fuller, 2006; Edwards et al., 2007a; Ajmone-Marsan et al., 2010; Chen et al., 2010). These migrations again generated a differential distribution of the mtDNA haplogroups I1 and I2, with eastern Asia populated almost exclusively by I1 haplotypes (Fig. 2.1).

Later migrations took zebu to tropical or subtropical zones of all inhabited continents, populating China, Indochina and Indonesia in the east and Africa and North and South America in the west. Pictures in Egypt at 3800 BP show the earliest African zebu, but large-scale introduction of zebu bulls occurred about 2000 BP (Epstein and Mason, 1984) and again following the Islamic invasions after AD 700 (Ajmone-Marsan et al., 2010; Payne and Hodges, 1997). Importation from India in the 19th century brought zebu to America.

These migrations led to various admixtures with taurine cattle as well as other bovine species (Lenstra and Bradley, 1999). Selembu is the offspring after terminal crossing of zebu with gayal (Bos frontalis), which is practised in Myanmar and Malaysia. Indonesian and south Chinese zebu breeds carry 10–30% autosomal alleles from domestic banteng (Bos javanicus) with the frequency of banteng mtDNA ranging from 0 to 100% (Fig. 2.1, Mohamad et al., 2009; Decker et al., 2014). Southwest Asian and Chinese cattle from mixed taurindicine descent may carry both taurine and indicine mtDNA or Y chromosomes (Mannen et al., 2004; Lai et al., 2006; Lei et al., 2006; Edwards et al., 2007b; Jia et al., 2010; Li et al., 2013). In contrast, only zebu bulls were exported to Africa and almost only bulls to America; in fact, indicine mtDNA is rare outside Asia (Bradley et al., 1998; Meirelles et al., 1999; Ginja et al., 2010).

African Sanga cattle descend from early taurindicine crosses and around AD 1500 they were the dominant type of cattle in eastern and central Africa (Payne and Hodges, 1997). Various degrees of zebu introgression in taurine and Sanga populations resulted in the present continuous spectrum of taurine, admixed taurine, Zenga (zebu × Sanga) and African zebu breeds. By crossbreeding of Indian bulls to American Criollo breeds, which are of Iberian maternal descent, overall frequency of the zebu-specific Y3 Y chromosomal haplotype is almost 51% (Ginja et al., 2010) and their ancestry became taurindicine (Martinez et al., 2012; McTavish et al., 2013). During the past century, several American and Australian synthetic breeds have been formed by planned crosses of taurine breeds with zebu (Buchanan and Lenstra, Chapter 3, this volume; Felius et al., 2014).

Several traits contribute to the adaptation of zebu to tropical and dry environments: a low metabolic rate, proliferation of large sweat glands, a large skin surface, a predominance of intramuscular instead of subcutaneous fat, a smooth coat, a low susceptibility to insects, ticks and protozoa and good utilization of low-quality fodder (Turton, 1991, see also Chapter 23). The heat tolerance often encouraged zebu introgression in spite of the higher productivity of European taurine cattle. Zebu did not develop the trypanotolerance of West African taurine breeds (see Chapter 9), but has a higher resistance to rinderpest and largely replaced the East African sanga after the epidemic of 1887–1897.

Several distinct indicine and taurindicine breeds have been described. However, selective breeding and genetic isolation of zebu have been less systematic and consequential than for taurine cattle.

Large Taurines during Classical Antiquity

The relatively peaceful Hellenistic and Roman societies and an increase of trading stimulated farming on a larger scale than in earlier pastoral societies. This also coincided with a reversal of the size reduction of taurine cattle in and around the Roman Empire. Already in the 8th century BC large cattle existed in the Greek Messenia region (Kron, 2004). It was probably the first type of cattle that spread over a large region because of its superior qualities: to Epiros, Greece, Anatolia, Sicily, from the 3rd century BC to south Italy and to north Italy and the

Mediterranean French coast even before the Roman occupation. If these so-called Epirote cattle are ancestral to the current Italian Podolian breeds, it would explain the maternal genetic link between Anatolia and Italy as evidenced by the high mtDNA in Italy (Pellecchia et al., 2007; Bonfiglio et al., 2010).

Export probably accounted for the large size of several cattle during the Roman period in the European part of the Roman Empire (Kron, 2004; Bökönyi, 1974; Schlumbaum et al., 2003). In the peripheral regions of the empire large cattle coexisted with small local short-horns (Jewell, 1962; Bökönyi, 1974; Riedel, 1985). One of these cattle found in the Alpine region appeared to carry the T2 mtDNA haplotype, which is rare in current Swiss cattle (Schlumbaum et al., 2006).

Written documentation from Greek and Roman sources provides the first contemporary reports of different types of cattle. White cattle were already reported in the Mycenean period on Pylos (Bökönyi, 1974; McInerney, 2010). Archimedes (287–212 BC) mentioned four coat colours or colour patterns on Sicily: creamy white, ebony black, yellow and spotted (Archimedes, 1999). Several Roman authors described a large variety of Italian cattle of different sizes and coat colours, which were used mainly for draught and played a role in religious rituals (Barker, 1985; MacKinnon, 2010). Beef was consumed, but fetched in AD 305 a lower price than pork (Diocletian Price Edict, Leake, 1826). Roman cattle were not milked in contrast to Alpine and Germanic cattle (Caesar, 50–40 BC; Strabo, 1969; MacKinnon, 2010).

Small Cattle in Medieval Europe

After the fall of the Roman Empire, cattle accompanied the migrations of various Germanic tribes. Fossil remains, which have outlasted any written evidence, show a swift disappearance of the large Roman cattle with the possible exception of Italian cattle (Zeuner, 1963; Barker, 1985; Kron, 2002). Thus most cattle found in the graves from the empire of the Azars in Balkan from the 6th and 9th century were small and short-horned (Bökönyi, 1974). Presumably, these small cattle were adequate for local demands and represent an adaptation to medieval farming practices and to the vicissitudes of the unruly societies (Jewell, 1962). This phenotype has been preserved in the Balkan Busha cattle and in American Criollo, which descend from 15th-century Iberian cattle.

Presumably, frequent depletion of local livestock populations during famines or by plundering maintained gene flow between neighbouring regions. For instance, the clear phylogenetic relationship of southern French and Swiss breeds (Buchanan and Lenstra, Chapter 3, this volume) may indicate import into France from the Alpine regions, which were less affected by the devastating Roman conquest, medieval wars and 14th-century famines. Nevertheless, it is plausible that already during the Middle Ages local developments generated geographic differentiation. At least the characteristic phenotypes found at the European periphery have medieval or even older origins: the small Nordic polled cattle, several long-horned or short-horned British types of cattle, the long-horned Podolic cattle in Italy and the Balkan steppe, and the large variety in coat colour and horns in Iberian cattle (Bishko, 1953; Felius, 1995). Present cattle from these regions also have high Y chromosomal diversity (Ginja et al., 2010; Edwards et al., 2011).

Iberian cattle expanded to the south during the Reconquista from AD 900–1492. Incorporation of south Iberian cattle introduced the T1 mtDNA haplogroup (Cymbron et al., 1999). On regained territories a beef cattle ranching economy with extensive management developed, which after 1492 was also introduced in the New World (Bishko, 1953). The high frequency of the mtDNA T1 haplogroup in Latin American Criollo cattle (Fig. 2.1, 29%) as well as SNP genotyping (Decker et al., 2014) indicate an Iberian ancestry. The increased frequency of the T1 subvariant T1c1a1 (originally named AA, Miretti et al., 2004; Ginja et al., 2010; Bonfiglio et al., 2012) reflects a founder effect, which very well may have taken place on one of the Caribbean islands before transport of cattle to the American continent.

The origin of the Podolian-type of cattle in Italy and the Balkans is unclear: a landrace with roots in the antiquity (Ciani and Matassino, 2001), which may have originated from the Epirote cattle (see above) and may in the 5th and 6th centuries have been influenced by cattle brought in by various German tribes; as suggested by their name, importation from Podolia in the Ukraine;

or, as proposed on the basis of fossil evidence, an emergence in Hungary since the 14th century by selective breeding (Bökönyi, 1974) stimulated by their large-scale export as beef animals (see below). Podolian cattle in Italy continued the Roman tradition of using cattle as draught animals, although Aragonian rule in south Italy encouraged sheep ranching at the expense of cattle (Kron, 2004). Remarkably Italian and east European Podolian mtDNA haplogroup distributions are clearly different (Fig. 2.1), emphasizing that gene flow was mostly male-mediated.

Intensive dairy farming was practised in at least two regions, in the Alps with the vertical transhumance between summer and winter pastures and on the rich pastures of the North Sea and Baltic coastal regions. This probably continued a tradition predating the Roman era. In both regions a single Y chromosome is now predominant, a northern Y1 haplotype and a central European Y2 haplotype (Edwards et al., 2011). These haplotypes are separated by a sharp genetic boundary that divides both France and Germany in northern and southern parts. This now reflects the contrast of specialized dairy cattle from the northwestern continental green lowland with the beef or dual-purpose cattle from the more hilly and mountainous regions (Comberg, 1984). However, it also coincides with historic cultural differences between northern and southern Europe, in France corresponding to the langue d'oïl and langue d'oc and Germany to the Niederdeutsch and Hochdeutsch, respectively.

Modernization of Cattle Husbandry

In post-medieval society higher literacy rates led to a more extensive documentation of agricultural history, so we can identify the several processes that led to the development of modern cattle, which shows a few clear parallels with Hellenistic and Roman agriculture.

- An increased demand for beef and dairy products stimulated international trade and large-scale movements of cattle. If the urban centres could not be directly supplied with locally produced beef, cattle were produced in extensive cow-calf operations on more distant breeding grounds and driven to the cities. Already in the Middle Ages Welsh cattle were moved along the so-called droves to London. From early 17th- to early 19th-century London and other English cities were supplied by cattle raised on Scottish grazing fields. Hungarian Grey cattle reared on the puzsta were driven westwards to Munich, Vienna and Venice in the 16th and 17th century (Bartosiewicz, 1997). During the same period The Netherlands and Germany were supplied by Danish cattle (Petersen, 1970). Both Danish and Hungarian cattle suffered competition from Ukrainian cattle driven via Krakow to the west (Carter, 1994). The cattle drives declined following improved husbandry practices that increased productivity of cattle bred near the urban centres. Most cattle were moved as meat-on-the-hoof, but in Italy Podolian cattle may have been recruited as breeding material (see above). In the 16th century Dutch, Holstein and Danish dairy cattle were imported in Sweden (Falk, 2012) and in the 17–18th century Dutch cattle contributed to the Shorthorn, Ayrshire, Hereford and other British breeds (Hall and Clutton-Brock, 1989; Felius, 1995).

- Improvements in agricultural practices allowed larger cattle to be kept, especially the increased availability of fodder during the winters. It is likely that selection contributed to the gradual size increase since the Middle Ages.
- A higher density of cattle probably played a role in the frequent occurrence of rinderpest epidemics (Broad, 1983; Spinage, 2003). In The Netherlands this necessitated the import of cattle from Jutland and Holstein, which were the ancestors of the Dutch black and red pied dairy cattle (Felius, 1995).
- From the 18th century, breed formation had a most profound influence on the appearance, productivity and genetic constitution of cattle. This was achieved by systematic selection of breeding bulls according to explicit breeding objectives within genetically isolated regions. Pedigrees and performance were documented in herd books and organized by breeding societies. Eventually this led to the development of hundreds of specialized breeds, which over time became integral parts of local tradition (Felius, 1995; Buchanan and Lenstra, Chapter 3, this volume). Inbreeding by genetic isolation was

for many breeds counteracted by introgression and crossbreeding. In the UK, the Nordic countries, south France and the Iberian Peninsula, local breeds with minimal crossbreeding to cattle from other regions have retained their common ancestry and constitute regional clusters of genetically related breeds (Felius et al., 2011). Distance analysis indicates also a cluster of Russian and Siberian breeds in spite of crossbreeding with west European breeds (Li and Kantanen, 2010; Felius et al., 2011).

- Other more international clusters of related breeds were created in the 19th century by systematic crossbreeding of local breeds with sires of successful breeds (Felius, 1995). In the first half of the 19th century, English Dairy Shorthorn bulls (Plate 19) were crossed into several Belgium and north French breeds (Béranger and Vissac, 1994). Hardy dairy Ayrshire cattle from Scotland were imported to Finland and Scandinavia and crossed into several local breeds, generating a Nordic Ayrshire breed cluster. Dairy Black-Pied cattle (Plate 11) and dual-purpose Red-Pied breeds were kept in most north European countries. Baltic Red dairy cattle (Plate 7) spread along the Baltic coasts and were also crossed into German Highland Red Cattle. Spotted dual-purpose cattle descending from the Swiss Simmental (Plate 20) became popular in Central Europe and the Balkans (Epstein and Mason, 1984; Averdunk and Krogmeier, 2011). Another dairy breed, the Swiss Brown, was crossed into mountain breeds in France, Germany, Italy and Spain. Finally, Podolian breeds (Plates 6 and 12) are now kept in the Balkan countries and Italy.
- Massive exports of European and Asian cattle to America, Australia and New Zealand replicated in the 19th and 20th centuries a large portion of the cattle genetic resources in the New World (Ajmone-Marsan et al., 2010). Holstein Friesians (Plate 11) originating from The Netherlands and the British Jersey (Plate 13) became the major dairy breeds. British Hereford and Angus (Plates 2 and 9) are still the dominating beef cattle, but since 1960 several European continental breeds have been imported as well. Zebus were imported in large numbers into Brazil, the USA

and Australia. Since then, many new taurine as well as taurindicine synthetic breeds have expanded considerably the cattle genetic repertoire (Buchanan and Lenstra, Chapter 3, this volume; Felius *et al.*, 2014).

These developments accelerated the evolution of cattle and expanded the phenotypic differentiation with regard to dairy or beef production, correspondingly influencing milk composition and meat quality (see Chapters 16–18). Dairy development in north or central Europe induced the typical wedge shape of dairy conformation (Plates 7, 11 and 17) in stark contrast to the appearance of muscular-hypertrophic beef cattle (Plate 3).

The continental lowland dairy cattle and the central European dairy cattle, carry different Y chromosomal haplotypes (Edwards *et al.*, 2011). Together with the separate development of the dairy island breeds Jersey and Guernsey, both free of crossbreeding since 1789, this indicates multiple origins of specialized dairy cattle. Therefore, different sets of genomic variants may confer the dairy productivity traits in cattle from the north European lowland, from central Europe and from the Channel Islands, respectively.

Autosomal DNA allows a monitoring of inbreeding and crossbreeding. By phylogenetic analysis it reveals local and international breed clusters mentioned above, which underlie a comprehensive classification of European breeds (Edwards *et al.*, 2011; Felius *et al.*, 2011; Buchanan and Lenstra, Chapter 3, this volume). However, the genetic surveys also demonstrate that breeds have considerable genetic overlap and still contain a large part of the total variety of the species (European Cattle Genetic Diversity Consortium, 2006; Decker *et al.*, 2009).

Industrial Cattle, Crossbreds and Local Breeds

Since World War II the development of cattle has been intensified and facilitated by modern reproductive techniques such as artificial insemination (AI) and multiple ovulation embryo transfer (MOET). The introduction of the tractor ended the requirement for draught power (Averdunk and Krogmeier, 2011). A growing role of the American cattle is illustrated by the allopatric

development of highly productive breeds on the basis of imported European breeds (Felius, 1995; Felius *et al.*, 2011). Several beef breeds have in recent decades been bred for solid black colour by crossbreeding with Angus cattle.

Genomic approaches now accelerate the identification of genetic signatures of selection and of sequence variants that are causative of phenotypic variation. In addition, genomic selection of favourable quantative trait locus (QTL) variants is expected to offer a viable alternative for traditional selection, which evaluates animals on the basis of the phenotypic characteristics of the individual or its immediate relatives (see Chapters 15 and 19).

To counteract negative consequences associated with selection for high productivity, crossbreeding is gaining popularity, with Holstein Friesian × Jersey becoming a usual combination for dairying (Freyer *et al.*, 2008; Sørensen *et al.*, 2008). In New Zealand 36% of dairy cattle are now crossbred. In Europe the Viking Red is being developed by combining animals from Danish, Norwegian and Swedish Red breeds and Finnish Ayrshire (http://www.genusbreeding. co.uk). Crossbreeding has also created several taurine or taurindicine synthetic beef breeds (Plate 18) both in America and Australia, several of which are suitable for extensive management.

Concerns about the growing focus on productivity have also led to a renewed interest in local breeds that are characterized by low productivity but are better suited for extensive management (FAO, 2007). Such traits would be useful for a new purpose of cattle: rewilding of uninhabited areas by the release of cattle and other megaherbivores (www.megafauna-foundation.org). The crossbred Heck cattle, a much disputed attempt to revive the aurochs (Felius, 1995), is often used for this purpose, but an alternative is being developed (www. taurosproject.com). The new feral cattle populations may very well develop new adaptations to their environment and reverse some of the changes associated with domestication.

Acknowledgements

We thank Drs Albano Beja-Pereira (Oporto), Dan Bradley (Dublin), Licia Colli (Piacenza) and Juha Kantanen (Jokioinen) for sending the data on haplogroup distributions used in Fig. 2.1.

References

Achilli, A., Bonfiglio, S., Olivieri, A., Malusa, A., Pala, M., Hooshiar Kashani, B., Perego, U.A., Ajmone-Marsan, P., Liotta, L., Semino, O., Bandelt, H.J., Ferretti, L. and Torroni, A. (2009) The multifaceted origin of taurine cattle reflected by the mitochondrial genome. *PLoS ONE* 4, e5753.

Ajmone-Marsan, P., Garcia, J.F. and Lenstra, J.A. (2010) On the origin of cattle: how aurochs became cattle and colonized the world. *Evolutionary Anthropology* 19, 148–157.

Allais-Bonnet, A., Grohs, C., Medugorac, I., Krebs, S., Djari, A., Graf, A., Fritz, S., Seichter, D., Baur, A., Russ, I. *et al.* (2013) Novel insights into the bovine polled phenotype and horn ontogenesis in *Bovidae*. *PLoS ONE* 8, e63512.

Allchin, B. and Allchin, R.A. (1974) *The Birth of Indian Civilization: India and Pakistan before 500 B.C.* Penguin, London.

Arbuckle, B.S. and Makarewicz, C.A. (2009) The early management of cattle (*Bos taurus*) in Neolithic central Anatolia. *Antiquity* 83, 669–686.

Archimedes (1999) *The Cattle Problem in English Verse* by S.J.P. Hillion & H.W. Lenstra Jr. Mercator, Santpoort, The Netherlands.

Armstrong, E., Iriarte, A., Martinez, A.M., Feijoo, M., Vega-Pla, J.L., Delgado, J.V. and Postiglioni, A. (2013) Genetic diversity analysis of the Uruguayan Creole cattle breed using microsatellites and mtDNA markers. *Genetics and Molecular Research* 12, 1119–1131.

Averdunk, G. and Krogmeier, D. (2011) Minor and dual-purpose *Bos taurus* breeds. In: Fuquay, J.W. (ed.) *Encyclopedia of Dairy Sciences*. Elsevier, Philadelphia, Pennsylvania, pp. 293–299.

Balasse, M. and Tresset, A. (2002) Early weaning of Neolithic domestic cattle (Bercy, France) revealed by intra-tooth variation in nitrogen isotope ratios. *Journal of Archaeological Science* 29, 853–859.

Bard, K.A. and Shubert, S.B. (1999) *Encyclopedia of the Archaeology of Ancient Egypt*. Routledge, Oxford.

Barker, G. (1985) *Prehistoric Farming in Europe*. Cambridge University Press, Cambridge.

Bartosiewicz, L. (1997) Hungarian Grey cattle: a traditional European breed. *Animal Genetic Resources* 21, 49–60.

Beja-Pereira,A., Alexandrino, P., Bessa, I., Carretero, Y., Dunner, S., Ferrand, N., Jordana, J., Laloe, D., Moazami-Goudarzi, K., Sanchez, A. and Canon, J. (2003) Genetic characterization of southwestern European bovine breeds: a historical and biogeographical reassessment with a set of 16 microsatellites. *Journal of Heredity* 94, 243–250.

Beja-Pereira, A., Caramelli, D., Lalueza-Fox, C., Vernesi, C., Ferrand, N., Casoli, A., Goyache, F., Royo, L.J., Conti, S., Lari, M. *et al.* (2006) The origin of European cattle: evidence from modern and ancient DNA. *Proceedings of the National Academy of Sciences of the United States of America* 103, 8113–8118.

Béranger, C. and Vissac, B. (1994) An holistic approach to livestock farming systems: theoretical and methodological aspects. In: Gibon, A. and Flamant, J.C. (eds) *The Study of Livestock Farming Systems in a Research and Development Framework. EAAP Publication 63.* Wageningen University Press, Wageningen, The Netherlands, pp. 5–17.

Bishko, C.J. (1953) The peninsular background of Latin American cattle ranching. *Hispanic Historical Review* 32, 491–515.

Bökönyi, S. (1974) *History of Domestic Mammals in Central and Eastern Europe.* Akadémiai Kiadó, Budapest.

Bollongino, R., Edwards, C.J., Alt, K.W., Burger, J. and Bradley, D.G. (2006) Early history of European domestic cattle as revealed by ancient DNA. *Biology Letters* 2, 155–159.

Bollongino, R., Elsner, J., Vigne, J.D. and Burger, J. (2008) Y-SNPs do not indicate hybridisation between European aurochs and domestic cattle. *PLoS ONE* 3, e3418.

Bollongino, R., Burger, J., Powell, A., Mashkour, M., Vigne, J.D. and Thomas, M.G. (2012) Modern taurine cattle descended from small number of Near-Eastern founders. *Molecular Biology and Evolution* 29, 2101–2104.

Bonfiglio, S., Achilli, A., Olivieri, A., Negrini, R., Colli, L., Liotta, L., Ajmone-Marsan, P., Torroni, A. and Ferretti, L. (2010) The enigmatic origin of bovine mtDNA haplogroup R: sporadic interbreeding or an independent event of *Bos primigenius* domestication in Italy? *PLoS ONE* 5, e15760.

Bonfiglio, S., Ginja, C., De Gaetano, A., Achilli, A., Olivieri, A., Colli, L., Tesfaye, K., Agha, S.H., Gama, L.T., Cattonaro, F. *et al.* (2012) Origin and spread of *Bos taurus*: new clues from mitochondrial genomes belonging to haplogroup T1. *PLoS ONE* 7, e38601.

Bradley, D.G. (2006) Genetics and origins of domestic cattle. In: Zeder, M.A., Bradley, D.G., Emshwiller, E. and Smith, B.D. (eds) *Documenting Domestication. New Genetic and Archaeological Paradigms.* University of California Press, Berkeley, California, pp. 317–328.

Bradley, D.G., Loftus, R.T., Cunningham, C. and MacHugh, D.E. (1998) Genetics and domestic cattle origin. *Evolutionary Anthropology* 6, 79–86.

Brass, M. (2012) Revisiting a hoary chestnut: the nature of early cattle domestication in North-East Africa. *Sahara* 24, 7–12.

Broad, J. (1983) Cattle plague in eighteenth-century England. *Agricultural History Review* 32, 104–115.

Caesar, J. (50–40 BC) Commentarii Rerum in Gallia Gestarum VI:22.

Carter, F.W. (1994) *Trade and Urban Development in Poland: An Economic Geography of Cracow, from its Origins to 1795.* Cambridge University Press, Cambridge.

Carvajal-Carmona, L.G., Bermudez, N., Olivera-Angel, M., Estrada, L., Ossa, J., Bedoya, G. and Ruiz-Linares, A. (2003) Abundant mtDNA diversity and ancestral admixture in Colombian criollo cattle (*Bos taurus*). *Genetics* 165, 1457–1463.

Chen, S., Lin, B.Z., Baig, M., Mitra, B., Lopes, R.J., Santos, A.M., Magee, D.A., Azevedo, M., Tarroso, P., Sasazaki, S. *et al.* (2010) Zebu cattle are an exclusive legacy of the South Asia Neolithic. *Molecular Biology and Evolution* 27, 1–6.

Ciani, F. and Matassino, D. (2001) Il bovino grigio allevato in Italia: origine ed evoluzione. Nota 2: il bovino macrocero [Grey cattle reared in Italy: origin and evolution. Note 1: long-horned cattle]. *Taurus Speciale* 13, 89–99.

Clutton-Brock, J. (1989) Five thousand years of livestock in Britain. *Biological Journal of the Linnean Society* 38, 31–37.

Comberg, G. (1984) *Die Deutsche Tierzucht Im 19. und 20. Jahrhundert [German Cattle Breeding in the 19th and 20th Centuries].* Eugen Ulmer, Stuttgart.

Copley, M.S., Berstan, R., Dudd, S.N., Docherty, G., Mukherjee, A.J., Straker, V., Payne, S. and Evershed, R.P. (2003) Direct chemical evidence for widespread dairying in prehistoric Britain. *Proceedings of the National Academy of Sciences of the USA* 100, 1524–1529.

Cortes, O., Tupac-Yupanqui, I., Dunner, S., Garcia-Atance, M.A., Garcia, D., Fernandez, J. and Canon, J. (2008) Ancestral matrilineages and mitochondrial DNA diversity of the Lidia cattle breed. *Animal Genetics* 39, 649–654.

Cymbron, T., Loftus, R.T., Malheiro, M.I. and Bradley, D.G. (1999) Mitochondrial sequence variation suggests an African influence in Portuguese cattle. *Proceedings of the Royal Society B: Biological Sciences* 266, 597–603.

Decker, J.E., Pires, J.C., Conant, G.C., McKay, S.D., Heaton, M.P., Chen, K., Cooper, A., Vilkki, J., Seabury, C.M., Caetano, A.R. *et al.* (2009) Resolving the evolution of extant and extinct ruminants with high-throughput phylogenomics. *Proceedings of the National Academy of Sciences of the USA* 106, 18644–18649.

Decker, J.E., McKay, S.D., Rolf, M.M., Kim, J., Molina Alcalá, A., Sonstegard, T.S., Hanotte, O., Götherström, A., Seabury, C.M. *et al.* (2014) Worldwide patterns of ancestry, divergence, and admixture in domesticated cattle. *PLoS Genetics* 10, e1004254.

Dunne, J., Evershed, R.P., Salque, M., Cramp, L., Bruni, S., Ryan, K., Biagetti, S. and di Lernia, S. (2012) First dairying in green Saharan Africa in the fifth millennium BC. *Nature* 486, 390–394.

Edwards, C.J., Baird, J.F. and MacHugh, D.E. (2007a) Taurine and zebu admixture in Near Eastern cattle: a comparison of mitochondrial, autosomal and Y-chromosomal data. *Animal Genetics* 38, 520–524.

Edwards, C.J., Bollongino, R., Scheu, A., Chamberlain, A., Tresset, A., Vigne, J.D., Baird, J.F., Larson, G., Ho, S.Y., Heupink, T.H. *et al.* (2007b) Mitochondrial DNA analysis shows a Near Eastern Neolithic origin for domestic cattle and no indication of domestication of European aurochs. *Proceedings of the Royal Society B: Biological Sciences* 274, 1377–1385.

Edwards, C.J., Ginja, C., Kantanen, J., Pérez-Pardal, L., Tresset, A., Stock, F., Gama, L.T., Penedo, M.C.T., Bradley, D.G., Lenstra, J.A., Nijman, I.J. and European Cattle Genetic Diversity Consortium (2011) Dual origins of dairy cattle farming – evidence from a comprehensive survey of European Y-chromosomal variation. *PLoS ONE* 6, e15922.

Epstein, H. and Mason, I.L. (1984) Cattle. In: Mason, I.L. (ed.) *Evolution of Domesticated Animals.* Longman Group, Harlow, UK.

European Cattle Genetic Diversity Consortium (2006) Marker-assisted conservation of European cattle breeds: an evaluation. *Animal Genetics* 37, 475–481.

Evershed, R.P., Payne, S., Sherratt, A.G., Copley, M.S., Coolidge, J., Urem-Kotsu, D., Kotsakis, K., Ozdogan, M., Ozdogan, A.E., Nieuwenhuyse, O., Akkermans, P.M., Bailey, D., Andeescu, R.R., Campbell, S., Farid, S., Hodder, I., Yalman, N., Ozbasaran, M. *et al.* (2008) Earliest date for milk use in the Near East and southeastern Europe linked to cattle herding. *Nature* 455, 528–531.

Falk, R.P. (2012) SRB – *The Economical Alternative to Holstein.* Available at: www.scanred.se (accessed 29 August 2014).

FAO (2007) *The State of the World's Animal Genetic Resources for Food and Agriculture*. FAO, Rome.

Felius, M. (1995) *Cattle Breeds, an Encyclopedia*. Misset Uitgeverij, Doetinchem, The Netherlands.

Felius, M., Koolmees, P.A., Theunissen, B., European Cattle Genetic Diversity Consortium and Lenstra, J.A. (2011) On the breeds of cattle – historic and current classlflcatlons. *Diversily* 3, 660–692.

Felius, M., Theunisse, B. and Lenstra, J.A. (2014) On the conservation of cattle – the role of breeds. *Journal of Agricultural Science* (in press).

Finlay, E.K., Gaillard, C., Vahidi, S.M., Mirhoseini, S.Z., Jianlin, H., Qi, X.B., El-Barody, M.A., Baird, J.F., Healy, B.C. and Bradley, D.G. (2007) Bayesian inference of population expansions in domestic bovines. *Biology Letters* 3, 449–452.

Freyer, G., Konig, S., Fischer, B., Bergfeld, U. and Cassell, B.G. (2008) Invited review: crossbreeding in dairy cattle from a German perspective of the past and today. *Journal of Dairy Science* 91, 3725–3743.

Fuller, D.Q. (2006) Agricultural origins and frontiers in South Asia: a working synthesis. *World Prehistory* 20, 1–86.

Gerbault, P., Liebert, A., Itan, Y., Powell, A., Currat, M., Burger, J., Swallow, D.M. and Thomas, M.G. (2011) Evolution of lactase persistence: an example of human niche construction. *Philosophical Transactions of the Royal Society of London B: Biological Sciences* 366, 863–877.

Ginja, C., Penedo, M.C., Melucci, L., Quiroz, J., Martinez Lopez, O.R., Revidatti, M.A., Martinez-Martinez, A., Delgado, J.V. and Gama, L.T. (2010) Origins and genetic diversity of New World Creole cattle: inferences from mitochondrial and Y chromosome polymorphisms. *Animal Genetics* 41, 128–141.

Hall, S.J.G. (2004) *Livestock Biodiversity. Genetic Resources for the Farming of the Future*. Blackwell Publishing, Oxford.

Hall, S.J.G. and Clutton-Brock, J. (1989) *Two Hundred Years of British Farm Livestock*. British Museum, London.

Ho, S.Y. and Shapiro, B. (2011) Skyline-plot methods for estimating demographic history from nucleotide sequences. *Molecular Ecology Resources* 11, 423–434.

Hongo, H., Pearson, J.Ö.B. and Ilgezdi, G. (2009) The process of ungulate domestication at Çayönü, south-eastern Turkey: a multidisciplinary approach focusing on *Bos* sp. and *Cervus elaphus*. *Antropozoologica* 44, 63–78.

Horsburgh, K.A., Prost, S., Gosling, A., Stanton, J.A., Rand, C. and Matisoo-Smith, E.A. (2013) The genetic diversity of the Nguni breed of African cattle (*Bos* spp.): complete mitochondrial genomes of haplogroup T1. *PLoS ONE* 8, e71956.

Jarrige, J.F. (2006) Mehrgarh Neolithic. In: *International Seminar on the 'First Farmers in Global Perspective', Lucknow, India, 18–20 January, 2006*. Lucknow, India, pp. 135–154.

Jewell, P.A. (1962) Changes in size and type of cattle from prehistoric to mediaeval times in Britain. *Journal of Animal Breeding and Genetics* 77, 159–167.

Jia, S., Zhou, Y., Lei, C., Yao, R., Zhang, Z., Fang, X. and Chen, H. (2010) A new insight into cattle's maternal origin in six Asian countries. *Journal of Genetics and Genomics* 37, 173–180.

Kantanen, J., Edwards, C.J., Bradley, D.G., Viinalass, H., Thessler, S., Ivanova, Z., Kiselyova, T., Cinkulov, M., Popov, R., Stojanovic, S., Ammosov, I. and Vilkki, J. (2009) Maternal and paternal genealogy of Eurasian taurine cattle (*Bos taurus*). *Heredity* 103, 404–415.

Kim, K.I., Lee, J.H., Lee, S.S. and Yang, Y.H. (2003) Phylogenetic relationship of northeast Asian cattle to other cattle populations determined using mitochindrial DNA D-loop sequence polymorphism. *Biochemical Genetics* 41, 91–98.

Komatsu, M., Yasuda, Y., Matias, J.M., Niibayashi, T., Abe-Nishimura, A., Kojima, T., Oshimna, K., Tajeda, H., Hasegawa, K., Abe, S., Yamamoto, N. and Shiraishi, T. (2004) Mitochondrial DNA polymorphisms of D-loop and four coding regions (ND2, ND4, ND5) in three Philippine native cattle: indicus and taurus maternal lineages. *Animal Science Journal* 75, 363–378.

Kron, G. (2002) Archaeozoological evidence for the productivity of Roman livestock farming. *Münstersche Beiträge zur Antiken Handelsgeschichte* 21, 53–73.

Kron, H. (2004) Roman livestock farming in southern Italy: the case against environmental determinism. In: Clavel-Lévêque, M. and Hermon, E. (eds) *Espaces Integrés et Ressources Naturelles Dans l'Empire Romain*. Presses Universitaires de Franche-Comté, Besançon, France, pp. 119–134.

Kyselý, R.A. and Hájek, M.B. (2012) MtDNA haplotype identification of aurochs remains originating from the Czech Republic. *Environmental Archaeology* 17, 118–125.

Lai, S.J., Liu, Y.P., Liu, Y.X., Li, X.W. and Yao, Y.G. (2006) Genetic diversity and origin of Chinese cattle revealed by mtDNA D-loop sequence variation. *Molecular Phylogenetics and Evolution* 38, 146–154.

Lari, M., Rizzi, E., Mona, S., Corti, G., Catalano, G., Chen, K., Vernesi, C., Larson, G., Boscato, P., De Bellis, G., Cooper, A., Caramelli, D. and Bertorelle, G. (2011) The complete mitochondrial genome of an 11,450-year-old aurochsen (*Bos primigenius*) from Central Italy. *BMC Evolutionary Biology* 11, 32.

Lauwerier, R. (2011) Het dominante hoornloze rund [The dominant hornless cattle]. *Vitruvius* 16, 26–31.

Leake, W.L. (1826) *An Edict of Diocletian Fixing a Maximum of Prices Throughout the Roman Empire*. John Murray, London.

Lei, C.Z., Chen, H., Zhang, H.C., Cai, X., Liu, R.Y., Luo, L.Y., Wang, C.F., Zhang, W., Ge, Q.L., Zhang, R.F. *et al.* (2006) Origin and phylogeographical structure of Chinese cattle. *Animal Genetics* 37, 579–582.

Lenstra, J.A. and Bradley, D.G. (1999) Systematics and phylogeny of cattle. In: Fries, R. and Ruvinsky, A. (eds) *The Genetics of Cattle*. CAB International, Wallingford, pp. 1–14.

Lenstra, J.A., Ajmone-Marsan, P., Beja-Pereira, A., Bollongino, R., Bradley, D.G., Colli, L., De Gaetano, A., Edwards, C.J., Felius, M., Ferretti, L. *et al.* (2014) Meta-analysis of mitochondrial DNA reveals several population bottlenecks during worldwide migrations of cattle. *Diversity* 6, 178–187.

Li, M.H. and Kantanen, J. (2010) Genetic structure of Eurasian cattle (*Bos taurus*) based on microsatellites: clarification for their breed classification. *Animal Genetics* 41, 150–158.

Li, R., Zhang, X.M., Campana, M.G., Huang, J.P., Chang, Z.H., Qi, X.B., Shi, H., Su, B., Zhang, R.F., Lan, X.Y. *et al.* (2013) Paternal origins of Chinese cattle. *Animal Genetics* 44, 446–449.

Ludwig, A., Alderson, L., Fandrey, E., Lieckfeldt, D., Soederlund, T.K. and Froelich, K. (2013) Tracing the genetic roots of the indigenous White Park Cattle. *Animal Genetics* 44, 383–386.

MacKinnon, M. (2010) Cattle 'breed' variation and improvement in Roman Italy: connecting the zooarchaeological and ancient textual evidence. *World Archaeology*, 55–73.

Magee, D.A., Meghen, C., Harrison, S., Troy, C.S., Cymbron, T., Gaillard, C., Morrow, A., Maillard, J.C. and Bradley, D.G. (2002) A partial African ancestry for the Creole cattle population of the Carribean. *Journal of Heredity* 93, 429–432.

Mannen, H., Kohno, M., Nagata, Y., Tsuji, S., Bradley, D.G., Yeo, J.S., Nyamsamba, D., Zagdsuren, Y., Yokohama, M., Nomura, K. and Amano, T. (2004) Independent mitochondrial origin and historical genetic differentiation in North Eastern Asian cattle. *Molecular Phylogenetics and Evolution* 32, 539–544.

Martinez, A.M., Gama, L.T., Canon, J., Ginja, C., Delgado, J.V., Dunner, S., Landi, V., Martin-Burriel, I., Penedo, M.C., Rodellar, C. *et al.* (2012) Genetic footprints of Iberian cattle in America 500 years after the arrival of Columbus. *PLoS ONE* 7, e49066.

McInerney, J. (2010) *The Cattle of the Sun: Cows and Culture in the World of the Ancient Greeks*. Princeton University Press, Princeton, New Jersey.

McTavish, E.J.D., Decker, J.E., Schnabel, R.D., Taylor, J.F. and Hillis, D.M. (2013) New World cattle show ancestry from multiple independent domestication events. *Proceedings of the National Academy of Sciences of the United States of America*, E1398–E1406.

Medugorac, I., Seichter, D., Graf, A., Russ, I., Blum, H., Gopel, K.H., Rothammer, S., Forster, M. and Krebs, S. (2012) Bovine polledness – an autosomal dominant trait with allelic heterogeneity. *PLoS ONE* 7, e39477.

Meirelles, F.V., Rosa, A.J.M., Lôbo, R.B., Garcia, J.M., Smith, L.C. and Duarte, F.A.M. (1999) Is the American zebu really *Bos indicus*? *Genetics and Molecular Biology* 22, 543–546.

Miretti, M.M., Pereira, H.A., Jr, Poli, M.A., Contel, E.P. and Ferro, J.A. (2002) African-derived mitochondria in South American native cattle breeds (*Bos taurus*): evidence of a new taurine mitochondrial lineage. *Journal of Heredity* 93, 323–330.

Miretti, M.M., Dunner, S., Naves, M., Contel, E.P. and Ferro, J.A. (2004) Predominant African-derived mtDNA in Caribbean and Brazilian Creole cattle is also found in Spanish cattle (*Bos taurus*). *Journal of Heredity* 95, 450–453.

Mohamad, K., Olsson, M., van Tol, H.T., Mikko, S., Vlamings, B.H., Andersson, G., Rodriguez-Martinez, H., Purwantara, B., Paling, R.W., Colenbrander, B. and Lenstra, J.A. (2009) On the origin of Indonesian cattle. *PLoS ONE* 4, e5490.

Mona, S., Catalano, G., Lari, M., Larson, G., Boscato, P., Casoli, A., Sineo, L., Di Patti, C., Pecchioli, E., Caramelli, D. and Bertorelle, G. (2010) Population dynamic of the extinct European aurochs: genetic evidence of a north–south differentiation pattern and no evidence of post-glacial expansion. *BMC Evolutionary Biology* 10, 83.

Murray, C., Huerta-Sanchez, E., Casey, F. and Bradley, D.G. (2010) Cattle demographic history modelled from autosomal sequence variation. *Philosophical Transactions of the Royal Society of London, B: Biological Sciences* 365, 2531–2539.

Payne, W.J.A. and Hodges, J. (1997) *Tropical Cattle: Origin, Breeding and Breeding Policies*. Blackwell Science, Oxford.

Pellecchia, M., Negrini, R., Colli, L., Patrini, M., Milanesi, E., Achilli, A., Bertorelle, G., Cavalli-Sforza, L.L., Piazza, A., Torroni, A. and Ajmone-Marsan, P. (2007) The mystery of Etruscan origins: novel clues from *Bos taurus* mitochondrial DNA. *Proceedings of the Royal Society B: Biological Sciences* 274, 1175–1179.

Perez-Pardal, L., Royo, L.J., Beja-Pereira, A., Chen, S., Cantet, R.J., Traore, A., Curik, I., Solkner, J., Bozzi, R., Fernandez, I. *et al.* (2010a) Multiple paternal origins of domestic cattle revealed by Y-specific interspersed multilocus microsatellites. *Heredity* 105, 511–519.

Perez-Pardal, L., Royo, L.J., Beja-Pereira, A., Curik, I., Traore, A., Fernandez, I., Solkner, J., Alonso, J., Alvarez, I., Bozzi, R., Chen, S., Ponce de Leon, F.A. and Goyache, F. (2010b) Y-specific microsatellites reveal an African subfamily in taurine (*Bos taurus*) cattle. *Animal Genetics* 41, 232–241.

Petersen, L. (1970) The Danish cattle trade during the sixteenth and seventeenth centuries. *Scandinavian Economic History Review* 18, 69–85.

Pinhasi, R., Thomas, M.G., Hofreiter, M., Currat, M. and Burger, J. (2012) The genetic history of Europeans. *Trends in Genetics* 28, 496–505.

Price, T.D. (2000) *Europe's First Farmers: an Introduction.* Cambridge University Press, Cambridge.

Rasteiro, R. and Chikhi, L. (2013) Female and male perspectives on the Neolithic transition in Europe: clues from ancient and modern genetic data. *PLoS ONE* 8, e60944.

Rawlinson, G. (1985) *The History of Herodotus* Vol. 3, Book 4. Appleton and Co, New York.

Riedel, A. (1985) Ergebnisse der Untersuchung einiger Südtiroler Faune [Investigations of South Tyrolean Fauna]. *Preistoria Alpina – Museo Tridentino di Scienze Naturali* 21, 113–177.

Salque, M., Bogucki, P.I., Pyzel, J., Sobkowiak-Tabaka, I., Grygiel, R., Szmyt, M. and Evershed, R.P. (2013) Earliest evidence for cheese making in the sixth millennium BC in northern Europe. *Nature* 493, 522–525.

Schlumbaum, A., Stopp, B., Breuer, G., Rehazek, A., Blatter, R., Turgay, M. and Schibler, J. (2003) Combining archaeozoology and molecular genetics: the reason behind the changes in cattle size between 150BC and 700AD in northern Switzerland. *Antiquity* 77. Available at: http://antiquity.ac.uk/ Projgall/schlumbaum/index.html (accessed 1 September 2014).

Schlumbaum, A., Turgay, M. and Schibler, J. (2006) Near East mtDNA haplotype variants in Roman cattle from Augusta Raurica, Switzerland, and in the Swiss Evolene breed. *Animal Genetics* 37, 373–375.

Sørensen, M.K., Norberg, E., Pedersen, J. and Christensen, L.G. (2008) Invited review: crossbreeding in dairy cattle: a Danish perspective. *Journal of Dairy Science* 91, 4116–4128.

Spinage, C.A. (2003) *Cattle Plague: a History.* Kluwer Academic, New York.

Stock, F. and Gifford-Gonzalez, D. (2013) Genetics and African cattle domestication. *African Archaeological Review* 30, 51–72.

Stock, F., Edwards, C.J., Bollongino, R., Finlay, E.K., Burger, J. and Bradley, D.G. (2009) Cytochrome *b* sequences of ancient cattle and wild ox support phylogenetic complexity in the ancient and modern bovine populations. *Animal Genetics* 40, 694–700.

Strabo (1969) *The Geography of Strabo in Eight Volumes with an English Translation by H.L. Jones.* Henry G. Bohn, London.

Strouhal, E. (1992) *Life of the Ancient Egyptians.* University of Oklahoma Press, Norman, Oklahoma.

Svensson, E. and Gotherstrom, A. (2008) Temporal fluctuations in Y-chromosomal variation in Bos taurus. *Biology Letters* 4, 752–754.

Teasdale, M.D. and Bradley, D.G. (2012) The origins of cattle. In: Womack, J. (ed.) *Bovine Genomics.* Wiley, Oxford, pp. 1–10.

Tresset, A. (2003) French connections II: Of cows and men. In: Armit, I., Murphy, E., Nelis, E. and Simpson, D. (eds) *Neolithic Settlement in Ireland and Western Britain.* Oxbow Books, Oxford, pp. 18–30.

Troy, C.S., MacHugh, D.E., Bailey, J.F., Magee, D.A., Loftus, R.T., Cunningham, P., Chamberlain, A.T., Sykes, B.C. and Bradley, D.G. (2001) Genetic evidence for Near-Eastern origins of European cattle. *Nature* 410, 1088–1091.

Turton, J.D. (1991) Modern needs for different genetic types. In: Hickman, C.G. (ed.) *Cattle Genetic Resources.* Elsevier, Amsterdam, pp. 21–49.

Vigne, J.D. (2011) The origins of animal domestication and husbandry: a major change in the history of humanity and the biosphere. *Comptes Rendus Biologies* 334, 171–181.

Wiedemar, N., Tetens, J., Jagannathan, V., Menoud, A., Neuenschwander, S., Bruggmann, R., Thaller, G. and Drögemüller, C. (2014) Independent polled mutations leading to complex gene expression differences in cattle. *PLoS One* 9, e93435.

Zeder, M.A., Emshwiller, E., Smith, B.D. and Bradley, D.G. (2006) Documenting domestication: the intersection of genetics and archaeology. *Trends in Genetics* 22, 139–155.

Zeuner, F.E. (1963) *A History of Domesticated Animals.* Harper and Row, New York.

3 Breeds of Cattle

D.S. Buchanan[1] and J.A. Lenstra[2]

[1]*North Dakota State University, Fargo, North Dakota, USA;*
[2]*Utrecht University, Utrecht, The Netherlands*

Introduction

Since domestication began more than 8000 years ago, cattle have become adapted to widely varying geographical areas and a multiplicity of breeding purposes (meat, dairy, draught, hides, ceremonial, etc.) (Ajmone-Marsan *et al.*, 2010). Since the 18th century, this diversification has been reinforced by systematic breeding of separate subpopulations, which we have come to refer to as 'breeds'.

In its most basic form, a breed is anything that is bred. This has often resulted in groups of animals with similar physical characteristics, such as colour, horns, body type, performance, etc. In Europe, breeds are developed in a highly directed fashion by organizations that protect the purity of the breed and pursue its further improvement. These 'breed societies' originated in the UK during the early part of the 19th century (Willham, 1987) and spread to other countries, most notably in Europe and the USA.

In contrast, many (sub)tropical populations are still managed by the owners of the animals and differ gradually from neighbouring populations. However, gene flow between breeds is also normal for the developed breeds. Many breeding societies manage similar types of cattle and exchange breeding sires. Animals from most breeds have a relatively recent common ancestry yet several herdbooks have allowed entry of animals from exotic ancestry or even large-scale crossbreeding to exotic sires from a highly productive breeds. One step further is the emergence, mainly in the New World, of several synthetic breeds that combine traits from widely divergent parental populations.

More than 1000 different breeds have been described (Mason, 1969, 1996; Felius, 1985), but this counts many national derivatives of a breed imported from its native country. A restricted number of cosmopolitan breeds with high census numbers account for a large part of the dairy and beef production. On the other hand, many local breeds with low numbers are important either historically or as a source of unique genetic material. For breeds kept under extensive management this may include an adaptation to environmental conditions such as a local disease resistance. Breeds may therefore be conserved for economic, scientific or cultural reasons.

Bos taurus and Bos indicus

All cattle are contained within the genus *Bos* (Hassanin, Chapter 1). Most breeds can be assigned to the species *Bos taurus* or *Bos indicus* (Felius, 1985) or are of mixed taurindicine ancestry. This assignment to two species is common in scientific circles and among cattle producers. This is in line with the divergence of their mitochondrial (Achilli *et al.*, 2009) and Y chromosomal (Nijman *et al.*, 2008) DNA and with their separate domestication events (Lenstra and Felius, Chapter 2). However, because of their cross-fertility they should formally be described as subspecies of the extinct ancestor *Bos promigenius* (Hassanin, Chapter 1).

Bos taurus cattle evolved in Mesopotamia ~10,000 years ago and migrated into Europe and Africa between 5500 and 7000 years ago, adapting to both temperate and (sub)tropical climates. The tropically adapted *Bos indicus* cattle, commonly denoted as zebu or humped cattle, emerged on the Indian subcontinent ~8000 years ago and migrated ~2500 years later to West and East Asia and then to eastern Africa. Crossbreeding of indicine bulls and taurine cows in Africa resulted in indicine admixture in the taurindicine 'Sanga', the zebu–Sanga intermediate Zenga and the African zebu (Hanotte *et al.*, 2000; Ajmone-Marsan *et al.*, 2010). Import of Iberian cattle into the Americas started late in the 15th century and adaptation for several hundred years resulted in the American Criollo cattle. Import of zebu cattle into America started in the 19th century and several Criollo breeds are now also taurindicine (Ajmone-Marsan *et al.*, 2010; Ginja *et al.*, 2010; Delgado *et al.*, 2012; McTavish *et al.*, 2013). In the last century, numerous American and Australian taurindicine breeds have been developed, which combine characteristics of *Bos taurus* and *Bos indicus* cattle.

The distinction between taurine and indicine cattle has been supported by genome-wide analysis of genetic variation as revealed by almost 50,000 single nucleotide polymorphisms (SNPs) (Bovine Hapmap Consortium, 2009). The indicine Brahman, Gir and Nelore breeds formed a group that was distinct from all taurine cattle, while the African taurine N'Dama (African, taurus) was separate from the European taurine breeds. The taurindicine

Beefmaster and Santa Gertrudis were intermediate between the taurine and indicine breeds. These results also indicated that many of these breeds are experiencing rapid decline in effective population size since their descent from large ancestral populations.

Categorization According to Utility and Mode of Origin

In addition to the species categorization, cattle breeds may be divided by utility. The introduction of tractors has ended the use of cattle as draught animals in most production systems, while selection for fighting abilities is restricted to the Iberian fighting bulls and the Swiss Valais or Italian Valdostana fighting cows. Systematic selection facilitated by artificial insemination has now caused many European and North American breeds to excel in either meat or milk production. Dairy breeds such as the Holstein produce much more milk than can be consumed by a calf and have become well adapted to being milked twice, or even thrice, daily. Other breeds give only enough milk to sustain a calf but have highly developed muscularity, possibly originating from their former use as a draught animal, and are now important for meat production. The different breeding histories of dairy and beef cattle were confirmed by genome-wide survey of SNP variation suggesting different dairy and beef selection signatures (Hayes *et al.*, 2008; Utsunomiya *et al.*, 2013).

Another categorization of breeds considers the mode of origin, interaction with other breeds and international status (FAO, 2007). Felius *et al.* (2014) have refined the categorization of the FAO, differentiating: (i) authentic local breeds from the 18th century or earlier; (ii) local breeds derived from 19th- or 20th-century imports; (iii) highly productive cosmopolitan breeds; and (iv) breeds maintained by crossbreeding.

Classification and Phylogeny of Breeds

A final categorization of cattle breeds is based on a comprehensive classification. Since the 19th century, several classifications have been proposed, focusing on European breeds and based on cranial and horn morphology, colour pattern, (supposed) history, geographical origin and molecular analysis (Alderson, 1992; Felius et al., 2011). Table 3.1 shows the integrative classification emphasizing geography and, within regions, history and morphology. Worldwide, 16 major groups are recognized comprising taurine, taurindicine and indicine breeds or cattle derived from other *Bos* species. Further subdivision of the major groups results in a comprehensive account of the global diversity of cattle (Felius 1995; Felius et al., 2011).

This classification is largely recapitulated by a molecular-genetic classification of European taurine cattle (Table 3.2). This is based on genetic distances and model-based clustering

of microsatellite genotyping (Laloe et al., 2011; Edwards et al., 2011; Felius et al., 2011), and is in good agreement with a classification based on protein polymorphisms (Baker and Manwell, 1980) and a clustering based on 50K SNP typing (Decker et al., 2009). Both classifications demonstrate for the European breeds a major subdivision in northern (integrative groups 1 and 2; genetic cluster I), central (3 and 4; II), Iberian, Podolian and Balkan cattle. Northern European cattle also differ from other European breeds owing to the predominance of the Y chromosomal haplogroup Y1 (Edwards et al., 2011).

Phylogenetic trees of breeds can be constructed as done for species, but with two major differences (Lenstra et al., 2012). First, differences between breeds are not based on

Table 3.1. Integrative classification of cattle breeds (Felius *et al.*, 2011).

Number	Group	Subgroups
1	Northern Polled, Celtic	Nordic Polled, Longhorned Dairy, British Polled, Celtic
2	North-Western Lowland	Lowland Red, Lowland Pied Dairy, Lowland Pied Dual-Purpose, British Shorthorn, English Lowland, Channel Island and Northwest French
3	Western-Central Highland	Vosges and Black Forest, Highland Red, Shortheaded Alpine, Central European, Yellow and Blonde, Broadheaded Spotted, Charolais
4	Highland solid-coloured	Middle French, Southwest French, Pyrenean Grey and Blonde, North Italian Fawn-Brown, Central European Brown and Grey, Illyric Shorthorn
5	Iberian	West Mediterranean isolates, Northwest Iberian and Balearic Blonde-Brown, Northwest Iberian Chestnut, Central and Southwest Iberian Black, Central and Southwestern Iberian Red and Southeastern Iberian
6	Podolian	Italian White, Italian and Croatian Podolian, East European, Balkan, Anatolian
7	Southwest Asian and Egyptian Shorthorn	Caucasian, Anatolian, Levant Shorthorns, Damascus type
8	Indo-Pakistani zebu	Central West Asian, Convex foreheaded, Shorthorned, Longhorned, Mysore, small Deshi, Himalayan
9	Central and Northeast Asian	Turuano-Mongolian, Northeast Asian, yak and yak hybrids
10	Central and South China, Southeast Asian	Chinese yellow, Chinese and Southwest Asian zebu, banteng, gayal and their hybrids
11	North and West African taurine	North African Shorthorn, Lake Chad Longhorn, N'Dama, West African Dwarf Shorthorn
12	West African zebu	Sahel Shorthorn, Sahel Longhorn
13	East African zebu	Northeast African, East African Shorthorn, Madagascar
14	Sanga and zenga	Northeast African, Central African Ankole, Southern African
15	Iberian-American	Texas Longhorn, Gulf Coast, Criollo
16	Modern American, New Zealand, Australian; bison	Original imports, taurine (British, continental, Japanese), indicine, taurindicine, crossbreds, dairy, beef, dual-purpose

Table 3.2. Molecular-genetic classification. (From Edwards *et al.*, 2011; Felius *et al.*, 2011, Table S35.)

Number	Cluster	Description
I	**North European breeds**	
I.1	British	British dairy and beef breeds, including also the Channel Island breed (Jersey, Guernsey), but not the Shorthorn. Jersey tends to be different from the other breeds in this group and may have apparent affinity to Podolian or Alpine grey breeds
I.2	Nordic	Authentic Norwegian, Swedish, Finnish and Baltic breeds, including both polled as long-horned (Døle, Telemark) breeds
I.3	Nordic Ayrshire	Imported Finnish Ayrshire and Ayrshire crossbreds: Norwegian Red, Swedish Red-and-White, Ringamåla, Väne
I.4	Lowland-Pied	Black- and red-pied dairy cattle originating from the northwestern European lowlands. Also includes the solid Red Flemish
I.5	Baltic Red	Solid red dairy cattle from the Baltic coasts and the German Highland. Also includes the Russian Suksun, Byelorussian Red and Ukrainian Red Steppe
I.6	Northwest Intermediate	Cattle from northwestern Europe that are not closely related to each other, but are influenced to different degrees by surrounding breeds: Shorthorn, Maine-Anjou (similar to Shorthorn), Bretonne-Pie Noir, Normande, Parthenaise (close to southern French breeds), Vosges, Charolais. Charolais clustered with south French breeds and Vosges with central western breeds in a 50K SNP analysis (Decker *et al.*, 2009; Gautier *et al.*, 2010)
1.7	Eastern crossbred	Russian breeds heavily influenced by western breeds: Istoben (influenced by Lowland Black- and Red-Pied), Kazakh Whiteheaded (influenced by Hereford), Ukranian Whitehead (influenced by Groningen Whiteheaded), Bestuzhev (influenced by several breeds)
I.8	Eastern	Russian and Siberian breeds: Kholmogory, Pechora (both influenced by I.4), Kalmyk, Yaroslav, Yakut
II	**Central European cattle**	
II.1	Central Western	Includes four subtypes (II.1.1–4) and Hinterwald; SNP data suggest inclusion of Charolais and Vosges
II.1.1	Central Spotted	Central European spotted dairy cattle with Simmentaler as prototype breed from which several other breeds have been derived
II.1.2	Central Blonde	Carinthian and Waldviertel Blonde, genetically close to the Central Spotted
II.1.3	West Alpine	French Alpine, Swiss Valais (Wallis) and Italian Valdostana breeds, AustrianTux-Zillertaler
II.1.4	Central Yellow	German Yellow breeds, Murbodner, Portuguese Minhota
II.2	South French	Southern French beef breeds and the Spanish Pirenaica, which is also influenced by Iberian cattle
II.3	Central Brown	Brown Swiss dairy cattle and derived breeds in Germany, Italy and Spain; including Murnau-Werdenfelder
II.3.1	Spanish Brown	Spanish breeds derived from Central Brown: Bruna dels Pirineus, Parda Montana and Serrana de Teruel

Continued

Table 3.2. Continued.

Number	Cluster	Description
II.4	Central Grey	Tyrolean Grey, Grigia Alpina
II.5	Central Eastern	Pinzgauer, Pustertaler, Cika
III	**Iberian cattle**	Authentic and morphologically diverse Spanish and Portuguese breeds. Relationships with the Mallorquina and Menorquina are unclear owing to the high degree of inbreeding of both Balearic breeds. The feral Betizu is genetically between the Iberian and southern French cattle. Contains regional clusters of breeds:
III.1.1	Cantabrian	Tudanca, Monchina, Betizu
III.1.2	Andalusian	Andalusian breeds: Berrenda, Cardena, Marismeña, Mostrenca, Pajuna, Fighting cattle (Lidia, Brava)
III.1.3	Iberian Black	Avileña, Morucha, Negra Andaluza, Preta
III.1.4	Morenas	Alistana, Barrosa, Cachena, Frieiresa, Caldelana, Limiana, Marinhoa, Maronesa, Mirandesa, Vianesa
III.1.5	South Portuguese Red	Alentejana, Garvonesa, Mertolenga
IV	**Podolian cattle**	Steppe cattle, presumed to originate from the Podolia region. Contains also Ukranian Grey, Turkish Grey and Chianina
V	**Balkan and Southwest Asian taurine cattle**	Authentic taurine cattle smaller and less developed than most European breeds; Busha, Anatolian and Caucasian cattle

substitutions, but on differences in frequencies of alleles, most of which are shared by the breeds. Second, unlike well diverged species, breeds may keep interacting long after their divergence, which invalidates the hierarchical, tree-like phylogenies by introducing reticulations. Nevertheless, neighbour-joining trees are still popular as convenient, if incomplete visualizations of the breed phylogeny. Reticulations may be visualized by phylogenetic networks as in the NeighborNet graphs (see Fig 3.1).

After an early tree based on protein polymorphisms (Baker and Manwell, 1980), phylogenetic trees and networks of microsatellite genetic distances have been published for representative Asian, African and European (Cymbron et al., 2005), British (Wiener et al., 2004), Danish (Withen et al., 2011), North Eurasian (Kantanen et al., 2000; Li and Kantanen, 2010), Polish (Grzybowski and Prusak, 2004), French (Moazami-Goudarzi et al.,1997; Maudet et al., 2003), Alpine (Del Bo et al., 2001), Iberian (Martín-Burriel et al., 2007, 2011), Italian (D'Andrea et al., 2011), Slovenian (Simcic et al., 2013) Balkan (Medugorac et al., 2009), Southwest Asian (Loftus et al., 1999), Indian

(Shah et al., 2012), Chinese (Zhang et al., 2007), African (Moazami-Goudarzi et al., 2001; Freeman et al., 2004; Ibeagha-Awemu et al., 2004; Zerabruk, 2012), Brazilian (Egito et al., 2007), Cuban (Acosta et al., 2013) and American (Delgado et al., 2012; Martínez et al., 2012) breeds. Combining several of these data with a Europe-wide database (Lenstra et al., 2008) yielded a network covering most European breeds (Felius et al., 2011).

Higher phylogenetic resolution has been achieved with a genome-wide analysis using 1536 SNPs for French and African breeds (Gautier et al., 2007) or 50 K SNPs for a global (Decker et al., 2009) and French (Gautier et al., 2010) set of breeds.

These phylogenies show close relationships of recently diverged breeds as, for example, black-pied and Baltic red breeds (Withen et al., 2011. However, recent genetic drift is reflected in long terminal branches (e.g. with the island breed Mallorquina and the Betizu A subpopulation, Martín-Burriel et al., 2007) and obscures deeper phylogenetic relationships of the clusters of related breeds. These may be approached by pooling breeds from the same cluster as

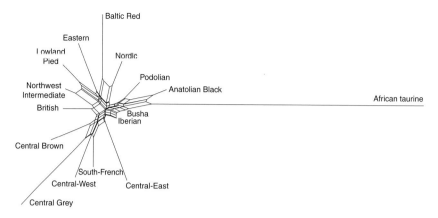

Fig. 3.1. NeighborNet graph illustrating microsatellite phylogenetic network of taurine cattle breeds.

reconstructions of the ancestral populations (Lenstra, 2008). Figure 3.1 shows a phylogeny of Eurasian breed clusters, visualizing a genetic cline from eastern (Anatolian) to Balkan, Mediterranean and Nordic cattle, and then to the more developed breeds from central and northwestern Europe. It also shows the major subdivision of European breeds (see above).

Description of Cattle Breeds

Tables 3.3 to 3.9 give, for a selection of the cattle breeds worldwide, species origin (taurine, indicine, etc.), integrative and, for European breeds, genetic classification, category of origin, main and morphological traits (Rouse, 1970a, 1970b; Briggs and Briggs, 1980; Felius, 1985, 1995; Walker, 1989; Committee on Managing Global Genetic Resources, 1993; see also http://dad.fao.org, www.ansi.okstate.edu/breeds/cattle and http://en.wikipedia.org/wiki/List_of_ cattle_breeds).

Where information is available, breeds are also described for the seven traits that have been used since the late 1960s for the Germ Plasm Evaluation (GPE) experiment: size, age at puberty, marbling, tenderness, lean to fat ratio, milk production and tropical adaptation (Cundiff et al., 1986, 1993, 1997; Cundiff, 2003a, 2003b). Descriptors for breeds other than those in the GPE are subjective and reflect a performance that is dependent upon the environment in which the breeds are used instead of what would be achieved in a uniform environment for all breeds.

Table 3.10 provides references to research information about the most common breeds, listing for each of the papers the breeds that have been evaluated.

Plates 2–25 contain colour images of several breeds representing various breed categories.

Perspectives

The concept of a breed is likely to remain rather fluid. Several beef cattle breeds developed in North America during the 20th century and this may be indicative of a general effort to identify combinations of germplasm for use in the varied environments in which cattle are raised. These developments are, apparently, continuing. It is tempting to assume that the important breeds of today will continue to be important in the future. One has only to examine the history of breeds during the 20th century, in cattle and in other species of livestock to see the fallacy of this assumption. The evolutionary pace in both beef and dairy cattle may even speed up due to improved techniques for identification of superior genetic material such as genomic selection. This may very well create new breeds by recombining gene variants from different genetic stocks.

Acknowledgement

We thank Marleen Felius for her expert advice.

Table 3.3. Breeds from Asia.

Breed	Classification[b]	Category of origin[c]	Species[d]	Country of origin	Global range	Colour	Size and growth	Milk production	Other traits, remarks
Bengali	8	1	I	Bangladesh and Bengal, India	Asia	White, blonde, brown	Small		Tropical triple-purpose, tolerates poor food
Chinese Yellow	10	1	T, T/I, I	China	Asia	Yellow	Small		Comprising many local varieties
Dhanni	8	1	I	Pakistan	Asia	White, black spotted	Moderate		Tropical, triple-purpose
Gir	8	1, 2	I	Gujarat, India	Asia, South America	Red, red with white spots	Moderate	High	Large population in Brazil, tropical dairy
Guzerat	8	1, 2	I	India	Asia, South America	Steel-grey, black markings	Moderate-large	Moderate	Kankrej in India, tropical
Hissar	8	1	I	India	Asia	White	Moderate-large		
Krishna Valley	8	1	I	India	Asia	White	Moderate-large		Tropical
Mongolian	9	1	T	Mongolia	Asia	Variable	Small		Adapted to extreme conditions
Nelore	8	2	I	India	South America, Australia	White, black markings	Moderate-large[a]	Moderate[a]	Derived from Ongole, tropical, low tenderness[a], late puberty[a], low marbling[a]
Ongole	8	1	I	India	Asia	White, black markings	Moderate-large	Moderate	Tropical, work-dairy
Red Sindhi	8	1, 2	I	Pakistan	Asia, Africa, Australia	Red	Small	Moderate-high	Tropical, triple purpose
Sahiwal	8	1, 2	I	India	Asia, Africa, America, Australia	Red	Small- moderate[a]	Moderate-high[a]	Tropical, low tenderness[a], late puberty[a], low marbling[a]
Tharparkar	8	1	I	India	Asia	Grey to white	Moderate-large		Tropical, draught

Continued

Table 3.3. Continued.

Breed	Classification[b]	Category of origin[c]	Species[d]	Country of origin	Global range	Colour	Size and growth	Milk production	Other traits, remarks
Wagyu	9	1, 2	T	Japan	Asia, North America Europe	Both black and red strains	Small[a]	Moderate[a]	Very high marbling[a], high tenderness[a], early puberty[a], low lean-to-fat[a]
Xinjiang Brown	9	1	T	China	Asia	Variable	Small-moderate	Low-moderate	Derived from Hazake × Alpine Brown Mountain
Yakut	9, I.8	1	T	Sakha Republic, Siberia	Asia	White, red or black spots	Small		Uniquely adapted to climate above polar circle

[a]For more details, see Cundiff, 2003. Breed names in bold refer to pictures in Plates 2–25.
[b]codes of integrative (Arabic numbers) and genetic (Roman followed by Arabic numbers) classifications, respectively, as in Tables 3.1 and 3.2
[c]1, authentic local; 2, imported local.
[d]Species: I, *Bos indicus*; T, *Bos taurus*; T/I, taurindicine.

Table 3.4. Breeds from Africa.

Breed	Classification[b]	Category of origin[c]	Species[d]	Country of origin	Global range	Colour	Size and growth	Milk production	Other traits, remarks
Abyssinian Shorthorned Zebu	13	1	I	Ethiopia	Africa	Variable	Small		Comprises several landraces adapted to the range of cool highlands to hot lowlands
Adamawa	12	1	I	Nigeria	Africa	Variable	Moderate		Bamenda in Nigeria, N'Gaoundéré in Cameroon, tropical, hanging hump
Africander	14	1, 2	S	South Africa	Africa, Australia	Red	Moderate		Tropical, moderate lean-to-fat, moderate to late age at puberty
Watusi	16	1, 2	I	DR Congo, Uganda, Rwanda, Burundi, Tanzania	Africa, North America	Dark brown, white spots	Moderate		Tropical, very large white horns, high fat milk, moderate-high lean-to-fat, tropical, in the US Texas Longhorn introgression
Arsi	13	1	I	Central Ethiopia	Africa	Variable	Small		Highland work-beef
Bonsmara	14	1, 2	T/I	South Africa	Africa, Australia, South America	Red	Moderate[a]	Moderate[a]	Afrikander–Hereford–Shorthorn hybrid, moderate lean-to-fat[a], moderate marbling[a]
Boran	13	1, 2	I	Ethiopia	Africa, South America	White, grey, fawn, red	Moderate[a]	Moderate[a]	Tropical, low tenderness[a], low marbling[a], late puberty[a], moderate lean-to-fat[a]
Brown Atlas	11	1	T	Algeria, Morocco	Africa	Brown	Small		Comprises several landraces
Butana	13	1	I	Sudan	Africa	Dark red	Moderate-large	Moderate-high	Tropical
Danakil	14	1	S	Ethiopia	Africa	Variable	Moderate-large	Moderate	Afar Sanga in Ethiopia, tropical
Dinka (Nilotic)	14	1	S	Sudan	Africa	White, also pied	Small-moderate		Tropical, very long horns
Fogera	14	1	S/I	Ethiopia	Asia	Variable	Moderate		Tropical

Continued

Table 3.4. Continued.

Breed	Classification[b]	Category of origin[c]	Species[d]	Country of origin	Global range	Colour	Size and growth	Milk production	Other traits, remarks
Kenana	13	1	I	Sudan	Africa	White	Small-moderate	Moderate	Tropical
Keteku	11	1	T/I	Nigeria	Africa	White, also black spotted	Small		Zebu–Shorthorn hybrid, tropical
Menufi	11	1	T	Egypt	Africa	Red	Small		Baladi variety, tropical
Muturu	11	1	T	Nigeria	Africa	Black and white spotted	Dwarf		Tropical, trypanotolerant
N'Dama	11	1	T	Guinea, West Africa	Africa	Fawn	Small		Tropical, trypanotolerant
Nguni	14	1	S	South Africa	Africa	Variable	Small-moderate		Amalgate of local varieties
Tuli	14	1,2	S	Zimbabwe	Africa, Australia	Yellow	Moderate[a]	Moderate[a]	Moderate marbling[a], low tenderness[a], moderate age at puberty[a]
White Fulani	12	1	I	Nigeria, Niger	Africa	White, also black spotted	Moderate		Tropical, triple purpose

[a]For more details, see Cundiff, 2003. Breed names in bold refer to pictures in Plates 2–25.
[b]Codes of integrative classification as in Table 3.1.
[c]1, authentic local.
[d]Species: I, *Bos indicus*; S, Sanga; S/I, Zenga; T/I, taurindicine.

Table 3.5. Breeds from continental Europe.

Breed	Classification[b]	Category of origin[c]	Species[d]	Country of origin	Global range	Colour	Size and growth	Milk production	Other traits, remarks
Alentejana	5, III.1.5	1	T	Portugal	Europe	Red	Moderate-large		
Aubrac	4, II.2	1	T	France	Europe	Brown, white muzzle band	Moderate		
Belgian Blue	2, I.6	1, 3	T	Belgium	North America, Europe	White, blue roan, black pied	Moderate-large[a]	Moderate[a]	High frequency double-muscled, high lean-to-fat[a], very low marbling[a], moderate tenderness[a]
Blonde d'Aquitaine	4, II.2	1, 3	T	France	North America, Europe	Yellow to red	Large	Low-moderate	Recent local amalgate
Braunvieh	4, II.3	1, 3	T	Switzerland	America, Europe, Africa	Grey-brown, white muzzle band	Moderate-large[a]	Moderate-high[a]	Early-moderate puberty[a], moderate-high lean-to-fat[a], moderate marbling[a]
Brown Swiss	4, II.3	3	T	Switzerland	America, Europe, Asia, Africa	Light brown, white muzzle band	Large	High	American derivate of Braunvieh
Charolais	3, I.6	1, 3	T	France	America, Europe	Creamy white	Large[a]	Low[a]	High lean-to-fat[a], late puberty[a], low marbling[a], low tenderness[a]
Chianina	6, IV	1, 3	T	Italy	North America, Europe	White, black points	Very large[a]	Low[a]	Podolian, high lean-to-fat[a], late puberty[a], low marbling[a], low tenderness[a], late puberty[a]
Danish Red	2, I.5	1	T	Denmark	Europe	Red	Moderate	Moderate to high	Brown Swiss, Red Holstein introgression
Danish Red and White	2, I.4	2	T	Denmark	Europe	Red and white	Moderate	Moderate	Shorthorn–Red-Pied composite

Continued

Table 3.5. Continued.

Breed	Classification[b]	Category of origin[c]	Species[d]	Country of origin	Global range	Colour	Size and growth	Milk production	Other traits, remarks
Dutch Belted	2, 1.4	1	T	Netherlands	Europe	Black and white belted	Moderate	Moderate	Hobby breed, Galloway introgression
Fighting Bull (Toro de Lidia)	5, III.1.2	1, 2	T	Spain	Europe, Latin American countries	Variable	Small-moderate		Bred for bullfighting several inbred lines
Flamande	2, I.6	1	T	France	Europe	Dark brown-black	Moderate-large	Moderate	Related to Simmental, dual-purpose or beef
Fleckvieh	3, II.1.1	2, 3	T	Germany	America, Europe	Red and white, white head	Moderate-large	Moderate	Amalgate, low marbling[a], low tenderness[a], early puberty[a]
Gelbvieh	3, II.1.2	1, 3	T	Bavaria, Germany	North America, Europe, Australia	Blond to red	Moderate-large[a]	Moderate-high[a]	Derived from Dutch Black-Pied, separate beef Friesian strain, moderate marbling[a], early puberty[a]
Holstein	2, 1.4	3	T	Netherlands	Global	Black and white, red-and-white	Large[a]	Very high[a]	
Hungarian Grey	6, IV	1	T	Hungary	Europe	Grey	Large	Low	Isolated since Viking import
Icelandic	1, I.2	1	T	Iceland	Iceland	Variable	Small-moderate	Moderate	High lean-to-fat, low marbling[a], late puberty[a], low tenderness[a]
Limousin	4, II.2	1, 3	T	France	America, Europe	Red	Moderate[a]	Low[a]	Influenced by British Shorthorn, high lean-to-fat[a], low marbling[a], low tenderness[a], moderate puberty age[a]
Maine-Anjou	4, I.6	2, 3	T	France	North America, Europe	Red and white	Large[a]	Moderate[a]	

Marchigiana	6, IV	1, 3	T	Marche, Italy	North America, Europe	White, black eye markings	Moderate-large		Podolian, moderate-high lean-to-fat, early-moderate age-puberty
Meuse-Rhine-Yssel	2, 1.4	1, 2	T	Netherlands	Europe	Red and white	Moderate-large		Ancestral to other lowland red-pied breeds
Montbéliarde	4, II.1.1	1	T	France	Europe	Red and white, white head	Moderate-large	Moderate-high	Related to Simmental
Normande	4, I.6	1, 3	T	Manche & Calvados, France	America, Europe	Red-brown spotted, pied, brindled, white face	Moderate-large	Moderate-high	Moderate-high lean-to-fat
Norwegian Red	1, I.3	2	T	Norway	Europe	Red-and-white	Moderate[a]	Moderate-high[a]	Ayrshire influence, moderate marbling[a], early puberty[a]
Piedmontese	6	1, 3	T	Italy	North America, Europe	Grey-white, black markings	Low[a]	Low-moderate[a]	High frequency double-muscled, very high lean-to-fat[a], low marbling[a], moderate tenderness[a], early puberty[a]
Pinzgauer	3, II.5	1, 3	T	Austria	North America, Europe, Africa	Red, defined white markings	Moderate[a]	Moderate[a]	Moderate lean-to-fat[a], moderate marbling[a], early age puberty[a]
Polish Red	2, I.5	2	T	Poland	Europe	Red	Small-moderate		Influenced by Danish Red
Romagnola	6, IV	1, 3	T	Italy	North America, Europe	Grey with black shades	Large		Podolian, high lean-to-fat, moderate age-puberty
Rotvieh (German Red)	2 or 4, I.5	1, 2	T	Germany	Europe	Red	Moderate		Influenced by Danish Red
Salers	4, II.2	1, 3	T	France	North America, Europe	Red	Moderate-large[a]	Moderate[a]	Moderate lean-to-fat[a], low marbling[a], moderate puberty age[a]

Continued

Table 3.5. Continued.

Breed	Classification[b]	Category of origin[c]	Species[d]	Country of origin	Global range	Colour	Size and growth	Milk production	Other traits, remarks
Simmental	3, II.1.1	1, 3	T	Switzerland	America, Europe, Asia	Red-and-white, white face, in America mostly black	Large[a]	Moderate-high[a]	Low marbling[a], moderate puberty age[a], moderate-high lean-to-fat[a], in America influenced by Angus
Swedish Red and White	1, I.3	2	T	Sweden	Europe	Red and white	Moderate-high[a]	Moderate-high[a]	Moderate-high lean-to-fat[a], moderate marbling[a], early puberty[a]
Swedish Red Polled	1, I.2	1	T	Sweden	Europe	Red	Small		
Tarentaise	4, II.1.3	1, 3	T	France	North America, Europe	Red, white muzzle band	Moderate[a]	Moderate[a]	Moderate lean-to-fat[a], low marbling[a], early puberty[a]

[a]For more details, see Cundiff, 2003. Breed names in bold refer to pictures in Plates 2–25.
[b]Codes of integrative (Arabic numbers) and genetic (Roman followed by Arabic numbers) classifications, respectively, as in Tables 3.1 and 3.2.
[c]1, authentic local; 2, imported local; 3, cosmopolitan.
[d]Species:T, *Bos taurus*.

Table 3.6. Breeds from Great Britain (and nearby islands).

Breed	Classification[b]	Category of origin[c]	Species[d]	Country of origin	Global range	Colour	Size and growth	Milk production	Other traits, remarks
Angus	2, I.1	1, 3	T	Scotland	America, Europe, Australia	Black, also red strain	Moderate-high[a]	Moderate[a]	Polled, high marbling[a], high tenderness[a], low-moderate lean-to-fat[a], early puberty[a]
Ayrshire	1, I.1	1, 3	T	Scotland	North America, Europe, Africa, Australia	Red-and-white	Moderate	Moderate-high	Exported to Finland, ancestral to several Scandinavian breeds
Belted Galloway	1, I.1	1, 3	T	Scotland	North America, Europe	Black with white belt	Small-moderate		Polled, curly hair
British White	1, I.1	1	T	England	Europe	White, black points	Moderate		Polled
Devon	2, I.1	1	T	England	North America, Europe	Red	Small-moderate[a]	Moderate[a]	Moderate marbling[a], moderate tenderness[a], moderate puberty age[a], moderate lean-to-fat[a]
Dexter	1, I.1	1, 3	T	Ireland	North America, Europe	Black, also dun or red	Dwarf	Moderate	Ancestral to American miniature breeds, milk high in butterfat, small size by heterozygozity of chondrodysplasia mutation, bulldog calves by homozygosity
Galloway	2, I.1	1, 3	T	Scotland	North America, Europe	Black, also dun	Small-moderate[a]	Low-moderate[a]	Long, curly hair, polled, moderate tenderness[a], moderate marbling[a], moderate puberty age[a]
Guernsey	2, I.1	1, 3	T	Guernsey	North America, Europe	Fawn and white	Moderate	Moderate	Moderately high in butterfat, genetically largely isolated since 1789

Continued

Table 3.6. Continued.

Breed	Classification[b]	Category of origin[c]	Species[d]	Country of origin	Global range	Colour	Size and growth	Milk production	Other traits, remarks
Hereford	2, I.1	1, 3	T	England	North America, Europe, Australia	Red, white face and markings	Moderate-high[a]	Low-moderate[a]	Moderate marbling[a], moderate tenderness[a], moderate lean to fat[a]
Jersey	1, I.1	1, 3	T	Jersey	Global	Fawn, also pied	Small[a]	High[a]	High in butterfat, low lean-to-fat, high marbling[a], moderate tenderness[a], early puberty[a], genetically largely isolated since 1789, used for dairy crossbreeding
Lincoln Red	2	1	T	England	Europe	Red	Large		Related to Shorthorn
Longhorn (English)	1, I.1	1	T	England	Europe	Red or grey colour-sided, also brindled or speckled	Moderate		Long downward horns
Red Angus	2, I.1	1, 3	T	Scotland	America, Europe, Australia	Red	Moderate-high[a]	Moderate[a]	Polled, moderate marbling[a], moderate tenderness[a], early puberty[a], low-moderate lean-to-fat[a]
Red Poll	1	1, 3	T	England	North America, Europe, Africa, Australia	Red	Moderate	Moderate-high	Polled, moderate tenderness, moderate-high marbling, low-moderate lean-to-fat, early puberty
Scottish Highland	1, I.1	1, 3	T	Scotland	North America, Europe	Brown, black or red	Small		Long hair, long horns, rustic, cold adaptation

Breed	[b]	[c]	Species[d]	Origin	Distribution	Colour			Traits
Shorthorn	2, I.6	1, 3	T	England	America, Europe, Australia	Dark red, red-and-white or roan	Moderate-high[a]	Moderate[a]	Low-moderate lean-to-fat[a], moderate-high marbling[a], moderate tenderness[a], early-moderate age puberty[a]
South Devon	2, I.1	1, 3	T	England	North America, Europe	Light-red	Moderate[a]	Moderate[a]	Moderate lean-to-fat[a], moderate-high marbling[a], moderate tenderness[a], early-moderate age puberty[a]
Sussex	2	1, 2	T	England	Europe, Africa	Blood-red	Small		
Welsh Black	1	1	T	Wales	Europe	Black	Moderate		
White Park	1, I.1	1, 3	T	England	Europe, North America	White with black points	Moderate		

[a]For more details, see Cundiff, 2003. Breed names in bold refer to pictures in Plates 2–25.
[b]codes of integrative (Arabic numbers) and genetic (Roman followed by Arabic numbers) classifications, respectively, as in Tables 3.1 and 3.2.
[c]1, authentic local; 2, imported local; 3, cosmopolitan.
[d]Species:T, *Bos taurus*.

Table 3.7. Breeds from North America.

Breed	Classification[b]	Category of origin[c]	Species[d]	Country of origin	Global range	Colour	Size and growth	Milk production	Other traits, remarks
American White Park	16	2	T	United States	North America	White with red points	Moderate	Moderate	Introgressed with beef and dairy cattle
Amerifax	16	4	T	United States	North America	Red or black	Moderate-large	Moderate-high	Polled, Angus–Beef Friesian hybrid
Barzona	16	2	T/I	United States	North America	Dark red	Moderate		Africander–Hereford–Angus–Santa Gertrudis hybrid
Beefmaster	16	2	T/I	United States	North America, Africa	Red and other colours	Moderate-high[a]	Moderate[a]	Moderate lean-to-fat[a], low marbling[a], low tenderness[a], moderate puberty age[a]
Braford	16	2	T/I	United States	North America, Australia	Red, white face and markings	Moderate	Moderate	Brahman–Hereford hybrid, tropical
Brah-Maine	16	2	T/I	United States	North America	Red and white markings	Moderate-large	Moderate	Brahman–Maine Anjou hybrid, tropical
Brahman	16	2	T/I	United States	America, Africa, Australia	Grey strains, red strains	Moderate-large[a]	Moderate-large[a]	Blending of Gir, Guzerat and Nelore, low tenderness[a], late puberty[a], moderate-high lean-to-fat[a]
Brahmousin	16	2	T/I	United States	North America	Red	Moderate	Moderate	Limousin–Brahman hybrid, tropical
Bralers	16	2	T/I	United States	North America	Red	Moderate	Moderate	Brahman–Salers hybrid, tropical
Brangus	16	2	T/I	United States	North America, Africa	Black	Moderate-large[a]	Moderate[a]	Angus–Brahman hybrid, moderate lean-to-fat[a], moderate marbling[a], low tenderness[a], moderate puberty age[a], tropical

Canadienne	16	1	T	Canada	North America, Europe	Red	Small		16/17th century Breton/Normandy derivative+ since 1990 in France
Charbray	16	2	T/I	United States	North America, Australia	White-tan	Moderate-large	Moderate	Charolais–Brahman hybrid, tropical
Chiangus	16	4	T	United States	North America	Black	Moderate-large		Former Ankina, polled
Corriente	15	1	T	Mexico	North America	Variable	Small		Criollo of northern Mexico
Florida Cracker	15	1	T	United States	North America	Variable	Small		Criollo, adapted to humid tropics
Gelbray	16	2	T/I	United States	North America	Red	Moderate-large	Moderate-high	Gelbvieh–Brahman crossbred, tropical
Hays Converter	16	2	T	Canada	North America	Black or red with white face and markings	Moderate-large	Moderate-high	Holstein–Hereford–Brown Swiss crossbred
Red Brangus	16	2	T/I	United States	North America	Red	Moderate	Moderate	Red Angus–Brahman crossbred, tropical
RX3	16	4	T	United States	North America	Red	Moderate	Moderate-high	Hereford–Holstein–Red Angus hybrid
Salorn	15	2	T	United States	North America	Red	Moderate		Salers–Texas Longhorn crossbred
Santa Cruz	16	4	T/I	United States	North America	Red	Moderate		Santa, Gertrudis, Red Angus, Gelbvieh hybrid crossbred
Santa Gertrudis	16	2	T/I	United States	North America, Africa	Red	Moderate[a]	Low-moderate[a]	Shorthorn–Brahman hybrid+, tropical
Senepol	16	2	T	Virgin Islands	North America	Red	Small-moderate		N'Dama/Red Poll crossbred, tropical

Continued

Table 3.7. Continued.

Breed	Classification[b]	Category of origin[c]	Species[d]	Country of origin	Global range	Colour	Size and growth	Milk production	Other traits, remarks
Simbrah	16	2	T/I	United States	North America	Red with white face and markings	Moderate-large		Simmental–Brahman hybrid, tropical
Texas Longhorn	15	1	T/I	Mexico and United States	North America	Variable	Small[a]	Small-moderate[a]	Criollo cattle of Southwestern United States, moderate lean-to-fat[a], low-moderate marbling[a], moderate puberty age[a]

[a]For more details, see Cundiff, 2003. Breed names in bold refer to pictures in Plates 2–25.
[b]Codes of integrative classification as in Table 3.1.
[c]1, authentic local; 2, imported local; 4, continuously crossbred.
[d]Species: T, *Bos taurus*; T/I, taurindicine.

Table 3.8. Breeds of cattle with origin in Australia, Southwest Asia and Oceania.

Breed	Classifi-cation[b]	Category of origin[c]	Species[d]	Country of origin	Global range	Colour	Size and growth	Milk production	Other traits, remarks
Bali Cattle	10	1	J	Bali	Australia, SW Asia Oceania	Dark-brown bulls, tan cows; white mirror and lower legs	Small		Tropical, work-beef, high fertility, adapted to extensive management
Grati	10	2	T	Indonesia	Oceania, tropical	Red or black-and-white	Moderate		Crossbred of Javanese, Black-pied and other dairy taurine breeds
Illawarra	16	1	T	Australia	Australia, Asia, America, Oceania	Red, some roans or white	Moderate		Crossbred of mainly Ayrshire and other dairy taurine breeds
Javanese	10	1	I/J	Indonesia	Oceania	Tan	Small-moderate		Banteng–Indochinese hybrid upgraded with Ongole
Kedah-Kelantan	10	1	I	Malaysia	Oceania	Tan-brown	Small		Indo-Chinese zebu
Local Indian Dairy	8	2	I	Malaysia	Asia, Oceania	White	Small		Kedah–Kelantan zebu crossbred, now × Holstein
Madura	10	1	I/J	Indonesia	Oceania	Tan	Small		Zebu (paternal)–Banteng cross, 'racing bull'
Mandalong Special	16	4	T/I	Australia	Oceania	Cream to red	Moderate-large		Charolais–Chianina-poll Shorthorn–British White–Brahman crossbred
Murray Grey	16	2	T	Australia	Australia, Europe, North America	Silver to dun-grey	Moderate		Shorthorn × Angus crossbred, polled

[a]For more details, see Cundiff, 2003.
[b]Codes of integrative classification as in Table 3.1.
[c]1, authentic local; 2, imported local; 4, continuously crossbred.
[d]Species: I, *Bos indicus*; J, *Bos javanicus*; T, *Bos taurus*; I/J, mixed *B. indicus–javanicus* origin; T/I, taurindicine.

Table 3.9. Breeds of cattle with origin in South America.

Breed	Classification[b]	Category of origin[c]	Species[d]	Country of origin	Global range	Colour	Size and growth	Milk production	Other traits, remarks
Blanco Orejinegro	15	1	T/I	Colombia	South America	White with black points	Moderate		Colombian Criollo, longhorned or polled, tropical
Caracu	15	1	T/I	Brazil	South America	Blonde to light red	Moderate		Criollo cattle of Brazil, tropical
Indo-Brazil	16	2	I	Brazil	America	White or grey	Large		Guzerá, Nelore, Hissar, Gir crossbred, tropical, leaflike hanging ears
Romosinuano	15	1	T/I	Colombia	America	Tan to red	Small[a]	Moderate[a]	Colombian Criollo, polled, moderate lean-to-fat[a], low marbling[a], low tenderness[a], moderate puberty age[a]

[a]For more details, see Cundiff, 2003.
[b]Codes of integrative classification, respectively, as in Table 3.1.
[c]1, authentic local; 2, imported local.
[d]Species: I, *Bos indicus*; T, *Bos taurus*; T/I, taurindicine.

Table 3.10. References for beef cattle breed comparison research.

Authors	Location	Breeds
Prayaga, 2003a,b, 2004	Australia	Africander, Boran, Brahman, Charolais, Hereford, Shorthorn, Simmental, Tuli
Chase *et al.*, 1997; Chenoweth *et al.*, 1996	Florida	Angus, Brahman, Hereford, Nelore, Romosinuano, Senepol
Koger, 1980	Florida	Zebu, Brahman, Santa Gertrudis, Beefmaster, Brangus, Braford, Barzona, Charbray, Simbrah, Bramousin
Arango, 2002b; Crouse *et al.*, 1975; Koch and Dikeman, 1977; Koch *et al.*, 1976, 1983; Laster *et al.*, 1972, 1976; Smith, 1976; Smith *et al.*, 1976c,d	Nebraska	Angus, Charolais, Hereford, Jersey, Limousin, Simmental, South Devon
Arango *et al.*, 2002a; Crouse *et al.*, 1989; Cundiff *et al.*, 1984; Gregory *et al.*, 1979a,b; Koch *et al.*, 1982a,b	Nebraska	Angus, Brahman, Hereford, Pinzgauer, Sahiwal, Tarentaise
Arango *et al.*, 2002c	Nebraska	Angus, Braunvieh, Chianina, Gelbvieh, Hereford, Maine Anjou, Red Poll
Arango *et al.*, 2004a; Cundiff *et al.*, 1990, 1998; Thallman *et al.*, 1999; Wheeler *et al.*, 1996, 1997	Nebraska	Angus, Charolais, Galloway, Gelbvieh, Hereford, Longhorn, Nelore, Piedmontese, Pinzgauer, Salers, Shorthorn
Arango *et al.*, 2004b; Cundiff *et al.*, 1986b; Cundiff *et al.*, 1993; Jenkins *et al.*, 1991b; Notter *et al.*, 1978a,b; Young *et al.*, 1978a,b	Nebraska	Angus, Brahman, Brangus, Braunvieh, Brown Swiss, Charolais, Chianina, Devon, Galloway, Gelbvieh, Hereford, Holstein, Jersey, Limousin, Longhorn, Maine-Anjou, Nelore, Piedmontese, Pinzgauer, Red Poll, Sahiwal, Salers, Santa Gertrudis, Shorthorn, Simmental, South Devon, Tarentaise
Casas and Cundiff, 2003; Casas *et al.*, 2011; Cundiff *et al.*, 1994; Freetly and Cundiff, 1997, 1998; Freetly *et al.*, 2011; Wheeler *et al.*, 2001	Nebraska	Angus, Belgian Blue, Boran, Brahman, Hereford, Piedmontese, Tuli
Casas and Cundiff, 2006; Casas *et al.*, 2007, 2012; Cundiff and Thallman, 2002; Wheeler *et al.*, 2004	Nebraska	Angus, Friesian, Hereford, Norwegian Red, Swedish Red and White, Wagyu
Casas *et al.*, 2010; Wheeler *et al.*, 2010	Nebraska	Angus, Beefmaster, Bonsmara, Brangus, Hereford, Romosinuano
Cundiff *et al.*, 1974a,b, 1992; Gregory *et al.*, 1965, 1966a,b,c; Long and Gregory 1974, 1975a,b; Núñez-Dominguez *et al.*, 1991, 1992; Olson *et al.*, 1978a,b,c; Smith and Cundiff, 1976; Smith *et al.*, 1976a,b; Wiltbank *et al.*, 1966, 1967	Nebraska	Angus, Hereford, Shorthorn
Cundiff *et al.*, 1981; Dearborn, 1986, 1987a,b; Gregory *et al.*, 1978a,b,c,d,e; Jenkins *et al.*, 1991a; Koch *et al.*, 1979, 1981; Laster *et al.*, 1979	Nebraska	Angus, Brown Swiss, Chianina, Gelbvieh, Hereford, Maine Anjou, Red Poll
Cundiff *et al.*, 1986a	Nebraska	Angus, Brown Swiss, Charolais, Chianina, Gelbvieh, Jersey, Limousin, Maine Anjou, Pinzgauer, South Devon, Tarentaise
Cushman *et al.*, 2007; Rodríguez-Almeida *et al.*, 1995a,b; Wheeler *et al.*, 2005	Nebraska	Angus, Charolais, Gelbvieh, Hereford, Limousin, Pinzgauer, Red Angus, Simmental

Continued

Table 3.10. Continued.

Authors	Location	Breeds
Ferrell, 1982; Laster et al., 1973a,b	Nebraska	Angus, Brown Swiss, Charolais, Hereford, Jersey, Limousin, Red Poll, Simmental, South Devon
Ferrell and Jenkins, 1998a,b; Jenkins and Ferrell, 2004	Nebraska	Angus, Boran, Brahman, Hereford, Tuli
Gregory and Cundiff, 1980	Nebraska	Angus, Brahman, Charolais, Hereford
Gregory et al., 1991a,b,c, 1992a,b,c, 1994; Jenkins and Ferrell, 1992, 1994, 1997	Nebraska	Angus, Braunvieh, Charolais, Gelbvieh, Hereford, Limousin, Pinzgauer, Red Poll, Simmental
Laster and Gregory, 1973	Nebraska	Angus, Charolais, Hereford, Jersey, Limousin, Red Poll, Simmental, South Devon
Laster et al., 1973a,b	Nebraska	Angus, Charolais, Hereford, Jersey, Limousin, Simmental, South Devon
Amer et al., 1992	Review	Angus, Charolais, Hereford, Limousin, Simmental
Cundiff, 1970	Review	Angus, Brahman, Brangus, Brown Swiss, Charolais, Hereford, Shorthorn
Franke, 1980	Review	Angus, Brahman, Brangus, Charolais, Devon, Shorthorn
Franke, 1997	Review	Beefmaster, Boran, Brahman, Brangus, Gir, Indu-Brazil, Nelore, Sahiwal, Santa Gertrudis, Tuli
Hetzel, 1988	Review	Africander, Angoni, Barotse, Boran, Mashona, Tuli
Long, 1980	Review	Angus, Blonde D'Aquitaine, Brahman, Brown Swiss, Charolais, Chianina, Gelbvieh, German Black and White, German Red and White, Hereford, Holstein, Jersey, Limousin, Maine Anjou, Normande, Marchigiana, Piedmontese, Pinzgauer, Red Poll, Romagnola, Sahiwal, Santa Gertrudis, Shorthorn, Simmental, South Devon, Tarentaise
Mason, 1971	Review	Angus, Blonde d'Aquitaine, Brown Swiss, Chianina, Charolais, Danish Red, Eastern Red Pied, French Brown, Friesian, Galloway, German Black Pied, German Brown, German Yellow, Hereford, Jersey, Limousin, Lincoln Red, Maine Anjou, Marchigiana, Montbeliard, Meuse-Rhine-Yssel, Normandy, Piedmontese, Red Poll, Romagnola, Shorthorn, Simmental, South Devon, Sussex, Swedish Red and White, Welsh Black
Plasse, 1983	Review	Brahman, Brown Swiss, Charolais, Criollo, Marchigiana, Red Poll, Simmental, Zebu
Roughsedge et al., 2001	Review	Angus, Belgian Blue, Braunvieh, Blonde d'Aquitaine, Brown Swiss, Charolais, Chianina, Devon, Friesian, Gelbvieh, Galloway, Hereford, Holstein, Jersey, Limousin, Longhorn, Maine Anjou, Pinzgauer, Piedmontese, Red Angus, Red Poll, Simmental, Salers, South Devon, Shorthorn, Tarentaise
Sanders, 1980	Review	Brahman, Gir, Guzerat, Indu-Brazil, Nelore, Zebu

Continued

Table 3.10. Continued.

Authors	Location	Breeds
Thrift, 1997	Review	Beefmaster, Boran, Braford, Brahman, Brangus, Gir, Indu-Brazil, Nelore, Romana Red, Sahiwal, Santa Gertrudis, Senepol, Simbrah, Tuli
Thrift et al., 2010	Review	Angus, Beefmaster, Bonsmara, Brahman, Brangus, Boran, Charolais, Gelbray, Gelbvieh, Gir, Hereford, Indu-Brazil, Nelore, Red Poll, Romosinuano, Sahiwal, Santa Gertrudis, Senepol, Simbrah, Tuli
Turner, 1980	Review	Boran, Brahman, other breeds
Turton, 1964	Review	Charolais and other breeds
Baker et al., 1984, 1989; Jenkins et al., 1981; Long et al., 1979a,b; Nelson et al., 1982a,b; Rohrer et al., 1988; Sacco et al., 1987, 1989a,b, 1990, 1991; Stewart et al., 1980; Talamantes et al., 1984	Texas	Angus, Brahman, Hereford, Holstein and Jersey
Baker et al., 2001	Texas	Angus, Brahman, Hereford, Tuli
Paschal et al., 1991, 1995	Texas	Angus, Brahman, Gir, Indu-Brazil, Nelore

References

Achilli, A., Bonfiglio, S., Olivieri, A., Malusà, A., Pala, M., Kashani, B.H., Perego, U.A., Ajmone-Marsan, P., Liotta, L., Semino, O., Bandelt, H.J., Ferretti, L. and Torroni, A. (2009) The multifaceted origin of taurine cattle reflected by the mitochondrial genome. *PLoS One* 4, e5753.

Acosta, A.C., Uffo, O., Sanz, A., Ronda, R., Osta, R., Rodellar, C., Martín-Burriel, I. and Zaragoza, P. (2013) Genetic diversity and differentiation of five Cuban cattle breeds using 30 microsatellite loci. *Journal of Animal Breeding and Genetics* 30, 79–86.

Ajmone-Marsan, P., Garcia, J.F. and Lenstra, J.A. (2010) On the origin of cattle: how aurochs became cattle and colonized the world. *Evolutionary Anthropology* 19, 148–157.

Alderson, L. (1992) The categorization of types and breeds of cattle in Europe. *Archives Zootechnia* 41, 325–334.

Amer, P.R., Kemp, R.A. and Smith, C. (1992) Genetic differences among the predominant beef cattle breeds in Canada: analysis of published results. *Canadian Journal of Animal Science* 72, 759–771.

Arango, J.A., Cundiff, L.V. and VanVleck, L.D. (2002a) Breed comparisons of Angus, Brahman, Hereford, Pinzgauer, Sahiwal, and Tarentaise for weight, weight adjusted for condition score, height, and body condition score. *Journal of Animal Science* 80, 3142–3149.

Arango, J.A., Cundiff, L.V. and VanVleck, L.D. (2002b) Breed comparisons of Angus, Charolais, Hereford, Jersey, Limousin, Simmental, and South Devon for weight, weight adjusted for body condition score, height, and body condition score of cows. *Journal of Animal Science* 80, 3123–3132.

Arango, J.A., Cundiff, L.V. and VanVleck, L.D. (2002c) Comparisons of Angus-, Braunvieh-, Chianina-, Hereford-, Gelbvieh-, Maine Anjou-, and Red Poll-sired cows for weight, weight adjusted for body condition score, height, and body condition score. *Journal of Animal Science* 80, 3133–3141.

Arango, J.A., Cundiff, L.V. and VanVleck, L.D. (2004a) Breed comparisons of weight, weight adjusted for condition score, height, and condition score of beef cows. *Professional Animal Scientist* 20, 15–26.

Arango, J.A., Cundiff, L.V. and VanVleck, L.D. (2004b) Comparisons of Angus, Charolais, Galloway, Hereford, Longhorn, Nellore, Piedmontese, Salers, and Shorthorn breeds for weight, weight adjusted for condition score, height, and condition score of cows. *Journal of Animal Science* 82, 74–84.

Baker, C.M. and Manwell, C. (1980) Chemical classification of cattle. 1. Breed groups. Anim Blood Groups. *Biochemical Genetics* 11, 127–150.

Baker, J.F., Long, C.R. and Cartwright, T.C. (1984) Characterization of cattle of a five breed diallel. V. Breed and heterosis effects on carcass merit. *Journal of Animal Science* 59, 922–933.

Baker, J. F.,Long, C.R. Posada, G.A., McElhenney, W.H. and Cartwright, T.C. (1989) Comparison of cattle of a five breed diallel: size, growth, condition and pubertal characters of second generation heifers. *Journal of Animal Science* 67, 1218–1229.

Baker, J.F., Williams, S.E. and Vann, R.C. (2001) Effects of Tuli, Brahman, Angus, and Polled Hereford sires on carcass traits of steer offspring. *Professional Animal Scientist* 17, 154–159.

Bovine Hapmap Consortium (2009) Genome-wide survey of SNP variation uncovers the genetic structure of cattle breeds. *Science* 324, 528–532.

Briggs, H.M. and Briggs, D.M. (1980) *Modern Breeds of Livestock*, 4th ed. Macmillan, New York.

Casas, E. and Cundiff, L.V. (2003) Maternal grandsire, granddam, and sire breed effects on growth and carcass traits of crossbred cattle. *Journal of Animal Science* 81, 904–911.

Casas, E. and Cundiff, L.V. (2006) Postweaning growth and carcass traits in crossbred cattle from Hereford, Angus, Norwegian Red, Swedish Red and White, Friesian, and Wagyu maternal grandsires. *Journal of Animal Science* 84, 305–310.

Casas, E., Lunstra, D.D., Cundiff, L.V. and Ford, J.J. (2007) Growth and pubertal development of F1 bulls from Hereford, Angus, Norwegian Red, Swedish Red and White, Friesian, and Wagyu sires. *Journal of Animal Science* 85, 2904–2909.

Casas, E., Thallman, R.M., Kuehn, L.A. and Cundiff, L.V. (2010) Postweaning growth and carcass traits in crossbred cattle from Hereford, Angus, Brangus, Beefmaster, Bonsmara, and Romosinuano maternal grandsires. *Journal of Animal Science* 88, 102–108.

Casas, E., Thallman, R.M. and Cundiff, L.V. (2011) Birth and weaning traits in crossbred cattle from Hereford, Angus, Brahman, Boran, Tuli, and Belgian Blue sires. *Journal of Animal Science* 89, 979–987.

Casas, E., Thallman, R.M. and Cundiff, L.V. (2012) Birth and weaning traits in crossbred cattle from Hereford, Angus, Norwegian Red, Swedish Red and White, Wagyu, and Friesian sires. *Journal of Animal Science* 90, 2916–2920.

Chase, C.C., Chenoweth, P.J., Larsen, R.E., Olson, T.A., Hammond, A.C., Menchaca, M.A. and Randel, R.D. (1997) Growth and reproductive development from weaning through 20 months of age among breeds of bulls in subtropical Florida. *Theriogenology* 47, 723–745.

Chenoweth, P.J., Chase Jr, C.C., Thatcher, M.J.D., Wilcox, C.J. and Larsen, R.E. (1996) Breed and other effects on reproductive traits and breeding soundness categorization in young beef bulls in Florida. *Theriogenology* 46, 1159–1170.

Committee on Managing Global Genetic Resources (1993) *Managing Global Genetic Resources – Livestock*. National Academy Press, Washington DC.

Crouse, J.D., Dikeman, M.E., Koch, R.M. and Murphey, C.E. (1975) Evaluation of traits in the U.S.D.A. yield grade equation for predicting beef carcass cutability in breed groups differing in growth and fattening characteristics. *Journal of Animal Science* 41, 548–553.

Crouse, J.D., Cundiff, L.V., Koch, R.M., Koohmaraie, M. and Seideman, S.C. (1989) Comparisons of *Bos indicus* and *Bos taurus* inheritance for carcass beef characteristics and meat palatability. *Journal of Animal Science* 67, 2661–2668.

Cundiff, L.V. (1970) Experimental results on crossbreeding cattle for beef production. *Journal of Animal Science* 30, 694–705.

Cundiff, L.V. (2003a) Beef cattle: breeds and genetics. In: Pond, W. and Bell, A. (eds) *Encyclopedia of Animal Science*. Taylor and Francis Online, Ithaca, New York.

Cundiff, L.V. (2003b) Implications of breed evaluations. In: *Proceedings Texas A&M Beef Cattle Shortcourse*. Department of Animal Science, Texas A&M University, College Station, Texas.

Cundiff, L.V. and Thallman, R.M. (2002) Reproduction and maternal performance of Angus, Hereford, Norwegian Red, Swedish Red and White, Friesian, and Wagyu sired F1 females. In: *Proceedings of the 7th World Congress on Genetics Applied to Livestock Production*, 1–4. INRA, Montpellier, France.

Cundiff, L.V., Gregory, K.E. and Koch, R.M. (1974a) Effects of heterosis on reproduction in Hereford, Angus and Shorthorn cattle. *Journal of Animal Science* 38, 711–727.

Cundiff, L.V., Gregory, K.E., Schwulst, F.J. and Koch, R.M. (1974b) Effects of heterosis on maternal performance and milk production in Hereford, Angus and Shorthorn cattle. *Journal of Animal Science* 38, 728–745.

Cundiff, L.V., Koch, R.M., Gregory, K.E. and Smith, G.M. (1981) Characterization of biological types of cattle – Cycle II. IV. Postweaning growth and feed efficiency of steers. *Journal of Animal Science* 53, 332–346.

Cundiff, L.V., Koch, R.M. and Gregory, K.E. (1984) Characterization of biological types of cattle (Cycle III). IV. Postweaning growth and feed efficiency. *Journal of Animal Science* 58, 312–323.

Cundiff, L.V., MacNeil, M.D., Gregory, K.E. and Koch, R.M. (1986a) Between and within breed genetic analysis of calving traits and survival to weaning in beef cattle. *Journal of Animal Science* 63, 27–33.

Cundiff, L.V., Gregory, K.E., Koch, R.M. and Dickerson, G.E. (1986b) Genetic diversity among cattle breeds and its use to increase beef production efficiency in a temperate environment. In: Dickerson, G.E. and Johnson, R.K. (eds) *Proceedings 3rd World Congress on Genetics Applied to Livestock Production: IX. Breeding Programs for Dairy and Beef Cattle, Water Buffalo, Sheep and Goats.* University of Nebraska, Lincoln, Nebraska, pp. 271–282.

Cundiff, L.V., Koch, R.M., Gregory, K.E., Crouse, J.D. and Dikeman, M.E. (1990) Preliminary results for carcass and meat characteristics of diverse breeds in cycle IV of the cattle germplasm evaluation program. In: Dickerson, G.E. (ed.) *Proceedings of the 4th World Congress on Genetics applied to Livestock Production.* University of Nebraska-Lincoln, Lincoln, Nebraska, pp. 291–294.

Cundiff, L.V., Núñez-Dominguez, R., Dickerson, G.E., Gregory, K.E. and Koch, R.M. (1992) Heterosis for lifetime production in Hereford, Angus, Shorthorn, and crossbred cows. *Journal of Animal Science* 70, 2397–2410.

Cundiff, L.V., Szabo, F., Gregory, K.E., Koch, R.M., Dikeman, M.E. and Crouse, J.D. (1993) Breed comparisons in the germplasm evaluation program at MARC. In: Bolze, R. (ed.) *Proceedings of Beef Improvement Federation Research Symposium and Annual Meeting.* Beef Improvement Federation, Colby, Kansas, pp. 124–138.

Cundiff, L.V., Gregory, K.E., Wheeler, T.L., Shackelford, S.D. and Koohmaraie, M. (1994) Carcass and meat characteristics of Tuli, Boran, Brahman, Belgian Blue, Piedmontese, Hereford and Angus breed crosses in the cattle germplasm evaluation program. In: Smith, C. (ed.) *Proceedings, 5th World Congress on Genetics Applied to Livestock Production.* University of Guelph, Guelph, Ontario, pp. 272–275.

Cundiff, L.V., Gregory, K.E., Wheeler, T.L., Shackelford, S.D., Koohmaraie, M., Freetly, H.C. and Lunstra, D.D. (1997) Preliminary results from Cycle V of the cattle germplasm evaluation program at the Roman L. Hruska U.S. Meat Animal Research Center. *Germplasm Evaluation Program Progress Report No. 16.* US Department of Agriculture, Clay Center, Nebraska, pp. 2–11.

Cundiff, L.V., Gregory, K.E. and Koch, R.M. (1998) Germplasm evaluation in beef cattle – cycle IV: birth and weaning traits. *Journal of Animal Science* 76, 2528–2535.

Cushman, R.A., Allan, M.F., Thallman, R.M. and Cundiff, L.V. (2007) Characterization of biological types of cattle (Cycle VII): influence of postpartum interval and estrous cycle length on fertility. *Journal of Animal Science* 85, 2156–2162.

Cymbron, T., Freeman, A.R., Isabel Malheiro, M., Vigne, J.D. and Bradley, D.G. (2005) Microsatellite diversity suggests different histories for Mediterranean and Northern European cattle populations. *Proceedings Biological Science* 272, 1837–1843.

D'Andrea, M., Pariset, L., Matassino, D., Valentini, A., Lenstra, J.A., Maiorano, G. and Pilla, F. (2011) Genetic characterization and structure of the Italian Podolian cattle breed and its relationship with some major European breeds. *Italian Journal of Animal Science* 10, e54.

Dearborn, D.D., Gregory, K.E., Cundiff, L.V. and Koch, R.M. (1986) Heterosis and breed maternal and transmitted effects in beef cattle. V. Weight, height and condition score of females. *Journal of Animal Science* 64, 706–713.

Dearborn, D.D., Gregory, K.E., Cundiff, L.V. and Koch, R.M. (1987a) Maternal heterosis and grandmaternal effects in beef cattle: preweaning traits. *Journal of Animal Science* 65, 33–41.

Dearborn, D.D., Gregory, K.E., Cundiff, L.V. and Koch, R.M. (1987b) Heterosis and breed maternal and transmitted effects in beef cattle. V. Weight, height and condition score of females. *Journal of Animal Science* 64, 706–713.

Decker, J.E., Pires, J.C., Conant, G.C., McKay, S.D., Heaton, M.P., Chen, K., Cooper, A., Vilkki, J., Seabury, C.M., Caetano, A.R. *et al.* (2009) Resolving the evolution of extant and extinct ruminants with high-throughput phylogenomics. *Proceedings National Academy of Science* 106, 18644–18649.

Del Bo, L., Polli, M., Longeri, M., Ceriotti, G., Looft, C., Barre-Dirie, A., Dolf, G. and Zanotti, M. (2001) Genetic diversity among some cattle breeds in the Alpine area. *Journal of Animal Breeding and Genetics* 118, 317–328.

Delgado, J.V., Martínez, A.M., Acosta, A., Alvarez, L.A., Armstrong, E., Camacho, E., Cañón, J., Cortés, O., Dunner, S., Landi, V. *et al.* (2012) Genetic characterization of Latin-American Creole cattle using microsatellite markers. *Animal Genetics* 43, 2–10.

Edwards, C.J., Ginja, C., Kantanen, J., Pérez-Pardal, L., Tresset, A., Stock, F., Gama, L.T., Penedo, M.C.T., Bradley, D.G., Lenstra, J.A. and Nijman, I.A. (2011) Dual origins of dairy cattle farming – evidence from a comprehensive survey of European Y-chromosomal variation. *PLoS One* 2011, e15922.

Egito, A.A., Paiva, S.R., Albuquerque Mdo, S., Mariante, A.S., Almeida, L.D., Castro, S.R. and Grattapaglia, D. (2007) Microsatellite based genetic diversity and relationships among ten Creole and commercial cattle breeds raised in Brazil. *BMC Genetics* 8, 83.

FAO (2007) *The State of the World's Animal Genetic Resources for Food and Agriculture.* FAO Communication Division, Rome.

Felius, M. (1985) *Cattle Breeds of the World.* Merck & Co., Inc., Rahway, New Jersey.

Felius, M. (1995) *Cattle Breeds: an Encyclopedia.* Misset, Doetinchem, The Netherlands.

Felius, M., Koolmees, P.A. and Theunissen, B. (2011) On the breeds of cattle – historic and current classification. *Diversity* 3, 660–692.

Felius, M., Theunissen, B. and Lenstra, J.A. (2014) On the conservation of cattle genetic resources – the role of breeds? *Journal of Agricultural Science* (in press).

Ferrell, C.L. (1982) Effect of postweaning rate of gain on onset of puberty and productive performance of heifers of different breeds. *Journal of Animal Science* 55, 1272–1283.

Ferrell, C.L. and Jenkins, T. (1998a) Body composition and energy utilization by steers of diverse genotypes fed a high-concentrate diet during the finishing period. I. Angus, Belgian Blue, Hereford, and Piedmontese sires. *Journal of Animal Science* 76, 637–646.

Ferrell, C.L. and Jenkins, T. (1998b) Body composition and energy utilization by steers of diverse genotypes fed a high-concentrate diet during the finishing period. II. Angus, Boran, Brahman, Hereford, and Tuli sires. *Journal of Animal Science* 76, 647–657.

Franke, D.E. (1980) Breed and heterosis effects of American Zebu cattle. *Journal of Animal Science* 50, 1206.

Franke, D.E. (1997) Postweaning performance and carcass merit of F1 steers sired by Brahman and alternative subtropically adapted breeds. *Journal of Animal Science* 75, 2604–2608.

Freeman, A.R., Meghen, C.M., MacHugh, D.E., Loftus, R.T., Achukwi, M.D., Bado, A., Sauveroche, B. and Bradley, D.G. (2004) Admixture and diversity in West African cattle populations. *Molecular Ecology* 13, 3477–3487.

Freetly, H.C. and Cundiff, L.V. (1997) Postweaning growth and reproduction characteristics of heifers sired by bulls of seven breeds and raised on different levels of nutrition. *Journal of Animal Science* 75, 2841–2851.

Freetly, H.C. and Cundiff, L.V. (1998) Reproductive performance, calf growth and milk production of first-calf heifers sired by seven breeds and raised on different levels of nutrition. *Journal of Animal Science* 76, 1513–1522.

Freetly, H.C., Kuehn, L.A. and Cundiff, L.V. (2011) Growth curves of crossbred cows sired by Hereford, Angus, Belgian Blue, Brahman, Boran, and Tuli bulls, and the fraction of mature body weight and height at puberty. *Journal of Animal Science* 89, 2373–2379.

Gautier, M., Faraut, T.,Moazami-Goudarzi, K., Navratil, V., Foglio, M., Grohs, C., Boland, A., Garnier, J.G., Boichard, D., Lathrop, G.M., Gut, I.G. and Eggen, A. (2007) Genetic and haplotypic structure in 14 European and African cattle breeds. *Genetics* 177, 1059–1070.

Gautier, M., Laloë, D. and Moazami-Goudarzi, K. (2010) Insights into the genetic history of French cattle from dense SNP data on 47 worldwide breeds. *PLoS One* 5, e13038.

Ginja, C., Penedo, M.C., Melucci, L., Quiroz, J., Martínez López, O.R., Revidatti, M.A., Martínez-Martínez, A., Delgado, J.V. and Gama, L.T. (2010) Origins and genetic diversity of New World Creole cattle: inferences from mitochondrial and Y chromosome polymorphisms. *Animal Genetics* 41, 128–141.

Gregory, K.E. and Cundiff, L.V. (1980) Crossbreeding in beef cattle: evaluation of systems. *Journal of Animal Science* 51, 1224–1242.

Gregory, K.E., Swiger, L.A., Koch, R.M., Sumption, L.J., Rowden, W.W. and Ingalls, J.E. (1965) Heterosis in preweaning traits of beef cattle. *Journal of Animal Science* 24, 21.

Gregory, K.E., Swiger, L.A., Koch, R.M. Sumption, L.J., Ingalls, J.E., Rowden, W.W. and Rothlisberger, J.A. (1966a) Heterosis effects on growth rate of beef heifers. *Journal of Animal Science* 25, 290.

Gregory, K.E., Swiger, L.A., Sumption, L.J., Koch, R.M., Ingalls, J.E., Rowden, W.W. and Rothlisberger, J.A. (1966b) Heterosis effects on growth rate and feed efficiency of beef steers. *Journal of Animal Science* 25, 299.

Gregory, K.E., Swiger, L.A., Sumption, L.J., Koch, R.M., Ingalls, J.E., Rowden, W.W. and Rothlisberger, J.A. (1966c) Heterosis effects on carcass traits of beef steers. *Journal of Animal Science* 25, 643.

Gregory, K.E., Cundiff, L.V., Koch, R.M., Laster, D.B. and Smith, G.M. (1978a) Heterosis and breed maternal and transmitted effects in beef cattle. I. Preweaning traits. *Journal of Animal Science* 47, 1031–1041.

Gregory, K.E., Laster, D.B., Cundiff, L.V., Koch, R.M. and Smith, G.M. (1978b) Heterosis and breed maternal and transmitted effects in beef cattle. II. Growth rate and puberty in females. *Journal of Animal Science* 47, 1042–1053.

Gregory, K.E., Koch, R.M., Laster, D.B., Cundiff, L.V. and Smith, G.M. (1978c) Heterosis and breed maternal and transmitted effects in beef cattle. III. Growth traits of steers. *Journal of Animal Science* 47, 1054–1062.

Gregory, K.E., Crouse, J.D., Koch, R.M., Laster, D.B., Cundiff, L.V. and Smith, G.M. (1978d) Heterosis and breed maternal and transmitted effects in beef cattle. IV. Carcass traits of steers. *Journal of Animal Science* 47, 1063–1079.

Gregory, K.E., Cundiff, L.V., Smith, G.M., Laster, D.B. and Fitzhugh, H.A., (1978e) Characterization of biological types of cattle – cycle II: I. Birth and weaning traits. *Journal of Animal Science* 47, 1022–1030.

Gregory, K.E., Laster, D.B., Cundiff, L.V., Smith, G.M. and Koch, R.M. (1979a) Characterization of biological types of cattle – cycle III: II. Growth rate and puberty in females. *Journal of Animal Science* 49, 461–471.

Gregory, K.E., Smith, G.M., Cundiff, L.V., Koch, R.M. and Laster, D.B. (1979b) Characterization of biological types of cattle – cycle III: I. Birth and weaning traits. *Journal of Animal Science* 48, 271–279.

Gregory, K.E., Cundiff, L.V. and Koch, R.M. (1991a) Breed effects and heterosis in advanced generations of composite populations for birth weight, birth date, dystocia, and survival as traits of dam in beef cattle. *Journal of Animal Science* 69, 3574–3589.

Gregory, K.E., Cundiff, L.V. and Koch, R.M. (1991b) Breed effects and heterosis in advanced generations of composite populations for growth traits in both sexes of beef cattle. *Journal of Animal Science* 69, 3202–3212.

Gregory, K.E., Cundiff, L.V. and Koch, R.M. (1991c) Breed effects and heterosis in advanced generations of composite populations for preweaning traits of beef cattle. *Journal of Animal Science* 69, 947–960.

Gregory, K.E., Cundiff, L.V. and Koch, R.M. (1992a) Breed effects and heterosis in advanced generations of composite populations for reproduction and maternal traits of beef cattle. *Journal of Animal Science* 70, 656–672.

Gregory, K.E., Cundiff, L.V. and Koch, R.M. (1992b) Breed effects and heterosis in advanced generations of composite populations on actual weight, adjusted weight, hip height, and condition score of beef cows. *Journal of Animal Science* 70, 1742–1754.

Gregory, K.E., Cundiff, L.V. and Koch, R.M. (1992c) Effects of breed and retained heterosis on milk yield and 200-day weight in advanced generations of composite populations of beef cattle. *Journal of Animal Science* 70, 2366–2372.

Gregory, K.E., Cundiff, L.V., Koch, R.M., Dikeman, M.E. and Koohmaraie, M. (1994) Breed effects and retained heterosis for growth, carcass and meat traits in advanced generations of composite populations of beef cattle. *Journal of Animal Science* 72, 833–850.

Grzybowski, G. and Prusak, B. (2004) Genetic variation in nine European cattle breeds as determined on the basis of microsatellite markers. III. Genetic integrity of the Polish Red cattle included in the breed preservation program. *Animal Science Papers and Reports* 22, 37–44.

Hanotte, O., Tawah, C.L., Bradley, D.G., Okomo, M., Verjee, Y., Ochieng, J. and Rege, J.E. (2000) Geographic distribution and frequency of a taurine *Bos taurus* and an indicine *Bos indicus* Y specific allele amongst sub-saharan African cattle breeds. *Molecular Ecology* 9, 387–396.

Hayes, B.J., Chamberlain, A.J., Maceachern, S., Savin, K., McPartlan, H., MacLeod, I., Sethuraman, L. and Goddard, M.E. (2008) A genome map of divergent artificial selection between *Bos taurus* dairy cattle and *Bos taurus* beef cattle. *Animal Genetics* 40, 176–184.

Hetzel, D.J.S. (1988) Comparative productivity of the Brahman and some indigenous Sanga and *Bos indicus* breeds of east and southern Asia. *Animal Breeding Abstracts* 56, 243–255.

Ibeagha-Awemu, E.M., Jann, O.C., Weimann, C. and Erhardt, G. (2004) Genetic diversity, introgression and relationships among West/Central African cattle breeds. *Genetic Selection Evolution* 36, 673–690.

Jenkins, T.G. and Ferrell, C.L. (1992) Lactation characteristics of nine breeds of cattle fed various quantities of dietary energy. *Journal of Animal Science* 70, 1652–1660.

Jenkins, T.G. and Ferrell, C.L. (1994) Productivity through weaning of nine breeds of cattle under varying feed availabilities. I. Initial evaluation. *Journal of Animal Science* 72, 2787–2797.

Jenkins, T.G. and Ferrell, C.L. (1997) Changes in proportions of empty body depots and constituents for nine breeds of cattle under various feed availabilities. *Journal of Animal Science* 75, 95–104.

Jenkins, T.G. and Ferrell, C.L. (2004) Preweaning efficiency for mature cows of breed crosses from tropically adapted *Bos indicus* and *Bos taurus* and unadapted *Bos taurus* breeds. *Journal of Animal Science* 82, 1876–1881.

Jenkins, T.G., Long, C.R., Cartwright, T.C. and Smith, G.C. (1981) Characterization of cattle of a five-breed diallel. IV. Slaughter and carcass characters of serially slaughtered bulls. *Journal of Animal Science* 53, 62–79.

Jenkins, T.G., Cundiff, L.V. and Ferrell, C.L. (1991a) Differences among breed crosses of cattle in the conversion of food energy to calf weight during the preweaning interval. *Journal of Animal Science* 69, 2762–2769.

Jenkins, T.G., Cundiff, L.V. and Ferrell, C.L. (1991b) Evaluation of between- and within-breed variation in measures of weight-age relationships. *Journal of Animal Science* 69, 3118–3128.

Kantanen, J., Olsaker, I., Holm, L.E., Lien, S., Vilkki, J., Brusgaard, K., Eythorsdottir, E., Danell, B. and Adalsteinsson, S. (2000) Genetic diversity and population structure of 20 North European cattle breeds. *Journal of Heredity* 91, 446–457.

Koch, R.M. and Dikeman, M.E. (1977) Characterization of biological types of cattle. V. Carcass wholesale cut composition. *Journal of Animal Science* 45, 30–42.

Koch, R.M., Dikeman, M.E., Allen, D.M., May, M., Crouse, J.D. and Campion, D.R. (1976) Characterization of biological types of cattle. III. Carcass composition, quality and palatability. *Journal of Animal Science* 43, 48–62.

Koch, R.M., Dikeman, M.E., Lipsey, J., Allen, D.M. and Crouse, J.D. (1979) Characterization of biological types of cattle – cycle II: III. Carcass composition, quality and palatability. *Journal of Animal Science* 49, 448–460.

Koch, R.M., Dikeman, M.E. and Cundiff, L.V. (1981) Characterization of biological types of cattle (cycle II). V. Carcass wholesale cut composition. *Journal of Animal Science* 53, 992–999.

Koch, R.M., Dikeman, M.E. and Crouse, J.D. (1982a) Characterization of biological types of cattle (cycle III). III. Carcass composition, quality and palatability. *Journal of Animal Science* 54, 35–45.

Koch, R.M., Dikeman, M.E. and Cundiff, L.V. (1982b) Characterization of biological types of cattle (cycle III). V. Carcass wholesale cut composition. *Journal of Animal Science* 54, 1160–1168.

Koch, R.M., Dikeman, M.E., Grodzki, H., Crouse, J.D. and Cundiff, L.V. (1983) Individual and maternal genetic effects for beef carcass traits of breeds representing diverse biological types (Cycle I). *Journal of Animal Science* 57, 1124–1132.

Koger, M. (1980) Effective crossbreeding systems utilizing Zebu cattle. *Journal of Animal Science* 50, 1215–1220.

Laloe, D.,Moazami-Goudarzi, K., Lenstra, J.A., Ajmone-Marsan, P., Azor, P., Baumung, R., Bradley, D.G., Bruford, M.W., Canon, J., Dolf, G., Dunner, S., Erhardt, G., Hewitt, G., Kantanen, J., Obexer-Ruff, G., Olsaker, I., Rodellar, C., Valentini, A., Wiener, P. and European Cattle Genetic Diversity Consortium and Econogene Consortium (2010) Spatial trends of genetic variation of domestic ruminants in Europe. *Diversity* 2, 932–945.

Laster, D.B. and Gregory, K.E. (1973) Factors influencing peri- and early postnatal calf mortality. *Journal of Animal Science* 37, 1092.

Laster, D.B., Glimp, H.A. and Gregory, K.E. (1972) Age and weight at puberty and conception in different breeds and breed-crosses of beef heifers. *Journal of Animal Science* 34, 1031–1036.

Laster, D.B., Glimp, H.A., Cundiff, L.V. and Gregory, K.E. (1973a) Factors affecting dystocia and the effects of dystocia on subsequent reproduction in beef cattle. *Journal of Animal Science* 36, 695–705.

Laster, D.B., Glimp, H.A. and Gregory, K.E. (1973b) Effects of early weaning on postpartum reproduction of cows. *Journal of Animal Science* 36, 734–740.

Laster, D.B., Smith G.M. and Gregory, K.E. (1976) Characterization of biological types of cattle. IV. Postweaning growth and puberty of heifers. *Journal of Animal Science* 43, 63–70.

Laster, D.B., Smith, G.M., Cundiff, L.V. and Gregory, K.E. (1979) Characterization of biological types of cattle (cycle II). II. Postweaning growth and puberty of heifers. *Journal of Animal Science* 48, 500–508.

Lenstra, J.A. (2008) What's in a breed. Criteria for conservation. In: *Proceeding of Genetic Diversity, Selection and Adaptation in Wildlife and Livestock—Molecular Approaches,* Salzburg, Austria, Available online: http://www.nas.boku.ac.at/fileadmin/_/H93/H932-NUWI/Kurse-Workshops-Tagungen/ESF-Workshop/Session1/Lenstra.pdf.

Lenstra, J.A., Groeneveld, L.F., Eding, H., Kantanen, J., Williams, J.L., Taberlet, P., Nicolazzi, E.L., Sölkner, J., Simianer, H., Ciani, E., Garcia, J.F., Bruford, M.W., Ajmone-Marsan, P. and Weigend, S. (2012) Molecular tools and analytical approaches for the characterisation of farm animal genetic diversity. *Animal Genetics* 43, 483–502.

Li, M.H. and Kantanen, J. (2010) Genetic structure of Eurasian cattle (*Bos taurus*) based on microsatellites: clarification for their breed classification. *Animal Genetics* 41, 150–158.

Li, M.H., Tapio, I., Vilkki, J., Ivanova, Z., Kiselyova, T., Marzanov, N., Cinkulov, M., Stojanović, S., Ammosov, I., Popov, R. and Kantanen, J. (2007) The genetic structure of cattle populations (*Bos taurus*) in northern Eurasia and the neighbouring Near Eastern regions: implications for breeding strategies and conservation. *Molecular Ecology* 16, 3839–3853.

Loftus, R.T., Ertugrul, O., Harba, A.H., El-Barody, M.A., MacHugh, D.E., Park, S.D. and Bradley, D.G. (1999) A microsatellite survey of cattle from a centre of origin: the Near East. *Molecular Ecology* 8, 2015–2022.

Long, C.R. and Gregory, K.E. (1974) Heterosis and breed effects in preweaning traits of Angus, Hereford and reciprocal cross calves. *Journal of Animal Science* 39, 11–17.

Long, C.R. and Gregory, K.E. (1975a) Heterosis and management effects in postweaning growth of Angus, Hereford and reciprocal cross cattle. *Journal of Animal Science* 41, 1563–1571.

Long, C.R. and Gregory, K.E. (1975b) Heterosis and management effects in carcass characters of Angus, Hereford and reciprocal cross cattle. *Journal of Animal Science* 41, 1572–1580.

Long, C.R. (1980) Crossbreeding for beef production: experimental results. *Journal of Animal Science* 51, 1197–1223.

Long, C.R., Stewart, T.S., Cartwright, T.C. and Baker, J.F. (1979a) Characterization of cattle of a five breed diallel: I. Measures of size, condition and growth in bulls. *Journal of Animal Science* 49, 418–431.

Long, C.R., Stewart, T.S., Cartwright, T.C. and Baker, J.F. (1979b) Characterization of cattle of a five breed diallel: II. Measures of size, condition and growth in heifers. *Journal of Animal Science* 49, 432–447.

McTavish, E.J., Decker, J.E., Schnabel, R.D., Taylor, J.F. and Hillis, D.M. (2013) New World cattle show ancestry from multiple independent domestication events. *Proceedings National Academy of Science USA* 110, E1398–E1406.

Martín-Burriel, I., Rodellar, C., Lenstra, J.A., Sanz, A., Cons, C., Osta, R., Reta, M., De Argüello, S., Sanz, A. and Zaragoza, P. (2007) Genetic diversity and relationships of endangered Spanish cattle breeds. *Journal of Heredity* 98, 687–691.

Martín-Burriel, I., Rodellar, C., Cañón, J., Cortés, O., Dunner, S., Landi, V., Martínez-Martínez, A., Gama, L.T., Ginja, C., Penedo, M.C., Sanz, A., Zaragoza, P. and Delgado, J.V. (2011) Genetic diversity, structure, and breed relationships in Iberian cattle. *Journal of Animal Science* 89, 893–906.

Martínez, A.M., Gama, L.T., Canon, J., Ginja, C., Delgado, J.V., Dunner, S., Landi, V., Martín-Burriel, I., Penedo, M.C.T., Rodellar, C. *et al.* (2012) Genetic footprints of Iberian cattle in America 500 years after the arrival of Columbus. *PLoS* 7, e49066.

Mason, I.L. (1969) *A World Dictionary of Livestock Breed Types and Varieties.* Commonwealth Agricultural Bureaux, Farnham Royal, UK.

Mason, I.L. (1971) Comparative beef performance of the large cattle breeds of Western Europe. *Animal Breeding Abstracts* 39, 1–29.

Mason, I.L. (1996) *A World Dictionary of Livestock Breeds, Types and Varieties*, 4th edn. CAB International, Wallingford, UK.

Maudet, C., Luikart, G. and Taberlet, P. (2002) Genetic diversity and assignment tests among seven French cattle breeds based on microsatellite DNA analysis. *Journal of Animal Science* 80, 942–945.

Medugorac, I., Medugorac, A., Russ, I., Veit-Kensch, C.E., Taberlet, P., Luntz, B., Mix, H.M. and Förster, M. (2009) Genetic diversity of European cattle breeds highlights the conservation value of traditional unselected breeds with high effective population size. *Molecular Ecology* 18, 3394–3341.

Moazami-Goudarzi, K., Laloë, D., Furet, J.P. and Grosclaude, F. (1997) Analysis of genetic relationships between 10 cattle breeds with 17 microsatellites, *Animal Genetics* 28, 338–345.

Moazami-Goudarzi, K., Belemsaga, D.M.A., Ceriotti, G., Laloë, D., Fagbohoun, F., Kouagou, N.T., Sidibé, I., Codjia, V., Crimella, M.C., Grosclaude, F. and Touré, S.M. (2001) Caractérisation de la race bovine Somba à l'aide de marqueurs moléculai. *Revue d'Elevage et de Médecine Vétérinaire des pays Tropicaux* 54, 1–1.

Nelsen, T.C., Long, C.R. and Cartwright, T.C. (1982a) Postinflection growth in straightbred and crossbred cattle. I. Heterosis for weight, height and maturing rate. *Journal of Animal Science* 55, 280–292.

Nelsen, T.C., Long, C.R. and Cartwright, T.C. (1982b) Postinflection growth in straightbred and crossbred cattle. II. Relationships among weight, height and pubertal characters. *Journal of Animal Science* 55, 293–304.

Nijman, I.J., Van Boxtel, D.C.J., Van Cann, L.M., Marnoch, Y., Cuppen, E. and Lenstra, J.A. (2008) Phylogeny of Y chromosomes from bovine species. *Cladistics* 24, 723–726.

Notter, D.R., Cundiff, L.V., Smith, G.M., Laster, D.B. and Gregory, K.E. (1978a) Characterization of biological types of cattle VI. Transmitted and maternal effects on birth and survival traits in progeny of young cows. *Journal of Animal Science* 46, 892–907.

Notter, D.R., Cundiff, L.V., Smith, G.M., Laster, D.B. and Gregory, K.E. (1978b) Characterization of biological types of cattle. VII. Milk production in young cows and transmitted and maternal effects on preweaning growth of progeny. *Journal of Animal Science* 46, 908–921.

Núñez-Dominguez, R., Cundiff, L.V., Dickerson, G.E., Gregory, K.E. and Koch, R.M. (1991) Heterosis for survival and dentition in Hereford, Angus, Shorthorn, and crossbred cows. *Journal of Animal Science* 69, 1885–1898.

Núñez-Dominguez, R., Dickerson, G.E., Cundiff, L.V., Gregory, K.E. and Koch, R.M. (1992) Economic evaluation of heterosis and culling policies for lifetime productivity in Hereford, Angus, Shorthorn, and crossbred cows. *Journal of Animal Science* 70, 2328–2337.

Olson, L.W., Dickerson, G.E., Cundiff, L.V. and Gregory, K.E. (1978a) Individual heterosis for postweaning growth efficiency in beef cattle. *Journal of Animal Science* 46, 1529–1538.

Olson, L.W., Dickerson, G.E., Cundiff, L.V. and Gregory, K.E. (1978b) Maternal heterosis effects on post-weaning growth and carcass traits in beef cattle. *Journal of Animal Science* 46, 1552–1562.

Olson, L.W., Dickerson, G.E., Cundiff, L.V. and Gregory, K.E. (1978c) Maternal heterosis for postweaning growth efficiency in beef cattle. *Journal of Animal Science* 46, 1539–1551.

Paschal, J.C., Sanders, J.O. and Kerr, J.L. (1991) Calving and weaning characteristics of Angus-, Gray Brahman-, Gir-, Indu-Brazil-, Nellore- and Red Brahman-sired F1 calves. *Journal of Animal Science* 69, 2395–2402.

Paschal, J.C., Sanders, J.O., Kerr, J.L., Lunt, D.K. and Herring, A.D. (1995) Postweaning and feedlot growth and carcass characteristics of Angus-, Gray Brahman-, Gir-, Indu-Brazil, Nellore- and Red Brahman-sired F1 calves. *Journal of Animal Science* 73, 373–380.

Plasse, D. (1983) Crossbreeding results from beef cattle in the Latin American tropics. *Animal Breeding Abstracts* 51, 779–797.

Prayaga, K.C. (2003a) Evaluation of beef cattle genotypes and estimation of direct and maternal genetic effects in a tropical environment. 1. Growth traits. *Australian Journal of Agricultural Research* 54, 1013–1025.

Prayaga, K.C. (2003b) Evaluation of beef cattle genotypes and estimation of direct and maternal genetic effects in a tropical environment. 2. Adaptive and temperament traits. *Australian Journal of Agricultural Research* 54, 1027–1038.

Prayaga, K.C. (2004) Evaluation of beef cattle genotypes and estimation of direct and maternal genetic effects in a tropical environment. 3. Fertility and calf survival traits *Australian Journal of Agricultural Research* 55, 811–824.

Rodríguez-Almeida, F.A., VanVleck, L.D., Cundiff, L.V. and Kachman, S.D. (1995a) Heterogeneity of variance by sire breed, sex, and dam breed in 200- and 365-day weights of beef cattle from a top cross experiment. *Journal of Animal Science* 73, 2579–2588.

Rodríguez-Almeida, F.A., VanVleck, L.D. and Cundiff, L.V. (1995b) Effect of accounting for different phenotypic variances by sire breed and sex on selection of sires based on expected progeny differences for 200- and 365-day weights. *Journal of Animal Science* 73, 2589–2599.

Rohrer, G.A., Baker, J.F., Long, C.R. and Cartwright, T.C. (1988) Productive longevity of first cross cows produced in a five breed diallel: I. Reasons for removal. *Journal of Animal Science* 66, 2826–2835.

Roughsedge, T., Thompson, R., Villanueva, B. and Simm, G. (2001) Synthesis of direct and maternal genetic components of economically important traits from beef breed-cross evaluations. *Journal of Animal Science* 79, 2307–2319.

Rouse, J.E. (1970a) *Cattle of Africa and Asia. World Cattle, Volume II.* University of Oklahoma Press, Norman, Oklahoma.

Rouse, J.E. (1970b) *Cattle of Europe, South America, Australia and New Zealand. World Cattle*, Volume I. University of Oklahoma Press, Norman, Oklahoma.

Rouse, J.E. (1973) *Cattle of North America. World Cattle, Volume III.* University of Oklahoma Press, Norman, Oklahoma.

Sacco, R.E., Baker, J.F. and Cartwright, T.C. (1987) Production characters of primiparous females of a five breed diallel. *Journal of Animal Science* 64, 1612–1618.

Sacco, R.E., Baker, J.F., Cartwright, T.C., Long, C.R. and Sanders, J.O. (1989a) Lifetime productivity of straightbred and F1 cows of a five breed diallel. *Journal of Animal Science* 67, 1964–1971.

Sacco, R.E., Baker, J.F., Cartwright, T.C., Long, C.R. and Sanders, J.O. (1989b) Production characters of straightbred F1 and F2 cows: birth and weaning characters of terminal cross calves. *Journal of Animal Science* 67, 1972–1979.

Sacco, R.E., Baker, J.F., Cartwright, T.C., Long, C.R. and Sanders, J.O. (1990) Measurements at calving for straightbred and crossbred cows of diverse types. *Journal of Animal Science* 68, 3103–3108.

Sacco, R.E., Baker, J.F., Cartwright, T.C., Long, C.R. and Sanders, J.O. (1991) Heterosis retention for birth and weaning characters of calves in the third generation of a five breed diallel. *Journal of Animal Science* 69, 4754–4762.

Sanders, J.O. (1980) History and development of Zebu cattle in the United States. *Journal of Animal Science* 50, 1188–1200.

Shah, T.M., Patel, J.S., Bhong, C.D., Doiphode, A., Umrikar, U.D., Parmar, S.S., Rank, D.N., Solanki, J.V. and Joshi, C.G. (2012) Evaluation of genetic diversity and population structure of west-central Indian cattle breeds. *Animal Genetics* 44, 442–445.

Simcic, M., Lenstra, J.A., Baumung, R., Bovc, P., Cepon, M. and Kompan, D. (2013) On the origin of the Slovenian Cika cattle. *Journal of Animal Breeding and Genetics* 130, 4877–4895.

Smith, G.M. (1976) Sire breed effects on economic efficiency of a terminal-cross beef production system. *Journal of Animal Science* 43, 1163–1170.

Smith, G.M. and Cundiff, L.V. (1976) Genetic analysis of relative growth rate in crossbreed and straightbred Hereford, Angus and Shorthorn steers. *Journal of Animal Science* 43, 1171–1175.

Smith, G.M., Fitzhugh, H.A., Jr, Cundiff, L.V., Cartwright, T.C. and Gregory, K.E. (1976a) Heterosis for maturing patterns in Hereford, Angus and Shorthorn cattle. *Journal of Animal Science* 43, 380–388.

Smith, G.M., Fitzhugh, H.A., Jr, Cundiff, L.V., Cartwright, T.C. and Gregory, K.E. (1976b) A genetic analysis of maturing patterns in straightbred and crossbred Hereford, Angus and Shorthorn cattle. *Journal of Animal Science* 43, 389–395.

Smith, G.M., Laster, D.B. and Gregory, K.E. (1976c) Characterization of biological types of cattle I. Dystocia and preweaning growth. *Journal of Animal Science* 43, 27–36.

Smith, G.M., Laster, D.B., Cundiff, L.V. and Gregory, K.E. (1976d) Characterization of biological types of cattle II. Postweaning growth and feed efficiency of steers. *Journal of Animal Science* 43, 37–47.

Stewart, T.S., Long, C.R. and Cartwright, T.C. (1980) Characterization of cattle of a five-breed diallel. III. Puberty in bulls and heifers. *Journal of Animal Science* 50, 808–820.

Talamantes, M.A., Long, C.R., Smith, G.C., Jenkins, T.G., Ellis, W.C. and Cartwright, T.C. (1986) Characterization of cattle of a five breed diallel: VI. Fat depositing patterns of serially slaughtered bulls. *Journal of Animal Science* 62, 1259–1266.

Thallman, R.M., Cundiff, L.V., Gregory, K.E. and Koch, R.M. (1999) Germplasm evaluation in beef cattle – Cycle IV: Postweaning growth and puberty of heifers. *Journal of Animal Science* 77, 2651–2659.

Thrift, F.A. (1997) Reproductive performance of cows mated to and preweaning performance of calves sired by Brahman vs alternative subtropically adapted breeds. *Journal of Animal Science* 75, 2597–2603.

Thrift, F.A., Sanders, J.O., Brown, M.A., Brown, A.H.,Jr, Herring, A.D., Riley, D.G., DeRouen, S.M., Holloway, J.W., Wyatt, W.E., Vann, R.C. *et al.* (2010) Review: preweaning, postweaning, and carcass trait comparisons for progeny sired by subtropically adapted beef sire breeds at various US locations. *Professional Animal Scientist* 26, 451–473.

Turner, J.W. (1980) Genetic and biological aspects of Zebu adaptability. *Journal of Animal Science* 50, 1201.

Turton, J.D. (1964) The Charolais and its use in crossbreeding. *Animal Breeding Abstracts* 32, 119.

Utsunomiya, Y.T., Pérez O'Brien, A.M., Sonstegard, T.S., Van Tassell, C.P., do Carmo, A.S., Mészáros, G., Sölkner, J. and Garcia, J.F. (2013) Detecting loci under recent positive selection in dairy and beef cattle by combining different genome-wide scan methods. *PLoS One* 16;8, e64280.

Walker, H.I. (1989) *Blue Book of Beef Breeds*. PAW Publishing, Allen, Kansas.

Wheeler, T.L., Cundiff, L.V., Koch, R.M. and Crouse, J.D. (1996) Characterization of biological types of cattle (cycle IV): carcass traits and longissimus palatability. *Journal of Animal Science* 74, 1023–1035.

Wheeler, T.L., Cundiff, L.V., Koch, R.M., Dikeman, M.E. and Crouse, J.D. (1997) Characterization of different biological types of steers (cycle IV): wholesale, subprimal, and retail product yields. *Journal of Animal Science* 75, 2389–2403.

Wheeler, T.L., Cundiff, L.V., Shackelford, S.D. and Koohmaraie, M. (2001) Characterization of biological types of cattle (Cycle V): carcass traits and longissimus palatability. *Journal of Animal Science* 79, 1209–1222.

Wheeler, T.L., Cundiff, L.V., Shackelford, S.D. and Koohmaraie, M. (2004) Characterization of biological types of cattle (Cycle VI): carcass, yield, and longissimus palatability traits. *Journal of Animal Science* 82, 1177–1189.

Wheeler, T.L., Cundiff, L.V., Shackelford, S.D. and Koohmaraie, M. (2005) Characterization of biological types of cattle (Cycle VII): carcass, yield, and longissimus palatability traits. *Journal of Animal Science* 83, 196–207.

Wheeler, T.L., Cundiff, L.V., Shackelford, S.D. and Koohmaraie, M. (2010) Characterization of biological types of cattle (Cycle VIII): carcass, yield, and longissimus palatability traits. *Journal of Animal Science* 88, 3070–3083.

Wiener, P., Burton, D. and Williams, J.L. (2004) Breed relationships and definition in British cattle: a genetic analysis. *Heredity (Edinb)*. 93, 597–602.

Willham, R.L. (1987) *Taking Stock*. Iowa State University, Ames, Iowa.

Wiltbank, J.N., Gregory, K.E., Swiger, L.A., Ingalls, J.E., Rothlisberger, J.A. and Koch, R.M. (1966) Effects of heterosis on age and weight at puberty in beef heifers. *Journal of Animal Science* 25, 744.

Wiltbank, J.N., Gregory, K.E., Rothlisberger, J.A., Ingalls, J.E. and Kasson, C.W. (1967) Fertility in beef cows bred to produce straightbred and crossbred calves. *Journal of Animal Science* 26, 1005–1010.

Withen, K.B., Brüniche-Olsen, A., Pedersen, B.V., European Cattle Genetic Diversity Consortium and Gravlund, P.J. (2011) The Agersoe cattle: the last remnants of the Danish island cattle (*Bos taurus*). *Animal Breeding and Genetics*. 128, 141–152.

Young, L.D., Cundiff, L.V., Crouse, J.D., Smith, G.M. and Gregory, K.E. (1978a) Characterization of biological types of cattle. VIII. Postweaning growth and carcass traits of three-way cross steers. *Journal of Animal Science* 46, 1178–1191.

Young, L.D., Laster, D.B., Cundiff, L.V., Smith, G.M. and Gregory, K.E. (1978b) Characterization of biological types of cattle. IX. Postweaning growth and puberty of three-breed cross heifers. *Journal of Animal Science* 47, 843–852.

Zerabruk, M., Li, M.H., Kantanen, J., Olsaker, I., Ibeagha-Awemu, E.M., Erhardt, G. and Vangen, O. (2012) Genetic diversity and admixture of indigenous cattle from North Ethiopia: implications of historical introgressions in the gateway region to Africa. *Animal Genetics* 43, 257–266.

Zhang, G.X., Wang, Z.G., Chen, W.S., Wu, C.X., Han, X., Chang, H., Zan, L.S., Li, R.L., Wang, J.H., Song, W.T., Xu, G.F., Yang, H.J. and Luo, Y.F. (2007) Genetic diversity and population structure of indigenous yellow cattle breeds of China using 30 microsatellite markers. *Animal Genetics* 38, 550–559.

4 Molecular Genetics of Coat Colour Variation

Anatoly Ruvinsky

University of New England, Armidale, New South Wales, Australia

Introduction

Studying genetics of coat colour in mammals commenced immediately after the rediscovery of Mendelian laws during the first years of the 20th century. The life-long work of C. Little (1958) created a solid understanding of colour variation in mammals and an impetus for further investigations. He identified several major loci influencing coat colour variation and described numerous mutations. A comparative approach developed by A. Searle (1968) convincingly showed that a great deal of knowledge obtained in laboratory mice and dogs was relevant to other mammals. A comprehensive description of coat colour genetics in cattle available at the time can be found in the first edition of *The Genetics of Cattle* (Olson, 1999).

This chapter is mainly focused on molecular genetic aspects of coat colour genetics.

Mammalian melanocytes produce only two types of pigments – black eumelanin and red pheomelanin. These two pigments are sufficient to produce the great variety of coat colours observed in cattle and other species. The understanding of molecular mechanisms responsible for making these 'combinatorial' colours has significantly accelerated during the past two decades. This chapter considers three major mechanisms producing numerous cattle coat colour variations due to: (i) pigment intensity and balance; (ii) switching genes affecting spatial and temporal distribution of pigments; and (iii) development, migration and survival of melanosomes, the pigment producing cells. Other mechanisms influencing colour variation include mRNA longevity and processing, which impacts translation and melanin production (Rouzaud et al., 2010). Despite a long and successful history of discoveries in genetics of colour variation one may expect further progress in this classical field of science.

The Basis of Pigmentation

Despite the deceptive simplicity of only two pigments generating all colour variations, the entire system of mammalian coat colour determination is complicated, comprising more than 150 coat colour-associated genes discovered to date (Cieslak et al., 2011). However, the genetic pathways influencing coat colour are still poorly understood and, hence, significant progress is anticipated in the future. Coat colour depends on type of melanin produced by melanocytes or a ratio of the two pigments. It also depends on intensity of melanin production and distribution of pigments along hairs. Distribution and activity of melanocytes across the body is another important factor affecting the final colour of an animal. The current knowledge of coat colour genes in cattle is much narrower than in better studied rodents and dogs. This might be explained by the history of mouse and dog domestication, during which fancy colours were particularly attractive, as well as by more practical aspects governing domestication of cattle. For instance,

there is some evidence that in tropical conditions characterized by high solar radiation, animals with light-coloured coats and darkly pigmented skin might be better adapted (Finch and Western, 1977; Finch et al., 1984). Interestingly most zebu breeds and Italian breeds like the Chianina (Plate 6) and some others have this type of coloration (Olson, 1999).

T. van Vuure (http://members.chello.nl/~t.vanvuure/oeros/uk/lutra.pdf) concluded that sexual dimorphism in domestic cattle has decreased markedly including coat colour comparatively to the known ancestral differences. The coat colour and pattern are rather uniform in many current cattle breeds. Nevertheless there are some breeds where segregating alleles in the colour determining gene(s) create several common colour and pattern types, all of which are acceptable within a breed.

A mammalian melanocyte can produce two types of pigments, eumelanin (black) and pheomelanin (red), which are incorporated in organelles called melanosomes. Synthesis of both pigments starts from tyrosine and includes several steps, the first of which is catalysed by tyrosinase (TYR gene). The major chemical difference in the formation of the two pigments is the involvement of cysteine in pheomelanin synthesis. The final steps of eumelanin synthesis are catalysed by dopachrome tautomerase (DCT gene) and tyrosinase related protein 1 (TRP1 gene). Melanosomes containing eumelanin are compact, dark and have an elliptical oblong shape, whereas those with pheomelanin are less compact, reddish and have a spherical shape. Depending on animal genotype, individual melanocytes may contain both types of melanosomes; the coat colour of Limousin (Plate 14) is an example (Renieri et al., 1993). However, production of melanosomes in a melanocyte is usually shifted towards one or other type. The solid black phenotype in numerous cattle breeds (Plate 2) is a result of predominant production of eumelanosomes, whereas breeds with light red (Plate 9) or yellow coloration exclusively produce pheomelanosomes. Eventually melanosomes are transferred to keratinocytes of growing hair at which time they determine the colour of the hair. A simplified illustration of the role that some major coat colour-related genes play in melanocyte biology is shown in Fig. 4.1. All genes shown

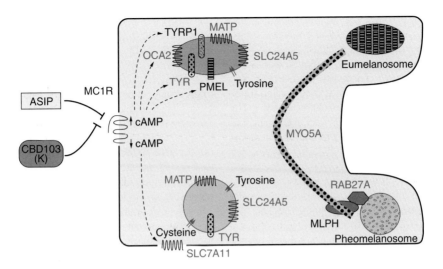

Fig. 4.1. The role of bovine coat colour genes in melanocyte cell biology. (This figure was adapted from Kaelin and Barsh (2012) with the kind permission of Chris Kaelin.) The diagram shows a melanocyte, with eumelanogenesis and pheomelanogenesis depicted in the upper and the lower sections, respectively. Protein names correspond to the genes, some of which are discussed in this chapter. The proteins with allelic variation in cattle are shown in black (except MLPH, which is monomorphic), and those that have been implicated in other systems or in different species are shown in grey. The type of pigment synthesized by melanocytes is controlled by the MC1R and its second messenger cAMP. High levels of basal MC1R signalling cause increased expression of TYR (tyrosinase), TYRP1 (tyrosinase-related protein 1), OCA2 (a membrane protein implicated in oculocutaneous albinism) and PMEL (a melanocyte-specific transmembrane glycoprotein), leading to increased eumelanin synthesis. Low levels of cAMP cause increased expression of the cysteine transporter SLC7A11, leading to increased pheomelanin synthesis. CBD103, which is named DEFB300 or beta-defensin 103B (an MC1R ligand, encoded by the K locus), prevents ASIP from inhibiting MC1R, thereby promoting eumelanin synthesis. This illustration is drawn to emphasize the differences between eumelanin and pheomelanin synthesis in the melanosomes; in reality, biogenesis of the different organelles is more complex and involves a common precursor organelle and several distinct protein trafficking steps. As melanosomes mature, they are transported to dendritic tips via a process that depends on the unconventional myosin (MYO5A), a GTP-binding protein (RAB27A) and an adapter protein (MLPH). MATP is a member-associated transporter protein; SLC24A5 is a solute carrier family 24 member 5 (also known as sodium/potassium/calcium exchanger 5 precursor), which has not yet been studied in cattle.

in the figure have been identified in cattle and mutations of several these genes have been studied.

The wild type

Colour variations occurring in different cattle breeds are easier to understand when the wild type is known. According to Olson (1999) the wild predecessor of domestic cattle, the auroch, was essentially reddish brown to brownish black with a tan muzzle. According to Hassanin (Chapter 1) 'the coat colour [of aurochs, *Bos primigenius*] was reddish-brown for calves and cows, and changed to a dark brown or black in bulls, with apparently a white dorsal stripe running down the spine. Both sexes had a pale muzzle'. A detailed description of the aurochs (http://members.chello.nl/~t.vanvuure/oeros/uk/lutra.pdf) was given by van Vuure. These views match well and provide a reasonable description of the wild type. Some Jersey (Plate 13), Brown Swiss purebreds as well as some crosses may have more or less similar phenotypes. 'Cattle with such brownish-black colour at maturity are born a reddish brown and darken when the calves shed out for the first time' (Olson, 1999). An opinion was expressed that the wild type of the ancestral cattle breeds might be caused by a genotype causing higher pheomelanin content, and that steady fixation of mutations led to the significant coat colour variations seen today (Seo *et al.*, 2007). Lenstra and Felius (Chapter 2) note that white, creamy white, ebony black, yellow and spotted cattle were already known in antiquity. The modern breeds represent an even wider variety of colours and patterns. Still the most commonly observed coat colour phenotypes in bovine are deviations from red or black.

Sometime in those genes where only dominant mutations have been observed so far, the wild type alleles are considered as recessive and designated by the lower case letter. This creates an uneven approach to different genes and here we follow a simple rule: all wild alleles are designated by the capital letter with superscripted + symbol. Obviously this kind of nomenclature does not influence any conclusion regarding dominance or recessiveness of the wild allele in question. Such an approach makes the rules uniform across all genes and is compatible with the genetic nomenclature requirements (Chapter 24).

Genes Causing Spatial and Temporal Pigment-type Switching

Introduction

In cattle, as in some other mammals, mutations of the *Extension* and *Agouti* loci have critical importance for switching between eumelanin and/or pheomelanin production over the entire coat or in a particular region and also during development (Searle, 1968; Silvers, 1979). Figure 4.1 shows that the synthesis of both melanins is regulated by *MC1R* and *ASIP* genes controlling an intercellular signalling pathway. The proteins produced by these genes are antagonists and the outcome of their interaction influences the passage of the activation signal into the melanocyte. A few mutant and wild type alleles have been described for both loci in cattle (Table 4.1). Until very recently nothing or very little was known about the contribution of the *K* locus

(*DEFB300*) in this signalling pathway. The latest available information regarding this locus is given below. The antagonistic interactions between these genes and proteins have been known for some time and create complex phenotypes.

The *Extension* locus (*MC1R*)

The three most common alleles were identified in the *Extension* locus. E^D dominant black causing uniformly distributed black colour at birth (Plate 2); E^1, which in homozygotes leads to a brown-black coat with darker extremities (wild type); and recessive e, which in homozygotes leads to a red coat without any dark pigmentation (Plate 7) (Olson, 1999). Thus, cattle with genotypes E^D- generate only eumelanin and with genotype ee only pheomelanin, while E^+E^+ cattle produce both pigments. Obviously such a description oversimplifies the situation as it ignores the effects of other genes and complex interactions.

As initially established in the mouse, the *Extension* locus corresponds to *Mc1r*, the gene coding for the melanocyte-stimulating hormone receptor (melanocortin 1 receptor) (Robbins *et al.*, 1993). In several mammalian species, some mutations in the *MC1R* gene led to the dominant black phenotype due to exclusive eumelanin production or the recessive red phenotype determined by pheomelanin. MC1R signalling controls switching between the production of eumelanin and pheomelanin. This is done by binding either melanocyte-stimulating hormone (α-MSH), acting as an agonist, or the agouti signalling protein acting as an antagonist (Fig. 4.1). As a result, synthesis of eumelanin and/or pheomelanin is stimulated or inhibited, respectively. In cattle three *MC1R* alleles were initially found. One that corresponds to the dominant E^D allele determining black colour is the result of a substitution. Another allele is caused by a frameshift mutation which generates a stop codon and leads to prematurely terminated MC1R protein in e/e red coat colour homozygotes. The third corresponds to wild-type allele E^+, generating a variety of colours. The *MC1R* gene has been mapped to bovine Chromosome 18 (Klungland *et al.*, 1995).

Table 4.1. Major bovine genes switching and diluting coat colours

Symbol	Gene name	Alleles	Classical view — Allele description	Representative breeds	Molecular genetic view — Location	Symbol	Name	Effect on pigmentation
A	Agouti	A^{bp}	Patterned blackish	Holstein, Jersey	Chr. 13	*ASIP*	Agouti-signalling protein	Switch between eumelanin and pheomelanin
		A^+	Wild type	Auroch				
		a^w	White-bellied	Brown Swiss, Hungarian Grey				
		a^f	Fawn	Limousin, Brahman, Chianina				
B	Brown	B^+	Wild type	Most breeds	Chr. 8	*TYRP1*	Tyrosinase-related protein 1	Change black eumelanin to brown
		b^{Dx}	Brown	Dexter				
C	Albino	C^+	Wild type (full colour)	Most breeds	Chr. 29	*TYR*	Tyrosinase precursor	Total or partial lack of pigmentation in hair, skin and eyes in mutant homozygotes
		c	Albino	Braunvieh (some animals)				
		c^{WH}	Himalayan type	White Highlands				
D	Dilution	D^C	Dilution Charolais	Charolais	Chr. 5	*PMEL17*	Melanocyte protein PMEL precursor	Dilution of pigment
		D^H	Dilution Highland	Highland, Galloway				
		D^S	Dilution Simmental	Simmental				
		D^+	Wild type	Most breeds				
E	Extension	E^D	Dominant black	Black Angus	Chr. 18	*MC1R*	Melanocortin 1 receptor	Relative production and distribution of eumelanin and pheomelanin
		E^+	Wild type	Jersey, Brown Swiss				
		e	Red	Red Angus				
K	(Black)	K^+	Wild type	Most breeds	Chr. 27	*DEFB300*	β-defensin 103B precursor	Switch between eumelanin and pheomelanin
		K^{br}	Brindle	Some animals in Icelandic cattle, Highland, etc., carry allele K^{br}				
		K^{vr}	'Variant red'	Holstein				

A deletion in the MC1R gene was discovered that is associated with red coat colour in Holstein cattle (Joerg et al., 1996). MC1R mutations creating premature stop codons are widespread in several cattle populations, which indicates that this gene may not have other critical for life functions (Klungland and Våge, 2003). A new allele, named E^1, was found in Aubrac and Gasconne breeds in hetero- and homozygotes. This allele has a duplication of 12 nucleotides, which generates four additional amino acids in the protein, located within a region known for interaction with G proteins transmitting signals from outside to inside a cell (Rouzaud et al., 2000). In this case the signal is α-MSH.

Further studies demonstrate that the MC1R mutation corresponding to the e allele is a non-functional receptor, as it was unresponsive to a wide range of α-MSH concentrations. Another recessive e^f allele found in the Simmental breed was more responsive but under much higher α-MSH concentrations. Two additional alleles were found in the Brown Swiss population (E^{D1} and E^{D2}), both of which act in a dose-dependent manner to stimulate α-MSH. This is in contrast to a common dominant E^D allele, which encodes constitutively activated MC1R receptor (Graphodatskaya et al., 2002). Allele E^{D2} contains the same 12 nucleotide duplication earlier described by Rouzaud et al. (2000), which does not affect colour in the Brown Swiss. This observation is supported by the study of Dreger and Schmutz (unpublished data cited by Schmutz, 2012). A simple Mendelian explanation of phenotype–genotype relations was put under some pressure when a bull with mosaic expression of red versus black pigment was found to be a carrier of the dominant E^D allele (Klungland and Våge, 1999). According to the International Genetic Nomenclature rules the Extension alleles should be named $MC1R^D$, $MC1R^+$ and $MC1R^r$.

The *Agouti* locus (*ASIP*)

The *Agouti* locus also has several alleles. The A^{bp} allele, so called *pattern blackish*, slightly modifies typical wild type by making it nearly black and is not influenced by sex (Olson, 1999).

This allele is dominant in the presence of E^+ but hypostatic to E^D. No molecular data are available at this stage for the A^{bp} allele and, until such information is published, one should accept it as a preliminary observation. A similar notion is relevant to other alleles, which fit into the same category. *White-bellied agouti*, a^w, is a recessive allele that in homozygotes removes red pigment and partially black pigment and distributes black pigment more uniformly across the sides of the animal. The belly is usually white or light. Another allele is the recessive *fawn*, a^f, which was postulated by Olson (1999) and causes removal of red and black pigment, particularly red, along the underline. No molecular data relevant to this allele are currently available.

As initially discovered in the mouse, the *Agouti* locus encodes a protein (agouti signalling protein, ASIP), which is a high-affinity antagonist of the MC1R and blocks α-MSH stimulation of adenylyl cyclase, the effector through which α-MSH induces eumelanin synthesis (Lu et al., 1994; Fig. 4.1). High levels of basal MC1R signalling cause elevated quantities of TYR, TYRP1, OCA2 and PMEL proteins, thus leading to increased eumelanin synthesis. Low levels of cAMP cause increased expression of the cysteine transporter SLC7A11 (as yet uncharacterized in cattle), leading to increased pheomelanin synthesis.

The bovine *ASIP* gene is located at chromosome 13 (Schläpfer et al., 2000) and has three coding and several non-coding exons (Fig. 4.2). A widespread expression of *ASIP* in different bovine tissues at mRNA (Girardot et al., 2005) and protein levels (Albrecht et al., 2012) was clearly demonstrated. Mutations of *ASIP* may affect switches of melanin synthesis from the black/brown eumelanin to red/yellow pheomelanin. Numerous mutant alleles of this gene have been described in the mouse. In cattle four alleles, mentioned in the introduction and causing similar phenotypes, have also been identified (Adalsteinsson et al., 1995; Olson, 1999; Table 4.1), although final confirmation might be desirable. Sequencing the coding exons of the *ASIP* gene in 20 animals belonging to six Spanish (Asturiana de los Valles, Asturiana de la Montana, Negra Serrana, Parda Alpina, Sayaguesa and Tudanca) and three French cattle breeds (Parthenais, Tarantaise and Normande) has not revealed a single mutation. This result suggests that the *ASIP* coding region does not play a central role in coat colour variation in cattle (Royo *et al.*, 2005); Girardot *et al.* (2005) also independently found 'no evidence of coding-region sequence variation within and between eight breeds representing a large panel of coat colour phenotypes'.

The temporal and spatial regulation of *ASIP* (*Agouti*) was not understood in molecular terms until two alternative promoters were discovered in mice that behave differently on the dorsal and ventral sides of the body (Vrieling *et al.*, 1994). Similar discoveries were made in dogs (Kaelin and Barsh, 2012) and other species. In cattle three mRNAs with the same coding region but different 5' untranslated regions were also discovered. 'Upstream regulatory sequences display two alternative promoters involved with the broad expression in tissues other than skin' (Girardot *et al.*, 2005). These sequences are highly homologous to upstream sequences of other studied mammals. Further investigations are warranted in order to check whether or how alternative promoters operate in cattle with different coat colours and hence genotypes.

Unlike in other studied mammals, in cattle *ASIP* is expressed in many tissues other than skin. 'ASIP mRNA was up-regulated more than ninefold in intramuscular fat of Japanese Black cattle compared to Holstein ($p < 0.001$). Further

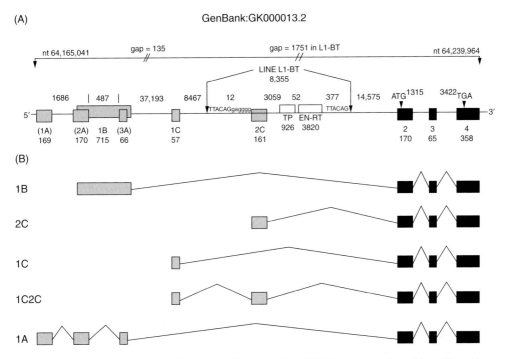

(A)   GenBank:GK000013.2

Fig. 4.2. Structure of the bovine *ASIP* gene and its transcripts. (This figure was adapted from Albrecht *et al.* (2012) with the kind permission of Steffen Maak.) (A) Non-coding (grey) and coding exons of *ASIP* (black) are given as boxes and are numbered below. Exons in parentheses were not observed in the study of Albrecht *et al.* (2012). Smaller numbers indicate exon and intron sizes in base pairs. A LINE element (L1-BT) is inserted between non-coding and coding exons. The underlying sequence (GenBank accession no. GK000013.2) contains two gaps. The size of the first gap was determined by sequencing, whereas the second gap was closed in silico by insertion of a partial sequence from DQ000238.1. (B) Transcripts of the bovine *ASIP* gene resulting from different use of non-coding exons. Transcript 2C recruits a non-coding exon from the LINE. Transcript 1A was not observed in the study of Albrecht *et al.* (2012). The figure was modified and supplemented by Albrecht *et al.* (2012) on the basis of data from Girardot *et al.* (2005, 2006).

analyses revealed that a transposon-derived transcript was solely responsible for the increased ASIP mRNA abundance' (Albrecht et al., 2012). However, these differences, are not related to the intramuscular fat content in cattle. According to the International Genetic Nomenclature rules the *Agouti* alleles could be named *ASIP⁺* and possibly *ASIPᵇᵖ*, *ASIPʷ* and *ASIPᶠ* after molecular identification of the expected alleles.

Extension–Agouti interactions

There should be a number of molecular interactions that influence α-MSH signal-activating melanogenesis. Despite the significant limitations in our current knowledge, two critically important participants are known: MC1R and ASIP proteins. ASIP antagonizes MC1R signalling and thus promotes pheomelanin synthesis (Fig. 4.1). As already mentioned mutations in *MC1R* or *ASIP* in cattle alter the timing and/or distribution of eumelanin and pheomelanin. As established in other species, sometimes *MC1R* gain-of-function or *ASIP* loss-of-function mutations cause exclusive production of eumelanin. Conversely, *MC1R* loss-of-function or *ASIP* gain-of-function mutations cause exclusive production of pheomelanin (Kaelin and Barsh, 2012). Whether the same rule is applicable to cattle is not yet known.

Many authors emphasize that red and black colours are the most common among numerous cattle breeds. It has to be taken into consideration that unidentified modifiers may cause significant variations around these two basic colours and thus complicate the general situation. Still it can be concluded that in cattle the red phenotype is predominantly generated by the genotype ($A⁺A⁺ee$), which indeed has non-functional (loss-of-function) *MC1R* alleles. Thus, it may represent an example of recessive epistasis, when functional ASIP protein determined by wild type alleles cannot further antagonize non-functional MC1R; a similar explanation may be true for A^{bp} allele. A special case of red with a different genetic determination is mentioned below. Alternatively, gain-of-function *MC1R* mutation, like E^D, which encodes constitutively activated MC1R

receptors (Graphodatskaya et al., 2002), usually leads to the black phenotype, and this is an example of dominant epistasis. Adalsteinsson et al. (1995) concluded that the *Agouti* alleles are only able to express their effect in $E⁺/–$ genotypes.

According to Schmutz (2009) '$E⁺$ appears to act as a "neutral" allele in most breeds and $E^D/E⁺$ cattle are typically black and $E⁺/e$ cattle are typically red. $E⁺/E⁺$ cattle can be almost any colour since other genes, such as the *Agouti* take over in dictating what pigments are produced.' Even if '$E⁺/e$ cattle are typically red' sounds a bit controversial, it may mean that the $E⁺$ allele does not exert epistatic influence on *Agouti* and even might be itself subject to epistatic influence, like in the case of white-bellied cattle. However, such influence has obvious dorso–ventral differences. The origin of such differences is not related to coding exons, which are conservative in animals of different colours but rather regulatory sequences (Fig. 4.2). Hopefully future molecular studies in cattle will bring more relevant information. Paradoxically Berge (1949) realized that inheritance of black, brindle, brown and red in cattle is really complicated. The data presented in several tables and particularly Table 5 of his paper demonstrate that even crosses of red × red produce not only red offspring (101 from total 125), but also 3 black, 6 brindle and 15 brown.

An observation described by Schmutz (2009) regarding age-related changes in coat colour adds further complexity: 'some Holstein cattle change from red to black or less commonly from black to red as they age from calf to adult.' In Highland cattle these age related changes are expressed particularly clearly (Glen and Karen Hastie; http://www.bairnsley.com/Colour%20-%20black.htm).

The *K* locus (*DEFB300*)

The phenotype of cattle carrying mutation *Brindle* (Plate 26) may hint at pigment switching in different sections of the coat. Indirect observations have been published (Girardot et al., 2006) which could indicate that allele *Brindle* belongs to the *Agouti* locus. However, as explained below, the latest publication negates this possibility (Albrecht *et al.*, 2012) and the *Brindle* is more likely to be a mutation belonging to the *K* locus.

This locus was discovered during the past several years owing to the efforts of canine geneticists (Candille *et al.*, 2007; Kaelin and Barsh, 2012). In mammals, beyond the *MC1R–ASIP* pair encoding ligand-receptor system controlling pigment type-switching, there is a third gene (named the *K* locus) which represents a previously unrecognized component of the melanocortin pathway. In dogs this gene was identified as beta-defensin 103 (CBD103). Its protein binds with high affinity to the MC1R and promotes eumelanin synthesis by inhibiting ASIP antagonism of MC1R (Candille *et al.*, 2007). Bovine beta-defensin 103B precursor, *DEFB300* gene, has ~70–80% identity with the orthologous genes from other studied mammals, like dog, mouse, human and others (*Ensembl* genome browser).

The so-called 'variant red' coat colour observed in some Holstein cattle is a good reminder that similar phenotypes can be produced by very different genotypes. Co-segregation analysis conducted by Dreger and Schmutz (2010) excluded melanocortin 1 receptor, agouti-signalling protein, attractin and melatonin receptor 1A as causative genes, but indicated that β-defensin, which is homologous to the canine gene and located at chromosome 27, seems to be responsible for the segregation pattern. According to Schmutz, this 'variant red' acts like it is dominant to black, but it is rather epistatic since it is not caused by a *MC1R* allele. Temporarily this allele may be named K^{vr}. Thus, proteins produced by *MC1R*, *ASIP* and *DEFB103* interact on the melanocyte entrance and depending on genotype pass a signal transmitted by α-MSH.

In the dog, the *brindle* mutation causes an irregular pattern of circular stripes. In cattle a similar phenotype with black and red stripes also can be observed. Sometimes such a phenotype is described as the 'tiger stripe'. As already explained, in the dog *brindle* is a mutation of *CBD103* gene. A study in the Normande cattle breed discovered insertion of a full-length LINE 1 transposon element in the 5′ regulatory sequence of the *ASIP* gene. Some of these animals had the brindle phenotype,

which generated a hypothesis that the insertion may explain the origin of the brindle coat colour in the Normande breed (Girardot *et al.*, 2006). A similar observation was made regarding Highland cattle, with one essential difference: non-brindle animals also had the allele with a LINE 1 element (Schmutz, 2009). Albrecht *et al.* (2012) tested this hypothesis, while studying expression of ASIP transcript 2C from a single allele in skin and found four animals carrying LINE 1 (Fig. 4.2). However, none of these animals displayed a brindle phenotype. The conclusion was drawn that the hypothesis does not work, at least in the original form. In other words no causative connections between the LINE1 insertion and brindle phenotype were found. 'Either a homozygous status for the insertion of the LINE 1 is required, which was not observed in the animals, or additional factors may be necessary to cause the brindle coat colour in cattle' (Albrecht *et al.*, 2012).

Similarity with the facts established in the dog investigations is very significant and allows us to assume that brindle in cattle is caused by a mutation of the *DEFB300* gene. Certainly only experimental testing will allow the verification of this assumption. At this stage this assumption looks to be the most parsimonious. If so, the appropriate symbol for bovine brindle should be K^{br}. Brindle cattle always have at least one E^+ and no E^D alleles (Schmutz, 2012). Thus, at the moment two rare mutant bovine alleles are assumed in the *K* locus: K^{vr} and K^{br}. Unavoidably a question about the wild type allele arises and such a wild allele could be named K^+. If the assertion regarding the nature of brindle and 'variant red' phenotypes is correct, then according to the International Genetic Nomenclature rules the *K* locus alleles could be named $DEFB^+$, $DEFB^{br}$ and $DEFB^{vr}$.

Genes Diluting Pigmentation

Introduction

So far three genes causing coat colour dilution in cattle have been identified. All these three genes operate within melanocytes: *TYR*, *TYRP1* and *PMEL* (Fig. 4.1). The first gene (*TYR*) encodes tyrosinase, a transmembrane melanosomal protein. The intramelanosomal domain of this protein catalyses a critical step in both eumelanin and pheomelanin synthesis. The second gene (*TYRP1*) encodes an enzyme (5,6-dihydroxyindole-2-carboxylic acid oxidase) that catalyses the oxidation of intermediates in eumelanin synthesis. The third gene (*PMEL*) encodes a melanocyte-specific type I transmembrane glycoprotein, which is involved in generating internal matrix fibres of melanosomes. These three proteins are involved in different but essential processes taking place during pigment biosynthesis in melanosomes. Not surprisingly mutations of these genes lead to similar phenotypic consequences in studied mammalian species. Independent actions of these genes and proteins create complex phenotypes.

The *Albino* locus (*TYR*)

Mutations of several major mammalian loci cause different types of pigment dilution. The *Albino* locus is one of them. The wild type allele (C^+) provides full colour, whatever it might be. A couple of classical forms of albinism are known in several mammalian species like the complete albino (*cc*) with red eyes, and partial forms of albinism (c^pc^p) when only the tip of muzzle, ears and lower limbs are coloured. Both forms are found in cattle and have been studied. No such forms of albinism are common for a cattle breed but these recessive alleles do exist in low frequencies in Braunvieh, Brown Swiss and other breeds. Recessive epistasis expressed in albino animals is caused by a deficiency of tyrosinase – a key enzyme for melanin biosynthesis.

A molecular basis for the *Albino* locus was understood long before the genomic era. Russell and Russell (1948) demonstrated that deficiency of tyrosinase enzyme was a reason for the albino phenotype. The *TYR* (tyrosinase) gene was mapped to bovine chromosome 29 (Schmutz and Cundiff, 1999). A study of the full-length protein-coding sequence of tyrosinase (*TYR*) in albino Braunvieh calves revealed an insertion that caused a premature stop codon at residue 316. All three studied calves were homozygous for this insertion and

their parents were heterozygotes. This condition is also called oculocutaneous albinism. However, an albino Holstein calf did not have this insertion, and the exact nature of the mutation that led to albinism in this case was not established (Schmutz et al., 2004). According to the International Genetic Nomenclature rules, this albino Braunvieh allele could be named TYR^{aBr} and the wild type should be TYR^+.

Another form of albinism is typical for White Galloway cattle, which have coloured muzzles, ears and lower limbs. The colours vary depending on the animal's genotype for other colour-determining alleles. A similar looking phenotype in other mammalian species is usually called Himalayan. Observed in homozygotes for c^h alleles, it served for decades as a classical example for genetics textbooks. The molecular cause for this kind of mutation in White Galloway cattle has not been studied as yet. However, information available from other mammals is useful. In mice for instance, the Himalayan allele contains an A→G substitution at nucleotide 1259 that alters a histidine residue to an arginine residue. Importantly this residue and the surrounding amino acids are conserved from mouse to human. It seems likely that the altered amino acid may play a role in stabilization of the tyrosinase molecule, or in interaction with other molecules (Kwon et al., 1989). The answer to the question of why only extremities are coloured in such Himalayan animals was found many years ago and is related to the temperature of the coloured areas. The 'Himalayan' tyrosinase has maximum activity at temperatures (15°C to 25°C) well below normal body temperature. In other areas higher temperature prevents normal functioning of this temperature-sensitive mutant tyrosinase (Kidson and Fabian, 1981). According to the International Genetic Nomenclature rules this albino White Galloway allele could be named TYR^{aWH}.

The *Brown* locus (*TYRP1*)

Variations of brown coat colour are quite often seen in cattle. However, the origin of these bovine colours seems to be entirely different from the well-studied brown colour in mice, where a wild type B and two recessive b and b^c alleles have been described. Coat colour of B-mice is determined by other genes, the mutant homozygotes however are brown. Until molecular techniques were introduced nothing was known about the *Brown* locus in cattle.

Our current knowledge about the *TYRP1* gene in cattle is based on two publications (Berryere *et al.*, 2002, 2003). This gene was mapped to chromosome 8. Brown coat colour in Dexter cattle is inherited as an autosomal recessive trait and, as became clear during the investigation, concomitant dun colour has nothing to do with this phenomenon. A mutation (H424Y) in the *TYRP1* gene that caused amino acid substitution was found in the homozygous condition in all 25 affected 'brown' Dexter animals regardless of shade of dun ranging from a pale golden to dark brown. Importantly, black Dexter animals had either one mutant allele or none. This mutation was not found in any of the 121 examined animals from other breeds.

In mice ultrastructural evaluation revealed that mutations in the *TYRP1* gene affected melanosome maturation and compromised tyrosinase activity within the organelle. Dihydroxyphenylalanine (DOPA) histochemistry also revealed differences in the melanosomal stages between black and brown melanocytes (Sarangarajan *et al.*, 2000). These findings confirmed the classical understanding of the phenomenon observed in mice (Searle, 1968; Silvers, 1979). It is reasonable to assume that similar changes in bovine TYRP1 protein might lead to the brown colour in the Dexter. According to the International Genetic Nomenclature rules this brown Dexter allele could be named $TYRP1^{bDx}$ and the wild type allele $TYRP1^+$.

The *Dilution* locus (*PMEL17*)

Dilution of a major colour is known in several cattle breeds like the Charolais, Highland, Galloway, Simmental and others. The mode of inheritance was described as dominant or semi-dominant. Olson (1999) indicated that Charolais and Simmental alleles are different, Dc and Ds correspondingly. Wild allele D^+, while behaving as recessive or rather less dominant in heterozygotes is unable to resist visible dilution of a colour, while in homozygotes it produces full colour. Mutant homozygotes show significant and uniform pigment dilution over the entire body. These types of dilutions affect black and red pigmentation. Only the introduction of molecular techniques allowed some progress in identification of different mutations causing dilution in cattle breeds.

Guibert *et al.* (2004) established that pheomelanin coat colour dilution in French cattle breeds is not correlated with *TYR*, *TYRP1* or *DCT* transcription levels. Then two papers indicated a connection between dilution of eumelanin and pheomelanin in cattle and the *SILV* gene on chromosome 5 (Gutiérrez-Gil *et al.*, 2007; Kühn and Weikard, 2007). A non-synonymous mutation in exon 1 of the *SILV* gene was detected ($c.64A \rightarrow G$) and associated with coat colour dilution in an F2-Backcross Charolais × Holstein population, where Charolais animals were mutant homozygotes. This mutation was initially described in Oulmouden, A., Julien, R., Laforet, J.-M. and Leveziel, H. Patent Publication in 2005 (WO2005/019473 cited by Gutiérrez-Gil *et al.*, 2007). Kühn and Weikard (2007) confirmed this finding and draw similar conclusions. Hence, the mutation observed in the Charolais is equivalent to the previously known *Dc* allele. Thus, the *SILV* gene was identified as the *PMEL17* gene encoding melanocyte protein PMEL precursor. This protein plays a central role in the early stages of melanosome biogenesis (Theos *et al.*, 2005). PMEL is found in pigment cells and forms fibrils during early stage melanosomes upon which eumelanins are deposited later. Murine melanosomes within $Pmel^{-/-}$ melanocytes, where both alleles have been inactivated, are spherical in contrast to the ellipsoid oblong shape typical for wild-type animals (Hellström *et al.*, 2011). Something similar can be expected in mutant homozygotes in cattle.

Hereford × Friesian crossbreds were used to investigate cases of coat colour dilution and hypotrichosis (abnormal hair patterns due to loss or reduction). An affected calf and its Hereford sire were heterozygous for a three-base deletion in exon 1 of the *PMEL17* gene (a loss of leucine codone, CTT, in position 18). The two animals were also heterozygous for a second mutation in exon 11 of the *PMEL17* gene (C→A substitution in 612 codon, replacing alanine with glutamic acid). It appears almost certain that the mutation in exon 11 is an independent event. Four other related animals also carried the same mutations (Jolly *et al.*, 2008). Authors of this study believe that a similar genetic disorder was previously described in Simmental crossbred calves (Jolly *et al.*, 2008). A conclusion could be drawn that the described Hereford × Friesian crossbreds were distinct from the above mentioned affected Charolais animals, which have a different mutation in the first exon.

The latest available study of *Dilution* in Highland and Galloway breeds described a similar, or rather likely the same, *PMEL17* deletion that also led to a loss of leucine in position 18 in the PMEL17 protein (Schmutz and Dreger, 2013). There are two shades of colour dilution in Highland and Galloway cattle, less and more intense. The first shade is known as dun and the second as silver dun. Schmutz and Drager (2013) have convincingly shown that heterozygotes for this deletion have significantly lighter colour regardless of the *MC1R* genotype. Homozygotes for this deletion have drastically reduced colour and can be categorized as silver dun, again regardless of *MC1R* genotype, and inheritance type is semidominant. The mutant allele found in Charolais was not observed in Highland or Galloway cattle.

A region on chromosome 28 influences the intensity of pigmentation and therefore may include a modifier of the *Dc/Dc* genotype. A candidate gene, *LYST*, was identified (Gutiérrez-Gil *et al.*, 2007). Kühn and Weikard (2007) suggested further investigation because the existing data indicate that a single mutation in the *PMEL17* gene may not be sufficient to explain all observed variations relevant to dilution colours.

According to the International Genetic Nomenclature rules the described *Dilution* alleles could be named as: *PMEL17^C* (Charolais type), Simmental/Hereford type *PMEL17^S*, *PMEL17^HiG* (Highland/Galloway type, which might be equivalent to Simmental/Hereford type) and *PMEL17^+*.

Genes Affecting Migration and Survival of Pigment Cells

Introduction

Genes and phenotypes described in this section do not affect coat colour as such but rather the pattern or distribution of coloured and white areas over the entire body or a particular part of it. These genes influence either migration of embryonic melanoblasts or development of functional melanocytes and their survival in relevant tissues. This is unlike the genes and phenotypes that are described in the previous sections. In most domestic animals, including cattle, all kinds of white spotting are very common, which is not observed in their wild ancestors. Genetic and developmental changes occurring during domestication significantly increase the frequency of spotted patterns (Belyaev, 1979; Trut *et al.*, 2012).

The variety of spotted patterns in cattle is significant and there are a number of genes, mutations of which cause the phenomenon. Some animals may have more than one spotted phenotype expressed simultaneously. This may complicate investigations and only some relevant genes have been studied deeply enough. Others still await their turn. *Blaze* is one such mutation. Some Simmental cattle as well as animals from other breeds (possibly Holstein and Groningen) may have a solid colour (i.e. black) phenotype with a white blaze on their forehead and face. The size of such a white blaze may vary from a wide stripe along the face to an almost entirely white head, except eyes and muzzle. It probably never or very rarely includes areas beyond the ears and may be accompanied by another type of white spotting, which is not obligatory. In some documented cases two solid black siblings differed, with one of them having a well-expressed blaze phenotype and the other a black face. According to Olson, the blaze phenotype is determined by incompletely dominant allele *Bl* (Olson, 1999; http://www.braunviehcenter.com/cattle_genetics_part2.html). As accepted in this chapter, the wild allele should be designated *Bl^+*. The phenotype of mutant homozygotes was not described with certainty but it is probably a stronger expressed blaze phenotype.

No molecular genetics data relevant to this mutation have been published so far. Another insufficiently studied mutation is *Brockling*, which causes pigmentation in areas of white spotting caused by mutations in other genes. *Brockling* (*Bc*) is considered to be dominant and found in several solid-coloured breeds as well as in Shorthorns, Ayrshire and Normande breeds (Olson, 1999).

Five genes generating different spotted patterns are described below. These include *Belted*, *Colour-sided*, *Piebald*, *Roan* and *White-spotting* genes (Table 4.2). The exact molecular nature of *Belted* may require further confirmation, but a candidate gene encoding for a transcription cofactor regulating cell differentiation was framed in a short DNA fragment on chromosome 3. *Colour-sided* represents a unique case, when translocations of the *KIT* gene from chromosome 6 to chromosome 29 and back create the alleles. *Piebald* (*MITF*) encodes microphthalmia-associated transcription factor, which is responsible for pigment cell-specific transcription of the genes involved in melanogenesis. The *Roan* gene (*KITLG*) encodes the ligand of the tyrosine-kinase receptor encoded by the *KIT* gene. The *White-spotting* (*KIT*) gene encodes tyrosine-protein kinase, which plays an essential role in the regulation of cell survival, proliferation, migration, melanogenesis and other functions. While all five genes are different, the unifying feature is their involvement in gene regulation and cell differentiation and development.

The *Belted* locus (BTA3, possibly *HES6*)

Belted is one the most striking white-spotting phenotypes commonly observed in Dutch Belted and Belted Galloway breeds (Plate 27). The dominant allele (*Bt*) causes this special phenotype and is possibly either widespread or fixed in these two breeds. Wild type animals (*Bt*+ *Bt*+) are most common in cattle generally.

The recent attempt to find a gene associated with the belted phenotype brought significant progress. During the first stage linkage mapping was performed using Brown Swiss animals, which identified the telomeric region

of bovine chromosome 3 as the point of interest. Then fine-mapping and haplotype analysis using 19 additional markers in this region refined the critical region of the belted locus to a 922-kb interval (Drögemüller *et al.*, 2009). Consequently two additional cattle breeds with the belted phenotype: Galloway and Dutch Belted (Lakenvelder) were investigated, which led to confirmation that this phenotype in Galloway is strongly associated with the same chromosomal locus as in Brown Swiss cattle. Eventually a single belt-associated haplotype was identified for each of the analysed breeds. These haplotypes share alleles in four blocks. The largest shared haplotype block incorporates nine SNPs along a 336-kb interval. A potential candidate gene within this interval, *HES6*, is a transcription co-factor playing a developmental role (Drögemüller *et al.*, 2010). So far no belt-associated polymorphisms have been found despite studying the complete *HES6* coding sequence. Whether a mutation in regulatory sequences is involved remains unknown. The pedigree data suggest a common founder for animals with the belt phenotype in different cattle breeds.

According to the International Genetic Nomenclature rules the described *Belted* alleles could be preliminarily named as: *HES6*^B (belted type) and *HES6*+ (wild type).

The *Colour-sided* locus (*KIT*; BTA6→BTA29→BTA6 translocations)

Another peculiar phenotype observed in a number of breeds, including Texas Longhorn, White Park, British White, Florida Cracker, English Longhorn, Belgian Blue, Dutch Belted, Brown Swiss and a few more, is the so-called colour-sided pattern. In heterozygotes (*CsCs*+) an irregular white strip along the dorsal and ventral parts is common. Olson (1999) provided a comprehensive description of typical phenotypes and suggested that the mutant allele is semi-dominant. As recently established, the nature of this genetic change is novel, unusual, complex and there is more than one 'allele' (Durkin *et al.*, 2012; see below).

Olson (1999) noticed that 'The spotting patterns produced by animals heterozygous for

Table 4.2. Major pattern creating bovine genes affecting pigment cells migration and survival.

		Classical view			Molecular genetic view			
Symbol	Name	Alleles	Allele description	Representative breeds	Location	Symbol	Name	Effect on pigmentation
Bt	Belted	Bt	White belt; dominant	Dutch Belted, Belted Galloway	Chr. 3, telomeric region	Possibly $HES6$	Transcription cofactor	Lack of pigmentation around midsection
		Bt^+	Wild type					
Cs	Colour sidedness	Cs_6	Brown Swiss type; semi-dominant	Several other breeds including domestic yak	Chr. 6, translocation Chr. 29 & Chr. 6, translocation	KIT modified	Modified mast/stem cell growth factor receptor	Irregular white stripes at the dorsal and ventral parts of animal
		Cs_{29}	Belgian Blue type; semi-dominant					
		Cs^+	Wild type					
R	Roan	R	Roan; semi-dominant	Belgian Blue	Chr. 5	$KITLG$	KIT ligand	Mixture of pigmented and white hairs
		R^+	Wild type					
P	Piebald	P^+	Wild type	Holstein and numerous other breeds	Chr. 22	$MITF$	Microphthalmia-associated transcription factor	Irregular white spots, piebaldness
		p	Piebald; recessive					
S	White-spotting	S^H	Hereford type; semi-dominant	Hereford	Chr. 6	KIT	Mast/stem cell growth factor receptor	Specifically patched depigmentation
		S^+	Wild type					

Cs may be differentiated from those produced by S^P (Pinzgauer type) in that the spotting produced by Cs generally has a ragged or roan-like edge, whereas the edge of spots produced by S^P are clearly defined.' Interestingly this observation indicates some similarity and difference between the genetic nature of both mutations. A recent publication (Durkin *et al.*, 2012) demonstrates the unique origin of colour-sidedness. 'Colour-sidedness is determined by a first allele on chromosome 29 (Cs_{29}), which results from the translocation of 492-kilobase chromosome 6 segment encompassing *KIT* to chromosome 29, and a second allele on chromosome 6 (Cs_6), derived from the first by repatriation of fused 575-kilobase chromosome 6 and 29 sequences to the *KIT* gene' (Durkin *et al.*, 2012).

According to the paper, initially a section of chromosome 6 that included the *KIT* gene (see the paragraph below on the *White-spotting* locus) and surrounding DNA was cut out or amplified. Then, this circular DNA intermediate was nicked in a different spot, linearized and inserted into chromosome 29 (Durkin *et al.*, 2012). This resulted in the new *KIT* allele (Cs_{29}), which while being homologous to the wild-type *KIT* allele located on chromosome 6, was nevertheless non-syntenic. In such an unusual case even the term 'allele' may not be appropriate. When this first set of translocation events occurred, there were probably three *KIT* alleles in the original animals. Then later a new version of chromosome 29 carrying Cs_{29} was fixed and such animals possess two copies of *KIT* on chromosome 6 and two modified copies on chromosome 29. This was observed in Belgian Blue animals (Durkin *et al.*, 2012).

In the next stage a part of the previously translocated DNA without the *KIT* gene, together with a fraction of the original chromosome 29, was cut out or amplified and inserted back into chromosome 6 close to the *KIT* gene. This second translocation somehow modified the *KIT* gene and a new allele (Cs_6), which can be found in Brown Swiss animals and causes colour-sidedness, was originated. The manner in which this translocation occurred led to duplication of a DNA fragment from chromosome 6 and the inclusion of an additional fragment from chromosome 29.

Testing for the presence of the two Belgian Blue-specific fusion points, the Brown Swiss-specific fusion point and the Belgian Blue/Brown Swiss-shared fusion point in several breeds where the colour-sidedness is known revealed the following. 'Colour-sided Dutch Witrik (Plate 28) and Ethiopian Fogera animals were shown to carry the Belgian Blue Cs_{29} allele; Austrian Pustertaler Sprinzen, Czech Red Spotted cattle and French Vosgienne the Brown Swiss Cs_6 allele, and Irish Moiled, Swedish Mountain and domestic yak carried both the Cs_{29} and Cs_6 alleles. Authors assume that Cs_{29} and Cs_6 alleles were introgressed in yak after its domestication via well-documented hybridization of *Bos taurus* and *Bos grunniens*. These findings indicate that the Cs_{29} and Cs_6 alleles account for most if not all colour-sidedness in cattle' (Durkin *et al.*, 2012).

This excellent research project provided very convincing explanation of genetic determination of colour-sidedness. There are other questions which probably will be addressed in the near future: What changes normal expression of *KIT* alleles and how is this change effected? How many *KIT* alleles exist in different breeds and animals? Do *KIT* alleles from chromosomes 6 and 29 interact and, if so, how?

The white-coloured variations in White Galloway cattle and White Park cattle are also caused by a *KIT* gene (chromosome 6) duplication and aberrant insertion on chromosome 29 (Cs_{29}) (Brenig *et al.*, 2013) as described for colour-sided Belgian Blue and some other breeds. White Galloway cattle show significant variation from fully black to animals without marks. Importantly all 27 studied fully black individuals were homozygotes for the wild-type chromosome 29; all 104 well and strongly marked individuals were heterozygotes (Cs_{29}/wt) and all 37 animals without marks were homozygotes (Cs_{29}/Cs_{29}).

According to the International Genetic Nomenclature rules the described *Colour-sided* alleles could be preliminarily named as: KIT^{Cs29} (Belgian Blue type) and KIT^{Cs6} (Brown Swiss type).

The *Piebald* locus (*MITF*)

Piebaldness or white spotting are commonly used descriptions of widespread phenotypes observed in many cattle breeds. In the past there was a tendency to describe all this great variety of phenotypes assuming the existence of one gene with many alleles. Olson (1999) suggests three mutant alleles: semi-dominant S^H – Hereford pattern with white face, belly, feet and tail (Plate 9); semi-dominant S^P – Pinzgauer pattern with variable amount of white along dorsal and ventral areas; and recessive s – piebald with irregular pigmented and white areas and usually white feet, belly and tail (Plate 15). Obviously there should be the wild type allele S^+. The limitations of hybridological analysis particularly in large animals did not help to establish allelic relationships in this and other cases. Hence, only direct molecular genetic evidence could advance this complex matter.

Use of molecular genetic methods has allowed distinguishing piebaldness and specific white-spotted phenotypes like the Hereford pattern (see below). Holstein–Friesian crossbred cows from an F_2 experimental design were used for the genome scan. Significant QTLs were found on chromosomes 6, 18 and 22. Haplotype data revealed the highly significant QTL on chromosome 22 in the interval covering Microphthalmia-associated transcription factor (*MITF*) gene (Liu *et al.*, 2009). This gene was proven to be associated with pigmentation traits in some other mammals and is a regulator of differentiation of the neural crest-derived melanoblasts (Hozumi *et al.*, 2012). Fontanesi *et al.* (2012) performed a candidate gene analysis, QTL mapping and genome-wide association study for piebaldness in Holstein cattle. These authors obtained clear evidence that the most likely gene causing/affecting the trait in question is the *MITF* gene located at chromosome 22. Sequencing *MITF* in numerous animals from breeds where the piebald animals are common led to identification of 17 SNPs. 'The allele frequencies of one polymorphism (g.32386957A→T) were clearly different between spotted (A = 0.875; T = 0.125) and non-spotted breeds (A = 0.125; T = 0.875) (P = 8.2E-12).' Altogether 21 different haplotypes were also inferred in the study. Although observed *MITF* variability explains the existence of piebald and solid coloured animals, other genetic factors also make a contribution (Fontanesi *et al.*, 2012). This conclusion

matches very well with numerous inconclusive studies of piebaldness made over a long time.

According to the International Genetic Nomenclature rules the described *Piebald* alleles could be preliminarily named as: *MITF*^{p} (piebald type) and *MITF*$^+$ (wild type).

The *Roan* locus (*KITLG*)

Belgian Blue cattle, Shorthorns, Texas Longhorns and several other breeds have a phenotypic trait called roan, caused by a mixture of pigmented and white hairs (Plate 29). This phenotype occurs in heterozygotes (*RR*$^+$) due to the presence of a semi-dominant allele. Mutant homozygotes (*RR*) are almost white with rare pigmented hairs mainly in the ears. At least in Shorthorns and Belgian Blue such female homozygotes suffer from so called 'White Heifer disease' affecting reproduction (Hanset, 1969). The colour of heterozygotes depends on other genes. The available information on the genes, which were studied using molecular methods, is given below and in Table 4.3.

Understanding the molecular nature of *Roan* was among the first discoveries of the molecular genetic era. *Roan* was mapped to bovine chromosome 5 and the interval, where it was located, included newly mapped gene coding for mast cell growth factor (Charlier *et al.*, 1996). This gene was proposed as the candidate gene for *Roan*. Further study undertaken by Seitz *et al.* (1999) identified a missense mutation at position 654 bp (amino acid 193, Ala→Asp) of *KITLG* (former *MGF*) gene, which created the *R* allele.

According to the International Genetic Nomenclature rules the described *Roan* alleles could be preliminarily named as: *KITLG*^{R} (roan) and *KITLG*$^+$ (wild type).

The *White-spotting* locus (*KIT*)

Various white spotting phenotypes in mice (Mackenzie *et al.*, 1997), pig (Andersson and Plastow, 2011) and other mammals are determined by mutations of the *KIT* gene encoding mast/stem cell growth factor receptor Kit (tyrosine kinase receptor). In cattle *KIT* is located on chromosome 6 and genome scanning found a QTL for the proportion of white coat with large effects in German Simmental and German Holstein cattle (Plate 11) is located within the same interval (Reinsch *et al.*, 1999). A similar result was obtained in a Hereford cross population, where the *S* locus was mapped to the interval between markers BM4528 and EL03 (Grosz and MacNeil 1999), thus suggesting a connection between specific Hereford (Plate 9) or white face types of spotting and *KIT*.

Fontanesi *et al.* (2010) investigated variability of the *KIT* gene and haplotype distribution in three breeds (Angus, Hereford and Holstein) with different anticipated alleles at the *S* locus (*S*+, *S*^{H} and *s*, respectively). Re-sequencing a large DNA section (0.485 Mb) including the *KIT* gene revealed 111 polymorphisms. 'The global nucleotide diversity was 0.087%. Tajima's D-values were negative for all breeds, indicating putative directional selection. Of the 28 inferred haplotypes, only five were observed in the Hereford breed, in which one was the most frequent. Coalescent simulation showed that it is highly unlikely (*P* < 10E-6) to obtain this low number of haplotypes conditionally on the observed number of segregating SNPs. Therefore, the neutral model could be rejected for the Hereford breed, suggesting that a selection sweep occurred at the *KIT* gene' (Fontanesi *et al.*, 2010).

The obtained data did not provide evidence in favour of selective sweeps in two other breeds. A conclusion can be drawn that the *S*^{H} allele determining the Hereford type white face phenotype is likely an allele of the *KIT* gene. Angus, having a solid colour without white marks, unsurprisingly showed no evidence of selective sweep. The same seems to be correct for the Holstein spotting, which, as described above, is caused by possible mutations in the *MITF* gene (Fontanesi *et al.*, 2012).

According to the International Genetic Nomenclature rules the described *White-spotting* alleles could be preliminarily named as: *KIT*^{H} (white face Hereford type) and *KIT*+ (wild type). Other possible alleles in this gene have not been studied using molecular methods so far.

Other Genes Influencing Coat Colour

As previous sections testify, the understanding of the molecular genetic basis of coat colour variation in cattle has advanced remarkably over the past 10–15 years. Nevertheless it seems very possible that other genes, which are proven to be involved in coat colour determination in other mammalian species, may be added to the list of currently identified genes. For instance, a whole genome Bayesian scan for adaptive genetic divergence in West African cattle revealed a number of candidate genes that are under strong pressure of natural selection (Gautier *et al.*, 2009). Among them were the *EDNRB* (Endothelin B receptor) gene located on chromosome 12, which is referred to as the *piebald* or *S* locus in the mouse (Shin *et al.*, 1997), and a null mutation induces a white coat colour in the rat (Gariepy *et al.*, 1996). Intensive sun radiation in tropical areas indeed could act as a powerful selective factor for coat colour in cattle. Such a notion has been expressed in the literature many times. Interestingly Olson (1999) mentioned that some African breeds 'appear to possess recessive (white) spotting'.

Another gene that might be potentially involved in coat colour development in cattle is *DCT*, encoding L-dopachrome tautomerase precursor (former *TRP2*), which is also mapped to bovine chromosome 12 (Hawkins *et al.*, 1996). This gene is known to be involved in eumelanin and pheomelanin synthesis in mouse melanoblasts (Lamoreux *et al.*, 2001; Hirobe *et al.*, 2006) and also is a factor affecting the development of neural crest-derived melanoblasts (Pavan and Tilghman, 1994). *DCT* actively interacts with *KIT* and with *MITF*, which are established as major coat colour genes in cattle (Opdecamp *et al.*, 1997). Gene *MYO5A* (myosin VA) mapped to bovine chromosome 10, causes dilution in the mouse and other mammals (Engle and Kennett, 1994). The list is probably longer, but the whole point is to stress that there is a multitude of key genes, which may affect coat colour in cattle. The only critical requirement is functional connection of these genes to a molecular or cellular process leading to wild type pigmentation.

An autosomal recessive mutation causing dilution of coat colour and a bleeding disorder has been reported in cattle and other mammals half a century ago (Padgett et al., 1964; Kuneida et al., 1999). This disorder was later named Chediak-Higashi syndrome (CHS) and was always considered separately from other kinds of dilution. The gene called LYST (lysosomal-trafficking regulator), which is located at bovine chromosome 28, causes CHS in cattle and a few other mammals (Prieur et al., 1976; Ensembl genome browser accessed 23 July 2013). In Japanese Black (Wagyu) cattle A→G transition led to substitution of histidine with arginine (H2015R). This particular mutation was found to be causative for CHS in Japanese Black (Wagyu) cattle (Kuneida et al., 1999; Yamakuchi et al., 2000). The exact reason for the lighter coat colour in affected animals is not known.

Pleiotropic Effects Caused by Coat Colour Mutations

Sometime presence or absence of coloration may impact productive traits like in Holstein cows, where percentage of white coat correlates with milk yield and reproductive traits are congruent with the intensity of solar radiation (King et al., 1988; Hansen, 1990; Becerril et al., 1994; Olson, 1999). Another association of a colour-related trait and an economically important trait is the relationship between eyelid pigmentation and the susceptibility to eye lesions leading to 'cancer eye' in Hereford and other cattle breeds (Anderson, 1991). Stronger eyelid pigmentation in Hereford cattle results in a decreased incidence of lesion development.

Several coat colour genes act during early development and affect basic molecular and cellular processes. Quite often this is sufficient to create significant pleotropic effects. The latest review of various pleiotropic effects of coat colour-associated mutations in mammals was recently published by Reissmann and Ludwig (2013). Several such effects have been studied in cattle on the molecular level (Table 4.3) and have negative pleiotropic effects, as in the case of German White Fleckvieh syndrome. A missense mutation (R210I) has been identified in the MITF gene (chromosome 22) as causative for the syndrome. This mutation affects the highly conserved basic region of the protein and causes a negative-dominant effect, which includes hypopigmentation, heterochromia irides, colobomatous eyes and bilateral hearing loss (Philipp et al., 2011). Another example is White Heifer Disease found among homozygotes for the semi-dominant missense mutation of the KITLG gene (causing amino acid substitution Ala→Asp), which is characterized by a loss of fertility (Charlier et al., 1996; Seitz et al., 1999).

Crossbred calves from Simmental × Angus as well as Hereford × Friesian crosses are known to develop some coat colour dilution and hypotrichosis, which is specifically expressed in coloured areas and, if the tail is affected, leads to the phenomenon of so-called 'rat-tailed' calves (Schalles and Cundiff, 1999; Jolly et al., 2008). Such affected calves are less efficient in gaining weight, and might be up to 36 kg lighter at slaughter. It was established that some Simmental and Hereford bulls are heterozygotes for a three-nucleotide deletion removing leucine codon from the first exon of the PMEL17 gene. The same Herefords also carry another mutation in PMEL17, which is the C→A mutation, causing alanine→glutamic acid substitution (Jolly et al., 2008). While further investigation of this phenomenon might be useful, the molecular cause of this phenomenon has been established. As follows from available pedigree data, not all animals that carry the mutations develop hypotrichosis and 'rat-tailed' syndrome. This observation indicates that penetrance of such a mutation is definitely below 100%. In case like this a molecular test is very useful for discovering all carriers of such a potentially undesirable trait. Such a test was developed and allowed identification of a common ancestor, which was responsible for bringing this mutation into the herd. A similar approach or direct DNA sequencing tests can be applied to mutations of other genes, particularly those with a recessive mode of inheritance or semi-dominant mutations with low penetrance.

Conclusion

There has been significant progress in understanding molecular causes of coat colour mutations in cattle during the past 10–15 years that lays the foundation for further research in this area. Several directions for future research seem

Table 4.3. Pleotropic effects of coat colour affecting genes in cattle with known molecular cause.

Gene symbol	Gene name	Chromosome	Inheritance	Syndrome	Pleoitropic effects	Coat colour	Breed where found	Reference
KITLG	KIT ligand (*Roan*)	Chr. 5	Semi-dominant, missense mutation	White Heifer Disease	Fertility loss	Intermingled coloured and white hairs in heterozygotes, white homozygotes	Belgian Blue	Charlier et al., 1996 Seitz et al., 1999
LYST	Lysosomal-trafficking regulator	Chr. 28	Recessive, missense mutation	Chediak-Hagashi syndrome (CHS)	Increased bleeding tendency, abnormal platelet granules	Light ccat colour	Japanese Black (Wagyu)	Kunieda et al., 1999 Yamakuchi et al., 2000
MITF	Microphthalmia-associated transcription factor (*Piebald*)	Chr. 22	Semi-dominant, missense mutation	German White Fleckvieh syndrome	Deafness, colobomatous eyes, heterochromia	White	German White, Fleckvieh	Philipp et al., 2011
PMEL17	Melanocyte protein PMEL precursor (*Dilution*)	Chr. 5	Semi-dominant, 3-nucleotide deletion exon 1 and mutation in exon 11	Hypotrichosis, 'rat-tailed' calves	Reduction of hair in pigmented areas, white areas are intact, reduced weight gain	Coat-colour dilution	Simmental and Hereford crossbred calves	Jolly et al., 2008
TYR	Tyrosinase precursor (*Albino*)	Chr. 29	Recessive frameshift insertion	Albinism	Multiple ocular abnormalities	White	Numerous	Schmutz et al., 2004

feasible. This may include investigation of those alleles whose molecular nature has not been described so far; studies of complex interactions between coat colour and other genes creating complex colour variations; and finally development and application of DNA-based methods for identification of animals carrying undesirable alleles. It seems that this classical field of genetics has successfully passed the test of time and has good options for further advancement.

Acknowledgements

The author is grateful to Chris Kaelin for permission to use the picture from Kaelin and Barsh (2012); as well as to Steffen Maak for the picture from Albrecht et al. (2012). I am also very grateful to Amanda Slater, Dave from Leicester, Henk Kosters and Malcolm Reedy for their kind permission to publish several colour plates.

References

Adalsteinsson, S., Bjarnadottir, S., Våge, D.I. and Jonmundsson, J.V. (1995) Brown coat color in Icelandic cattle produced by the loci *Extension* and *Agouti*. *Journal of Heredity* 86, 395–398.

Albrecht, E., Komolka, K., Kuzinski, J. and Maak, S. (2012) Agouti revisited: transcript quantification of the *ASIP* gene in bovine tissues related to protein expression and localization. *PLoS One* 7, e35282.

Anderson, D.E. (1991) Genetic study of eye cancer in cattle. *Journal of Heredity* 82, 21–26.

Andersson, L. and Plastow, G. (2011) Molecular genetics of coat colour variation. In: Rothschild, M.F. and Ruvinsky A. (eds) *The Genetics of the Pig*. CAB International, Wallingford, UK, pp. 38–50.

Becerril, C.M., Wilcox, C.J., Wiggins, G.R. and Sigmon, K.N. (1994) Transformation of measurements of percentage of white coat colour for Holsteins and estimation of heritability. *Journal of Dairy Science* 77, 2651–2667.

Belyaev, D.K. (1979) Destabilizing selection as a factor in domestication. *Journal of Heredity* 70, 301–308.

Berge, S. (1949) Inheritance of dun, brown and brindle colour in cattle. *Heredity Pt. 2* 3, 195–204. Available at: http://www.nature.com/hdy/journal/v3/n2/pdf/hdy194911a.pdf (accessed 21 March 2014).

Berryere, T.G., Schmutz, S.M., Cowan C.M. and Potter, J. (2002) Effects of the brown locus (TYRP1) on coat color in cattle. International Society of Animal Genetics meeting, August, Goettingen, Germany, pp. 47–48.

Berryere, T.G., Schmutz, S.M., Schimpf, R.J., Cowan, C.M. and Potter, J. (2003) TYRP1 is associated with dun coat colour in Dexter cattle or how now brown cow? *Animal Genetics* 34, 169–175.

Brenig, B., Beck, J., Floren, C., Bornemann-Kolatzki, K., Wiedemann, I., Hennecke, S., Swalve, H. and Schütz, E. (2013) Molecular genetics of coat colour variations in White Galloway and White Park cattle. *Animal Genetics* 44, 450–453.

Candille, S.I., Kaelin, C.B., Cattanach, B.M., Yu, B., Thompson, D.A., Nix, M.A., Kerns, J.A., Schmutz, S.M., Millhauser, G.L. and Barsh, G.S. (2007). A b-defensin mutation causes black coat color in domestic dogs. *Science* 318, 1418–1423.

Charlier, C., Denys, B., Belanche, J.I., Coppieters, W., Grobet, L., Mni, M., Womack, J., Hanset, R. and Georges, M. (1996) Microsatellite mapping of the bovine roan locus: a major determinant of White Heifer disease. *Mammalian Genome* 7, 138–142.

Cieslak, M., Reissmann, M., Hofreiter, M. and Ludwig, A. (2011) Colours of domestication. *Biological Reviews of the Cambridge Philosophical Society* 86, 885–899.

Dreger, D.L. and Schmutz, S.M. (2010) The variant red coat colour phenotype of Holstein cattle maps to BTA27. *Animal Genetics* 41, 109–112.

Drögemüller, C., Engensteiner, M., Moser, S., Rieder, S. and Leeb, T. (2009) Genetic mapping of the belt pattern in Brown Swiss cattle to BTA3. *Animal Genetics* 40, 225–229.

Drögemüller, C., Demmel, S., Engensteiner, M., Rieder, S. and Leeb, T. (2010) A shared 336 kb haplotype associated with the belt pattern in three divergent cattle breeds. *Animal Genetics* 41, 304–307.

Durkin, K., Coppieters, W., Drögemüller, C., Ahariz, N., Cambisano, N., Druet, T., Fasquelle, C., Haile, A., Horin, P., Huang, L., Kamatani, Y., Karim, L., Lathrop, M., Moser, S., Oldenbroek, K., Rieder, S., Sartelet, A., Sölkner, J., Stålhammar, H., Zelenika, D., Zhang, Z., Leeb, T., Georges, M. and Charlier, C. (2012) Serial translocation by means of circular intermediates underlies colour sidedness in cattle. *Nature* 482, 81–84.

Engle, L.J. and Kennett, R.H. (1994) Cloning, analysis, and chromosomal localization of myoxin (MYH12), the human homologue to the mouse dilute gene. *Genomics* 19, 407–416.

Finch, V.A. and Western, D. (1977) Cattle colours in pastoral herds: natural selection or social preference. *Ecology* 58, 1384.

Finch, V.A., Bennett, I.L. and Holmes, C.R. (1984) Coat colour in cattle: effect on thermal balance, behaviour and growth, and relationship with coat type. *Journal of Agricultural Science (Cambridge)* 102, 141.

Fontanesi, L., Tazzoli, M., Russo, V. and Beever, J. (2010) Genetic heterogeneity at the bovine KIT gene in cattle breeds carrying different putative alleles at the spotting locus. *Animal Genetics* 41, 295–303.

Fontanesi, L., Scotti, E. and Russo, V. (2012) Haplotype variability in the bovine MITF gene and association with piebaldism in Holstein and Simmental cattle breeds. *Animal Genetics* 43, 250–256.

Gariepy, C.E., Cass, D.T. and Yanagisawa, M. (1996) Null mutation of endothelin receptor type B gene in spotting lethal rats causes aganglionic megacolon and white coat color. *Proceedings of the National Academy of Sciences USA* 93, 867–872.

Gautier, M., Flori, L., Riebler, A., Jaffrézic, F., Laloé, D., Gut, I., Moazami-Goudarzi, K. and Foulley, J.-L. (2009) A whole genome Bayesian scan for adaptive genetic divergence in West African cattle. *BMC Genomics* 10, 550.

Girardot, M., Martin, J., Guibert, S., Leveziel, H., Julien, R. and Oulmouden, A. (2005) Widespread expression of the bovine Agouti gene results from at least three alternative promoters. *Pigment Cell Research* 18, 34–41.

Girardot, M., Guibert, S., Laforet, M.P., Gallard, Y., Larroque, H. and Oulmouden, A. (2006) The insertion of a full-length Bos taurus LINE element is responsible for a transcriptional deregulation of the Normande Agouti gene. *Pigment Cell Research* 19, 346–355.

Graphodatskaya, D., Joerg, H. and Stranzinger, G. (2002) Molecular and pharmacological characterisation of the MSH-R alleles in Swiss cattle breeds. *Journal of Receptors and Signal Transduction Research* 22, 421–430.

Grosz, M.D. and MacNeil, M.D. (1999) The 'spotted' locus maps to bovine chromosome 6 in a Hereford-Cross population. *Journal of Heredity* 90, 233–236.

Guibert, S., Girardot, M., Leveziel, H., Julien, R. and Oulmouden, A. (2004) Pheomelanin coat colour dilution in French cattle breeds is not correlated with the TYR, TYRP1 and DCT transcription levels. *Pigment Cell Research* 17, 337–345.

Gutiérrez-Gil, B., Wiener, P. and Williams, J.L. (2007) Genetic effects on coat colour in cattle: dilution of eumelanin and phaeomelanin pigments in an F2-Backcross Charolais x Holstein population. *BMC Genetics* 8, 56.

Hansen, P.J. (1990) Effects of coat colour on physiological responses to solar radiation in Holsteins. *Veterinary Record* 127, 333.

Hanset, R. (1969) La White Heifer Disease dans la race bovine de Moyenne et Haute Belgique: un bilan de dix années. *Annales de Medécine Vétérinaire* 113, 12–21.

Hawkins, G.A., Eggen, A., Hayes, H., Elduque, C. and Bishop, M.D. (1996) Tyrosinase-related protein-2 (DCT; TYRP2) maps to bovine chromosome 12. *Mammalian Genome* 7, 474–475.

Hellström, A.R., Watt, B., Fard, S.S., Tenza, D., Mannström, P., Narfström, K., Ekesten, B., Ito, S., Wakamatsu, K., Larsson, J., Ulfendahl, M., Kullander, K., Raposo, G., Kerje, S., Hallböök, F., Marks, M.S. and Andersson, L. (2011) Inactivation of Pmel alters melanosome shape but has only a subtle effect on visible pigmentation. *PLoS Genetics* 7, e1002285.

Hirobe, T., Wakamatsu, K., Ito, S., Kawa, Y., Soma, Y. and Mizoguchi, M. (2006) The slaty mutation affects eumelanin and pheomelanin synthesis in mouse melanocytes. *European Journal of Cell Biology* 85, 537–549.

Hozumi, H., Takeda, K., Yoshida-Amano, Y., Takemoto, Y., Kusumi, R., Fukuzaki-Dohi, U., Higashitani, A., Yamamoto, H. and Shibahara, S. (2012) Impaired development of melanoblasts in the black-eyed white Mitf (mi-bw) mouse, a model for auditory-pigmentary disorders. *Genes to Cells* 17, 494–508.

Joerg, H., Fries, H.R., Meijerink, E. and Stranzinger, G.F. (1996) Red coat color in Holstein cattle is associated with a deletion in the MSHR gene. *Mammalian Genome* 7, 317–318.

Jolly, R.D., Wills, J.L., Kenny, J.E., Cahill, J.I. and Howe, L. (2008) Coat-colour dilution and hypotrichosis in Hereford crossbred calves. *New Zealand Veterinary Journal* 56, 74–77.

Kaelin, C. and Barsh, G.S. (2012) Molecular genetics of coat colour, texture and length in the dog. In: Ostrander, E. and Ruvinsky, A. (eds) *The Genetics of the Dog.* CAB International, Wallingford, UK, pp. 57–82.

Kidson, S.H. and Fabian, B.C. (1981) The effect of temperature on tyrosinase activity in Himalayan mouse skin. *Journal of Experimental Zoology* 215, 91–97.

King, V.L., Denise, S.K., Armstrong, D.V., Torabi, M. and Weirsma, F. (1988) Effects of a hot climate on the performance of first lactation cows grouped by coat colour. *Journal of Dairy Science* 71, 1093–1096.

Klungland, H. and Våge, D.I. (1999) Presence of the dominant extension allele E(D) in red and mosaic cattle. *Pigment Cell Research* 12, 391–393.

Klungland, H. and Våge, D.I. (2003) Pigmentary switches in domestic animal species. *Annals of New York Academy of Sciences* 994, 331–338.

Klungland, H., Våge, D.I., Gomez-Raya, L., Adalsteinsson, S. and Lien, S. (1995) The role of melanocyte-stimulating hormone (MSH) receptor in bovine coat color determination. *Mammalian Genome* 6, 636–639.

Kühn, C. and Weikard, R. (2007) An investigation into the genetic background of coat colour dilution in a Charolais x German Holstein F2 resource population. *Animal Genetics* 38, 109–113.

Kuneida, T., Nakagiri, M., Takami, M., Ide, H. and Ogawa, H. (1999) Cloning of bovine LYST gene and identification of a missense mutation associated with Chediak-Higashi syndrome of cattle. *Mammalian Genome* 10, 1146–1149.

Kwon, B.S., Halaban, R. and Chintamaneni, C. (1989) Molecular basis of mouse Himalayan mutation. *Biochemistry Biophysics Research Communication* 161, 252–260.

Lamoreux, M.L., Wakamatsu, K. and Ito, S. (2001) Interaction of major coat color gene functions in mice as studied by chemical analysis of eumelanin and pheomelanin. *Pigment Cell Research* 14, 23–31.

Little, C.C. (1958) Coat colour genes in rodents and carnivores. *Quarterly Review of Biology* 33, 103–137.

Liu, L., Harris, B., Keehan, M. and Zhang, Y. (2009) Genome scan for the degree of white spotting in dairy cattle. *Animal Genetics* 40, 975–977.

Lu, D., Willard, D., Patel, I.R., Kadwell, S., Overton, L., Kost, T., Luther, M., Chen, W., Woychik, R.P., Wilkison, W.O. *et al.* (1994) Agouti protein is an antagonist of the melanocyte-stimulating-hormone receptor. *Nature* 371, 799–802.

Mackenzie, M.A., Jordan, S.A., Budd, P.S. and Jackson, I.J. (1997) Activation of the receptor tyrosine kinase Kit is required for the proliferation of melanoblasts in the mouse embryo. *Developmental Biology* 192, 99–107.

Olson, T.A. (1999) Genetics of coat colour variation. In: Fries, R. and Ruvinsky, A. (eds) *The Genetics of Cattle.* CAB International, Wallingford.

Opdecamp, K., Nakayama, A., Nguyen, M.T., Hodgkinson, C.A., Pavan, W.J. and Arnheiter, H. (1997) Melanocyte development *in vivo* and in neural crest cell cultures: crucial dependence on the Mitf basic-helix-loop-helix-zipper transcription factor. *Development* 124, 2377–2386.

Padgett, G.A., Leader, R.W., Gorham, J.R. and O'Mary, C.C. (1964) The familial occurrence of the Chediak-Higashi syndrome in mink and cattle. *Genetics* 49, 505–512.

Pavan, W.J. and Tilghman, S.M. (1994) Piebald lethal (sl) acts early to disrupt the development of neural crest-derived melanocytes. *Proceedings of the National Academy of Sciences USA* 91, 7159–7163.

Philipp, U., Lupp, B., Mömke, S., Stein, V., Tipold, A., Eule, J.C., Rehage, J. and Distl, O. (2011) A MITF mutation associated with a dominant white phenotype and bilateral deafness in German Fleckvieh cattle. *PLoS One* 6, e28857.

Prieur, D.J., Holland, J.M., Bell, T.G. and Young, D.M. (1976) Ultrastructural and morphometric studies of platelets from cattle with the Chediak-Higashi syndrome. *Laboratory Investigations* 35, 197–204.

Reinsch, N., Thomsen, H., Xu, N., Brink, M., Looft, C., Kalm, E., Brockmann, G.A., Grupe, S., Kühn, C., Schwerin, M., Leyhe, B., Hiendleder, S., Erhardt, G., Medjugorac, I., Russ, I., Förster, M., Reents, R. and Averdunk, G. (1999) A QTL for the degree of spotting in cattle shows synteny with the KIT locus on chromosome 6. *Journal of Heredity* 90, 629–634.

Reissmann, M. and Ludwig, A. (2013) Pleiotropic effects of coat colour-associated mutations in humans, mice and other mammals. *Seminars in Cell and Developmental Biology* 24, 576–586.

Renieri, C., Ceccarelli, P., Gargiulo, A.M., Lauvergne, J.J. and Monacelli, G. (1993) Chemical and electron microscopic studies of cattle (*Bos taurus*) with four types of phenotypic pigmentation. *Pigment Cell Research* 6, 165–170.

Robbins, L.S., Nadeau, J.H., Johnson, K.R., Kelly, M.A., Roselli-Rehfuss, L., Baack, E., Mountjoy, K.G. and Cone, R.D. (1993) Pigmentation phenotypes of variant extension locus alleles result from point mutations that alter MSH receptor function. *Cell* 72, 827–834.

Rouzaud, F., Martin, J., Gallet, P.F., Delourme, D., Goulemot-Leger, V., Amigues, Y., Ménissier, F., Levéziel, H., Julien, R. and Oulmouden, A. (2000) A first genotyping assay of French cattle breeds based on a new allele of the extension gene encoding the melanocortin-1 receptor (MC1R). *Genetics Selection Evolution* 32, 511–520.

Rouzaud. F., Oulmouden, A. and Kos, L. (2010) The untranslated side of hair and skin mammalian pigmentation: beyond coding sequences. *IUBMB Life* 62, 340–346.

Royo, L.J., Alvarez, I., Fernández, I., Arranz, J.J., Gómez, E. and Goyache, F. (2005) The coding sequence of the ASIP gene is identical in nine wild-type coloured cattle breeds. *Journal of Animal Breeding and Genetics* 122, 357–360.

Russell, L.B. and Russell, W.L. (1948) A study of the physiological genetics of coat color in the mouse by means of the dopa reaction in frozen sections of skin. *Genetics* 33, 237–262.

Sarangarajan, R., Zhao, Y., Babcock, G., Cornelius, J., Lamoreux, M.L. and Boissy, R.E. (2000) Mutant alleles at the brown locus encoding tyrosinase-related protein-1 (TRP-1) affect proliferation of mouse melanocytes in culture. *Pigment Cell Research* 13, 337–344.

Schalles, R.R. and Cundiff, L.V. (1999) Inheritance of the 'rat-tail' syndrome and its effect on calf performance. *Journal of Animal Science* 77, 1144–1147.

Schläpfer, J., Stahlberger-Saitbekova, N., Womack, J.E., Gaillard, C. and Dolf, G. (2000) Assignment of six genes to bovine chromosome 13. *Journal of Animal Breeding and Genetics* 118, 189–196.

Schmutz, S.M. (2009) Red and Black. Series of web pages *Genetics of Coat Color in Cattle.* http://homepage.usask.ca/~schmutz/Variant%20Red.html (accessed 12 July 2013).

Schmutz, S.M. (2012) Genetics of coat color in cattle. In: Womack, J. (ed.) *Bovine Genomics.* John Wiley & Sons, Oxford, pp. 20–33.

Schmutz, S.M. and Dreger, D.L. (2013) Interaction of MC1R and PMEL alleles on solid coat colors in Highland cattle. *Animal Genetics* 44, 9–13.

Schmutz, S.M. and Moker, J.S. (1999) In situ hybridization mapping of TYR and CCND1 to cattle chromosome 29. *Animal Genetics* 30, 241–242.

Schmutz, S.M., Berryere, T.G., Ciobanu, D.C., Mileham, A.J., Schmidtz, B.H. and Fredholm, M. (2004) A form of albinism in cattle is caused by a tyrosinase frameshift mutation. *Mammalian Genome* 15, 62–67.

Searle, A.G. (1968) *Comparative Genetics of Coat Colour in Mammals.* Logos Press, London.

Seitz, J.J., Schmutz, S.M., Thue, T.D. and Buchanan, F.C. (1999) A missense mutation in the bovine MGF gene is associated with the roan phenotype in Belgian Blue and Shorthorn cattle. *Mammalian Genome* 10, 710–712.

Seo, K., Mohanty, T.R., Choi, T. and Hwang, I. (2007) Biology of epidermal and hair pigmentation in cattle: a mini-review. *Veterinary Dermatology* 18, 392–400.

Shin, M.B., Russel, L.B. and Tilghman, M. (1997) Molecular characterization of four induced alleles of the Ednrb locus. *Proceedings of the National Academy of Sciences USA* 94, 13105–13110.

Silvers, W.K. (1979) *The Coat Colour of Mice: A Model for Mammalian Gene Action and Interaction.* Springer-Verlag, New York.

Theos, A.C., Truschel, S.T., Raposo, G., and Marks, M.S. (2005) The Silver locus product Pmel17/gp100/Silv/ME20: controversial in name and in function. *Pigment Cell Research* 18, 322–336.

Trut, L.N., Oskina, I.N. and Kharlamova, A.V. (2012) Experimental studies of early canid domestication. In: Ostrander, E. and Ruvinsky. A. (eds) *The Genetics of the Dog.* CAB International, Wallingford, pp. 12–37.

Vrieling, H., Duhl, D.M., Millar, S.E., Miller, K.A. and Barsh, G.S. (1994) Differences in dorsal and ventral pigmentation result from regional expression of the mouse agouti gene. *Proceedings of the National Academy of Sciences USA*, 91, 5667–5671.

Yamakuchi, H., Agaba, M., Hirano, T., Hara, K., Todoroki, J., Mizoshita, K., Kubota, C., Tabara, N. and Sugimoto, Y. (2000) Chediak-Higashi syndrome mutation and genetic testing in Japanese black cattle (Wagyu). *Animal Genetics* 31, 13–19.

5 Genetics of Morphological Traits and Inherited Disorders

F.W. Nicholas

University of Sydney, Sydney, New South Wales, Australia

Introduction

The spectrum of morphological traits and inherited disorders ranges from those that are definitely due to the action of just one gene, to those that are due to the combined action of many genes and many non-genetic (environmental) factors. In between these two extremes are many traits and disorders that appear to run in families, but for which there is insufficient information to enable a conclusion to be drawn about whether one or more genes are involved. Unfortunately, the literature abounds with examples of traits and disorders that have been claimed to be due to just one gene, despite the data being so sparse that such a claim cannot be justified. Similar problems exist with claims of inheritance being recessive or dominant; in most cases, there is insufficient information to justify the claims that have been made. In the fullness of time, of course, additional data might support the initial claims. But we must be careful not to jump the gun.

This scarcity of reliable data on the inheritance of traits and disorders poses a challenge to those who are asked to compile lists of such traits – as required for this chapter. Since no two reviewers will interpret the evidence in exactly the same way, we must expect that lists

of single-locus traits and disorders compiled by different authors will differ at the margins. In the fullness of time, as more data become available, these differences will be resolved.

Previous Reviews

Many reviews of inherited traits and disorders in cattle have been published over the years. The first major summary specifically for cattle was by Shrode and Lush (1947). Since then, there have been surveys of inherited disorders by Gilmore (1957), Lauvergne (1968), Leipold et al. (1972), Jolly and Leipold (1973), Leipold and Schalles (1977), Herzog (1992), Kuhn (1997), Millar et al. (2000), Gentile and Testoni (2006), Agerholm (2007), Whitlock et al. (2008), Windsor and Agerholm (2009), Windsor et al. (2011a,b) and Leeb (2012).[1]

No discussion of inherited disorders in cattle can be complete without a special mention of the pioneering work of Dr Horst Leipold, whose name appears often in the list of reviews. His pioneering research into the inheritance of disorders, and his encylopedic knowledge of inherited disorders, will continue to earn him the gratitude of those who follow in his footsteps. Mention must also be made of the mammoth 'retirement' project of Dr Keith Huston, a former colleague of Dr Leipold. In reviewing all the published information on inherited disorders, Dr Huston compiled an annotated list of inherited disorders in cattle (Millar *et al.*, 2000) that will be a very important source of information for many years to come.

Current Sources of Information

While a list of reviews is useful, it is even more useful to have a single catalogue of morphological traits and inherited disorders that is regularly updated, and which is made freely available on the internet. Human geneticists have long had access to such a resource, namely Dr Victor McKusick's Online Mendelian Inheritance in Man (OMIM), which is freely accessible at http://www.omim.org. OMIM contains a wealth of information on thousands of morphological traits and inherited disorders in humans. It also contains a surprising quantity of information on cattle, because McKusick was always interested in potential animal models of human disorders.

In 1978, the present author commenced compiling a catalogue of inherited traits and disorders in a wide range of animal species. Being modelled on, and complementary to, McKusick's catalogue, this animal catalogue is called Online Mendelian Inheritance in Animals (OMIA). It is freely accessible on the internet at http://omia.angis.org.au.

OMIA includes entries for all inherited disorders in cattle, together with other traits in cattle for which single-locus inheritance has been claimed, however dubiously. Each entry includes a list of references arranged chronologically, so as to present a convenient history of knowledge about each disorder or trait. For some entries, there is additional information on inheritance or molecular genetics. If the disorder or trait has a human homologue, the relevant OMIM number is included, providing a direct hyperlink to the relevant entry in McKusick's online catalogue OMIM.

OMIA is updated regularly and therefore makes it possible for readers throughout the world to obtain freely the latest information on any single-locus trait or inherited disorder in cattle.

An Overview

When the first edition of this book was published in 1999, the chapter corresponding to the present chapter stated 'With the molecular revolution now in full swing, and, in particular, with the development of gene markers covering all regions of all bovine chromosomes..., knowledge of morphological traits and inherited disorders in cattle will increase rapidly in the decades ahead.' The extent to which this prediction has come to pass is illustrated by the number of single-locus bovine morphological traits and inherited disorders with a known causal mutation, which has risen from 12 in 1999 to 81 at the time of writing in early 2013; a nearly sevenfold increase in 13 years!

Because of this revolution, there is no longer sufficient space available in a single chapter to provide all the relevant textual material for all relevant traits. Indeed, with the increasing use of the internet, there is really no need to provide all the textual information in a chapter such as this one. Instead, it is better to devote the available space to a listing of the relevant traits, with pointers to the relevant internet entry. With so many single-locus morphological traits and inherited disorders now characterized at the DNA level, and with quite high-quality sequence assemblies of the bovine genome now available, it makes sense for this chapter to concentrate on presenting a list of bovine single-locus morphological traits and inherited disorders with known causal mutations, arranged according to their map position in the genome. This is done in Table 5.1. Such a table is best called a mostly-morbid map, by analogy with McKusick's long-established morbid map of the human genome, which includes only disorders.[2] Table 5.2 presents bovine morphological traits and inherited disorders for which there is reasonable evidence of single-locus inheritance but for which the causal mutation has not yet been determined. Table 5.3 lists bovine morphological traits and inherited

Table 5.1. A mostly-morbid map of the bovine genome, incorporating all Mendelian morphological traits and inherited disorders that have been characterized at the DNA level, as at 1 March 2013.[a] (Full details for each entry are available at http://omia.angis.org.au/.)

Name of trait/disorder	OMIA number[b]	Gene	Location in bovine genome assembly UMD 3.1		
			Chromosome	Nucleotide start	Nucleotide end
Horns/polled	000483-9913	?[c]	1	1,168,000	2,049,000
Deficiency of uridine monophosphate synthase	000262-9913	UMPS	1	69,732,777	69,782,823
Renal dysplasia	001135-9913	CLDN16	1	77,492,293	77,469,356
Leukocyte adhesion deficiency, type I	000595-9913	ITGB2	1	145,133,580	145,104,426
Muscular hypertrophy (double muscling)	000683-9913	MSTN	2	6,213,565	6,220,195
Polled and multisystemic syndrome	001736-9913	ZEB2	2	49,422,588	53,130,732
Ichthyosis congenita	000547-9913	ABCA12	2	103,720,886	103,520,023
Complex vertebral malformation	001340-9913	SLC35A3	3	43,418,922	43,404,022
Dwarfism, proportionate, with inflammatory lesions	001686-9913	RNF11	3	95,601,694	95,598,416
Scurs, type 2	001593-9913	TWIST1	4	27,855,319	27,853,325
Osteopetrosis	000755-9913	SLC4A2	4	114,438,014	114,450,606
Coat colour, roan	001216-9913	KITLG	5	18,377,443	18,317,747
Ehlers-Danlos syndrome, Holstein variant	001716-9913	EPYC	5	20,950,210	20,909,662
Epidermolysis bullosa	000340-9913	KRT5	5	27,541,427	27,547,278
Arachnomelia, BTA5	000059-9913	SUOX	5	57,643,833	57,639,564
Coat colour, dilution	001545-9913	PMEL	5	57,669,834	57,677,940
Hypotrichosis with coat colour dilution	001544-9913	PMEL	5	57,669,834	57,677,940
Abortion due to mutation in APAF1	000001-9913	APAF1	5	63,125,176	63,207,284
Mannosidosis, beta	000626-9913	MANBA	6	23,390,301	23,541,418
Coat colour, dominant white	000209-9913	KIT	6	71,796,317	71,917,430
Coat colour, colour-sided	001576-9913	KIT	6	71,796,317	71,917,430
Dwarfism, Angus	001485-9913	PRKG2	6	97,735,626	97,652,568
Chondrodysplasia	000187-9913	EVC2	6	105,291,555	105,437,261
Ehlers-Danlos syndrome, type VII (Dermatosparaxis)	000328-9913	ADAMTS2	7	1,956,351	2,165,241
Mannosidosis, alpha	000625-9913	MAN2B1	7	13,954,084	13,969,420
Myoclonus	000689-9913	GLRA1	7	65,112,635	65,025,010
Coat colour, brown	001249-9913	TYRP1	8	31,726,908	31,710,696
Marfan syndrome	000628-9913	FBN1	10	61,877,807	62,142,170
Spinal dysmyelination	001247-9913	SPAST	11	14,714,303	14,769,811

Disorder/Trait	Accession	Gene	Chr		
Citrullinaemia	000194-9913	ASS1	11	100,791,338	100,843,336
Beta-lactoglobulin, aberrant low expression	001437-9913	LGB	11	103,301,663	103,306,380
Neuronal ceroid lipofuscinosis, 5	001482-9913	CLN5	12	52,453,835	52,461,656
Coat colour, agouti	000201-9913	ASIP	13	64,213,311	64,239,963
Acrodermatitis enteropathica	000593-9913	SLC39A4	14	1,719,731	1,724,220
Goitre, familial	000424-9913	TG	14	9,278,155	9,281,191
Abortion due to mutation in CWC15	001697-9913	CWC15	15	15,705,459	15,713,949
Yellow fat	001079-9913	BCO2	15	22,838,239	22,905,944
Syndactyly	000963-9913	LRP4	15	77,701,217	77,663,789
Trimethylaminuria	001360-9913	FMO3	15	39,505,887	39,531,948
Axonopathy	001106-9913	MFN2	16	42,587,220	42,560,146
Lethal multi-organ developmental dysplasia	001722-9913	KDM2B	17	55,898,977	56,018,629
Multiple ocular defects	000733-9913	WFDC1	18	10,558,019	10,587,118
Coat colour, extension	001199-9913	MC1R	18	14,757,331	14,759,081
Maple syrup urine disease	000627-9913	BCKDHA	18	50,819,364	50,838,368
Cardiomyopathy and woolly haircoat syndrome	000161-9913	PPP1R13L	18	53,447,530	53,432,741
Cardiomyopathy, dilated	000162-9913	OPA3	18	53,612,019	53,579,091
Abortion and stillbirth	001565-9913	MIMT1	18	64,325,122	64,431,506
Myasthenic syndrome, congenital	000685-9913	CHRNE	19	27,118,516	27,123,113
Spherocytosis	001228-9913	SLC4A1	19	44,708,380	44,692,186
Tail, crooked	001452-9913	MRC2	19	47,689,028	47,748,129
Dwarfism, growth-hormone deficiency	001473-9913	GH1	19	48,772,013	48,768,617
Glycogen storage disease II	000419-9913	GAA	19	53,113,263	53,100,964
Dwarfism, Dexter	001271-9913	ACAN	21	20,800,157	20,868,836
Brachyspina	000151-9913	FANCI	21	21,137,917	21,198,617
Coat colour, white spotting	000214-9913	MITF	22	31,769,465	31,735,989
Dominant white with bilateral deafness	001680-9913	MITF	22	31,769,465	31,735,989
Epidermolysis bullosa, dystrophic	000341-9913	COL7A1	22	51,859,651	51,889,953
Arachnomelia, BTA23	001541-9913	MOCS1	23	13,866,949	13,832,464
Myopathy of the diaphragmatic muscles	001319-9913	HSPA1B	23	27,333,869	27,331,771
Protoporphyria	000836-9913	FECH	24	57,333,272	57,298,433
Spinal muscular atrophy	000939-9913	KDSR	24	62,180,437	62,118,138
Congenital muscular dystonia 1	001450-9913	ATP2A1	25	26,204,651	26,187,386
Pseudomyotonia, congenital	001464-9913	ATP2A1	25	26,204,651	26,187,386
Forelimb-girdle muscular anomaly	001442-9913	GFRA1	26	37,020,528	36,789,222

Continued

Table 5.1. Continued.

Name of trait/disorder	OMIA number[b]	Gene	Location in bovine genome assembly UMD 3.1		
			Chromosome	Nucleotide start	Nucleotide end
Mucopolysaccharidosis IIIB	001342-9913	NAGLU	26	43,258,821	43,265,801
Factor XI deficiency	000363-9913	F11	27	15,350,936	15,370,081
Chediak-Higashi syndrome	000185-9913	LYST	28	8,567,654	8,423,714
Hypotrichosis	000540-9913	HEPHL1	29	744,426	653,015
Coat colour, albinism	000202-9913	TYR	29	6,462,239	6,351,876
Congenital muscular dystonia 2	001451-9913	SLC6A5	29	24,618,069	24,564,841
Glycogen storage disease V	001139-9913	PYGM	29	43,617,816	43,606,016
Thrombopathia	001003-9913	RASGRP2	29	43,602,498	43,590,192
Haemophilia A	000437-9913	F8	X	38,838,454	38,982,286
Anhidrotic ectodermal dysplasia	000543-9913	EDA	X	86,099,972	85,708,002
Ovotesticular DSD (disorder of sexual development)	001230-9913	SRY	Y	42,225,120	42,225,990
Epilepsy[d]	000344-9913	?	?	?	?
Hydrocephalus[d]	000487-9913	?	?	?	?
Tibial hemimelia[d]	001009-9913	?	?	?	?
Arthrogryposis multiplex congenital[d]	001465-9913	?	?	?	?
Contractural arachnodactyly (Fawn calf syndrome)[d]	001511-9913	?	?	?	?
Pulmonary hypoplasia with anasarca[d]	001562-9913	?	?	?	?

[a]This table is a revised and updated version of a table that first appeared in Nicholas, F.W. (2012) Mendelian inheritance in cattle. In: Womack, J. (ed.) *Bovine Genomics*. Wiley-Blackwell, Ames, Iowa, pp. 11–19.

[b]xxxxxx-9913, where xxxxxx is the unique six-digit OMIA ID for a trait/disorder, and 9913 is the NCBI taxonomy ID for cattle.

[c]The polled locus is an enigma: two alleles that are completely associated with polledness in European cattle map to a region of chromosome BTA1 with no known functional sequences.

[d]Causal mutations for these disorders have been discovered, and in some cases are being offered for DNA testing to control the disorder. However, for intellectual-property reasons, the causal mutation has not yet been published.

Table 5.2. Bovine morphological traits and inherited disorders for which there is reasonable published evidence for single-locus inheritance, but for which no causal mutations have been reported, as at 1 March 2013. (Full details for each entry are available at http://omia.angis.org.au/.)

OMIA number[a]	Name of trait/disorder
000004-9913	Achondroplasia
000010-9913	Acroteriasis congenita
000036-9913	Amputated
000991-9913	Androgen insensitivity syndrome (AIS)
000047-9913	Ankylosis, generalized
001556-9913	Ankylosis, jaw
000083-9913	Atresia ani
000117-9913	Blood group system A
000120-9913	Blood group system B
000121-9913	Blood group system C
000124-9913	Blood group system F
000130-9913	Blood group system J
000132-9913	Blood group system L
000133-9913	Blood group system M
001629-9913	Blood group system N'
001630-9913	Blood group system R'
000139-9913	Blood group system S
001631-9913	Blood group system T'
001632-9913	Blood group system Z
001502-9913	Caprine-like Generalized Hypoplasia Syndrome
000160-9913	Cardiomyopathy
000168-9913	Cataract, generic
001585-9913	Cleft lip and jaw, right-sided
000204-9913	Coat colour, albinism, incomplete
001320-9913	Coat colour, spotted
001529-9913	Coat colour, variant red
001469-9913	Coat colour, white belt
000313-9913	Congenital dyserythropoietic anaemia with dyskeratosis and progressive alopecia
001659-9913	Dwarfism, dominant
001294-9913	Dwarfism, growth-hormone-receptor deficiency
000308-9913	Dwarfism, proportionate
000310-9913	Dwarfism, snorter
000311-9913	Dwarfism, stumpy
000317-9913	Ears, crop
000321-9913	Ears, notched
000343-9913	Epididymal aplasia
000348-9913	Epitheliogenesis imperfecta
000402-9913	Gangliosidosis, GM1
000468-9913	Heterochrirides
000493-9913	Hydrops foetalis
000495-9913	Hyperbilirubinaemia, unclassified
000542-9913	Hypotrichosis, streaked
000601-9913	Limber legs
000603-9913	Limbs, curved
001407-9913	Lipofuscinosis, renal
001557-9913	Ljutikow's lethal
001558-9913	Micromelia, achondroplastic
000656-9913	Molars, impacted
000664-9913	Mucopolysaccharidosis I

Continued

Table 5.2. Continued.

OMIA number[a]	Name of trait/disorder
000673-9913	Mummified foetus
000674-9913	Muscle contracture
000181-9913	Neuronal Ceroid Lipofuscinosis, generic
000754-9913	Osteogenesis imperfecta
001175-9913	Porphyria, congenital erythropoietic
000827-9913	Progressive degenerative myeloencephalopathy (Weaver syndrome)
000834-9913	Protamine-2 deficiency
000850-9913	Rectovaginal constriction
000894-9913	Scurs
000911-9913	Short spine
001372-9913	Slick hair
000927-9913	Spastic lethal
000973-9913	Tail, kinky
001510-9913	Tail, wry
000992-9913	Testicular hypoplasia
001048-9913	Vertical fibre hide defect

[a]xxxxxx-9913, where xxxxxx is the unique six-digit OMIA ID for a trait/disorder, and 9913 is the NCBI taxonomy ID for cattle.

Table 5.3. Bovine morphological traits and inherited disorders for which there is insufficient published evidence for single-locus inheritance, as at 1 March 2013. (Full details for each entry are available at http://omia.angis.org.au/.)

OMIA number[a]	Name of trait/disorder
001119-9913	Abomasum, displaced
000002-9913	Abrachia
000005-9913	Achondroplasia foetalis
000012-9913	Adactyly
000014-9913	Adenohypophyseal aplasia
000021-9913	Agenesis of corpus callosum
000022-9913	Agnathia
000027-9913	Allergic rhinitis
000030-9913	Alopecia
001702-9913	Alopecia areata
000040-9913	Amyloidosis, renal
000044-9913	Anencephaly
000049-9913	Anophthalmos
000050-9913	Anophthalmos microphthalmos
001411-9913	Anotia
000056-9913	Aplasia segmentalis ductus wolffii
000058-9913	Aprosopia
000061-9913	Arnold-Chiari malformation
000069-9913	Arthrogryposis
000070-9913	Arthrogryposis and palatoschisis syndrome
001169-9913	Ascites
000077-9913	Ataxia
000078-9913	Ataxia, cerebellar
001091-9913	Ataxia, progressive
000085-9913	Atresia coli
000086-9913	Atresia ilei

Continued

Table 5.3. Continued.

OMIA number[a]	Name of trait/disorder
000087-9913	Atresia intestinal
000089-9913	Atrial septal defect
000091-9913	Atrophic rhinitis
000110-9913	Black hair follicle dysplasia
000112-9913	Bleeding diathesis
000113-9913	Bleeding disorder
000116-9913	Blood group systems, generic
000147-9913	Brachygnathia
000149-9913	Brachygnathia superior
000150-9913	Brachygnathia superior and degenerative joint disease
000159-9913	Cardiac anomaly
000515-9913	Cardiomyopathy, hypertrophic
000175-9913	Cerebellar abiotrophy
000177-9913	Cerebellar cortical atrophy
000178-9913	Cerebellar disease
000179-9913	Cerebellar hypoplasia
000189-9913	Chondrodystrophy
000195-9913	Claw defects
000197-9913	Cleft palate
000214-9913	Coat colour, white spotting
000200-9913	Coat colours, generic
000219-9913	Coloboma
001689-9913	Congenital hydranencephaly and cerebellar hypoplasia
001270-9913	Convulsions and ataxia, familial
001118-9913	Corkscrew penis
000231-9913	Corneal opacity
000235-9913	Cranial duplication
000237-9913	Cranioschisis
000243-9913	Cryptorchidism
000245-9913	Curly coat
000246-9913	Curly hair, karakul-type
000249-9913	Cyclopia
000254-9913	Cystic ovary
000260-9913	Debility, congenital
000261-9913	Defective keratogenesis of hooves and mouth
000272-9913	Dermoid sinus
000274-9913	Dermoid, ocular, congenital
000279-9913	Diabetes mellitus
000283-9913	Diabetes mellitus, type I
001410-9913	Diphallus
000290-9913	Diprosopus
000291-9913	Doddler
000293-9913	Double cervix
000295-9913	Duck-legged
000299-9913	Dwarfism
001323-9913	Dwarfism, Laron
000318-9913	Ears, double
000324-9913	Ectrodactyly
001126-9913	Ectromelia
001488-9913	Encephalomyelopathy, multifocal symmetrical necrotizing, Angus
001489-9913	Encephalomyelopathy, multifocal symmetrical necrotizing, Limousin
001490-9913	Encephalomyelopathy, multifocal symmetrical necrotizing, Simmental

Continued

Table 5.3. Continued.

OMIA number[a]	Name of trait/disorder
000338-9913	Epidermal dysplasia
000339-9913	Epidermolysis
000342-9913	Epidermolysis bullosa, junctionalis
001239-9913	Epiphora
000346-9913	Epistaxis
000353-9913	Exophthalmos with strabismus
000355-9913	Extra ear lobes
000358-9913	Eye defects
000360-9913	Facial digital syndrome
001276-9913	Facial eczema
000366-9913	Fanconi syndrome
000391-9913	Fragile X
000393-9913	Freemartin
000401-9913	Gangliosidosis
000406-9913	Genital hypoplasia
000410-9913	Glandular aplasia
000426-9913	Gonadal hypoplasia
000428-9913	Haemochromatosis
000430-9913	Haemolytic anaemia
000435-9913	Haemolytic uremic syndrome
000440-9913	Hair, long
000444-9913	Harelip
000446-9913	Heart defect, congenital
000726-9913	Hemeralopia
001191-9913	Hemivertebrae
000454-9913	Hepatic fibrosis, idiopathic
000457-9913	Hernia, brain
000462-9913	Hernia, inguinal
000464-9913	Hernia, scrotal
000465-9913	Hernia, umbilical
000466-9913	Hernia, ventral
000470-9913	High lysozyme activity
000473-9913	Hip dysplasia
000474-9913	Histocytosis
000475-9913	Hocks, straight
001197-9913	Horner syndrome
000484-9913	Horse rump
000486-9913	Hydranencephaly
000489-9913	Hydrocephalus, internal
001107-9913	Hymen, imperforate
001412-9913	Hyperextension of fetlock joints
001231-9913	Hyperhidrosis
000506-9913	Hypermetria
000513-9913	Hypertrichosis
000527-9913	Hypomyelinogenesis, congenital
000530-9913	Hypoplasia of sex organs
001187-9913	Hypospadias
000541-9913	Hypotrichosis, semi
000555-9913	Immunoglobulin G2 deficiency
000558-9913	Imperforate anus
000560-9913	Impotentia cocundi
000562-9913	Infertility

Continued

Table 5.3. Continued.

OMIA number[a]	Name of trait/disorder
000563-9913	Interdigital tissue pachydermia
000564-9913	Intersex
001227-9913	Intussusception
000570-9913	Joint laxity and dwarfism, congenital
001125-9913	Ketosis
000576-9913	Knobbed acrosome
001693-9913	Lacrimal fistula, bilateral congenital
000579-9913	Lameness
001225-9913	Laminitis
001206-9913	Laryngeal paralysis
000584-9913	Leg defects
000585-9913	Leg weakness
001404-9913	Leptin concentration
000590-9913	Lethal gene
000599-9913	Limb deformity, congenital
000605-9913	Lipidosis, hepatic
000606-9913	Lipomatosis, multiple
000611-9913	Luxating patella
000613-9913	Lymphoedema
000621-9913	Malignant hyperthermia
000629-9913	Megacolon
000631-9913	Megaoesophagus
000644-9913	Micrencephaly
000646-9913	Micrognathia
000647-9913	Microhydranencephalus
000649-9913	Microphthalmia
000675-9913	Muscle contracture and chondrodysplasia
000690-9913	Myoclonus epilepsy of Lafora
000697-9913	Myositis ossificans
001097-9913	Necrotizing encephalopathy, subacute, of Leigh
000714-9913	Neuraxial oedema
000716-9913	Neurofibromatosis
001351-9913	Neuromuscular disease, degenerative
000728-9913	Nipples, depressed
000729-9913	Nipples, inverted
000735-9913	Ocular squamous cell carcinoma
000741-9913	Omphalocele
000747-9913	Osteoarthritis
000750-9913	Osteochondrosis
000753-9913	Osteodystrophy
001277-9913	Otitis interna, susceptibility to
001266-9913	Otitis media, susceptibility to
001127-9913	Otocephaly
000759-9913	Ovarian aplasia
000761-9913	Ovarian hypoplasia
001539-9913	Pancytopenia, neonatal
001147-9913	Papillomatosis, cutaneous
000775-9913	Pasterns, bowed
000776-9913	Pasterns, flexed
000778-9913	Patellar luxation
000779-9913	Patent ductus arteriosus
001188-9913	Pemphigus

Continued

Table 5.3. Continued.

OMIA number[a]	Name of trait/disorder
001453-9913	Periodic spasticity, inherited
000789-9913	Perosomus elumbis
000792-9913	Persistent frenulum praeputii
000795-9913	Persistent truncus arteriosus
001218-9913	Persistent truncus arteriosus with ventricular septal defect and patent foramen ovale
000798-9913	Phalanges, reduced
001337-9913	Platelet aggregation disorder
000803-9913	Platelet function defect
000809-9913	Polycythemia
000810-9913	Polydactyly
001226-9913	Polymelia
001100-9913	Polymicrogyria
000817-9913	Portosystemic shunt
001265-9913	Preputial prolapse
000823-9913	Prognathism
000824-9913	Progressive alopecia
000832-9913	Progressive spinal myelopathy
000833-9913	Prolonged gestation
000840-9913	Pulmonary adenomatosis
000841-9913	Pulmonary hypertension
001691-9913	Recombination rate
001533-9913	Resistance to infectious bovine keratoconjunctivitis
001744-9913	Resistance to mastitis
001663-9913	Retinal dysplasia and internal hydrocephalus
001413-9913	Retinitis pigmentosa
000890-9913	Schistosomus reflexus
000896-9913	Seminal defect
000898-9913	Serum cholesterol level
000908-9913	Sheep's head
001102-9913	Situs inversus
000917-9913	Skeletal deformity
000919-9913	Skin defect
000922-9913	Smooth tongue
000926-9913	Spastic lameness
000928-9913	Spastic paresis
000929-9913	Spastic syndrome
000930-9913	Sperm defect
000932-9913	Sperm, dag defect of
001662-9913	Sperm, decapitated
001334-9913	Sperm, short tail
000933-9913	Spina bifida
000935-9913	Spina bifida with myelomeningocele
000938-9913	Spinal dysraphism
000944-9913	Spongiform encephalopathy
001156-9913	Spongiform myelopathy
000947-9913	Stenosis, spinal
000950-9913	Stringhalt
000965-9913	Syringomyelia
000975-9913	Tail, short
000977-9913	Taillessness
000984-9913	Teat injury
000985-9913	Teat number

Continued

Table 5.3. Continued.

OMIA number[a]	Name of trait/disorder
000986-9913	Teats, bottle
000987-9913	Teats, fused
000990-9913	Tendons, contracted, congenital
000994-9913	Tetralogy of fallot
000997-9913	Thoracic limb, angular deformity of
001001-9913	Thrombocytopaenia
001491-9913	Tomaculous neuropathy
001012-9913	Tongue-wagging
001022-9913	Twinning
001023-9913	Twinning, conjoined
001026-9913	Udder abnormality
001033-9913	Urolithiasis
001039-9913	Various disorders
001041-9913	Ventricular septal defect
001043-9913	Ventricular septal defect with atrioventricular valvular anomaly
001055-9913	Vitiligo
001056-9913	Von Willebrand disease
001060-9913	Warts between hooves
001142-9913	Wilms tumour
001071-9913	Wilson disease
001194-9913	Wolff-Parkinson-White syndrome
001769-9913	Y anomaly in low reproductive females

[a]xxxxxx-9913, where xxxxxx is the unique six-digit OMIA ID for a trait/disorder, and 9913 is the NCBI taxonomy ID for cattle.

disorders for which there is insufficient evidence of single-locus inheritance. Details for each entry in all three tables, including comprehensive reference lists, are available from http://omia.angis.org.au.

It is readily acknowledged that the information in OMIA is incomplete, and that it includes errors of omission and commission. One of the advantages of having this type of information stored in a database is that errors can be rectified easily as soon as they are spotted. The author would therefore be very grateful to any readers who identify errors in the information supplied in this chapter or on the website.

Conclusions

The lists of inherited morphological traits and disorders presented in this chapter provide an indication of the range of such traits and disorders that have been observed and studied in cattle. The molecular and gene-mapping revolutions now underway have already led to an explosion of knowledge in this area, and there is much more to come! To exploit fully the genetic variation that does occur, breeders and researchers need to be continually on the look-out for unusual animals, saving DNA from them where possible. The power of modern genomic technologies is such that only a handful of affected animals are required in order to map the disorder (if it is single-locus) and (in many, but not all, cases) to determine the causal mutation.

Notes

[1] For a full list of reviews pertaining to cattle, see the *Bos taurus* section of http://omia.angis.org.au/key_articles/reviews/. It should be noted that some of these reviews are concerned with congenital traits and disorders, i.e. traits and disorders that are present at birth. Not all such traits and disorders are inherited.

[2] Available for download from http://www.omim.org/downloads/.

References

Agerholm, J.S. (2007) Inherited disorders in Danish cattle. APMIS Supplement s122, 1 76.

Gentile, A. and Testoni, S. (2006) Inherited disorders of cattle: a selected review. *Slovenian Veterinary Research* 43, 17–29.

Gilmore, L.O. (1957) Inherited defects in cattle. *Journal of Dairy Science* 40, 593–595.

Herzog, A. (1992) Genetic defects in cattle and the possibilities of control. *Wiener Tierarztliche Monatsschrift* 79, 142–148.

Jolly, R.D. and Leipold, H.W. (1973) Inherited diseases of cattle – a perspective. *New Zealand Veterinary Journal* 21, 147–155.

Kuhn, C. (1997) [Molecular genetic background of inherited defects in cattle] [German]. *Archiv fur Tierzucht-Archives of Animal Breeding* 40, 121–127.

Lauvergne, J.J. (1968) Catalogue des anomalies hereditaires des bovines. *Bulletin Technique du Département de Génétique Animale*, Number 1, Institut national de la recherche agronomique (INRA), Paris, France.

Leeb, T. (2012) Animal DNA diagnostics – personal genomics for our pets and livestock is at the Horizon. *Molecular and Cellular Probes* 26, 223.

Leipold, H.W. Dennis, S.M. and Huston, K. (1972) Congenital defects of cattle: nature, cause and effect. *Advances in Veterinary Science and Comparative Medicine* 16, 103–150.

Leipold, H.W. and Schalles, R. (1977) Genetic defects in cattle, transmission and control. *Veterinary Medicine and Small Animal Clinician* 45, 80–85.

Millar, P., Lauvergne, J.J. and Dolling, C.H.S. 2000. *Mendelian Inheritance in Cattle.* EAAP Publication No 101, Wageningen Academic Publishers, Wageningen, The Netherlands. 450 pp. The catalogue on which this book is based is available at http://dga.jouy.inra.fr/lgbc/mic2000/

Shrode, R.R. and Lush, J.L. (1947) The genetics of cattle. *Advances in Genetics* 1, 209–261.

Windsor, P. and Agerholm, J. (2009) Inherited diseases of Australian Holstein-Friesian cattle. *Australian Veterinary Journal* 87, 193–199.

Windsor, P., Kessell, A. and Finnie, J. (2011a) Review of neurological diseases of ruminant livestock in Australia. V: congenital neurogenetic disorders of cattle. *Australian Veterinary Journal* 89, 394–401.

Windsor, P., Kessell, A. and Finnie, J. (2011b) Review of neurological diseases of ruminant livestock in Australia. VI: postnatal bovine, and ovine and caprine, neurogenetic disorders. *Australian Veterinary Journal* 89, 432–438.

Whitlock, B.K., Kaiser, L. and Maxwell, H.S. (2008) Heritable bovine fetal abnormalities. *Theriogenology* 70, 535–549.

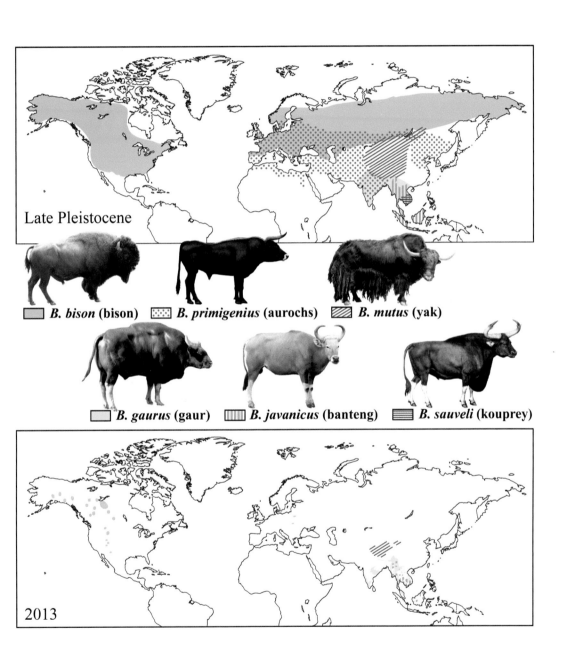

Plate 1. Former range (Late Pleistocene) and current geographic distribution of wild species of *Bos* (from Alexandre Hassanin, Chapter 1). The maps were interpreted from various references, including Olsen (1990), Van Vuure (2005), Louys *et al.* (2007), Sanderson *et al.* (2008), Sipko (2009), IUCN (2012) and Kerley *et al.* (2012). The illustration of aurochs was drawn by Daniel Foidl. The illustration of kouprey was modified from Coolidge (1940). Photo of American bison by Nicolas Puillandre. Photos of yak, gaur and banteng by Alexandre Hassanin.

2

3

4

5

6

7

Plate 2. Angus, Scotland, UK. (Photo: Wikimedia Commons.)
Plate 3. Belgian Blue, Belgium. (Photo: Wikimedia Commons.)
Plate 4. Brahman, USA. (Photo: Wikimedia Commons.)
Plate 5. Charolais, France. (Photo: Wikimedia Commons.)
Plate 6. Chianina, Italy. (Photo: Wikimedia Commons.)
Plate 7. Danish Red, Denmark. (Photo: Wikimedia Commons.)

8

9

10

11

12

13

Plate 8. Dexter, Ireland. (Photo: Wikimedia Commons.)
Plate 9. Hereford, England, UK. (Photo: Wikimedia Commons.)
Plate 10. Highland, Scotland, UK. (Photo: Wikimedia Commons.)
Plate 11. Holstein, Netherlands. (Photo: Wikimedia Commons.)
Plate 12. Hungarian Grey, Hungary. (Photo: Wikimedia Commons.)
Plate 13. Jersey, Jersey Island. (Photo: Wikimedia Commons.)

14

15

16

17

18

19

Plate 14. Limousin, France. (Photo: Treftz Limousin, Wetonka, South Dakota, USA.)
Plate 15. Maine Anjou, France. (Photo: American Maine-Anjou Association, Platte City, Missouri, USA.)
Plate 16. N'Dama, Guinea, West Africa. (Photo: Wikimedia Commons.)
Plate 17. Normande, France. (Photo: New Hope Normande, Scandinavia, Wisconsin, USA.)
Plate 18. Santa Gertrudis, USA. (Photo: Santa Gertrudis Breeders International, Kingsville, Texas, USA.)
Plate 19. Shorthorn, England, UK. (Photo: Wikimedia Commons.)

20

21

22

23

24

25

Plate 20. Simmental, Switzerland. (Photo: Roberts Country Simbrah, Mountainair, New Mexico, USA.)
Plate 21. Spanish Fighting Bull, Spain. (Photo: Wikimedia Commons.)
Plate 22. Texas Longhorn, USA. (Photo: Wikimedia Commons.)
Plate 23. Wagyu, Japan. (Photo: Rob Cumine, Natural Wagyu Beef, Pembrokeshire, UK.)
Plate 24. Watusi, DR Congo. (Photo: Wikimedia Commons.)
Plate 25. Yakutian, East Siberia, Russia. (Photo: Anu Osva, Helsinki, Finland.)

26

27

28

29

Plate 26. English Longhorn bull with brindle phenotype. (Photo: Wikimedia Commons, by Dave, Leicester, UK.)
Plate 27. A belted Galloway at Gretna Green. (Photo: Wikimedia Commons, by Amanda Slater, Coventry, UK.)
Plate 28. Colour-sided Dutch Witrik cows. (Photo: Henk Kosters, Kwadendamme, Zeeland, The Netherlands.)
Plate 29. Belgian Blue roan calf. (Permission is kindly granted by Malcolm Reedy on behalf of The Australian Belgian Blue Cattle Society.)

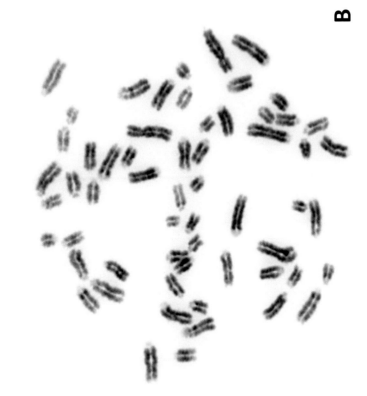

Plate 30. Cattle metaphase chromosomes stained with DAPI (A) and black and white version of the same image (B). From Rebecca O'Connor, School of Biosciences, University of Kent.

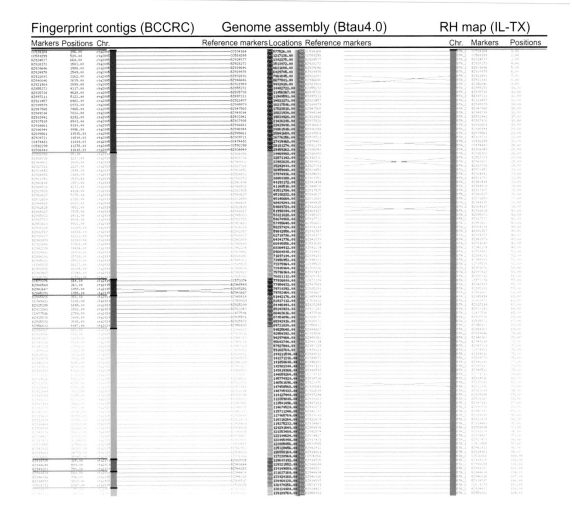

Plate 31. Comparison of cattle chromosome 2 genome assembly (centre panel) with the BCCRC fingerprint (left panel) and IL-TX radiation hybrid (right panel) maps. Comparison of the three maps demonstrates differences between the two maps (physical and RH) and the genome assembly. The BCCRC fingerprint map consists of multiple (quite long) contigs indicated by different colours in the left panel, while the IL-TX radiation hybrid map of the chromosome is represented by a single linkage group due to the high quality of the radiation panel used (Womack *et al.*, 1997) and marker selection strategy (see the main text). However, the BCCRC map has fewer inconsistencies in marker order with the genome assembly than the RH map because local flips of markers are a common problem in RH maps when resolution of a panel is not high enough to resolve the order of closely located markers (see the main text for details). The graphic was generated using AutoGRAPH web server (Derrien *et al.*, 2007).

6 Cytogenetics and Chromosome Maps

Denis M. Larkin and Marta Farré

Royal Veterinary College, London, UK

Introduction

In 2009 the cattle genome sequencing and assembly was completed. The success of this effort was the result of an international collaboration between six countries. An assembly of the genome became possible due to numerous projects started in the 1970s to understand the organization of cattle chromosomes (Heuertz and Hors-Cayla, 1978; Womack and Moll, 1986), to perform the microsatellite (Barendse et al., 1997) and gene mapping (Itoh et al., 2003), and to construct high resolution physical (Everts-van der Wind et al., 2005; Snelling et al., 2007) and linkage maps (Ihara et al., 2004). With the availability of the genome sequence and accurate assembly it became feasible to perform analysis of the genome at a level that was not possible before (Elsik et al., 2009).

One of the major drivers of cattle genome studies is an attempt to understand the genetic nature of quantitative traits and diseases affecting economically important traits, such as milk or meat quality. Significant progress has been achieved in this area leading to identification of several quantitative trait nucleotides (QTNs) that contribute to such traits in cattle (Grisart et al., 2002; Cohen-Zinder et al., 2005). The availability of the whole-genome annotation in conjunction with less expensive sequencing and genotyping techniques opens new exciting opportunities for identification and genotyping of all single-nucleotide mutations in any breed. Together with genome-wide association studies (GWAS), this will lead to detection of all common and some individual QTNs (Larkin et al., 2012).

The cattle genome is a great resource for studying mammalian genome evolution. Unique genome features formed in the processes of speciation and adaptation are reflected in the genome by gene mutations, sequence losses, duplications and repositions due to multiple chromosomal rearrangements that distinguish the cattle genome from other mammalian genomes and a putative mammalian ancestor (Murphy *et al.*, 2005; Larkin *et al.*, 2009; Elsik *et al.*, 2009). However, when compared to other sequenced mammalian genomes, the cattle genome in some chromosomal regions represents an ancestral organization, allowing for the detection of evolutionary events that happened in the course of genome evolution in other species (Murphy *et al.*, 2005).

We briefly summarize results of the cattle genome mapping efforts, annotation and the evolutionary history analysis. We start with earlier efforts in cattle genome analysis, including cattle cytogenetic and somatic cell hybrid mapping, linkage mapping, and later present advances achieved with the use of radiation hybrid mapping and fingerprint map construction. Together these efforts have gradually built a basis for understanding Mendelian traits and some cattle quantitative trait loci (QTLs), facilitated genome assembly, and made possible functional and evolutionary study of the cattle genome.

The Cattle Chromosome Nomenclature

Domestic cattle (*Bos taurus* and *Bos indicus*) have 30 chromosome pairs: 58 acrocentric autosomes and 2 submetacentric sex chromosomes. Using a uniform staining on metaphase chromosomes, the cattle karyotype can be presented as a decreasing series of arbitrarily divided chromosome groups, using the relative length of each chromosome as the only criterion (Fig. 6.1).

Early in the 1970s, cytogenetists used different banding techniques, such as C-bands, G-bands, Q-bands with Hoechst 33258 or quinacrine and R-bands, to differentiate cattle chromosomes. Contemporary cytogenetics uses 4′,6- diamidino-2-phenylindole (DAPI) as a

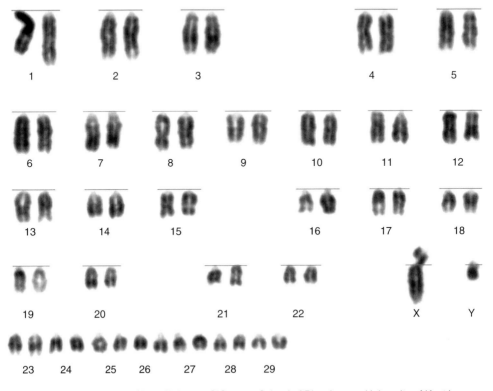

Fig. 6.1. A cattle karyotype. (From Rebecca O'Connor, School of Biosciences, University of Kent.)

fluorescent stain (Plate 30). Different staining methods would produce slightly different cytogenetic nomenclatures of the cattle chromosomes, leading to several major disagreements in nomenclatures. It was during the Ninth North American Colloquium on Domestic Animal Cytogenetics and Gene Mapping, held at Texas A&M University in 1995, that the cattle karyotype was standardized. This nomenclature ('Texas standard') has merged the previous nomenclature's attempts with the data on somatic cell and *in situ* hybridization gene mapping for each cattle chromosome. It also has indicated rough homologies between cattle, human and sheep chromosomes (Table 6.1) (Popescu *et al.*, 1996). This nomenclature has failed to resolve differences between cattle chromosomes 25, 27 and 29. To resolve this issue Hayes *et al.* (2000) unambiguously localized 31 marker genes on to the 31 cattle chromosomes using fluorescent *in situ* hybridization (FISH) technique (Hayes *et al.*, 2000).

While G-, Q- and C-banding techniques produce the characteristic banding pattern of all chromosomes; silver staining is used to reveal the active nuclear organizer regions (NORs). In cattle, NORs can be found in telomeres of chromosomes 2, 3, 4 and 11 (Henderson and Bruere, 1979; Di Berardino *et al.*, 1981; Mayr and Czaker, 1981). Moreover, chromosomes 25 and 28 probably contain NORs, although their assignment is difficult due to problems with chromosome identification and polymorphism detection.

Chromosome Abnormalities

Chromosome abnormalities can be classified into numerical or structural aberrations. Numerical abnormalities affect the diploid number of the cell, whereas structural mutations affect the arrangement of chromosomes. Chromosome abnormalities have been broadly studied since they are associated with fertility problems or reproductive failure in humans and domestic animals.

Numerical chromosome aberrations

There are two types of numerical chromosome aberrations: polyploidy and aneuploidy.

Table 6.1. The Texas standard chromosome nomenclature, showing homologies with human and sheep chromosomes. (From Popescu *et al.*, 1996.)

Texas standard	% length	Human chromosome	Sheep chromosome
1	5.87	3, 21	1q
2	5.12	1p, 2q	2q
3	4.71	1p, 2[a]	1p
4	4.67	7p	4
5	4.48	1q[b], 12, 22	3q
6	4.33	4	6
7	4.18	1[a], 5q, 19p	5
8	4.13	4[a], 8p, 9q	2p
9	3.86	6q	8
10	3.67	5q, 14, 15	7
11	3.94	2, 9q	3p
12	3.29	13	10
13	3.09	10p, 20	13
14	3.15	8q	9
15	3.11	5[b], 11p	15
16	3.07	1q	12
17	2.83	4q, 12q, 22	17
18	2.60	16q, 19q	14
19	2.54	17	11
20	2.75	5	16
21	2.72	14, 15	18
22	2.51	3, 7[a]	19
23	2.09	6p	20
24	2.37	18	23
25	1.97	7q, 16p	24
26	1.96	10q	22
27	1.83	3[a], 4[a], 8	26
28	1.73	1[a], 10q	25
29	1.99	11	21
X	5.45	X	X
Y	2.13	Y	Y

Length of each chromosome is expressed as relative length of the haploid genome.
[a] Indicates additional homologies with human chromosomes detected using the alignment of human and cattle genome sequences.
[b] Indicates homologies with human not confirmed with the whole genome sequence alignment.

Polyploidy is the result of abnormal fertilization (polyandry or polygyny): suppression of the first cleavage division in embryogenesis or fusion of embryonic cells. Aneuploidy arises from nondisjunction of homologous chromosomes during meiosis.

Aneuploidy is the only numerical chromosome aberration found in mammals, with monosomies and trisomies the most common abnormalities. Compared to humans, fewer cases of aneuploidy have been found in the cattle breeds analysed so far. This is probably due to the elimination of the embryos carrying these mutations prior to implantation or to low levels of sperm aneuploidies (Nicodemo *et al.*, 2009; Pauciullo *et al.*, 2012). Some well-documented cases include autosomal trisomies for chromosome 17 (Herzog *et al.*, 1977), chromosome 18 (Herzog *et al.*, 1982) and chromosome 28 (Iannuzzi *et al.*, 2001). However, several sex chromosome trisomies, including XXX, XXY and XYY, have been documented in cattle (Citek *et al.*, 2009), but with negligible or no effect on normal development, due to the gene dosage inactivation mechanism of the mammalian X chromosome.

Numerical chromosome aberrations have limited economic consequences, since they reduce the fertility of the affected animal. Therefore, spreading of these mutations at a population level does not represent a problem.

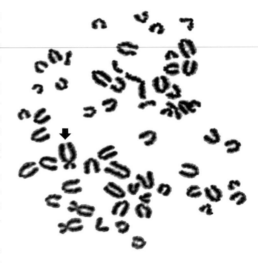

Fig. 6.2. A cattle cell containing 1/29 Robertsonian translocation. (From Dr Pietro Parma, Department of Agricultural and Environmental Sciences, University of Milan.) Chromosome containing the translocation is indicated by a black arrow.

Freemartin syndrome

The freemartin condition represents the most frequent form of intersexuality found in cattle, and occasionally other species. Freemartins are females born co-twin to a male. Vascular connections form between the placentae of developing twin feti, XX/XY chimerism develops, and ultimately there is masculinization of the female tubular reproductive tract to varying degrees (Padula, 2005). From a cytogenetic point of view, freemartins are chimerical organisms, carrying XX and XY cells in blood and haematopoietic organs. However, the ratio of XX/XY cells present in a freemartin is not an indicator of the severity of the masculinization that has occurred. The male co-twin to a freemartin female is also a chimera, but the genital organs are normal. However, they usually have poor semen quality and are subfertile (Dunn *et al.*, 1979; Peretti *et al.*, 2008).

Other chromosomal abnormalities occasionally coincident with XX/XY chimeras have been reported such as 4/21 tandem fusion (Pinheiro *et al.*, 1995), 1/29 Robertsonian translocation (Fig. 6.2) (Zhang *et al.*, 1994; Guanti and Minola, 1978), 6/1 translocation and XXY trisomy (Zhang *et al.*, 1994), undetermined centric fusion (Zhang *et al.*, 1994) and mixoploid chromosome constitution (Hare, 1976).

Structural chromosome aberrations

Structural chromosome aberrations can be defined as a change in the integrity of the chromosome, affecting part of its length or its entirety. They occur after a misrepair of breaks during meiosis. They can be classified as balanced or unbalanced depending on the modifications of the genome. Balanced chromosome aberrations do not alter the DNA content of the cell. Typically, balanced reorganizations include fusions, fissions, translocations (where a fragment of one chromosome breaks and fuses to a different chromosome) or inversions (being pericentromeric or paracentromeric, whether the centromere is affected or not, respectively). Balanced aberrations are often associated with reproductive failure because of the possible formation of unbalanced gametes during meiosis. These unbalanced gametes are able to participate in fertilization but will give rise to a non-viable zygote. However, if a balanced gamete is formed, it will contribute to a viable zygote with reduced fertility.

Deletions and duplications are unbalanced aberrations, since they reduce or increase the DNA content of the cell. These aberrations can produce balanced gametes, and therefore contribute to offspring formation.

In cattle, the most commonly detected structural chromosome change is the so-called Robertsonian translocation or centric fusion, where two acrocentric chromosomes break and fuse at the centromere region. So far, 44 Robertsonian translocations have been described in cattle breeds, affecting almost all chromosomes (Table 6.2). From these centric fusions, the 1/29 translocation (Fig. 6.2) is the most widely spread across different breeds and environments. Its frequency varies considerably from one breed to another, reaching up to 60% in British White (Eldridge, 1975) and Corsican breeds (Hari et al., 1984). This translocation causes a 5–10% reduction of fertility due to the formation of unbalanced gametes and an increase in embryonic mortality (Dyrendahl and Gustavsson, 1979), reaching values of up to 2.76% of unbalanced sperm and 4.06% of unbalanced oocytes (Bonnet-Garnier et al., 2008).

Other types of structural chromosome aberrations have been described in cattle, with reciprocal translocations the second most commonly identified. To date, only 19 reciprocal translocations have been described (Table 6.3), representing a frequency of 0.03% (Ducos et al., 2008). This low frequency could be due to low occurrence of such abnormalities in cattle or due to difficulties in detecting smaller rearrangements by routine cytogenetics using Giemsa staining. In a recent study, De Lorenzi et al. (2012) showed that only 16% of reciprocal translocations could be detected using these techniques; therefore, the frequency of these rearrangements would be underestimated (De Lorenzi et al., 2012). This was exemplified in a recent paper, where using a combination of cytogenetic and sequencing techniques, Durkin et al. (2012) described a new serial translocation led by circular intermediates responsible for colour sidedness in cattle breeds (Durkin et al., 2012).

Structural chromosome aberrations are responsible for significant economic losses in cattle breeding, and thus, their identification in animals intended for reproduction represents an important step in cytogenetic studies. Therefore, accurate and fast cytogenetic techniques, such as array Comparative Genomic Hybridization (aCGH), could be applied to screen chromosome aberrations in these animals.

Chromosome Maps

Overview of chromosome mapping

At early stages of genome mapping the maps contained 'synteny' or 'linkage' groups of genes and other markers that were not assigned to specific chromosomes. The number of groups could be significantly larger than the haploid number of chromosomes in a mapped species due to limitations in resolution of mapping panels and populations or insufficient number of markers to detect linkage or synteny. Such maps produced with somatic cell hybrid or linkage mapping techniques would contain a few markers and often would fail to resolve the order of closely or even distantly located genes. However, even sparse linkage maps were powerful enough to roughly identify chromosome regions that control economically important traits in cattle (Heyen et al., 1999). With advances in human and mouse genomics, mapping techniques and development of molecular markers with single orthologues in various mammalian genomes, a 'comparative mapping' approach has become widely utilized in livestock genomics. Comparative genomics has been able to efficiently couple information about the association of markers and phenotypes produced by genetic studies in 'map poor' species (e.g. cattle) with information on functional 'candidate genes' from 'map rich' human and mouse genomes. This has resulted in the detection of multiple candidate genes and actual mutations controlling genetic disorders and some economically important traits in cattle (Grobet et al., 1997). This approach became especially effective when linkage mapping was able to be supported with radiation hybrid maps of hundreds or even thousands of ordered genes and microsatellites. The radiation hybrid maps were integrated with linkage maps by enabling positioning of the same markers (e.g. microsatellite) and simultaneous integration with physical maps (cytogenetic or fingerprint) providing a strong link between the genetic and physical maps (Everts-van der Wind et al., 2005). Radiation hybrid maps have also provided high enough resolution to compare patterns of chromosome evolution in multiple mammals and have been used as a basis for a

Table 6.2. Robertsonian translocations described in cattle breeds.

Robertsonian translocation	Breed	Reference
1;4		Lojda et al., 1976
1;7		Frank and Robert, 1981
1;21	Holstein Friesian	Miyaket et al., 1991
1;23		Lojda et al., 1976
1;25	Piebald	Stranzinger and Forster, 1976
1;26	Holstein Friesian	Miyake and Kaneda, 1987
1;28		Lojda et al., 1976
1;29	Different breeds	See text for details
2;8	Friesian	Pollock, 1974
2;27		Yu and Xin, 1991
2;28	Vietnamese	Tanaka et al., 2000
3;4	Limousin	Popescu, 1977
3;27	Friesian	Samarineanu et al., 1977
4;4		Lojda et al., 1975
4;8	Chianina	De Giovanni et al., 1988
4;10	Blonde d'Aquitaine	Bahri-Darwich et al., 1993
5;18	Simmental	Papp and Kovacs, 1980
5;21	Japanese Black	Masuda et al., 1978
5;22	Polish Red	Sysa and Slota, 1992
5;23	Brune Roumaine	Samarineanu et al., 1977
6;16	Dexter	Loghe and Harvey, 1978
6;28		Lojda et al., 1976
7;21	Japanese Black	Hanada et al., 1981
8;9	Brown Swiss	Tschudi et al., 1977
8;23	Ukrainian Grey	Biltueva et al., 1994
9;23	Blonde d'Aquitaine	Cribiu et al., 1989
11;16	Simmental	Kovacs, 1975
11;22		Lodja et al., 1976
12;12	Simmental	Herzog and Hohn, 1984
12;15	Holstein Friesian	Roldan et al., 1984
13;19		Molteni et al., 1998
13;21	Holstein Friesian	Kovacs et al., 1973
13;24	Red and White	Slota et al., 1988
14;19	Braunvieh	Stranzinger, 1989
14;20	Simmental	Logue and Harvey, 1978
14;21	Simmental	Kovacs and Szepeshlyi, 1977
14;24	Podolian	Di Berardino et al., 1979
14;28	Holstein Friesian	Ellsworth et al., 1979
15;25	Barrosa	Iannuzzi et al., 1992
16;18	Barrosa	Iannuzzi et al., 1993
16;19	Marchigiana	Malerba, 1997
16;20	Ger. Red Pied × Czech. Red Pier	Rubes et al., 1996
16;21	Ger. Red Pied × Czech. Red Pier	Rubes et al., 1992
19;21	Holstein Friesian	Pinton et al., 1997
20;20	Simmental	Herzog and Hohn, 1984
21;27	Blonde d'Aquitaine	Berland et al., 1988
24;27	Holstein hybrid	Mahrous et al., 1994
25;27	Grey Alpine	De Giovanni et al., 1979

series of important discoveries. Murphy et al. (2005) reported that evolutionary breakpoints in mammalian chromosomes are often reused in evolution, and identified such regions in mammalian chromosomes with the use of the data from sequenced human, mouse and rat genomes and radiation hybrid maps from five additional species, including cattle. The same

Table 6.3. Reciprocal translocations (RCPs) in cattle. (Modified from De Lorenzi *et al.*, 2012.)

RCP	Reference
1;5	Iannuzzi *et al.*, 2001
1;8;9	Kovacs *et al.*, 1992
1;15	Ducos *et al.*, 2008
2;4	Switonski *et al.*, 2008
2;5	Pinton *et al.*, 2003
2;20	De Schepper *et al.*, 1982
4;7	De Lorenzi *et al.*, 2010
8;13	Ansari *et al.*, 1993
8;21	Ducos *et al.*, 2008
8;27	De Schepper *et al.*, 1982
9;11	De Lorenzi *et al.*, 2007
9;12	Ducos *et al.*, 2008
11;21	Molteni *et al.*, 2007
12;17	Ducos *et al.*, 2000
20;24	Villagomez *et al.*, 1993
Y;9	Iannuzzi *et al.*, 2001
Y;17	Vallenzasca *et al.*, 1990
Y;21	Switonski *et al.*, 2011
X;23	Basrur *et al.*, 2001

group reported that rates of chromosomal rearrangement in mammals were not equal throughout evolutionary time and have increased significantly after the Cretaceous–Tertiary boundary about 65 MYA at the time of active speciation of mammals (Murphy *et al.*, 2005).

Since the completion of sequencing of the cattle genome, high resolution radiation hybrid (Everts-van der Wind *et al.*, 2005) or integrated (Snelling *et al.*, 2007) maps have become invaluable to build whole genome assembly at chromosomal level (Plate 31). As has been demonstrated by the cattle genome sequencing initiative, different maps used to assemble exactly the same sequence data lead to different enough assemblies (Elsik *et al.*, 2009; Zimin *et al.*, 2009) to contain a number of large-scale differences in the chromosome structure.

Somatic cell hybrid maps

Somatic cell hybrid mapping utilizes an ability of cultured mammalian cells from different species to fuse forming heterokaryons in the presence of some viruses (Barski *et al.*, 1961) or polyethylene glycol (Ahkong *et al.*, 1975). In the next rounds of division heterokaryons randomly lose chromosomes from the donor but retain chromosomes of the recipient species. A panel of 20–30 independent hybrid cell clones containing different combinations of donor chromosomes is used for synteny mapping. These clones are analysed for the presence/absence of donor markers that are distinguishable from the recipient orthologues. A concordance between the presence of a donor chromosome and a marker suggests location of the marker in the donor chromosome. If multiple markers are found in the clones containing the same donor chromosome, these markers are syntenic. In this manner somatic cell hybrid mapping identifies groups of markers co-located on chromosomes ('syntenic groups'). Somatic cell hybrids normally contain complete donor chromosome(s); therefore they can only be used to identify marker synteny in the donor genome. The order of markers within syntenic groups in the majority of cases remains unresolved.

The first work using interspecies hybrids of somatic cells for establishing synteny between cattle genes reported genes *G6PD*, *PGK*, *GALA* and *HPRT* being located on cattle chromosome X (Heuertz and Hors-Cayla, 1978). Later, after the construction of a rodent-cattle somatic cell hybrid panel (Womack and Moll, 1986), containing 31 independent clones, a large number of cattle markers were mapped using this approach. Even at the early stages of somatic cell hybrid mapping in cattle the map has demonstrated a higher level of synteny conservation between human and cattle than between human and mouse chromosomes (Womack and Moll, 1986). Now the cattle somatic cell hybrid map contains over 2700 genes, 1400 of which were genotyped on the cattle–hamster somatic cell hybrid panel by a Japanese research group (Itoh *et al.*, 2003). The somatic cell hybrid map was integrated with the USDA-MARC linkage map (Kappes *et al.*, 1997) by the genotyping of over 200 microsatellite markers from the linkage map on the somatic cell hybrid panel providing a detailed integration of physical and linkage data. Another interesting attempt to improve the cattle somatic cell hybrid map was made by Laurent *et al.* (2000), who used 233 human Expressed Sequence Tag (EST) PCR primers that amplified cattle sequences to map additional gene markers on cattle chromosomes. Eventually they assigned 60 human ESTs to cattle chromosomes, of which 46 ESTs had assignments consistent with human–cattle chromosome painting results (Laurent *et al.*, 2000).

Somatic cell hybrid maps provided information about the chromosomal assignments of gene markers or microsatellites, but the information about the order of markers in chromosomes was very limited because the majority of clones would contain complete or nearly complete cattle chromosomes. However, these maps were invaluable for the assignment of ordered markers within linkage groups to cattle chromosomes and for some pioneering studies of chromosome evolution in mammals.

Linkage maps

Since the first linkage map, built by A. Sturtevant in T. Morgan's laboratory more than 100 years ago, construction of meiotic or linkage maps has become an essential genetic procedure (Barendse and Fries, 1999; Moran and James, 2005). Genetic linkage was initially revealed as a deviation from Mendel's law of independent assortment. Genes that are located close to each other on the same chromosome do not assort independently at meiosis, which is explained by the linkage. The exchanges or crossovers between homologous chromosomes, which occur at meiosis during formation of gametes, break the linkage with certain frequency. The proportion of recombinant haplotypes is a measure of crossing over frequency. In general, the further apart two loci are on a chromosome the greater the chance a crossover event will have taken place between them and so the greater will be the proportion of recombinants. The recombination rate can be used for measuring distance between two loci on a chromosome. There are two important requirements for basic linkage mapping – large pedigrees, in which the relationships are known, and availability of polymorphic genetic loci. Both these requirements were satisfied for cattle in the early 1990s when polymorphic DNA markers like microsatellites became available. Modern genomic tools provide practically endless sources of polymorphic loci.

It should be emphasized that physical distances between loci on DNA are constant and can be expressed as the number of nucleotides or other common metrics. On the contrary, the recombination rate or linkage between two genes or markers varies for the same physical distance depending on type of cross, genotype, region of a chromosome, sex and other factors. Despite this well-known 'volatility' in measuring recombination distances between loci, linkage maps remain a unique instrument in genetic research and selection even in the post-genomic era. While physical/genomic maps allow the highest possible accuracy, linkage maps provide a valuable link between genomes and phenotypes.

Cattle are hardly the best choice for building good linkage genetic maps due to large size, slow growth and usually a single offspring in each parity. Also, as the chapter testifies, cattle have 30 pairs of chromosomes, which adds complications caused by the necessity to construct 30 linkage maps for females and 31 for males. The difficulties in building a linkage map are usually compounded by a lack of knowledge of the relative position of alleles on homologous chromosomes. The major solution to the problem was calculating a likelihood ratio that takes into account alternative phases. This procedure can be quite complex, particularly with large and convoluted pedigrees. Fortunately several computer programs were developed in the late 1980s including LINKAGE (Lathrop and Lalouel, 1988) and CRI-MAP (Green *et al.*, 1990), which have been widely used for resolving these problems in most cases. The theoretical solution of these problems and the corresponding computer programs were major advancements that eventually led to construction of multilocus linkage maps. For cattle, such linkage maps were built by the late 1990s and included nearly all polymorphic genes and microsatellites available at that time (for details see Barendse and Fries, 1999). The latest release of bovine maps can be found at the website of Roslin Bioinformatics Group (UK) (http://www.thearkdb.org/arkdb/).

The first two whole-genome linkage maps for cattle were published in 1994 (Barendse *et al.*, 1994; Bishop *et al.*, 1994). These maps contained about 200 and 300 polymorphic markers, respectively, with an average interval between markers >10 cM. Individual linkage groups were assigned to cattle chromosomes using an overlapping set of markers placed on the somatic cell hybrid or cytogenetic physical maps. Significant progress in cattle linkage mapping has been achieved with the construction of the USDA-MARC linkage map containing ~1200 polymorphic markers with an average spacing of 2.5 cM and a total genome length of 2990 cM (Kappes et al., 1997). This map has provided the basis for integration of four linkage maps (Barendse et al., 1994; Bishop et al., 1994, Georges et al., 1995; Ma et al., 1996), which significantly increased the power of QTL detection. The next significant improvement in the USDA-MARC map was an addition of 2277 microsatellite markers, resulting in the generation of a 3802 microsatellite map with an average interval between markers of 1.4 cM (Ihara et al., 2004). Later BAC end sequences (BESs) and EST-based SNPs were added to the linkage map resulting in a 4585 marker map (Snelling et al., 2005). After the cattle genome sequence became available, the cattle linkage maps were enriched for biallelic SNP markers. A high-density bovine linkage map was recently constructed using 294 microsatellites, 3 milk protein haplotypes and 6769 SNPs. This map was built by combining genetic and physical information in an iterative mapping process. Markers were mapped to 3155 unique positions; the 6924 autosomal markers were mapped to 3078 unique positions and the 123 non-pseudoautosomal and 19 pseudoautosomal sex chromosome markers were mapped to 62 and 15 unique positions, respectively (Arias et al., 2009).

Linkage maps, besides their significant theoretical value in several fields of genetics, are essential for locating quantitative trait loci (QTLs) and can be used in marker assisted selection. During the last 10–15 years the underlying genetic architecture of critically important bovine traits like growth, disease resistance, milk production, meat and carcass quality and behavioural characteristics became more accessible for investigation. Further study of QTLs and their interactions will continue to be of considerable interest. However, a link between phenotype and genotype for quantitative traits is usually not very strong as these traits are polygenic and individual genes involved in development of these traits do not have large effects;

there is always significant influence of environmental factors as well as the unavoidable contribution of developmental randomness (Ruvinsky, 2009).

Nevertheless tracking the inheritance of markers in cattle populations with well recorded performance data should allow some of the QTLs to be detected and the genetic control of production traits to be at least partially identified. The general principle of such an approach is simple; as soon as significant associations between the inheritance of a particular chromosomal region (as determined by marker inheritance) and trait variation is detected in a sufficiently large population, this suggests existence of a gene or genes affecting the traits in question. Efforts of numerous research groups and particularly from Iowa State University led to the creation of a QTL database for different agricultural animals including cattle (www.animalgenome.org/cgi-bin/QTLdb/BT/index). Once a QTL has been mapped to an interval between two arbitrary markers, there is a need to identify markers that are as close to the QTL as possible. Tightly linked markers are rarely involved in meiotic recombination and will continue to frame the QTL for a long time. As outlined in this and the following chapters, syntenic or linkage relationships over short distances (<3 cM) are often conserved across species and the cattle genome is no exception. Now when the cattle genome is resequenced (Elsik et al., 2009), genes in many QTL regions can be examined for causative mutations. For example a whole-genome scan for QTLs affecting milk production traits in Holstein cattle (Georges et al., 1995; Heyen et al., 1999; Keele et al., 1999) was performed. In addition, 31 chromosomal regions affecting milk production QTLs were detected using Finnish Ayrshire dairy cattle (Viitala et al., 2003). Several monogenic disorders were identified using the genetic linkage map information and genome wide association analysis. A missense mutation in the bovine ATP2A1 gene was found to be associated with congenital pseudomyotonia of Chianina cattle and potentially can be used as an animal model of human Brody disease (Drögemüller et al., 2008). A deletion of the myostatin gene causes the double-muscled phenotype in cattle (Grobet et al., 1997).

Radiation hybrid maps

Based on an observation of Goss and Harris (1975) and subsequent application by Cox et al. (1990) for the human genome high-resolution mapping (Cox et al., 1990, Goss and Harris, 1975), radiation hybrid (RH) mapping has been widely used for mammalian species to build ordered maps with marker spacing between millions and as few as thousands of base pairs. The approach is based on a random segregation of irradiated chromosomal fragments from donor cells used for the construction of somatic cell hybrids. As a result heterokaryons from a donor and recipient cell fusions contain a complete recipient genome and a random set of donor chromosomal fragments. Sizes of the donor chromosomal fragments correlate negatively with the dose or radiation applied to the donor cell. Therefore the principle of RH clone construction is equivalent to that of somatic cell hybrids with the addition of irradiation of the donor cells step. In contrast the principle behind RH mapping is similar to that of genetic mapping, where instead of estimating distances between markers based upon frequencies of recombination in a population of related individuals, these distances are estimated based on the frequency of physical DNA breakage. The closer markers are in a chromosome the higher is the frequency of their co-appearance in DNA fragments found in independent RH clones. RH panels generated using higher doses of radiation allow for estimating distances and the order of markers located closer to one another, but often fail, producing long linkage groups, while RH panels generated with lower radiation doses produce longer linkage groups but often are non-informative to resolve the order of closely located markers. Because RH markers are genotyped on the fragmented chromosomes of the same donor individual, there is no need for within-species polymorphism to estimate the order of markers. However interspecies differences between the donor and recipient marker counterparts are important. Therefore unlike linkage maps, RH maps could be built with markers that lack within-species polymorphism (e.g. genes) and there is no need for a large mapping population, making this type of map ideal for mapping genes and

other molecular markers in mammals that have a limited number of offspring (e.g. cattle). It worth mentioning that unlike linkage maps RH maps could not be used to find an association between specific chromosomal interval and QTLs. However, RH maps could be integrated with linkage maps to enrich candidate intervals of a linkage map with gene markers.

In cattle, an RH mapping approach was first applied by Yang and Womack (1998) for the creation of a comparative map of cattle chromosome 19 and human chromosome 17 (Yang and Womack, 1998). A 5000 Rad radiation hybrid panel constructed by Womack et al. (1997) was used to build three generations of Illinois-Texas (IL-TX) whole-genome cattle radiation hybrid maps containing 1087, 1913 and 3484 markers, respectively (Womack et al., 1997). The first generation medium-resolution IL-TX RH map contained 768 gene markers and 319 microsatellites, which were used to link RH linkage groups to the USDA-MARC linkage map (Band et al., 2000). A total of 638 markers on this RH map had known orthologues in the human genome, and an estimated comparative coverage of the human genome was ~50%. Regardless of the relatively small number of markers, this map provided a great resource for predicting positions of cattle BAC end sequences (BESs) using the 'comparative mapping by annotation and sequence similarity' (COMPASS) approach that utilizes comparative maps of cattle and human genomes for the prediction of positions of cattle genomic sequences on cattle chromosomes (Ma et al., 1998; Rebeiz and Lewin, 2000; Larkin et al., 2003).

To generate a higher resolution cattle IL-TX RH map, 870 new markers with predicted positions in gaps of cattle–human comparative coverage were selected for a new mapping project. As a result, 1913 markers were placed on a new version of the cattle RH map. This provided ~66% comparative coverage between human and cattle genomes and almost maximum resolution and coverage of the cattle genome that could be achieved using EST markers because of uneven distribution of genes in mammalian genomes. Most of the large gaps in the comparative coverage between the human and cattle genomes were located in gene-poor regions. To build the third generation whole-genome IL-TX RH map of the cattle genome, a set of genomic markers rather than ESTs was used (Everts-van der Wind et al., 2005). This set of markers was generated by the International Cattle BAC Mapping Consortium and represented ~500 bp terminal end sequences of cattle BAC clones (Larkin et al., 2003; Snelling et al., 2007). These sequences were compared to the human genome sequence and over 3000 cattle sequences with evenly spaced (~1 Mb apart) unique BLASTn hits in human chromosome sequences were placed on the cattle RH map. The resulting map, containing 2516 ordered BESs, 736 ESTs and 232 microsatellites, was integrated with a physical fingerprint map. The third generation IL-TX cattle radiation hybrid map had ~91% comparative coverage of the human genome and demonstrated ~93% agreement in the order of markers with the cattle physical fingerprint map containing the same BAC clones (Plate 31). Whereas the focus of IL-TX radiation hybrid maps was on mapping markers with known orthologues in the human genome, other groups have built RH maps that contained a significant number of microsatellite and SNP markers. For example, a 3966 marker map built using Roslin 3000 Rad panel contained 1072 microsatellite markers, 1999 genes, BESs and amplified fragment length polymorphism (AFLP) markers (Jann et al., 2006). Another map (SUNbRH, 7000 Rad) contained 5593 markers, of which 3216 markers were microsatellites and 2377 were ESTs (Itoh et al., 2005). An attempt for bioinformatics-based integration of several RH and linkage maps into a single integrated map resource has been made (Snelling et al., 2007). The resulting composite map contained 17,254 markers and was integrated with the cattle fingerprint map. The latest version of the map was used to assign scaffolds to chromosomes and to establish their order on the Maryland cattle genome assembly (UMD2.0) (Zimin et al., 2009).

Fingerprint maps

A 'fingerprint' physical map contains contigs of cloned DNA fragments (often YAC or BAC clones) combined and ordered based on similarities in their patterns of digestion with

endonucleases of restriction. To increase resolution often a combination of two enzymes is used to generate digests that are separated on agarose gels and compared with thousands of other clone digestion patterns using specially written software (e.g. FPC; Soderlund et al., 2000). The length and quality of contigs depends on the representativeness of the clone library used to build the map as well as on the actual number of overlapping clones from the library that were included in the mapping experiment. Physical fingerprint maps provide the highest resolution among other physical maps (except for the recently introduced optical maps and genome assembly) and are very useful to build maps for specific regions for QTLs or disease gene mining or for whole genomes for selection of the minimum 'tiling paths' of clones for the whole-genome sequencing using a clone-by-clone sequencing approach.

A British Columbia Cancer Research Center (BCCRC) fingerprint physical map of the cattle genome has been built by the International Bovine BAC Mapping Consortium (Snelling et al., 2007). This map contains 290,797 BAC clones from free cattle BAC libraries generated from different breeds, 200,064 clones from CHORI-240 (Hereford male), 94,848 from RPCI-42 (Holstein male) and 44,948 from TAMBT (Angus male). The initial set of ~13,000 contigs has been merged by FPC software into a set of 655 large contigs, containing 257,914 clones. Comparative data obtained from the alignment of cattle BES with the human genome allowed for selection of probable merge points between contigs that were examined by an FPC program using relaxed threshold criteria. The use of a comparative information and a high number of fingerprinted clones allowed for significant decrease in the number and increase in the length of contigs compared to another cattle fingerprint map constructed at INRA. The INRA fingerprint map contained 6615 contigs designed from ~105,000 clones from the INRA BAC library and ~27,000 clones from the CHORI-240 library (Schibler et al., 2004).

The BCCRC physical map has been integrated with the third generation IL-TX RH map by 3400 BESs. These independent maps show about 93% agreement in the order or clones indicating high quality of these resources

(Plate 31). Several hundred RH and linkage map markers were assigned to BAC clones from the BCCRC fingerprint map using PCR analysis or *in silico* comparative mapping against the human genome. This whole-genome contig map provided the highest level of resolution of the cattle genome until the sequence assembly became available. The human–cattle comparative map based on the fingerprint map and BES hits in the human genome has been used for discovery of long regions of amniote chromosomes that are non-randomly maintained during chromosomal evolution (Larkin *et al.*, 2009). A skim of ~19,600 overlapping BAC clones from the CHORI-240 library from this map has been selected to complement the whole-genome shotgun (WGS) sequence for genome sequencing (Elsik *et al.*, 2009).

Genome assembly

The cattle genome assembly (~7.1× Sanger reads) has been generated at Baylor College of Medicine combining BAC sequencing from a Hereford male CHORI-240 library and the whole-genome shotgun sequences of DNA taken from a Hereford dam, L1 Dominette 01449 (Elsik *et al.*, 2009). The overlapping set of BAC clones for sequencing has been selected from the BCCRC fingerprint map. Combining BAC and shotgun sequences, a set of scaffolds with N50 of 1.9 Mb was generated. The published build of the cattle genome (Btau 4.0) has ~90% of the cattle genome sequence placed on 29 autosomes and chromosome X. The estimated cattle genome size based on that map is 2.87 Gb (Elsik *et al.*, 2009).

Another assembly of the cattle genome was built at the University of Maryland (UMD2.0) (Zimin *et al.*, 2009). For this assembly the same set of raw sequences was used as for Baylor assembly, however a different assembly approach and software were used. This allowed 5% more of the sequence to be placed on chromosomes compared to Btau 4.0 and resulted in larger N50 size of the sequence contigs. There are significant discrepancies between the Btau 4.0 and UMD 2.0 assemblies. Additional efforts were required to resolve the discrepancies case by case.

An annotation of the cattle genome was performed by the Bovine Genome Sequencing and Analysis Consortium on the Baylor version of the cattle genome assembly. The number of genes in the cattle genome was estimated as >22,000 protein-coding and 496 miRNA genes. The genome contains a large number of ruminant-specific transposable elements that comprise 27% of the genome. Some transposable elements from the BOV-B group have intact open reading frames and could still be active. An analysis of orthologous gene pairs between human, cattle, dog, mouse, rat, opossum and platypus genomes has revealed 1217 genes that could be placental-specific because they are not present in opossum and platypus genomes. About 3.1% of the cattle genome is in segmental duplications. Seventy-six per cent of segmental duplications contain complete or partial gene duplications. This set is enriched for genes involved in interactions of the organism with its external environment, e.g. immune proteins and olfactory receptors.

A comparison of the cattle chromosome architecture to the chromosomes of other mammals has revealed 124 evolutionary breakpoint regions in the cattle lineage of which 24 are shared by cattle and pig chromosomes (artiodactyl-specific). The remaining 100 were found only in cattle chromosomes. Our current studies show that only about 50% of these evolutionary breakpoints have occurred in the cattle lineage, while the rest of them are ancestral Pecora or even ruminant-specific events. Nine additional breakpoints were shared by all ferungulate species (cattle, pig, dog) and represent events that originated in the ferungulate ancestor. Interestingly, there is a strong negative correlation between the positions of cattle and artiodactyl-specific breakpoints and transposable elements, e.g. some long interspersed elements (LINEs) and short interspersed elements (SINEs), whereas more recent LINE-L1 and LINE-RTE elements are significantly enriched in these breakpoint regions. Another group of repeats, tRNA^{Glu}-derived SINEs originating in the common ancestor of all artiodactyls has a higher than expected density in artiodactyl-specific breakpoint regions, but not in the cattle-specific breakpoints. This suggests that evolutionary breakpoints tend to happen in genome regions with a high density of repetitive elements that are active (Fig. 6.3), and therefore have high sequence similarity between different copies required for non-allelic homologous recombination. In confirmation of this observation, an analysis revealed a high density of large (>10 kb) segmental duplications in cattle and artiodactyl-specific breakpoint regions, phenomena previously reported for primate (Murphy *et al.*, 2005) and murine rodent (Armengol *et al.*, 2005) genomes.

Comparative Studies

Among the livestock species, cattle have one of the best and most detailed sets of chromosome comparative maps available mostly due to its global economic importance. Whereas somatic cell hybrid maps and cross-species chromosome painting with human and other species DNA probes have provided an important but patched correspondence between cattle, human, mouse and pig genomes (Womack and Moll, 1986; Hayes, 1995; Chowdhary *et al.*, 1996; Schmitz *et al.*, 1998), the real breakthrough in cattle comparative studies started with the introduction of high-resolution ordered radiation hybrid maps (Band *et al.*, 2000; Everts-van der Wind *et al.*, 2004, 2005) and a COMPASS-based approach of marker selection for mapping (Rebeiz and Lewin, 2000; Larkin *et al.*, 2003). The whole-genome high-resolution ordered chromosome comparative maps identify approximately 201–211 large blocks of homologous synteny between human and cattle chromosomes (Fig. 6.4). These blocks have two or more genes found in human and cattle on the same chromosome and most likely represent the gene order inherited by both human and cattle from their common ancestor. Comparable numbers are reported for the comparison of completely sequenced cattle and human genomes with the highest number of 268 homologous synteny blocks being reported by Zimin and coworkers (2009). The variation in the number of conserved blocks could

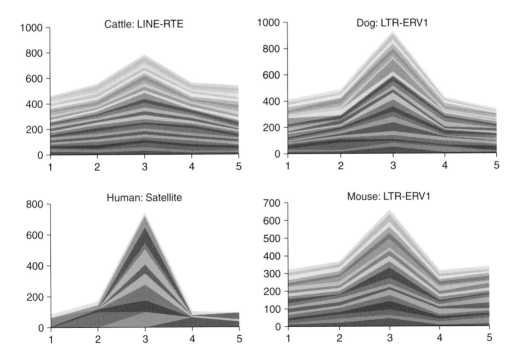

Fig. 6.3. Average density of lineage-specific repetitive elements in lineage-specific evolutionary breakpoint regions (EBRs, point 3) and two immediately adjacent chromosomal intervals (points 1,2 and 3–5). Each line represents data for individual chromosomes. The data demonstrate that EBRs are preferably located in the regions of genomes enriched for repeat families that were recently active in species evolution.

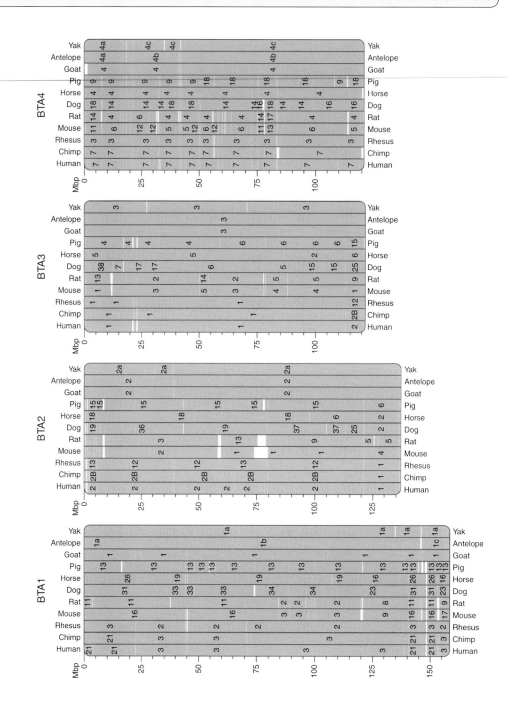

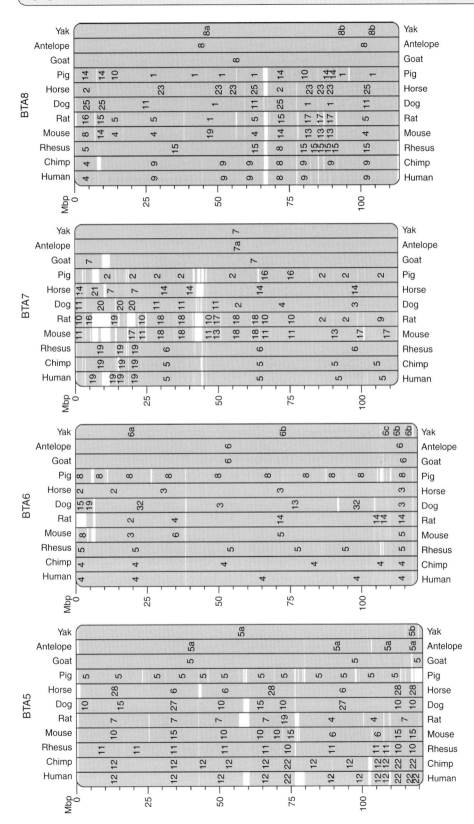

Fig. 6.4. Complete comparative map of the cattle genome. See page 121 for details.

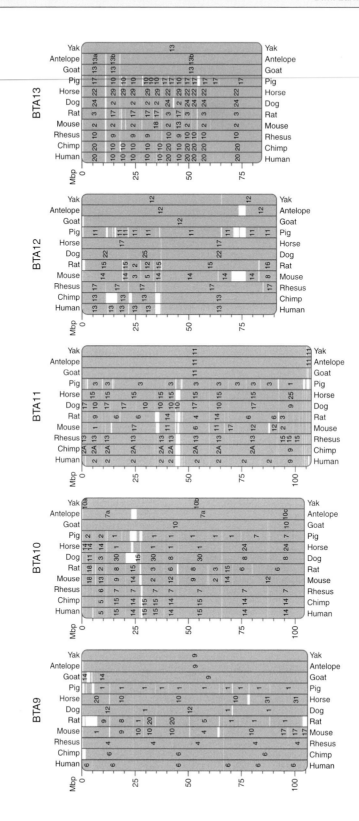

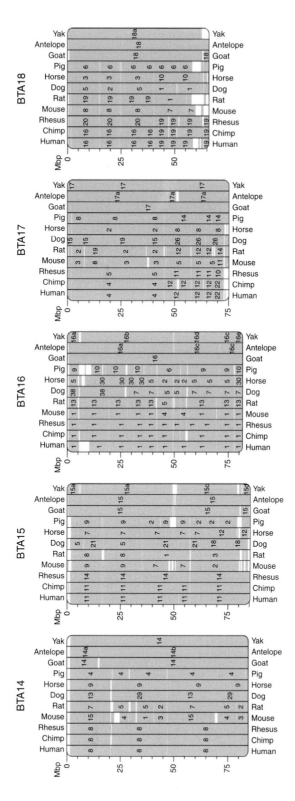

Fig. 6.4. Complete comparative map of the cattle genome. See page 121 for details.

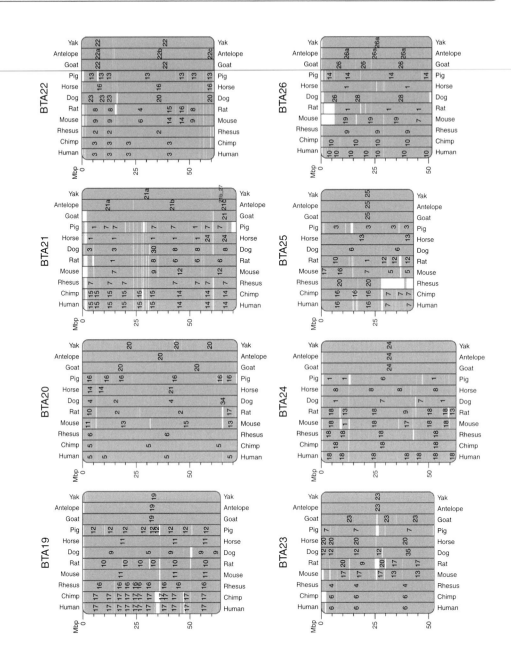

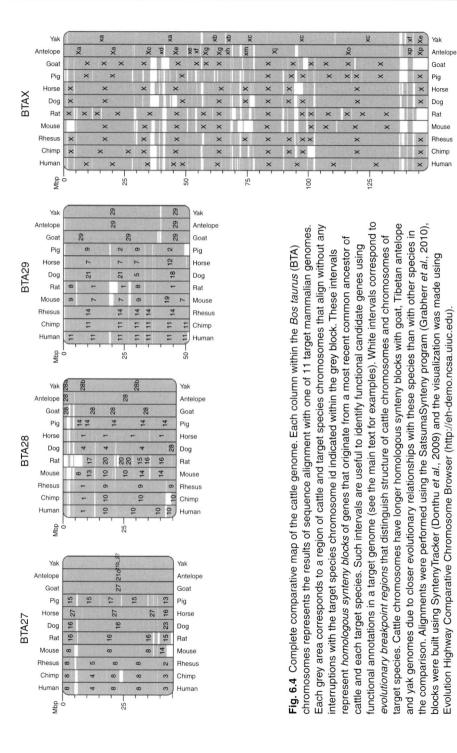

Fig. 6.4 Complete comparative map of the cattle genome. Each column within the *Bos taurus* (BTA) chromosomes represents the results of sequence alignment with one of 11 target mammalian genomes. Each grey area corresponds to a region of cattle and target species chromosomes that align without any interruptions with the target species chromosome id indicated within the grey block. These intervals represent *homologous synteny blocks* of genes that originate from a most recent common ancestor of cattle and each target species. Such intervals are useful to identify functional candidate genes using functional annotations in a target genome (see the main text for examples). White intervals correspond to *evolutionary breakpoint regions* that distinguish structure of cattle chromosomes and chromosomes of target species. Cattle chromosomes have longer homologous synteny blocks with goat, Tibetan antelope and yak genomes due to closer evolutionary relationships with these species than with other species in the comparison. Alignments were performed using the SatsumaSynteny program (Grabherr *et al.*, 2010), blocks were built using SyntenyTracker (Donthu *et al.*, 2009) and the visualization was made using Evolution Highway Comparative Chromosome Browser (http://eh-demo.ncsa.uiuc.edu).

be explained by differences in the rule sets used by different groups for the identification of evolutionary breakpoints and conserved parts of chromosomes. An identification of evolutionary breakpoints that distinguish cattle chromosome architecture from the chromosomal architecture of other species has played an important role in finding the major patterns in the cattle and mammalian chromosome evolution. For example, Everts-van der Wind and co-workers (2004) reported that evolutionary breakpoints between the cattle and human genomes are enriched in genes, at least in the human genome (Everts-van der Wind et al., 2004). Lately this observation has been confirmed by multi-species genome comparisons (Murphy et al., 2005; Larkin et al., 2009). In addition, it has been demonstrated that in cattle gene families coding milk proteins have been significantly rearranged compared to other mammals. One example is histatherin (*HSTN*), a gene in the casein cluster on BTA6. In cattle *HSTN* was moved to a regulatory element important for β-casein expression, and as a probable consequence, *HSTN* is regulated like the casein genes during the lactation cycle (Elsik et al., 2009). Also, it was demonstrated that the cluster of β-defensin genes coding antimicrobial peptides was significantly reorganized and expanded in the cattle genome due to a large segmental duplication located in a cattle-specific evolutionary breakpoint region. The analysis of cattle segmental duplications has confirmed the previous observation that evolutionary breakpoint regions in different mammalian species are enriched for this type of sequence. Therefore, one important lesson learned from the cattle-whole genome analysis and its comparison to other mammalian genomes is that evolutionary breakpoint regions could play an important role in adaptation of the genome to the environment because they are reorganizing the genes that contribute to the species' response to external stimuli, immune response and other functions related to the lineage-specific features (Larkin et al., 2009, Lemay et al., 2009).

Future of Cattle Genome Mapping

With the introduction of next generation sequencing (NGS) platforms genome sequencing became

a trivial task. However, *de novo* assembly of sequences generated from a large number of short or pair-end reads is still a difficult and resource-consuming endeavour especially if a genome needs to be assembled to the chromosome level. Next generation resequencing projects in cattle will be focused on assembly using a reference genome as the basis for contig construction. As with humans (Levy et al., 2007; Wheeler et al., 2008) resequencing projects in cattle will become the major source of polymorphic markers (mostly SNPs and indels) for QTL association studies and QTN discoveries (Eck et al., 2009). For example in a recent study Larkin et al. (2012) reported *de novo* sequencing and reference-based assembly of the two most influential bulls in the history of the Holstein Friesian breed in the USA (Larkin et al., 2012). These two bulls (sire and son) have contributed about 16% of their genes to the population and both had multiple prominent offspring. By using Roche-454 sequencing technology and a reference-based assembly approach the authors were able to reconstruct genome-wide phases of alleles for both individuals' chromosomes and trace these chromosomes in seven generations of descendants. The phasing and haplotype-reconstruction step would be impossible without a high-quality reference genome that was assembled with the use of radiation hybrid and other types of chromosome maps described in this chapter. By following the frequency of alleles originating from one of the bulls in the Holstein Friesian population subjected to strong artificial selection the authors identified 49 chromosomal intervals affected by selection in Holsteins including 11 candidate point mutations for traits related to milk production and disease resistance. This final example demonstrates how a combination of contemporary chromosome mapping techniques and NGS facilitates efficient and high-throughput detection of mutations controlling economically important traits in cattle.

Conclusions

In several decades the study of cattle chromosomes has advanced from Giemsa-stained metaphase spreads, low resolution mapping of microsatellites and genes to whole-genome studies involving thousands of genes and millions of polymorphisms. Altogether these studies have drastically improved our understanding of the genetic basis of economically important phenotypes in cattle and have successfully connected many phenotypes to chromosomes. Due to high quality resources available for cattle chromosomes, the cattle genome was also successfully utilized to study chromosomal evolution. A success of the cattle genome studies could not have been achieved without a combination of cytogenetic and various genomic techniques. While a traditional cytogenetics could be used on a daily basis to screen for chromosome aberrations in cattle in a cost-effective manner, it also provides a link between chromosomes, phenotypes mapped in linkage studies, RH maps and genome assemblies. Coupled together these techniques have already led to detection of intervals, genes and mutations in cattle chromosomes related to multiple QTLs like milk production, fertility and disease resistance. With the introduction of the NGS technologies and the continuing hunt for rare or breed-specific QTLs, the role of high-quality chromosome resources will stay important because of a need to detect long-range haplotypes and linkage disequilibrium.

Acknowledgements

The authors would like to thank Dr Pietro Parma, Department of Agricultural and Environmental Sciences, University of Milan and Rebecca O'Connor, School of Biosciences, University of Kent for providing images of cattle chromosomes. This work was funded in part by the BBSRC grant BB/J010170/1.

References

Ahkong, Q.F., Fisher, D., Tampion, W. and Lucy, J.A. (1975) Mechanisms of cell-fusion. *Nature* 253, 194–195.

Ansari, H.A., Jung, H.R., Hediger, R., Fries, R., König, H. and Stranzinger, G. (1993) A balanced autosomal reciprocal translocation in an azoospermic bull. *Cytogenetics Cell Genetics* 62, 117–123.

Arias, J.A., Keehan, M., Fisher, P., Coppieters, W. and Spelman, R. (2009) A high density linkage map of the bovine genome. *BMC Genetics* 10, 18.

Armengol, L., Marques-Bonet, T., Cheung, J., Khaja, R., Gonzalez, J.R., Scherer, S.W., Navarro, A. and Estivill, X. (2005) Murine segmental duplications are hot spots for chromosome and gene evolution. *Genomics* 86, 692–700.

Bahri-Darwich, I., Cribiu, E.P., Berland, H.M. and Darre, R. (1993) A new Roberstonian translocation in Blonde d'Aquitaine cattle, rob(4:10). *Genetics, Selection, Evolution* 25, 413–419.

Band, M.R., Larson, J.H., Rebeiz, M., Green, C.A., Heyen, D.W., Donovan, J., Windish, R., Steining, C., Mahyuddin, P., Womack, J.E. and Lewin, H.A. (2000) An ordered comparative map of the cattle and human genomes. *Genome Research* 10, 1359–1368.

Barendse, W. and Fries, R. (1999) *Genetic Linkage Mapping, The Gene Maps of Cattle and the List of Loci.* CAB International, Wallingford.

Barendse, W., Armitage, S.M., Kossarek, L.M., Shalom, A., Kirkpatrick, B.W., Ryan, A.M., Clayton, D., Li, L., Neibergs, H.L., Zhang, N. *et al.* (1994) A genetic linkage map of the bovine genome. *Nature Genetics* 6, 227–235.

Barendse, W., Vaiman, D., Kemp, S.J., Sugimoto, Y., Armitage, S.M., Williams, J.L., Sun, H.S., Eggen, A., Agaba, M., Aleyasin, S.A. *et al.* (1997) A medium-density genetic linkage map of the bovine genome. *Mammalian Genome* 8, 21–28.

Barski, G., Sorieul, S. and Cornefert, F. (1961) 'Hybrid' type cells in combined cultures of two different mammalian cell strains. *Journal Natural Cancer Institute* 26, 1269–1291.

Basrur, P.K., Reyes, E.R., Farazmand, A., King, W.A. and Popescu, P.C. (2001) X-autosome translocation and low fertility in a family of crossbred cattle. *Animal Reproduction Sciences* 67, 1–16.

Berland, H.M., Sharma, A., Cribiu, E.P., Darre, R., Boscher, J. and Popescu, C.P. (1988) A new case of Robertsonian translocation in cattle. *Journal of Heredity* 79, 33–36.

Biltueva, L., Sharshova, S., Sharshov, A., Ladygina, T., Borodin, P. and Graphodatski, A. (1994) A new Robertsonian translocation, 8;23, in cattle. *Genetics, Selection, Evolution* 26, 159–165.

Bishop, M.D., Kappes, S.M., Keele, J.W., Stone, R.T., Sunden, S.L., Hawkins, G.A., Toldo, S.S., Fries, R., Grosz, M.D., Yoo, J. *et al.* (1994) A genetic linkage map for cattle. *Genetics* 136, 619–639.

Bonnet-Garnier, A., Lacaze, S., Beckers, J.F., Berland, H.M., Pinton, A., Yerle, M. and Ducos, A. (2008) Meiotic segregation analysis in cows carrying the t(1;29) Robertsonian translocation. *Cytogenetics Genome Research* 120, 91–96.

Chowdhary, B.P., Fronicke, L., Gustavsson, I. and Scherthan, H. (1996) Comparative analysis of the cattle and human genomes: detection of ZOO-FISH and gene mapping-based chromosomal homologies. *Mammalian Genome* 7, 297–302.

Citek, J., Rubes, J. and Hajkova, J. (2009) Short communication: Robertsonian translocations, chimerism, and aneuploidy in cattle. *Journal of Dairy Sciences* 92, 3481–3483.

Cohen-Zinder, M., Seroussi, E., Larkin, D.M., Loor, J.J., Everts-van der Wind, A., Lee, J.H., Drackley, J.K., Band, M.R., Hernandez, A.G., Shani, M., Lewin, H.A., Weller, J.I. and Ron, M. (2005) Identification of a missense mutation in the bovine ABCG2 gene with a major effect on the QTL on chromosome 6 affecting milk yield and composition in Holstein cattle. *Genome Research* 15, 936–944.

Cox, D.R., Burmeister, M., Price, E.R., Kim, S. and Myers, R.M. (1990) Radiation hybrid mapping: a somatic cell genetic method for constructing high-resolution maps of mammalian chromosomes. *Science* 250, 245–250.

Cribiu, E.P., Matejka, M., Darre, R., Durand, V., Berland, H.M. and Bouvet, A. (1989) Identification of chromosomes involved in a Robertsonian translocation in cattle. *Annales de Génétique et Selection Animale* 21, 555–560.

De Giovanni, A.M. Succi, G., Molteni, L. and Castiglioni, M. (1979) A new autosomal translocation in 'Alpine grey cattle'. *Annales de Genetique et Selection Animal* 11, 115–120.

De Giovanni, A.M., Molteni, L., Succi, G., Galliani, C., Boscher, J. and Popescu, C.P. (1988) A new type of Robertsonian translocation in cattle. In: Long, S. (ed.) *Proceedings of the 8th European Colloquium on Cytogenetics of Domestic Animals*, Bristol, UK, pp. 53–59.

De Lorenzi, L., De Giovanni, A., Molteni, L., Denis, C., Eggen, A. and Parma, P. (2007) Characterization of a balanced reciprocal translocation, rcp(9;11)(q27;q11) in cattle. *Cytogenetics and Genome Research* 119, 231–234.

De Lorenzi, L., Kopecna, O., Gimelli, S., Cernohorska, H., Zannotti, M., Béna, F., Molteni, L., Rubes, J. and Parma P. (2010) Reciprocal translocation t(4;7)(q14;q28) in cattle: molecular characterization. *Cytogenetics and Genome Research* 129, 298–304.

De Lorenzi, L., Morando, P., Planas, J., Zannotti, M., Molteni, L. and Parma, P. (2012) Reciprocal translocations in cattle: frequency estimation. *Journal Animal Breeding Genetics* 129, 409–416.

De Schepper, G.G., Aalbers, J.G. and Te Brake, J.H. (1982) Double reciprocal translocation heterozygosity in a bull. *Veterinary Research* 110, 197–199.

Derrien, T., Andre, C., Galibert, F. and Hitte, C. (2007) AutoGRAPH: an interactive web server for automating and visualizing comparative genome maps. *Bioinformatics* 23, 498–499.

Di Berardino, D., Iannuzzi, L., Ferrara, L. and Matassino, D. (1979) A new case of Robertsonian translocation in cattle. *Journal of Heredity* 70, 436–438.

Di Berardino, D., Iannuzzi, L., Bettini, T.M. and Matassino, D. (1981) Ag-NORs variation and banding homologies in two species of Bovidae: *Bubalus bubalis* and *Bos taurus*. *Candian Journal of Genetics and Cytology* 23, 89–99.

Donthu, R., Lewin, H.A. and Larkin, D.M. (2009) SyntenyTracker: a tool for defining homologous synteny blocks using radiation hybrid maps and whole-genome sequence. *BMC Research Notes* 2, 148.

Drögemüller, C., Drögemüller, M., Leeb, T., Mascarello, F., Testoni, S., Rossi, M., Gentile, A., Damiani, E. and Sacchetto, R. (2008) Identification of a missense mutation in the bovine ATP2A1 gene in congenital pseudomyotonia of Chianina cattle: an animal model of human Brody disease. *Genomics* 92, 474–477.

Ducos, A., Séguéla, A., Pinton, A., Berland, H., Brun-Baronnat, C., Darré, A., Manesse, M. and Darré R. (2000) Trisomy 26 mosaicism in a sterile Holstein-Friesian heifer. *Veterinary Research* 146, 163–164.

Ducos, A., Revay, T., Kovacs, A., Hidas, A., Pinton, A., Bonnet-Garnier, A., Molteni, L., Slota, E., Switonski, M., Arruga, M. V., Van Haeringen, W.A., Nicolae, I., Chaves, R., Guedes-Pinto, H., Andersson, M. and Iannuzzi, L. (2008) Cytogenetic screening of livestock populations in Europe: an overview. *Cytogenetics Genome Research* 120, 26–41.

Dunn, H.O., McEntee, K., Hall, C.E., Johnson, R.H., Jr and Stone, W.H. (1979) Cytogenetic and reproductive studies of bulls born co-twin with freemartins. *Journal Reproduction and Fertility* 57, 21–30.

Durkin, K., Coppieters, W., Drögemüller, C., Ahariz, N., Cambisano, N., Druet, T., Fasquelle, C., Haile, A., Horin, P., Huang, L. *et al.* (2012) Serial translocation by means of circular intermediates underlies colour sidedness in cattle. *Nature* 482, 81–84.

Dyrendahl, I. and Gustavsson, I. (1979) Sexual functions, semen characteristics and fertility of bulls carrying the 1/29 chromosome translocation. *Hereditas* 90, 281–289.

Eck, S.H., Benet-Pages, A., Flisikowski, K., Meitinger, T., Fries, R. and Strom, T.M. (2009) Whole genome sequencing of a single *Bos taurus* animal for single nucleotide polymorphism discovery. *Genome Biology* 10, R82.

Eldridge, F.E. (1975) High frequency of a Robertsonian translocation in a herd of British White cattle. *Veterinary Research* 97, 71–73.

Ellsworth, S.M., Paul, S.R. and Bunch, T.D. (1979) A 14/28 dicentricrobertsonian translocation in a Holstein cow. *Theriogenology* 11, 165–170.

Elsik, C.G., Tellam, R.L., Worley, K.C., Gibbs, R.A., Muzny, D.M., Weinstock, G.M., Adelson, D.L., Eichler, E.E., Elnitski, L., Guigo, R. *et al.* (2009) The genome sequence of taurine cattle: a window to ruminant biology and evolution. *Science* 324, 522–528.

Everts-van der Wind, A., Kata, S.R., Band, M.R., Rebeiz, M., Larkin, D.M., Everts, R.E., Green, C.A., Liu, L., Natarajan, S., Goldammer, T. *et al.* (2004) A 1463 gene cattle-human comparative map with anchor points defined by human genome sequence coordinates. *Genome Research* 14, 1424–1437.

Everts-van der Wind, A., Larkin, D.M., Green, C.A., Elliott, J.S., Olmstead, C.A., Chiu, R., Schein, J.E., Marra, M.A., Womack, J.E. and Lewin, H.A. (2005) A high-resolution whole-genome cattle-human comparative map reveals details of mammalian chromosome evolution. *Proceedings of the National Academy of Sciences USA* 102, 18526–18531.

Frank, M. and Robert, J.M. (1981) La pathologiechromosomique. Etude chez Bos Taurus. *Revue de Sciences Vétérinaires* 132, 405–411.

Georges, M., Nielsen, D., Mackinnon, M., Mishra, A., Okimoto, R., Pasquino, A.T., Sargeant, L.S., Sorensen, A., Steele, M.R., Zhao, X. *et al.* (1995) Mapping quantitative trait loci controlling milk production in dairy cattle by exploiting progeny testing. *Genetics* 139, 907–920.

Goss, S.J. and Harris, H. (1975) New method for mapping genes in human chromosomes. *Nature* 255, 680–684.

Grabherr, M.G., Russell, P., Meyer, M., Mauceli, E., Alfoldi, J., Di Palma, F. and Lindblad-Toh, K. (2010) Genome-wide synteny through highly sensitive sequence alignment: satsuma. *Bioinformatics* 26, 1145–1151.

Green, P., Falls, K. and Crooks, S. (1990) *Documentation for CRI-MAP Version 2.4.* Washington University School of Medicine, St. Louis, Missouri.

Grisart, B., Coppieters, W., Farnir, F., Karim, L., Ford, C., Berzi, P., Cambisano, N., Mni, M., Reid, S., Simon, P., Spelman, R., Georges, M. and Snell, R. (2002) Positional candidate cloning of a QTL in dairy cattle: identification of a missense mutation in the bovine DGAT1 gene with major effect on milk yield and composition. *Genome Research* 12, 222–231.

Grobet, L., Martin, L.J., Poncelet, D., Pirottin, D., Brouwers, B., Riquet, J., Schoeberlein, A., Dunner, S., Menissier, F., Massabanda, J., Fries, R., Hanset, R. and Georges, M. (1997) A deletion in the bovine myostatin gene causes the double-muscled phenotype in cattle. *Nature Genetics* 17, 71–74.

Guanti, G. and Minola, P. (1978) A Robertsonian translocations in the female cells of a bull, co-twin to a freemartin. *Cornell Veterinary* 68, 94–98.

Hanada, M., Muramatsu, S., Abe, T. and Fukushima, T. (1981) Robertsonian chromosome polymorphism found in a local herd of the Japanese Black cattle. *Annales de Génétique et Selection Animale* 13, 205–211.

Hare, W.C. (1976) Congenital retroflexion of the penis and inguinal cryptorchidism in a presumptive bovine twin with a 60,XY/60,XX/61,XX,+cen chromosome constitution. *Canadian Journal of Comparative Medicine* 40, 429–433.

Hari, J.J., Franceschi, P., Casabianca, F., Bosher, J. and Popescu, C.P. (1984) Etude cytogénétique d'une population de bovins corses. *Compte Rendu de l'Académie d'Agriculture de France* 70, 8.

Hayes, H. (1995) Chromosome painting with human chromosome-specific DNA libraries reveals the extent and distribution of conserved segments in bovine chromosomes. *Cytogenetics Cell Genetics* 71, 168–174.

Hayes, H., Di Meo, G.P., Gautier, M., Laurent, P., Eggen, A. and Iannuzzi, L. (2000) Localization by FISH of the 31 Texas nomenclature type I markers to both Q- and R-banded bovine chromosomes. *Cytogenetics Cell Genetics* 90, 315–320.

Henderson, L.M. and Bruere, A.N. (1979) Conservation of nucleolus organizer regions during evolution in sheep, goat, cattle and aoudad. *Canadian Journal of Genetics and Cytology* 21, 1–8.

Herzog, A. and Hohn, H. (1984) Two new translocation type trisomies in calves, 60,XX,t(12;12),+12 and 60,XX,t(20;20),+20. In: *Proceedings of the 6th European Colloquium on Cytogenetics of Domestic Animals*, Zurich, Switzerland, pp.313–317.

Herzog, A., Hohn, H. and Rieck, G. (1977) Survey of recent situation of chromosome pathology in different breeds of German cattle. *Annales de Génétique et Selection Animale* 9, 471–491.

Herzog, A., Hohn, H. and Oyschlager, F. (1982) Autosomal trisomy in calves with dwarfism. *Deutche Tierärztliche Wochenschrift* 3.

Heuertz, S. and Hors-Cayla, M.C. (1978) Carte genetique des bovins por la technique d'hybridation cel-lulaire. Localisation sur le chromosome X de la glucose-6-phosphate deshydrogenose, la phospho-glycerate kinase, l'd-galactosidose a et l'hypoxanthine guanine phosphoribosyl transferase. *Annale Génétique* 21, 197–202.

Heyen, D.W., Weller, J.I., Ron, M., Band, M., Beever, J.E., Feldmesser, E., Da, Y., Wiggans, G.R., Vanraden, P.M. and Lewin, H. A. (1999) A genome scan for QTL influencing milk production and health traits in dairy cattle. *Physiology Genomics* 1, 165–175.

Iannuzzi, L., Rangel-Figueredo, T., Di Meo, G.P. and Ferrara, L. (1992) A new Robertsonian translocation in cattle, rob(15;25). *Cytogenetics and Cell Genetics* 59, 280–283.

Iannuzzi, L., Di Meo, G.P., Rangel-Figueredo, T. and Ferrara, L. (1993) A new case of centric fusion trans-location in cattle, rob(16;18). In: *Proceedings of the 8th North America Colloquium on Domestic Animal Cytogenetics and Gene Mapping*, Guelph, Canada, pp.127–128.

Iannuzzi, L., Di Meo, G.P., Leifsson, P.S., Eggen, A. and Christensen, K. (2001) A case of trisomy 28 in cattle revealed by both banding and FISH-mapping techniques. *Hereditas* 134, 147–151.

Ihara, N., Takasuga, A., Mizoshita, K., Takeda, H., Sugimoto, M., Mizoguchi, Y., Hirano, T., Itoh, T., Watanabe, T., Reed, K.M. *et al.* (2004) A comprehensive genetic map of the cattle genome based on 3802 micro-satellites. *Genome Research* 14, 1987–1998.

Itoh, T., Takasuga, A., Watanabe, T. and Sugimoto, Y. (2003) Mapping of 1400 expressed sequence tags in the bovine genome using a somatic cell hybrid panel. *Animal Genetics* 34, 362–370.

Itoh, T., Watanabe, T., Ihara, N., Mariani, P., Beattie, C.W., Sugimoto, Y. and Takasuga, A. (2005) A com-prehensive radiation hybrid map of the bovine genome comprising 5593 loci. *Genomics* 85, 413–424.

Jann, O.C., Aerts, J., Jones, M., Hastings, N., Law, A., McKay, S., Marques, E., Prasad, A., Yu, J., Moore, S.S. *et al.* (2006) A second generation radiation hybrid map to aid the assembly of the bovine genome sequence. *BMC Genomics* 7, 283.

Kappes, S.M., Keele, J.W., Stone, R.T., McGraw, R.A., Sonstegard, T.S., Smith, T.P., Lopez-Corrales, N.L. and Beattie, C.W. (1997) A second-generation linkage map of the bovine genome. *Genome Research* 7, 235–249.

Keele, J.W., Shackelford, S.D., Kappes, S.M., Koohmaraie, M. and Stone, R.T. (1999) A region on bovine chro-mosome 15 influences beef longissimus tenderness in steers. *Journal of Animal Science* 77, 1364–1371.

Kovacs, A. (1975) Uber eine neue autosomale Translokation beim ungarischen Fleckviehrind. In: *Proceedings of the 2nd European Colloquium on Cytogenetics of Domestic Animals*, Giessen, Germany, pp. 162–167.

Kovacs, A. and Szepeshelyi, F. (1977) Chromosomal screening of breeding bulls in Hungary. *Journal of Dairy Science* 70, 236.

Kovacs, A., Meszaros, I., Sellyei, M. and Vass, I. (1973) Mosaic centromeric fusion in a Holstein-Friesian bull. *ActaBiologica Academia Sciencia Hungaria* 24, 215–220.

Kovács, A., Villagómez, D.A., Gustavsson, I., Lindblad, K., Foote, R.H. and Howard, T.H. (1992) Synaptonemal complex analysis of a three-breakpoint translocation in a subfertile bull. *Cytogenetics Cell Genetics* 61, 195–201.

Larkin, D.M., Everts-van der Wind, A., Rebeiz, M., Schweitzer, P.A., Bachman, S., Green, C., Wright, C.L., Campos, E.J., Benson, L.D. *et al.* (2003) A cattle-human comparative map built with cattle BAC-ends and human genome sequence. *Genome Research* 13, 1966–1972.

Larkin, D.M., Pape, G., Donthu, R., Auvil, L., Welge, M. and Lewin, H.A. (2009) Breakpoint regions and homologous synteny blocks in chromosomes have different evolutionary histories. *Genome Research* 19, 770–777.

Larkin, D.M., Daetwyler, H.D., Hernandez, A.G., Wright, C.L., Hetrick, L.A., Boucek, L., Bachman, S.L., Band, M.R., Akraiko, T.V., Cohen-Zinder, M. *et al.* (2012) Whole-genome resequencing of two elite sires for the detection of haplotypes under selection in dairy cattle. *Proceedings of Natural Academy of Science USA* 109, 7693–7698.

Lathrop, G.M. and Lalouel, J.M. (1988) Efficient computations in multilocus linkage analysis. *American Journal Human Genetics* 42, 498–505.

Laurent, P., Elduque, C., Hayes, H., Saunier, K., Eggen, A. and Leveziel, H. (2000) Assignment of 60 human ESTs in cattle. *Mammialn Genome* 11, 748–754.

Lemay, D.G., Lynn, D.J., Martin, W.F., Neville, M.C., Casey, T.M., Rincon, G., Kriventseva, E.V., Barris, W.C., Hinrichs, A.S., Molenaar, A.J. *et al.* (2009) The bovine lactation genome: insights into the evolution of mammalian milk. *Genome Biology* 10, R43.

Levy, S., Sutton, G., Ng, P.C., Feuk, L., Halpern, A.L., Walenz, B.P., Axelrod, N., Huang, J., Kirkness, E.F., Denisov, G. *et al.* (2007) The diploid genome sequence of an individual human. *PLoS Biology* 5, e254.

Loghe, D.N. and Harvey, M.J.A. (1978) Meiosis and spermatogenesis in bulls heterozygous for a presumptive 1/29 Robertsonian translocation. *Journal of Reproduction and Fertility* 54, 159–165.

Lojda, L., Mikulas, L., and Rubes, L. (1975) Einige Ergnisse der Chromosomenuntersuchungen im Rahmen der staatlichen Erbgesundheitskintrolle beim Rin. In: *Proceedings of the 2nd European Colloquium of Cytogenetics of Domestic Animals*, Giessen, Germany, pp. 269–276.

Lojda, L., Rubes, J., Staiskskova, M. and Havrandsova, J. (1976) Chromosomal findings in some reproductive disorders in bulls. *Proceedings of the 8th International Congress of Animal Reproduction and Artificial Insemination*, Krakow, Poland, pp. 151–158.

Ma, R.Z., Beever, J.E., Da, Y., Green, C.A., Russ, I., Park, C., Heyen, D.W., Everts, R.E., Fisher, S.R., Overton, K.M. *et al.* (1996) A male linkage map of the cattle (*Bos taurus*) genome. *Journal of Heredity* 87, 261–271.

Ma, R.Z., Van Eijk, M.J., Beever, J.E., Guerin, G., Mummery, C.L. and Lewin, H.A. (1998) Comparative analysis of 82 expressed sequence tags from a cattle ovary cDNA library. *Mammalian Genome* 9, 545–549.

Mahrous, K.F., Hassanane, M.S. and El-Kholy, A.F. (1994) Robertsonian translocation and freemartin cases in hybrid Friesian cows raised in Egypt. *Egyptian Journal of Animal Production* 31, 213–220.

Marleba, F. (1997) Individuazione di una nuova translocazione Robertsoniana in un torello di razza Marchigiana: sua determinazione e studio dei suoi effetti sulla fertilita. Corso di Laurea in Scienze Agrarie, Universita degli Studi di Milano, Milano, Italy.

Masuda, H., Takahaschi, T., Soejima, A. and Waido, Y. (1978) Centric fusion of chromosome in a Japanese black bull and its offspring. *Japanese Journal of Zootechnological Sciences* 49, 853–858.

Mayr, B. and Czaker, R. (1981) Variable positions of nucleolus organizer regions in Bovidae. *Experientia* 37, 564–565.

Miyake, Y.I. and Kaneda, Y. (1987) A new type of Roberstonian translocation (1/26) in a bull with unilateral cryptorchidism, probably occurring de novo. *Japanese Journal of Veterinary Science* 49, 1015–1019.

Miyake, Y.I., Murakami, R.K. and Kaneda, Y. (1991) Inheritance of the Robertsonian translocation (1/21) in the Holstein-Friesian Cattle I. Chromosome analysis. *Journal of Veterinary Medical Science* 53, 113–116.

Molteni, L., De Giovanni-Macchi, A., Succi, G., Cremonesi, F., Stacchezzini, S., Di Meo, G.P. and Iannuzzi, L. (1998) A new centric fusion translocation in cattle: rob (13;19). *Hereditas* 129, 177–180.

Molteni, L., Perucatti, A., Iannuzzi, A., Di Meo, G.P., De Lorenzi, L., De Giovanni, A., Incarnato, D., Succi, G., Cribiu, E., Eggen, A. *et al.* (2007) A new case of reciprocal translocation in a young bull: rcp(11;21) (q28;q12). *Cytogenetics and Genome Research* 116, 80–84.

Moran, C. and James, J.W. (2005) *Linkage Mapping.* CAB International, Wallingford.

Murphy, W.J., Larkin, D.M., Everts-van der Wind, A., Bourque, G., Tesler, G., Auvil, L., Beever, J.E., Chowdhary, B.P., Galibert, F., Gatzke, L. *et al.* (2005) Dynamics of mammalian chromosome evolution inferred from multispecies comparative maps. *Science* 309, 613–617.

Nicodemo, D., Pauciullo, A., Castello, A., Roldan, E., Gomendio, M., Cosenza, G., Peretti, V., Perucatti, A., Di Meo, G.P., Ramunno, L. *et al.* (2009) X-Y sperm aneuploidy in 2 cattle (*Bos taurus*) breeds as determined by dual color fluorescent in situ hybridization (FISH). *Cytogenetics Genome Research* 126, 217–225.

Padula, A.M. (2005) The freemartin syndrome: an update. *Animal Reproduction Sciences* 87, 93–109.

Papp, M. and Kovacs, A. (1980) 5/18 dicentricRobertsonian translocation in a Simmental bull. In: *Proceedings of the 4th European Colloquium on Cytogenetics of Domestic Animals*, Uppsala, Sweden, pp. 51–54.

Pauciullo, A., Nicodemo, D., Cosenza, G., Peretti, V., Iannuzzi, A., Di Meo, G.P., Ramunno, L., Iannuzzi, L., Rubes, J. and Di Berardino, D. (2012) Similar rates of chromosomal aberrant secondary oocytes in two indigenous cattle (*Bos taurus*) breeds as determined by dual-color FISH. *Theriogenology* 77, 675–683.

Peretti, V., Ciotola, F., Albarella, S., Paciello, O., Dario, C., Barbieri, V. and Iannuzzi, L. (2008) XX/XY chimerism in cattle: clinical and cytogenetic studies. *Sex Development* 2, 24–30.

Pinheiro, L.E., Carvalho, T.B., Oliveira, D.A., Popescu, C.P. and Basrur, P.K. (1995) A 4/21 tandem fusion in cattle. *Hereditas* 122, 99–102.

Pinton, A., Ducos, A., Berland, H.M., Seguela, A., Blanc, M.F., Darre, A., Mimar, S. and Darre, R. (1997) A new Robertsonian translocation in Holstein-Friesian cattle. *Genetics, Selection, Evolution* 29, 523–526.

Pinton, A., Ducos, A. and Yerle, M. (2003) Chromosomal rearrangements in cattle and pigs revealed by chromosome microdissection and chromosome painting. *Genetics, Selection, Evolution* 35, 685–696.

Pollock, D.L. (1974) Chromosome studies in artificial insemination sires in Great Britain. *Veterinary Records* 95, 266–277.

Popescu, C.P. (1977) A new type of Robertsonian translocation in cattle. *Journal of Heredity* 68, 139–142.

Popescu, C.P., Long, S., Riggs, P., Womack, J., Schmutz, S., Fries, R. and Gallagher, D.S. (1996) Standardization of cattle karyotype nomenclature: report of the committee for the standardization of the cattle karyotype. *Cytogenetics Cell Genetics* 74, 259–261.

Rebeiz, M. and Lewin, H.A. (2000) Compass of 47,787 cattle ESTs. *Animal Biotechnology* 11, 75–241.

Roldan, E.R.S., Merani, M.S. and Von Lawzewitsch, I. (1984) Two abnormal chromosomes found in one cell line of mosaic cow with low fertility. *Genetics, Selection, Evolution* 16, 135–142.

Rubes, J., Borkovec, L., Borkovcova, Z. and Urbanova, J. (1992) A new Robertsonian translocation in cattle, rob(16;21). In: *Proceedings of the 10th European Colloquium on Cytogenetics of Domestic Animals*, Utrecht, The Netherlands, pp. 201–205.

Rubes, J., Musilova, P., Borkovec, L., Borkovcova, Z., Svecova, D. and Urbanova, J. (1996) A new Robertsonian translocation in cattle, rob(16;20). *Hereditas* 124, 275–279.

Ruvinsky, A. (2009) *Genetics and Randomness*. CRC Press, Boca Raton, Florida.

Samarineanu, N.E., Livescu, B., and Granciu, I. (1977) Identification of Robertsonian translocation in some breeds of cattle. *Lucrarile Stiintifce ale Institutului de Cercetari si Cresterea Taurinelor Corbeanca* 3, 53–60.

Schibler, L., Roig, A., Mahe, M.F., Save, J.C., Gautier, M., Taourit, S., Boichard, D., Eggen, A. and Cribiu, E.P. (2004) A first generation bovine BAC-based physical map. *Genetics, Selection, Evolution* 36, 105–122.

Schmitz, A., Oustry, A., Vaiman, D., Chaput, B., Frelat, G. and Cribiu, E.P. (1998) Comparative karyotype of pig and cattle using whole chromosome painting probes. *Hereditas* 128, 257–263.

Slota, E., Danielak, B. and Kozubska-Sobocinska, A. (1988) The Robertsonian translocation in the cattle quintuplet. In: *Proceedings of the 8th European Colloquium on Cytogenetics of Domestic Animals*, Bristol, UK, pp. 122–123.

Snelling, W.M., Casas, E., Stone, R.T., Keele, J.W., Harhay, G.P., Bennett, G.L. and Smith, T.P. (2005) Linkage mapping bovine EST-based SNP. *BMC Genomics* 6, 74.

Snelling, W.M., Chiu, R., Schein, J.E., Hobbs, M., Abbey, C.A., Adelson, D.L., Aerts, J., Bennett, G.L., Bosdet, I.E., Boussaha, M. *et al.* (2007) A physical map of the bovine genome. *Genome Biology* 8, R165.

Soderlund, C., Humphray, S., Dunham, A. and French, L. (2000) Contigs built with fingerprints, markers, and FPC V4.7. *Genome Research* 10, 1772–1787.

Stranzinger, G. (1989) Zytogenetische Kontrolluntersuchungen an Nutztieren. *Landwirtschaf Schweiz* 2, 355–362.

Stranzinger, G.F. and Forster, M. (1976) Autosomal chromosome translocation of piebald cattle and brown cattle. *Experimenta* 32, 24–27.

Switonski, M., Andersson, M., Nowacka-Woszuk, J., Szczerbal, I., Sosnowski, J., Kopp, C., Cernohorska, H. and Rubes J. (2008) Identification of a new reciprocal translocation in an AI bull by synaptonemal complex analysis, followed by chromosome painting. *Cytogenetics Genome Research* 121, 245–248.

Switonski, M., Szczerbal, I., Krumrych, W. and Nowacka-Woszuk, J. (2011) A case of Y-autosome reciprocal translocation in a Holstein-Friesian bull. *Cytogenetics Genome Research* 132, 22–25.

Sysa, P.S. and Slota, E. (1992) The investigation of karyotype in cattle within the new system of cytogenetic control of bulls in Poland. In: *Proceedings of the 10th European Colloquium on Cytogenetics of Domestic Animals*, Utrecht, The Netherlands, pp. 248–249.

Tanaka, K., Yamamoto, Y., Amano, T., Yamagata, T., Dang, V.B., Matsuda, Y. and Namikawa, T. (2000) A Robertsonian translocation, rob(2;28), found in Vietnamese cattle. *Hereditas* 133, 19–23.

Tschudi, P., Zahner, B., Kupfer, U. and Stamfli, G. (1977) Chromosomen untersuchungen an Schweizerischen Rinderrasen. *Schweizerisches Archiv fur Tierheilkunde* 119, 329–336.

Viitala, S.M., Schulman, N.F., De Koning, D.J., Elo, K., Kinos, R., Virta, A., Virta, J., Maki-Tanila, A. and Vilkki, J.H. (2003) Quantitative trait loci affecting milk production traits in Finnish Ayrshire dairy cattle. *Journal Dairy Sciences* 86, 1828–1836.

Villagómez, D.A., Andersson, M., Gustavsson, I. and Plöen, L. (1993) Synaptonemal complex analysis of a reciprocal translocation, rcp(20;24) (q17;q25), in a subfertile bull. *Cytogenetics Cell Genetics* 62, 124–130.

Wheeler, D.A., Srinivasan, M., Egholm, M., Shen, Y., Chen, L., McGuire, A., He, W., Chen, Y.J., Makhijani, V., Roth, G.T. *et al.* (2008) The complete genome of an individual by massively parallel DNA sequencing. *Nature* 452, 872–876.

Womack, J.E. and Moll, Y.D. (1986) Gene map of the cow: conservation of linkage with mouse and man. *Journal of Heredity* 77, 2–7.

Womack, J.E., Johnson, J.S., Owens, E.K., Rexroad, C.E., 3rd, Schlapfer, J. and Yang, Y.P. (1997) A whole-genome radiation hybrid panel for bovine gene mapping. *Mammalian Genome* 8, 854–856.

Yang, Y.P. and Womack, J.E. (1998) Parallel radiation hybrid mapping: a powerful tool for high-resolution genomic comparison. *Genome Research* 8, 731–736.

Yu, R. and Xin, C.L.I. (1991) The 2/27 Robertsonian translocation in Wenling hump cattle. *Hereditas (Beijing)* 13, 17–18.

Zhang, T., Buoen, L.C., Seguin, B.E., Ruth, G.R. and Weber, A.F. (1994) Diagnosis of freemartinism in cattle: the need for clinical and cytogenic evaluation. *Journal American Veterinary Medicine Associated* 204, 1672–1675.

Zimin, A.V., Delcher, A.L., Florea, L., Kelley, D.R., Schatz, M.C., Puiu, D., Hanrahan, F., Pertea, G., Van Tassell, C.P., Sonstegard, T.S. *et al.* (2009) A whole-genome assembly of the domestic cow, *Bos taurus*. *Genome Biology* 10, R42.

7 Bovine Genomics

David L. Adelson, Zhipeng Qu, Joy M. Raison and Sim L. Lim

The University of Adelaide, Adelaide, South Australia, Australia

Introduction

Because of the economic importance of cattle as a source of dairy and meat products, there has been sustained interest in and extensive resources applied to genetic improvement of both dairy and beef cattle. The availability of high-resolution genetic maps and millions of potential genetic markers in the bovine genome sequence data have been critical to advances in the study of associations between genotype and phenotype.

Whilst the origin and influence of bovine gene mapping and its importance to the assembly of the bovine genome has been recently reviewed (Womack, 2012), for the purposes of this review we consider the beginning of bovine genomics to be the first release of comprehensive microsatellite-based linkage maps of the bovine genome in 1994 (Barendse *et al.*, 1994; Bishop *et al.*, 1994). The elaboration of these maps, along with the availability of radiation hybrid panels and physical maps (Schlapfer *et al.*, 1997; Womack *et al.*, 1997) provided the first resources for comparative mapping of the bovine genome (Everts-van der Wind *et al.*, 2004), and set the stage for the sequencing of the bovine genome. The first release of the first preliminary bovine genome assembly, Btau_1.0 in 2004 (www.hgsc.bcm.edu/content/bovine-genome-project) was produced by the Baylor College of Medicine Human Genome Sequencing Center, was contig-based and incorporated no chromosomal scaffolding. Subsequent assemblies incorporating tiled bovine bacterial artificial chromosome (BAC) clones that had been physically mapped using restriction fragment fingerprinting (Schibler *et al.*, 2004) led to the release of the first draft assembly, Btau_3.1 in 2006. The culmination of the mapping era of bovine genomics occurred with the release of the integrated genome map (Snelling *et al.*, 2007), which was the foundation for high-resolution long-range assembly of the bovine genome sequence, and contributed to the Btau_4.0 draft assembly. The improved and annotated draft genome of the cow was published a few years later (Elsik *et al.*, 2009), and has served as the basis for subsequent rapid growth in bovine genomics. A unique feature of bovine genomics compared to other species is the availability of two independent draft genome sequence assemblies based on the same primary sequence data. Shortly after the publication of the annotated draft genome, an alternative assembly was released (Zimin *et al.*, 2009). Since then, the field of bovine genomics has grown rapidly, with improved draft genome assemblies and >10 million mapped single nucleotide polymorphisms (SNPs) for genotyping and trait association studies.

At present, the existence of two genome assemblies poses a challenge for researchers wishing to map traits or carry out whole genome analyses based on a reference genome sequence. While the reference genome is Hereford-based, additional reference quality assemblies are a certainty as the cost of whole genome sequencing and assembly drops. The recent publication of a *Bos indicus* genome aligned to both reference assemblies (Canavez *et al.*, 2012) can be viewed as the prototype new cattle genome aligned and anchored to the cattle reference genome. As whole genome sequencing technologies improve, become cheaper and more widely available to the scientific community, newer consensus assemblies will be produced reducing some of the current challenges. The likely eventual outcome will be a world where reference genomes from multiple breeds will be aligned and annotated to each other. Currently the 1000 bull genomes project (http://1000bullgenomes. com) aims to collect additional bull genome assemblies for use in genotype imputation and is anchoring those assemblies to existing reference assemblies.

Bovine Genome Sequence

Sequencing strategy

The Baylor College of Medicine Human Genome Sequencing Center sequenced the bovine genome by using a hybrid approach pioneered for the rat genome sequence (Gibbs *et al.*, 2004). That approach combined both whole genome shotgun (WGS) sequencing and shotgun sequencing of physically mapped BAC clones to generate two assemblies: a WGS assembly and a series of assembled BACs. These two assemblies were then combined to take advantage of the high confidence local assembly from the BACs and the combined assembly contigs were then positioned on chromosome scaffolds. The long-range marker information used for scaffolding was obtained from the integrated map of the bovine genome, which included the BAC physical map (Snelling *et al.*, 2007). This approach was used to produce the Btau_3.1 assembly. In theory this strategy should provide an improved contig assembly compared to shotgun sequencing alone. However shortcomings in the Btau_3.1 assembly required reassembly using a high-resolution bovine–human comparative map (Everts-van der Wind *et al.*, 2005), the BAC physical map (Snelling *et al.*, 2007), and the inclusion of bovine–ovine comparative genome information (Dalrymple *et al.*, 2007).

Baylor assembly

The result of the above assembly was named Btau_4.0, the assembly process and outcome are described in detail in Liu *et al.* (2009). It was used as the basis for the published bovine genome analysis (Elsik *et al.*, 2009). The inclusion and merging of multiple marker sets for long range assembly required additional development of the ATLAS assembly program (Havlak *et al.*, 2004) and in the process relied heavily on comparative genome mapping between bovine and human, ovine and human and bovine and ovine. The resulting assembly had all the hallmarks of a high-quality draft assembly in terms of contig length distribution and proportion of reads incorporated into contigs, but it still had a significant amount

(10%) of sequence that could not be placed on to chromosomes.

Since 2009, the Baylor assembly has been improved and the current assembly release is known as Btau_4.6.1 and differs from the published assembly by the inclusion of more finished BAC sequences and some paired-end sequencing by synthesis (SOLiD) data to provide additional long-range links between scaffolds. This current assembly differs in terms of overall sequence content from Btau_4.0 because it includes a Y chromosome sequence derived from BACs. In addition to the Y chromosome sequence, Btau_4.6.1 has fewer/smaller gaps and is an improvement in terms of chromosome level assembly compared to Btau_4.0/4.2. However, the current bovine assembly still has a significant amount of unplaced sequence, with 11% of the assembly not placed on to chromosomes. This is in stark contrast to the rat genome, where only 4% of the contig sequences could not be mapped to chromosomes. Details of the assembly process and assembly statistics for Btau_4.6.1 can be obtained from ftp://ftp.hgsc.bcm.edu/Btaurus/fasta/Btau20101111/readme.Btau20101111.txt.

The reasons for the relatively poor assembly of the bovine genome compared to the rat genome are unclear. The bovine genome benefited from improvements in sequencing technology and the ATLAS assembler compared to the rat, and had a greater number of markers available for long-range assembly.

University of Maryland (UMD) assembly

Shortly after publication of the bovine genome analysis, an alternative assembly (UMD2) for the bovine genome was released (Zimin *et al.*, 2009). This assembly used the same sequence reads used for Btau_4.0 as input to the assembly process. The first released UMD assembly was produced using the Celera assembler first developed for the *Drosophila* genome project (Myers *et al.*, 2000) and then used for the Celera human genome sequence (Venter *et al.*, 2001). In addition to using a different assembler, the UMD assembly was carried out as a pure WGS assembly, incorporating the BAC shotgun sequences along with the WGS sequences

as input into the Celera assembler. Placing of contigs on to chromosomes used the same integrated bovine map as the Btau_4.0 assembly and conserved synteny with human was used to orient contigs on scaffolds as required. The UMD2 assembly was compared to the Btau_4.0 assembly and found to have contigs of greater length and incorporated ~2% more sequence into chromosomes. This improvement in sequence assembly is a direct result of using a WGS assembly method instead of first assembling BACs and incorporating WGS reads as done with the ATLAS assembler. WGS assembly includes reads that would otherwise be missed because of gaps between BACs. Comparison of the two assemblies showed a number of discrepancies, mostly inversions or deletions. A striking difference between UMD2 and Btau_4.0 assemblies was the inclusion of 53 Mbp of additional sequence in the UMD2 assembly of the X chromosome. The UMD2 assembly was comparable to the Btau_4.0 assembly in terms of unplaced contigs, with about 8.5% of assembled sequence not placed on chromosome scaffolds. As a result of these comparisons, the UMD2 assembly was believed to be superior to the Btau_4.0 assembly, particularly for chromosome X. Subsequently UMD3.1 was released and compared to Btau_4.2 (Zimin *et al.*, 2012) revealing that Btau_4.2 contained significantly more misassemblies contributing to spurious segmental duplications. The UMD3.1 assembly included ~3% more placed contig sequence than the UMD2 assembly and so represents a significant improvement.

Since the release of the UMD3.1 assembly, the Baylor assembly has been further improved, and the current version is Btau_4.6.1. However no direct comparison of these two assemblies has been published. We have carried out simple pairwise alignments of chromosome assemblies between UMD3.1 and Btau_4.6.1 in order to display the current gross differences between the two assemblies described below.

Differences in the Current Baylor and UMD Assemblies

Pairwise alignments of chromosomes from the two current bovine assemblies were carried out using Mummer (Delcher *et al.*, 2003) (Fig. 7.1).

These large-scale alignments show significant differences such as inversions and indels in ten autosomes, and as reported previously (Zimin *et al.*, 2012) the two X chromosome assemblies aligned very poorly. In addition to the significant differences there were many smaller regions, perhaps segmental duplications in one assembly not present in the other.

While the UMD3.1 assembly may be significantly better than Btau_4.6.1 it is still a draft assembly and contains known gaps, unknown misassemblies and missing sequences. However,

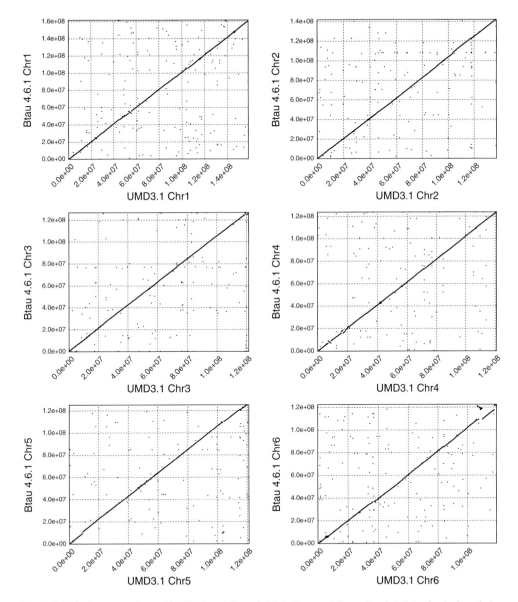

Fig. 7.1. Pairwise comparison of UMD3.1 and Btau_4.6.1 draft assemblies using dot plots of pairwise whole chromosome alignments. Deviations from the diagonal indicate discrepancies between the two assemblies. Alignments carried out using Mummer 3.2.3, following procedure for aligning two draft sequences as described in section 4.2 of the Mummer 3 manual (http://mummer/sourceforge.net/manual/#mummer).

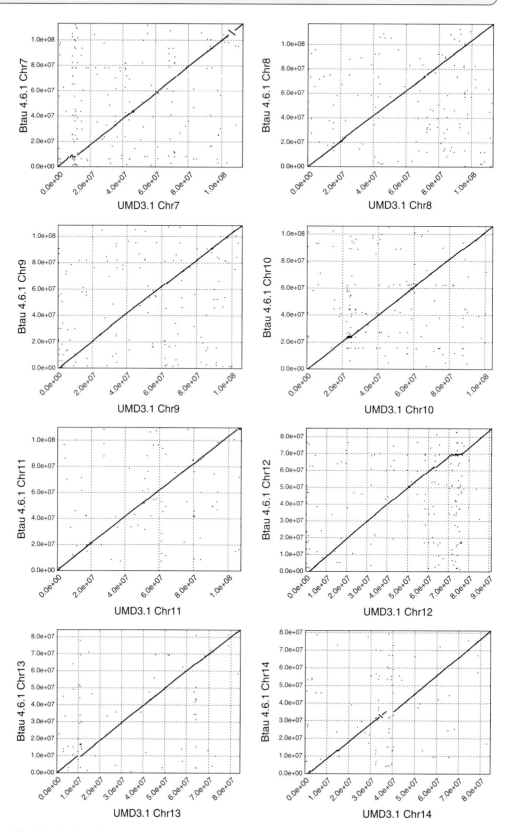

Fig. 7.1. Continued.

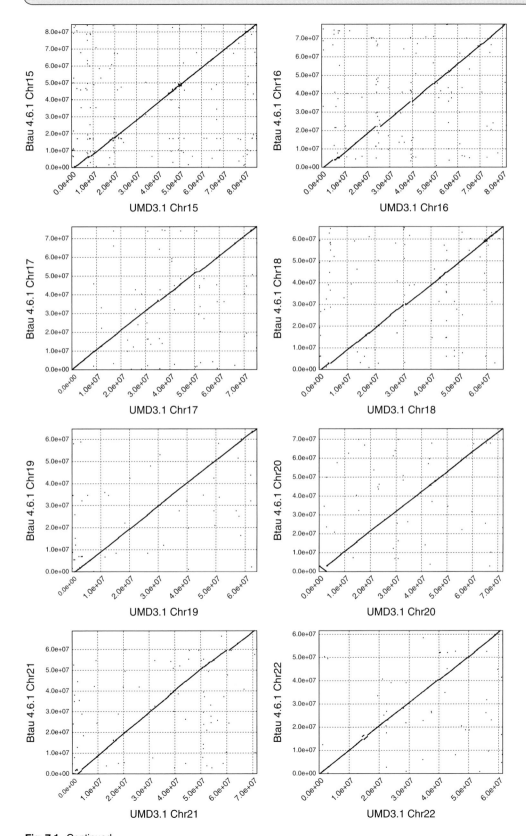

Fig. 7.1. Continued.

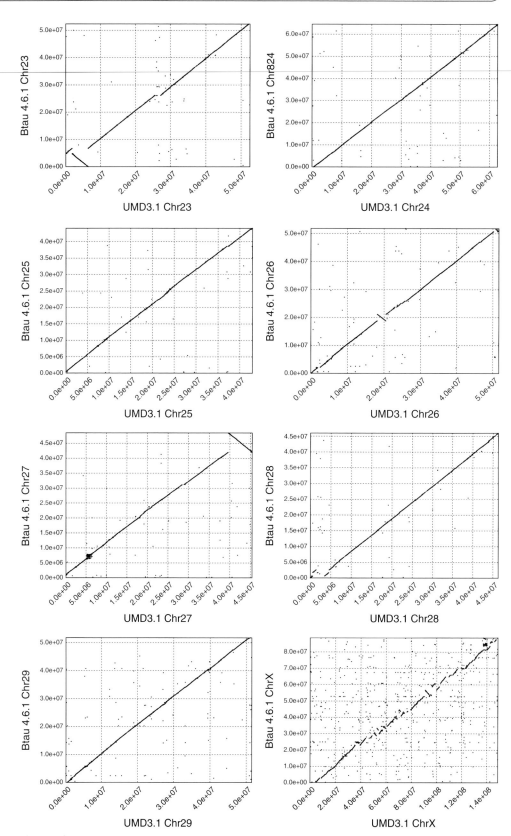

Fig. 7.1. Continued.

it is clear that the Btau_4.6.1 draft assembly is very likely much less accurate overall than the UMD3.1 assembly. In spite of this, investigators should be wary of complete reliance on any draft assembly and should be cognizant of the fact that there are misassembled regions on UMD3.1 that are likely correctly assembled in Btau_4.6.1, particularly in regions lying within a single sequenced BAC clone or Bactig. For quantitative trait locus (QTL) mapping we expect that the better long-range assembly of UMD3.1 should be a better resource, but for shorter range information, such as identifying candidate mutations in a critical interval the Btau_4.6.1 assembly information might be more reliable. In future, polymorphism discovery and genotyping for trait mapping may be done simultaneously, by genome sequencing. At present, cost restricts such analysis to a sequencing depth of 4–10× for whole genome sequencing or 50–100× for exome sequencing. For either of these, a reference genome on which to map reads is required. Even when mapping 10× read-depth resequenced genomes to a finished genome, there are reads that do not map. For example, in the mouse we mapped ~10× average sequencing depth 100 bp paired-end Illumina reads from C57BL/6J or knock out mutants derived from a related 129S strain to the reference genome. We found that ~18% of raw reads or ~2.5% of the de-duplicated (non-redundant) reads did not map at all using the Burroughs–Wheeler Aligner (BWA) (Li and Durbin, 2010). When mapping whole genome resequencing reads to the cattle draft genome assembly we can expect more unmapped reads and we can expect different mapping between the UMD3.1 and Btau_4.6.1 assemblies.

Genome Annotation

The contig level (short range) assembly of a genome is the primary determinant of annotation quality, because many, if not most, gene models will fall within a single contig. Higher order assembly may alter the order and orientation of contigs, and hence the coordinates of the annotated features, but will not alter their contents. Because the contigs for UMD3.1 are on average longer than for Btau_4.6.1, there are some small differences in the gene models between the two assemblies. Bovine genome annotation is available from the National Center for Biotechnology Information (NCBI) (www.ncbi.nlm.nih.gov), Ensembl (www.ensembl.org), the UCSC Genome Browser (http://genome.ucsc.edu) and Bovinegenome.org (http://bovinegenome.org). The last website is a bovine-centric, comprehensive one-stop shop for bovine genome assembly data and annotations (Reese et al., 2010; Childers et al., 2011).

Sequence polymorphism

As of February 2013, there were 13,146,622 cattle single nucleotide polymorphisms (SNPs) deposited in dbSNP (Sherry et al., 2001), of which 66,994 are coding and non-synonymous. These constitute an enormous resource for genotyping and trait mapping. SNPs selected to be evenly spaced across the genome and possessing useful minor allele frequencies in economically important cattle breeds have been included in various SNP genotyping platforms. At present there are two high density SNP genotyping arrays available, one from Illumina (BovineHD Genotyping BeadChip with 777,962 SNPs) and one from Affymetrix (Axiom Genome-Wide BOS 1 Array with 648,874 SNPs). The overlap in SNP content between these two platforms is only ~100,000 SNPs, based on sequence coordinates. The performance of these two SNP genotyping arrays has been evaluated in dairy cattle (Rincon et al., 2011) and is roughly comparable for most uses, with some degree of superiority for Copy Number Variation (CNV) calling from the BovineHD array. In addition, the limited overlap between the SNPs on the two high-density platforms provides an advantage when fine-mapping traits through the use of both arrays. In addition to genotyping for trait mapping and selection, the BovineHD array can also be used to impute microsatellite genotypes currently used for bovine parentage testing (McClure et al., 2012). This allows the BovineHD to be used for more accurate, lower cost genotyping on living animals, while providing an avenue for parentage verification using ancestors with microsatellite genotypes.

In addition, a low density SNP array designed for imputation in dairy cattle is also available from Illumina (BovineLD Genotyping BeadChip; 6909 SNPs) based on the Bovine LD Consortium design (Boichard *et al.*, 2012). That low density chip is effective for implementing genomic selection in dairy cattle at greatly reduced cost compared to high density genotyping alternatives, allowing genotyping to be extended beyond elite sires, into entire herds. Other custom Illumina products based on the BovineLD are available from GeneSeek, the GGP-LD (8762 SNPs) and GGP-HD customized for either *Bos taurus* (77k) or *Bos indicus* (80k). Those products also include causal mutations for diseases and other traits.

Gene models

Identifying all the genes in an organism has long been viewed as the principal justification for sequencing genomes. Because most genes deduced from genome sequences have not been validated, they are referred to as gene models, which correctly conveys the uncertainty with which most of them should be viewed. The major genome data repositories, such as NCBI and Ensembl have developed their gene annotation pipelines to use a variety of data to best predict gene models (Hubbard *et al.*, 2005; Maglott *et al.*, 2007). In broad terms these pipelines integrate *in silico* gene predictions, aligned cDNA and expressed sequence tags (ESTs) and comparative alignment of gene models from other organisms. Because NCBI and Ensembl have independent annotation pipelines and procedures, the sets of gene models they produce are never identical, although they largely overlap.

NCBI, as of February 2013 had entries for 29,754 genes for the Btau_4.6.1 assembly and 27,155 genes for the UMD3.1 assembly. Ensembl had entries for 26,740 genes for the UMD3.1 assembly, but does not host data for Btau_4.6.1. So investigators not only have to choose between two assemblies, but between gene predictions from two different sources.

This produces a conundrum for researchers who have to choose which model to base their analyses or experiments on. The solution has been to create a consensus gene set that merges the NCBI, Ensembl and other gene prediction data to provide gene models with accompanying probabilistic confidence scores (Elsik *et al.*, 2007). The cow is unique among mammals in having a consensus gene set that integrates the NCBI, Ensembl and other gene prediction data that has been annotated and reviewed (Elsik *et al.*, 2006; Reese *et al.*, 2010) by many in the bovine genomics community. This approach provides both a single source of information and a measure of confidence in the gene models. The consensus gene set for cattle is called the Bovine Official Gene Set (OGSv2) and has entries for 26,835 gene models.

Repetitive elements

In addition to SNPs and gene models, other genomic features deserve annotation for various reasons. Non-protein coding or non-gene elements such as repetitive elements and ncRNAs are also annotated because of their significance from either an experimental viewpoint or because they are potentially regulatory elements. Repetitive sequences account for at least 40–50% of mammalian genomes, but probably account for more, as our detection threshold based on sequence similarity searches is effectively ~65% identity, depending on the length of the element. As most repetitive elements are short, this means that most old, divergent elements cannot be detected by sequence similarity searching.

Repeats that we can detect, usually with tools such as RepeatMasker (Smit *et al.*, 1996–2010) fall into several broad classes such as DNA transposons, which use a cut and paste mechanism to jump throughout the genome. DNA transposons were first identified in maize as the causal mechanism for colour spotting in kernels (Fedoroff *et al.*, 1983; McClintock, 1950). Other types of repeats include retrotransposons, which move via a copy and paste mechanism using an RNA intermediate and re-insertion into the genome via reverse transcription (Esnault *et al.*, 2000; Babushok *et al.*, 2006; Kubo *et al.*, 2006). Retrotransposons include long terminal repeat (LTR)-containing elements, which include retroviruses or retrovirus like elements, or non-LTR elements, which include Long INterspersed Elements (LINEs) or Short INterspersed Elements (SINEs) (reviewed by Jurka *et al.*, 2007).

Bovine repetitive elements were characterized *de novo* for the analysis of the bovine genome sequence (Adelson *et al.*, 2009) and as with most mammals are primarily composed of retrotransposons. However, cattle and other ruminants and Afrotheria have a different repetitive element profile from most other Eutheria, in that they include a class of non-LTR LINEs known as BovB, which makes up a substantial proportion of their genomes and appears to have been horizontally transferred between reptiles, ticks and ruminants (Walsh *et al.*, 2013). BovB retrotransposons are also of interest because they contain a short region (~50 bp) of sequence identity with another ruminant specific LTR-containing repeat known as BTLTR1 (Fig. 7.2). It is not clear if this short, shared sequence is functionally significant and has resulted in transfer from

BovB to BTLTR1 in ruminants, or if it is just a chance sequence similarity.

Of all the eutheria harbouring BovB, cattle are the best model system to study how the horizontal transfer of a retrotransposon can remodel a genome and illuminate some of the dynamics of genome evolution. Because BovB were incorporated into the ruminant ancestor genome <50 MYA, they can be considered recent or modern repeats compared to ancestral repeats such as LINE L2 and SINE MIR, both of which are present in the mammalian common ancestor, but have been inactive for >100 MYA (Jurka *et al.*, 1995). This allows us to determine which portions of the genome have been permissive to BovB invasion and which have been protected from BovB invasion.

To identify these regions each of the chromosomes in the genome was divided into a series 1.5 Mbp bins down its length. For each bin where the number of unknown (N) bases was less than 0.5 Mbp (≥1 Mbp called bases) the following information was obtained: the number of each of 18 interspersed repeat types that were contained within the bin, the number of

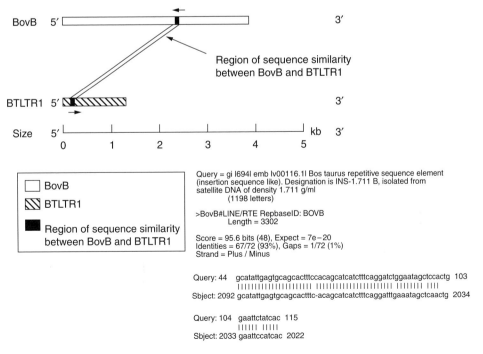

Fig. 7.2. Sequence similarity between BovB and BTLTR1 as determined by BLASTN alignments. The ~70 bp region of sequence identity is present on opposite strands of the two repeats.

genes that started within the bin, the number of CpG Islands contained within the bin, the CpG Island coverage of the bin, and the GC content of the bin. Previously bins had been identified as ancestral if the sum of ancestral repeats (LINE L2/SINE MIR) ranks exceeded that expected in 5% of the genome (Adelson *et al.*, 2009). This method has been refined so that it does not rely on an arbitrary cutoff of the expected coverage of the genome by ancestral regions. In the current method a principal components analysis is performed on the data, transformed by taking the square root of the counts, and an arcsine transformation taken of the CpG Island coverage to stabilize the variance. The principal component appropriate for identifying ancient areas – that is the one with high weights in the same direction for SINE/MIR, LINE/L2 and LINE/CR1 – is identified. For the identified principal component (PC2), the running average from a window size of five bins, and its 95% confidence band was obtained over each chromosome. Ancient bins were identified as those whose running average's lower confidence limit was >0, and new bins, those in which ancient repeats are significantly under-represented, as those whose upper confidence limit was <0. Results of this analysis are shown in Fig. 7.3, and it is clear that the genome bins enriched in ancient repeats tend to cluster into regions. New bins also tend to cluster into regions.

The previous method could identify which bins in the genome were most enriched in ancient repeats. The new method identifies all regions of the genome that statistically, are enriched in ancient repeats. There is evidence to support the view that ancient repeats have either been exapted into genes or recruited as enhancer/promoter elements for genes (Lowe *et al.*, 2007). PC2 ancient regions cover 24.8% of the UMD3.1 assembly, and we believe this method is an alternative approach to substitution rate-based methods for identifying evolutionarily conserved regions, particularly in non-coding regions of the genome that may have regulatory functions.

Long non-coding RNAs

While the protein coding portion of the bovine genome only occupies about 2% of the sequence, there are also many non-coding transcripts. The existence, distribution and abundance of ncRNAs have been the subject of debate (Mattick and Makunin, 2006; van Bakel *et al.*, 2010), but it is clear that long non-coding RNAs (lncRNA) can regulate the expression of protein coding genes through a variety of mechanisms (Qu and Adelson, 2012b). Analysis of bovine transcripts has shown that cattle are no exception when it comes to lncRNAs, particularly long intergenic non-coding RNAs (lincRNA) (Qu and Adelson, 2012a). Based on very stringent criteria, we have previously shown that 23,060 ncRNAs are transcribed from the bovine genome, with 12,614 of these lincRNAs (Qu and Adelson, 2012a). The lincRNAs are primarily (8500 of 12,614) located within 20 Kbp of protein coding genes and show evidence of both negative and positive selection (Qu and Adelson, 2012a). These are not unusual results and similar results have been described in human, mouse and zebra fish (Qu and Adelson, 2012c). Because gain and loss of function experiments involving lincRNAs in other organisms have shown they can regulate genes in *cis* and in *trans* (Martens *et al.*, 2004; Costa, 2008; Leeb *et al.*, 2009), it is likely that ncRNAs, particularly lincRNAs in cattle may have important regulatory functions and thus be critical to our understanding of beef and dairy cattle traits of economic importance. One line of evidence for regulatory functions of ncRNAs stems from their highly tissue-specific expression patterns. Below, evidence is provided for tissue-specific co-expression of both coding and non-coding RNAs in a number of bovine tissues.

Gene Expression Sub-networks

We have used existing public domain Massively Parallel Signature Sequencing (MPSS) data from a wide selection of bovine tissues. Bovine MPSS data were downloaded from the NCBI GEO database (GSE21544). The samples were collected from 92 adult, juvenile and fetal cattle tissues and three cattle cell lines (Harhay *et al.*, 2010). The sequence tags were mapped to bovine RefSeqs and ncRNAs using GSNAP with maximum one mismatch.

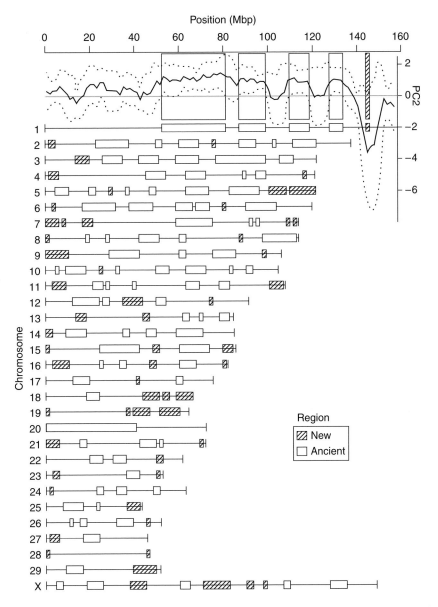

Fig. 7.3. Location of ancient and new regions in the bovine genome. Plotted above chromosome 1 is its running average of the second principal component with its 95% confidence interval. Ancient regions are those where the lower confidence limit is >0, and new regions where the upper confidence limit <0.

The digital expression value for each transcript (RefSeq or ncRNA) was calculated by normalizing sequence counts mapped to the 3′ ends of transcripts with library size. Expression values for all transcripts in 95 samples were combined as the expression profile, which was used to reconstruct protein-coding and non-coding co-expression networks. The co-expression networks were reconstructed using R package 'WGCNA' (Langfelder and Horvath, 2008). The visualization of co-expression networks was performed with Power Graph in Cytoscape (Royer et al., 2008).

There were 15,829 protein-coding genes and 4012 ncRNAs with digital expression in at least 3 out of 95 samples. Among these 19,841 transcripts, 4945 transcripts were reconstructed into 33 modules, which represent transcripts showing highly correlated expression patterns across 95 samples. We have used one of these modules, containing 41 protein-coding genes and 11 intergenic ncRNAs as a simple example of a specific sub-network reconstruction. All 52 transcripts showed high expression in three samples compared to all other 92 samples. These three samples were 'rumen', 'duodenum' and 'ventricle'. Thirty-four of the 41 protein-coding genes had human orthologues, and their annotation based on GeneCards (www. genecards.org) is shown in Table 7.1. The other seven protein-coding genes are 'LOC615250', 'MGC128175', 'LOC787094', 'LOC784776', 'LOC521764', 'LOC790435' and 'LOC615365'.

The co-expression network for our example module is shown in Fig. 7.4. Ten ncRNAs are in the two big power nodes, sharing highly correlated co-expression with multiple transcripts, including protein-coding genes. In particular 'NCRNA_DT885051' and 'NCRNA_EE220881' show significantly highly correlated co-expression with other transcripts in the network. This co-expression is consistent with a regulatory role for these ncRNAs, perhaps by acting as decoys for miRNAs.

GO (Gene ontology) classification of all 41 protein-coding genes in the example module is shown in Table 7.2. These GO annotations are consistent with this sub-network being specifically associated with cardiac physiology. Interestingly this sub-network is also found in two gut-specific tissues, indicating shared genetic regulation in these somewhat disparate tissues. The top three significantly over-represented terms are 'regulation of system process', 'regulation of the force of heart contraction' and 'circulatory system process'. We expect that using this approach in combination with gene mapping for quantitative traits associated with particular tissues such as milk composition or intramuscular fat might prove useful in resolving QTL to the gene or

ncRNA level and in identifying epistatic interactions difficult to map with current resources.

Conclusions

We are now able to use the bovine genome sequence and associated annotations as an unparalleled resource for bovine genetics. Genetic mapping is now done routinely at single nucleotide resolution and genetic intervals containing QTLs can now be mined for information associated with both protein coding and non-coding genes. The use of transcriptome profiling as a tool to narrow down potential candidate genes is likely to increase dramatically, as it provides the means to identify interacting genes, both coding and non-coding, that are specific to economically important tissues/traits.

However the existence of two draft assemblies that differ significantly in some regions is a complication for the applications of bovine genomics. Whilst the UMD3.1 draft assembly is currently superior for many mapping based applications, it is still far from perfect. As more and more genomes are sequenced it is apparent that individual genomes vary in their content, and that this individual content will have to be taken into account when determining the genetic basis of traits of interest. The ultimate solution to this problem will be *de novo* whole genome sequencing and assembly, with comparisons carried out between assemblies, rather than mapping sequence reads to a reference assembly.

For genomic selection in the short term, it is likely that low-density SNP arrays tailored to maximize imputation of genome sequence/structure will become more and more significant because of their cost advantages and ease of use.

Acknowledgement

The authors wish to acknowledge Dan Kortschak for helpful discussions and his willingness to challenge the obvious.

Table 7.1. Gene annotation of 34 of 41 protein-coding genes in a sub-network expression module based on human orthologues in GeneCards.

Gene symbol	Gene description	Aliases and descriptions	Diseases/disorders
MYBPC3	Myosin binding protein C, cardiac	Cardiac MyBP-C | MYBP-C | C-protein, cardiac muscle isoform | myosin binding protein C, cardiac | myosin-binding protein C, cardiac-type | FHC | myosin-binding protein C, cardiac | CMH4	Hypertrophic cardiomyopathy | familial hypertrophic cardiomyopathy | dilated cardiomyopathy | heart failure
NEBL	Nebulette	Actin-binding Z-disk protein | LIM-nebulette | nebulette | LNEBL |LASP2 | Actin-binding Z-disk protein	Endocardial fibroelastosis | Ebstein anomaly
CCDC141	Coiled-coil domain containing 141	CAMDI | coiled-coil protein associated with myosin II and DISC1 | coiled-coil domain containing 141 | FLJ39502 | coiled-coil domain-containing protein 141	
PDE3A	Phosphodiesterase 3A,cGMP-inhibited	Cyclic GMP-inhibited phosphodiesterase A | cAMP phosphodiesterase, myocardial cGMP-inhibited | CGI-PDE | cGMP-inhibited 3′,5′-cyclic phosphodiesterase A | CGI-PDE A | CGI-PDE-A | Cyclic GMP-inhibited phosphodiesterase A | phosphodiesterase 3A, cGMP-inhibited | EC 3.1.4.17	
TRIM55	Tripartite motif containing 55	RNF29 | MURF-2 | tripartite motif containing 55 | MuRF2 | Muscle-specific RING finger protein 2 | tripartite motif-containing 55 | muscle-specific RING finger protein 2 | muscle specific ring finger 2 | tripartite motif-containing protein 55 | MURF2 | MuRF-2 | ring finger protein 29 | RING finger protein 29 | muRF2	
SYT15	Synaptotagmin XV	Synaptotagmin XV | CHR10SYT | SytXV | Chr10Syt | synaptotagmin XV | synaptotagmin XV-a | sytXV | chr10 synaptotagmin | synaptotagmin-15	
SOX11	SRY (sex determining region Y)-box 11	SRY-related HMG-box gene 11 | SRY (sex determining region Y)-box 11 | SRY (sex-determining region Y)-box 11 | transcription factor SOX-11	Lymphoma
LMO7	LIM domain 7	FBXO20 | zinc-finger domain-containing protein | LMO-7 | LIM domain only 7 protein | LIM domain only protein 7 | F-box protein Fbx20 | LIM domain only 7 | F-box only protein 20 | FBX20 | LOMP | LIM domain 7 | KIAA0858	Emery–Dreifuss muscular dystrophy

Continued

Table 7.1. Continued.

Gene symbol	Gene description	Aliases and descriptions	Diseases/disorders
KCNE1	Potassium voltage-gated channel, IsK-related family, member 1	IsK producing slow voltage-gated potassium channel subunit beta Mink	voltage gated potassium channel accessory subunit	delayed rectifier potassium channel subunit IsK	potassium voltage-gated channel, IsK-related family, member 1	minimal potassium channel	cardiac delayed rectifier potassium channel protein	potassium voltage-gated channel subfamily E member 1	delayed rectifier potassium channel subunit IsK	JLNS2	JLNS	MinK	minK	ISK	LQT2/5	Minimal potassium channel	potassium voltage-gated channel, IsK-related subfamily, member 1	human cardiac delayed rectifier potassium channel protein	LQT5	Long QT syndrome	Jervell–Lange Nielsen syndrome	Brugada syndrome	sensorineural hearing loss	congenital heart block	hypokalaemia
TNNI3K	TNNI3 interacting kinase	Serine/threonine-protein kinase TNNI3K	TNNI3 interacting kinase	cardiac ankyrin repeat kinase	cardiac troponin I-interacting kinase	TNNI3-interacting kinase	CARK	cardiac troponin I-interacting kinase	cardiac ankyrin repeat kinase	EC 2.7.11.1	
SCN5A	Sodium channel, voltage-gated, type V, alpha subunit	HB1	HBBD	sodium channel, voltage-gated, type V, alpha subunit	SSS1	ICCD	sodium channel protein cardiac muscle subunit alpha	sodium channel protein cardiac muscle subunit alpha	IVF	voltage-gated sodium channel subunit alpha Nav1.5	CMD1E	sodium channel protein type 5 subunit alpha	voltage-gated sodium channel subunit alpha Nav1.5	PFHB1 |cardiac tetrodotoxin-insensitive voltage-dependent sodium channel alpha subunit	sodium channel protein type V subunit alpha	Nav1.5	LQT3	CMPD2	sodium channel, voltage-gated, type V, alpha (long QT syndrome 3)	HH1	CDCD2	VF1	HB2	Brugada syndrome	long QT syndrome	sudden infant death syndrome	heart disease	congenital heart block	Jervell–Lange Nielsen syndrome	congenital epilepsy
ACTC1	Actin, alpha, cardiac muscle 1	Actin, alpha cardiac muscle 1	CMH11	LVNC4	Alpha-cardiac actin	ACTC	actin, alpha, cardiac muscle	ASD5	CMD1R	actin, alpha, cardiac muscle 1	alpha-cardiac actin	Hypertrophic cardiomyopathy	dilated cardiomyopathy	nemaline myopathy	familial hypertrophic cardiomyopathy	restrictive cardiomyopathy

Continued

Table 7.1. Continued.

Gene symbol	Gene description	Aliases and descriptions	Diseases/disorders
TNNT2	Troponin T type 2 (cardiac)	RCM3 | cardiomyopathy, dilated 1D (autosomal dominant) | cTnT | TnTc | TnTC | troponin T type 2 (cardiac) | Cardiac muscle troponin T | troponin T, cardiac muscle | CMPD2 | cardiac muscle troponin T | LVNC6 | CMD1D | CMH2 | troponin T2, cardiac | cardiomyopathy, hypertrophic 2	Myocardial infarction | intermediate coronary syndrome | familial hypertrophic cardiomyopathy | hypertrophic cardiomyopathy | heart failure | dilated cardiomyopathy | kidney failure | pulmonary embolism | myocarditis | coronary heart disease | restrictive cardiomyopathy | cerebrovascular accident | gas gangrene | myopathy | hypertension | congenital heart defect | diabetes mellitus
TECRL	Trans-2,3-enoyl-coa reductase-like	DKFZp313D0829 | GPSN2L | SRD5A2L2 | TERL | steroid 5-alpha-reductase 2-like 2 protein | trans-2,3-enoyl-CoA reductase-like | DKFZp313B2333 | steroid 5-alpha-reductase 2-like 2 protein | glycoprotein, synaptic 2-like | steroid 5 alpha-reductase 2-like 2 | EC 1.3.1.-	
PLN	Phospholamban	PLB | CMD1P |CMH18 | phospholamban | cardiac phospholamban	Heart failure | dilated cardiomyopathy | familial hypertrophic cardiomyopathy | myocardial infarction | hypertrophic cardiomyopathy
CASQ2	Calsequestrin 2 (cardiac muscle)	Calsequestrin 2 (cardiac muscle) | calsequestrin 2, fast-twitch, cardiac muscle | calsequestrin, cardiac muscle isoform | Calsequestrin, cardiac muscle isoform | calsequestrin-2 | PDIB2	Heart failure | malignant hyperthermia | gas gangrene | myopathy | dilated cardiomyopathy | hyperthyroidism | autoimmune thyroiditis
FGF16	Fibroblast growth factor 16	FGF-16 | FGF-16 | fibroblast growth factor 16	
SLC8A1	Solute carrier family 8 (sodium/calcium exchanger), member 1	NCX1 | Na(+)/Ca(2+)-exchange protein 1 | CNC | sodium/calcium exchanger 1 | solute carrier family 8 (sodium/calcium exchanger), member 1 | Na+/Ca++ exchanger | Na+/Ca2+ exchanger	Heart failure | vascular disease
RNF207	Ring finger protein 207	RING finger protein 207 | FLJ32096 | chromosome 1 open reading frame 188 | C1orf188 | ring finger protein 207 | FLJ46380	Long QT syndrome

Continued

Table 7.1. Continued.

Gene symbol	Gene description	Aliases and descriptions	Diseases/disorders
HSPB3	Heat shock 27 kDa protein 3	HspB3 | HSP 17 | heat shock 17 kDa protein | protein 3 | HMN2C | heat shock 27 kDa protein 3 | HSP27 | DHMN2C | heat shock 17 kDa protein | HSPL27 | protein 3 | heat shock protein beta-3 | heat shock 27 kDa protein 3	
SH3RF2	SH3 domain containing ring finger 2	Protein phosphatase 1, regulatory subunit 39 | HEPP1 | putative E3 ubiquitin-protein ligase SH3RF2 | POSH-eliminating RING protein | protein phosphatase 1 regulatory subunit 39 | PPP1R39 | RNF158 | POSHER | Hepp1 | SH3 domain containing ring finger 2 | heart protein phosphatase 1-binding protein | FLJ23654 | protein phosphatase 1 regulatory subunit 39 | RING finger protein 158 | SH3 domain-containing RING finger protein 2 | heart protein phosphatase 1-binding protein | EC 6.3.2.-	
MYL3	Myosin, light chain 3, alkali; ventricular, skeletal, slow	Ventricular/slow twitch myosin alkali light chain | ventricular/slow twitch myosin alkali light chain | CMH8 | cardiac myosin light chain 1 | CMLC1 | myosin light chain 3 | myosin, light chain 3, alkali; ventricular, skeletal, slow | cardiac myosin light chain 1 | myosin light chain 1, slow-twitch muscle B/ ventricular isoform | MLC1SB | myosin, light polypeptide 3, alkali; ventricular, skeletal, slow | myosin light chain 1, slow-twitch muscle B/ventricular isoform | MLC1V | VLC1	Hypertrophic cardiomyopathy
MYLK3	Myosin light chain kinase 3	MLCK | cardiac-MyBP-C-associated Ca/CaM kinase | putative myosin light chain kinase 3 | caMLCK | cardiac-MyBP-C-associated Ca/CaM kinase | myosin light chain kinase 3 | cardiac-MyBP-C associated Ca/CaM kinase | MLC kinase | MLCK2 |EC 2.7.11.18	Anthrax disease
ADPRHL1	ADP-ribosylhydrolase like 1	[Protein ADP-ribosylarginine] hydrolase-like protein 1 | ADP-ribosylhydrolase 2 | ARH2 | ADP-ribosyl-hydrolase | ADP-ribosylhydrolase like 1	

Continued

Table 7.1. Continued.

Gene symbol	Gene description	Aliases and descriptions	Diseases/disorders
ADRB1	Adrenoceptor beta 1	Beta-1 adrenoceptor | beta-1 adrenergic receptor | adrenoceptor beta 1 | ADRB1R | Beta-1 adrenoreceptor | adrenergic, beta-1-, receptor | beta-1 adrenoceptor | RHR | BETA1AR | beta-1 adrenoreceptor | B1AR	Heart failure | dilated cardiomyopathy | hypertension | congenital aortic valve stenosis | myocardial infarction
PKP2	Plakophilin 2	plakophilin-2 | ARVD9 | plakophilin 2	Arrhythmogenic right ventricular dysplasia | palmoplantar keratosis
KCNJ8	Potassium inwardly rectifying channel, subfamily J, member 8	Inwardly rectifying potassium channel KIR6.1 | Inward rectifier K(+) channel Kir6.1 | Kir6.1 | ATP-sensitive inward rectifier potassium channel 8 | potassium channel, inwardly rectifying subfamily J member 8 | KIR6.1 | inward rectifier K(+) channel Kir6.1 | potassium inwardly rectifying channel, subfamily J, member 8 | potassium channel, inwardly rectifying subfamily J member 8 | uKATP-1	Vascular disease
RBM20	RNA binding motif protein 20	Probable RNA-binding protein 20 | RNA-binding motif protein 20 | RNA-binding protein 20	Dilated cardiomyopathy | heart failure
RYR2	Ryanodine receptor 2 (cardiac)	ARVD2 | cardiac-type ryanodine receptor | type 2 ryanodine receptor | ryanodine receptor 2 (cardiac) | RYR-2 | cardiac muscle ryanodine receptor-calcium release channel | cardiac muscle ryanodine receptor-calcium release channel | cardiac muscle ryanodine receptor | hRYR-2 | kidney-type ryanodine receptor | type 2 ryanodine receptor | cardiac-type ryanodine receptor | ryanodine receptor 2 | RyR2 | VTSIP | islet-type ryanodine receptor | cardiac muscle ryanodine receptor | ARVC2 | arrhythmogenic right ventricular dysplasia 2 |RyR	Heart failure | malignant hyperthermia | central core myopathy | arrhythmogenic right ventricular dysplasia | long QT syndrome | Brugada syndrome | dilated cardiomyopathy

Continued

Table 7.1. Continued.

Gene symbol	Gene description	Aliases and descriptions	Diseases/disorders
TNNI3	Troponin I type 3 (cardiac)	CMD2A | CMD1FF | troponin I, cardiac muscle | CMH7 | RCM1 | TNNC1 | troponin I type 3 (cardiac) | troponin I, cardiac | cardiac troponin I |cTnI	Myocardial infarction | intermediate coronary syndrome | hypertrophic cardiomyopathy | heart failure | familial hypertrophic cardiomyopathy | restrictive cardiomyopathy | myocarditis | pulmonary embolism | dilated cardiomyopathy | kidney failure | congenital heart defect | gas gangrene | pericardial effusion | coronary heart disease | pericarditis
LRRC10	Leucine rich repeat containing 10	Leucine rich repeat containing 10 | LRRC10A | HRLRRP | leucine-rich repeat-containing protein 10	
POPDC2	Popeye domain containing 2	Popeye protein 2 | popeye protein 2 | popeye domain containing 2 | popeye domain-containing protein 2 | POP2	
ANKRD1	Ankyrin repeat domain 1 (cardiac muscle)	Cytokine-inducible gene C-193 protein | cardiac ankyrin repeat protein | ankyrin repeat domain-containing protein 1 | CVARP | cytokine-inducible nuclear protein | MCARP | liver ankyrin repeat domain 1 | C-193 | bA320F15.2 | C193 | cytokine-inducible gene C-193 protein | ALRP | CARP | ankyrin repeat domain 1 (cardiac muscle) | HA1A2 | Cardiac ankyrin repeat protein | Cytokine-inducible nuclear protein	Scimitar syndrome
FHOD3	Formin homology 2 domain containing 3	KIAA1695 | FHOS2 | formin homolog overexpressed in spleen 2 | FLJ22297 | formactin-2 | hFHOS2 | formin homology 2 domain containing 3 | formin homologue overexpressed in spleen 2 | FH1/FH2 domain-containing protein 3 | formactin2 | FLJ22717 | formactin-2	

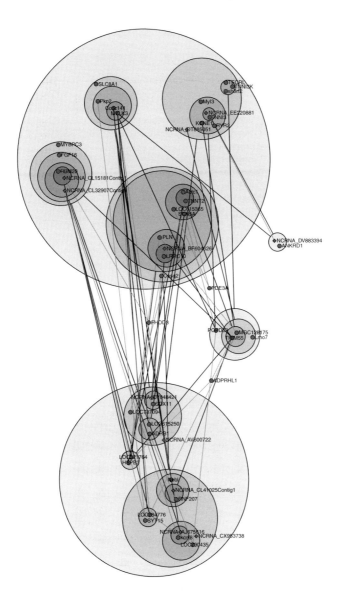

Fig. 7.4. Power graph representation of co-expression network of 52 transcripts, including 41 protein-coding genes and 11 ncRNAs in example module. Small circles filled with darker grey colours represent protein-coding gene nodes. Small diamonds filled with darker grey colours represent ncRNA nodes. Bigger circles represent power nodes, which are sets of nodes that are connected by power edges. A power edge between two power nodes signifies that all nodes in the first set are connected to all nodes in the second set. The line width of power edges represents the strength of the correlation of expression profiles of transcripts between power nodes.

Table 7.2. Over-represented GO terms for 41 protein-coding genes in example sub-network module.

GO term category	Term	Count	P value	Fold enrichment
BP_3	GO:0044057~regulation of system process	6	4.51E-07	33.2
BP_3	GO:0002026~regulation of the force of heart contraction	4	1.22E-06	164.1
BP_3	GO:0003013~circulatory system process	4	2.95E-04	27.7
BP_3	GO:0051239~regulation of multicellular organismal process	6	5.49E-04	7.7
BP_3	GO:0048513~organ development	7	9.03E-04	5.2
BP_3	GO:0048731~system development	7	0.0024	4.3
BP_3	GO:0055082~cellular chemical homeostasis	4	0.0028	12.9
BP_3	GO:0003012~muscle system process	3	0.0036	30.8
BP_3	GO:0042592~homeostatic process	5	0.00367.0	
BP_3	GO:0009888~tissue development	4	0.0124	7.5
BP_3	GO:0006811~ion transport	5	0.0134	4.8
MF_3	GO:0005246~calcium channel regulator activity	2	0.0143	132.8
BP_3	GO:0002027~regulation of heart rate	2	0.0208	89.5
MF_3	GO:0043169~cation binding	11	0.0415	1.8
BP_3	GO:0009266~response to temperature stimulus	2	0.0557	32.8
BP_3	GO:0009887~organ morphogenesis	3	0.0616	6.9
BP_3	GO:0030154~cell differentiation	4	0.0999	3.3

References

Adelson, D.L., Raison, J.M.and Edgar, R.C. (2009) Characterization and distribution of retrotransposons and simple sequence repeats in the bovine genome. *Proceedings of the National Academy of Sciences USA* 106, 12855–12860.

Babushok, D.V., Ostertag, E.M., Courtney, C.E., Choi, J.M. and Kazazian, H.H. Jr (2006) L1 integration in a transgenic mouse model. *Genome Research* 16, 240–250.

Barendse, W., Armitage, S.M., Kossarek, L.M., Shalom, A., Kirkpatrick, B.W., Ryan, A.M., Clayton, D., Li, L., Neibergs, H.L. and Zhang, N. (1994) A genetic-linkage map of the bovine genome. *Nature Genetics* 6, 227–235.

Bishop, M.D., Kappes, S.M., Keele, J.W., Stone, R.T., Sunden, S.L., Hawkins, G.A., Toldo, S.S., Fries, R., Grosz, M.D., Yoo, J. *et al.* (1994) A genetic linkage map for cattle. *Genetics* 136, 619–639.

Boichard, D., Chung, H., Dassonneville, R., David, X., Eggen, A., Fritz, S., Gietzen, K.J., Hayes, B.J., Lawley, C.T., Sonstegard, T.S. *et al.* (2012) Design of a bovine low-density SNP array optimized for imputation. *PLoS ONE* 7, e34130.

Canavez, F.C., Luche, D.D., Stothard, P., Leite, K.R.M., Sousa-Canavez, J.M., Plastow, G., Meidanis, J., Souza, M.A., Feijao, P., Moore, S.S. and Camara-Lopes, L.H. (2012) Genome sequence and assembly of *Bos indicus. Journal of Heredity*, 103, 342–348.

Childers, C.P., Reese, J.T., Sundaram, J.P., Vile, D.C., Dickens, C.M., Childs, K.L., Salih, H., Bennett, A.K., Hagen, D.E., Adelson, D.L. and Elsik, C.G. (2011) Bovine Genome Database: integrated tools for genome annotation and discovery. *Nucleic Acids Research* 39, D830–834.

Costa, F.F. (2008) Non-coding RNAs, epigenetics and complexity. *Gene* 410, 9–17.

Dalrymple, B.P., Kirkness, E.F., Nefedov, M., McWilliam, S., Ratnakumar, A., Barris, W., Zhao, S., Shetty, J., Maddox, J.F., O'Grady, M. *et al.* (2007) Using comparative genomics to reorder the human genome sequence into a virtual sheep genome. *Genome Biology* 8, R152.

Delcher, A.L., Salzberg, S.L. and Phillippy, A.M. (2003) Using MUMmer to identify similar regions in large sequence sets. *Current Protocols in Bioinformatics*, Chapter 10, 10.13.11–10.13.18.

Elsik, C.G., Worley, K.C., Zhang, L., Milshina, N.V., Jiang, H., Reese, J.T., Childs, K.L., Venkatraman, A., Dickens, C.M., Weinstock, G.M. and Gibbs, R.A. (2006) Community annotation: procedures, protocols, and supporting tools. *Genome Research* 16, 1329–1333.

Elsik, C.G., Mackey, A.J., Reese, J.T., Milshina, N.V., Roos, D.S. and Weinstock, G.M. (2007) Creating a honey bee consensus gene set. *Genome Biology* 8, R13, doi:10.1186/gb-2007-8-1-r13.

Elsik, C.G., Tellam, R.L., Worley, K.C., Gibbs, R.A., Muzny, D.M., Weinstock, G.M., Adelson, D.L., Eichler, E.E., Elnitski, L., Guigo, R. *et al.* (2009) The genome sequence of taurine cattle: a window to ruminant biology and evolution. *Science* 324, 522–528.

Esnault, C., Maestre, J. and Heidmann, T. (2000) Human LINE retrotransposons generate processed pseudogenes. *Nature Genetics* 24, 363–367.

Everts-van der Wind, A., Kata, S.R., Band, M.R., Rebeiz, M., Larkin, D.M., Everts, R.E., Green, C.A., Liu, L., Natarajan, S., Goldammer, T. *et al.* (2004) A 1463 gene cattle-human comparative map with anchor points defined by human genome sequence coordinates. *Genome Research* 14, 1424–1437.

Everts-van der Wind, A., Larkin, D.M., Green, C.A., Elliott, J.S., Olmstead, C.A., Chiu, R., Schein, J.E., Marra, M.A., Womack, J.E.and Lewin, H.A. (2005) A high-resolution whole-genome cattle-human comparative map reveals details of mammalian chromosome evolution. *Proceedings of the National Academy of Sciences USA* 102, 18526–18531.

Fedoroff, N., Wessler, S. and Shure, M. (1983) Isolation of the transposable maize controlling elements Ac and Ds. *Cell* 35, 235–242.

Gibbs, R.A., Weinstock, G.M., Metzker, M.L., Muzny, D.M., Sodergren, E.J., Scherer, S., Scott, G., Steffen, D., Worley, K.C., Burch, P.E. *et al.* (2004) Genome sequence of the Brown Norway rat yields insights into mammalian evolution. *Nature* 428, 493–521.

Harhay, G.P., Smith, T.P., Alexander, L.J., Haudenschild, C.D., Keele, J.W., Matukumalli, L.K., Schroeder, S.G., Van Tassell, C.P., Gresham, C.R., Bridges, S.M. *et al.* (2010) An atlas of bovine gene expression reveals novel distinctive tissue characteristics and evidence for improving genome annotation. *Genome Biology* 11, R102.

Havlak, P., Chen, R., Durbin, K.J., Egan, A., Ren, Y., Song, X.-Z., Weinstock, G.M.and Gibbs, R.A. (2004) The atlas genome assembly system. *Genome Research* 14, 721–732.

Hubbard, T., Andrews, D., Caccamo, M., Cameron, G., Chen, Y., Clamp, M., Clarke, L., Coates, G., Cox, T., Cunningham, F. *et al.* (2005) Ensembl 2005. *Nucleic Acids Research* 33, D447–453.

Jurka, J., Zietkiewicz, E. and Labuda, D. (1995) Ubiquitous mammalian-wide interspersed repeats (MIRs) are molecular fossils from the mesozoic era. *Nucleic Acids Research* 23, 170–175.

Jurka, J., Kapitonov, V.V., Kohany, O. and Jurka, M.V. (2007) Repetitive sequences in complex genomes: structure and evolution. *Annual Review of Genomics and Human Genetics* 8, 241–259.

Kubo, S., Seleme, M.C., Soifer, H.S., Perez, J.L., Moran, J.V., Kazazian, H.H., Jr and Kasahara, N. (2006) L1 retrotransposition in nondividing and primary human somatic cells. *Proceedings of the National Academy of Sciences USA* 103, 8036–8041.

Langfelder, P. and Horvath, S. (2008) WGCNA: an R package for weighted correlation network analysis. *BMC Bioinformatics* 9, 559.

Leeb, M., Steffen, P.A. and Wutz, A. (2009) X chromosome inactivation sparked by non-coding RNAs. *RNA Biology* 6, 94–99.

Li, H. and Durbin, R. (2010) Fast and accurate long-read alignment with Burrows-Wheeler transform. *Bioinformatics* 26, 589–595.

Liu, Y., Qin, X., Song, X.Z., Jiang, H., Shen, Y., Durbin, K.J., Lien, S., Kent, M.P., Sodeland, M., Ren, Y. *et al.* (2009) *Bos taurus* genome assembly. *BMC Genomics* 10, 180.

Lowe, C.B., Bejerano, G. and Haussler, D. (2007) Thousands of human mobile element fragments undergo strong purifying selection near developmental genes. *Proceedings of the National Academy of Sciences USA* 104, 8005–8010.

McClintock, B. (1950) The origin and behavior of mutable loci in maize. *Proceedings of the National Academy of Sciences USA* 36, 344–355.

McClure, M., Sonstegard, T., Wiggans, G. and Van Tassell, C.P. (2012) Imputation of microsatellite alleles from dense SNP genotypes for parental verification. *Front Genetics* 3, 140.

Maglott, D., Ostell, J., Pruitt, K.D. and Tatusova, T. (2007) Entrez Gene: gene-centered information at NCBI. *Nucleic Acids Research* 35, D26–31.

Martens, J.A., Laprade, L. and Winston, F. (2004) Intergenic transcription is required to repress the Saccharomyces cerevisiae SER3 gene. *Nature* 429, 571–574.

Mattick, J.S.and Makunin, I.V. (2006) Non-coding RNA. *Human Molecular Genetics* 15, R17–R29.

Myers, E.W., Sutton, G.G., Delcher, A.L., Dew, I.M., Fasulo, D.P., Flanigan, M.J., Kravitz, S.A., Mobarry, C.M., Reinert, K.H., Remington, K.A. *et al.* (2000) A whole-genome assembly of Drosophila. *Science* 287, 2196–2204.

Qu, Z. and Adelson, D.L. (2012a) Bovine ncRNAs are abundant, primarily intergenic, conserved and associ-
    ated with regulatory genes. *PLoS ONE* 7, e42638.
Qu, Z. and Adelson, D.L. (2012b) Evolutionary conservation and functional roles of ncRNA. *Frontier
    Genetics* 3, 205.
Qu, Z. and Adelson, D.L. (2012c) Identification and comparative analysis of ncRNAs in human, mouse and
    zebrafish indicate a conserved role in regulation of genes expressed in brain. *PLoS ONE* 7, e52275.
Reese, J.T., Childers, C.P., Sundaram, J.P., Dickens, C.M., Childs, K.L., Vile, D.C.and Elsik, C.G. (2010)
    Bovine Genome Database: supporting community annotation and analysis of the *Bos taurus* genome.
    BMC Genomics 11, 645.
Rincon, G., Weber, K.L., Eenennaam, A.L., Golden, B.L.and Medrano, J.F. (2011) Hot topic: performance
    of bovine high-density genotyping platforms in Holsteins and Jerseys. *Journal of Dairy Science* 94,
    6116–6121.
Royer, L., Reimann, M., Andreopoulos, B. and Schroeder, M. (2008) Unraveling protein networks with
    power graph analysis. *PLoS Computational Biology* 4, e1000108.
Schibler, L., Roig, A., Mahe, M.F., Save, J.C., Gautier, M., Taourit, S., Boichard, D., Eggen, A. and
    Cribiu, E.P. (2004) A first generation bovine BAC-based physical map. *Genetics, Selection, Evolution*
    36, 105–122.
Schlapfer, J., Yang, Y., Rexroad, C., 3rd and Womack, J.E. (1997) A radiation hybrid framework map of
    bovine chromosome 13. *Chromosome Research* 5, 511–519.
Sherry, S.T., Ward, M.H., Kholodov, M., Baker, J., Phan, L., Smigielski, E.M.and Sirotkin, K. (2001) dbSNP:
    the NCBI database of genetic variation. *Nucleic Acids Research* 29, 308–311.
Smit, A.F.A., Hubley, R. and Green, P. (1996–2010) RepeatMasker Open-3.0. Available at: http://www.
    repeatmasker.org.
Snelling, W.M., Chiu, R., Schein, J.E., Hobbs, M., Abbey, C.A., Adelson, D.L., Aerts, J., Bennett, G.L.,
    Bosdet, I.E., Boussaha, M. *et al.* (2007) A physical map of the bovine genome. *Genome Biology* 8, 17.
van Bakel, H., Nislow, C., Blencowe, B.J.and Hughes, T.R. (2010) Most 'dark matter' transcripts are associ-
    ated with known genes. *PLoS Biology* 8, e1000371.
Venter, J.C., Adams, M.D., Myers, E.W., Li, P.W., Mural, R.J., Sutton, G.G., Smith, H.O., Yandell, M., Evans, C.A.,
    Holt, R.A. *et al.* (2001) The sequence of the human genome. *Science* 291, 1304–1351.
Walsh, A.M., Kortschak, R.D., Gardner, M.G., Bertozzi, T. and Adelson, D.L. (2013) Widespread horizontal
    transfer of retrotransposons. *Proceedings of the National Academy of Sciences USA* 110,
    1012–1016.
Womack, J.E. (2012) First steps: bovine genomics in historical perspective. *Animal Genetics* 43 2–8.
Womack, J.E., Johnson, J.S., Owens, E.K., Rexroad, C.E., 3rd, Schlapfer, J. and Yang, Y.P. (1997) A whole-
    genome radiation hybrid panel for bovine gene mapping. *Mammalian Genome* 8, 854–856.
Zimin, A.V., Delcher, A.L., Florea, L., Kelley, D.R., Schatz, M.C., Puiu, D., Hanrahan, F., Pertea, G.,
    Van Tassell, C.P., Sonstegard, T.S. *et al.* (2009) A whole-genome assembly of the domestic cow, *Bos
    taurus*. *Genome Biology* 10, R42.
Zimin, A.V., Kelley, D.R., Roberts, M., Marçais, G., Salzberg, S.L.and Yorke, J.A. (2012) Mis-assembled
    'segmental duplications' in two versions of the *Bos taurus* genome. *PLoS ONE* 7, e42680.

8 Bovine Immunogenetics

Yoko Aida,[1] Shin-nosuke Takeshima,[1] Cynthia L. Baldwin[2] and Azad K. Kaushik[3]

[1]*RIKEN Brain Science Institute, Saitama, Japan;*
[2]*University of Massachusetts, Amherst, Massachusetts, USA;*
[3]*University of Guelph, Guelph, Ontario, Canada*

Introduction

The immune system of jawed vertebrates evolved to provide innate and adaptive immunity against a diverse array of potentially harmful antigens. The adaptive immune effector cells are B and T lymphocytes (also known as B cells and T cells), while innate system cells include those of the myeloid lineage (monocytes, macrophages, eosinophils, basophils, mast cells, neutrophils and dendritic cells) as well as primitive lymphoid cells known as natural killer (NK) cells. The cells of the innate immune system not only play their own direct role in immunity, for example, killing infectious microbes following phagocytosis, but in the case of macrophages and dendritic cells function as accessory cells for T cells by presenting antigenic peptides on major histocompatibility complex (MHC) molecules and producing cytokines that direct T cell functional responses. The immune response has been historically broken into two aspects known as humoral and cell-mediated immunity. While B cells produce antibodies, which are mediators of humoral immunity, T cells can promote the B cell response through their production of specific soluble molecules known as cytokines, thereby facilitating humoral immunity. Alternatively, T cells mediate cellular immunity by killing infected host cells and by their production of cytokines that activate macrophages to more effectively kill phagocytosed infectious organisms or inhibit viral replication. The host's immune system must differentiate between self and non-self antigens but still recognize a diverse array of potentially harmful antigens, estimated to be between 10^8 and 10^{11}. Significant advances have been made in describing the genetics of the bovine immune system receptors and MHC molecules that are involved in presenting peptides to T cells to engage their so-called T cell receptor (TCR) and will be reviewed here. We describe in detail the genes that code for the T and B cell antigen-specific receptors (TCR and B cell receptor (BCR)) and the immunoglobulins (antibodies) that are secreted by B cells and which mirror the BCR of the secreting cell. These receptors and antibodies are formed by somatic gene rearrangements. In addition we describe germline encoded multigene receptor families that are expressed by both innate and adaptive immune system cells and which interact with pathogen-associated molecular patterns (PAMP), host cell-derived damage-associated molecule patterns (DAMP), as well as classical and non-classical MHC molecules.

The Major Histocompatibility Complex

The major histocompatibility complex (MHC) plays a crucial role in determining immune responsiveness (Klein, 1986). The MHC is referred to as a 'complex' because its genes are clustered together within a single genomic region or locus. This region of tightly linked genes encodes the proteins responsible for presenting self and non-self-antigens to T cells and is, therefore, fundamental to immune recognition and regulation. Classical MHC class I and class II molecules bind antigen-derived peptides and present them to T cells. Antigen presentation is one aspect of a complex series of events that results in activation of T cells with the ultimate objectives being elimination of invading parasites, microorganisms and infected host cells, but can also result in rejection of transplanted organs. Diversity of antigenic peptides presented is achieved in several ways: (i) the coding of the MHC is polygenic, meaning that a single characteristic is controlled by two or more genes; (ii) MHC genes are highly polymorphic; and (iii) MHC genes are expressed from both inherited alleles.

Our knowledge of MHC genetics, gene function, disease associations, protein structure, phylogeny and genomic organization has (mainly) been gained through human and mouse studies. The murine MHC was first described 75 years ago by Peter Gorer (Gorer, 1937). George Snell then identified the MHC locus by selectively breeding two different mouse strains in order to create a new mouse strain that was almost genetically identical to one of the original parental strains, differing only in the locus that governs histocompatibility (the ability for a tumour to be accepted) (Snell and Higgins, 1951). The first human MHC antigen, initially called Mac, is now known as human leukocyte antigen (HLA)-A2, was identified 21 years later (Dausset, 1958). Since then, the study of MHC genes has undergone a marked expansion, driven by the identification and characterization of many MHC class I and class II genes in humans, mice, chickens and other jawed vertebrates, although not in jawless vertebrates (e.g. hagfish, lamprey) or invertebrates (Flajnik and Kasahara, 2001).

Bovine MHC genes were first identified by Amorena and Stone (1978) and Spooner *et al.* (1978). The genetic regions they defined are referred to as the bovine leukocyte antigen (BoLA) system (Spooner *et al.*, 1979). The BoLA is encoded on chromosome 23, within a 4000 kb stretch of DNA that contains more than 154 tightly linked genes (Elsik *et al.*, 2009). The current BoLA nomenclature, as detailed in the Immuno Polymorphism Database (IPD-MHC; www.ebi.ac.uk/ipd/mhc/bola) (Robinson *et al.*, 2010), has been in place since 2004 and is modelled on the HLA nomenclature system

(Marsh *et al.*, 2005). The BoLA section of the IPD-MHC database currently contains sequences assigned to class I genes and class II *DRB3*, *DQA* and *DQB* genes. An overview of the essential features of the most studied MHC molecules in both humans and mice is presented below, but the main focus of the remaining sections will be on the structural and functional features and the disease associations of the BoLA genes.

Molecular structure and function of MHC molecules

The MHC gene family is divided into three subgroups: class I, class II and class III. The class I and II regions comprise genes that encode both classical and non-classical MHC proteins (Painter and Stern, 2012). The 'classical MHC' molecules, encoded by highly polymorphic class I and class II genes, serve to present antigen-derived peptides of varying sizes to conventional αβ T cells and play an integral role in vertebrate adaptive immunity. There are also non-polymorphic representatives-termed 'non-classical MHC molecules' for those encoded by genes located within the MHC region, and 'MHC-like molecules' when encoded by genes located outside the MHC; these exist for both class I and class II MHC-related proteins, with the most diverse being related to the MHC class I molecules. Some αβ T cells, including the so-called natural killer (NK) T cells, recognize protein antigens that have lipid components and which are presented by particular MHC-class I like molecules in the CD1 family.

Class I MHC molecules

Classical MHC class I molecules have an α-chain (molecular mass, 45 kDa) comprising three domains, α1, α2 and α3. The α-chain is non-covalently associated with the non-MHC molecule, β2 microglobulin (β2m; 12 kDa) (Fig. 8.1A, left panel). The α1 and α2 domains form a groove that accommodates antigen-derived peptides (Bjorkman *et al.*, 1987). Classical MHC class I molecules are expressed on the surface of all nucleated cells (Adams and Luoma, 2013). Proteins within the cytosol are degraded by the proteasome, which is encoded

by *PSMB8* and *PSMB9* located within the MHC region; liberated peptides are internalized by the MHC-encoded transporter associated with antigen processing (TAP) channel in the endoplasmic reticulum (ER), where they are loaded on to MHC class I molecules. This complex traffics through the Golgi apparatus and fuses with the cell membrane where it can interact with the TCR on cytotoxic T cells that co-express the CD8 molecule. CD8 docks with the classical MHC class I molecule, thereby anchoring the T cell to the antigen-presenting cell and facilitating signal transduction. If the TCR interacts strongly with this complex the T cell is stimulated to proliferate, produce cytokines and/or kill the cell.

A recent study showed the high-resolution crystal structure of the bovine MHC class I *N*01301* allele bound to an immunodominant 11-amino acid fragment ($Tp1_{214-224}$) derived from *Theileria parva* (Macdonald *et al.*, 2010). As shown in Fig. 8.1, the peptide is presented in a distinctive raised conformation rather than being buried in the groove, which is likely to have a significant impact on TCR recognition. This unconventional structure results from a hydrophobic ridge within the MHC peptide binding groove, which is found in a set of BoLA alleles. This feature is extremely rare in other species, although it occurs in a small group of murine MHC class I molecules.

Non-classical MHC class I and class I MHC-like molecules (Table 8.1) have limited polymorphism, expression patterns and antigen presentation abilities but are important for their ability to act as ligands for the killer receptors. Non-classical MHC proteins are those molecules encoded within the MHC loci (e.g. *HLA-E*, *-F* and *-G* and *MIC* in humans and *M3* and *Qa* in mice), while MHC class I-like molecules are encoded across chromosomes. Some of the non-classical MHC class I molecules present peptides or act as ligands for killer receptors as mentioned above. MHC class I-like molecules include the stress ligands that interact with Natural Killer Group 2D (NKG2D) receptors (described below). They are known as UL16-binding proteins (ULBPs) in humans and in mice retinoic acid early inducible gene 1 (Rae1), histocompatibility 60 (H60) and murine UL16-binding protein-like transcript 1 (MULT1). These molecules interact

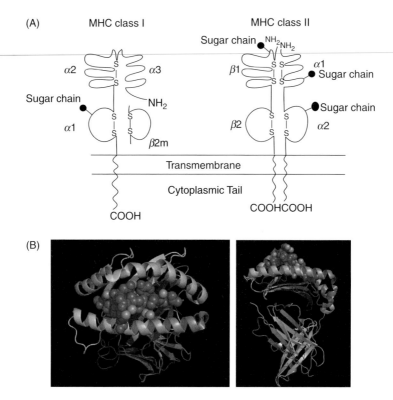

Fig. 8.1. Structure of the MHC molecules. (A) The MHC class I molecule (left) and the MHC class II molecule (right). (B) 3D structure of a bovine MHC class I molecule (PDB: 2xfx) as reported by Macdonald *et al.* (2010). Left panel: looking down from the T cell receptor; right panel, side view. The peptide is presented by the bovine class I molecule in a distinctive raised conformation, which is extremely rare in other species.

with CD8[+] T cells, NKT cells (which are a unique type of αβ T cell), γδ T cell and NK cells to exert functions that can be immune or non-immune related through a variety of receptors (described below in this chapter). Other MHC-like molecules such as those in the CD1 family, like their classical relatives, are able to bind and present peptides, although distinct structural features within the peptide binding groove and their low polymorphism distinguish them from classical MHC molecules and allow them to present molecules with lipid tails.

Class II MHC molecules

Classical MHC class II molecules are heterodimeric transmembrane glycoproteins of approximately 50 kDa, which are formed by the non-covalent association between α- and β-chains encoded by distinct genes within the MHC (Fig. 8.1A, right panel): the α1 and β1 domains form the peptide binding groove (Brown *et al.*, 1993). Classical class II molecules are expressed by so-called 'professional antigen-presenting cells' (APCs), which include dendritic (DCs) cells, macrophages and B cells (Painter and Stern, 2012). These cells process peptides derived from self and foreign proteins and display them at the cell surface in conjunction with MHC class II molecules (described below). The MHC class II/peptide complex is recognized by CD4[+] T cells. Newly synthesized classical MHC class II α- and β-chains are translocated to the lumen of the ER where they associate with a trimeric chaperone protein known as the class II-associated invariant chain, which directs them to endosomal compartments. Endosomal proteases cleave the invariant chain to yield a small peptide called the class II-associated invariant chain peptide (CLIP), which is presented in the class II peptide binding groove.

Table 8.1. Non-classical and MHC class I-like genes.

Known as	Gene map	Cattle	Humans	Mice	Function of these gene products
Non-classical MHC class I genes	MHC class I locus	17 alleles of 5 genes (e.g. NC1–NC5)	HLA-E, HLA-F, HLA-G	Q, M gene clusters (e.g. M3, Qa)	Present peptides to T cells; HLA-E is ligand for NKG2A and NKG2C
		BoLA MIC1, BoLA MIC2, BoLA MIC3	MICA, MICB	Not found in mice	Ligands for NKG2D receptors on NK and T cells
MHC class I-like molecules	Non-MHC loci encoded	ULBP1– ULBP30	10 ULBP genes (aka Rae)	Mult1, Rae, H60	Ligand for NKG2D receptors on NK and T cells
		CD1a, CD1b (2 copies), CD1d pseudogene	CD1a–e	CD1a–c	Present lipid-containing antigens to αβ T cells

The non-classical MHC class II molecules, DM or DO, act as a catalytic peptide exchange factor, which releases CLIP and promotes the binding of self or foreign antigen peptides to MHC class II molecules; this complex goes to the cell surface, where it interacts with TCRs expressed by CD4$^+$ T cells (Kropshofer et al., 1999). The CD4 molecules themselves bind to the MHC class II molecules to stabilize the interaction and this results in signal transduction to activate the T cell. CD4$^+$ T cells are known as helper T cells (Th), which are divided into functional subpopulations known as type 1 (Th1), type 2 (Th2), type 17 (Th17) (Harrington et al., 2005) and regulatory (Treg) (Sakaguchi et al., 1995). The crystal structures of DM and DO show that these accessory proteins are structural homologues of classical MHC class II proteins and differ mainly within the MHC II peptide binding groove. However they are restricted to the membrane of lysosomes.

Class III molecules

The MHC class III locus resides between the class I and class II loci, and the molecules coded for have a different physiological role. Class III molecules include several secretory proteins that have immune functions other than antigen processing and presentation or interaction with killer receptors such as NKG2D. Examples include components of the complement system (such as C2, C4 and B factor), cytokines involved in immune signalling (such as tumour necrosis factor-α (TNF-α), lymphotoxin-α (LTA) and lymphotoxin-β (LTB)), and heat shock proteins (HSP), which buffer cells from stress.

Genomic organization of the bovine MHC

The genome of a Hereford cow has been sequenced and assembled ('Bos_taurus_UMD_3.1'; www.hgsc.bcm.tmc.edu/projects/bovine; http://www.ncbi.nlm.nih.gov/projects/mapview). Although there is some discordance with other published results, we have placed the MHC genes in order and predicted the size of the MHC region in cattle (Fig. 8.2).

The BoLA and structural genes have been mapped to bovine autosome 23 (BTA23). The MHC of most mammalian taxa comprises tightly linked genes with the general structure relatively conserved, being divided into three main regions: class I, class II and class III. However, the organization of the BoLA complex differs in that the class II loci are found within two regions, designated IIa and IIb (Fig. 8.2). Linkage (Andersson et al., 1988) and cytogenetic

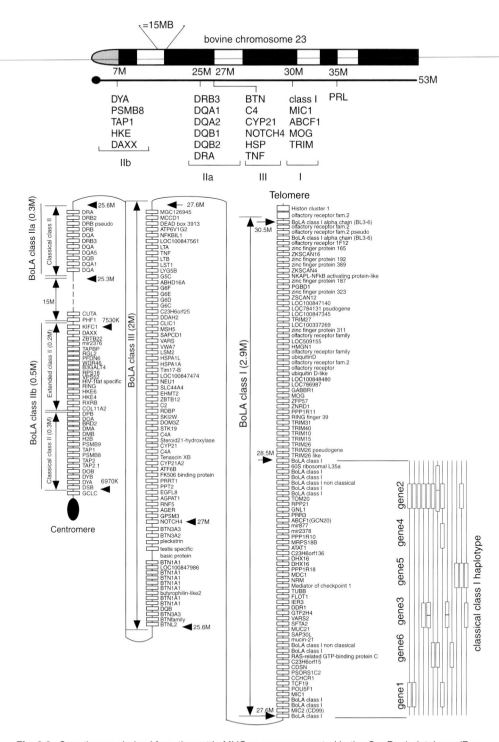

Fig. 8.2. Genetic map derived from the cattle MHC sequence reported in the GenBank database (Bos_taurus_UMD_3.1). The upper panel provides an overview of bovine chromosome 23; the lower panel shows the MHC region in detail. Genes are ordered from centromere to telomere (not to scale). The order of classical class I genes in the class I region was identified by Codner *et al.* (2012), and is shown on the right of the class I scheme.

analyses (McShane *et al*., 2001) show that the class IIb is located within a region in close proximity to the centromere at BTA23q12, and that the IIa region is located in a region near the class I and III regions at BTA23q23. As shown in Fig. 8.2, the BoLA region contains 0.5 megabases (Mb) of centromeric DNA encoding class IIb (comprising classical class II (0.3 Mb) and extended class II (0.2 Mb)), 0.3 Mb encoding class IIa, 2.0 Mb encoding class III and 2.9 Mb encoding class I. Detailed mapping of BTA23 using radiation hybrid analysis techniques (Band *et al*., 1998; Itoh *et al*., 2005; Snelling *et al*., 2005) revealed that the ancestral MHC was probably disrupted by a large inversion, which produced the BoLA IIa and IIb regions. Comparative studies in other species show that the inversion is likely to be a common feature of the MHC in all ruminants (Skow *et al*., 1996).

The annotated genomic sequence of BoLA IIb clearly shows that the class IIb region spans roughly 450 kb and includes 20 potential genes (Childers *et al*., 2006). Comparative sequence analysis of classical class II molecules encoded by the human MHC shows that the proximal inversion breakpoint occurred approximately 2.5 kb from the 3′ end of the glutamate-cysteine ligase catalytic subunit (*GCLC*), and that the distal breakpoint occurred about 2 kb from the 5′ end of a divergent class II DRβ-like sequence, designated *DSB* (Fig. 8.2). The *DSB* gene is followed by the divergent class II *DYA* and *DYB* loci. The remainder of the assembly contains most of the genes typically found in classical class II regions, including the class II *DOB* gene, proteasome genes (*PSMB8* and *PSMB9*), transporter genes (*TAP2.1*, *TAP1* and *TAP2*), the CLIP-releasing class II gene pair *DMB* and *DMA*, and the class II gene *DOA*. Although a short fragment showing 78% sequence identity to exon 3 of the HLA class II *DPB* gene was identified approximately 9.4 kb from *DOA*, analysis of BoLA IIb failed to identify any additional sequences that were similar to *DPB* or *DPA*. This region also contains non-MHC class II genes, including histone H2B-like (*H2B*) and bromodomain-containing 2 (*BRD2*). This sequence extends into the extended class II region, which includes apparently intact genes that encode *COL11A2*, *RXRB*, *SLC39A7* (old name: *HKE4*) and *HSD17B8* (old name: *HKE6*). In cattle the extended MHC class II region goes from *KIFC1*

to *COLA11A2*. Thus, the BoLA class IIb regions are thought to result from genetic transposition, with the order running in the opposite direction.

The class IIa sub-region has two gene clusters: *BoLA-DR* and *BoLA-DQ* (Fig. 8.2). The most centromeric gene within the BoLA class IIa region is *BoLA-DQA* and the most telomeric is *BoLA-DRA*. All class IIa genes are classical class II genes, whose products are expressed on the cell surface and function to present antigen to CD4+ T cells.

Shiina *et al*. (2004) showed that the most centromeric gene within the MHC class III region is *BTNL2*. BoLA sequencing also showed that *BTNL2* is next to *DRA*. *BTNL2* belongs to the butyrophilin-like (Btnl) family of genes, which regulate milk droplet secretion and T cell activation and proliferation (Abeler-Dorner *et al*., 2012) (Fig. 8.2). The centromeric region of BoLA class III (1.4 Mb) contains several genes that encode butyrophilins (BTN) and butyrophilin-like (BTNL) molecules. Other typical class III genes, such as those that encode complement *C2* and *C4*, *TNF*, *HSPs*, *NF-kB* and *CYP21* are embedded within the remaining 0.6 Mb of the class III region.

The BoLA class I genes, including the *BoLA-A* genes (which are the most commonly expressed class I genes), are embedded within the most centromeric portion of the class I region, along with the non-classical MHC class I chain-related (*MIC*) genes (Fig. 8.2). Class I genes are located within the most centromeric (0.9 Mb) stretch of the class I region (apart from one class I gene, which is located 2 Mb closer to the telomere; indicated by the arrow in Fig. 8.2). The 2 Mb telomeric region of class I contains the *TRIM* gene cluster and the genes encoding the olfactory receptor family. The order of the genes is similar to that in BoLA regions; however, some genes such as *TRIM* and the olfactory receptor family genes show extensive duplication.

Bovine class I gene products and polymorphisms

Classical bovine class I molecules

The bovine section of IPD-MHC (www. ebi.ac.uk/ipd/mhc/bola; update 18 November 2011) currently contains 107 classical BoLA class I sequences, which fall into 94 allele groups; thus, the runs in a single series are numbered from 1 to 94 (prefixed 'N'). Previous studies used Southern blotting and restriction fragment length polymorphism (RFLP) analysis to show that the bovine class I region contains about 10–20 different class I genes (Lindberg and Andersson, 1988). A recent phylogenetic analysis of all classical BoLA class I cDNA sequences (*n* = 80) in the IPD-MHC, using nucleotide sequences from exon four to the stop codon, showed that all the classical alleles cluster into six groups: *Genes 1–6* (Codner *et al.*, 2012) (Table 8.2). Phylogenetic analyses (Birch *et al.*, 2006; Holmes *et al.*, 2003), mapping data (Di Palma *et al.*, 2002) and functional studies (Gaddum *et al.*, 2003; Guzman *et al.*, 2008, 2010b; MacHugh *et al.*, 2009) suggest that these highly polymorphic classical BoLA class I genes are present in cattle. To date, 16 alleles have been assigned to *Gene 1*, 33 to *Gene 2*, 30 to *Gene 3*, 3 to *Gene 4*, 4 to *Gene 5* and 9 to *Gene 6* (Hammond *et al.*, 2012). Table 8.2 lower right column lists 29 BoLA class I haplotypes relating to *Genes 1–6* (Codner *et al.*, 2012). The order of *Genes 1, 2, 4* and *5* shown in Table 8.2 and Fig. 8.2 was determined from the genome and a BAC contig containing the BoLA class I region of one known bovine haplotype, *A14* (Di Palma *et al.*, 2002). *Genes 3* and *6* are placed arbitrarily between *Genes 1* and *5*, although we do not know their true location; however, it is thought that they are located within this extended 'classical' region.

What do we know about the expression of the six classical class I genes? Humans express three highly polymorphic classical MHC class I genes (*HLA-A, HLA-B* and *HLA-C*). By contrast, some species, for example rats and rhesus macaques, maintain diversity by generating haplotypes that show considerable variation in terms of the number and combination of transcribed genes. Interestingly, cattle appear to use both strategies. Further characterization of BoLA class I haplotypes *A19, A17, A11, A20, A33, RSCA2, A10, A14, A15, A12 (W12B), A12 (A13), A10/KN104, A13, A18* and *A18v* suggests that between one and three of the six classical genes are usually transcribed in a particular haplotype (Birch *et al.*, 2006;

Ellis *et al.*, 1999). For example, highly polymorphic genes, such as *Gene 1, Gene 2* and *Gene 3*, are transcribed in different haplotypes, and almost all haplotypes have a transcribed allele assigned to *Gene 2*, together with either *Gene 1-* or *Gene 3*-derived alleles. Interestingly, Codner *et al.* (2012) examined the frequency with which all genes and haplotype structures occur in a cohort of Holstein–Friesian animals. BoLA class I haplotypes that express two classical class I genes are the most frequently observed, and *Gene 2* is almost always expressed although it is usually in combination with one or two of the other five genes. The frequency data support the dominance of *Gene 2*. Haplotype frequency in cattle populations is likely to impact on disease susceptibility. Indeed, a previous study identified foot and mouth disease (FMDV)-specific CD8+ T cell responses in cattle of a known BoLA class I phenotype (Gerner *et al.*, 2009; Guzman *et al.*, 2008, 2010b). Therefore, characterization of diversity of classical BoLA class I haplotypes is the current approach for various cattle breeds worldwide.

Non-classical bovine class I molecules

To date, 17 non-classical (NC) BoLA class I sequences have been assigned to mapped genes in the IPD-MHC (*NC1–NC4*; Birch *et al.*, 2008a; Shu *et al.*, 2012). These three, *NC2, NC3* and *NC4*, are close to the *MIC* genes (described below in this chapter), and one is close to the classical class I genes. *NC1* currently has eight alleles plus a number of splice variants, *NC2* has three alleles, *NC3* has a single allele and *NC4* has four alleles. A recently submitted sequence was named *NC5*00101*, since it is not closely related to any existing non-classical sequence and is presumed to be encoded at a fifth locus; however, to date, there is no evidence for an additional gene. In contrast to classical BoLA class I genes, non-classical BoLA class I sequences show little, if any, polymorphism and have restricted tissue distribution (Davies *et al.*, 2006). For example, all four non-classical BoLA class I genes, *NC1, NC2, NC3* and *NC4*, are expressed at higher levels in trophoblasts than in peripheral blood mononuclear cells (PBMCs), although the relative levels vary considerably. A recent study reported that the

Table 8.2. Predicted haplotypes of the BoLA class I and class II regions.

Class I haplotype[a]	Gene 1	Gene 6	Gene 3	Gene 5	Gene 4	Gene 2
A12(W12B)	N*01901					N*00801
A12(A30)	N*02001					N*00801
A31	N*02101					N*02201
A13	N*03101					N*03201N
BF1						N*05401
BF5						N*01601
A11			N*01701			N*01801
A20			N*02701			N*02601
A20v			N*02702			N*02602
A33			N*00401			N*00501
H1			N*03801			4960.1a
A10			N*00201			N*01201
BF2			N*01702			N*05501
BF3			N*05101			N*04402
BF7			N*00402/01101			N*04801
BF8			N*03601/03701			N*05601
A19		N*01401				N*01601
BF4		N*01501				N*04501
A17		N*01502				N*00601/00801
A17v		N*01501				N*00602/00802
A14	N*02301				N*02401	N*02501
A15	N*00901				N*02401	N*02501
A15v	N*00902				N*02401	N*02501
H5			N*03601/03701			
A18		N*01301				
A18v		N*01302				
H2				N*03901		
A10/KN104			N*00101	N*00301		

Class II haplotypes

Miyasaka *et al.* (2012)[b]	ISAG[c]	*DRB3*	*DQA1*	*DQA2*	*DQA4*	*DQA5*	*DQB*	*DQB*
0101A	DH24A	*0101	*0101				*0101	
0201A	DH07A	*0201	*0203(1)				*0201	
0501A	DH01A	*0501		*22021			*1301	
0502A		*0502		*2901	*2703		*3101	
0503A		*0503	*0101				*2702	
0504A		*0504						
0601A		*0601					*0302	
0701A	DH28A	*0701	*0101	*22031			*0103	
0801A		*0801	*0801	*22031		*2801	*0801	
0801B	DH21A	*0801	*0801			*2801	*0801	*2001
0801C		*0801	*1203	*2201				
0801D		*0801		*22031		*2801		
0901A	DH11C	*0901	*1203	*2201			*1006	*0901
0902A	DH11A	*0902	*0204				*0301	
0902B		*0902	*0204				*1803	
0902C		*0902	*0204				*1807	
1001A		*1001	*10012	*2101			*10021	*0901
1001B		*1001	*10012	*2206			*10021	*1402
1001C	DH03A	*1001	*10012				*1003	*0902
1101A	DH22H	*1101	*10011	*2206			*10021	*1402

Continued

Table 8.2. Continued.

Class II haplotypes

Miyasaka et al. (2012)[b]	ISAG[c]	DRB3	DQA1	DQA2	DQA4	DQA5	DQB	DQB
1101B	DH22E	*1101	*10012	*2101			*10021	*2902
1101C		*1101	*1301				*1802	
1103A		*1103	*0203(2)				*3602	
1201A		*1201	*10011	*2206			*10021	*1402
1201B	DH08A	*1201	*12011	*2201			*1005	*1201
1301A		*1301			*27011			
1302A		*1302			*27011			
14011A		*14011	*10012	*22021			*10021	*1301
14011B	DH27A	*14011	*1401				*1401	
1501A		*1501	*10011	*2101			*10021	*2903
1501B	DH16A	*1501	*10011	*22021			*0102	*1101
1501C	DH16A	*1501	*10011	*22021			*10021	*1301
1501D		*1501	*10012	*2101			*10021	*0901
1601A	DH10C	*1601	*10011	*2002				*1302
1601B		*1601	*10012	*22021			*10021	*1302
1601C		*1601	*12021	*22021			*1001	*1301
1701A		*1701			*3001			
1801A	DH18A	*1801		*2201			*0601	*1702
1902A		*1902		*2206			*1402	
2002A	DH15B	*2002	*0101				*0101	
2703A		*2703	*0101	*22031			*0103	
2703B	DH23A	*2703	*0101	*22031			*1803	
2703C		*2703	*0203(1)				*0201	
2703D		*2703	*0204				*1808	
2703E		*2703	*12012	*2201			*1001	*0501
3401A		*3401	*0103	*22021			*2501	
3401B		*3401	*0203(1)				*0201	
3401C		*3401	*0301				*1501	
4401A		*4401	*0103				*2701	
nd	DH09B		*0301					
nd	DH11B		*0301					
nd	DH12B		*0802	*22023				
nd	DH15A		*0103					
nd	DH17A		*1101					
nd	DH22A		*0401					
nd	DH22F						*0401	

[a]Class I haplotypes identified by Codner *et al.*, 2012.
[b]Class II haplotype identified by Miyasaka *et al.*, 2011.
[c]Class II haplotype identified by Lewin *et al.*, 1999.

expression of three non-classical BoLA class I genes, *NC1*, *NC2* and *NC3*, by bovine PBMCs varied according to the stage of pregnancy; the same genes were upregulated in fetal ear tissue, but were downregulated in the placenta (Shu *et al.*, 2012). These studies show that non-classical BoLA class I molecules play a crucial role during reproduction in dairy cows.

Bovine MHC class I-like molecules

In addition to non-classical BoLA class I, bovine MHC class I-like molecules that are coded outside the MHC locus include a CD1 family (Nguyen *et al.*, 2013; Van Rhijn *et al.*, 2006) and a ULBP (described below in this chapter as ligands for killer and stress receptors) in cattle as demonstrated by study of the genomic organization, expression and function of these genes. In the CD1 family, the five known CD1 isoforms have been divided into two subsets: Group 1 CD1 molecules (CD1a, CD1b and CD1c) have been shown in humans to present mycobacterial lipid antigens; and Group 2 CD1 molecules (CD1d) are known to present antigen to NKT cells. It is interesting that the evidence to date indicates that cattle express CD1a, CD1e and multiple CD1b molecules, but no CD1c and CD1d molecules; the gene for CD1c is not present in the genome and both *CD1D* genes are pseudogenes in cattle, suggesting CD1d-restricted NKT cells are absent in cattle (Van Rhijn *et al.*, 2006). By contrast, it has been reported that the bovine *CD1D* gene has an unusual gene structure and is expressed but cannot present α-galactosylceramide variants with shorter fatty acids (Nguyen *et al.*, 2013).

Bovine class II gene products and polymorphisms

A major chromosome rearrangement within the class II region led to the division of the BoLA region on chromosome 23 into two distinct sub-regions: class IIa and class IIb (Fig. 8.2).

The bovine class IIa region

The BoLA class IIa region contains the functionally expressed classical class II genes, *BoLA-DR* and *BoLA-DQ*. As shown in Fig. 8.2, unlike in humans, a functional *DP* gene was not identified in the class IIa region of the bovine genome, although a fragment from exon 3 of the *DPB* was identified in the class IIb region (Childers *et al.*, 2006).

The genetic structure of the DR sub-region is well conserved across species. It contains a single, almost non-polymorphic, *DRA* gene, and a number of pseudogenes, gene fragments and expressed *DRB* genes. For example, humans predominantly express *HLA-DRA1* and *HLA-DRB1*, although *HLA-DRB3*, *-DRB4* and *-DRB5* are functional in some haplotype groups (Schreuder *et al.*, 2005). By contrast, cattle have only one known functional *BoLA-DRB* gene. In addition, only one *BoLA-DRA* allele has been identified from sequence data. Unlike for HLA, there are at least three *BoLA-DRB* loci, but only

one *DRB* gene (*BoLA-DRB3*) is functional. The *BoLA-DRB1* gene is a pseudogene containing multiple stop codons. The *BoLA-DRB2* gene is expressed poorly (Burke *et al.*, 1991; Russell *et al.*, 1994), although it does show some polymorphism (Muggli-Cockett and Stone, 1991). The *BoLA-DRB3* gene is strongly expressed and is the most polymorphic class II locus in cattle. The gene regulates both antigen recognition and the magnitude of the antigen-specific T cell response mounted upon exposure to infectious diseases (Lewin *et al.*, 1999). Indeed, population-based studies of *BoLA-DRB3* polymorphisms have been performed in many breeds including European breeds, zebu breeds and native breeds of South America and Asia (Giovambattista *et al.*, 1996, 2013; Gilliespie *et al.*, 1999; Maillard *et al.*, 1999; Miretti *et al.*, 2001; da Mota *et al.*, 2002; Takeshima *et al.*, 2002, 2003, 2009a,b, 2011; Ripoli *et al.*, 2004; Behl *et al.*, 2007; Miyasaka *et al.*, 2011, 2012; Lee *et al.*, 2012; Lei *et al.*, 2012). To date, techniques such as sequencing cloned genomic DNA, cDNA or cloned polymerase chain reaction (PCR) products, PCR-RFLP analysis, PCR-sequence-based typing (SBT), and next generation sequencing have identified 131 *BoLA-DRB3* alleles in various breeds of cattle (Sigurdardottir *et al.*, 1991; Ammer *et al.*, 1992; van Eijk *et al.*, 1992; Aida *et al.*, 1995; Mikko and Andersson, 1995; Russell *et al.*, 1997, 2000; Maillard *et al.*, 1999, 2001; Takeshima *et al.*, 2001, 2002, 2003, 2009a, 2011; Baxter *et al.*, 2008; Miyasaka *et al.*, 2011; Baltian *et al.*, 2012; Lee *et al.*, 2012; Giovambattista *et al.*, 2013). These genes are listed in the IPD-MHC. Genetic variations in *BoLA-DRB3* influence resistance and susceptibility to a wide variety of infectious diseases, e.g. bovine leukaemia virus (BLV)-induced B-cell lymphoma (Aida, 2001) and lymphocytosis (Xu *et al.*, 1993; Sulimova *et al.*, 1995; Starkenburg *et al.*, 1997; Juliarena *et al.*, 2008), BLV proviral load (Miyasaka *et al.*, 2013), mastitis (Dietz *et al.*, 1997a; Sharif *et al.*, 1998; Takeshima *et al.*, 2008; Yoshida *et al.*, 2009a,b; Baltian *et al.*, 2012) and dermatophilosis (Maillard *et al.*, 2002). In addition, *BoLA-DRB3* polymorphisms are associated with differences in susceptibility to immunological conditions using 20 indicator traits of innate and adaptive immunity (Dietz *et al.*, 1997b), posterior spinal paresis, ketosis (Mejdell *et al.*, 1994) and retained placenta (Joosten *et al.*, 1991; Park *et al.*, 1993; Mejdell *et al.*, 1994). Such polymorphisms also affect responses to foot and mouth disease virus (FMDV) (Baxter *et al.*, 2009) and *Theileria parva* (Ballingall *et al.*, 2004b) vaccines.

The *DQ* genes within the mouse, rat, pig and rabbit MHC class II regions are single copy genes. Multiple *DQ* genes have been identified in humans and dogs, but only one DQ molecule is actually expressed (Kappes and Strominger, 1988; Ando *et al.*, 1989). Some individual cows harbour a single copy of the *BoLA-DQA* and *BoLA-DQB* genes, whereas others harbour duplicate copies; in the latter case, both BoLA-DQ molecules appear to be expressed (Table 8.2). Sequence comparisons, Southern blotting and phylogenetic analyses indicate that there are at least five different *BoLA-DQA* genes (Sigurdardottir *et al.*, 1991; Morooka *et al.*, 1995; Nishino *et al.*, 1995; Ballingall *et al.*, 1997, 1998; Gelhaus *et al.*, 1999b; Miyasaka *et al.*, 2013). In contrast to the *BoLA-DQA4* and *-DQA5* genes, the *-DQA1*, *-DQA2* and *-DQA3* genes are highly polymorphic. Likewise, five different *BoLA-DQB* genes have been identified by sequences analysis (Sigurdardottir *et al.*, 1992; Xu *et al.*, 1994; Dikiniene and Aida, 1995; Marello *et al.*, 1995; Gelhaus *et al.*, 1999a). The *BoLA-DQB1* gene is the most common, whereas the *BoLA-DQB2*, *-DQB3*, *-DQB4* and *-DQB5* genes are only found in duplicated haplotypes. Polymorphisms in *BoLA-DQA* and *BoLA-DQB* have been studied in Holstein, Jersey, Japanese Shorthorn, Japanese Black and African cattle (Ballingall *et al.*, 1997; Maillard *et al.*, 2001; Wang *et al.*, 2005, 2007; Takeshima *et al.*, 2007, 2008; Miyasaka *et al.*, 2011, 2012). To date, 54 *BoLA-DQA* and 78 *BoLA-DQB* alleles have been registered in the IPD-MHC database.

The particular set of alleles present on a chromosome is referred to as the 'MHC haplotype'. Almost all of the reported BoLA class II DR-DQ-linked haplotypes have focused on Holstein cattle and have used a relatively small sample population (Lewin *et al.*, 1999; Glass *et al.*, 2000; Russell *et al.*, 2000; Park *et al.*, 2004; Norimine and Brown, 2005; Staska *et al.*, 2005). A recent study genotyped populations

of Japanese Black and Holstein cattle and analysed the BoLA class II haplotypes, the *BoLA-DRB3* locus, five *BoLA-DQA* loci and five *BoLA-DQB* loci. The newly designated *BoLA-DRB3-DQA-DQB* haplotypes are shown in Table 8.2, together with previously published haplotype data.

Recent evidence suggests that there are more than 56 *BoLA-DRB3-DQA-DQB* haplotypes (Miyasaka *et al.*, 2012). Furthermore, 39 *DRB3-DQA1* haplotypes were identified, including 29 in Japanese Black and 22 in Holstein cattle. The majority of these haplotypes were identified in both breeds, although several were identified in only a single breed. Interestingly, two *DRB3-DQA1* haplotypes, namely *0902B* or *C* (*DRB3*0902-DQA1*0204*) and *1101A* (*DRB3*1101-DQA1*10011*), were associated with a low BLV proviral load of BLV-infected cattle, whereas one haplotype *1601B* (*DRB3*1601-DQA1*10012*) was associated with a high BLV proviral load. This is the first report to identify an association between the *DRB3-DQA1* haplotype and differences in BLV proviral load (Miyasaka *et al.*, 2013).

Thus, unlike HLA, many BoLA haplotypes have duplicated *DQ* genes and only one functional *DRB3* gene, suggesting that BoLA-DR molecules alone cannot present a sufficiently broad spectrum of antigens and that BoLA-DQ molecules are equally important for priming CD4[+] T cells. Several studies support this hypothesis: when cells are transiently transfected with cDNAs corresponding to MHC class II A and B genes, they express BoLA-DQ molecules that function as effectively as BoLA-DR molecules (Aida, 1995; Aida *et al.*, 1994, 1995; Dikiniene and Aida, 1995; Morooka *et al.*, 1995; Nishino *et al.*, 1995). The surface expression of BoLA-DQ molecules has been demonstrated using locus-specific monoclonal antibodies and isoelectric focusing (IEF) (Davies *et al.*, 1992; Bissumbhar *et al.*, 1994; Escayg *et al.*, 1996). Furthermore, Glass *et al.* (2000) used monoclonal antibody-blocking assays to show that BoLA-DQ molecules presented FMDV-peptides to CD4[+] T cells. Moreover, Norimine and Brown (2005) showed that functional BoLA-DQ molecules are generated by both intra-haplotype and inter-haplotype pairing of A and B chains, and play a similar role to that of BoLA-DR during priming.

Bovine class IIb region

The BoLA class IIb locus is divided into two regions. These are the 'extended class II region', which contains non-MHC genes, and the 'classical class II region', which contains genes of unknown function (e.g. *DSB, DYA, DYB*), class II genes involved in antigen presentation (*DMA, DMB, DOB, DOA*) and class I genes such as *TAP1, TAP2, TAP2.1, PSMB8* and *PSMB9*.

Unlike BoLA class IIa region genes, *BoLA-DY* genes are found only in ruminants, show a low level of polymorphism and are transcribed only in dendritic cells (Ballingall *et al.*, 2001). An analysis of the first full-length *BoLA-DYA* and *BoLA-DYB* transcripts identified in cattle shows that they contained open reading frames with coding potential for proteins of 253 and 259 amino acids, respectively (Ballingall *et al.*, 2004a). Expression analysis of tagged constructs showed for the first time that the *BoLA-DY* genes of cattle can encode distinct class II MHC α and β polypeptide chains. In humans, the *DMA* and *DMB* genes encode a molecule that plays a role in assembling the complexes of peptides with class II molecules, whereas *DOA* and *DOB* encode a protein that may regulate the function of the DM molecule. In cattle, the *BoLA-DM* (Niimi *et al.*, 1995) and *BoLA-DO* genes (Takeshima and Aida, unpublished data) have been sequenced from cDNA using primers derived from a human sequence. These cDNAs are closely related to human genes, an observation that supports the hypothesis that the corresponding genes might be expressed and functional.

T Cell Antigen Receptors

To provide protection from pathogens, T cells recognize a large set of antigens in a very specific way through their diverse set of TCRs. Each lymphocyte has only one of the myriad of possible TCRs expressed, but it is estimated that well over a million different possible receptors can be constructed by the mechanisms described below. Ligation of the TCR during the induction of an immune response leads to clonal expansion of that particular lymphocyte, thereby generating large numbers of cells with identical receptors specific for the particular antigen. Since this process of receptor generation through somatic gene rearrangements is somewhat random and the product so diverse it can result in receptors that react with non-harmful antigens including self antigens as well as receptors that react with foreign antigens derived from infectious agents.

The TCR is coded for by a group of germline genes (previously referred to as 'gene segments' because all parts are needed to create a functional transcript) that are 'rearranged' in a variety of possible combinations during lymphocyte maturation to give rise to the two polypeptide chains needed to make a TCR (α and β chains or γ and δ chains). Those genes are the so-called variable (*V*), diversity (*D*), joining (*J*) and constant (*C*) genes with one to several hundred occurring in each group. The TCRα and TCRγ chains are formed from rearrangement of *V-J-C* genes, while the TCRδ and TCRβ chains are coded for by *V-D-J-C* genes making the latter chains potentially more variable and complex. A gene is chosen from each group in a variety of combinations such that lymphocytes have the ability to recognize a nearly unlimited array of antigens (Davis and Bjorkman, 1988; Rock *et al.*, 1994) especially when the additional mechanisms that contribute to diversity beyond the *V-(D)-J-C* combinations are considered. These include imprecise joining of genes, and deletion and addition of nucleotides during recombination of *V-(D)-J* genes (Schatz, 2004).

Sites of greatest sequence variability within the TCR genes are localized to three distinct regions designated complementarity-determining regions (CDR)1, CDR2 and CDR3. The CDR1s and CDR2s are coded for by the *V* genes, while CDR3s are the site with the highest level of variability; CDR3s are formed by the combinations of *V-(D)-J* genes with imprecise joining of genes during recombination, and untemplated nucleotide additions and deletions in this area (Schatz, 2004). The transcript for a single TCRδ chain of cattle also can incorporate multiple *D* genes, each of which can be edited by nucleotide deletions and untemplated nucleotide additions (Herzig *et al.*, 2010a).

T cells are defined as αβ vs γδ based on their TCR gene usage

With regard to T cells, there are two main populations in cattle and other mammals defined by their TCR gene usage, known as αβ T cells and γδ T cells. Although both types of TCRs perceive antigens, they differ in the types of antigens with which they react: T cells using α and β genes react with antigenic peptides in the context of self major histocompatibility complex (MHC) molecules (discussed above in this chapter), while those expressing the γ and δ TCR genes have been shown to react with self molecules on cells including non-classical MHC molecules and molecules on stressed macrophages (Havran *et al.*, 1991; Chien *et al.*, 1996; Okragly *et al.*, 1996; Groh *et al.*, 1998; Egan and Carding, 2000; Sathiyaseelan *et al.*, 2002) as well as non-proteinaceous molecules, none of which involve MHC presentation (Morita *et al.*, 1994). Thus for those cells expressing α and β TCR genes, their CDR1 and CDR2 primarily interact with MHC molecules presenting the antigenic peptide, although the CDR1 also interacts with the peptide. The CDR3 interacts with the antigenic peptide only. For those expressing the γ and δ TCR genes there is a different interaction of the TCR with MHC molecules and thus the role of the CDRs is also different. In one instance where an MHC-like molecule (known as T22/T10) acts as an antigen for γδ TCR in mice (Chien *et al.*, 1996) it is the CDR3 of the TCR δ chain that is interacting. It is important to stress that the T22 is acting as an antigen in this case and not an MHC molecule presenting peptide.

Cattle are described as a γδ T cell high species, meaning at birth their γδ T cells may comprise up to 60% of the mononuclear cell population in the blood in contrast with healthy humans where regardless of the age the γδ T cells generally do not exceed 20%. In both mice and humans there is canonical pairing of TCRγ and TCRδ chains incorporating a particular Vγ gene with that incorporating the product of a particular Vδ gene and the γδ T cells localize to particular tissues based on the TCR V-gene segments expressed (Carding and Egan, 2002; Hayday, 2000). However, in cattle, this pairing or tissue local-

ization does not seem to occur according to the particular TCR genes used (Van Rhijn *et al.*, 2007).

Nomenclature for the TCR

Traditionally and still in widespread use, the designation of TCR genes was as written above, with the gene type (e.g. V for variable) followed by the Greek letter of the TCR chain name (e.g. γ) followed by the individual gene's number (e.g. 5): this therefore is Vγ5. However ImmunoGenetics (IMGT; www.imgt.org) has suggested a nomenclature that has been widely accepted including for cattle immunogenetics. Using IMGT naming, the Vγ5 gene would be *TRGV5* or T cell receptor gamma variable gene number 5. In subgroups where more than one gene exists according to the IMGT rules they are indicated as a hyphen followed by a number. For example duplicated genes within TRGV5 are further designated as *TRGV5-1* and *TRGV5-2* because the subgroup comprises two mapped genes. This does not correspond with the general rules as discussed in Chapter 24 but is accepted by the greater immunogenetics community.

TCR gamma chain immunogenetics

In ruminants, the TRG genes occur at two distinct loci, named T cell receptor gamma @ locus 1 and locus 2 (e.g. *TRG@1*) (Miccoli *et al.*, 2003) (Fig. 8.3). *TRG1@* is homologous to the single locus where TRG genes are found on human chromosomes. This contrasts with *TRB* genes, which are at a single locus, and the *TRA* and *TRD* genes, which are at a combined single locus. Between the two loci, there are a total of 11 bovine TRGV genes, which fall into eight subgroups, and six TRGC genes (Fig. 8.3). Homologous TRG genes in cattle and sheep were assigned, using four accepted criteria (Herzig *et al.*, 2006), although the names are not necessarily identical, since in the IMGT system new genes are sequentially assigned a number as they are discovered (Lefranc *et al.*, 2003). For example, bovine

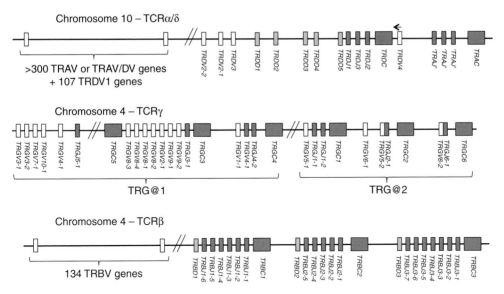

Fig. 8.3. General proposed organization of the bovine TCR loci. Based on the literature this is a schematic presentation of the genes that code for the four bovine TCR chains. This is not to scale and the exact number of Variable (*V*) genes is not certain for the TCRα/δ locus or the number of '*TRAJ*' genes and thus they appear in quotes. Open rectangles are for V genes, lightly shaded rectangles for D genes, deeply shaded small rectangles are for J genes and large deeply shaded rectangles are for C genes. The arrow for *TRDV4* indicates it is in reverse orientation.

genes described by Herzig *et al.* included the bovine so-named *TRGC6*, *TRGV2* and *TRGV4*, which are homologues of the earlier described ovine *TRGC4*, *TRGV2* and *TRGV4*, respectively. Also the newer IMGT nomenclature may supersede the previous names. For example the bovine Vγ7 and BTGV1 clones (previously *TRGV4* and *TRGV2*, respectively) were reassigned to new subgroups *TRGV7* and *TRGV8*, respectively.

The Bovine Genome Sequencing and Annotation Consortium (Elsik *et al.*, 2009) found all previously reported *TRG* genes (Herzig *et al.*, 2006; Conrad *et al.*, 2007) had been sequenced and assembled correctly except for *TRGJ5-1* and *TRGJ6-1*, which are missing in the assembly. It also confirmed previous information (Conrad *et al.*, 2007) that three cassettes were found at each locus with each cassette composed of one to seven TRGV genes, one or two TRGJ and a single *TRGC*. It has been shown experimentally that somatic recombination is largely restricted to within a cassette (Herzig *et al.*, 2006). The expression

of TRG genes differs for γδ T cells according to whether they do or do not express the WC1 co-receptor (described below). Those bovine γδ T cells that express the WC1 co-receptor only express genes in the cassette containing *TRGC5*. In contrast WC1-non-expressing γδ T cells in peripheral blood may use TRG genes found in any of the six cassettes (Blumerman *et al.*, 2006).

TCR delta chain immunogenetics

For mammals the *TRD* genes are embedded within the *TRA* locus as shown for bovine chromosome 10 in Fig. 8.3. Although this region has not been fully assembled for the bovine chromosome, in humans and mice the combined *TRA/TRD* locus is over 1 Mb (Glusman *et al.*, 2001). For those species, the *TRD* locus comprises a cluster of TRDV genes, followed by TRDD genes, TRDJ genes, the single *TRDC* and an additional *TRDV* gene that is located 3' of the *TRDC* in an inverted transcriptional orientation (Lefranc, 2001). In addition, five and ten bi-functional TRAV/DV genes (that are capable of recombining with either *TRDD* or *TRAJ* and presumably giving rise to a mature TCRδ or TCRα chain, respectively) have been identified for human and mice, respectively. These V genes are upstream of the *TRD* locus (Lefranc, 2001). The situation is similar for cattle: the *TRD* genes are found within the *TRA* locus (Fig. 8.3); however, cattle are unique in that there is a profound expansion of the number of genes in the *TRDV1* subfamily (Takeuchi *et al.*, 1992; Ishiguro *et al.*, 1993; Van Rhijn *et al.*, 2007; Herzig *et al.*, 2010a).

The genomic sequence for previously reported TRD genes (Herzig *et al.*, 2010a) was identified during the Bovine Genome Sequencing and Annotation project (Elsik *et al.*, 2009), but due to insufficient scaffolding it was not possible to determine their genomic organization completely. Exceptions to this include the five TRDD genes, three TRDJ genes and the single *TRDC* gene, which have been mapped and are all located on chromosome 10 (Fig. 8.3) (Herzig *et al.*, 2010a), and, in addition, the single *TRDV4* was found to be located downstream of the *TRDC* but upstream of *TRAC* in an inverted orientation, as is the case with orthologous human and mouse genes. Two *TRDV2* genes and one *TRDV3* were identified but their placement is unknown. Finally, 52 TRDV1 genes were identified (Herzig *et al.*, 2010a) and thought to be functional based on their structure, but their organization within the genome could not be determined. This large number of genes within the TRDV1 family had been predicted previously based on cDNA evidence (see Accession numbers of cDNA sequences in Table 4 of Herzig *et al.*, 2010a), and they were broken into 11 subgroups using phylogenetic trees (Herzig *et al.*, 2010a). Others have identified 107 TRDV1 genes by annotating a later assembly (Van Rhijn *et al.*, 2007; Herzig *et al.*, 2010a). Bovine TRDV1 genes are co-mingled with the TRAV genes and there is apparent dual usage of some V genes since some V gene sequences are found rearranged with either *TRAC* and *TRDC* with the intervening *TRAJ* or *TRDJ* plus *TRDD* as appropriate (Van Rhijn *et al.*, 2007; Herzig *et al.*, 2010a).

The CDRs of the TRD genes are of interest on several fronts. First there is cDNA evidence indicating that between one and five TRDD sequences can be incorporated into a single transcript (Herzig *et al.*, 2010a). This can result in a very long CDR3 region that upon translation would have 8 to 20 amino acids. By contrast it is interesting that this does not occur for TRB for cattle or other mammalian species despite the fact that TRB also incorporates *D* genes. The CDR2-coding region of bovine TRDV genes is also of interest. The CDR2 are the loops of the αβ TCR proteins that bind to the MHC portion of the MHC/antigen peptide complex. However bovine *TRDV1* CDR2s have sequence that would code for only three amino acids or which is absent completely in some genes (e.g. *TRDV1f*, *TRDV1ae*, *TRDV1ar* and *TRDV1o*) (Herzig *et al.*, 2010a) contrasting with *TRBV* CDR2 regions, which have coding sequence for five to seven amino acids in mammals. Since γδ T cells are not MHC-restricted (i.e. their TCR does not see antigenic peptides presented on MHC molecules), it may be logical that the CDR2 lengths are gone or abbreviated. Transcripts that appear complete but lack coding sequence for CDR2 have been found so it is presumed that these genes are not pseudogenes (Reinink and Van Rhijn, 2009). Finally, the CDR1 loops of the TCR α and β chains interact with the antigenic peptide as well as the MHC presenting; it was found that the *TRDVs'* CDR1s in cattle exhibited diverse lengths coding for between five and ten amino acids consistent with what occurs for mouse and human *TRDV*.

TCR alpha chain immunogenetics

As indicated above, the TRA locus is complicated by the insertion of the TRD genes within it, and further by the fact that some *V* gene sequences are incorporated into transcripts coding for either TCRδ or TCRα chains, thereby being designated TRAV/DV genes. The TRA locus is found on bovine chromosome 10 (Fig. 8.3) and spans a 2.4 mb region (Van Rhijn *et al.*, 2007). The ability of a V gene to be bi-functional, serving as either *TRAV* or *TRDV*, is not dictated by its position among other V genes. That is, they are not closest or adjacent to V genes that solely serve as *TRDV*. Because the assembly of the bovine genome in this region is still fragmented, it has not been possible to say definitively what the total number of TRAV genes is. However there are over 300 TRAV or TRAV/DV genes reported (Ishiguro *et al.*, 1993; Van Rhijn *et al.*, 2007). This total of more than 300 genes is substantially greater than the 53 found in humans and 104 in mice, not all of which are functional. However it parallels the expansion of the TRDV1 gene family in cattle. While the number of TRAJ genes is still unknown, a single *TRAC* has been described, and both the TRAJ genes and *TRAC* are found downstream of *TRCV4* (Herzig *et al.*, 2010a) as in humans and mice.

TCR beta chain immunogenetics

For the bovine TRB locus, Connelley and colleagues (2008a, 2008b, 2009) identified 134 TRBV genes from the third assembly of the Bovine Genome Annotation and Sequencing on chromosome 4 (Connelley *et al.*, 2009). This number of V genes is more than twice the number for humans and four times that of mice despite the fact that the bovine genome assembly is incomplete and thus even more V genes may be identified in the future. It was shown that the TRBD, TRBJ and TRBC genes of the locus are interestingly organized into three cassettes (Fig. 8.3) each containing *D-J-C* (Conrad *et al.*, 2002; Connelley *et al.*, 2009). One cassette is an apparent duplication. This is somewhat akin to the bovine TRG gene organization at the two loci but differs in that the extensive number of TRBV genes are outside the D-J-C cassettes, while for the TRG loci the TRGV genes are together with the TRGJ and TRGC genes belonging to that particular cassette. Each TRB cassette has a single *TRBD* gene and a single *TRBC* gene, and either five or seven TRBJ genes. The order of the genes within the cassettes and the order of the cassettes are conserved with that of humans and mice (Connelley *et al.*, 2009).

Of those 134 V genes identified to date, 79 are predicted to be functional. They were subdivided into 24 subgroups based on sequence characteristics (Connelley *et al.*, 2009). The large number of V genes can be attributed to expansion in several TRBV subgroups. So again, this expansion of bovine TRBV genes mirrors the expansion of bovine TRDV and bovine TRAV genes described above.

Multigene Families of Germline Encoded Receptors

Pattern recognition receptors (PRR) and natural killer receptors (both described below) differ from the TCR and BCR since they are not requiring somatic recombination. Thus, they do not have the same level of diversity as the TCR and BCR. Nevertheless, they empower the expressing cell with the ability to immediately recognize and respond to pathogens or stressed or altered/infected host cells. Because PRRs have restricted diversity they react with conserved structures (i.e. 'patterns') of foreign infectious agents known as pathogen-associated molecular patterns (PAMPs) including those of bacteria, viruses, fungi, protozoa and helminths. This does not mean the particular structure (e.g. lipopolysaccharide (LPS)) is identical among all infectious agents but rather that the rules of engagement of PRRs are less stringent than that of TCR and BCR and thus variations in the basic ligand structure can be accommodated by the receptor. Moreover, a single PRR or NK receptor may have a large number of different ligands among PAMPs and/or danger-associated molecular patterns (DAMPs) as do some of the toll-like receptors (TLR) and killer receptors within the NKG2 family (Champsaur and Lanier, 2010), respectively. Further adding to the complexity is the 'dual receptor' paradigm. This indicates that the identical extracellular structures of a receptor may be paired with a choice of inhibitory or activating intracytoplasmic tails, thus conveying opposing signals following ligand engagement (Fourmentraux-Neves *et al.*, 2008).

The role of these receptors for activation of different subpopulations of lymphoid cells (NK cells vs. T cells) and myeloid cells is complex, since depending upon the cell type on which it is expressed, the same receptor may be either the primary regulator of cellular response, or have a supporting role as a co-receptor to augment responses mediated through the TCR (Snyder *et al.*, 2004). The current working paradigm is that they send fully activating signals when engaging their ligand on innate immune system cells such as NK cells, but act as co-receptors on T cells. This difference in outcome occurs because of differences in the intracellular signalling molecules with which the receptors associate in the different cell types (Snyder *et al.*, 2004). However, it may be even more complex, since for γδ T cells there is evidence that a combination of PRRs and TCRs gives the most profound activation even though either alone may give some lower level of activation (Bonneville *et al.*, 2010). This contrasts with adaptive immune system αβ T cells, in which the role of PRR is clearly subservient to that of TCR and is not thought to be independent (Snyder *et al.*, 2004), thereby relegating it to the role of co-receptor.

Pattern recognition receptors

Toll-like receptors

Probably the most well-known of the PRRs are those of the toll-like receptor (TLR) gene family, first identified in flies but since shown to occur in mammals; thus the TLR family members are highly conserved PRRs. TLR genes are most often expressed by innate immune system cells of the myeloid lineage, such as macrophages and dendritic cells. Crystal structures of the TLR gene products indicate that the extracellular leucine rich repeat region (LRR) is a horseshoe shaped structure (Bell *et al.*, 2003, 2006), which encompasses many ligand binding sites. It may occur as a homodimer or a heterodimer, which increases its ligand diversity. TLR engagement by the ligand stimulates cytokine and chemokine production as well as promotes phagocytosis. Some TLR genes are also expressed by lymphoid lineage cells including NK cells, B cells and γδ T cells.

While the total number of TLR genes identified in some species is 13, only ten TLR genes have been found for cattle as in human, pigs and mice (Werling and Coffey, 2007). Moreover, the ten TLR genes in cattle are homologues of the ten found in humans and

have 83–90% similarity at the nucleotide level between corresponding cattle and human genes (Menzies and Ingham, 2006). The evolutionary trees of Dubey *et al.* (2013) revealed the clustering of major TLR gene subfamilies from several species including cattle, humans and mice as follows: *TLR7*, *TLR8* and *TLR9* of different species clustered together under a single clade, while the other gene subfamily included *TLR1*, *TLR2*, *TLR6* and *TLR10*. *TLR3*, *TLR4* and *TLR5* of different species were found to be clustered individually as separate clades. There have been reports that polymorphisms in TLR genes exist in cattle and that the polymorphisms contribute to disease susceptibility and resistance for *Mycobacterium avium paratuberculosis* (Fisher *et al.*, 2011) and *M. bovis* (Sun *et al.*, 2012a). Some TLR gene products are on the outer membrane of cells (*TLR1*, *TRL2*, *TLR4*, *TLR5*, *TLR6* and *TLR10*), where they can interact with bacterial and fungal components, while others are in the intracellular endosomal compartment membranes (*TLR3*, *TRL7*, *TRL8* and *TRL9*), where they tend to react with viral components.

NOD-like receptors

Nucleotide-binding oligomerization domain, leucine rich repeat and pyrin domain containing proteins (NLRP) are members of the NOD-like receptors. Like the TLRs they are important in sensing microbes but they function intracellularly only. Genes coding for a total of eight NOD-like receptors are found on bovine chromosomes 7, 15, 18 and 19 (Table 8.3) (Tian *et al.*, 2009). Their placement on the chromosomes indicates that the duplication of this large gene family occurred prior to the divergence of mammals.

WC1 co-receptors

The WC1 gene family codes for a PRR family of co-receptors that are predominant in ruminants and pigs and whose products are expressed uniquely by γδ T cells. In sheep, WC1 is also known as T19 (Mackay *et al.*, 1986). WC1 stands for 'Workshop Cluster 1', which is actually a placeholder name (Morrison and Davis, 1991) from international workshops comparing monoclonal antibodies and making 'clusters of differentiation', or more commonly CD designations. WC1 molecules act as co-receptors to the TCR, in that their co-ligation with the TCR potentiates T cell activation (Hanby-Flarida et al., 1996) dependent on the phosphorylation of a tyrosine in the intracellular tail sequence (Hanby-Flarida et al., 1996; Wang et al., 2009). SRCR superfamily members closely related to WC1 bind to yeast and bacteria and thus can function as PRRs (Bikker et al., 2004; Sarrias et al., 2005, 2007; Matthews et al., 2006; Ligtenberg et al., 2007; Fabriek et al., 2009; Vera et al., 2009). Similar proof is available for the WC1 molecules that are unique to ruminants and pigs (Wang et al., 2011). Also the response of bovine γδ T cells to specific bacteria varies according to the WC1 genes expressed by the lymphocytes (Rogers et al., 2005; Lahmers et al., 2006).

Based on results from the Bovine Sequencing and Annotation project (Herzig and Baldwin, 2009) and subsequent analyses (Chen et al., 2012), 13 WC1 genes are found at two loci on bovine chromosome 5. When quantitative PCR was used to determine gene number in a number of breeds of cattle it was established as 13, and little or no polymorphism among WC1 genes from various animals was found (Chen et al., 2012). The genes are WC1-1 through WC1-13, and all but WC1-11 have coding sequence for 11 extracellular SRCR domains. WC1-11 is most similar to that previously described for swine WC1 (Kanan et al., 1997). WC1-1, WC1-2, WC1-3, WC1-4, WC1-5, WC1-6, WC1-7, WC1-8 and WC1-13 all have four exons coding for their intracytoplasmic tails while five exons exist for the coding sequence of the intracytoplasmic tails of WC1-9, WC1-10 and WC1-12 and six exons for WC1-11. They have multiple scavenger receptor cysteine-rich (SRCR) domains (Sarrias et al., 2004; Herzig et al., 2010b) and because they share the gene sequences that code for domains with the pattern b, c, d, e, d (Sarrias et al., 2004) with other CD163 family members (Herzig et al., 2010b), WC1s should be re-named 'CD163d'. Cattle also have CD163A, which is located in the middle of the two WC1 loci (Herzig and Baldwin, 2009), as well as CD163c-α (Herzig et al., 2010b) as do humans and mice, but neither humans nor mice have an exact equivalent of WC1 (Herzig et al., 2010b). CD designation is dependent upon human homologues and thus they remain as 'WC1'.

Killer receptors

As described above, many receptor gene families play a pivotal role in innate immune responses, particularly those innate immune responses mediated by NK cells. However, members of these gene families may be expressed by γδ T cells and memory αβ T cells (Parham, 2004). The majority of receptors classically associated with NK cells are in the killer cell lectin-like receptor (KLR) group. The KLR include multi-gene families of Ly49 (aka KLRA) and natural killer group 2 (NKG2, aka KLRC & K) that form heterodimers with CD94 (aka KLRD by its gene symbol), as well as other KLR that have only one or two genes including some found in cattle (Table 8.3; Dissen et al., 2008). NK receptors are also coded for by the killer cell Ig-like receptor (KIR) genes whose products are functionally analogous to Ly49 gene products even though they are structurally unrelated and located on different chromosomes. That is, Ly49 molecules are homodimers with type II C-type lectin-like domains, while KIRs are members of the superimmunoglobulin gene family. They are summarized in Table 8.3.

Receptor engagement may result in cell inhibition or activation if the intracytoplasmic tail sequence has an immunoreceptor tyrosine-based inhibitory motif (ITIM) sequence or codes for a protein that associates with the intracellular adaptor protein DAP-12, respectively. In some cases the gene sequence coding for the extracellular portion of the receptor may be paired with either an inhibitory or activating intracytoplasmic tail sequence resulting in the dual receptor paradigm (Fourmentraux-Neves et al., 2008). The ligands for these receptors are classical MHC class I molecules for Ly49 and KIR, non-classical MHC class I molecules for NKG2 receptors and MHC class I-like molecules for NKG2D described above (Tables 8.1 and 8.3).

KIR

Cattle are unique outside of primates as they have an expanded KIR gene family along with

Table 8.3. Bovine pattern recognition receptors and killer receptors.[a]

Receptor family name	Gene symbols	Receptor's protein name	No. of receptor genes in cattle	Ligands for receptors shown in other species
Killer cell immunoglobulin-like receptor (KIR)	*BotaKIR2DL1, BotaKIR3DS1, BotaKIR3DL1, BotaKIR3DL1-like, BotaKIR2DS1*	KIR	≥5–8	MHC class I
Killer cell lectin-like receptors (KLR)A	*btLy49*02, btLy49*03, btLy49*01*	Ly49	1[b]	MHC class I
KLRC	*NKG2A-01* to *NKG2A-07*	NKG2A	7	Non-classical MHC class I
KLRK	*NKG2D*	NKG2D	possibly 7	MHC class I-like molecules (e.g. ULBP) and non-classical class I MHC (e.g. MIC)
KLRC	*NKG2C*	NKG2C	1+1 pseudogene	Non-classical MHC class I
KLRD	*CD94*	CD94	2	Pairs with NKG2s, so no ligand alone
KLRB	*NKR-P1*	NKR-P1	1	?
KLRF	*NKp80*	NKp80	1	?
KLRG	*MAFA*	MAFA	1	?
KLRH	*KLRH1*	KLRH1	1	?
KLRI	*KLRI1, KLRI2*	KLRI1, KLRI2	2	?
KLRJ	*KLRJ1*	KLRJ1	1	?
KLRE	*KLRE1*	KLRE1	1	?
Toll-like receptors	*TLR1, TLR2, TLR3, TLR4, TLR5, TLR6, TLR7, TLR8, TLR9, TLR10*	TLR	10	Microbial components
NOD-like receptors family	*NLRP1, NLRP3, NLRP4, NLRP5, NLRP6, NLRP8, NLRP9, NLRP13*	Nucleotide-binding oligomerization domain leucine rich repeat and pyrin domain containing proteins (NLRP)	8	Microbial components
CD163	*WC1-1, WC1-2, WC1-3, WC1-4, WC1-5, WC1-6, WC1-7, WC1-8, WC1-9, WC1-10, WC1-11, WC1-12, WC1-13*	WC1	13	Ligands include microbes

[a]The number of genes in each family is indicated for cattle and humans. Table after that in Dissen *et al.* (2008), but with additional information as cited in the text.
[b]This gene is polymorphic.

a single functional Ly49 gene that is polymorphic (btLy49*01, btLy49*02, btLy49*03) (McQueen et al., 2002; Storset et al., 2003; Dobromylskyj and Ellis, 2007). The bovine KIR gene family has two lineages, with the lineage related to the primate KIR3DX1 lineage having a large number of genes (BotaKIR3DS1, BotaKIR3DL1, BotaKIR3DL1-like, Bota-KIR2DS1), although primates have only the single gene KIR3DX1. In addition, cattle have only one complete gene from KIR3DL-lineage (BotaKIR2DL1), while in primates it is a variable and expanded group (Guethlein et al., 2007). Like humans, cattle apparently have both activating and inhibitory KIR, the latter characterized by the intracytoplasmic tail having an ITIM motif. The second build of the bovine genome revealed seven putative KIR genes (McQueen et al., 2002; Storset et al., 2003; Dobromylskyj and Ellis, 2007). The sequences of these genes differed from the four previously described by Storset et al. (McQueen et al., 2002; Storset et al., 2003; Dobromylskyj and Ellis, 2007) which are BtKIR2DL1, BtKIR3DL1, BtKIR2DS1 and BtKIR3DS1. However, it should be noted that these genomic regions are notoriously difficult to assemble due to stretches of high sequence identity and long repeat elements.

Ly49

The single bovine Ly49 gene btLy49 was originally thought to be monomorphic, but recently sequences representing polymorphisms have been found in cattle and designated btLy49*01, btLy40*02 and btLy49*03 (Table 8.3). The gene products are predicted to differ from one another by up to 16 amino acids (Dobromylskyj et al., 2009). In addition, splice variants of this gene were found when transcripts were analysed.

NKG2

Cattle have a multigenic NKG2 family with more than seven members including multiple NKG2A genes (NKG2A-01 to NKG2A-07), which code for two tyrosine-based inhibitory motifs in the intracytoplasmic domain, and at least two NKG2C genes whose homologue in humans is activating (Birch and Ellis, 2007;

bovine genome build Btau 4.6.1). The NKG2A genes are similar to that for mouse 'short NKG2D' (Fikri et al., 2007). The bovine CD94 genes, of which there are two and which are polymorphic, whose products associate with NKG2 chains to form a heterodimer, have also been cloned (Storset et al., 2003; Birch and Ellis, 2007) along with the genes coding for both bovine adaptor proteins involved in NKG2D signalling, which are DAP10 and DAP12 (Fikri et al., 2007).

It has been shown in humans that of this family, only the NKG2D gene product interacts with non-classical MHC class I MIC gene products and MHC class I-like ULBP gene products, while the other NKG2 receptors react with different non-classical MHC class I molecules. In addition to the 4 non-classical MHC class I genes defined for cattle and whose products may interact with NKG2 family members as in humans, both MIC (Birch et al., 2008b; Guzman et al., 2010a) and ULBP (Larson et al., 2006) genes also have been defined in cattle (Table 8.3). For MIC genes there are definitively three in the bovine genome, although it is possible that four exist (Birch et al., 2008). They have been mapped to bovine chromosome 23 and provisionally named BoLA MIC1, BoLA MIC2 and BoLA MIC3. There are four bovine ULBP genes in the minor cluster and 26 in the major (named ULBP1 through ULBP30) (Larson et al., 2006).

Immunoglobulins

Immunoglobulins are composed of two identical heavy (H) and two identical light (L) polypeptide chains in cattle. The heavy chains are known as μ, δ, γ, ϵ and α, while the light chains are known as κ or λ, so-named for the genes that code for a portion of the chains referred to as the constant domains (IGHC for the heavy chain; IGKC and IGLC for the κ and λ light chain, respectively). The standard IMGT nomenclature for immunoglobulin heavy and light chain genes has been used and explained and takes into consideration the historical gene designations widely cited in the literature. The designation IGHD as per IMGT nomenclature must be viewed in the proper context as it might refer to a heavy chain (*Cδ*) or diversity (*D*) mini-gene involved in encoding the variable-region of the heavy chain. Functionally, immunoglobulins are known as antibodies when secreted by B cells or as the BCR when bound to the membranes of B cells. Antibodies are the main effector molecules produced by B cells, while the BCR allows the cell to interact with antigens thereby becoming activated. The immunoglobulin chains have terms for specific parts of the molecule: the part responsible for interacting with antigens is known as the 'variable domain' and occurs in both the heavy and light chains. The other parts of these chains are the constant domains and some of those in the heavy chains convey the functional differences among antibodies. The part of the antibody composed of heavy chain constant domains that convey function is known as the fragment-crystallizable or Fc piece. The variety of functions mediated by it include the ability of the antibody to interact with specific receptors on other cells (known as Fc receptors) or to activate an enzyme system in blood and interstitial fluids known as the complement system. Thus, immunoglobulins are divided into various classes (previously termed isotypes) according to their heavy constant regions as follows: IgM, IgD, IgG, IgA and IgE. For example, IgM means it is an immunoglobulin with a μ-heavy chain encoded by the *IGHM* gene.

Immunoglobulins are coded for by a set of germline genes (previously referred to as 'gene segments' or exons because all parts are needed to create a functional transcript) that are 'rearranged' in a variety of possible combinations during lymphocyte development to give rise to the two polypeptide chains (known as heavy and light chains). Those genes are the so-called variable (*V or IGHV*), diversity (*D or IGHD*) and joining (*J or IGHJ*) genes, with one to several hundred occurring in each set. The heavy chains are coded for by *IGHV-IGHD-IGHJ-IGHC* genes, where the variable domain, encoded by *V-D-J* gene recombinations, is potentially more variable and complex (Tonegawa, 1983; Kaushik and Lim, 1996; Jones and Simkus, 2009). The κ light chains are formed from rearrangement of *IGKV-IGKJ* and the constant (*C or IGKC*) gene, while λ light chains

are formed from rearrangement of *IGLV-IGLJ* and the constant (*C or IGLC*) gene. A gene is chosen from each group in a variety of combinations such that lymphocytes have the ability to recognize a nearly unlimited array of antigens especially when the additional mechanisms that contribute to diversity beyond the *V-(D)-J-C* recombinations are considered (Schatz, 2004).

When the BCR engages the appropriate antigen, the B cell is activated and undergoes two genetic processes known as somatic hypermutation, which affects the variable domains of both chains, and class switch recombination, which affects the constant domains of the heavy chain. These processes are mediated by activation-induced deaminase enzyme (Neuberger and Scott, 2000; Hackney *et al.*, 2009; Verma *et al.*, 2010). Somatic hypermutation means that additional random changes occur in the coding sequence for the variable domain, concentrated in regions known as 'complementarity determining regions' (CDR); some of these changes in coding sequence will make the interaction with the antigen stronger and, as a result, those B cells will be selected and stimulated to replicate and survive more efficiently. This phenomenon is known as affinity maturation during the development of the antibody response. In contrast, class switching affects the constant domains of the heavy chain and means that the genes that code for those regions of the protein are changed or 'switched' leaving the variable region intact but making the class of antibody different. For example, as indicated above, IgM is an immunoglobulin with a μ heavy chain since the constant domains are coded for by the *IGHM* gene, but its variable region genes could become associated with genes that code for a different constant region, e.g. *IGHA* gene making it now an IgA class of antibody.

Immunoglobulin heavy chain immunogenetics

The functional immunoglobulin heavy chain locus, called IGH, is located on chromosome 21q23-q24 (Gu *et al.*, 1992; Tobin-Janzen and Womack, 1992; Zhao *et al.*, 2003) where it spans approximately 150 kb (Zhao *et al.*, 2003). Complete characterization of this locus on chromosomes 8 and 21 (Fig. 8.4A) must await full assembly of the bovine genome, since in Btau_4.2 version it is incomplete (Zimin *et al.*, 2009; Niku *et al.*, 2012; Walther *et al.*, 2013). An additional IGH indicating a partial duplication has a functional *IGHM* and a pseudo *IGHD* gene and is identified on chromosome 11q23 (Hayes and Petit, 1993), now assigned to chromosome 8. No other mammalian species is known to have two functional IGH loci on two chromosomes (Das *et al.*, 2008).

Variable-region heavy chain genes

A polymorphic bovine *IGHV1* (designated earlier as *BovVH1*) gene family that includes 13–15 genes based on Southern analysis, with a significant similarity to human *IGHV4*

(67.4–69.8%) genes, encodes the entire cattle antibody repertoire (Berens *et al.*, 1997; Saini *et al.*, 1997; Sinclair *et al.*, 1997). A total of 36 *IGHV* genes, 10 being functional, were identified upon analysis of Btau_4.2 and UMD_3.1 bovine genome assemblies (Niku *et al.*, 2012; Walther *et al.*, 2013). Another study suggested 11 functional and 6 pseudogenes (Das *et al.*, 2008) in accord with genomic complexity of 13–15 genes in the bovine *IGHV1* gene family (Saini *et al.*, 1997). Consistent with the presence of other genes detected in a Southern blot (Saini *et al.*, 1996), another bovine *IGHV2* (earlier designated as *BovVH2*) gene family has been identified, mostly containing pseudogenes (Walther *et al.*, 2013). The limited germline encoded combinatorial diversity in cattle (Table 8.4) differs from humans and rodents, but is similar to other domestic species, such as sheep,

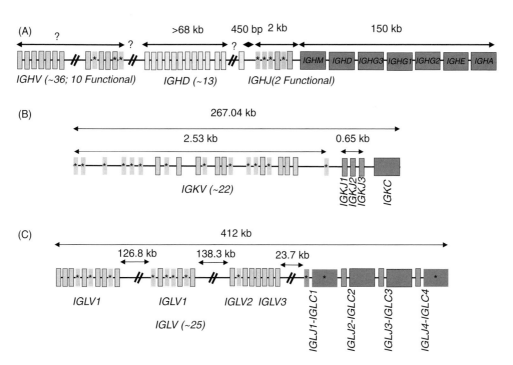

Fig. 8.4. Diagrammatic organization of bovine IGH, IGK and IGL loci. The annotation of the bovine IGH, IGK and IGL loci is based on available literature. (A) Organization of bovine IGH locus based on published data (Zhao *et al.*, 2003; Koti *et al.*, 2010; Niku *et al.*, 2012) but its organization on two chromosomes, 8 and 21, requires complete assembly of the cow genome. (B) Organization of bovine IGK locus on chromosome 11 (Ekman *et al.*, 2009). (C) Organization of bovine IGL locus on chromosome 17 (Pasman *et al.*, 2010). Asterisks indicate pseudogenes.

Table 8.4. Genetic elements and combinatorial antibody diversity in cattle.

		Light chain	
Germline genes	Heavy chain[a]	Kappa	Lambda
V	36 (10 functional)	22 (8 functional)	25 (17 functional)
D	10–13	–	–
J	6 (2 functional)	3	4 (2 functional)
Potential recombinational diversity	260	24	34
Potential H + L pairings	$260 \times (24 + 34) = 0.15 \times 10^5$		

[a]Total germline genes identified on chromosome 8 and 21.

rabbit, camel, pig and horse. Sites of greatest sequence variability within the immunoglobulins are localized to three distinct regions designated complementarity-determining regions 1, 2 and 3 (CDR1, CDR2 and CDR3). The *IGHV* gene codes for the CDR1 and CDR2 of the variable-heavy region are often referred as CDR1H and CDR2H. The CDR3 of the heavy-variable region (CDR3H) is the site with the highest level of variability formed by the combinatorial joining of the *V-D-J* genes, as well as further diversification generated by nucleotide additions and deletions at the junctions because of imprecise joining of genes during recombination (Schatz, 2004). The bovine *IGHV1* genes encode a CDR1H that has conserved five codons, while the CDR2H is strictly 16 codons long. The *V-D-J* encoded CDR3H size ranges from 3 to 66 codons that include characteristic GGT and TAT repetitive codons (Saini *et al.*, 1999; Larsen and Smith, 2012; Wang *et al.*, 2013). Such an exceptionally long CDR3H is the largest known to exist in a species to date where multiple even numbered cysteine residues permit intra-CDR3H disulphide bridging (Saini *et al.*, 1999; Wang *et al.*, 2013). The atypical CDR3H, first noted in IgM, is observed in all antibody classes including IgG, IgA and IgE (Larsen and Smith, 2012; Walther *et al.*, 2013).

A total of 10 to 13 bovine *IGHD* (earlier designated as *BovDH*) genes, flanked by a recombination signal sequence (RSS) comprising 9 bp nonamer and 7 bp heptamer with intervening 12 bp spacers are identified (Shojaei *et al.*, 2003; Koti *et al.*, 2008, 2010). The J-proximal conserved *IGHDQ52* gene is distinct from the majority of bovine *IGHD* genes that characteristically have repetitive GGT and

TAT codons encoding glycine and tyrosine, respectively. A single unusually long *IGHD2* (earlier designated as *BovD*$_H$*2*) gene is identified with the potential to directly encode ≤49 codons. Other bovine *IGHD* genes range from 14–48 bp in size, with the conserved *IGHDQ52* gene (14 bp) being the shortest. The bovine *IGHD* genes are classified into four *IGHD* families known as A through D as follows: *IGHD-A* (earlier designated as *BovD*$_H$*A*; *IGHD1* and *IGHD6*), *IGHD-B* (earlier designated as *BovD*$_H$*B*; *IGHD2, IGHD3, IGHD5, IGHD7* and *IGHD8*), *IGHD-C* (earlier designated as *BovD*$_H$*C*; *IGHD4*) and *IGHD-D* (earlier designated as *BovD*$_H$*D*; *IGHDQ52*). Two new *IGHD* genes, *IGHD-U16* and *IGHD-U31*, are yet to be classified. Six bovine *IGHJ* genes, 130–500 bp apart, span 18 kb 7 kb upstream of the *IGHM* exons (Fig. 8.4A). Only two *IGHJ* genes, *IGHJ1* (previously called *BovJ*$_H$*1*) and *IGHJ2* (earlier called *BovJ*$_H$*2*), are expressed and encode typical amino acid VTVSS motifs at the 3′ end. Other *IGHJ* genes either lack RSS or the splice site due to which their expression is affected. The *IGHJ* genes on chromosome 21 are found duplicated, together with *IGHM* exons, on chromosome 8, earlier assigned to chromosome 11 (Hosseini *et al.*, 2004). The extent of bovine heavy chain locus duplication on chromosome 8 will be known upon complete annotation of the fully assembled bovine genome.

Constant-region heavy chain genes

As indicated above, immunoglobulins are named as classes according to the expression of heavy chain genes that code for their constant domains. The genes that code for the constant domains, μ, δ, γ, ε and α of the heavy chains are designated as *IGHM, IGHD, IGHG IGHE* and *IGHA*, respectively, as per IGMT nomenclature. The *IGHG* class is composed of several genes that code for so-called subclasses of IgG antibodies such as IgG1. In cattle, the heavy chain constant region genes are found arranged as *IGHM, IGHD, IGHG3, IGHG1, IGHG2, IGHE* and *IGHA* (Fig. 8.4A), spanning approximately 150 kb on chromosome 21 (Zhao *et al.*, 2003). The additional heavy chain locus on chromosome 11q23 (now assigned to chromosome 8) contains a functional *IGHM* and a pseudo *IGHD* gene (Hayes and Petit, 1993). Each heavy chain has three or four constant domains, with each being coded for by a single constant (C) gene, but with separate exons encoding the individual domains. Individual classes of various heavy chain genes are considered below.

The *IGHM* constant gene is composed of four exons (*CH1–4*), each one coding for one of the four constant domains, whereas two other exons (*M1* and *M2*) encode the transmembrane domains. Similar to other species, the *M1* exon is spliced on to the *CH4* (Mousavi *et al.*, 1998) and this results in the immunoglobulin being expressed on the B cell surface where it acts as the BCR. With deletion of these exons during RNA processing, the B cells produce immunoglobulin in secreted form. The antigen-binding function of bovine IgM seems to be influenced by relative inflexibility of *CH2*, which acts as a hinge, because of fewer proline amino acids (Saini and Kaushik, 2001). The rigidity in *CH2* may facilitate exposure of the C1q-binding site subsequent to antigen binding and enhance IgM's complement fixing ability. Three bovine IgM allotypes (IgMa, IgMb and IgMc) are described based on nucleotide substitutions in all the *CH* exons resulting in amino acid replacements (Saini and Kaushik, 2001). Additional IgM variants may originate via alternative splicing where, for example, three in-frame codons are inserted at the *CH1–CH2* junction (Saini and Kaushik, 2001).

The bovine *IGHD* gene encodes three domains (*CH1–CH3*) similar to other species. However, the *CH1* exon has a high level of identity at the nucleotide (96.6%) and protein (93.5%) levels with *CH1* of *IGHM* gene (Zhao *et al.*, 2002). Unlike other species, the *IGHD* gene has a short switch region (Sδ) between

the *IGHM* and *IGHD* genes that may permit class switch recombination (Sun *et al.*, 2012b). Evidence in support of possible class switch recombination from IgM to IgD is not available, however. Consistent with the observation that IgD could not be serologically detected on bovine B cells, the germline *IGHD* gene is found to be transcriptionally active at a low level (Zhao *et al.*, 2002).

Constant *IGHG* genes, *IGHG1, IGHG2* and *IGHG3* coding for three subclasses of IgG are found in cattle and give rise to the IgG1, IgG2 and IgG3 (Knight *et al.*, 1988; Kacskovics and Butler, 1996; Rabbani *et al.*, 1997). This differs from humans, sheep and rabbits that have four, two or one IgG subclass, respectively. The bovine *IGHG1* gene is most probably the homologue of the *IGHG2* and *IGHG3* genes since the first gene duplication led to the *IGHG2* gene followed by the second duplication event that gave rise to the *IGHG3* gene. Indeed, a high nucleotide sequence identity exists between *IGHG3* and *IGHG1* genes (85.1%) as compared to the *IGHG2* gene (83.4%) (Zhao *et al.*, 2003). Allelic variation is found for all the bovine IgG subclasses. Four allotypic variants of IgG1 (IgG1a, IgG1b, IgG1c and IgG1d) have been described with amino acid replacements in the hinge and all *CH* exons (Symons *et al.*, 1989; Saini *et al.*, 2007). The unique *Pro–Ala–Ser–Ser* motifs in the CH1 (positions 189–192 and 205–208) domain of the IgG1c allotype seem to confer a novel cellular adhesion and migration function (Saini *et al.*, 2007). The role of IgG1 class in protection of mucosal surfaces is yet to be fully understood from a functional perspective. The nucleotide sequences of two IgG2 allotypes, designated as IgG2A1 and IgG2A2, have been described, and may differ in four regions including hinge, and *CH1* and *CH3* exons encoded domains (Kacskovics and Butler, 1996). Two allotypes (IgG3a and IgG3b) of IgG3 differ by six amino acids in the coding region and an 84 base pair insertion in the intron between the *CH2* and *CH3* exons (Rabbani *et al.*, 1997).

A single copy of the *IGHE* constant gene has been identified in cattle (Knight *et al.*, 1988). It has four exons (*CH1–4*) similar to other species and shares 87% sequence identity with the sheep *IGHE* gene. Bovine IgE has heat labile skin sensitizing ability analogous to human IgE (Hammer *et al.*, 1971).

A single IGHA constant gene has been identified in the bovine genome (Knight *et al.*, 1988) with three bovine *CH* exons (*CH1–CH3*) separated by two introns (Brown *et al.*, 1997). Bovine IgA is closest to swine IgA at the protein level but shares an additional N-linked glycosylation site at position 282 with rabbit IgA3 and IgA4. RFLP (Brown *et al.*, 1997) and serological (De Benedictis *et al.*, 1984) analysis have suggested two allelic variants of bovine IgA, but genomic DNA analysis of 50 Swedish cattle did not support it.

Immunoglobulin light chain immunogenetics

Significant differences exist across species with regard to expression of kappa (κ) and lambda (λ) light chains (reviewed by Saini and Kaushik, 2002b). Similar to other ruminant species, λ light chains are predominantly expressed by B cells in cattle (estimated at up to 98% of B cells) (Butler, 1997, 1998) in contrast to κ light chains (≤9%) (Arun *et al.*, 1996; Beyer *et al.*, 2002). The IGK (κ) and IGL (λ) loci are located on different chromosomes where *IGKV*, *IGKJ* and *IGKC* genes encode κ, while *IGLV*, *IGLJ* and *IGLC* genes encode λ light chains. The immunoglobulin light chain constant (*C*) genes provide the light chain with its name of κ or λ.

The bovine κ light chain locus spans 267 Kb on chromosome 11 (Fig. 8.4B) (Ekman *et al.*, 2009). Twenty-two *IGKV* (eight of which are functional), three *IGKJ* and one *IGKC* genes have been identified in proper transcriptional orientation that permits recombination via deletion of intervening sequences (Ekman *et al.*, 2009). The low level of κ-light chain gene expression in cattle is not because of either genomic complexity or recombinational potential of the IGK locus (Table 8.4), but seems related to factors intrinsic to light chain repertoire selection.

The bovine λ light chain (IGL) locus spans 412 kb on chromosome 17 (Fig. 8.4C). It includes 25 *IGLV* genes (17 being functional) organized in three sub-clusters (Pasman *et al.*, 2010)

followed by four *IGLJ-IGLC* recombination units of which two are functional (*IGLJ2-IGLC2* and *IGLJ3-IGLC3*) (Chen *et al.*, 2008). Three bovine *IGLV* (earlier called *BovV$_\lambda$*) gene families, *IGLV1*, *IGLV2* and *IGLV3* (Saini *et al.*, 2003), exist in cattle where pseudogenes in the *IGLV1* and *IGLV2* families (Pasman *et al.*, 2010) could potentially diversify the λ-light chain repertoire via gene conversion (Parng *et al.*, 1996). The *IGLV1* genes recombined with the *IGLJ3-IGLC3* unit are predominantly expressed in the primary antibody repertoire of cattle (Chen *et al.*, 2008; Pasman *et al.*, 2010). Specific recombined *IGLV1* genes (*IGLV1d*, *IGLV1e* and *IGLV1x*) pair with immunoglobulin heavy chains that have exceptionally long CDR3 segments, coded for by up to 61 base pairs (Saini *et al.*, 2003). Overall, *IGLV1-IGLJ3-IGLC3* recombination encodes most of the λ-light chains in cattle (Pasman *et al.*, 2010).

Surrogate light chains are known to pair with the nascent heavy chain in B cell development. In cattle four surrogate light chain genes, *Vpreb1*, *Vpreb2*, *Vpreb3* and *IgLL1* (Ekman *et al.*, 2009, 2012), are identified. The *Vpreb1* and *IgLL1*, but not *Vpreb2* and *Vpreb3* gene products, seem to function as a surrogate light chain during pre-B cell development in cattle. The genes coding for Vpreb1, Vpreb3 and IgLL1 are found on chromosome 17 (Ekman *et al.*, 2009).

Development of B cells and antibody repertoires

The IgM-bearing B cells have been detected in the bovine fetus as early as 59 days into gestation (Schultz *et al.*, 1973). However, *V-D-J* and *V-J* recombinations were observed in splenic B cells at 125 days of gestation and serum immunoglobulin was detectable in a 145-day-old fetus (Saini and Kaushik, 2002a). At this developmental stage, some splenic B-cells may express *V-D-J* recombinations alone, while others may secrete λ light chain only because of non-productive *V-D-J* recombinations.

In cattle, perinatal immunoglobulin diversification occurs in the ileal Peyer's patches, suggesting that the ileal Peyer's patches serve as the primary lymphoid organ in ruminants (Yasuda *et al.*, 2004, 2006). The lymphoid follicles of ileal Peyer's patches consist mostly of IgM-bearing B cells that develop and expand oligoclonally (David *et al.*, 2003), similar to bursal follicles in chicken. Nevertheless, *IGLV-IGLJ* recombination-associated λ light chain diversification has been noted in bovine fetal spleen prior to the establishment of a diverse repertoire in the ileum (Lucier *et al.*, 1998). B lymphopoiesis (as shown by the presence of so-called pre-B like cells that had intracellular μ heavy chains) also has been observed in bovine fetal bone marrow and lymph node in parallel to ileal Peyer's patches (Ekman *et al.*, 2012). Thus, ileal Peyer's patches may not be the sole primary lymphoid organ in cattle. In general, variations with regard to B cell development across species seem to exemplify an outcome of divergent evolution (Yasuda *et al.*, 2006; Alitheen *et al.*, 2010).

There are some known differences between immunoglobulin gene usage in fetal development versus the adult. Two *IGHV* genes (*gl.110.20* and *BF2B5*) are preferentially used in the fetal *V-D-J* recombinations (Saini and Kaushik, 2002a). In contrast to J-proximal conserved *IGHDQ52* gene, *IGHD7* and *IGHD5* genes are favourably expressed in both fetal and adult B cells (Koti *et al.*, 2010). The bovine *IGHJ1* gene (*IGHJpB7S2*) expression is also predominant in both fetal and adult *V-D-J* recombinations (Saini *et al.*, 1997). Analysis of somatic hypermutations in the CDRs revealed that transition nucleotide substitutions predominate over transversions (Kaushik *et al.*, 2009). Further, somatic hypermutations result in higher diversification in the third framework region of IgG as compared to IgM antibodies in cattle (Kaushik *et al.*, 2009).

Mechanisms of antibody diversification

In species where immunoglobulins can be transferred across the placenta and into colostrum as well (e.g. mice and humans) significant germline *IGHV*, *IGHD* and *IGHJ* gene sequence divergence and combinatorial diversity exists. In contrast, the primary antibody repertoire of cattle is composed of limited combinatorial diversity (1.5×10^4) because of restricted germline sequence divergence both at IGH and IGK or IGL loci (Table 8.4). For example, while in mice and humans there are over 200 *IGHV* genes for the heavy chain, cattle have only 36 of which 10 are functional. Thus, several other mechanisms compensate for this restricted combinatorial diversity in cattle including somatic hypermutations (Kaushik *et al.*, 2009; Verma and Aitken, 2012), insertion of conserved short nucleotide sequences (CSNS) specifically at *V-D* junctions (Koti *et al.*, 2010) and extensive junctional flexibility in *V-D-J* recombination involving deletions and templated or untemplated nucleotide additions at the junctions (Koti *et al.*, 2010).

While no evidence exists for gene conversion for the heavy chain (Kaushik *et al.*, 2009), it has been suggested to occur at the λ light chain variable region (Parng *et al.*, 1996; Lucier *et al.*, 1998). Activation induced cytidine deaminase (AID), an enzyme crucial to somatic hypermutation, has been characterized in cattle (Verma *et al.*, 2010). AID gene, located on chromosome 5, is expressed in neonatal and adult lymphoid tissue of cattle. The biased 'hot spot' triplets in the CDRs of bovine *V-D-J* recombinations predispose them to somatic hypermutations (Kaushik *et al.*, 2009) similar to other species. Somatic hypermutations are also involved in diversifying the *V-J* recombinations encoding λ-light chains (Lucier *et al.*, 1998). Cattle have been shown to use somatic hypermutation without exposure to exogenous antigen to diversify the developing antibody repertoire during B cell ontogeny (Koti *et al.*, 2010), with somatic hypermutations evident in the heavy chain CDR1 and CDR2 of 125-day-old fetus. Finally, extensive size heterogeneity (3 to 66 codons) in the heavy chain CDR3 together with disulphide bridging between multiple even numbered cysteines leads to significant configurational diversity of this region, which constitutes the antigen-combining site (Saini and Kaushik, 2002a; Wang *et al.*, 2013).

In general, the CDR3 of the bovine heavy chains has an average length of 22.7 ± 3.2 amino acids (Almagro *et al.*, 2006), although it varies by class, with IgMs having 21.7 ± 1.8 and IgGs having 18.2 ± 1.3 (Kaushik *et al.*, 2009).

However, cattle antibodies can express exceptionally long heavy chain CDR3s (>50 amino acids) with multiple even numbered cysteine residues, both in fetal and adult B cells (Saini *et al.*, 1999; Saini and Kaushik, 2002a; Wang *et al.*, 2013). The exceptionally long CDR3H occurs in 8–10% of circulating B cells and, while initially observed in IgM (Saini *et al.*, 1999), it occurs in IgG (Larsen and Smith, 2012), IgA and IgE classes of immunoglobulins (Walther *et al.*, 2013). Recent crystallization of bovine antibodies with exceptionally long heavy chain CDR3 has revealed a unique 'stalk and knob' structure where configurational diversity is generated via creation of mini-domains through intra-CDR3H disulphide bridges between the cysteine amino acids (Wang *et al.*, 2013). Such a structural diversity via mini-domains in the antigen-binding site is not yet known to exist in other species. Both fetal and adult antibodies with exceptionally long CDR3H originate from unique recombinations of the germline *IGHV-gl.110.20*, longest *IGHD2* and *IGHJ1-pB7S2* genes (Saini *et al.*, 1999; Saini and Kaushik, 2002a; Koti *et al.*, 2010). An insertion of 13–18 nucleotide long CSNS of unknown origin in adult *V-D-J* recombinations, which has a disproportionate number of adenines, specifically at the *V-D* junction increases the CDR3 size to ~61 codons following encounter with antigen in the periphery, providing a novel mechanism of antibody diversification (Koti *et al.*, 2010). Such insertions at the *V-D* junction (Koti *et al.*, 2010) that contribute to the stalk structure of the antigen-combining site (Wang *et al.*, 2013) are absent in *V-D-J* recombinations in fetal B cells. Thus, the structure of the antigen-combining site of exceptionally long CDR3H encoded by fetal *V-D-J* recombinations is likely to be different due to a relatively shorter or non-existent stalk. The B cells expressing immunoglobulin with exceptionally long heavy chain CDR3 undergo affinity maturation via somatic mutations upon antigen encounter (Kaushik *et al.*, 2002, 2009) and these heavy chains with unusually long CDR3s exclusively pair with λ light chains with *Ser90* conserved in the light chain CDR3, which provide minimal structural support without making contact with antigen (Saini *et al.*, 1999, 2003). In conclusion, these exceptionally long heavy chain CDR3s found in all bovine antibody classes provide a distinct novel mechanism of antibody diversification.

Acknowledgements

The studies on bovine MHC were supported by Grants-in-Aid for Scientific Research (A, B and C) from the Japan Society for the Promotion of Science (JSPS), and by a grant from the Program for the Promotion of Basic and Applied Research for Innovations in Bio-oriented Industry (Yoko Aida). The studies on bovine immunoglobulin genetics were supported by a research grant from NSERC Canada (Azad K. Kaushik). Some of the studies on TCR gamma and TCR delta genes and the WC1 genes were funded by grants from the USDA-NIFA to Cynthia Baldwin. We also thank Dr John Hammond, UK, for editorial comments. The section on MHC was written by Drs Aida and Takeshima, the section on TCR and Multigene Receptor Families by Dr Baldwin and the section on Immunoglobulins by Dr Kaushik.

References

Abeler-Dorner, L., Swamy, M., Williams, G., Hayday, A.C. and Bas, A. (2012) Butyrophilins: an emerging family of immune regulators. *Trends in Immunology* 33, 34–41.

Adams, E.J. and Luoma, A.M. (2013) The adaptable major histocompatibility complex (MHC) fold: structure and function of nonclassical and MHC class I-like molecules. *Annual Review of Immunology* 31, 529–561.

Aida, Y. (1995) Characterization and expression of bovine MHC class II genes. *Bulletin de la Société Franco-Japonaise des Sciences Vétérinaires* 6, 17–24.

Aida, Y. (2001) Influence of host genetic differences on leukemogenesis induced by bovine leukemia virus. *AIDS Research and Human Retroviruses* 17, S12.

Aida, Y., Kohda, C., Morooka, A., Nakai, Y., Ogimoto, K., Urao, T. and Asahina, M. (1994) Cloning of cDNAs and the molecular evolution of a bovine MHC class II DRA gene. *Biochemical and Biophysical Research Communications* 204, 195–202.

Aida, Y., Niimi, M., Asahina, M., Okada, K., Nakai, Y. and Ogimoto, K. (1995) Identification of a new bovine MHC class II DRB allele by nucleotide sequencing and an analysis of phylogenetic relationships. *Biochemical and Biophysical Research Communications* 209, 981–988.

Alitheen, N.B., McClure, S. and McCullagh, P. (2010) B-cell development: one problem, multiple solutions. *Immunology and Cell Biology* 88, 445–450.

Almagro, J.C., Martinez, L., Smith, S.L., Alagon, A., Estevez, J. and Paniagua, J. (2006) Analysis of the horse V(H) repertoire and comparison with the human IGHV germline genes, and sheep, cattle and pig V(H) sequences. *Molecular Immunology* 43, 1836–1845.

Ammer, H., Schwaiger, F.W., Kammerbauer, C., Gomolka, M., Arriens, A., Lazary, S. and Epplen, J.T. (1992) Exonic polymorphism vs intronic simple repeat hypervariability in MHC-DRB genes. *Immunogenetics* 35, 332–340.

Amorena, B. and Stone, W.H. (1978) Serologically defined (SD) locus in cattle. *Science* 201, 159–160.

Andersson, L., Lunden, A., Sigurdardottir, S., Davies, C.J. and Rask, L. (1988) Linkage relationships in the bovine MHC region. High recombination frequency between class II subregions. *Immunogenetics* 27, 273–280.

Ando, A., Kawai, J., Maeda, M., Tsuji, K., Trowsdale, J. and Inoko, H. (1989) Mapping and nucleotide sequence of a new HLA class II light chain gene, DQB3. *Immunogenetics* 30, 243–249.

Arun, S.S., Breuer, W. and Hermanns, W. (1996) Immunohistochemical examination of light-chain expression (lambda/kappa ratio) in canine, feline, equine, bovine and porcine plasma cells. *Journal of Veterinary Medicine Series A* 43, 573–576.

Ballingall, K.T., Luyai, A. and McKeever, D.J. (1997) Analysis of genetic diversity at the DQA loci in African cattle: evidence for a BoLA-DQA3 locus. *Immunogenetics* 46, 237–244.

Ballingall, K.T., Marasa, B.S., Luyai, A. and McKeever, D.J. (1998) Identification of diverse BoLA DQA3 genes consistent with non-allelic sequences. *Animal Genetics* 29, 123–129.

Ballingall, K., MacHugh, N., Taracha, E., Mertens, B. and McKeever, D. (2001) Transcription of the unique ruminant class II major histocompatibility complex-DYA and DIB genes in dendritic cells. *European Journal of Immunology* 31, 82–86.

Ballingall, K.T., Ellis, S.A., MacHugh, N.D., Archibald, S.D. and McKeever, D.J. (2004a) The DY genes of the cattle MHC: expression and comparative analysis of an unusual class II MHC gene pair. *Immunogenetics* 55, 748–755.

Ballingall, K.T., Luyai, A., Rowlands, G.J., Sales, J., Musoke, A.J., Morzaria, S.P. and McKeever, D.J. (2004b) Bovine leukocyte antigen major histocompatibility complex class II DRB3*2703 and DRB3*1501 alleles are associated with variation in levels of protection against *Theileria parva* challenge following immunization with the sporozoite p67 antigen. *Infection and Immunity* 72, 2738–2741.

Baltian, L.R., Ripoli, M.V., Sanfilippo, S., Takeshima, S.N., Aida, Y. and Giovambattista, G. (2012) Association between BoLA-DRB3 and somatic cell count in Holstein cattle from Argentina. *Molecular Biology Reports* 39, 7215–7220.

Band, M., Larson, J.H., Womack, J.E. and Lewin, H.A. (1998) A radiation hybrid map of BTA23: identification of a chromosomal rearrangement leading to separation of the cattle MHC class II subregions. *Genomics* 53, 269–275.

Baxter, R., Hastings, N., Law, A. and Glass, E.J. (2008) A rapid and robust sequence-based genotyping method for BoLA-DRB3 alleles in large numbers of heterozygous cattle. *Animal Genetics* 39, 561–563.

Baxter, R., Craigmile, S.C., Haley, C., Douglas, A.J., Williams, J.L. and Glass, E.J. (2009) BoLA-DR peptide binding pockets are fundamental for foot-and-mouth disease virus vaccine design in cattle. *Vaccine* 28, 28–37.

Behl, J.D., Verma, N.K., Behl, R., Mukesh, M. and Ahlawat, S.P. (2007) Characterization of genetic polymorphism of the bovine lymphocyte antigen DRB3.2 locus in Kankrej cattle (*Bos indicus*). *Journal of Dairy Science* 90, 2997–3001.

Bell, J.K., Mullen, G.E., Leifer, C.A., Mazzoni, A., Davies, D.R. and Segal, D.M. (2003) Leucine-rich repeats and pathogen recognition in Toll-like receptors. *Trends in Immunology* 24, 528–533.

Bell, J.K., Askins, J., Hall, P.R., Davies, D.R. and Segal, D.M. (2006) The dsRNA binding site of human Toll-like receptor 3. *Proceedings of the National Academy of Sciences USA* 103, 8792–8797.

Berens, S.J., Wylie, D.E. and Lopez, O.J. (1997) Use of a single VH family and long CDR3s in the variable region of cattle Ig heavy chains. *International immunology* 9, 189–199.

Beyer, J., Kollner, B., Teifke, J.P., Starick, E., Beier, D., Reimann, I., Grunwald, U. and Ziller, M. (2002) Cattle infected with bovine leukaemia virus may not only develop persistent B-cell lymphocytosis but also persistent B-cell lymphopenia. *Journal of Veterinary Medicine B Infectious Diseases, Veterinary Public Health* 49, 270–277.

Bikker, F.J., Ligtenberg, A.J., End, C., Renner, M., Blaich, S., Lyer, S., Wittig, R., van't Hof, W., Veerman, E.C., Nazmi, K. *et al.* (2004) Bacteria binding by DMBT1/SAG/gp-340 is confined to the VEVLXXXXW motif in its scavenger receptor cysteine-rich domains. *Journal of Biological Chemistry* 279, 47699–47703.

Birch, J. and Ellis, S.A. (2007) Complexity in the cattle CD94/NKG2 gene families. *Immunogenetics* 59, 273–280.

Birch, J., Murphy, L., MacHugh, N.D. and Ellis, S.A. (2006) Generation and maintenance of diversity in the cattle MHC class I region. *Immunogenetics* 58, 670–679.

Birch, J., Codner, G., Guzman, E. and Ellis, S.A. (2008a) Genomic location and characterisation of non-classical MHC class I genes in cattle. *Immunogenetics* 60, 267–273.

Birch, J., De Juan Sanjuan, C., Guzman, E. and Ellis, S.A. (2008b) Genomic location and characterisation of MIC genes in cattle. *Immunogenetics* 60, 477–483.

Bissumbhar, B., Nilsson, P.R., Hensen, E.J., Davis, W.C. and Joosten, I. (1994) Biochemical characterization of bovine MHC DQ allelic variants by one-dimensional isoelectric focusing. *Tissue Antigens* 44, 100–109.

Bjorkman, P.J., Saper, M.A., Samraoui, B., Bennett, W.S., Strominger, J.L. and Wiley, D.C. (1987) Structure of the human class I histocompatibility antigen, HLA-A2. *Nature* 329, 506–512.

Blumerman, S.L., Herzig, C.T., Rogers, A.N., Telfer, J.C. and Baldwin, C.L. (2006) Differential TCR gene usage between WC1– and WC1+ ruminant gammadelta T cell subpopulations including those responding to bacterial antigen. *Immunogenetics* 58, 680–692.

Bonneville, M., O'Brien, R.L. and Born, W.K. (2010) Gammadelta T cell effector functions: a blend of innate programming and acquired plasticity. *Nature Reviews in Immunology* 10, 467–478.

Brown, J.H., Jardetzky, T.S., Gorga, J.C., Stern, L.J., Urban, R.G., Strominger, J.L. and Wiley, D.C. (1993) Three-dimensional structure of the human class II histocompatibility antigen HLA-DR1. *Nature* 364, 33–39.

Brown, W.R., Rabbani, H., Butler, J.E. and Hammarstrom, L. (1997) Characterization of the bovine C alpha gene. *Immunology* 91, 1–6.

Burke, M.G., Stone, R.T. and Muggli-Cockett, N.E. (1991) Nucleotide sequence and northern analysis of a bovine major histocompatibility class II DR beta-like cDNA. *Animal Genetics* 22, 343–352.

Butler, J.E. (1997) Immunoglobulin gene organization and the mechanism of repertoire development. *Scandanavian Journal of Immunology* 45, 455–462.

Butler, J.E. (1998) Immunoglobulin diversity, B-cell and antibody repertoire development in large farm animals. *Reviews in Science and Techology* 17, 43–70.

Carding, S.R. and Egan, P.J. (2002) Gammadelta T cells: functional plasticity and heterogeneity. *Nature Reviews in Immunology* 2, 336–345.

Champsaur, M. and Lanier, L.L. (2010) Effect of NKG2D ligand expression on host immune responses. *Immunological Reviews* 235, 267–285.

Chen, C., Herzig, C.T., Alexander, L.J., Keele, J.W., McDaneld, T.G., Telfer, J.C. and Baldwin, C.L. (2012) Gene number determination and genetic polymorphism of the gamma delta T cell co-receptor WC1 genes. *BMC Genetics* 13, 86.

Chen, L., Li, M., Li, Q., Yang, X., An, X. and Chen, Y. (2008) Characterization of the bovine immunoglobulin lambda light chain constant IGLC genes. *Veterinary Immunology and Immunopathology* 124, 284–294.

Chien, Y.H., Jores, R. and Crowley, M.P. (1996) Recognition by gamma/delta T cells. *Annual Reviews in Immunology* 14, 511–532.

Childers, C.P., Newkirk, H.L., Honeycutt, D.A., Ramlachan, N., Muzney, D.M., Sodergren, E., Gibbs, R.A., Weinstock, G.M., Womack, J.E. and Skow, L.C. (2006) Comparative analysis of the bovine MHC class IIb sequence identifies inversion breakpoints and three unexpected genes. *Animal Genetics* 37, 121–129.

Codner, G.F., Birch, J., Hammond, J.A. and Ellis, S.A. (2012) Constraints on haplotype structure and variable gene frequencies suggest a functional hierarchy within cattle MHC class I. *Immunogenetics* 64, 435–445.

Connelley, T., Burrells, A., MacHugh, N.D. and Morrison, W.I. (2008a) Use of a Pan-Vbeta primer permits the amplification and sequencing of TCRbeta chains expressed by bovine T-cell clones following a single semi-nested PCR reaction. *Veterinary Immunology and Immunopathology* 126, 156–162.

Connelley, T., MacHugh, N.D., Burrells, A. and Morrison, W.I. (2008b) Dissection of the clonal composition of bovine alphabeta T cell responses using T cell receptor Vbeta subfamily-specific PCR and hetero-duplex analysis. *Journal of Immunological Methods* 335, 28–40.

Connelley, T., Aerts, J., Law, A. and Morrison, W.I. (2009) Genomic analysis reveals extensive gene duplication within the bovine TRB locus. *BMC Genomics* 10, 192.

Conrad, M.L., Pettman, R., Whitehead, J., McKinnel, L., Davis, S.K. and Koop, B.F. (2002) Genomic sequencing of the bovine T cell receptor beta locus. *Veterinary Immunology and Immunopathology* 87, 439–441.

Conrad, M.L., Mawer, M.A., Lefranc, M.P., McKinnell, L., Whitehead, J., Davis, S.K., Pettman, R. and Koop, B.F. (2007) The genomic sequence of the bovine T cell receptor gamma TRG loci and localization of the TRGC5 cassette. *Veterinary Immunology and Immunopathology* 115, 346–356.

da Mota, A.F., Gabriel, J.E., Martinez, M.L. and Coutinho, L.L. (2002) Distribution of bovine lymphocyte antigen (BoLA-DRB3) alleles in Brazilian dairy Gir cattle (*Bos indicus*). *European Journal of Immunogenetics* 29, 223–227.

Das, S., Nozawa, M., Klein, J. and Nei, M. (2008) Evolutionary dynamics of the immunoglobulin heavy chain variable region genes in vertebrates. *Immunogenetics* 60, 47–55.

Dausset, J. (1958) Iso-leuco-anticorps. *Acta Haematologica* 20, 156–166.

David, C.W., Norrman, J., Hammon, H.M., Davis, W.C. and Blum, J.W. (2003) Cell proliferation, apoptosis, and B- and T-lymphocytes in Peyer's patches of the ileum, in thymus and in lymph nodes of preterm calves, and in full-term calves at birth and on day 5 of life. *Journal of Dairy Science* 86, 3321–3329.

Davies, C.J., Andersson, L., Joosten, I., Mariani, P., Gasbarre, L.C. and Hensen, E.J. (1992) Characterization of bovine MHC class II polymorphism using three typing methods: serology, RFLP and IEF. *European Journal of Immunogenetics* 19, 253–262.

Davies, C.J., Eldridge, J.A., Fisher, P.J. and Schlafer, D.H. (2006) Evidence for expression of both classical and non-classical major histocompatibility complex class I genes in bovine trophoblast cells. *American Journal of Reproductive Immunology* 55, 188–200.

Davis, M.M. and Bjorkman, P.J. (1988) T-cell antigen receptor genes and T-cell recognition. *Nature* 334, 395–402.

De Benedictis, G., Capalbo, P. and Dragone, A. (1984) Identification of an allotypic IgA in cattle serum. *Comparative Immunology Microbiology and Infectious Diseases* 7, 35–42.

Di Palma, F., Archibald, S.D., Young, J.R. and Ellis, S.A. (2002) A BAC contig of approximately 400 kb contains the classical class I major histocompatibility complex (MHC) genes of cattle. *European Journal of Immunogenetics* 29, 65–68.

Dietz, A.B., Cohen, N.D., Timms, L. and Kehrli, M.E., Jr (1997a) Bovine lymphocyte antigen class II alleles as risk factors for high somatic cell counts in milk of lactating dairy cows. *Journal of Dairy Science* 80, 406–412.

Dietz, A.B., Detilleux, J.C., Freeman, A.E., Kelley, D.H., Stabel, J.R. and Kehrli, M.E., Jr (1997b) Genetic association of bovine lymphocyte antigen DRB3 alleles with immunological traits of Holstein cattle. *Journal of Dairy Science* 80, 400–405.

Dikiniene, N. and Aida, Y. (1995) Cattle cDNA clones encoding MHC class II DQB1 and DQB2 genes. *Immunogenetics* 42, 75.

Dissen, E., Fossum, S., Hoelsbrekken, S.E. and Saether, P.C. (2008) NK cell receptors in rodents and cattle. *Seminars in Immunology* 20, 369–375.

Dobromylskyj, M. and Ellis, S. (2007) Complexity in cattle KIR genes: transcription and genome analysis. *Immunogenetics* 59, 463–472.

Dobromylskyj, M.J., Connelley, T., Hammond, J.A. and Ellis, S. (2009) Cattle Ly49 is polymorphic. *Immunogenetics* 61, 789–795.

Dubey, P.K., Goyal, S., Kathiravan, P., Mishra, B.P., Gahlawat, S.K. and Kataria, R.S. (2013) Sequence characterization of river buffalo Toll-like receptor genes 1-10 reveals distinct relationship with cattle and sheep. *International Journal of Immunogenetics* 40, 140–148.

Egan, P.J. and Carding, S.R. (2000) Downmodulation of the inflammatory response to bacterial infection by gammadelta T cells cytotoxic for activated macrophages. *Journal of Experimental Medicine* 191, 2145–2158.

Ekman, A., Niku, M., Liljavirta, J. and Iivanainen, A. (2009) *Bos taurus* genome sequence reveals the assortment of immunoglobulin and surrogate light chain genes in domestic cattle. *BMC Immunology* 10, 22.

Ekman, A., Ilves, M. and Iivanainen, A. (2012) B lymphopoiesis is characterized by pre-B cell marker gene expression in fetal cattle and declines in adults. *Developmental and Comparative Immunology* 37, 39–49.

Ellis, S.A., Holmes, E.C., Staines, K.A., Smith, K.B., Stear, M.J., McKeever, D.J., MacHugh, N.D. and Morrison, W.I. (1999) Variation in the number of expressed MHC genes in different cattle class I haplotypes. *Immunogenetics* 50, 319–328.

Elsik, C.G., Tellam, R.L., Worley, K.C., Gibbs, R.A., Muzny, D.M., Weinstock, G.M., Adelson, D.L., Eichler, E.E., Elnitski, L., Guigo, R. *et al.* (2009) The genome sequence of taurine cattle: a window to ruminant biology and evolution. *Science* 324, 522–528.

Escayg, A.P., Hickford, J.G., Montgomery, G.W., Dodds, K.G. and Bullock, D.W. (1996) Polymorphism at the ovine major histocompatibility complex class II loci. *Animal Genetics* 27, 305–312.

Fabriek, B.O., van Bruggen, R., Deng, D.M., Ligtenberg, A.J., Nazmi, K., Schornagel, K., Vloet, R.P., Dijkstra, C.D. and van den Berg, T.K. (2009) The macrophage scavenger receptor CD163 functions as an innate immune sensor for bacteria. *Blood* 113, 887–892.

Fikri, Y., Nyabenda, J., Content, J. and Huygen, K. (2007) Cloning, sequencing, and cell surface expression pattern of bovine immunoreceptor NKG2D and adaptor molecules DAP10 and DAP12. *Immunogenetics* 59, 653–659.

Fisher, C.A., Bhattarai, E.K., Osterstock, J.B., Dowd, S.E., Seabury, P.M., Vikram, M., Whitlock, R.H., Schukken, Y.H., Schnabel, R.D., Taylor, J.F. *et al.* (2011) Evolution of the bovine TLR gene family and member associations with *Mycobacterium avium* subspecies paratuberculosis infection. *PLoS One* 6, e27744.

Flajnik, M.F. and Kasahara, M. (2001) Comparative genomics of the MHC: glimpses into the evolution of the adaptive immune system. *Immunity* 15, 351–362.

Fourmentraux-Neves, E., Jalil, A., Da Rocha, S., Pichon, C., Chouaib, S., Bismuth, G. and Caignard, A. (2008) Two opposite signaling outputs are driven by the KIR2DL1 receptor in human CD4+ T cells. *Blood* 112, 2381–2389.

Gaddum, R.M., Cook, R.S., Furze, J.M., Ellis, S.A. and Taylor, G. (2003) Recognition of bovine respiratory syncytial virus proteins by bovine CD8+ T lymphocytes. *Immunology* 108, 220–229.

Gelhaus, A., Forster, B. and Horstmann, R.D. (1999a) Evidence for an additional cattle DQB locus. *Immunogenetics* 49, 879–885.

Gelhaus, A., Forster, B., Wippern, C. and Horstmann, R.D. (1999b) Evidence for an additional cattle DQA locus, BoLA-DQA5. *Immunogenetics* 49, 321–327.

Gerner, W., Hammer, S.E., Wiesmuller, K.H. and Saalmuller, A. (2009) Identification of major histocompatibility complex restriction and anchor residues of foot-and-mouth disease virus-derived bovine T-cell epitopes. *Journal of Virology* 83, 4039–4050.

Gilliespie, B.E., Jayarao, B.M., Dowlen, H.H. and Oliver, S.P. (1999) Analysis and frequency of bovine lymphocyte antigen DRB3.2 alleles in Jersey cows. *Journal of Dairy Science* 82, 2049–2053.

Giovambattista, G., Golijow, C.D., Dulout, F.N. and Lojo, M.M. (1996) Gene frequencies of DRB3.2 locus of Argentine Creole cattle. *Animal Genetics* 27, 55–56.

Giovambattista, G., Takeshima, S.N., Ripoli, M.V., Matsumoto, Y., Franco, L.A., Saito, H., Onuma, M. and Aida, Y. (2013) Characterization of bovine MHC DRB3 diversity in Latin American Creole cattle breeds. *Gene* 519, 150–158.

Glass, E.J., Oliver, R.A. and Russell, G.C. (2000) Duplicated DQ haplotypes increase the complexity of restriction element usage in cattle. *Journal of Immunology* 165, 134–138.

Glusman, G., Rowen, L., Lee, I., Boysen, C., Roach, J.C., Smit, A.F., Wang, K., Koop, B.F. and Hood, L. (2001) Comparative genomics of the human and mouse T cell receptor loci. *Immunity* 15, 337–349.

Gorer, P.A. (1937) The genetic and antigenic basis of tumour transplantation. *Journal of Pathology and Bacteriology* 44, 691–697.

Groh, V., Steinle, A., Bauer, S. and Spies, T. (1998) Recognition of stress-induced MHC molecules by intestinal epithelial gammadelta T cells. *Science* 279, 1737–1740.

Gu, F., Chowdhary, B.P., Andersson, L., Harbitz, I. and Gustavsson, I. (1992) Assignment of the bovine immunoglobulin gamma heavy chain (IGHG) gene to chromosome 21q24 by in situ hybridization. *Hereditas* 117, 237–240.

Guethlein, L.A., Abi-Rached, L., Hammond, J.A. and Parham, P. (2007) The expanded cattle KIR genes are orthologous to the conserved single-copy KIR3DX1 gene of primates. *Immunogenetics* 59, 517–522.

Guzman, E., Taylor, G., Charleston, B., Skinner, M.A. and Ellis, S.A. (2008) An MHC-restricted CD8+ T-cell response is induced in cattle by foot-and-mouth disease virus (FMDV) infection and also following vaccination with inactivated FMDV. *Journal of General Virology* 89, 667–675.

Guzman, E., Birch, J.R. and Ellis, S.A. (2010a) Cattle MIC is a ligand for the activating NK cell receptor NKG2D. *Veterinary Immunology and Immunopathology* 136, 227–234.

Guzman, E., Taylor, G., Charleston, B. and Ellis, S.A. (2010b) Induction of a cross-reactive CD8(+) T cell response following foot-and-mouth disease virus vaccination. *Journal of Virology* 84, 12375–12384.

Hackney, J.A., Misaghi, S., Senger, K., Garris, C., Sun, Y., Lorenzo, M.N. and Zarrin, A.A. (2009) DNA targets of AID evolutionary link between antibody somatic hypermutation and class switch recombination. *Advances in Immunology* 101, 163–189.

Hammer, D.K., Kickhofen, B. and Schmid, T. (1971) Detection of homocytotropic antibody associated with a unique immunoglobulin class in the bovine species. *European Journal of Immunology* 1, 249–257.

Hammond, J.A., Marsh, S.G., Robinson, J., Davies, C.J., Stear, M.J. and Ellis, S.A. (2012) Cattle MHC nomenclature: is it possible to assign sequences to discrete class I genes? *Immunogenetics* 64, 475–480.

Hanby-Flarida, M.D., Trask, O.J., Yang, T.J. and Baldwin, C.L. (1996) Modulation of WC1, a lineage-specific cell surface molecule of gamma/delta T cells augments cellular proliferation. *Immunology* 88, 116–123.

Harrington, L.E., Hatton, R.D., Mangan, P.R., Turner, H., Murphy, T.L., Murphy, K.M. and Weaver, C.T. (2005) Interleukin 17-producing CD4+ effector T cells develop via a lineage distinct from the T helper type 1 and 2 lineages. *Nature Immunology* 6, 1123–1132.

Havran, W.L., Chien, Y.H. and Allison, J.P. (1991) Recognition of self antigens by skin-derived T cells with invariant gamma delta antigen receptors. *Science* 252, 1430–1432.

Hayday, A.C. (2000) [gamma][delta] cells: a right time and a right place for a conserved third way of protection. *Annual Reviews in Immunology* 18, 975–1026.

Hayes, H.C. and Petit, E.J. (1993) Mapping of the beta-lactoglobulin gene and of an immunoglobulin M heavy chain-like sequence to homoeologous cattle, sheep, and goat chromosomes. *Mammalian Genome* 4, 207–210.

Herzig, C.T. and Baldwin, C.L. (2009) Genomic organization and classification of the bovine WC1 genes and expression by peripheral blood gamma delta T cells. *BMC Genomics* 10, 191.

Herzig, C., Blumerman, S., Lefranc, M.P. and Baldwin, C. (2006) Bovine T cell receptor gamma variable and constant genes: combinatorial usage by circulating gammadelta T cells. *Immunogenetics* 58, 138–151.

Herzig, C.T., Lefranc, M.P. and Baldwin, C.L. (2010a) Annotation and classification of the bovine T cell receptor delta genes. *BMC Genomics* 11, 100.

Herzig, C.T., Waters, R.W., Baldwin, C.L. and Telfer, J.C. (2010b) Evolution of the CD163 family and its relationship to the bovine gamma delta T cell co-receptor WC1. *BMC Evolutionary Biology* 10, 181.

Holmes, E.C., Roberts, A.F., Staines, K.A. and Ellis, S.A. (2003) Evolution of major histocompatibility complex class I genes in Cetartiodactyls. *Immunogenetics* 55, 193–202.

Hosseini, A., Campbell, G., Prorocic, M. and Aitken, R. (2004) Duplicated copies of the bovine JH locus contribute to the Ig repertoire. *International Immunology* 16, 843–852.

Ishiguro, N., Aida, Y., Shinagawa, T. and Shinagawa, M. (1993) Molecular structures of cattle T-cell receptor gamma and delta chains predominantly expressed on peripheral blood lymphocytes. *Immunogenetics* 38, 437–443.

Itoh, T., Watanabe, T., Ihara, N., Mariani, P., Beattie, C.W., Sugimoto, Y. and Takasuga, A. (2005) A comprehensive radiation hybrid map of the bovine genome comprising 5593 loci. *Genomics* 85, 413–424.

Jones, J.M. and Simkus, C. (2009) The roles of the RAG1 and RAG2 'non-core' regions in V(D)J recombination and lymphocyte development. *Archivum Immunologiae et Therapiae Experimentalis* 57, 105–116.

Joosten, I., Sanders, M.F. and Hensen, E.J. (1991) Involvement of major histocompatibility complex class I compatibility between dam and calf in the aetiology of bovine retained placenta. *Animal Genetics* 22, 455–463.

Juliarena, M.A., Poli, M., Sala, L., Ceriani, C., Gutierrez, S., Dolcini, G., Rodriguez, E.M., Marino, B., Rodriguez-Dubra, C. and Esteban, E.N. (2008) Association of BLV infection profiles with alleles of the BoLA-DRB3.2 gene. *Animal Genetics* 39, 432–438.

Kacskovics, I. and Butler, J.E. (1996) The heterogeneity of bovine IgG2-VIII. The complete cDNA sequence of bovine IgG2a (A2) and an IgG1. *Molecular Immunology* 33, 189–195.

Kanan, J.H., Nayeem, N., Binns, R.M. and Chain, B.M. (1997) Mechanisms for variability in a member of the scavenger-receptor cysteine-rich superfamily. *Immunogenetics* 46, 276–282.

Kappes, D. and Strominger, J.L. (1988) Human class II major histocompatibility complex genes and proteins. *Annual Review of Biochemistry* 57, 991–1028.

Kaushik, A. and Lim, W. (1996) The primary antibody repertoire of normal, immunodeficient and autoimmune mice is characterized by differences in V gene expression. *Research in Immunology* 147, 9–26.

Kaushik, A., Shojaei, F. and Saini, S.S. (2002) Novel insight into antibody diversification from cattle. *Veterinary Immunology and Immunopathology* 87, 347–350.

Kaushik, A.K., Kehrli, M.E., Jr, Kurtz, A., Ng, S., Koti, M., Shojaei, F. and Saini, S.S. (2009) Somatic hypermutations and isotype restricted exceptionally long CDR3H contribute to antibody diversification in cattle. *Veterinary Immunology and Immunopathology* 127, 106–113.

Klein, J. (1986) *Natural History of the Major Histocompatibility Complex*, 99th Edition. John Wiley & Sons Inc, London.

Knight, K.L., Suter, M. and Becker, R.S. (1988) Genetic engineering of bovine Ig. Construction and characterization of hapten-binding bovine/murine chimeric IgE, IgA, IgG1, IgG2, and IgG3 molecules. *Journal of Immunology* 140, 3654–3659.

Koti, M., Kataeva, G. and Kaushik, A.K. (2008) Organization of D(H)-gene locus is distinct in cattle. *Developmental Biology (Basel)* 132, 307–313.

Koti, M., Kataeva, G. and Kaushik, A.K. (2010) Novel atypical nucleotide insertions specifically at VH-DH junction generate exceptionally long CDR3H in cattle antibodies. *Molecular Immunology* 47, 2119–2128.

Kropshofer, H., Hammerling, G.J. and Vogt, A.B. (1999) The impact of the non-classical MHC proteins HLA-DM and HLA-DO on loading of MHC class II molecules. *Immunological Reviews* 172, 267–278.

Lahmers, K.K., Hedges, J.F., Jutila, M.A., Deng, M., Abrahamsen, M.S. and Brown, W.C. (2006) Comparative gene expression by WC1+ gammadelta and CD4+ alphabeta T lymphocytes, which respond to Anaplasma marginale, demonstrates higher expression of chemokines and other myeloid cell-associated genes by WC1+ gammadelta T cells. *Journal of Leukocyte Biology* 80, 939–952.

Larsen, P.A. and Smith, T.P. (2012) Application of circular consensus sequencing and network analysis to characterize the bovine IgG repertoire. *BMC Immunology* 13, 52.

Lee, B.Y., Hur, T.Y., Jung, Y.H. and Kim, H. (2012) Identification of BoLA-DRB3.2 alleles in Korean native cattle (Hanwoo) and Holstein populations using a next generation sequencer. *Animal Genetics* 43, 438–441.

Lefranc, M.P. (2001) IMGT, the international ImMunoGeneTics database. *Nucleic Acids Research* 29, 207–209.

Lefranc, M.P., Pommie, C., Ruiz, M., Giudicelli, V., Foulquier, E., Truong, L., Thouvenin-Contet, V. and Lefranc, G. (2003) IMGT unique numbering for immunoglobulin and T cell receptor variable domains and Ig superfamily V-like domains. *Developmental and Comparative Immunology* 27, 55–77.

Lei, W., Liang, Q., Jing, L., Wang, C., Wu, X. and He, H. (2012) BoLA-DRB3 gene polymorphism and FMD resistance or susceptibility in Wanbei cattle. *Molecular Biology Reports* 39, 9203–9209.

Lewin, H.A., Russell, G.C. and Glass, E.J. (1999) Comparative organization and function of the major histocompatibility complex of domesticated cattle. *Immunological Reviews* 167, 145–158.

Ligtenberg, A.J., Veerman, E.C., Nieuw Amerongen, A.V. and Mollenhauer, J. (2007) Salivary agglutinin/glycoprotein-340/DMBT1: a single molecule with variable composition and with different functions in infection, inflammation and cancer. *Biological Chemistry* 388, 1275–1289.

Lindberg, P.G. and Andersson, L. (1988) Close association between DNA polymorphism of bovine major histocompatibility complex class I genes and serological BoLA-A specificities. *Animal Genetics* 19, 245–255.

Lucier, M.R., Thompson, R.E., Waire, J., Lin, A.W., Osborne, B.A. and Goldsby, R.A. (1998) Multiple sites of V lambda diversification in cattle. *Journal of Immunology* 161, 5438–5444.

Macdonald, I.K., Harkiolaki, M., Hunt, L., Connelley, T., Carroll, A.V., MacHugh, N.D., Graham, S.P., Jones, E.Y., Morrison, W.I., Flower, D.R. and Ellis, S.A. (2010) MHC class I bound to an immunodominant *Theileria parva* epitope demonstrates unconventional presentation to T cell receptors. *PLoS Pathogens* 6, e1001149.

MacHugh, N.D., Connelley, T., Graham, S.P., Pelle, R., Formisano, P., Taracha, E.L., Ellis, S.A., McKeever, D.J., Burrells, A. and Morrison, W.I. (2009) CD8+ T-cell responses to *Theileria parva* are preferentially directed to a single dominant antigen: Implications for parasite strain-specific immunity. *European Journal of Immunology* 39, 2459–2469.

Mackay, C.R., Maddox, J.F. and Brandon, M.R. (1986) Three distinct subpopulations of sheep T lymphocytes. *European Journal of Immunology* 16, 19–25.

McQueen, K.L., Wilhelm, B.T., Harden, K.D. and Mager, D.L. (2002) Evolution of NK receptors: a single Ly49 and multiple KIR genes in the cow. *European Journal of Immunology* 32, 810–817.

McShane, R.D., Gallagher, D.S., Jr, Newkirk, H., Taylor, J.F., Burzlaff, J.D., Davis, S.K. and Skow, L.C. (2001) Physical localization and order of genes in the class I region of the bovine MHC. *Animal Genetics* 32, 235–239.

Maillard, J.C., Renard, C., Chardon, P., Chantal, I. and Bensaid, A. (1999) Characterization of 18 new BoLA-DRB3 alleles. *Animal Genetics* 30, 200–203.

Maillard, J.C., Chantal, I. and Berthier, D. (2001) Sequencing of four new BoLA-DRB3 and six new BoLA-DQB alleles. *Animal Genetics* 32, 44–46.

Maillard, J.C., Chantal, I., Berthier, D., Thevenon, S., Sidibe, I. and Razafindraibe, H. (2002) Molecular immunogenetics in susceptibility to bovine dermatophilosis: a candidate gene approach and a concrete field application. *Annals of the New York Academy of Science* 969, 92–96.

Marello, K.L., Gallagher, A., McKeever, D.J., Spooner, R.L. and Russell, G.C. (1995) Expression of multiple DQB genes in *Bos indicus* cattle. *Animal Genetics* 26, 345–349.

Marsh, S.G., Albert, E.D., Bodmer, W.F., Bontrop, R.E., Dupont, B., Erlich, H.A., Geraghty, D.E., Hansen, J.A., Hurley, C.K., Mach, B.*et al.* (2005) Nomenclature for factors of the HLA system, 2004. *Tissue Antigens* 65, 301–369.

Matthews, K.E., Mueller, S.G., Woods, C. and Bell, D.N. (2006) Expression of the hemoglobin-haptoglobin receptor CD163 on hematopoietic progenitors. *Stem Cells Development* 15, 40–48.

Mejdell, C.M., Lie, O., Solbu, H., Arnet, E.F. and Spooner, R.L. (1994) Association of major histocompatibility complex antigens (BoLA-A) with AI bull progeny test results for mastitis, ketosis and fertility in Norwegian cattle. *Animal Genetics* 25, 99–104.

Menzies, M. and Ingham, A. (2006) Identification and expression of Toll-like receptors 1–10 in selected bovine and ovine tissues. *Veterinary Immunology and Immunopathology* 109, 23–30.

Miccoli, M.C., Antonacci, R., Vaccarelli, G., Lanave, C., Massari, S., Cribiu, E.P. and Ciccarese, S. (2003) Evolution of TRG clusters in cattle and sheep genomes as drawn from the structural analysis of the ovine TRG2@ locus. *Journal of Molecular Evolution* 57, 52–62.

Mikko, S. and Andersson, L. (1995) Extensive MHC class II DRB3 diversity in African and European cattle. *Immunogenetics* 42, 408–413.

Miretti, M.M., Ferro, J.A., Lara, M.A. and Contel, E.P. (2001) Restriction fragment length polymorphism (RFLP) in exon 2 of the BoLA-DRB3 gene in South American cattle. *Biochemical Genetics* 39, 311–324.

Miyasaka, T., Takeshima, S.N., Matsumoto, Y., Kobayashi, N., Matsuhashi, T., Miyazaki, Y., Tanabe, Y., Ishibashi, K., Sentsui, H. and Aida, Y. (2011) The diversity of bovine MHC class II DRB3 and DQA1 alleles in different herds of Japanese Black and Holstein cattle in Japan. *Gene* 472, 42–49.

Miyasaka, T., Takeshima, S.N., Sentsui, H. and Aida, Y. (2012) Identification and diversity of bovine major histocompatibility complex class II haplotypes in Japanese Black and Holstein cattle in Japan. *Journal of Dairy Science* 95, 420–431.

Miyasaka, T., Takeshima, S.N., Jimba, M., Matsumoto, Y., Kobayashi, N., Matsuhashi, T., Sentsui, H. and Aida, Y. (2013) Identification of bovine leukocyte antigen class II haplotypes associated with variations in bovine leukemia virus proviral load in Japanese Black cattle. *Tissue Antigens* 81, 72–82.

Morita, C.T., Parker, C.M., Brenner, M.B. and Band, H. (1994) TCR usage and functional capabilities of human gamma delta T cells at birth. *Journal of Immunology* 153, 3979–3988.

Morooka, A., Asahina, M., Kohda, C., Tajima, S., Niimi, M., Nishino, Y., Sugiyama, M. and Aida, Y. (1995) Nucleotide sequence and the molecular evolution of a new A2 gene in the DQ subregion of the bovine major histocompatibility complex. *Biochemical and Biophysical Research Communications* 212, 110–117.

Morrison, W.I. and Davis, W.C. (1991) Individual antigens of cattle. Differentiation antigens expressed predominantly on CD4- CD8- T lymphocytes (WC1, WC2). *Veterinary Immunology and Immunopathology* 27, 71–76.

Mousavi, M., Rabbani, H., Pilstrom, L. and Hammarstrom, L. (1998) Characterization of the gene for the membrane and secretory form of the IgM heavy-chain constant region gene (C mu) of the cow (*Bos taurus*). *Immunology* 93, 581–588.

Muggli-Cockett, N.E. and Stone, R.T. (1991) Restriction fragment length polymorphisms in bovine major histocompatibility complex class II beta-chain genes using bovine exon-containing hybridization probes. *Animal Genetics* 22, 123–136.

Neuberger, M.S. and Scott, J. (2000) Immunology. RNA editing AIDs antibody diversification? *Science* 289, 1705–1706.

Nguyen, T.K., Koets, A.P., Vordermeier, M., Jervis, P.J., Cox, L.R., Graham, S.P., Santema, W.J., Moody, D.B., van Calenbergh, S., Zajonc, D.M. et al. (2013) The bovine CD1D gene has an unusual gene structure and is expressed but cannot present alpha-galactosylceramide with a C26 fatty acid. *International Immunology* 25, 91–98.

Niimi, M., Nakai, Y. and Aida, Y. (1995) Nucleotide sequences and the molecular evolution of the DMA and DMB genes of the bovine major histocompatibility complex. *Biochemistry and Biophysics Research Communications* 217, 522–528.

Niku, M., Liljavirta, J., Durkin, K., Schroderus, E. and Iivanainen, A. (2012) The bovine genomic DNA sequence data reveal three IGHV subgroups, only one of which is functionally expressed. *Developmental and Comparative Immunology* 37, 457–461.

Nishino, Y., Tajima, S. and Aida, Y. (1995) Cattle cDNA clone encoding a new allele of the MHC class II DQA1 gene. *Immunogenetics* 42, 306–307.

Norimine, J. and Brown, W.C. (2005) Intrahaplotype and interhaplotype pairing of bovine leukocyte antigen DQA and DQB molecules generate functional DQ molecules important for priming CD4(+) T-lymphocyte responses. *Immunogenetics* 57, 750–762.

Okragly, A.J., Hanby-Flarida, M., Mann, D. and Baldwin, C.L. (1996) Bovine gamma/delta T-cell proliferation is associated with self-derived molecules constitutively expressed *in vivo* on mononuclear phagocytes. *Immunology* 87, 71–79.

Painter, C.A. and Stern, L.J. (2012) Conformational variation in structures of classical and non-classical MHCII proteins and functional implications. *Immunological Reviews* 250, 144–157.

Parham, P. (2004) Killer cell immunoglobulin-like receptor diversity: balancing signals in the natural killer cell response. *Immunological Letters* 92, 11–13.

Park, C.A., Hines, H.C., Monke, D.R. and Threlfall, W.T. (1993) Association between the bovine major histocompatibility complex and chronic posterior spinal paresis – a form of ankylosing spondylitis – in Holstein bulls. *Animal Genetics* 24, 53–58.

Park, Y.H., Joo, Y.S., Park, J.Y., Moon, J.S., Kim, S.H., Kwon, N.H., Ahn, J.S., Davis, W.C. and Davies, C.J. (2004) Characterization of lymphocyte subpopulations and major histocompatibility complex haplotypes of mastitis-resistant and susceptible cows. *Journal of Veterinary Science* 5, 29–39.

Parng, C.L., Hansal, S., Goldsby, R.A. and Osborne, B.A. (1996) Gene conversion contributes to Ig light chain diversity in cattle. *Journal of Immunology* 157, 5478–5486.

Pasman, Y., Saini, S.S., Smith, E. and Kaushik, A.K. (2010) Organization and genomic complexity of bovine lambda-light chain gene locus. *Veterinary Immunology and Immunopathology* 135, 306–313.

Rabbani, H., Brown, W.R., Butler, J.E. and Hammarstrom, L. (1997) Polymorphism of the IGHG3 gene in cattle. *Immunogenetics* 46, 326–331.

Reinink, P. and Van Rhijn, I. (2009) The bovine T cell receptor alpha/delta locus contains over 400 V genes and encodes V genes without CDR2. *Immunogenetics* 61, 541–549.

Ripoli, M.V., Liron, J.P., De Luca, J.C., Rojas, F., Dulout, F.N. and Giovambattista, G. (2004) Gene frequency distribution of the BoLA-DRB3 locus in Saavedreno Creole dairy cattle. *Biochemical Genetics* 42, 231–240.

Robinson, J., Mistry, K., McWilliam, H., Lopez, R. and Marsh, S.G. (2010) IPD – the Immuno Polymorphism Database. *Nucleic Acids Research* 38, D863–869.

Rock, E.P., Sibbald, P.R., Davis, M.M. and Chien, Y.H. (1994) CDR3 length in antigen-specific immune receptors. *Journal of Experimental Medicine* 179, 323–328.

Rogers, A.N., Vanburen, D.G., Hedblom, E.E., Tilahun, M.E., Telfer, J.C. and Baldwin, C.L. (2005) Gammadelta T cell function varies with the expressed WC1 coreceptor. *Journal of Immunology* 174, 3386–3393.

Russell, G.C., Marello, K.L., Gallagher, A., McKeever, D.J. and Spooner, R.L. (1994) Amplification and sequencing of expressed DRB second exons from *Bos indicus*. *Immunogenetics* 39, 432–436.

Russell, G.C., Davies, C.J., Andersson, L., Mikko, S., Ellis, S.A., Hensen, E.J., Lewin, H.A., Muggli-Cockett, N.E. and van der Poel, J.J. (1997) BoLA class II nucleotide sequences, 1996: report of the ISAG BoLA Nomenclature Committee. *Animal Genetics* 28, 169–180.

Russell, G.C., Fraser, D.C., Craigmile, S., Oliver, R.A., Dutia, B.M. and Glass, E.J. (2000) Sequence and transfection of BoLA-DRB3 cDNAs. *Animal Genetics* 31, 219–222.

Saini, S.S. and Kaushik, A. (2001) Origin of bovine IgM structural variants. *Molecular Immunology* 38, 389–396.

Saini, S.S. and Kaushik, A. (2002a) Extensive CDR3H length heterogeneity exists in bovine foetal VDJ rearrangements. *Scandinavian Journal of Immunology* 55, 140–148.

Saini, S.S. and Kaushik, A.K. (2002b) Immunoglobulin genes and their diversification in vertebrates. *Current Trends in Immunology* 4, 161–176.

Saini, S., Teo, K., Nangpal, A., Mallard, B.A. and Kaushik, A. (1996) Homologues of murine Vh11 gene are conserved during evolution. *Experimental and Clinical Immunogenetics* 13, 154–160.

Saini, S.S., Hein, W.R. and Kaushik, A. (1997) A single predominantly expressed polymorphic immunoglobulin VH gene family, related to mammalian group, I, clan, II, is identified in cattle. *Molecular Immunology* 34, 641–651.

Saini, S.S., Allore, B., Jacobs, R.M. and Kaushik, A. (1999) Exceptionally long CDR3H region with multiple cysteine residues in functional bovine IgM antibodies. *European Journal of Immunology* 29, 2420–2426.

Saini, S.S., Farrugia, W., Ramsland, P.A. and Kaushik, A.K. (2003) Bovine IgM antibodies with exceptionally long complementarity-determining region 3 of the heavy chain share unique structural properties conferring restricted VH + Vlambda pairings. *International Immunology* 15, 845–853.

Saini, S.S., Farrugia, W., Muthusamy, N., Ramsland, P.A. and Kaushik, A.K. (2007) Structural evidence for a new IgG1 antibody sequence allele of cattle. *Scandinavian Journal of Immunology* 65, 32–38.

Sakaguchi, S., Sakaguchi, N., Asano, M., Itoh, M. and Toda, M. (1995) Immunologic self-tolerance maintained by activated T cells expressing IL-2 receptor alpha-chains (CD25). Breakdown of a single mechanism of self-tolerance causes various autoimmune diseases. *Journal of Immunology* 155, 1151–1164.

Sarrias, M.R., Gronlund, J., Padilla, O., Madsen, J., Holmskov, U. and Lozano, F. (2004) The Scavenger Receptor Cysteine-Rich (SRCR) domain: an ancient and highly conserved protein module of the innate immune system. *Critical Reviews in Immunology* 24, 1–37.

Sarrias, M.R., Rosello, S., Sanchez-Barbero, F., Sierra, J.M., Vila, J., Yelamos, J., Vives, J., Casals, C. and Lozano, F. (2005) A role for human Sp alpha as a pattern recognition receptor. *Journal of Biological Chemistry* 280, 35391–35398.

Sarrias, M.R., Farnos, M., Mota, R., Sanchez-Barbero, F., Ibanez, A., Gimferrer, I., Vera, J., Fenutria, R., Casals, C., Yelamos, J. and Lozano, F. (2007) CD6 binds to pathogen-associated molecular patterns and protects from LPS-induced septic shock. *Proceedings of the National Academy of Sciences USA* 104, 11724–11729.

Sathiyaseelan, T., Naiman, B., Welte, S., MacHugh, N., Black, S.J. and Baldwin, C.L. (2002) Immunological characterization of a gammadelta T-cell stimulatory ligand on autologous monocytes. *Immunology* 105, 181–189.

Schatz, D.G. (2004) Antigen receptor genes and the evolution of a recombinase. *Seminars in Immunology* 16, 245–256.

Schreuder, G.M., Hurley, C.K., Marsh, S.G., Lau, M., Fernandez-Vina, M.A., Noreen, H.J., Setterholm, M. and Maiers, M. (2005) HLA dictionary 2004: summary of HLA-A, -B, -C, -DRB1/3/4/5, -DQB1 alleles and their association with serologically defined HLA-A, -B, -C, -DR, and -DQ antigens. *Human Immunology* 66, 170–210.

Schultz, R.D., Dunne, H.W. and Heist, C.E. (1973) Ontogeny of the bovine immune response. *Infection and Immunity* 7, 981–991.

Sharif, S., Mallard, B.A., Wilkie, B.N., Sargeant, J.M., Scott, H.M., Dekkers, J.C. and Leslie, K.E. (1998) Associations of the bovine major histocompatibility complex DRB3 (BoLA-DRB3) alleles with occurrence of disease and milk somatic cell score in Canadian dairy cattle. *Animal Genetics* 29, 185–193.

Shiina, T., Inoko, H. and Kulski, J.K. (2004) An update of the HLA genomic region, locus information and disease associations: 2004. *Tissue Antigens* 64, 631–649.

Shojaei, F., Saini, S.S. and Kaushik, A.K. (2003) Unusually long germline DH genes contribute to large sized CDR3H in bovine antibodies. *Molecular Immunology* 40, 61–67.

Shu, L., Peng, X., Zhang, S., Deng, G., Wu, Y., He, M., Li, B., Li, C. and Zhang, K. (2012) Non-classical major histocompatibility complex class makes a crucial contribution to reproduction in the dairy cow. *Journal of Reproductive Development* 58, 569–575.

Sigurdardottir, S., Borsch, C., Gustafsson, K. and Andersson, L. (1991) Cloning and sequence analysis of 14 DRB alleles of the bovine major histocompatibility complex by using the polymerase chain reaction. *Animal Genetics* 22, 199–209.

Sigurdardottir, S., Borsch, C., Gustafsson, K. and Andersson, L. (1992) Gene duplications and sequence polymorphism of bovine class II DQB genes. *Immunogenetics* 35, 205–213.

Sinclair, M.C., Gilchrist, J. and Aitken, R. (1997) Bovine IgG repertoire is dominated by a single diversified VH gene family. *Journal of Immunology* 159, 3883–3889.

Skow, L.C., Snaples, S.N., Davis, S.K., Taylor, J.F., Huang, B. and Gallagher, D.H. (1996) Localization of bovine lymphocyte antigen (BoLA) DYA and class I loci to different regions of chromosome 23. *Mammalian Genome* 7, 388–389.

Snell, G.D. and Higgins, G.F. (1951) Alleles at the histocompatibility-2 locus in the mouse as determined by tumor transplantation. *Genetics* 36, 306–310.

Snelling, W.M., Casas, E., Stone, R.T., Keele, J.W., Harhay, G.P., Bennett, G.L. and Smith, T.P. (2005) Linkage mapping bovine EST-based SNP. *BMC Genomics* 6, 74.

Snyder, M.R., Weyand, C.M. and Goronzy, J.J. (2004) The double life of NK receptors: stimulation or co-stimulation? *Trends in Immunology* 25, 25–32.

Spooner, R.L., Leveziel, H., Grosclaude, F., Oliver, R.A. and Vaiman, M. (1978) Evidence for a possible major histocompatibility complex (BLA) in cattle. *Journal of Immunogenetics* 5, 325–346.

Spooner, R.L., Millar, P. and Oliver, R.A. (1979) The production and analysis of antilymphocyte sera following pregnancy and skin grafting of cattle. *Animal Blood Groups and Biochemical Genetics* 10, 99–105.

Starkenburg, R.J., Hansen, L.B., Kehrli, M.E., Jr and Chester-Jones, H. (1997) Frequencies and effects of alternative DRB3.2 alleles of bovine lymphocyte antigen for Holsteins in milk selection and control lines. *Journal of Dairy Science* 80, 3411–3419.

Staska, L.M., Davies, C.J., Brown, W.C., McGuire, T.C., Suarez, C.E., Park, J.Y., Mathison, B.A., Abbott, J.R. and Baszler, T.V. (2005) Identification of vaccine candidate peptides in the NcSRS2 surface protein of *Neospora caninum* by using CD4+ cytotoxic T lymphocytes and gamma interferon-secreting T lymphocytes of infected Holstein cattle. *Infection and Immunity* 73, 1321–1329.

Storset, A.K., Slettedal, I.O., Williams, J.L., Law, A. and Dissen, E. (2003) Natural killer cell receptors in cattle: a bovine killer cell immunoglobulin-like receptor multigene family contains members with divergent signaling motifs. *European Journal of Immunology* 33, 980–990.

Sulimova, G.E., Udina, I.G., Shaikhaev, G.O. and Zakharov, I.A. (1995) [DNA polymorphism of the BoLA-DRB3 gene in cattle in connection with resistance and susceptibility to leukemia]. *Genetika* 31, 1294–1299.

Sun, L., Song, Y., Riaz, H., Yang, H., Hua, G., Guo, A. and Yang, L. (2012a) Polymorphisms in toll-like receptor 1 and 9 genes and their association with tuberculosis susceptibility in Chinese Holstein cattle. *Veterinary Immunology and Immunopathology* 147, 195–201.

Sun, Y., Liu, Z., Ren, L., Wei, Z., Wang, P., Li, N. and Zhao, Y. (2012b) Immunoglobulin genes and diversity: what we have learned from domestic animals. *Journal of Animal Science and Biotechnology* 3, 18.

Symons, D.B., Clarkson, C.A. and Beale, D. (1989) Structure of bovine immunoglobulin constant region heavy chain gamma 1 and gamma 2 genes. *Molecular Immunology* 26, 841–850.

Takeshima, S.N., Ikegami, M., Morita, M., Nakai, Y. and Aida, Y. (2001) Identification of new cattle BoLA-DRB3 alleles by sequence-based typing. *Immunogenetics* 53, 74–81.

Takeshima, S.N., Nakai, Y., Ohta, M. and Aida, Y. (2002) Short communication: characterization of DRB3 alleles in the MHC of Japanese shorthorn cattle by polymerase chain reaction-sequence-based typing. *Journal of Dairy Science* 85, 1630–1632.

Takeshima, S.N., Saitou, N., Morita, M., Inoko, H. and Aida, Y. (2003) The diversity of bovine MHC class II DRB3 genes in Japanese Black, Japanese Shorthorn, Jersey and Holstein cattle in Japan. *Gene* 316, 111–118.

Takeshima, S.N., Miki, A., Kado, M. and Aida, Y. (2007) Establishment of a sequence-based typing system for BoLA-DQA1 exon 2. *Tissue Antigens* 69, 189–199.

Takeshima, S.N., Matsumoto, Y., Chen, J., Yoshida, T., Mukoyama, H. and Aida, Y. (2008) Evidence for cattle major histocompatibility complex (BoLA) class II DQA1 gene heterozygote advantage against clinical mastitis caused by *Streptococci* and *Escherichia* species. *Tissue Antigens* 72, 525–531.

Takeshima, S.N., Matsumoto, Y. and Aida, Y. (2009a) Short communication: Establishment of a new poly-merase chain reaction-sequence-based typing method for genotyping cattle major histocompatibility complex class II DRB3. *Journal of Dairy Science* 92, 2965–2970.

Takeshima, S.N., Sarai, Y., Saitou, N. and Aida, Y. (2009b) MHC class II DR classification based on antigen-binding groove natural selection. *Biochemical and Biophysical Research Communication* 385, 137–142.

Takeshima, S.N., Matsumoto, Y., Miyasaka, T., Arainga-Ramirez, M., Saito, H., Onuma, M. and Aida, Y. (2011) A new method for typing bovine major histocompatibility complex class II DRB3 alleles by combining two established PCR sequence-based techniques. *Tissue Antigens* 78, 208–213.

Takeuchi, N., Ishiguro, N. and Shinagawa, M. (1992) Molecular cloning and sequence analysis of bovine T-cell receptor gamma and delta chain genes. *Immunogenetics* 35, 89–96.

Tian, X., Pascal, G. and Monget, P. (2009) Evolution and functional divergence of NLRP genes in mam-malian reproductive systems. *BMC Evolutionary Biology* 9, 202.

Tobin-Janzen, T.C. and Womack, J.E. (1992) Comparative mapping of IGHG1, IGHM, FES, and FOS in domestic cattle. *Immunogenetics* 36, 157–165.

Tonegawa, S. (1983) Somatic generation of antibody diversity. *Nature* 302, 575–581.

van Eijk, M.J., Stewart-Haynes, J.A. and Lewin, H.A. (1992) Extensive polymorphism of the BoLA-DRB3 gene distinguished by PCR-RFLP. *Animal Genetics* 23, 483–496.

Van Rhijn, I., Koets, A.P., Im, J.S., Piebes, D., Reddington, F., Besra, G.S., Porcelli, S.A., van Eden, W. and Rutten, V.P. (2006) The bovine CD1 family contains group 1 CD1 proteins, but no functional CD1d. *Journal of Immunology* 176, 4888–4893.

Van Rhijn, I., Spiering, R., Smits, M., van Blokland, M.T., de Weger, R., van Eden, W., Rutten, V.P. and Koets, A.P. (2007) Highly diverse TCR delta chain repertoire in bovine tissues due to the use of up to four D segments per delta chain. *Molecular Immunology* 44, 3155–3161.

Vera, J., Fenutria, R., Canadas, O., Figueras, M., Mota, R., Sarrias, M.R., Williams, D.L., Casals, C., Yelamos, J. and Lozano, F. (2009) The CD5 ectodomain interacts with conserved fungal cell wall components and protects from zymosan-induced septic shock-like syndrome. *Proceedings of the National Academy of Sciences USA* 106, 1506–1511.

Verma, S. and Aitken, R. (2012) Somatic hypermutation leads to diversification of the heavy chain immunoglobulin repertoire in cattle. *Veterinary Immunology and Immunopathology* 145, 14–22.

Verma, S., Goldammer, T. and Aitken, R. (2010) Cloning and expression of activation induced cytidine deaminase from *Bos taurus*. *Veterinary Immunology and Immunopathology* 134, 151–159.

Walther, S., Czerny, C.P. and Diesterbeck, U.S. (2013) Exceptionally long CDR3H are not isotype restricted in bovine immunoglobulins. *PLoS One* 8, e64234.

Wang, F., Herzig, C., Ozer, D., Baldwin, C.L. and Telfer, J.C. (2009) Tyrosine phosphorylation of scavenger receptor cysteine-rich WC1 is required for the WC1-mediated potentiation of TCR-induced T-cell proliferation. *European Journal of Immunology* 39, 254–266.

Wang, F., Herzig, C.T., Chen, C., Hsu, H., Baldwin, C.L. and Telfer, J.C. (2011) Scavenger receptor WC1 contributes to the gammadelta T cell response to Leptospira. *Molecular Immunology* 48, 801–809.

Wang, F., Ekiert, D.C., Ahmad, I., Yu, W., Zhang, Y., Bazirgan, O., Torkamani, A., Raudsepp, T., Mwangi, W., Criscitiello, M.F. *et al.* (2013) Reshaping antibody diversity. *Cell* 153, 1379–1393.

Wang, K., Sun, D., Xu, R. and Zhang, Y. (2005) Identification of 19 new BoLA-DQB alleles. *Animal Genetics* 36, 166–167.

Wang, K., Sun, D.X. and Zhang, Y. (2007) Identification of genetic variations of exon 2 of BoLA-DQB gene in five Chinese yellow cattle breeds. *International Journal of Immunogenetics* 34, 115–118.

Werling, D. and Coffey, T.J. (2007) Pattern recognition receptors in companion and farm animals – the key to unlocking the door to animal disease? *Veterinary Journal* 174, 240–251.

Xu, A., van Eijk, M.J., Park, C. and Lewin, H.A. (1993) Polymorphism in BoLA-DRB3 exon 2 correlates with resistance to persistent lymphocytosis caused by bovine leukemia virus. *Journal of Immunology* 151, 6977–6985.

Xu, A., Park, C. and Lewin, H.A. (1994) Both DQB genes are expressed in BoLA haplotypes carrying a duplicated DQ region. *Immunogenetics* 39, 316–321.

Yasuda, M., Fujino, M., Nasu, T. and Murakami, T. (2004) Histological studies on the ontogeny of bovine gut-associated lymphoid tissue: appearance of T cells and development of IgG+ and IgA+ cells in lymphoid follicles. *Developmental and Comparative Immunology* 28, 357–369.

Yasuda, M., Jenne, C.N., Kennedy, L.J. and Reynolds, J.D. (2006) The sheep and cattle Peyer's patch as a site of B-cell development. *Veterinary Research* 37, 401–415.

Yoshida, T., Mukoyama, H., Furuta, H., Kondo, Y., Takeshima, S.N., Aida, Y., Kosugiyama, M. and Tomogane, H. (2009a) Association of BoLA-DRB3 alleles identified by a sequence-based typing method with mastitis pathogens in Japanese Holstein cows. *Animal Science Journal* 80, 498–509.

Yoshida, T., Mukoyama, H., Furuta, H., Kondo, Y., Takeshima, S.N., Aida, Y., Kosugiyama, M. and Tomogane, H. (2009b) Association of the amino acid motifs of BoLA-DRB3 alleles with mastitis pathogens in Japanese Holstein cows. *Animal Science Journal* 80, 510–519.

Zhao, Y., Kacskovics, I., Pan, Q., Liberles, D.A., Geli, J., Davis, S.K., Rabbani, H. and Hammarstrom, L. (2002) Artiodactyl IgD: the missing link. *Journal of Immunology* 169, 4408–4416.

Zhao, Y., Kacskovics, I., Rabbani, H. and Hammarstrom, L. (2003) Physical mapping of the bovine immunoglobulin heavy chain constant region gene locus. *Journal of Biological Chemistry* 278, 35024–35032.

Zimin, A.V., Delcher, A.L., Florea, L., Kelley, D.R., Schatz, M.C., Puiu, D., Hanrahan, F., Pertea, G., Van Tassell, C.P., Sonstegard, T.S., Marcais, G., Roberts, M., Subramanian, P., Yorke, J.A. and Salzberg, S.L. (2009) A whole-genome assembly of the domestic cow, *Bos taurus*. *Genome Biology* 10, R42.

9 Genetics of Disease Resistance

Anatoly Ruvinsky[1] and Alan Teale[2]

[1]University of New England, Armidale, New South Wales, Australia; [2]Retired, UK

Introduction

Ongoing attempts to prevent, control and eradicate the most significant cattle diseases caused by different biological agents have been undertaken in many countries. A recent review of challenges and opportunities in USA shows the high complexity and variability of the situation (Miller *et al.*, 2013). The bulk of these efforts involve management of cattle production systems and/or veterinary interventions. The general question relevant to this chapter is how knowledge of genetic resistance to certain diseases can be used in practice to complement the other approaches. The focus here is resistance to the deleterious consequences of the infected state, and more particularly, on the genetic basis of diversity in resistance within domestic cattle. Non-infectious diseases were not considered.

Genetics of Disease Resistance in Cattle: Relevant Notes

Studies in this field have been driven by two principal objectives. The first of these, as with all scientific endeavour, is to increase knowledge and understanding. In this regard the genomic revolution dramatically widens opportunities for obtaining previously unavailable information on genes influencing resistance to different diseases. The second objective for research into disease resistance in cattle is the prospect of useful applications in agriculture to improve animal productivity, improve animal welfare or reduce risk of zoonoses.

There is significant variation in cattle in terms of resistance to diseases, and this variation is of economic importance. Nevertheless, the application of selection for resistance in the field has been slow to develop for several reasons.

Alternative options for disease control are common, such as efficient management practices including test and slaughter or isolation and quarantine, as are veterinary treatments, vaccination and control of infections. In contrast the genetic route to improving the disease resistance of entire breeds is slow and arduous due to long generation intervals, can be compromised by genetic change in the pathogen, and is difficult in many parts of the world that lack adequate animal breeding expertise and required infrastructure. Extra effort is usually undertaken to minimize the disease exposure of elite animals and their herdmates, thereby reducing the opportunity for direct selection and limiting the information available for conventional prediction of breeding values. For many diseases, there are not obvious phenotypic traits that are reliably correlated to the level of disease, such that the observed phenotypes are often categorical rather than continuous. Disease incidence can vary between herds and years, reducing the amount of information available for prediction when the incidence is low. Considering all pros and cons, one should not miss possible negative correlations between some productivity traits and disease resistance as well as cost of selection. Finally, the logistics of experimentation in disease resistance in cattle can pose a considerable challenge.

However, the situation is changing and the stimulus to undertake research that will provide new options for disease control in cattle seems to be increasing. This is so for three major reasons. First, resistance among pathogens to chemotherapeutic and chemoprophylactic drugs is apparently increasing. Compelling examples are resistance to anthelminthics (Waller, 1997; Sutherland and Leathwick, 2011) and to trypanocidal compounds (Peregrine, 1994; Delespaux et al., 2010). In the case of trypanocide resistance, it can be argued that development of the livestock sector in some of the poorest countries of the world is jeopardized. Second, safe, effective and inexpensive vaccines have not been developed yet for some economically important diseases. The comparative costs of non-genetic disease control options are also a consideration. Third, growing volumes of information relevant to genetic resistance to diverse diseases in cattle should provide a background for breeding and selection.

Nevertheless a realistic outlook is necessary. Obviously parasite and pathogen genomes will not remain unchanged while cattle genomes are modified by ongoing selection for resistance. Still the fact that some livestock populations are relatively resistant to certain diseases, and have remained so for thousands of years in some cases, suggests that 'agreements' between pathogens and hosts can be brokered at various levels. This view of the genetic option raises another important point. It implies that selection will usually be a means of disease control rather than a means of infection or parasite control per se.

A subjective comparison of different disease control options, in terms of a variety of features, is given in Fig. 9.1. As the figure shows, most options contain weak and negative features. Some options like vaccination and movement control look particularly attractive and are used in cattle populations very regularly. The option of selecting for disease resistance has one major problem, namely difficulty in creating such cattle. Except for a few rare examples, some of which will be mentioned below, this option was not widely used in the past despite existence of genetic variability in different breeds to a variety of diseases. There are indications that the situation may change in the future due to new knowledge generated by genomics. However, even in the most advanced cases the final verdict will be written in economic terms.

Host resistance can operate at different levels:

- physical barrier, be it an epidermis, mucosal or serosal surface;
- innate or acquired immune responses (see Chapter 8); or
- a range of innate non-immunological factors such as lack of essential nutrients/ substrates, lack of receptors for potential intracellular parasites, incompatible intracellular processing mechanisms, etc.

The last group of factors reflects fundamental incompatibilities in what would otherwise be a host–parasite or host–pathogen relationship. At this third level, potential pathogens may establish, but not cause a significant illness. A good example is provided by *Trypanosoma congolense* infection in resistant cattle types (Murray et al., 1984; Trail et al., 1989).

One may speculate that disease control achieved by exploiting resistance to the

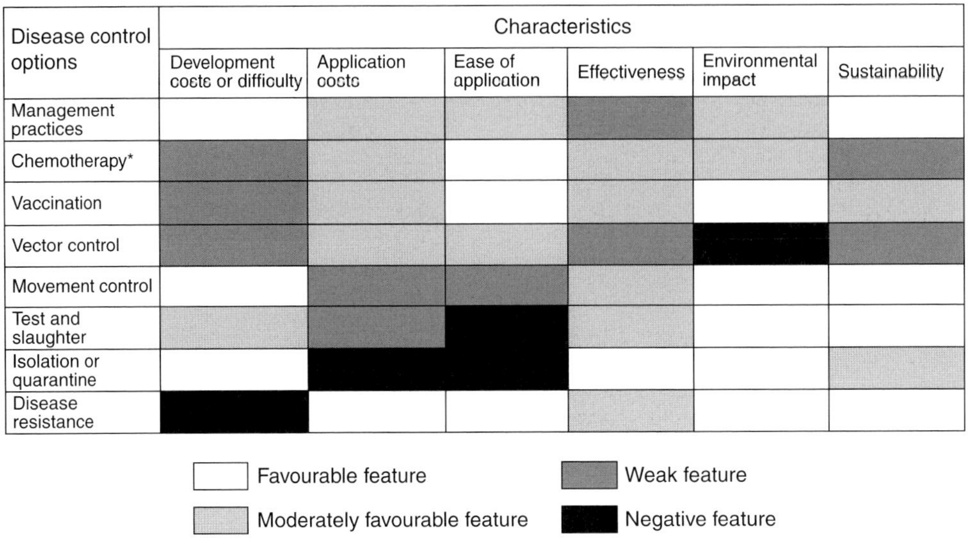

Disease control options	Characteristics					
	Development costs or difficulty	Application costs	Ease of application	Effectiveness	Environmental impact	Sustainability
Management practices	Favourable	Moderately favourable	Moderately favourable	Weak	Favourable	Favourable
Chemotherapy*	Weak	Moderately favourable	Favourable	Favourable	Weak	Weak
Vaccination	Weak	Moderately favourable	Favourable	Moderately favourable	Favourable	Moderately favourable
Vector control	Moderately favourable	Moderately favourable	Moderately favourable	Weak	Negative	Weak
Movement control	Favourable	Moderately favourable	Moderately favourable	Favourable	Favourable	Favourable
Test and slaughter	Moderately favourable	Moderately favourable	Negative	Favourable	Favourable	Favourable
Isolation or quarantine	Favourable	Negative	Negative	Favourable	Favourable	Moderately favourable
Disease resistance	Negative	Favourable	Favourable	Moderately favourable	Favourable	Favourable

Legend: ☐ Favourable feature   ▨ Weak feature   ▨ Moderately favourable feature   ■ Negative feature

Fig. 9.1. A subjective comparison of different disease control options in cattle in terms of some essential characteristics.

consequences of the infected or parasitized state may be the most sustainable, because it allows a continuing host–parasite relationship. As soon as genetic resistance is practically achievable at a sufficient level and financially affordable it probably should become an integral part of the general strategy of protecting cattle from the most damaging diseases. The general constraints and targets of this chapter can be formulated as follows:

- It is expected that many disease resistance traits are complex in nature due to involvement of numerous genes and their multiple interactions, which form dynamic systems operating in changing environmental conditions.
- The extent of genetic variation is a critical factor in dictating the opportunities for conferring resistance or tolerance to a wide range of diseases.
- Unlike productivity, disease resistance such as that found in some rare breeds has largely been the product of natural selection over very long time periods. With the increasing knowledge in genetic resistance to major cattle diseases this process can be radically accelerated.
- The ultimate objective of these efforts to select for disease resistance is to create animals that also exhibit efficiency and productivity.

- Resistance to disease is a particularly important attribute of livestock in low input production systems typically in developing tropical countries and is often the key factor in the sustainability of such agricultural systems (Gibson and Bishop, 2005).

The choice of diseases, resistance to which is discussed below, is based on three criteria: the damaging consequences of the disease to agriculture and human health; the degree of disease virulence and currently available knowledge. Our preference was given to diseases which inflict the most significant economic losses and may present some danger for humans; possess moderate to considerable virulence and have been intensively investigated to characterize genetic resistance.

Genetic Aspects of Resistance Traits for the Most Significant Cattle Diseases

Brucellosis

Brucellosis in cattle is caused by *Brucella abortus* and may cause abortion during the last 3 months of pregnancy. While the abortion rate might not be high, animals suffer, productivity is affected and there is a significant risk of human infection. Human incidence rate is highly correlated to cattle rate, $r = 0.82$ (Lee et al., 2013). Vaccination and slaughter are often used for control and eradication of brucellosis in cattle.

Since Thimm (1973) reported evidence of natural resistance to brucellosis in East African shorthorn zebu cattle, a considerable amount of research has been undertaken to confirm genetic control of variation in resistance to brucellosis in cattle, and to identify underlying cellular and molecular mechanisms. Breeding experiments confirming heritable variation in brucellosis resistance as a basis for numerous subsequent studies were initiated by Templeton and Adams and colleagues in the 1970s (Templeton and Adams, 1995). In these studies, males and females received a standard challenge with *Brucella abortus* S2308 in mid-gestation. Resistant cows did not abort and *Brucella* cells could not be recovered from these cows or calves. Similarly bulls were classified resistant if *Brucella* cells could not be recovered from semen at slaughter. Candidate immune responses were examined in these animals, and of these, two stand out. First, resistant and susceptible animals were found to differ with respect to anti-LPS IgG2a allotypes, with the A allotype over-represented in susceptible animals (Estes et al., 1990). Second, significant differences emerged in macrophage function in terms of respiratory burst in response to *B. abortus* (Harmon et al., 1989) and ability to control bacterial growth (Price et al., 1990; Campbell and Adams, 1992; Qureshi et al., 1996). A detailed characterization of immune response to brucellosis in mice and human was given by Ko and Splitter (2003).

With the realization that the differences in antibacterial activity of macrophages derived from resistant and susceptible cattle may extend to other intracellular pathogens, like *Mycobacterium bovis* and *Salmonella dublin* (Qureshi et al., 1996), close parallels with resistance to intracellular pathogens in mice under the control of the *Bcg/Ity/Lsh* gene were drawn. The murine gene had been positionally cloned and designated *NRAMP1* (natural resistance-associated macrophage protein 1) (Vidal et al., 1993), and its product

was postulated to function in the phagolyso-somal membrane to concentrate oxidation products with antibacterial activity. Subsequently, Feng et al. (1996) reported cloning and analysis of the bovine homologue that revealed significant genetic variability for resistance or susceptibility to brucellosis in cattle. Part of this resistance in cattle has been associated with a 3' untranslated polymorphism in the gene (Adams and Templeton, 1998; Barthel et al., 2001). Thus, there is some evidence that resistance to brucellosis is at least partially under the control of the bovine gene formerly called *NRAMP1* but currently known as *SLC11A1* (solute carrier family 11) and located at bovine chromosome 2. Studies suggest that in cattle other genes might also be involved (Adams and Templeton, 1998).

It was established that this gene is associated with resistance or susceptibility to more than one bacterial species and likely in a number of mammalian species. A polymorphism in some introns and exons of *SLC11A1* gene was described. 'A substitution at nucleotide position 1202 in exon 5 of the Japanese black, Angus, Philippine and Bangladesh swamp-type buffaloes which coded for Thr, while the Korean cattle, Holstein, African N'dama, Indonesian swamp-type buffalo and the Bangladesh river-type buffalo had Ile. All the breeds of cattle and buffaloes tested in this study coded for Gly at the position in exon 6, which corresponds to the same amino acid of the murine Nramp1-resistant phenotype at position 169' (Ables et al., 2002). Additional polymorphism in this bovine locus was described by Coussens et al. (2004).

The resistance/susceptibility of two cattle breeds (Blanco Orejinegro Creole and Brahman) to brucellosis was evaluated in an F1 population generated by all possible crosses between purebred resistant and susceptible animals based on challenges *in vitro* and *in vivo*. 'The association between single nucleotide polymorphisms identified in the coding region of the *SLC11A1* gene and resistance/susceptibility was estimated. The trait resistance or susceptibility to brucellosis, evaluated by a challenge *in vitro*', showed a high heritability ($h^2 = 0.54 \pm 0.11$; Table 9.1). 'In addition, there was a significant association ($P < 0.05$) between the control of bacterial survival and two polymorphisms (a 3'UTR and SNP4 located in exon 10). The antibody response in animals classified as resistant to *Brucella abortus* infection differed significantly ($P < 0.05$) from that of susceptible animals. However, there was no significant association between single nucleotide polymorphisms located in *SLC11A1* gene and the antibody response stimulated by a challenge *in vivo*' (Martínez *et al.*, 2010).

Importantly, all studies conducted during the last two decades have demonstrated the possibility of selection for brucellosis resistance in cattle and identified at least one strong candidate gene.

Dermatophilosis

The bacterial infection dermatophilosis (Ambrose, 1996) is caused by *Dermatophilus congolensis*, and most commonly manifests as a skin infection that can vary significantly. The disease occurs sporadically throughout the world, and in a broad range of species, including humans. The principal economic importance of dermatophilosis arises from the disease in cattle, which affects productivity. The most severe bovine form was found in the humid tropical regions, especially in West Africa and the Caribbean. Concurrent tick infestation, particularly with the bont tick *Amblyomma variegatum*, is associated with dramatically increased severity of lesions and

Table 9.1. Heritability (additive genetic variance) values for different traits relevant to some infection diseases of cattle.

Breeds	Disease-related traits	Heritability	Authors
Creole breeds, Brahman breeds	Brucellosis susceptibility	0.54 ± 0.11	Martínez *et al.*, 2010
Dutch Holstein Friesian	Faecal egg count	0.07 to 0.21 variable	Coppieters *et al.*, 2009
Angus	Faecal egg count	029 ± 0.18	Leighton *et al.*, 1989
Angus calves 7–12 months old	log FEC (faecal egg count)	0.28 ± 0.05	Morris and Amyes, 2012
Angus cows 2–3 years old	Ave. anti-nematode antibody concentration	0.30	Morris *et al.*, 2003
Angus	Keratoconjunctivitis susceptibility	0.06	Kizilkaya *et al.*, 2013
Hereford	Keratoconjunctivitis susceptibility	0.28 ± 0.05	Snowder *et al.*, 2005
Dairy cattle	Leukosis susceptibilty	0.07 to 0.5 highly variable	Ernst and Petukhov, 1978
Dairy cattle	Mastitis susceptibility somatic cell score	0.11 ± 0.04	Mrode and Swanson, 1996
German Holstein cows	Liability to clinical mastitis	0.08 to 0.15 depending on models	Henrichs *et al.*, 2011
Austrian Fleckvieh dual-purpose cows	Clinical mastitis rate	0.02 to 0.06	Koeck *et al.*, 2010
Average over many breeds	Tick counts	0.34 ± 0.06	Davis, 1993
Taurine and zebu breeds	Tick burden	~0.30	Porto Neto *et al.*, 2011b
Black and White	Bovine tuberculosis susceptibility	0.06 to 0.08	Petukhov *et al.*, 1998
Holstein Friesian Irish and British	Bovine tuberculosis susceptibility	0.12 to 0.18 depending on models	Bermingham *et al.*, 2009, 2011
West African taurine breeds	Trypanosomiasis anaemia control	0.09 to 0.22 depending on models	Dayo *et al.*, 2012

continuing disease progression in most cattle breeds (Ambrose, 1996).

Resistance to dermatophilosis has been described in West Africa, where it is most apparent in the N'Dama breed (Morrow et al., 1996) and it is also a characteristic of the Creole cattle of Guadeloupe (Maillard et al., 1993a). The N'Dama is a West African long-horn B. taurus breed descended from the earliest domestic cattle on the African continent. The Caribbean Creole cattle of Guadeloupe are an admixture of B. taurus and zebu types and, interestingly, there is some evidence that at least some of the B. taurus genetic component originated in West Africa (Maillard et al., 1993a) and includes N'Dama genes.

As far as the aetiology of dermatophilosis is concerned, it seems clear that immunosuppression occurs in animals with severe and progressive skin lesions (Koney et al. 1994, 1996), and its extent correlates with susceptibility. Thus, while D. congolensis can induce skin lesions in the absence of tick infestation, the disease only becomes a significant problem where ticks occur on animals. This in turn, where tick A. variegatum is involved, leads to generalized immunosuppression.

The first studies on the possible underlying genetic mechanism responsible for variation in dermatophilosis resistance took a candidate gene approach, which has resulted in identification of a strong association with polymorphisms in exon 2 of the BoLA-DRB3 gene in Brahman cattle of Martinique (Maillard et al, 1993b, 1996). A subsequent functional candidate gene approach clearly confirmed that the polymorphism within the major histocompatibility complex (MHC), where the BoLA-DRB3 and DQB genes encode molecules involved in the antigen presentation to T cell receptors, is very relevant to resistance/susceptibility to infection agents causing dermatophilosis. A unique BoLA class II haplotype, containing DRB3 exon 2 allele and a specific DQB allele, highly correlates with the susceptibility trait ($P < 0.001$). This particular haplotype was also found in several bovine populations. A marker-assisted strategy was developed in order to remove animals carrying this unique haplotype from the population. The end result of the project was impressive: in a Martinique cattle population under such selection the disease prevalence was reduced from 0.76 to 0.02 over 5 years (Maillard et al., 2002). Introgressing the resistance haplotype to those breeds where it is rare or absent can be achieved by crossbreeding.

Helminthiasis

Helminthiasis is caused by infestation of animals by parasitic worms, which are broadly classified into three major groups: tapeworms, flukes and round worms. The problem is so multifaceted and unevenly studied that it cannot be comprehensively covered in this chapter. Rather we shall concentrate on a group of round worms, nematodes.

Infestation of mainly grazing cattle by different species of nematodes or nematodiasis is a global problem. While usually nematodiasis does not create dramatic consequences for animals, economic losses are very significant, as it is 'only second to mastitis in terms of health costs to dairy farmers in developed countries' (Coppieters et al., 2009). Different management strategies and anthelmintic drugs were developed and introduced in regular practice decades ago in order to fit local conditions and a variety of nematode species (i.e. Trichostrongylus axei, Cooperia punctata, Ostertagia ostertagi, Haemonchus placei, Oesophagostomum radiatum and Trichuris spp.). The success of such complex efforts, and anthelmintics in particular, possibly limited attention to genetic aspects of the disease. However, as far as anthelminthic use is concerned, resistance to the drugs rapidly became an issue in grazing cattle (El-Abdellati et al., 2011; Sutherland and Leathwick, 2011).

Differences in resistance to haemonchosis were reported in Nigerian zebu more than 35 years ago (Ross et al., 1959, 1960). Subsequently significant sire effects for resistance to other helminths were also published (Kloosterman et al., 1978; Leighton et al., 1989; Gasbarre et al. 1990). Kloosterman and colleagues (1978) focused on Dutch Friesian cattle and infection with Cooperia spp. Their results concerning genetic influences were somewhat equivocal and suggested that the possibility for useful selection was remote. More encouraging facts were obtained subsequently in studies in Angus cattle where heritability of 0.29 ± 0.18 for faecal egg count was observed (Leighton et al., 1989). The complexity of this resistance began to emerge in a comprehensive follow-up study (Gasbarre et al., 1990). These initial heritability estimates were supported by recent results obtained in different breeds and countries (Morris et al., 2003; Coppieters et al., 2009; Morris and Amyes, 2012; Table 9.1). A good example of a breed difference in nematodiasis resistance has been reported in cattle under natural challenge in village herds of Gambia. N'Dama cattle appeared to shed lower numbers of eggs (during high challenge periods) and carry smaller worm burdens than zebu cattle (Claxton and Leperre, 1991). This breed difference applied to infections at all levels of the gut, including in the abomasum where Haemonchus contortus was the major worm species involved. These and other data create confidence that there is sufficient genetic variability in regard to nematode resistance and selection programmes can be implemented (Gasbarre et al., 1993, 2001; Sonstegard and Gasbarre, 2001; Morris, 2007). This conclusion is supported by an experiment carried out at CSIRO (Rockhampton, Australia) starting as early as 1966, during which Herefords and Shorthorns were crossed. Crossbred bulls were selected for high weight gain in a stressful environment including nematode parasite challenge (Frisch, 1981). Eventually a new strain Adaptur with genetic resistance to nematodes was developed (O'Neill et al., 1998).

The underlying mechanisms responsible for nematode resistance in cattle are still unknown but relevant facts are steadily emerging. From heritability estimates and other data it was expected that many genes contribute to this set of traits. Such expectation was confirmed by the mapping of quantitative trait loci (QTLs) influencing gastrointestinal nematode burden in Dutch Holstein Friesian dairy cattle (Coppieters et al., 2009). Two genome-wide significant QTLs were identified on chromosomes 9 and 19, in an across-family analysis, coinciding with previous findings in orthologous chromosome regions in sheep. It should be mentioned that some MHC-related genes are mapped to these two chromosomes. Additionally six possible QTLs were

identified as well as 73 informative single nucleotide polymorphisms (SNPs) on chromosome 19.

During earlier periods of investigation into nematodiasis resistance, serum antibody levels received significant attention (Gasbarre et al., 1990), although their role in genetically determined resistance is unclear. Gasbarre et al. (1993) reported a high heritability for serum antibodies, but suggested that this trait was under separate genetic control than control of faecal egg output. Subsequently evidence has been reported to support the role of antibody response in terms of mediating genetically determined variation in egg output (Hammond et al., 1997).

More recent molecular results provide insights into the development of host immunity to gastrointestinal nematode infection and will facilitate understanding of mechanism underlying host resistance (Li et al., 2011). The response of the abomasal transcriptome to gastrointestinal parasites was evaluated in parasite-susceptible and parasite-resistant Angus cattle. Gene (Bt.14427) with unknown function, which produced the most abundant transcript, accounting for 10.4% of sequences in the transcriptome, was identified (Li et al., 2011). Location of this gene is not exactly known. The authors indicated bovine chromosome 29 as a possible location. As of May 2013 both UniGene (GenBank) and Ensembl indicate chromosome 14 as a potential location. Additionally, PIGR (polymeric immunoglobulin receptor precursor, chr. 16); Complement C3 gene (BT.19562 located on chr. 7) and Immunoglobulin J chain (IGJ gene located on chr. 6) were among other abundant transcripts in this transcriptome (Li et al., 2011).

Study of Nelore cattle with different degrees of resistance to Cooperia punctata natural infection suggested that immune response to this nematode was probably mediated by Th2 cytokines in the resistant group and by Th1 cytokines in the susceptible group (Bricarello et al., 2008). Cytokine gene expression in response to Haemonchus placei infections was also studied in Nelore cattle (Zaros et al., 2010). 'The seven most resistant and the eight most susceptible animals were selected based on nematode faecal egg counts (FEC) and worm burden.' Gene expression analysis in the abomasal tissue indicated that *IL4* and *IL13* (TH2 cytokines) genes (chr. 7) were up-regulated in the resistant group, whereas *TNF* (TH1/TH2 cytokine) gene (chr. 23) was up-regulated in the susceptible group. The authors suggested 'a protective TH2-mediated immune response against *H. placei* in the resistant group and a less protective TH1 response in the susceptible group' (Zaros *et al.*, 2010).

Microarray technology was used to delineate gene expression patterns in cattle selected for resistance or susceptibility to intestinal nematodes and the expression patterns of 381 genes with known association to host immune responses (Araujo *et al.*, 2009). The 'results confirmed that in the small intestine mucosa, susceptible animals showed significantly higher levels of expression in the genes encoding IGHG1, CD3E, ACTB, IRF1, CCL5 and C3, while in the mesenteric lymph node of resistant animals, higher levels of expression were confirmed for PTPRC, CD1D and ITGA4'. Genes corresponding to the above proteins are located on chromosomes 21, 15, 11, 7, 19, 7, 16, 3 and 2, respectively. 'Combined, the results indicate that immune responses against gastrointestinal nematode infections involve multiple response pathways. Higher levels of expression for IGE receptor, integrins, complement, monocyte/macrophage and tissue factors are related to resistance. In contrast, higher levels of expression for immunoglobulin chains and TCRs are related to susceptibility' (Araujo *et al.*, 2009).

This accumulated knowledge creates an opportunity for further analysis of possible genomic changes during selection for resistance or susceptibility to intestinal nematodes. Comparative genomic hybridization of Angus cattle with extreme faecal egg count phenotypes and pepsinogen levels identified 20 loci with copy number variations, of which 12 were located within known chromosomes harbouring or adjacent to gains or losses of DNA fragments (Liu *et al.*, 2011). 'These variable regions are particularly enriched for immune function affecting receptor activities, signal transduction, and transcription.' The authors also show that common transcription factors are probably involved in parasite resistance. Further investigations may bring new information, which might be important for improving genetic resistance to intestinal nematodes.

Keratoconjunctivitis (IBK)

Infectious bovine keratoconjunctivitis (IBK) is caused by virulent strains of *Moraxella bovis*, which colonizes and affects the corneal surface. Environmental conditions as well as *Mycoplasma* sp. or infectious bovine rhinotracheitis virus may enhance or hasten the disease process. The intensity of IBK may range from mild conjunctivitis to severe ulceration, corneal perforation and blindness (Brown *et al.*, 1998). IBK imposes a significant economic effect on the cattle industry worldwide (Alexander, 2010). *Moraxela bovis* can be controlled by predative Gram-negative, obligate aerobic bacteria; *Bdellovibrio bacteriovorus*, strain 109J was shown to be quite effective in achieving biological control over *Moraxela bovis* (Boileau *et al.*, 2011).

Significant breed differences regarding IBK incidence in the USA have been reported (Snowder *et al.*, 2005). Herefords were significantly more susceptible (22.4%) than the average of 11 other breeds and crosses compared (4.6%). Channel Island breeds seems to be also highly susceptible to IBK (Slatter *et al.*, 1982). Heritability of IBK susceptibility has been measured several times, most recent estimates for Angus cattle was rather low ~0.06 (Kizilkaya *et al.*, 2013) compared to the Hereford breed 0.28 ± 0.05 (Snowder *et al.*, 2005; Table 9.1). Eyelid pigmentation may have a protective effect from IBK (Ward and Nielson, 1979) and large heritability estimates for eyelid pigmentation of 0.64–0.83 have been reported (French, 1959).

Whole genome analysis of IBK in Angus cattle showed that 'magnitudes of genetic variances estimated in localized regions across the genome indicated that SNPs within the most informative regions accounted for much of the genetic variance of IBK and pointed out some degree of association to IBK. There are many candidate genes in these regions which could include a gene or group of genes' somehow connected with susceptibility to IBK in cattle (Kizilkaya *et al.*, 2011). Polymorphism in Toll-like receptor 4 (*TLR4*, chromosome 8) that typically binds to Gram-negative bacteria was studied in American Angus cattle (Kataria *et al.*, 2011). 'Animals with previously calculated breeding values for the IBK susceptibility were used to identify two SNPs in *TLR4*'; one of which observed in intron1 (A→G) had a significant effect on IBK infection rates. This SNP alone could account for 2.1% of phenotypic variation in IBK infection during the disease season and 3.0% of phenotypic variation in IBK infection at the time of weaning. Other genes involved in IBK susceptibility might be uncovered in the near future. Such expectation was confirmed in the latest report, which is aiming to identify other essential genetic polymorphisms (Kizilkaya *et al.*, 2013). This study revealed 11 candidate genes located on chromosomes 2, 12, 13 and 21. The authors draw a conclusion confirming the polygenic nature of IBK, where many loci with small effects are expected and disease incidence is affected by the environment.

A study of genetic parameters of IBK and its relationship with weight and parasite infestations in Australian tropical *Bos taurus* cattle was recently conducted (Ali *et al.*, 2012). The main conclusion is that selection for low IBK rate looks feasible and that genetically susceptible calves could be genetically predisposed to a slower growth.

Leukosis (BL)

Bovine leukosis caused by bovine leukaemia virus (BLV) is of considerable economic importance in the dairy industries of North America and Europe in particular. Affected animals suffer from persistent lymphocytosis (PL) and have reduced milk and fat yields by comparison with infected PL-negative peers (Da *et al.*, 1993). The loss in the American dairy industry due to the effects of BLV infection is measured in hundreds of millions of dollars annually. 'Besides the lethal form of BLV-induced leukemia, persistent lymphocytosis (PL) is characterized by a permanent and relatively stable increase in the number of B lymphocytes in the peripheral blood. The PL stage, which affects approximately one third of infected animals, is considered to be a benign form of the disease resulting from the accumulation of untransformed B lymphocytes. Finally, viral infection

is asymptomatic in the majority of BLV-infected animals; in these settings, small % of peripheral blood cells in animals are found to be infected by virus' (Gillet *et al.*, 2007).

Eradication of BL has been on the agenda for a long time and the first positive results have begun to emerge. Denmark seems to be the first country where the virus has been eradicated through systematic destruction of infected herds. The identification of infected animals was performed on the basis of peripheral blood cell counts without specific serological tests (Gillet *et al.*, 2007).

An attempt to estimate heritability of resistance or susceptibility to BL was undertaken many years ago. A large-scale study of dairy cattle in Russia (~14,000 animals) revealed that daughters of affected bulls were more likely BL sufferers than the average for the populations. A similar observation was true for daughters of affected cows, which got ill more often than those of healthy animals. The heritability for BL was variable from 0.07 to 0.50 (Ernst and Petukhov, 1978; Table 9.1). A lack of other heritability estimates makes it difficult to draw a conclusion regarding the nature of such unusual variation. One may speculate that if some pedigrees carry certain alleles of a major gene affecting susceptibility to BLV, this could significantly affect resistance to BLV in some large families.

That susceptibility of cattle to BLV is influenced by genetic factors was suggested by Burridge and colleagues in 1979 (Burridge *et al.*, 1979). A significant component of the genetic influence was associated with bovine lymphocyte antigen (BoLA) types (Lewin and Bernoco, 1986). The major gene cluster of BoLA or MHC genes is located at bovine chromosome 23 (see Chapter 8 for details). A study conducted in a single herd of Shorthorn cattle revealed that resistance and susceptibility (in terms of virus-dependent B-cell proliferation and lymphocytosis in seropositive animals) is associated with alternative BoLA haplotypes defined by class I antigens. Importantly, Lewin and Bernoco (1986) extended the association study at herd level to the offspring of an individual bull carrying the alternative haplotypes associated with resistance or susceptibility. These authors were able to demonstrate co-segregation of the haplotypes and resistance/susceptibility in 33 offspring examined. The following studies confirmed that several BoLA class I types could be associated with BL resistance and susceptibility as defined by lymphocyte numbers in seropositive animals (Lewin et al. 1988; Stear et al., 1988).

Then attention shifted to the BoLA class II region, which led to discovery of an association between development of persistent lymphocytosis in cattle and BoLA-DRB2 (van Eijk et al., 1992). The effort focused on DRB3 gene (Xu et al., 1993), and specifically on amino acid residues 70 and 71 in the corresponding protein. It was reported that Glu-Arg at positions 70 and 71 was a feature of haplotypes previously found to be associated with resistance to persistent lymphocytosis. Since the original report linking DRB3 with resistance/susceptibility to persistent lymphocytosis in American Holstein Friesians, several DRB3 types (alleles) have been associated with the trait in Black Pied cattle (Sulimova et al., 1995; Ernst et al., 1997).

A study of two Holstein Friesian herds in Italy suggested association with BoLA haplotypes defined by A-locus serology and restriction fragment length polymorphisms in the class II region, including the DRB3 gene (Zanotti et al., 1996). It is notable that the resistant haplotype in the Italian study carries one of the DRB3 types associated with resistance in the studies of Black Pied cattle (Sulimova et al., 1995; Ernst et al., 1997). Likewise, the susceptible haplotype in the Italian Holstein Friesians carries a DRB3 allele associated with susceptibility in the Black Pied cattle.

Most recently BLV proviral load and polymorphism of BoLA class II haplotypes was investigated in Japanese Black cattle (Miyasaka et al., 2013). BoLA-DRB3*0902 and BoLA-DRB3*1101 alleles were associated with a low proviral load (LPVL), and BoLA-DRB3*1601 was associated with a high proviral load (HPVL). Two other alleles BoLA-DQA1*0204 and BoLA-DQA1*10012 were related to LPVL and HPVL, respectively. Furthermore, the correlation was confirmed between the DRB3-DQA1 haplotype and BLV proviral load. Two haplotypes, namely 0902B or C (DRB3*0902-DQA1*0204) and 1101A (DRB3*1101-DQA1*10011), were associated

with a low BLV proviral load, whereas one haplotype 1601B (DRB3*1601-DQA1*10012) was associated with a high BLV proviral load. The authors conclude that resistance is a dominant trait and susceptibility is a recessive trait. The same resistant alleles were common between Japanese Black and Holstein cattle, but susceptible alleles differed. This report identifies an association between the DRB3-DQA1 haplotype and variations in BLV proviral load (Miyasaka et al., 2013). A useful tool for evaluating BLV infection status was developed (Jimba et al., 2012).

The mechanism(s) underlying resistance to the effects of BLV infection still remain unresolved. It was suggested that factors affecting viral spread are important, and that they are likely under genetic control (Mirsky et al., 1996). It is also not well understood why BLV infection is spread much wider in dairy than in beef cattle (Murakami et al., 2011). Interestingly selection of cattle 'carrying alleles of the bovine leukocyte antigen BoLA-DRB3.2 gene associated with BLV-infection resistance, like *0902, emerges as the best additional tool toward controlling virus spread' (Juliarena et al., 2009). Perhaps this selection experiment conducted in Argentina might be in a reasonable agreement with the latest results obtained in Japanese Black and Holstein cattle (Miyasaka et al., 2013).

Mastitis

Mastitis is inflammation of the mammary gland and udder tissue. It is usually a result of immune response to bacterial invasion of the teat canal by a variety of bacteria present on the farm. Mastitis treatment and control are among the largest costs in dairy farms of many countries (Heringstad et al., 2000). Not surprisingly the genetic basis of resistance and susceptibility to mastitis has received significant attention.

Heritability of resistance and susceptibility to mastitis has been estimated in several studies in various cattle breeds (Stanik and Vasil, 1986; Stavikova et al., 1990; Philipsson et al., 1995; Kelm et al., 1997; Koeck et al., 2010; Hinrichs et al., 2011). The utility of somatic cell scores (SCS) in milk samples as a selection parameter has also been established and was based on the high correlation between SCS and mastitis occurrence (Shook and Schutz, 1994; Philipsson et al., 1995). The genetic correlations between clinical mastitis (CM) and somatic cell count traits range from 0.64 to 0.77. As CM and SCS describe 'different aspects of udder health, information on both traits should be considered for selection of bulls. A favourable significant genetic trend in decreasing clinical mastitis in Norwegian Dairy Cattle was estimated as 0.19% per year for cows born after 1990. This trend was considered against several key traits and the conclusion was drawn' that selection for increased milk production will result in an unfavourable correlated increase in mastitis incidence, if mastitis is ignored in the breeding program (Heringstad et al., 2003).

The search for MHC markers of mastitis resistance led to reports that associations with MHC class I type (Oddgeirsson et al., 1988; Lundén et al., 1990; Simpson et al., 1990; Weigel et al., 1990; Aarestrup et al., 1995; Simon et al., 1995) and with class II type (Lundén et al., 1990; Dietz et al., 1997; Kelm et al., 1997) do exist. Schukken et al. (1994) reported a serologically defined class I specificity associated with susceptibility to intramammary challenge with Staphylococcus aureus in artificial challenge experiments. However, associations between mastitis and MHC alleles and haplotypes were not always confirmed (Våge et al., 1992). A large number of reports coming from different sources brought numerous and convincing confirmations that BoLA-DB3 alleles indeed are relevant to resistance or susceptibility to mastitis. Presence of glutamine at position 74 of pocket 4 in the BoLA-DR antigen binding groove is associated with occurrence of clinical mastitis caused by Staphylococcus species (Sharif et al., 2000). In Japanese Holstein cows DRB3.2*8(DRB3*1201) and DRB3.2*16(DRB3*1501) alleles were found to be associated with susceptibility, while DRB3.2*22(DRB3*1101), DRB3.2*23 (DRB3*2703) and DRB3.2*24(DRB3*0101) alleles were found to be associated with resistance (Yoshida et al., 2012). Similarly in Argentinean Holstein cattle a significant association was revealed between BoLA-DRB3.2*23 and DRB3.2*27 alleles and protective or susceptibility effects, respectively, in

regard to mastitis. The phenotypic trait used in this case was somatic cell count (Baltian et al., 2012). BoLA-DRB3.2 alleles' association with resistance to persistent lymphocytosis caused by the bovine leukaemia virus and to various forms of mastitis were identified in the Mongolian and Kalmyk cattle (Ruzina et al., 2010). Other MHC genes are also involved in the mastitis-related immune response. It was found that two BoLA-DQA1 alleles promote susceptibility to Streptococci-induced mastitis, namely BoLA-DQA1*0101 and BoLA-DQA1*10012. The homozygous BoLA-DQA1*0101/0101 and BoLA-DQA1*10011/10011 genotypes promote susceptibility to mastitis caused by Streptococci and Escherichia spp., respectively (Takeshima et al., 2008).

Numerous QTLs located on at least 15 bovine chromosomes contributing to development of mastitis resistance/susceptibility have been identified so far (www.animalgenome.org/cgi-bin/QTLdb/BT/search). However, these QTLs are not equally influential. A genome-wide association study for milk production and somatic cell score in Irish Holstein Friesian cattle among other associations indicates prominence of a region on chromosome 13 associated with milk yield (Meredith et al., 2012). Pearson correlations between somatic cell score and milk, fat and protein yield are 0.12, 0.16 and 0.17, respectively. There are many publications pointing to different traits that could be involved in development of resistance to mastitis; unfortunately only in rare cases were genes behind these traits identified.

A few recent publications have brought new important information. For example, high-mobility group box protein 1 (HMGB1) gene located at bovine chromosome 12 'has universal sentinel function for nucleic acid-mediated innate immune responses and acts as a pathogenic mediator in the inflammatory disease' (Li et al., 2012). A novel SNP (g. +2776 A>G) in the 3'-UTR region of HMGB1 gene, altering the binding of the UTR sequence with microRNA (bta-miR-223), was found to be associated with somatic count scores in cows. 'The expression of bta-miR-223 is significantly upregulated in the bovine mastitis-infected mammary gland tissues.' Expression of HMGB1 mRNA in cows with GG genotype (homozygous for this SNP) is significantly higher than in animals with other genotypes. This new allele is a promising functional marker for mastitis resistance (Li *et al.*, 2012).

Toll-like receptor 1 (*TLR1*) gene, putatively located on bovine chromosome 6, has an integral role in initiation and regulation of the immune response to microbial pathogens, and has been known to be involved in numerous inflammatory diseases in several species (see Chapter 8 for details). Two SNPs in the *TLR1* gene were found within a Holstein Friesian herd (the tagging SNP -79 T → G and the 3′UTR SNP +2463 C → T) that were associated with CM. There was a favourable relationship between reduced CM incidents with increased milk fat and protein production. Cows 'with the GG genotype (from the tag SNP -79 T → G) had significantly lower *TLR1* expression in milk-producing somatic cells when compared with TT or TG animals. In addition, stimulation of leucocytes from GG animals with the TLR1-ligand PAM3CSK4 resulted in significantly lower levels of CXCL8 mRNA and protein.' These SNPs are significantly associated with CM (Russell *et al.*, 2012).

The latest contribution is detection of a QTL for CM status on bovine chromosome 11. This large-scale study used exome and genome sequence data to examine the QTL region in more detail. It reveals association with markers encompassing the interleukin-1 (*IL1*) gene cluster and prioritizes the *IL1* gene cluster for further analysis in the genetic control of mammary infection resistance (Littlejohn *et al.*, 2014, unpublished; D.J. Garrick, personal communication). Hopefully further research in this field may combine existing data and generate new practically important knowledge leading to higher genetic resistance to mastitis particularly in dairy cattle.

Paratuberculosis (Johne's disease)

Paratuberculosis or Johne's disease in cattle is caused by *Mycobacterium avium* ssp. *paratuberculosis* and is highly infectious. Resistance to paratuberculosis in Dutch dairy cattle has low heritability (0.06; Table 9.1) as estimated by Koets *et al.* (2000). Minozzi *et al.* (2010) and independently Kirkpatrick *et al.* (2011) undertook genome-wide association studies in

order to find QTLs relevant to resistance and susceptibility to paratuberculosis in cattle, and identified several potential QTLs. Meta-analysis of two genome-wide association studies of bovine paratuberculosis (Minozzi *et al.*, 2012) identified contributions of loci on chromosomes 1, 12 and 15, from which associations on chromosome 12 had earlier been identified in an Italian population. Meanwhile Ruiz-Larrañaga *et al.* (2010) investigated *SP110* gene as potential candidate, which was earlier implicated in tuberculosis resistance/susceptibility. This group was able to identify a SNP that causes amino acid change in codon 196 (Asn→Ser), which was considered as a putative causal variant for susceptibility to paratuberculosis. This result may constitute a step forward towards the implementation of marker-assisted selection in breeding programmes aimed at controlling paratuberculosis. Very recently another SNP was described in the bovine *CD209* candidate gene (Ruiz-Larrañaga *et al.*, 2012). However, resistance/susceptibility to bovine paratuberculosis might be controlled on a number of levels and further studies should add clarity to this important matter.

Tick resistance

Resistance or susceptibility to numerous tick species, particularly in tropical countries, has a tremendous economic impact on cattle production systems (Machado *et al.*, 2010). In Australia alone despite active veterinary measures, ticks and tick-borne diseases result in estimated losses of livestock production of around AUS$200 million per year (Porto Neto *et al.*, 2012). Tick resistance in cattle has been the subject of study over many decades in different breeds in a variety of locations. It has involved studies of infestations with several major tick species, e.g. *Boophilus microplus* (O'Kelly and Spiers, 1976; Stear, *et al.*, 1984), *B. decoloratus* (Rechav and Kostrzewski, 1991; Ali and de Castro, 1993), *Amblyomma americanum* (George *et al.*, 1985; Barnard, 1990a), *A. hebraeum* (Rechav *et al.*, 1991; Norval *et al.*, 1996), *A. variegatum* (Claxton and Leperre, 1991; Mattioli *et al.*, 1993; Morrow *et al.*, 1996), *Rhipicephalus appendiculatus* (Latif *et al.*, 1991), *Haemphysalis longicornis* (Dicker and Sutherst, 1981) and *Ixodes rubicundus* (Fourie and Kok, 1995). These studies have involved both natural and artificial challenges, and resistance has generally been measured on the basis of an overall visual assessment of tick burdens (de Castro *et al.*, 1991), most commonly by tick counts, and survival (Utech and Wharton, 1982), weight and fecundity (Barnard, 1990b; Latif *et al.*, 1991). Effects on tick population growth rates have also been determined (Barnard, 1990a) and have served to underscore the value of this disease resistance trait.

In some cases, attempts have been made to correlate host morphological characteristics (de Castro *et al.*, 1991), and especially aspects of immune response, with tick resistance (George *et al.*, 1985; Rechav *et al.*, 1990; Claxton and Leperre, 1991), although causal relationships have not been established unequivocally. MHC class I typing of cattle of defined resistance/susceptibility to *B. microplus* revealed only weak associations (Stear *et al.*, 1984).

Nevertheless some general conclusions can be drawn from the results of studies of tick resistance reported to date. First, there is evidence that in situations where cattle are challenged by multiple tick species, resistance is generated to all of them (de Castro *et al.*, 1991; Mattioli *et al.*, 1993; Rechav *et al.*, 1990; Ali and de Castro, 1993; Solomon and Kaaya, 1996). Proportions of ticks belonging to different species tend to be the same on susceptible and resistant animals. Second, tick resistance tends to be acquired with exposure (O'Kelly and Spiers, 1976; George *et al.*, 1985; Spickett *et al.*, 1989; Morrow *et al.*, 1996).

The most commonly reported observation on tick resistance is relatively higher resistance of *B. indicus* (zebu) cattle in comparison with *Bos taurus* types (Dicker and Sutherst, 1981; Rechav *et al.*, 1990). The N'Dama cattle (*B. taurus*) represent a rare exception from the general rule. The crossbreds have generally been found to be intermediate in terms of resistance in comparison with pure *B. indicus* and *B. taurus* types (O'Kelly and Spiers, 1976; Dicker and Sutherst, 1981; Utech and Wharton, 1982; George *et al.*, 1985; Barnard, 1990a,b; Ali and de Castro, 1993; Fourie and Kok, 1995; Norval *et al.*,

1996). This is an indirect argument that resistance to ticks has to be described as a classical quantitative trait. Heritability for tick resistance has been estimated more than once (Table 9.1). While depending on circumstances the heritability values vary (Port Neto *et al.*, 2010), an average estimate came close to 0.30. This moderate value indicates the presence of a genetic factor affecting traits relevant to the resistance. Studies of the past few years have led to the conclusion that individual 'genetic markers identified for tick burden explain a relatively small proportion of the variance, which is typical of markers for quantitative traits' (Porto Neto *et al.*, 2011b). A QTL affecting tick burden was located on chromosome 10 in the vicinity of the *ITGA11* gene encoding integrin α11. Several haplotypes were also identified on the same chromosome that may individually explain between 1.3% and 1.5% of the residual variance in tick burden (Porto Neto *et al.*, 2010) as well as on chromosome 3 with an effect of less than 1% (Porto Neto *et al.*, 2011a). In a bovine F2 population derived from the Gyr (*B. indicus*) × Holstein (*B. taurus*) cross, several tick burden related QTLs significant in certain conditions were also found on chromosomes 2 and 10 as well as on chromosomes 5, 11 and 27. Furthermore, a highly significant QTL was identified on chromosome 23 (Machado *et al.*, 2010).

Gene expression studies clearly show that both immune and non-immune mechanisms are involved in tick resistance (Porto Neto *et al.*, 2011b). For instance, sequencing of the gene-encoding heavy chain of IgG2 in tick-resistant Nelore (*Bos t. indicus*) and tick-susceptible Holstein breeds revealed several SNPs and haplotypes, of which three were exclusive for Nelore and five for Holstein breeds (Carvalho *et al.*, 2011). In a similar way SNPs and haplotypes from the region on chromosome 14, where *RIPK2* (receptor-interacting serine/threonine-protein kinase 2) gene is located, are associated with tick burden in both dairy and beef cattle. This gene is possibly involved in modulation of IgG production, which is supported by the results obtained using knock-out mice. Namely IgG production in the *RIPK2* −/− mouse was significantly reduced (Porto Neto *et al.*, 2012).

Genes involved in inflammation processes and immune responsiveness to tick infestation were up-regulated in more susceptible Holstein Friesian cattle; however no significant changes were observed in resistant Brahman cattle. Tick-susceptible *B. taurus* cattle display an increased cellular response (Piper *et al.*, 2010). At least one innate immune receptor is involved in reaction to tick in cattle. Toll-like receptor 4 is associated with tick infestation rates and blood histamine concentration (Zhao *et al.*, 2013). There is a significant negative correlation between blood histamine concentration and intensity of tick infestation in BMY cattle (½ Brahman, ¼ Murray Grey and ¼ Yunnan Yellow cattle). Two alleles A and B were identified in the *TLR4* gene (chromosome 8). Homozygotes *BB* were significantly less tick infested and had significantly higher concentration of histamine in blood than two other genotypes, particularly *AA*. It seems that allele *B* has a potential to serve as a marker for zebu-originated tick-resistance (Zhao *et al.*, 2013). Current data on tick resistance testify in favour of numerous genes with small effects. Existence of numerous tick species and a great variety of ecological conditions in which cattle live put in an additional layer of complexity on an already complex situation.

Trypanosomiasis

Resistance to African tsetse-transmitted trypanosomiasis in cattle, which is commonly referred to as trypanotolerance, is among actively researched traits in livestock species due to significant economic losses. The disease was 'exported' to South and Central America, where it continues to spread, in the 19th century (Cadioli *et al.*, 2012). Trypanosomiasis in cattle is caused by infection with *Trypanosoma brucei brucei*, *T. congolense* and *T. vivax*. These haemoprotozoa are the essence of extracellular infections in a range of wild and domestic species being commonly transmitted by the bites of infected tsetse flies (*Glossina* spp.). The disease is characterized by anaemia, lymphadenopathy, weight loss and abortion, and in susceptible animals advancing cachexia eventually leads to death after months or even years of infection. Control of trypanosomiasis by conventional means is difficult. This fact, and the very serious impacts on livestock agriculture, has focused attention on trypanotolerance as an option for disease control (Trail *et al.*, 1989).

It has been recognized for many decades that some *B. taurus* breeds of West African cattle appear to be resistant to trypanosomiasis by comparison with susceptible *B. indicus* breeds (Stewart, 1937). The trypanotolerance trait is particularly evident in the longhorn N'Dama and the small West African shorthorn breeds (Roberts and Gray, 1973; Roelants, 1986; Doko *et al.*, 1991). Notably, in the small West African shorthorn breeds, there is evidence of considerable variation in resistance, especially at higher challenge levels (Roelants, 1986). There is also some evidence of variation in resistance within the generally susceptible Boran zebu breed of East Africa (Njogu *et al.*, 1985). Moreover, N'Dama cattle appear to be as susceptible as Boran cattle to challenge with an unusual *T. vivax*, which causes a particularly acute and highly haemorrhagic syndrome (Williams *et al.*, 1992). However, in laboratory challenge experiments, the majority of animals recover from a severe anaemic episode, irrespective of whether they are N'Dama or Boran. Controlled challenge experiments in animals reared in a non-challenge environment (Logan *et al.*, 1988; Paling *et al.*, 1991a, 1991b) confirmed the trypanotolerance of the N'Dama breed, and the detailed comparisons of responses to challenge with *T. congolense* established the key parameters that have been used subsequently to detect and measure trypanotolerance and to study genetic control.

Studies of field challenges have demonstrated genetic control of variation in trypanotolerance within the N'Dama breed (Trail *et al.*, 1991) and at the same time emphasized the importance of resistance to anaemia development as a component of the trait (Trail *et al.*, 1990a). Indeed, a field challenge test, in which animals are ranked for selection purposes based on their ability to control anaemia, has been described (Trail *et al.*, 1990b). Depending on the method used, the heritability of anaemia caused by trypanosomiasis ranged from 0.09 to 0.22 (Dayo *et al.*, 2012). Orenge *et al.* (2012) have shown that females have higher trypanotolerance as well as F1, backcross and pure Kenya–Boran animals ranked in that order with respect to trypanotolerance. Controlled challenge experiments with resistant N'Dama and susceptible Boran cattle revealed superior control of parasites in peripheral blood as a key feature of trypanotolerant animals (Logan *et al.*, 1988; Paling *et al.*, 1991a,b). These studies have demonstrated breed differences in response to challenge with *T. congolense* in terms of antibody responses to cryptic and invariant trypanosome antigens, in early development of co-stimulatory cytokines, in T cell responses and in circulating leukocyte populations (Williams *et al.*, 1991, 1996; Flynn *et al.*, 1992; Authié *et al.*, 1993a,b; Sileghem *et al.*, 1993). Despite this effort, which in terms of advancing knowledge of trypanosomiasis immunology in cattle has been very valuable, a causal relationship between immunological responses and trypanotolerance has not been clearly established (Taylor, 1998).

A genetic approach, involving a genome scan, to understanding the basis of trypanotolerance suggested by Soller and Beckman (1988) was a very demanding proposition at that time. In the following years the mouse model was used quite often. Genome scans of independent murine F2 populations revealed three regions of the C57BL/6 genome harbouring resistance genes (Kemp *et al.*, 1996, 1997; Kemp and Teale, 1998). The most recent investigation shows that in the mouse 'the mapped QTL regions encompass genes that are vital to innate immune response and can be potential candidate genes for the underlying QTL' (Nganga *et al.*, 2010). Despite the significant value of the murine model for bovine trypanosomiasis resistance, it is no more than indicative of potential homologous bovine genetic regions that might be involved in trypanotolerance in cattle.

The first attempt to identify QTL controlling trypanotolerance in hybrids between tolerant West African N'Dama and susceptible East African Boran cattle revealed nine QTL of the N'Dama origin and five QTLs from the Kenya Boran; there were also four QTLs consistent with overdominant mode of inheritance (Hanotte *et al.*, 2003). Further analysis identified polymorphisms in *ARHGAP15* (Rho GTPase-activating protein 15) gene in the trypanotolerance QTL

located at chromosome 2. This polymorphism affects the gene function *in vitro* and could contribute to the observed differences in expression of the mitogen-activated protein kinase (MAPK) signalling pathway *in vivo*. 'The expression data showed that TLR (toll-like receptor) and MAPK pathways responded to infection, and the former contained *TICAM1* (TIR domain-containing adapter molecule 1) gene, which is within a QTL on chromosome 7. Genetic analyses showed that selective sweeps had occurred at *TICAM1* and *ARHGAP15* genes in African taurine cattle, making them strong candidates for the genes underlying the QTL. Candidate genes were identified in other QTL by their expression profile and the pathways in which they participate' (Noyes *et al.*, 2011). Dayo *et al.* (2012) identified several QTLs affecting traits related to trypanotolerance within an experimentally infected F2 population. Finally, an association analysis identified an allele of the MNB42 marker on chromosome 4 as being strongly associated with anaemia control. A candidate gene, *INHBA* (Inhibin beta A chain), is closely located to this marker and awaits detailed investigation.

'Laboratory studies, comparing *Trypanosoma congolense* infections in trypanotolerant N'Dama cattle (*Bos taurus*) and in more susceptible Boran cattle (*Bos indicus*), confirmed the field observations. Experiments using haemopoietic chimeric twins, composed of a tolerant and a susceptible co-twin, and T cell depletion studies suggested that trypanotolerance is composed of two independent traits. The first is a better capacity to control parasitaemia and is not mediated by haemopoietic cells, T lymphocytes or antibodies. The second is a better capacity to limit anaemia development and is mediated by haemopoietic cells, but not by T lymphocytes or antibodies' (Naessens, 2006). 'Thus, mortality and morbidity in trypanosome-infected cattle are primarily due to self-inflicted damage by disproportionate immune and/or innate responses. These features of bovine trypanotolerance differ greatly from those in murine models.' This study posed the question of whether bovine trypanotolerance is based on natural ability to prevent severe anaemia and haemophagocytic syndrome (Naessens, 2006). Despite great scientific efforts, many questions regarding bovine trypanotolerance currently remain unanswered, which is a clear demonstration of the complexity of this phenomenon.

Tuberculosis

The principal agent of bovine tuberculosis (TB) is *Mycobacterium bovis*. TB has not been able to be eradicated in many countries, causes considerable economic losses and, being transmissible to humans, represents a threat (Humblet *et al.*, 2009). Heritability of susceptibility to TB estimated in Black and White cattle is not high (h^2 = 0.06–0.08) (Petukhov *et al.*, 1998; Table 9.1). Similar estimates made for Irish Holstein Friesian cattle depended on model and were also modest (h^2 = 0.12–0.18) (Bermingham *et al.*, 2009, 2011). Analysis of five dairy breeds in Russia found positive genetic correlations between susceptibilities to TB and leukosis (Kulikova and Petukhov, 1994). Bermingham *et al.* (2010) concluded that 'selection for increased survival may indirectly reduce susceptibility to *M. bovis* infection, whereas selection for reduced somatic cell count and increased fat production and body condition score may increase susceptibility to *M. bovis* infection'.

In mice *NRAMP1* gene may create resistance not only in the case of *B. abortus* but also *M. bovis*; the initial expectation that the same gene (*SLC11A1*) in cattle has a similar resistance spectrum was not confirmed for cattle in the case of *M. bovis* (Barthel *et al.*, 2000). A real-time quantitative polymerase chain reaction (qRT-PCR) led to identification of several genes expressed at lower levels in cattle infected with *M. bovis*. 'In total, 378 gene features were differentially expressed at the $P \leq 0.05$ level in bovine tuberculosis (BTB)-infected and control animals, of which 244 were expressed at lower levels (65%) in the infected group'. Relatively lower expression of key innate immune genes, including the Toll-like receptor 2 (*TLR2*) and *TLR4* genes, lack of differential expression of indicator adaptive immune genes (*IFNG, IL2, IL4*) and lower expression of major histocompatibility complex class I (*BoLA*) and class II (*BoLA-DRA*) genes was consistent with innate immune gene

repression in the infected animals. Statistical analysis identified a panel of 15 genes predictive of disease status (Meade *et al.*, 2007); further investigations are desirable.

Meanwhile it was discovered that in humans *IPR1* gene (the current symbol is *SP110* nuclear body protein) mediates innate immunity to tuberculosis (Pan *et al.*, 2005). The protein produced by the gene 'might have a previously undocumented function in integrating signals generated by intracellular pathogens with mechanisms controlling innate immunity, cell death and pathogenesis'. The same gene *SP110*, located at chromosome 2 not very far from *SLC11A1*, also mediates RAW 264.7 macrophage cell line resistance to *M. bovis* in cattle. He *et al.* (2011) demonstrated that when 'RAW 264.7 macrophage was transduced with lentiviral vector carrying *IPR1* (named Lenti-IPR1); transgenic cells were identified by RT-PCR and western blotting. Transgenic positive cells (R-IPR1) were then infected with an *M. bovis* virulent strain, with non-transduced cells used as control. When cell proliferation, viability and apoptosis of the two groups were investigated, it was found that infected RAW 264.7 cells died by necrosis whereas R-IPR1 underwent apoptosis. Furthermore, the numbers of intracellular bacteria in R-IPR1 were lower than those in control cells ($P < 0.05$)' (He *et al.*, 2011). To identify the roles of *IPR1* (*SP110*), *CASP3*, *MCL1* and *NOS2A* genes, which are associated with macrophage activation and apoptosis, transcriptions were measured by qRT-PCR. The results demonstrated that *IPR1* (*SP110*) gene expression can enhance anti-*M. bovis* activity of macrophages. This finding establishes a basis for potential production of IPR1-transgenic cattle to strengthen tuberculosis resistance (He *et al.*, 2011). Another possible option is discovering a polymorphism in *SP110* that can be used for selection purposes.

Concluding Remarks

The past decade has brought a lot of new information about genetic factors determining resistance or susceptibility to major cattle diseases caused by very different infectious agents. In rare cases this knowledge is already successfully applied to development of natural resistance in cattle. However, the major work in this direction is still ahead and its progress will depend not only on our understanding of the genetic nature of the resistance but also on economic, veterinary and technological requirements. Hopefully a basic trend towards sustainable and clean agriculture will stimulate further investigations of the problem and will eventually lead to the incorporation of more disease resistance traits into breeding programmes.

References

Aarestrup, F.M., Jensen, N.E. and Ostergard, H. (1995) Analysis of associations between major histocompatibility complex (BoLA) class I haplotypes and subclinical mastitis of dairy cows. *Journal of Dairy Science* 78, 1684–1692.

Ables, G., Nishibori, M., Kanemaki, M. and Watanabe, T. (2002) Sequence analysis of the NRAMP1 genes from different bovine and buffalo breeds. *Journal of Veterinary Medical Science* 64, 1081–1083.

Adams, L.G. and Templeton, J.W. (1998) Genetic resistance to bacterial diseases of animals. *Revue Scientifique et Technique de l'Office International des Epizooties* 17(1), 200–219.

Alexander, D. (2010) Infectious bovine keratoconjunctivitis: a review of cases in clinical practice. Veterinary Clinics of North America: Food Animal Practice 26, 487–503.

Ali, A.A., O'Neill, C.J., Thomson, P.C. and Kadarmideen, H.N. (2012) Genetic parameters of infectious bovine keratoconjunctivitis and its relationship with weight and parasite infestations in Australian tropical *Bos taurus* cattle. *Genetics, Selection and Evolution* 44, 22.

Ali, M. and de Castro, J.J. (1993) Host resistance to ticks (Acari: Ixodidae) in different breeds of cattle at Bako, Ethiopia. *Tropical Animal Health and Production* 25, 215–222.

Ambrose, N.C. (1996) The pathogenesis of dermatophilosis. *Tropical Animal Health and Production* 28, 29S–37S.

Araujo, R.N., Padilha, T., Zarlenga, D., Sonstegard, T., Connor, E.E., Van Tassel, C., Lima, W.S., Nascimento, E. and Gasbarre, L.C. (2009) Use of a candidate gene array to delineate gene expression patterns in cattle selected for resistance or susceptibility to intestinal nematodes. *Veterinary Parasitology* 162, 106–115.

Authié, É., Muteti, D.K. and Williams, D.J. (1993a) Antibody responses to invariant antigens of *Trypanosoma congolense* in cattle of differing susceptibility to trypanosomiasis. *Parasite Immunology* 15, 101–111.

Authié, É., Duvallet, G., Robertson, C. and Williams, D.J.L. (1993b) Antibody responses to a 33 kDa cysteine protease of *Trypanosoma congolense*: relationship to 'trypanotolerance' in cattle. *Parasite Immunology* 15, 465–474.

Baltian, L.R., Ripoli, M.V., Sanfilippo, S., Takeshima, S.N., Aida, Y. and Giovambattista, G. (2012) Association between BoLA-DRB3 and somatic cell count in Holstein cattle from Argentina. *Molecular Biology and Reproduction* 39, 7215–7220.

Barnard, D.R. (1990a) Population growth rates for *Amblyomma americanum* (Acari: Ixodidae) on *Bos indicus, B. taurus* and *B. indicus x B. taurus* cattle. *Experimental and Applied Acarology* 9, 259–265.

Barnard, D.R. (1990b) Cattle breed alters reproduction in *amblyomma americanum* (Acari: Ixodidae) *Experimental and Applied Acarology* 10, 105–109.

Barthel, R., Piedrahita, J.A., McMurray, D.N., Payeur, J., Baca, D., Suárez Güemes, F., Perumaalia, V.S., Ficht, T.A., Templeton, J.W. and Adams, L.G. (2000) Pathologic findings and association of *Mycobacterium bovis* infection with the bovine NRAMP1 gene in cattle from herds with naturally occurring tuberculosis. *American Journal of Veterinary Research* 61, 1140–1144.

Barthel, R., Feng, J., Piedrathia, J.A., McMurray, D.N., Templeton, J.W. and Adams, G. (2001) Stable transfection of the bovine *NRAMP1* gene into murine RAW264.7 cells: effect on *Brucella abortus* survival. *Infection and Immunity* 69, 3110–3119.

Bermingham, M.L., More, S.J., Good, M., Cromie, A.R., Higgins, I.M., Brotherstone, S. and Berry, D.P. (2009) Genetics of tuberculosis in Irish Holstein-Friesian dairy herds. *Journal of Dairy Science* 92, 3447–3456.

Bermingham, M.L., More, S.J., Good, M., Cromie, A.R., Higgins, I.M. and Berry, D.P. (2010) Genetic correlations between measures of *Mycobacterium bovis* infection and economically important traits in Irish Holstein-Friesian dairy cows. *Journal of Dairy Science* 93, 5413–5422.

Bermingham, M.L., Brotherstone, S., Berry, D.P., More, S.J., Good, M., Cromie, A.R., White, I.M., Higgins, I.M., Coffey, M., Downs, S.H. *et al.* (2011) Evidence for genetic variance in resistance to tuberculosis in Great Britain and Irish Holstein-Friesian populations. *BMC Proceedings* 5 (Suppl 4), S15.

Boileau, M.J., Clinkenbeard, K.D. and Iandolo, J.J. (2011) Assessment of Bdellovibrio bacteriovorus 109J killing of *Moraxella bovis* in an *in vitro* model of infectious bovine keratoconjunctivitis. *Canadian Journal of Veterinary Research* 75, 285–291.

Bricarello, P.A., Zaros, L.G., Coutinho, L.L., Rocha, R.A., Silva, M.B., Kooyman, F.N., De Vries, E., Yatsuda, A.P. and Amarante, A.F. (2008) Immunological responses and cytokine gene expression analysis to *Cooperia punctata* infections in resistant and susceptible Nelore cattle. *Veterinary Parasitology* 155, 95–103.

Brown, M.H., Brightman, A.H., Fenwick, B.W. and Rider, M.A. (1998) Infectious bovine keratoconjunctivitis: a review. *Journal of Veterinary Internal Medicine* 12, 259–266.

Burridge, M.J., Wilcox, C.J. and Hennemann, J.M. (1979) Influence of genetic factors on the susceptibility of cattle to bovine leukaemia virus infection. *European Journal of Cancer* 15, 1395–1400.

Cadioli, F.A., Barnabé, Pde. A., Machado, R.Z., Teixeira, M.C., André, M.R., Sampaio, P.H., Fidélis Junior, O.L., Teixeira, M.M. and Marques, L.C. (2012) First report of Trypanosoma vivax outbreak in dairy cattle in São Paulo state, Brazil. *Revista Brasileira de Parasitologia Veterinária* 21, 118–124.

Campbell, G.A. and Adams L.G. (1992) The long-term culture of bovine monocyte-derived macrophages and their use in the study of intracellular proliferation of *Brucella abortus*. *Veterinary Immunology and Immunopathology* 34, 291–305.

Carvalho, W.A., Ianella, P., Arnoldi, F.G., Caetano, A.R., Maruyama, S.R., Ferreira, B.R., Conti, L.H., da Silva, M.R., Paula, J.O., Maia, A.A. and Santos, I.K. (2011) Haplotypes of the bovine IgG2 heavy gamma chain in tick-resistant and tick-susceptible breeds of cattle. *Immunogenetics* 63, 319–324.

Claxton, J. and Leperre, P. (1991) Parasite burdens and host susceptibility of Zebu and N'Dama cattle in village herds in Gambia. *Veterinary Parasitology* 40, 293–304.

Coppieters, W., Mes, E.H.M., Druet, T., Farnir, F., Tamma, N., Schrooten, C., Cornelissen, A.W.C.A., Georges, M. and Ploeger, H.W. (2009) Mapping QTL influencing gastrointestinal nematode burden in Dutch Holstein-Friesian dairy cattle. *BMC Genomics* 10, 96.

Coussens, P.M., Coussens, M.J., Tooker, B.C. and Nobis, W. (2004) Structure of the bovine natural resistance associated macrophage protein (NRAMP 1) gene and identification of a novel polymorphism. *DNA Sequence* 15, 15–25.

Da, Y., Shanks, P.D., Stewart, J.A. and Lewin, H.A. (1993) Milk and fat yields decline in leukaemia virus infected Holstein cattle with persistent lymphocytosis. *Proceedings of the National Academy of Sciences USA* 90, 6538–6541.

Dayo, G.K., Gautier, M., Berthier, D., Poivey, J.P., Sidibe, I., Bengaly, Z., Eggen, A., Boichard, D. and Thevenon, S. (2012) Association studies in QTL regions linked to bovine trypanotolerance in a West African crossbred population. *Animal Genetics* 43, 123–132.

de Castro, J.J., Capstick, P.B., Nokoe, S., Kiara, H., Rinkanya, F., Slade, R., Okello, O. and Bennun, L. (1991) Towards the selection of cattle for tick resistance in Africa. *Experimental and Applied Acarology* 12, 219–227.

Delespaux, V., Vitouley, H.S., Marcotty, T., Speybroeck, N., Berkvens, D., Roy, K., Geerts, S. and Van den Bossche, P. (2010) Chemosensitization of Trypanosoma congolense strains resistant to isometamidium chloride by tetracyclines and enrofloxacin. *PLoS Neglected Tropical Diseases* 4, e828.

Dicker, R.W. and Sutherst, R.W. (1981) Control of the bush tick (*Haemaphysalis longicornis*) with Zebu x European cattle. *Australian Veterinary Journal* 57, 66–68.

Dietz, A.B., Cohen, N.D., Timms, L. and Kehrli, M.E. Jr (1997) Bovine lymphocyte antigen class II alleles as risk factors for high somatic cell counts in milk of lactating dairy cows. *Journal of Dairy Science* 80, 406–412.

Doko, A., Guedegbe, B., Baelmans, R., Demey, F., N'Diaye, A., Pandey, V.S. and Verhulst, A. (1991) Trypanosomiasis in different breeds of cattle from Benin. *Veterinary Parasitology* 40, 1–7.

El-Abdellati, A., De Graef, J., Van Zeveren, A., Donnan, A., Skuce, P., Walsh, T., Wolstenholme, A., Tait, A., Vercruysse, J., Claerebout, E. and Geldhof, P. (2011) Altered avr-14B gene transcription patterns in ivermectin-resistant isolates of the cattle parasites, *Cooperia oncophora* and *Ostertagia ostertagi*. *International Journal of Parasitology* 41, 951–957.

Ernst, L.K. and Petukhov, V.L. (1978) Problems in the genetics of bovine leukemia. II. The effect of fathers and mothers on the leukemia disease frequency in the progeny. *Genetika* 14, 1247–1256. [In Russian.]

Ernst, L.K., Sulimova, G.E., Orlova, A.R., Udina, I.G. and Pavlenko, S.K. (1997) Features of the distribution of BoLA-A antigens and alleles of the BoLA-DRB3 gene in Black Pied cattle in relation to association with leukaemia. *Genetika* 33, 87–95.

Estes, D.M., Templeton, J.W., Hunter, D.M. and Adams. L.G. (1990) Production and use of murine monoclonal antibodies reactive with bovine 1gM isotype and IgG subisotypes (IgGI, IgG2a and IgG2b) in assessing immunoglobulin levels in serum of cattle. *Veterinary Immunology and Immunopathology* 25, 61–72.

Feng, J., Li, Y., Hashad, M., Schurr, E., Gros, P., Adams, L.G. and Templeton, J.W. (1996) Bovine natural resistance associated macrophage protein I (*Nramp* I) gene. *Genome Research* 6, 956–964.

Flynn, J.N., Sileghem, M. and Williams, D.J.L. (1992) Parasite-specific T-cell responses of trypanotolerant and trypanosusceptible cattle during infection with *Trypanosoma congolense*. *Immunology* 75, 639–645.

Fourie, L.J. and Kok, D.J. (1995) A comparison of *Ixodes rubicundus* (Acari: Ixodidae) infestations on Friesian and Bonsmara cattle in South Africa. *Experimental and Applied Acarology* 19, 529–531.

French, G.T. (1959) A clinical and genetic study of eye cancer in Hereford cattle. *Australian Veterinary Journal* 35, 474–481.

Frisch, J.E. (1981) Changes occurring in cattle as a consequence of selection for growth rate in a stressful environment. *Journal of Agricultural Science* (Cambridge) 96, 23–38.

Gasbarre, L.C., Leighton, E.A. and Davies, C.J. (1990) Genetic control of immunity to gastrointestinal nematodes of cattle. *Veterinary Parasitology* 37, 257–272.

Gasbarre, L.C., Leighton, E.A. and Davies, C.J. (1993) Influence of host genetics upon antibody responses against gastrointestinal nematode infections in cattle. *Veterinary Parasitology* 46, 81–91.

Gasbarre, L.C., Leighton, E.A. and Sonstegard, T. (2001) Role of the bovine immune system and genome in resistance to gastrointestinal nematodes. *Veterinary Parasitology* 98, 51–64.

George, J.E., Osburn, R.L. and Wikel, S.K. (1985) Acquisition and expression of resistance by *Bos indicus* and *Bos indicus* x *Bos taurus* calves to *Amblyomma americanum* infestation. *Journal of Parasitology* 71, 174–182.

Gibson, J.P. and Bishop, S. (2005) Use of molecular markers to enhance resistance of livestock to disease: a global approach. *Revue scientifique et technique* (International Office of Epizootics), 24, 343–353

Gillet, N., Florins, A., Boxus, M., Burteau, C., Nigro, A., Vandermeers, F., Balon, H., Bouzar, A.B., Defoiche, J., Burny, A., Reichert, M., Kettmann, R. and Willems, L. (2007) Mechanisms of leukemogenesis induced by bovine leukemia virus: prospects for novel anti-retroviral therapies in human. *Retrovirology* 4, 18.

Hammond, A.C., Williams, M.J., Olson, T.A., Gasbarre, L.C., Leighton, E.A. and Menchaca, M.A. (1997) Effect of rotational vs continuous intensive stocking of bahiagrass on performance of Angus cows and calves and interaction with sire type on gastrointestinal nematode burden. *Journal of Animal Science* 75, 2291–2299.

Hanotte, O., Ronin, Y., Agaba, M., Nilsson, P., Gelhaus, A., Horstmann, R., Sugimoto, Y., Kemp, S., Gibson, J., Korol, A., Soller, M. and Teale, A. (2003) Mapping of quantitative trait loci controlling trypanotolerance in a cross of tolerant West African N'Dama and susceptible East African Boran cattle. *Proceedings of the National Academy of Sciences USA* 100, 7443–7448.

Harmon, B.G., Adams, L.G., Templeton, J.W. and Smith, R. III (1989) Macrophage function in mammary glands of *Brucella abortus*-infected cows and cows that resisted infection after inoculation of *Brucella abortus*. *American Journal of Veterinary Research* 50, 459–465.

He, X.N., Su, F., Lou, Z.Z., Jia, W.Z., Song, Y.L., Chang, H.Y., Wu, Y.H., Lan, J., He, X.Y. and Zhang, Y. (2011) Ipr1 gene mediates RAW 264.7 macrophage cell line resistance to *Mycobacterium bovis*. *Scandinavian Journal of Immunology* 74, 438–444.

Heringstad, B., Klemetsdal, G. and Ruane, J (2000) Selection for mastitis resistance in dairy cattle: a review with focus on the situation in the Nordic countries. *Livestock Production Science* 64, 95–106.

Heringstad, B., Klemetsdal, G. and Steine, T. (2003) Selection responses for clinical mastitis and protein yield in two Norwegian dairy cattle selection experiments. *Journal of Dairy Science* 86, 2990–2999.

Hinrich, D., Bennewitz, J., Stamer, E., Junge, W., Kalm, E. and Thaller, G. (2011) Genetic analysis of mastitis data with different models. *Journal of Dairy Science* 94, 471–478.

Humblet, M.-F., Boschiroli, M.L. and Saegerman, C. (2009) Classification of worldwide bovine tuberculosis risk factors in cattle: a stratified approach. *Veterinary Research* 40, 50.

Jimba, M., Takeshima, S.-N., Murakabi, H., Kohara, J., Kobayashi, N., Matsuhashi, T., Ohmori, T., Nunoya, T. and Aida, Y. (2012) BLV-CoCoMo-qPCR: a useful tool for evaluating bovine leukemia virus infection status. *BMC Veterinary Research* 8, 167.

Juliarena, M.A., Poli, M., Ceriani, C., Sala, L., Rodríguez, E., Gutierrez, S, Dolcini, G., Odeon, A. and Esteban, E.N. (2009) Antibody response against three widespread bovine viruses is not impaired in Holstein cattle carrying bovine leukocyte antigen DRB3.2 alleles associated with bovine leukemia virus resistance. *Journal of Dairy Science* 92, 375–381.

Kataria, R.S., Tait, R.G. Jr, Kumar, D., Ortega, M.A., Rodiguez, J. and Reecy, J.M. (2011) Association of toll-like receptor four single nucleotide polymorphisms with incidence of infectious bovine keratoconjunctivitis (IBK) in cattle. *Immunogenetics* 63, 115–119.

Kelm, S.C., Detilleux, J.C., Freeman, A.E., Kehrli, M.E. Jr, Dietz, A.B., Fox, L.K., Butler, J.E., Kasckovics, I. and Kelly, D.H. (1997) Genetic association between parameters of innate immunity and measures of mastitis in periparturient Holstein cattle. *Journal of Dairy Science* 80, 1767–1775.

Kemp, S.J. and Teale, A.J. (1998) The genetic basis of trypanotolerance in mice and cattle. *Parasitology Today* 14, 450–454.

Kemp, S.J., Iraqi, F., Darvasi, A., Soller, M. and Teale, A.J. (1996) Genetic control of resistance to trypanosomiasis. *Veterinary Immunology and Immunopathology* 54, 239–243.

Kemp, S.J., Iraqi, F., Darvasi, A., Soller, M. and Teale, A.J. (1997) Localisation of genes controlling resistance to trypanosomiasis in mice. *Nature Genetics* 16, 194–196.

Kirkpatrick, B.W., Shi, X., Shook, G.E. and Collins, M.T. (2011) Whole-genome association analysis of susceptibility to paratuberculosis in Holstein cattle. *Animal Genetics* 42, 149–160.

Kizilkaya, K., Tait, R.G., Garrick, D.J., Fernando, R.L. and Reecy, J.M. (2011) Whole genome analysis of infectious bovine keratoconjunctivitis in Angus cattle using Bayesian threshold models. *BMC Proceedings* 5 Suppl 4, S22.

Kizilkaya, K., Tait, R.G., Garrick, D.J., Fernando, R.L. and Reecy, J.M. (2013) Genome-wide association study of infectious bovine keratoconjunctivitis in Angus cattle. *BMC Genetics* 14, 23.

Kloosterman, A., Albers, G.A.A. and Van den Brink, R. (1978) Genetic variation among calves in resistance to nematode parasites. *Veterinary Parasitology* 4, 353–368.

Ko, J. and Splitter, G.A. (2003) Molecular host-pathogen interaction in brucellosis: current understanding and future approaches to vaccine development for mice and humans. *Clinical Microbiology Reviews* 16, 65–78.

Koeck, A., Heringstad, B., Egger-Danner, C., Fuerst, C., Winter, P. and Fuerst-Waltl, B. (2010) Genetic analysis of clinical mastitis and somatic cell count traits in Austrian Fleckvieh cows. *Journal of Dairy Science* 93, 5987–5995.

Koets, A.P., Adugna, G., Janss, L.L.G., van Weering, H.J., Kalis, C.H.J., Wentink, G.H., Rutten, V.P.M.G. and Schukken, Y.H. (2000) Genetic variation of susceptibility to *Mycobacterium avium* subsp. paratuberculosis infection in dairy cattle. *Journal of Dairy Science* 83, 2702–2708.

Koney, E.B., Morrow, A.N., Heron, I., Ambrose, N.C. and Scott, G.R. (1994) Lymphocyte proliferative responses and the occurrence of dermatophilosis in cattle naturally infested with *Amblyomma variegatum*. *Veterinary Parasitology* 55, 245–256.

Koney, E.B., Morrow, A.N. and Heron, I.D. (1996) The association between *Amblyomma variegatum* and dermatophilosis: epidemiology and immunology. *Tropical Animal Health and Production* 28, 18S–25S.

Kulikova, S.G. and Petukhov, V.L. (1994) Genetic correlation of cattle resistance to tuberculosis and leucosis. *Proceedings of the 5th World Congress on Genetics Applied to Livestock Production* 20, 300–301.

Latif, A.A., Nokoe, S., Punyua, D.K. and Capstick, P.B. (1991) Tick infestations on Zebu cattle in Western Kenya: quantitative assessment of host resistance. *Journal of Medical Entomology* 28, 122–126.

Lee, H.S., Her, M., Levine, M. and Moore, G.E. (2013) Time series analysis of human and bovine brucellosis in South Korea from 2005 to 2010. *Preventive Veterinary Medicine* 110, 190–197.

Leighton, E.A., Murrell, K.D. and Gasbarre, L.C. (1989) Evidence for genetic control of nematode egg-shedding rates in calves. *Journal of Parasitology* 75, 498–504.

Lewin, H.A. and Bernoco, D. (1986) Evidence for BoLA-linked resistance and susceptibility to subclinical progression of bovine leukaemia virus infection. *Animal Genetics* 17, 197–207.

Lewin, H.A., Wu, M.C., Stewart, J.A. and Nolan, T.J. (1988) Association between *BoLA* and subclinical bovine leukaemia virus infection in a herd of Holstein-Friesian cows. *Immunogenetics* 27, 338–344.

Li, L., Huang, J., Zhang, X., Ju, Z., Qi, C., Zhang, Y., Li, Q., Wang, C., Miao, W., Zhong, J. *et al.* (2012) One SNP in the 3'-UTR of HMGB1 gene affects the binding of target bta-miR-223 and is involved in mastitis in dairy cattle. *Immunogenetics* 64, 817–824.

Li, R.W., Rinaldi, M. and Capuco, A.V. (2011) Characterization of the abomasal transcriptome for mechanisms of resistance to gastrointestinal nematodes in cattle. *Veterinary Research* 42, 114.

Liu, G.E., Brown, T., Hebert, D.A., Cardone, M.F., Hou, Y., Choudhary, R.K., Shaffer, J., Amazu, C., Connor, E.E., Ventura, M. and Gasbarre, L.C. (2011) Initial analysis of copy number variations in cattle selected for resistance or susceptibility to intestinal nematodes. *Mammalian Genome* 22, 111–121.

Logan, L.L., Paling, R.W., Moloo, S.K. and Scott, J.R. (1988) Comparative studies on the responses of N'Dama and Boran cattle to experimental challenge with tsetse-transmitted *Trypanosoma congolense*. In: ILCA and ILRAD (ed.) *Livestock Production in Tsetse Affected Areas of Africa*. ILCA/ILRAD, Nairobi, pp. 152–167.

Lundén, A., Sigurdardóttir, S., Edfors-Lilja, I., Danell, B., Rendel, J. and Andersson, L. (1990) The relationship between bovine major histocompatibility complex class II polymorphism and disease studied by use of bull breeding values. *Animal Genetics* 21, 221–232.

Machado, M.A., Azevedo, A.L., Teodoro, R.L., Pires, M.A., Peixoto, M.G., de Freitas, C., Prata, M.C., Furlong, J., da Silva, M.V., Guimarães, S.E. *et al.* (2010) Genome wide scan for quantitative trait loci affecting tick resistance in cattle (*Bos taurus* x *Bos indicus*). *BMC Genomics* 11, 280.

Maillard, J.C., Kemp, S.J., Naves, M., Palin, C., Demangel, C., Accipe, A., Maillard, N. and Bensaid, A. (1993a) An attempt to correlate cattle breed origins and diseases associated with or transmitted by the tick *Amblyomma variegatum* in the French West Indies. *Revue d'Élevage et de Médecine Vétérinaire des Pays Tropicaux* 46, 283–290.

Maillard, J.C., Palin, C., Trap, I. and Bensaid, A. (1993b) An attempt to identify genetic markers of resistance or susceptibility to dermatophilosis in the zebu Brahman population of Martinique. *Revue d'Élevage et de Médecine Vétérinaire des Pays Tropicaux* 46, 291–295.

Maillard, J.C., Martinez, D. and Bensaid, A. (1996) An amino acid sequence coded by the exon 2 of the BoLA DRB3 gene associated with a BoLA class I specificity constitutes a likely genetic marker of resistance to dermatophilosis in Brahman zebu cattle of Martinique (FWI) *Annals of the New York Academy of Sciences* 791, 185–197.

Maillard, J.C., Chantal, I., Berthier, D., Thevenon, S., Sidibe, I. and Razafindraibe, H. (2002) Molecular immunogenetics in susceptibility to bovine dermatophilosis: a candidate gene approach and a concrete field application. *Annals of the New York Academy of Sciences* 969, 92–96.

Martínez, R., Dunner, S., Toro, R., Tobón, J., Gallego, J. and Cañón, J. (2010) Effect of polymorphisms in the Slc11a1 coding region on resistance to brucellosis by macrophages *in vitro* and after challenge in two Bos breeds (Blanco Orejinegro and Zebu). *Genetics and Molecular Biology* 33, 463–470.

Mattioli, R.C., Bah, M., Faye, J., Kora, S. and Cassama, M. (1993) A comparison of field tick infestation on N'Dama, Zebu and N'Dama x Zebu crossbred cattle. *Veterinary Parasitology* 47, 139–148.

Meade, K.G., Gormley, E., Doyle, M.B., Fitzsimons, T., O'Farrelly, C., Costello, E., Keane, J., Zhao, Y. and MacHugh, D.E. (2007) Innate gene repression associated with *Mycobacterium bovis* infection in cattle: toward a gene signature of disease. *BMC Genomics* 8, 400.

Meredith, B.K., Kearney, F.J., Finlay, E.K., Bradley, D.G., Fahey, A.G., Berry, D.P. and Lynn, D.J. (2012) Genome-wide associations for milk production and somatic cell score in Holstein-Friesian cattle in Ireland. *BMC Genetics* 13, 21.

Miller, R.S., Farnsworth, M.L. and Malmberg, J.L. (2013) Diseases at the livestock-wildlife interface: status, challenge and opportunities in the United States. *Preventive Veterinary Medicine* 11, 119–132.

Minozzi, G., Buggiotti, L., Stella, A., Strozzi, F., Luini, M. and Williams, J.L. (2010) Genetic loci involved in antibody response to *Mycobacterium avium* ssp. *paratuberculosis* in cattle. *PLoS ONE* 5, e11117.

Minozzi, G., Williams, J.L., Stella, A., Strozzi, F., Luini, M., Settles, M.L., Taylor, J.F., Whitlock, R.H., Zanella, R. and Neibergs, H.L. (2012) Meta-analysis of two genome-wide association studies of bovine paratuberculosis. *PLoS ONE* 7, e32578.

Mirsky, M.L., Olmstead, C.A., Da, Y. and Lewin, H.A. (1996) The prevalence of proviral bovine leukaemia virus in peripheral blood mononuclear cells at two subclinical stages of infection. *Journal of Virology* 70, 2178–2183.

Miyasaka, T., Takeshima, S.N., Jimba, M., Matsumoto, Y., Kobayashi, N., Matsuhashi, T., Sentsui, H., Aida, Y. (2013) Identification of bovine leukocyte antigen class II haplotypes associated with variations in bovine leukemia virus proviral load in Japanese Black cattle. *Tissue Antigens* 81, 72–82.

Morris, C.A. (2007) A review of genetic resistance to disease in *Bos taurus* cattle. *Veterinary Journal* 174, 481–491.

Morris, C.A. and Amyes, N.C. (2012) *Proceedings of the New Zealand Society of Animal Production* 72, 236–239.

Morris, C.A., Green, R.S., Cullen, N.G. and Hickey, S.M. (2003) Genetic and phenotypic relationships among faecal egg count, anti-nematode antibody level and live weight in Angus cattle. *Animal Science* 76, 167–174.

Morrow, A.N., Koney, E.B. and Heron, I.D. (1996) Control of *Amblyomma variegatum* and dermatophilosis on local and exotic breeds of cattle in Ghana. *Tropical Animal Health and Production* 28, 44S–49S.

Murakami, K., Kobayashi, S., Konishi, M., Kameyama, K., Yamamoto, T. and Tsutsui, T. (2010) The recent prevalence of bovine leukemia virus (BLV) infection among Japanese cattle. *Veterinary Microbiology* 148, 84–88.

Murray, M., Trail, J.C.M., Davies, C.E. and Black, S.J. (1984) Genetic resistance to African trypanosomiasis. *Journal of Infectious Diseases* 149, 311–319.

Naessens, J. (2006) Bovine trypanotolerance: a natural ability to prevent severe anaemia and hae-mophagocytic syndrome? *International Journal of Parasitology* 36, 521–528.

Nganga, J.K., Soller, M. and Iraqi, F.A. (2010) High resolution mapping of trypanosomosis resistance loci Tir2 and Tir3 using F12 advanced intercross lines with major locus Tir1 fixed for the susceptible allele. *BMC Genomics* 11, 394.

Njogu, A.R., Dolan, R.B., Wilson, A.J. and Sayer, P.D. (1985) Trypanotolerance in East African Orma Boran cattle. *Veterinary Record* 117, 632–636.

Norval, R.A.I., Sutherst, R.W. and Kerr, J.D. (1996) Infestations of the bont tick *Amblyomma hebraeum* (Acari: Ixodidae) on different breeds of cattle in Zimbabwe. *Experimental and Applied Acarology* 20, 599–605.

Noyes, H., Brass, A., Obara, I., Anderson S., Archibald, A.L., Bradley, D.G., Fisher, P., Freeman, A., Gibson, J., Gicheru, M. *et al.* (2011) Genetic and expression analysis of cattle identifies candidate genes in path-ways responding to *Trypanosoma congolense* infection. *Proceedings of the National Academy of Sciences USA* 108, 9304–9309.

Oddgeirsson, O., Simpson, S.P., Morgan, A.L., Ross, D.S. and Spooner, R.L. (1988) Relationship between the bovine major histocompatibility complex (BoLA), erythrocyte markers and susceptibility to mastitis in Icelandic cattle. *Animal Genetics* 19, 11–16.

O'Kelly, J.C. and Spiers, W.G. (1976) Resistance to *Boophilus microplus* (Canestrini) in genetically different types of calves in early life. *Journal of Parasitology* 62, 312–317.

O'Neill, C.J., Weldon, G.J., Hill, R.A. and Thomas, J.E. (1998) Adaptaur Association of Australia: a breed association based on the performance recording of Adaptaur cattle. *Animal Production in Australia* (Proceedings of the Australian Society of Animal Production) 22, 229–232.

Orenge, C.O., Munga, L., Kimwele, C.N., Kemp, S., Korol, A., Gibson, J.P., Hanotte, O. and Soller, M. (2012) Trypanotolerance in N'Dama x Boran crosses under natural trypanosome challenge: effect of test-year environment, gender, and breed composition. *BMC Genetics* 13, 87.

Paling, R.W., Moloo, S.K., Scott, J.R., McOdimba, F.A., Logan-Henfrey, L.L., Murray, M. and Williams, D.J.L. (1991a) Susceptibility of N'Dama and Boran cattle to tsetse-transmitted primary and rechallenge infec-tions with a homologous serodeme of *Trypanosoma congolense*. *Parasite Immunology* 13, 413–425.

Paling, R.W., Moloo, S.K., Scott, J.R., Gettinby, G., McOdimba, F.A. and Murray, M. (1991b) Susceptibility of N'Dama and Boran cattle to sequential challenges with tsetse-transmitted clones of *Trypanosoma congolense*. *Parasite Immunology* 13, 427–445.

Pan, H., Yan, B.S., Rojas, M., Shebzukhov, Y.V., Zhou, H., Kobzik, L., Higgins, D.E., Daly, M.J., Bloom, B.R. and Kramnik, I. (2005) Ipr1 gene mediates innate immunity to tuberculosis. *Nature* 434, 767–772.

Peregrine, A.S. (1994) Chemotherapy and delivery systems: haemoparasites. *Veterinary Parasitology* 54, 223–248.

Petukhov, V.L., Kochnev, N.N., Panov, B.L., Korotkevich, O.S., Kulikova, S.G. and Marenkov, V.G. (1998) Genetics of cattle resistance to tuberculosis. *Proceedings of the 6th World Congress on Genetics Applied to Livestock Production* 27, 365–366.

Philipsson, J., Ral, G. and Berglund, B. (1995) Somatic cell count as a selection criterion for mastitis resist-ance in dairy cattle. *Livestock Production Science* 41, 195–200.

Piper, E.K., Jackson, L.A., Bielefeldt-Ohmann, H., Gondro, C., Lew-Tabor, A.E. and Jonsson, N.N. (2010) Tick-susceptible *Bos taurus* cattle display an increased cellular response at the site of larval *Rhipicephalus* (*Boophilus*) *microplus* attachment, compared with tick-resistant *Bos indicus* cattle. *International Journal of Parasitology* 40, 431–441.

Porto Neto, L.R., Bunch, R.J., Harrison, B.E., Prayaga, K.C. and Barendse, W. (2010) Haplotypes that include the integrin alpha 11 gene are associated with tick burden in cattle. *BMC Genetics* 11, 55.

Porto Neto, L.R., Bunch, R.J., Harrison, B.E. and Barendse, W. (2011a) DNA variation in the gene ELTD1 is associated with tick burden in cattle. *Animal Genetics* 42, 50–55.

Porto Neto, L.R., Jonsson, N.N., D'Occhio, M.J. and Barendse, W. (2011b) Molecular genetic approaches for identifying the basis of variation in resistance to tick infestation in cattle. *Veterinary Parasitology* 180, 165–172.

Porto Neto, L.R., Jonsson, N.N., Ingham, A., Bunch, R.J., Harrison, B.E., Barendse, W. and Cooperative Research Centre for Beef Genetic Technologies (2012) The RIPK2 gene: a positional candidate for tick burden supported by genetic associations in cattle and immunological response of knockout mouse. *Immunogenetics* 64, 379–388.

Price, R.E., Templeton, J.W., Smith, R. III and Adams. L.G. (1990) Ability of mononuclear phagocytes from cattle naturally resistant or susceptible to brucellosis to control *in vitro* intracellular survival of *Brucella abortus*. *Infection and Immunity* 58, 879–886.

Qureshi, T., Templeton, J.W. and Adams, L.G. (1996) Intracellular survival of *Brucella abortus*, *Mycobacterium bovis* (BCG), *Salmonella dublin* and *Salmonella typhimurium* in macrophages from cattle genetically resistant to *Brucella abortus*. *Veterinary Immunology and Immunopathology* 50, 55–66.

Rechav, Y. and Kostrzewski, M.W. (1991) Relative resistance of six cattle breeds to the tick *Boophilus decoloratus* in South Africa. *Onderstepoort Journal of Veterinary Research* 58, 181–186.

Rechav, Y., Dauth, J. and Els, D.A. (1990) Resistance of Brahman and Simmentaler cattle to southern African ticks. *Onderstepoort Journal of Veterinary Research* 57, 7–12.

Rechav, Y., Kostrzewski, M.W. and Els, D.A. (1991) Resistance of indigenous African cattle to the tick *Amblyomma hebraeum*. *Experimental and Applied Acarology* 12, 229–241.

Roberts, C.J. and Gray, A.R. (1973) Studies on trypanosome-resistant cattle II. The effects of trypanosomiasis on N'Dama, Muturu and Zebu cattle. *Tropical Animal Health and Production* 5, 220–233.

Roelants, G.E. (1986) Natural resistance to African trypanosomiasis. *Parasite Immunology* 8, 1–10.

Ross, J.G., Lee, R.P. and Armour, J. (1959) Haemonchosis in Nigeria zebu cattle: the influence of genetical factors in resistance. *Veterinary Record* 71, 27–31.

Ross, J.G., Armour, J. and Lee, R.P. (1960) Further observations on the influence of genetical factors in resistance to helminthiasis in Nigeria zebu cattle. *Veterinary Parasitology* 72, 119–122.

Ruiz-Larrañaga, O., Garrido, J.M., Iriondo, M., Manzano, C., Molina, E., Montes, I., Vazquez, P., Koets, A.P., Rutten, V.P., Juste, R.A. and Estonba, A. (2010) SP110 as a novel susceptibility gene for *Mycobacterium avium* subspecies paratuberculosis infection in cattle. *Journal of Dairy Science* 93(12), 5950–5958.

Ruiz-Larrañaga, O., Iriondo, M., Manzano, C., Agirre, M., Garrido, J.M., Juste, J.A. and Estonba, A. (2012) Single-nucleotide polymorphisms in the bovine CD209 candidate gene for susceptibility to infection by Mycobacterium avium subsp. Paratuberculosis. *Animal Genetics* 43, 646–647.

Russell, C.D., Widdison, S., Leigh, J.A. and Coffey, T.J. (2012) Identification of single nucleotide polymorphisms in the bovine Toll-like receptor 1 gene and association with health traits in cattle. *Veterinary Research* 43, 17.

Ruzina, M.N., Shtyfurko, T.A., Mohammadabadi, M.R., Gendzhieva, O.V., Tsedev, T. and Sulimova, G.E. (2010) Polymorphism of the BoLA-DRB3 gene in the Mongolian, Kalmyk, and Yakut cattle breeds. *Genetika* 46, 517–525. [Article in Russian]

Schukken, Y.H., Mallard, B.A., Dekkers, J.C.M., Leslie, K.E. and Stear, M.J. (1994) Genetic impact on the risk of intramammary infection following *Staphylococcus aureus* challenge. *Journal of Dairy Science* 77, 639–647.

Sharif, S., Mallard, B.A. and Sargeant, J.M. (2000) Presence of glutamine at position 74 of pocket 4 in the BoLA-DR antigen binding groove is associated with occurrence of clinical mastitis caused by *Staphylococcus* species. *Veterinary Immunology and Immunopathology* 76, 231–238.

Shook, G.E. and Schutz, M.M. (1994) Selection on somatic cell score to improve resistance to mastitis in the United States. *Journal of Dairy Science* 77, 648–658.

Sileghem, M.R., Flynn, J.N., Saya, R. and Williams, D.J. (1993) Secretion of co-stimulatory cytokines by monocytes and macrophages during infection with *Trypanosoma* (Nannomonas) *congolense* in susceptible and tolerant cattle. *Veterinary Immunology and Immunopathology* 37, 123–134.

Simon, M., Dusinsky, R. and Stavikova, M. (1995) Association between BoLA antigens and bovine mastitis. *Veterinary Medicine* 40, 7–10.

Simpson, S.P., Oddgeirsson, O., Jonmundsson, J.V. and Oliver, R.A. (1990) Associations between the bovine major histocompatibility complex (BoLA) and milk production in Icelandic dairy cattle. *Journal of Dairy Research* 57, 437–440.

Slatter, D.H., Edwards, M.E., Hawkins, C.D. and Wilcox, G.E. (1982) A national survey of the occurrence of infectious bovine keratoconjunctivitis. *Australian Veterinary Journal* 59, 65–68.

Snowder, G.D., van Vleck, L.D., Cundiff, L.V. and Bennett, G.L. (2005) Genetic and environmental factors associated with incidence of infectious bovine keratoconjunctivitis in preweaned beef calves. *Journal of Animal Science* 83, 507–518.

Soller, M. and Beckman, J.S. (1988) *Maping Trypanotolerance Loci of the N'Dama Cattle of West Africa*. Consultation Report, FAO, Rome.

Solomon, G. and Kaaya, G.P. (1996) Comparison of resistance in three breeds of cattle against African ixodid ticks. *Experimental and Applied Acarology* 20, 223–230.

Sonstegard, T.S. and Gasbarre, L.C. (2001) Genomic tools to improve parasite resistance. *Veterinary Parasitology* 101, 387–403.

Spickett, A.M., De Klerk, D., Enslin, C.B. and Scholtz, M.M. (1989) Resistance of Nguni, Bonsmara and Hereford cattle to ticks in a Bushveld region of South Africa. *Onderstepoort Journal of Veterinary Research* 56, 245–250.

Stanik, J. and Vasil, M. (1986) Mastitis in dairy cows in large-scale farming operating from the genetic aspect. *Veterinary Medicine* 31, 21–26.

Stavikova, M., Lojda, L. Zakova, M., Mach, P., Prikryl, S. and Pospisil, J. (1990) The genetic contribution of cows to the prevalence of mastitis in the following generation. *Veterinarni Medicina* 35, 257–265.

Stear, M.J., Newman, M.J., Nicholas, F.W., Brown, S.C. and Holroyd, R.G. (1984) Tick resistance and the major histocompatibility system. *Australian Journal of Experimental Biology and Medical Science* 62, 47–52.

Stear, M.J., Dimmock, C.K., Newman, M.J. and Nicholas, F.W. (1988) BoLA antigens are associated with increased frequency of persistent lymphocytosis in bovine leukaemia virus infected cattle and with increased incidence of antibodies to bovine leukaemia virus. *Animal Genetics* 19, 151–158.

Stewart, J.L. (1937) The cattle of the Gold Coast. *Veterinary Record* 49, 1289–1297.

Sulimova, G.E., Udina, I.G., Shaikhaev, G.O. and Zakharov, I.A. (1995) DNA polymorphism of the BoLA-DRB3 gene in cattle in connection with resistance and susceptibility to leukaemia. *Genetika* 9, 1294–1299.

Sutherland, I.A. and Leathwick, D.M. (2011) Anthelmintic resistance in nematode parasites of cattle: a global issue? *Trends in Parasitology* 27, 176–181.

Takeshima, S., Matsumoto, Y., Chen, J., Yoshida, T., Mukoyama, H. and Aida, Y. (2008) Evidence for cattle major histocompatibility complex (BoLA) class II DQA1 gene heterozygote advantage against clinical mastitis caused by *Streptococci* and *Escherichia* species. *Tissue Antigens* 72, 525–531.

Taylor, K.A. (1998) Immune responses of cattle to African trypanosomes: protective or pathogenic? *International Journal for Parasitology* 28, 219–240.

Templeton, J.W. and Adams, L.G. (1995) Natural resistance to bovine brucellosis. In: Adams, L.G. (ed.) *Advances in Brucellosis Research*. Texas A&M University Press, College Station, Texas, pp. 144–150.

Thimm, B. (1973) Hypothesis of a higher natural resistance against brucellosis in East African shorthorn zebu. *Zentralblatt für Veterinärmedizin* [B] 20, 490–494.

Trail, J.C.M., d'Ieteren, G. and Teale, A.J. (1989) Trypanotolerance and the value of conserving livestock genetic resources. *Genome* 31, 805–812.

Trail, J.C., d'Ieteren, G.D.M., Feron, A., Kakiese, O., Mulungo, M. and Pelo, M. (1990a) Effect of trypanosome infection, control of parasitaemia and control of anaemia development on productivity of N'Dama cattle. *Acta Tropica* 48, 37–45.

Trail, J.C.M., d'Ieteren, G.D.M., Colardelle, C., Maille, J.C., Ordner, G., Sauveroche, B. and Yangari, G. (1990b) Evaluation of a field test for trypanotolerance in young N'Dama cattle. *Acta Tropica* 48, 47–57.

Trail, J.C.M., d'Ieteren, G.D.M., Maille, J.C. and Yangari, G. (1991) Genetic aspects of control of anaemia development in trypanotolerant N'Dama cattle. *Acta Tropica* 48, 285–291.

Utech, K.B. and Wharton, R.H. (1982) Breeding for resistance to *Boophilus microplus* in Australian Illawarra Shorthorn and Brahman x Australian Illawarra Shorthorn cattle. *Australian Veterinary Journal* 58, 41–46.

Våge, D.I., Lingaas, F., Spooner, R.L., Arnet, E.F., and Lie, Ø. (1992) A study on association between mastitis and serologically defined class I bovine lymphocyte antigens (BoLA-A) in Norwegian cows. *Animal Genetics* 23, 533–536.

van Eijk, M.J., Stewart-Haynes, J.A., Beever, J.E., Fernando, R.L. and Lewin, M.A. (1992) Development of persistent lymphocytosis in cattle is closely associated with DRB2. *Immunogenetics* 37, 64–68.

Vidal, S.M., Malo, D., Vogan, K., Skamene, E. and Gros, P. (1993) Natural resistance to infection with intracellular parasites: isolation of a candidate for *Bcg*. *Cell* 73, 469–485.

Waller, P.J. (1997) Nematode parasite control of livestock in the tropics/subtropics: the need for novel approaches. *International Journal of Parasitology* 27, 1193–1201.

Ward, J.K. and Nielson, M.K. (1979) Pinkeye (bovine infectious keratoconjunctivitis) in beef cattle. *Journal of Animal Science* 49, 361–366.

Weigel, K.A., Freeman, A.E., Kehrli, M.E. Jr, Stear, M.J. and Kelley, D.H. (1990) Association of class I bovine lymphocyte antigen complex alleles with health and production traits in dairy cattle. *Journal of Dairy Science* 73, 2538–2546.

Williams, D.J.L., Naessens, J., Scott, J.R. and McOdimba, F.A. (1991) Analysis of peripheral leucocyte populations in N'Dama and Boran cattle following a rechallenge infection with *Trypanosoma congolense. Parasite Immunology* 13, 171–185.

Williams, D.J.L., Logan-Henfrey, L.L., Authié, E., Seely, C. and McOdimba, F. (1992) Experimental infection with a haemorrhage-causing *Trypanosoma vivax* in N'Dama and Boran cattle. *Scandinavian Journal of Immunology* 36, 34–36.

Williams, D.J.L., Taylor, K., Newson, J., Gichuki, B. and Naessens, J. (1996) The role of anti-variable surface glycoprotein antibody responses in bovine trypanotolerance. *Parasite Immunology* 18, 209–218.

Xu, A., van Eijk, M.J., Park, C. and Lewin, H.A. (1993) Polymorphism in *BoLA-DRB3* exon 2 correlates with resistance to persistent lymphocytosis caused by bovine leukaemia virus. *Journal of Immunology* 151, 6977–6985.

Yoshida, T., Furuta, H., Kondo, Y. and Mukoyama, H. (2012) Association of BoLA-DRB3 alleles with mastitis resistance and susceptibility in Japanese Holstein cows. *Animal Science Journal* 83, 359–366.

Zanotti, M., Poli, G., Ponti, W., Polli, M., Rocchi, M., Bolzani, E., Longeri, M., Russo, S., Lewin, H.A. and van Eijk, M.J. (1996). Association of BoLA class II haplotypes with subclinical progression of bovine leukaemia virus infection in Holstein-Friesian cattle. *Animal Genetics* 27, 337–341.

Zaros, L.G., Bricarello, P.A., Amarante, A.F., Rocha, R.A., Kooyman, F.N., De Vries, E. and Coutinho, L.L. (2010) Cytokine gene expression in response to *Haemonchus placei* infections in Nelore cattle. *Veterinary Parasitology* 171, 68–73.

Zhao, G., Yu, M., Cui, Q.W., Zhou, X., Zhang, J.C., Li, H.X., Qu, K.X., Wang, G.L. and Huang, B.Z. (2013) Association of bovine toll-like receptor 4 with tick infestation rates and blood histamine concentration. *Genetics and Molecular Research* 12, 2783–2793.

10 Molecular Biology and Genetics of Bovine Spongiform Encephalopathy

N. Hunter

University of Edinburgh, Easter Bush, Midlothian, UK

Introduction

Bovine spongiform encephalopathy (BSE), although a disease currently in decline, is still a subject of much debate concerning its aetiology, epidemiology, mode of transmission and genetics. BSE was first recognized as a new neurological disease in cattle in the UK in 1986 and since then there have been over 180,000 UK cases. The economic impact of BSE and the associated control measures of culling of healthy but at-risk animals has been enormous but the result is that BSE has all but disappeared from the UK (seven cases in 2011, three in 2012). It is present at similarly low levels in many other countries, for example Spain, Portugal and Poland (www.oie.int), and without continued vigilance it could re-emerge as a major problem in later years.

BSE is one of a group of related diseases known as transmissible spongiform encephalopathies (TSEs) or prion diseases, the oldest known of which is scrapie, which occurs in sheep and goats, but there are also human forms of TSEs including variant Creutzfeldt–Jakob disease (vCJD), which was mostly likely the result of consumption of BSE-contaminated cattle meat. The TSEs are all slowly progressive, inevitably fatal, neurodegenerative disorders characterized by vacuolated brain neurones and the deposition of an abnormal form of a host protein, PrP, or prion protein. TSEs are experimentally transmissible, and are usually studied in laboratory rodents. The most likely source of infection in cattle was the use of a dietary protein supplement, meat and bone meal (MBM), which was regularly fed, particularly to dairy cattle, and contained the rendered remains of animal offal and carcases, principally from ruminants (Wilesmith *et al.*, 1991; Nathanson *et al.*, 1997). Such feeding practices were made illegal in July 1998; however, because BSE cases are now occurring less and less frequently there is discussion amongst government policy-makers about relaxing the surveillance and control measures (Budka, 2011). However, the heavy surveillance of cattle globally has inadvertently revealed other similar diseases of cattle – atypical BSE forms – and the origin of these is less clear (Tranulis *et al.*, 2011). Atypical BSE is very rare – by 2010 only 52 cases had been reported worldwide (Seuberlich *et al.*, 2010). Although unlikely to be related to MBM, these novel diseases are of unpredictable risk to animal welfare and human health.

There is a strong genetic component in the patterns of disease incidence of scrapie in sheep and of some forms of human TSE and there is overwhelming evidence that the genetic component is the gene that encodes the PrP protein, *PRNP*. (Many papers still refer to the gene as 'the PrP gene' however it is now more usual to use the term *PRNP*.) In mice, sheep, goats and humans, there are polymorphisms and mutations of the *PRNP* gene linked to TSE disease incidence, but such linkage has not so far been demonstrated for cattle.

This chapter describes *PRNP* genetics in cattle with respect to BSE and sets this against the background of what is known about *PRNP* genetics in sheep and humans.

Clinical Signs and Pathology

BSE affected cattle become very difficult to handle and show increasing signs of ataxia, altered behaviour with fear and/or aggression and oversensitivity to noises and to touch. Affected animals spend less time ruminating than healthy cattle (Austin and Pollin, 1993), although their physiological drive to eat appears to remain normal. Several studies have noted that BSE

cattle have low heart rates (brachycardia), which may be related to the low food intake associated with reduced rumination or which may indicate that there is some damage to the vagus during disease development (Austin *et al.*, 1997). BSE affected cattle show significant neuronal loss in the brain (Jeffrey and Halliday, 1994), and the appearance of vacuolar lesions in brain sections is very similar to that seen in sheep scrapie. BSE was confirmed as a TSE by demonstration of diagnostic TSE-related PrP protein fibrils in brain extracts (Hope *et al.*, 1988) and by transmission of the disease to mice (Bruce *et al.*, 1994).

Cattle affected by the atypical forms of BSE have mostly been found during rapid high-throughput testing in cattle from abattoirs when brain samples have been positive for disease-related PrP protein. Atypical BSE cattle are usually much older than those affected by BSE and occasionally neurological clinical signs have been reported such as difficulty standing (Hagiwara *et al.*, 2007).

The Importance of PrP Protein

The PrP protein is a normal host protein found in every mammal so far examined and consists of approximately 250 amino acids (exact length depends on the species). PrP is glycosylated at either one, or both, of two possible glycosylation sites and is attached to the outside of the neuronal cell membrane by a glycophosphatidylinositol anchor (Hope, 1993). The protein, in a conformationally altered form (PrP^{Sc}) that is relatively resistant to protease digestion, is the major constituent of scrapie-associated fibrils (SAFs), now known to be a hallmark of TSEs in general, and has a characteristic triple-banded pattern when visualized on electrophoresis gels. Western blots from BSE-affected cattle show three bands at around 29 kDa, 24 kDa and 20 kDa in apparent molecular weight. Forms of atypical BSE give different patterns, particularly with the lowest (unglycosylated) band, which in some cases appeared to be slightly higher molecular weight (H-type BSE) and in some cases slightly lower molecular weight (L-type BSE).

The normal protein is designated PrP^{C} and is fully sensitive to protease digestion. It is thought that PrP^{Sc} is formed directly from

PrP^C by a poorly understood self-propagation mechanism that induces a change in the three-dimensional structure of the molecule. The main physical differences between PrP^C and PrP^{Sc} are shown in Table 10.1. Analysis suggests that PrP^C protein has a structured C-terminus made up of three α-helices and two small β-sheets (Chen and Thirumalai, 2012), whereas PrP^{Sc} has much more β-sheet. Molecules with high β-sheet content are more resistant to protease enzymatic digestion, probably because the structure gives regions of the protein protection from physical exposure to the enzyme.

The nature of the TSE infectious agents was an unsettled question for decades, but the most widely held view is the prion hypothesis, which proposes that the diseases are caused by an infectious self-replicating protein. PrP^{Sc} is so closely associated with TSE infectivity that it can be considered as a reliable marker for infection, and indeed, the prion hypothesis proposes that it is PrP^{Sc} that is itself the infectious agent, causing disease by acting as a seed for the conversion of the normal endogenous PrP^C into new PrP^{Sc} and thus appearing to have replicated the infectivity (Prusiner et al., 1990). In order for this previously heretical idea to be accepted, it was necessary for the process to be demonstrated to work in vitro with defined constituents free of any active live cells. In recent years this has been achieved via a process known as protein misfolding cyclic amplification (PMCA) and similar techniques (Soto, 2011), however it is still unclear whether additional molecules other than PrP polypeptides are required to convey strain variation information.

Table 10.1. Differences between normal PrP (PrP^c) and its disease associated isoform (PrP^{Sc}).

	PrP^c	PrP^{Sc}
Proteinase K (PK)	Sensitive	Partially resistant
Molecular mass (−PK)	33–35 kDa	33–35 kDa
Molecular mass (+PK)	Degraded	27–30 kDa
Detergent	Soluble	Insoluble
Location	Cell surface	Aggregates
Turnover	Rapid	Slow
Prion infectivity	Does not copurify	Copurifies

Different variant forms (allotypes) of the PrP protein are associated with differences in incubation period of experimental scrapie both in laboratory mice (Carlson et al., 1986) and in sheep (Goldmann et al., 1991a). In addition, some of the human TSEs appear to be familial and present excellent linkage between PrP gene mutations and the incidence of disease. The prion hypothesis therefore also accommodates the idea that TSEs can sometimes be simply genetic in origin, with the mutant protein being more likely spontaneously to adopt the disease-associated conformation both causing disease and producing a seed for a new infection should a transmission to another individual occur (Collinge and Palmer, 1994). There are still proponents of alternative views, however, as the biology of TSEs, for example natural sheep scrapie, resembles that of viral infections in how they spread between animals and within their body tissues. An alternative view is that variant forms of PrP^C control susceptibility to an infecting agent and that PrP^{Sc} is a by-product of the infection perhaps the result of a 'hit and run' virus, long gone by the time the clinical signs emerge in the animal (Miyazawa et al., 2012).

Whatever the nature of the infectious agent, BSE, like many other TSEs, is very resistant to heat and to chemical methods of inactivation (Taylor et al., 1994, 1995) making it difficult and expensive to decontaminate farms, abattoirs and laboratories.

Pathogenesis of BSE in Cattle

Natural BSE in cattle

A single major strain of BSE predominated throughout much of the epizootic, although since 2004 high levels of surveillance resulted in the discovery of rare atypical forms (L-BSE and H-BSE, see below) in older cattle. BSE is sometimes also known as Classical BSE (C-BSE) to distinguish it from the atypical forms, however, in this chapter the term BSE will be used to refer to the original and predominant form of the disease.

BSE is transmissible to laboratory mice, both inbred lines expressing the mouse PRNP gene, and transgenic mice, which express the bovine *PRNP* instead of, or in addition to, the endogenous mouse gene. The inbred line RIII gives the shortest incubation period and a characteristic pattern of brain pathology following injection with BSE (Fraser *et al.*, 1992) and were used in the first major investigations of BSE-affected cattle. RIII bioassay detected no infectivity in any fluid (including milk) (Taylor *et al.*, 1995), or tissue other than brain and spinal cord (Fraser and Foster, 1993). The mouse bioassay is not as sensitive as cattle-to-cattle transmission, however it did pick up BSE from the spleens of experimentally infected sheep (Foster *et al.*, 1996), providing an early suggestion that pathogenesis of BSE in sheep was similar to scrapie in sheep and not at all like pathogenesis of BSE in cattle. The very much lower levels of infectivity in peripheral tissues of cattle are confirmed by analysis of PrP^{Sc} protein, which is detected in central nervous system (CNS) tissues in cattle but in both CNS and peripheral tissues in experimental BSE in sheep and scrapie in sheep (Somerville *et al.*, 1997).

Transgenic mice expressing bovine *PRNP* genes produced more sensitive models for bioassay of BSE cattle tissues. The TgbovXV line was shown to be 10,000-fold more sensitive than RIII mice but even TgbovXV mice did not detect BSE infection in cattle lymphatic tissues, with the exception of Peyer's patches in the distal ileum of the intestine. There is now a considerable amount of data supporting the view that in cattle, unlike sheep and mice, BSE travels through the body from the site of infection to the brain only via neuronal cells (Buschmann and Groschup, 2005). More recently, careful study of late-stage clinically affected cattle has found low levels of BSE in the tongue and nasal mucosa, thought to have spread out from the brainstem via facial nerves as the animals became terminally ill, a time when they would be very unlikely to be acceptable for human consumption (Balkema-Buschmann *et al.*, 2011).

Atypical BSE in cattle

Due to the usual means of detection of atypical BSE (H- and L-type) from abattoir brain sampling, access to a whole carcass to test which body tissues are infected is rare in natural cattle cases. However transmission of atypical BSEs to mice was achieved using affected cattle brain (Baron *et al.*, 2006), confirming the infectious nature of atypical BSE. Using rare samples available from Italian atypical BSE cases, no infection was found in lymphoid and kidney tissues although skeletal muscle was positive (Suardi *et al.*, 2010) suggesting a potential risk to cattle-meat consumers. Brain pathology is different in classical BSE and atypical BSE as, in the former, lesions are predominantly in the brain stem, whereas L- and H-type BSE both have more lesions in cortical areas. L-type BSE also has large PrP protein deposits (amyloid plaques) which give rise to its alternative name bovine amyloidotic spongiform encephalopathy or BASE, however, technically this term is only correctly used if brain sections are available to identify the plaques and that is not always possible.

Experimental BSE in cattle

Because BSE cattle are detected only at the clinical stage, experiments were set up to study development of BSE from the point of inoculation to find out how the infection spreads throughout the animal's body till it reaches the CNS, particularly of course the brain. In one large study, calves were dosed orally with 100 g BSE brain resulting in clinical signs in the cattle from about 36 months after inoculation. Infectivity was found in the distal ileum of cattle killed at 6 and 10 months after inoculation (Wells *et al.*, 1994, 1998). Infectivity was also demonstrated (by inbred mouse bioassay) in the peripheral nervous system: in the cervical and dorsal root ganglia at 32–40 months after infection and trigeminal ganglia at 36 and 38 months after inoculation but in no other tissues examined. These tissues were negative in the naturally infected BSE cases but this may be related to the initial dose of infection, which is likely to have been greater than in naturally infected cattle. Using more sensitive techniques, signs of infection were found in vagus nerve and adrenal gland of these challenged cattle (Masujin *et al.*, 2007). Further studies using transgenic mouse bioassay have shown in some detail the path taken by BSE infection from the gut via the peripheral nerves to the brain (Kaatz et al., 2012). Infection was first seen in distal ileum and enteric nervous system, spreading then through the sympathetic and parasympathetic nervous system to the brain stem. It is clear that BSE cattle, even those receiving a huge initial dose, are far less likely to contain the relatively high levels of infectivity seen outside the CNS in scrapie-affected sheep.

An attempt was made to establish the smallest amount of BSE brain that would produce disease in inoculated cattle. Even exposure to as little as 1 mg brain homogenate from clinically affected field cases of BSE was enough to produce BSE so the limiting dose for calves must be lower than 1 mg (Wells et al., 2007). However there were found to be fewer signs of infection in the small intestine in the lowest dose animals, similar to field cases of BSE, supporting the hypothesis that field cases must have received very low amounts of infection (Stack et al., 2011).

Experimental atypical BSE in cattle

Experimental challenge studies have been set up to study pathogenesis of atypical BSE development in cattle. In one experiment, six cattle were inoculated intracerebrally with L-type BSE and five with H-type BSE (Balkema-Buschmann et al., 2013) resulting in clinical disease in the cattle after around 15 months' incubation. The main early clinical sign was depression but hyperesthesia prevented more detailed clinical analysis in later stages. The western blot patterns expected of L-type and H-type BSE were reproduced in the inoculated animals. In a separate study (Konold et al., 2012), two groups of four cattle were inoculated with L-type or H-type BSE and the predominant clinical sign was that of difficulty rising. Here L-type was distinguishable from H-type in immunohistochemical staining of disease-related PrP with different antibodies. On western blot examination, the pattern expected of H-type BSE was reproduced in the inoculated animals but the L-type pattern was less obvious. As is often the case, two laboratories using slightly different techniques produce differences in detail but the main conclusion is clear that both forms of atypical BSE are transmissible to other cattle.

Preclinical Diagnosis of BSE in Cattle

Many research groups are looking for markers that would diagnose any of the TSEs in tests which could be carried out on live animals or humans. TSEs are remarkably difficult diseases for which to find specific markers and BSE in cattle is no exception. There is not so much interest specifically in BSE since the control measures seem to be working very well in reducing the incidence, however, mention of some projects is warranted. Cerebrospinal fluid (CSF) has been found to have elevated levels of apolipoprotein E (Hochstrasser et al., 1997) and another protein called 14-3-3, which is of some use as a screening method for CJD in humans (Hsich et al., 1996), may also be informative in cattle with BSE (Lee and Harrington, 1997a, 1997b). An additional marker erythroid associated factor (ERAF), which looked promising as a blood test for scrapie-infected laboratory mice, unfortunately turned out not to be so useful when studies were extended to BSE cattle and scrapie sheep (Brown et al., 2007).

Simply detecting PrP^{Sc} in affected individuals is also potentially of interest, although much less of the disease-associated form of PrP is found in peripheral cattle tissues than in sheep with scrapie or experimental BSE (Somerville et al., 1997). Humans with the new variant form of CJD (vCJD) have been found to have PrP^{Sc} deposits in tonsil biopsies, something which is not the case with the more common sporadic CJD (Arya, 1997; Collinge et al., 1997) although it does occur in sheep scrapie (Schreuder et al., 1996). Such tests would simply not work in cattle due to the very low levels of PrP^{Sc} in their lymphoid tissues. Atypical BSE has a similar lack of lymphoid tissue involvement to that seen with BSE, and so diagnostic tests relying on biopsy sampling in cattle are likely to be uninformative.

BSE Transmission Characteristics

In order to try to understand the role of genetics in control of susceptibility to BSE, it is necessary to understand how, and to what species, it transmits and whether that aetiology bears a resemblance to any of the other transmissible spongiform encephalopathies, such as scrapie in sheep and CJD in humans.

Within cattle most cases were singletons (one case per herd) and, although it is thought unlikely that BSE can spread between cattle via close contact, there has been some argument about whether there is maternal transmission of disease from affected cow to calf. This is an important issue because if maternal transmission does occur in cattle, it could suggest there is inheritance of susceptibility. However, in a study on the offspring of BSE-affected pedigree beef suckler cows, much less likely than dairy cattle to be fed meat and bone meal-derived protein concentrates, none of 219 calves which had been suckled for at least a month went on to develop BSE themselves (Wilesmith and Ryan, 1997). As these animals would have consumed 111,500 l of milk, it suggests that either cattle milk is not a potential source of infection, or that inheritance of susceptibility has not occurred in the study group. A large-scale cohort study has also been carried out comparing animals born to BSE-affected cattle with animals whose mothers were healthy (Wilesmith et al., 1997). Of the offspring from BSE-affected mothers, 42 out of 301 (14%) developed BSE, whereas only 13 out of 301 (4.3%) offspring of BSE-unaffected mothers developed BSE. This places the calves from BSE-affected cows at greater risk of developing disease themselves ($P < 0.0001$) but does not distinguish between inheritance of susceptibility and true maternal transmission of disease. Re-analysis of the data provided support for a genetic component (Ferguson et al., 1997), but the fact that a calf is even more likely to go on to develop BSE if it is born after the onset of symptoms in its mother argues for an element of direct maternal transmission of infection (Donnelly et al., 1997a). However, such a low frequency of maternal transmission, if it occurs at all, was not thought to be able to sustain the epizootic in the UK beyond 2001 (Anderson et al., 1996). BSE has however continued to occur in a trickle of cases known collectively as BARBs (born after the reinforced feed ban). There is no satisfactory explanation for these cases and no systematic genetic study is available.

Atypical BSE, both L- and H-types, is believed to be a sporadic disease due to the occurrence of single cases that have been found globally, including a single case in Brazil where cattle are grass fed. Nevertheless there is interest in looking for evidence of natural transmission and genetic markers and in one study from Japan, the offspring of a beef cow affected by L-type BSE was retained and observed for 4 years before being culled and examined for disease-related PrP protein; however, none was detected in brain or spinal cord (Yokoyama et al., 2011). There is also concern about the potential risks for humans from consumption of meat from atypical BSE cattle and transmissions to primates indicate that L-type BSE will cause disease in macaques by the intracerebral route (Ono et al., 2011) and lemurs by the oral route (Mestre-Frances et al., 2012) and therefore it is potentially a risk for transmission to humans. This has prompted studies of infection with atypical BSE of transgenic mice encoding the human PRNP gene with variable results of very low or negative transmission rates suggesting there is a substantial barrier to human infection with atypical BSE (Kong et al., 2008; Wilson et al., 2012).

Genetics of TSEs in Sheep and Humans

Studies of natural scrapie in sheep have confirmed the importance of three codons in the sheep PRNP gene (136, 154 and 171) (Belt et al., 1995; Clouscard et al., 1995; Hunter et al., 1996) originally shown to be associated with differing incubation periods following experimental challenge of sheep with different sources of scrapie and BSE (Goldmann et al., 1991a, 1994) and, although there are breed differences in PRNP allele frequencies and in disease-associated alleles, some clear rules have emerged from this work. The usual way to describe sheep genotypes is to use the single letter amino acid code, each codon in turn and each allele in turn. The genotype most resistant to natural scrapie in all sheep breeds is thought to be ARR/ARR. This genotype is also resistant to experimental oral challenge with both scrapie and BSE (Goldmann et al., 1994) although is susceptible if the intracerebral route is used (Houston et al., 2003) and so such sheep could potentially act as non-clinical carriers of infection. Other homozygous genotypes encoding glutamine (Q) at codon 171 are more susceptible to scrapie. For example in Suffolk sheep the genotype ARQ/ARQ is most susceptible, although not all animals of this genotype succumb to disease and it is a relatively common genotype amongst healthy animals (Westaway et al., 1994; Hunter et al., 1997b). The PRNP genetic variation in Suffolk sheep is much less than in some other breeds, the so-called 'valine breeds'. Breeds such as Cheviots, Swaledales and Shetlands encode PRNP gene alleles with valine at codon 136 and the genotype VRQ/VRQ is the most susceptible to scrapie (Hunter et al., 1994a, 1996). VRQ/VRQ is a rare genotype and when it does occur, is almost always in scrapie-affected sheep and so it has been suggested that scrapie may be simply a genetic disease (Ridley and Baker, 1995). However healthy animals of this genotype can live up to 8 years of age, well past the usual age-at-death from scrapie (2–4 years) (Hunter et al., 1996, 1997a) and can be easily found in scrapie-free countries (Australia and New Zealand) and so the genetic disease hypothesis seems less likely than an aetiology that involves host genetic control of susceptibility to an infecting agent. Other codons in the sheep PRNP gene are also now known to be linked to, or associated with, differences in survival time, incubation period and/or susceptibility to a range of different strains of scrapie, but the 'three codon' genotype remains the most usual one seen in selection for resistance, for example in the UK National Scrapie Plan (Dawson et al., 2008).

In humans, sporadic forms of CJD are associated with PRNP gene codon 129 polymorphism (methionine/valine) in that homozygous individuals (either MM_{129} or VV_{129}) are over-represented in CJD cases and heterozygosity seems to confer some protection (Lloyd et al., 2011). Variant CJD (vCJD), which unlike sporadic CJD is caused by an infectious agent indistinguishable from BSE, has also been confirmed so far only in MM_{129} genotypes. Other forms of TSEs in humans appear to be genetic diseases, for example GSS which is linked to a codon 102 proline to leucine mutation (Hsiao et al., 1989). There are many other human PRNP gene mutations associated with disease,

for example one familial form of CJD is linked to an insert of 144 bp coding for six extra octapeptide repeats at codon 53 (Poulter et al., 1992) and a codon 200 mutation (glutamic acid to lysine) that is linked to CJD in Israeli Jews of Libyan origin, Slovaks in north central Slovakia, a family in Chile and a German family in the USA (Prusiner and Scott, 1997). However the fact that sheep scrapie, which also demonstrates excellent linkage with PrP genotype, has been shown to be unlikely to be a genetic disease also has implications for interpretation of the human data (Hunter et al., 1997a).

The Bovine PRNP Gene

When BSE was found in cattle, it was an obvious step to study the bovine PRNP gene for markers of resistance or susceptibility to disease similar to those that had been found in sheep and humans. There is a great deal of allelic complexity in both the PRNP coding region and its flanking regions in the sheep (Hunter et al., 1989, 1993; Goldmann et al., 1990; Muramatsu et al., 1992; Laplanche et al., 1993; Bossers et al., 1996) and human PRNP genes (Collinge and Palmer, 1994; Prusiner and Scott, 1997). In contrast, the bovine PRNP gene is remarkably invariant with very few polymorphisms described.

The bovine PRNP gene coding region, which was originally mapped to bovine syntenic group U11 (Ryan and Womack, 1993), was first sequenced in 1991 (Goldmann et al., 1991b) and this sequence was subsequently confirmed by two other groups (Yoshimoto et al., 1992; Prusiner et al., 1993). Allowing for various polymorphic forms of the gene in each species, there is very little difference (>90% identity) between the cattle and sheep PRNP gene. The bovine PRNP gene has so far revealed a limited number of polymorphisms of the coding region (Goldmann et al., 1991b; Goldmann, 2008). These fall into four groups, first the single DNA nucleotide polymorphisms (SNPs) in the PRNP gene coding region, which sometimes, not always, result in an amino acid change in the protein. SNPs in cattle are relatively few compared with the numbers found in sheep and humans. Bos taurus, Bos indicus, Bos javanicus and *Bos mutus* have been studied and only eight different alleles have been found, based on single amino acid changes (Goldmann, 2008). It is noticeable that in Europe and America, domesticated cattle have very little genetic variation, compared to African and Asian cattle. This may be due to differences in effective population sizes as a result of wider use of practices such as artificial insemination in the more industrially developed countries.

The second group of polymorphic variants, again in the *PRNP* coding region, involves a series of glycine-rich repeats encoded by 24 or 27 nucleotide G-C-rich elements (Goldmann *et al.*, 1991b; Prusiner *et al.*, 1993) forming octapeptide repeats in the protein (Fig. 10.1). This region produces variants with different numbers of these repeats, a phenomenon also seen in *PRNP* genes from other species including humans (Goldfarb *et al.*, 1992) where many variations in the PrP octapeptide repeat number have been described, some of which have clear linkage to the incidence of human TSE (Poulter *et al.*, 1992). In cattle, *PRNP* alleles have been described with four to seven repeats (Goldmann, 2008).

The coding region may be rather invariant but there are polymorphisms in the control regions of the *PRNP* gene in cattle. In the promoter and first intron (Fig. 10.1) two short insertion/deletions (indels) of 23 and 12 bp have

been described (Haase *et al.*, 2007). These are potentially of interest as levels of PrP protein are directly linked to susceptibility to scrapie in mice (Bueler *et al.*, 1993) and the same might be true for cattle.

Cattle *PRNP* Gene Polymorphisms and Susceptibility to BSE

PRNP gene

Two early studies addressed the question of association of *PRNP* genotype with incidence of BSE in cattle hoping to find similarly clear linkage with disease to that seen in sheep scrapie. One study is discussed in the next section on family studies but in the other (Hunter *et al.*, 1994b) *PRNP* genotypes of BSE-affected cattle were compared with healthy animals and a case-control study of a single BSE-affected herd. Genotype frequencies of the 5.6 octapeptide repeat polymorphism are presented for 172 histopathologically confirmed BSE-affected cattle in Table 10.2. The majority of the cattle (91%) were of the 6:6 genotype with 9% 6:5 and no 5:5 animals. For convenience cattle were separated into five breed groups: Friesian (92), Friesian × Holstein (14), other Friesian crosses (20), Ayrshire (16) and others (30).

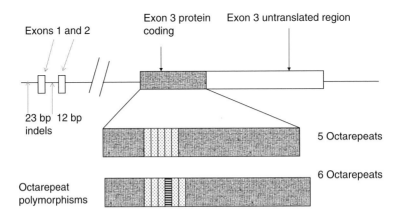

Fig. 10.1. Bovine *PRNP* gene polymorphisms. Octapeptide repeats are indicated in the protein coding open reading frame in Exon 3 of the gene. Each octapeptide repeat is distinguishable on the basis of DNA sequence, the extra repeat in the six-repeat encoding allele is therefore indicated in stripes rather than dots. Insertion/deletion polymorphisms (indels) are indicated in promoter region and the intron between Exons 1 and 2. Direction of transcription is from left to right.

Friesian × Holsteins and Ayrshires had higher frequencies of the 6:5 genotype (21% and 31%, respectively) than other breed groups (ranging from 0–10% but these differences were not significant.

In the case-control single herd study cattle (total 90 animals) shown in Table 10.3 the octapeptide repeat frequencies were 89% 6:6 and 11% 6:5 in the 85 healthy cattle. All five BSE cases were 6:6. (BSE case study frequencies are also given in Table 10.3 for comparison.) There were no significant differences between BSE affected and healthy cattle in this herd in frequencies of the octarepeat *PRNP* polymorphisms.

The healthy (or unaffected) cattle group represented 108 animals from three herds with no history of BSE. Again the majority of animals were of the genotype 6:6 (82%) with 17% being 6:5. A single animal of 5:5 genotype (representing 1% of this sample) was also found. The age of onset was also examined. The youngest animal, an Ayrshire, was 29 months (m) and the oldest, an Ayrshire cross, was 121 m. Table 10.4

gives the age/octarepeat genotype comparisons for all 172 cattle in the BSE case study and for three breed groups large enough to analyse: Friesian, Ayrshire and Friesian × Holstein. There was no association between genotype and age. All the 6:5 genotype animals fell well within the age range set by the greater numbers of 6:6 animals.

Whether or not the cattle were home bred (born on the same farm where they later became BSE-affected) or purchased (born elsewhere and transferred at some later date to the affected farm) may give an indication of where the animals contracted BSE. Of the BSE-affected cattle samples collected in 1991 (146 animals, all female) the majority (67%) were home bred. There was no evidence that this frequency was related to genotype, for instance the ten 6:5 animals in the 1991 group were 80% home bred. The frequency was breed dependent, however, in that in this sample, more than 80% of Friesian, Ayrshire and Friesian × Holstein cattle were home bred, but other Friesian crosses,

Table 10.2. *PRNP* gene octarepeat genotypes in a BSE case study. (Data taken from Hunter *et al.,* 1994b.)

Breed group	Genotype frequency (%)			
	6:6[a]	6:5[b]	5:5[c]	Number
All	91	9	0	172
Friesian	95	5	0	92
Friesian × Holstein	79	21	0	14
Friesian crosses (others)	100	0	0	20
Ayrshire	69	31	0	30
Others	90	10	0	

[a]6:6 – homozygous for the 6-octapeptide repeat-encoding allele.
[b]6:5 – heterozygous for the 6- and the 5-octapeptide repeat-encoding alleles.
[c]5:5 – homozygous for the 5-octapeptide repeat-encoding allele.

Table 10.3. Cattle *PRNP* octapeptide genotype frequencies. (Data taken from Hunter *et al.,* 1994b.)

Cattle group	No. of cattle	Frequency (%)		
		6:6[a]	6:5[b]	5:5[c]
Case study BSE	172	91	9	0
Herd study				
Healthy	85	89	11	0
BSE	5	100	0	0
Unaffected	108	82	17	1

[a]6:6 – homozygous for the 6-octapeptide repeat-encoding allele.
[b]6:5 – heterozygous for the 6- and the 5-octapeptide repeat-encoding alleles.
[c]5:5 – homozygous for the 5-octapeptide repeat-encoding allele.

Table 10.4. BSE case-study: genotype comparison with age of onset of BSE. (Data taken from Hunter et al., 1994b.)

Breed	Genotype[a]	Number	Mean age (months)	SD (months)	Range (months)
All	All	172	60	13	29–121
	6:6	156	60	13	29–121
	6:5	16	59	10	47–79
Friesian	All	92	59	12	38–110
	6:6	87	59	12	38–110
	6:5	5	60	11	49–72
Ayrshire	All	16	58	16	29–105
	6:6	11	61	19	29–105
	6:5	5	53	5	47–60
Friesian × Holstein	All	14	63	12	49–81
	6:6	11	62	12	49–81
	6:5	3	69	12	56–79

[a]Genotype designation as in Tables 10.2 and 10.3.

Aberdeen Angus crosses, Simmental, Limousin, Herefords and other crosses were much more likely to have been purchased animals.

The genotype frequencies found in the above study (Hunter et al., 1994b) were similar to frequencies found in studies of healthy Belgian (Grobet et al., 1994) and US cattle (McKenzie et al., 1992) and in 210 Holstein and 46 Hereford bulls used actively in artificial insemination programmes in the US, the frequency of 6:6 was 97% and 99%, respectively (Brown et al., 1993). The other bulls were 6:5 reducing the frequency of the 5:5 genotype to less than 0.5% in these US Holsteins.

According to a review of the published data in 2008, in a total of 1250 cattle PRNP alleles, 16% (representing 200 alleles) showed variation in the protein coding sequence (Goldmann, 2008). Although none of these was conclusively linked to BSE incidence, the two 23- and 12-bp indel polymorphisms in the upstream region of the PRNP gene from cattle of Holstein and Simmental-related breeds are at higher risk of developing BSE (Sander et al., 2004, 2005). The main link to BSE incidence was from the deletion of the 12-bp sequence, but highest risk of BSE in the tested cattle was associated with PRNP genotypes that had the 23-bp and the 12-bp deletions on both alleles. Around 450 BSE affected and 430 control cattle were included in this study. However, this relationship was not seen in all breeds as German Brown and Swiss Brown cattle did not show the association despite having good frequencies of the deletion alleles. It seems that these polymorphisms may modulate disease outcome, but that the main controller of BSE incidence is whether or not the animal is exposed to infection in the first place (Sander et al., 2004; Gedermann et al., 2006; Juling et al., 2006; Kashkevich et al., 2007).

Unfortunately the evidence suggests that the upstream region polymorphisms are not associated with atypical BSE incidence (Brunelle et al., 2007), and indeed there is no definitive evidence of any PRNP link with these diseases. However, there are tantalizing pieces of information, one of which concerns a variant at codon 211 (glutamate to lysine) that was found in a US case of atypical BSE. Further study showed this polymorphism to be very rare (<1 in 2000) in US cattle so if it was linked to disease it presents a low risk (Heaton et al., 2008). Further more detailed analysis has suggested there is a haplotype, straddling the PRNP gene that may be a genetic determinant of susceptibility to atypical BSE, however, it is a frequent haplotype in healthy cattle so would not be straightforward to use for selection breeding in cattle (Clawson et al., 2008).

Family studies

Given the paucity of evidence for the involvement of the PRNP gene in cattle susceptibility to BSE, is there any sign that offspring of

BSE-affected cows are at more risk of contracting BSE themselves? Several studies attempting to find evidence of inheritance of genetic control of susceptibility, rather than maternal transmission of disease, have not conclusively ruled out some element of genetic control of susceptibility in cattle (Curnow and Hau, 1996; Wilesmith and Ryan, 1997; Donnelly et al., 1997b; Fergusson et al., 1997; Wilesmith et al., 1997).

The information for atypical BSE is lacking, however, as mentioned previously, one reported instance of following the offspring of a cow affected by L-type BSE in Japan found no evidence of disease after 4 years (Yokoyama et al., 2011).

Influence of other genes

Various researchers have tried to find other genes that might show linkage with BSE. One study (Neibergs et al., 1994) used a technique known as single strand polymorphism analysis (SSCP), which is designed to reveal the presence of a polymorphism within stretches of DNA but does not provide details of exactly what that polymorphism is. The SSCP analysis revealed three possible alleles (designated A, B and C) in the PRNP gene region (Neibergs et al., 1994). The source of the changes in DNA that resulted in each allele was unknown, however BSE-affected animals and their relatives were found to be more likely to have the AA genotype than the other animals analysed, with BSE-affected animals giving AA frequency of 48%, their relatives 58% and unrelated healthy animals 29%. Although the AA genotype cannot be regarded as a marker for BSE susceptibility in these cattle, it is suggestive that there may be some genetic linkage with disease incidence within the PrP gene coding region itself. It is interesting that in this study, non-UK cattle (Boran and N'Dama from Kenya, Friesian Sahiwal, Brahman and Brahman crosses from Australia and Brangus from the USA) had extremely low frequencies of AA (5%), suggesting that something is indeed genetically different about UK cattle, however this has never been followed up.

In recent years the development of sophisticated and powerful methods allowing whole genome study has shown that there are likely to be several possible loci involved in control of different aspects of BSE susceptibility to cattle. Different techniques have pointed to different chromosomal regions, for example, on bovine chromosomes BTA 5, 10 and 20 (Hernandez-Sanchez et al., 2002) and using a larger dataset BTA 17 and BTAX/Yps and some additional evidence for BTA 1, 13 and 19 (Zhang et al., 2004). More detailed study of BTA 10 revealed a candidate gene, HEXA, a gene associated in humans with a neurodegenerative disorder Tay-Sachs disease (Juling et al., 2008). The mRNA from this gene is also elevated in mice inoculated with CJD and the product of the gene is the alpha subunit of β hexosaminidase A, which in the lysosomal catabolic pathway catalyses nerve cell membrane components. HEXA has SNPs in intron regions that were associated with absence of BSE in UK cattle. A comparison with German Holstein cattle revealed the opposite effect such that it was over-represented in BSE-affected animals. This linkage disequilibrium may be significant and could tell us more about the development of BSE, however, it is unlikely at present to be as useful a marker of resistance as PRNP gene polymorphisms are in sheep. There are undoubtedly many gene products that will have an influence on the progress of disease and on the pathology and degeneration once infection is established and this may be what is being picked up in these studies.

Is BSE not subject to host genetic control by the PRNP gene?

Despite the lack of evidence that the PRNP gene coding region controls incidence of BSE and atypical BSE in cattle, transmission studies of classical BSE to mice, to sheep and to goats strongly suggest that BSE infectivity does 'select' animals of certain PRNP genotypes in these experimental models. In the mouse transmission studies of Fraser et al. (1992), BSE-affected cattle brain homogenate was injected into strains of mice that differed at amino acids 108 and 189 of the PrP gene (Hunter et al., 1992) giving shorter incubation periods in the mice of $Prnp^{a/a}$ genotype (leucine and threonine at codons 108 and 189) than in mice of $Prnp^{b/b}$ genotype (phenylalanine and valine). Transmission of BSE to Cheviot sheep (Goldmann *et al.*, 1994) also reveals an association of disease with the *PRNP* genotype homozygous for glutamine at codon 171, and when BSE is injected into goats, animals with isoleucine rather than methionine at codon 142 have longer incubation periods (Goldmann *et al.*, 1996). BSE in mice, sheep and goats does associate with *PRNP* variants so why not in cattle?

It may be that in cattle the 6 octarepeat *PRNP* allele is dominant in conferring susceptibility as most of the BSE cases so far described have been 6:6 or 6:5. This could be tested by direct challenge of cattle and/or transgenic mice carrying different octarepeat alleles however if the 6 allele does confer susceptibility, why have there not been more BSE cases in a cattle population that has apparently very high frequencies of this allele? The evidence from variants in the promoter region that controls expression levels of PrP protein may point to differences in susceptibility controlled via this route, but cattle are clearly different from sheep and humans in the lack of a definite marker for prediction of the risk of development of prion disease.

Using BSE occurrence as sole measure of the frequency of susceptible cattle and 'absence of BSE' to estimate the numbers of resistant cattle could simply be wrong. John Wilesmith (Wilesmith *et al.*, 1988), writing at the height of the epizootic and suggesting that BSE resulted from feeding cattle infected ruminant material, also described the difficulties of carrying out case-control studies to confirm this (Wilesmith *et al.*, 1992). The difficulties still remain in epidemiological studies of BSE cattle data. Of particular interest now in the UK are the BSE cases that continue to occur well after the banning (in 1988) of feeding of ruminant derived protein to ruminants (BARB cases). In BARB animals a link with home-made feed mixes has been noted but no association with environmental sources of contamination or indeed of waste on grassland or the presence of other species on the holding (Ortiz-Pelaez *et al.*, 2012).

It remains true that most UK dairy cattle would have been fed potentially contaminated concentrates before 1988, however most did not develop BSE. There may therefore have been uneven distribution of infection in the food and only those cattle ingesting a large enough dose went on to develop BSE. This problem, which affects the epidemiology, may also apply to the *PRNP* genetics. With essentially one form of the cattle *PRNP* gene predominating and if this is the 'susceptible' allele, most cattle may have the potential to develop BSE if given a sufficiently high dose of infection.

Conclusions

In cattle, unlike sheep, the option to control TSE disease by breeding for resistance is not available – there are no genetic markers linked in in a straightforward way with BSE. BSE in the UK is in decline as a result of the physical measures taken to control cattle food along with the slaughter of any animal considered at risk of disease. Because of this, it may be thought that there is no point trying to understand the genetics of BSE, however BSE has in the past apparently spread to other species including humans and it has the potential to do so again if control measures lapse. In particular the BARB cases in the UK, although few in number, represent a risk for a future source of infection should control measures be relaxed beyond a safe point. Our knowledge of BSE may therefore protect us from similar new disease outbreaks in the future.

References

Anderson, R.M., Donnelly, C.A., Ferguson, N.M., Woolhouse, M.E.J., Watt, C.J., Udy, H.J., Mawhinney, S., Dunstan, S.P., Southwood, T.R.E., Wilesmith, J.W. *et al.* (1996) Transmission dynamics and epidemiology of BSE in British cattle. *Nature* 382, 779–788.

Arya, S.C. (1997) Diagnosis of new variant Creutzfeldt-Jakob disease by tonsil biopsy. *The Lancet* 349, 1322–1323.

Austin, A. and Pollin, M. (1993) Reduced rumination in bovine spongiform encephalopathy and scrapie. *Veterinary Record* 132, 324–325.

Austin, A.R., Pawson, L., Meek, S. and Webster, S. (1997) Abnormalities of heart rate and rhythm in bovine spongiform encephalopathy. *Veterinary Record* 141, 352–357.

Balkema-Buschmann, A., Eiden, M., Hoffmann, C., Kaatz, M., Ziegler, U., Keller, M. and Goschup, M. (2011) BSE infectivity in the absence of detectable PrP^{Sc} accumulation in the tongue and nasal mucosa of terminally diseased cattle. *Journal of General Virology* 92, 467–476.

Balkema-Buschmann, A., Ziegler, U., McIntyre, L., Keller, M., Hoffmann, C., Rogers, R., Hills, B. and Groschup, M. (2013) Experimental challenge of cattle with German atypical bovine spongiform encephalopathy (BSE) isolates. *Journal of Toxicology and Environmental Health, Part A: Current Issues* 74, 103–109.

Baron, T.G.M., Biacabe, A.-G., Bencsik, A. and Langeveld, J.P.M. (2006) Transmission of new bovine prion to mice. *Emerging Infectious Diseases* 12, 1125–1128.

Belt, P.B.G.M., Muileman, I.H., Schreuder, B.E.C., Bos-De Ruijter, J., Gielkens, A.L.J. and Smits, M.A. (1995) Identification of five allelic variants of the sheep PrP gene and their association with natural scrapie. *Journal of General Virology* 76, 509–517.

Bossers, A., Schreuder, B.E.C., Muileman, I.H., Belt, P.B.G.M. and Smits, M.A. (1996) PrP genotype contributes to determining survival times of sheep with natural scrapie. *Journal of General Virology* 77, 2669–2673.

Brown, A.R., Alejo Blanco, A.R., Miele, G., Hawkins, S.A., Hopkins, J., Fazakerley, J.K., Manson, J. and Clinton, M. (2007) Differential expression of erythroid genes in prion disease. *Biochemical and Biophysical Research Communications* 64.

Brown, D.R., Zhang, H.M., Denise, S.K. and Ax, R.L. (1993) Bovine prion gene allele frequencies determined by AMFLP and RFLP analysis. *Animal Biotechnology* 4, 47–51.

Bruce, M., Chree, A., McConnell, I., Foster, J., Pearson, G. and Fraser, H. (1994) Transmission of bovine spongiform encephalopathy and scrapie to mice – strain variation and the species barrier. *Philosophical Transactions of the Royal Society of London Series B: Biological Sciences* 343, 405–411.

Brunelle, B.W., Hamir, A.N., Baron, T., Biacabe, A.-G., Richt, J., Kunkle, R.A., Cutlip, R.C., Miller, J.M. and Nicholson, E.M. (2007) Polymorphisms of the prion gene promoter region that influence classical bovine spongiform encephalopahy susceptibility are not applicable to other transmissible spongiform encephalopathies in cattle. *Journal of Animal Science* 85, 3142–3147.

Budka, H. (2011) The European response to BSE: a success story. *EFSA Journal* 9, doi: 10.2903/j.efsa.2011.e991.

Bueler, H., Aguzzi, A., Sailer, A., Greiner, R.A., Autenried, P., Aguet, M. and Weissmann, C. (1993) Mice devoid of PrP are resistant to scrapie. *Cell* 73, 1339–1347.

Buschmann, A. and Groschup, M. (2005) Highly bovine spongiform encephalopathy-sensitive transgenic mice confirm the essential restriction of infectivity to the nervous system in clinically diseased cattle. *Journal of Infectious Disease* 192, 934–942.

Carlson, G.A., Kingsbury, D.T., Goodman, P.A., Coleman, S., Marshall, S.T., Dearmond, S., Wesetaway, D. and Prusiner, S.B. (1986) Linkage of prion protein and scrapie incubation time genes. *Cell* 46, 503–511.

Chen, J. and Thirumalai, D. (2012) Helices 2 and 3 are the initiation sites in the PrPC to PrPSc transition. *Biochemistry* 52, 310–319.

Clawson, M.L., Richt, J.A., Baron, T., Biacabe, A.-G., Czub, S., Heaton, M.P., Smith, T.P.L. and Laegeid, W.W. (2008) Association of a bovine prion gene haplotype with atypical BSE. *Plos One* 3, e1830 doi:10.1371/journal.pone.001830.

Clouscard, C., Beaudry, P., Elsen, J.M., Milan, D., Dussaucy, M., Bouneau, C., Schelcher, F., Chatelain, J., Launay, J.M. and Laplanche, J.L. (1995) Different allelic effects of the codons 136 and 171 of the prion protein gene in sheep with natural scrapie. *Journal of General Virology* 76, 2097–2101.

Collinge, J. and Palmer, M.S. (1994) Human prion diseases. *Baillieres Clinical Neurology* 3, 241–255.

Collinge, J., Hill, A., Ironside, J. and Zeidler, M. (1997) Diagnosis of new variant Creutzfeldt-Jakob disease by tonsil biopsy – Reply. *Lancet* 349, 1323.

Curnow, R.N. and Hau, C.M. (1996) The incidence of bovine spongiform encephalopathy in the progeny of affected sires and dams. *Veterinary Record* 138, 407–408.

Dawson, M., Moore, R.C. and Bishop, S.C. (2008) Progress and limits of PrP gene selection policy. *Veterinary Research* 39, 25,doi:10.1051/vetres:2007064.

Donnelly, C.A., Ferguson, N.M., Ghani, A.C., Wilesmith, J.W. and Anderson, R.M. (1997a) Analysis of dam–calf pairs of BSE cases: confirmation of a maternal risk enhancement. *Proceedings of the Royal Society of London Series B: Biological Sciences* 264, 1647–1656.

Donnelly, C.A., Ghani, A.C., Ferguson, N.M., Wilesmith, J.W. and Anderson, R.M. (1997b) Analysis of the bovine spongiform encephalopathy maternal cohort study: evidence for direct maternal transmission. *Applied Statistics–Journal of the Royal Statistical Society Series C* 46, 321–344.

Ferguson, N.M., Donnelly, C.A., Woolhouse, M.E.J. and Anderson, R.M. (1997) Genetic interpretation of heightened risk of BSE in offspring of affected dams. *Proceedings of the Royal Society of London Series B: Biological Sciences* 264, 1445–1455.

Foster, J.D., Bruce, M., McConnell, I., Chree, A. and Fraser, H. (1996) Detection of BSE infectivity in brain and spleen of experimentally infected sheep. *Veterinary Record* 138, 546–548.

Fraser, H. and Foster, J.D. (1993) Transmission of BSE to mice, sheep and goats and bioassay of bovine tissues. In: Bradley, R. and Marchant, B. (eds) *Transmissible Spongiform Encephalopathies: Proceedings of a Consulation on BSE with the Scientific Veterinary Committee of the Commission of the European Communities*. European Commission, Brussels.

Fraser, H., Bruce, M.E., Chree, A., McConnell, I. and Wells, G.A. (1992) Transmission of bovine spongiform encephalopathy and scrapie to mice. *Journal of General Virology* 73, 1891–1897.

Gedermann, H., He, H., Bobal, P., Bartenschlager, H. and Peuss, S. (2006) Comparison of DNA variants in the *PRNP* and *NF1* regions between bovine spongiform encephalopathy and control cattle. *Animal Genetics* 37, 469–474.

Goldfarb, L.G., Brown, P.B. and Gajdusek, D.C. (1992) The molecular genetics of human transmissible spongiform encephalopathies. In: Prusiner, S.B., Collinge, J., Powell, J. and Anderton, B. (eds) *Prion Diseases of Humans and Animals*. Ellis Horwood, Chichester.

Goldmann, W. (2008) PrP genetics in ruminat transmissible spongiform encephalopathies. *Veterinary Research* 39, 30, doi:10.1051/vetres:2008010.

Goldmann, W., Hunter, N., Foster, J.D., Salbaum, J.M., Beyreuther, K. and Hope, J. (1990) Two alleles of a neural protein gene linked to scrapie in sheep. *Procedings of the National Academy of Sciences USA* 87, 2476–2480.

Goldmann, W., Hunter, N., Benson, G., Foster, J.D. and Hope, J. (1991a) Different scrapie-associated fibril proteins (PrP) are encoded by lines of sheep selected for different alleles of the *Sip* gene. *Journal of General Virology* 72, 2411–2417.

Goldmann, W., Hunter, N., Martin, T., Dawson, M. and Hope, J. (1991b) Different forms of the bovine PrP gene have five or six copies of a short, G-C-rich element within the protein coding exon. *Journal of General Virology* 72, 201–204.

Goldmann, W., Hunter, N., Smith, G., Foster, J. and Hope, J. (1994) PrP genotype and agent effects in scrapie – change in allelic interaction with different isolates of agent in sheep, a natural host of scrapie. *Journal of General Virology* 75, 989–995.

Goldmann, W., Martin, T., Foster, J., Hughes, S., Smith, G., Hughes, K., Dawson, M. and Hunter, N. (1996) Novel polymorphisms in the caprine PrP gene: a codon 142 mutation associated with scrapie incubation period. *Journal of General Virology* 77, 2885–2891.

Grobet, L., Vandevenne, S., Charlier, C., Pastoret, P.P. and Hanset, R. (1994) Polymorphism of the prion protein gene in Belgian cattle. *Annales De Médecine Vétérinaire* 138, 581–586.

Haase, B., Doherr, M.G., Seuberlich, T., Drogemuller, C., Dolf, G., Nicken, P., Schiebel, K., Ziegler, U., Groschup, M., Zurbriggen, A. and Leeb, T. (2007) *PRNP* promoter polymorphisms are associated with BSE susceptibility in Swiss and German cattle. *BMC Genetics* 8, 15.

Hagiwara, K., Yamakawa, Y., Sato, Y., Nakamura, Y., Tobiume, M., Shinagawa, M. and Sata, T. (2007) Accumulation of mono-glycosylated form-rich, plaque-forming PrP^{Sc} in the second atypical bovine spongiform encephalopathy case in Japan. *Japanese Journal of Infectious Disease* 60, 305–308.

Heaton, M.P., Keele, J.W., Harhay, G.P., Richt, J.A., Koohmaraie, K., Wheeler, T.J., Shackelfod, S.D., Casas, E., King, D.A., Sonstegard, T.S., Van Tassell, C.P., Neibergs, H.L., Chase, C.C., Kalbfleisch, T.S., Smith, T.P.L., Clawson, M.L. and Laegeid, W.W. (2008) Prevalence of the prion protein gene E211K variant in US cattle. *BMC Veterinary Research* 4, 25, doi:10.1186/1746-6148-4-25.

Hernandez-Sanchez, J., Waddington, D., Wiener, P., Haley, C.S. and Williams, J.L. (2002) Genome-wide search for markers associaed with bovine spongiform encephalopathy. *Mammalian Genome* 13, 164–168.

Hochstrasser, D.F., Frutiger, S., Wilkins, M.R., Hughes, G. and Sanchez, J.C. (1997) Elevation of apolipoprotein E in the CSF of cattle affected by BSE. *Febs Letters* 416, 161–163.

Hope, J. (1993) The biology and molecular biology of scrapie-like diseases. *Archives of Virology* 7, 201–214.

Hope, J., Reekie, L.J.D., Hunter, N., Multhaup, G., Beyreuther, K., White, H., Scott, A.C., Stack, M.J., Dawson, M. and Wells, G.A.H. (1988) Fibrils from brains of cows with new cattle disease contain scrapie-associated protein. *Nature* 336, 390–392.

Houston, F., Goldmann, W., Chong, A., Jeffrey, M., Gonzalez, L., Foster, J., Parnham, D. and Hunter, N. (2003) BSE in sheep bred for resistance to infection. *Nature* 23, 98.

Hsiao, K., Baker, H.F., Crow, T.J., Poulter, M., Owen, F., Terwilliger, J.D., Westaway, D., Ott, J. and Prusiner, S.B. (1989) Linkage of a prion protein missense variant to Gerstmann Sraussler Syndrome. *Nature* 338, 342–345.

Hsich, G., Kenney, K., Gibbs, C.J., Lee, K.H. and Harrington, M.G. (1996) The 14-3-3 brain protein in cerebrospinal fluid as a marker for transmissible spongiform encephalopathies. *New England Journal of Medicine* 335, 924–930.

Hunter, N., Foster, J.D., Dickinson, A.G. and Hope, J. (1989) Linkage of the gene for the scrapie-associated fibril protein (PrP) to the *Sip* gene in Cheviot sheep. *Veterinary Record* 124, 364–366.

Hunter, N., Dann, J.C., Bennett, A.D., Somerville, R.A., McConnell, I. and Hope, J. (1992) Are Sinc and the PrP gene congruent? Evidence from PrP gene analysis in *Sinc* congenic mice. *Journal of General Virology* 73, 2751–2755.

Hunter, N., Goldmann, W., Benson, G., Foster, J.D. and Hope, J. (1993) Swaledale sheep affected by natural scrapie differ significantly in PrP genotype frequencies from healthy sheep and those selected for reduced incidence of scrapie. *Journal of General Virology* 74, 1025–1031.

Hunter, N., Goldmann, W., Smith, G. and Hope, J. (1994a) The association of a codon 136 PrP gene variant with the occurrence of natural scrapie. *Archives of Virology* 137, 171–177.

Hunter, N., Goldmann, W., Smith, G. and Hope, J. (1994b) Frequencies of PrP gene variants in healthy cattle and cattle with BSE in Scotland. *Veterinary Record* 135, 400–403.

Hunter, N., Foster, J., Goldmann, W., Stear, M., Hope, J. and Bostock, C. (1996) Natural scrapie in a closed flock of Cheviot sheep occurs only in specific PrP genotypes. *Archives of Virology* 141, 809–824.

Hunter, N., Cairns, D., Foster, J., Smith, G., Goldmann, W. and Donnelly, K. (1997a) Is scrapie a genetic disease? Evidence from scrapie-free countries. *Nature* 386, 137.

Hunter, N., Moore, L., Hosie, B., Dingwall, W. and Greig, A. (1997b) Natural scrapie in a flock of Suffolk sheep in Scotland is associated with PrP genotype. *Veterinary Record* 140, 59–63.

Jeffrey, M. and Halliday, W.G. (1994) Numbers of neurons in vacuolated and non-vacuolated neuroanatomical nuclei in bovine spongiform encephalopathy-affected brains. *Journal of Comparative Pathology* 110, 287–293.

Juling, K.H.S., Williams, J.L. and Fries, R. (2006) A major genetic component of BSE susceptibility. *BMC Biology* 4, 33.

Juling, K., Schwarzenbacher, H., Frankenberg, U., Ziegler, U., Groschup, M., Williams, J.L. and Fries, R. (2008) Characterization of a 320-bb region containing the *HEXA* gene on bovine chromosome 10 and analysis of its association with BSE susceptibility. *Animal Genetics* 39, 400–406.

Kaatz, M., Fast, C., Ziegler, U., Balkema-Buschmann, A., Hammerschmidt, B., Keller, M., Oelschlegel, A., McIntyre, L. and Groschup, M. (2012) Spread of classic BSE prions from the gut via the peripheral nervous system to the brain. *American Journal of Pathology* 181, 515–524.

Kashkevich, K., Humeny, A., Zeigler, U., Groschup, M., Nicken, P., Leeb, T., Fischer, C., Becker, C.M. and Schiebel, K. (2007) Functional relevance of DNA polymorphisms within the promoter region of the prion protein gene and their association to BSE infection. *FASEB Journal* 21, 1547–1555.

Kong, Q., Zheng, M., Casalone, C., Qing, L., Huang, S., Chakaborty, B., Wang, P., Chen, J., Cali, I., Corona, C., Martucci, F., Iulini, B., Acutis, P., Wang, L., Liang, J., Wang, M., Li, X., Monaco, S., Zanusso, G., Zou, W.Q., Caramelli, M. and Gambetti, P. (2008) Evaluation of the human transmission risk of an atypical bovine spongiform encephalopathy prion strain. *Journal of Virology* 82, 3697–3701.

Konold, T., Lee, Y.H., Stack, M.J., Horrocks, C., Green, R.B., Chaplin, M., Simmons, M.M., Hawkins, S.A.C., Lockey, R., Spiopoulos, J., Wilesmith, J.W. and Wells, G.A.H. (2006) Different prion disease phenotypes result from inoculations of cattle with two temporallly separated sources of sheep scrapie from Great Britain. *BMC Veterinary Research* 2, 31.

Konold, T., Bone, G.E., Clifford, D., Chaplin, M.J., Cawthraw, S., Stack, M.J. and Simmons, M.M. (2012) Experimental H-type and L-type bovine spongiform encephalopathy in cattle: observation of two clinical syndromes and diagnostic challenges. *BMC Veterinary Research* 8, 22.

Laplanche, J.L., Chatelain, J., Westaway, D., Thomas, S., Dussaucy, M., Brugerepicoux, J. and Launay, J.M. (1993) PrP polymorphisms associated with natural scrapie discovered by denaturing gradient gel-electrophoresis. *Genomics* 15, 30–37.

Lee, K. and Harrington, M. (1997a) 14-3-3 and BSE. *Veterinary Record* 140, 206–207.

Lee, K.H. and Harrington, M.G. (1997b) The assay development of a molecular marker for transmissible spongiform encephalopathies. *Electrophoresis* 18, 502–506.

Lloyd, S., Mead, S. and Collinge, J. (2011) Genetics of prion disease. *Topics in Current Chemistry* 305, 1–22.

McKenzie, D.I., Cowan, C.M. and Marsh, R.F. (1992) PrP gene variability in the US cattle population. *Animal Biotechnology* 3, 309–315.

Masujin, K., Matthews, D., Wells, G.A.H., Mohri, S. and Yokohama, T. (2007) Prions in the peripheral nerves of bovine spongiform encephalopathy-affected cattle. *Journal of General Virology* 88, 1850–1858.

Mestre-Frances, N., Nicot, S., Rouland, S., Biacabe, A.-G., Quadrio, I., Peret-Liaudet, A., Baron, T. and Verdier, J.-M. (2012) Oral transmission of L-type bovine spongiform encephalopathy in primate model. *Emerging Infectious Diseases* 18, 142–145.

Miyazawa, K., Kipkorir, T., Tittman, D. and Manuelidis, L. (2012) Continuous production of prions after infectious particles are eliminated: implications for Alzheimer's disease. *PLoS ONE* 7, doi:10.1371/journal.pone.0035471.

Muramatsu, Y., Tanaka, K., Horiuchi, M., Ishiguro, N., Shinagawa, M., Matsui, T. and Onodeera, T. (1992) A specific RFLP type associated with the occurrence of sheep scrapie in Japan. *Archives in Virology* 127, 1–9.

Nathanson, N., Wilesmith, J. and Griot, C. (1997) Bovine spongiform encephalopathy (BSE): causes and consequences of a common source epidemic. *American Journal of Epidemiology* 145, 959–969.

Neibergs, H.L., Ryan, A.M., Womack, J.E., Spooner, R.L. and Williams, J.L. (1994) Polymorphism analysis of the prion gene in BSE-affected and unaffected cattle. *Animal Genetics* 25, 313–317.

Ono, F., Tase, N., Kurosawa, A., Hiyaoka, A., Ohyama, A., Tezuka, Y., Wada, N., Sato, Y., Tobiume, M., Hagiwara, K., Yamakawa, Y., Terao, K. and Sata, T. (2011) Atypical L-type bovine spongiform encephalopathy (L-BSE) transmission to cynomolgus macaques, a non-human primate. *Japanese Journal of Infectious Disease* 64, 81–84.

Ortiz-Pelaez, A., Steven, M.A., Wilesmith, J.W., Ryan, J.B.M. and Cook, A.J.C. (2012) Case-control study of cases of bovine spongiform encephalopathy born after July 31, 1996 (BARB cases) in Great Britain. *Veterinary Record* 170, 389.

Poulter, M., Baker, H.F., Frith, C.D., Leach, M., Lofthouse, R., Ridley, R.M., Shah, T., Owen, F., Collinge, J. and Brown, J. (1992) Inherited prion disease with 144 base pair gene insertion. 1. Genealogical and molecular studies. *Brain* 115, 675–685.

Prusiner, S.B. and Scott, M.R. (1997) Genetics of prions. *Annual Review of Genetics* 31, 139–175.

Prusiner, S.B., Scott, M., Foster, D., Pan, K.-M., Groth, D., Mirenda, C., Torchia, M., Yang, S.-L., Serban, D., Carlson, G.A., Hoppe, P.C., Westaway, D. and Dearmond, S.J. (1990) Transgenetic studies implicate interactions between homologous PrP isoforms in scrapie prion replication. *Cell* 63, 673–686.

Prusiner, S.B., Miklos, F., Scott, M., Serban, H., Taraboulos, A., Gabriel, J.-M., Wells, G.A.H., Wilesmith, J.W., Bradley, R., Dearmond, S.J. and Kristensson, K. (1993) Immunologic and molecular biologic studies of prion proteins in bovine spongiform encephalopathy. *Journal of Infectious Diseases* 136, 602–613.

Ridley, R.M. and Baker, H.F. (1995) The myth of maternal transmission of spongiform encephalopathy. *British Medical Journal* 311, 1071–1075.

Ryan, A.M. and Womack, J.E. (1993) Somatic cell mapping of the bovine prion protein gene and restriction fragment length polymorphism studies in cattle and sheep. *Animal Genetics* 24, 23–26.

Sander, P., Hamann, H., Preiffer, I., Wemeuuer, W., Brenig, B., Groschup, M., Ziegler, U., Distl, O. and Leeb, T. (2004) Analysis of sequence variability of the bovine prion protein gene (*PRNP*) in German cattle breeds. *Neurogenetics* 5, 19–25.

Sander, P., Hamann, H., Drogemuller, C., Kashkevich, K., Sschiebel, K. and Leeb, T. (2005) Bovine prion protein gene (*PRNP*) promoter polymorphisms modulate *PRNP* expression and may be responsible for differences in bovine spongiform encephalopathy susceptibility. *Journal of Biological Chemistry* 280, 37408–37414.

Schreuder, B.E.C., Van Keulen, L.J.M., Langeveld, J.P.M. and Smits, M.A. (1996) Pre clinical test for prion diseases. *Nature* 381, 563.

Seuberlich, T., Heim, D. and Zurbriggen, A. (2010) Atypical transmissible spongiform encephalopathies in ruminants: a challenge for disease surveillance and control. *Journal of Veterinary Diagnostic Investigation* 22, 823–842.

Somerville, R.A., Birkett, C.R., Farquhar, C.F., Hunter, N., Goldmann, W., Dornan, J., Gover, D., Hennion, R., Percy, C., Foster, J. and Jeffrey, M. (1997) Immunodetection of PrPSC in spleens of some scrapie-infected sheep but not BSE-infected cows. *Journal of General Virology* 78, 2389–2396.

Soto, S. (2011) Prion hypothesis: the end of the controversy? *Trends in Biochemical Science* 36, 151–158.

Stack, M., Moore, S.J., Vidal-Diez, A., Arnold, M.E., Jones, E.M., Spencer, Y.I., Webb, P., Spiropoulos, J., Powell, L., Bellerby, P. *et al.* (2011) Experimental bovine spongiform encephalopathy: detection of PrPSc in the small intestine relative to exposure dose and age. *Journal of Comparative Pathology* 145, 289–310.

Suardi, S., Vimercati, C., Casalone, C., Gelmetti, D., Corona, C., Iulini, B., Mazza, M., Lombardi, G., Moda, F., Ruggerone, M. *et al.* (2012) Infectivity in skeletal muscle of cattle with atypical bovine spongiform encephalopathy. *PLoS ONE* 7, doi:10.1371/journal.pone.0031449.

Taylor, D.M., Fraser, H., McConnell, I., Brown, D.A., Brown, K.L., Lamza, K.A. and Smith, G.R.A. (1994) Decontamination studies with the agents of bovine spongiform encephalopathy and scrapie. *Archives of Virology* 139, 313–326.

Taylor, D.M., Ferguson, C.E., Bostock, C.J. and Dawson, M. (1995) Absence of disease in mice receiving milk from cows with bovine spongiform encephalopathy. *Veterinary Record* 136, 592.

Tranulis, M.A., Benestad, S.L., Baron, T. and Kretzschmar, H. (2011) Atypical prion diseases in humans and animals. *Topics in Current Chemistry* 305, 23–50.

Wells, G.A.H., Dawson, M., Hawkins, S.A.C., Green, R.B., Dexter, I., Francis, M.E., Simmons, M.M., Austin, A.R. and Horigan, M.W. (1994) Infectivity in the ileum of cattle challenged orally with bovine spongiform encephalopathy. *Veterinary Record* 135, 40–41.

Wells, G.A.H., Hawkins, S.A.C., Green, R.B., Austin, A.R., Dexter, I., Spencer, Y.I., Chaplin, M.J., Stack, M.J. and Dawson, M. (1998) Preliminary observations on the pathogenesis of experimental bovine spongiform encephalopathy (BSE): an update. *Veterinary Record* 142, 103–106.

Wells, G.A.H., Konold, T., Arnold, M.E., Austin, N.A.R., Hawkins, S.A.C., Stack, M., Simmons, M.M., Lee, Y.H., Gavier-Widen, D., Dawson, M. and Wilesmith, J.W. (2007) Bovine spongiform encephalopathy: the effect of oral exposure dose on attack rate and incubation period in cattle. *Journal of General Virology* 88, 1363–1373.

Westaway, D., Zuliani, V., Cooper, C.M., Dacosta, M., Neuman, S., Jenny, A.L., Detwiler, L. and Prusiner, S.B. (1994) Homozygosity for prion protein alleles encoding glutamine-171 renders sheep susceptible to natural scrapie. *Genes & Development* 8, 959–969.

Wilesmith, J. and Ryan, J. (1997) Absence of BSE in the offspring of pedigree suckler cows affected by BSE in Great Britain. *Veterinary Record* 141, 250–251.

Wilesmith, J.W., Wells, G.A.H., Cranwell, M.P. and Ryan, J.B.M. (1988) Bovine spongiform encephalopathy – epidemiological Studies. *Veterinary Record* 123, 638–644.

Wilesmith, J.W., Ryan, J.B.M. and Atkinson, M.J. (1991) Bovine spongiform encephalopathy – epidemiologic studies on the origin. *Veterinary Record* 128, 199–203.

Wilesmith, J.W., Ryan, J.B. and Hueston, W.D. (1992) Bovine spongiform encephalopathy: case-control studies of calf feeding practices and meat and bonemeal inclusion in proprietary concentrates. *Research in Veterinary Science* 52, 325–331.

Wilesmith, J., Wells, G., Ryan, J., Gavier-Widen, D. and Simmons, M. (1997) A cohort study to examine maternally-associated risk factors for bovine spongiform encephalopathy. *Veterinary Record* 141, 239–243.

Wilson, R., Plinston, C., Hunter, N., Casalone, C., Corona, C., Tagliavini, F., Suardi, S., Ruggerone, M., Moda, F., Graziano, S., Sbriggoli, M., Cardone, F., Pocchiari, M., Ingrosso, L., Baron, T., Richt, J., Andreoletti, O., Simmons, M., Lockey, R., Manson, J.C. and Barron, R.M. (2012) Chronic wasting disease and atypical forms of bovine spongiform encephalopathy and scrapie are not transmissible to mice expressing wild-type levels of human prion protein. *Journal of General Virology* 93, 1624–1629.

Yokoyama, T., Okada, H., Murayama, Y., Masujin, K., Iwamaru, Y. and Mohri, S. (2011) Examination of the offspring of a Japanese cow affected wtih L-type bovine spongiform encephalopathy. *Journal of Veterinary and Medical Science* 73, 121–123.

Yoshimoto, J., Iinumo, T., Ishiguro, N., Imamura, M. and Shiniagawa, M. (1992) Comparative sequence analysis and expression of bovine PrP gene in mouse-L929 cells. *Virus Genes* 6, 343–356.

Zhang, C., De Koning, D.-J., Hernandez-Sanchez, J., Haley, C.S., Williams, J.L. and Wheeler, P. (2004) Mapping of multiple quantitative trait loci affecting bovine spongiform encephalopathy. *Genetics* 167, 1863–1872.

11 Genetics of Behaviour in Cattle

Pamela Wiener

University of Edinburgh, Easter Bush, Midlothian, UK

Introduction

Cattle behaviour has been altered both by natural and artificial selection, the latter following domestication by humans. As with other traits, some characteristics that were favoured in a wild-living situation may have been undesirable under domestication. Albright and Arave (1997) (based on Hale, 1969) outline a number of behavioural characteristics that are beneficial in domesticated animals and could therefore have been favoured during cattle domestication. These include tendency to get close to humans, calmness and unspecialized dietary habits, among others. Genetic variation for these traits would of course be required for selection to succeed, and thus to understand the domestication of cattle requires knowledge of the level of genetic variation underlying behavioural traits.

There are a number of motivations for improving understanding of the genetics of cattle behaviour in addition to basic scientific interest. One major issue regards welfare and ethical considerations. There are increasing concerns from both governmental regulatory institutions and consumers for the welfare of farm animals. As livestock production has become more intensive and commercialized, disparities between animal and environment have become apparent (Phillips, 2002). While some welfare issues can be dealt with by modification of management, it is not clear that all can be without substantially

© CAB International 2015. *The Genetics of Cattle*, 2nd Edn (eds D.J. Garrick and A. Ruvinsky)

reducing productivity (Phillips, 2002) and thus, selection for improved 'suitability' for domesticated life could provide an alternative approach. While improved animal welfare has an indirect benefit on humans, there are also potential direct benefits from selective breeding for altered behaviour; these may derive from the improved safety for farm workers working with these large and potentially dangerous animals as well as the possible associations between behavioural and production characteristics (dairy or meat) of cattle.

Compared to other traits, the application of genomic technologies to the study of cattle behaviour has not been well developed. This is possibly due to a number of factors including the many difficulties inherent in measuring behavioural traits and the substantial environmental component of these traits (Buchenauer, 1999). In order to distinguish the role of genetic from other factors in the determination of behaviour, it is necessary to develop measurements that are well defined and objective. If possible, traits should also be consistent across time; however, if an individual's change in response over time follows a consistent pattern and variation between animals exists, the traits can still be used for genetic analysis (Gibbons et al., 2009a). An important focus for research in cattle behaviour has been refinement of the behavioural tests, with the aim of developing measurements that are well defined, objective and consistent, and can thus be used in genetic and other studies (e.g. Curley et al., 2006; Gibbons et al., 2009b, 2010). Another focus is to characterize the relationship between different behavioural tests and determine how well individual behavioural measures capture the overall characteristics of the animal (Boissy and Bouissou, 1995; Grignard et al., 2001; Gibbons et al., 2011). As for any trait, the interpretation of results from genetic studies will be limited by the quality of the behavioural data.

This chapter details several aspects of cattle behaviour where genetic factors have been investigated. These include temperament, social behaviour, feeding behaviour and mating behaviour. A particular focus regards the substantial developments related to genetic improvement of cattle in terms of their manageability.

Temperament

Burrow (1997) defined temperament as 'the animal's behavioural response to handling by humans' and this chapter will adopt that definition. Accordingly, animals with 'better' or 'favourable' temperament are those that are less agitated by the presence of or handling by humans. Temperament is generally measured using some form of handling test, which Burrow (1997) categorizes as either restrained or non-restrained, in which the former involves placing the animal in a small, confined space. The most common type of restrained test is the chute or crush test, which involves placing the animal in a chute or crush (with or without restraining the head) and assessing the animal's level of agitation (Kuehn et al., 1998; Beckman et al., 2007). Tests of non-restrained behaviour include flight speed (or flight distance), the speed of exit (or distance travelled) when the animal is released from a weighing scale, and exit score (a categorical assessment of speed) where low scores are associated with more favourable temperaments. Alternatively, flight time is inversely related to flight speed or distance and high scores are associated with more favourable temperament. Other examples include social separation or yard tests, which measure how easy it is to corner an animal in a pen after separating it from other animals. Fordyce et al. (1982) found moderate and significant phenotypic correlations (0.34–0.51) between crush test and yard test measurements in a group of various Bos taurus and Bos indicus cross beef cattle; Grignard (2001) found significant and moderate correlations (0.28–0.37) between scores from a crush test and a social separation test in Limousin heifers, and Vetters et al. (2013) found a moderate correlation (0.30) between flight speed and exit score in mixed-breed B. taurus (which increased to 0.45 when the former trait was corrected for body weight). However, Gibbons et al. (2011) did not find a significant correlation between crush score and either flight speed or response to human approach in a study of Holstein Friesian dairy cattle. Flight speed/distance/time give quantitative measures, while other implementations of temperament tests give an overall score (e.g. 1–5, 1–6) of the animal's level of flightiness; other approaches divide temperament into different components (e.g. amount of time spent running or in the corner during the yard test).

Beef cattle

A substantial genetic contribution to beef cattle temperament has been clearly demonstrated from several lines of evidence. First, *B. indicus* and *indicus*-derived breeds have been shown to be more excitable than *B. taurus* breeds raised under the same management regime (Burrow, 1997; Voisinet *et al.*, 1997b). Significant differences between *B. taurus* breeds have also been documented, with breeds of British ancestry (Hereford and German Angus) showing more favourable temperaments than those with continental European ancestry (Charolais and Limousin) (Hoppe *et al.*, 2010). Differences have also been shown between *B. indicus*-derived breed-groups, where Brahman heifers were more docile than those of the Tropical Composite breed (Prayaga *et al.*, 2009). Finally, a large number of studies have found significant heritability for either overall temperament scores or component traits, with most (60%) estimates between 0.11 and 0.40 and some as high as 0.70, varying with test (or component), age of testing, sex, country and model used (Table 11.1). For example, published heritability estimates for the chute test in North American Limousins are generally higher (0.29–0.40; Kuehn *et al.*, 1998; Beckman *et al.*, 2007) than yard test heritabilities in French Limousins (0.18–0.22; Le Neindre *et al.*, 1995; Phocas *et al.*, 2006). Benhajali *et al.* (2010) also estimated heritabilities for crush scores of French Limousins lower than those in the North American studies, indicating that the differences between the North American and French estimates may be explained by management regimes or genetic backgrounds of the animals rather than differences between the tests. An analysis of the heritability estimates for *B. taurus* breeds shown in Table 11.1 revealed that

those based on dam–offspring regression were significantly greater than those estimated using other methods ($P < 0.01$), which is likely to be due to confounding of maternal genetic and maternal environmental effects (Falconer and Mackay, 1996). There were no significant differences between *B. taurus* heritability estimates from restrained and unrestrained tests.

Breeding for docility in Limousins

One of the most important developments over recent years in genetics of cattle behaviour has been the application of research on temperament to breeding programmes in several countries, particularly for the Limousin breed. Among *B. taurus* beef cattle breeds, Limousins have the reputation of being particularly volatile and there is some evidence to support this from studies of Limousin and Limousin-cross animals (Burrow and Corbet, 2000; Hoppe *et al.*, 2010). In addition to overall differences between Limousin and other breeds, a number of studies (Table 11.1) have demonstrated moderate heritabilities for temperament-related traits in Limousins, and breed societies report substantial genetic improvement in 'docility' in Australian/New Zealand and North American Limousins since they began publishing estimated breeding values (EBVs) (or estimated progeny differences, EPD) in the 1990s (Beitia and Epperly, 2011; www.limousin.com.au/genetictrends%20for%20docility.pdf, retrieved 14 March 2013). Estimated breeding values for docility have also been published for Irish and British Limousins (www.irishlimousin.com/html/docility_improvement.html, retrieved 14 March 2013; www.signetfbc.co.uk/news/index.aspx?section=67anditem=201, retrieved 14 March 2013). American Angus breeders have also recently begun to publish EPDs for this trait (www.angus.org/Nce/Documents/ByThenumbersDocility.pdf, retrieved 14 March 2013).

Dairy cattle

Studies by Murphey *et al.* (1980, 1981) indicated that dairy cattle are generally more

Table 11.1. Estimates of heritability of cattle temperament traits.

Measurement	h^2 (s.e.)	No. animals	Means (SD)	No. sires	Breed	Sex	Age	Model used[a]	Country	Reference
Non-restrained tests										
Flight speed (weaning)	0.54 (0.16)	561	1.17 s	42	*B. indicus-* derived	M&F	6 months	Paternal half-sib	Australia	Burrow et al., 1988
Flight speed (18 months)	0.26 (0.13)	558	1.10 s	38	*B. indicus -* derived	M&F	18 months	Paternal half-sib	Australia	Burrow et al., 1988
Flight distance (6 months)	0.40 (0.15)	485	3.30 m	49	Brahman cross	M	6 months	Paternal half-sib	Australia	O'Rourke, 1989
Flight distance (12 months)	0.32 (0.14)	485	2.78 m	49	Brahman cross	M	12 months	Paternal half-sib	Australia	O'Rourke, 1989
Flight distance (24 months)	0.70 (0.23)	485	2.57 m	49	Brahman cross	M	24 months	Paternal half-sib	Australia	O'Rourke, 1989
Docility score	0.22	904	13.73 units	34	Limousin	F	10–11 months	Sire model	France	Le Neindre et al., 1995
Docility criterion	0.18	904	2.13 units	34	Limousin	F	10–11 months	Mixed threshold	France	Le Neindre et al., 1995
Average flight time (2.2 m)	0.35	851	~1.88 s (0.10)	79	Various B. indicus & B. indicus– B. taurus crosses	M&F	12–30 months	Animal	Australia	Burrow and Corbet, 2000
Flight time (1.7 m)	0.40	≤1,871	1.04 s (0.29)	139	B. taurus composite & B. indicus– B. taurus composite	M&F	Various	Animal repeatability	Australia	Burrow, 2001
Temperament score (1–5) during handling	0.61 (0.17)	259	2.05 units (0.97)	n.a.	German Angus	M&F	8 months	Animal	Germany	Gauly et al., 2001
Temperament score (1–5) during handling	0.55 (0.15)	209	2.19 units (1.00)	102	Simmental	M&F	8 months	Animal	Germany	Gauly et al., 2001
Flight time (1.7 m)	0.31 (0.05)	5,204	~1.33 s (~0.56)	n.a.	Various B. indicus & B. indicus- derived	M&F	8–18 months	Animal	Australia	Kadel et al., 2006

Table 11.1. Continued.

Measurement	h^2 (s.e.)	No. animals	Means (SD)	No. sires	Breed	Sex	Age	Model used[a]	Country	Reference
Docility score (6.5–17)	0.18	2,781	13.0 units (2.1)	n.a.	Limousin	F	Weaned	Sire	France	Phocas et al., 2006
Flight speed (2.44 m)	0.49 (0.18)	302	2.52 m/s (0.73)	n.a.	Various B. taurus & B. taurus hybrid	M	8 months	Animal	Canada	Nkrumah et al., 2007
Flight speed (exit velocity)	0.39 (0.08)	~917	~1.70 m/s (~0.77)	n.a.	Simmental	n.a.	n.a.	Animal	USA	Weaber et al., 2010
Flight score (1–4)	0.20 (0.08)	706	~1.49 units (0.06)	40	German Angus	M&F	6–9 months	Animal	Germany	Hoppe et al., 2010
Flight score (1–4)	0.25 (0.10)	556	~1.73 units (0.06)	32	Charolais	M&F	6–9 months	Animal	Germany	Hoppe et al., 2010
Flight score (1–4)	0.36 (0.06)	697	~1.46 units (0.06)	40	Hereford	M&F	6–9 months	Animal	Germany	Hoppe et al., 2010
Flight score (1–4)	0.11 (0.07)	424	~1.76 units (0.07)	56	Limousin	M&F	6–9 months	Animal	Germany	Hoppe et al., 2010
Flight score (1–4)	0.28 (0.07)	667	~1.81 units (0.07)	45	Simmental	M&F	6–9 months	Animal	Germany	Hoppe et al., 2010
Flight speed	0.26 (0.05)	7,402	2.26 m/s (1.00)	706	Nelore	M&F	16 months	Animal	Brazil	Sant'Anna et al., 2012
Restrained tests										
Behaviour score	0.40 (0.30)	144	2.52 units	n.a.	Angus	M&F	12 months	Paternal half-sib	USA	Shrode and Hammack, 1971
Temperament score	0.45	191	n.a.	5	Japanese Black	M&F	Various	Paternal half-sib	Japan	Sato, 1981
Temperament score	0.67	191	n.a.	5	Japanese Black	M&F	Various	Dam–offspring reg	Japan	Sato, 1981
Bail test (movement score)	0.67 (0.26)	957	3.03 units	n.a.	Various B. indicus & B. taurus crosses	M&F	10/22 months	Paternal half-sib	Australia	Fordyce et al., 1982

Trait	h^2 (SE)	n	Phenotypic value	No. sires	Breed	Sex	Age	Method	Country	Reference
Race test (movement score)	0.17 (0.21)	957	2.03 units	n.a.	Various B. indicus & B. taurus crosses	M&F	10/22 months	Paternal half-sib	Australia	Fordyce et al., 1982
Race test (audible respirat.)	0.57 (0.22)	957	1.43 units	n.a	Various B. indicus & B. taurus crosses	M&F	10/22 months	Paternal half-sib	Australia	Fordyce et al., 1982
Crush test (movement score)	0.25 (0.20)	957	1.82 units	n.a.	Various B. indicus & B. taurus crosses	M&F	10/22 months	Paternal half-sib	Australia	Fordyce et al., 1982
Crush test (audible respirat.)	0.20 (0.16)	957	1.44 units	n.a	Various B. indicus & B. taurus crosses	M&F	10/22 months	Paternal half-sib	Australia	Fordyce et al., 1982
Crush test (movement score)	0	1,852	2.08 units	63	Droughtmaster	F	Mature cows	Paternal half-sib	Australia	Fordyce and Goddard, 1984
Crush test (audible respirat.)	0	1,852	0.31 units	63	Droughtmaster	F	Mature cows	Paternal half-sib	Australia	Fordyce and Goddard, 1984
Crush test (movement score)	0.09	1,852	2.08 units	63	Droughtmaster	F	Mature cows	Dam–daughter reg	Australia	Fordyce and Goddard, 1984
Crush test (audible respirat.)	0.05	1,852	0.31 units	63	Droughtmaster	F	Mature cows	Dam–daughter reg	Australia	Fordyce and Goddard, 1984
Temperament score	0	1,852	2.45 units	63	Droughtmaster	F	Mature cows	Paternal half-sib	Australia	Fordyce and Goddard, 1984
Temperament score	0.09	1,852	2.45 units	63	Droughtmaster	F	Mature cows	Dam–daughter reg	Australia	Fordyce and Goddard, 1984
Temperament score	0.03 (0.28)	209	1.05 units	~60	Various B. taurus	M&F	6–9 months	Paternal half-sib	New Zealand	Hearnshaw and Morris, 1984

Continued

Table 11.1. Continued.

Measurement	h² (s.e.)	No. animals	Means (SD)	No. sires	Breed	Sex	Age	Model used[a]	Country	Reference
Temperament score	0.46 (0.37)	358	1.96 units	~90	Various B. indicus–B. taurus crosses	M&F	6–9 months	Paternal half-sib	New Zealand	Hearnshaw and Morris, 1984
Crush test (movement score)	0.10 (0.11)	485	2.6 units	49	Brahman cross	M	6 months	Paternal half-sib	Australia	O'Rourke, 1989
Crush test (movement score)	0.23 (0.13)	435	2.1 units	49	Brahman cross	M	12 months	Paternal half-sib	Australia	O'Rourke, 1989
Crush test (movement score)	0.11 (0.11)	485	2.2 units	49	Brahman cross	M	24 months	Paternal half-sib	Australia	O'Rourke, 1989
Temperament score	0.14 (0.11)	485	3.11 units	49	Brahman cross	M	6 months	Paternal half-sib	Australia	O'Rourke, 1989
Temperament score	0.12 (0.11)	485	2.42 units	49	Brahman cross	M	12 months	Paternal half-sib	Australia	O'Rourke, 1989
Temperament score	0.08 (0.10)	485	2.65 units	49	Brahman cross	M	24 months	Paternal half-sib	Australia	O'Rourke, 1989
Docility score (1–3)	0.40 (0.034)	24,960	n.a.	n.a.	Limousin	M&F	Weaning	Animal	N. America	Kuehn et al., 1998
Crush score (1–15)	0.15 (0.05)	3,416	~6.2 units (2.1)	1,878	Various B. indicus & B. indicus-derived	M&F	8–18 months	Animal	Australia	Kadel et al., 2006
Chute test (1–6)	0.34 (0.01)	21,932	1.93 (0.85)	~73	Limousin	M&F	Weaning	Animal	N. America	Beckman et al., 2007
No. movements during weighing	0.31 (0.10)	1,439	28.2 (18.8)	~73	Limousin	M&F	7 months	Animal	France	Benhajali et al., 2010
No. rush movements during human exposure (1–6)	0.16 (0.07)	1,440	2.0 (1.3)	40	Limousin	M&F	7 months	Animal	France	Benhajali et al., 2010

Trait	h^2 (SE)	n	Mean (units)	n	Breed	Sex	Age	Method	Country	Reference
Chute score (1–5)	0.15 (0.06)	706	~2.52 units (0.07)	32	German Angus	M&F	6–9 months	Animal	Germany	Hoppe et al., 2010
Chute score (1–5)	0.17 (0.07)	556	~2.78 units (0.07)	40	Charolais	M&F	6–9 months	Animal	Germany	Hoppe et al., 2010
Chute score (1–5)	0.33 (0.10)	697	~2.08 units (0.08)	56	Hereford	M&F	6–9 months	Animal	Germany	Hoppe et al., 2010
Chute score (1–5)	0.11 (0.08)	424	~2.95 units (0.08)	45	Limousin	M&F	6–9 months	Animal	Germany	Hoppe et al., 2010
Chute score (1–5)	0.18 (0.07)	667	~2.33 units (0.08)	n.a.	Simmental	M&F	6–9 months	Animal	Germany	Hoppe et al., 2010
Dairy temperament score										
Temperament score	0.40 (0.09)	n.a.	n.a.	n.a.	Holstein	F	Mature cows	Dam–daughter reg	USA	O'Bleness et al., 1960
Temperament score	0.16	1,400	n.a.	133	Holstein	F	<35 months	Sire	USA	Van Vleck, 1964
Temperament score	0	4,080	n.a.	209	Holstein	F	>35 months	Sire	USA	Van Vleck, 1964
Temperament score	0.53	1,017	1.9 units (0.8)	31	Holstein	F	Mature cows	Paternal half-sib	USA	Dickson et al., 1970
Disposition	<0.15 (<0.09)	11,106	2.8 units (0.7)	n.a.	Holstein	F	Mature cows	Paternal half-sib	USA	Aitchison et al., 1972
Temperament score	0.11	4,891	n.a.	157	Friesian	F	Mature cows	Paternal half-sib	New Zealand	Wickham, 1979
Temperament score	0.11	4,171	n.a.	135	Jersey	F	Mature cows	Paternal half-sib	New Zealand	Wickham, 1979
Temperament score	0.19 (0.19)	319	1.71 units (0.11)	n.a.	B. indicus-cross	F	Mature cows	Paternal half-sib	India	Sharma and Khanna, 1980
Disposition	0.07 (0.02)	8,977	2.00 units (0.52)	125	Holstein	F	Mature cows	Sire	USA	Thompson et al., 1981
Disposition	<0.08	5,601	n.a.	187	Holstein	F	Mature cows	Sire	USA	Agyemang et al., 1982
Temperament score	0.12 (0.02)	9,646	28.8 units (7.6)	208	Holstein	F	Mature cows	Sire	USA	Lawstuen et al., 1988

Continued

Table 11.1. Continued.

Measurement	h^2 (s.e.)	No. animals	Means (SD)	No. sires	Breed	Sex	Age	Model used[a]	Country	Reference
Disposition (1–50)	0.08 (0.02)	43,428	28.9 (9.07)	228	Holstein	F	Mature cows	Paternal half-sib	USA	Foster et al., 1988
Disposition (1–3)	0.10 (~0.08)	880	1.30 (0.51)	121	Holstein	F	Mature cows	Sire	USA	Erf et al., 1992
Temperament (1–5)	0.22 (0.03)	14,596	2.6 (0.9)	334	Holstein	F	Mature cows	Sire	Australia	Visscher and Goddard, 1995
Temperament (1–5)	0.25 (0.06)	4,695	2.5 (0.8)	125	Jersey	F	Mature cows	Sire	Australia	Visscher and Goddard, 1995
Shed temperament	0.137 (0.015)	59,623	5.696 (1.485)	1,116	Holstein	F	Mature cows (first lactation)	Sire	New Zealand	Cue et al., 1996
Shed temperament	0.172 (0.015)	45,396	6.024 (1.428)	773	Jersey	F	Mature cows (first lactation)	Sire	New Zealand	Cue et al., 1996
Shed temperament	0.333 (0.06)	6,599	5.995 (1.530)	210	Ayrshire	F	Mature cows (first lactation)	Sire	New Zealand	Cue et al., 1996
Temperament score (1–5)	0.128 (0.014)	1,940,092	~3.31 units	28,837	Holstein	F	Mature cows	Animal	Canada	Sewalem et al., 2011

[a]Table is an updated version of that published by Burrow 1997.
reg, regression.

approachable than beef cattle in comparisons of adult females from both *B. taurus* and *B. indicus* breeds. Murphey *et al.* (1980) also demonstrated that raising dairy breeds for meat and vice versa had little effect on their approachability, suggesting that this is a fixed characteristic of these breeds. These findings suggest that selection for milking ability may have involved greater coincident selection for docility compared to selection for meat production. However, somewhat surprisingly, there have not been follow-up studies that apply the same tests to a range of dairy and beef breeds, and thus the ability to make general conclusions is limited.

The speed and ease with which a dairy cow is milked is potentially an important production trait. Burnside *et al.* (1971) compared various breeds and found differences in percentage of animals culled for bad temperament. Among the studied *B. taurus* breeds, the greatest percentage of animals culled for bad temperament was in Ayrshires, followed by Jerseys, Guernseys and Holsteins. In traditional milking systems, estimates of heritabilities for milking temperament generally range between 0.07 and 0.35 (Table 11.1), although some more extreme values have also been reported (e.g. 0.53 by Dickson *et al.*, 1970). The largest published study (Sewalem *et al.*, 2011) reported a heritability of 0.13 (standard error 0.014). Published estimates of heritability for temperament in dairy breeds are generally lower than those seen in beef breeds (Table 11.1, Fig. 11.1). However, these values may not be directly comparable as the tests are usually different, with beef breeds often measured for docility in a chute/crush or in flight from the chute, as described above, whereas dairy cows are generally assessed during milking. Furthermore, estimates in dairy cattle are measured on females, while estimates in beef cattle are often on males or both sexes. If the lower heritability of temperament in dairy cattle is real, this may again reflect greater selection for docility.

Physiological correlates

A large number of studies have examined the effects of various stressors on cattle physiological

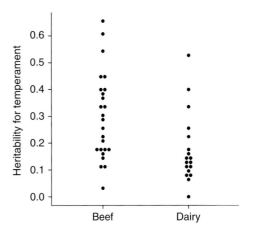

Fig. 11.1. Comparison of heritability estimates for temperament traits in *Bos taurus* beef and dairy cattle.

traits, including hormone levels. Hypothalamic–pituitary–adrenal (HPA) function is the standard approach to assessing stress and welfare in cattle and other livestock (Mormède *et al.*, 2007). While interpretation of HPA levels is complicated by the normal variation of these hormones (due to periodic secretion, diurnal and seasonal rhythms, animal age and numerous environmental factors), it is currently the best available method for evaluating animal stress levels, at least for acute forms of stress (Mormède *et al.*, 2007). In cattle, the main active hormone of the HPA axis is cortisol, a cholesterol-derived steroid synthesized in the adrenal cortex under the control of the pituitary-derived adrenocorticotropic hormone (ACTH), the release of which is triggered by two neuropeptides synthesized in the hypothalamus, corticotropin-releasing hormone (CRH) and vasopressin (AVP) (Mormède *et al.*, 2007). Blood levels of cortisol have been shown to increase in animals subjected to painful procedures like castration, but also when isolated from peers, restrained in a crush, transported and in various other situations; increases in ACTH levels tend to show a more graded response relative to stimulus intensity (Mormède *et al.*, 2007). Chronic stress has been shown in many studies to not affect basal levels of cortisol or ACTH, however, in some cases, the activity of the HPA system appears to have changed such that responses to stimulation are altered.

In order to test these effects, 'stimulation tests' are performed in which CRH, ACTH or other factors are injected into the animal and the effect on cortisol or ACTH levels is subsequently monitored (Mormède et al., 2007).

With regard to the underlying basis of behavioural characteristics, the question is whether animals that differ in temperament also differ in terms of HPA function. Studies of both B. indicus and B. taurus cattle have demonstrated that baseline cortisol and/or ACTH levels are higher in temperamental animals relative to those in calmer ones (Curley et al., 2006, 2008, 2010; King et al., 2006; Cafe et al., 2011a; Fazio et al., 2012; Sánchez-Rodríguez et al., 2013) (Fig. 11.2). However, mixed results have been seen regarding cortisol responses to stimulation tests: Curley et al. (2008) saw a greater cortisol response following ACTH or CRH challenge in calm animals compared to more temperamental ones, while Curley et al. (2010) saw the opposite effect following challenge with vasopressin, a stimulant of ACTH secretion, and Cafe et al. (2011a) did not observe an association between temperament and cortisol response to ACTH injection.

In addition to differences in the HPA axis, more temperamental cattle have been shown to differ from calm animals in other physiological traits: higher baseline rectal temperatures in temperamental animals were observed in beef breeds (Fazio et al., 2012; Sánchez-Rodríguez et al., 2013), and higher heart rates before and during milking were observed in more temperamental dairy cattle (Sutherland et al., 2012).

Associations with production traits

A number of studies have examined associations between temperament and cattle production traits in both dairy and beef breeds, with fairly consistent results. The general finding is that selecting for improved temperament is not expected to have undesirable correlated effects on production traits because either there is little or no association with these traits, or because there is a favourable correlation. The latter has been documented in several studies: Burrow and Dillon (1997) found a significant positive relationship between growth rate and

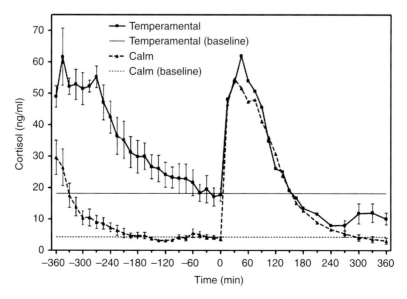

Fig. 11.2. Mean serum cortisol concentrations over a 12-h sampling period before and after ACTH challenge for both calm (triangles) and temperamental (squares) Brahman heifers. Mean baseline cortisol concentrations represented by the horizontal lines. Standard error bars partially omitted to enhance clarity. ACTH administered at time 0. Previously published by Curley *et al.* (2008) and reprinted with permission from Elsevier Limited.

temperament (measured as 'flight speed score', inversely related to flight speed) for *B. indicus* crosses raised on northern Australian feedlots and Sant'Anna *et al.* (2012) also found low but significant negative genetic correlation between growth traits and flight speed in Nelore cattle raised on Brazilian farms. Further support for such favourable associations comes from other studies of *B. indicus*-derived cattle: Voisinet *et al.* (1997a) showed a lower proportion of dark-cutting carcasses (where beef does not turn its normal bright red colour when exposed to air) in animals with desirable temperament and Kadel *et al.* (2006) found a positive genetic correlation between flight time and meat tenderness in several *B. indicus* breeds. Cafe *et al.* (2011b) found a strong favourable association between temperament and several production traits (e.g. growth rate, carcass weight, meat quality) in Brahman cattle. Positive or neutral associations between temperament and growth rate have also been seen in *B. taurus* breeds and *B. indicus*–*B. taurus* cross cattle. Favourable genetic correlations between temperament and production traits were described by Phocas *et al.* (2006) for heifer breeding traits in Limousins, by Hoppe *et al.* (2010) for growth rate in several European breeds, and either favourable or neutral correlations were seen by Turner *et al.* (2011) for growth rate in Limousin and Aberdeen Angus cattle. Müller and von Keyserlingk (2006) saw a less clear relationship between flight speed and average daily gain in crossbred Aberdeen Angus, but the animals with highest flight score did show low growth. Voisinet *et al.* (1997b) found a favourable correlation between temperament score and growth rate in both *B. indicus*–*B. taurus* crosses and crossbred *B. taurus*. Cafe *et al.* (2011b) found favourable associations between temperament and production traits in Angus cattle, although these were much weaker than those observed in Brahman. However, not all studies on beef cattle have found favourable associations. Burrow and Prayaga (2004) found no correlated response for flight time in *B. taurus*–*B. indicus* cross cattle selected for either growth rate or heat resistance and Burrow (2001) found very low phenotypic and genetic correlations between flight score and growth rate in composite *B. taurus* and *B. taurus*–*B. indicus* breeds.

King *et al.* (2006) did not find consistent associations between temperament and beef tenderness in a study of Angus and a cross of mixed *B. taurus*–*B. indicus* ancestry. With respect to dairy cattle, Lawstuen *et al.* (1988) and Sewalem *et al.* (2011) found moderate favourable genetic correlations between milking speed and temperament (0.36 and 0.24, respectively).

The association between temperament and meat quality is thought to be a function of preslaughter stress, which leads to depletion of muscle glycogen and higher ultimate pH in meat. Thus various aspects of meat quality may be compromised from animals that experience such stress (Turner *et al.*, 2011). While further investigation is required, it has been suggested that the association between temperament and milk production may be due to a greater inhibition of milk letdown (measured as residual milk volume following an oxytocin challenge) in cows with worse temperaments (Sutherland *et al.*, 2012). Inhibition of milk letdown has also been associated with reduced adrenal responsiveness to ACTH challenge (Macuhova *et al.*, 2002; Weiss *et al.*, 2004).

QTL studies

Compared to other traits in cattle, there has been much less progress in dissecting the genetic basis of behaviour by applying genomic technologies. However, the revolution in genomic technology has not completely bypassed the study of cattle behaviour, and several genome-wide quantitative trait locus (QTL) studies incorporating temperament-related traits were published between 1999 and 2012 (Spelman *et al.*, 1999; Schrooten *et al.*, 2000; Schmutz *et al.*, 2001; Hiendleder *et al.*, 2003; Gutiérrez-Gil *et al.*, 2008), the latter three of which reported QTLs for these traits. These studies used a similar approach to identifying genetic regions associated with temperament: the study populations were genotyped for 162–264 microsatellite markers across the genome and associations between genotype and temperament-related phenotypes were then assessed using regression- or maximum-likelihood-based analyses. However, the studies differed in a number of important ways. In terms of design, Schmutz et al. (2001) studied full-sib families of male and female calves produced by embryo transfer from the Canadian Reference Herd (developed from five beef breeds). Hiendleder et al. (2003) analysed paternal half-sib families of German Holstein cows in a granddaughter design. Gutiérrez-Gil et al. (2008) analysed a mixture of F2 and backcross heifers generated from a Holstein and Charolais cross in an experiment designed mainly to detect loci that were fixed for different alleles in the two breeds.

The phenotypes that were measured in the studies were also different. Schmutz et al. (2001) electronically measured the amount of movement made by each calf when kept for 1 min on a platform scale without the ability to see other calves. They also analysed the difference in levels of movement from the initial test and one carried out ~2–6 months later ('habituation'). Hiendleder et al. (2003) analysed deregressed EBVs of temperament during milking (graded 1–9). Gutiérrez-Gil et al. (2008) analysed data from two tests: flight from feeder (FF) and social separation (SS). In the FF test, a human approached the calf while at an automatic feeder. The score was determined by the distance between the observer and the calf when the animal moved away. In the SS test, the animal moved from a shared pen to its home pen on its own. The amount of time the animal spent engaged in various activities was recorded over a 5-min period. The change in these activities (habituation) between the original test and a later test (carried out within a few days of the first) was also analysed separately.

There were some consistent results across these studies as shown in Table 11.2. A QTL for temperament and habituation was found at the proximal end of BTA1 in the Schmutz et al. (2001) and Gutiérrez-Gil et al. (2008) studies, a QTL for habituation of two social separation activities (standing alert, vocalization) was also seen in that region. A later study on German Simmental and Angus breeds (Glenske et al. 2010b) tested associations between temperament and several microsatellites in this region and found an association between one marker and nervousness during weighing in Simmental cattle.

A QTL for temperament during milking (Hiendleder et al. 2003) was found close to one reported for habituation of a social separation trait (Gutiérrez-Gil, 2008) on BTA29. In a follow-up study of that chromosome, Glenske et al. (2011) also found a nearly significant association for nervousness during weighing very close to the Hiendleder et al. (2003) QTL. Another region at the distal end of BTA29 was shown by Gutiérrez-Gil et al. (2008) to be associated with flight response. Glenske et al. (2011) found a significant association between DRD4, a candidate gene for curiosity behaviour and novelty-seeking located at the distal end of BTA29, and nervousness during the separation and restraint test. However, they did not find any QTLs in this region in a chromosome-wide analysis.

A QTL for a social separation activity (vocalization) found by Gutiérrez-Gil et al. (2008) was in the same region of BTA9 as a QTL for temperament and habituation found by Schmutz et al. (2001) and a QTL for habituation of another social separation activity (standing alert) was found in the same region of BTA11 as a QTL for temperament and habituation found by Schmutz et al. (2001).

Although there are some consistencies between these few studies for temperament QTLs, the identified regions have large confidence intervals and the overlap in position is not entirely convincing. It is clear that much larger study populations, genotyped for high-density markers, are needed to further research into the genetics of temperament.

Candidate gene analyses

Despite the evidence for physiological differences between calm and temperamental animals described above, there have been relatively few studies that examine the association between cattle temperament and genes related to hormone production. Pugh et al. (2011) found a significant effect of the combined genotypes of the CRH and leptin genes, which are involved in mediating the HPA axis, on response to handling in beef steers. Glenske et al. (2010a) studied the cholecystokinin B receptor gene and Lüehken et al. (2010) studied the monoamine

Table 11.2. Comparison of QTL positions reported for cattle temperament.

Chromosome	Study 1				Study 2			
	Trait	Linked marker(s)	Position (cM)	Reference	Trait	Linked marker(s)	Position (cM)	Reference
1	Temperament + habituation	BMS574	15	Schmutz et al., 2001	SS test: time spent standing alert (habituation)	BM6438	2	Gutierrez-Gil et al., 2008
9	Temperament + habituation	ILSTS013	48	Schmutz et al., 2001	SS test: time spent vocalizing (test 2)	BM2504-UWCA9	31–50	Gutierrez-Gil et al., 2008
11	Temperament + habituation	ILSTS036	62	Schmutz et al., 2001	SS test: time spent standing alert (habituation)	ILSTS100-IDVGA-3	59–82	Gutierrez-Gil et al., 2008
29	Temperament score	BMS764-MBC8012	11–21	Hiendleder et al., 2003	SS test: time spent vocalizing (habituation)	RM044-MNB166	24–33	Gutierrez-Gil et al., 2008

SS, social separation.

oxidase A gene, but neither found significant associations with cattle temperament despite these genes having been associated with various behavioural traits in humans and other mammals. While heifer reactions to a surprise test (blast of air) were significantly associated with cortisol and oxytocin reactivity in blood samples (Yayou et al., 2010), whether this can be explained by variation in related genes is not yet known.

Social Interactions

The ancestors of B. taurus cattle lived in social units whose composition changed with the seasons (Buchenauer, 1999). Maternal-based groups of 20–30 animals lived separately during spring and summer while in autumn mature males joined the group for mating and remained through winter. Similar social structures have been observed in modern wild and semi-wild cattle, including Camargue, Highland and Donana cattle (Buchenauer, 1999).

Group structure and dominance

Cattle living in groups establish social hierarchies, which reduces the level of aggression in the herd (Buchenauer, 1999). Dominance has been shown to depend on a number of factors, including size, presence of horns, age, sex and breed (Buchenauer, 1999). In two studies of mixed-breed herds (Wagnon et al., 1966; Stricklin et al., 1983), Angus individuals were found to be dominant to Hereford cows, despite their smaller body size. Stricklin et al. (1983) also found that Angus showed greater group cohesiveness in that physical distances between individuals were lower; Angus individuals also tended to be located closer to the centre of the group while Herefords were located on the periphery. Additional studies have also demonstrated breed differences in dominance, some of which did correspond to body size differences (see Buchenauer, 1999, for references).

Dominance within breeds has also been suggested to have a genetic basis, although most studies have used small sample sizes, so results may not be statistically conclusive. Purcell and Arave (1991) found a very high intraclass correlation (0.93) for dominance rank between Holstein monozygotic twin pairs produced by transfer of split embryos to separate recipient dams. Heritability of dominance was also estimated as ~0.4 in Holsteins, based on pairs of monozygotic and dizygotic twins (Beilharz et al., 1966), but a very high standard error (>1.0) indicated that the estimate had low precision due to small sample size.

There also appear to be genetic effects on response to isolation although few studies have been carried out on this characteristic, and again sample sizes tend to be small. Watts et al. (2001) found either breed or family effects (these could not be distinguished in this study) for the number and quality of vocalizations when calves were isolated from peers for short periods. In a similar study of short-term isolation, Aubrac heifers were found to show greater levels of distress than Friesians, including more struggling and greater heart rates and cortisol levels (although not a greater number of vocalizations) (Boissy and Le Neindre, 1997).

Aggression

Maternal aggression is probably the most important cattle behavioural trait from the point of view of cattle handlers because of the risk of injury. Buddenberg et al. (1986) compared the behaviour of dams during calf ear-tagging from several beef breeds and found that Angus cows were more aggressive to farm staff than Charolais, Hereford or Red Poll. Morris et al. (1994) also found breed differences in maternal response score, with Angus cows again showing greater response than Herefords. Limousin cows were similarly noted to show greater aggressiveness compared to Charolais, Simmental and Gelbvieh (Buchenauer, 1999). German Angus showed greater maternal protective behaviour during calf ear-tagging than Simmental in a study by Hoppe et al. (2008). Despite these breed differences, heritability estimates for maternal behaviour traits are generally low: Buddenberg et al. (1986) estimated 0.06 (standard error 0.01) for aggressiveness in a mixed group of beef cattle, Morris et al. (1994) estimated 0.09 (standard error 0.03) for maternal responsiveness score in a mixed group of purebred and crossbred beef cattle and Hoppe et al. (2008) estimated 0.14 (standard error 0.08) for maternal responsiveness in German Angus (although they also estimated high maternal response heritability (0.42, standard error 0.05) in Simmental). The heritability of maternal response traits appears to be lower than that seen for temperament traits (Turner and Lawrence, 2007), which may reflect strong natural as well as artificial selection for maternal responsiveness.

While most farmers would prefer to work with non-aggressive animals for their own safety, some cattle breeds have been selected for aggressiveness in competition situations, either against humans or other animals. While this topic has not received a great deal of interest, the genetics of fighting behaviour has been studied in a few cattle breeds and for two different fighting situations. Bullfighting has a long history in Spain and its former colonies. Significant heritabilities of bullfighting ability, where the bull attempts to attack someone who provokes them with a lure, have been estimated for Spanish and Colombian bullfighting breeds. Silva et al. (2006) estimated heritabilities greater than 0.29 (standard error 0.02–0.09) for subjective characteristics (aggressiveness, ferocity and mobility) measured during fights involving Spanish bullfighting cattle. González Caicedo et al. (1994) estimated a somewhat lower heritability (0.19, standard error 0.03) for fighting ability, again scored subjectively, in a Colombian bullfighting breed. These heritabilities suggest that fighting ability should respond to selection in these breeds.

The lesser-known event of cow fighting is a traditional Swiss event that also takes place in Italy, France and Turkey. Two cows push each other until one backs away and the other is declared the winner. Plusquellec and Bouissou (2001) compared behavioural characteristics of a fighting breed (Herens, H) and a related non-fighting breed (Brune des Alpes, BA) and found that H cows showed dominance to BA cows, H cows were less reactive in situations devised to engender fear, and H cows maintained greater social distances than BA when on pasture. However, H individuals were not more aggressive than BA towards an unfamiliar cow introduced to their group and were less aggressive in undisturbed groups, suggesting that a very specific form of aggressiveness has been selected for in this breed. Sartori and Mantovani (2010) studied heritability of success in fighting tournaments for Italian Valdostana cattle. Heritability estimates ranged from 0.068 to 0.148 (standard error 0.04), depending on the model used and whether they analysed results from all competitions or only the best yearly performances of each cow. When Sartori and Mantovani (2013) included indirect genetic effects of social partners in their models of fight outcome, they found substantially higher heritabilities (estimates 0.27/0.38) than in models without indirect effects. It is not clear whether the higher heritability estimates for the bullfighting traits reflects a true difference in the genetic architecture of these traits. Sartori and Mantovani (2012) also found a negative relationship between the EBVs of placement scores and the level of inbreeding within lineages over a 19-year period, suggesting that inbreeding depression affects fighting ability in these cattle. In addition to cattle selected for fighting ability, there is evidence for genetic control of aggression under normal farm conditions; Buchenauer (1999) reviewed studies showing breed differences in aggression and moderate heritability in Holstein Friesians for aggressive activities at automatic feeders.

Feeding Behaviour

Cattle graze in a wide range of different environments and feed on a range of plant species (Buchenauer, 1999). They tend to graze between 4 and 14 h per day, preferably during daylight, with distinct periods of intense grazing (Buchenauer, 1999).

Evidence of genetic effects on feeding behaviour derives from breed comparisons and estimates of heritability. Differences in grazing selectivity between Hereford, Angus and Brangus cattle were reported in the Chihuahuan desert range (Winder et al., 1996) and between native breeds in southern Italy (Braghieri et al., 2011). Breed differences have also been noted for the number of feeding sessions and time spent eating/grazing in a large number of studies comparing different breeds (Alencar et al., 1995, 2009; Senn et al., 1995; Schutt et al., 2009; also see earlier references in Buchenauer, 1999). O'Driscoll et al. (2009) found that Holstein Friesian cows had a higher bite rate and fewer mastications during housed feeding than Norwegian Red cows, which may reflect stronger selection for feed intake in Holsteins (Houpt, 2011). Aharoni et al. (2009) used a global positioning system (GPS) and activity meters to compare the grazing behaviour of Baladi cattle, a small-sized breed native to the Middle East, and larger crossbred cattle with primarily European ancestry. The Baladi walked longer distances and appeared to graze more hours per day than the larger cattle in an experiment carried out on Israeli pasture, which may be related to the Baladi's adaptation to the Mediterranean climate. Senepol cows (derived from N'Dama and Red Poll breeds) were found to graze longer than Hereford cows in a tropical environment, again possibly due to climatic adaptation (Hammond and Olson, 1994). Comparisons have also been made within the Holstein breed where North American (NA) cattle were found to spend less time grazing than New Zealand (NZ) cattle, although NA also had greater dry matter intake (and produced more milk) (Sheahan et al., 2011). Heritabilities have been reported for various grazing traits, including the proportion of certain plants ingested during grazing and time spent grazing (Buchenauer, 1999), however, sample sizes were too small to be reliable. Buchenauer (1999) also reported a high heritability for the number of visits to automatic feeders by housed dairy cattle, although these may be over-estimated due to the relatively small sample sizes. A larger study also estimated high heritability (0.56–0.59, standard error ~0.18) for feeding frequency of crossbred feedlot cattle (Durunna et al., 2011).

Compared to other behavioural traits, there is substantial information regarding genetic control of feed intake in housed cattle, due to its importance as a production trait and the ease of measurement. Moderate heritability has been estimated for feed intake in many studies (e.g. MacNeil et al., 1991; Rolfe et al., 2011;

Liinamo et al., 2012). A number of QTL and association studies have been carried out on this trait and the roles of several candidate genes have been investigated. Snelling et al. (2011) identified a large number of SNPs associated with dry matter intake, and follow-up studies suggested several candidate genes in genomic regions on BTA6 and BTA14 (Lindholm-Perry et al., 2011, 2012), including lysophospholipase 1 (LYPLA1), transmembrane protein 68 (TMEM68) and condensin complex subunit 3 (NCAPG). Other genes have been investigated based on known functional roles. The leptin gene was discovered by cloning the obese gene from a mutant mouse that overeats to the point of hyperobesity (Zhang et al., 1994). Connections between sequence variation in leptin and feeding characteristics have been reported in a number of livestock species (Wylie, 2011), including cattle where several studies have identified SNPs associated with feed intake and also with carcass and milk production traits (Lagonigro et al., 2003; Nkrumah et al., 2005; Banos et al., 2008). The ghrelin precursor (GHRL) gene, another functional candidate for traits related to appetite and feeding, has also been investigated for its role in feed intake in cattle but results have not been consistent: Sherman et al. (2008) reported significant statistical associations while Sun et al. (2011) did not. The physiological effects of other hormones on feed intake have been demonstrated, for example, intracerebroventricular administration of CRH inhibited feed intake in steers (Yayou et al., 2011). However, an association between SNPs in the CRH gene and feed-related traits was not found (Sherman et al., 2008).

Very little is known about the genetics of suckling behaviour but there is a suggestion that suckling time of calves increases with proportion of B. indicus ancestry (Alencar et al., 1995; Das et al., 2000; de Vargas Junior, 2010).

Mating Behaviour

Traits related to mating behaviour are likely to have been under strong natural selection in wild species, however, following domestication of cattle and more recently the industrialization of livestock production, the relationship between such traits and reproduction has changed dramatically. Dairy cattle are primarily bred by artificial insemination (AI), where a major goal has been to sire as many daughters as possible from elite bulls (Buchenauer, 1999). Use of AI is also increasing in major beef breeds.

As previously reviewed (Buchenauer, 1999), bulls differ substantially in characteristics of their sexual behaviour in the presence of a cow. These traits can be measured in various ways: quantitative measures include number of services, time to first mount or time of sexual inactivity (Chenoweth et al., 1996), while qualitative measures include 'libido score', which assesses the level of sexual activity. As with other behavioural traits, data from breed comparisons as well as within-breed analyses provide evidence for a genetic component to sexual behaviour traits. Chenoweth et al. (1996) found that B. indicus breeds showed lower sexual activity (based on libido scores and number of services) than B. taurus breeds. Houpt (1991) also reported that B. taurus bulls were more likely than B. indicus to mount an inappropriate object (a male or anoestrous female). Differences have also been seen among B. indicus-derived breeds, where Belmont Red was shown to have greater serving capacity than Santa Gertrudis, Brahman and Brahman-cross cattle (Bertram et al., 2002). A study of bulls from two B. taurus breeds, Belgian Blue and Holstein Friesian, found no differences in libido (Hoflack et al., 2006) while Jezierski et al. (1989) found that Black and White cattle performed more bull–bull mounts than Hereford cattle, which could be viewed as support for Houpt's (2011) suggestion that due to extensive use of artificial insemination, dairy bulls have been selected (among other traits) for their willingness to mount inappropriate objects.

Relatively few studies have estimated heritability of male sexual behaviour traits. Morris et al. (1992) analysed two measures of male serving capacity (numbers of serves in a standard time period and a subjective 'serving grade') in Herefords and estimated heritabilities for both as approximately 0.14 (standard error 0.07). Quirino et al. (2004) estimated heritability for

libido (scored 0–10) in Nelore cattle as 0.34 (standard error ~0.11). They also found favourable correlations between libido and other fertility traits. Studies of large, multiple-sire ranch populations reveal substantial variation in reproductive output of bulls in natural mating situations (Fordyce et al., 2002; Van Eenennaam et al., 2007). Non-genetic factors such as age had a strong influence over reproductive success (Van Eenennaam et al., 2007); however, some of the variation may also be explained by behavioural differences. In the study of Fordyce et al. (2002), high reproductive output was associated with restricted movement, grazing with females and social dominance, which may vary genetically.

Regarding female fertility, cows in oestrus are known to show specific behaviours, particularly involving mounting or being mounted by other cows (or bulls, if present). They also increase their walking or restlessness activity and the frequency of flehmen (curling of the upper lip and inhaling), sniffing, rubbing, licking and urination (Buchenauer, 1999; Kommadath et al., 2010; Houpt, 2011). Oestrus behaviour is an important production trait especially for dairy cattle, as it allows the farmer to optimally time inseminations (Roelofs et al., 2010); silent heat (absence of oestrus behaviour) is a continuing problem on dairy farms (Heringstad, 2010). While it is clear that oestrus behaviour is strongly influenced by environmental and social factors (Landaeta-Hernandez et al., 2002), there is also evidence for genetic influences. Buchenauer (1999) reported that breed differences were noted for sexual behaviour, with Jersey cows showing higher mounting frequencies than Holstein Friesian cattle. B. indicus cattle have been reported to show less overt oestrous behaviour than B. taurus breeds but this is not a universal observation (Orihuela, 2000). Rottensten and Touchberry (1957) reported heritability of oestrus behaviour (scored 1–4) as 0.21, although Roxström et al. (2001) estimated heritability for heat intensity (scored 1–3) as much lower (0.01–0.03, standard error 0.001–0.005). Oestrus behaviour has also been shown to be associated with genetic merit for fertility traits in Holsteins, providing further evidence of genetic influence on these traits and suggesting a genetic correlation between fertility and oestrus behaviour.

A greater proportion of 'high fertility' cows showed behavioural oestrus than 'low fertility' cows, although there was no difference in mounting behaviour of the two groups (Cummins et al., 2012). Hormonal control of female reproduction in cattle has been studied intensively, but associations between oestrus behaviour and specific hormones are not consistent and direct causal relationships are not clear (van Eerdenburg, 2008). A specific example is progesterone: circulating progesterone levels were greater in the cows with high fertility and greater behavioural oestrus studied by Cummins et al. (2012), and van Eerdenburg (2008) noted positive correlations between levels of progesterone and oestrus intensity (total score of oestrus behaviours divided by duration of period showing these behaviours). However, other studies have reported either no association or suggested that high levels of progesterone can inhibit oestrus behaviour (Duchens et al., 1995; also see Orihuela, 2000).

The expression levels of several genes in the brain have been associated with levels of oestrus behaviour in Holsteins, which again suggests genomic influences on these traits. Wyszynska-Koko et al. (2011) found significant differences for expression levels of genes encoding immunoglobulins and genes related to cell adhesion and cytoskeleton structure. Kommadath et al. (2011) found expression associated with oestrus behaviour for genes related to other behavioural processes, neurotransmission and signalling, and ion regulation, including the oxytocin (OXT) and vasopressin (AVP) genes, which encode the related hormones oxytocin and vasopressin, respectively. Oxytocin has established roles in birth and lactation, while both hormones have been implicated in maternal care and aggression in rats (Bosch and Neumann, 2012).

Milking Tendency

On some dairy farms, automated (robotic) milking systems have been implemented to save on labour costs. In this situation, cows decide when they will go to the milking area, so a key production variable is milking frequency. König

et al. (2006) reported heritabilities of milking frequency between 0.16 and 0.28 (standard error 0.04–0.05), depending on the statistical model and stage of lactation. Heritabilities increased over the lactation period. As milking frequency had a substantial positive genetic correlation with milk yield, selection for the former trait should not have unfavourable effects on production.

Conclusions

Although there has been considerable progress in several areas of behavioural genetics of cattle, the genetic dissection of behavioural traits has not progressed rapidly compared to that for production phenotypes. There are very few studies carried out on a large scale and with both high-density genotype data and reliable behavioural measurements that would allow genetic inferences to be drawn. Genotyping costs are coming down (although are still substantial multiplied over large numbers of samples), but the primary costs and effort involve development of appropriate phenotypic measurements and implementation on large numbers of animals. Greater collaboration between behavioural scientists and those groups of scientists conducting large-scale genetic studies of cattle production traits would greatly facilitate progress in this area. An obvious first step would be a genome-wide association study on temperament in either beef or dairy cattle that have been genotyped using high-density SNPs and characterized for both temperament and production traits. Greater attention also needs to be focused on characterizing how genetic variation influences the physiological traits associated with behavioural characteristics, as has been initiated for temperament.

Acknowledgements

I would like to thank R. Pong-Wong and S. Turner for helpful discussions and D. Garrick, M. Haskell and A. Ruvinsky for comments on the manuscript.

References

Agyemang, K., Clapp, E. and Van Vleck, L.D (1982) Components of variance of dairymen's workability traits among Holstein cows. *Journal of Dairy Science* 65, 1334–1338.

Aharoni, Y., Henkin, Z., Ezra, A., Dolev, A., Shabtay, A., Orlov, A., Yehuda, Y. and Brosh, A. (2009) Grazing behavior and energy costs of activity: a comparison between two types of cattle. *Journal of Animal Science* 87, 2719–2731.

Aitchison, T.E., Freeman, A.E. and Thomson, G.M. (1972) Evaluation of a type appraisal program in Holsteins. *Journal of Dairy Science* 55, 840–844.

Albright, J.L. and Arave, C.W. (1997) *The Behaviour of Cattle*. CAB International, Wallingford, UK.

Alencar, M.M.D., Cruz, G.M.D., Tullio, R.R. and Correa, L.D.A. (1995) Suckling traits of straightbred Canchim and crossbred Canchim X Nelore calves. *Revista da Sociedade Brasileira de Zootecnia* 24, 706–714.

Alencar, M.M.D., Tullio, R.R., Cruz, G.M.D. and Correa, L.D.A. (1996) Grazing behavior of beef cows. *Revista da Sociedade Brasileira de Zootecnia* 25, 13–21.

Banos, G., Woolliams, J.A., Woodward, B.W., Forbes, A.B. and Coffey, M.P. (2008) Impact of single nucleotide polymorphisms in leptin, leptin receptor, growth hormone receptor, and diacylglycerol acyltransferase (DGAT1) gene loci on milk production, feed, and body energy traits of UK dairy cows. *Journal of Dairy Science* 91, 3190–3200.

Beckman, D.W., Enns, R.M., Speidel, S.E., Brigham, B.W. and Garrick, D.J. (2007) Maternal effects on docility in Limousin cattle. *Journal of Animal Science* 85, 650–657.

Beilharz, R.G., Butcher, D.F. and Freeman, A.E. (1966) Social dominance and milk production in Holsteins. *Journal of Dairy Science* 49, 887–892.

Beitia, J. and Epperly, J. (2011) Rediscover the value of efficiency. *Bottom Line* 13, 1, 15.

Benhajali, H., Boivin, X., Sapa, J., Pellegrini, P., Boulesteix, P., Lajudie, P. and Phocas, F. (2010) Assessment of different on-farm measures of beef cattle temperament for use in genetic evaluation. *Journal of Animal Science* 88, 3529–3537.

Bertram, J.D., Fordyce, G., McGowan, M.R., Jayawardhana, G.A., Fitzpatrick, L.A., Doogan, V.J., De Faveri, J. and Holroyd, R.G. (2002) Bull selection and use in northern Australia. 3. Serving capacity tests. *Animal Reproduction Science* 71, 51–66.

Boissy, A. and Bouissou, M.F. (1995) Assessment of individual differences in behavioural reactions of heifers exposed to various fear-eliciting situations. *Applied Animal Behaviour Science* 46, 17–31.

Boissy, A. and Le Neindre, P. (1997) Behavioral, cardiac and cortisol responses to brief peer separation and reunion in cattle. *Physiology and Behavior* 61, 693–699.

Bosch, O.J. and Neumann, I.D. (2012) Both oxytocin and vasopressin are mediators of maternal care and aggression in rodents: from central release to sites of action. *Hormones and Behavior* 61, 293–303.

Braghieri, A., Pacelli, C., Girolami, A. and Napolitano, F. (2011) Time budget, social and ingestive behaviours expressed by native beef cows in Mediterranean conditions. *Livestock Science* 141, 47–52.

Buchenauer, D. (1999) Genetics of behaviour in cattle. In: Fries, R. and Ruvinsky, A. (eds) *The Genetics of Cattle*. CAB International, Wallingford, UK, pp. 365–390.

Buddenberg, B.J., Brown, C.J., Johnson, Z.B. and Honea, R.S. (1986) Maternal behavior of beef cows at parturition. *Journal of Animal Science* 62, 42–46.

Burnside, E.B., Kowalchuck, S.B., Lambroughton, D.B. and Macleod, N.M. (1971) Canadian dairy cow disposals. I. Differences between breeds, lactation numbers and seasons. *Canadian Journal of Animal Science* 51, 75–83.

Burrow, H.M. (1988) A new technique for measuring temperament in cattle. *Proceedings of the Australian Society of Animal Production* 17, 154–157.

Burrow, H.M. (1997) Measurements of temperament and their relationships with performance traits of beef cattle. *Animal Breeding Abstracts* 65, 477–495.

Burrow, H.M. (2001) Variances and covariances between productive and adaptive traits and temperament in a composite breed of tropical beef cattle. *Livestock Production Science* 70, 213–233.

Burrow, H.M. and Corbet, N.J. (2000) Genetic and environmental factors affecting temperament of zebu and zebu-derived beef cattle grazed at pasture in the tropics. *Australian Journal of Agricultural Research* 51, 155–162.

Burrow, H.M. and Dillon, R.D. (1997) Relationships between temperament and growth in a feedlot and commercial carcass traits of Bos indicus crossbreds. *Australian Journal of Experimental Agriculture* 37, 407–411.

Burrow, H.M. and Prayaga, K.C. (2004) Correlated responses in productive and adaptive traits and temperament following selection for growth and heat resistance in tropical beef cattle. *Livestock Production Science* 86, 143–161.

Cafe, L.M., Robinson, D.L., Ferguson, D.M., Geesink, G.H. and Greenwood, P.L. (2011a) Temperament and hypothalamic-pituitary-adrenal axis function are related and combine to affect growth, efficiency, carcass, and meat quality traits in Brahman steers. *Domestic Animal Endocrinology* 40, 230–240.

Cafe, L.M., Robinson, D.L., Ferguson, D.M., McIntyre, B.L., Geesink, G.H. and Greenwood, P.L. (2011b) Cattle temperament: persistence of assessments and associations with productivity, efficiency, carcass and meat quality traits. *Journal of Animal Science* 89, 1452–1465.

Chenoweth, P.J., Chase, C.C., Larsen, R.E., Thatcher, M.J.D., Bivens, J.F. and Wilcox, C.J. (1996) The assessment of sexual performance in young *Bos taurus* and *Bos indicus* beef bulls. *Applied Animal Behaviour Science* 48, 225–235.

Cue, R.I., Harris, B.L. and Rendel, J.M. (1996) Genetic parameters for traits other than production in purebred and crossbred New Zealand dairy cattle. *Livestock Production Science* 45, 123–135.

Cummins, S.B., Lonergan, P., Evans, A.C.O. and Butler, S.T. (2012) Genetic merit for fertility traits in Holstein cows: II. Ovarian follicular and corpus luteum dynamics, reproductive hormones, and estrus behavior. *Journal of Dairy Science* 95, 3698–3710.

Curley, K.O., Jr, Paschal, J.C., Welsh, T.H., Jr and Randel, R.D. (2006) Technical note: exit velocity as a measure of cattle temperament is repeatable and associated with serum concentration of cortisol in Brahman bulls. *Journal of Animal Science* 84, 3100–3103.

Curley, K.O., Jr, Neuendorff, D.A., Lewis, A.W., Cleere, J.J., Welsh, T.H., Jr and Randel, R.D. (2008) Functional characteristics of the bovine hypothalamic-pituitary-adrenal axis vary with temperament. *Hormones and Behavior* 53, 20–27.

Curley, K.O., Jr, Neuendorff, D.A., Lewis, A.W., Rouquette, F.M., Jr, Randel, R.D. and Welsh, T.H., Jr (2010) The effectiveness of vasopressin as an ACTH secretagogue in cattle differs with temperament. *Physiology and Behavior* 101, 699–704.

Das, S.M., Redbo, I. and Wiktorsson, H. (2000) Effect of age of calf on suckling behaviour and other behavioural activities of Zebu and crossbred calves during restricted suckling periods. *Applied Animal Behaviour Science* 67, 47–57.

de Vargas Junior, F.M., Wechsler, F.S., Rossi, P., Morais de Oliveira, M.V. and Schmidt, P. (2010) Ingestive behavior of Nellore cows and their straightbred or crossbred calves. *Revista Brasileira de Zootecnia – Brazilian Journal of Animal Science* 39, 648–655.

Dickson, D.P., Barr, G.R., Johnson, L.P. and Wieckert, D.A. (1970) Social dominance and temperament of Holstein cows. *Journal of Dairy Science* 53, 904–907.

Duchens, M., Maciel, M., Gustafsson, H., Forsberg, M., Rodriguez-Martinez, H. and Edqvist, L.E. (1995) Influence of perioestrous suprabasal progesterone levels on cycle length, estrous behavior and ovulation in heifers. *Animal Reproduction Science* 37, 95–108.

Durunna, O.N., Wang, Z., Basarab, J.A., Okine, E.K. and Moore, S.S. (2011) Phenotypic and genetic relationships among feeding behavior traits, feed intake, and residual feed intake in steers fed grower and finisher diets. *Journal of Animal Science* 89, 3401–3409.

Erf, D.F., Hansen, L.B. and Lawstuen, D.A. (1992) Inheritance and relationships of workability traits and yield for Holsteins. *Journal of Dairy Science* 75, 1999–2007.

Falconer, D.S. and Mackay, T.F.C. (1996) *Introduction to Quantitative Genetics*. Addison Wesley Longman Limited, Harlow, UK.

Fazio, E., Medica, P., Cravana, C., Cavaleri, S. and Ferlazzo, A. (2012) Effect of temperament and prolonged transportation on endocrine and functional variables in young beef bulls. *Veterinary Record* 171, 644.

Fordyce, G. and Goddard, M.E. (1984) Maternal influence on the temperament of *Bos indicus* cross cows. *Proceedings of the Australian Society of Animal Production* 15, 345–348.

Fordyce, G., Goddard, M.E. and Seifert, G.W. (1982) The measurement of temperament in cattle and the effect of experience and genotype. *Proceedings of the Australian Society of Animal Production* 14, 329–332.

Fordyce, G., Fitzpatrick, L.A., Cooper, N.J., Doogan, V.J., De Faveri, J. and Holroyd, R.G. (2002) Bull selection and use in northern Australia. 5. Social behaviour and management. *Animal Reproduction Science* 71, 81–99.

Foster, W.W., Freeman, A.E., Berger, P.J. and Kuck, A. (1988) Linear type trait analysis with genetic parameter estimation. *Journal of Dairy Science* 71, 223–231.

Gauly, M., Mathiak, H., Hoffmann, K., Kraus, M. and Erhardt, G. (2001) Estimating genetic variability in temperamental traits in German Angus and Simmental cattle. *Applied Animal Behaviour Science* 74, 109–119.

Gibbons, J., Lawrence, A. and Haskell, M. (2009a) Responsiveness of dairy cows to human approach and novel stimuli. *Applied Animal Behaviour Science* 116, 163–173.

Gibbons, J.M., Lawrence, A.B. and Haskell, M.J. (2009b) Consistency of aggressive feeding behaviour in dairy cows. *Applied Animal Behaviour Science* 121, 1–7.

Gibbons, J.M., Lawrence, A.B. and Haskell, M.J. (2010) Measuring sociability in dairy cows. *Applied Animal Behaviour Science* 122, 84–91.

Gibbons, J.M., Lawrence, A.B. and Haskell, M.J. (2011) Consistency of flight speed and response to restraint in a crush in dairy cattle. *Applied Animal Behaviour Science* 131, 15–20.

Glenske, K., Brandt, H., Gauly, M. and Erhardt, G. (2010a) Analysis of association of two SNP in cholecystokinin B receptor gene with behaviour scores in German Angus and German Simmental cattle (Brief Report). *Archiv für Tierzucht – Archives of Animal Breeding* 53, 494–496.

Glenske, K., Brandt, H., Prinzenberg, E.-M., Gauly, M. and Erhardt, G. (2010b) Verification of a QTL on BTA1 for temperament in German Simmental and German Angus calves (Short Communication). *Archiv für Tierzucht – Archives of Animal Breeding* 53, 388–392.

Glenske, K., Prinzenberg, E.M., Brandt, H., Gauly, M. and Erhardt, G. (2011) A chromosome-wide QTL study on BTA29 affecting temperament traits in German Angus beef cattle and mapping of DRD4. *Animal* 5, 195–197.

González Caicedo, E., Durán Castro, C.V. and Dominguez Cadavid, J.F. (1994) Heritability and repeatability of the Nota De Trienta and Nota De Lidia traits of a herd of bullfighting cattle. *Archivos de Zootecnia* 43, 225–237.

Grignard, L., Boivin, X., Boissy, A. and Le Neindre, P. (2001) Do beef cattle react consistently to different handling situations? *Applied Animal Behaviour Science* 71, 263–276.

Gutiérrez-Gil, B., Ball, N., Burton, D., Haskell, M., Williams, J.L. and Wiener, P. (2008) Identification of quantitative trait loci affecting cattle temperament. *Journal of Heredity* 99, 629–638.

Hale, E.B. (1969) Domestication and the evolution of behaviour. In: Hafez, E.S.E. (ed.) *The Behavior of Domestic Animals,* 2nd edition. Williams and Wilkins, Baltimore, Maryland, pp. 23–42.

Hammond, A.C. and Olson, T.A. (1994) Rectal temperature and grazing time in selected beef-cattle breeds under tropical summer conditions in subtropical Florida. *Tropical Agriculture* 71, 128–134.

Hearnshaw, H. and Morris, C.A. (1984) Genetic and environment effects on a temperament score in beef cattle. *Australian Journal of Agricultural Research* 35, 723–733.

Heringstad, B. (2010) Genetic analysis of fertility-related diseases and disorders in Norwegian Red cows. *Journal of Dairy Science* 93, 2751–2756.

Hiendleder, S., Thomsen, H., Reinsch, N., Bennewitz, J., Leyhe-Horn, B., Looft, C., Xu, N., Medjugorac, I., Russ, I., Kuhn, C. *et al.* (2003) Mapping of QTL for body conformation and behavior in cattle. *Journal of Heredity* 94, 496–506.

Hoflack, G., Van Soom, A., Maes, D., de Kruif, A., Opsomer, G. and Duchateau, L. (2006) Breeding soundness and libido examination of Belgian blue and Holstein Friesian artificial insemination bulls in Belgium and the Netherlands. *Theriogenology* 66, 207–216.

Hoppe, S., Brandt, H.R., Erhardt, G. and Gauly, M. (2008) Maternal protective behaviour of German Angus and Simmental beef cattle after parturition and its relation to production traits. *Applied Animal Behaviour Science* 114, 297–306.

Hoppe, S., Brandt, H.R., Koenig, S., Erhardt, G. and Gauly, M. (2010) Temperament traits of beef calves measured under field conditions and their relationships to performance. *Journal of Animal Science* 88, 1982–1989.

Houpt, K.A. (1991) *Domestic Animal Behavior for Veterinarians and Animal Scientists,* 2nd Edn. Iowa State University Press, Ames, Iowa.

Houpt, K.A. (2011) *Domestic Animal Behavior for Veterinarians and Animal Scientists,* 5th Edn. Wiley-Blackwell, Ames, Iowa.

Jezierski, T.A., Koziorowski, M., Goszczynski, J. and Sieradzka, I. (1989) Homosexual and social behaviors of young bulls of different genotypes and phenotypes and plasma concentrations of some hormones. *Applied Animal Behaviour Science* 24, 101–113.

Kadel, M.J., Johnston, D.J., Burrow, H.M., Graser, H.-U. and Ferguson, D.M. (2006) Genetics of flight time and other measures of temperament and their value as selection criteria for improving meat quality traits in tropically adapted breeds of beef cattle. *Australian Journal of Agricultural Research* 57, 1029–1035.

King, D.A., Pfeiffer, C.E.S., Randel, R.D., Welsh, T.H., Jr, Oliphint, R.A., Baird, B.E., Curley, K.O., Jr, Vann, R.C., Hale, D.S. and Savell, J.W. (2006) Influence of animal temperament and stress responsiveness on the carcass quality and beef tenderness of feedlot cattle. *Meat Science* 74, 546–556.

Kommadath, A., Mulder, H.A., de Wit, A.A.C., Woelders, H., Smits, M.A., Beerda, B., Veerkamp, R.F., Frijters, A.C.J. and Pas, M.F.W.T. (2010) Gene expression patterns in anterior pituitary associated with quantitative measure of oestrous behaviour in dairy cows. *Animal* 4, 1297–1307.

Kommadath, A., Woelders, H., Beerda, B., Mulder, H.A., de Wit, A.A.C., Veerkamp, R.F., Pas, M.F.W.T. and Smits, M.A. (2011) Gene expression patterns in four brain areas associate with quantitative measure of estrous behavior in dairy cows. *BMC Genomics* 12, 200.

König, S., Koehn, F., Kuwan, K., Simianer, H. and Gauly, M. (2006) Use of repeated measures analysis for evaluation of genetic background of dairy cattle behavior in automatic milking systems. *Journal of Dairy Science* 89, 3636–3644.

Kuehn, L.A., Golden, B.L., Comstock, C.R. and Andersen, K.J. (1998) Docility EPD for Limousin cattle. *Journal of Dairy Science* 81, 85.

Lagonigro, R., Wiener, P., Pilla, F., Woolliams, J.A. and Williams, J.L. (2003) A new mutation in the coding region of the bovine leptin gene associated with feed intake. *Animal Genetics* 34, 371–374.

Landaeta-Hernandez, A.J., Yelich, J.V., Lemaster, J.W., Fields, M.J., Tran, T., Chase, C.C., Rae, D.O. and Chenoweth, P.J. (2002) Environmental, genetic and social factors affecting the expression of estrus in beef cows. *Theriogenology* 57, 1357–1370.

Lawstuen, D.A., Hansen, L.B., Steuernagel, G.R. and Johnson, L.P. (1988) Management traits scored linearly by dairy producers. *Journal of Dairy Science* 71, 788–799.

Le Neindre, P., Trillat, G., Sapa, J., Menissier, F., Bonnet, J.N. and Chupin, J.M. (1995) Individual differences in docility in Limousin cattle. *Journal of Animal Science* 73, 2249–2253.

Liinamo, A.E., Mantysaari, P. and Mantysaari, E.A. (2012) Short communication: Genetic parameters for feed intake, production, and extent of negative energy balance in Nordic Red dairy cattle. *Journal of Dairy Science* 95, 6788–6794.

Lindholm-Perry, A.K., Sexten, A.K., Kuehn, L.A., Smith, T.P.L., King, D.A., Shackelford, S.D., Wheeler, T.L., Ferrell, C.L., Jenkins, T.G., Snelling, W.M. and Freetly, H.C. (2011) Association, effects and validation of polymorphisms within the NCAPG-LCORL locus located on BTA6 with feed intake, gain, meat and carcass traits in beef cattle. *BMC Genetics* 12, 103.

Lindholm-Perry, A.K., Kuehn, L.A., Smith, T.P.L., Ferrell, C.L., Jenkins, T.G., Freetly, H.C. and Snelling, W.M. (2012) A region on BTA14 that includes the positional candidate genes LYPLA1, XKR4 and TMEM68 is associated with feed intake and growth phenotypes in cattle. *Animal Genetics* 43, 216–219.

Lüehken, G., Glenske, K., Brandt, H. and Erhardt, G. (2010) Genetic variation in monoamine oxidase A and analysis of association with behaviour traits in beef cattle. *Journal of Animal Breeding and Genetics* 127, 411–418.

MacNeil, M.D., Bailey, D.R.C., Urick, J.J., Gilbert, R.P. and Reynolds, W.L. (1991) Heritabilities and genetic correlations for postweaning growth and feed intake of beef bulls and steers. *Journal of Animal Science* 69, 3183–3189.

Macuhova, J., Tancin, V., Kraetzl, W.D., Meyer, H.H.D. and Bruckmaier, R.M. (2002) Inhibition of oxytocin release during repeated milking in unfamiliar surroundings: the importance of opioids and adrenal cortex sensitivity. *Journal of Dairy Research* 69, 63–73.

Mormède, P., Andanson, S., Auperin, B., Beerda, B., Guemene, D., Malmkvist, J., Manteca, X., Manteuffel, G., Prunet, P., van Reenen, C.G., Richard, S. and Veissier, I. (2007) Exploration of the hypothalamic-pituitary-adrenal function as a tool to evaluate animal welfare. *Physiology and Behavior* 92, 317–339.

Morris, C.A., Baker, R.L., Cullen, N.G. and Boyd, P. (1992) Genetic parameters for body weight, scrotal circumference, and serving capacity in beef cattle. *New Zealand Journal of Agricultural Research* 35, 195–198.

Morris, C.A., Cullen, N.G., Kilgour, R. and Bremner, K.J. (1994) Some genetic factors affecting temperament in *Bos taurus* cattle. *New Zealand Journal of Agricultural Research* 37, 167–175.

Müller, R. and von Keyserlingk, M.A.G. (2006) Consistency of flight speed and its correlation to productivity and to personality in *Bos taurus* beef cattle. *Applied Animal Behaviour Science* 99, 193–204.

Murphey, R.M., Duarte, F.A.M. and Penedo, M.C.T. (1980) Approachability of bovine cattle in pastures – breed comparisons and a breed × treatment analysis. *Behavior Genetics* 10, 171–181.

Murphey, R.M., Duarte, F.A.M. and Penedo, M.C.T. (1981) Responses of cattle to humans in open spaces – breed comparisons and approach-avoidance relationships. *Behavior Genetics* 11, 37–48.

Nkrumah, J.D., Li, C., Yu, J., Hansen, C., Keisler, D.H. and Moore, S.S. (2005) Polymorphisms in the bovine leptin promoter associated with serum leptin concentration, growth, feed intake, feeding behavior, and measures of carcass merit. *Journal of Animal Science* 83, 20–28.

Nkrumah, J.D., Crews, D.H., Jr, Basarab, J.A., Price, M.A., Okine, E.K., Wang, Z., Li, C. and Moore, S.S. (2007) Genetic and phenotypic relationships of feeding behavior and temperament with performance, feed efficiency, ultrasound, and carcass merit of beef cattle. *Journal of Animal Science* 85, 2382–2390.

O'Bleness, G.V., Van Vleck, L.D. and Henderson, C.R. (1960) Heritabilities of some type appraisal traits and their genetic and phenotypic correlations with production. *Journal of Dairy Science* 43, 1490–1498.

O'Driscoll, K., Boyle, L. and Hanlon, A. (2009) The effect of breed and housing system on dairy cow feeding and lying behaviour. *Applied Animal Behaviour Science* 116, 156–162.

Orihuela, A. (2000) Some factors affecting the behavioural manifestation of oestrus in cattle: a review. *Applied Animal Behaviour Science* 70, 1–16.

O'Rourke, P.K. (1989) Validation of genetic parameters for breeding *Bos indicus* cross cattle in the dry tropics.(ed. by Industries, Q.D.o.P.), Brisbane.

Phillips, C. (2002) *Cattle Behaviour and Welfare*. Blackwell Science Ltd, Oxford.

Phocas, F., Boivin, X., Sapa, J., Trillat, G., Boissy, A. and Le Neindre, P. (2006) Genetic correlations between temperament and breeding traits in Limousin heifers. *Animal Science* 82, 805–811.

Plusquellec, P. and Bouissou, M.F. (2001) Behavioural characteristics of two dairy breeds of cows selected (Herens) or not (Brune des Alpes) for fighting and dominance ability. *Applied Animal Behaviour Science* 72, 1–21.

Prayaga, K.C., Corbet, N.J., Johnston, D.J., Wolcott, M.L., Fordyce, G. and Burrow, H.M. (2009) Genetics of adaptive traits in heifers and their relationship to growth, pubertal and carcass traits in two tropical beef cattle genotypes. *Animal Production Science* 49, 413–425.

Pugh, K.A., Stookey, J.M. and Buchanan, F.C. (2011) An evaluation of corticotropin-releasing hormone and leptin SNPs relative to cattle behavior. *Canadian Journal of Animal Science* 91, 567–572.

Purcell, D. and Arave, C.W. (1991) Isolation vs. group rearing in monozygous twins. *Applied Animal Behaviour Science* 31, 147–156.

Quirino, C.R., Bergmann, J.A.G., Vale, V.R., Andrade, V.J., Reis, S.R., Mendonca, R.M. and Fonseca, C.G. (2004) Genetic parameters of libido in Brazilian Nellore bulls. *Theriogenology* 62, 1–7.

Roelofs, J., Lopez-Gatius, F., Hunter, R.H.F., van Eerdenburg, F.J.C.M. and Hanzen, C. (2010) When is a cow in estrus? Clinical and practical aspects. *Theriogenology* 74, 327–344.

Rolfe, K.M., Snelling, W.M., Nielsen, M.K., Freetly, H.C., Ferrell, C.L. and Jenkins, T.G. (2011) Genetic and phenotypic parameter estimates for feed intake and other traits in growing beef cattle, and opportunities for selection. *Journal of Animal Science* 89, 3452–3459.

Rottensten, K. and Touchberry, R.W. (1957) Observations on the degree of expression of estrus in cattle. *Journal of Dairy Science* 40, 1457–1465.

Roxström, A., Strandberg, E., Berglund, B., Emanuelson, U. and Philipsson, J. (2001) Genetic and environmental correlations among female fertility traits, and between the ability to show oestrus and milk production in dairy cattle. *Acta Agriculturae Scandinavica Section a: Animal Science* 51, 192–199.

Sánchez-Rodríguez, H.L., Vann, R.C., Youngblood, R.C., Baravik-Munsell, E., Christiansen, D.L., Willard, S. and Ryan, P.L. (2013) Evaluation of pulsatility index and diameter of the jugular vein and superficial body temperature as physiological indices of temperament in weaned beef calves: Relationship with serum cortisol concentrations, rectal temperature, and sex. *Livestock Science* 151, 228–237.

Sant'Anna, A.C., Paranhos da Costa, M.J.R., Baldi, F., Rueda, P.M. and Albuquerque, L.G. (2012) Genetic associations between flight speed and growth traits in Nellore cattle. *Journal of Animal Science* 90, 3427–3432.

Sartori, C. and Mantovani, R. (2010) Genetics of fighting ability in cattle using data from the traditional battle contest of the Valdostana breed. *Journal of Animal Science* 88, 3206–3213.

Sartori, C. and Mantovani, R. (2012) Effects of inbreeding on fighting ability measured in Aosta Chestnut and Aosta Black Pied cattle. *Journal of Animal Science* 90, 2907–2915.

Sartori, C. and Mantovani, R. (2013) Indirect genetic effects and the genetic bases of social dominance: evidence from cattle. *Heredity* 110, 3–9.

Sato, S. (1981) Factors associated with temperament of beef cattle. *Japanese Journal of Zootechnical Science* 52, 595–605.

Schmutz, S.M., Stookey, J.M., Winkelman-Sim, D.C., Waltz, C.S., Plante, Y. and Buchanan, F.C. (2001) A QTL study of cattle behavioral traits in embryo transfer families. *Journal of Heredity* 92, 290–292.

Schrooten, C., Bovenhuis, H., Coppieters, W. and Van Arendonk, J.A.M. (2000) Whole genome scan to detect quantitative trait loci for conformation and functional traits in dairy cattle. *Journal of Dairy Science* 83, 795–806.

Schutt, K.M., Arthur, P.F. and Burrow, H.M. (2009) Brahman and Brahman crossbred cattle grown on pasture and in feedlots in subtropical and temperate Australia. 3. Feed efficiency and feeding behaviour of feedlot-finished animals. *Animal Production Science* 49, 452–460.

Senn, M., Durst, B., Kaufmann, A. and Langhans, W. (1995) Feeding patterns of lactating cows of 3 different breeds fed hay, corn silage, and grass silage. *Physiology and Behavior* 58, 229–236.

Sewalem, A., Miglior, F. and Kistemaker, G.J. (2011) Short communication: genetic parameters of milking temperament and milking speed in Canadian Holsteins. *Journal of Dairy Science* 94, 512–516.

Sharma, J.S. and Khanna, A.S. (1980) Note on genetic group and parity differences in dairy temperament score of crossbred cattle. *Indian Journal Animal Research* 14, 127–128.

Sheahan, A.J., Kolver, E.S. and Roche, J.R. (2011) Genetic strain and diet effects on grazing behavior, pasture intake, and milk production. *Journal of Dairy Science* 94, 3583–3591.

Sherman, E.L., Nkrumah, J.D., Murdoch, B.M., Li, C., Wang, Z., Fu, A. and Moore, S.S. (2008) Polymorphisms and haplotypes in the bovine neuropeptide Y, growth hormone receptor, ghrelin, insulin-like growth factor 2, and uncoupling proteins 2 and 3 genes and their associations with measures of growth, performance, feed efficiency, and carcass merit in beef cattle. *Journal of Animal Science* 86, 1–16.

Shrode, R.R. and Hammack, S.P. (1971) Chute behavior of yearling beef cattle. *Journal of Animal Science* 33, 193.

Silva, B., Gonzalo, A. and Canon, J. (2006) Genetic parameters of aggressiveness, ferocity and mobility in the fighting bull breed. *Animal Research* 55, 65–70.

Snelling, W.M., Allan, M.F., Keele, J.W., Kuehn, L.A., Thallman, R.M., Bennett, G.L., Ferrell, C.L., Jenkins, T.G., Freetly, H.C., Nielsen, M.K. and Rolfe, K.M. (2011) Partial-genome evaluation of postweaning feed intake and efficiency of crossbred beef cattle. *Journal of Animal Science* 89, 1731–1741.

Spelman, R.J., Huisman, A.E., Singireddy, S.R., Coppieters, W., Arranz, J., Georges, M. and Garrick, D.J. (1999) Quantitative trait loci analysis on 17 nonproduction traits in the New Zealand dairy population. *Journal of Dairy Science* 82, 2514–2516.

Stricklin, W.R. (1983) Matrilinear social dominance and spatial relationships among Angus and Hereford cows. *Journal of Animal Science* 57, 1397–1405.

Sun, J., Jin, Q., Zhang, C., Fang, X., Gu, C., Lei, C., Wang, J. and Chen, H. (2011) Polymorphisms in the bovine ghrelin precursor (GHRL) and Syndecan-1 (SDC1) genes that are associated with growth traits in cattle. *Molecular Biology Reports* 38, 3153–3160.

Sutherland, M.A., Rogers, A.R. and Verkerk, G.A. (2012) The effect of temperament and responsiveness towards humans on the behavior, physiology and milk production of multi-parous dairy cows in a familiar and novel milking environment. *Physiology and Behavior* 107, 329–337.

Turner, S.P. and Lawrence, A.B. (2007) Relationship between maternal defensive aggression, fear of handling and other maternal care traits in beef cows. *Livestock Science* 106, 182–188.

Turner, S.P., Navajas, E.A., Hyslop, J.J., Ross, D.W., Richardson, R.I., Prieto, N., Bell, M., Jack, M.C. and Roehe, R. (2011) Associations between response to handling and growth and meat quality in frequently handled *Bos taurus* beef cattle. *Journal of Animal Science* 89, 4239–4248.

Van Eenennaam, A.L., Weaber, R.L., Drake, D.J., Penedo, M.C.T., Quaas, R.L., Garrick, D.J. and Pollak, E.J. (2007) DNA-based paternity analysis and genetic evaluation in a large, commercial cattle ranch setting. *Journal of Animal Science* 85, 3159–3169.

van Eerdenburg, F.J.C.M. (2008) Possible causes for the diminished expression of oestrous behaviour. *Veterinary Quarterly* 30, 79–100.

Van Vleck, L.D. (1964) Variation in type appraisal scores due to sire + herd effects. *Journal of Dairy Science* 47, 1249–1256.

Vetters, M.D.D., Engle, T.E., Ahola, J.K. and Grandin, T. (2013) Comparison of flight speed and exit score as measurements of temperament in beef cattle. *Journal of Animal Science* 91, 374–381.

Visscher, P.M. and Goddard, M.E. (1995) Genetic parameters for milk yield, survival workability, and type traits for Australian dairy cattle. *Journal of Dairy Science* 78, 205–220.

Voisinet, B.D., Grandin, T., O'Connor, S.F., Tatum, J.D. and Deesing, M.J. (1997a) *Bos indicus* cross feedlot cattle with excitable temperaments have tougher meat and a higher incidence of borderline dark cutters. *Meat Science* 46, 367–377.

Voisinet, B.D., Grandin, T., Tatum, J.D., O'Connor, S.F. and Struthers, J.J. (1997b) Feedlot cattle with calm temperaments have higher average daily gains than cattle with excitable temperaments. *Journal of Animal Science* 75, 892–896.

Wagnon, K.A., Loy, R.G., Rollins, W.C. and Carroll, F.D. (1966) Social dominance in a herd of Angus, Hereford and Shorthorn cows. *Animal Behaviour* 14, 474–479.

Watts, J.M., Stookey, J.M., Schmutz, S.M. and Waltz, C.S. (2001) Variability in vocal and behavioural responses to visual isolation between full-sibling families of beef calves. *Applied Animal Behaviour Science* 70, 255–273.

Weaber, R.L., Taxis, T.M., Shafer, W.R., Berger, L.L., Faulkner, D.B., Rolf, M.M., Dow, D.L., Taylor, J.F. and Lorenzen, C.L. (2010) Heritabilities, genetic and phenotypic correlations among Warner-Bratzler shear force and repeated objective measurements of temperament in fed cattle. *Journal of Dairy Science* 93, 744–745.

Weiss, D., Helmreich, S., Mostl, E., Dzidic, A. and Bruckmaier, R.M. (2004) Coping capacity of dairy cows during the change from conventional to automatic milking. *Journal of Animal Science* 82, 563–570.

Wickham, B.W. (1979) Genetic parameters and economic values of traits other than production for dairy cattle. *Proceedings of the New Zealand Society of Animal Production* 39, 180–193.

Winder, J.A., Walker, D.A. and Bailey, C.C. (1996) Effect of breed on botanical composition of cattle diets on Chihuahuan desert range. *Journal of Range Management* 49, 209–214.

Wylie, A.R.G. (2011) Leptin in farm animals: where are we and where can we go? *Animal* 5, 246–267.

Wyszynska-Koko, J., de Wit, A.A.C., Beerda, B., Veerkamp, R.F. and Pas, M.F.W.T. (2011) Gene expression patterns in the ventral tegmental area relate to oestrus behaviour in high-producing dairy cows. *Journal of Animal Breeding and Genetics* 128, 183–191.

Yayou, K.-I., Ito, S., Yamamoto, N., Kitagawa, S. and Okamura, H. (2010) Relationships of stress responses with plasma oxytocin and prolactin in heifer calves. *Physiology and Behavior* 99, 362–369.

Yayou, K.-I., Kitagawa, S., Ito, S., Kasuya, E. and Sutoh, M. (2011) Effect of oxytocin, prolactin-releasing peptide, or corticotropin-releasing hormone on feeding behavior in steers. *General and Comparative Endocrinology* 174, 287–291.

Zhang, Y.Y., Proenca, R., Maffei, M., Barone, M., Leopold, L. and Friedman, J.M. (1994) Positional cloning of the mouse obese gene and its human homolog. *Nature* 372, 425–432.

12 Genetics of Reproduction in Cattle

Brian W. Kirkpatrick

University of Wisconsin-Madison, Madison, Wisconsin, USA

Introduction

Since publication of the previous edition of this book there has been a phenomenal increase in information about the bovine genome and development of genomic tools. As a result there have been a multitude of genomic studies published and a plethora of genomic associations with reproductive traits. This chapter attempts to be exhaustive in identifying those published studies, but cannot be similarly exhaustive in describing the reported genomic associations for a variety of reasons including differences between studies in statistical thresholds employed and the sheer number of reported associations. As a compromise, the most significant results from the various studies are reported and discussed here. The interested reader can consult the primary sources for information on marginally significant results and their potential correspondence across studies. A helpful resource in that regard are cattle quantitative trait locus (QTL) databases, which are more comprehensive in the listing of results (Polineni *et al.*, 2006; Hu and Reecy, 2007; Hu *et al.*, 2013).

Genetics of Ovulation Rate and Related Traits

The primary source of bovine ovulation rate data is the long-running selection experiment for bovine twinning rate conducted by the USDA Meat Animal Research Center (USMARC) (Gregory *et al.*, 1990). Several studies have reported evidence of ovulation rate QTLs in cattle using data from this population (Blattman *et al.*, 1996; Kappes *et al.*, 2000; Kirkpatrick *et al.*, 2000; Arias and Kirkpatrick, 2004; Gonda *et al.*, 2004). These studies have taken various approaches that have resulted in identification of different QTLs in some cases. Scientists at the University of Wisconsin-Madison examined three elite families from within the herd, which were also the largest families available, and tracked inheritance of patriarch's alleles through three-generation pedigrees. Scientists at USMARC utilized paternal half-sib families of sons in the equivalent of a granddaughter design (Kappes *et al.*, 2000). Initial results from sons were subsequently tested using data from daughters. While this granddaughter design provided a broader search of the population, power was limited by small family size. The Wisconsin studies provided evidence for QTLs on chromosomes 7, 10, 14 and 19 (chromosome-wise $P < 0.01$) and lesser evidence on chromosome 23. USMARC scientists reported strong evidence for a QTL on chromosome 5, and lesser evidence for locations on chromosomes 1, 7 and 23.

While ovulation rate data is not routinely recorded in industry, records of single versus multiple births often are, enabling mapping of QTLs for twinning rate. Most twin births (i.e. ~90%) in cattle are dizygotic (Cady, 1978; Nielen et al., 1989; Ryan and Boland, 1991; del Rio et al., 2006), resulting from the ovulation of two eggs. Consequently, the genetic correlation between ovulation rate and twinning rate is very high with estimates from 0.75 to >0.90 (Van Vleck et al., 1991; Gregory et al., 1997).

A genome-wide search for twinning rate QTLs in the Norwegian cattle population revealed evidence for twinning rate QTLs (5% genome-wise significance level) on chromosomes 5, 7, 12 and 23 (Lien et al., 2000). The most likely location of the chromosome 5 QTL from this study was distal to that reported for ovulation rate in studies with the USMARC population, in contrast the chromosome 7 QTL corresponded well in location (Kappes et al., 2000). A genome-wide search using microsatellite markers and Holstein families provided evidence of twinning rate QTLs on chromosomes 8, 10 and 14 at a chromosome-wise $P < 0.01$ (Cobanoglu et al., 2005), with the reported QTL location on chromosome 14 showing great similarity to a previously reported ovulation rate QTL (Gonda et al., 2004). A genome-wide analysis of the Israeli Holstein population using a granddaughter design (Weller et al., 2008b) provided only marginal evidence of twinning rate QTLs (nominal $P < 0.01$ in three cases), though the results for chromosomes 14 and 23 displayed similar locations in comparisons to previous reports of ovulation rate and twinning rate QTLs (Blattman et al., 1996; Gonda et al., 2004; Cobanoglu et al., 2005). Subsequent genome-wide analysis (Kim et al., 2009a) of the North American Holstein population with an early, low density single nucleotide polymorphism (SNP) panel (10 k) suggested evidence for twinning

rate QTLs on multiple chromosomes, including some of the chromosomal regions identified in previous analyses.

Further analysis has validated several initial QTL reports described above. Subsequent analysis in North American Holsteins of the chromosome 5 region initially identified by Lien et al. (2000) provided corroborating evidence of a twinning rate QTL (Cruickshank et al., 2004), and re-analysis of the Norwegian data provided an example of the utility of combined linkage–linkage disequilibrium analysis for QTL fine-mapping (Meuwissen et al., 2002). A narrow region containing the insulin-like growth factor-1 (IGF1) gene was identified as a positional candidate gene region containing two SNPs in strong linkage disequilibrium with the QTL (Kim et al., 2009b).

The chromosome 5 QTL reported by Kappes et al. (2000) was re-examined in a subsequent analysis (Allan et al., 2009) of more extensive data from the same population (USMARC twinning herd, subsequently referred to as the Production Efficiency Population), with both twinning rate and ovulation rate phenotypes considered. SNP associations were observed over a region from roughly 29–40 Mb, with the strongest association observed for a SNP located at 28.9 Mb. This is not the same location as the QTL reported by Lien et al. (2000) and subsequently validated by Cruickshank et al. (2004) and Kim et al. (2009b). Interestingly, one of the IGF1 SNPs identified in the latter study has been validated in the Production Efficiency Population with an association significant at $P < 0.0005$ for twinning rate (L.A. Kuehn, US Meat Animal Research Center, 2011, personal communication), suggesting there are multiple genes affecting ovulation and twinning rate on chromosome 5.

Multiple locations on chromosome 14 (Bierman et al., 2010b) were identified as significant in validation analyses, suggesting that the chromosome 14 QTLs previously identified were the result of multiple, linked loci. This validation work went on to develop a prediction equation for twinning rate genomic breeding value based on SNPs identified as significant in initial analyses with the North American Holstein population (AI sires) and subsequently validated in an independent set of sires (Bierman et al., 2010a).

While work in sheep has identified several major genes for ovulation rate (Galloway et al., 2000; Wilson et al., 2001; Hanrahan et al., 2004; Drouilhet et al., 2009), until recently no evidence for major genes had been presented for cattle. However, a cow ('Treble') with an exceptional record of prolificacy (Morris et al., 2010) is a matriarch that produced three sets of triplets in her lifetime, and a son from one of the triplet sets was used as a sire and produced several daughters that had either twin or triplet births. Semen from this son ('Trio') was used in artificial insemination matings at the University of Wisconsin-Madison, where daughters born in 2008 and 2009 (n = 88) have been evaluated for ovulation rate (count of corpora lutea) over an average of four oestrous cycles from 12–16 months of age. A within-family linkage analysis provided strong evidence ($P < 1 \times 10^{-19}$) of segregation of a single gene with large effect on ovulation rate (1.02 ± 0.08 additional corpora lutea per cycle). The gene and mutation causing the phenotype have not yet been elucidated.

The identification of single genes for fecundity in sheep provides candidate genes for ovulation and twinning rate in cattle, though no candidate gene studies for twinning rate or naturally occurring ovulation rate have been reported. However, these or other genes with known roles in reproductive endocrinology have been the basis for candidate gene studies of hormone-induced ovulation rate (i.e. response to superovulatory treatment) and assessments of embryo fertilization and transferability. Association of polymorphisms in growth and differentiation factor 9 (GDF9), luteinizing hormone/choriogonadotropin receptor (LHCGR), follicle stimulating hormone receptor (FSHR), inhibin alpha (INHA) and progesterone receptor genes with response to superovulation in Chinese Holsteins have been examined in a series of studies (Yang et al., 2010, 2011, 2012; Tang et al., 2011, 2013; Yu et al., 2012; Cory et al., 2013). In all cases authors reported associations between polymorphisms in these genes and various traits (total number of ova, number of unfertilized ova, number of degenerate embryos, number of transferable embryos), albeit typically at marginal levels of significance ($P < 0.05$). Power in these studies is low given relatively few animals with records

of superovulation response. In only one case has a genome-wide search for hormone-induced ovulation rate in cattle been reported, resulting subsequently in a positional candidate gene analysis of the ionotropic glutamate receptor AMPA 1 (GRIA1) gene (Sugimoto et al., 2010). The authors identified a missense mutation in GRIA1 that was associated with variation in gonadotropin releasing hormone release for hypothalamic cells in vitro, drawing a strong link between the polymorphism and phenotype at whole animal and cellular levels.

Recent estimates of heritability for twinning rate in cattle are consistent with earlier work. Analysis of twinning in Iranian Holsteins yielded heritability estimates ranging from 0.01 to 0.02 when analysed separately by parity with linear animal or sire models, and from 0.05 to 0.11 when analysed with a threshold model (Ghavi Hossein-Zadeh et al., 2009). Similarly for US Holsteins, heritability of twinning rate was reported as 0.02 and 0.09 when estimated with linear and threshold models (Johanson et al., 2001). Reported frequencies of twin birth in these reports for the Iranian and US Holstein populations were 3.01 and 5.02%, respectively. Analysis of data from the USMARC Production Efficiency Population yielded estimates of heritability of 0.04 and 0.13 for twinning rate, and 0.07 and 0.17 for ovulation rate with linear and Bayesian threshold models, respectively (Van Tassell et al., 1998).

Genetics of Puberty and Related Traits

Measurement of age at puberty is complicated by ambiguity in males, owing to the absence of a signal event, and the cost and effort required in females where the first oestrous cycle clearly signals attainment of puberty but requires assessment of animal behaviour and/or circulating progesterone levels. As a consequence, easily recorded correlated traits such as scrotal circumference at a year of age in males or age at first calving in females are often the subject of study. In the latter case, imposed management often constrains the biological variation observed, such as limiting the earliest age at which heifers are exposed to a bull when evaluating age at first calving.

In spite of these limitations, heritable variation in traits related to age at puberty has been observed. In a study utilizing pregnancy records from 11,487 Nelore heifers from three Brazilian farms, where heifers were exposed to bulls starting at 14 months of age, a heritability of 0.57 ± 0.01 for probability of pregnancy was reported (Eler *et al.*, 2002). Given the well-documented later age at puberty for *Bos taurus indicus* cattle (Nogueira, 2004), the initial exposure of heifers to bulls at 14 months of age in this study was likely not a constraint on biological variation. In contrast, the necessity of maintaining an annual calving season and initiating calving at 2 years of age in many production systems utilizing *Bos taurus taurus* germplasm likely limits the range of biological variation evaluated with consequences for estimates of heritability from field data. In an evaluation of reproductive records from 3144 purebred Angus heifers in the USA, heritability estimates of 0.13 ± 0.07 and 0.03 ± 0.03 were reported for pregnancy rate and first-service conception rate, respectively (Bormann *et al.*, 2006). While likelihood of conception and pregnancy is expected to increase as a heifer moves from pubertal to later oestrous cycles, typical management in these herds would likely lessen the opportunity to observe variation between animals (heifers not bred until ~14 months of age, oestrus synchronization methods employed in some cases that initiate cycling in peripubertal heifers). A similar estimate of heritability for heifer pregnancy, 0.17 ± 0.01, was reported using data from 37,802 animals of the Red Angus breed in the US (McAllister *et al.*, 2011). Likewise, in an analysis of 12 *B. taurus taurus* breeds, age at puberty, age at first calving, heifer pregnancy rate and heifer calving rate had estimated heritabilities of 0.16 ± 0.04, 0.08 ± 0.04, 0.14 ± 0.03 and 0.14 ± 0.03 (Martinez-Velazquez *et al.*, 2003).

A negative genetic correlation between scrotal circumference in males and age at puberty in females, together with a moderate heritability for scrotal circumference (Martinez-Velazquez *et al.*, 2003; Kealey *et al.*, 2006; Araujo Neto *et al.*, 2011; McAllister *et al.*, 2011; Santana *et al.*, 2012) has led to interest in the use of scrotal circumference as an indicator trait and a component of heifer pregnancy

expected progeny difference (EPD) calculation. However, because of a low genetic correlation and non-linear relationship between scrotal circumference and heifer pregnancy and moderate heritability estimate for heifer pregnancy, direct selection for heifer pregnancy has been suggested instead (Evans *et al.*, 1999). Estimates of genetic correlation between scrotal circumference and heifer pregnancy reported for *B. taurus taurus* cattle have not been significantly different from zero in most cases. Evans *et al.* (1999) reported a genetic correlation estimate of 0.002 ± 0.45 using heifer pregnancy data for 986 females from a large Hereford herd in New Mexico, USA. The genetic correlation between these two traits estimated using the more substantial Red Angus data mentioned above produced an estimate of 0.05 ± 0.09.

Efforts to identify the contribution of specific genomic regions to variation in age at puberty in females have been conducted in population studies with Brahman (Hawken *et al.*, 2012), Holstein (Daetwyler *et al.*, 2008) and indicine × taurine composite populations (Fortes *et al.*, 2012b; Hawken *et al.*, 2012) as well as an F2 Jersey × Limousin cross (Morris *et al.*, 2009). Actual age at puberty, as determined by presence of the first corpus luteum, has been recorded in some studies, while in others a correlated trait such as age at first service or first service conception rate has been used as a proxy. Correspondence between results across studies has been low (Table 12.1), which is not surprising given the genetic diversity between the populations examined and limitations in statistical power. One exception is the identification of a common region on chromosome 14 contributing to age at puberty in both Brahman and indicine × taurine composite populations (Fortes *et al.*, 2012b; Hawken *et al.*, 2012). Another limitation in making such comparisons, however, is the inadequacy of simply comparing SNP associations across studies.

Fortes *et al.* (2010) developed an approach based on a so-called association weight matrix to use results from a genome-wide association study (GWAS) of multiple related traits to identify gene networks and candidate genes. Age at puberty in cattle was used as the test case for this methodology (Fortes *et al.*, 2010) and was also examined in subsequent studies with the same approach (Fortes *et al.*, 2011, 2012b). Comparison of gene networks identified from the association weight matrix analysis of Brahman and indicine × taurine composite cattle found five significant transcription factor-associated networks in common between the populations (Fortes *et al.*, 2011). The power of this approach is that this commonality was identified but would have been overlooked by simple comparison of SNP associations with age at puberty between studies.

Genetics of Conception or Pregnancy Rate

Recent studies in which heritability of pregnancy rate or conception rate has been estimated have typically reported estimates in the low single digits (range 0.005 to 0.13) (Weigel and Rekaya, 2000; Oseni *et al.*, 2004; Bormann *et al.*, 2006; Kuhn *et al.*, 2006; Chang *et al.*, 2007; Aguilar *et al.*, 2011; Berry *et al.*, 2012; Tiezzi *et al.*, 2012). None the less, several studies have been conducted examining genomic associations with conception rate and related traits, most commonly in dairy cattle. In most cases the information utilized is field data collected through national dairy herd data recording schemes. Specific traits evaluated include daughter pregnancy rate (percentage of non-pregnant cows that become pregnant during each 21-day period after a voluntary waiting period following calving), days open (number of days from calving to conception), number of days from calving to first insemination, first service conception rate, non-return rate (proportion of cows not seen to come back into oestrus within a specified period after breeding), number of inseminations per conception, interval from first to last insemination and heifer pregnancy rate in beef cattle (proportion of heifers pregnant following the conclusion of the breeding season).

Using the information available from the Cattle QTL database (www.animalgenome.org/cgi-bin/QTLdb/BT/index) as well as that listed here (Table 12.1), clusters of genomic associations with conception rate and related traits can be found on chromosomes 1, 2, 4, 5, 7, 9, 10 and 20, suggesting greater confidence that gene(s) in those genomic regions contribute to variation in conception rate and related traits. Such strongly supported regions are the best targets for efforts aimed at progressing from SNP association or QTL mapping to identification of functional polymorphisms, however such efforts are challenging. Glick *et al.* (2011) used medium density (50 k) SNP genotype data in a concordance analysis of Holstein sires with known QTL genotype (based on daughter design analysis) to narrow the proximal QTL location on chromosome 7 from 27 cM to 270 kb. Subsequent positional candidate gene analysis within this region revealed a copy number variant (CNV) in strong association with fertility, but a definitive determination that this or some other polymorphism was causative was not achieved.

Another source of genetic variation in pregnancy rate, separate from variation in conception, is loss of pregnancy due to deleterious alleles that cause early embryonic lethality. One of the first examples of this type of genetic variant in cattle was the disorder deficiency of uridine monophosphate synthase or DUMPS (Shanks *et al.*, 1992). Embryos homozygous for deleterious mutation in the uridine monophosphate synthase gene causing DUMPS lack functional uridine monophosphate synthase and were not viable past approximately day 40 of gestation. Since the embryonic loss cannot be directly observed, the disorder manifested as cows (carriers) that appeared difficult to breed. Other examples of embryonic lethal alleles have been more recently identified including mutations in Fanconi anaemia, complementation group I (*FANCI*) (Charlier *et al.*, 2012), signal transducer and activator of transcription 5A (*STAT5A*) (Khatib *et al.*, 2008b, 2009b), apoptotic peptidase activating factor 1 (*APAF1*) (Adams *et al.*, 2012) and spliceosome-associated protein homologue (*CWC15*) (Sonstegard *et al.*, 2013) genes. Identification of the *APAF1* mutation in the US Holstein and the *CWC15* mutation in the US Jersey population followed comprehensive genome screening at the population level in which SNP haplotypes were identified that had an apparent absence of homozygotes, implying embryonic lethality (VanRaden *et al.*, 2011). In addition to these, one deleterious haplotype was identified in the US Brown

Table 12.1. Chromosomal regions most significantly associated with reproductive traits.

Trait	Type of study	Chromosome and location *gene*	Population	Citation
Daughter pregnancy rate	Interval mapping	1, 9 cM	US Holstein	Schnabel et al., 2005
Embryonic lethal	Haplotype analysis	1, 0–9.3 Mb, HH4 haplotype (*GART*, 24.9 Mb)	French Holstein	Fritz et al., 2013
Non return rate	Interval mapping	1, 78.9–132.5 Mb	French Prim'Holstein, Normande and Montbeliarde	Ben Jemaa et al., 2008
Conception rate	Interval mapping	1, 62 cM	French dairy breeds	Boichard et al., 2003
First service conception rate	GWAS and association weight matrix	1, 88.6 Mb (*ZMAT3*)	US Brangus	Fortes et al., 2012b
Number of inseminations	GWAS	1, 88.1 Mb	Finnish Ayrshire	Schulman et al., 2011
Embryonic lethal	Haplotype analysis	1, 92–97 Mb, HH2 haplotype	US Holstein	VanRaden et al., 2011
Daughter pregnancy rate	GWAS	1, 129–141 Mb	US Holstein	Cole et al., 2011
Interval from calving to first insemination	Interval mapping	1, 140.8 cM	Danish and Swedish Holstein	Hoglund et al., 2009a,b
Days open	Interval mapping	1, 146 cM	Finnish Ayrshire	Schulman et al., 2008
Days open	Interval mapping	2, 2 cM	Finnish Ayrshire	Schulman et al., 2008
Non return rate	Interval mapping	2, 3.9 cM	Danish and Swedish Holstein	Hoglund et al., 2009b
Non return rate	GWAS	2, 12.4 Mb	Finnish Ayrshire	Schulman et al., 2011
Sire conception rate	GWAS	2, 24.8 Mb	US Holstein	Penagaricano et al., 2012a
Commencement of luteal activity	GWAS	2, 130.1–134.6 Mb	Irish, British, Swedish and Dutch Holstein	Berry et al., 2012
Non return rate	Interval mapping, LDLA	3, 19 cM	French Holstein	Druet et al., 2008
Non return rate	Interval mapping	3, 26 cM	French Holstein	Guillaume et al., 2007
Daughter pregnancy rate	GWAS	3, 90 Mb	US Holstein	Cole et al., 2011
Stillbirth, direct	Interval mapping	3, BM7225–BM2924 (96.0–111.8 Mb)	Danish Holstein	Thomasen et al., 2008
Days from calving to first ovulation postpartum	GWAS	3, 112.3 Mb	Australian Brahman	Hawken et al., 2012
Gestation length	Interval mapping	4, 17 cM	Dutch Holstein	Schrooten et al., 2000
Non return rate	GWAS	4, 34.9 Mb	Finnish Ayrshire	Schulman et al., 2011
Interval from first to last insemination	Interval mapping	4, 43.2 cM	Danish and Swedish Holstein	Hoglund et al., 2009b
Maternal calving ease	Interval mapping	4, 74 cM	German Holstein	Seidenspinner et al., 2009
Scrotal circumference	Interval mapping	4, 105.7 Mb	US Angus	McClure et al., 2010

Continued

Table 12.1. Continued.

Trait	Type of study	Chromosome and location gene	Population	Citation
Twinning rate, ovulation rate	Interval mapping, association analysis	5, 46 cM	Composite, Bos taurus taurus	Kappes et al., 2000; Allan et al., 2009
Postpartum anoestrous interval	GWAS	5, 46.0 Mb	Bos taurus taurus x Bos taurus inducus composite	Hawken et al., 2012
First service conception rate	GWAS and association weight matrix	5, 56.7 Mb (STAT6)	US Brangus	Fortes et al., 2012b
Embryonic lethal	Haplotype analysis	5, 58–66 Mb, HH1 haplotype (APAF1, 63.2 Mb)	US Holstein	VanRaden et al., 2011; Adams et al., 2012
Twinning rate	Interval mapping	5, 66.6 Mb	Composite, Bos taurus taurus	Lien et al., 2000; Meuwissen et al., 2002
Twinning rate	Interval mapping	5, 71.6 Mb	US Holstein	Cruickshank et al., 2004
First service conception rate	GWAS and association weight matrix	5, 74.8 Mb (RFX4)	US Brangus	Fortes et al., 2012b
Retained placenta	GWAS	5, 89.6 Mb	Norwegian Red	Olsen et al., 2011
Age at first corpus luteum	GWAS	5, 96.3 Mb	Bos taurus taurus x Bos taurus inducus composite	Hawken et al., 2012; Fortes et al., 2010
Days open	Interval mapping	5, 108 cM	Finnish Ayrshire	Schulman et al., 2008
Sire conception rate	GWAS	5, 112.8 Mb	US Holstein	Penagaricano et al., 2012a
Number of inseminations per conception	GWAS	5, 116.3 Mb	Swedish Holstein	Sahana et al., 2010
Calving ease	Interval mapping	6, BM1329 BM143 (26–44 Mb)	Swedish Holstein	Holmberg and Andersson-Eklund, 2006
Stillbirth direct	Interval mapping, haplotype analysis	6, 38.2 Mb (NCAPG)	German Holstein	Kuhn et al., 2003; Eberlein et al., 2009
Stillbirth direct, calving ease direct	LDLA	6, 37.6–38.4 Mb	Norwegian Red	Olsen et al., 2010
Age at first service	Variance component linkage analysis	6, 59 cM, 68 cM and 100 cM	Canada Holstein	Daetwyler et al., 2008
Postpartum anoestrous interval	GWAS	6, 118.4 Mb	Australian Brahman	Hawken et al., 2012
Pregnancy rate	Interval mapping	6, 122 cM	US Holstein	Ashwell et al., 2004
Stillbirth, direct	Interval mapping	7, BMS2258–OARAE129 (64.2–96.6 Mb)	Danish Holstein	Thomasen et al., 2008

Stillbirth	Interval mapping	7, BM6105 (25.4 Mb)	Swedish Holstein	Holmberg and Andersson-Eklund, 2006
Ovulation rate	Interval mapping	7, 0.03 Mb	Composite, *Bos taurus taurus*	Blattman *et al.*, 1996
Number of inseminations to conception	Association analysis	7, 4.9 Mb	Israeli Holstein	Glick *et al.*, 2011
Number of inseminations to conception	Association analysis	7, 11 cM	Israeli Holstein	Weller *et al.*, 2008a
Gestation length	Interval mapping	7, 15 cM	US Holstein–Jersey × Holstein backcross	Maltecca *et al.*, 2009
Daughter pregnancy rate	GWAS	7, 15.4 Mb	US Holstein	Cole *et al.*, 2011
Sperm motility	Interval mapping	7, 25–71 cM	French Holstein	Druet *et al.*, 2009
Ovulation rate	Interval mapping	7, 39.6 Mb	Composite, *Bos taurus taurus*	Blattman *et al.*, 1996
Embryonic lethal	Haplotype analysis	7, 41–47 Mb, BH1 haplotype	US Brown Swiss	VanRaden *et al.*, 2011
Ovulation rate, hormone induced	Interval mapping	7, 66.9 Mb	Japanese Black	Sugimoto *et al.*, 2010
Twinning rate	Interval mapping	7, 96.7 Mb	Composite, *Bos taurus taurus*	Lien *et al.*, 2000
Number of inseminations per conception	Interval mapping	7, 111.6	Danish and Swedish Holstein	Hoglund *et al.*, 2009b
Maternal calving ease	Interval mapping	7, 126 cM	US Angus	McClure *et al.*, 2010
Interval from first to last insemination	GWAS	8, 21.5 Mb	Finnish Ayrshire	Schulman *et al.*, 2011
Embryonic lethal	Haplotype analysis	8, 90–95 Mb, HH3 haplotype	US Holstein	VanRaden *et al.*, 2011
Maternal calving ease	Interval mapping	8, 93 cM	German Holstein	Kuhn *et al.*, 2003
Twinning rate	Interval mapping	8, 104.2 Mb	US Holstein	Cobanoglu *et al.*, 2005
Commencement of luteal activity	GWAS	8, 108.0 Mb	Irish, British, Swedish and Dutch Holstein	Berry *et al.*, 2012
Maternal calving ease	Interval mapping	8, 116 cM	US Holstein	Ashwell *et al.*, 2005
Interval from first to last insemination	Interval mapping	9, 4.9 cM	Danish and Swedish Holstein	Hoglund *et al.*, 2009b
Maternal calving ease	Interval mapping	9, 50.0 cM	US Angus	McClure *et al.*, 2010
Non return rate, Number of inseminations	Interval mapping	9, 34 cM; 9, 40 cM	Swedish Holstein	Holmberg and Andersson-Eklund, 2006; Holmberg *et al.*, 2007
Retained placenta	GWAS	9, 52.6 Mb	Norwegian Red	Olsen *et al.*, 2011
Maternal calving ease	Interval mapping	9, 58 cM	US Angus	McClure *et al.*, 2010
Non return rate	GWAS	9, 63.1 Mb	Norwegian Red	Olsen *et al.*, 2011
Twinning rate	Interval mapping	10, 31.1 Mb	US Holstein	Cobanoglu *et al.*, 2005

Continued

Table 12.1. Continued.

Trait	Type of study	Chromosome and location gene	Population	Citation
Calving ease	Interval mapping	10, 74 cM	US Holstein	Schnabel et al., 2005
Maternal calving ease and stillbirth	Association analysis	10, 86.6 Mb (PGF)	German Holstein	Seidenspinner et al., 2011
Maternal calving ease	Interval mapping	10, 87 cM	German Holstein	Kuhn et al., 2003
Interval from first to last insemination	Interval mapping	10, 90.8 cM	Danish and Swedish Holstein	Hoglund et al., 2009b
Sperm motility	Interval mapping	11, 11–58 cM	French Holstein	Druet et al., 2009
Non return rate	Interval mapping	11, INRA177 (24.6 Mb)	Swedish Holstein	Holmberg and Andersson-Eklund, 2006
Stillbirth	Interval mapping	11, BMS7169 (38.0 Mb)	Swedish Holstein	Holmberg and Andersson-Eklund, 2006
Non return rate	GWAS	11, 39.6 Mb	Norwegian Red	Olsen et al., 2011
Number of inseminations	Interval mapping	11, ILSTS036 (46.7 Mb)	Swedish Holstein	Holmberg and Andersson-Eklund, 2006
Retained placenta	GWAS	11, 89.3 Mb	Norwegian Red	Olsen et al., 2011
Interval from calving to first insemination	Interval mapping	11, 92.5 cM	Danish and Swedish Holstein	Hoglund et al., 2009b
First service conception rate	GWAS and association weight matrix	11, 95.6 Mb (NR6A1)	US Brangus	Fortes et al., 2012b
Non return rate	GWAS	12, 6.3–8.5 Mb	Norwegian Red	Olsen et al., 2011
Twinning rate	Interval mapping	12, 7.3 Mb	Composite, Bos taurus taurus	Lien et al., 2000
Non return rate	GWAS	12, 11.9–18.1 Mb	Norwegian Red	Olsen et al., 2011
Interval from first to last insemination	GWAS	12, 24.3 Mb	Finnish Ayrshire	Schulman et al., 2011
Number of inseminations	GWAS	12, 28.7 Mb	Finnish Ayrshire	Schulman et al., 2011
Non return rate	Interval mapping	12, 40.6 cM	Danish and Swedish Holstein	Hoglund et al., 2009b
Interval from calving to first insemination	GWAS	13, 18.1 Mb	Finnish Ayrshire	Schulman et al., 2011
Calving ease	Interval mapping	13, BMS1352, 28.8 Mb	Swedish Holstein	Holmberg and Andersson-Eklund, 2006
Interval from calving to first insemination	GWAS	13, 33.5 Mb	Swedish Holstein	Sahana et al., 2010
Interval from calving to first insemination	Interval mapping	13, 89.7 cM	Danish and Swedish Holstein	Hoglund et al., 2009b

Trait	Method	Position	Population	Reference
Calving ease, direct	Association analysis	14, 2.7 Mb	German Holstein	Kaupe et al., 2007
Pregnancy rate	Interval mapping	14, 11 cM	US Holstein	Ashwell et al., 2004
Gestation length	Interval mapping	14, 20 cM	US Holstein–Jersey × Holstein backcross	Maltecca et al., 2009
Calving ease, direct	Association analysis	14, 24.1 Mb	German Fleckvieh	Pausch et al., 2011
Age at first corpus luteum	GWAS	14, 22–25 Mb	Australian Brahman	Hawken et al., 2012
Scrotal circumference	GWAS	14, 21–28 Mb	Australian Brahman	Fortes et al., 2012a
First service conception rate	GWAS and association weight matrix	14, 25.0 Mb (PLAG1)	US Brangus	Fortes et al., 2012b
Postpartum anoestrous interval	GWAS	14, 26–28 Mb	Australian Brahman	Hawken et al., 2012
Twinning rate	Interval mapping	14, 51.3 Mb	US Holstein	Cobanoglu et al., 2005
Daughter pregnancy rate	Interval mapping	14, 60 cM	US Holstein	Schnabel et al., 2005
Ovulation rate	Interval mapping	14, 61.1 Mb	Composite, Bos taurus taurus	Gonda et al., 2004
Age at first service	Variance component linkage analysis	14, 3 cM and 62 cM	Canada Holstein	Daetwyler et al., 2008
Embryonic lethal	Haplotype analysis	15, 13–18 Mb, JH1 haplotype, (CWC15, 15.7 Mb)	US Jersey	VanRaden et al., 2011; Sonstegard et al., 2013
Sperm volume	Interval mapping	15, 14–45 cM	French Holstein	Druet et al., 2009
Non return rate	Interval mapping	15, 24.2 Mb	Swedish Holstein	Holmberg and Andersson-Eklund, 2006
Age at first corpus luteum	GWAS	15, 31–38 Mb	Australian Brahman	Hawken et al., 2012
Stillbirth	Interval mapping	15, 53 cM	German Holstein	Seidenspinner et al., 2009
Daughter stillbirth	GWAS	15, 75.5 Mb	US Holstein	Cole et al., 2011
Stillbirth	Interval mapping	16, BM1706–DIK4982 (16.0–69.2 Mb)	Danish Jersey	Mai et al., 2010
Age at first corpus luteum	GWAS and association weight matrix	16, 21.2 Mb (ESRRG)	Bos taurus taurus × Bos taurus indicus composite	Fortes et al., 2010
Postpartum anoestrous interval	GWAS	16, 40–45 Mb	Bos taurus taurus × Bos taurus indicus composite	Hawken et al., 2012
Calving ease	Interval mapping	16, 63 cM	US Holstein	Schnabel et al., 2005
Pregnancy rate	Interval mapping	16, 81 cM	US Holstein	Ashwell et al., 2004
Maternal calving ease	Interval mapping	17, 69 cM	US Holstein	Ashwell et al., 2005
Scrotal circumference	Interval mapping	17, 70.4 Mb	US Angus	McClure et al., 2010
Pregnancy rate	Interval mapping	18, 14 cM	US Holstein	Ashwell et al., 2004
Daughter calving ease	GWAS	18, 15.8 Mb	US Holstein	Cole et al., 2011
Age at first oestrus	Interval mapping	18, 21 cM	Jersey × Limousin F_2	Morris et al., 2009

Continued

Table 12.1. Continued.

Trait	Type of study	Chromosome and location gene	Population	Citation
Non return rate	Interval mapping	18, BMS2639. 43.7 Mb	Swedish Holstein	Holmberg and Andersson-Eklund, 2006
Maternal stillbirth and direct calving ease	LDLA	18, 53.9 Mb	German Holstein	Brand et al., 2010
Pregnancy rate	Interval mapping	18, 54 cM	US Holstein	Ashwell et al., 2004; Muncie et al., 2006
Sire conception rate	GWAS	18, 55.0 Mb	US Holstein	Penagaricano et al., 2012a
Calving ease	Interval mapping	18, BMS2785, 56.8 Mb	Swedish Holstein	Holmberg and Andersson-Eklund, 2006
Gestation length, direct	GWAS	18, 57.1 Mb	US Holstein, Italian Brown Swiss	Maltecca et al., 2011
Direct and maternal calving ease	GWAS	18, 57.1 Mb	US Holstein	Cole et al., 2009
Direct calving ease	GWAS	18, 57.1 Mb	Danish Holstein	Hoglund et al., 2012
Pregnancy with first 42 days of mating	GWAS, haplotype analysis	18, 63 Mb	Australian Holstein and Jersey	Pryce et al., 2010
Calving ease, direct	Interval mapping	18, BM6507–TGLA227, 63.1–65.4 Mb	Danish Holstein	Thomasen et al., 2008
Stillbirth	Interval mapping	18, 75 cM	German Holstein	Kuhn et al., 2003
Non return rate	Interval mapping	18, 111 cM	German Holstein	Kuhn et al., 2003
Embryonic lethal	Haplotype analysis	19, 21.6–35.4 Mb, MH1 haplotype (*SHBG*, 28.0 Mb)	French Montbeliarde	Fritz et al., 2013
Ovulation rate	Interval mapping	19, 42.4 Mb	Composite, *Bos taurus taurus*	Arias and Kirkpatrick, 2004
Interval calving to first insemination	GWAS	20, 4.9 Mb	Finnish Ayrshire	Schulman et al., 2011
Non return rate	Interval mapping	20, BMS1282, 10.8 Mb	Swedish Holstein	Holmberg and Andersson-Eklund, 2006
Percentage of sperm with abnormal cytoplasmic droplet	Interval mapping	20, 40–78 cM	French Holstein	Druet et al., 2009
Calving ease, direct	Association analysis	21, 2.2 Mb	German Fleckvieh	Pausch et al., 2011
Percentage of sperm with abnormal cytoplasmic droplet	Interval mapping	21, 38–49 cM	French Holstein	Druet et al., 2009

Trait	Method	Position	Breed	Reference
Commencement of luteal activity	GWAS	21, 93.8–94.1 Mb	Irish, British, Swedish and Dutch Holstein	Berry et al., 2012
Sperm volume	Interval mapping	22, 0–28 cM	French Holstein	Druet et al., 2009
Non return rate	Interval mapping	22, 43.5 cM	Danish and Swedish Holstein	Hoglund et al., 2009b
Scrotal circumference	Interval mapping	22, 61.1 Mb	US Angus	McClure et al., 2010
Daughter stillbirth	GWAS	23, 0.2 Mb	US Holstein	Cole et al., 2011
Percentage of live sperm after thaw	Interval mapping	23, 8–48 cM	French Holstein	Druet et al., 2009
Percentage of live sperm after osmotic stress	Interval mapping	23, 8–35 cM	French Holstein	Druet et al., 2009
Twinning rate	Interval mapping	23, 25.5 Mb	Composite, Bos taurus taurus	Lien et al., 2000
Female fertility	Interval mapping	23, MB026–MB019, 26.2–27.8 Mb	Danish Red	Mai et al., 2010
Ovulation rate	Interval mapping	23, 27.2 Mb	Composite, Bos taurus taurus	Blattman et al., 1996
Stillbirth	Interval mapping	23, 59 cM	German Holstein	Seidenspinner et al., 2009
Number of inseminations per conception	Interval mapping	24, 6.2 cM	Danish and Swedish Holstein	Hoglund et al., 2009b
Interval from calving to first insemination	GWAS	24, 25.5–26.0 Mb	Finnish Ayrshire	Schulman et al., 2011
Interval from calving to first insemination	Interval mapping	24, 30.4 cM	Danish and Swedish Holstein	Hoglund et al., 2009b
Sire conception rate	GWAS	25, 0.9–4.3 Mb	US Holstein	Penagaricano et al., 2012a
Ability to recycle after calving	GWAS	25, 1.1–1.4 Mb	International Brown Swiss	Guo et al., 2012
Non return rate	Interval mapping	25, 17.3 cM	Danish and Swedish Holstein	Hoglund et al., 2009b
Sperm path velocity	Interval mapping	25, 24–63 cM	French Holstein	Druet et al., 2009
Calving ease, direct	Interval mapping	25, BMS1353–AF5, 31.4–40.3 Mb	Danish Holstein	Thomasen et al., 2008
Days open	Interval mapping	25, 47 cM	Finnish Ayrshire	Schulman et al., 2008
Scrotal circumference	Interval mapping	25, 60.6 Mb	US Angus	McClure et al., 2010
Stillbirth, maternal	Interval mapping	26, BM804–BM7237, 45.5–47.4 Mb	Danish Holstein	Thomasen et al., 2008
Non return rate	Interval mapping	26, 45.7 cM	Danish and Swedish Holstein	Hoglund et al., 2009b
Interval from first to last insemination	Interval mapping	26, 53.7 cM	Danish and Swedish Holstein	Hoglund et al., 2009b
Non return rate	GWAS	27, 6.1–15.1 Mb	Finnish Ayrshire	Schulman et al., 2011
Interval from first to last insemination, number of inseminations	GWAS	27, 21.6 Mb	Finnish Ayrshire	Schulman et al., 2011

Continued

Table 12.1. Continued.

Trait	Type of study	Chromosome and location gene	Population	Citation
Maternal calving ease	Interval mapping	27, 36 cM	US Holstein	Ashwell et al., 2005
Sperm motility	Interval mapping	27, 46–58 cM	French Holstein	Druet et al., 2009
Percentage of live sperm after thaw	Interval mapping	27, 49–68 cM	French Holstein	Druet et al., 2009
Pregnancy rate	Interval mapping	27, 62 cM	US Holstein	Ashwell et al., 2004
Scrotal circumference	Interval mapping	28, 22.8 Mb	US Angus	McClure et al., 2010
Pregnancy rate	Interval mapping	28, 48 cM	US Holstein	Ashwell et al., 2004
Sire conception rate	GWAS	29, 14.3 Mb	US Holstein	Penagaricano et al., 2012a
Embryonic lethal	Haplotype analysis	29, 21.9–35.1 Mb, MH2 haplotype (SLC37A2, 28.9 Mb)	French Montbeliarde	Fritz et al., 2013
Non return rate	Interval mapping	29, BMS1600–ILSTS81	Swedish Holstein	Holmberg and Andersson-Eklund, 2006
Daughter pregnancy rate	GWAS	X, 0.4 Mb (ATP1B4)	US Holstein	Cole et al., 2011
Non return rate	Interval mapping	X, 5 cM	German Holstein	Kuhn et al., 2003
Maternal calving ease	Interval mapping	X, 7 cM	German Holstein	Kuhn et al., 2003
Daughter pregnancy rate	GWAS	X, 7.3 Mb (GRIA3)	US Holstein	Cole et al., 2011
Scrotal circumference	GWAS	X, 62–92 Mb	Australian Brahman	Fortes et al., 2012a
Daughter pregnancy rate	GWAS	X, 106 Mb	US Holstein	Cole et al., 2011
Fertility	GWAS	Y	US Bos taurus taurus composite, Bos taurus taurus × Bos taurus indicus composite	McDaneld et al., 2012

Swiss population and two additional deleterious haplotypes were identified in the US Holstein population. Subsequently, a similar analysis was performed in three French dairy breeds leading to the identification of 34 candidate haplotypes which showed a deficit relative to expectation for homozygote frequency (Fritz et al., 2013). Some of the identified haplotypes corresponded to previously identified loci (VanRaden et al., 2011; Adams et al., 2012; Charlier et al., 2012), while three unique candidate mutations associated with embryonic lethality were identified in the genes GART, SHBG and SLC37A2.

While most of the studies mentioned above have considered phenotypes at the whole animal level (cow pregnant vs non-pregnant), researchers have in some cases considered fertility at a more basic level using in vitro systems to study genetic contributions to variation in fertilization and embryo survival. This approach has identified potential associations of either fertilization rate or embryo survival with polymorphisms in signal transducer and activator of transcription 1 and 3 (STAT1, STAT3) (Khatib et al., 2009a), the fibroblast growth factor 2 (FGF2) (Khatib et al., 2008a) and STAT5A (Khatib et al., 2008b, 2009b) in US Holstein cattle. It is noteworthy that the genomic location of STAT3 and STAT5A is internal to the genomic region of a deficit haplotype (HH14) identified in the French study examining potential embryonic lethals (Fritz et al., 2013). In addition to these a priori candidate gene studies, the same in vitro model system has been used in GWAS with selectively pooled DNA samples (Huang et al., 2010), identifying significant associations on chromosomes 2, 8, 18 and 22.

Perhaps one of the most interesting results recently reported from a GWAS of cattle fertility was the association of a chromosome, as opposed to an individual marker, with female fertility. McDaneld et al. (2012) used selective, pooled genotyping by high density (800 k) beadchip of pregnant and non-pregnant beef cows. A surprising result of their study was the identification of Y chromosomal DNA in a considerable proportion of cows that failed to get pregnant. The proportion of such cows with Y chromosomal DNA was far in excess of that which might be attributed to accidental inclusion of freemartin females in the non-pregnant samples. Y-specific SNPs with significant association with fertility were distributed across the chromosome, though only limited segments of the Y chromosome were present in any given non-pregnant female. Whether the presence of the Y chromosomal DNA in low fertility females occurs because of an aberrant meiotic recombination event or some other mechanism is currently unknown.

Genetics of Gestation Length

Gestation length in cattle is well known to vary by breed (Gregory et al., 1979; McElhenney et al., 1985; Cundiff et al., 1998; Norman et al., 2009; Casas et al., 2011) and estimates of heritability have generally been moderate for direct effects, ranging from 0.04 to 0.77, and low for maternal effects, ranging from 0.01 to 0.10 (Bourdon and Brinks, 1982; MacNeil et al., 1984; Cundiff et al., 1986; Wray et al., 1987; Silva et al., 1992; Gregory et al., 1995; Crews, 2006; Mujibi and Crews, 2009; Norman et al., 2009; Yague et al., 2009; Cervantes et al., 2010, Eghbalsaied, 2011; Johanson et al., 2011; Maltecca et al., 2011). While a moderate heritability for gestation length suggests the opportunity to change trait levels through selection, analyses in dairy cattle suggest intermediate trait levels may be optimal (Norman et al., 2011). Commercially, sires with direct genetic effects causing reduced gestation length are sometimes used selectively on late-calving cows in annual breeding systems to maintain shorter calving seasons and annual calving intervals (Livestock Improvement Corporation, 2008). The even shorter gestation length of the yak (Bos brunniens) species, shorter than cattle by 25 days, has led to the consideration of their use for this purpose (Livestock Improvement Corporation, 2007).

QTL or SNP associations for gestation length (Table 12.1) have been reported in three studies (Schrooten et al., 2000; Maltecca et al., 2009, 2011). The first two studies employed microsatellite genotyping and interval mapping to identify QTLs on chromosomes 4, 7, 14, 15, 26 and 29. A subsequent GWAS in US Holstein and Italian Brown Swiss provided confirmatory SNP associations for the QTL on chromosome 7 and pointed to additional chromosomal regions with significant effect. The largest among these was a SNP on chromosome 18 that has been strongly associated in other studies with the calving complex, meaning a group of calving-related traits including gestation length, calf weight and dystocia (Cole et al., 2009; Hoglund et al., 2012).

Genetics of Dystocia and Related Traits

Factors affecting dystocia (difficult birth), both genetic and non-genetic have been the topic of multiple review articles (Hickson et al., 2006; Mee, 2008; Zaborski et al., 2009). Feto-pelvic disproportion is considered the primary cause of dystocia in primiparous females with fetal malpresentation a more common cause in multiparous females. Breed differences in both direct (calf) and maternal (cow, apart from maternal contribution to calf genotype) contributions to dystocia are well documented and have been reviewed in the previous version of this chapter (Kirkpatrick, 1999). In the previous review, reported heritability estimates for direct and maternal contributions to dystocia ranged from 0.03–0.42 and 0.03–0.47, respectively; more recent heritability estimates fall in the same range (Varona et al., 1999; Carnier et al., 2000; Bennett and Gregory, 2001; Gutierrez et al., 2007; Eaglen and Bijma, 2009; Mujibi and Crews, 2009; Cervantes et al., 2010; Tarres et al., 2010).

Efforts to map QTLs or identify SNP associations with calving ease have identified genomic contributions on chromosome 4, 6, 7, 8, 9, 10, 13, 14, 16, 17, 18, 21, 25, 27 and X (Table 12.1). Of these associations, the most commonly repeated result is that for SNP rs109478645 located on chromosome 18, or the genomic region in close proximity to it, which has been associated with calving ease (Holmberg and Andersson-Eklund, 2006; Cole et al., 2009; Hoglund et al., 2012) or gestation length (Maltecca et al., 2011) in multiple studies. rs109478645 is a base substitution in an intron of the sialic acid binding IG-like lectin 5 (SIGLEC5) gene, and other members of the gene family are known to be placentally expressed and to bind the hormone leptin.

Cole et al. (2009) have suggested the possibility that the unknown causative mutation may affect expression of SIGLEC5, with consequent effects on timing of parturition, calf size and dystocia. Another result that appears to correspond across studies is an association with calving ease and/or stillbirth in the ~38 Mb region of chromosome 6 (Kuhn et al., 2003; Holmberg and Andersson-Eklund, 2006; Eberlein et al., 2009; Olsen et al., 2010). Eberlein et al. (2009) have recently reported both functional and genetic evidence that implicates a mis-sense mutation in the non-SMC condensin I complex subunit G (NCAPG) gene as the polymorphism underlying this QTL/genomic association. The NCAPG product is a subunit of condensin I, a protein complex involved in mitotic chromosome condensation.

Stillbirth has been variously defined as a calf born dead or dying within 24 (Philipsson, 1979) or 48 h (Meyer et al., 2001) after birth. Given a definition including calf death up to 24–48 h following parturition, stillbirth and dystocia are intrinsically linked. The genetic correlation between the traits are positive and large having been recently estimated as 0.67 and 0.45 for direct and maternal components in Holstein cattle (Johanson et al., 2011). Estimates of heritability for stillbirth range from 0.01 to 0.12 for direct, and 0.02 to 0.13 for maternal effects, respectively, in recent reports (Meyer et al., 2001; Steinbock et al., 2003; Hansen et al., 2004). QTL or SNP associations for stillbirth have been reported on chromosomes 3, 6, 7, 10, 11, 15, 16, 18, 23 and 26. The positional candidate genes (SIGLEC5, NCAPG) mentioned in the preceding paragraph for dystocia are likewise candidates for stillbirth given the strong genetic correlation between these traits and the correspondence in QTL and GWAS results for these two traits.

Genetics of Male Reproductive Traits

Studies of the genetics of male reproduction are largely limited to analyses of scrotal circumference, for its potential value as an indicator trait for age at puberty, and direct sire contributions to conception rate. As discussed above, scrotal circumference is a moderately heritable trait, leading to the expectation that genomic associations with this trait should be identifiable. Examination of sires from the US Angus population identified five loci on chromosomes 4, 17, 22, 25 and 28 that exceeded thresholds for genome-wise significance, with additional loci on other chromosomes exceeding suggestive chromosome-wise significance levels (McClure et al., 2010). GWAS for testicular and related hormonal traits in Australian Brahman cattle identified significant associations for both scrotal circumference at a year of age and age at a scrotal circumference of 26 cm for SNPs on chromosomes 14 and X. Interestingly, the SNP associations on chromosome 14 were coincident in location with SNP associations for serum levels of insulin-like growth factor 1 (IGF1) at 6 months of age (Fortes et al., 2012a). While these GWAS results with Brahman cattle suggest a correlation between IGF1 and scrotal circumference, no direct association of the IGF1 gene was suggested. In contrast, a candidate gene study in Angus cattle did identify an IGF1 SNP with significant association with scrotal circumference (Liron et al., 2012).

Regarding direct sire contributions to conception rate, one definition of the latter is the proportion of successful matings in artificial insemination when viable sperm number is not a limiting factor, termed non-compensatory fertility. This phenotype was used in a preliminary GWAS with selectively chosen US Holstein sires (10 high and 10 low fertility sires) and a 10k SNP chip (Feugang et al., 2009), and a subsequent analysis with a more comprehensive set of sires and 50k genotyping (Blaschek et al., 2011). Results from the latter did not support findings of the preliminary study but did identify associations on chromosomes 6, 8, 10, 12, 13, 22 and X. A more recent GWAS examined a closely related trait, sire conception rate (Kuhn and Hutchison, 2008), using US Holstein sires, and found the strongest evidence for SNP associations on chromosomes 2, 5, 18, 25 and 29 (Penagaricano et al., 2012a). A subsequent gene set enrichment analysis using the same data found evidence of overrepresentation for 20 of 662 gene ontology categories, suggesting several candidate gene pathways (Penagaricano et al., 2012b). Other studies have examined

sperm characteristics directly rather than fertility per se. In a study examining French Holstein cattle, multiple sperm characteristics were evaluated including sperm motility, morphology and concentration. Heritabilities ranged from moderate to high for most traits, and several QTLs were identified in a genome-wide search utilizing microsatellite marker genotypes in an interval mapping analysis (Druet et al., 2009).

Candidate gene analyses of sire conception rate have been conducted as a follow-up to the in vitro studies of fertilization rate and embryo survival mentioned earlier. Significant associations between STAT5A and FGF2 polymorphisms and sire conception rate for US Holstein sires confirmed associations initially observed with in vitro analysis of Holstein embryos (Khatib et al., 2010).

Conclusions

With the development of tools for rapid, higher density genotyping in cattle there has been a proliferation of genomic studies aimed at dissecting the genetic basis for reproductive variation within cattle. While these tools are already being beneficially used for genomic selection to accelerate genetic improvement of reproduction in cattle, there remains a need and desire to advance these results from genomic association to an understanding of the specific genes and polymorphisms that underlie genetic variation at the population level. This objective has been achieved in relatively few cases to date; however, opportunities exist to advance that agenda. One would be the application of meta-analysis to results from studies examining similar traits in independent populations, an approach that has been beneficial in human studies for more clearly resolving and identifying additional relevant loci (Jostins et al., 2012; Theodoratou et al., 2012). Another opportunity is applying genomic information to gene pathway analysis to identify relevant groups of interacting genes and likely candidate genes (Wang et al., 2007; Luo et al., 2010). Additional benefit will likely be derived by applying this approach to combined analysis of genomic data from closely related traits (Fortes et al., 2010).

References

Adams, H.A., Sonstegard, T., Vanraden, P.M., Null, D.J., Van Tassell, C.P. and Lewin, H. (2012) Identification of a nonsense mutation in APAF1 that is causal for a decrease in reproductive efficiency in dairy cattle. *Plant and Animal Genome Meeting*, Poster P0555, 14–18 January 2012, San Diego, California.

Aguilar, I., Misztal, I., Tsuruta, S., Wiggans, G.R. and Lawlor, T.J. (2011) Multiple trait genomic evaluation of conception rate in Holsteins. *Journal of Dairy Science* 94, 2621–2624.

Allan, M.F., Kuehn, L.A., Cushman, R.A., Snelling, W.M., Echternkamp, S.E. and Thallman, R.M. (2009) Confirmation of quantitative trait loci using a low-density single nucleotide polymorphism map for twinning and ovulation rate on bovine chromosome 5. *Journal of Animal Science* 87, 46–56.

Araujo Neto, F.R., Lobo, R.B., Mota, M.D. and Oliveira, H.N. (2011) Genetic parameter estimates and response to selection for weight and testicular traits in Nelore cattle. *Genetics and Molecular Research* 10, 3127–3140.

Arias, J. and Kirkpatrick, B. (2004) Mapping of bovine ovulation rate QTL; an analytical approach for three generation pedigrees. *Animal Genetics* 35, 7–13.

Ashwell, M.S., Heyen, D.W., Sonstegard, T.S., Van Tassell, C.P., Da, Y., Vanraden, P.M., Ron, M., Weller, J.I. and Lewin, H.A. (2004) Detection of quantitative trait loci affecting milk production, health, and reproductive traits in Holstein cattle. *Journal of Dairy Science* 87, 468–475.

Ashwell, M.S., Heyen, D.W., Weller, J.I., Ron, M., Sonstegard, T.S., Van Tassell, C.P. and Lewin, H.A. (2005) Detection of quantitative trait loci influencing conformation traits and calving ease in Holstein-Friesian cattle. *Journal of Dairy Science* 88, 4111–4119.

Ben Jemaa, S., Fritz, S., Guillaume, F., Druet, T., Denis, C., Eggen, A. and Gautier, M. (2008) Detection of quantitative trait loci affecting non-return rate in French dairy cattle. *Journal of Animal Breeding and Genetics* 125, 280–288.

Bennett, G.L. and Gregory, K.E. (2001) Genetic (co)variances for calving difficulty score in composite and parental populations of beef cattle: II. Reproductive, skeletal, and carcass traits. *Journal of Animal Science* 79, 52–59.

Berry, D.P., Bastiaansen, J.W., Veerkamp, R.F., Wijga, S., Wall, E., Berglund, B. and Calus, M.P. (2012) Genome-wide associations for fertility traits in Holstein-Friesian dairy cows using data from experimental research herds in four European countries. *Animal* 6, 1206–1215.

Bierman, C.D., Kim, E., Shi, X.W., Weigel, K., Jeffrey Berger, P. and Kirkpatrick, B.W. (2010a) Validation of whole genome linkage-linkage disequilibrium and association results, and identification of markers to predict genetic merit for twinning. *Animal Genetics* 41, 406–416.

Bierman, C.D., Kim, E., Weigel, K., Berger, P.J. and Kirkpatrick, B.W. (2010b) Fine-mapping quantitative trait loci for twinning rate on *Bos taurus* chromosome 14 in North American Holsteins. *Journal of Animal Science* 88, 2556–2564.

Blaschek, M., Kaya, A., Zwald, N., Memili, E. and Kirkpatrick, B.W. (2011) A whole-genome association analysis of noncompensatory fertility in Holstein bulls. *Journal of Dairy Science* 94, 4695–4699.

Blattman, A.N., Kirkpatrick, B.W. and Gregory, K.E. (1996) A search for quantitative trait loci for ovulation rate in cattle. *Animal Genetics* 27, 157–162.

Boichard, D., Grohs, C., Bourgeois, F., Cerqueira, F., Faugeras, R., Neau, A., Rupp, R., Amigues, Y., Boscher, M.Y. and Leveziel, H. (2003) Detection of genes influencing economic traits in three French dairy cattle breeds. *Genetics Selection Evolution* 35, 77–101.

Bormann, J.M., Totir, L.R., Kachman, S.D., Fernando, R.L. and Wilson, D.E. (2006) Pregnancy rate and first-service conception rate in Angus heifers. *Journal of Animal Science* 84, 2022–2025.

Bourdon, R.M. and Brinks, J.S. (1982) Genetic, environmental and phenotypic relationships among gestation length, birth weight, growth traits and age at first calving in beef cattle. *Journal of Animal Science* 55, 543–553.

Brand, B., Baes, C., Mayer, M., Reinsch, N., Seidenspinner, T., Thaller, G. and Kuhn, C. (2010) Quantitative trait loci mapping of calving and conformation traits on *Bos taurus* autosome 18 in the German Holstein population. *Journal of Dairy Science* 93, 1205–1215.

Cady, R.A. and Van Vleck, L.D. (1978) Factors affecting twinning and effects of twinning in Holstein dairy cattle. *Journal of Animal Science* 46, 950, 950–956.

Carnier, P., Albera, A., Dal Zotto, R., Groen, A.F., Bona, M. and Bittante, G. (2000) Genetic parameters for direct and maternal calving ability over parities in Piedmontese cattle. *Journal of Animal Science* 78, 2532–2539.

Casas, E., Thallman, R.M. and Cundiff, L.V. (2011) Birth and weaning traits in crossbred cattle from Hereford, Angus, Brahman, Boran, Tuli, and Belgian Blue sires. *Journal of Animal Science* 89, 979–987.

Cervantes, I., Gutierrez, J.P., Fernandez, I. and Goyache, F. (2010) Genetic relationships among calving ease, gestation length, and calf survival to weaning in the Asturiana de los Valles beef cattle breed. *Journal of Animal Science* 88, 96–101.

Chang, Y.M., Gonzalez-Recio, O., Weigel, K.A. and Fricke, P.M. (2007) Genetic analysis of the twenty-one-day pregnancy rate in US Holsteins using an ordinal censored threshold model with unknown voluntary waiting period. *Journal of Dairy Science* 90, 1987–1997.

Charlier, C., Agerholm, J.S., Coppieters, W., Karlskov-Mortensen, P., Li, W., De Jong, G., Fasquelle, C., Karim, L., Cirera, S., Cambisano, N.*et al.* (2012) A deletion in the bovine FANCI gene compromises fertility by causing fetal death and brachyspina. *PLoS One* 7, e43085.

Cobanoglu, O., Berger, P.J. and Kirkpatrick, B. W. (2005) Genome screen for twinning rate QTL in four North American Holstein families. *Animal Genetics* 36, 303–308.

Cole, J.B., Vanraden, P.M., OConnell, J.R., Van Tassell, C.P., Sonstegard, T.S., Schnabel, R.D., Taylor, J.F. and Wiggans, G.R. (2009) Distribution and location of genetic effects for dairy traits. *Journal of Dairy Science* 92, 2931–2946.

Cole, J.B., Wiggans, G.R., Ma, L., Sonstegard, T.S., Lawlor, T.J., Crooker, B.A., Van Tassell, C.P., Yang, J., Wang, S., Matukumalli, L.K. and Da, Y. (2011) Genome-wide association analysis of thirty one production, health, reproduction and body conformation traits in contemporary U.S. Holstein cows. *BMC Genomics* 12, 408.

Cory, A.T., Price, C.A., Lefebvre, R. and Palin, M.F. (2013) Identification of single nucleotide polymorphisms in the bovine follicle-stimulating hormone receptor and effects of genotypes on superovulatory response traits. *Animal Genetics* 44, 197–201.

Crews, D.H., Jr (2006) Age of dam and sex of calf adjustments and genetic parameters for gestation length in Charolais cattle. *Journal of Animal Science* 84, 25–31.

Cruickshank, J., Dentine, M.R., Berger, P.J. and Kirkpatrick, B.W. (2004) Evidence for quantitative trait loci affecting twinning rate in North American Holstein cattle. *Animal Genetics* 35, 206–212.

Cundiff, L.V., MacNeil, M.D., Gregory, K.E. and Koch, R.M. (1986) Between- and within-breed genetic analysis of calving traits and survival to weaning in beef cattle. *Journal of Animal Science* 63, 27–33.

Cundiff, L.V., Gregory, K.E. and Koch, R.M. (1998) Germplasm evaluation in beef cattle – cycle IV: birth and weaning traits. *Journal of Animal Science* 76, 2528–2535.

Daetwyler, H.D., Schenkel, F.S., Sargolzaei, M. and Robinson, J.A. (2008) A genome scan to detect quantitative trait loci for economically important traits in Holstein cattle using two methods and a dense single nucleotide polymorphism map. *Journal of Dairy Science* 91, 3225–3236.

Del Rio, N.S., Kirkpatrick, B.W. and Fricke, P.M. (2006) Observed frequency of monozygotic twinning in Holstein dairy cattle. *Theriogenology* 66, 1292–1299.

Drouilhet, L., Lecerf, F., Bodin, L., Fabre, S. and Mulsant, P. (2009) Fine mapping of the FecL locus influencing prolificacy in Lacaune sheep. *Animal Genetics* 40, 804–812.

Druet, T., Fritz, S., Boussaha, M., Ben-Jemaa, S., Guillaume, F., Derbala, D., Zelenika, D., Lechner, D., Charon, C., Boichard, D., Gut, I.G., Eggen, A. and Gautier, M. (2008) Fine mapping of quantitative trait loci affecting female fertility in dairy cattle on BTA03 using a dense single-nucleotide polymorphism map. *Genetics* 178, 2227–2235.

Druet, T., Fritz, S., Sellem, E., Basso, B., Gerard, O., Salas-Cortes, L., Humblot, P., Druart, X. and Eggen, A. (2009) Estimation of genetic parameters and genome scan for 15 semen characteristics traits of Holstein bulls. *Journal of Animal Breeding and Genetics* 126, 269–277.

Eaglen, S.A. and Bijma, P. (2009) Genetic parameters of direct and maternal effects for calving ease in Dutch Holstein-Friesian cattle. *Journal of Dairy Science* 92, 2229–2237.

Eberlein, A., Takasuga, A., Setoguchi, K., Pfuhl, R., Flisikowski, K., Fries, R., Klopp, N., Furbass, R., Weikard, R. and Kuhn, C. (2009) Dissection of genetic factors modulating fetal growth in cattle indicates a substantial role of the non-SMC condensin I complex, subunit G (NCAPG) gene. *Genetics* 183, 951–964.

Eghbalsaied, S. (2011) Estimation of genetic parameters for 13 female fertility indices in Holstein dairy cows. *Tropical Animal Health and Production* 43, 811–816.

Eler, J.P., Silva, J.A., Ferraz, J.B., Dias, F., Oliveira, H.N., Evans, J.L. and Golden, B.L. (2002) Genetic evaluation of the probability of pregnancy at 14 months for Nellore heifers. *Journal of Animal Science* 80, 951–954.

Evans, J.L., Golden, B.L., Bourdon, R.M. and Long, K.L. (1999) Additive genetic relationships between heifer pregnancy and scrotal circumference in Hereford cattle. *Journal of Animal Science* 77, 2621–2628.

Feugang, J.M., Kaya, A., Page, G.P., Chen, L., Mehta, T., Hirani, K., Nazareth, L., Topper, E., Gibbs, R. and Memili, E. (2009) Two-stage genome-wide association study identifies integrin beta 5 as having potential role in bull fertility. *BMC Genomics* 10, 176.

Fortes, M.R., Reverter, A., Zhang, Y., Collis, E., Nagaraj, S.H., Jonsson, N.N., Prayaga, K.C., Barris, W. and Hawken, R.J. (2010) Association weight matrix for the genetic dissection of puberty in beef cattle. *Proceedings of the National Academy of Sciences USA* 107, 13642–13647.

Fortes, M.R., Reverter, A., Nagaraj, S.H., Zhang, Y., Jonsson, N.N., Barris, W., Lehnert, S., Boe-Hansen, G.B. and Hawken, R.J. (2011) A single nucleotide polymorphism-derived regulatory gene network underlying puberty in 2 tropical breeds of beef cattle. *Journal of Animal Science* 89, 1669–1683.

Fortes, M.R., Reverter, A., Hawken, R.J., Bolormaa, S. and Lehnert, S.A. (2012a) Candidate genes associated with testicular development, sperm quality, and hormone levels of inhibin, luteinizing hormone, and insulin-like growth factor 1 in Brahman bulls. *Biology of Reproduction* 87, 58.

Fortes, M.R., Snelling, W.M., Reverter, A., Nagaraj, S.H., Lehnert, S.A., Hawken, R.J., Deatley, K.L., Peters, S.O., Silver, G.A., Rincon, G., Medrano, J.F., Islas-Trejo, A. and Thomas, M.G. (2012b) Gene network analyses of first service conception in Brangus heifers: use of genome and trait associations, hypothalamic-transcriptome information, and transcription factors. *Journal of Animal Science* 90, 2894–2906.

Fritz, S., Capitan, A., Djari, A., Rodriguez, S.C., Barbat, A., Baur, A., Grohs, C., Weiss, B., Boussaha, M., Esquerre, D. *et al.* (2013) Detection of haplotypes associated with prenatal death in dairy cattle and identification of deleterious mutations in GART, SHBG and SLC37A2. *PLoS ONE* 8, e65550.

Galloway, S.M., McNatty, K.P., Cambridge, L.M., Laitinen, M.P., Juengel, J.L., Jokiranta, T.S., Mclaren, R.J., Luiro, K., Dodds, K.G., Montgomery, G.W. *et al.* (2000) Mutations in an oocyte-derived growth factor gene (BMP15) cause increased ovulation rate and infertility in a dosage-sensitive manner. *Nature Genetics* 25, 279–283.

Ghavi Hossein-Zadeh, N., Nejati-Javaremi, A., Miraei-Ashtiani, S.R. and Kohram, H. (2009) Estimation of variance components and genetic trends for twinning rate in Holstein dairy cattle of Iran. *Journal of Dairy Science* 92, 3411–3421.

Glick, G., Shirak, A., Seroussi, E., Zeron, Y., Ezra, E., Weller, J.I. and Ron, M. (2011) Fine mapping of a QTL for fertility on BTA7 and its association with a CNV in the Israeli Holsteins. *G3 (Bethesda)* 1, 65–74.

Gonda, M.G., Arias, J.A., Shook, G.E. and Kirkpatrick, B.W. (2004) Identification of an ovulation rate QTL in cattle on BTA14 using selective DNA pooling and interval mapping. *Animal Genetics* 35, 298–304.

Gregory, K.E., Smith, G.M., Cundiff, L.V., Koch, R.M. and Laster, D.B. (1979) Characterization of biological types of cattle-cycle III: I. Birth and weaning traits. *Journal of Animal Science* 48, 271–279.

Gregory, K.E., Echternkamp, S.E., Dickerson, G.E., Cundiff, L.V., Koch, R.M. and Van Vleck, L.D. (1990) Twinning in cattle: I. Foundation animals and genetic and environmental effects on twinning rate. *Journal of Animal Science* 68, 1867–1876.

Gregory, K.E., Cundiff, L.V. and Koch, R.M. (1995) Genetic and phenotypic (co)variances for production traits of female populations of purebred and composite beef cattle. *Journal of Animal Science* 73, 2235–2242.

Gregory, K.E., Bennett, G.L., Van Vleck, L.D., Echternkamp, S.E. and Cundiff, L.V. (1997) Genetic and environmental parameters for ovulation rate, twinning rate, and weight traits in a cattle population selected for twinning. *Journal of Animal Science* 75, 1213–1222.

Guillaume, F., Gautier, M., Ben Jemaa, S., Fritz, S., Eggen, A., Boichard, D. and Druet, T. (2007) Refinement of two female fertility QTL using alternative phenotypes in French Holstein dairy cattle. *Animal Genetics* 38, 72–74.

Guo, J., Jorjani, H. and Carlborg, O. (2012) A genome-wide association study using international breeding-evaluation data identifies major loci affecting production traits and stature in the Brown Swiss cattle breed. *BMC Genetics* 13, 82.

Gutierrez, J.P., Goyache, F., Fernandez, I., Alvarez, I. and Royo, L.J. (2007) Genetic relationships among calving ease, calving interval, birth weight, and weaning weight in the Asturiana de los Valles beef cattle breed. *Journal of Animal Science* 85, 69–75.

Hanrahan, J.P., Gregan, S.M., Mulsant, P., Mullen, M., Davis, G.H., Powell, R. and Galloway, S.M. (2004) Mutations in the genes for oocyte-derived growth factors GDF9 and BMP15 are associated with both increased ovulation rate and sterility in Cambridge and Belclare sheep (Ovis aries). *Biology of Reproduction* 70, 900–909.

Hansen, M., Lund, M.S., Pedersen, J. and Christensen, L.G. (2004) Genetic parameters for stillbirth in Danish Holstein cows using a Bayesian threshold model. *Journal of Dairy Science* 87, 706–716.

Hawken, R.J., Zhang, Y.D., Fortes, M.R., Collis, E., Barris, W.C., Corbet, N.J., Williams, P.J., Fordyce, G., Holroyd, R.G., Walkley, J.R. *et al.* (2012) Genome-wide association studies of female reproduction in tropically adapted beef cattle. *Journal of Animal Science* 90, 1398–1410.

Hickson, R.E., Morris, S.T., Kenyon, P.R. and Lopez-Villalobos, N. (2006) Dystocia in beef heifers: a review of genetic and nutritional influences. *New Zealand Veterinary Journal* 54, 256–264.

Hoglund, J.K., Buitenhuis, A.J., Guldbrandtsen, B., Su, G., Thomsen, B. and Lund, M.S. (2009a) Overlapping chromosomal regions for fertility traits and production traits in the Danish Holstein population. *Journal of Dairy Science* 92, 5712–5719.

Hoglund, J.K., Guldbrandtsen, B., Su, G., Thomsen, B. and Lund, M.S. (2009b) Genome scan detects quantitative trait loci affecting female fertility traits in Danish and Swedish Holstein cattle. *Journal of Dairy Science* 92, 2136–2143.

Hoglund, J.K., Guldbrandtsen, B., Lund, M.S. and Sahana, G. (2012) Analyzes of genome-wide association follow-up study for calving traits in dairy cattle. *BMC Genetics* 13, 71.

Holmberg, M. and Andersson-Eklund, L. (2006) Quantitative trait loci affecting fertility and calving traits in Swedish dairy cattle. *Journal of Dairy Science* 89, 3664–3671.

Holmberg, M., Sahana, G. and Andersson-Eklund, L. (2007) Fine mapping of a quantitative trait locus on chromosome 9 affecting non-return rate in Swedish dairy cattle. *Journal of Animal Breeding and Genetics* 124, 257–263.

Hu, Z.L. and Reecy, J.M. (2007) Animal QTLdb: beyond a repository. A public platform for QTL comparisons and integration with diverse types of structural genomic information. *Mammalian Genome* 18, 1–4.

Hu, Z.L., Park, C.A., Wu, X.L. and Reecy, J.M. (2013) Animal QTLdb: an improved database tool for livestock animal QTL/association data dissemination in the post-genome era. *Nucleic Acids Research* 41, D871–D879.

Huang, W., Kirkpatrick, B.W., Rosa, G.J. and Khatib, H. (2010) A genome-wide association study using selective DNA pooling identifies candidate markers for fertility in Holstein cattle. *Animal Genetics* 41, 570–578.

Johanson, J.M., Bergert, P.J., Kirkpatrick, B.W. and Dentine, M.R. (2001) Twinning rates for North American Holstein sires. *Journal of Dairy Science* 84, 2081–2088.

Johanson, J.M., Berger, P.J., Tsuruta, S. and Misztal, I. (2011) A Bayesian threshold-linear model evaluation of perinatal mortality, dystocia, birth weight, and gestation length in a Holstein herd. *Journal of Dairy Science* 94, 450–460.

Jostins, L., Ripke, S., Weersma, R.K., Duerr, R.H., McGovern, D.P., Hui, K.Y., Lee, J.C., Schumm, L.P., Sharma, Y., Anderson, C. A. *et al.* (2012) Host–microbe interactions have shaped the genetic architecture of inflammatory bowel disease. *Nature* 491, 119–124.

Kappes, S.M., Bennett, G.L., Keele, J.W., Echternkamp, S.E., Gregory, K.E. and Thallman, R.M. (2000) Initial results of genomic scans for ovulation rate in a cattle population selected for increased twinning rate. *Journal of Animal Science* 78, 3053–3059.

Kaupe, B., Brandt, H., Prinzenberg, E.M. and Erhardt, G. (2007) Joint analysis of the influence of CYP11B1 and DGAT1 genetic variation on milk production, somatic cell score, conformation, reproduction, and productive lifespan in German Holstein cattle. *Journal of Animal Science* 85, 11–21.

Kealey, C.G., MacNeil, M.D., Tess, M.W., Geary, T.W. and Bellows, R.A. (2006) Genetic parameter estimates for scrotal circumference and semen characteristics of Line 1 Hereford bulls. *Journal of Animal Science* 84, 283–290.

Khatib, H., Maltecca, C., Monson, R.L., Schutzkus, V., Wang, X. and Rutledge, J.J. (2008a) The fibroblast growth factor 2 gene is associated with embryonic mortality in cattle. *Journal of Animal Science* 86, 2063–2067.

Khatib, H., Monson, R.L., Schutzkus, V., Kohl, D.M., Rosa, G.J. and Rutledge, J.J. (2008b) Mutations in the STAT5A gene are associated with embryonic survival and milk composition in cattle. *Journal of Dairy Science* 91, 784–793.

Khatib, H., Huang, W., Mikheil, D., Schutzkus, V. and Monson, R.L. (2009a) Effects of signal transducer and activator of transcription (STAT) genes STAT1 and STAT3 genotypic combinations on fertilization and embryonic survival rates in Holstein cattle. *Journal of Dairy Science* 92, 6186–6191.

Khatib, H., Maltecca, C., Monson, R.L., Schutzkus, V. and Rutledge, J.J. (2009b) Monoallelic maternal expression of STAT5A affects embryonic survival in cattle. *BMC Genetics* 10, 13.

Khatib, H., Monson, R.L., Huang, W., Khatib, R., Schutzkus, V., Khateeb, H. and Parrish, J.J. (2010) Short communication: validation of *in vitro* fertility genes in a Holstein bull population. *Journal of Dairy Science* 93, 2244–2249.

Kim, E.S., Berger, P.J. and Kirkpatrick, B.W. (2009a) Genome-wide scan for bovine twinning rate QTL using linkage disequilibrium. *Animal Genetics* 40, 300–307.

Kim, E.S., Shi, X., Cobanoglu, O., Weigel, K., Berger, P.J. and Kirkpatrick, B.W. (2009b) Refined mapping of twinning-rate quantitative trait loci on bovine chromosome 5 and analysis of insulin-like growth factor-1 as a positional candidate gene. *Journal of Animal Science* 87, 835–843.

Kirkpatrick, B.W. (1999) Genetics and biology of reproduction in cattle. In: Fries, R. and Ruvinsky, A. (eds) *The Genetics of Cattle*. CAB International, Wallingford, UK, pp. 391–410.

Kirkpatrick, B.W., Byla, B.M. and Gregory, K.E. (2000) Mapping quantitative trait loci for bovine ovulation rate. *Mammalian Genome* 11, 136–139.

Kuhn, C., Bennewitz, J., Reinsch, N., Xu, N., Thomsen, H., Looft, C., Brockmann, G.A., Schwerin, M., Weimann, C., Hiendleder, S. *et al.* (2003) Quantitative trait loci mapping of functional traits in the German Holstein cattle population. *Journal of Dairy Science* 86, 360–368.

Kuhn, M.T., Hutchison, J.L. and Wiggans, G.R. (2006) Characterization of Holstein heifer fertility in the United States. *Journal of Dairy Science* 89, 4907–4920.

Kuhn, M.T. and Hutchison, J.L. (2008) Prediction of dairy bull fertility from field data: use of multiple services and identification and utilization of factors affecting bull fertility. *Journal of Dairy Science* 91, 2481–2492.

Lien, S., Karlsen, A., Klemetsdal, G., Vage, D.I., Olsaker, I., Klungland, H., Aasland, M., Heringstad, B., Ruane, J. and Gomez-Raya, L. (2000) A primary screen of the bovine genome for quantitative trait loci affecting twinning rate. *Mammalian Genome* 11, 877–882.

Liron, J.P., Prando, A.J., Fernandez, M.E., Ripoli, M.V., Rogberg-Munoz, A., Goszczynski, D.E., Posik, D.M., Peral-Garcia, P., Baldo, A. and Giovambattista, G. (2012) Association between GNRHR, LHR and IGF1 polymorphisms and timing of puberty in male Angus cattle. *BMC Genetics* 13, 26.

Livestock Improvement Corporation (2007) It's a cow Jim but not as we know it. Available at: www.lic.co.nz/lic_News_Archive.cfm?nid=44 (accessed 27 February 2014).

Livestock Improvement Corporation (2008) Short gestation. Available at: www.lic.co.nz/lic_Short_gestation.cfm (accessed 17 June 2013).

Luo, L., Peng, G., Zhu, Y., Dong, H., Amos, C.I. and Xiong, M. (2010) Genome-wide gene and pathway analysis. *European Journal of Human Genetics* 18, 1045–1053.

McAllister, C.M., Speidel, S.E., Crews, D.H., Jr and Enns, R.M. (2011) Genetic parameters for intramuscular fat percentage, marbling score, scrotal circumference, and heifer pregnancy in Red Angus cattle. *Journal of Animal Science* 89, 2068–2072.

McClure, M.C., Morsci, N.S., Schnabel, R.D., Kim, J.W., Yao, P., Rolf, M.M., McKay, S.D., Gregg, S.J., Chapple, R.H., Northcutt, S.L. and Taylor, J.F. (2010) A genome scan for quantitative trait loci influencing carcass, post-natal growth and reproductive traits in commercial Angus cattle. *Animal Genetics* 41, 597–607.

McDaneld, T.G., Kuehn, L.A., Thomas, M.G., Snelling, W.M., Sonstegard, T.S., Matukumalli, L.K., Smith, T.P., Pollak, E.J. and Keele, J.W. (2012) Y are you not pregnant: identification of Y chromosome segments in female cattle with decreased reproductive efficiency. *Journal of Animal Science* 90, 2142–2151.

McElhenney, W.H., Long, C.R., Baker, J.F. and Cartwright, T.C. (1985) Production characters of first-generation cows of a five-breed diallel: reproduction of young cows and preweaning performance of inter se calves. *Journal of Animal Science* 61, 55–65.

MacNeil, M.D., Cundiff, L.V., Dinkel, C.A. and Koch, R.M. (1984) Genetic correlations among sex-limited traits in beef cattle. *Journal of Animal Science* 58, 1171–1180.

Mai, M.D., Rychtarova, J., Zink, V., Lassen, J. and Guldbrandtsen, B. (2010) Quantitative trait loci for milk production and functional traits in two Danish Cattle breeds. *Journal of Animal Breeding and Genetics* 127, 469–473.

Maltecca, C., Weigel, K.A., Khatib, H., Cowan, M. and Bagnato, A. (2009) Whole-genome scan for quantitative trait loci associated with birth weight, gestation length and passive immune transfer in a Holstein x Jersey crossbred population. *Animal Genetics* 40, 27–34.

Maltecca, C., Gray, K.A., Weigel, K.A., Cassady, J.P. and Ashwell, M. (2011) A genome-wide association study of direct gestation length in US Holstein and Italian Brown populations. *Animal Genetics* 42, 585–591.

Martinez-Velazquez, G., Gregory, K.E., Bennett, G.L. and Van Vleck, L.D. (2003) Genetic relationships between scrotal circumference and female reproductive traits. *Journal of Animal Science* 81, 395–401.

Mee, J.F. (2008) Prevalence and risk factors for dystocia in dairy cattle: a review. *Veterinary Journal* 176, 93–101.

Meuwissen, T.H., Karlsen, A., Lien, S., Olsaker, I. and Goddard, M.E. (2002) Fine mapping of a quantitative trait locus for twinning rate using combined linkage and linkage disequilibrium mapping. *Genetics* 161, 373–379.

Meyer, C.L., Berger, P.J., Thompson, J.R. and Sattler, C.G. (2001) Genetic evaluation of Holstein sires and maternal grandsires in the United States for perinatal survival. *Journal of Dairy Science* 84, 1246–1254.

Morris, C.A., Pitchford, W.S., Cullen, N.G., Esmailizadeh, A.K., Hickey, S.M., Hyndman, D., Dodds, K.G., Afolayan, R.A., Crawford, A.M. and Bottema, C.D. (2009) Quantitative trait loci for live animal and carcass composition traits in Jersey and Limousin back-cross cattle finished on pasture or feedlot. *Animal Genetics* 40, 648–654.

Morris, C.A., Wheeler, M., Levet, G.L. and Kirkpatrick, B.W. (2010) A cattle family in New Zealand with triplet calving ability. *Livestock Science* 128, 193–196.

Mujibi, F.D. and Crews, D.H., Jr (2009) Genetic parameters for calving ease, gestation length, and birth weight in Charolais cattle. *Journal of Animal Science* 87, 2759–2766.

Muncie, S.A., Cassady, J.P. and Ashwell, M.S. (2006) Refinement of quantitative trait loci on bovine chromosome 18 affecting health and reproduction in US Holsteins. *Animal Genetics* 37, 273–275.

Nielen, M., Schukken, Y.H., Scholl, D.T., Wilbrink, H.J. and Brand, A. (1989) Twinning in dairy cattle: a study of risk factors and effects. *Theriogenology* 32, 845–862.

Nogueira, G.P. (2004) Puberty in South American *Bos indicus* (Zebu) cattle. *Animal Reproduction Science* 82–83, 361–372.

Norman, H.D., Wright, J.R., Kuhn, M.T., Hubbard, S.M., Cole, J.B. and Vanraden, P.M. (2009) Genetic and environmental factors that affect gestation length in dairy cattle. *Journal of Dairy Science* 92, 2259–2269.

Norman, H.D., Wright, J.R. and Miller, R. (2011) Potential consequences of selection to change gestation length on performance of Holstein cows. *Journal of Dairy Science* 94, 1005–1010.

Olsen, H.G., Hayes, B.J., Kent, M.P., Nome, T., Svendsen, M. and Lien, S. (2010) A genome wide association study for QTL affecting direct and maternal effects of stillbirth and dystocia in cattle. *Animal Genetics* 41, 273–280.

Olsen, H.G., Hayes, B.J., Kent, M.P., Nome, T., Svendsen, M., Larsgard, A.G. and Lien, S. (2011) Genome-wide association mapping in Norwegian Red cattle identifies quantitative trait loci for fertility and milk production on BTA12. *Animal Genetics* 42, 466–474.

Oseni, S., Tsuruta, S., Misztal, I. and Rekaya, R. (2004) Genetic parameters for days open and pregnancy rates in US Holsteins using different editing criteria. *Journal of Dairy Science* 87, 4327–4333.

Pausch, H., Flisikowski, K., Jung, S., Emmerling, R., Edel, C., Gotz, K.U. and Fries, R. (2011) Genome-wide association study identifies two major loci affecting calving ease and growth-related traits in cattle. *Genetics* 187, 289–297.

Penagaricano, F., Weigel, K.A. and Khatib, H. (2012a) Genome-wide association study identifies candidate markers for bull fertility in Holstein dairy cattle. *Animal Genetics* 43 Suppl 1, 65–71.

Penagaricano, F., Weigel, K.A., Rosa, G.J. and Khatib, H. (2012b) Inferring quantitative trait pathways associated with bull fertility from a genome-wide association study. *Frontiers in Genetics* 3, 307.

Philipsson, J., Foulley, J.L., Lederer, J., Liboriussen, T. and Osinga, A. (1979) Sire evaluation standards and breeding strategies for limiting dystocia and stillbirth. Report of an EEC/E.A.A.P working group. *Livestock Production Science* 6, 111–127.

Polineni, P., Aragonda, P., Xavier, S.R., Furuta, R. and Adelson, D.L. (2006) The bovine QTL viewer: a web accessible database of bovine quantitative trait loci. *BMC Bioinformatics* 7, 283.

Pryce, J.E., Bolormaa, S., Chamberlain, A.J., Bowman, P.J., Savin, K., Goddard, M.E. and Hayes, B.J. (2010) A validated genome-wide association study in 2 dairy cattle breeds for milk production and fertility traits using variable length haplotypes. *Journal of Dairy Science* 93, 3331–3345.

Ryan, D.P. and Boland, M.P. (1991) Frequency of twin births among Holstein-Friesian cows in a warm dry climate. *Theriogenology* 36, 1–10.

Sahana, G., Guldbrandtsen, B., Bendixen, C. and Lund, M.S. (2010) Genome-wide association mapping for female fertility traits in Danish and Swedish Holstein cattle. *Animal Genetics* 41, 579–588.

Santana, M.L., Eler, J.P., Ferraz, J.B. and Mattos, E.C. (2012) Genetic relationship between growth and reproductive traits in Nellore cattle. *Animal* 6, 565–570.

Schnabel, R.D., Sonstegard, T.S., Taylor, J.F. and Ashwell, M.S. (2005) Whole-genome scan to detect QTL for milk production, conformation, fertility and functional traits in two US Holstein families. *Animal Genetics* 36, 408–416.

Schrooten, C., Bovenhuis, H., Coppieters, W. and Van Arendonk, J.A. (2000) Whole genome scan to detect quantitative trait loci for conformation and functional traits in dairy cattle. *Journal of Dairy Science* 83, 795–806.

Schulman, N.F., Sahana, G., Lund, M.S., Viitala, S.M. and Vilkki, J.H. (2008) Quantitative trait loci for fertility traits in Finnish Ayrshire cattle. *Genetics Selection Evolution* 40, 195–214.

Schulman, N.F., Sahana, G., Iso-Touru, T., McKay, S.D., Schnabel, R.D., Lund, M.S., Taylor, J.F., Virta, J. and Vilkki, J.H. (2011) Mapping of fertility traits in Finnish Ayrshire by genome-wide association analysis. *Animal Genetics* 42, 263–269.

Seidenspinner, T., Bennewitz, J., Reinhardt, F. and Thaller, G. (2009) Need for sharp phenotypes in QTL detection for calving traits in dairy cattle. *Journal of Animal Breeding and Genetics* 126, 455–462.

Seidenspinner, T., Tetens, J., Habier, D., Bennewitz, J. and Thaller, G. (2011) The placental growth factor (PGF)–a positional and functional candidate gene influencing calving ease and stillbirth in German dairy cattle. *Animal Genetics* 42, 22–27.

Shanks, R.D., Popp, R.G., McCoy, G.C., Nelson, D.R. and Robinson, J.L. (1992) Identification of the homozygous recessive genotype for the deficiency of uridine monophosphate synthase in 35-day bovine embryos. *Journal of Reproduction and Fertility* 94, 5–10.

Silva, H.M., Wilcox, C.J., Thatcher, W.W., Becker, R.B. and Morse, D. (1992) Factors affecting days open, gestation length, and calving interval in Florida dairy cattle. *Journal of Dairy Science* 75, 288–293.

Sonstegard, T.S., Cole, J.B., Vanraden, P.M., Van Tassell, C.P., Null, D.J., Schroeder, S.G., Bickhart, D. and McClure, M.C. (2013) Identification of a nonsense mutation in CWC15 associated with decreased reproductive efficiency in Jersey cattle. *PLoS ONE* 8, e54872.

Steinbock, L., Nasholm, A., Berglund, B., Johansson, K. and Philipsson, J. (2003) Genetic effects on stillbirth and calving difficulty in Swedish Holsteins at first and second calving. *Journal of Dairy Science* 86, 2228–2235.

Sugimoto, M., Sasaki, S., Watanabe, T., Nishimura, S., Ideta, A., Yamazaki, M., Matsuda, K., Yuzaki, M., Sakimura, K., Aoyagi, Y. and Sugimoto, Y. (2010) Ionotropic glutamate receptor AMPA 1 is associated with ovulation rate. *PLoS One* 5, e13817.

Tang, K.Q., Li, S.J., Yang, W.C., Yu, J.N., Han, L., Li, X. and Yang, L.G. (2011) An MspI polymorphism in the inhibin alpha gene and its associations with superovulation traits in Chinese Holstein cows. *Molecular Biology Reports* 38, 17–21.

Tang, K.Q., Yang, W.C., Li, S.J. and Yang, L.G. (2013) Polymorphisms of the bovine growth differentiation factor 9 gene associated with superovulation performance in Chinese Holstein cows. *Genetics and Molecular Research* 12, 390–399.

Tarres, J., Fina, M. and Piedrafita, J. (2010) Parametric bootstrap for testing model fitting of threshold and grouped data models: an application to the analysis of calving ease of Bruna dels Pirineus beef cattle. *Journal of Animal Science* 88, 2920–2931.

Theodoratou, E., Montazeri, Z., Hawken, S., Allum, G.C., Gong, J., Tait, V., Kirac, I., Tazari, M., Farrington, S.M., Demarsh, A. *et al.* (2012) Systematic meta-analyses and field synopsis of genetic association studies in colorectal cancer. *Journal of the National Cancer Institute* 104, 1433–1457.

Thomasen, J.R., Guldbrandtsen, B., Sorensen, P., Thomsen, B. and Lund, M.S. (2008) Quantitative trait loci affecting calving traits in Danish Holstein cattle. *Journal of Dairy Science* 91, 2098–2105.

Tiezzi, F., Maltecca, C., Cecchinato, A., Penasa, M. and Bittante, G. (2012) Genetic parameters for fertility of dairy heifers and cows at different parities and relationships with production traits in first lactation. *Journal of Dairy Science* 95, 7355–7362.

Van Tassell, C.P., Van Vleck, L.D. and Gregory, K.E. (1998) Bayesian analysis of twinning and ovulation rates using a multiple-trait threshold model and Gibbs sampling. *Journal of Animal Science* 76, 2048–2061.

Van Vleck, L.D., Gregory, K.E. and Echternkamp, S.E. (1991) Ovulation rate and twinning rate in cattle: heritabilities and genetic correlation. *Journal of Animal Science* 69, 3213–3219.

Vanraden, P.M., Olson, K.M., Null, D.J. and Hutchison, J.L. (2011) Harmful recessive effects on fertility detected by absence of homozygous haplotypes. *Journal of Dairy Science* 94, 6153–6161.

Varona, L., Misztal, I. and Bertrand, J.K. (1999) Threshold-linear versus linear-linear analysis of birth weight and calving ease using an animal model: I. Variance component estimation. *Journal of Animal Science* 77, 1994–2002.

Wang, K., Li, M. and Bucan, M. (2007) Pathway-based approaches for analysis of genome-wide association studies. *American Journal of Human Genetics* 81, 1278–1283.

Weigel, K.A. and Rekaya, R. (2000) Genetic parameters for reproductive traits of Holstein cattle in California and Minnesota. *Journal of Dairy Science* 83, 1072–1080.

Weller, J.I., Golik, M., Reikhav, S., Domochovsky, R., Seroussi, E. and Ron, M. (2008a) Detection and analysis of quantitative trait loci affecting production and secondary traits on chromosome 7 in Israeli Holsteins. *Journal of Dairy Science* 91, 802–813.

Weller, J.I., Golik, M., Seroussi, E., Ron, M. and Ezra, E. (2008b) Detection of quantitative trait loci affecting twinning rate in Israeli Holsteins by the daughter design. *Journal of Dairy Science* 91, 2469–2174.

Wilson, T., Wu, X.Y., Juengel, J.L., Ross, I.K., Lumsden, J.M., Lord, E.A., Dodds, K.G., Walling, G.A., McEwan, J.C., O'Connell, A.R., McNatty, K.P. and Montgomery, G.W. (2001) Highly prolific Booroola sheep have a mutation in the intracellular kinase domain of bone morphogenetic protein IB receptor (ALK-6) that is expressed in both oocytes and granulosa cells. *Biology of Reproduction* 64, 1225–1235.

Wray, N.R., Quaas, R.L. and Pollak, E.J. (1987) Analysis of gestation length in American Simmental cattle. *Journal of Animal Science* 65, 970–974.

Yague, G., Goyache, F., Becerra, J., Moreno, C., Sanchez, L. and Altarriba, J. (2009) Bayesian estimates of genetic parameters for pre-conception traits, gestation length and calving interval in beef cattle. *Animal Reproduction Science* 114, 72–80.

Yang, W.C., Li, S.J., Tang, K.Q., Hua, G.H., Zhang, C.Y., Yu, J.N., Han, L. and Yang, L.G. (2010) Polymorphisms in the 5' upstream region of the FSH receptor gene, and their association with super-ovulation traits in Chinese Holstein cows. *Animal Reproduction Science* 119, 172–177.

Yang, W.C., Tang, K.Q., Li, S.J. and Yang, L.G. (2011) Association analysis between variants in bovine progesterone receptor gene and superovulation traits in Chinese Holstein cows. *Reproduction in Domestic Animals* 46, 1029–1034.

Yang, W.C., Tang, K.Q., Li, S.J., Chao, L.M. and Yang, L.G. (2012) Polymorphisms of the bovine luteinizing hormone/choriogonadotropin receptor (LHCGR) gene and its association with superovulation traits. *Molecular Biology Reports* 39, 2481–2487.

Yu, Y., Pang, Y., Zhao, H., Xu, X., Wu, Z., An, L. and Tian, J. (2012) Association of a missense mutation in the luteinizing hormone/choriogonadotropin receptor gene (LHCGR) with superovulation traits in Chinese Holstein heifers. *Journal of Animal Science* 3, 35.

Zaborski, D., Grzesiak, W., Szatkowska, I., Dybus, A., Muszynska, M. and Jedrzejczak, M. (2009) Factors affecting dystocia in cattle. *Reproduction in Domestic Animals* 44, 540–551.

13 Modern Reproductive Technologies and Breed improvement

Habib A. Shojaei Saadi and Claude Robert

Université Laval, Québec, Canada

Introduction

Well-defined objectives are essential for success in livestock breeding and will usually evolve over time in relation to societal concerns as well as technological and economic developments. In the case of cattle, higher output per unit input has been the main breeding objective, but other issues including well-being, longevity and reproduction have been raised (Flint and Woolliams, 2008). Dairy cattle breeding is undergoing a paradigm shift to genomic selection of sires and dams (Amann and DeJarnette, 2012). Genetic improvement relies on two different physiological applications, namely genetic selection and selective mating. The first requires estimation of genetic potential, the second pertains to the selection of mating pairs and reproductive technologies used to disseminate high merit genetics and is the focus of this chapter. Reproductive technologies have a crucial role in improvement of livestock reproduction to meet the rising challenges to the food supply of the 21st century (Murphy, 2012). Despite major advances associated with reproduction in dairy cattle (Moore and Thatcher, 2006), reproductive decline in elite dairy cattle is a serious concern of farmers and the dairy

© CAB International 2015. *The Genetics of Cattle*,
2nd Edn (eds D.J. Garrick and A. Ruvinsky)

industry worldwide (Lucy, 2001, 2007; Pryce et al., 2004; Dobson et al., 2007; Walsh et al., 2011). Fertility has emerged as a growing genetic and management challenge in high-producing dairy herds (Funk, 2006). Fertility requires both paternal and maternal contributions to produce an embryo which eventually leads to live birth. The following aims at portraying the current and emerging reproductive technologies that are destined to increase the rate of genetic improvement with a focus on dairy herds.

The Paternal Contribution

Male fertility is an important factor in cattle breeding, as bulls in service are used to breed numerous cows and defective semen quality can have a profound contribution to reproductive failure (DeJarnette et al., 2004; Kathiravan et al., 2011). It was estimated that sperm from the majority of sires is not able to fertilize 100% of oocytes (>90% of oocytes in heifers or non-lactating cows), and male-associated deficiencies account for approximately 5–20% of embryos dying by day 8 of development (Amann and DeJarnette, 2012). In cattle, primarily, post-thaw semen evaluation and analysis of several sperm characteristics are routinely performed in breeding centres to assess bull fertility.

Sperm is the ultimate example of a structural and functionally differentiated cell in which its genetic component located in the highly condensed nucleus results in gene expression silencing (Eddy, 2006; Johnson et al., 2011). This high level of compaction is achieved through chromatin remodelling during spermatogenesis by removing and replacing most somatic histone cores with transition proteins and then protamines (arginine-rich, nuclear proteins that have higher DNA affinity), leading to extreme chromatin compaction in the late haploid phase of spermatogenesis (Ward, 2010). The extent of histone replacement differs between species, and varies between 85 and 98% in mammals (Balhorn, 2007; Ward, 2010). Interestingly, recent studies imply that retained histones in mature sperm genome are believed to be gene-specific and non-randomly distributed and have unique

histone modifications, indicating their potential epigenetic regulatory mechanisms in early embryo development (Arpanahi et al., 2009; Hammoud et al., 2009; Miller et al., 2010). In the sperm nucleus, interaction of the protamines and DNA make unique coiling structures of sperm DNA called toroidal subunits, or doughnut loops, that contain roughly 50 kb of DNA (D'Occhio et al., 2007).

Mammalian spermatozoa are endowed with a unique cytoskeleton (Fouquet and Kann, 1994) that varies in morphology, biochemical and physiological characteristics among the species (Pesch and Bergmann, 2006). Substantial differences in sperm head shape and size have been found within and between breeds in livestock (Phetudomsinsuk et al., 2008). Bull sperm has a very flat and highly condensed sperm head due to the extent and/or efficiency of disulphide bonding in its nucleus (Perreault et al., 1988). Sperm DNA is enclosed in a vestigial nuclear envelope that is protected by several outer membranes formed when the typical elongated spermatozoa take shape during spermiogenesis (Hess and Franca, 2009). These membranes are known to delineate subcellular compartments and different regions that are essential for the complex physiological transformation which sperm must undergo to achieve successful fertilization (Brewis and Gadella, 2010). The acrosome is a unique membranous secretory organelle located in the sperm head (Yanagimachi, 1994; Mayorga et al., 2007; Berruti and Paiardi, 2011) that contains enzymes to digest and interact with the zona pellucida of the oocyte (Buffone et al., 2008). During the early stages of sperm–egg interaction, the acrosome undergoes an exocytotic process known as the acrosome reaction (AR), which is an irreversible step and functions as a behavioural switch, converting sperm into a state in which they are competent to interact with oocytes (Florman et al., 2008). Prior to the AR in the female reproductive tract, mammalian sperm must undergo capacitation, which is a complex series of biochemical and physiological modifications followed by changes in sperm motility pattern (hyperactivation), which altogether are crucial changes in sperm for successful fertilization (Ickowicz *et al.*, 2012). Depending on the species, about 100 mitochondria are localized in the sperm mid-piece, having several key roles in sperm pathology and physiology such as providing energy (ATP) for sperm metabolism, membrane function and motility (Pena *et al.*, 2009).

Some of these structural, physiological and biochemical characteristics of sperm have been routinely examined through techniques to evaluate and determine the quality of the semen and characterize the most elite bulls in terms of fertility potential. These sperm attributes can be evaluated *in vitro* at molecular and cellular levels. Semen evaluation using light microscopy provides useful information at the cellular level (morphology and motility), but is subjective, which limits its prognostic value for the reproductive performance of males or the outcome of assisted fertilization. However, at molecular level, sperm characteristics (mainly abnormalities) can be detected by an array of biomarkers including fluorescent markers for sperm plasmalema integrity, permeability and stability, acrosomal status, sperm mitochondrial integrity and activity, capacitation status and membrane fluidity of sperm, altered sperm chromatin or DNA integrity, apoptotic, oxidative stress and lipid peroxidation events in sperm, and antibodies detecting proteins that are either up- or down-regulated in defective spermatozoa (Fig. 13.1) (Sutovsky and Lovercamp, 2010).

Using flow cytometry (FC), the majority of these biomarkers can be efficiently evaluated (Martínez-Pastor *et al.*, 2010). Flow cytometry enables objective measurement of these biomarkers (individually or simultaneously) in semen through automated, high-throughput and rapid methods (Gillan *et al.*, 2005; Hossain *et al.*, 2011).

Breeding soundness evaluation (BSE) and estimation of sire fertility

Recently, bull fertility has received increasing attention as the results of artificial insemination (AI) success rates are declining in highly selected dairy cattle populations (Karoui *et al.*, 2011). One of the most important challenges in the AI industry is to identify bulls producing large numbers of fertile sperm and accurately predict the fertility of dairy bulls with apparently normal semen (Foote, 2003). Breeding soundness, which refers to a bull's ability to get cows pregnant, can be classically evaluated to identify bulls with substantial deficits in fertility, but does not consistently identify sub-fertile bulls (Kastelic and Thundathil, 2008). There is an urgent need for accurate biomarkers of fertility to complement traditional breeding soundness evaluation, identifying and eliminating bulls with inferior fertility/semen quality (Shojaei *et al.*, 2012;

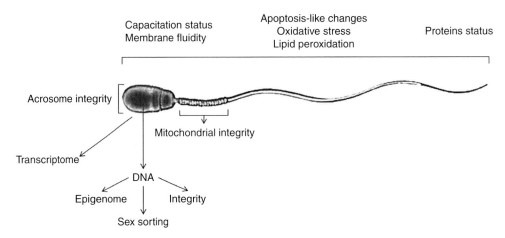

Fig. 13.1. Different spermatic characteristics that can be evaluated at the molecular level to either assess sperm quality or gender selection.

Sutovsky and Kennedy, 2013). Currently, relative sire fertility can be retrospectively estimated based on AI records for the calculation of the very general parameter of non-return rate (NRR), which is based on whether or not the female had a second insemination (indicative of a failed gestation causing a return to oestrus) within a certain period (either 28, 56 or 128 days post-insemination). Usually 56 days after the first insemination (day of AI = day 0) is currently the most commonly used time point for recording fertility in Canada and Europe (Rodríguez-Martínez, 2003). While, NRR provides reliable results over longer periods of time, it is not completely accurate and is highly prone to bias due to several factors including unexplained variation. (Rodríguez-Martínez, 2003; Foote, 2003; Amann and DeJarnette, 2012). The NRR value represents the combined contribution of sire and dam to successfully sustain gestation. Female factors such as age and lactation status are known to impact fertility as heifers are more fertile than cows. In USA, the estimated relative conception rate (ERCR) was used from 1986 to 2008. It was based on the 70-day NRR of an AI service sire relative to service sires of herd-mates and provided more reliable fertility estimates for bulls through exclusion of services to cows that had left the herd (Cornwell et al., 2006; Amann and DeJarnette, 2012). Since 2008, sire conception rate (SCR), a phenotypic predictor of bull fertility, has been made available to dairy producers by the US Department of Agriculture (USDA) as a new and more accurate evaluation of AI bull (service-sire) fertility in USA compared to ERCR. The SCR is based on 70 day post-AI confirmed pregnancy (or not pregnant) after ≥ 300 total inseminations (≥ 100 in last 12 months) and is reported as a sire's fertility deviation from the average fertility of a population (>10 herds). (Kuhn and Hutchison, 2008; Kuhn et al., 2008; Norman et al., 2010).

Sperm attributes

Sperm count

The most traditional and 'gold standard' method to count sperm is microscopic manual chamber counting using a haemocytometer, which is a time-consuming and has several limitations such as variation between operators and haemocytometer designs. Several methods have been developed to precisely count the sperm, amongst which the flow cytometer approach was found to be the most precise, however, high cost and overestimation and the need for appropriate semen dilution and flow rate adjustment are its drawbacks (Petrunkina and Harrison, 2010). Overall, frequency of use, size of sample required, the number of samples routinely assessed, species and cost are important factors to consider for standardizing a laboratory procedure for sperm counts (Prathalingam et al., 2006).

Sperm morphometry

Computer automated sperm head morphology analysis (ASMA) was introduced to objectively evaluate several sperm morphometric parameters (Gravance et al., 1996) and identify sperm subpopulations. Due to some technical challenges, inconsistency of results and low predictive value for bull fertility assessment, so far it has received little attention especially in dairy bulls (Hoflack et al., 2005; Gravance et al., 2009; Jenkins and Carrell, 2012). However, recent attempts have been done to improve its performance (Vicente-Fiel et al., 2013).

Sperm motility

Motility is considered to be one of the most important characteristics of sperm indicating their viability and structural integrity as well as being a good indication of their fertilizing ability (Gaffney et al., 2011). Traditionally, assessment of sperm motility has been based on subjective optical microscopic evaluation of parameters, such as population and individual motility, which has resulted in 30–60% variations in motility parameters of the same ejaculates (Verstegen et al., 2002). Computer-assisted sperm analysis (CASA) is a precise, automated and objective method to evaluate sperm head motion characteristics, which provides several sperm kinematic parameters. A combination of some of these parameters has been found to be correlated with fertility, as well as classification of different sperm sub-populations with specific patterns of movement in a given semen based on path velocity or hyperactivated motility in bulls (Muiño et al., 2008; Kathiravan et al., 2011; Shojaei et al., 2012).

Adding 'new' parameters to sperm quality evaluation

Although the aforementioned standard seminal parameters such as motility, morphology and sperm concentration measurements are the most commonly implemented methods to evaluate the quality of a bull's ejaculate, they are insufficient as some service bulls with good semen quality parameters still provide low NRR values (Petrunkina et al., 2007; Kastelic and Thundathil, 2008). It is not known if this is caused by genetically based incompatibility or is associated with other spermatic characteristics. As such, several groups are currently looking into integrating different assays to complement the current set of parameters. Most of these additional semen quality parameters are based on older assays developed to study spermatic physiology (Table 13.1). They mainly target the assessment of the integrity of membranes as they play crucial roles in the spermatic functions for capacitation and AR, or of the mitochondrial potential needed to sustain motility and also of the integrity of genetic material itself as DNA fragmentation could lead to embryonic failure (Fig. 13.1).

Sperm sorting

Using flowcytometric approaches, sperm can be sorted based on the sex or several parameters and markers (Martínez-Pastor et al., 2010).

Sex sorting

The availability of sexed semen in dairy cattle has been eagerly anticipated for many years and is growing in adoption (Weigel, 2004) such that sex-selected sperm straws have become a new product offered by many AI centres (Garner and Seidel Jr, 2008). Sex sorting separates semen into fractions enriched with either X- or Y-bearing sperm. One of the most efficient and widely used methods for sex sorting sperm is based on fluorescence-activated cell sorting (FACS) and staining sperm with a DNA-binding fluorescent dye to separate sperm based on the quantity of the DNA content. The bovine X chromosome-bearing sperm contain 3.8% more DNA than Y chromosome-bearing sperm, which allows their separation by FACS (Weigel, 2004; Garner, 2006; Seidel, 2007). The demand is for sperm containing an X chromosome for the production of replacement heifers. Despite current advances (Sharpe and Evans, 2009), sex sorting is not completely accurate (with 85–95% of the sperm containing the desired chromosome), and the slow procedure compromises viability, motility, lifespan and fertility potential of the sex-sorted sperm (Wheeler et al., 2006; Vazquez et al., 2008; Seidel, 2012). However, this technology is continuously being improved and is expected to be more widely available and used in the near future (De Vries et al., 2008; Garner and Seidel Jr, 2008; Sharpe and Evans, 2009).

Marker-based sorting of fertile sperm

Many techniques have been developed to isolate sperm capable of fertilizing oocytes, especially in the context of medically assisted reproduction (Ickowicz et al., 2012). Recent studies on human sperm have proved that using different fluorochromes and FACS, sperm can be sorted based on apoptosis (Annexin-V) (Hoogendijk et al., 2009) as well as apoptotic and dead sperm (YO-PRO) (Ribeiro et al., 2013). Magnet-activated (or magnetic-bead activated) cell sorting (MACS) technique is another sorting method (immunomagnetic), which eliminates apoptotic sperm from a sperm suspension through Annexin V-conjugated paramagnetic microbeads (Grunewald et al., 2001; Said et al., 2008; Dirican, 2012). Overall, these testify the possibility that sperm can be physically sorted depending on one or several parameters providing new possibilities for research and for future practical use in breeding (Martínez-Pastor et al., 2010).

Table 13.1. Summaries of current possible methods of sperm quality assessments.

	Classic		Modern		References
	Non-FC	FC-based	Non-FC	FC-based	
Sperm characteristics					
Enumerating	Haemocytometry, spectrophotometry and microcells	–	Fluorescent plate reading and image analysis	Flow cytometry	Prathalingam et al., 2006
Morphometry	Eosin–nigrosin staining	–	ASMA	–	Gravance et al., 1996
Motility	Microscopic	–	CASA	–	
Sperm intactness Integrity (plasma membrane)	Eosin–nigrosin staining, HOST	CFDA, CFDA/PI, CMFDA, CAM	SYBR-14/PI	SYBR-14/PI	Resli et al., 1983; Jeyendran et al., 1984; Garner et al., 1986; 1994; PeÑA et al., 2005; Hallap et al., 2006
Permeability and stability	–	–	–	Annexin V/PI, Hoechst 33342, YO-PRO-1, Merocyanine 540, SNARF-1, ethidium homodimer	
Acrosome integrity	Phase-contrast and DIC microscopy; dyes for bright-field microscopy; fluorescent labels	–	FITC-PSA/PNA	SYBR-14/PE-PNA/PI, FITC-PNA/PI, FITC-PSA/PI, LysoTracker™	Cross and Meizel, 1989; Hinsch et al., 1997; Thomas et al., 1997a,b
Mitochondrial status		R123-EtBr		JC-1, Mitotracker Green, Mitotracker Deep Red	Evenson et al., 1982; Garner et al., 1997; Garner and Thomas, 1999; Boe-Hansen et al., 2005; Hallap et al., 2005
Chromatin intactness	AOT	–	Comet assay, CMA3	SCSA, TUNEL, CMA3	Evenson and Wixon, 2006; Simoes et al., 2009; Tavalaee et al., 2010; Rahman et al., 2011; Evenson, 2013

Continued

Table 13.1. Continued.

	Classic		Modern		References
	Non-FC	FC-based	Non-FC	FC-based	
Sperm functionality					
Changes induced during capacitation	–	–	CTC	CTC, M540/Yo-Pro 1/Hoechst 33342, Fluo, Indo-1	Thundathil et al., 1999; Fathi et al., 2001; Hallap et al., 2006; Piehler et al., 2006; Pons-Rejraji et al., 2009
Apoptotic-like changes	–	–	–	Annexin V/PI, YO-PRO-1	Anzar et al., 2002; Martin et al., 2004
Detection of oxidative stress and lipid peroxidation	TBARS assay	–	–	H2DCFDA, HE, MitoSOX, BODIPY probes	Bansal and Bilaspuri, 2011; Hossain et al., 2011
Sperm surface targets	–	–	–	PR, ubiquitin-PNA, pY	Gadkar et al., 2002; Piehler et al., 2006; Odhiambo et al., 2011

AOT, Acridine orange test; BODIPY, 4-bora-3a,4a-diaza-s-indacene; CAM, calcein acetomethyl ester; CASA, Computer Assisted Sperm Analysis; CFDA, 6-carboxyfluorescein diacetate; CMA3, chromomycin A3; CTC, chlortetracycline; EtBr, ethidium bromide; FC, Flow cytometry; FITC, Fluorescein isothiocyanate; H2DCFDA, 2′,7′-dichlorodihydroflucrescein diacetate; HE, hydroethidine; HOST, hypo-osmotic swelling test; JC-1, 5,5′,6,6′-tetrachloro-1,1′,3,3′-tetraethylbenzimidazolylcarbocyanine iodide; M540, merocyanine 540; PE, phycoerythrin; PI, propidium iodide; PNA, peanut agglutinin; PR, progesterone receptor; PSA, *Pisum sativum* agglutinin; pY, phosphotyrosine; R123, rodamine 123; SCSA, sperm chromatin structure assay; SNARF-1, seminaphthorhodafluor-1; SYBR14, membrane-permeant fluorescent nucleic acid stain; TBARS, thiobarbituric acid reactive substances; ~UNEL, terminal deoxynucleotidyl transferase dUTP nick end labelling.

Improving male fertility through integration of the 'omics'

The recent development, implementation and acceptance of genomic evaluations (based first on the BovineSNP50 Bead Chip) have had great impacts on AI industry and breeding, such that currently all young bulls entering the major AI centres are preselected based on genomic evaluations resulting in extensive marketing of 2-year-old bulls (Wiggans et al., 2011). However, genomics advances in terms of male fertility have not been significant, and currently there is limited information on genes associated with bull fertility (Peñagaricano et al., 2012). Several studies using genome-wide analysis, comparative genomics, single nucleotide polymorphism (SNP) identification (whole-genome SNP Chip using BovineSNP50 Bead Chip or other methods plus Imputation) and candidate pathway approaches have shown that bull fertility is influenced by genetic factors (Feugang et al., 2009; Khatib et al., 2010; Blaschek et al., 2011; Fortes et al., 2012; Li et al., 2012; Peñagaricano et al., 2012, 2013). In spite of the prominent place of AI in animal breeding where a single male has greater impact on genetic improvement than any female, study of the genetic variation behind the expression of male reproductive traits has received less attention than that of the cow (Li et al., 2012). The few studies targeting the identification of quantitative trait loci (QTLs) or of molecular markers for male reproductive traits only found weak association, indicating the importance of and need for replications and further studies (Druet et al., 2009; Casas et al., 2010). The difficulty in finding genetic markers of fertility is likely attributable to the complex nature of its aetiology in addition to the challenging nature of phenotypic data collection, where gestational failure alone is actually the combined result of different phenotypes. Hence, it is expected that industry-wide data collection and analysis to evaluate genetic merit would be the most important tool for genetic progress in the future (Powell and Norman, 2006). Increasing the level of precision at which the genome is mined using high-density or whole-genome SNP chips or even specially designed 'fertility chips' might be future tools to better identify genomic fertility markers and characterize the fertility of elite sires (Amann and DeJarnette, 2012). Alternate perspectives include the study of gene products such as RNA and proteins.

Recently, sperm transcriptome studies in dairy bulls have received increasing attention but they remain controversial (Lalancette et al., 2008). Some studies have found sperm mRNA (Gilbert et al., 2007; Feugang et al., 2010; Arangasamy et al., 2011; Kasimanickam et al., 2012; Card et al., 2013) and microRNAs (Govindaraju et al., 2012) may have potential to be applied as molecular biomarkers for male gamete quality and bull fertility.

Recently, proteomics approaches, especially 2D-PAGE coupled with mass spectrometry (MS) technique, have been increasingly used on fertile and sub-fertile dairy bulls to identify candidate proteins that are differentially expressed in sperm membrane as biomarkers associated with bull fertility potential (Moura et al., 2006a; D'Amours et al., 2010; Gaviraghi et al., 2010; Park et al., 2012), seminal plasma (Killian et al., 1993; Kumar et al., 2012), accessory sex gland fluid (Moura et al., 2006b, 2007) or epididymal fluid (Moura et al., 2006a).

Epigenetics is the study of potential heritable changes in gene function that occur independently of alterations to primary DNA sequence (Bernstein et al., 2007; Kiefer, 2007). Epigenetics may help find missing causality and missing heritability of complex traits and diseases, which genomics approaches have to date been unable to address (Gonzalez-Recio, 2012). Among all the epigenetic effectors DNA methylation is the most intensively studied and the most stable type of epigenetic modification modulating the transcriptional plasticity of mammalian genomes (Eckhardt et al., 2006; Suzuki and Bird, 2008; McGraw et al., 2013). Recent studies have shown that sperm is heavily methylated compared to oocytes, with approximately 85% vs 30% global cytosine-phosphate-guanine (CG) methylation levels, respectively (Seisenberger et al., 2013). Sperm DNA methylation mainly occurs outside of the promoter regions (intergenic) similar to embryonic stem cells and embryonic germ cells (Schagdarsurengin et al., 2012). Developing embryos require appropriate epigenetic marks from sperm (Jenkins and Carrell, 2012). This emphasizes the importance and crucial role of the sperm epigenome in successful fertilization, embryo development and full term pregnancies. Recently there is an emerging interest in studying the sperm epigenome in livestock to improve breeding through identification of potential fertility biomarkers, environmental effects and transgenerational epigenome inheritance.

Artificial insemination (AI)

Due to the ease of access and rate of production of male compared to female gametes, reproductive technologies have initially focused on maximizing benefits from AI. More than two centuries ago (1784) the first successful insemination was performed in a dog, however, 100 years later it had been used for studies in rabbits, dogs and horses. In the early 20th century, Ivanow established AI as a practical procedure in farm animals and later was followed by other scientists mainly in Europe. Modern development of AI in dairy cattle was initiated over 70 years ago in the USA by establishing AI cooperatives and, in the 1980s, tremendous growth occurred because of strong export markets, especially for Holstein semen (Funk, 2006). AI has been considered as the first remarkable and most widely used biotechnology applied to improve reproduction and genetics of farm animals, and because of its worldwide acceptance it opened the doors for other reproductive biotechnologies such as semen evaluation techniques, semen freezing and sexing, bull sexual behaviour and sire power, genetic selection of bulls for milk and detection of oestrus, synchronization and timing of insemination (Foote, 2002; Cardellino, 2003). Currently, based on all the collected and analysed data, the trend in AI industries would be the following: breeding for long-lasting, durable and profitable dairy cattle through shifting emphasis from production (i.e. yield) to non-production traits; selection based on fertility potential; and inbreeding for purebred dairy breeds and crossbreeding to overcome inbreeding depression within the pure breeds (Funk, 2006).

Sperm cryopreservation

Cryopreservation in general is a procedure in which cells (gametes, embryos and somatic cells) are suspended in a solution of salts and a low-molecular-weight (low-MW) organic compound, cooled to very low subzero temperatures (usually $-196°C$ in liquid nitrogen), stored for some theoretically unlimited period of time, then warmed and recovered to resume their normal function (Leibo and Pool, 2011). Semen cryopreservation is extensively used by the dairy cattle AI industry for conservation and distribution of animal genetic resources and was the main reason for the rapid growth of the bovine AI industry since the early 1950s (Funk, 2006). In contrast to many other species, bull sperm are characterized by superior cryoresistance because of their physiology, biochemistry and structure, which enable them to efficiently survive cryopreservation (Holt, 2000a; Słowińska et al., 2008). This cryotolerance and the anatomy and physiology of sperm transport in the bovine female reproductive tract (Holt, 2000b) have led to remarkable success with cryopreservation of bull semen enabling worldwide distribution of selected male genetics as cryopreserved semen straws. Currently several methods are available for semen cryopreservation (Vishwanath and Shannon, 2000; Barbas and Mascarenhas, 2009). Despite all the advances, improvements in the field of cryoprotection are still ongoing due to the low survival rate (30–50% viability) as well as the approximately seven-fold reduction in fertilizing ability of post-thaw sperm using the current methods (Watson, 2000; Sullivan, 2004; Muiño et al., 2007; Hu et al., 2011; Benson et al., 2012). The post-thaw survival of sperm depends on several factors such as physiology, biochemistry and structure of sperm, storage temperature, cooling rate, chemical composition of the extender, cryoprotectant concentration, reactive oxygen species (ROS), seminal plasma composition and hygienic control (Barbas and Mascarenhas, 2009). The ejaculated semen is composed of a heterogenous cohort. This heterogeneity is believed to be an important aspect of fertility, where a more heterogeneous gametic population will offer a larger window of opportunity for successful fertilization, since oestrus, insemination and ovulation occur over a variable time window (Curry, 2000). Hence, there has been always a sub-population of sperm that are more sensitive than others to environmental conditions and various stresses such as cryopreservation, leading to survival, cell death or functional impairment, which can be detected in several assays (Watson, 1995). In addition to within-species variations, sensitivity to cryopreservation is also known to exhibit some variance within the individual bull population (Lisa and Paul, 2002). Cryo-damage in sperm has been shown to cause premature capacitation and induced DNA damage that would compromise embryonic development (Bailey et al., 2000; Barbas and Mascarenhas, 2009). In AI centres, the ultimate safeguard in a quality control programme is the post-thaw semen evaluation programme (Amann and DeJarnette, 2012).

Conclusions

Accurately identifying bulls that produce large numbers of fertile sperm is critically important for the AI industry. From an AI centre standpoint, profits are associated with the lineup of bulls that are under contract which dictates the demands but also the quality of the semen that is produced to maximize the number of straws per ejaculate in order to efficiently provide the offer to meet the demands (DeJarnette et al., 2004). Hence, the AI centres are trying to define characteristics on the sperm population to estimate fertility potential before putting the sire in service. Simple assessment of mating ability and physical examination of a sire cannot predict his potential fertility. Rather, based on the current available knowledge and modern technologies, there is strong consensus that a combination of laboratory methods testing a large, heterogeneous sperm population for characteristics relevant to both fertilization and embryo development should be applied (Rodríguez-Martínez and Barth, 2007; Kastelic and Thundathil, 2008; Rodríguez-Martínez, 2003).

The Maternal Contribution

A lowering of dairy herd fertility especially in high producing cows has been reported worldwide. The exact nature of this decline is still not clear but some mechanisms have been suggested (Lucy, 2001, 2007; Pryce et al., 2004; Diskin et al., 2006; Dobson et al., 2007; Evans and Walsh, 2011; Walsh et al., 2011). Despite the central role of the male testes and sperm attributes in fertility assessments (Krawetz, 2005), the female oocyte (maternal gamete), ovaries and endocrine function are of great importance in physiology and management of cow reproduction.

Female endocrine function and communication between the hypothalamic–pituitary–ovaries axis in cows play a crucial role in normal oocyte development, ovulation, fertilization, early embryo development, implantation, fetal development and parturition. Their manipulations have been of great importance in dairy cow reproductive management. With current advances in AI some proven sires can produce up to 50,000 offspring in 1 year (Funk, 2006), while the best females have much more limited impact, with generally fewer than 100 offspring. In addition, the current important limiting factor for the efficiency of dairy production systems is failure of cows to successfully establish pregnancy after AI (Evans and Walsh, 2011). Together, these factors emphasize the importance of maternal contribution in fertility and breeding.

Oocyte quality is instrumental for fertility. Oocytes are the largest cells in the body and are produced through oogenesis, which is a complex process regulated by a vast number of intra- and extra-ovarian factors in the ovaries (Sánchez and Smitz, 2012). Unlike sperm, oocytes are limited resources and cattle normally release only one oocyte prior to fertilization. Bovine follicular development occurs through two or three consecutive waves of follicle growth during the oestrous cycle (lasting approximately 21–28 days) (Fig. 13.2). Each follicular wave includes the initial recruitment of a group of follicles each containing a single oocyte arrested at the germinal vesicle (GV) stage, from which one is selected to pursue its growth, resume meiosis and eventually ovulate, while the others undergo atresia (Shoubridge and Wai, 2007; Adams et al., 2008; Mapletoft et al., 2009). The bovine follicle initiates growth following recruitment from the ovarian reserves several months before the observable antral stage development that can be monitored

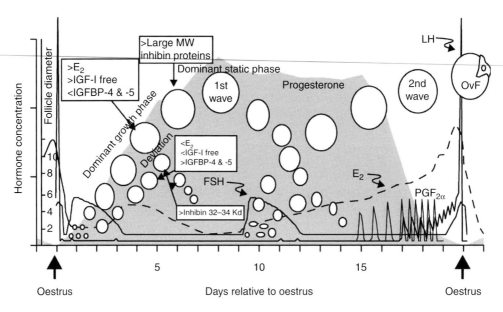

Fig. 13.2. Ovarian follicular and corpus luteum development correlated with endocrine changes during the bovine oestrous cycle. E_2, Oestradiol; IGFBP-4 and -5, insulin-like growth factor binding proteins 4 and 5; OvF, ovulatory follicle. (Reproduced from Moore and Thatcher, 2006, with permission.)

by ultrasonography. During antral growth, the follicle enormously increases in size and volume (300- to 400-fold in diameter from the primary (50 μm) to the pre-ovulatory (15–20 mm) stage (Rajakoski, 1960). The bovine oocyte and follicle grow in parallel until the follicle reaches a diameter of 3 mm, at which point the oocyte is fully grown (diameter plateaus at about 120–130 μm) and already harbours transcriptionally inactive, highly condensed chromatin (GV3 stage) (Macaulay et al., 2011). The follicle can continue to grow up to 15–20 mm in diameter before ovulation (Fair et al., 1995; Fair, 2003).

In addition to providing half of the nuclear genetic material, oocytes endow the embryo with almost all membrane and cytoplasmic determinants such as maternal RNAs, mitochondria and other organelles required for successful fertilization and embryo development and quality (Sirard et al., 2006; Ferreira et al., 2009). The prime example of such asymmetric contribution to embryogenesis is represented by mitochondrial inheritance. While one sperm has less than 100 mitochondria, an oocyte has up to 100,000–400,000 mitochondria. The existence of a mitochondrial bottleneck has been established but still remains to be explained: as

few as 0.01% of these mitochondria in the oocyte actually contribute to the offspring of the next generation, while selective destruction of paternal mitochondria leads to the maternally exclusive lineage of mitochondria (Krawetz, 2005; Shoubridge and Wai, 2007). Hence the oocyte has a crucial role in embryogenesis compared to sperm, and in pregnancy establishment and maintenance through follicular determinants (Pohler et al., 2012; Geary et al., 2013).

Bovine oocytes, similar to those in other mammals, initiate meiotic maturation during fetal life then arrest at the diplotene stage of the first meiotic prophase, also called the germinal vesicle stage (GV) oocyte. Only fully grown female gametes individually enclosed in a large antral follicle will resume meiosis in vivo following the luteinizing hormone (LH) surges that initiates at puberty (Mehlmann, 2005). Oocyte maturation involves complex and distinct, although linked, events of nuclear (chromosomal segregation) and cytoplasmic maturation (organelle reorganization and storage of mRNAs and proteins) (Ferreira et al., 2009). Over the years, nuclear maturation has been the most studied process. During follicular growth, the oocyte remains at the GV stage but its chromatin undergoes remodelling, which is concomitant with transcriptional silencing. The bovine oocyte exhibits unique patterns of chromatin configurations (from GV0 to GV3) distinct from the mouse (Liu et al., 2006; Lodde et al., 2008). Following the endogenous LH surge or following oocyte extraction from the follicle, meiotic resumption is morphologically characterized by germinal vesicle breakdown (GVBD), which is followed by progression to metaphase-I (MI) manifested by extrusion of the first polar body, and then the meiotic cell cycle is arrested a second time at metaphase-II (MII) until fertilization. Cytoplasmic maturation is more discreet and less understood. It consists of three main events: (i) redistribution of cytoplasmic organelles; (ii) dynamics of the cytoskeletal filaments; and (iii) molecular maturation (Ferreira et al., 2009).

It is the sequence of all these cellular and molecular events that leads to the proper preparation of the oocyte to successfully sustain early development. The acquisition of the intrinsic developmental potential by the oocyte is referred to as its developmental competence. The nature and underlying mechanisms of oocyte developmental competence have yet to be elucidated (Duranthon and Renard, 2001). It is known that during oogenesis the growing antral follicle oocyte's cytoplasm enriches with stabilized transcripts (Gilbert et al., 2009), which can be stored for several days before they are used to support early embryonic development at least until embryonic genome activation (EGA) (8–16-cell stage embryo in bovine) after three or four cell cycles (Macaulay et al., 2011). Considering that within an in vitro context where hundreds of collected cumulus–oocytes complexes can be submitted to the same developmental opportunity and that most embryonic losses occur before the developmental stage at which the embryonic genome is known to activate, many studies are looking at the oocyte's transcriptome to identify the potentially lacking transcripts that fail to support protein synthesis before the embryonic take-over (Assidi et al., 2008, 2010; Nivet et al., 2012, 2013; O'Shea et al., 2012; Bunel et al., 2013).

As mentioned, the mechanisms by which an oocyte acquires its potential for embryonic development are not known, but it is clear that the oocyte and its surrounding somatic cells have a series of paracrine and junctional interactions (through transzonal projections), which allow for the exchange of many regulatory signals that control oocyte metabolism, cytoskeletal remodelling, cell cycle progression and fertilization, all of which are key events for initiating and sustaining early embryogenesis (Albertini et al., 2003; Li and Albertini, 2013). In spite of remarkable advancements in reproductive biology and the array of techniques for dairy cattle reproduction (Moore and Thatcher, 2006), oocyte quality in cattle is poorly defined, and the effects of metabolic disorders and disease in the post-partum period on oocyte quality are not well understood (Walsh et al., 2011).

Over several decades, a number of therapies have been developed that manipulate ovarian follicle growth to improve oocyte quality and conception rates in cattle (Baruselli et al., 2012a). Many of the current reproductive technologies in application or being developed are aiming at increasing and maximizing the maternal contribution to genetic improvement efforts. The following aims at portraying the current and emerging technologies that might increase the rate of genetic improvement in dairy herds.

Synchronization

Ovarian follicular growth and development is of great importance as there are a number of intriguing aspects of reproductive physiology that differ somewhat in lactating dairy cows and which may be related to the control of follicular wave dynamics. Most of those changes become more dramatic as milk production increases. Synchronization (mainly on ovulation) has been applied currently to overcome reproductive inefficiencies in dairy cows (Wiltbank et al., 2011). Furthermore, synchronization programmes (including synchronization of oestrus and/or ovulation) have been successfully used in many dairy cow management systems to reduce the intervals from calving to conception (Macmillan, 2010). Any method that will synchronize oestrus will also synchronize the time of ovulation; however, synchronization may not be sufficient to yield good success with timed AI. Synchrony programmes such as Ovsynch (Fig. 13.3) and its

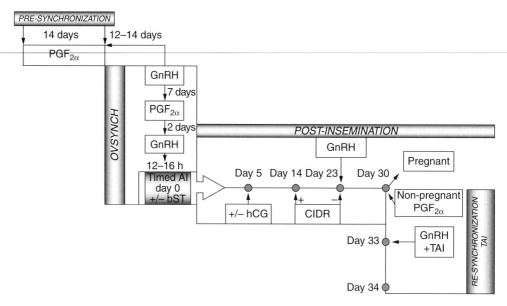

Fig. 13.3. Reproductive management alternatives to improve reproductive performance of lactating dairy cows with the use of presynchronization. Ovsynch for timed artificial insemination (TAI), post-insemination endocrine treatments and resynchronization for TAI. Endocrine treatments involve injection of bovine somatotropin (bST) at TAI and injection of human chorionic gonadotropin (hCG) at day 5 after TAI. Resynchronization of non-pregnant cows involves the insertion of an intravaginal progesterone device (CIDR) between days 14 and 23 after artificial insemination and injection of GnRH at day 23 at the time of CIDR withdrawal. Cows diagnosed non-pregnant at day 30 receive an injection of PGF2α, and on day 33 are injected with GnRH and concurrently inseminated. (Reproduced from Moore and Thatcher, 2006, with permission.)

various modified protocols have been developed for dairy cattle, and some have shown more efficiency than others (Rabiee *et al.*, 2005).

Oestrus/ovulation synchronization

In using AI, oestrus synchronization becomes an important tool, mainly because of its application to improve the efficiency and accuracy of oestrus detection (heat) or because it can reduce the work involved in oestrus detection (Xu, 2011a). However, using oestrus synchronization, oestrus detection is problematic in modern herds of high producing cows as they may display less obvious behavioural symptoms of oestrus and shortened oestrus period (Macmillan, 2010; Roelofs *et al.*, 2010; Evans and Walsh, 2011; Walsh *et al.*, 2011; Wiltbank *et al.*, 2011).

Therefore, due to all the challenges associated with efficient oestrous detection, the current trend is towards ovulation synchronization

or induction, which allows timed breeding without the need for oestrus detection (Macmillan, 2010; Xu, 2011a). Ovulation synchronization (synchronization of follicle wave emergence and growth, which is deliberately terminated by induction of ovulation through administration of luteinizing hormone) is the key component of recent protocols for AI and embryo transfer (ET), namely fixed-time artificial insemination (FTAI) or fixed-time embryo transfer (FTET) (Mapletoft *et al.*, 2009). These technologies provide organized approaches to enhance the use of AI or ET.

Fixed-timed artificial insemination (FTAI)

Currently, FTAI programmes are widely used as an integrated part of reproductive management strategies with satisfactory pregnancy rates in many parts of the world with various protocols (Baruselli *et al.*, 2012b). The key point is synchronization of the follicular wave at the initiation of the programme. GnRH-based (Ovsynch programme) (Fig. 13.3) and oestradiol/progesterone-based programmes for ovulation synchronization are the most common programmes currently used (Fig. 13.4) (Wiltbank *et al.*, 2011). The GnRH-based programme is the original FTAI protocol introduced in 1995 (Pursley *et al.*, 1995). It starts with GnRH administration (to synchronize a new follicular wave and to ensure the presence of a corpus luteum (CL) during the programme), followed 7 days later by prostaglandin $F_{2\alpha}$ (PGF) treatment (to regress the CL and allow the dominant follicle to proceed toward ovulation), 56 h later by a second GnRH injection to synchronize ovulation, and finally timed artificial insemination (TAI) at 16 h after the second GnRH injection (Wiltbank *et al.*, 2011). Presynch–Ovsynch, Co-synch, Heatsynch and Selectsynch are recent variations of Ovsynch, which keep the interval of 7 days between the first GnRH injection and the one following PGF used in the Ovsynch programme (Macmillan, 2010).

(A) Ovynch programme (GnRH-based programme)

(B) Oestradiol/progesterone-based programme

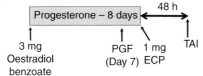

Fig. 13.4. Typical programmes designed to synchronize ovulation of the dominant follicle and facilitate success with a timed AI (TAI) protocol. (A) Representation of the typical Ovsynch programme that begins with GnRH treatment, followed 7 days later by PGF treatment, 56 h later by a second GnRH treatment to synchronize ovulation, and finally TAI at 16 h after the second GnRH. (B) Representation of a typical E2/P4-based programme which begins with insertion of a P4-releasing device and treatment with E2 (in this case 3 mg E2-benzoate), followed 7 days later by PGF treatment. One day later, cows are given 1 mg of oestradiol cypionate (ECP), with TAI 48 h later. (Reproduced from Wiltbank *et al.*, 2011 with permission.)

Recently, there have been some advancements in reproductive management strategies for timed insemination (Xu, 2011b). A representative typical estradiol (E2)/progesterone (P4)-based programme begins with a P4-releasing device being inserted into the vagina (mainly controlled intravaginal drug release (CIDR) inserts) and treatment with E2 (to regress follicles that are present on the ovaries, and approximately 3 to 5 days later, there is initiation of a new follicular wave) followed 7 days later by PGF treatment (to regress any CL). One day later, cows are given E2 cypionate and the CIDR is removed (to synchronize ovulation), with FTAI 48 h later (Wiltbank *et al.*, 2011).

Fixed-time embryo transfer (FTET)

The widespread application of embryo transfer has been limited due to inefficiency of oestrus detection in dairy herds, especially in recipients, although it has been used commercially for many years. FTET allows ET independent of oestrus detection and has been applied as the most useful alternative to increase the sufficient number of well-synchronized recipient cattle utilized in an ET programme (Rodrigues *et al.*, 2010). Such an increase in proportion of recipients is reported to lead to higher pregnancy rates without compromising the conception rate in the case of a single FTET (Baruselli *et al.*, 2010). The basic protocol for FTET is similar to that of FTAI (Ovsynch), with the exception that the embryo transfer is carried out 8 days after the second GnRH (day 17) instead of timed-AI at 16–20 h after the second GnRH injection. Improvements in protocols are ongoing (Baruselli *et al.*, 2011; Bó *et al.*, 2011).

Occurrence of embryonic death and strategies to improve embryo survival

While fertilization rates (>80%) are not considered to be a main contributor to the poor fertility seen in dairy cows, embryo and fetal mortality are (Sartori *et al.*, 2009; Evans and Walsh, 2011; Xu, 2011b). It is estimated that based on fertilization rates of 90%, embryonic and fetal loss (from fertilization to birth) may be up to 60% (Fig. 13.5), while calving rates may only be 30–40% in high-yielding dairy cows (Moore and Thatcher, 2006; Diskin and Morris, 2008). Furthermore, fertility depreciation in dairy cows is believed to be caused by abnormal early embryo development rather than fertilization failure (Lucy, 2007; Bamber et al., 2009) or failure of the cow to maintain the pregnancy (Bamber et al., 2009). In high-producing dairy cows, embryo mortality is the single biggest factor reducing calving rates. Poor oocyte quality (probably caused by the adverse metabolic environment) and poor maternal uterine environment (probably caused by carry-over effects of uterine infection and low circulating progesterone concentrations) are possible reasons for embryo mortality (Evans and Walsh, 2011). Hence, embryonic mortality is a big challenge and can be distinguished as being early embryo or late embryo/fetal loss (Santos et al., 2004; Diskin et al., 2006).

Early embryo mortality occurs between fertilization (day 0) and day 24 of gestation, which includes the developmental stages of cleavage, compaction, blastulation, expansion, hatching and elongation, termed together as the pre-attachment period (Fig. 13.6) (Peippo et al., 2011). Late embryo mortality occurs between day 25 and 45, at which time embryonic differentiation is mostly completed, while fetal mortality occurs after this and up to parturition (Nomenclature, 1972). In high-producing dairy cows (very) early embryo loss occurs mainly within the first 8 days post-fertilization (accounts for ~35% of embryonic mortality in the first week post-fertilization) (Courot et al., 1985; Santos et al., 2004; Sartori et al., 2009). The causes of early embryonic failure are not well understood mainly due to the fact that it is generally detected as a return into oestrus on the following cycle. It has been proposed to have a paternal origin due to inadequacies in the fertilizing sperm. However, maternal failure to recognize pregnancy (estimated to be up to 25% of failures of conception in dairy cows) and gross chromosomal abnormalities (approximately 5% of embryonic mortalities) are other important factors leading to the same phenotype of early embryo loss (Walsh et al., 2011). The extent of late embryo and early fetal mortality is relatively low (~7% on day 24 and 80 for lactating cows) and similarly to early embryo losses the causes are numerous ranging from genetic, physiological, endocrinological to environmental factors (Silke et al., 2002; Van Soom et al., 2007; Diskin et al., 2011; Evans and Walsh, 2011; Diskin and Morris, 2008). Overall, embryo viability is currently a big challenge and is the topic of many research programs aiming at developing means to define and increase embryonic quality.

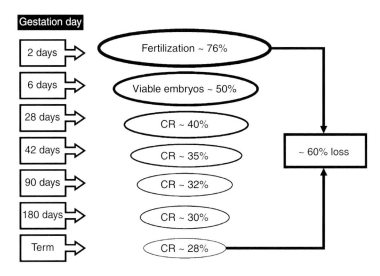

Fig. 13.5. Timing and extent of pregnancy losses in the high producing lactating dairy cow. CR, conception rate. (Reproduced from Santos et al., 2004, with permission.)

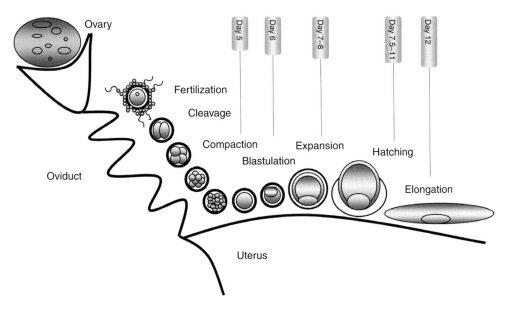

Fig. 13.6. Terms and concepts in the pre-attachment period in a bovine embryo. (Reproduced from Machatya *et al.*, 2012, with permission.)

Multiple ovulation and embryo transfer (MOET)

Being a mono-ovulant, a cow generally produces one offspring in each 9-month gestation. To increase the dissemination of high genetic merit females, it is common to manipulate the endogenous hormonal response and supplement it with exogenous hormones to bypass the natural negative hormonal feedback that limits ovulation to exactly one follicle. Through hormonal manipulation of cow reproductive cycles, a number of oocytes can be simultaneously ovulated and fertilized by AI. Embryos may then be collected and graded according to their quality to either transfer into synchronized recipients of lesser genetic merit (which allows dairy producers to obtain multiple offspring from genetically superior females) or be frozen for later transfer or sale (Moore and Thatcher, 2006). Generally, using superovulation the mean ovulation rate will be about 15 (ovulated oocytes) of which about 10 unfertilized ova/embryos will be recovered. From such a cohort, 60–70% of the embryos will be of good quality and can be stored or transferred to recipients (Lonergan and Boland, 2011). MOET breeding schemes have been widely established (~80% of commercial ET) in

cattle (Rodríguez-Martínez, 2012). This technology has not reached optimality due to some limitations (mainly inefficiency of superovulation protocols in generating large homogenous cohorts of high developmental potential oocytes), however, currently several approaches to improve and simplify superovulation protocols in cattle have been developed (Blondin *et al.*, 2002; Bó *et al.*, 2010; Mapletoft and Bó, 2011; Nivet *et al.*, 2012).

Compared to natural breeding, the use of MOET substantially increases the number of descendants per animal thus increases genetic improvement rates when applied to elite animals by producing more embryos from genetically superior cows in shorter time periods. The standard approach involves the aforementioned hormonal regimen to induce and support follicular growth that will lead to multiple ovulations. The eggs are fertilized *in vivo* by artificial insemination and the embryos are allowed to develop for 6–7 days until their exit from the oviducts to the uterine environment where they are collected by flushing the uterine lumen. The current most efficient embryo production strategy involves superovulation and collecting the immature oocytes by ultrasound-guided transvaginal follicular puncture and aspiration (termed ovum pick-up (OPU)). The oocytes are then matured and fertilized *in vitro* and the embryos are kept in *in vitro* culture until they reach the early blastocyst stage (about 7 days post-fertilization). The combination of OPU and *in vitro* embryo production (IVP) can produce 3.4 more embryos per cow per year than conventional ET (Bousquet *et al.*, 1999). Using OPU up to 1000 oocytes may be collected and up to 300 *in vitro*-produced embryos can be obtained annually per cow (Baruselli *et al.*, 2011; Rodríguez-Martínez, 2012; Wu and Zan, 2012). Future improvements in standard MOET or in the OPU-IVP based MOET aim at reducing the number of animal handlings without compromising embryo production and pregnancy rates, but most importantly aim for better control of the inter-animal variance in ovarian response to the hormonal regimen. The extent of ovarian response is a repeatable phenotype where low yielding animals are consistently providing low numbers of oocytes in response to a specific hormonal regimen. This exemplifies the need to adapt hormonal stimulation protocols on a per animal basis (Keller and Teepker, 1990; Yaakub *et al.*, 1999).

In vitro production of embryos

Since production of the first live calf by *in vitro* fertilization (IVF) in 1981 (Brackett *et al.*, 1982), and from an *in vitro* matured (IVM) and IVF oocyte in 1986 (Hanada *et al.*, 1986), and IVM/IVF/*in vitro* culture (IVC) in 1988 of an oocyte/embryo (Sirard *et al.*, 1988), significant advancements have been achieved in nonsurgical OPU techniques, protocols, materials, culture media, etc. None the less, the efficacy of *in vitro* embryo production is still suboptimal. Typically under normal IVP conditions, >90% of retrieved oocytes are able to complete nuclear maturation (transiting from an immature oocyte to a fertilizable egg), of which 80% can be fertilized and will reach the first cell division (cleavage stage). However, about a third of the collected oocytes will reach the late developmental stages (morula and blastocyst) *in vitro* (Mermillod, 2011). In cattle, it is at these later developmental stages, preferably at compacted morula or early blastocyst, that embryos are transferred into synchronized recipients. For breeders who want to generate as many offspring as possible from a single elite cow, IVF is currently the most efficient approach. Hence, a combination of MOET and IVF is applied by breeders for the genetically superior females such as those identified via genomic testing (Schefers and Weigel, 2012).

To maximize embryo production, oocyte developmental competence plays a critical role (Sirard *et al.*, 2006; Pohler *et al.*, 2012; Geary *et al.*, 2013). This developmental potential intrinsic to the oocyte differs between species: while ovulated mouse oocytes express about 80% developmental potential to reach the blastocyst stage *in vitro*, in bovines, the developmental competence of immature oocytes collected from medium size follicles is seldom higher than 33%. So far, improving embryonic yield through better control of oocyte maturation has proved to be complex since most of the compounds tested so far have yielded marginal results or are affected by a lack of consistency, where blastocyst rates greatly fluctuate from one run to the other without modification to the procedures. Strategies to modulate oocytes' *in vivo* quality prior to follicular aspiration have been developed. One involves withdrawal of growth stimulation about 2 days prior to oocyte aspiration. The rationale is that this 'coasting' period permits follicular differentiation where growth signal alone cannot. It has been successfully used in commercial embryo production settings (Blondin *et al.*, 2002). Using this stimulation regimen, blastocyst yields have been increased from 50% up to 80% (Blondin *et al.*, 2002; Nivet *et al.*, 2012). In addition to the quality of the gamete, culture conditions have been shown to affect embryonic development and quality expressed as post-cryopreservation survival and/or post-transfer survival rates (Hansen *et al.*, 2010).

On the down side, this higher rate of embryonic production through IVP is currently plagued by concerns regarding the quality of the produced embryos. Reports are indicating that IVP embryos: (i) lead to lower conception rates (30–40%) compared to AI or embryos recovered non-surgically from donor animals (50–70%); (ii) lead to higher rates of early embryonic mortality (mainly during the first 30 days); and (iii) are more prone to developmental aberrations such as the large offspring syndrome (LOS), abortions and dystocia (Moore and Thatcher, 2006).

Similar afflictions have been reported for human embryos produced *in vitro*, and recently there have been attempts and promising advances in reproductive medicine on embryo culture platforms to improve developmental rates and embryonic quality through modifications of the microenvironment (Swain and Smith, 2011; Smith *et al.*, 2012). There are also great incentives to select the best gametes and embryos to increase pregnancy success rates, which will lower costs. Some of the promising approaches are based on metabolomics by assessing the excreted metabolites in the culture medium or through oxygen respiration measurement (Montag *et al.*, 2013). Timing of the first zygotic cleavage is a valuable, non-invasive marker of embryonic competence in cattle (Lechniak *et al.*, 2008). Recently, development in a time-lapse cinematography (TLC) system has been considered to be a highly promising, simple and non-invasive technology to objectively classify IVP embryos according to developmental competence. The TLC system enables simultaneous monitoring of embryo morphology and measurement of the length of each developmental stage, together with morphokinetics of the embryo (Somfai *et al.*, 2010; Sugimura *et al.*, 2010; Kirkegaard *et al.*, 2012).

Embryo sexing

Sex determination of embryos has been performed on-site for more than a decade. It is routinely performed by veterinarians following collection of embryos. Embryonic biopsies are performed by taking away several cells from the trophectoderm (extra-embryonic tissue), which are destroyed to detect the presence of a specific sequence found on the Y chromosome by DNA amplification (Fig. 13.7). Biopsied embryos can be kept alive in a portable incubator or frozen for later transfer. The most common hurdles are false negatives as many protocols do not include a positive control of amplification and when the reaction fails, it is by default determined to be a female embryo. The recurrent amplification of a single target can also lead to amplification carry over, where the template from previous amplifications contaminates stock solutions or equipment leading to false positives where all embryos are detected as males (Machaty *et al.*, 2012). It is common in laboratories sexing embryos to have only female technicians perform genotyping to avoid exogenous contamination from human sources. The choice of candidate locus located on the Y chromosome can have an impact on PCR efficiency. Traditionally, the sex-determining region on the Y chromosome (SRY) locus has been targeted for sexing. A multiplexed approach targeting an X chromosome sequence as a positive control and a Y-specific sequence has been shown to lead to more precise results. Sensitivity can be improved by targeting a repeated sequence found on the Y chromosome. Different means of DNA amplification and detection have been proposed. The loop-mediated isothermal amplification (LAMP) reaction does not require any expensive equipment and generates results in under an hour. The amplified product can be run on a gel for detection but as the amplification of the target is very efficient, it leads to solution turbidity, which can be used to detect the presence of amplification using a turbidity meter (Hirayama *et al.*, 2004). In our hands, this method is efficient but has sometimes given ambiguous results when turbidity is low. However, recent improvement permits detection with the naked eye without electrophoresis or a turbidity meter (Zoheir and Allam, 2010).

Oocyte/embryo cryopreservation

The desire of breeders to take full advantage of elite cows drives the development and increases the usage of assisted reproductive methodologies. The exports of cryopreserved semen represent a significant market, and to take the full advantage of both genders, the trade in cryopreserved embryos is expanding. Freezing bovine embryos is now common, and with improved methodologies and cryoprotectant to limit embryonic damage, pregnancy rates can only be slightly lower than those achieved with fresh embryos (Mapletoft and Hasler, 2005).

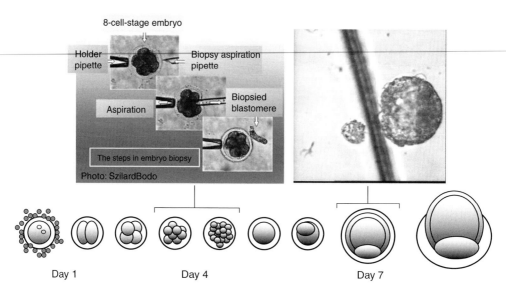

Fig. 13.7. Biopsy of bovine embryo for sex diagnosis. (Reproduced from Machatya *et al.*, 2012, with permission.)

Controlled-rate freezing and vitrification are two basic cryopreservation techniques. Slow freezing commonly leads to intracellular ice crystallization and cell damage. Vitrification is the alternative method of cryopreservation (a non-equilibrium method), which uses an ultra-rapid cooling rate, using high concentrations of cryoprotectants, which avoids water precipitation, preventing intracellular ice crystal formation. Vitrification is becoming more popular, but has yet to achieve convincing results capable of widespread application. Oocyte and embryo cryopreservation can be performed for animal genetic conservation (Prentice and Anzar, 2011; Saragusty and Arav, 2011; Youngs, 2011).

Applications

Cloning

One of the main incentives driving the need to generate genetically identical copies of dairy cattle is to increase the potential semen offer from the next top of the breed bulls. Producing one or more copies of an individual either naturally, artificially (embryo splitting) or by nuclear transfer is called 'cloning' (Moore and Thatcher, 2006). Two general methods can be used to obtain two and/or multiple identical embryos

from one embryo in cattle: (i) cleavage-stage (2–4 and 8-cell embryos) blastomere separation technique to produce multiple (twins, triplets, quadruplet monozygotic) calves; and (ii) embryo bisection technique on post-compacted embryos (morula or blastocysts) to produce mainly identical twins (hemi-embryos) using microsurgery (Rho *et al.*, 1998; Tagawa *et al.*, 2008). Both techniques (separation and bisection) have proven efficient to yield monozygotic twins after laparoscopic transfer to recipient cows, with satisfactory pregnancy rates and up to 60% overall efficiency of cow embryo splitting (number of calves born per embryos bisected and transferred). Somatic (adult or fetal cells) nuclear transfer (SCNT) has been successfully used, however, it is quite inefficient (due to aberrant DNA reprogramming) and very costly. Such drawbacks preclude widespread application at commercial level. In addition to generating genetic copies of genetically superior animals, SCNT can be used for production of transgenic animals (Rodríguez-Martínez, 2012).

Transgenesis

Several methods are available to produce transgenic animals. Transgenesis is used to genetically modify cattle for production improvements (altered production characteristics and quality), better health and welfare (increased disease resistance) (Lewis et al., 2004) or for use as bioreactors (Rudolph, 1999; van Berkel et al., 2002; Yang et al., 2008) for biopharmaceutical applications. Due to extremely low efficiencies (<1%) and ethical considerations, transgenesis remains far from being accepted for widespread commercial use (Lewis et al., 2004; Moore and Thatcher, 2006; Murphy, 2012).

Integrating genomic selection and reproductive technologies

Reproductive technologies have traditionally been applied to cattle for two main objectives: (i) to treat low fertility; (ii) to increase the dissemination rate of elite genetic merit. Nowadays, breeding companies are using the latest genetic selection tools to improve all traits of economic importance including fertility (Veerkamp and Beerda, 2007; Berglund, 2008). It is a means to address the problem of reduced fertility from a genetic perspective with long-term consequences, while the application of reproductive technologies is destined to help producers in managing the problem on a daily basis. Genetic selection is used through a number of approaches and technologies such as conventional breeding (e.g. EBVs, PT), marker-gene assisted selection (e.g. microsatellites, DNA polymorphisms including single nucleotides), and, more recently, genomic selection (e.g. GEBV) and functional genomics. The main hurdles facing genetic improvement of fertility traits are the low accuracy of fertility breeding values, its low heritability and the fact that animals must be selected as parents in order to allow individual observations on their fertility. The fact fertility traits are difficult to grasp is partly explained by the numerous aetiologies that may result in reduced fertility. As with any other biological system, the reproductive system is complex, but it is not a vital system as individuals can live without procreating. Furthermore, it is an accessory system that is highly responsive to environmental conditions such as nutrition, stress, temperature or presence of pathogens. As such, phenotypic recording of infertility is therefore very complex

and confounding. Current fertility evaluations are based on recording unsuccessful induction and maintenance of gestation, which can have numerous causes. None the less, genome-wide association studies (GWASs) and the candidate gene approach have been successfully used in identification of QTLs associated with traits related to dairy cow fertility (e.g. ovulation rate, multiple ovulations, twinning) (Veerkamp and Beerda, 2007; Huang et al., 2010; Pryce et al., 2010; Sahana et al., 2010).

Research in functional genomics is conducted to identify biomarkers to better understand the physiology of reduced fertility, but also as a means to improve reproductive technologies and to better phenotype and classify fertility (Veerkamp and Beerda, 2007; Robert, 2008). More precise phenotyping would translate into more precise genetic evaluations and increase genetic improvement. Functional genomics includes transcriptomics and proteomics studies aiming at identifying biomarkers that are expressed in relation to the studied physiology. It involves every aspect of the reproductive system from oocyte developmental competence and its associated follicular environment (Misirlioglu et al., 2006; Berendt et al., 2009; Gilbert et al., 2011, 2012; Mondou et al., 2012; Walsh et al., 2012; Christenson et al., 2013; Nivet et al., 2013); the mechanisms supporting early embryo development (Bhojwani et al., 2006; Massicotte et al., 2006; Memili et al., 2007; Chaze et al., 2008; Kues et al., 2008; Vigneault et al., 2009; Huang and Khatib, 2010; Robert et al., 2011; Clemente et al., 2011; Driver et al., 2012; Chitwood et al., 2013) and conceptus–maternal interactions that will lead to a successful gestation (Berendt et al., 2005; Forde et al., 2009; Ledgard et al., 2009; Mamo et al., 2011, 2012; Forde et al., 2012a,b; Walker et al., 2012).

Recently, an extra layer of biological complexity has been added that can affect reproductive performances: environmental conditions can modulate gene expression by modifying the epigenome, which represents the sum of the alterations to the chromatin structure, not affecting the DNA sequence itself but modulating how genes are expressed. Simplified, the epigenome represents a layer of chemical factors (from simple elements such as phosphate to more complex ones such as ubiquitin and even binding of RNA molecules) that is added to the chromatin structure (the DNA and the proteins that package it) and can influence how a gene is expressed to the extent of even silencing it long term. (For a review see Faulk and Dolinoy, 2011; Zaidi et al., 2011; van Montfoort et al., 2012; Inbar-Feigenberg et al., 2013; Kohda and Ishino, 2013.) It is believed that the epigenome is a means through which biological systems adapt to environmental conditions. The reproductive system is of particular interest since it has been shown that some stress applied early on during development can have long-term consequences. Examples of the impact of the epigenome are actually numerous. Some of the most striking ones are phenotypes derived from the application of reproductive technologies where a stress applied in the first days of development carries post-birth consequences. For example, although SCNT-derived clones are genetically identical copies (aside from the mitochondrial background), some clones exhibit distinct phenotypic performances although raised in the same environment. In cattle, in vitro-derived calves can suffer from fetal overgrowth (LOS), with in vitro culture conditions such as the presence of serum in the medium proposed to be a causative agent (Young et al., 1998; Hiroyuki et al., 2004).

An important research focus is now directed at better understanding how early events can shape later performances. Several studies are focusing on the first embryonic cell divisions since this time is known to be a window of sensitivity to epigenetic perturbations, as the paternal and maternal genomes must reprogramme to erase their ultra-specialized gametic programme to become embryonic stem cells that will give rise to all cell types to form a complete individual. Another point of focus is the time during gestation where maternal cues can lead to fetal programming. It has been recently proposed that reduced fertility in dairy cattle is attributable to metabolic stress provided by negative energy balance during early lactation, which leads to the birth of heifers with lower ovarian reserves and thus lower fertility (Walsh et al., 2011). These reports are now raising numerous questions pertaining to the long-term impacts of the uterine environment, which would translate into better selection of recipients for embryo transfer. Other impacts of the epigenome could also be modulated from the rearing conditions provided to heifers and young bulls especially around the time of puberty in the latter.

Studying the (epi)-genome

Whole-genome sequencing allows complete discovery of SNPs without any ascertainment bias, but also permits identification of structural variants like copy number variants (CNVs). Current sequencing platforms have not yet reached a low-enough cost per genome to permit routine sequencing of large cohorts of animals within the context of academic research funding. It is anticipated that sequencing the genome of sires or dams will soon become available at the bench in every laboratory and will eventually revolutionize livestock breeding (Pérez-Enciso and Ferretti, 2010; Murphy, 2012). The same impact can be expected in terms of epigenetics study, especially for DNA methylation and the study of RNA populations (de Montera et al., 2013; McGraw et al., 2013). Although currently the study of livestock epigenomes is challenging, particularly for limited samples such as early embryo, platforms are being developed (Shojaei Saadi et al., 2014). Similarly, while third generation platforms for genome sequencing (using single-molecule templates, limiting sample amplification, using less starting material and being less error prone) are being used and improved, fourth generation platforms such as Oxford Nanopore (Oxford, UK) technology offer future promise (Pennisi, 2012; Ku and Roukos, 2013).

Until then, all high-throughput platforms performing genomic profiling are array based and are thus limited to surveying the information represented on the arrays. This information can range from several thousand up to >1.2 M loci (Mardis, 2008; Metzker, 2010; Zhang et al., 2011). Overall, the wealth of data generated from all these high-throughput platforms is currently generating a bottleneck as data processing and analysis is becoming challenging and requires dedicated expertise in bioinformatics.

Conclusions

Modern genetic improvement is becoming increasingly intertwined with the application of reproductive technologies. Traditionally it has been used for the dissemination of semen of elite bulls with a clear focus on improving AI and semen cryopreservation. Although research is still ongoing to improve sire conception, the maternal contribution is now being considered in the application and development of reproductive technologies, with a clear aim of increasing the number of descendants per cow and reducing the generation intervals. From the improvements in the reliability and efficacy of reproductive technologies, breeding companies are now looking into combining the latest genetic selection tools (genomic selection) with the most advanced reproductive technologies (ovarian stimulation, OPU, IVP) to increase the rate at which the next generations of high genetic merit animals

reach market. Genomic selection bypasses the need for progeny testing and thus offers the potential to look for the next top elite bull readily at birth and even earlier when determining the genetic value of the embryos pre-transfer through an embryonic biopsy. Furthermore, improvements in the control of ovarian and testicular functions combined with optimized *in vitro* embryo production could foreseeably lead to the production of offspring from prepubertal parents.

The awareness that pre- and post-conception as well as gestational and pre- and post-puberty conditions can modulate fertility of the animals in the long term (and even in the next generation) is opening up an entirely new field of research opportunities to improve and maximize cattle fertility.

References

Adams, G.P., Jaiswal, R., Singh, J. and Malhi, P. (2008) Progress in understanding ovarian follicular dynamics in cattle. *Theriogenology* 69, 72–80.

Albertini, D.F., Sanfins, A. and Combelles, C.M. (2003) Origins and manifestations of oocyte maturation competencies. *Reprod Biomed Online* 6, 410–415.

Amann, R.P. and DeJarnette, J.M. (2012) Impact of genomic selection of AI dairy sires on their likely utilization and methods to estimate fertility: a paradigm shift. *Theriogenology* 77, 795–817.

Anzar, M., He, L., Buhr, M.M., Kroetsch, T.G. and Pauls, K.P. (2002) Sperm apoptosis in fresh and cryopreserved bull semen detected by flow cytometry and its relationship with fertility. *Biology of Reproduction* 66, 354–360.

Arangasamy, A., Kasimanickam, V.R., DeJarnette, J.M. and Kasimanickam, R.K. (2011) Association of CRISP2, CCT8, PEBP1 mRNA abundance in sperm and sire conception rate in Holstein bulls. *Theriogenology* 76, 570–577.

Arpanahi, A., Brinkworth, M., Iles, D., Krawetz, S.A., Paradowska, A., Platts, A.E., Saida, M., Steger, K., Tedder, P. and Miller, D. (2009) Endonuclease-sensitive regions of human spermatozoal chromatin are highly enriched in promoter and CTCF binding sequences. *Genome Research* 19, 1338–1349.

Assidi, M., Dufort, I., Ali, A., Hamel, M., Algriany, O., Dielemann, S. and Sirard, M.A. (2008) Identification of potential markers of oocyte competence expressed in bovine cumulus cells matured with follicle-stimulating hormone and/or phorbol myristate acetate *in vitro*. *Biology of Reproduction* 79, 209–222.

Assidi, M., Dieleman, S.J. and Sirard, M.A. (2010) Cumulus cell gene expression following the LH surge in bovine preovulatory follicles: potential early markers of oocyte competence. *Reproduction* 140, 835–852.

Bailey, J.L., Blodeau, J.-F. and Cormier, N. (2000) Semen cryopreservation in domestic animals: a damaging and capacitating phenomenon minireview. *Journal of Andrology* 21, 1–7.

Balhorn, R. (2007) The protamine family of sperm nuclear proteins. *Genome Biology* 8, 227.

Bamber, R.L., Shook, G.E., Wiltbank, M.C., Santos, J.E.P. and Fricke, P.M. (2009) Genetic parameters for anovulation and pregnancy loss in dairy cattle. *Journal of Dairy Science* 92, 5739–5753.

Bansal, A.K. and Bilaspuri, G.S. (2011) Impacts of oxidative stress and antioxidants on semen functions. *Veterinary Medicine International* 2011, article 686137.

Barbas, J.P. and Mascarenhas, R.D. (2009) Cryopreservation of domestic animal sperm cells. *Cell and Tissue Banking* 10, 49–62.

Baruselli, P.S., Ferreira, R.M., Filho, M.F., Nasser, L.F., Rodrigues, C.A. and Bo, G.A. (2010) Bovine embryo transfer recipient synchronisation and management in tropical environments. *Reproduction, Fertility, and Development* 22, 67–74.

Baruselli, P.S., Ferreira, R.M., Sales, J.N., Gimenes, L.U., Sá Filho, M.F., Martins, C.M., Rodrigues, C.A. and Bó, G.A. (2011) Timed embryo transfer programs for management of donor and recipient cattle. *Theriogenology* 76, 1583–1593.

Baruselli, P.S., Sá Filho, M.F., Ferreira, R.M., Sales, J.N.S., Gimenes, L.U., Vieira, L.M., Mendanha, M.F. and Bó, G.A. (2012a) Manipulation of follicle development to ensure optimal oocyte quality and conception rates in cattle. *Reproduction in Domestic Animals* 47, 134–141.

Baruselli, P.S., Sales, J.N.S., Sala, R.V., Vieira, L.M. and Sá Filho, M.F. (2012b) History, evolution and perspectives of timed artificial insemination programs in Brazil. *Animal Reproduction, Belo Horizonte* 9, 139–152.

Benson, J., Woods, E., Walters, E. and Critser, J. (2012) The cryobiology of spermatozoa. *Theriogenology* 78, 1682–1699.

Berendt, F.J., Fröhlich, T., Schmidt, S.E.M., Reichenbach, H.-D., Wolf, E. and Arnold, G.J. (2005) Holistic differential analysis of embryo-induced alterations in the proteome of bovine endometrium in the pre-attachment period. *Proteomics* 5, 2551–2560.

Berendt, F.J., Frohlich, T., Bolbrinker, P., Boelhauve, M., Gungor, T., Habermann, F.A., Wolf, E. and Arnold, G.J. (2009) Highly sensitive saturation labeling reveals changes in abundance of cell cycle-associated proteins and redox enzyme variants during oocyte maturation *in vitro*. *Proteomics* 9, 550–564.

Berglund, B. (2008) Genetic improvement of dairy cow reproductive performance. *Reproduction in Domestic Animals* 43, 89–95.

Bernstein, B.E., Meissner, A. and Lander, E.S. (2007) The mammalian epigenome. *Cell* 128, 669–681.

Berruti, G. and Paiardi, C. (2011) Acrosome biogenesis: revisiting old questions to yield new insights. *Spermatogenesis* 1, 95–98.

Bhojwani, M., Rudolph, E., Kanitz, W., Zuehlke, H., Schneider, F. and Tomek, W. (2006) Molecular analysis of maturation processes by protein and phosphoprotein profiling during *in vitro* maturation of bovine oocytes: a proteomic approach. *Cloning and Stem Cells* 8, 259–274.

Blaschek, M., Kaya, A., Zwald, N., Memili, E. and Kirkpatrick, B.W. (2011) A whole-genome association analysis of noncompensatory fertility in Holstein bulls. *Journal of Dairy Science* 94, 4695–4699.

Blondin, P., Bousquet, D., Twagiramungu, H., Barnes, F. and Sirard, M.-A. (2002) Manipulation of follicular development to produce developmentally competent bovine oocytes. *Biology of Reproduction* 66, 38–43.

Bó, G.A., Guerrero, D.C., Tribulo, A., Tribulo, H., Tribulo, R., Rogan, D. and Mapletoft, R.J. (2010) New approaches to superovulation in the cow. *Reproduction, Fertility, and Development* 22, 106–112.

Bó, G.A., Peres, L.C., Cutaia, L.E., Pincinato, D., Baruselli, P.S. and Mapletoft, R.J. (2011) Treatments for the synchronisation of bovine recipients for fixed-time embryo transfer and improvement of pregnancy rates. *Reproduction, Fertility and Development* 24, 272–277.

Boe-Hansen, G.B., Morris, I.D., Ersboll, A.K., Greve, T. and Christensen, P. (2005) DNA integrity in sexed bull sperm assessed by neutral Comet assay and sperm chromatin structure assay. *Theriogenology* 63, 1789–1802.

Bousquet, D., Twagiramungu, H., Morin, N., Brisson, C., Carboneau, G. and Durocher, J. (1999) *In vitro* embryo production in the cow: an effective alternative to the conventional embryo production approach. *Theriogenology* 51, 59–70.

Brackett, B.G., Bousquet, D., Boice, M.L., Donawick, W.J., Evans, J.F. and Dressel, M.A. (1982) Normal development following *in vitro* fertilization in the cow. *Biology of Reproduction* 27, 147–158.

Brewis, I.A. and Gadella, B.M. (2010) Sperm surface proteomics: from protein lists to biological function. *Molecular Human Reproduction* 16, 68–79.

Buffone, M.G., Foster, J.A. and Gerton, G.L. (2008) The role of the acrosomal matrix in fertilization. *International Journal of Developmental Biology* 52, 511–522.

Bunel, A., Nivet, A.L., Blondin, P., Vigneault, C., Richard, F.J. and Sirard, M.A. (2013) Cumulus cell gene expression associated with pre-ovulatory acquisition of developmental competence in bovine oocytes. *Reproduction, Fertility, and Development* doi:10.1071/RD13061.

Card, C.J., Anderson, E.J., Zamberlan, S., Krieger, K.E., Kaproth, M. and Sartini, B.L. (2013) Cryopreserved bovine spermatozoal transcript profile as revealed by high-throughput ribonucleic acid sequencing. *Biology of Reproduction* 88, Article 49, 41–49.

Cardellino, R., Hoffmann, I. and Tempelman, K.A. (2003) First report on the state of the world's animal genetic resources: views on biotechnology as expressed in country reports. In: FAO/IAEA International Symposium on the Applications of Gene-based Technologies for Improving Animal Production and Health in Developing Countries. FAO/IAEA, Vienna, Austria, pp. 12–14.

Casas, E., Ford, J.J. and Rohrer, G.A. (2010) Quantitative genomics of male reproduction. In: Jiang, Z. and Ott, T.L. (eds) *Reproductive Genomics in Domestic Animals*. Wiley-Blackwell, Oxford, pp. 53–66.

Chaze, T., Meunier, B., Chambon, C., Jurie, C. and Picard, B. (2008) *In vivo* proteome dynamics during early bovine myogenesis. *Proteomics* 8, 4236–4248.

Chitwood, J., Rincon, G., Kaiser, G., Medrano, J. and Ross, P. (2013) RNA-seq analysis of single bovine blastocysts. *BMC Genomics* 14, 350.

Christenson, L.K., Gunewardena, S., Hong, X., Spitschak, M., Baufeld, A. and Vanselow, J. (2013) Research resource: pre-ovulatory LH surge effects on follicular theca and granulosa transcriptomes. *Molecular Endocrinology* 27, 1153–1171.

Clemente, M., Lopez-Vidriero, I., O'Gaora, P., Mehta, J.P., Forde, N., Gutierrez-Adan, A., Lonergan, P. and Rizos, D. (2011) Transcriptome changes at the initiation of elongation in the bovine conceptus. *Biology of Reproduction* 85, 285–295.

Cornwell, J.M., McGilliard, M.L., Kasimanickam, R. and Nebel, R.L. (2006) Effect of sire fertility and timing of artificial insemination in a Presynch + Ovsynch protocol on first-service pregnancy rates. *Journal of Dairy Science* 89, 2473–2478.

Courot, M., Colas, G. and Scaramuzzi, C.D. (1985) The role of the male in embryonic mortality (cattle and sheep). In: Sreenan, J.M. and Diskin, M.G. (eds) *Embryonic Mortality in Farm Animals*, Vol. 34. Springer Netherlands, pp. 195–206.

Cross, N.L. and Meizel, S. (1989) Methods for evaluating the acrosomal status of mammalian sperm. *Biology of Reproduction* 41, 635–641.

Curry, M. (2000) Cryopreservation of semen from domestic livestock. *Reviews in Reproduction* 5, 46–52.

D'Amours, O., Frenette, G., Fortier, M., Leclerc, P. and Sullivan, R. (2010) Proteomic comparison of detergent-extracted sperm proteins from bulls with different fertility indexes. *Reproduction* 139, 545–556.

D'Occhio, M., Hengstberger, K. and Johnston, S. (2007) Biology of sperm chromatin structure and relationship to male fertility and embryonic survival. *Animal Reproduction Science* 101, 1–17.

de Montera, B., Fournier, E., Shojaei Saadi, H.A., Gagne, D., Laflamme, I., Blondin, P., Sirard, M.A. and Robert, C. (2013) Combined methylation mapping of 5mC and 5hmC during early embryonic stages in bovine. *BMC Genomics* 14, 406.

De Vries, A., Overton, M., Fetrow, J., Leslie, K., Eicker, S. and Rogers, G. (2008) Exploring the impact of sexed semen on the structure of the dairy industry. *Journal of Dairy Science* 91, 847–856.

DeJarnette, J., Marshall, C., Lenz, R., Monke, D., Ayars, W. and Sattler, C. (2004) Sustaining the fertility of artificially inseminated dairy cattle: the role of the artificial insemination industry. *Journal of Dairy Science* 87, E93–E104.

Dirican, E. (2012) Magnetic-activated cell sorting of human spermatozoa. In: Nagy, Z.P., Varghese, A.C. and Agarwal, A. (eds) *Practical Manual of In Vitro Fertilization*. Springer, New York, pp. 265–272.

Diskin, M.G. and Morris, D.G. (2008) Embryonic and early foetal losses in cattle and other ruminants. *Reproduction in Domestic Animals* 43, 260–267.

Diskin, M.G., Murphy, J.J. and Sreenan, J.M. (2006) Embryo survival in dairy cows managed under pastoral conditions. *Animal Reproduction Science* 96, 297–311.

Diskin, M.G., Parr, M.H. and Morris, D.G. (2011) Embryo death in cattle: an update. *Reproduction, Fertility and Development* 24, 244–251.

Dobson, H., Smith, R.F., Royal, M.D., Knight, C.H. and Sheldon, I.M. (2007) The high-producing dairy cow and its reproductive performance. *Reproduction in Domestic Animals* 42, 17–23.

Driver, A., Penagaricano, F., Huang, W., Ahmad, K., Hackbart, K., Wiltbank, M. and Khatib, H. (2012) RNA-Seq analysis uncovers transcriptomic variations between morphologically similar *in vivo*- and *in vitro*-derived bovine blastocysts. *BMC Genomics* 13, 118.

Druet, T., Fritz, S., Sellem, E., Basso, B., Gérard, O., Salas-Cortes, L., Humblot, P., Druart, X. and Eggen, A. (2009) Estimation of genetic parameters and genome scan for 15 semen characteristics traits of Holstein bulls. *Journal of Animal Breeding and Genetics* 126, 269–277.

Duranthon, V. and Renard, J.P. (2001) The developmental competence of mammalian oocytes: a convenient but biologically fuzzy concept. *Theriogenology* 55, 1277–1289.

Eckhardt, F., Lewin, J., Cortese, R., Rakyan, V.K., Attwood, J., Burger, M., Burton, J., Cox, T.V., Davies, R., Down, T.A. *et al.* (2006) DNA methylation profiling of human chromosomes 6, 20 and 22. *Nature Genetics* 38, 1378–1385.

Eddy, E.M. (2006) The spermatozoon. In: Neill, J.D., Plant, T.M., Pfaff, D.W., Challis, J.R.G., de Kretser, D.M., Richards, J.S. and Wassarman, P.M. (eds). *Knobil and Neill's Physiology of Reproduction (Third Edition)*. Academic Press, St Louis, Missouri, pp. 3–54.

Evans, A.C.O. and Walsh, S.W. (2011) The physiology of multifactorial problems limiting the establishment of pregnancy in dairy cattle. *Reproduction, Fertility and Development* 24, 233–237.

Evenson, D.P. (2013) Sperm chromatin structure assay (SCSA(R)). *Methods in Molecular Biology* 927, 147–164.

Evenson, D.P. and Wixon, R. (2006) Clinical aspects of sperm DNA fragmentation detection and male infertility. *Theriogenology* 65, 979–991.

Evenson, D.P., Darzynkiewicz, Z. and Melamed, M.R. (1982) Simultaneous measurement by flow cytometry of sperm cell viability and mitochondrial membrane potential related to cell motility. *Journal of Histochemistry and Cytochemistry* 30, 279–280.

Fair, T. (2003) Follicular oocyte growth and acquisition of developmental competence. *Animal Reproduction Science* 78, 203–216.

Fair, T., Hyttel, P. and Greve, T. (1995) Bovine oocyte diameter in relation to maturational competence and transcriptional activity. *Molecular Reproduction and Development* 42, 437–442.

Faulk, C. and Dolinoy, D.C. (2011) Timing is everything: the when and how of environmentally induced changes in the epigenome of animals. *Epigenetics* 6, 791–797.

Ferreira, E.M., Vireque, A.A., Adona, P.R., Meirelles, F.V., Ferriani, R.A. and Navarro, P.A.A.S. (2009) Cytoplasmic maturation of bovine oocytes: Structural and biochemical modifications and acquisition of developmental competence. *Theriogenology* 71, 836–848.

Feugang, J.M., Kaya, A., Page, G.P., Chen, L., Mehta, T., Hirani, K., Nazareth, L., Topper, E., Gibbs, R. and Memili, E. (2009) Two-stage genome-wide association study identifies integrin beta 5 as having potential role in bull fertility. *BMC Genomics* 10, 176.

Feugang, J.M., Rodriguez-Osorio, N., Kaya, A., Wang, H., Page, G., Ostermeier, G.C., Topper, E.K. and Memili, E. (2010) Transcriptome analysis of bull spermatozoa: implications for male fertility. *Reproductive BioMedicine Online* 21, 312–324.

Flint, A.P.F. and Woolliams, J.A. (2008) Precision animal breeding. *Philosophical Transactions of the Royal Society B: Biological Sciences* 363, 573–590.

Florman, H.M., Jungnickel, M.K. and Sutton, K.A. (2008) Regulating the acrosome reaction. *International Journal of Developmental Biology* 52, 503–510.

Foote, R. (2002) The history of artificial insemination: selected notes and notables. *Journal of Animal Science* 80, 1–10.

Foote, R.H. (2003) Fertility estimation: a review of past experience and future prospects. *Animal Reproduction Science* 75, 119–139.

Forde, N., Carter, F., Fair, T., Crowe, M.A., Evans, A.C.O., Spencer, T.E., Bazer, F.W., McBride, R., Boland, M.P., O'Gaora, P. *et al.* (2009) Progesterone-regulated changes in endometrial gene expression contribute to advanced conceptus development in cattle. *Biology of Reproduction* 81, 784–794.

Forde, N., Duffy, G.B., McGettigan, P.A., Browne, J.A., Prakash Mehta, J., Kelly, A.K., Mansouri-Attia, N., Sandra, O., Loftus, B.J., Crowe, M.A. *et al.* (2012a) Evidence for an early endometrial response to pregnancy in cattle: both dependent upon and independent of interferon tau. *Physiological Genomics* 44, 799–810.

Forde, N., Mehta, J.P., Minten, M., Crowe, M.A., Roche, J.F., Spencer, T.E. and Lonergan, P. (2012b) Effects of low progesterone on the endometrial transcriptome in cattle. *Biology of Reproduction* 87, 124, 121–111.

Fortes, M.R.S., Reverter, A., Hawken, R.J., Bolormaa, S. and Lehnert, S.A. (2012) Candidate genes associated with testicular development, sperm quality, and hormone levels of inhibin, luteinizing hormone, and insulin-like growth factor 1 in Brahman bulls. *Biology of Reproduction* 87, 58, 51–58.

Fouquet, J.-P. and Kann, M.-L. (1994) The cytoskeleton of mammalian spermatozoa. *Biology of the Cell* 81, 89–93.

Funk, D.A. (2006) Major advances in globalization and consolidation of the artificial insemination industry. *Journal of Dairy Science* 89, 1362–1368.

Gadkar, S., Shah, C.A., Sachdeva, G., Samant, U. and Puri, C.P. (2002) Progesterone receptor as an indicator of sperm function. *Biology of Reproduction* 67, 1327–1336.

Gaffney, E.A., Gadêlha, H., Smith, D.J., Blake, J.R. and Kirkman-Brown, J.C. (2011) Mammalian sperm motility: observation and theory. *Annual Review of Fluid Mechanics* 43, 501–528.

Garner, D.L. (2006) Flow cytometric sexing of mammalian sperm. *Theriogenology* 65, 943–957.

Garner, D.L. and Seidel, G.E., Jr (2008) History of commercializing sexed semen for cattle. *Theriogenology* 69, 886–895.

Garner, D.L. and Thomas, C.A. (1999) Organelle-specific probe JC-1 identifies membrane potential differences in the mitochondrial function of bovine sperm. *Molecular Reproduction and Development* 53, 222–229.

Garner, D.L., Pinkel, D., Johnson, L.A. and Pace, M.M. (1986) Assessment of spermatozoal function using dual fluorescent staining and flow cytometric analyses. *Biology of Reproduction* 34, 127–138.

Garner, D.L., Johnson, L.A., Yue, S.T., Roth, B.L. and Haugland, R.P. (1994) Dual DNA staining assessment of bovine sperm viability using SYBR-14 and propidium iodide. *Journal of Andrology* 15, 620–629.

Garner, D.L., Thomas, C.A., Joerg, H.W., DeJarnette, J.M. and Marshall, C.E. (1997) Fluorometric assessments of mitochondrial function and viability in cryopreserved bovine spermatozoa. *Biology of Reproduction* 57, 1401–1406.

Gaviraghi, A., Deriu, F., Soggiu, A., Galli, A., Bonacina, C., Bonizzi, L. and Roncada, P. (2010) Proteomics to investigate fertility in bulls. *Veterinary Research Communications* 34, 33–36.

Geary, T.W., Smith, M.F., MacNeil, M.D., Day, M.L., Bridges, G.A., Perry, G.A., Abreu, F.M., Atkins, J.A., Pohler, K.G., Jinks, E.M. and Madsen, C.A. (2013) Triennial Reproduction Symposium: influence of follicular characteristics at ovulation on early embryonic survival. *Journal of Animal Science* 91, 3014–3021.

Gilbert, I., Bissonnette, N., Boissonneault, G., Vallee, M. and Robert, C. (2007) A molecular analysis of the population of mRNA in bovine spermatozoa. *Reproduction* 133, 1073–1086.

Gilbert, I., Scantland, S., Sylvestre, E.-L., Gravel, C., Laflamme, I., Sirard, M.-A. and Robert, C. (2009) The dynamics of gene products fluctuation during bovine pre-hatching development. *Molecular Reproduction and Development* 76, 762–772.

Gilbert, I., Robert, C., Dieleman, S., Blondin, P. and Sirard, M.-A. (2011) Transcriptional effect of the LH surge in bovine granulosa cells during the peri-ovulation period. *Reproduction* 141, 193–205.

Gilbert, I., Robert, C., Vigneault, C., Blondin, P. and Sirard, M.-A. (2012) Impact of the LH surge on granulosa cell transcript levels as markers of oocyte developmental competence in cattle. *Reproduction* 143, 735–747.

Gillan, L., Evans, G. and Maxwell, W.M.C. (2005) Flow cytometric evaluation of sperm parameters in relation to fertility potential. *Theriogenology* 63, 445–457.

Gonzalez-Recio, O. (2012) Epigenetics: a new challenge in the post-genomic era of livestock. *Frontiers in Genetics* 2, 1–4.

Govindaraju, A., Uzun, A., Robertson, L., Atli, M., Kaya, A., Topper, E., Crate, E., Padbury, J., Perkins, A. and Memili, E. (2012) Dynamics of microRNAs in bull spermatozoa. *Reproductive Biology and Endocrinology* 10, 82.

Gravance, C.G., Vishwanath, R., Pitt, C. and Casey, P.J. (1996) Computer automated morphometric analysis of bull sperm heads. *Theriogenology* 46, 1205–1215.

Gravance, C.G., Casey, M.E. and Casey, P.J. (2009) Pre-freeze bull sperm head morphometry related to post-thaw fertility. *Animal Reproduction Science* 114, 81–88.

Grunewald, S., Paasch, U. and Glander, H.J. (2001) Enrichment of non-apoptotic human spermatozoa after cryopreservation by immunomagnetic cell sorting. *Cell and Tissue Banking* 2, 127–133.

Hallap, T., Nagy, S., Jaakma, Ü., Johannisson, A. and Rodríguez-Martínez, H. (2005) Mitochondrial activity of frozen-thawed spermatozoa assessed by MitoTracker Deep Red 633. *Theriogenology* 63, 2311–2322.

Hallap, T., Nagy, S., Jaakma, Ü., Johannisson, A. and Rodríguez-Martínez, H. (2006) Usefulness of a triple fluorochrome combination Merocyanine 540/Yo-Pro 1/Hoechst 33342 in assessing membrane stability of viable frozen-thawed spermatozoa from Estonian Holstein AI bulls. *Theriogenology* 65, 1122–1136.

Hammoud, S.S., Nix, D.A., Zhang, H., Purwar, J., Carrell, D.T. and Cairns, B.R. (2009) Distinctive chromatin in human sperm packages genes for embryo development. *Nature* 460, 473–478.

Hanada, A., Enya, Y. and Suzuki, T. (1986) Birth of calves by non-surgical transfer of *in vitro* fertilized embryos obtained from oocytes matured *in vitro*. *Japan Journal of Animal Reproduction* 32, 208.

Hansen, P.J., Block, J., Loureiro, B., Bonilla, L. and Hendricks, K.E. (2010) Effects of gamete source and culture conditions on the competence of *in vitro*-produced embryos for post-transfer survival in cattle. *Reproduction, Fertility, and Development* 22, 59–66.

Hess, R. and Franca, L. (2009) Spermatogenesis and cycle of the seminiferous epithelium. In: Cheng, C.Y. (ed.) *Molecular Mechanisms in Spermatogenesis, Vol. 636*. Springer, New York, pp. 1–15.

Hinsch, E., Ponce, A.A., Hagele, W., Hedrich, F., Muller-Schlosser, F., Schill, W.B. and Hinsch, K.D. (1997) A new combined *in-vitro* test model for the identification of substances affecting essential sperm functions. *Human Reproduction* 12, 1673–1681.

Hirayama, H., Kageyama, S., Moriyasu, S., Sawai, K., Onoe, S., Takahashi, Y., Katagiri, S., Toen, K., Watanabe, K., Notomi, T. *et al.* (2004) Rapid sexing of bovine preimplantation embryos using loop-mediated isothermal amplification. *Theriogenology* 62, 887–896.

Hiroyuki, A., Hitoshi, S., Shigeo, A. and Hiroyoshi, H. (2004) *In vitro* culture and evaluation of embryos for production of high quality bovine embryos. *Journal of Mammalian Ova Research* 21, 22–30.

Hoflack, G., Rijsselaere, T., Maes, D., Dewulf, J., Opsomer, G., De Kruif, A. and Van Soom, A. (2005) Validation and usefulness of the sperm quality analyzer (SQA II-C) for bull semen analysis. *Reproduction in Domestic Animals* 40, 237–244.

Holt, W.V. (2000a) Basic aspects of frozen storage of semen. *Animal Reproduction Science* 62, 3–22.

Holt, W.V. (2000b) Fundamental aspects of sperm cryobiology: the importance of species and individual differences. *Theriogenology* 53, 47–58.

Hoogendijk, C.F., Kruger, T.F., Bouic, P.J. and Henkel, R.R. (2009) A novel approach for the selection of human sperm using annexin V-binding and flow cytometry. *Fertility and Sterility* 91, 1285–1292.

Hossain, M., Johannisson, A., Wallgren, M., Nagy, S., Siqueira, A. and Rodríguez-Martínez, H. (2011) Flow cytometry for the assessment of animal sperm integrity and functionality: state of the art. *Asian Journal of Andrology* 13, 406–419.

Hu, J.-H., Jiang, Z.-L., Lv, R.-K., Li, Q.-W., Zhang, S.-S., Zan, L.-S., Li, Y.-K. and Li, X. (2011) The advantages of low-density lipoproteins in the cryopreservation of bull semen. *Cryobiology* 62, 83–87.

Huang, W. and Khatib, H. (2010) Comparison of transcriptomic landscapes of bovine embryos using RNA-Seq. *BMC Genomics* 11, 711.

Huang, W., Kirkpatrick, B.W., Rosa, G.J.M. and Khatib, H. (2010) A genome-wide association study using selective DNA pooling identifies candidate markers for fertility in Holstein cattle. *Animal Genetics* 41, 570–578.

Ickowicz, D., Finkelstein, M. and Breitbart, H. (2012) Mechanism of sperm capacitation and the acrosome reaction: role of protein kinases. *Asian Journal of Andrology* 14, 816–821.

Inbar-Feigenberg, M., Choufani, S., Butcher, D.T., Roifman, M. and Weksberg, R. (2013) Basic concepts of epigenetics. *Fertility and Sterility* 99, 607–615.

Jenkins, T.G. and Carrell, D.T. (2012) The sperm epigenome and potential implications for the developing embryo. *Reproduction* 143, 727–734.

Jeyendran, R.S., Van der Ven, H.H., Perez-Pelaez, M., Crabo, B.G. and Zaneveld, L.J. (1984) Development of an assay to assess the functional integrity of the human sperm membrane and its relationship to other semen characteristics. *Journal of Reproduction and Fertility* 70, 219–228.

Johnson, G.D., Lalancette, C., Linnemann, A.K., Leduc, F., Boissonneault, G. and Krawetz, S.A. (2011) The sperm nucleus: chromatin, RNA, and the nuclear matrix. *Reproduction* 141, 21–36.

Karoui, S., Díaz, C., Serrano, M., Cue, R., Celorrio, I. and Carabaño, M.J. (2011) Time trends, environmental factors and genetic basis of semen traits collected in Holstein bulls under commercial conditions. *Animal Reproduction Science* 124, 28–38.

Kasimanickam, V., Kasimanickam, R., Arangasamy, A., Saberivand, A., Stevenson, J.S. and Kastelic, J.P. (2012) Association between mRNA abundance of functional sperm function proteins and fertility of Holstein bulls. *Theriogenology* 78, 2007–2019.

Kastelic, J.P. and Thundathil, J.C. (2008) Breeding soundness evaluation and semen analysis for predicting bull fertility. *Reproduction in Domestic Animals* 43, 368–373.

Kathiravan, P., Kalatharan, J., Karthikeya, G., Rengarajan, K. and Kadirvel, G. (2011) Objective sperm motion analysis to assess dairy bull fertility using computer-aided system – a review. *Reproduction in Domestic Animals* 46, 165–172.

Keller, D.S. and Teepker, G. (1990) Effect of variability in response to superovulation on donor cow selection differentials in nucleus breeding schemes. *Journal of Dairy Science* 73, 549–554.

Khatib, H., Monson, R.L., Huang, W., Khatib, R., Schutzkus, V., Khateeb, H. and Parrish, J.J. (2010) Short communication: validation of *in vitro* fertility genes in a Holstein bull population. *Journal of Dairy Science* 93, 2244–2249.

Kiefer, J.C. (2007) Epigenetics in development. *Developmental Dynamics* 236, 1144–1156.

Killian, G.J., Chapman, D.A. and Rogowski, L.A. (1993) Fertility-associated proteins in Holstein bull seminal plasma. *Biology of Reproduction* 49, 1202–1207.

Kirkegaard, K., Agerholm, I.E. and Ingerslev, H.J. (2012) Time-lapse monitoring as a tool for clinical embryo assessment. *Human Reproduction* 27, 1277–1285.

Kohda, T. and Ishino, F. (2013) Embryo manipulation via assisted reproductive technology and epigenetic asymmetry in mammalian early development. *Philosophical Transactions of the Royal Society B: Biological Sciences*, 368, doi:10.1098/rstb.2012.0353.

Krawetz, S.A. (2005) Paternal contribution: new insights and future challenges. Nature reviews. *Genetics* 6, 633–642.

Ku, C.S. and Roukos, D.H. (2013) From next-generation sequencing to nanopore sequencing technology: paving the way to personalized genomic medicine. *Expert Review of Medical Devices* 10, 1–6.

Kues, W.A., Sudheer, S., Herrmann, D., Carnwath, J.W., Havlicek, V., Besenfelder, U., Lehrach, H., Adjaye, J. and Niemann, H. (2008) Genome-wide expression profiling reveals distinct clusters of transcriptional regulation during bovine preimplantation development *in vivo. Proceedings of the National Academy of Sciences* 105, 19768–19773.

Kuhn, M.T. and Hutchison, J.L. (2008) Prediction of dairy bull fertility from field data: use of multiple services and identification and utilization of factors affecting bull fertility. *Journal of Dairy Science* 91, 2481–2492.

Kuhn, M.T., Hutchison, J.L. and Norman, H.D. (2008) Modeling nuisance variables for prediction of service sire fertility. *Journal of Dairy Science* 91, 2823–2835.

Kumar, P., Kumar, D., Singh, I. and Yadav, P.S. (2012) Seminal plasma proteome: promising biomarkers for bull fertility. *Agricultural Research* 1, 78–86.

Lalancette, C., Miller, D., Li, Y. and Krawetz, S.A. (2008) Paternal contributions: new functional insights for spermatozoal RNA. *Journal of Cellular Biochemistry* 104, 1570–1579.

Lechniak, D., Pers-Kamczyc, E. and Pawlak, P. (2008) Timing of the first zygotic cleavage as a marker of developmental potential of mammalian embryos. *Reproductive Biology* 8, 23–42.

Ledgard, A.M., Lee, R.S. and Peterson, A.J. (2009) Bovine endometrial legumain and TIMP-2 regulation in response to presence of a conceptus. *Molecular Reproduction and Development* 76, 65–74.

Leibo, S.P. and Pool, T.B. (2011) The principal variables of cryopreservation: solutions, temperatures, and rate changes. *Fertility and Sterility* 96, 269–276.

Lewis, I.M., French, A.J., Tecirlioglu, R.T., Vajta, G., McClintock, A.E., Nicholas, K.R., Zuelke, K.A., Holland, M.K. and Trounson, A.O. (2004) Commercial aspects of cloning and genetic modification in cattle. *Australian Journal of Experimental Agriculture* 44, 1105–1111.

Li, G., Peñagaricano, F., Weigel, K.A., Zhang, Y., Rosa, G. and Khatib, H. (2012) Comparative genomics between fly, mouse, and cattle identifies genes associated with sire conception rate. *Journal of Dairy Science* 95, 6122–6129.

Li, R. and Albertini, D.F. (2013) The road to maturation: somatic cell interaction and self-organization of the mammalian oocyte. Nature reviews. *Molecular Cell Biology* 14, 141–152.

Lisa, M.T. and Paul, F.W. (2002) Semen cryopreservation: a genetic explanation for species and individual variation? *Cryoletters* 23, 255–262.

Liu, Y., Sui, H.-S., Wang, H.-L., Yuan, J.-H., Luo, M.-J., Xia, P. and Tan, J.-H. (2006) Germinal vesicle chromatin configurations of bovine oocytes. *Microscopy Research and Technique* 69, 799–807.

Lodde, V., Modina, S., Maddox-Hyttel, P., Franciosi, F., Lauria, A. and Luciano, A.M. (2008) Oocyte morphology and transcriptional silencing in relation to chromatin remodeling during the final phases of bovine oocyte growth. *Molecular Reproduction and Development* 75, 915–924.

Lonergan, P. and Boland, M.P. (2011) Gamete and embryo technology: multiple ovulation and embryo transfer. In: John, W. F. (ed.) *Encyclopedia of Dairy Sciences (Second Edition).* Academic Press, San Diego, California, pp. 623–630.

Lucy, M.C. (2001) Reproductive loss in high-producing dairy cattle: where will it end? *Journal of Dairy Science* 84, 1277–1293.

Lucy, M.C. (2007) Fertility in high-producing dairy cows: reasons for decline and corrective strategies for sustainable improvement. *Society of Reproduction and Fertility Supplement* 64, 237–254.

Macaulay, A., Scantland, S. and Rober, C. (2011) RNA processing during early embryogenesis: managing storage, utilisation and destruction. In: Grabowski, P. (ed.) *RNA Processing.* InTech, DOI:10.5772/20375.

Machaty, Z., Peippo, J. and Peter, A. (2012) Production and manipulation of bovine embryos: techniques and terminology. *Theriogenology* 78, 937–950.

Macmillan, K.L. (2010) Recent advances in the synchronization of estrus and ovulation in dairy cows. *Journal of Reproduction and Development* 56 Suppl, S42–47.

McGraw, S., Shojaei Saadi, H.A. and Robert, C. (2013) Meeting the methodological challenges in molecular mapping of the embryonic epigenome. *Molecular Human Reproduction* 19, 809–827.

Mamo, S., Mehta, J.P., McGettigan, P., Fair, T., Spencer, T.E., Bazer, F.W. and Lonergan, P. (2011) RNA sequencing reveals novel gene clusters in bovine conceptuses associated with maternal recognition of pregnancy and implantation. *Biology of Reproduction* 85, 1143–1151.

Mamo, S., Mehta, J.P., Forde, N., McGettigan, P. and Lonergan, P. (2012) Conceptus-endometrium crosstalk during maternal recognition of pregnancy in cattle. *Biology of Reproduction* 87, 6, 1–9.

Mapletoft, R.J. and Bó, G.A. (2011) The evolution of improved and simplified superovulation protocols in cattle. *Reproduction, Fertility, and Development* 24, 278–283.

Mapletoft, R.J. and Hasler, J.F. (2005) Assisted reproductive technologies in cattle: a review. *Revue Scientifique et Technique* 24, 393–403.

Mapletoft, R.J., Bó, G.A. and Baruselli, P.S. (2009) Control of ovarian function for assisted reproductive technologies in cattle. *Animal Reproduction, Belo Horizonte* 6, 114–124.

Mardis, E.R. (2008) Next-generation DNA sequencing methods. *Annual Review of Genomics and Human Genetics* 9, 387–402.

Martin, G., Sabido, O., Durand, P. and Levy, R. (2004) Cryopreservation induces an apoptosis-like mechanism in bull sperm. *Biology of Reproduction* 71, 28–37.

Martínez-Pastor, F., Mata-Campuzano, M., Álvarez-Rodríguez, M., Álvarez, M., Anel, L. and De Paz, P. (2010) Probes and techniques for sperm evaluation by flow cytometry. *Reproduction in Domestic Animals* 45, 67–78.

Massicotte, L., Coenen, K., Mourot, M. and Sirard, M.A. (2006) Maternal housekeeping proteins translated during bovine oocyte maturation and early embryo development. *Proteomics* 6, 3811–3820.

Mayorga, L.S., Tomes, C.N. and Belmonte, S.A. (2007) Acrosomal exocytosis, a special type of regulated secretion. *IUBMB Life* 59, 286–292.

Mehlmann, L.M. (2005) Stops and starts in mammalian oocytes: recent advances in understanding the regulation of meiotic arrest and oocyte maturation. *Reproduction* 130, 791–799.

Memili, E., Peddinti, D., Shack, L.A., Nanduri, B., McCarthy, F., Sagirkaya, H. and Burgess, S.C. (2007) Bovine germinal vesicle oocyte and cumulus cell proteomics. *Reproduction* 133, 1107–1120.

Mermillod, P. (2011) Gamete and embryo technology: *in vitro* fertilization. In: John, W.F. (ed.) *Encyclopedia of Dairy Sciences (Second Edition)*. Academic Press, San Diego, California, pp. 616–622.

Metzker, M.L. (2010) Sequencing technologies – the next generation. *Nature Reviews. Genetics* 11, 31–46.

Miller, D., Brinkworth, M. and Iles, D. (2010) Paternal DNA packaging in spermatozoa: more than the sum of its parts? DNA, histones, protamines and epigenetics. *Reproduction* 139, 287–301.

Misirlioglu, M., Page, G.P., Sagirkaya, H., Kaya, A., Parrish, J.J., First, N.L. and Memili, E. (2006) Dynamics of global transcriptome in bovine matured oocytes and preimplantation embryos. *Proceedings of the National Academy of Sciences* 103, 18905–18910.

Mondou, E., Dufort, I., Gohin, M., Fournier, E. and Sirard, M.-A. (2012) Analysis of microRNAs and their precursors in bovine early embryonic development. *Molecular Human Reproduction* 18, 425–434.

Montag, M., Toth, B. and Strowitzki, T. (2013) New laboratory techniques in reproductive medicine. *Journal für Reproduktionsmed und Endokrinologie* 10, 33–37.

Moore, K. and Thatcher, W.W. (2006) Major advances associated with reproduction in dairy cattle. *Journal of Dairy Science* 89, 1254–1266.

Moura, A.A., Chapman, D.A., Koc, H. and Killian, G.J. (2006a) Proteins of the cauda epididymal fluid associated with fertility of mature dairy bulls. *Journal of Andrology* 27, 534–541.

Moura, A.A., Koc, H., Chapman, D.A. and Killian, G.J. (2006b) Identification of proteins in the accessory sex gland fluid associated with fertility indexes of dairy bulls: a proteomic approach. *Journal of Andrology* 27, 201–211.

Moura, A.A., Chapman, D.A., Koc, H. and Killian, G.J. (2007) A comprehensive proteomic analysis of the accessory sex gland fluid from mature Holstein bulls. *Animal Reproduction Science* 98, 169–188.

Muiño, R., Fernández, M. and Peña, A. (2007) Post-thaw survival and longevity of bull spermatozoa frozen with an egg yolk-based or two egg yolk-free extenders after an equilibration period of 18 h. *Reproduction in Domestic Animals* 42, 305–311.

Muiño, R., Tamargo, C., Hidalgo, C.O. and Peña, A.I. (2008) Identification of sperm subpopulations with defined motility characteristics in ejaculates from Holstein bulls: effects of cryopreservation and between-bull variation. *Animal Reproduction Science* 109, 27–39.

Murphy, B.D. (2012) Research in animal reproduction: quo vadimus? *Animal Reproduction* 9, 217–222.

Nivet, A.L., Bunel, A., Labrecque, R., Belanger, J., Vigneault, C., Blondin, P. and Sirard, M.A. (2012) FSH withdrawal improves developmental competence of oocytes in the bovine model. *Reproduction* 143, 165–171.

Nivet, A.-L., Vigneault, C., Blondin, P. and Sirard, M.-A. (2013) Changes in granulosa cells' gene expression associated with increased oocyte competence in bovine. *Reproduction* 145, 555–565.

Nomenclature, C.O.B.R. (1972) Recommendations for standardising bovine reproductive terms. *Cornell Veterinarian* 216–237.

Norman, H.D., Hutchison, J.L. and Wright, J.R. (2010) Sire conception rate: new national AI bull fertility evaluation. AIPL Research Report, available at: http://aipl.arsusda.gov/reference/arr-scr1.htm (accessed 4 March 2014).

Odhiambo, J.F., Sutovsky, M., DeJarnette, J.M., Marshall, C. and Sutovsky, P. (2011) Adaptation of ubiquitin-PNA based sperm quality assay for semen evaluation by a conventional flow cytometer and a dedicated platform for flow cytometric semen analysis. *Theriogenology* 76, 1168–1176.

O'Shea, L.C., Mehta, J., Lonergan, P., Hensey, C. and Fair, T. (2012) Developmental competence in oocytes and cumulus cells: candidate genes and networks. *Systems Biology in Reproductive Medicine* 58, 88–101.

Park, Y.-J., Kwon, W.-S., Oh, S.-A. and Pang, M.-G. (2012) Fertility-related proteomic profiling bull spermatozoa separated by percoll. *Journal of Proteome Research* 11, 4162–4168.

Peippo, J., Machaty, Z. and Peter, A. (2011) Terminologies for the pre-attachment bovine embryo. *Theriogenology* 76, 1373–1379.

Peña, F.J., Saravia, F., Johannisson, A., Walgren, M. and Rodríguez-Martínez, H. (2005) A new and simple method to evaluate early membrane changes in frozen–thawed boar spermatozoa. *International Journal of Andrology* 28, 107–114.

Peña, F.J., Rodríguez-Martínez, H., Tapia, J.A., Ortega Ferrusola, C., Gonzalez Fernandez, L. and Macias Garcia, B. (2009) Mitochondria in mammalian sperm physiology and pathology: a review. *Reproduction in Domestic Animals* 44, 345–349.

Peñagaricano, F., Weigel, K.A. and Khatib, H. (2012) Genome-wide association study identifies candidate markers for bull fertility in Holstein dairy cattle. *Animal Genetics* 43, 65–71.

Peñagaricano, F., Weigel, K., Rosa, G.J.M. and Khatib, H. (2013) Inferring quantitative trait pathways associated with bull fertility from a genome-wide association study. *Frontiers in Genetics* 3, 307.

Pennisi, E. (2012) Going solid-state. *Science* 336, 536.

Pérez-Enciso, M. and Ferretti, L. (2010) Massive parallel sequencing in animal genetics: wherefroms and wheretos. *Animal Genetics* 41, 561–569.

Perreault, S.D., Barbee, R.R., Elstein, K.H., Zucker, R.M. and Keefer, C.L. (1988) Interspecies differences in the stability of mammalian sperm nuclei assessed *in vivo* by sperm microinjection and *in vitro* by flow cytometry. *Biology of Reproduction* 39, 157–167.

Pesch, S. and Bergmann, M. (2006) Structure of mammalian spermatozoa in respect to viability, fertility and cryopreservation. *Micron* 37, 597–612.

Petrunkina, A.M. and Harrison, R.A. (2010) Systematic misestimation of cell subpopulations by flow cytometry: a mathematical analysis. *Theriogenology* 73, 839–847.

Petrunkina, A.M., Waberski, D., Günzel-Apel, A.R. and Töpfer-Petersen, E. (2007) Determinants of sperm quality and fertility in domestic species. *Reproduction* 134, 3–17.

Phetudomsinsuk, K., Sirinarumitr, K., Laikul, A. and Pinyopummin, A. (2008) Morphology and head morphometric characters of sperm in Thai native crossbred stallions. *Acta Veterinaria Scandinavica* 50, 41.

Piehler, E., Petrunkina, A.M., Ekhlasi-Hundrieser, M. and Töpfer-Petersen, E. (2006) Dynamic quantification of the tyrosine phosphorylation of the sperm surface proteins during capacitation. *Cytometry Part A* 69A, 1062–1070.

Pohler, K., Geary, T., Atkins, J., Perry, G., Jinks, E. and Smith, M. (2012) Follicular determinants of pregnancy establishment and maintenance. *Cell Tissue Research* 349, 649–664.

Pons-Rejraji, H., Bailey, J.L. and Leclerc, P. (2009) Cryopreservation affects bovine sperm intracellular parameters associated with capacitation and acrosome exocytosis. *Reproduction, Fertility, and Development* 21, 525–537.

Powell, R.L. and Norman, H.D. (2006) Major advances in genetic evaluation techniques. *Journal of Dairy Science* 89, 1337–1348.

Prathalingam, N.S., Holt, W.W., Revell, S.G., Jones, S. and Watson, P.F. (2006) The precision and accuracy of six different methods to determine sperm concentration. *Journal of Andrology* 27, 257–262.

Prentice, J.R. and Anzar, M. (2011) Cryopreservation of mammalian oocyte for conservation of animal genetics. *Veterinary Medicine International* 2011, doi:10.4061/2011/146405.

Pryce, J.E., Royal, M.D., Garnsworthy, P.C. and Mao, I.L. (2004) Fertility in the high-producing dairy cow. *Livestock Production Science* 86, 125–135.

Pryce, J.E., Bolormaa, S., Chamberlain, A.J., Bowman, P.J., Savin, K., Goddard, M.E. and Hayes, B.J. (2010) A validated genome-wide association study in 2 dairy cattle breeds for milk production and fertility traits using variable length haplotypes. *Journal of Dairy Science* 93, 3331–3345.

Pursley, J.R., Mee, M.O. and Wiltbank, M.C. (1995) Synchronization of ovulation in dairy cows using PGF2alpha and GnRH. *Theriogenology* 44, 915–923.

Rabiee, A.R., Lean, I.J. and Stevenson, M.A. (2005) Efficacy of ovsynch program on reproductive performance in dairy cattle: a meta-analysis. *Journal of Dairy Science* 88, 2754–2770.

Rahman, M.B., Vandaele, L., Rijsselaere, T., Maes, D., Hoogewijs, M., Frijters, A., Noordman, J., Granados, A., Dernelle, E., Shamsuddin, M., Parrish, J.J. and Van Soom, A. (2011) Scrotal insulation and its relationship to abnormal morphology, chromatin protamination and nuclear shape of spermatozoa in Holstein-Friesian and Belgian Blue bulls. *Theriogenology* 76, 1246–1257.

Rajakoski, E. (1960) The ovarian follicular system in sexually mature heifers with special reference to seasonal, cyclical, end left-right variations. *Acta Endocrinologica* 34(Suppl 52), 1–68.

Rathi, R., Colenbrander, B., Bevers, M.M. and Gadella, B.M. (2001) Evaluation of *in vitro* capacitation of stallion spermatozoa. *Biology of Reproduction* 65, 462–470.

Resli, I., Gaspar, R., Szabo, G., Matyus, L. and Damjanovich, S. (1983) Biophysical analysis of fertility of sperm cells. *Magy Allatorvosok Lapja* 38, 38–41.

Rho, G.-J., Johnson, W.H. and Betteridge, K.J. (1998) Cellular composition and viability of demi- and quarter-embryos made from bisected bovine morulae and blastocysts produced *in vitro*. *Theriogenology* 50, 885–895.

Ribeiro, S.C., Sartorius, G., Pletscher, F., de Geyter, M., Zhang, H. and de Geyter, C. (2013) Isolation of spermatozoa with low levels of fragmented DNA with the use of flow cytometry and sorting. *Fertility and Sterility* 100, 686–694.

Robert, C. (2008) Challenges of functional genomics applied to farm animal gametes and pre-hatching embryos. *Theriogenology* 70, 1277–1287.

Robert, C., Nieminen, J., Dufort, I., Gagné, D., Grant, J.R., Cagnone, G., Plourde, D., Nivet, A.-L., Fournier, É., Paquet, É., Blazejczyk, M., Rigault, P., Juge, N. and Sirard, M.-A. (2011) Combining resources to obtain a comprehensive survey of the bovine embryo transcriptome through deep sequencing and microarrays. *Molecular Reproduction and Development* 78, 651–664.

Rodrigues, C.A., Teixeira, A.A., Ferreira, R.M., Ayres, H., Mancilha, R.F., Souza, A.H. and Baruselli, P.S. (2010) Effect of fixed-time embryo transfer on reproductive efficiency in high-producing repeat-breeder Holstein cows. *Animal Reproductive Science* 118, 110–117.

Rodríguez-Martínez, H. (2003) Laboratory semen assessment and prediction of fertility: still utopia?. *Reproduction in Domestic Animals* 38, 312–318.

Rodríguez-Martínez, H. (2012) Assisted reproductive techniques for cattle breeding in developing countries: a critical appraisal of their value and limitations. *Reproduction in Domestic Animals* 47, 21–26.

Rodríguez-Martínez, H. and Barth, A.D. (2007) *In vitro* evaluation of sperm quality related to *in vivo* function and fertility. *Society of Reproduction and Fertility, Supplement* 64, 39–54.

Roelofs, J., López-Gatius, F., Hunter, R.H.F., van Eerdenburg, F.J.C.M. and Hanzen, C. (2010) When is a cow in estrus? Clinical and practical aspects. *Theriogenology* 74, 327–344.

Rudolph, N.S. (1999) Biopharmaceutical production in transgenic livestock. *Trends in Biotechnology* 17, 367–374.

Sahana, G., Guldbrandtsen, B., Bendixen, C. and Lund, M.S. (2010) Genome-wide association mapping for female fertility traits in Danish and Swedish Holstein cattle. *Animal Genetics* 41, 579–588.

Said, T.M., Agarwal, A., Zborowski, M., Grunewald, S., Glander, H.-J. and Paasch, U. (2008) ANDROLOGY LAB CORNER*: utility of magnetic cell separation as a molecular sperm preparation technique. *Journal of Andrology* 29, 134–142.

Sánchez, F. and Smitz, J. (2012) Molecular control of oogenesis. *Biochimica et Biophysica Acta (BBA) – Molecular Basis of Disease* 1822, 1896–1912.

Santos, J.E.P., Thatcher, W.W., Chebel, R.C., Cerri, R.L.A. and Galvão, K.N. (2004) The effect of embryonic death rates in cattle on the efficacy of estrus synchronization programs. *Animal Reproduction Science* 82–83, 513–535.

Saragusty, J. and Arav, A. (2011) Current progress in oocyte and embryo cryopreservation by slow freezing and vitrification. *Reproduction* 141, 1–19.

Sartori, R., Bastos, M.R. and Wiltbank, M.C. (2009) Factors affecting fertilisation and early embryo quality in single- and superovulated dairy cattle. *Reproduction, Fertility, and Development* 22, 151–158.

Schagdarsurengin, U., Paradowska, A. and Steger, K. (2012) Analysing the sperm epigenome: roles in early embryogenesis and assisted reproduction. *Nature Reviews. Urology* 9, 609–619.

Schefers, J.M. and Weigel, K.A. (2012) Genomic selection in dairy cattle: integration of DNA testing into breeding programs. *Animal Frontiers* 2, 4–9.

Seidel, G.E., Jr (2007) Overview of sexing sperm. *Theriogenology* 68, 443–446.

Seidel, J.G.E. (2012) Sexing mammalian sperm – where do we go from here? *Journal of Reproduction and Development* 58, 505–509.

Seisenberger, S., Peat, J., Hore, T., Santos, F., Dean, W. and Reik, W. (2013) Reprogramming DNA methylation in the mammalian life cycle: building and breaking epigenetic barriers. *Philosophical Transactions of the Royal Society of London B: Biological Sciences* 368, 20110330.

Sharpe, J.C. and Evans, K.M. (2009) Advances in flow cytometry for sperm sexing. *Theriogenology* 71, 4–10.

Shojaei, H., Kroetsch, T., Wilde, R., Blondin, P., Kastelic, J.P. and Thundathil, J.C. (2012) Moribund sperm in frozen-thawed semen, and sperm motion end points post-thaw and post-swim-up, are related to fertility in Holstein AI bulls. *Theriogenology* 77, 940–951.

Shojaei Saadi, H.A., O'Doherty, A.M., Gagné, D., Fournier, E., Grant, J.R., Sirard, M.A. and Robert, C. (2014) An integrated platform for bovine DNA methylome analysis suitable for small samples. *BMC Genomics* 15, 451.

Shoubridge, E.A. and Wai, T. (2007) Mitochondrial DNA and the mammalian oocyte. *Current Topics in Developmental Biology* 77, 87–111.

Silke, V., Diskin, M.G., Kenny, D.A., Boland, M.P., Dillon, P., Mee, J.F. and Sreenan, J.M. (2002) Extent, pattern and factors associated with late embryonic loss in dairy cows. *Animal Reproduction Science* 71, 1–12.

Simoes, R., Feitosa, W.B., Mendes, C.M., Marques, M.G., Nicacio, A.C., de Barros, F.R., Visintin, J.A. and Assumpcao, M.E. (2009) Use of chromomycin A3 staining in bovine sperm cells for detection of protamine deficiency. *Biotechnic and Histochemistry* 84, 79–83.

Sirard, M.A., Parrish, J.J., Ware, C.B., Leibfried-Rutledge, M.L. and First, N.L. (1988) The culture of bovine oocytes to obtain developmentally competent embryos. *Biology of Reproduction* 39, 546–552.

Sirard, M.A., Richard, F., Blondin, P. and Robert, C. (2006) Contribution of the oocyte to embryo quality. *Theriogenology* 65, 126–136.

Słowińska, M., Karol, H. and Ciereszko, A. (2008) Comet assay of fresh and cryopreserved bull spermatozoa. *Cryobiology* 56, 100–102.

Smith, G.D., Takayama, S. and Swain, J.E. (2012) Rethinking *in vitro* embryo culture: new developments in culture platforms and potential to improve assisted reproductive technologies. *Biology of Reproduction* 86, 62, 61–10.

Somfai, T., Inaba, Y., Aikawa, Y., Ohtake, M., Kobayashi, S., Konishi, K. and Imai, K. (2010) Relationship between the length of cell cycles, cleavage pattern and developmental competence in bovine embryos generated by *in vitro* fertilization or parthenogenesis. *Journal of Reproduction and Development* 56, 200–207.

Sugimura, S., Akai, T., Somfai, T., Hirayama, M., Aikawa, Y., Ohtake, M., Hattori, H., Kobayashi, S., Hashiyada, Y., Konishi, K. and Imai, K. (2010) Time-lapse cinematography-compatible polystyrene-based microwell culture system: a novel tool for tracking the development of individual bovine embryos. *Biology of Reproduction* 83, 970–978.

Sullivan, R. (2004) Male fertility markers, myth or reality. *Animal Reproduction Science* 82–83, 341–347.

Sutovsky, P. and Kennedy, C.E. (2013) Biomarker-based nanotechnology for the improvement of reproductive performance in beef and dairy cattle. *Industrial Biotechnology* 9, 24–30.

Sutovsky, P. and Lovercamp, K. (2010) Molecular markers of sperm quality. *Society of Reproduction and Fertility, Supplement* 67, 247–256.

Suzuki, M.M. and Bird, A. (2008) DNA methylation landscapes: provocative insights from epigenomics. *Nature Reviews, Genetics* 9, 465–476.

Swain, J.E. and Smith, G.D. (2011) Advances in embryo culture platforms: novel approaches to improve preimplantation embryo development through modifications of the microenvironment. *Human Reproduction Update* 17, 541–557.

Tagawa, M., Matoba, S., Narita, M., Saito, N., Nagai, T. and Imai, K. (2008) Production of monozygotic twin calves using the blastomere separation technique and Well of the Well culture system. *Theriogenology* 69, 574–582.

Tavalaee, M., Kiani, A., Arbabian, M., Deemeh, M.R. and Esfahani, M.H.N. (2010) Flow cytometry: a new approach for indirect assessment of sperm protamine deficiency. *International Journal of Fertility and Sterility* 3, 177–184.

Thomas, C.A., Garner, D.L., DeJarnette, J.M. and Marshall, C.E. (1997) Fluorometric assessments of acrosomal integrity and viability in cryopreserved bovine spermatozoa. *Biology of Reproduction* 56, 991–998.

Thundathil, J., Gil, J., Januskauskas, A., Larsson, B., Soderquist, L., Mapletoft, R. and Rodríguez-Martínez, H. (1999) Relationship between the proportion of capacitated spermatozoa present in frozen-thawed bull semen and fertility with artificial insemination. *International Journal of Andrology* 22, 366–373.

van Berkel, P.H., Welling, M.M., Geerts, M., van Veen, H.A., Ravensbergen, B., Salaheddine, M., Pauwels, E.K., Pieper, F., Nuijens, J.H. and Nibbering, P.H. (2002) Large scale production of recombinant human lactoferrin in the milk of transgenic cows. *Nature Biotechnology* 20, 484–487.

van Montfoort, A.P., Hanssen, L.L., de Sutter, P., Viville, S., Geraedts, J.P. and de Boer, P. (2012) Assisted reproduction treatment and epigenetic inheritance. *Human Reproduction Update* 18, 171–197.

Van Soom, A., Vandaele, L., Goossens, K., de Kruif, A. and Peelman, L. (2007) Gamete origin in relation to early embryo development. *Theriogenology* 68, Supplement 1, S131–S137.

Vazquez, J.M., Parrilla, I., Gil, M.A., Cuello, C., Caballero, I., Vazquez, J.L., Roca, J. and Martínez, E.A. (2008) Improving the efficiency of insemination with sex-sorted spermatozoa. *Reproduction in Domestic Animals* 43, 1–8.

Veerkamp, R.F. and Beerda, B. (2007) Genetics and genomics to improve fertility in high producing dairy cows. *Theriogenology* 68, Supplement 1, S266–S273.

Verstegen, J., Iguer-Ouada, M. and Onclin, K. (2002) Computer assisted semen analyzers in andrology research and veterinary practice. *Theriogenology* 57, 149–179.

Vicente-Fiel, S., Palacín, I., Santolaria, P. and Yániz, J.L. (2013) A comparative study of sperm morphometric subpopulations in cattle, goat, sheep and pigs using a computer-assisted fluorescence method (CASMA-F). *Animal Reproduction Science* 139, 182–189.

Vigneault, C., Gravel, C., Vallée, M., McGraw, S. and Sirard, M.-A. (2009) Unveiling the bovine embryo transcriptome during the maternal-to-embryonic transition. *Reproduction* 137, 245–257.

Vishwanath, R. and Shannon, P. (2000) Storage of bovine semen in liquid and frozen state. *Animal Reproduction Science* 62, 23–53.

Walker, C.G., Littlejohn, M.D., Mitchell, M.D., Roche, J.R. and Meier, S. (2012) Endometrial gene expression during early pregnancy differs between fertile and subfertile dairy cow strains. *Physiological Genomics* 44, 47–58.

Walsh, S.W., Williams, E.J. and Evans, A.C.O. (2011) A review of the causes of poor fertility in high milk producing dairy cows. *Animal Reproduction Science* 123, 127–138.

Walsh, S.W., Mehta, J.P., McGettigan, P.A., Browne, J.A., Forde, N., Alibrahim, R.M., Mulligan, F.J., Loftus, B., Crowe, M.A., Matthews, D., Diskin, M., Mihm, M. and Evans, A.C. (2012) Effect of the metabolic environment at key stages of follicle development in cattle: focus on steroid biosynthesis. *Physiological Genomics* 44, 504–517.

Ward, W. (2010) Function of sperm chromatin structural elements in fertilization and development. *Molecular Human Reproduction* 16, 30–36.

Watson, P.F. (1995) Recent developments and concepts in the cryopreservation of spermatozoa and the assessment of their post-thawing function. *Reproduction, Fertility, and Development* 7, 871–891.

Watson, P.F. (2000) The causes of reduced fertility with cryopreserved semen. *Animal Reproduction Science* 60–61, 481–492.

Weigel, K.A. (2004) Exploring the role of sexed semen in dairy production systems. *Journal of Dairy Science* 87, E120–E130.

Wheeler, M.B., Rutledge, J.J., Fischer-Brown, A., VanEtten, T., Malusky, S. and Beebe, D.J. (2006) Application of sexed semen technology to *in vitro* embryo production in cattle. *Theriogenology* 65, 219–227.

Wiggans, G.R., VanRaden, P.M. and Cooper, T.A. (2011) The genomic evaluation system in the United States: past, present, future. *Journal of Dairy Science* 94, 3202–3211.

Wiltbank, M.C., Sartori, R., Herlihy, M.M., Vasconcelos, J.L.M., Nascimento, A.B., Souza, A.H., Ayres, H., Cunha, A.P., Keskin, A., Guenther, J.N. and Gumen, A. (2011) Managing the dominant follicle in lactating dairy cows. *Theriogenology* 76, 1568–1582.

Wu, B. and Zan, L. (2012) Enhance beef cattle improvement by embryo biotechnologies. *Reproduction in Domestic Animals* 47, 865–871.

Xu, Z.Z. (2011a) Reproduction, events and management. Control of estrous cycles: synchronization of estrus. In: John, W.F. (ed.) *Encyclopedia of Dairy Sciences (Second Edition)*. Academic Press, San Diego, California, 448–453.

Xu, Z.Z. (2011b) Reproduction, events and management. Control of estrous cycles: synchronization of ovulation and insemination. In: John, W.F. (ed.) *Encyclopedia of Dairy Sciences (Second Edition)*. Academic Press, San Diego, California, pp. 454–460.

Yaakub, H., O'Callaghan, D. and Boland, M.P. (1999) Effect of type and quantity of concentrates on super-ovulation and embryo yield in beef heifers. *Theriogenology* 51, 1259–1266.

Yanagimachi, R. (1994) Fertility of mammalian spermatozoa: its development and relativity. *Zygote* 2, 371–372.

Yang, P., Wang, J., Gong, G., Sun, X., Zhang, R., Du, Z., Liu, Y., Li, R., Ding, F., Tang, B., Dai, Y. and Li, N. (2008) Cattle mammary bioreactor generated by a novel procedure of transgenic cloning for large-scale production of functional human lactoferrin. *PLoS ONE* 3, e3453.

Young, L.E., Sinclair, K.D. and Wilmut, I. (1998) Large offspring syndrome in cattle and sheep. *Reviews in Reproduction* 3, 155–163.

Youngs, C.R. (2011) Cryopreservation of preimplantation embryos of cattle, sheep, and goats. *Journal of Visualized Experiments* 54, 2764.

Zaidi, S.K., Young, D.W., Montecino, M., van Wijnen, A.J., Stein, J.L., Lian, J.B. and Stein, G.S. (2011) Bookmarking the genome: maintenance of epigenetic information. *Journal of Biological Chemistry* 286, 18355–18361.

Zhang, J., Chiodini, R., Badr, A. and Zhang, G. (2011) The impact of next-generation sequencing on genomics. *Journal of Genetics and Genomics* 38, 95–109.

Zoheir, K.M.A. and Allam, A.A. (2010) A rapid method for sexing the bovine embryo. *Animal Reproduction Science* 119, 92–96.

14 Developmental Genetics

Anatoly Ruvinsky

University of New England, Armidale, New South Wales, Australia

© CAB International 2015. *The Genetics of Cattle*,
2nd Edn (eds D.J. Garrick and A. Ruvinsky)

Introduction

Just a decade ago the vast bulk of developmental genetics data was generated mainly by studies of murine models. In the past few years, investigation of bovine embryos has increasingly contributed a significant volume of developmental information. The success of the Bovine Genome Project, and introduction of the latest molecular methods and advances in bovine embryo manipulations are among the causes that have stimulated this progress. In spite of this, any review of genetic aspects of development in cattle cannot avoid significant gaps and should refer to other species, mainly to the mouse. Fortunately high similarity in the genetic regulation of mammalian development makes such an approach acceptable and useful.

Nevertheless, numerous distinctions between mammalian species, resulting from differences in embryo morphology, placental structures, longevity and schedule of development processes should be taken into consideration. These diverse features of development are based on genetic differences, many of which are still awaiting better understanding. In this chapter we present bovine data where available and also use information obtained in other mammalian species.

Developmental Stages of the Cattle Embryo

Gametogenesis

The ovaries of newly born females contain a lifetime supply of oocytes stored in quiescent primordial follicles, which can be successfully used for production of full-term pregnancies (Kauffold et al., 2005). Primordial germ-like cells can be identified as early as 40 days of development, and oogonia and oocytes have been recognized in ovaries of 80–130-day-old fetuses (Lavoir et al., 1994). As in many other mammals, oocytes are arrested at prophase of the first meiotic division (i.e. germinal vesicle stage). At puberty and continuing throughout the reproductive life of the female, under appropriate hormonal control, pools of primordial follicles are recruited to grow. In the cow >100,000 restricted primordial follicles are present at birth. Follicular development follows a series of events characterized by follicular and oocyte growth, as well as cell proliferation and differentiation. Global gene expression analysis during bovine oocyte maturation revealed that 209 transcripts were significantly up-regulated and 612 down-regulated (Fair et al., 2007).

Spermatogenesis is a process that begins with spermatogonial stem cells, which in contrast to oogenesis, continually replenish the testicular seminiferous tubules with a virtually unlimited number of gametes. The spermatogonia proceed through two meiotic divisions, which are followed by spermiogenesis in which haploid spermatids develop into spermatozoa. All types of male germ cells are found in a single section of the seminiferous tubule epithelium. The cycle of the seminiferous epithelium refers to complete progression through these series of cellular stages and is unique for each species, as is the duration of spermatogenesis. The length of a seminiferous cycle in bull is about 13.5 days (França and Russell, 1998), while the complete process of spermatogenesis from spermatogonia A to a fully formed spermatozoa requires 61 days. Due to the complexity and duration of the spermatogenic process, the underlying genetic mechanisms are not fully understood in domestic animals, however, according to Yan (2009) 20 different genes, when deleted, adversely affect male fertility.

During both oogenesis and spermatogenesis epigenetic mechanisms that regulate gene transcription are active, and DNA methylation or demethylation are commonly involved. The primordial germ cells which are the starting point of gametogenesis in the early embryo have highly methylated DNA (Reik et al., 2001; Hajkova et al., 2002). However, once these cells have populated the developing gonads they are generally hypomethylated. During mammalian gametogenesis there is a de novo methylation of the gametic genomes catalysed by DNA (cytosine-5)-methyltransferases (DNMTs), which regulate genes during development and play a role in genomic imprinting. Genomic methylation patterns are erased and reacquired differentially in the developing male and female gametes, further modified in the early embryo, but become relatively stable by late embryogenesis. Sex-specific DNA methylation occurring in particular

DNA sequences forms the basis for paternally and maternally imprinted genes. It seems possible that the erasure and resetting of DNA methylation occurring during gametogenesis might be essential in preventing DNA methylation defects from being passed from one generation to the next.

Fertilization and embryonic development within the zona pellucida

Fertilization occurs within several hours after ovulation in an inseminated cow. Following fertilization of an oocyte the arrested meiotic process resumes, the second polar body is extruded into the perivitelline space and then the male and female pronuclei are formed. The completion of meiosis is facilitated by various signal transduction pathways. The mitogen-activated protein kinase (MAPK) cascade is essential in regulating the meiotic cell cycle of oocytes. MAPK is involved in the regulation of microtubule organization and meiotic spindle assembly after germinal vesicle breakdown. The activation of this kinase is essential for the maintenance of metaphase II arrest, while its inactivation is a prerequisite for pronuclear formation after fertilization (Fan and Sun, 2004). The Mos/MAP kinase pathway (the c-mos proto-oncogene product, Mos, is a serine/threonine kinase that can activate two MAP kinases) is also considered to be responsible for the phosphorylation of spindlin protein (Oh et al., 1997; Vallée et al., 2006), which is associated with the meiotic spindle and changing of the metaphase spindle into an anaphase configuration that requires the presence of calcium/calmodulin-dependent protein kinase II (CaM) and has been conserved in oocytes of mammalian species (Fan et al., 2003; Vallée et al. 2006; Xu et al., 2009). Oocytes of c-mos knockout mice undergo spontaneous activation due to a lack of spindlin phosphorylation, which destabilizes metaphase II arrest (Hashimoto et al., 1994). Fertilization also triggers waves of increased calcium concentration passing through the cytoplasm, which is thought to be mediated by the soluble sperm protein oscillin (Parrington et al., 1996). This calcium wave, also referred to as calcium transient, leads to a remodelling of the

cytoplasm and nuclear compartments suggesting it plays a role in the initiation of transcription. Study of the calcium transient in pigs has generally been related to the refining of in vitro fertilization procedures (Funahashi et al., 1995; Ito et al., 2003) although it would appear that this is only one of the regulatory events that may proceed in waves through the cytoplasm at the time of fertilization. Additional factors are being discovered that promote oocyte cytoplasmic and nuclear maturation, including the Sonic hedgehog signalling (Shh) pathway (Nguyen et al., 2009).

There are few gene expression studies of bovine gametes around the time of fertilization. Study of the kinetics of gene expression and signalling that affects oocyte maturation has revealed a number of significant changes (Salhab et al., 2011). Transcription and protein abundance of glutathione-S-transferase A1 (GSTA1), FSH receptor and aromatase (CYP19A1) steroidogenic enzymes (steroidogenic acute regulatory protein, cytochrome P450scc and 3-beta-hydroxysteroid dehydrogenase) are significantly decreased during oocyte maturation. In the meantime the expression of progesterone receptor (PGR) and clusterin (CLU) mRNA and phosphorylations of protein kinases AKT, MAPK P38 and SMAD2 increased. Such an expression pattern might indicate involvement of these factors in the oocyte metaphase-I check point of meiosis (Salhab et al., 2011). In mammalian spermatozoa transcripts for clusterin (CLU), protamine 2 (PRM2), calmegin (CLGN), cAMP-response element modulator protein (CREM), methyltransferase 1 (DNMT1), linker histone 1 (H1), protamine 1 (PRM1), TATA box-binding protein associated factor 1 (TAF1) and TATA box-binding protein (TBP) in porcine spermatozoa were detected (Kempisty et al., 2008). The oocytes contained only CREM, H1, TAF1, and TBP mRNAs. The zygote and two-cell stage embryos have the set of transcripts found in sperm cells for the CLU, CREM, H1, PRM1, PRM2, TAF1 and TBP genes. This suggests that spermatozoa may deliver CLU, PRM1 and PRM2 mRNAs into the oocyte, which probably contributes to zygotic and early embryonic development.

The initial cleavage to a two-celled embryo occurs within the first day after fertilization. The bovine conceptus then develops through a series of stages prior to implantation[1]. These steps and timing of critical events during the prenatal bovine fetal development are summarized in Table 14.1. Within 48 h of development the bovine embryo is transformed from the two- to four-cell stage and, at the end of the third cell cycle, transition from maternal to embryonic genome expression commences. Subsequently, the embryo development continues and at the 8- to 16-cell stage it reaches morula stage as early as by days 2–3 of the development, which is followed by transfer of the morula from the oviduct to the upper uterine horn. At this point a compaction process causes flattening of the blastomeres and cell-to-cell contacts become more pronounced; internal and external cells steadily differentiate and obtain some degree of polarity (Rossant and Tam, 2009). Tight intercellular junctions develop and this provides a condition for accumulation of fluid within the central cavity (the blastocoele). The next stage of development is blastocyst, which attains typical structure and forms two distinct cell lineages: the inner cell mass (ICM) and trophectoderm (TE). As a part of this process, the outer layers, closest to the zona pellucida, become connected by tight junctions and desmosomes to seal the expanding blastocyst cavity in which the ICM forms a tight cluster of cells, until the blastocyst ultimately hatches from the zona pellucida by days 9–10.

Post-hatching development

At the time of hatching (days 9–10), the TE makes up the majority of the external cells of the blastocyst. It will develop an epithelial phenotype, form much of the extra-embryonic tissue, and play a critical role during implantation and formation of the trophoblast layers of the placenta. The ICM makes up the remaining cells and differentiates into the epiblast and the hypoblast (primitive endoderm). At day 12 of development there is a disintegration of the TE polar region covering the epiblast to create the

---

[1]The term implantation is commonly used in literature relevant to bovine embryonic development, however in cattle there is no invasion of the uterine epithelium typical for primates and rodents but rather attachment (Melton *et al.*, 1951).

structure known as the embryonic disc. At this time the embryos have two cell populations of the epiblast: one constituting a distinctive basal layer apposing the hypoblast, and one arranged inside or above the former layer, including cells apposing the Rauber layer (Vejlsted *et al.*, 2005). By day 13 the conceptus still has the ovoid shape and continues to enlarge. The embryo may be referred to as being at the pre-streak 1 stage at this point (Vejlsted *et al.*, 2006). During days 14–15, the embryonic disc transforms from a circle into an oval structure and at one pole of the disc, a prominent crescent-shaped thickening appears. This represents the pre-streak 2 stage and the first signs of anterior–posterior polarity of the embryo. At this point a dramatic more than 1000-fold elongation of the conceptus begins (Clemente *et al.*, 2011). The conceptus will also signal its presence during this period to allow for maternal recognition of pregnancy to occur in order that the corpus luteum (CL) is maintained and the uterine environment remains such that it will support and promote pregnancy (in rare cases there is more than one CL). In the cow, blastocysts begin to produce oestrogens at some point after days 12–13 of development, which through a series of processes (discussed below) prevent secretion of the uterine luteolytic factor (PGF2 alpha) in an endocrine direction, while allowing secretion in an exocrine direction (i.e. into the uterine lumen), thereby protecting the CL from luteolysis or regression (Spencer *et al.*, 2004). Then, by approximately days 20–21 of development, the primitive streak appears at the posterior end of the embryonic disc, corresponding to the onset of gastrulation.

Gastrulation involves a complex sequence of cellular differentiation events and movement that ultimately facilitates the generation of uniquely distinct structures and tissues within the conceptus. The major result is formation of the three primary germ layers: endoderm, mesoderm and ectoderm. The ectoderm will eventually give rise to the nervous system and epidermis; the mesoderm develops into the cardiovascular, urogenital and muscular systems; while the endoderm is the starting point for the digestive, pulmonary and endocrine systems. In addition to these somatic germ layers, the primordial germ cells are also formed later at days 24–25 (Wrobel and Süss, 1998). The

Table 14.1. Essential events and timing of bovine prenatal development.

Days after fertilization	Developmental stage/event
Day 1	Cleavage to two-cell stage
Days 1–2	Cleavage to four-cell stage
Days 2–3	Development to four- to eight-cell stage; embryo genome activation
Day 4	Transition of embryo from oviduct to uterus
Days 5–6	Blastomere compaction; morula development
Days 7–8	Blastocyst formation and development of internal cell mass (ICM) and trophectoderm (TE)
Day 9	Blastocyst hatches from zona pellucida
Days 12–13	Ovoid conceptus and initiation of trophoblast elongation; embryonic disc forms
Days 14–15	Gastrulation begins; trophoblast cells elongate; amnion folding
Days 15–16	Notochord develops; chorion extension
Days 16–17	Maternal recognition of pregnancy; first somite pair; neural tube developing; enormous filamentous chorion occupies uterine horn
Day 18	Primordial germ cell formation; elongation of trophoblast is complete
Day 19	Implantation begins
Days 20–21	Primitive streak develops; gastrulation continues; five somites; open neural tube forms; head fold
Days 22–23	Placentation begins, migration of binucleate cells and formation of syncytium in euteraine epitelium; neural tube closed; heart beating, otic and optic vesicle forming; up to 14 somites; allantois emerges
Days 24–25	The embryonic membrane can cover both uterine horns; putative primordial germ cells form; three brain vesicles; bud of forelimb
Day 26	Gastro-intestinal structures, mesonephros develops, amnionic vesicle visible; bud of hindlimb
Days 27–28	Visible forelimb and hindlimb buds; gonadal ridge develops
Days 32–33	Somite formation accomplished, allantochorion formed
Days 32–36	Placentomas are detectable
Day 39	Sexually indifferent gonadal fold exists
Day 40	Sexual differentiation begins, optic lens, complex interdigitating villi of placentomes formed
Day 45	Split hooves
Days 55–60	Eyelids cover eyes; further head development, ribs formed
Days 60–70	Further limb development
Days 65–70	Prepuce, scrotum or labia and clitoris present
Day 90	Hair follicles
Day 100	Horn pits
Day 110	Fetal teeth begin to erupt
Day 140	Testicles migration completed
Day 230	Hair covers the whole fetus
Days 276–290	Birth

Compiled from: Assis Neto *et al.* (2010); Bazer *et al.* (1993); Blomberg *et al.* (2008); Clemente *et al.* (2011); Cruz and Pedersen (1991); Curran *et al.* (1986); Degrelle *et al.* (2005); Forde and Lonergan (2012); Greenstein and Foley (1958); Guillomot *et al.* (1993, 1995); Jainudeen and Hafez (1993); Maddox-Hyttel *et al.* (2003); Mamo *et al.* (2012a, b); Ménézoand Renard (1993); Silva and Ginther (2010); Winters *et al.* (1942); Wrobel and Süss (1998). Depending on breed, sex and individual differences the timing of events may vary by 12–24 h during the first 30–40 days of development. The length of the entire developmental process may vary up to 10–14 days.

initiation of gastrulation as a developmental phase precedes neurulation, but its completion overlaps with this later process. The first sign of neurulation is a thickening of the anterior ectoderm as the primitive streak regresses and the formation of neural plate folds to become the neural groove (Maddox-Hyttel *et al.*, 2003). In the bovine embryo the neural groove develops at approximately days 16–17, which also coincides with initiation of segmentation (Maddox-Hyttel *et al.*, 2003). Segmentation is the developmental process that subdivides the body into a series of subunits. In vertebrates the earliest form of segmentation is the development of somites, which result from a thickening of the mesoderm in the midline of the embryo to form blocks of mesodermal cells. In parallel to somite formation, the neural tube progressively develops and begins to close by the five- to seven-somite stage and the process is complete by the 28-somite stage (days 32–33) (Maddox-Hyttel et al., 2003; Table 14.1). Soon after gastrulation the endoderm layer forms a primitive gut tube, which subsequently leads to organ specification (foregut, midgut and hindgut), then formation of organ buds, and finally to more specialized cell lineages.

Genetic Control of Pre-implantation Development

Expression of maternal genes

Despite unovulated mammalian oocytes being arrested at prophase of the first meiotic division, both transcription and translation are very active and under 'maternal command'. Mouse oocytes express about 5400 genes and transposable elements, many of which are conserved in chordates (Evsikov et al., 2006). Numerous, newly synthesized mRNAs are stored and used later during oocyte maturation and up until embryonic genome activation, which occurs at the two-cell stage in the mouse (Hamatani et al., 2006) and the four to eight-cell stage during bovine development. Meanwhile, depletion of maternal mRNA intensifies prior to fertilization and continues until activation of the embryonic genome. By this time, nearly 90% of maternal mRNA is degraded and the majority of such transcripts are exclusively expressed from the oocyte genome (Bettegowda et al., 2008). In the mouse, and likely other mammals including cattle, 'housekeeping' genes are underrepresented in the oocyte and early embryo transcriptomes. It has been suggested that this unique feature indicates that the core function of the oocyte is acting like a 'reprogramming machine' in order to create a totipotent embryo (Evsikov and Marín de Evsikova, 2009a). In bovine oocytes reprogramming is connected with changes in three groups of transcripts.

Endogenous retrotransposons (with LTR) and mitochondrial transcripts are up-regulated, while genes encoding ribosomal proteins were down-regulated (Bui et al., 2009).

While the current understanding of transition towards mature oocyte and embryonic development in mammals is only emerging, and some species-specific deviations are possible, it is useful to form a more general view. Several genes, some identified recently, guide this process (Fig. 14.1). Among such genes is $Eif4lb$, which is involved in translational repression of maternal mRNAs. In the mouse, an oocyte-specific mammalian form of eukaryotic translation initiation factor 4E coded by $Eif4lb$ gene may influence the speed of oocyte maturation (Evsikov et al., 2006; Evsikov and Marín de Evsikova, 2009b). Another example is the inhibitory phosphorylation of CDC2 protein with kinase activity, which is catalysed by pig Wee1B protein. This involves meiotic arrest of porcine oocytes (Shimaoka et al., 2009). The inactivation of $Wee1B$ gene, in combination with other factors, leads to the resumption of meiosis and exit of mouse oocytes from metaphase II (Shimaoka et al., 2009; Oh et al., 2011). The regulatory mechanisms in porcine and murine oocytes are different (Shimaoka et al., 2011). In mature oocytes, degradation of maternal transcripts becomes more prominent and seems to be nearly completed by the two-cell stage when the so called minor zygotic genome activation takes place. In fact, the $ZAR1$ gene (zygote arrest 1) is one of the few known oocyte-specific maternal-effect genes essential for the beginning of embryo development (Wu et al. 2003). Surprisingly some $Zar1$ (–/–) mice are viable and look normal. However, $Zar1$ (–/–) females are infertile, probably due to the arrest of embryonic development in the majority of zygotes at the one-cell stage and the fact that maternal and paternal genomes remain separate in such zygotes. These $Zar1$ (–/–) embryos show a marked reduction in the synthesis of the transcription-requiring complex, with fewer than 20% of them progressing to the two-cell stage, and none develop to the four-cell stage (Wu et al., 2003). This gene is evolutionarily conserved and the protein plays a role in transcription regulation during oocyte maturation and early post-fertilization development (Uzbekova et al., 2006). Several additional

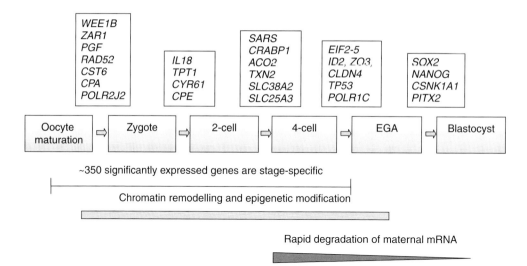

Fig. 14.1. Current knowledge on genetical regulation of oocyte-to-embryo transition in bovine development. (Redrawn from Bettegowda *et al.* (2008) with modifications compiled from several sources including Magnani and Cabot (2008, 2009); Shimaoka *et al.* (2009); Evsikov and Marín de Evsikova (2009a); Kues *et al.* (2008); Khan *et al.* (2012).) Only some genes with stage-specific expression are shown on the diagram.

maternal genes active during transition from oocyte to zygote are depicted in Fig. 14.1. The control of oocyte maturation, the maternal-to-embryonic transition as well as the first steps of embryo differentiation might involve several HOX genes (Paul *et al.*, 2011). Studies of developmental changes in gene expression indicated resistance to proapoptotic signals until the 8- to 16-cell stage in bovine pre-implantation embryos (Fear and Hansen, 2011).

The first investigation of transcriptomes from bovine metaphase II oocytes up to the blastocyst stages using the Affymetrix GeneChip Bovine Genome Array described approximately 23,000 transcripts (Kues *et al.*, 2008). The data show that bovine oocytes and embryos transcribed a significantly higher number of genes than somatic cells. Several hundred genes were transcriptionally active well before the eight-cell stage, at which the major activation of gene expression occurs. Microarray analysis of bovine oocyte cytoplasm fractions discovered expression of 4320 annotated genes. Most of these genes were associated with RNA processing, translation, and RNA binding. The content of mRNAs expressed in metaphase II oocytes influences activation of the embryonic genome and enables further development (Biase *et al.*,

2012). Pre-implantation development includes four major events: the transition from maternal transcripts to zygotic transcripts, compaction and the first lineage differentiation into inner cell mass and trophectoderm, and implantation (Hamatani *et al.*, 2006). Zygote genome activation (ZGA) in mice follows two stages: a minor prior to cleavage and a major at the two-cell stage and later (Hamatani *et al.*, 2006). In the bovine embryo ZGA occurs slightly later, at the four to eight-cell stage. Then the nucleoli develop, which are essential for ribosomal RNA (rRNA) and ribosome production. After fertilization structures resembling the nucleolar remnant are established in pronuclei, they are engaged in re-establishment of fibrillo-granular nucleoli during the major activation of the embryonic genome (Maddox-Hyttel *et al.*, 2007). The first divisions of the mammalian embryo are largely controlled by proteins and transcripts stored during oogenesis and oocyte maturation; bovine embryonic development is no different in that sense.

It is well known that in *Drosophila melanogaster* and *Caenorhabditis elegans*, gradients of morphogens in the zygote and early embryo are crucial for establishing positional information (St Johnston and Nüsslein-Volhard,

1992; Nüsslein-Volhard, 1996). These gradients are essentially products of maternal gene expression. The extent that similar gradients and elements of the cytoskeleton are important during the earliest stage of mammalian development is unclear. Increasing cell polarity was described at the eight-cell stage of mouse and rat development (Reeve, 1981; Gueth-Hallonet and Maro, 1992). Cell fate, controlled by positional information, seems reversible and provides the developing embryo with a certain degree of flexibility. In cattle, cellular polarization occurred in some blastomeres between the 8- and 16-cell stages, but distinct polarity was possibly observed after the 16-cell stage, with approximately 40% polar cells per embryo (Koyama et al., 1994). Chimeric murine embryos, constructed from two-cell stage blastomeres from which the animal or the vegetal poles have been removed, can develop into normal fertile adult mice. Although polarity of the post-implantation embryo can be traced back to the eight-cell stage and in turn to the organization of the oocyte, its role is not entirely clear (Ciemerych et al., 2000). It seems that mammalian axis specification during oogenesis and through to the early stages of cleavage is under strong regulation. This is unlike what is observed in other metazoans and

may be related to viviparity (Evsikov and Marín de Evsikova, 2009b). If so, then the gradients which are so important in insects and worms may not be crucial for very early stages of mammalian development. The establishment of axial polarity during cleavage and blastocyst formation is considered later in this chapter.

Activation of the embryonic genome

Bovine transcriptome analysis allowed detection of stage-specific expression patterns starting from the two-cell stage until blastocysts (Kues et al., 2008). The number of detected transcripts on different stages of development from oocyte to blastocyst varies from 12,000 to 14,500 and these numbers are considerably higher than the average 8000 transcripts typical for somatic cells. This fact might be an indirect indication of the uncommitted state of earlier pluripotent blastomeres. About 35 genes were found to be significantly up-regulated only in oocytes, some of which are depicted in Fig. 14.2. From oocyte to four-cell stage the great majority of these transcripts are of maternal origin; then a significant drop in quantity of such transcripts is observed

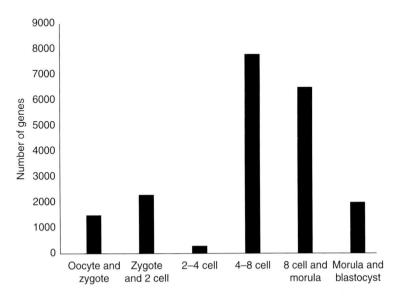

Fig. 14.2. Global analysis of transcriptome features in naturally developing bovine embryos. Number of differentially regulated genes of subsequent stages. (Redrawn from Kues et al., 2008, with modifications. Copyright (2008) The National Academy of Sciences, USA was granted as well as the kind permission of H. Lehrach.)

between the four- and eight-cell stages. According to published data the first and minor wave of embryonic genome activation (EGA) in the bovine conceptus occurs between the two- and four-cell stages (Kues *et al.*, 2008); the second and major wave follows the eight-cell stage, when the highest number of differentially regulated genes was observed. Approximately 2473 genes are significantly up-regulated in eight-cell embryos, morula and blastocysts, compared to earlier stages. Rapid degradation of maternal mRNAs occurs after the four-cell stage and thus coincides or slightly precedes major EGA, supporting an assumption of coordination between the events (Kues *et al.*, 2008). Quite expectedly dynamic changes in the transcriptome and in transiently active genes were typical. More than 120 biochemical pathways were found to be active during the early pre-implantation stage. A proportion of stage-specific transcripts is low during the oocyte and zygote stages (~2–3%) but increases significantly at the two- to eight-cell stages (~22–25%); transcription falls during morula (~7%) and rises again in the blastocyst (~15%).

Compared to zygotes in the two-cell stage, 321 genes (13%) were up-regulated and in the four-cell stage 285 genes (11.5%); 197 (8%) genes were common to both stages and remained up-regulated until the blastocyst stage (Kues *et al.*, 2008). A group of 48 genes that show highly coordinated behaviour is up-regulated in the two-cell stage, down-regulated in the four-cell stage, and finally up-regulated again, remaining highly transcribed to the blastocyst stage. Two genes from the group deserve particular notice as they are strongly up-regulated: *IL18* (Interleukin 18) and *TPT1* (tumour protein, translationally-controlled protein 1) (Adjaye *et al.*, 2007). Approximately 350 highly expressed genes show a stage-specific pattern before the major onset of EGA at the eight-cell stage.

Comparisons of the mouse and cattle data show a number of differences. In the mouse *Eif4b* gene, coding for translation initiation factor is actively expressed on the way from oocyte to zygote (Evsikov and Marín de Evsikova, 2009b), while in bovine embryo similar *EIF2-5* genes became selectively up-regulated at the eight-cell stage (Kues *et al.*, 2008). Three genes (*POU5F1* (formerly *OCT4*), *SOX2* and *NANOG*), which are considered as pluripotency genes in the

mouse and implicated in EGA, in the early bovine embryos have a different developmental dynamic and are not implicated in EGA (Duranthon *et al.*, 2008; Khan *et al.*, 2012). However, *SOX2* and *NANOG* are pertinent candidates for bovine pluripotent lineage specification occurring later in ICM (Khan *et al.*, 2012). According to the latest data microRNAs are possibly involved in transition from maternal to embryonic expression. Mature and precursor forms of miR-21 and miR-130a were quantified at several stages from oocyte to blastocyst. A linear increase during the one- to eight-cell stage for the mature forms of miR-130a and miR-21 and for the precursor form of miR-130a was observed in bovine embryos (Mondou *et al.*, 2012). Another microRNA precursor family, miR-196a, is a likely negative regulator of the *NOBOX* gene during bovine early embryogenesis (Tripurani *et al.*, 2011).

Koehler *et al.* (2009) demonstrated temporal correlation between transcriptional activation and major rearrangement of chromatin topography in blastomere nuclei at the time of major genome activation in bovine embryos. After the eight-cell stage so-called gene-dense chromosome territories are localized more internally and gene-poor territories more peripherally.

Reprogramming, methylation patterns and genomic imprinting

During the first 24 h or so after fertilization the mammalian oocyte and sperm undergo natural reprogramming that gives rise to a totipotent zygote (de Vries *et al.*, 2008). Genomic reprogramming is a complex process involving numerous mechanisms (Seisenberger *et al.*, 2013). Protein and mRNA molecules accumulated in the oocyte facilitate reprogramming through chromosome remodelling as well as differential utilization and degradation of mRNA. DNA methylation is erased from chromatin very early during development thus creating a critically important condition for the next cycle of life. In the bovine zygote an intense DNA demethylation process is followed by rapid *de novo* methylation, which restores the paternal level of methylation (Park *et al.*, 2007). The bovine paternal genome undergoes rapid demethylation within 10 h after IVF and 6 h after intracyto-plasmic sperm injection effect (Abdalla et al., 2009). Beside changes in DNA methylation level, extensive chromatin remodelling includes post-translational histone tail modifications (Canovas et al., 2012). A novel type of epigenetic modi-fication found in bovine and some other mam-malian zygotes, that is 5-hydroxymethylcytosine (5hmC), suggests an additional role for DNA methylation reprogramming processes during early embryonic development (Wossidlo et al., 2011). Kues et al. (2008) indicated that tran-scripts related to DNA methylation, histone modification and chromatin remodelling could be clustered based on maximal or minimal expres-sion at the eight-cell stage, which may reflect molecular remodelling at this time. Such epige-netic modifications are important during pre-implantation development in particular, and spurious epigenetic marks may have long-lasting consequences for developing embryos. Incom-plete epigenetic reprogramming was described for embryos generated by nuclear transfer and contributes to the low efficiency of cloning (Dean et al., 2001).

Modification of epigenetic features and gene expression patterns is the essence of genome reprogramming, and this process is compromised after somatic cell nuclear transfer (SCNT) in oocytes. Comparative experiments show that three groups of transcripts are mostly affected during somatic reprogramming. This includes LTR retrotransposons and mitochondrial tran-scripts, which are up-regulated, as well as down-regulated genes encoding ribosomal proteins (Bui et al., 2009). In contrast to normal embryos, which significantly increase DNA methylation levels during the elongation stage, in embryos derived by somatic cell nuclear transfer, methyl-ation was significantly lower in the trophectoderm compared to the normal level in the embryonic disc. DNMT1 expression in similarly derived blastocysts was significantly reduced compared to that in naturally produced embryos (Sawai et al., 2010).

The consequent developmental stages may generate pluripotent cell types. Gene expression programmes operating in these pluripotent cells steadily become more defined, production of core transcription factors begins, and expression of pluripotency-associated genes commences. At least three so-called pluripotency genes,

POU5F1, NANOG and SOX2, influencing transcription factors during mammalian devel-opment have been identified. These genes are responsible for activation of gene cascades essential for maintenance of pluripotency and temporary repression of genes required for fur-ther differentiation (de Vries et al., 2008). Computational comparison of the influence of POU5F1 on transcription factor networks shows a significant level of conservation in human, mouse and cow (Xie et al., 2011). NANOG also plays a key role in stable induction of pluri-potency in bovine adult fibroblasts and prob-ably during embryonic development (Sumer et al., 2011).

Genes that are required later in develop-ment are repressed by histone marks, which confer short-term, and therefore flexible, epi-genetic silencing (Reik, 2007). As soon as dem-ethylation is accomplished, a new wave of DNA methylation begins and it leads to stable and long-term epigenetic silencing of certain genetic elements like transposons, imprinted genes and pluripotency-associated genes. Evidence is accumulating that such DNA methylation epigenetic marks play a key role in deter-mining cell and lineage commitment (Senner et al., 2012).

Thus bovine development shows typical features discovered in other mammalian spe-cies including reprogramming and epigenetic modification. For instance, methylation of the lysine residue 27 of histone H3 (H3K27me3), which is established by polycomb group genes, renders an essential epigenetic mark associ-ated with stable and heritable gene silencing in early bovine embryos (Ross et al., 2008a). Transcripts of EED, EZH2 and SUZ12 genes, which regulate the H3K27 methylation process, were high at early stages of development. Nuclear expression of EZH2 was detected in oocytes, while EED and SUZ12 were only evident at the morula and blastocyst stages (Ross et al., 2008a). As shown in pig development H3K27me3 tri-methylation is an epigenetic marker of mater-nally derived chromatin that undergoes global remodelling (Park et al., 2009). The paternal bovine genome is actively demethylated within several hours after fertilization, while, on the contrary, the maternal genome is demethyl-ated passively by a replication-dependent mechanism after the two-cell embryo stage

(Abdalla *et al.*, 2009). Such demethylation is related to DNA repair processes (Reik, 2007).

Gametic or genomic imprinting is a developmental phenomenon typical for eutherian mammals and based on differential expression of maternal and paternal alleles in certain genes. These genes are essential for regulation of embryonic and placental growth. In genes like *IGF2* only the paternal allele is expressed (maternal imprinting). However, in genes like *H19* only the maternal allele is expressed (paternal imprinting). Imprint acquisition occurs before fertilization and imprint propagation extends up until the morula–blastocyst stage (Shemer *et al.*, 1996). In *H19* the 2-kb region is methylated on the paternal allele during spermatogenesis. The maternal allele has a different methylation pattern (Davis *et al.*, 1999). Bovine *H19* is exclusively expressed from the maternal allele in all major organs similarly to other species (Robbins *et al.*, 2012). The bovine intergenic IGF2-H19 imprinting control region was recently investigated in some detail (Hansmann *et al.*, 2011).

The molecular mechanisms of gametic imprinting are still under investigation. It seems possible that primary genomic signals are not simply copied from the gametes, but rather the methylation pattern typical for imprinted genes establishes gradually during early development (Shemer *et al.*, 1996). The regulatory elements that control genomic imprinting have differential epigenetic marking in oogenesis and spermatogenesis, which results in the parental allele-specific expression of imprinted genes during development and after birth (Feil, 2009). Both DNA and histone methylation are essential for imprinting. DNA methylation is involved in the acquisition and/or maintenance of histone methylation at imprinting control regions (Henckel *et al.*, 2009). The developmental function of genomic imprinting is also not absolutely clear, but an explanation proposed by Moore and Haig (1991) is widely accepted. It is based on the concept of genetic conflict arising during pregnancy between maternally and paternally inherited genes. Thus, it is likely that genomic imprinting evolved in mammals to regulate intrauterine growth and to increase the safety of embryonic development. A lack or presence of an extra copy of maternally or paternally derived alleles or abnormal expression of such alleles in a zygote or later during development

may lead to embryonic mortality and impose strict requirements on the stability of imprinting signals.

According to the database of imprinted genes (http://igc.otago.ac.nz/Search.html) the number of bovine imprinted genes so far described is 32. The latest genome scan identified 24 imprinted genes/DNA sections or, in other words, parent-of-origin effects quantitative trait loci (QTLs) on 15 *Bos taurus* autosomes of which six were significant at 5% genome-wide (GW) level and 18 at the 5% chromosome-wide (CW) significance level (Imumorin *et al.*, 2011). These QTLs affect mainly bovine growth and carcass traits. There is a genuine need to identify and validate other bovine orthologues of imprinted genes already identified in other mammals.

Gene expression during blastocyst formation, expansion and elongation

During early embryonic development some proteins are produced from maternal mRNA templates and then nearly without interruption from reactivated embryonic genes. Among them are centromere protein F (*CENPF* gene) and arginine/serine-rich 3 splicing factor (*SRSF3* gene), which show high mRNA content during the two- to four-cell and then from late eight-cell stages (Kanka *et al.*, 2009). A similar pattern was observed for the high mobility group nucleosomal binding domain 2 protein coded by the *HMGN2* gene whose mRNA level was high at the two- to four-cell stage and later at morula. *SRSF3* is also expressed during bovine minor genome activation (Kanka *et al.*, 2009). The morula-to-blastocyst transition (days 6–7) is sensitive to disruption of methionine metabolism caused in part by S-adenosylmethionine deficiency, which may lead to DNA hypomethylation and altering expression of genes affecting normal blastocyst development (Ikeda *et al.*, 2012). Transcriptional activity of many genes changes as the bovine embryo transforms from early spherical blastocyst (day 7) to an ovoid conceptus at the initial stage of elongation (day 13). Comparison of day 7 and day 13 embryos revealed significant temporal changes in transcription profile of 1806 genes (Clemente *et al.*, 2011). These researchers identified genes and pathways crucial for the transition from a spherical blastocyst to an ovoid conceptus.

It was shown that lamin B appeared as a constitutive component of nuclei at all pre-implantation stages but lamins A/C had a stage-related distribution (Shehu *et al.*, 1996). The nuclei from the early cleavage stages contain lamins A/C, which generally disappeared later, with a few possible exceptions in the morula and blastocyst. Many other proteins are produced at this stage including several cytoskeletal and cytoskeleton-related components such as F-actin, alpha-catenin and E-cadherin. These proteins appear from day 6 and their polarized distribution in blastomeres seems to be relevant to morula compaction (Shehu *et al.*, 1996). Pig data suggest that several other morphogenetically important proteins appear during early cleavage and compaction like alpha-fodrin, vinculin and others (Reima *et al.*, 1993).

The crucial differentiation event occurs during the blastocyst stage when totipotent blastomeres differentiate into either pluripotent ICM or multipotent TE. Analysis revealed 870 genes that were differentially expressed between ICM and TE in bovine embryos. A number of key genes like *NANOG*, *SOX2* and *STAT3* in ICM and *ELF5*, *GATA3*, and *KRT18* in TE behave similarly in distant mammalian species like mouse, human and cattle. Not all genes however demonstrate such uniformity. Analysis of gene expression made by Ozawa *et al.* (2012) shows

that differentiation of blastomeres into the ICM and TE is accompanied by gene expression differences between the two cell lineages controlling metabolic processes like endocytosis, hatching from the zona pellucida, paracrine and endocrine signalling with the mother, and genes supporting development of the trophoblast. *NANOG* and *SOX2* are considered as pertinent candidates for mammalian as well as bovine pluripotent lineage specification occurring in ICM (Khan *et al.*, 2012). Ozawa *et al.* (2012) have demonstrated that in bovine embryo 'expression of two genes important for ICM commitment, *NANOG* and *SOX2*, was significantly higher for ICM than TE while expression of two other genes important for ICM commitment, *POU5F1* and *SALL4*, did not differ significantly between ICM and TE' (Fig. 14.3). These authors also identified 'four genes important for TE commitment – *CDX2*, *GATA3*, *TEAD4*, and *YAP1*'. Expression of *GATA3* was significantly higher for TE but there were no significant differences in expression between ICM and TE for the other three genes. Another gene essential for TE differentiation in later development, *ELF5*, was highly expressed unlike *EOMES,* which surprisingly was barely detectable in ICM and TE (Ozawa *et al.*, 2012). Thus *NANOG* and *GATA3* play a central role in lineage commitment during bovine ICM/TE split, where *NANOG* expression is essential for ICM and *GATA3* for TE differentiation (Ozawa *et al.*, 2012).

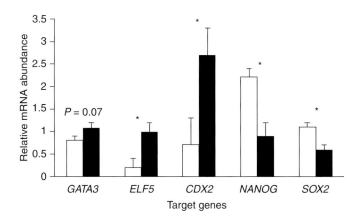

Fig. 14.3. Differences between inner cell mass (ICM) and trophectoderm (TE) in expression of six genes as determined by quantitative PCR. Data represent least-squares means ± SEM of results from six biological replicates. Open bars represent ICM and filled bars TE. *$P < 0.05$. (Reprinted from Ozawa *et al.*, 2012, with the kind permission of the corresponding author, P.J. Hansen.)

Although TE and ICM are affected by reciprocal expression of *Cdx2* and *Pou5f1* in murine late blastocysts, similar developmental processes are somewhat delayed and modified in cattle where *CDX2* is required later for TE maintenance and it does not repress *POU5F1* expression to the same degree. The comparative data suggest that the regulatory circuitry determining ICM/TE identity has been rewired in mice, to allow rapid TE differentiation and early blastocyst implantation, which is different from bovine embryos (Berg *et al.* 2011). The differentiation of the trophoblast in ruminants is characterized by the formation of binucleated cells (BNCs), which appear from pre-implantation onwards. The expression pattern of *DLX3* and *PPARG* regulating early placenta formation as well as expression of related *SP1* were studied by Degrelle *et al.* (2011a), who demonstrated co-expression of these genes in bovine BNC nuclei.

Despite the simplicity of the blastocyst structure, the mechanisms of its formation are still elusive. Three models: mosaic, positional and polarization have been suggested and extensively studied (Johnson, 2009). The ICM differentiates into the epiblast and primitive endoderm. The epiblast gives rise to the embryo itself and also to some extra-embryonic tissues. The TE is responsible for development of remaining extra-embryonic tissues and plays a critical role during implantation and formation of the trophoblast layers of the placenta. It has been found that *Cdx2*, which encodes a caudal-related homeodomain protein, is a key regulator of the TE lineage (Rossant and Tam, 2009), even though expression of this gene begins earlier. Mouse gene *Tead4* producing transcription enhancer is tentatively considered as an upstream factor relevant to *Cdx2*; *Eomes*, on the contrary, is a downstream-located factor (Rossant and Tam, 2009). Cdx2 and Eomes proteins are restricted to outer layer cells. The genes that specify the pluripotent cells like *Pou5f1/Oct4* (Nichols *et al.*, 1998), *Sox2* (Nichols *et al.*, 1998; Avilion *et al.*, 2003) and *Nanog* (Chambers *et al.*, 2003; Mitsui *et al.*, 2003) are initially expressed in all blastomeres but progressively become restricted to ICM cells after blastocyst formation. Adjaye *et al.* (2005) identified in developing human blastocysts marker transcripts specific to ICM (e.g. *POU5F1*, *NANOG*, *HMGB1* and *DPPA5*) and TE (e.g. *CDX2*, *ATP1B3*, *SFN* and *IPL*). The emergence of pluripotent ICM and TE cell lineages starting from morula is controlled by metabolic and signalling pathways, which include *inter alia*, Wnt, mitogen-activated protein kinase, transforming growth factor-beta, Notch, integrin-mediated cell adhesion, phosphatidylinositol 3-kinase and apoptosis. Several studies of bovine development indicate that the mouse model is not always applicable.

Experiments recently conducted by Kuijk *et al.* (2012) demonstrate high plasticity of the initial steps in establishing epiblast and hypoblast cell lines in bovine embryos. For instance FGF4 and heparin have a powerful influence and in experimental conditions ICM might be composed only of hypoblast cells and no epiblast cells. On the contrary, inhibition of dual-specificity threonine and tyrosine recognition kinase (MEK) causes a significant increase of epiblast precursors and decrease of hypoblast precursors in bovine embryos. Surprisingly human embryos behave differently, which proves intrinsic differences in early development between mammalian species. The authors concluded that bovine embryonic cells are heterogeneous in response to MEK inhibition and that in some cells GATA6 expression is independent of activated MEK. This paper also indicates that such critically important transcription factors like NANOG and GATA6 have 'pepper-and-salt' distribution in the ICM of day 8 embryos and are expressed in the epiblast and hypoblast precursors, respectively (Kuijk *et al.*, 2012).

Polarity of cells in the blastocyst increases due to an accumulation of protein kinase3, polarity protein Par3 and ezrin in the apical domain of blastomeres and apical membrane; other proteins like Lg1 and Par1 are exclusively found in the basal portion of murine blastomeres (Rossant and Tam, 2009). Connexin proteins are differently expressed both temporally and spatially in the pig embryo, influencing formation of gap junctions in the trophoblast and later controlling the exponential growth of the trophoblast in pre-implantation pig blastocysts (Fléchon *et al.*, 2004a).

Although the rate of embryonic development differs in cow and mouse, there is a correlation between the developmental stage and cytoskeletal organization in both species. Likewise, in the expanded bovine blastocyst, the distribution of several cytoskeletal and cytoskeleton-related proteins appeared similar (Shehu *et al.*, 1996). Extracellular fibronectin was first detected in the early blastocyst before differentiation of the primitive endoderm and, at this stage, was localized at the interface between the trophectoderm and the extra-embryonic endoderm (Shehu *et al.*, 1996). Cingulin, the tight junction peripheral membrane protein, also contributes to morphological differentiation in early mouse development, and other mammals including cattle have the same gene. Its synthesis is tissue-specific in blastocysts, is up-regulated in the TE, and down-regulated in ICM (Javed *et al.*, 1993). The bovine epiblast (day 12) consists of at least two cell subpopulations and each of them exclusively contains POU5F1/OCT4 (the POU class 5 homeobox 1); other cell populations of the blastocyst do not show presence of this developmentally critical protein (Vejlsted *et al.*, 2005). Colocalization of vimentin and POU5F1 was also shown. The proliferation marker Ki-67 was localized to most nuclei throughout the epiblast as well. Trophectodermal cells on the contrary exclusively contain alkaline phosphatase and a basement membrane covering epiblast has laminin (Vejlsted *et al.* 2005). These molecular distinctions provide additional evidence for the ongoing morphological and functional differentiation of cellular populations.

The following elongation stage of bovine development (days 12–18) coincides with a large proportion of embryonic losses, which are particularly prevalent after artificial insemination (~30% of embryonic losses). During this developmental stage physiological interactions between the conceptus and the uterus are steadily established, and are essential for successful implantation. However identification of many genes and biological networks playing key roles in elongation processes specific for *Bos taurus* still requires further efforts (Hue *et al.*, 2012). The transformation of blastocysts from ovoid to filamentous stages is initiated by day 12 and the elongation process begins (Clemente *et al.*, 2011). During days 12–19 of bovine development, when elongation and transformation to a thin filamentous structure is in progress, the pattern of gene expression becomes rather complex. Elongation increases the size of the bovine conceptus more than 1000-fold mainly due to an increase in cell number and cellular growth (Thompson *et al.*, 1998; Maddox-Hyttel *et al.*, 2003; Degrelle *et al.*, 2005). In bovine embryos the mononucleate trophoblast cells differentiate into a cell type important for implantation, the binucleate trophoblast. A gene cluster associated with this rapid proliferation and differentiation of the trophoblast was initially described by Blomberg *et al.* (2008). The latest comprehensive review of transcriptomic changes in the bovine conceptus from the early blastocyst until initiation of implantation refers to more than 18,500 transcripts shared among the five conceptus development stages (days 7, 10, 13, 16, 19), which were described, identified and compared (Mamo *et al.*, 2012b). During this period several critically important events take place including the formation of an ovoid conceptus, initiation of elongation, maternal recognition of pregnancy and initiation of implantation. Mamo *et al.* (2012b) identified 20 genes with the highest expression at each stage and demonstrated similarities and differences in gene expression patterns. The most prevalent genes include those that code for various trophoblast Kunitz domain proteins, pregnancy-associated glycoproteins, cytoskeletal transcripts, heat shock proteins, calcium binding proteins, as well as *APOA1*, *AHSG*, *BOP1*, *TMSB10*, *CALR*, *APOE*, *TPT1*, *BSG*, *FETUB*, *MYL6*, *GNB2L1*, *PRDX1*, *PRF1*, *IFNT* and *FTH1*. The comparison study of Clemente *et al.* (2011) shows that of 909 genes that were differentially expressed in *in vivo* embryos on day 7 and day 13, 408 genes were up-regulated and 501 down-regulated. Some microarray findings were confirmed by quantitative real-time PCR, which opened up the opportunity for mapping the differentially expressed genes (*CYP51A1*, *FADS1*, *TDGF1*, *HABP2*, *APOA2* and *SLC12A2*) into relevant functional groups. It was concluded that between day 7 and day 13 465 differentially expressed genes play a role in elongation of the bovine conceptus (Clemente *et al.*, 2011).

Genetic control of gastrulation

Two major processes occur during gastrulation: the development of three germ layers and the establishment of anterio-posterior and dorso-ventral axes. At the stage preceding gastrulation (pre-streak) in pigs, the rate of cell proliferation in the posterior section of the epiblast is heightened. Migration of the cells, which are the precursors of the developing primitive streak, begins (Fléchon et al., 2004b). As even the smallest bovine embryos displayed a primitive streak at day 21 (Maddox-Hyttell et al., 2003), one can safely assume that the pre-streak stage begins at least a day or two earlier and, while gastrulation starts earlier, it overlaps with implantation (Table 14.1). Elongation of bovine trophoblasts starts prior to gastrulation and influences this critically important process. A set of six genes that discriminate stages of bovine embryonic development (CALM1, CPA3, CITED1, DLD, HNRNPDL and TGFB3) has been identified (Degrelle et al., 2011a).

It has been known since 1924 that in vertebrates the Spemann organizer, which forms at mid blastula, plays a crucial role as signalling centre for the dorso-ventral axis specification. The Spemann organizer blocks action of BMP-4 by secreting several proteins like Noggin, Chordin, Nodal-related and Cerberus. Wnt-signalling is strongly involved in the formation of the organizer. The signal transduction cascade is a complex system of interactions of several proteins, which prevents degradation of β-catenin essential for the following gene activation (Sokol, 1999). Extra-embryonic cells, known as the anterior visceral endoderm (AVE), migrating from the distal to a more proximal region of the embryo specify the anterior–posterior body axis (Migeotte et al., 2010). AVE secretes inhibitors of the Wnt and Nodal pathways. Other essential regulators of cell migration are Rac proteins, which play a role in AVE migration. Rac1 mutant murine embryos fail to specify an anterior–posterior axis. AVE cells extend long lamellar protrusions that span several cell diameters and are polarized in the direction of cell movement. This represents a critical step in the establishment of the mammalian body (Migeotte et al., 2010). Cdx2 seems to be significantly involved in the integration of the pathways controlling embryonic axial elongation and anterior–posterior patterning (Chawengsaksophak et al., 2004). A number of other players in the Wnt signalling pathway have been discovered including Axin: mutations that affect development of axial skeleton and the tail in particular (Zeng et al., 1997; Fagotto et al., 1999).

At the present time information about the gene expression pattern from pre-streak stage to gastrulation is limited in mammals. Fortunately studies of other vertebrates like chickens, have recently defined three genes as markers for early responses of epiblast cells to signals from Hensen's node inducing the cascade of following events (Pinho et al., 2011). There are two groups of genes during the earliest hours of signalling: a 'pre-streak group' and a 'streak group' (expressed in the later gastrula stage). Three genes that belong to the second group are: Asterix, TrkC and Obelix. Interactions of these genes with hormones and transcription factors operating in mammals are not yet known.

Development of the three-germ-layer embryo with ectoderm, endoderm and mesoderm is the major result of gastrulation. Brachyury is a T-box-containing transcription factor involved in mesoderm formation during vertebrate gastrulation as well as in tissue specification, morphogenesis and organogenesis (Müller and Herrmann, 1997; Hue et al., 2001; Blomberg et al., 2008). Mutations in this gene may lead to serious morphological abnormalities. It has been found that the normal expression pattern of Brachyury is temporarily reduced in the anterior part of the primitive streak in bovine embryos (Hue et al., 2001).

In mammals Brachyury interacts with the Goosecoid gene (GSC) encoding a homeobox protein. In the pig embryo the intensive expression becomes more pronounced in differentiating mesodermal cells that ingress from the epiblast via the Hensen's node (van de Pavert et al., 2001). This process finally leads to formation of the mesoderm and embryonic endoderm. Goosecoid over-expression may repress the Brahyury gene and affect normal development (Boucher et al., 2000). In porcine embryos at the expanding hatched blastocyst stage, POU5F1 is confined to the ICM. Following separation of the hypoblast, and formation of the embryonic disc, this marker of pluripotency was selectively observed in the epiblast. Progressive differentiation of germ layers and tissues leads to the silencing of this gene with the exception of the primordial germ cells (Vejlsted et al., 2006).

At the pre-streak stage prior to gastrulation and migration of extra-embryonic mesoderm the embryonic disc becomes polarized (Fléchon et al., 2004b). The early primitive streak is characterized by both high pseudostratified epithelium with an almost continuous but unusually thick basement membrane, and Brachyury expression. Brachyury is crucial for notochord development in all examined chordates, and at least 44 notochord-expressed genes are its transcriptional targets (Hue et al., 2001; Capellini et al., 2008). Expression of the NODAL gene is essential for axial patterning during early mammalian gastrulation as well as induction of the dorso-anterior and ventral mesoderm (Jones and Clemmons, 1995). As gastrulation proceeds, the primitive streak extends anteriorly and at its distal end the Hensen's node is developed, which is composed of a mass of epithelium-like cells but without cilia (Blum et al., 2007). Expression of the Goosecoid gene is typical for these cells, which originate the notochord.

The notochord is a flexible fibro-cellular cord lying ventral to the developing central nervous system and represents the major axial structure of the embryo, playing a very important role in induction of the neural plate, chondrogenesis and somite formation (Gomercić et al., 1991). Glycoproteins compose a core of the notochord with cells encased in a sheath of collagen fibres. Two genes controlling notochord formation encode laminin β1 and laminin γ1, which are essential for building the scaffold on which individual cells organize the rod-like structure typical for the notochord in vertebrates (Pollard et al., 2006). There is higher protein production of integrin subunits that regulate interactions with collagens and laminin in notochordal cells (Chen et al., 2006). In vertebrates the notochord is replaced during development by the vertebrate column. The notochord grows anteriorly from the Hensens's node below the embryonic disc and is composed of cells derived from differentiating mesodermal cells that ingress from the epiblast. Three key mammalian genes, SOX17, NODAL and Brachyury (T), are involved in the early development of the axial structure during gastrulation (Blomberg et al., 2008; Hassoun et al., 2009). According to Zorn and Wells (2009) the Nodal signalling pathway is necessary and sufficient for initiation of endoderm and mesoderm

development and it is required for proper gastrulation and axial patterning. Nodal ligands are members of the TGFβ family of secreted growth factors. NOTO is another gene that is required for the formation of the caudal part of the notochord as well as for ciliogenesis in the posterior notochord. The data also show that Noto acts during murine development as a transcription factor upstream of Foxj1 and Rfx3. According to Beckers et al. (2007) this genetic cascade is important for the expression of multiple proteins required for cilia formation and function. Later these processes influence dorsal and ventral axis specification, neural tube and spinal cord patterning. Complex interactions between Noto and Foxj1 were studied recently. Foxj1 expression from the Noto locus is functional and restores the formation of structurally normal motile cilia in the absence of Noto, however, Foxj1 alone is not sufficient (Alten et al., 2012). Genes for cytokeratins 8, 18 and 19 are active in the notochord and later in the vertebrate discs of adult mammals (Shapiro and Risbud, 2010).

Clearly, activation of nuclear genes responsible for basic morphogenetic rearrangements is the prerequisite for notochord formation and development. The T gene, which was first described as the Brachyury mutation in mice 80 years ago, is the key regulator of events required for differentiation of the notochord and formation of mesoderm during posterior development. The T protein is located in the cell nuclei and acts as a tissue-specific transcription factor (Kispert et al., 1995). Cloning and sequencing of the T gene led to the discovery of the T-box gene family, which is characterized by a conserved sequence called T-box (Bollag et al., 1994). This ancient family of transcription factors, which underwent duplication around 400 million years ago, is common to all vertebrates (Ruvinsky and Silver, 1997). There are indications that several murine T-box genes are essential for formation of different mesodermal cell sub-populations and one of the T-box genes is essential for the development of early endoderm occurring during gastrulation (Papaioannou, 1997). Involvement of other T-box genes in the developmental process is diverse. For instance Tbx2–Tbx5 genes are involved in vertebrate limb specification and development (Gibson-Brown et al., 1998).

Tbx5 is also influential in heart development, and mutations in these genes are associated with congenital heart defects (Hariri *et al.*, 2012). During the evolution of heart in amniotes ventricular septation is steadily established by a steep and correctly positioned gradient of *Tbx5* activity (Koshiba-Takeuchi *et al.*, 2009).

Brachyury also plays a key role in the process activated in the blastopore region, which might be a 'primary' function of the gene (Satoh *et al.*, 2012). In the course of chordates' evolution, *Brachyury* gained an additional expression domain at the dorsal midline region of the blastopore, where *Brachyury* performed a 'secondary' function, recruiting another set of target genes to form the notochord (Satoh *et al.*, 2012). Formation of the notochord leads to several key ontogenetic events including induction of the neural tube and then the central nerve system. Shh protein (Sonic hedgehog signalling) secreted by the floor plate and notochord, specifies the fate of multiple cell types in the dorsoventral axis of the neural tube and thus affects formation of the vertebrate nervous system (Gray and Dale, 2010). Shh in turn induces expression of *Gli1*, which affects later development of dorsal midbrain and hindbrain (Hynes *et al.*, 1997). It was also found that Notch signalling operates like a balance for Shh by promoting axial progenitor cells to the floor plate and inhibiting contribution to the notochord (Gray and Dale, 2010). These authors suggested that 'Notch regulates the allocation of appropriate numbers of progenitor cells from Hensen's node of vertebrate embryo to the notochord and the floor plate'. Recent results show that a certain combination of transcription factors is influential in determining Nodal pathway output during mesendoderm patterning (Slagle *et al.*, 2011).

Specification of the germ layers and hence definitive body axes occur prior to primitive streak formation. The Wnt/β-catenin, BMP/Nodal and FGF-signalling pathways are also involved in the transcriptional activation of *Brachyury*. These processes are emerging as decisive steps in the initial patterning of the pre-gastrulation embryo. Their ensuing signalling leads to the specification of axial epiblast and hypoblast compartments through cellular migration and differentiation and, in particular, the specification of the early germ layer tissues in the epiblast via gene expression characteristic of endoderm and mesoderm precursor cells (Blomberg *et al.*, 2008). Unlike the notochord cells, other emerging mesodermal cells spread out more or less uniformly and give rise to numerous organs and structures.

Establishment of axial identity

The early blastocyst and even possibly the late morula have some degree of polarization, which later may influence axial identity. Several genes significantly contributing to the emerging polarity have been identified so far. Genes encoding ezrin, PAR family proteins and CDX2 are likely the key regulators of the process. Other proteins like CDC42, E-cadherin, β-catenin and Hippo are strongly involved in the process; laminin and integrins also play a role (Johnson, 2009). Development of the primitive streak and the notochord is the convincing demonstration that both anterior–posterior (AP) and dorso-ventral (DV) axes are strictly determined.

The left–right (LR) axis may look like an automatic consequence of the earlier-defined AP and DV axes, as it is perpendicular to both (Levin, 2004). However, the cause of LR asymmetry in vertebrates, and mammals in particular, is a complicated question. Levin (2004) compiled a long list of genes that may affect the symmetry. More recent findings show that in the developing mouse embryo, leftward fluid flow on the ventral side of the Hensen's node determines LR asymmetry. Morphological analyses of the node cilia demonstrated that the cilia stand, not perpendicular to the node surface, but tilted posteriorly (Nonaka *et al.*, 2005). This morphological asymmetry can produce leftward flow. A genetic cause of LR asymmetries of the internal organs in vertebrates is steadily becoming clearer. Gros *et al.* (2009) considered two possibilities. The initial asymmetric cell rearrangements in chick embryos create a leftward movement of cells around the Hensen's node. This is relevant to expression of *Shh* and *Fgf8* (fibroblast growth factor 8). The alternative is a passive effect of cell movements. It was also shown that a Nodal-BMP signalling cascade drives LR heart morphogenesis by regulating the speed and direction of cardiomyocyte movement (Medeiros de Campos-Baptista *et al.*, 2008). Interplay between two TGFβ ligands, GDF1 and Nodal together with inhibitors Lefty and Cerl2 provide the signals for the establishment of laterality. By blocking TGFβ signalling, APOBEC2 protein also regulates LR specification (Vonica *et al.*, 2010). The mouse transcription factor *Noto*, as mentioned earlier, is expressed in the node and influences formation of nodal cilia and hence LR asymmetry. There is a synergy between *Noto* and *Foxj1*. However, *Foxj1* alone is not sufficient for the correct positioning of cilia on the cell surface within the plane of the nodal epithelium, and other factors are involved (Alten *et al.*, 2012).

The three germ layers and their derivates

By the end of gastrulation three germ layers are established: endoderm, mesoderm and ectoderm. Molecular mechanisms driving this highly complex combination of processes began to emerge relatively recently. Zorn and Wells (2009) published one of the first reviews covering the entire endoderm development and organ formation. Here we can highlight only the major regulatory systems influencing the variety of genes and processes involved in endoderm morphogenesis and formation of certain organs. The Nodal signalling pathway is necessary and sufficient to initiate ectoderm and mesoderm development and itself is influenced by the canonic WNT/β-catenin pathway (Zorn and Wells 2009). High-level Nodal signalling supports endoderm development and lower activity specifies mesoderm identity. The activity of the Nodal pathway is controlled by an auto regulatory loop. Several genes in vertebrates, involved in the pathway like *Nodal*, have conserved *Foxh1* DNA-binding sites in their first introns, sustaining the high activity essential for endoderm development. On the other hand, in developing ectoderm a negative feedback of Nodal activity is caused via transcriptional target *Lefty2* (Shen, 2007). Soon after gastrulation, the endoderm germ layer forms a primitive gut tube, which leads to organ specification, then formation of organ buds and finally to more specialized cell lineages (Zorn and Wells, 2009).

Developmental events in the mesoderm and ectoderm progress simultaneously but independently with significant interactions. As is well known, many organs have cellular components originating from different germ layers. Certain genes play a key role in the earliest stages of germ layer development. For instance the *Eed* gene, initially identified in mice, is critical for embryonic ectoderm development (Sharan *et al.*, 1991) as deletion of this gene prevents formation of ectoderm. As already mentioned, *T* gene is crucial for mesoderm development. Mice homozygous for mutant alleles of the *T* gene do not generate enough mesoderm and show severe disruption in morphogenesis of mesoderm-derived structures, in particular the notochord (Wilkinson *et al.*, 1990). One of the T-box genes, *Tbx6*, in mice is implicated in the development of paraxial mesoderm (Chapman *et al.*, 1996; White and Chapman, 2005). *Tbx6* transcripts are first detected in the gastrulation stage embryo in the primitive streak and in newly recruited paraxial mesoderm.

Genetic Regulation of Implantation and Placentation

Endometrial and trophoblast gene expression

Both endometrial and trophoblast cells prepare for implantations in advance. Conceptus–maternal communication becomes increasingly important for successful establishment and maintenance of pregnancy (Mamo, *et al.*, 2012b). Forde and Lonergan (2012) argue that in cows around days 13–16 of pregnancy 'carefully orchestrated spatio-temporal alterations in the transcriptomic profile of the endometrium are required to drive conceptus elongation, via secretions from the endometrium and establish uterine receptivity to implantation'. Progesterone (P4) and the pregnancy recognition signal interferon tau (IFNT) are particularly important. It has been shown that modulation of circulating P4 affects endometrial expression of genes that can cause beneficial (when P4 is supplemented) or detrimental (when P4 is reduced) influences on the developing conceptus (Forde and Lonergan, 2012). These critical factors promote a sequence of molecular signals required for successful establishment of pregnancy in cattle. A comparison of transcriptome analysis in bovine, porcine and equine endometrium revealed not only some differences 2–3 days preceding implantation, but also provided strong indication of the importance of interferons during the establishment of pregnancy (Bauersachs and Wolf, 2012). Several changes of transcriptome profile are similar between studied domestic mammals. *CLDN4* (claudin 4) coding for a cell adhesion molecule essential in tight junctions and *DKK1* inhibiting WNT signalling are among such 'universally' important genes in these species. The list of such genes is significantly longer and depends on species and statistical criteria. The overview of processes and signalling pathways involved in conceptus–endometrium interaction demonstrates not only the role of interferons, but also oestrogens, progesterone and prostaglandins between days 14–16 of bovine development (Bauersachs and Wolf, 2012).

Progesterone plays the central role preparing endometrium for embryo implantation and maintenance of pregnancy (Bazer *et al.*, 2008). Progesterone-induced blocking factor (PIBF) and galectins are likely involved in uterine receptivity and possibly implantation acting as positive and negative regulators. As reported by Okumu *et al.* (2012) galectins *LGALS9* and *LGALS3BP* were expressed at low levels in both cyclic and pregnant endometrial from day 7 until day 15; however, on day 16 expression increased in the pregnant heifers. It seems likely that PIBF and galectins are produced by different cell types of endometrium. Forde *et al.* (2012b) discovered that the expression of *MEP1B*, *NID2* and *PRSS23* genes increased in bovine endometrium from day 13 onwards. These genes are possibly regulated by changes in P4 concentrations which probably impact conceptus elongation.

Trophoblast cells (called TE after the onset of gastrulation) forming the outer layer of the bovine blastocyst, expand dramatically from day 14 and elongate >1000-fold along the villous folds of the uterus and thus create the functional point of the fetal–maternal contact. Differentiation of the trophoblast begins early in embryonic development and ultimately results in functionally diverse cells. Significant work

on trophoblast gene expression has been carried out in the murine model whose placentation is quite different from the cow. Roberts *et al.* (2004) describe this process and consider that the first key step in trophoblast differentiation is down-regulation of *POU5F1*, which normally acts as a negative regulator of genes required for the next stage of differentiation (de Vries *et al.*, 2008). *POU5F1* acts in the pluriplotent ICM to silence genes related to differentiation, but once this restraint is removed the below-mentioned genes can come under the control of transcriptional activators. Knofler *et al.* (2001) provide a view of the regulatory factors involved in trophoblast development and differentiation. Of these, the T-box gene *Eomes*, which is considered to be one of the earliest trophoblast determining factors in the pre-implantation embryo, is required for trophoblast differentiation (Russ *et al.*, 2000). Both Eomes and the homeodomain protein CDX2 are absent in ICM, but present in TE (Beck *et al.*, 1995). *Cdx2* and *Eomes* murine knockout embryos fail to implant and only develop to the blastocyst stage (Chawengsaksopak *et al.*, 1997; Russ *et al.*, 2000).

As reviewed by Blomberg *et al.* (2008) the global gene expression profiling data of the elongating bovine embryo indicate two main points. 'First, trophoblast mononucleate cell (TMC) proliferation and functional differentiation may be induced by specific genes, such as interferon-τ (*IFNT*), trophoblast Kunitz domain proteins (TKDPs), and the transcription factors POU-domain class 5 transcription factor (POU5F1), ERG, and CDX2 that can regulate IFNT and TKDPs. Second, the specific expression of transcripts encoding pregnancy-associated glycoproteins (PAGs), prolactin-related proteins (PRPs), and placental lactogen (CSH1) in the trophoblast giant cell (TGC) lineage may not be related to the differentiation of TGCs from TMCs. The activating enhancer-binding protein 2 (AP2) family and endogenous retroviruses (ERVs) may be relevant to trophoblast cell differentiation.' Ushizawa *et al.* (2004) detected ~80 up-regulated genes on days 17–19, i.e. prior to implantation, and show that the population of trophoblast binucleate cells consists of two cell types; one of which expresses CSH1 protein, and the other does not. The latest list of the most prevalent 20 up-regulated genes observed at days 13, 16 and 19 of developing bovine conceptus includes various trophoblast Kunitz domain proteins, pregnancy associated glycoproteins, cytoskeletal transcripts, heat shock proteins, calcium binding proteins, APOA1, AHSG, BOP1, TMSB10, CALR, APOE, TPT1, BSG, FETUB, MYL6, GNB2L1, PRDX1, PRF1, IFNT and FTH1 (Mamo et al., 2012b). GATA1 has high expression level in conceptuses and endometrium prior to and at the time of trophoblast attachment playing a potentially important role in regulating antagonistic GATA2 (Bai et al., 2012).

A family of transcription factors of basic helix-loop-helix (bHLH) proteins is important in trophoblast development. In the mouse this includes Mash2, whose expression is crucial in the specification of the trophoblast lineage and particularly sphongiotrophoblast development. Bovine MASH2 is maternally expressed after implantation, but the paternal allele is silenced after implantation (Arnold et al., 2006). This family of factors also includes HAND1, which is important for trophoblast giant cell formation in the mouse. Mice lacking the HAND1 gene show defects in the development of these cells (Riley et al., 1998), also HAND1 expression may be related to the regulation of Mash-2 (Scott et al., 2000). HAND1 is actively involved in bovine trophoblast development (Arnold et al., 2006) and recent data mining has confirmed that HAND1 as well as other transcriptions factors affect elongation of bovine embryos (Turenne et al., 2012). As knockout studies show, various transcription factors widely expressed in embryonic, fetal and adult tissues are necessary for placental development. These factors include ETS2, involved in regulation of bovine interferon-tau gene expression (Yamamoto et al., 1998; Erashi et al., 2008) and AP1 (Schorpp-Kistner et al., 1999; Schreiber et al., 2000; Das et al., 2008). A general pattern of gene expression and signalling during bovine trophoblast development was recently discussed by Pfeffer and Pearton (2012).

Maternal recognition of pregnancy

Both the conceptus and the uterine luminal epithelium have to be prepared for implantation. On days 13–14, which coincides with the beginning of conceptus elongation, endometrial gene expression undergoes the first measurable changes suggesting commencement of an interactive stage during bovine pregnancy (Forde et al., 2011). The genes that show activation of transcription include: pregnancy associated glycoprotein 8 (PAG8), TDGF1, the trophoblast Kunitz domain protein 1 (TKDP1) as well as TKDP4 and TKDP5. A correlation between activity of these genes and production of interferon tau was found (Clemente et al., 2011). An interaction network analysis (from day 13) revealed a number of genes involved in cellular development, lipid metabolism and small-molecular biochemistry including: MYC, SLC25A12, HSPH1, LXN, ALDH18A1, PMP22, PEG3 and CDH2 (Clemente et al., 2011). MYC gene encoding influential transcription factor possibly regulating global chromatin structure could have a profound effect on many other genes (Cotterman et al., 2008).

Reciprocal signalling between the maternal endometrium and the developing conceptus is necessary for successful implantation and placentation in eutherian mammals and the cow is no exception (Bazer, 1992). The elongation of the conceptus requires a specific uterine environment (Mamo et al., 2012b) and signals from the developing bovine conceptus are essential for the developing receptivity of maternal endometrium (Bauersachs et al., 2009; Sandra et al., 2011). The elongating ruminant conceptus secretes interferon tau and inhibits the luteolytic mechanism by repressing transcription of the oestrogen receptor alpha gene (ESR1) in uterine epithelial cells. This prevents oestrogen-induced production of oxytocin receptors (OXTR) and generation of luteolytic prostaglandin F2-alpha pulses (Spencer et al., 2007; Blomberg et al., 2008).

The chain of events continues with the down-regulation of progesterone receptor gene (PGR) in endometrium epithelial cells, which is coupled with a reduction of anti-adhesive MUC1 (mucin glycoprotein 1) and induction of secreted LGALS15 (galectin 15) and SPP1 (secreted phosphoprotein 1). According to the current understanding these molecules regulate trophectoderm growth and adhesion (Spencer et al., 2007). By day 16 bovine trophectoderm normally secretes sufficient quantities of interferon tau (IFNT) to inhibit the luteolytic pulses of prostaglandin F2α produced by endometrial epithelial cells (Forde and Lonergan, 2012). Significant changes in the transcriptome profiles of endometria in pregnant and cycling cows become detectable by about day 16, when the endometrium of the pregnant cow responds to the increasing IFNT level produced by the filamentous conceptus (Forde *et al.*, 2011).

IFNT acts on the epithelial cells (days 15–19) and induces WNT7A protein (wingless-type MMTV integration site family member 7A), which stimulates production of LGALS15, CTSL1 (cathepsin L1) and CST3 (cystatin C), thus regulating further conceptus development and implantation. At the same time several interferon-stimulated genes and interleukin-10 (IL10) are up-regulated in peripheral blood (Shirasuna *et al.*, 2012). Prior to implantation, newly formed trophoblast giant binucleate cells migrate and fuse, forming multinucleated syncytial plaques, and secrete chorionic somatotropin (CSH1 or placental lactogen), which influences the endometrial glands and stimulates their differentiation. A steadily coordinated set of actions leads to essential fetal–maternal interactions and finally to maternal recognition of pregnancy (Spencer *et al.*, 2007).

Concerted efforts have been made to comprehensively describe the molecular essence of bovine conceptus–endometrium cross-talk based on RNA sequencing of more than 287 million reads and detection of more than 22,700 unique transcripts (Mamo *et al.*, 2011). Using bioinformatic tools 2261 and 2505 transcripts have been identified as conceptus and endometrium specific. Further efforts led to identification of 133 conceptus ligands interacting with the corresponding receptors on bovine endometrium and 121 endometrium ligands that interact with the conceptus receptors. There are also 87 ligands which are commonly observed, 46 of which are conceptus specific and 34 are endometrium specific (Mamo *et al.*, 2012a). These data obviously demonstrate the very complex nature of conceptus–endometrium communication.

Implantation and placental development

The formation of extra-embryonic membranes is an obligatory step in establishing the ability of the conceptus to attach and interact with the uterus. The membranes that originate from the primary germ layers include the yolk sac, chorion (serosa), amnion and allantois. The yolk sac is formed from the ICM. The amnion and chorion are both formed from the primitive endoderm and mesoderm. The amnion is a fluid-filled membrane that surrounds the developing embryo, while the chorion is the outer most extra-embryonic membrane that interacts with the uterine endometrium. The amnionic folds first appear shortly after the primitive streak stage; then the structure quickly develops into a fluid-filled membrane that encases the developing embryo. Amnion formation is complete by days 26–27 of development.

As the amnion is developing, the allantois emerges (days 22–23) as a sac-like invagination from the primitive gut. While the embryo grows, the allantois fills and eventually contacts the chorion. Increasing contact between the chorion and allantois, and rapid angiogenesis, results in the fusion of these two membranes and infiltration of the chorion by allantoic vessels. By day 32 the allantochorion enters into the contralateral horn and within the next 10 days the allantochorion fills both horns (Silva and Ginther, 2010). The vascular perfusion in each uterine horn during early pregnancy is mediated by direct contact between conceptus and uterus.

Differentiation of trophoblast mononucleate cells (TMCs) to trophoblast giant cells (TGCs) takes place during transition from tubular to filamentous conceptus (Blomberg *et al.*, 2008). Then some TGCs undergo binucleation (BNC) and eventually comprise ~20% of the trophoblast cell population in cow (Wooding, 1992). The fetal binucleate cells, which have chorionic origin, migrate between adjacent cell tight junctions and fuse with uterine luminal epithelium to form multinucleated cells or syncytia and a synepitheliochorial placenta (Bazer *et al.*, 2009). It has been shown that fusion molecules called fertilin and CD9 are involved in bovine binucleate cell migration and fusion (Xiang *et al.*, 2002). TGCs migrate along a matrix of laminin and are involved in cell–cell contact with mononuclear trophoblast cells via protein $\alpha_2\beta_1$ heterodimers (Pfarrer *et al.*, 2003).

Cellular transformations and the molecular basis supporting these differentiation events are highly complex and are based on distinct molecular substrates in different species (Pfeffer and Pearton, 2012). Differentiation into TGCs always requires the bHLH transcription factor HAND1 (Scott et al., 2000). Differentiation of trophoblast cells in cattle involves expression of endogenous retrovirus envelope genes. This might be particularly relevant to bovine BNC-specific genes and the progression of binucleation in trophoblast cells (Koshi et al., 2012). The BNCs produce placental lactogen (CSH1) (Bazer et al., 2009).

As soon as embryonic trophoblast and endometrium epithelium are well prepared, implantation in the pregnant cow starts at day 20 (Wathes and Wooding, 1980). At this time expression of mesenchymal-related genes by the bovine trophectoderm seems to be necessary for the conceptus attachment to the endometrial epithelium (Yamakoshi et al., 2012). Then (days 22–23, Table 14.1) placentation commences from development of minute papillae of embryonic membranes that penetrate the vestibule of the uterine glands. These papillae disappear before day 30 of pregnancy (Guillomot and Guay, 1982) and are replaced with interdigitating microvilli connections of the placental allantochorion to maternal caruncular crypts of the uterine endometrium (King et al., 1979). The placenta microvilli grow and become vascularized, as do the associated caruncular areas, causing formation of placentomes, which have a convex shape in cattle. No syncytia is formed in the intercaruncular epithelium, however abundant giant cells are present during the fourth week of pregnancy, which steadily decline thereafter (King et al., 1981). Thus implantation leads eventually to formation of extra-embryonic membranes and placenta, which are vital for normal embryonic growth and development.

Attachment of chorioallantoic membranes of bovine placenta with uterine caruncles promotes development of placental cotyledons, which together with maternal caruncles create the placentome, the major site for transfer of nutrients and gases (Bazer et al., 2009). The bovine placenta is classified as epitheliochorial type and is much less invasive than in other mammals because it relies on the elongated conceptus establishing maximal surface contact with the uterine endometrium (Schlafer et al., 2000). The placenta is growing in the course of pregnancy and growth hormone is involved in placental metabolism and embryonic devel-opment from the early beginning of pregnancy until birth (Kölle et al., 1997).

At the beginning of implantation the expression of more than 20 genes is intensified in mainly TGCs (Ushizawa et al., 2004) including PAGs, PRPs, CSH1, BCL2A1 and CTSL1. A subtractive bioinformatic analysis of sequences expressed in cattle placenta also revealed eight novel genes and two previously known, TKDP1 and a splice variant of TKDP4 (Larson et al., 2006). More recently gene expression profiles were described for bovine caruncular (C) and intercaruncular endometrium (IC) at the points of implantation, revealing 446 and 1295 differentially expressed genes in C and IC areas, respectively (Mansouri-Attia et al., 2009). This study indicates that the impact of the conceptus is greater on the immune response function in C but more prominent in the regulation of metabolism function in IC. Genomic imprinting in mammals, as mentioned earlier, might be a natural response to a genetic conflict arising during pregnancy between mother and fetus. Obviously the placenta is a particularly important point for such interactions. The first data on genomic imprinting in bovine placenta have begun to emerge. For instance, imprinted gene PHLDA2, expressed from the maternal allele in several mammals, remains practically silent in extra-embryonic tissues until day 32 of gestation, but then its expression in placenta increases throughout the pregnancy (Sikora et al., 2012). Interestingly abnormal expression in the fetal villi from the oversized bovine placenta from somatic cell nuclear transfer experiments provide some explanation for numerous problems well known from cloning (Guillomot et al., 2010). Information regarding epigenetic modifications of trophoblast-specific genes in the bovine embryo is still limited (Kremenskoy et al., 2006). However, there is significant literature demonstrating that placental and embryonic abnormalities result from abnormal epigenetic regulation caused by some artificial reproduction procedures, but this is beyond the scope of this chapter.

Molecular signals affecting implantation and placentation

It is obvious that natural reproduction in mammals is not possible without sperm and egg cells, nevertheless even such basic requirements have to be supported by an exchange of molecular signals. It has been shown that inflammatory byproducts caused by penetration of sperm cells into the uterus must be counter-balanced by the effective elimination of excessive numbers of sperm cells and subsequent rapid return of the endometrium to a normal state prerequisite for successful pregnancy (Katila, 2012). Starting from this 'entry point' the whole process of embryonic developments is a complex net of numerous signalling. Another recently discovered confirmation of this general rule of interactions is pre-implantation factor (PIF), a novel peptide secreted by embryos, which is essential for implantation. PIF is capable of modulating local immune reactions, 'promotes decidual pro-adhesion molecules and enhances trophoblast invasion' (Stamatkin et al., 2011).

Despite the fact that mammalian species have several different implantation strategies, the common feature is that progesterone causes down-regulation of expression of its own receptors in uterine epithelia prior to implantation events (Bazer et al., 2009, 2010). Uterine receptivity to implantation involves expression of interferon stimulated genes (ISGs) that have numerous roles (Bazer et al., 2011). A comprehensive list of ISGs is available (http://interferome.its.monash.edu.au/interferome/home.jspx). Interferons (IFNs) are pro-inflammatory cytokines that are also secreted in the uterus during early pregnancy. They are cell-signalling proteins and, in most cases, the action of IFNs on ISGs is preceded by induction of the genes by progesterone, which is elevated during pregnancy. This process has been well described in several species (Bazer et al., 2008). In ruminants, IFN tau (IFNT) is well established as the factor expressed by the conceptus that acts as the signal for recognition of pregnancy and therefore impedes regression of the corpora luteum until the end of pregnancy (Forde et al., 2011). The predominant effect of the bovine conceptus is to elicit a classical type 1 IFN response in the endometrium. In other words the conceptus through IFNT activates expression of several genes including *MX2*, *BST2*, *IFITIRSAD2* and *OAS1* (Forde and Lonergan, 2012). These events lead to production of proteins and expression of genes involved in establishment of uterine receptivity in cattle: fatty acid binding protein 3, muscle and heart (mammary-derived growth inhibitor), *FABP3*; serpin peptidase inhibitor, clade A (alpha-1 antiproteinase, antitryosin), member 14, *SERPINA14*, *CA2* (Forde and Lonergan, 2012). The most recent study identified several novel endometrial pregnancy-associated genes during early bovine pregnancy that are not regulated by IFNT *in vivo* (Forde et al., 2012a). This study also identified a number of other genes that are directly regulated by IFNT *in vivo*. The role of these pregnancy-specific IFN-stimulated genes active *in uterine* is unclear but could involve: (i) protecting the conceptus from immune rejection; (ii) limiting the ability of the conceptus to invade the endometrium; and (iii) stimulating uterine/placental angiogenesis. There are other factors affecting the peri-implantation period. Among them are BMP2 and BMP4 (bone morphogenetic proteins) ligands involved in the receptor system that is active in bovine trophoblast cells prior to uterine attachment. BMP4 negatively impacts trophoblast cell growth and both BMPs affect IFNT mRNA abundance (Pennington and Ealy, 2012).

Implantation of an embryo into endometrium carries a risk of rejection caused by the reaction of the maternal immune system. Several major loci and cellular types might be involved in such an immune response. Human and mouse data indicate that successful pregnancy requires some modification of the function and secretion profile of certain types of immune cells. In cow there is a marked increase in the population of CD14 (+) cells and CD172a–CD11c (+) cells in the endometrium as a response to pregnancy. At the same time production of some interleukins is affected: IL12B and IL15 are up-regulated and IL18 down-regulated. In addition, several novel IFNT- and progesterone-regulated factors, like IL12B, MCP1, MCP2, PTX3, RSAD2 and TNFA have been identified that might be essential for normal pregnancy (Mansouri-Attia et al., 2012). It has also been found that class I of major histocompatability complex (MHC) at the point of implantation may contribute significantly to establishment of pregnancy in cattle. MHC class I genes *NC2*, *NC3* and *NC4* in blastocysts were a subject of a preferential immunomodulation by interleukins (IL1B, IL3, IL4, IL10) during pre-implantation embryo development

(Al Naib *et al.*, 2012). Tuning the maternal immune system during bovine pregnancy is achieved using several mechanisms of uterine immune suppression (Oliveira *et al.*, 2012). The latest includes reduced expression of MHC proteins by the trophoblast; recruitment of macrophages to the pregnant endometrium; and modulation of immune-related genes in response to the presence of the conceptus. Fetal DNA has been detected in cow blood at the time of penetration of trophoblast fragments into the placenta and endometrium. As confirmed by global transcriptome studies of bovine endometrium the maternal immune system is aware of the presence of the fetus even in the early stages of pregnancy (Oliveira *et al.*, 2012).

Insulin-like growth factors (IGFs) are members of a large superfamily found in many vertebrate species. IGF1 and IGF2 are small polypeptides that promote cellular differentiation, proliferation and migration, and inhibit apoptosis. These factors are required for uterine and embryonic growth and generate pleiotrophic effects during development (Velazquez *et al.*, 2009). Both are involved in the remodelling that occurs during development of the placenta and its endometrial attachment site. The IGFs bind with high affinity to their receptors, namely IGF1R and IGF2R. IGF1R is a member of the tyrosine kinase family and is structurally related to the insulin receptor (Jones and Clemmons, 1995). IGF1R binds with equal affinity to both IGF1 and IGF2, whereas IGF2R binds with high affinity only to IGF2 (Pollak, 2008). The bioavailability and biological actions of IGFs are regulated by at least six IGF-binding proteins (Clemmons, 1997). IGF-binding proteins may augment or inhibit IGF action and several proteases cleave IGFBPs, reducing or eliminating their ability to bind IGFs. Bovine IGF1 is essential for follicular growth, oocyte competence and embryo viability, and has mitogenic and anti-apoptotic activities (Velazquez *et al.*, 2009). Decreased IGF1 and IGFBP6 expression together with an increase in IGF1R and IGFBP2 are necessary for the establishment of a uterine environment promoting embryonic growth and development prior to implantation (McCarthy *et al.*, 2012). In mammalian development, *IGF2* is characterized by a complex transcriptional regulation involving multiple promoters, alternative splicing and genomic imprinting, which is also common for cattle and effects development of the placenta (Dindot *et al.*, 2004; Curchoe *et al.*, 2005). The latest data suggest that imprinted genes might be implicated in the emergence, maintenance and function of trophoblast glycogen cells (Lefebvre, 2012).

The pregnancy-associated glycoproteins (PAGs) belong to a multigenic family of aspartic peptidases (Davies, 1990; Touzard *et al.*, 2012). The production of PAGs has been identified in various species during pregnancy and their expression consistently initiates at the time of implantation and continues in the TE as pregnancy proceeds (Szafranska *et al.*, 1995). In *B. taurus* this family consists of 21 identified genes divided into two phylogenetic groups (modern and ancient). Modern PAGs (PAG I) are produced by binucleate trophoblastic cells of the cotyledons, whereas ancient PAGs (PAG II) are produced by both mononucleate and binucleate trophoblastic cells (Hashizume, 2007; Touzard *et al.*, 2012). PAG II are specific to non-villous tissues, whereas PAG I are mainly synthesized within the villous part of the cotyledon. The PAG II glycoproteins seem to be an exception, belonging to the ancient group, but expressed as a modern PAG and synthesized only in the cotyledon (Touzard *et al.*, 2012). *In vitro* studies have revealed the potential role of the PAG family as chorionic signalling ligands that interact with gonadotropin receptors in cyclic pigs and cows (Szafranska *et al.*, 2007). However, the overall role of PAGs is still under investigation. These secretory proteins are easily detectable in maternal blood circulation and are used for pregnancy diagnosis (Touzard *et al.* 2012).

Placentogenesis is a unique biological process caused by communication and interaction between fetal and maternal tissues. Hormonal and cytokine signals produced by the placenta organize communication between cotyledonary villi and the maternal caruncle. Investigation of gestational-stage-specific gene expression profiles in bovine placentomes has indicated that the AP-2 family of transcription factors may serve as a consensus regulator for the gene cluster that characteristically appears in bovine placenta as gestation progresses (Ushizawa *et al.*, 2007). Two members of the family, TFAP2A and TFAP2B, seem to be involved in regulation of bovine binucleate cell-specific genes such as *CSH1*, some *PAG* or *SULT1E1*. Ushizawa *et al.* (2007) suggest that the AP-2 family acts a specific transcription factor for clusters of crucial placental genes. A detailed review of fetal–maternal signalling during implantation and genetic aspects of bovine placentome morphogenesis can be found elsewhere (Hashizume, 2007).

Angiogenesis

Angiogenesis, the process by which new blood vessels are generated from an existing vascular system, occurs extensively during pregnancy to support the conceptus. A variety of factors support angiogenesis, but vascular endothelial growth factor (VEGF) and its receptors (VEGFR1, VEGFR2, VEGFR3) appear to be of primary importance (Loges *et al.*, 2009). Among these receptors VEGFR2 is dominant and placenta growth factor (PGF) is a high-affinity ligand for VEGFR1 (Carmeliet *et al.*, 2001).

In bovine interplacentomal areas, VEGFs have been found in luminal and glandular epithelia as well as in trophoblast, particularly in TGCs. VEGFR1 was initially observed in trophoblast and uterine epithelium around implantation. Later, in definite placentomes, VEGFR1 was localized in TGCs in the centre of the placentome (Pfarrer *et al.*, 2006). VEGFR1 and VEGFR2 were co-localized in uterine epithelium and trophoblast as well as in blood vessel tissue and uterine glands. Pfarrer *et al.* (2006) concluded that 'The presence of VEGF, VEGFR1 and VEGFR2 at the feto-maternal interface and in vasculature indicates that in the bovine VEGF may have (1) classic functions in angiogenesis and vascular permeability, (2) growth factor properties, facilitating feto-maternal exchange via paracrine action, (3) chemotactic activity on capillary endothelium, and (4) an autocrine influence on TGC migratory activity'.

Although information on angiogenic gene profiles in cattle is lacking, porcine endometrial tissues have been studied for *VEGF*, *PCF*, *VEGFR1* and *VEGFR2* expression in the trophoblast associated with healthy conceptus, and those experiencing arrested development (Tayade *et al.*, 2006; Linton *et al.*, 2008). At day 20 and 50 of development fewer *VEGF*

transcripts were detected in the endometrial tissues associated with the arrested conceptuses compared to the healthy, but the amounts of *VEGF* transcripts in the trophoblast were not altered. In addition, between days 15 and 28 of porcine development there is a dramatic onset of angiogenic activity that coincides with elevated numbers of a unique lymphocyte type referred to as uterine natural killer cells (uNK) (Engelhardt *et al.*, 2002). When endometrial lymphocytes from these same day 20 and 50 conceptuses were screened for angiogenic gene expression, the endometrial lymphocytes were found to have a greater abundance of the *VEGF* transcript than the endometrial endothelium or the trophoblasts. However, for the conceptuses demonstrating arrested development, their attachment sites showed severely reduced VEGF expression and an increase in *PGF* expression by the lymphocytes. The uterine lymphocytes preferentially expressed *VEGFR1*, while the trophoblasts were abundant in *VEGF2* transcripts indicating that mechanisms which regulate angiogenesis differ between the maternal and embryonic/fetal compartments (Tayade *et al.*, 2007). This molecular evaluation of the porcine conceptus attachment sites shows a clear role for immune cells in the acceleration of angiogenesis in these tissues.

Genes Involved in Post-implantation Development

Development of segment identity and HOX genes

Segmentation observed in different groups of animals, and particularly in vertebrates, has deep evolutionary roots. Segments with a common origin remain relatively separate during development causing diversification and specialization. This evolutionary–developmental strategy has been commonly used for the creation of morphological structures or groups of cells with distinct features. For instance, development of two major structures, the ectodermal neural tube and the paraxial mesoderm, depends on segmentation. The first is critical for development of the hindbrain, the head process and the spinal cord. The second is essential in generation of somites, which give rise to the axial skeleton and skeletal muscles. The first five somites are developed by the time of neural fold closure in the mid-region on about day 20 (Table 14.1; Asis Neto *et al.*, 2010); the following somites develops anteriorly and their number increases. The genetic and cellular processes driving segmentation depend on expression patterns of HOX genes (Alexander *et al.*, 2009).

The homeotic genes, which encode helix-turn-helix transcription factors, were first described in *Drosophila* as the primary determinants of segment identity. They all contain a conservative 180-bp DNA sequence motif named the homeobox. Comparative analysis of the *Drosophila* homeotic gene complex, called HOM-C, and the mammalian homeobox genes, called the HOX complex, demonstrates a striking case of evolutionary conservation. The HOX genes family determines a set of transcription factors crucial for development of axial identity in a wide range of animal species (Maconochie *et al.*, 1996). Figure 14.4 shows the remarkable similarity and collinearity existing in the molecular anatomy of the insect and mammalian HOX complexes. The main difference is the number of complexes per genome. In insects there is only one, while mammals and other higher vertebrates have four separate chromosome clusters (Alexander *et al.*, 2009). There are 39 HOX genes in mammalian genomes, which belong to 13 paralogous groups. The HOX genes are expressed in a segmental fashion in the developing somites and central nervous system and each HOX gene acts from a particular anterior limit in a posterior direction. The anterior and posterior limits are distinct for different HOX genes (Fig. 14.4). A hallmark of HOX genes is the correlation between their linear arrangement along the chromosome and their timing and anterior–posterior limits of expression during development (Alexander *et al.* 2009). HOX genes determine anterior–posterior positional identity within the paraxial and lateral mesoderm, neuroectoderm, neural crest and endoderm.

Thus, the vertebrate body plan is, at least partially, a result of the interactions of HOX genes that provide cells with essential positional and functional information. Signals from HOX genes force embryonic cells to migrate to the appropriate destination and generate certain structures. Major signalling pathways like fibroblast growth factor (FGF), Wnt and retinoic acid (RA) play their roles in affecting expression of different HOX genes in different developmental conditions. The expression of RA and its protein binding ability as well as their other functions during development of the mammalian conceptus have been described earlier (Yelich *et al.*, 1997). RA can affect the expression of HOX genes and there is a 5′ to 3′ gradient in responsiveness of the genes to retinoids (Marshall *et al.*, 1996). RA acts via its receptors, which comprise two families, RAR and RXR, that are members of the ligand-activated nuclear receptor superfamily. The receptors interact to form complexes that in turn regulate target gene binding to retinoic acid response elements (RAREs). These RAREs are found in the 5′ regulatory regions of the murine Hox genes and other mammals. HOX genes have a profound influence on the whole array of developmental processes and establishment of segment identity. As was found recently, proper activation of HOX genes and establishment of anterior–posterior identity require active involvement of the *UTX* gene (ubiquitously transcribed tetratricopeptide repeat X) (Canovas *et al.*, 2012).

Pattern formation

Early pattern formation is similar for all vertebrates. There are two essential patterning processes: gastrulation, forming the trilaminar embryo; and axis formation visualized in developing the notochord. These processes establish embryo patterns for intermediate structures that are transformed during the following embryonic development characterized by overt differentiation of tissues and organs known as organogenesis. During the further embryonic period musculoskeletal patterning commences, as well as formation of the head and limbs. A few examples of embryo patterning in musculoskeletal, neural and renal development are given below, identifying common mechanisms used throughout the embryo.

The T-box transcription factor is essential for formation of the posterior mesoderm and the

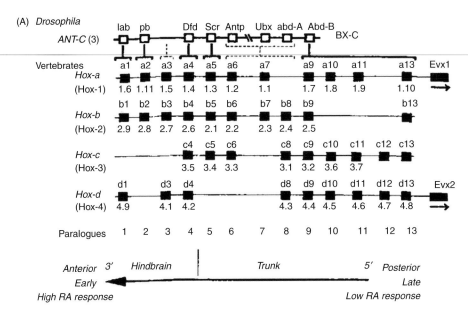

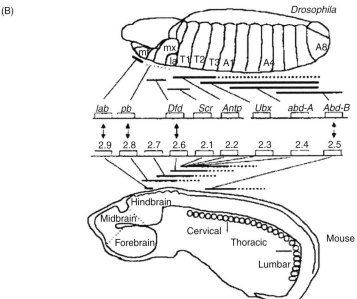

Fig. 14.4. (A) Alignment of the four mouse *Hox* complexes with that of *Hox* gene cluster from *Drosophila*. The vertical shaded boxes indicate related genes. The 13 paralogous groups are noted at the bottom of alignment. The colinear properties of the *Hox* complexes with respect to timing of expression, anteroposterior (a–P) level, and retinoic acid (RA) response are also noted at the bottom. (From Maconochie *et al.*, 1996, with the author's permission.) (B) Summary of *HOM-C* and *Hox-2* expression patterns. The upper part of the figure is a diagram of a 10-h-old *Drosophila* embryo with projections of expression patterns of different genes from *Hox* gene cluster to particular body segments. The lower part of the figure is a diagram of a 12-day-old mouse embryo with projections of expression patterns of different genes from *Hox-2* complex to particular body segments. (From McGinnis *et al.*, 1992; with the kind permission of the corresponding author R. Krumlauf.)

notochord in vertebrate embryos (Evans *et al.*, 2012). A highly homologous gene was found in cattle. Whole mount *in situ* hybridization reveals a normal expression pattern except for a transiently reduced expression in the anterior part of the primitive streak. According to these results, gastrulation in mammals is implemented and regulated irrespective of implantation (Hue *et al.*, 2001). Bovine mutations of this gene have not been discovered as yet. Mice homozygous for mutant alleles of the *T* gene do not generate enough mesoderm, and show severe disruption in morphogenesis of mesoderm-derived structures, in particular the notochord (Wilkinson *et al.*, 1990). One of the T-box genes, *Tbx6*, in mice is implicated in development of paraxial mesoderm (Chapman *et al.*, 1996). *Tbx6* transcripts are first detected in the gastrulation stage embryo in the primitive streak and newly recruited paraxial mesoderm.

Limb patterning begins with establishing position on the trunk and limb identity by T-box genes (Rodriguez-Esteban *et al.*, 1999). It has been shown that murine *Tbx5* and *Tbx4* expression is primarily restricted to the developing fore- and hindlimb buds, respectively (Agarwal *et al.*, 2003). These two genes appear to have been divergently selected in vertebrate evolution to play a role in the differential specification of fore- (pectoral) versus hind- (pelvic) limb identity (Gibson-Brown *et al.*, 1998). Mutations in the human *TBX3* gene cause the ulnar-mammary syndrome characterized by posterior limb deficiencies or duplications, mammary gland dysfunction and genital abnormalities. It has been suggested that *TBX3* and *TBX5* evolved from a common ancestral gene and each has acquired specific, yet complementary, roles in patterning the mammalian upper limb (Bamshad *et al.*, 1997). *Tbx4* and *Tbx5* contribute to regulation of limb outgrowth and their roles seem to be linked to the activity of signalling proteins that are required for initial limb outgrowth, the fibroblast growth factors Fgf4 and Fgf8 (Boulet *et al.*, 2004). Fore- and hindlimb bud musculoskeleton patterning occurs by similar mechanisms and both are structurally mesoderm (somatic) proliferating cells within a covering ectoderm. The ectoderm at the limb tip is thickened by *Wnt7a* forming an apical ectodermal ridge (AER) that expresses FGF encoding genes (*Fgf4*, *Fgf8*, *Fgf9*, *Fgf17*)

stimulating underlying mesoderm proliferation and establishing the limb proximodistal axis (reviewed by Towers and Tickle, 2009). A number of mutations causing abnormal limb development are known in cattle, and some such as Bulldog dwarfism in Dexter cattle have been mapped and identified (Chapter 5; OMIA). Affected embryos display the trait caused by abnormal cartilage development (chondrodysplasia). Heterozygotes show a milder form of dwarfism, most noticeably having shorter legs. Aggrecan (*ACAN*) is the most likely candidate gene, which has one of two so far observed insertions (Cavanagh *et al.*, 2007).

The ventral mesoderm forms a region known as 'the zone of polarizing activity' (ZPA) secreting SHH protein, competing with bone morphogenetic protein (BMP4) and establishing the dorsoventral axis (Kicheva and Briscoe, 2010). Abnormalities such as canine preaxial polydactyly (PPD), a developmental trait that restores the missing digit lost during canine evolution, is due to a ZPA-related change in an intronic sequence of *LMBR1* gene (Park *et al.*, 2008).

Differentiation of mesoderm-derived musculoskeletal tissues (like cartilage, bone and muscle) is similar throughout the embryo and is regulated by tissue-specific pathways. Chondrogenesis of mesoderm forms cartilage template structures in the position of, and replaced by, future bone formation described as endochondral ossification (Goldring *et al.*, 2006). In the adult, this cartilage remains only on the articular surfaces of the bony skeleton. Chrondrogenesis transcription factors determined by *Znf219*, *Sox9* and *Runx2* genes interact with secreted factors (Indian hedgehog, parathyroid hormone-related peptide, FGFs) to determine whether the differentiated chondrocytes remain within cartilage elements in articular joints or undergo hypertrophic maturation prior to ossification (Cheng and Genever, 2010). Endochondral ossification occurs in all bones except the cranial vault and scapula. Death of chondrocytes releases VEGF stimulating vascular growth and deposition of osteoclasts that further erode cartilage and osteoblasts that ossify the previous cartilage template region (Mackie *et al.*, 2008). The transcription factors coded by *Runx2* and *Runx3* are essential for chondrocyte maturation, while *Runx2* and *Osterix* are essential for osteoblast differentiation. Osteogenesis is inhibited by the Wnt signalling pathway antagonists including genes like *DKK1*, *SOST* and *SFRP1* (Fujita and Janz, 2007).

Patterning of the ectoderm-derived neural tube is initially affected by the notochord secreted Shh, dorsally secreted bone morphogenic protein (Bmp) and Wnt signalling (Ulloa and Martí, 2010). Factors affecting spread of Shh include its receptor patched 1 (Ptc1), Hedgehog interacting protein (Hhip1), and the proteins Cdo, Boc and Gas1 (Ribes and Briscoe, 2009). The rostrocaudal pattern is mediated by the homeobox (HOX) gene family, differentially expressed along the neural tube and within the neural crest (Mallo *et al.*, 2010). At about the same time two genes, *Otx2*, expressed in the forebrain and midbrain, and *Gbx2*, expressed in the anterior hindbrain, play an essential and interactive role in positioning of the mid/hindbrain junction. This junction acts as an organizer, directing development of midbrain and anterior hindbrain (Millet *et al.*, 1999).

Muscle development and gene regulation

The musculoskeletal system develops from the paraxial presomitic mesoderm (PSM) cells (Tam and Beddington, 1987). Once these cells reach a specified position, gene expression changes significantly and a segmentation process begins. New somites appear approximately every few hours and they are separated from the anterior PSM (Dunty *et al.*, 2008). The bHLH transcription factor encoded by the *Mesp2* gene and controlled by the Notch signalling pathway is essential in the segmentation programme (Saga *et al.*, 1997). *Ripply2* is another identified gene involved in segment boundary regulation and it is also under the influence of Notch pathway genes as well as the mesodermal transcription factors *T* and *Tbx6* (Oginuma *et al.*, 2008).

Somitogenesis is probably controlled by a segmentation clock, which consists of molecular oscillators in the Wnt3a, Fgf8 and Notch pathways (Pourquie, 2003; Aulehla and Herrmann, 2004; Rida *et al.*, 2004). Alternatively directed gradients of fibroblast growth factor 8 (Fgf8) and/or Wnt3a and RA establish a boundary front in the anterior PSM. Dunty *et al.* (2008) demonstrated that the canonical Wnt3a/ β-catenin pathway is necessary for molecular oscillations in all three signalling pathways, but does not function as an integral component of the oscillator. On the contrary, Notch pathway genes continue to oscillate in the presence of stabilized β-catenin and thus drive periodic expression of the target genes lunatic fringe (*Lfng*) and *Hes7* (Bessho *et al.*, 2001; Morimoto *et al.*, 2005). Further investigations should reach a deeper understanding of this sophisticated dynamic system.

Myogenesis forms skeletal muscle from somite myotome-derived mesoderm. Muscle regulatory transcription factors like MyoD (MYOD1), Myf6 (MYF6) and Pax7 (PAX7) control differentiation (Ropka-Molik *et al.*, 2011). The first two of them are bHLH transcription factors that initiate formation of muscle fibres and regulate transcription of muscle-specific genes. MyoD needs to form a dimer to be active and is maintained in an inactive state by binding of the inhibitor Id. *Pax7* as a member of the paired-box transcription factors is also required for muscle growth and both renewal and maintenance of muscle stem cells. Studies have shown that *Lbx1h* gene is also involved in regulation of muscle precursor cell migration and is necessary for acquisition of dorsal identities of forelimb muscles (Schafer and Braun, 1999). The genetic mechanisms of muscle development have been reviewed (Firulli and Olson, 1997; Bentzinger *et al.*, 2012). Skeletal, cardiac and smooth muscle cells express overlapping sets of muscle-specific genes, however there are genes expressed in one particular muscle type. So called modules or independent cis-regulatory regions are required to direct the complete developmental pattern of expression of individual muscle-specific genes, even within a single muscle cell type. The temporospatial specificity of these myogenic regulatory modules is established by unique combinations of transcription factors (Firulli and Olson, 1997).

Several factors (LIF, p21, caspase 3/ CASP3, MEK and possibly others) are involved in complex mechanisms of inhibition of differentiation of skeletal muscle cells (Hunt *et al.*, 2011); SHH and IGF1 stimulate myogenic differentiation and proliferation (Madhala-Levy *et al.*, 2012). An inhibitor of MAPK reduces

IGF-stimulated proliferation of L6A1 myoblasts and associated events, such as phosphorylation of the MAPKs and elevation of c-fos mRNA and cyclin D protein. This inhibitor causes a dramatic enhancement of differentiation, evident at both morphological and biochemical levels. In sharp contrast, an inhibitor of phosphatidylinositol 3-kinase and p70 S6 kinase completely abolished IGF stimulation of L6A1 differentiation. These data demonstrate that the MAPK pathway plays a primary role in the mitogenic response and is inhibitory to the myogenic response in L6A1 myoblasts, while activation of the phosphatidylinositol 3-kinase/ p70 (S6k) pathway is essential for IGF-stimulated differentiation. Thus, it appears that signalling from the IGF1 receptor utilizes two distinct pathways leading either to proliferation or differentiation of muscle cells (Coolican et al., 1997). IGF1 and IGF2, their binding proteins and members of the transforming growth factor (TGF) superfamily (myostatin/MSTN and TGFß1) are key regulators of proliferation and differentiation of myogenic cells.

Obviously understanding the development of muscle tissue in bovine embryo has a great practical interest, particularly in beef cattle. Two groups of cells have been identified during fetal development in cattle. The first appears at a very early stage and gives rise to fast adult type I fibres. The second group of cells gives rise to adult fast type IIA and IIB fibres, and to type IIC. The beginning of myogenesis is characterized by expression of transitory miosin forms that are not found in adult cattle (Picard et al., 1994). Insulin-like growth factor IGF2 modulates myogenesis and between 60 and 162 days of gestation the bulk of IGF2 molecules are localized within the developing muscle cells. It seems likely that IGF2 acts as an autocrine-acting growth factor during myogenesis (Listrat et al., 1994). Expressions of IGF1 and IGF2 and their receptors IGF1R and IGF2R in fetal skeletal muscle are influenced by maternal nutrient intake and probably have permanent and sex-specific effect on post-natal skeletal muscle growth (Micke et al., 2011). In addition it has to be reiterated that IGF2 is undergoing genomic imprinting in mammalian species including cattle (Dindot et al., 2004). Bovine cathepsin B, a lysosomal cystein proteinase, is involved in fetal muscle development and is

encoded by two different transcripts resulting from alternative polyadenilation. These two mRNAs decline in a similar fashion from 80 to 250 days of fetal age (Bechet et al, 1996). The gene expression of bovine fetal skeletal muscle myosin heavy chain is influenced by genotype and indirectly by external factors (Gore et al., 1995; Cassar-Malek et al., 2007).

Molecular and developmental studies of double-muscling (DM) in cattle, caused by recessive mutations located at chromosome 2, shed additional light on skeletal muscle development (Charlier et al., 1995; Grobet et al., 1997). It was shown that an 11-bp deletion in the coding sequence of myostatin gene (MSTN), which belongs to transforming growth factor-beta (TGF-beta) superfamily causes muscular hypertrophy (the double muscled phenotype) in Belgian Blue (Grobet et al., 1997; Kambadur et al., 1997). In Peidmontese cattle G→A transition in the same region of the gene is responsible for the phenotype (Kambadur et al., 1997). Thus myostatin acts a negative regulator of muscle growth in cattle and other mammals, and loss of MSTN function caused by amino acid substitution leads to double muscling in fetuses and adults (Oldham et al., 2001). Several genes have been identified that affect regulation of myostatin during development of skeletal muscle (Potts et al., 2003). The latest results support activation of the cell survival pathway (PI3K/Akt) and loss of the apoptosis pathway within the muscles of DM animals. Alteration of both pathways may increase myonuclear or satellite cell survival, which is crucial for protein synthesis and could contribute to muscle hypertrophy (Chelh et al., 2011). The differentiation of muscle fibres occurs at a slower rate in DM fetuses particularly during the first two-thirds of gestation (Picard et al., 1995).

Description of normal bovine fetal growth and development from day 90 to day 255 is presented elsewhere (Prior and Laster, 1979). Growth hormone (GH), IGF1 and IGF2, their associated binding proteins and transmembrane receptors (GHR, IGF1R and IGF2R) play an important role in the physiology of mammalian growth. The effects of allele substitutions in microsatellites surrounding GHR were significant for weight gain and body weight in several studied cattle breeds (Curi et al., 2005). Polymorphisms in MYH3 (myosin 3), a major contractile protein predominantly found in skeletal and heart muscles in developing embryos and calves, is associated with growth and carcass traits in Qinchuan cattle (Wang *et al.*, 2013). The complexity of muscle development was demonstrated by studying a two-step mechanism for UTX-mediated demethylation at muscle-specific genes during myogenesis (Seenundun *et al.*, 2010). It was shown that the removal of repressive histone H3K27me3 trimethylation marks by UTX demethylase occurs through targeted recruitment of specific genomic regions and requires RNA polymerase II (Pol II) elongation (Seenundun *et al.*, 2010). Changes in expression of IGF genes in progeny caused by variations of maternal protein intake during gestation may have long-lasting and sex-specific effects on postnatal muscle growth (Micke *et al.*, 2011).

Organogenesis: T-box and PAX genes

Numerous genes are involved in organogenesis, and current knowledge covers only a small fraction of this variety. Only a few genes, with multiple effects on organ formation, are briefly mention here. The T-box family of transcription factor genes is a good example of such massive involvement in organogenesis due to their contribution to embryonic cell fate decisions, control of extra-embryonic structures, embryonic patterning and numerous aspects of organogenesis (Naiche *et al.*, 2005). Some T-box genes are involved in limb morphogenesis and specification of forelimb/hindlimb identity. It has been shown that *Tbx5* and *Tbx4* expression is primarily restricted to the developing fore- and hindlimb buds, respectively. These two genes appear to have been divergently selected in vertebrate evolution to play a role in the differential specification of fore- (pectoral) versus hind- (pelvic) limb identity (Gibson-Brown *et al.*, 1998). Mutations in the human *TBX3* gene cause the ulnar-mammary syndrome characterized by limb deficiencies or duplications, mammary gland dysfunction and genital abnormalities. It has been suggested that *TBX3* and *TBX5* evolved from a common ancestral gene and each has acquired specific yet complementary roles in patterning the mammalian upper limb (Bamshad *et al.*, 1997).

At least seven T-box genes, including *Tbx1*, *Tbx2*, *Tbx3*, *Tbx5*, *Tbx18* and *Tbx20*, are involved in heart development. Each T-box gene has a unique expression profile in specific heart regions (Naiche *et al.*, 2005). The study of transcriptional activation and repression of several T-box genes in three-dimensional space provides essential knowledge on the development of heart morphology, function and pathology (Stennard and Harvey, 2005). T-box genes are not only important players in the vertebrate heart development, but they also cause alternative outcomes. For instance *Tbx5* and *Tbx20* promote, while *Tbx2* and *Tbx3* inhibit the genetic programme responsible for early heart tube development (Singh and Kispert, 2010). T-box genes interact with the Bmp2/Smad signalling pathway. Alternative splicing seems to be common among the genes of the Tbx/TBX family and this may provide an additional explanation to their diversified functions (DeBenedittis and Jiao, 2011).

Formation of the intermediate mesoderm-derived kidney occurs through a series of epithelial to mesenchymal inductive interactions, with patterning signals by Wnts, BMPs, FGFs, Shh, Ret/glial cell-derived neurotrophic factor and Notch pathways. Renal overt differentiation requires the specific transcription factors Odd1, Eya1, Pax2, Lim1 and Wt-1 (Reidy and Rosenblum, 2009). PAX nuclear transcription factors contain the so-called 'paired domain', a highly conserved amino acid motif with DNA binding activity. PAX genes regulate development of organs and structures like kidney, eye, ear, nose, limb muscles, vertebral column and brain. Vertebrate PAX genes are involved in pattern formation possibly by determining the time and place of organ initiation or morphogenesis (Dahl *et al.*, 1997). Murine *Pax1*, for instance, is a mediator of notochord signals during the dorso-ventral specification of vertebrae (Koseki *et al.*, 1993) and together with *Pax9* regulates vertebral column development (Peters *et al.*, 1999). The *Pax3* gene may mediate activation of *MyoD* and *Myf5*, the myogenic regulatory factors, in response to muscle-inducing signals from either axial tissues or overlying ectoderm and may act as a regulator of somitic myogenesis (Maroto *et al.*, 1997). The temporal and spatial expressions of these genes are tightly regulated throughout early fetal development and organogenesis; the genes usually switch off during terminal differentiation of most structures (Wang *et al.*, 2008).

The *Pax2* gene is involved in optic nerve formation and *Pax6* is considered as a master gene for eye development as well as some other ectodermic tissues (Schimmenti, 2011; Shaham *et al.*, 2012). Mutations in *Pax6* result in eye malformation, known as *Aniridia* in humans and small eye syndrome in mice (Dahl *et al.*, 1997). Relevant data for cattle embryonic development are not available. Several other genes such as *Bmp4*, *Msx1* and *Msx2*, which encode bone morphogenetic proteins and are expressed before and after neural tube closure, interact with *Pax2* and *Pax3* (Monsoro-Burq *et al.*, 1996).

Sex Differentiation

The major steps in gonad differentiation

Complex epigenetic modifications play an important role in defining cell fate in general and primordial germ cells (PGCs) in particular. There are two waves of reprogramming DNA methylation. The first one takes place after fertilization in the zygote and the second in the PGCs (Seisenberger *et al.*, 2013). It is not clear how the epigenetic processes are connected to the differences between male and female bovine blastocysts which have been observed in the mean mtDNA copy number, in telomere length, methylation level and transcription intensity for *Dnmt3a*, *Dnmt3b*, *Hmt1* and *Ilf3* (Bermejo-Alvarez *et al.*, 2008). One can only suggest that epigenetic events may cause some measurable differences observed between embryos with alternative sexual genetic determination during pre-implantation development prior to gonadal differentiation. Similarly transcription of *SRY* located on the Y chromosome in potential male embryos has been detected as early as from the four to eight-cell stage up until the blastocyst stage (Gutiérrez-Adán *et al.*, 1997). It has also been known for some time that XY blastocysts develop faster than XX blastocysts during the first days, long before the undifferentiated gonads even appear (Xu *et al.*, 1992).

In mice, expression of the *Bmp4* gene (bone morphogenetic protein 4) in the trophectoderm layer, which is in the closest contact with the epiblast, is responsible for differentiation of both the primordial germ cells and the allantois (Lawson *et al.*, 1999). If a similar mechanism operates in cattle, then BMP4 protein would be also produced and thus precedes cellular migration. Due to ongoing proliferation a significant number of germ cells reach the genital ridge, which consists of a thin layer of mesenchymal cells located between the coelomic epithelium and the mesonephros. Two genes, *Sf1* and *Wt1*, are particularly important in development of the murine genital ridge (McLaren, 1998). Eventually, four different cell lines comprise the genital ridge: primordial germ cells, somatic steroidogenic cells, supporting cells and connective tissue. The fate of each lineage depends on the sexual determination of the embryo in which they develop, and patterns of genetic activity are quite different in testes and ovaries. During bovine sexual differentiation *DDX4* plays a role and it has homologues in other mammalian species (Lee *et al.*, 2005; Bartholomew and Parks, 2007). In adult tissues, *DDX4* transcription is only in the ovary or testis. In porcine fetuses the transcript has been found at all stages, except for days 17–18. DDX4 protein has been observed in proliferating primordial germ cells but not in embryonic germ cells.

The earliest steps of gonadal development in mammals commence at a similar period in XX and XY embryos. Primordial germ cells, which differentiate relatively late in mammals, migrate into the gonad area of either presumptive sex indiscriminately and may function even across species barriers (McLaren, 1998, 1999). Wrobel and Süss (1998) described gonad development in bovine embryos in sufficient detail: 'During … folding process primordial germ cells located in the proximal yolk sac area are incorporated into the embryo when this portion of the yolk sac becomes the hind- and mid-gut. Consequently, in D23–25 embryos putative primordial germ cells (alkaline phosphatase- and lection-positive) are situated predominantly in the axial body region at the level of the mesonephros. When the gonadal ridge develops in this region (about day 27) it contains a certain number of primordial germ cells… Within the gonadal ridge (days 27–31) and later in the still sexually indifferent gonadal fold (32–39 days) the primordial germ cells are unevenly distributed.'

In the majority of mammalian species, the sex determination mechanism is based on presence or absence of the Y chromosome. The discovery of SRY paved the way for molecular understanding of sex determination and differentiation in mammals (Goodfellow and Lovell-Badge, 1993). Since this discovery the testis-determining role of SRY in mammals is widely accepted. In mammals gonadal differentiation starts relatively late in embryonic development, and morphological changes in XY embryos appear earlier than in XX embryos. A similar situation is true for cattle where androstenedione metabolism commences in male embryos at about 25–27 days of gestation and several days later in female embryos (Juarez-Oropeza et al., 1995). At this stage gonads still look similar for both types of embryos. Visible morphological differences between sexes start to emerge after day 40. Embryos carrying a Y chromosome develop testicular cords and interstitium, and begin testosterone production, whereas at this time undifferentiated gonadal blastema can be seen in XX embryos. Further events in male embryos include formation of Sertoli cells, production of anti-Müllerian hormone (AMH) and 3β-HSD (hydroxysteroid dehydrogenase), which leads to the appearance of Leydig cells. Müllerian ducts disappear and Wolfian ducts are transformed into epididymides, vas deferens, seminal vesicles, prostate and other structures. Germ cells with a high proliferation rate are observed from day 50 to day 80. These cells are in transition from primordial germ cells to prespermatogonia. Between 18 and 27 weeks the second phase of prepubertal germ cell multiplication coincides with the period when the pre-Sertoli cells transform into adult-type Sertoli cells (Wrobel, 2000).

Entry of bovine fetal oocytes into meiotic prophase occurs before day 140. Then expression of YBX2 increases and after day 141 the majority of oocytes had been arrested at the diplotene stage (Yang and Fortune, 2008). Morphologically observable sex differentiation of bovine female gonads takes place at about day 45 (Kurilo et al., 1987). In females, Müllerian ducts develop, no Leydig cells form in the gonads, no testosterone is produced and development steadily moves towards a female phenotype.

The genes involved in sex differentiation

The bovine SRY gene has been cloned and sequenced. The bovine SRY protein consists of 229 amino acid residues. The conserved high-mobility group (HMG) box consists of a 79-amino-acid motif which confers DNA-binding ability and act as a transcription factor (Goodfellow and Lovell-Badge, 1993; Daneau et al., 1995). Outside the HMG box, the bovine SRY structure shows a greater resemblance to human SRY than to mouse Sry (Daneau et al., 1995). A comparative study of SRY structures including bovine sequence demonstrated a high degree of variability between mammalian species (Ross et al., 2008b).

Expression of SRY was confirmed in bovine embryos both for a short time at the sex-determining stage of development around the period of the primitive undifferentiated gonad, and in adult testes (Gutiérrez-Adán et al., 1997). Expression of bovine SRY was reported as early as the four- to eight-cell stage and through to the blastocyst stage. The intensive SRY expression in male bovine embryos begins at about day 37 and peaks 2 days later, activating a cascade of regulatory, signalling and steroidogenic gene expression (Ross et al., 2009). The SRY gene is expressed within the cells of the genital ridge in developing male embryos. At this time the primitive gonads are still bipotential, however after day 40 testis determination is morphologically evident (Wrobel and Süss, 1998). In mammalian genetic males genital ridge cells express not only Y chromosome-located SRY or ZFY genes, but X chromosome-located DAX1 (a nuclear receptor protein encoded by the NR0B1 gene) and DFX genes, but also autosomal AMH, WT1, SOX2 and SOX9, SF-1 (steroidogenic factor-1) and several other genes (Ramkissoon and Goodfellow, 1996; Lahbib-Mansais et al., 1997; Parma et al., 1997; Pevny and Lovell-Badge, 1997). SF1 (steroidogenic factor 1) protein possibly transactivates the SRY promoter (Pilon et al., 2003). Sox9 has an essential function in sex determination, possibly downstream of Sry in mammals and critical for Sertoli cell differentiation (Morais da Silva et al., 1996). Despite significant efforts many questions concerning interactions of these genes with SRY remains. As Sekido and Lovell-Badge (2013) wrote recently: 'even at the adult stage, some of these genes retain a key role in maintaining testicular fate because conditional ablation of the genes leads to adult testis dysgenesis or transdifferentiation into an ovary. This sheds light on mammalian sex-reprogramming, despite the prevailing dogma that postnatal sex change does not occur in mammals.'

Figure 14.5 shows a simplified genetic model describing the major events in sex determination pathways. The activation of *SRY* in normal XY embryos shifts the balance in favour of testis development and male pathways through up-regulation of the *SOX9* gene and

signalling of *FGF9* (fibroblast growth factor 9), which promotes secretion of prostaglandin D$_2$ (Nef and Vassilli, 2009). FGF9 and PGD$_2$ form a positive feedback loop and intensify *SOX9* expression thus directing differentiation of supporting cells to Sertoli cells, which in turn down-regulate female signals like WNT4 and FOXL2 (Forkhead transcription factor; Lamba *et al.*, 2009). Thus, testicular development in mammals is triggered by *SRY*. In genetic males, this factor induces differentiation of Sertoli cells (reviewed by McLaren, 1991) and secretion of anti-Müllerian hormone (AMH). AMH, which

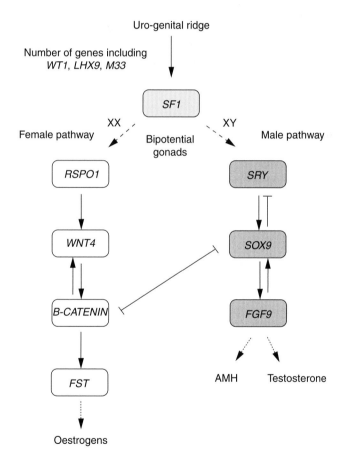

Fig. 14.5. A genetic model for sex determination, controlled by a balance of antagonistic pathways. In XY gonads, *Sry* triggers upregulation of *Sox9*, leading to Sertoli cell commitment and testicular differentiation. Sertoli cell differentiation is a result of the establishment of a positive feedback loop between Sox9 and secretion of Fgf9 (and also PGD$_2$; not shown), which act in a paracrine manner to recruit additional Sertoli cells. In XX gonads, two independent signalling pathways involving the Rspo1/Wnt4/β-catenin pathway and Foxl2 tilt the balance towards the female side and silence *Sox9* and *Fgf9*. Arrows indicate stimulation; T bars indicate inhibition. (Redrawn from Edson *et al.* (2009) with modifications.)

belongs to the transforming growth factors β family, causes regression of Müllerian ducts, promotes development of Wolffian ducts and the differentiation of Leydig cells, which secrete the male steroid hormone, testosterone (Behringer, 1995). Testosterone binds to androgen receptors, which act as transcription factors.

In the absence of an *SRY* gene, which is typical for XX embryos, gonads develop into ovaries. Two independent signalling pathways, the R-spondin1/WNT4/β-catenin pathway and FOXL2 transcription factor, support this developmental sequence of events. R-spondin1 has been recently recognized as a key female-determining factor (Nef and Vassilli, 2009). Quite often the female developmental programme is considered as the 'default', while the male programme requires 'switching on' the *SRY* gene followed by activation of other genes. The comparison of four regulatory regions located upstream of *SRY* shows high conservation between human, bovine, pig and goat. These regions of homology share transcription factor binding sites that appear to be subject to strong evolutionary pressure for conservation and may therefore be important for correct regulation of *SRY* (Ross *et al.*, 2008b). In contrast, the structure of the *SRY* region in the mouse is significantly different.

Cycle of the X chromosome

As proposed by Lyon (1961) and now widely accepted, one of the X chromosomes in eutherian females undergoes inactivation during early embryonic development as a dosage compensation measure. Numerous investigations have shed light on different aspects of this phenomenon including preferential inactivation of the paternal X chromosome in the trophoblast of female conceptuses, random inactivation of the X chromosome in the developing female embryos and molecular mechanisms of the inactivation (Goto and Monk, 1998). The cycle of the X chromosome inactivations and reactivations in cattle follows a similar scenario with species-specific timing.

In mature oocytes prior and immediately after fertilization the X chromosome is active. The paternal X chromosome, however, enters the zygote being inactive but soon after fertilization,

it reactivates. The expression of X chromosome-located monoamine oxidase A alleles has been confirmed in 4-, 8-, 16-cell, blastocyst and expanded blastocyst embryos, from both the maternal and paternal X chromosomes. However in morula only the maternal allele was expressed. Ferreira *et al.* (2010) concluded that in cattle the paternal allele was inactive at the morula stage and reactivated later at the blastocyst stage.

In XX embryos both X chromosomes are expected to be active for a relatively short time. In the next stage preferential inactivation of the paternal X chromosome in the trophoblast of bovine XX embryos probably occurs after days 7–8 and at about the same time random inactivation of one X chromosome takes place in the embryo proper during later blastocyst development (Xue *et al.*, 2002). Such random inactivation of one X chromosome in every somatic cell makes cows natural mosaics. However, uncertainty about the exact time of inactivation remains (Bermejo-Alvarez *et al.*, 2011). A great deal of transcriptional sexual dimorphism has been discovered in pre-implantational bovine embryos and a number of possible causes have been considered and investigated (Bermejo-Alvarez *et al.*, 2011). It is not known whether *Xist* gene expression is affected by the maternal metabolic state at the time of ovulation as has been shown in pigs (Vinsky *et al.*, 2007), but the fact that bovine *Xist* displays monoallelic expression patterns in bovine embryos has been established (Cruz *et al.*, 2008).

X chromosome inactivation is an essential condition for gene dose compensation. The mechanisms of silencing the X chromosome are complex (Ohhata and Wutz, 2013). Several chromatin modifications are necessary in order to form stable facultative chromatin, capable of propagating through numerous cell divisions. The so-called *X-inactivation center* located on the X chromosome contains the *Xist* gene and cis regulatory genetic elements. The *Xist* gene encodes an RNA molecule that plays a key role in silencing the inactive X chromosome (Plath *et al.*, 2002). *Xist* is negatively regulated by its antisense transcript Tsix. It seems, however, that *Tsix* (the reverse spelling of *Xist*) is not the only regulator, and additional transcription factors are involved in this complex process (Senner and Brockdorff, 2009). Different heterochromatin regions on the bovine inactive X chromosome can be identified by their histone isoform composition (Coppola *et al.*, 2008).

Anomalies of sex development

The *SRY* gene is among relatively few genes located on the Y chromosome. If this gene is missing from the Y chromosome, XY bovine embryos develop gonadal hypoplasia (Kawakura *et al.*, 1997). Mosaicism in a heifer carrying predominantly XY cells seems to be a cause of non-developed gonads, malformation of Müllerian ducts and infertility (Pinheiro *et al.*, 1990). Detection of bovine fetal SRY sequence from peripheral blood of pregnant cows by PCR analysis allows sex prediction (Yang *et al.*, 1996).

Usually only embryos carrying the Y chromosome possess *SRY*. However, *SRY* can become non-functional or be transferred from the Y chromosome to the X chromosome by a rare recombination event. Such events can cause sex reversal, whereby XY individuals become females and XX individuals become males. In either case, intersex individuals can arise (Cattanach *et al.*, 1982). In much better studied species like humans, XY sex reversals are rather frequent (about 1 in 3000 newborns) and are genetically heterogeneous. XX sex reversals on the contrary are rare (about 1 in 20,000 newborns) and are usually caused by the translocation of *SRY*. In the majority of human cases (~75%) the cause of sex reversal cannot be precisely identified (Nef and Vassilli, 2009). Less information is available on sex reversal in cattle (Favetta *et al.*, 2012), but excluding freemartins, it is not high. Kawakura *et al.* (1997) described three cases of bovine XY females that lacked the *SRY* gene (PCR confirmed) and possessed a structurally abnormal Y chromosome. Two unrelated cases of bovine male pseudohermaphrodites have been described recently (Moriyama *et al.*, 2010). One calf was a XX/XY chimera and another had low plasma testosterone, which was not caused by a lack of *SRY* and should have a different genetic defect. The comparative study of several mammalian species revealed upstream located *SRY* regulatory elements, mutations of which might lead to XY sex reversal (Ross *et al.*, 2008b).

Abnormalities of sex development may have different aetiology. Belgian Blue and Shorthorn cattle give a good example of such an interaction, being involved in determination of the White Heifer Disease. This disorder is characterized by a group of anomalies of the female genital tract, which stem from abnormal development of the Müllerian ducts. This includes numerous anatomical lesions in the vagina, cervix, uterus and consequently disturbances in their function. Interestingly, no disorders were found in the ovaries. The *Roan* gene was mapped to bovine chromosome 5, and the candidate murine *Steel* gene coding for the mast cell growth factor (MGF) was mapped in the same interval (Charlier *et al.*, 1996). As shown later the bovine *MGF* gene is indeed responsible for the roan phenotype (Seitz *et al.*, 1999). The current symbol of the gene is *KITLG* (KIT ligand).

Freemartinism in Cattle

Phenomenology

A freemartin is a sterile heifer (genetic female, XX) born co-twin to a bull. The frequency of male–female twin pairs is about 50% of non-identical twins. The freemartin condition represents the most frequent form of intersexuality found in cattle (Padula, 2005). Freemartinism results from the sexual modification of a female twin by *in utero* exchange of blood with its male co-twin. The dramatic reduction in size of the uterus and oviducts in freemartin cattle (Khan and Foley, 1994) is due in part to the female fetus being exposed to AMH from the male fetus (Vigier *et al.*, 1984). Between days 50 and 80, Müllerian ducts regression occurs simultaneously in males and freemartins and positively correlates with serum AMH concentrations. However, gonadal production of AMH in freemartins is very low (Vigier *et al.*, 1984). Beyond day 70 gubernacula development in freemartins shows male characteristics (van der Schoot *et al.*, 1995), and the ovaries produce abnormally high amounts of testosterone and little or no oestradiol (Shore and Shemesh, 1981). There are indications that the testes of the male co-twin to a freemartin display abnormal steroidogenesis, being for instance responsive to LH stimulation (Shore et al., 1984). However the majority of these males do not later differ from normal bulls. Interestingly, transgenic female murine fetuses expressing the human AMH gene underwent regression of Müllerian ducts and had reduced ovarian aromatase activity (Lyet et al., 1995). The transcription levels of several X-linked genes (HPRT, MECP2, RPS4X, SLC25A6, XIAP, XIST and ZFX), especially XIST, had abnormal patterns in the freemartin SCNT embryos, suggesting aberrant X chromosome inactivation in freemartins (Jeon et al., 2012).

There is a variation among freemartin females ranging from a more typical female phenotype (but usually with a blind-ending vagina and primordial ovarian and uterus structures) to a nearly male phenotype (body conformations with male traits and presence of primordial prepuce, penis and testicles) (Peretti et al., 2008). Harikae et al. (2012) also report a case of a freemartin calf with almost complete XX sex reversal and expressed the view that this observation strongly supports the idea of a high sexual plasticity of mammalian gonads.

Specific features of the bovine placenta and chimerism

The placenta in cattle is composed of specialized areas on the fetal chorion called cotyledons, which are in direct contact with the uterine epithelium of the dam at specialized areas called caruncles. Oxygen and nutrients pass from the maternal blood to the blood of the fetus, and waste products pass from the fetal blood into the blood of the dam without exchange of fetal and maternal blood. Shortly after implantation of twin pregnancies in cattle, chorionic fusion and vascular anastomoses of the two foetuses usually occur (Mellor, 1969). The vascular connections are more common for bovine species and much less typical for other ruminants. This allows exchange of blood cells between fetuses and leads to chimerism (60, XX/XY) in peripheral blood mononuclear leukocytes (Dunn et al., 1968; Basrur et al., 1970).

About half of the investigated freemartin males, despite chimerism, had normal body conformation, male-like external genitalia and some of them were fertile. In addition to the

cytogenetic diagnosis, chromosome fragility was described (CA: aneuploidy, gaps, chromatid breaks, chromosome breaks and fragments) as well as sister chromatid exchange (SCE). Freemartins show a higher percentage of aneuploid cells and significant statistical differences in mean values of gaps, chromatid and chromosome breaks when compared with control animals (Peretti et al., 2008).

Diagnosis of freemartins and practical aspect

Several methods have been described for detection of freemartinism (van Haeringen and Hradil, 1993), including sex chromatin karyotyping (Khan and Foley, 1994; Zhang et al., 1994; Bhatia and Shanker, 1985) and using PCR (Schellander et al., 1992; Ennis et al., 1999), which is rapid, highly sensitive and suitable for routine testing (Justi et al., 1995). The PCR method is also used for detection of sex in bovine pre-implantation embryos (Kirkpatrick and Monson, 1993; Ennis and Gallagher, 1994). A more recent approach based on fluorescence in situ hybridization on interphase nuclei using a bovine Y chromosome-specific DNA probe has been proven to be a 'rapid and reliable procedure that can be used for early-life diagnosis of bovine freemartinism' (Sohn et al., 2007).

Some breeding programmes aim to reduce the costs of rearing calves. This encourages selection for twins in cattle and eventually increases the number of sterile freemartin heifers (Kastli and Hall, 1978). Introduction of modern diagnostics has suggested that ~17.5% of heifers born with male co-twins may not be freemartins (Zhang et al., 1994), but further confirmations in different breeds are desirable. The freemartin syndrome limits the number of reproductively normal female offspring per dam unless sex predetermination methods are in place, and this may affect breeding systems that rely on increased twinning (Padula, 2005). As twinning rates in some dairy breeds are increasing, the economic importance of early diagnosis of the freemartin might be more significant. However, from a production point of view, slaughter-age freemartins do not differ from normal females in growth traits (Hallford *et al.*, 1976; Gregory *et al.*, 1996) but freemartins may show higher marbling scores than normal females (Gregory *et al.*, 1996).

Conclusions

Fifteen years ago, when the first edition of this book was published, very little was known about genetics of bovine development. Advancement in genomics, molecular methods, artificial reproduction and bioinformatics has changed the situation dramatically. Major progress has been achieved in molecular description of the first 15–20 days of development. This chapter is an attempt to collect and present in one location relevant data. Perhaps this kind of review may be helpful for those researchers and graduate students who are interested in bovine developmental genetics.

References

Abdalla, H., Hirabayashi, M. and Hochi, S. (2009) Demethylation dynamics of the paternal genome in pronuclear-stage bovine zygotes produced by *in vitro* fertilization and ooplasmic injection of freeze-thawed or freeze-dried spermatozoa. *Journal of Reproduction and Development* 55, 433–439.

Adjaye, J., Huntriss, J., Herwig, R., BenKahla, A., Brink, T.C., Wierling, C., Hultschig, C., Groth, D., Yaspo, M.-Y., Picton, H.R., Gosden, R.G. and Lehrach, H. (2005) Primary differentiation in the human blastocyst: comparative molecular portraits of inner cell mass and trophectoderm cells. *Stem Cells* 23, 1514–1525.

Adjaye, J., Herwig, R., Brink, T.C., Herrmann, D., Greber, B., Sudheer, S., Groth, D., Carnwath, J.W., Lehrach, H. and Niemann, H. (2007) Conserved molecular portraits of bovine and human blastocysts as a consequence of the transition from maternal to embryonic control of gene expression. *Physiological Genomics* 31, 315–327.

Agarwal, P., Wylie, J.N., Galceran, J., Arkhitko, O., Li, C., Deng, C., Grosschedl, R. and Bruneau, B.G. (2003) Tbx5 is essential for forelimb bud initiation following patterning of the limb field in the mouse embryo. *Development* 130(3), 623–633.

Al Naib, A., Mamo, S., O'Gorman, G.M., Lonergan, P., Swales, A. and Fair, T. (2012) Regulation of non-classical major histocompatability complex class I mRNA expression in bovine embryos. *Journal of Reproductive Immunology* 91, 31–40.

Alexander, T., Nolte, C.A. and Krumlauf, R. (2009) *Hox* genes and segmentation of the hindbrain and axial skeleton. *Annual Review of Cell and Developmental Biology* 25, 431–456.

Alten, L., Schuster-Gossler, K., Beckers, A., Groos, S., Ulmer, B., Hegermann, J., Ochs, M. and Gossler, A. (2012) Differential regulation of node formation, nodal ciliogenesis and cilia positioning by Noto and Foxj1. *Development* 139, 1276–1284.

Arnold, D.R., Lefebvre, R. and Smith, L.C. (2006) Characterization of the placenta specific bovine mammalian achaete scute-like homologue 2 (Mash2) gene. *Placenta* 27, 1124–1131.

Assis Neto, A.C., Pereira, F.T., Santos, T.C., Ambrosio, C.E., Leiser, R. and Miglino, M.A. (2010) Morpho-physical recording of bovine conceptus (*Bos indicus*) and placenta from days 20 to 70 of pregnancy. *Reproduction in Domestic Animals* 45, 760–772.

Aulehla, A. and Herrmann, B.G. (2004) Segmentation in vertebrates,clock and gradient finally joined. *Genes and Development* 18, 2060–2067.

Avilion, A.A., Nicholis, S.K., Pevney, L.H., Perez, L., Vivian, N. and Lowell-Badge, R. (2003) Multipotent cell lineages in early mouse development depend on SOX 2 function. *Genes and Development* 17, 126–140.

Bai, H., Sakurai, T., Konno, T., Ideta, A., Aoyagi, Y., Godkin, J.D. and Imakawa, K. (2012) Expression of GATA1 in the ovine conceptus and endometrium during the peri-attachment period. *Molecular Reproduction and Development* 79, 64–73.

Bamshad, M., Lin, R.C., Law, D.J., Watkins, W.C. Krakowiak, P.A., Moore, M.E., Franceschini, P., Lala, R., Holmes, L.B., Gebuhr, T.C., Bruneau, B.G., Schinzel, A., Seidman, C.E. and Jorde, L.B. (1997) Mutations in human *TBX3* alter limb, apocrine and genital development in ulnar-mammary syndrome. *Nature Genetics* 16, 311–315.

Bartholomew, R.A. and Parks, J.E. (2007) Identification, localization, and sequencing of fetal bovine VASA homolog. *Animal Reproductive Science* 101, 241–251.

Basrur, P.K., Kosaka, S. and Kanagawa, H. (1970) Blood cell chimerism and freemartinism in heterosexual bovine quadruplets. *Journal of Heredity* 61, 15–18.

Bauersachs, S. and Wolf, E. (2012) Transcriptome analyses of bovine, porcine and equine endometrium during the pre-implantation phase. *Animal Reproduction Science* 134, 84–94.

Bauersachs, S., Ulbrich, S.E., Zakhartchenko, V., Minten, M., Reichenbach, M., Reichenbach, H.D., Blum, H., Spencer, T.E. and Wolf, E. (2009) The endometrium responds differently to cloned versus fertilized embryos. *Proceedings of the National Academy of Sciences USA* 106, 5681–5686.

Bazer, F.W. (1992) Mediators of maternalrecognition of pregnancy in mammals. *Proceedings of the Society for Experimental Biology and Medicine* 199, 373–384.

Bazer, F.W., Geisert, R.D. and Zavy, M.T. (1993) Fertilization, cleavage, and implantation. In: Hafez, E.S.E. (ed.) *Reproduction in Farm Animals*, 6th Edn. Lea and Febiger, Philadelphia, Pennsylvania, pp. 188–212.

Bazer, F.W., Burghardt, R.C., Johnson, G.A., Spencer, T.E. and Wu, G. (2008) Interferons and progesterone for establishment and maintenance of pregnancy: interactions among novel cell signaling pathways. *Reproductive Biology* 8, 179–211.

Bazer, F.W., Spencer, T.E., Johnson, G.A., Burghardt, R.C. and Wu, G. (2009) Comparative aspects of implantation. *Reproduction* 138, 195–209.

Bazer, F.W., Wu, G., Spencer, T.E., Johnson, G.A., Burghardt, R.C. and Bayless, K. (2010) Novel pathways for implantation and establishment and maintenance of pregnancy in mammals. *Molecular Human Reproduction* 16, 135–152.

Bazer, F.W., Spencer, T.E., Johnson, G.A. and Burghardt, R.C. (2011) Uterine receptivity to implantation of blastocysts in mammals. *Frontiers in Bioscience (Scholar edition)* 3, 745–767.

Bechet, D.M., Deval, C., Robelin, J., Ferrara, M.J. and Obled, A. (1996) Developmental control of cathepsin B expression in bovine fetal muscles. *Archive Biochemistry and Biophysics* 334, 362–368.

Beck, F., Erler, T., Russell, A. and James, R. (1995) Expression of Cdx-2 in the mouse embryo and placenta: possible role in patterning of the extra-embryonic membranes. *Developmental Dynamics* 204, 219–227.

Beckers, A., Alten, L., Viebahn, C., Andre, P. and Gossler, A. (2007) The mouse homeobox gene Noto regulates node morphogenesis, notochordal ciliogenesis, and left right patterning. *Proceedings of the National Academy of Sciences USA* 104, 15765–15770.

Bentzinger, C.F., Wang, Y.X. and Rudnicki, M.A. (2012) Building muscle: molecular regulation of myogenesis. *Cold Spring Harbor Perspective in Biology* 4, pii:a008342.

Berg, D.K., Smith, C.S., Pearton, D.J., Wells, D.N., Broadhurst, R., Donnison, M. and Pfeffer, P.L. (2011) Trophectoderm lineage determination in cattle. *Developmental Cell* 20, 244–255.

Bermejo-Alvarez, P., Rizos, D., Rath, D., Lonergan, P. and Gutiérrez-Adán, A. (2008) Epigenetic differences between male and female bovine blastocysts produced *in vitro*. *Physiological Genomics* 32, 264–272.

Bermejo-Alvarez, P., Rizos, D., Lonergan, P. and Gutiérrez-Adán, A. (2011) Transcriptional sexual dimorphism in elongating bovine embryos: implications for XCI and sex determination genes. *Reproduction* 141, 801–808.

Bessho, Y., Sakata, R., Komatsu, S., Shiota, K., Yamada, S. and Kageyama, R. (2001) Dynamic expression and essential functions of Hes7 in somite segmentation. *Genes and Development* 15, 2642–2647.

Behringer, R.R. (1995) The mullerian inhibitor and mammalian sexual development. *Philosopical Transactions of Royal Society B: Biological Sciences* 350, 285–288; discussion 289.

Bettegowda, A., Lee, K.-B. and Smith, G.W. (2008) Cytoplasmic and nuclear determinants of the maternal-to-embryonic transition. *Reproduction, Fertility and Development* 20, 45–53.

Biase, F.H., Everts, R.E., Oliveira, R., Santos-Biase, W.K., Fonseca Merighe, G.K., Smith, L.C., Martelli, L., Lewin, H. and Meirelles, F.V. (2012) Messenger RNAs in metaphase II oocytes correlate with successful embryo development to the blastocyst stage. *Zygote* 10, 1–11.

Bhatia, S. and Shanker, V. (1985) Sex chromatin as a useful tool for detection of freemartinism in bovine twins. *British Veterinary Journal* 141, 42–48.

Blomberg, L., Hashizume, K. and Viebahn, C. (2008) Blastocyst elongation, trophoblastic differentiation, and embryonic pattern formation. *Reproduction* 135, 181–195.

Blum, M., Andre, P., Muders, K., Schweickert, A., Fischer, A., Bitzer, E., Bogusch, S., Beyer, T., van Straaten, H.W. and Viebahn, C. (2007) Ciliation and gene expression distinguish between node and posterior notochord in the mammalian embryo. *Differentiation* 75, 133–146.

Bollag, R.J., Siegfried, Z., Cebra-Thomas, J.A., Davison, E.M. and Silver, L.M. (1994) An ancient family of embryonically expressed mouse genes sharing a conserved protein motif with the *T* locus. *Nature Genetics* 7, 383–389.

Boucher, D.M., Schäffer, M., Deissler, K., Moore, C.A., Gold, J.D., Burdsal, C.A., Meneses, J.J., Pedersen, R.A. and Blum, M. (2000) Goosecoid expression represses Brachyury in embryonic stem cells and affects craniofacial development in chimeric mice. *International Journal of Developmental Biology* 44, 279–288.

Boulet, A.M., Moon, A.M., Arenkiel, B.R. and Capecchi, M.R. (2004) The roles of Fgf4 and Fgf8 in limb bud initiation and outgrowth. *Developmental Biology* 273, 361–372.

Bui, L.C., Evsikov, A.V., Khan, D.R., Archilla, C., Peynot, N., Hénaut, A., Le Bourhis, D., Vignon, X., Renard, J.P. and Duranthon, V. (2009) Retrotransposon expression as a defining event of genome reprograming in fertilized and cloned bovine embryos. *Reproduction* 138, 289–299.

Canovas, S., Cibelli, J.B., Ross, P.J. (2012) Jumonji domain-containing protein 3 regulates histone 3 lysine 27 methylation during bovine preimplantation development. *Proceedings of the National Academy of Sciences USA* 109, 2400–2405.

Capellini, T.D., Dunn, M.P., Passamaneck, Y.J., Selleri, L. and Di Gregorio, A. (2008) Conservation of notochord gene expression across chordates: insights from the Leprecan gene family. *Genesis* 46, 683–696.

Carmeliet, P., Moons, L., Luttun, A., Vincenti, V., Compernolle, V., De Mol, M., Wu, Y., Bono, F., Devy, L., Beck, H. *et al.* (2001) Synergism between vascular endothelial growth factor and placental growth factor contributes to angiogenesis and plasma extravasation in pathological conditions. *Nature Medicine* 7, 575–583.

Cassar-Malek, I., Picard, B., Kahl, S. and Hocquette, J.F. (2007) Relationships between thyroid status, tissue oxidative metabolism, and muscle differentiation in bovine foetuses. *Domestic Animals Endocrinology* 33, 91–106.

Cattanach, B.M., Evans, E.P., Burtenshaw, M.D. and Barlow, J. (1982) Male, female and intersex development in mice of identical chromosome constitution. *Nature* 300, 445–446.

Cavanagh, J.A., Tammen, I., Windsor, P.A., Bateman, J.F., Savarirayan, R., Nicholas, F.W. and Raadsma, H.W. (2007) Bulldog dwarfism in Dexter cattle is caused by mutations in ACAN. *Mammalian Genome* 218, 808–814.

Chambers, I., Colby, D., Robertson, M., Nichols, J., Lee. S., Tweedie, S. and Smith, A. (2003) Functional expression cloning of Nanog, a pluripotentcy sustaining factor in embryonic stem cells. *Cell* 113, 643–665.

Chapman, D.L., Agulnik, I., Hancock, S., Silver, L.M. and Papaioannou, V.E. (1996) Tbx6, a mouse T-Box gene implicated in paraxial mesoderm formation at gastrulation. *Developmental Biology* 180, 534–542.

Charlier, C., Coppieters, W., Farnir, F., Grobet, L., Leroy, P.L., Michaux, C., Mni, M., Schwers, A., Vanmanshoven, P., Hanset, R. and Georges, M. (1995) The *mh* gene causing double-muscling in cattle maps to bovine Chromosome 2. *Mammalian Genome* 6, 788–792.

Charlier, C., Denys, B., Belanche, J.I., Coppieters, W., Grobet, L., Mni, M., Womack, J., Hanset, R. and Georges, M. (1996) Microsatellite mapping of the bovine roan locus: a major determinant of White Heifer disease. *Mammalian Genome* 7, 138–142.

Chawengsaksophak, K., de Graaff, W., Rossant, J., Deschamps, J. and Beck, F. (2004) Cdx2 is essential for axial elongation in mouse development. *Proceedings of the National Academy of Sciences USA* 101, 7641–7645.

Chelh, I., Picard, B., Hocquette, J.F. and Cassar-Malek, I. (2011) Myostatin inactivation induces a similar muscle molecular signature in double-muscled cattle as in mice. *Animal* 5, 278–286.

Chen, J., Yan, W. and Setton, L.A. (2006) Molecular phenotypes of notochordal cells purified from immature nucleus pulposus. *European Spine Journal* 15(Suppl 3), 303–311.

Cheng, A. and Genever, P.G. (2010) SOX9 determines RUNX2 transactivity by directing intracellular degradation. *Journal of Bone and Mineral Research* 25, 2404–2413.

Ciemerych, M.A., Mesnard, D. and Zernicka-Goetz, M. (2000) Animal and vegetal poles of the mouse egg predict the polarity of the embryonic axis, yet are nonessential for development. *Development.* 127, 3467–3474.

Clemente, M., Lopez-Vidriero, I., O'Gaora, P., Mehta, J.P., Forde, N., Gutiérrez-Adán, A., Lonergan, P. and Rizos, D. (2011) Transcriptome changes at the initiation of elongation in the bovine conceptus. *Biology of Reproduction* 85, 285–295.

Clemmons, D.R. (1997) Insulin-like growth factor binding proteins and their role in controlling IGF actions. *Cytokine Growth Factor Reviews* 8, 45–62.

Coolican, S.A., Samuel, D.S., Ewton, D.Z., McWade, F.J. and Florini, J.R. (1997) The mitogenic and myogenic actions of insulin-like growth factors utilize distinct signaling pathways. *Journal of Biological Chemistry* 272, 6653–6662.

Coppola, G., Pinton, A., Joudrey, E.M., Basrur, P.K. and King, W.A. (2008) Spatial distribution of histone isoforms on the bovine active and inactive X chromosomes. *Sexual Development* 2, 12–23.

Cotterman, R., Jin, V.X., Krig, S.R., Lemen, J.M., Wey, A., Farnham, P.J. and Knoepfler, P.S. (2008) N-Myc regulates a widespread euchromatic program in the human genome partially independent of its role as a classical transcription factor. *Cancer Research* 68, 9654–9662.

Cruz, N.T., Wilson, K.J., Cooney, M.A., Tecirlioglu, R.T., Lagutina, I., Galli, C., Holland, M.K. and French, A.J. (2008) Putative imprinted gene expression in uniparental bovine embryo models. *Reproduction Fertility and Development* 20, 589–597.

Cruz, Y.P. and Pedersen, R.A. (1991) Origin of embryonic and extraembryonic cell lineages in mammalian embryos. In: Pedersen, R.A., McLaren, A. and First, N.L. (eds) *Animal Applications of Research in Mammalian Development. Current Communications in Cell and Molecular Biology*, 4. Cold Spring Harbor Laboratory Press, Cold Spring Harbor, New York, pp. 147–204.

Curchoe, C., Zhang, S., Bin, Y., Zhang, X., Yang, L., Feng, D., O'Neill, M. and Tian, X.C. (2005) Promoter-specific expression of the imprinted IGF2 gene in cattle (*Bos taurus*). *Biology of Reproduction* 73, 1275–1281.

Curi, R.A., Oliveira, H.N., Silveira, A.C. and Lopes, C.R. (2005) Effects of polymorphic microsatellites in the regulatory region of IGF1 and GHR on growth and carcass traits in beef cattle. *Animal Genetics* 36, 58–62.

Curran, S., Pierson, R.A. and Ginther, O.J. (1986) Ultrasonographic appearance of the bovine conceptus from days 20 through 60. *Journal of the American Veterinary Medical Association* 189, 1295–1302.

Dahl, E., Koseki, H. and Balling, R. (1997) *Pax* genes and organogenesis. *Bioessays* 19, 755–765.

Daneau, I., Houde, A., Ethier, J.F., Lussier, J.G. and Silversides, D.W. (1995) Bovine SRY gene locus: cloning and testicular expression. *Biology of Reproduction* 52, 591–599.

Das, P., Ezashi, T., Gupta, R. and Roberts, R.M. (2008) Combinatorial roles of protein kinase A, Ets2, and 3',5'-cyclic-adenosine monophosphate response element-binding protein-binding protein/p300 in the transcriptional control of interferon-tau expression in a trophoblast cell line. *Molecular Endocrinology* 22, 331–343.

Davies, D.R. (1990) The structure and function of the aspartic proteinases. *Annual Review of Biophysics and Biophysical Chemistry* 19, 189–215.

Davis, T.L., Trasler, J.M., Moss, S.B., Yang, G.J. and Bartolomei, M.S. (1999) Acquisition of the *H19* methylation imprint occurs differentially on the parental alleles during spermatogenesis. *Genomics* 58, 18–28.

de Vries, W.N., Evsikov, A.V., Brogan, L.J., Anderson, C.P., Graber, J.H. and Solter, D. (2008) Reprogramming and differentiation in mammals: motifs and mechanisms. *Cold Spring Harbor Symposium on Quantitative Biology* 73, 33–38.

Dean, W., Santos, F., Stojkovic, M., Zakhartchenko, V., Walter, J., Wolf, E. and Reik, W. (2001) Conservation of methylation reprogramming in mammalian development: aberrant reprogramming in cloned embryos. *Proceedings of the National Academy of Sciences USA* 98, 13734–13738.

DeBenedittis, P. and Jiao, K. (2011) Alternative splicing of T-box transcription factor genes. *Biochemical and Biophysical Research Communication* 412, 513–517.

Degrelle, S.A., Campion, E., Cabau, C., Piumi, F., Reinaud, P., Richard, C., Renard, J.P. and Hue, I. (2005) Molecular evidence for a critical period in mural trophoblast development in bovine blastocysts. *Developmental Biology* 288, 448–460.

Degrelle, S.A., Lê Cao, K.A., Heyman, Y., Everts, R.E., Campion, E., Richard, C., Ducroix-Crépy, C., Tian, X.C., Lewin, H.A., Renard, J.P. *et al.* (2011a) A small set of extra-embryonic genes defines a new landmark for bovine embryo staging. *Reproduction* 141, 79–89.

Degrelle, S.A., Murthi, P., Evain-Brion, D., Fournier, T. and Hue, I. (2011b) Expression and localization of DLX3, PPARG and SP1 in bovine trophoblast during binucleated cell differentiation. *Placenta* 32, 917–920.

Dindot, S.V., Kent, K.C., Evers, B., Loskutoff, N., Womack, J. and Piedrahita, J.A. (2004) Conservation of genomic imprinting at the XIST, IGF2, and GTL2 loci in the bovine. *Mammalian Genome* 15, 966–974.

Dunn, H.O., Kenney, R.M. and Lein D.H. (1968) XX-XY chimerism in a bovine true hermaphrodite: an insight into the understanding of freemartinism. *Cytogenetics* 7, 390–402.

Dunty, W.C. Jr, Biris, K.K., Chalamalasetty, R.B., Taketo, M.M., Lewandoski, M. and Yamaguchi, T.P. (2008) Wnt3a/β-catenin signaling controls posterior body development by coordinating mesoderm formation and segmentation. *Development* 135, 85–94.

Duranthon, V., Watson, A.J. and Lonergan, P. (2008) Preimplantation embryo programming: transcription, epigenetics, and culture environment. *Reproduction* 135, 141–150.

Edson, M.A., Nagaraja, A.K. and Matzuk, M.M. (2009) The mammalian ovary from genesis to revelation. *Endocrine Reviews* 30, 624–712.

Engelhardt, H., Croy, B.A. and King, G.J. (2002) Evaluation of natural killer cell recruitment to embryonic attachment sites during early porcine pregnancy. *Biology of Reproduction* 66, 1185–1192.

Ennis, S. and Gallagher, T.F. (1994) A PCR-based sex-determination assay in cattle based on the bovine amelogenin locus. *Animal Genetics* 25, 425–427.

Ennis, S., Vaughan. L. and Gallagher, T.F. (1999) The diagnosis of freemartinism in cattle using sex-specific DNA sequences. *Research in Veterinary Science* 67, 111–112.

Evans, A.L., Faial, T., Gilchrist, M.J., Down, T., Vallier, L., Pedersen, R.A., Wardle, F.C. and Smith, J.C. (2012) Genomic targets of brachyury (T) in differentiating mouse embryonic stem cells. *PLoS ONE* 7, e33346.

Evsikov, A.V. and Marín de Evsikova, C. (2009a) Gene expression during oocyte-to-embryo transition in mammals. *Molecular Reproduction and Development* 76, 805–818.

Evsikov, A.V. and Marín de Evsikova, C. (2009b) Evolutionary origin and phylogenetic analysis of the novel oocyte-specific eukaryotic translation initiation factor 4E in Tetrapoda. *Development Genes and Evolution* 219, 111–118.

Evsikov, A.V., Graber, J.H., Brockman, J.M., Hampl, A., Holbrook, A.E., Singh, P., Eppig, J.J., Solter, D. and Knowles, B.B. (2006) Cracking the egg: molecular dynamics and evolutionary aspects of the transition from the fully grown oocyte to embryo. *Genes and Development* 20, 2713–2727.

Fagotto, F., Jho Eh, Zeng, L., Kurth, T., Joos, T., Kaufmann, C. and Costantini, F. (1999) Domains of axin involved in protein-protein interactions, Wnt pathway inhibition, and intracellular localization. *Journal of Cellular Biology* 145, 741–756.

Fair, T., Carter, F., Park, S., Evans, A.C. and Lonergan, P. (2007) Global gene expression analysis during bovine oocyte *in vitro* maturation. *Theriogenology* 68 (Suppl 1), S91–S97.

Fan, H.Y. and Sun, Q.Y. (2004) Involvement of mitogen-activated protein kinase cascade during oocyte maturation and fertilization in mammals. *Biology of Reproduction.* 70, 535–547.

Fan, H.Y., Huo, L.J., Meng, X.Q., Zhong, Z.S., Hou, Y., Chen, D.Y. and Sun, Q.Y. (2003) Involvement of calcium/calmodulin-dependent protein kinase II (CaMKII) in meiotic maturation and activation of pig oocytes. *Biology of Reproduction* 69, 1552–1564.

Favetta, L.A., Villagómez, D.A., Iannuzzi, L., Di Meo, G., Webb, A., Crain, S. and King, W.A. (2012) Disorders of sexual development and abnormal early development in domestic food-producing mammals: the role of chromosome abnormalities, environment and stress factors. *Sexual Develeopment.* 6, 18–32.

Fear, J.M. and Hansen, P.J. (2011) Developmental changes in expression of genes involved in regulation of apoptosis in the bovine preimplantation embryo. *Biology of Reproduction* 84, 43–51.

Feil, R. (2009) Epigenetic asymmetry in the zygote and mammalian development. *International Journal of Developmental Biology* 53, 191–201.

Ferreira, A.R., Machado, G.M., Diesel, T.O., Carvalho, J.O., Rumpf, R., Melo, E.O., Dode, M.A. and Franco, M.M. (2010) Allele-specific expression of the MAOA gene and X chromosome inactivation in *in vitro* produced bovine embryos. *Molecular Reproduction and Development* 77, 615–621.

Firulli, A.B. and Olson, E.N. (1997) Modular regulation of muscle gene transcription; a mechanism for muscle cell diversity. *Trends in Genetics* 13, 364–369.

Fléchon, J.E., Degrouard, J., Fléchon, B., Lefèvre, F. and Traub, O. (2004a) Gap junction formation and connexin distribution in pig trophoblast before implantation. *Placenta* 25, 85–94.

Fléchon, J.E., Degrouard, J. and Fléchon, B. (2004b) Gastrulation events in the prestreak pig embryo: ultrastructure and cell markers. *Genesis* 38, 13–25.

Forde, N. and Lonergan, P. (2012) Transcriptomic analysis of the bovine endometrium: what is required to establish uterine receptivity to implantation in cattle? *Journal of Reproduction and Development* 58, 189–195.

Forde, N., Carter, F., Spencer, T.E., Bazer, F.W., Sandra, O., Mansouri-Attia, N., Okumu, L.A., McGettigan, P.A., Mehta, J.P., McBride, R., O'Gaora, P., Roche, J.F. and Lonergan, P. (2011) Conceptus-induced changes in the endometrial transcriptome: how soon does the cow know she is pregnant? *Biology of Reproduction* 85, 144–156.

Forde, N., Mehta, J.P., Minten, M., Crowe, M.A., Roche, J.F., Spencer, T.E., Lonergan, P. (2012a) Effects of low progesterone on the endometrial transcriptome in cattle. *Biology of Reproduction* 87, 124.

Forde, N., Duffy, G.B., McGettigan, P.A., Browne, J.A., Mehta, J.P., Kelly, A.K., Mansouri-Attia, N., Sandra, O., Loftus, B.J., Crowe, M.A. *et al.* (2012b) Evidence for an early endometrial response to pregnancy in cattle: both dependent upon and independent of interferon tau. *Physiological Genomics* 44, 799–810.

França, L.R. and Russell, L.D. (1998) The testis of domestic animals. In: Martinez, F. and Regadera, J. (eds) *Male Reproduction. A Multidisciplinary Over-view.* Churchill Livingstone, Madrid, pp. 197–219.

Fujita, K. and Janz, S. (2007) Attenuation of WNT signaling by DKK-1 and -2 regulates BMP2-induced osteoblast differentiation and expression of OPG, RANKL and M-CSF. *Molecular Cancer* 6, 71.

Funahashi, H., Stumpf, T.T., Cantley, T.C., Kim, N.H. and Day, B.N. (1995) Pronuclear formation and intracellular glutathione content of *in vitro*-matured porcine oocytes following *in-vitro* fertilization and/or electrical activation. *Zygote* 3, 273–281.

Gibson-Brown, J.J., Agulnik, S.I., Silver, L.M., Niswander, L. and Papaioannou, V.E. (1998) Involvement of T-box genes Tbx2-Tbx5 in vertebrate limb specification and development. *Development* 125, 2499–509.

Goldring, M.B., Tsuchimochi, K. and Ijiri, K. (2006) The control of chondrogenesis. *Journal of Cellular Biochemistry* 97, 33–44.

Gomercic, H., Vuković, S., Gomercić, V. and Skrtić, D. (1991) Histological and histochemical characteristics of the bovine notochord. *International Journal of Developmental Biology* 35, 353–358.

Goodfellow, P.N. and Lovell-Badge, R. (1993) SRY and sex determination in mammals. *Annual Review of Genetics* 27, 71–92.

Gore, M.T., Young, R.B., Bird, C.R., Rahe, C.H., Marple, D.N. Griffin, J.L. and Mulvaney, D.R. (1995) Myosin heavy chain gene expression in bovine fetuses and neonates representing genotypes with contrasting patterns of growth. *Proceedings of the Society for Experimental Biology and Medicine* 209, 86–91.

Goto, T. and Monk, M. (1998) Regulation of X-chromosome inactivation in development in mice and humans. *Microbiology and Molecular Biology Reviews* 62, 362–378.

Gray, S.D. and Dale, J.K. (2010) Notch signalling regulates the contribution of progenitor cells from the chick Hensen's node to the floor plate and notochord. *Development* 137, 561–568.

Greenstein, J.S. and Foley, R.C. (1958) Early embryology of the cow. I. Gastrula and primitive streak stages. *Journal of Dairy Science* 41, 409–421.

Gregory, K.E., Echternkamp, S.E. and Cundiff, L.V. (1996) Effects of twinning on dystocia, calf survival, calf growth, carcass traits, and cow productivity. *Journal of Animal Science* 74, 1223–1233.

Grobet, L., Martin, L.J., Poncelet, D., Pirottin, D., Brouwers, B., Riquet, J., Schoeberlein, A., Dunner, S., Menissier, F., Massabanda, J., Fries, R., Hanset, R. and Georges, M. (1997) A deletion in the bovine myostatin gene causes the double-muscled phenotype in cattle. *Nature Genetics* 17, 71–74.

Gros, J., Feistel, K., Viebahn, C., Blum, M. and Tabin, C.J. (2009) Cell movements at Hensen's node establish left/right asymmetric gene expression in the chick. *Science* 15, 941–944.

Gueth-Hallonet, C. and Maro, B. (1992) Cell polarity and cell diversification during early mouse embryogenesis. *Trends in Genetics* 8, 274–279.

Guillomot, M. and Guay, P. (1982) Ultrastructural features of the cell surfaces of uterine and trophoblastic epithelia during embryo attachment in the cow. *Anatomical Record* 204, 315–322.

Guillomot, M., Fléchon, J.-E., and Leroy, F. (1993) Blastocyst development and implantation. In: Thibault, C., Levasseur, M.C. and Hunter, R.H.F. (eds) *Reproduction in Mammals and Man.* Ellipses, Paris, pp. 387–411.

Guillemot, F., Caspary, T., Tilghman, S.M., Copeland, N.G., Gilbert, D.J., Jenkins, N.A., Anderson, N.A., Joyner, A.L., Rossant, J. and Nagy, A. (1995) Genomic imprinting of Mash-2, a mouse gene required for trophoblast development. *Nature Genetics* 9, 235–242.

Guillomot, M., Taghouti, G., Constant, F., Degrelle, S., Hue. I., Chavatte-Palmer, P. and Jammes, H. (2010) Abnormal expression of the imprinted gene Phlda2 in cloned bovine placenta. *Placenta* 31, 482–490.

Gutiérrez-Adán, A., Behboodi, E., Murray, J.D. and Anderson, G.B. (1997) Early transcription of the SRY gene by bovine preimplantation embryos. *Molecular Reproduction and Development* 48, 246–250.

Hajkova, P., Erhardt, S., Lane N., Haaf, T., El-Maarri, O., Reik, W., Walter, J. and Surani, M.A. (2002) Epigenetic reprogramming in mouse primordial germ cells. *Mechanics of Development* 117, 15–23.

Hallford, D.M., Turman, E.J., Selk, G.E., Walters, L.E. and Stephens, D.F. (1976) Carcass composition in single and multiple birth cattle. *Journal of Animal Science* 42, 1098–1103.

Hamatani, T., Ko, M., Yamada, M., Kuji, N., Mizusawa, Y., Shoji, M., Hada, T., Asada, H., Maruyama, T. and Yoshimura, Y. (2006) Global gene expression profiling of preimplantation embryos. *Human Cell* 19, 98–117.

Hansmann, T., Heinzmann, J., Wrenzycki, C., Zechner. U., Niemann, H. and Haaf, T. (2011) Characterization of differentially methylated regions in 3 bovine imprinted genes: a model for studying human germ-cell and embryo development. *Cytogenetic and Genome Research* 132, 239–247.

Harikae, K., Tsunekawa, N., Hiramatsu, R., Toda, S., Kurohmaru, M. and Kanai, Y. (2012) Evidence for almost complete sex-reversal in bovine freemartin gonads: formation of seminiferous tubule-like structures and transdifferentiation into typical testicular cell types. *Journal of Reproduction and Development* 58, 654–660.

Hariri, F., Nemer, M. and Nemer, G. (2012) T-box factors: insights into the evolutionary emergence of the complex heart. *Annals of Medicine* 44, 680–693.

Hashimoto, N., Watanabe, N., Furuta, Y., Tamemoto, H., Sagata, N., Yokoyama, M., Okazaki, K., Nagayoshi, M., Takeda, N. and Ikawa, Y. (1994) Parthenogenetic activation of oocytes in c-mos-deficient mice. *Nature* 370, 68–71.

Hashizume, K. (2007) Analysis of uteroplacental-specific molecules and their functions during implantation and placentation in the bovine. *Journal of Reproduction and Development* 53, 1–11.

Hassoun, R., Schwartz, P., Feistel, K., Blum, M.and Viebahn, C. (2009) Axial differentiation and early gastrulation stages of the pig embryo. *Differentiation* 78, 301–311.

Henckel, A., Nakabayashi, K., Sanz, L.A., Feil, R., Hata, K. and Arnaud, P. (2009) Histone methylation is mechanistically linked to DNA methylation at imprinting control regions in mammals. *Human Molecular Genetics* 18, 3375–3383.

Hue, I., Renard, J.P. and Viebahn, C. (2001) Brachyury is expressed in gastrulating bovine embryos well ahead of implantation. *Development Genes and Evolution* 211, 157–159.

Hue, I., Degrelle, S.A. and Turenne, N. (2012) Conceptus elongation in cattle: genes, models and questions. *Animal Reproduction Science* 134, 19–28.

Hunt, L.C., Upadhyay, A., Jazayeri, J.A., Tudor, E.M. and White, J.D. (2011) Caspase-3, myogenic transcription factors and cell cycle inhibitors are regulated by leukemia inhibitory factor to mediate inhibition of myogenic differentiation. *Skeletal Muscle* 1, 17.

Hynes, M., Stone, D.M., Dowd, M., Pitts-Meek, S., Goddard, A., Gurney, A. and Rosenthal, A. (1997) Control of cell pattern in the neural tube by the zinc finger transcription factor and oncogene *Gli-1*. *Neuron* 19, 15–26.

Ikeda, S., Sugimoto, M. and Kume, S. (2012) Importance of methionine metabolism in morula-to-blastocyst transition in bovine preimplantation embryos. *Journal of Reproduction and Development* 58, 91–97.

Imumorin, I.G., Kim, E.H., Lee, Y.M., De Koning, D.J., van Arendonk, J.A., De Donato, M., Taylor, J.F. and Kim, J.J. (2011) Genome scan for Parent-of-Origin QTL effects on bovine growth and carcass traits. *Frontiers of Genetics* 2, 44.

Ito, J., Shimada, M. and Terada, T. (2003) Effect of protein kinase C activator on mitogen-activated protein kinase and p34(cdc2) kinase activity during parthenogenetic activation of porcine oocytes by calcium ionophore. *Biology of Reproduction*. 69, 1675–1682.

Jainudeen, M.R. and Hafez, E.S.E. (1993) Gestation, prenatal physiology, and parturition. In: Hafez, E.S.E. (ed.) *Reproduction in Farm Animals*, 6th Edn. Lea and Febiger, Philadelphia, Pennsylvania, pp. 213–236.

Javed, Q., Fleming, T.P., Hay, M. and Citi, S. (1993) Tight junction protein cingulin is expressed by maternal and embryonic genomes during early mouse development. *Development* 117, 1145–1151.

Jeon, B.G., Rho, G.J., Betts, D.H., Petrik, J.J., Favetta, L.A. and King, W.A. (2012) Low levels of X-inactive specific transcript in somatic cell nuclear transfer embryos derived from female bovine freemartin donor cells. *Sexual Development* 6, 151–159.

Johnson, M.H. (2009) From mouse egg to mouse embryo: polarities, axes, and tissues. *Annual Review of Cell Developmental Biology* 25, 483–512.

Jones, J.I. and Clemmons, D.R. (1995) Insulin-like growth factors and their binding proteins: biological actions. *Endocrine Reviews* 16, 3–34.

Juarez-Oropeza, M.A., Lopez, V., Alvarez-Fernandez, G., Gomez, Y. and Pedernera, E. (1995) Androstenedione metabolism in the indifferent stage of bovine gonad development. *Journal of Experimental Zoology* 271, 373–378.

Justi, A., Hecht, W., Herzog, A. and Speck J. (1995) Comparison of different methods for the diagnosis of freemartinism blood group serology, cytology and polymerase chain reaction. *Deutsche Tierarztliche Wochenschrift* 102, 471–474.

Kambadur, R., Sharma, M., Smith, T.P. and Bass, J.J. (1997) Mutation in myostatin (GDF8) in double-muscled Belgian Blue and Piedmontese cattle. *Genome Research* 7, 910–916.

Kanka, J., Kepková, K. and Nemcová, L. (2009) Gene expression during minor genome activation in preimplantation bovine development. *Theriogenology* 724, 572–583.

Kastli, F. and Hall, J.G. (1978) Cattle twins and freemartin diagnosis. *Veterinary Record* 102, 80–83.

Katila, T. (2012) Post-mating inflammatory responses of the uterus. *Reproduction in Domestic Animals* 47 (Suppl 5), 31–41.

Kauffold, J., Amer, H.A., Bergfeld, U., Müller, F., Weber, W. and Sobiraj, A. (2005) Offspring from non-stimulated calves at an age younger than two months: a preliminary report. *Journal of Reproduction and Development* 51, 527–532.

Kawakura, K., Miyake, Y., Murakami, R.K., Kondoh, S., Hirata, T.I. and Kaneda, Y. (1997) Abnormal structure of the Y chromosome detected in bovine gonadal hypoplasia (XY female) by FISH. *Cytogenetics and Cell Genetics* 76, 36–38.

Kempisty, B., Antosik, P., Bukowska, D., Jackowska, M., Lianeri, M., Jaskowski, J.M. and Jagodzinski, P.P. (2008) Analysis of selected transcript levels in porcine spermatozoa, oocytes, zygotes and two-cell stage embryos. *Reproduction Fertility Development* 20, 513–518.

Khan, D.R., Dubé, D., Gall, L., Peynot, N., Ruffini, S., Laffont, L., Le Bourhis, D., Séverine Degrelle, S., Jouneau, A. and Duranthon, V. (2012) Expression of pluripotency master regulators during two key developmental transitions: EGA and early lineage specification in the bovine embryo. *PLoS One* 7, e34110.

Khan, M.Z. and Foley, G.L. (1994) Retrospective studies on the measurements, karyotyping and pathology of reproductive organs of bovine freemartins. *Journal of Comparative Pathology* 110, 25–36.

Kicheva, A. and Briscoe, J. (2010) Limbs made to measure. *PLoS Biology* 8, e1000421.

King, G.J., Atknison, B.A. and Robertson, H.A. (1979) Development of the bovine placentome during the second month of gestation. *Journal of Reproduction and Fertility* 55, 173–180.

King, G.J., Atkinson, B.A. and Robertson, H.A. (1981) Development of the intercaruncular areas during early gestation and establishment of the bovine placenta. *Journal of Reproduction and Fertility* 61, 469–474.

Kirkpatrick, B.W. and Monson, R.L. (1993) Sensitive sex determination assay applicable to bovine embryos derived from IVM and IVF. *Journal of Reproduction and Fertility* 98, 335–340.

Kispert, A., Koschorz, B. and Herrmann, B.G. (1995) The protein encoded by Brachyury is a tissue-specific transcription factor. *European Molecular Biology Organization Journal* 14, 4763–4772.

Knofler, M., Vasicek, R. and Schreiber, M. (2001) Key regulatory transcription factors involved in placental trophoblast development–a review. *Placenta* 22 (Suppl A), S83–S92.

Koehler, D., Zakhartchenko, V., Lutz Froenicke, L., Stone, G., Stanyon, R., Wolf, E., Cremer, T. and Brero, A. (2009) Changes of higher order chromatin arrangements during major genome activation in bovine preimplantation embryos. *Experimental Cell Research* 315, 2053–2063.

Kölle, S., Sinowatz, F., Boie, G., Lincoln, D. and Waters, M.J. (1997) Differential expression of the growth hormone receptor and its transcript in bovine uterus and placenta. *Molecular and Cellular Endocrinology* 131, 127–136.

Koseki, H., Wallin, J., Wilting, J., Mizutani, Y., Kispert, A., Ebensperger, C., Herrmann, B.G., Christ, B. and Balling, R. (1993) A role for *Pax-1* as a mediator of notochordal signals during the dorsoventral specification of vertebrae. *Development* 119, 649–660.

Koshi, K., Suzuki, Y., Nakaya, Y., Imai, K., Hosoe, M., Takahashi, T., Kizaki, K., Miyazawa, T. and Hashizume, K. (2012) Bovine trophoblastic cell differentiation and binucleation involves enhanced endogenous retrovirus element expression. *Reproductive Biology and Endocrinology* 10, 41.

Koshiba-Takeuchi, K., Mori, A.D., Kaynak, B.L., Cebra-Thomas, J., Sukonnik, T., Georges, R.O., Latham, S., Beck, L., Henkelman, R.M., Black, B.L. *et al.* (2009) Reptilian heart development and the molecular basis of cardiac chamber evolution. *Nature* 461, 95–98.

Koyama, H., Suzuki, H., Yang, X., Jiang, S. and Foote, R.H. (1994) Analysis of polarity of bovine and rabbit embryos by scanning electron microscopy. *Biology of Reproduction* 50, 163–170.

Kremenskoy, M., Kremenska, Y., Suzuki, M., Imai, K., Takahashi, S., Hashizume, K., Yagi, S. and Shiota, K. (2006) Epigenetic characterization of the CpG islands of bovine Leptin and POU5F1 genes in cloned bovine fetuses. *Journal of Reproduction and Development* 52, 277–285.

Kues, W.A., Sudheer, S., Herrmann, D., Carnwath, J.W., Havlicek, V., Besenfelder, U., Lehrach, H., Adjaye, J. and Niemann, H. (2008) Genome-wide expression profiling reveals distinct clusters of transcriptional regulation during bovine preimplantation development *in vivo*. *Proceedings of the National Academy of Sciences USA* 105, 19768–19773.

Kuijk, E.W., van Tol, L.T.A., Van de Velde, H., Wubbolts, R., Welling, M., Geijsen, N. and Roelen, B.A.J. (2012) The roles of FGF and MAP kinase signaling in the segregation of the epiblast and hypoblast cell lineages in bovine and human embryos. *Development* 139, 871–882.

Kurilo, L.F., Tepliakova, N.P. and Lavrikova, G.V. (1987) Development of ovaries in bovine fetuses. *Ontogenez* 18, 500–506 (in Russian).

Lahbib-Mansais, Y., Barbosa, A., Yerle, M., Parma, P., Milan, D., Pailhoux, E., Gellin, J. and Cotinot, C. (1997) Mapping in pig of genes involved in sexual differentiation: AMH, WT1, FTZF1, SOX2, SOX9, AHC, and placental and embryonic CYP19. *Cytogenetics and Cell Genetics* 76, 109–114.

Lamba, P., Fortin, J., Tran, S., Wang, Y. and Bernard, D.J. (2009) A novel role for the forkhead transcription factor foxl2 in activin a-regulated follicle-stimulating hormone β subunit transcription. *Molecular Endocrinology* 23, 1001–1013.

Larson, J.H., Kumar, C.G., Everts, R.E., Green, C.A., Everts-van der Wind, A., Band, M.R. and Lewin, H.A. (2006) Discovery of eight novel divergent homologs expressed in cattle placenta. *Physiological Genomics* 25, 405–413.

Lawson, K.A., Dunn, N.R., Roelen, B.A.J., Zeinstra, L.M., Davis, A.M., Wright, C.V.E., Korving, J.P.W.F.M. and Hogon, B.L.M. (1999) Bmp4 is required for the generation of primordial germ cells in the mouse embryo. *Genes and Development* 13, 424–436.

Lavoir, M.C., Basrur, P.K. and Betteridge, K.J. (1994) Isolation and identification of germ cells from fetal bovine ovaries. *Molecular Reproduction and Development* 37, 413–424.

Lee, G.S., Kim, H.S., Hwang, W.S. and Hyun, S.H. (2008) Characterization of porcine growth differentiation factor-9 and its expression in oocyte maturation. *Molecular Reproduction and Development* 75, 707–714.

Lefebvre, L. (2012) The placental imprintome and imprinted gene function in the trophoblast glycogen cell lineage. *Reproductive BioMedicine Online* 25, 44–57.

Levin, M. (2004) Left–right asymmetry in embryonic development: a comprehensive review. *Mechanisms of Development* 122, 3–25.

Linton, N.F., Wessels, J.M., Cnossen, S.A., Croy, B.A. and Tayade, C. (2008) Immunological mechanisms affecting angiogenesis and their relation to porcine pregnancy success. *Immunological Investigations* 37, 611–629.

Listrat, A., Gerrard, D.E., Boulle, N., Groyer, A. and Robelin, J. (1994) *In situ* localization of muscle insulin-like growth factor-II m RNA in developing bovine fetuses. *Journal of Endocrinology* 140, 179–187.

Lyet, L., Louis, F., Forest, M.G., Jasso, N., Behringer, R.R. and Vigier, B. (1995) Ontogeny of reproductive abnormalities induced by deregulation of anti-mullerian hormone expression in transgenic mice. *Biology of Reproduction* 52, 444–454.

Lyon, M.F. (1961) Gene action in the X-chromosome of the mouse (*Mus musculus* L.). *Nature* 190, 372–373.

Mackie, E.J., Ahmed, Y.A., Tatarczuch, L., Chen, K-S. and Mirams, M. (2008) Endochondral ossification: how cartilage is converted into bone in the developing skeleton. *International Journal of Biochemistry and Cell Biology* 40, 46–62.

Maconochie, M., Nonchev, S., Morrison, A. and Krumlauf, R. (1996) Paralogous *Hox* genes: function and regulation. *Annual Review of Genetics* 30, 529–556.

Maddox-Hyttel, P., Alexopoulos, N.I., Vajta, G., Lewis, I., Rogers, P., Cann, L., Callesen, H., Tveden-Nyborg, P. and Trounson, A. (2003) Immunohistochemical and ultrastructural characterization of the initial post-hatching development of bovine embryos. *Reproduction* 125, 607–623.

Maddox-Hyttel, P., Svarcova, O. and Laurincik, J. (2007) Ribosomal RNA and nucleolar proteins from the oocyte are to some degree used for embryonic nucleolar formation in cattle and pig. *Theriogenology* 68 (Suppl 1), S63–S70.

Madhala-Levy, D., Williams, V.C., Hughes, S.M., Reshef, R. and Halevy, O. (2012) Cooperation between Shh and IGF-I in promoting myogenic proliferation and differentiation via the MAPK/ERK and PI3K/Akt pathways requires Smo activity. *Journal of Cellular Physiology* 227, 1455–1464.

Magnani, L. and Cabot, R.A. (2008) *In vitro* and *in vivo* derived porcine embryos possess similar, but not identical, patterns of Oct4, Nanog, and Sox2 mRNA expression during cleavage development. *Molecular Reproduction and Development* 75, 1726–1735.

Magnani, L. and Cabot, R.A. (2009) Manipulation of *SMARCA2* and *SMARCA4* transcript levels in porcine embryos differentially alters development and expression of *SMARCA1*, *SOX2*, *NANOG*, and *EIF1*. *Reproduction.* 137, 23–33.

Mamo, S., Mehta, J.P., McGettigan, P., Fair, T., Spencer, T.E., Bazer, F.W. and Lonergan, P. (2011) RNA sequencing reveals novel gene clusters in bovine conceptuses associated with maternal recognition of pregnancy and implantation. *Biology of Reproduction* 85, 1143–1151.

Mamo, S., Mehta, J.P., Forde, N., McGettigan, P. and Lonergan, P. (2012a) Conceptus-endometrium cross-talk during maternal recognition of pregnancy in cattle. *Biology of Reproduction* 87, 1–9.

Mamo, S., Rizos, D. and Lonergan, P. (2012b) Transcriptomic changes in the bovine conceptus between the blastocyst stage and initiation of implantation. *Animal Reproduction Science* 134, 56–63.

Mansouri-Attia, N., Aubert, J., Reinaud, P., Giraud-Delville, C., Taghouti, G., Galio, L., Everts, R.E., Degrelle, S., Richard, C. *et al.* (2009) Gene expression profiles of bovine caruncular and intercaruncular endometrium at implantation. *Physiological Genomics* 39, 14–27.

Mansouri-Attia, N., Oliveira, L.J., Forde, N., Fahey, A.G., Browne, J.A., Roche, J.F., Sandra, O., Reinaud, P., Lonergan, P. and Fair, T. (2012) Pivotal role for monocytes/macrophages and dendritic cells in maternal immune response to the developing embryo in cattle. *Biology of Reproduction* 87, 123.

Maroto, M., Reshef, R., Munsterbergh, A.E., Koester, S., Goulding, M. and Lassar, A.B. (1997) Ectopic Pax-3 activates MyoD and Myf-5 expression in embryonic mesoderm and neural tissue. *Cell* 89, 139–148.

Marshall, H., Morrison, A., Studer, M., Popperl, H. and Krumlauf, R. (1996) Retinoids and *Hox* genes. *Federation of American Societies for Experimental Biology Journal* 10, 969–978.

McCarthy, S.D., Roche, J.F. and Forde, N. (2012) Temporal changes in endometrial gene expression and protein localization of members of the IGF family in cattle: effects of progesterone and pregnancy. *Physiological Genomics* 44, 130–140.

McGinnis, W. and Krumlauf, R. (1992) Homeobox genes and axial patterning. *Cell* 68, 238–302.

McLaren, A. (1991) Development of the mammalian gonad: the fate of the supporting cell lineage. *BioEssays* 13, 151–156.

McLaren, A. (1998) Assembling the mammalian testis. *Current Biology* 8, R175–R177.

McLaren, A. (1999) Signaling for germ cells. *Genes and Development* 13, 373–376.

Medeiros de Campos-Baptista, M.I., Glickman Holtzman, N., Yelon, D. and Schier, A.F. (2008) Nodal signaling promotes the speed and directional movement of cardiomyocytes in zebrafish. *Developmental Dynamics* 237, 3624–3633.

Mellor, D.J. (1969) Chorioinic fusion and the occurrence of free-martins: a brief review. *British Veterinary Journal* 125, 442–444.

Melton, A.A., Berry, R.O. and Butler, O.D. (1951) The interval between the time of ovulation and attachment of the bovine embryo. *Journal of Animal Science* 10, 993–1005.

Ménézo, Y. and Renard, J.-P. (1993) The life of the egg before implantation. In: Thibault, C., Levasseur, M.C., and Hunter, R.H.F. (eds) *Reproduction in Mammals and Man*. Ellipses, Paris, pp. 349–367.

Micke, G.C., Sullivan, T.M., McMillen, I.C., Gentili, S. and Perry, V.E. (2011) Protein intake during gestation affects postnatal bovine skeletal muscle growth and relative expression of IGF1, IGF1R, IGF2 and IGF2R. *Molecular and Cellular Endocrinology* 332, 234–241.

Migeotte, I., Omelchenko, T., Hall, A. and Anderson, K.V. (2010) Rac1-dependent collective cell migration is required for specification of the anterior-posterior body axis of the mouse. *PLoS Biology* 8, e1000442.

Millet, S., Campbell, K., Epstein, D.J., Losos, K., Harris, E. and Joyner, A.L. (1999) A role for Gbx2 in repression of Otx2 and positioning the mid/hindbrain organizer. *Nature* 401, 161–164.

Mitsui, K., Tokuzawa, Y., Itoh, H., Segawa, K., Murakami, M., Takahashi, K., Maruyama, M., Maeda, M. and Yamanaka, S. (2003) The homeoprotein Nanog is required for maintenance of pluripotency in mouse epiblast and ES cells. *Cell* 113, 631–642.

Mondou, E., Dufort, I., Gohin, M., Fournier, E. and Sirard, M.A. (2012) Analysis of microRNAs and their precursors in bovine early embryonic development. *Molecular Human Reproduction* 18, 425–434.

Monsoro-Burq, A.H., Duprez, D., Watanabe, Y., Bontoux, M., Vincent, C., Brickel, P. and Le Dourian, N. (1996) The role of bone morphogenetic proteins in vertebral development. *Development* 122, 3607–3616.

Moore, T. and Haig, D. (1991) Genomic imprinting in mammalian development: a parental tug-of-war. *Trends in Genetics* 7, 45–49.

Morais da Silva, S., Hacker, A., Harley, V., Goodfellow, P., Swain, A. and Lovell-Badge, R. (1996) *Sox9* expression during gonadal development implies a conserved role for the gene in testis differentiation in mammals and birds. *Nature Genetics* 14, 62–68.

Morimoto, M., Takahashi, Y., Endo, M. and Saga, Y. (2005) The Mesp2 transcription factor establishes segmental borders by suppressing Notch activity. *Nature* 435, 354–359.

Moriyama, C., Tani, M., Nibe, K., Kitahara, G., Haneda, S., Matsui, M., Miyake, Y.I. and Kamimura, S. (2010) Two cases of bovine male pseudohermaphrodites with different endocrinological and pathological findings. *Journal of Veterinary Medical Science* 72, 507–510.

Müller, C.W. and Herrmann, B.G. (1997) Crystallographic structure of the T domain-DNA complex of the Brachury transcription factor. *Nature* 389, 884–888.

Naiche, L.A., Harrelson, H., Kelly, R.G. and Papaioannou, V.P. (2005) T-Box genes in vertebrate development. *Annual Review of Genetics* 39, 219–239.

Nef, S. and Vassulli, J.-D. (2009) Complementary pathways in mammalian female sex determination. *Journal of Biology* 8, 74.

Nguyen, N.T., Lin, D.P.-C., Yen, S.-Y., Tseng, J.-K., Chuang, J.-F., Chen, B.-Y., Lin, T.-A., Chang, H.-H. and Ju, J.-C. (2009) Sonic hedgehog promotes porcine oocyte maturation and early embryo development. *Reproduction, Fertility and Development* 21, 805–815.

Nichols, J., Zevnik, B., Anastassiadis, K., Niwa, H., Klewe-Nebenius, D., Chambers, I., Scholer, H. and Smith, A. (1998) Formation of pluripotent stem cells in the mammalian embryo depends on the POU transcription factor Oct4. *Cell* 95, 379–391.

Nonaka, S., Yoshiba, S., Watanabe, D., Ikeuchi, S., Goto, T., Marshall, W.F. and Hamada, H. (2005) De novo formation of left–right asymmetry by posterior tilt of nodal cilia. *PLoS Biology* 3, e268.

Nüsslein-Volhard, C. (1996) Gradients that organize embryo development. *Scientific American* 275, 54–55.

Oginuma, M., Niwa, Y., Chapman, D.L. and Saga, Y. (2008) Mesp2 and Tbx6 cooperatively create periodic patterns coupled with the clock machinery during mouse somitogenesis. *Development* 135, 2555–2562.

Oh, B., Hwang, S.Y., Solter, D. and Knowles, B.B. (1997) Spindlin, a major maternal transcript expressed in the mouse during the transition from oocyte to embryo. *Development* 124, 493–503.

Oh, J.S., Susor, A. and Conti, M. (2011) Protein tyrosine kinase Wee1B is essential for metaphase II exit in mouse oocytes. *Science* 332, 462–465.

Ohhata, T. and Wutz, A. (2013) Reactivation of the inactive X chromosome in development and reprogramming. *Cellular and Molecular Life Science* 70, 2443–2461.

Okumu, L.A., Fair, T., Szekeres-Bartho, J., O'Doherty, A.M., Crowe, M.A., Roche, J.F., Lonergan, P. and Forde, N. (2011) Endometrial expression of progesterone-induced blocking factor and galectins-1, -3, -9, and -3 binding protein in the luteal phase and early pregnancy in cattle. *Physiological Genomics* 43, 903–910.

Oldham, J.M., Martyn, J.A., Sharma, M., Jeanplong, F., Kambadur, R. and Bass, J.J. (2001) Molecular expression of myostatin and MyoD is greater in double-muscled than normal-muscled cattle foetuses. *American Journal of Physiology: Regulatory, Integrating and Comparative Physiology* 280, R1488–R1493.

Oliveira, L.J., Barreto, R.S., Perecin, F., Mansouri-Attia, N., Pereira, F.T. and Meirelles, F.V. (2012) Modulation of maternal immune system during pregnancy in the cow. *Reproduction in Domestic Animals* 47 (Suppl 4), 384–393.

Ozawa, M., Sakatani, M., Yao, J., Shanker, S., Yu, F., Yamashita, R., Wakabayashi, S., Nakai, K., Dobbs, K.B., Sudano, M.J., Farmerie, W.G. and Hansen, P.J. (2012) Global gene expression of the inner cell mass and trophectoderm of the bovine blastocyst. *BMC Developmental Biology* 12, 33.

Padula, A.M. (2005) The freemartin syndrome: an update. *Animal Reproduction Science* 87, 93–109.

Papaioannou, V.E. (1997) T-box family reunion. *Trends in Genetics* 13, 212–213.

Park, J.S., Jeong, Y.S., Shin, S.T., Lee, K.K. and Kang, Y.K. (2007) Dynamic DNA methylation reprogramming: active demethylation and immediate remethylation in the male pronucleus of bovine zygotes. *Developmental Dynamics* 236, 2523–2533.

Park, K., Kang, J., Subedi, K.P., Ha, J.H. and Park, C. (2008) Canine polydactyl mutations with heterogeneous origin in the conserved intronic sequence of LMBR1. *Genetics* 179, 2163–2172.

Park, K.E., Magnani, L. and Cabot, R.A. (2009) Differential remodeling of mono- and trimethylated H3K27 during porcine embryo development. *Molecular Reproduction and Development* 76, 1033–1042.

Parma, P., Pailhoux, E., Puissant, C. and Cotinot, C. (1997) Porcine Dax-1 gene: isolation and expression during gonadal development. *Molecular and Cellular Endocrinology* 135, 49–58.

Parrington, J., Swann, K., Shevchenko, V.I., Sesay, A.K. and Lai, F.A. (1996) Calcium oscillations in mammalian eggs triggered by a soluble protein. *Nature* 379, 364–368.

Paul, D., Bridoux, L., Rezsöhazy, R. and Donnay, I. (2011) HOX genes are expressed in bovine and mouse oocytes and early embryos. *Molecular Reproduction and Development* 78, 436–449.

Pennington, K.A. and Ealy, A.D. (2012) The expression and potential function of bone morphogenetic proteins 2 and 4 in bovine trophectoderm. *Reproductive Biology and Endocrinology* 10, 12.

Peretti, V., Ciotola, F., Albarella, S., Paciello, O., Dario, C., Barbieri, V. and Iannuzzi, L. (2008) XX/XY chimerism in cattle: clinical and cytogenetic studies. *Sexual Development* 2, 24–30.

Peters, H., Wilm, B., Sakai, N., Imai, K., Maas, R. and Balling, R. (1999) Pax1 and Pax9 synergistically regulate vertebral column development. *Development* 126, 5399–5408.

Pevny, L.H. and Lovell-Badge, R. (1997) Sox genes find their feet. *Current Opinions in Genetics and Development* 7, 338–344.

Pfarrer, C., Hirsch, P., Guillomot, M. and Leiser, R. (2003) Interaction of integrin receptors with extracellular matrix is involved in trophoblast giant cell migration in bovine placentomes. *Placenta* 24, 588–597.

Pfarrer, C.D., Ruziwa, S.D., Winther, H., Callesen, H., Leiser, R., Schams, D. and Dantzer, V. (2006) Localization of vascular endothelial growth factor (VEGF) and its receptors VEGFR-1 and VEGFR-2 in bovine placentomes from implantation until term. *Placenta* 27, 889–898.

Pfeffer, P.L. and Pearton, D.J. (2012) Trophoblast development. *Reproduction* 143, 231–246.

Picard, B. Robelin, J., Pons, F. and Geay, Y. (1994) Comparison of the foetal development of fibre types in four bovine muscles. *Journal of Muscle Research and Cell Motility* 15, 473–486.

Picard, B., Cagniere, H., Robelin, J. and Geay, Y. (1995) Comparison of the foetal development in normal and double-muscled cattle. *Journal of Muscle Research and Cell Motility* 16, 629–639.

Pilon, N., Daneau I., Paradis, V., Hamel, G., Lussier, J., Viger, R.S. and Silversides, D.W. (2003) Porcine *SRY* promoter is a target for steroidogenic factor 1. *Biology of Reproduction* 68,1098–1106.

Pinheiro, L.E., Mikich, A.B., Bechara, G.H., Almeida, I.L. and Basrur, P.K. (1990) Isochromosome Y in an infertile heifer. *Genome* 33, 690–695.

Pinho, S., Simonsson, P.R., Trevers, K.E., Stower, M.J., Sherlock, W.T., Khan, M., Streit, A., Sheng, G. and Stern, C.D. (2011) Distinct steps of neural induction revealed by *Asterix, Obelix* and *TrkC*, genes induced by different signals from the organizer. *PLoS One* 6, e19157.

Plath, K., Mlynarczyk-Evans, S., Nusinow, D.A. and Panning, B. (2002) *Xist* RNA and the mechanism of chromosome inactivation. *Annual Review of Genetics* 36, 233–278.

Pollak, M. (2008) Insulin and insulin-like growth factor signalling in neoplasia. *Nature Reviews Cancer* 8, 915–928.

Pollard, S.M., Parsons, M.J., Kamei, M., Kettleborough, R.N., Thomas, K.A., Pham, V.N., Bae, M.K., Scott, A., Weinstein, B.M. and Stemple, D.L. (2006) Essential and overlapping roles for laminin alpha chains in notochord and blood vessel formation. *Developmental Biology* 289, 64–76.

Potts, J.K., Echternkamp, S.E., Smith, T.P. and Reecy, J.M. (2003) Characterization of gene expression in double-muscled and normal-muscled bovine embryos. *Animal Genetics* 34, 438–444.

Pourquie, O. (2003) The segmentation clock: converting embryonic time into spatial pattern. *Science* 301, 328–330.

Prior, R.L. and Laster, D.B. (1979) Development of the bovine foetus. *Journal of Animal Science* 48, 1546–1553.

Ramkissoon, Y. and Goodfellow, P. (1996) Early steps in mammalian sex determination. *Current Opinions in Genetics and Development* 6, 316–321.

Reeve, W.J. (1981) Cytoplasmic polarity develops at compaction in rat and mouse embryos. *Journal of Experimental Morphology* 62, 351–367.

Reidy, K.J. and Rosenblum, N.D. (2009) Cell and molecular biology of kidney development. *Seminars in Nephrology* 29, 321–337.

Reik, W. (2007) Stability and flexibility of epigenetic gene regulation in mammalian development. *Nature* 447, 425–432.

Reik, W., Dean, W. and Walter, J. (2001) Epigenetic reprogramming in mammalian development. *Science* 293, 1089–1093.

Reima, I., Lehtonen, E., Virtanen, I. and Flechon, J.E. (1993) The cytoskeleton and associated proteins during cleavage, compaction and blastocyst differentiation in the pig. *Differentiation* 54, 35–45.

Ribes, V. and Briscoe, J. (2009) Establishing and interpreting graded Sonic Hedgehog signaling during vertebrate neural tube patterning: the role of negative feedback. *Cold Spring Harbor Perspectives in Biology* 1, a002014.

Riley, P., Anson-Cartwright, L. and Cross, J.C. (1998) The Hand1 bHLH transcription factor is essential for placentation and cardiac morphogenesis. *Nature Genetics* 18, 271–275.

Rida, P.C., Le Minh, N. and Jiang, Y.J. (2004) A Notch feeling of somite segmentation and beyond. *Developmental Biology* 265, 2–22.

Robbins, K.M., Chen, Z., Wells, K.D. and Rivera, R.M. (2012) Expression of KCNQ1OT1, CDKN1C, H19, and PLAGL1 and the methylation patterns at the KvDMR1 and H19/IGF2 imprinting control regions is conserved between human and bovine. *Journal of Biomedical Science* 19, 95.

Roberts, R.M., Ezashi, T. and Das, P. (2004) Trophoblast gene expression: transcription factors in the specification of early trophoblast. *Reproductive Biology and Endocrinology* 2, 47–49.

Rodriguez-Esteban, C., Tsukui, T., Yonei, S., Magallon, J., Tamura, K. and Izpisua Belmonte, J.C. (1999) The T-box genes *Tbx4* and *Tbx5* regulate limb outgrowth and identity. *Nature* 398, 814–818.

Ropka-Molik, K., Eckert, R. and Pirkowska, K. (2011) The expression pattern of myogenic regulatory factors MyoD, Myf6 and Pax7 in postnatal porcine skeletal muscles. *Gene Expression Patterns* 11, 79–83.

Ross, P.J., Ragina, N.P., Rodriguez, R.M., Iager, A.E., Siripattarapravat, K., Lopez-Corrales, N. and Cibelli, J.B. (2008a) Polycomb gene expression and histone H3 lysine 27 trimethylation changes during bovine preimplantation development. *Reproduction* 136, 777–785.

Ross, D.G., Bowles, J., Koopman, P. and Lehnert, S. (2008b) New insights into SRY regulation through identification of 5′ conserved sequences. *BMC Molecular Biology* 9, 85.

Ross, D.G.F., Bowles, J., Hope, M., Lehnert, S. and Koopman, P. (2009) Profiles of gonadal gene expression in the developing bovine embryo. *Sexual Development* 3, 273–283.

Rossant, J. and Tam, P.P.L. (2009) Blastocyst lineage formation, early embryonic asymmetries and axis patterning in the mouse. *Development* 136, 701–713.

Russ, A.P., Wattler, S., Colledge, W.H., Aparicio, A., Carlton, M.B., Pearce, J.J., Barton, S.C., Surani, M.A., Ryan, K., Nehls, M.C., Wilson, V. and Evans, M.J. (2000) Eomesodermin is required for mouse trophoblast development and mesoderm formation. *Nature* 404, 95–99.

Ruvinsky, I. and Silver, L.M. (1997) Newly identified paralogous groups on mouse chromosomes 5 and 11 reveal the age of a T-box cluster duplication. *Genomics* 40, 262–266.

Saga, Y., Hata, N., Koseki, H. and Taketo, M.M. (1997) Mesp2: a novel mouse gene expressed in the presegmented mesoderm and essential for segmentation initiation. *Genes and Development* 11, 1827–1839.

Salhab, M., Tosca, L., Cabau, C., Papillier, P., Perreau, C., Dupont, J., Mermillod, P. and Uzbekova, S. (2011) Kinetics of gene expression and signaling in bovine cumulus cells throughout IVM in different mediums in relation to oocyte developmental competence, cumulus apoptosis and progesterone secretion. *Theriogenology* 75, 90–104.

Sandra, O., Mansouri-Attia, N. and Lea, R.G. (2011) Novel aspects of endometrial function: a biological sensor of embryo quality and driver of pregnancy success. *Reproduction Fertility and Development* 24, 68–79.

Satoh, N., Tagawa, K. and Takahashi, H. (2012) How was the notochord born? *Evolution and Development* 14, 56–75.

Sawai, K., Takahashi, M., Moriyasu, S., Hirayama, H., Minamihashi, A., Hashizume, T. and Onoe, S. (2010) Changes in the DNA methylation status of bovine embryos from the blastocyst to elongated stage derived from somatic cell nuclear transfer. *Cell Reprogramming* 12, 15–22.

Schafer, K. and Braun, T. (1999) Early specification of limb muscle precursor cells by the homeobox gene Lbx1h. *Nature Genetics* 23, 213–216.

Schellander, K., Peli, J., Taha, T.A., Kopp, E. and Mayr, B. (1992) Diagnosis of bovine freemartinism by the polymerase chain reaction method. *Animal Genetics* 23, 549–551.

Schimmenti, L.A. (2011) Renal coloboma syndrome. *European Journal of Human Genetics* 19, 1207–1212.

Schlafer, D.H., Fisher, P.J. and Davies, C.J. (2000) The bovine placenta before and after birth: placental development and function in health and disease. *Animal Reproduction Science* 60–61, 145–160.

Schorpp-Kistner, M., Wang, Z.Q., Angel, P. and Wagner, E.F. (1999) JunB is essential for mammalian placentation. *European Molecular Biology Organization Journal* 18, 934–948.

Schreiber, M., Wang, Z.Q., Jochum, W., Fetka, I., Elliott, C. and Wagner, E.F. (2000) Placental vascularisation requires the AP-1 component fra1. *Development* 127, 4937–4948.

Scott, I.C., Anson-Cartwright, L., Riley, P., Reda, D. and Cross, J.C. (2000) The HAND1 basic helix-loop-helix transcription factor regulates trophoblast differentiation via multiple mechanisms. *Molecular Cellular Biology* 20, 530–541.

Sekido, R. and Lovell-Badge, R. (2013) Genetic control of testis development. *Sexual Development* 7, 21–32.

Seenundun, S., Rampalli, S., Liu, Q.C., Aziz, A., Palii, C., Hong, S., Blais, A., Brand, M., Ge, K. and Dilworth, F.J. (2010) UTX mediates demethylation of H3K27me3 at muscle-specific genes during myogenesis. *EMBO Journal* 29, 1401–1411.

Seisenberger, S., Peat, J.R., Hore, T.A., Santos, F., Dean, W. and Reik, W. (2013) Reprogramming DNA methylation in the mammalian life cycle: building and breaking epigenetic barriers. *Philosophical Transactions of Royal Society B: Biological Sciences* 368, 2011.0330.

Seitz, J.J., Schmutz, S.M., Thue, T.D. and Buchanan, F.C. (1999) A missense mutation in the bovine MGF gene is associated with the roan phenotype in Belgian Blue and Shorthorn cattle. *Mammalian Genome* 10, 710–712.

Senner, C.E. and Brockdorff, N. (2009) Xist gene regulation at the onset of X inactivation. *Current Opinions in Genetics and Development* 19, 122–126.

Senner, C.E., Krueger, F., Oxley, D., Andrews, S. and Hemberger, M. (2012) DNA methylation profiles define stem cell identity and reveal a tight embryonic-extraembryonic lineage boundary. *Stem Cells* 30, 2732–2745.

Shaham, O., Menuchin, Y., Farhy, C. and Ashery-Padan, R. (2012) Pax6: a multi-level regulator of ocular development. *Progress in Retinal and Eye Research* 31, 351–376.

Shapiro, I.M. and Risbud, M.V. (2010) Transcriptional profiling of the nucleus pulposus: say yes to notochord. *Arthritis Research and Therapy* 12, 117.

Sharan, S.K., Holdener-Kenny, B., Ruppert, S., Schedl, A., Kelsey, G., Rinchik, E.M. and Magnuson, T. (1991) The albino-deletion complex of the mouse: molecular mapping of deletion breakpoints that define regions necessary for development of the embryonic and extraembryonic ectoderm. *Genetics* 129, 825–832.

Shehu, D., Marsicano, G., Flechon, J.E. and Gali, C. (1996) Developmentally regulated markers of *in vitro*-produced preimplantaion bovine embryos. *Zygote* 4, 109–121.

Shemer, R., Birger, Y., Dean, W.L., Reik, W., Riggs, A.D. and Razin, A. (1996) Dynamic methylation adjustment and counting as part of imprinting mechanisms. *Proceedings of the National Academy of Sciences USA* 93, 6371–6376.

Shen, M.M. (2007) Nodal signalling: developmental roles and regulations. *Development* 134, 1023–1034.

Shore, L. and Shemesh, M. (1981) Altered steroidogenesis by the fetal bovine freemartin ovary. *Journal Reproduction and Fertility* 63, 309–314.

Shore, L.S., Shemesh, M. and Mileguir, F. (1984) Foetal testicular steroidogenesis and responsiveness to LH in freemartins and thier male co-twins. *International Journal of Andrology* 7, 87–93.

Shimaoka, T., Nishimura, T., Kano, K. and Naito, K. (2009) Critical effect of pigWee1B on the regulation of meiotic resumption in porcine immature oocytes. *Cell Cycle* 8, 2375–2384.

Shimaoka, T., Nishimura, T., Kano, K. and Naito, K. (2011) Analyses of the regulatory mechanism of porcine WEE1B: the phosphorylation sites of porcine WEE1B and mouse WEE1B are different. *Journal of Reproduction and Development* 57, 223–228.

Shirasuna, K., Matsumoto, H., Kobayashi, E., Nitta, A., Haneda, S., Matsui, M., Kawashima, C., Kida, K., Shimizu, T. and Miyamoto, A. (2012) Upregulation of interferon-stimulated genes and interleukin-10 in peripheral blood immune cells during early pregnancy in dairy cows. *Journal of Reproduction and Development* 58, 84–90.

Sikora, K.M., Magee, D.A., Berkowicz, E.W., Lonergan, P., Evans, A.C., Carter, F., Comte, A., Waters, S.M., MacHugh, D.E. and Spillane, C. (2012) PHLDA2 is an imprinted gene in cattle. *Animal Genetics* 43, 587–590.

Silva, L.A. and Ginther, O.J. (2010) Local effect of the conceptus on uterine vascular perfusion during early pregnancy in heifers. *Reproduction* 139,453–463.

Singh, R. and Kispert, A. (2010) Tbx20, Smads, and the atrioventricular canal. *Trends in Cardiovascular Medicine* 20, 109–114.

Slagle, C.E., Aoki, T. and Burdine, R.D. (2011) Nodal-dependent mesendoderm specification requires the combinatorial activities of FoxH1 and Eomesodermin. *PLoS Genetics* 7, e1002072.

Spencer, T.E., Burghardt, R.C., Johnson, G.A. and Bazer, F.W. (2004) Conceptus signals for establishment and maintenance of pregnancy. *Animal Reproduction Science* 82, 537–550.

Spencer, T.E., Johnson, G.A., Bazer, F.W. and Burghardt, R.C. (2007) Fetal-maternal interactions during the establishment of pregnancy in ruminants. *Society for Reproduction and Fertility* 64 (Suppl), 379–396.

Sohn, S.H., Cho, E.J., Son, W.J. and Lee, C.Y. (2007) Diagnosis of bovine freemartinism by fluorescence in situ hybridization on interphase nuclei using a bovine Y chromosome-specific DNA probe. *Theriogenology* 68, 1003–1011.

Sokol, S.Y. (1999) Wnt signalling and dorso-ventral axix specification in vertebrates. *Current Opinions in Genetics and Development* 9, 405–410.

Stamatkin, C.W., Roussev, R.G., Stout, M., Absalon-Medina, V., Ramu. S., Goodman, C., Coulam, C.B., Gilbert, R.O., Godke, R.A. and Barnea, E.R. (2011) PreImplantation Factor (PIF) correlates with early mammalian embryo development-bovine and murine models. *Reproductive Biology and Endocrinology* 15, 63.

Stennard, F.A. and Harvey, R.P. (2005) T-box transcription factors and their roles in regulatory hierarchies in the developing heart. *Development* 132, 4897–4910.

St Johnston, D. and Nüsslein-Volhard, C. (1992) The origin of pattern and polarity in the Drosophila embryo. *Cell* 68, 201–219.

Sumer, H., Liu, J., Malaver-Ortega, L.F, Lim, M.L., Khodadadi, K. and Verma, P.J. (2011) NANOG is a key factor for induction of pluripotency in bovine adult fibroblasts. *Journal of Animal Science* 89, 2708–2716.

Szafranska, B., Xie, S., Green, J. and Roberts, R.M. (1995) Porcine pregnancy-associated glycoproteins: new members of the aspartic proteinase gene family expressed in trophectoderm. *Biology of Reproduction* 53, 21–28.

Szafranska, B., Panasiewicz, G., Majewska, M., Romanowska, A. and Dajnowiec, J. (2007) Pregnancy-associated glycoprotein family (PAG) – as chorionic signaling ligands for gonadotropin receptors of cyclic animals. *Animal Reproduction Science* 99, 269–284.

Tam, P.P. and Beddington, R.S. (1987) The formation of mesodermal tissues in the mouse embryo during gastrulation and early organogenesis. *Development* 99, 109–126.

Tayade, C., Black, G.P., Fang, Y. and Croy, B.A. (2006) Differential gene expression in endometrium, endometrial lymphocytes, and trophoblasts during successful and abortive embryo implantation. *Journal of Immunology* 176, 148–156.

Tayade, C., Fang, Y. and Croy, B.A. (2007) A review of gene expression in porcine endometrial lymphocytes, endothelium and trophoblast during pregnancy success and failure. *Journal of Reproduction and Development* 53, 455–463.

Thompson, J.G., Sherman, A.N., Allen, N.W., McGowan, L.T. and Tervit, H.R. (1998) Total protein content and protein synthesis within pre-elongation stage bovine embryos. *Molecular Reproduction and Development* 50, 139–145.

Towers, M. and Tickle, C. (2009) Growing models of vertebrate limb development. *Development* 136, 179–190.

Touzard, E., Reinaud, P., Dubois, O., Joly-Guyader, C., Humblot, P., Ponsart, C. and Charpigny, G. (2012) 100 bovine pregnancy-associated glycoproteins are allocated to cotyledonary or intercotyledonary trophoblast according to their phylogenetic origin. *Reproduction, Fertility and Development* 25, 197–198.

Tripurani, S.K., Lee, K.-B., Wee, G., Smith, G.W. and Yao, J. (2011) MicroRNA-196a regulates bovine newborn ovary homeobox gene (NOBOX) expression during early embryogenesis. *BMC Developmental Biology* 11, 25.

Turenne, N.N., Tiys, E.E., Ivanisenko, V.V., Yudin, N.N., Ignatieva, E.E., Valour, D.D., Degrelle, S.S. and Hue, I.I. (2012) Finding biomarkers in non-model species: literature mining of transcription factors involved in bovine embryo development. *BioData Mining* 5, 12.

Ulloa, F. and Martí, E. (2010) Wnt won the war: antagonistic role of Wnt over Shh controls dorso-ventral patterning of the vertebrate neural tube. *Developmental Dynamics* 239, 69–76.

Ushizawa, K., Takahashi, T., Hosoe, M., Ishiwata, H., Kaneyama, K., Kizaki, K. and Hashizume, K. (2007) Global gene expression analysis and regulation of the principal genes expressed in bovine placenta in relation to the transcription factor AP-2 family. *Reproductive Biology and Endocrinology* 5, 17.

Ushizawa, K., Herath, C.B., Kaneyama, K., Shiojima, S., Hirasawa, A., Takahashi, T., Imai, K., Ochiai, K., Tokunaga, T., Tsunoda, Y., Tsujimoto, G. and Hashizume, K. (2004) cDNA microarray analysis of bovine embryo gene expression profiles during the preimplantation period. *Reproductive Biology and Endocrinology* 2, 77.

Uzbekova, S., Roy-Sabau, M., Dalbiès-Tran, R., Perreau, C., Papillier, P., Mompart, F., Thelie, A., Pennetier, S., Cognie, J., Cadoret, V., Royere, D., Monget, P. and Mermillod, P. (2006) Zygote arrest 1 gene in pig, cattle and human: evidence of different transcript variants in male and female germ cells. *Reproduction Biology and Endocrinology* 4, 12.

Vallée, M., Robert, C., Méthot, S., Palin, M.F. and Sirard, M.A. (2006) Cross-species hybridizations on a multi-species cDNA microarray to identify evolutionarily conserved genes expressed in oocytes. *BMC Genomics* 7, 113.

van de Pavert, S.A., Schipper, H., de Wit, A.A.C., Soede, N.M., van den Hurk, R., Taverne, M.A.M., Boerjan, M.L., Henri, W.J. and Stroband, H.W.J. (2001) Comparison of anterior–posterior development in the porcine versus chicken embryo, using *goosecoid* expression as a marker. *Reproduction, Fertility and Development* 13, 177–185.

van der Schoot, P., Vigier, B., Prepin, J., Perchellet, J.P. and Gittenberger-de Groot, A. (1995) Development of the gubernaculum and processus vaginalis in freemartinism:further evidence in support of a specific fetal testis hormone governing male-specific gubernacular development. *Anatomical Record* 241, 211–224.

van Haeringen, H. and Hradil, R. (1993) Twins in cattle. Freemartin or not? Current aspects. *Tijschrift Voor Diergeneeskunde* 118, 648–649.

Vejlsted, M., Avery, B., Schmidt, M., Greve, T., Alexopoulos, N. and Maddox-Hyttel, P. (2005) Ultrastructural and immunohistochemical characterization of the bovine epiblast. *Biology of Reproduction* 72, 678–686.

Vejlsted, M., Du, Y., Vajta, G. and Maddox-Hyttel, P. (2006) Post-hatching development of the porcine and bovine embryo – defining criteria for expected development *in vivo* and *in vitro*. *Theriogenology* 65, 153–165.

Velazquez, M.A., Zaraza, J., Oropeza, A., Webb, R. and Niemann, H. (2009) The role of IGF1 in the *in vivo* production of bovine embryos from superovulated donors. *Reproduction*. 137, 161–180.

Vigier, B., Tran, D., Legeai, L., Bezard, J. and Josso, N. (1984) Origin of anti-Müllerian hormone in bovine freemartin fetuses. *Journal of Reproduction and Fertility* 70, 473–479.

Vinsky, M.D., Murdoch, G.K. Dixon, W.T., Dyck, M.K. and Foxcroft, G.R. (2007) Altered epigenetic variance in surviving litters from nutritionally restricted lactating primiparous sows. *Reproduction Fertility and Development* 19, 430–435.

Vonica, A., Rosa, A., Arduini, B. and Brivanlou, A.H. (2011) APOBEC2, a selective inhibitor of TGFβ signaling, regulates left-right axis specification during early embryogenesis. *Developmental Biology* 350, 13–23.

Wang, L., Liu, X., Niu, F., Wang, H., He, H. and Gu, Y. (2013) Single nucleotide polymorphisms, haplotypes and combined genotypes in MYH3 gene and their associations with growth and carcass traits in Qinchuan cattle. *Molecular Biology Reports* 40, 417–426.

Wang, Q., Fang, W.H., Kruplnski, J., Kumar, S., Slevin, M. and Kumar, P. (2008) Paxgenes in embryogenesis and oncogenesis. *Journal of Cellular and Molecular Medicine* 12, 2281–2294.

Wathes, D.C. and Wooding, F.B. (1980) An electron microscopic study of implantation in the cow. *American Journal of Anatomy* 159, 285–306.

White, P.H. and Chapman, D.L. (2005) Dll1 is a downstream target of Tbx6 in the paraxial mesoderm. *Genesis* 42, 193–202.

Wilkinson, D.G., Bhatt, S. and Herrmann, B.G. (1990) Expression pattern of the mouse T gene and its role in mesoderm formation. *Nature* 343, 657–659.

Winters, L.M., Green, W.W. and Comstock, R.E. (1942) Prenatal development of the Bovine. *Technical Bulletin* 151, 1–44. University of Minnesota, Agricultural Experimental Station. Reprinted December 1953.

Wooding, F.B. (1992) Current topic: the synepitheliochorial placenta of ruminants: binucleate cell fusions and hormone production. *Placenta* 13 101–113.

Wossidlo, M., Nakamura, T., Lepikhov, K., Marques, C.J., Zakhartchenko, V., Boiani, M., Arand, J., Nakano, T., Reik, W. and Walter, J. (2011) 5-Hydroxymethylcytosine in the mammalian zygote is linked with epigenetic reprogramming. *Nature Communications* 2, 241. doi:10.1038/ncomms1240.

Wrobel, K.H. (2000) Prespermatogenesis and spermatogoniogenesis in the bovine testis. *Anatomy and Embryology (Berl)* 202, 209–222.

Wrobel, K.H. and Süss, F. (1998) Identification and temporospatial distribution of bovine primordial germ cells prior to gonadal sexual differentiation. *Anatomy and Embryology (Berl)* 197, 451–467.

Wu, X., Viveiros, M.M., Eppig, J.J., Bai, Y., Fitzpatrick, S.L. and Matzuk, M.M. (2003) Zygote arrest 1 (Zar1) is a novel maternal-effect gene critical for the oocyte-to-embryo transition. *Nature Genetics* 33, 187–191.

Xiang, W. and MacLaren, L.A. (2002) Expression of fertilin and CD9 in bovine trophoblast and endometrium during implantation. *Biology of Reproduction.* 66, 1790–1796.

Xie, D., Chen, C.C., He, X., Cao, X. and Zhong, S. (2011) Towards an evolutionary model of transcription networks. *PLoS Computational Biology* 7, doi:10.1371/journal.pcbi.1002064.

Xu, B.Z., Li, M., Xiong, B., Lin, S.L., Zhu, J.Q., Hou, Y., Chen, Y.D. and Sun, Q.Y. (2009) Involvement of calcium/calmodulin-dependent protein kinase kinase in meiotic maturation of pig oocytes. *Animal Reproduction Science* 111, 17–30.

Xu, K.P., Yadav, B.R., King, W.A. and Betteridge, K.J. (1992) Sex-related differences in developmental rates of bovine embryos produced and cultured *in vitro*. *Molecular Reproduction and Development* 31, 249–252.

Xue, F., Tian, X.C., Du, F., Kubota, C., Taneja, M., Dinnyes, A., Dai, Y., Levine, H., Pereira, L.V. and Yang, X. (2002) Aberrant patterns of X chromosome inactivation in bovine clones. *Nature Genetics* 31, 216–220.

Yamakoshi, S., Bai, R., Chaen, T., Ideta, A., Aoyagi, Y., Sakurai, T., Konno, T. and Imakawa, K. (2012) Expression of mesenchymal-related genes by the bovine trophectoderm following conceptus attachment to the endometrial epithelium. *Reproduction* 143, 377–387.

Yamamoto, H., Flannery, M.L., Kupriyanov, S., Pearce, J., McKercher, S.R., Henkel, G.W.,Maki, R.A., Werb, Z. and Oshima, R.G. (1998) Defective trophoblast function in mice with a targeted mutation of Ets2. *Genes and Development* 12, 1315–1326.

Yan, W. (2009) Male infertility caused by spermiogenic defects: lessons from gene knockouts. *Molecular and Cellular Endocrinology* 306, 24–32.

Yang, J., Wang, L., Jiang, X., Jiang, Y. and Liu, L. (1996) Detection of bovine fetal Y-specific Sry sequence from maternal blood. *Chinese Journal of Biotechnology* 12, 185–188.

Yang, M.Y. and Fortune, J.E. (2008) The capacity of primordial follicles in fetalbovine ovaries to initiate growth *in vitro* develops during mid-gestation and is associated with meiotic arrest of oocytes. *Biology of Reproduction* 78, 1153–1161.

Yelich, J.V., Pomp, D. and Geisert, R.D. (1997) Detection of transcripts for retinoic acid receptors, retinol-binding protein, and transforming growth factors during rapid trophoblastic elongation in the porcine conceptus. *Biology of Reproduction* 57, 286–294.

Zeng, L., Fagotto, F., Zhang, T., Hsu, W., Vasicek, T.J., Perry, W.L. 3rd, Lee, J.J., Tilghman, S.M., Gumbiner, B.M. and Costantini, F. (1997) The mouse Fused locus encodes Axin, an inhibitor of the Wnt signaling pathway that regulates embryonic axis formation. *Cell* 90, 181–192.

Zhang, T., Buoen, L.C., Seguin, B.E., Ruth, G.R. and Weber, A.F. (1994) Diagnosis of freemartinism in cattle: the need for clinical and cytogenic evaluation. *Journal of the American Veterinary Association* 204, 1672–1675.

Zorn, A.M. and Wells, J.M. (2009) Vertebrate endoderm development and organ formation. *Annual Review of Cell and Development* 25, 10.1–10.31.

15 Genetic Improvement of Dairy Cattle

Vincent Ducrocq[1] and George Wiggans[2]

[1]UMR 1313 Génétique Animale et Biologie Intégrative (GABI)
INRA, Jouy-en-Josas, France; [2]United States Department of Agriculture,
Beltsville, Maryland, USA

Introduction

As for any species or population, genetic improvement of dairy cattle involves determining a desirable direction for improvement, identifying traits that provide information to move in that direction, quantifying their heritability, deciding how to evaluate them and designing a breeding programme to achieve the goals. With regard to these issues, dairy cattle are one of the most highly studied of all domesticated species.

This chapter describes how to determine which goals should be established to emphasize profit or efficiency as the ultimate goal of the dairy enterprise. The traits typically measured are listed along with how they are related and the genetic parameters utilized in the selection process. Evaluation procedures used to establish genetic rankings based on observations on related animals (genetic evaluations) or on genomic information (genomic evaluations) are reviewed. Their incorporation into breeding programmes is outlined.

Breeding Objectives

Derivation of a breeding objective

The first step in the design of a breeding programme is to specify its goal: the breeding objective. The usual purpose of a breeding programme is assumed to be the increase of profitability by modifying the genetic mean of key traits. Often, this increase must be performed under an uncertain future economic environment, under diverse management systems and under some constraints (e.g. quota on overall production or constant feed supply or pasture area at farm level). Therefore, the definition of the breeding objective starts with an inventory of representative management systems and of likely future scenarios as well as the description of specific constraints to satisfy.

The derivation of the breeding objective involves a profit function that shows how a change in each relevant trait influences profitability (Goddard, 1998). This profit function is often based on a bioeconomic model of the farm and obviously depends on the prices the farmer receives for milk and other products, and the prices paid for inputs (Groen et al., 1997). Typical illustrations can be found in Visscher et al. (1994) for pasture-based dairy farming in Australia, or Steine et al. (2008) for Norwegian Red cattle. Other required characteristics are the genetic parameters, the phenotypic means and the age structure of the herd at demographic equilibrium. When the performance

level for one trait is modified by one unit under the specified constraints, a new equilibrium is reached and the economic efficiency of the herd changes. The economic weight of this trait is the monetary difference between the two situations, or mathematically, the value of the partial derivative of the profit function with respect to the trait. This weight will be used in the construction of a total merit index (TMI), i.e. a linear combination of estimated breeding values (EBVs) that will serve as selection criterion to generate genetic progress on the breeding objective.

When future economic scenarios are too vague or when the economic impact of some traits is too difficult to determine, it may be preferred to derive a breeding goal that induces genetic gain in a direction of general consensus (Olesen et al., 1999). One approach involves finding weights for the traits in the breeding objective that lead to desired or restricted genetic gains. This is also a way to incorporate farmer or consumer opinion. For example, continuous decline in fertility or resistance to mastitis may be regarded as no longer admissible, while solely economic consideration would tolerate the deterioration. On the other hand, constraints or restrictions must be included with care because they can have a strong negative impact on overall benefits. A common practice is a two-step approach where a bioeconomic model is first developed to derive reference weights for the traits in the breeding goal. Expected genetic gains under a typical value for selection intensity are computed and then relative weights are empirically modified to get a more acceptable response, while controlling its overall cost compared with the initial situation.

In practice, the economic weight for a given trait depends on the other traits that are included in the profit function. For instance, if feed intake is not included in the profit function, the economic weight for cow body weight is positive because increasing body weight increases income from sale of cull cows. However, if feed intake is included in the profit function, but not as a measured trait, the economic weight of body weight may be negative because larger cows have greater feed requirements for maintenance.

National breeding programmes are often compared by contrasting the (relative) economic weights of the traits of interest (Miglior et al., 2005). However, such a comparison may be misleading (Cunningham and Taubert, 2009; Ducrocq, 2010). First, the relative weights may be attached to traits expressed in different units, such as phenotypic, genetic or even average EBV standard deviations. Furthermore, traits are not independent: for example, productive life and fertility are genetically correlated, and the weight ascribed to productive life differs strongly when the cost of culling due to sterility is assigned to productive life, or instead, to fertility, even though the expected responses are similar. When some traits receive a negative weight, the meaning of a relative weight assuming a sum of 100% is questionable. Finally, when the average reliabilities of the EBVs included in TMI vary a lot, traits with high EBV variability may contribute more to the overall ranking of animals than their economic weight indicates. This is clearly the case for production compared to fertility and disease traits.

Recent evolutions in breeding objectives

For decades, breeding goals in dairy cattle included few traits worldwide. These were mainly production and type traits, with a strong emphasis on production. Exceptions included the Scandinavian countries, with far-sighted focus on udder health, and fertility and dual-purpose breeds for which growth and beef traits were also valued. Hence in most countries, *functional traits*, i.e. traits related to the ability to remain productive, fertile and in good health with minimum human intervention were basically ignored in breeding programmes, except indirectly through some morphological (type) predictors. As a result, the overall robustness of dairy cows has been decreasing along with the continuous and successful increase in performance for those production traits under selection. This was the consequence of the nearly universal negative genetic correlation between production and fitness (Jorjani, 2007). Functional traits are difficult to select because of their low heritability but they often have large genetic variability and, therefore,

they can also easily deteriorate. Attention towards more sustainable breeding schemes has increased tremendously over the past 20 years, following the path paved by Scandinavian countries. So breeding objectives are now broader, more complex but also more balanced in many countries. Nowadays, the relative weight given to production in breeding objectives is generally between 25 and 50%, and functional traits receive larger attention, in order to improve long term sustainability of dairy production.

Traits to consider

The yields of milk, fat and protein are the major determinants of income to dairy farmers and the most important traits in the objective. Their relative economic weights depend on the pricing formula by which farmers are paid. If the milk is used for manufacturing, protein is generally most valuable, fat has some value, but milk volume has a negative value because it must be transported from farm to factory and evaporated to make particular products.

Other traits commonly included in breeding objectives are health (in particular udder health), fertility, calving ease, body weight, milking speed, temperament and length of productive life.

Resistance to mastitis is the health trait of major concern in dairy cattle. It represents the ability to avoid udder infection or to quickly recover after infection. In some cases, resistance to a particular pathogen is considered but the latter is usually unknown. Negative economic consequences of a mastitis event are numerous: lower milk production, discarded milk because of the presence of antibiotics or inadequate composition, lower milk payment, increased veterinary and labour costs, and increased cow replacement. Indeed, other health traits (lameness, metabolic or reproductive disorders) share most of these negative impacts.

Cow fertility influences AI and veterinary costs, the interval between calvings and hence the pattern and yield from current and later lactations. In countries relying heavily on pasture (New Zealand, Ireland) or where male calves have a higher value and can be channelled towards meat production, fertility has always been an important trait. In contrast, in countries where dairy calves are of low value and where farmers can manage cows with long calving intervals so that those cows have long persistent lactations, the economic weight of cow fertility used to be low. However, even in such a situation, the degradation of fertility in Holsteins is an issue. This has led to an increase in the economic weight on fertility traits everywhere.

Calving ease is valuable because dystocia has potentially severe consequences on stillbirth, production, fertility and general health, leading to veterinary costs, extra labour costs, lost calves and cows, reduced milk yield and infertility. In dairy cattle, losses due to difficult calvings mostly occur at first calving. The economic weight depends heavily on the average incidence of dystocia. Calving ease is affected by the genetic merit of both the calf and its dam; therefore, selection needs to consider calving ease as a maternal trait (of the cow) and a direct trait of the calf.

Cows are culled when they are no longer economically or physically sustainable. Length of productive life (LPL) from first calving to culling can be seen as an overall measure of her ability to stay productive. If LPL is corrected for the major source of voluntary culling (production), the resulting functional longevity depicts her ability to elude involuntary culling related to fertility, health or workability problems. In most selected dairy breeds, the proportion of involuntary culling has been increasing and voluntary culling on production traits has been declining, leading to closer convergence between true and functional longevity.

Milking speed is of economic value because slow milkers increase the labour cost of milking. Good temperament, while it may be difficult to assign a monetary value to it, is valued highly by dairy farmers in Australia and New Zealand who milk large numbers of cows and want to avoid the disruption and danger caused by cows with poor temperament.

Genetic Variation

Genetic parameters quantify the rate of genetic change that it is possible to achieve.

They are required for estimation of genetic merit. Of these parameters, the heritability describes what portion of the variation (variance) in a trait is of genetic origin, and correlations among these traits indicate how genetic change in one trait can affect the others. When multiple traits are evaluated, covariances indicate to what degree the information from one trait influences the others. If an animal has more than one observation for a trait, the repeatability describes the expected similarity among those observations. Other genetic parameters include the effects of dominance, individual genes, breed, inbreeding and heterosis (crossbreeding).

Breed differences

The world's dairy cattle include *Bos taurus* and *Bos indicus* breeds. The *B. taurus* cattle are dominant in temperate regions and are noted for high production. The *B. indicus* are prevalent in hotter climates and subsistence farming. The breeds of the *B. taurus* population mostly arose in Europe. Globally, most animals are purebred but crossbreeding programmes have been proposed as a way of upgrading indigenous cattle to a high producing breed, or as a way to obtain the benefits of complementarity and heterosis.

Registry organizations maintain pedigree records, which enable animals to be traced to the origin of the breed, or its importation. With globalization, selection goals around the world have converged, as have the technologies to support high yields, particularly in temperate regions. In those environments, the Holstein breed has become dominant because if its high yield per cow. The Jersey has emerged as the primary alternative dairy breed, because of high component yields and smaller size, along with a collection of so-called Red breeds. Other breeds, in particular dual-purpose breeds, have regional importance, such as the Simmental/Montbéliarde breed(s) in Europe. Table 15.1 illustrates the differences in yields for the most common dairy breeds. A more complete overview can be obtained from the International Committee for Animal Recording (ICAR, 2013).

Table 15.1. 305-day lactation averages by breed and country in 2011. (From www.icar.org.)

Breed	Country	Cows (10^6)	Milk (kg)	Fat (%)	Protein (%)
Holstein	Canada	0.72	9,975	3.79	3.19
	France	1.72	7,873	3.97	3.38
	USA	3.82	10,607	3.66	3.07
Friesian	New Zealand	0.96	5,600	4.22	3.50
Simmental	Germany	0.89	6,922	4.11	3.48
Brown Swiss	Germany	0.17	7,002	4.22	3.59
Jersey	USA	0.23	7,626	4.75	3.63
	New Zealand	0.35	3,946	5.52	4.00

Within-breed variation

Yield traits

Milk, fat and protein yields are usually defined in a standard manner representing production in kilogrammes during the 305 days (10 months) following calving. Fat and protein concentrations derived from yields are also of interest because they often condition milk price. In practice, 305 day yields are obtained from periodic (most often monthly) measurements of daily production or 'test-day' yields.

For simplification, lactation yields are often regarded as repeated measurements of the same genetic trait. Genetic correlations among successive lactations provide an indication of the appropriateness of this assumption. Indeed, these correlations are high (>0.85 between first and later lactations, close to 1 between later lactations, e.g. Druet et al., 2005). One reason for the lower correlation with first and later lactations compared to between later lactations is that cows reach their mature production level at different rates.

Genetic parameters for lactation yields are remarkably similar across countries, with heritabilities from 0.25 to 0.35 for yields, with lower values in extensive or harsh environments; repeatabilities of 0.50 to 0.60; and much higher heritabilities (at least 0.50) for fat and protein concentrations. Genetic correlations are high between lactation yields. A review of 22 studies in Holstein in different countries over the past 20 years indicates correlations of 0.62 ± 0.17 between milk and fat yields, 0.84 ± 0.14 between milk and protein yields, 0.72 ± 0.12 between fat and protein yields and 0.48 ± 0.25 between fat and protein concentrations in the first lactation, with similar values in later lactations. Dominance variation – due to interactions among genes at a specific locus – is most often ignored but can reach 5% of the total variance (Van Tassell et al., 2000).

Test-day yields are measurements specific to a particular testing day, with such tests usually being distributed over the whole lactation. Longitudinal analyses of such data are particularly interesting compared to analysis of whole lactation yields because they allow a more precise description of how genetic and non-genetic factors affect production over the lactation. For example, the specific effect of a herd on a given test day can be accounted for, and the effects of month or age at calving or stage of gestation and the additive genetic merit of the animal on the level *and* shape of the lactation can be accurately modelled. Such test-day models have gained popularity since the mid-1990s, as special cases of random regression models. Heritability estimates of test-day production are typically lower at the beginning and end of the lactation but can be high in mid lactation. Genetic correlations between production traits at different stages of lactation are usually high to very high (close to 1), except for the beginning and the end of lactation. As a result, the overall heritability estimates over the lactation are definitely higher (up to 0.50) than when the total 305 day lactation yields are directly analysed. Persistency, which describes how steeply the production decreases during the lactation, and maturity, which expresses how production evolves between first and later lactations can be specifically evaluated using test-day models.

Their heritability is generally low (e.g. 0.09–0.16 for milk persistency according to Jakobsen et al., 2002).

Conformation traits

Visual appraisals of cows for conformation (also known as type) traits have been collected for many years. Improving type traits has been advocated as a way to improve fitness, longevity and workability. This view has been altered in the recent past: size (or height) is receiving considerable attention worldwide in most breeds, while its relationship with fitness is often uncertain, and in some production systems, clearly unfavourable (Pryce et al., 2009). It is now well established that dairy character – or dairy form, dairyness or angularity – is unfavourably associated with body condition score, fertility and mastitis resistance (Lassen et al., 2003). Indeed, the objectives of elite breeders regarding type traits often diverge from those of most commercial dairymen. Traits such as angularity, body condition score (Pryce et al., 2001), body depth or rump angle are useful, but mainly as predictors of poor fertility. In contrast, udder traits have unambiguous beneficial impact on functional longevity, resistance to mastitis and milking speed. Udder depth is certainly the most important udder trait in that respect, together with fore and rear attachment, suspensory ligament and teat length and placement. A few feet and leg traits (rear leg set, foot angle, locomotion) are routinely collected and evaluated nationally and internationally, but generally have low heritability and a disappointingly low correlation with, for example, actual longevity. Some countries are now investigating other relevant traits better related to lameness and longevity, such as information on claw disorders collected by hoof trimmers (e.g. Van der Linde et al., 2010). Other traits such as muscling are recorded in dual-purpose breeds (Simmental, Montbéliarde, Normande).

Conformation traits are most often scored on linear scales, e.g. on a scale from 1 to 9. Heritability is usually relatively low (0.05–0.20) for feet and leg traits, moderate (0.20–0.35) for udder traits and moderate to high (0.25–0.60) for traits related to size (Interbull, 2013). Cows usually get a final score to summarize overall conformation. The final score is a combination of scores characterizing udder, body or feet and legs quality. Because genetic parameters vary between type traits as well as the weights used to combine them into a final score, composite indices combining genetic merit of the elementary traits in a formal way are preferable to direct evaluation of final scores.

Workability traits

Workability traits include milking speed and temperament. They are often recorded at the same time as type traits, on linear scales (e.g. 1 to 5 in a within herd comparison) or with actual measure (milk flow). Except in the latter case where larger estimates were found, heritability estimates are moderate (0.20–0.25) for milking speed and low (0.10) for temperament.

Calving traits

Birth weight is seldom recorded in dairy herds, whereas dystocia is commonly recorded as a calving code (e.g. 1 = no assistance, 2 = easy pull, 3 = hard pull, 4 = caesarean). Stillbirth is recorded as an all or none trait (alive or dead within 24 or 48 h after birth). Calving traits are under the influence of the genetic and non-genetic characteristics of both the calf (direct effect, ease of birth) and its dam (maternal effect, ease of calving). Heritabilities are usually quite low (<0.10), especially when adult cows or maternal effects are considered (Interbull, 2013).

Fertility traits

Female fertility has been neglected in breeding programmes for decades. As a result, it has been notably compromised by intensive selection for production. Initially, female fertility traits were limited to crude measures such as calving intervals or days open, which can be directly extracted from milk recording data. However, fertility is a composite phenotype that can be broken down into various basic traits requiring joint analysis of insemination and calving data. Records corresponding to natural services are usually ignored in analysis of fertility data. Most fertility traits are considered as genetically different between heifers and adult cows, the latter being challenged by concomitant production. Jorjani (2007) classified female fertility traits into four groups: ability to conceive (non-return rates, conception rate, number of inseminations) for heifers; ability to conceive for adult cows; ability to recycle after calving (interval from calving to first AI); and interval measures of ability to conceive (interval from first to successful (or last) AI). Calving intervals and days open are pooled measures of these abilities. Gestation length is moderately heritable but does not vary much within breed and is rarely considered in breeding programmes.

Health traits

Milk samples collected to determine fat and protein content are also analysed for somatic cell counts (SCCs), which, after a normalizing transformation, become somatic cell scores (SCSs), an indicator of udder health (Ali and Shook, 1980). High SCSs are associated with clinical or sub-clinical mastitis and depressed milk yield. Scandinavian countries have a long history of systematic disease data collection (Aamand, 2006). In particular, the actual occurrence of clinical mastitis is routinely recorded. More countries are following this track, especially in Europe (Austria, France). Other health traits include feet and legs, reproductive or digestive disorders (e.g. Egger-Danner et al., 2012). Those health traits are characterized by a low incidence, a low heritability but a large genetic variance.

Longevity

A typical measure of longevity is length of productive life (LPL), defined as the number of days between first calving and culling. For cows still alive, only the current LPL (i.e. a lower bound of their 'final' LPL) is known: the observation is said to be censored. Another characteristic of LPL measures is that environmental factors influencing risk of being culled (season, parity, herd size, etc.) are changing at the same time as LPL is measured. Any statistical analysis of LPL should take these features into account. It is also possible to predict LPL of cows still in the herd so they can be analysed with a standard linear model. An alternative simplified trait is survival (0/1) to the next lactation, also called stayability. Heritability estimates for longevity are around 0.10 or less. Because of its low heritability and its relatively late availability, LPL information is generally combined with early predictors such as type traits or SCS to improve longevity evaluations.

Correlation between trait groups

Table 15.2 reports genetic correlation estimates between traits included in the breeding goal in France for the Holstein breed. Udder health traits (somatic cell count and clinical mastitis) are strongly correlated (0.70) but clearly correspond to distinct traits, themselves related to udder conformation traits. Fast milking Holstein cows have higher SCCs. The situation is reversed in other breeds such as the Montbéliarde, where selection for milking ease

Table 15.2. Estimated genetic correlations between some traits included in the Holstein total merit index in France[a].

Trait name and abbreviation		Milk yield	SCC	CIM	CRate	IC-1AI	MEase	UddD
Somatic cell count	SCC	*[b]						
Clinical mastitis	CIM	−0.26	0.70					
Conception rate	CRate	−0.22	0.25	0.24				
Interval calving – 1st AI	IC-1AI[c]	−0.42	*	0.23	0.16			
Milking ease	MEase	*	−0.37	−0.18	*	*		
Udder depth	UddD	−0.22	0.27	0.30	0.15	*	0.28	
Functional longevity	FLong	0.17	0.48	0.47	0.47	*	0.17	0.41

[a]Trait scales are transformed: positive values indicate favourable values, e.g. positive SCC means lower SCC.
[b]Absolute value of genetic correlation less than 0.15.
[c]Interval between calving and first insemination.

has not been as strong. Ability to conceive and ability to recycle are two poorly correlated fertility traits. Functional longevity exhibits a relatively high genetic correlation (close to 0.5) with a number of functional traits related to udder health (somatic cell count, clinical mastitis, udder depth) and fertility (conception rate).

Inbreeding, genetic variability and heterosis

An animal is inbred if its parents are related. The inbreeding coefficient is the probability that an animal receives from both parents the same ancestral copy of any particular allele or chromosome fragment. The intense selection of bulls in most breeds, each with (tens of) thousands of daughters, and the use of a reduced number of sires of sons at the international level have led to a continuous increase in inbreeding. The use of close family information in genetic evaluation tends to further increase inbreeding because the consideration of all relationships tends to make the evaluations of family members similar, i.e. more likely to be selected together.

Systematic calculation of inbreeding relative to a base population that is assumed unrelated and non-inbred is feasible in very large populations. Inbreeding can be strongly underestimated when pedigree data is incomplete, or pedigree depth, i.e. the equivalent number of generations of known parents is low, but methods have been proposed to account for missing ancestors (VanRaden, 1992). Other measures of genetic variability less sensitive to missing data exist. They are related to the probability of gene origin, e.g. the effective number of founders or ancestors, or the number of ancestors accounting for 50% of the genes (Boichard et al., 1997). They show that actual population size is not at all representative of genetic variability. This is demonstrated by some values obtained in Holstein in France: 8 bulls contributed 50% of the genes in females born between 2004 and 2007 and the effective number of ancestors was 21 (Danchin-Burge et al., 2012). This situation is observed in all Holstein populations. The evolution of inbreeding for the Holstein population in the USA is shown in Fig. 15.1. The base population consisted of animals born before 1960. For 20 years until 1980, inbreeding increased slowly at about 0.044%/year. For the next 15 years it rose rapidly at 0.275%/year. More recently, during the period from 2000, the rate of increase has decreased to 0.11%/year.

A consequence of receiving the same genes identical by descent from both ancestors

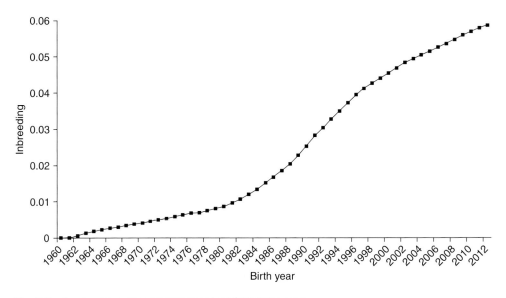

Fig. 15.1. Average inbreeding by birth year for US Holstein cows.

is that the likelihood of homozygosity among recessive alleles increases. Homozygosity can lead to embryo failure, Mendelian diseases and decreased productivity related to inbreeding depression (Wiggans et al., 1995). Table 15.3 presents estimates of inbreeding depression for Holstein cows in the USA.

Heterosis can be viewed as the opposite of inbreeding depression, and results from an increase in heterozygosity, reducing the likelihood of deleterious homozygous recessive genes. Heterosis measures the degree that offspring exceed the average of the performance of their parents, the magnitude of which depends on the genetic distance between the parents. Some estimates of heterosis are presented in Table 15.3. Heterosis is most apparent in breed crosses. If parental breeds are quite different in the trait of interest, the benefit of heterosis is unlikely to make the progeny competitive with the higher producing parental breed. However, heterosis may contribute a significant advantage in fitness. In New Zealand where most milk is used in manufacturing, cows from the Jersey breed (less milk, but high in fat) managed at higher stocking rates than their larger Holstein counterparts (more milk, but less fat) are perceived as financially competitive on a per hectare basis, and the progeny of crosses between these breeds are also highly regarded. In the USA, concern about reduced fertility and survival of Holstein cows is leading some farmers to crossbreed with Montbéliarde, Normande and Swedish Red or Norwegian Red bulls to improve fitness traits (Heins and Hansen, 2012).

Heterosis and a related concept – recombination loss – have been of interest in Europe since the introduction of Holstein semen and embryos from North America. Crossbred progeny were backcrossed to North American bulls, so the performance of generations beyond the F1 was important. If epistatic gene combinations in the parent breeds have a positive effect on yield, often breaking these combinations up in subsequent generations will reduce yield. The loss of these epistatic effects is called recombination loss. Estimates of recombinant loss are usually negative, meaning that the segregating generations perform worse than expected from the performance of the parent breeds and the F1. Although epistasis is a possible cause of these results, there are also other possible explanations, e.g. preferential treatment of the F1.

QTLs and individual genes affecting traits of economic importance

The traits considered for genetic evaluation are generally quantitative, affected by many genes as well as environmental factors. However, for many traits, there are some individual genes with a moderate effect that are worth studying more precisely: when these genes are identified, it increases our understanding of the mechanism of genetic control and this knowledge may be useful in improving accuracy of breeding value predictions.

In the 1990s, a lot of work was devoted to detection of quantitative trait loci (QTLs), i.e. polymorphic chromosomal segments that have an impact on quantitative traits. Mapping QTLs by linkage to genetic markers chosen to represent sparse coverage of the entire genome was a first step towards the identification of actual genes affecting traits. It was also a prerequisite to the implementation of marker

Table 15.3. Estimates of inbreeding depression and heterosis in US Holstein cattle.

Statistic	Milk yield (kg)	Fat yield (kg)	Protein yield (kg)	Somatic cell score	Daughter pregnancy rate (%)	Productive life (months)
Inbreeding depression[a]	−30	−1.1	−0.9	0.0045	−0.071	−0.27
Heterosis[b]	205	12	8	0.010	1.5	0.0

[a]Effect per 1% inbreeding.
[b]Effect for animals with 100% heterosis, i.e. in F1 crosses.

assisted selection (MAS). As for the more recent approach known as genomic selection, which will be covered later, QTL detection requires a reference population with phenotypic records and associated marker genotypes. Special designs were proposed to increase the power of QTL detection, in particular the daughter and grand-daughter designs (Weller et al., 1990), which are well suited to dairy cattle. The grand-daughter design traces the transmission of genetic markers in families comprising a sire and his progeny-tested sons, who all have accurate phenotypes (average adjusted performances of a large number of daughters). Initially, microsatellites were used as markers because they are highly polymorphic. However, microsatellite genotyping was rather expensive and QTL detection studies were typically restricted to the use of only dozens or hundreds of microsatellites per animal. With the availability of high density marker coverage, the prospects of detecting QTLs have improved considerably (Weller et al., 2013).

Large QTL detection programmes were implemented, in particular in the Holstein breeds (e.g. Ashwell et al., 2001; Spelman et al., 2001; Boichard et al., 2003). Dozens of QTLs were discovered (see Khatkar et al., 2004, for a review). Unfortunately, they usually explained at most a small percentage of genetic variance, and the confidence interval of their location remained large (>10 centimorgans). Nevertheless, these discoveries were used in some MAS programmes, in particular in New Zealand (Spelman, 2002) and in France, to pre-select young candidates before progeny testing. These programmes were complex and costly because association between alleles at the marker and at the QTL had to be established within each sire family. In France, 70,000 animals were genotyped between 2001 and 2008 for 14 chromosome regions traced by 45 microsatellites markers (Boichard et al., 2012b). The efficiency of this programme was retrospectively shown to be essentially as expected (Guillaume et al., 2008), i.e. small compared with what can be achieved nowadays with denser sets of markers. In practice, MAS programmes have been replaced by genomic selection strategies.

With QTL detection based on microsatellite markers, finding the gene(s) involved and the causal mutation(s) was an extremely difficult and time-consuming task, unless good positional candidates were identified by comparative mapping. The most prominent discoveries were the ATP-binding cassette, subfamily G, member 2 gene (ABCG2), the acylCoA:diacyglycerol acyltransferase (DGAT1) and the growth hormone receptor (GHR) genes (Grisart et al., 2002; Winter et al., 2002; Blott et al., 2003; Cohen-Zinder et al., 2005) on chromosomes 6, 14 and 20, respectively, which have very significant effects on milk yield and/or composition in Holsteins.

QTL detection became much more effective with the development of assays that can be used to genotype large numbers of another class of genomic markers at low cost: the single nucleotide polymorphisms (SNPs) (Van Tassel et al., 2008). The much denser coverage of the whole genome with markers makes the linkage disequilibrium between SNPs and QTLs extend beyond families to breed level. As a typical example, a QTL affecting dystocia, conformation and economic merit was discovered on chromosome 18 (Cole et al., 2009). It appears to be related to calf size or birth weight and may be the result of longer gestation lengths.

Genetic Evaluation

The goal of a genetic evaluation system is to produce rankings of animals that will generate progress on a breeding objective when selection decisions are based on the rankings. Genetic evaluation systems have been developed over the past 60 years, progressively making more efficient use of national databases that are growing in nature and scope, while exploiting sophisticated statistical techniques and fast-growing computing power.

Evaluation models

Currently, animal models are used in all major dairy countries for most of the traits evaluated. In an animal model, the phenotype of a particular cow is described as a function of her own additive genetic merit, in contrast with sire models for which only the genetic contribution of her sire is considered. Animal and sire models are special cases of mixed linear models (Henderson, 1984) where effects such as environmental groupings are typically treated as fixed effects while genetic effects are considered as random. Best linear unbiased prediction (BLUP) is the favourite estimation procedure (Henderson, 1963, 1984). From a Bayesian perspective, this means that environmental effects are estimated only from the data, while genetic effects are estimated (or predicted) combining information from data with prior knowledge of pedigree relationships between animals. The BLUP approach most commonly used involves a system of so-called mixed model equations, which can be constructed and solved relatively easily. The solutions include estimates of fixed effects corrected for all other factors and EBVs for every cow or bull in the pedigree. The EBV of any individual is a function of the EBVs of its parents and its progeny as well as its own records (VanRaden and Wiggans, 1991). Because all equations are solved simultaneously, data from one animal influences EBVs of all its relatives and herdmates. Together with EBVs, reliabilities are also computed or approximated from the mixed model equations, the reliabilities reflecting the accuracy of each EBV.

There has been considerable work in the recent past to adapt evaluation models, methods and software to the specific characteristics of each trait evaluated. These adaptations are based on relaxation of the underlying assumptions of simple linear mixed model analyses. For example, evaluations for production traits now often account for heterogeneous residual variances across herds and years (Wiggans et al., 1991; Robert-Granié et al., 1999). Random regression models for test-day evaluations include sophisticated modelling of fixed effects describing the lactation curves and description of the correlation structure between breeding values at different stages of lactation and different lactations (Jamrozik et al., 1997). Such modelling allows for a better correction for environmental factors, provides a higher heritability and allows evaluation of underlying traits such as production persistency. Type traits are generally analysed altogether in multiple trait animal model evaluations, where the high correlations particularly within trait groups (udder traits, body traits, feet and legs traits) allow data from one trait to contribute to the accuracy of the evaluation of the others (Misztal et al., 1992). Genetic evaluations of discrete data such as calving scores have been implemented using threshold models (Gianola and Foulley, 1983), possibly accounting for heterogeneous residual variances (Ducrocq, 2000; Kizilkaya and Tempelman, 2005). Survival analysis, i.e. the methodology to account for censored observations and time-dependent explanatory variables (Kalbfleisch and Prentice, 2002), is routinely used for longevity evaluations (e.g. Ducrocq, 2005). However, analyses with threshold and survival models often use sire (or sire and maternal grand-sire) models because animal models are either computationally demanding or lead to other complications.

Genotype by environment interaction

An interaction between genome and environment (G × E) exists when the effect of genes is different in different environments. Then, the best animals in one environment for a given trait are not necessarily the best ones in another environment. To account for G × E interactions, the same trait measured under two different environments can be considered as two distinct traits and the genetic correlation between these trait-by-environment combinations can be estimated. When the correlation is less than 1, animals do not rank equally in each environment. Correlations among countries for protein yield are in Table 15.4 for some countries. The correlation is highest within each continent and for countries using similar evaluation models (lactation vs test-day) to analyse their data. The correlations are lowest between New Zealand, Australia and Ireland on the one hand and all other countries on the other hand (0.75–0.82). Hence, somewhat different genes are required for high performance in the grazing-based management systems predominant in those countries compared to confinement feeding and greater reliance on concentrate feeds in North America and Europe.

More extreme situations exist. For example, Ojango and Pollott (2002) found a genetic correlation of 0.49 for milk yield of Holstein

Table 15.4. Estimated sire standard deviations (bold on diagonal) and genetic correlations (below diagonal) considered in the international evaluation of protein yield for Holstein cattle in December 2012 for a subset of countries; sire standard deviation estimates reflect the scale after the standardizations applied by the individual countries and are expressed in kg (lb in USA). (For a complete table, see www.interbull.org.)

	DEU	DFS	FRA	USA	NZL	AUS	IRL
DEU	**17.70**						
DFS	0.90	**10.50**					
FRA	0.85	0.90	**18.64**				
USA	0.87	0.91	0.90	**19.68**			
NZL	0.75	0.75	0.75	0.75	**9.18**		
AUS	0.75	0.75	0.78	0.75	0.85	**10.66**	
IRL	0.75	0.75	0.82	0.75	0.85	0.85	**4.74**

DEU, Germany; DFS, Denmark–Finland–Sweden; FRA, France; USA, United States; NZL, New Zealand; AUS, Australia; IRL, Ireland.

cows in the UK and Kenya. In the extreme, G × E is a major concern when introducing high producing cattle into marginal environments to upgrade indigenous cattle. Native cattle are well adapted to harsh conditions and will survive in environments where the improved cattle do not. Thus the native cattle are superior in that environment and the import of foreign animals or semen will produce disappointing results.

Even if the genetic ranking of sires does not change across environments, a smaller or larger response in one environment is still indicative of an interaction. A typical example is a higher genetic variance for production traits in more intensive or specialized environments (Huquet et al., 2012).

National vs international evaluations

Each country has adapted its evaluation system to model the structure of its data. Some of the ways in which systems differ include trait definition, trait measurement or calculation, parameter estimates, definition of environmental groups, definition of base population through unknown parent groups, account of inbreeding and heterosis, and reporting scale for evaluations.

The extensive marketing of bull semen and embryos internationally has generated enormous interest in international comparison of bulls. Since 1994, the Interbull Centre (Interbull, 2013) in Uppsala, Sweden, has

combined bull evaluations from participating countries on all continents to generate rankings that include bulls from all those countries, reported on each country's own evaluation scale. This evaluation first involves the estimation of genetic correlations between performances in different countries (as in Table 15.4). These are less than 1 because of genotype by environment interactions and also because of differences in national evaluations. As a consequence, the average daughter performance of each bull is considered as describing a trait that is different in each country, and a multiple trait, across country evaluation (MACE; Schaeffer, 1994) is performed. Interbull had 32 members in 2013 and performs MACE evaluations three times a year. Six (groups of) international breeds are considered: Holstein (with 29 populations participating to the production evaluations), Red dairy breeds (14), Jersey (11), Simmental (11), Brown Swiss (10) and Guernsey (6). Evaluated traits are grouped into production, type, udder health, fertility, calving, longevity and workability traits. Not all countries and breeds are considered for each trait group, but there is a continuous trend towards the inclusion of more breed × country × trait combinations.

Total merit indexes

Optimal selection for a particular breeding goal supposes that EBVs on different traits are combined into a TMI (Philipsson *et al.*, 1994). But EBVs on important functional traits are often available only on males and with a satisfactory reliability obtained too late to be used efficiently at young ages. Early predictors can be added to increase this reliability, especially for low heritability traits, such as longevity. This is frequently done using selection index theory leading to weights that depend on the reliability of each EBV, and the genetic and residual correlations between traits. In practice, this step is often simplified and the same coefficients are used in all situations (VanRaden, 2001). Another strategy consists of approximating a BLUP multiple trait animal model evaluation (MTAM). MTAM has a number of desirable features: it effectively merges all information sources, properly accounting for residual correlations and differences in reliabilities. MTAM also prevents biases in genetic trends due to selection on correlated traits. A full-scale implementation of MTAM for all traits of interest is not feasible in large populations, in particular because genetic evaluation models greatly differ from one group of traits to the other (e.g. test-day models for production traits vs survival analysis for longevity). Good two-step approximations exist (Lassen *et al.*, 2007): first single trait evaluations are carried out and average performances corrected for all non-genetic effects and their associated weight are computed for all recorded cows. Then, all ingredients for calculation of TMI are available for all males and females and a simpler MTAM evaluation is easily implemented. The optimal weights of the resulting EBVs in TMI are simply the economic weights of the traits. Other essential outcomes are EBVs for each functional trait optimally combining direct and indirect information from early predictors (e.g. longevity or fertility).

Genetic trends

Genetic trends reflect the historical progress achieved. Average breeding value of cows by birth year is a common measure of genetic trend and indication of the success of a national breeding programme. However, the evaluation model and adjustments for age effects can affect trend estimates. Table 15.5 shows the trend in breeding values for yield traits of US Holstein cows born during two periods. A comparison of the two periods indicates changes in selection goals over time. Recent trends in milk and protein were lower than earlier trends, but more favourable progress was made in productive life, SCS and daughter pregnancy rate.

Across-breed analysis

Historically, genetic evaluation was done within breed to limit the complexity of genetic evaluations and avoid the need to consider heterosis and differences in scale between breeds.

Table 15.5. Genetic trend for US Holstein cows.

Trait	1993–1997		2003–2007	
	Annual trend	% of phenotypic mean	Annual trend	% of phenotypic mean
Milk (kg)	94	0.9	76	0.6
Fat (kg)	2.6	0.7	2.9	0.7
Protein (kg)	2.8	0.9	2.5	0.7
Productive life (months)	0.13	0.5	0.25	0.9
Somatic cell score[a]	0.01	0.3	−0.02	−0.6
Daughter pregnancy rate (%)	−0.13[b]	−0.6[b]	0.05	0.2

[a]Score obtained after a normalizing transformation of somatic cell count (SCC): SCS = log base 2 (SCC/100,000) + 3.
[b]For the period 1998–2002.

In countries such as New Zealand where cross-breeding has become popular because of benefits from heterosis and complementarity, there is interest in using information from crossbred daughters in evaluating bulls, and having genetic evaluations of the crossbreds themselves. New Zealand implemented an across-breed analysis (Garrick et al., 1997; Harris et al., 2006) with TMI reported across breed in terms of profit per unit dry matter consumed. The USA began an across-breed analysis in 2007 for yield and fitness traits (VanRaden et al., 2007). In the USA, evaluations are calculated on an all-breed base but converted to traditional within-breed genetic bases for publication. The effect of heterosis is subtracted from each trait in the all-breed model, but when evaluations of crossbred animals are converted to the pure breed evaluation, the heterosis expected when crossbreds are mated to purebreds is included in the predicted transmitting ability.

Genomic Selection

The genomic era for dairy cattle began in 2007 with the development of assays that can be used to genotype large numbers of SNPs at low cost (Matukumalli et al., 2011). Since then it has become possible to obtain genomic evaluations of adequate accuracy as soon as a DNA sample is processed (Meuwissen et al., 2001).

Principles and methods

Although SNPs are only biallelic (two states), their large number allows tracking the inheritance of short chromosomal segments. The BovineSNP50® BeadChip (Illumina, 2011) with 54,001 approximately evenly spaced SNPs was a major innovation and its adoption in dairy cattle was extremely fast. Genomic selection requires the definition of a reference population of animals for which both phenotypes and genotypes are available. Because male phenotypes (defined as average daughter performances corrected for all non-genetic effects estimated in classical genetic evaluations) are much more precise than individual cow

records, reference populations mainly comprise genotyped bulls with progeny test results. However, some countries, in particular the USA, also include cows with individual performance in their reference population, after a particular standardization of the phenotypes to buffer the impact of potential preferential treatment of bull dams (Wiggans et al., 2011a).

Genomic evaluation comprises the random multiple regression of average daughter performances of each bull in the reference population on its SNP genotypes expressed as the number of one of the alleles. The substitution effect of this allele summarizes the effect of the surrounding chromosomal segment. Mixed linear or non-linear models are assumed. Many genomic evaluation methods have been proposed (Meuwissen et al., 2001; Hayes et al., 2009a; VanRaden et al., 2009; Verbyla et al., 2009, Croiseau et al., 2011). They vary in the proportion of SNPs actually contributing to the phenotype (from a small fraction to all of them), the underlying distribution of the SNP effects (most often normal with constant or heterogeneous variance) and the estimation method (GBLUP, i.e. a genomic extension of BLUP, Bayesian methods relying on Markov chain Monte Carlo methods, methods adapted to situations when there are many more unknowns (SNP effects) than observations (phenotypes) and machine learning approaches). In France, selected SNPs are grouped into haplotypes of four to six SNPs that are included in a QTL-BLUP evaluation (Boichard et al., 2012b). Using haplotypes improves SNP informativeness by increasing linkage disequilibrium between each group of SNPs and the neighbouring chromosomal region. Comparisons between methods show a modest advantage to Bayesian methods in most cases. This advantage becomes more substantial for traits characterized by a small number of larger QTLs, such as fat content.

Figure 15.2 illustrates the SNP effects over the whole genome obtained for protein yield in the US reference population. This figure illustrates a general feature of most traits: many chromosomal regions are involved in determination of the genetic effect. This is consistent with the observed success of selection previously based on an infinitesimal model that assumed a very large number of genes

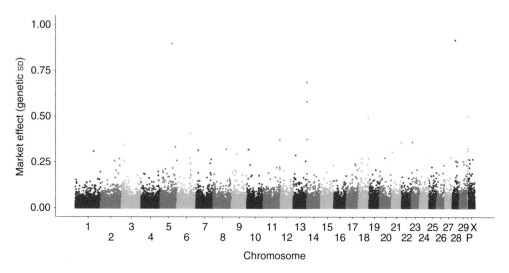

Fig. 15.2. Manhattan plot of estimated SNP effects for protein yield in the USA.

involved in a quantitative trait, each one with a small effect. Most genomic analyses also include a residual polygenic effect to account for that proportion of the additive genetic variance not explained by the SNPs.

Whatever the genomic evaluation approach, estimates of relevant SNPs or haplotypes can be used to predict the additive genetic value of any animal, e.g. at birth, without waiting for individual or progeny performances. In most cases, a final evaluation combines the sum of the SNP effects applied to the genotype with the residual polygenic effect and information blended from the classical genetic evaluation (VanRaden et al., 2009). The quality of these genomically enhanced breeding values (GEBVs, also known as genomic estimated breeding values) is usually assessed in a validation population of bulls by comparing their traditional parent average or GEBV computed when they were young, with their actual daughter performances a few years later. The increase in reliability above parent average due to the inclusion of genomic information varies greatly between traits (in absolute value, from 3% to 48% in Holstein according to Wiggans et al., 2011b). The gain is low for low heritability traits and maximum for fat content. For production traits, values of 25 to 30% are common in Holsteins. Other essential measures of the quality of GEBVs are the slope and intercept

of the regression of performance on prediction in the validation population. When these parameters deviate from 1 and 0, respectively, GEBVs of young bulls and EBVs of proven bulls are not consistent (Mäntysaari et al., 2011).

SNP chips

After the very successful BovineSNP50 BeadChip®, the Illumina company released in 2010 two additional genotyping chips: a low-density chip (Bovine3K®) with 2900 SNPs and a high-density chip (BovineHD®) with 777,962 SNPs. The Bovine3K reduced the cost of genotyping, thereby increasing application to females.

The BovineHD® chip (Matukumalli et al., 2009) has been used primarily for research, because only small increases in the accuracy of within-breed genomic evaluations have been reported (VanRaden et al., 2011b), insufficient to justify its more than twofold higher cost.

SNP chips have since become more diversified: in 2011, the Bovine3K was replaced by the BovineLD® chip (Boichard et al., 2012a) with 6909 SNPs, which uses the same Infinium® chemistry as the BovineSNP50®, with better performance than the GoldenGate® technology used in the Bovine3K. The BovineLD®

supports custom addition of up to 80,000 SNPs. Examples of custom chips include the 8000 SNP GGP-LD and 77,000 SNP GGP-HD from GeneSeek, and a 10,000 SNP chip from Eurogenomics, which all include individual gene tests as well as markers spread throughout the genome.

With this increasing variety of genotyping chips, it has been necessary to develop tools to simultaneously include all densities of genotypes in genomic evaluations. Provided SNP markers are spaced throughout the genome, it is possible to use statistical methods to predict (or 'impute') the missing markers, thus transforming lower density genotypes into higher density ones (e.g. LD into 50K or 50K into HD genotypes), albeit with some uncertainty. Imputation and phasing (i.e. the construction of haplotypes of paternal and maternal origin) basically use the same statistical approaches (Druet and George, 2010). The accuracy of imputation is often very good, but depends on the number of genotypes available at higher density, on the imputation method implemented and, for a particular animal, on the availability of genotypes of its parents, other direct ancestors and progeny. Imputation is also able to create the entire genotype of a non-genotyped animal if it has sufficient genotyped progeny (e.g. at least five).

Implementation

Genomic evaluation is particularly beneficial in dairy cattle because of the possibility to dramatically reduce the generation interval. Now, bulls can be used heavily at 1 or 2 years of age instead of waiting until 5 years when their progeny test information is available. The cost of genotyping is modest compared to the value of a dairy animal and is negligible compared to the cost of progeny testing. Furthermore, this cost has declined since genotyping first became available. This has resulted in the widespread genotyping of young calves (Table 15.6). Genotyping of females is used to select calves to be kept as replacements as well as the ones to be used as bull dams. Tens of thousands of animals are genotyped every month and most countries

with large dairy populations have a genomic evaluation system.

Before being included in genomic evaluations, the genotypes should be checked to determine if they are associated with the correct animal and if the individual SNP genotypes appear accurate. Because most sires have been genotyped, sire conflicts can be detected and the correct sire discovered in many cases. Conflicts may be due to sample ID error as well as errors in the pedigree. The quality of SNP genotypes is assessed by call rate, proportion of heterozygous calls, and parent–progeny conflicts on a SNP basis. Some SNPs are excluded because of low call rate, poor calling properties, high correlation with other SNPs, excessive parent–progeny conflicts or low minor allele frequency across all breeds of interest. Genotyping laboratories are able to improve quality by adjusting the clustering when making the genotype calls.

The gender of the animal can be confirmed from the SNPs on the Y chromosome if present on the chip, or the presence of heterozygous calls among the X-specific SNPs. Breed can be validated by checking SNPs that are usually homozygous for a specific allele in the designated breed and not in the other breeds.

Consequences on the international scene

The accuracy of genomic evaluations is determined primarily by the number of bulls with progeny test evaluations and genotypes included in the reference population (VanRaden et al., 2009; Liu et al., 2011). This has pushed countries to share the genotypes of their bulls to increase their numbers, in particular in Holsteins. In North America, the USA and Canada share the genotypes of both males and females and share bulls with Italy and the UK. In Europe, genotypes are shared in the EuroGenomics consortium, which includes various breeding organizations: Viking Genetics (Denmark/Finland/Sweden), UNCEIA (France), DHV and VIT (Germany), CRV (Netherlands/Belgium), Conafe (Spain) and Genomika Polska (Poland). These associations give each group over 20,000 Holstein bulls in their

Table 15.6. Key features of genomic selection programmes implemented in selected countries in 2012 (updated from Pryce and Daetwyler, 2011).

Feature	Australia	Ireland	NZ	France	Germany	Netherlands	Denmark–Sweden–Finland	USA–Canada
Year when genomic evaluation started	2008	2009	2008	2008	2009	2007	2008	2008
Year when genomic evaluation became official	2011	2009	2008	2009	2010	2010	2011	2009
Size of reference population (males; production traits)	4,364	5,000	5,503	25,000	25,050	24,504	25,000	20,822
Reliability (total merit index) (%)[a]	55	54	55	65	67	62	55–60	77
Reliability (protein yield) (%)[a]	65	61	55	65	73	68	63	72
Females included in reference population	13,851	Not yet	Not yet	Not yet	0	0	Not yet	34,008/0[b]
Number of young bulls genotyped per year	455	4,000	2,000	8,300	13,000	2,500	1,800	18,744
Number of bulls progeny-tested	271	50	200	0	400	140	175	2,000
Age at which young bulls are widely used (months)	16	24	14	16	15	18	17	12
Semen price relative to proven bulls	Same	Same	More	Less	Same	Same	Same	Same
Number of young genomically tested bulls in the top 20 bulls ranked on country's index	15	10	20	18	20	16	18	20
Market-share of genomically tested bulls (bulls without milking daughters) (%)	9	50	35	60	50	35–40	70	47

[a]Several methods exist for calculating the reliabilities of genomic breeding values; so in some cases, the reliabilities between countries are not directly comparable.
[b]The USA and Canada have essentially the same male reference population but Canada does not include females in the reference.

predictor populations and a substantial lift in terms of genomic reliabilities (Lund *et al.*, 2011). Other breeds took similar initiatives: the Red breeds in Scandinavia and the Brown Swiss, with eight countries in the Intergenomics consortium (Jorjani, 2011).

As for genetic evaluations, Interbull plays a role in making genomic evaluations comparable worldwide (Dürr and Philipsson, 2011). First of all, genomic evaluations must go through validation tests to be recognized at international level (Mäntysaari *et al.*, 2011). For the Intergenomics members, Interbull implemented a service to calculate genomic evaluations of Brown Swiss bulls. Interbull is also developing a genomic extension of MACE evaluations for young Holstein bulls from countries with domestic genomic evaluation systems.

Perspectives and challenges

Table 15.6 presents some key parameters in the Holstein breed for some of the countries using genomic selection. The countries mostly vary in reference population size with a large advantage to the two consortia, in inclusion of females in the reference population, in relative price paid for semen of young sires and in use of semen from young bulls. Reliabilities are variable even between countries with similar reference population size, in particular because there is no standard method to compute them. The USA and Australia are the only two countries to include females in their reference population. Some countries no longer have organized progeny testing programmes (e.g. France) or progeny test many fewer young bulls than before (e.g. Germany). But the most remarkable feature is the high (and fast-growing) market share of semen from young bulls with only genomic information.

Since 2010 in North America with the Bovine3K chip and 2011 in Europe with the BovineLD chip, genotyping of commercial females has become an affordable option at herd level and the number of genotyped cows is rapidly increasing worldwide. Benefits include better within-herd selection (especially when combined with use of sexed semen), management of replacement heifers, parentage

verification and increase of beef production through terminal crossbreeding for the below average cows. For potential buyers, access to GEBVs avoids unknown or biased information (due to preferential treatment) on the animals available in the market.

Currently, only widespread international breeds have the potential to create reference populations large enough for efficient genomic selection. In the USA and Canada, the Jersey, Brown Swiss and Ayrshire breeds also receive genomic evaluations, however, particularly for Ayrshires the gain in reliability is small for most traits. Smaller breeds must hope for across-breed evaluations to benefit from more accurate genomic evaluations. The popular BovineSNP50 BeadChip is unfortunately not dense enough (1 SNP for every 49,400 DNA bases) to guarantee that linkage disequilibrium between a marker and neighbouring QTL observed within breed will be conserved across breeds. Indeed, most across-breed genomic evaluation using the 50K chip gave disappointing results, except when crossbred animals were also available in the reference population (de Roos *et al.*, 2009; Hayes *et al.*, 2009b; Harris and Johnson, 2010). The higher density (1 SNP for every 3430 DNA bases) of the BovineHD chip is theoretically high enough to find identical-by-descent chromosomal segments across breeds. First attempts to use real or imputed HD genotypes for across-breed genomic evaluation have shown modest gains in accuracy (Erbe *et al.*, 2012; VanRaden *et al.*, 2013), but estimating the effects of conserved haplotypes rather than SNPs may give better results in the future.

The success of genomic selection introduces bias in classical genetic evaluations because the only animals that will have progeny with phenotypic records are those that were preselected at a young age based on their GEBV (Patry and Ducrocq, 2011). Their additive genetic value substantially deviates from the average of their parents. BLUP evaluations cannot account for this deviation unless the genomic information used in selection is suitably included in the evaluation. The computational challenges of making such a combined system with all pedigree, genomic and phenotypic information in large populations are considerable, but current progress in this direction is fast (Aguilar *et al.*, 2009; Christensen and Lund, 2010). Such so-called single step approaches offer a number of conceptual advantages making their development in the near future a priority.

Genomic information may be used to improve mating decisions. Discovery of haplotype segments that never occur in the homozygous state has led to the identification of abnormalities that cause early embryonic death (VanRaden *et al.*, 2011a). Mating allocation software using genomic information can avoid carrier by carrier matings.

As technology has improved, it has become affordable to obtain full sequence information on important bulls (1000 genome consortium, 2013). These data will enable discovery of millions of genetic variants. Research will in time associate these with diseases and variation in performance. As an increasing number of the causative variants are discovered, the accuracy of genomic evaluation should increase, along with improved likelihood of evaluation across breeds (Meuwissen and Goddard, 2010).

Design of Breeding Programmes

EBV and GEBV provide cattle breeders with tools for identifying the best bulls and cows for breeding. Obtaining maximum genetic gain through selection requires the design of efficient breeding programmes. Assuming that a fraction α of the candidates are selected in the top EBV list for a given trait, the theoretical annual genetic gain per year can be computed using the following formula (Rendel and Robertson, 1950):

$$\Delta G_y = (\iota \rho \sigma_G)/L \quad (15.1)$$

This involves four parameters: (i) the selection intensity (ι), a function of α, which measures the superiority of selected animals compared to all candidates to selection; (ii) the EBV accuracy (ρ); (iii) the generation interval (L), the average age of parents when their progeny are born; and (iv) the genetic standard deviation (σ_G) of the trait of interest. In dairy cattle, research into the design of breeding programmes has focused for decades on finding

the best combinations of these parameters, taking advantage of reproductive technology such as artificial insemination (AI) and multiple ovulation and embryo transfer (MOET).

AI and MOET

The availability of AI led to the development of breeding programmes based on progeny testing: a group of young bulls is progeny tested by producing a number of daughters each. When these daughters get recorded performances included in genetic evaluations, bulls selected for widespread use are mated with the best cows available to produce replacement heifers and a new generation of young bulls. Selection of cows to produce replacement heifers is also practised, but is of limited value because the low reproductive rate of cows means that a large proportion of each batch of new heifers is needed to maintain the herd size. Thus selection decisions concern four types of combinations: bulls to breed bulls (bb), bulls to breed cows (bc), cows to breed bulls (cb) and cows to breed cows (cc), but the selection intensity on the fourth pathway is low. The formula to compute annual genetic gain is then extended to the following (Rendel and Robertson, 1950):

$$\Delta G_y = \frac{\iota_{bb}\rho_{bb} + \iota_{bc}\rho_{bc} + \iota_{cb}\rho_{cb} + \iota_{cc}\rho_{cc}}{L_{bb} + L_{bc} + L_{cb} + L_{cc}} \sigma_G \quad (15.2)$$

Table 15.7 demonstrates such a calculation for a typical progeny test of the pre-genomic era. The parameters that have attracted the most attention in breeding programmes are the proportion of cows to be mated to young bulls, the number of young bulls progeny tested per year and the number of daughters per young bull. The optimum values of these parameters varied widely between studies, from 15% to almost 100% of cows mated to young bulls and 20 to 400 daughters per young bull. For a cow population of a given size, these parameters are used to determine the number of young bulls to be progeny tested (Dekkers *et al.*, 1996).

Different optima were found depending on the objective (maximum rate of genetic progress, maximum monetary outcome from the programme), the economic horizon, the population size and the breeding goal (inclusion of traits of low heritability or observed late in life). As the population size increased, the optimum proportion of matings to young bulls decreased, the number of daughters per bull increased, the number of bulls progeny tested increased and the rate of genetic gain increased. When economic benefit was the objective, 80 to 150 daughters per young bull was close to optimum, but in practice, these optima were relatively flat so there was little cost associated with departing slightly from the optimum value of a parameter.

When MOET became possible, new designs were proposed to increase genetic gain despite the associated extra cost. Nicholas and Smith (1989) suggested nucleus breeding herds with selection of bulls based on the performance of their sibs and older relatives. In MOET designs, the generation interval is reduced but at the expense of a less accurate selection than progeny testing.

Breeding programmes with genomic selection

The genomic revolution has imposed a complete revision of the previous golden standards and traditional breeding schemes are being disrupted. Genomic selection relies on the same feature as MOET schemes (reduced generation interval) but at a much lower global cost – especially when the genotyping of the reference population is financed through public funds and/or shared within consortia – and with only a modest decrease in reliability compared with progeny testing. Breeding parameters along the same four pathways are adapted to optimize breeding programmes. Schaeffer (2006) showed that under genomic selection, the annual genetic gain for the current breeding goals can be doubled even when keeping selection intensity unchanged: compared with a typical progeny testing scheme, the average generation interval can be more than halved and the higher reliability of dams of bulls compensates for the lower reliability of sires of bulls. Table 15.7 includes expected figures from Schaeffer (2006) for a typical breeding scheme fully based on GEBVs. His hypotheses may not all be realistic but they clearly illustrate the huge impact of the generation interval reduction on annual genetic gain. Yet changes induced by genomic selection are not limited to a faster global genetic gain. GEBVs of functional traits (often with low heritability) are available at birth with reliabilities comparable to traditional progeny testing without the requirement for large progeny groups. Furthermore, GEBVs of females and males have the same reliabilities because they are based on exactly the same sources of information: selection of dams of bulls or even dams of cows is possible with accuracy previously unachievable even for old cows with many daughters. As a consequence, genetic gains should become more balanced (Ducrocq, 2010). This is highly desirable because neither farmers nor processing

Table 15.7. An example of calculation of annual genetic gain for a breeding scheme only based on progeny test (upper line) or on genomic information (lower line, bold). (From Schaeffer, 2006.)

Pathway	Selected fraction (%)	Selection intensity (ι)	Selection accuracy (ρ)	Generation interval (L in years)	Annual genetic gain (ΔGy in genetic standard deviations)
Bulls to breed bulls	5	2.06	0.99	6.50	
			0.75	**1.75**	
Bulls to breed cows	20	1.40	0.75	6.00	
			0.75	**1.75**	
Cows to breed bulls	2	2.42	0.60	5.00	
			0.75	**2.00**	
Cows to breed cows	85	0.27	0.50	4.25	
					0.22
					0.47

plants can readily cope with much larger annual increases in production.

In contrast to progeny testing, genomic selection does not require exhaustive data collection. Specific reference (female) populations can be created where finer, i.e. detailed and more precise phenotypes can be collected in a reduced number of herds on new traits. Genomic selection offers new opportunities to include in breeding programmes selection on new traits related to production, such as detailed milk composition in fatty acids and protein measured using the mid-infrared spectra (Soyeurt *et al.*, 2006), health traits (feet and leg disorders, metabolic diseases), feed efficiency or even methane emission (Boichard and Brochard, 2012).

Minimizing inbreeding

Intense selection under progeny testing schemes implies a small number of parents for the next generation of bulls and in time this causes inbreeding. Small effective population size and inbreeding trigger inbreeding depression, increased incidence of recessive abnormalities, reduced genetic variation and random fluctuations in the mean of the population. Long-term selection requires maintenance of within-breed genetic variability. Therefore, a compromise needs to be found between minimizing inbreeding and maximizing genetic gain. Several categories of mating plans exist to limit the rate of inbreeding (Sonneson and Meuwissen, 2000): factorial matings, compensatory matings, minimum coancestry matings and strict limitation of number of progeny per sire. In efficient breeding schemes, the best option is to calculate the optimum contribution of each selection candidate (Sonneson and Meuwissen, 2000), which maximizes genetic gain while limiting the rate of inbreeding in the progeny at a given level by restricting relationships between selected parents. A challenging question is then how to choose a suitable maximum rate of inbreeding. An alternative is to choose an acceptable reduction of genetic gain and organize matings to minimize the average coancestry of future animals (Colleau *et al.*, 2004). A reduction of about 20% of this

parameter can be obtained with very little loss in genetic gain. To be efficient, these approaches require a strong and centralized control of the planned matings, which may be difficult in practice.

At first glance, minimizing inbreeding under genomic selection appears more critical: with a reduction in generation interval, the same increase in rate of inbreeding per generation results in a faster increase per year. Indeed, Colleau (according to Boichard *et al.*, 2012) showed that in a scenario where half of the cow population was bred to young bulls preselected on their GEBVs and the other half was bred to the best 25% of these bulls returning to AI after the performance of their progeny becomes available, the annual rate of inbreeding was 69% higher than in the progeny test reference situation. However, when the whole cow population was bred to young bulls, nearly the same increase in genetic gain was observed (>80%), but with a decrease in rate of inbreeding (−23%). The reason is that many more bulls are used, each one contributing a few thousand inseminations over a short period of time before being replaced. In France, this has been implemented and has led to the complete end of planned progeny testing. Of course, genetic evaluations of older bulls based on progeny performances are (and will be) still available. Another trend that is beneficial to control inbreeding is the large increase in the number of sires of bulls with no damaging impact on genetic gain: only their best sons are eventually selected based on their genomic information.

Conclusions

Rapid progress in the genetic improvement of dairy cattle has been achieved and the fast adoption of genomic selection portends an even more rapid increase. In recent years, the focus has shifted somewhat from yields of milk and components, to fitness and fertility traits to better track the total economic value. The investment of producers in milk recording and artificial insemination organizations in genotyping a large number of young bulls each year have been important contributions to this success. Data collection is somewhat easier with dairy cattle than some other farm species because of the intensive nature of production and the relatively high value of the individual animals. This situation has led to a highly developed system of data collection, genotyping and genetic evaluation. Further developments in genomic evaluation methods, mating programmes and breeding plans hold promise for further increases in the rate of a more sustainable genetic improvement.

References

1000 genome consortium (2013) Available at: http://www.1000bullgenomes.com (accessed 13 March 2014).

Aamand, G.P. (2006) Use of health data in genetic evaluation and breeding. *EAAP Scientific Series* 121, 275–282.

Aguilar, I., Misztal, I., Johnson, D.L., Legarra, A., Tsuruta, S. and Lawlor, T.J. (2010) A unified approach to utilize phenotypic, full pedigree, and genomic information for genetic evaluation of Holstein final score. *Journal of Dairy Science* 93, 743–752.

Ali, A.K.A. and Shook, G.E. (1980) An optimum transformation for somatic cell concentration in milk. *Journal of Dairy Science* 62, 487–490.

Ashwell, M.S., Van Tassell, C.P. and Sonstegard, T.S. (2001) A genome scan to identify quantitative trait loci affecting economically important traits in a US Holstein population. *Journal of Dairy Science* 84, 2535–2542.

Blott, S., Kim, J.J., Moisio, S., Schmidt-Küntzel, A., Cornet, A., Berzi, P., Cambiaso, N., Ford, C., Grisart, B., Johnson, D. *et al.* (2003) Molecular dissection of a quantitative trait locus: a phenylalanine-to-tyrosine substitution in the transmembrane domain of the bovine growth hormone receptor is associated with a major effect on milk yield and composition. *Genetics* 163, 253–266.

Boichard, D. and Brochard, M. (2012) New phenotypes for new breeding goals in dairy cattle. *animal* 6, 544–550.

Boichard, D., Maignel, L. and Verrier, E. (1997) Value of using probabilities of gene origin to measure genetic variabilitiy in a population. *Genetics Selection Evolution* 29, 5–23.

Boichard, D., Grohs, C., Bourgeois, F., Cerqueira, F. Faugeras, R., Neau, A., Rupp, R., Amigues, Y., Boscher, M.Y. and Leveziel, H. (2003) Detection of genes influencing economic traits in three French dairy cattle breeds. *Genetics Selection Evolution* 35, 77–101.

Boichard, D., Chung, H., Dassonneville, R., David, X., Eggen, A., Fritz, S., Gietzen, K.J., Hayes, B.J., Lawley, C.T., Sonstegard, T.S. *et al.* (2012a) Design of a bovine low-density SNP array optimized for imputation. *PLoS ONE* 7, e34130.

Boichard, D., Guillaume, F., Baur, A., Croiseau, P., Rossignol, M.N., Boscher, M.Y., Druet, T., Genestout, L., Colleau, J.J., Journaux, L., Ducrocq, V. and Fritz, S. (2012b) Genomic selection in French dairy cattle. *Animal Production Science* 52, 115–120.

Christensen, O.F. and Lund, M.S. (2010) Genomic prediction when some animals are not genotyped. *Genetics Selection Evolution* 42, 2.

Cohen-Zinder, M., Seroussi, E., Larkin, D.M., Loor, J.J., Everts-van der Wind, A. and Lee, J.-H. (2005) Identification of a missense mutation in the bovine ABCG2 gene with a major effect on the QTL on chromosome 6 affecting milk yield and composition in Holstein cattle. *Genome Research* 15, 936–944.

Cole, J.B., VanRaden, P.M., O'Connell, J.R., Van Tassell, C.P., Sonstegard, T.S., Schnabel, R.D., Taylor, J.F. and Wiggans, G.R. (2009) Distribution and location of genetic effects for dairy traits. *Journal of Dairy Science* 92, 2931–2946.

Colleau, J.J., Moureaux, S., Briend, M. and Béchu, J. (2004) A method for the dynamic management of genetic variability in dairy cattle. *Genetics Selection Evolution* 36, 373–394.

Croiseau, P., Legarra, A., Guillaume, F., Fritz, S., Baur, A., Colombani, C., Robert-Granié, C., Boichard, D. and Ducrocq, V. (2011) Fine tuning genomic evaluations in dairy cattle through SNP pre-selection with Elastic-Net algorithm. *Genetic Research* 93, 409–417.

Cunningham, E.P. and Täubert, H. (2009) Measuring the effect of change in selection indices. *Journal of Dairy Science* 92, 6192–6196.

Danchin-Burge, C., Leroy, G., Brochard, M., Moureaux, S. and Verrier, E. (2012) Evolution of genetic variability of eight French dairy cattle breeds assessed by pedigree analysis. *Journal of Animal Breeding and Genetics* 129, 206–217.

De Roos, A.P.W., Hayes, B.J. and Goddard, M.E. (2009) Reliability of genomic predictions across multiple populations. *Genetics* 183, 1545–1553.

Dekkers, J.C.M., Vandervoort, G.E. and Burnside, E.B. (1996) Optimal size of progeny groups for progeny-testing programs by artificial insemination firms. *Journal of Dairy Science* 79, 2056–2070.

Druet, T. and George, M. (2010) A hidden Markov model combining linkage and linkage disequilibrium information for haplotype reconstruction and quantitative trait locus fine mapping. *Genetics* 184, 789–798.

Druet, T., Jaffrezic, F. and Ducrocq, V. (2005) Estimation of genetic parameters for test day records of dairy traits in the first three lactations. *Genetics Selection Evolution* 37, 257–271.

Ducrocq, V. (2000) Calving ease evaluation of French dairy bulls with a heteroskedastic threshold model with direct and maternal effects. *Interbull Bulletin* 25, 123–130.

Ducrocq, V. (2005) An improved model for the French genetic evaluation of dairy bulls on length of productive life of their daughters. *Animal Science* 80, 249–256.

Ducrocq, V. (2010) Sustainable dairy cattle breeding: illusion or reality? *Proceedings of the 9th World Congress on Genetics Applied to Livestock Production.* Leipzig, Germany, 1–6 August, Communication 66.

Dürr, J. and Philipsson, J. (2011) International cooperation: the pathway for cattle genomics. *Animal Frontiers* 2, 16–21.

Egger-Danner, C., Fuerst-Waltl, B., Obritzhauser, W., Fürst, C., Schwarzenbacher, H., Grassauer, B., Mayerhofer, M. and Koeck, A. (2012) Recording of direct health traits in Austria – Experience report with emphasis on aspects of availability for breeding purposes. *Journal of Dairy Science* 95, 2765–2777.

Erbe, M., Hayes, B.J., Matukumalli, L.K., Goswami, S., Bowman, P.J., Reich, C.M., Mason, B.A. and Goddard, M.E. (2012) Improving accuracy of genomic predictions within and between dairy cattle breeds with imputed high-density single nucleotide polymorphism panels. *Journal of Dairy Science* 95, 4114–4129.

Garrick, D.J., Harris, B.L. and Johnson, D.L. (1997) The across-breed evaluation of dairy cattle in New Zealand. *Proceedings of the Association for the Advancement of Animal Breeding and Genetics 12th Conference* 12, 611–615.

Gianola, D. and Foulley, J.L. (1983) Sire evaluation for ordered categorical data with a threshold model. *Genetics Selection Evolution* 15, 201–224.

Goddard, M.E. (1998) Consensus and debate in the definition of breeding objectives. *Journal of Dairy Science* 81, 6–16.

Grisart, B., Coppieters, W., Farnir, F., Karim, L., Ford, C., Berzi, P., Cambisano, N., Mni, M., Reid, S., Simon, P. *et al.* (2002) Positional candidate cloning of a QTL in dairy cattle: identification of a missense mutation in the bovine DGAT1 gene with major effect on milk yield and composition. *Genome Research* 12, 222–231.

Groen, A., Steine, T., Colleau, J., Pedersen, J., Pribyl, J. and Reinsch, N. (1997) Economic values in dairy cattle breeding, with special reference to functional traits – report of an EAAP working group (review). *Livestock Production Science* 49, 1–21.

Guillaume, F., Fritz, S., Boichard, D. and Druet, T. (2008) Correlations of marker-assisted breeding values with progeny-test breeding values for eight hundred ninety-nine French Holstein bulls. *Journal of Dairy Science* 91, 2520–2522.

Harris, B.L.and Johnson, D.L. (2010) Genomic predictions for New Zealand dairy bulls and integration with national genetic evaluation. *Journal of Dairy Science* 93, 1243–1252.

Harris, B.L., Winkelman, A.M., Johnson, D.L. and Montgomerie W.A. (2006) Development of a national production testday model for New Zealand. *Interbull Bulletin* 35, 27–32.

Hayes, B.J., Bowman, P.J., Chamberlain, A.J. and Goddard, M.E. (2009a) Genomic selection in dairy cattle: progress and challenges. *Journal of Dairy Science* 92, 433–443.

Hayes, B.J., Bowman, P.J., Chamberlain, A.C., Verbyla, K. and Goddard, M.E. (2009b) Accuracy of genomic breeding values in multi-breed dairy cattle populations. *Genetics Selection Evolution* 24, 41–51.

Heins, B.J. and Hansen, L.B. (2012) Fertility, somatic cell score, and production of Normande × Holstein, Montbéliarde × Holstein, and Scandinavian Red × Holstein crossbreds versus pure Holsteins during their first 5 lactations. *Journal of Dairy Science* 95, 918–924.

Henderson, C.R. (1963). Selection index and expected genetic advance. In: Hanson, W.D. and Robinson, H.F. (eds) *Statistical Genetics and Plant Breeding*, Publ. 982. National Academy of Science, National Research Council, Washington, DC, pp. 141–163.

Henderson, C.R. (1984) *Applications of Linear Models in Animal Breeding*. University of Guelph, Guelph, Ontario, Canada.

Huquet, B., Leclerc, H. and Ducrocq, V. (2012) Modelling and estimation of genotype by environment interactions for production traits in French dairy cattle. *Genetics Selection Evolution* 44, 35.

Illumina (2011) Data sheet: DNA analysis. BovineSNP50 genotyping BeadChip. Publ. No. 370-2007-029. Illumina Inc., San Diego, California.

International Committee for Animal Recording (2013) Available at: http://www.icar.org (accessed 13 March 2014).

Interbull (2013) Available at: http://www.interbull.org (accessed 13 March 2014).

Jakobsen, J.H., Madsen, P., Jensen, J., Pedersen, J., Christensen, L.G. and Sorensen, D.A. (2002) Genetic parameters for milk production and persistency for Danish Holsteins estimated in random regression models using REML. *Journal of Dairy Science* 85, 1607–1616.

Jamrozik, J., Schaeffer, L.R. and Dekkers, J.C.M. (1997) Genetic evaluation of dairy cattle using test day yields and random regression model. *Journal of Dairy Science* 80, 1217–1226.

Jorjani, H. (2007) International genetic evaluation of female fertility traits in five major breeds. *Interbull Bulletin* 37, 151.

Jorjani, H. (2011) Genomic evaluation of BSW populations, intergenomics: results and deliverables. *Interbull Bulletin* 43, 5–8.

Kalbfleisch, J.D. and Prentice, R.L. (2002) *The Statistical Analysis of Failure Time Data*, second edition. Wiley series in probability and statistics. John Wiley and Sons, New York, 462pp.

Khatkar, M.S., Thomson P.C., Tammen, I. and Raadsmaa, H.W. (2004) Quantitative trait loci mapping in dairy cattle: review and meta-analysis. *Genetics Selection Evolution* 36, 163–190.

Kizilkaya, K. and Tempelman, R.J. (2005) A general approach to mixed effects modeling of residual variances in generalized linear mixed models. *Genetics Selection Evolution* 37, 31–56.

Lassen, J., Hansen, N., Sørensen, M.K., Aamand, G.P., Christensen, L.G. and Madsen, P. (2003) Genetic relationship between body condition score, dairy character, mastitis, and diseases other than mastitis in first-parity Danish Holstein cows. *Journal of Dairy Science* 86, 3730–3735.

Lassen, J., Sørensen, M.K., Madsen, P. and Ducrocq, V. (2007) A stochastic simulation study on validation of an approximate multitrait model using preadjusted data for prediction of breeding values. *Journal of Dairy Science* 90, 3002–3011.

Liu, Z., Seefried, F.R., Reinhardt, F., Rensing, S., Thaller, G. and Reents, R. (2011) Impacts of both reference population size and inclusion of a residual polygenic effect on the accuracy of genomic prediction. *Genetics Selection Evolution* 43, 19.

Lund, M.S., de Roos, A.P.W., de Vries, A.G., Druet, T., Ducrocq, V., Fritz, S., Guillaume, F., Guldbrandtsen, B., Liu, Z., Reents, R. *et al.* (2011) Common reference of four European Holstein populations increase reliability of genomic predictions. *Genetics Selection Evolution* 43, 43.

Mäntysaari, E., Liu, Z. and VanRaden, P. (2011) Interbull validation test for genomic evaluations. *Interbull Bulletin* 41, 17–21.

Matukumalli, L.K., Lawley, C.T., Schnabel, R.D., Taylor, J.F., Allan, M.F., Heaton, M.P., O'Connell, J., Moore, S.S., Smith, T.P., Sonstegard, T.S. and Van Tassell, C.P. (2009) Development and characterization of a high density SNP genotyping assay for cattle. *PLoS ONE* 4, e5350.

Meuwissen, T.H.E. (1991) Expectation and variance of genetic gain in open and closed nucleus and progeny testing schemes. *Animal Production* 53, 133–141.

Meuwissen, T.H.E. and Goddard, M.E. (2010) Accurate prediction of genetic values for complex traits by whole-genome resequencing. *Genetics* 185, 623–631.

Meuwissen, T.H.E., Hayes, B.J. and Goddard, M.E. (2001) Prediction of total genetic value using genome-wide dense marker maps. *Genetics* 157, 1819–1829.

Miglior, F., Muir, B.L. and Van Doormaal, B.J. (2005) Selection indices in Holstein cattle of various countries. *Journal of Dairy Science* 88, 1255–1263.

Misztal, I., Lawlor, T.J., Short, T.H. and VanRaden, P.M. (1992) Multiple-trait estimation of variance components of yield and type traits using an animal model. *Journal of Dairy Science* 75, 544–551.

Nicholas, F.W. and Smith, C. (1989) Increased rates of genetic change in dairy cattle by embryo transfer and splitting. *Animal Production* 36, 341–353.

Ojango, J.M.K. and Pollot, G.E. (2002) The relationship between Holstein bull breeding values for milk yield derived in both the UK and Kenya. *Livestock Production Science* 74, 1–12.

Olesen, I., Gjerde, B. and Groen, A.F. (1999) Methodology for deriving non-market trait values in animal breeding goals for sustainable production systems. *Interbull Bulletin* 23, 13–22.

Patry, C. and Ducrocq, V. (2011) Evidence of biases in genetic evaluations due to genomic pre-selection in dairy cattle. *Journal of Dairy Science* 94, 1011–1020.

Philipsson, J., Banos, G. and Arnason, T. (1994) Present and future uses of selection index methodology in dairy cattle. *Journal of Dairy Science* 77, 3252–3261.

Pryce, J.E. and Daetwyler, H.D. (2011) Designing dairy cattle breeding schemes under genomic selection: a review of international research. *Animal Production Science* 52, 107–114.

Pryce, J.E., Coffey, M.P. and Simm, G. (2001) The relationship between body condition score and reproductive performance. *Journal of Dairy Science* 84, 1508–1515.

Pryce, J.E., Harris, B.L., Bryant, J.R. and Montgomerie, W.A. (2009) Do robust dairy cows already exist? *EAAP Scientific Series* 126, 99–112.

Rendel, J.M. and Robertson, A. (1950) Estimation of genetic gain in milk yield by selection in a closed herd of dairy cattle. *Journal of Genetics* 50, 1–10.

Robert-Granié, C., Bonaiti, B., Boichard, D. and Barbat, A. (1999) Accounting for variance heterogeneity in French dairy cattle genetic evaluation. *Livestock Production Science* 60, 343–357.

Schaeffer, L.R. (1994) Multiple-country comparison of dairy sires. *Journal of Dairy Science* 77, 2671–2678.

Schaeffer, L.R. (2006) Strategy for applying genome-wide selection in dairy cattle. *Journal of Animal Breeding and Genetics* 123, 218–223.

Sonneson, A.K. and Meuwissen, T.H.E. (2000) Mating schemes for optimum contribution selection with constrained rate of inbreeding. *Genetics Selection Evolution* 32, 231–248.

Soyeurt, H., Dardenne, P., Dehareng, F., Lognay, G., Veselko, D., Marlier, M., Bertozzi, C., Mayeres, P. and Gengler, N. (2006) Estimating fatty acid content in cow milk using mid-infrared spectrometry. *Journal of Dairy Science* 89, 3690–3695.

Spelman, R.J. (2002) Utilisation of molecular information in dairy cattle breeding. *Proceedings of the 7th World Congress on Genetics Applied to Livestock Production* 33, 1–17.

Spelman R.J., Coppieters W., Grisart B., Blott, S. and Georges, M. (2001) Review of QTL mapping in the New Zealand and Dutch dairy cattle populations. *Proceedings of the Association for Advancement of Animal Breeding and Genetics* 14, 11–16.

Steine, G., Kristofersson, D. and Guttormsen, A.G. (2008) Economic evaluation of the breeding goal for Norwegian Red dairy cattle. *Journal of Dairy Science* 91, 418–428.

Van der Linde, C., De Jong, G., Koenen, E.P.C. and Eding, H. (2010) Claw health index for Dutch dairy cattle based on claw trimming and conformation data. *Journal of Dairy Science* 93, 4883–4891.

Van Tassell, C.P., Misztal, I. and Varona, L. (2000) Method R estimates of additive genetic, dominance genetic, and permanent environmental fraction of variance for yield and health traits of Holsteins. *Journal of Dairy Science* 83, 1873–1877.

Van Tassell, C.P., Smith, T.P.L., Matukumalli, L.K., Taylor, J.F., Schnabel, R.D., Taylor Lawley, C., Haudenschild, C.D., Moore, S.S., Warren, W.C. and Sonstegard, T.S. (2008) SNP discovery and allele frequency estimation by deep sequencing of reduced representation libraries. *Nature Methods* 5, 247–252.

VanRaden, P.M. (1992) Accounting for inbreeding and crossbreeding in genetic evaluation of large populations. *Journal of Dairy Science* 75, 3136.

VanRaden, P.M. (2001) Method to combine estimated breeding values obtained from separate sources. *Journal of Dairy Science* 84 (E. Suppl), E47–E55.

VanRaden, P.M. (2008) Efficient methods to compute genomic predictions. *Journal of Dairy Science* 91, 4414–4423.

VanRaden, P.M. and Wiggans, G.R. (1991) Derivation, calculation, and use of national animal model information. *Journal of Dairy Science* 74, 2737–2746.

VanRaden, P.M., Tooker, M.E., Cole, J.B., Wiggans, G.R. and Megonigal, J.H. (2007) Genetic evaluations for mixed-breed populations. *Journal of Dairy Science* 90, 2434–2441.

VanRaden, P.M., Van Tassell, C.P., Wiggans, G.R., Sonstegard, T.S., Schnabel, R.D., Taylor, J.F. and Schenkel, F.S. (2009) Reliability of genomic predictions for North American Holstein bulls. *Journal of Dairy Science* 92, 16–24.

VanRaden, P.M., O'Connell, J.R., Wiggans, G.R. and Weigel, K.A. (2011a) Genomic evaluations with many more genotypes. *Genetics Selection Evolution* 43, 10.

VanRaden, P.M., Olson, K.M. Null, D.J. and Hutchison, J.L. (2011b) Harmful recessive effects on fertility detected by absence of homozygous haplotypes. *Journal of Dairy Science* 94, 6153–6161.

VanRaden, P.M., Null, D.J., Sargolzaei, M., Wiggans, G.R., Tooker, M.E., Cole, J.B., Connor, E.E., Winters, M., Van Kaam, J.B.C.H.M., Sonstegard, T.S. *et al.* (2013) Genomic imputation and evaluation using high-density Holstein genotypes. *Journal of Dairy Science* 96, 668–678.

Verbyla, K.L., Hayes, B.J., Bowman, P.J. and Goddard, M.E. (2009) Accuracy of genomic selection using stochastic search variable selection in Australian Holstein Friesian dairy cattle. *Genetics Research* 91, 307–311.

Visscher, P.M., Bowman, P.J. and Goddard, M.E. (1994) Breeding objectives for pasture based dairy production systems. *Livestock Production Science* 40, 123–137.

Weller, J.I., Kashi, Y. and Soller, M. (1990) Power of daughter and granddaughter designs for determining linkage between marker loci and quantitative trait loci in dairy cattle. *Journal of Dairy Science* 73, 2525–2537.

Weller, J.I., VanRaden, P.M., and Wiggans, G.R. (2013) Application of a posteriori granddaughter and modified granddaughter designs to determine Holstein haplotype effects. *Journal of Dairy Science* 96, 5376–5387.

Wiggans, G.R. and VanRaden, P.M. (1991) Method and effect of adjustment for heterogeneous variance. *Journal of Dairy Science* 74, 4350–4357.

Wiggans, G.R., VanRaden, P.M. and Zuurbier, J. (1995) Calculation and use of inbreeding coefficients for genetic evaluation of United States dairy cattle. *Journal of Dairy Science* 78, 1584–1590.

Wiggans, G.R., Cooper, T.A.,VanRaden, P.M. and Cole, J.B. (2011a) Adjustment of traditional cow evaluations to improve accuracy of genomic predictions. *Journal of Dairy Science* 94, 6188–6193.

Wiggans, G.R., VanRaden, P.M. and Cooper, T.A. (2011b) The genomic evaluation system in the United States: past, present, future. *Journal of Dairy Science* 94, 3202–3221.

Winter, A., Kramer, W., Werner, F.A.O., Kollers, S., Kata, S., Durstewitz, G., Buitkamp, J., Womack, J.E., Thaller, G. and Fries, R. (2002) Association of a lysine-232/alanine polymorphism in a bovine gene encoding acyl-CoA: diacylglycerol acyltransferase (DGAT1) with variation at a quantitative trait locus for milk fat content. *Proceedings of the National Academy of Science* 99, 9300–9305.

16 Molecular Genetics of Milk Protein Production

Julie A. Sharp[1] and Kevin Nicholas[2]

[1]Deakin University, Geelong, Victoria, Australia;
[2]Monash University, Clayton, Victoria, Australia

Introduction

Increasing international competition for export and import replacement of primary produce has resulted in the adoption of a wide range of technologies to improve productivity in the dairy industry. Improved milk production, and more particularly, improved milk composition for value-added processing is a major goal of the industry. Protein is the most valuable component of milk and the industry would benefit considerably from cattle with increased protein yield without an additional increase in volume (Bulter and McGarry Wolf, 2010). Therefore, an understanding of the molecular genetics of the milk proteins and of mammary gland function is central to these goals.

Appropriate nutrition is essential for maximal milk production and to limit seasonal variation in milk composition, but significant economic gains in improved milk protein content are not likely by manipulation of diet alone (Duff and Galyean, 2007). Breeding programmes have used genetic selection based on progeny testing to identify superior sires for widespread use and the result has been small, but cumulative improvements in milk production and composition. Conventional selection based on production traits can be enhanced by the use of gene- or marker-assisted selection, which requires the identification of chromosomal markers associated with improved production. These markers can be used to identify superior animals for breeding and in the longer term may direct researchers to the genes that cause improved production.

The availability of the *Bos taurus* genome assembly has allowed bovine milk and lactation data to be used to increase our understanding of bovine milk in comparison with the milk of other species. This is particularly important as bovine milk is a major human food uniquely tied in with human nutrition. The arrival of the bovine genome sequence provides unique opportunities to study *in silico* milk proteome data, milk production quantitative trait loci (QTLs) and over 100,000 mammary-related bovine expressed sequence tags (ESTs) (Lemay *et al.*, 2009). MicroRNA technology has also advanced the field of bovine milk production by cloning of animals with knockout genes that produce low allergenic milk more suitable for human consumption (Formigli *et al.*, 2000). Unfortunately, given the current laws surrounding genetic modification, it is not likely that this milk will be available for human consumption in the near future.

A necessary prerequisite to successfully achieving improvements in milk composition in cattle is a substantial understanding of the regulation of milk protein genes. Transgene expression must be tissue specific, developmentally regulated and preferably copy number-dependent. This chapter reviews the current status of our knowledge of the structure, function and control of the milk protein genes in cattle.

Bovine Milk Composition

Typical bovine milk consists of a complex mixture of water (87.3%), lactose (4.8%), fat (3.7%), protein (3.5%) and a number of other minor components (Bedo *et al.*, 1995). Approximately 80% of the protein content of milk consists of α_{S1}-, α_{S2}-, β- and κ-casein, which comprise a group of acidic phosphoproteins that precipitate from skimmed milk at pH 4.6 and 20°C (Jolles, 1975). The whey proteins, including β-lactoglobulin (Palmer, 1934), α-lactalbumin (Jenness, 1974), lactoferrin (Blanc and Isliker, 1961) and various enzymes constitute the remaining fraction. Approximately 95% of milk proteins are synthesized in the mammary gland. The remaining 5% originate from the blood, including immunoglobulins IgG and IgM, as well as transferrin and serum albumin (Jenness, 1974).

The protein content of mammalian milk can range from 10 to 200 g kg^{-1} between various species (Murphy and O'Mara, 1993). In addition, the relative proportion of individual milk proteins can vary markedly from that found in the cow. For example, whey acidic protein (WAP), the major whey protein of mice and rats (Hennighausen and Sippel, 1982; Hobbs and Rosen, 1982), is not present in the milk of cattle or humans (Jenness, 1979). The protein content also varies significantly between cattle breeds (Murphy and O'Mara, 1993).

The *in silico* analysis of the bovine genome allows examination of its content and organization. Utilizing the genomes of seven mammals (bovine, dog, human, mouse, rat, opossum and platypus) gene loss and duplication events were investigated in addition to phylogeny, sequence conservation, and evolution of milk and mammary genes (Lemay *et al.*, 2009). Past analyses have suggested that these species have many differences in milk composition as revealed by the absence of some known abundant proteins, such as β-lactoglobulin and whey acidic protein, in the milk of some species (Mercier and Vilotte, 1993). This led to the hypothesis that variation in milk composition resides in part in variation in the milk genome. Comparative genome analysis shows that genomic rearrangements, such as those in the transcriptional regulatory region of the major milk genes (Kishore *et al.*, 2013) or gene duplication (Lemay *et al.*, 2009; Rullo *et al.*, 2010) contribute to changes in milk protein gene composition between species. Although the casein proteins are highly divergent across mammalian milks (Rijnkels, 2002; Lemay *et al.*, 2009), it is now known that other genes expressed in the mammary gland are more highly conserved and evolving more slowly, on average, than other genes in the bovine genome. Among mammals, it appears that the milk protein genes that are the most divergent have nutritional or immunological functions (e.g. caseins), whereas the most conserved milk protein genes have functions important for formation and secretion of mammalian milk (Lemay et al., 2009). Casein proteins appear the most divergent and provide the suckling neonate with a source of amino acids and calcium, while peptides derived from partially digested caseins have potential antimicrobial, immune-modulating and other bioactive properties. The fact that the caseins are the most divergent of the milk proteins suggests that the nutritional and immunological functions of these proteins do not particularly constrain their amino acid sequence and structure. In contrast, genes associated with the milk fat globule membrane proteome are highly conserved, suggesting that this mechanism is a highly specialized process by which the milk fat globule is secreted or that the exocytotic and lipid secretion pathways meet at some point during the secretion process. Such high conservation of milk fat globule membrane protein genes among the mammalian genomes suggests that the secretory process for milk production was established more than 160 million years ago (Lemay et al., 2009). This supports the theory that mammary secretion evolved from pre-existing tissue, presumably glandular tissue associated with the skin, which was co-opted for a new function of nutrient secretion in order to feed their young (Oftedal, 2012).

The caseins and their genes

The caseins are predominantly present in milk as a colloidal aggregation complexed with calcium phosphate (7% dry weight) to form casein micelles (Waugh, 1971). The calcium-sensitive α_{S1}-, α_{S2}- and β-casein contain clusters of phosphoseryl residues and when isolated will precipitate out of solution at low calcium concentration (Waugh, 1971). In contrast, κ-casein contains only one or two phosphoseryl residues and remains in solution over a broad range of calcium concentrations (Waugh and von Hippel,

1956). The structure of casein micelles has been the subject of extensive studies over past decades but details at the molecular level still remain elusive (Horne, 2006). While the internal structure of the casein micelles is a subject of debate, it is widely accepted that the stability of the micelles is provided by an outer layer of κ-caseins and the internal structure consists of a protein matrix in which calcium phosphate nanoclusters are dispersed (de Kruif et al., 2012). Despite their ability to associate into micelles, the caseins possess limited α-helix and β-sheet secondary structure and all attempts to crystallize them have failed. Nevertheless, computer-aided molecular models of the caseins have been reported (Kumosinski et al., 1991, 1993a, 1993b; Farrell et al., 2009).

The casein loci

The genomic sequences for all four casein genes have been determined (Lemay et al., 2009). These genes have been mapped to chromosome 6 (6q31) in cattle within a 250–350 kb region arranged in the order α_{S1}-β-α_{S2}-κ (Rijnkels, 2002) (Fig. 16.1).

The calcium-sensitive caseins possess similar gene structure, containing many small exons and low exon to intron ratio. Both α_{S1}- and α_{S2}-casein genes (CSN1S2 and CSN1S2, respectively) have relatively large transcriptional units, 17.5 and 18.5 kb, respectively, with a similar number of exons (Koczan et al., 1991; Groenen et al., 1993). The β-casein gene (CSN2) is approximately half the size of the other calcium-sensitive caseins and contains half the number of exons (Bonsing et al., 1988). Fine mapping of the casein locus has also demonstrated that the CSN2 is divergently transcribed with respect to the other casein genes (Rijnkels et al., 1997). These genes are characterized by multiple copies of repeat elements, many of which are Alu-like artiodactyla retroposons (Watanabe et al., 1982; Skowronski et al., 1984). The calcium-sensitive caseins are clustered within a 140-kb region, while κ-casein gene (CSN3) is positioned 95–120 kb downstream of the α_{S2}-casein transcriptional unit (Rijnkels et al., 1997).

Recent advances in genome sequencing of a number of mammalian species have provided invaluable resources for the comparative

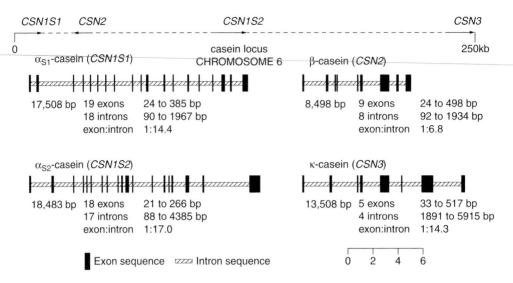

Fig. 16.1. The bovine casein genes. Schematic representation of the structural organization of the transcription units encoding the bovine casein proteins depicting their exon structure (including number and size range of exons and introns, exon to intron ratio) as well as their organization within the casein locus. (Derived from sequence data reported by Alexander *et al.*, 1988; Bonsing *et al.*, 1988; Koczan *et al.*, 1991; Groenen *et al.*, 1993; Rijnkels *et al.*, 1997 and genome sequencing data.)

evolutionary analysis of genes involved in lactation. Comparison of the casein cluster shows a physical proximity of the casein genes of all mammals examined so far (Rijnkels, 2002). The cluster has expanded during the radiation of mammals, while lineage-specific differences can be seen, such as a *CSN2* duplication in monotremes (Lefevre *et al.*, 2009) and *CSN1S2* in rodents (Rijnkels, 2002). The conserved gene structure indicates that the molecular diversity of the casein genes is provided through variation of the use of exons in different species and sequence divergence (Fig. 16.2). It has been shown that all genes identified in this region encode, or are predicted to encode, secreted proteins involved in nutrition, mineral homeostasis, and/or host defence, and are mostly expressed in the mammary and/or salivary glands (Rijnkels, 2002).

Sequence alignments of casein genes from various species reveal evidence of a high mutation rate, including major insertions and deletions in addition to sequence rearrangements. As discussed earlier, and in contrast to other mammary gland genes, the caseins appear to be a rapidly evolving gene family, presumably due to the minimal functional and

structural constraints on their amino acid sequence (Williams and Lovell, 2009). This is demonstrated by the fact that silent and replacement mutations are found with similar frequency within the *CSN1* cDNAs. Homologous proteins that possess more defined structural requirements for function, such as enzymes, exhibit a greater abundance of silent mutations. This lack of conservation of the majority of the coding region is consistent with the perceived loose structure/function relationship of the caseins in forming stable micelles (Bonsing and Mackinlay, 1987).

Despite their highly divergent nature there are regions of strong similarity between the calcium-sensitive casein genes. For example, the first exon always consists of the 5' non-coding sequence, which is highly conserved presumably due to important secondary structure interactions necessary for post-transcriptional regulation (Blackburn *et al.*, 1982; Stewart *et al.*, 1984). In addition, exon 2 is uniformly 63 bp in length and encodes the remaining 5' non-coding region, the entire signal peptide plus two additional amino acid residues of the mature casein. Another interesting feature is that all the

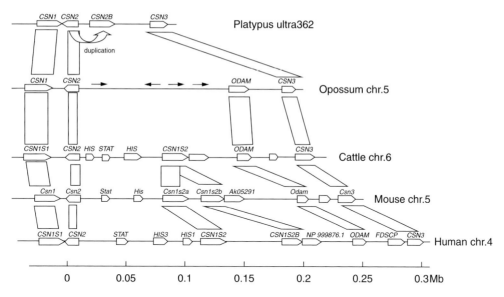

Fig. 16.2. Comparison of mammalian casein gene clusters. Casein locus organization in platypus, opossum, cattle, mouse and human genomes. Genes are represented by a box with a tail arrow pointing in the direction of gene transcription. Note the close proximity of α- (*CSN1*) and β- (*CSN2*) casein genes in reverse orientation on the left and the expansion of the region between β- and κ- (*CSN3*) casein on the right. Except for *CSN2*, all genes are transcribed from left to right. In monotremes, a recent duplication of *CSN2* has led to *CSN2B*, whereas in eutherians, an ancient duplication produced *CSN1S2*, which has been duplicated in some species to produce *CSN1S2B*, now a pseudogene in human but not in mouse. In the marsupial locus, there is no casein duplication and the spacing region contains several copies of an invading repetitive element (black arrows). (Modified from Lefevre *et al.*, 2009.)

protein coding exons end in a complete codon so that none of the codons are interrupted by splice sites (Bonsing and Mackinlay, 1987; Mercier and Vilotte, 1993). These similarities suggest calcium-sensitive caseins constitute a multigene family that is thought to have arisen from an ancestral gene via intra- and intergenic duplication (Yu-Lee *et al.*, 1986; Bonsing and Mackinlay, 1987). Evidence for this theory can be seen within *CSN1S1* where a 154-bp region spanning exon 10 differs by only four bases from a similar region comprising exon 13 (Koczan *et al.*, 1991). Exons 7–11 and 12–16 of *CSN1S2* may also have resulted from internal duplication (Stewart *et al.*, 1987). Although *CSN1S1* and *CSN1S2* appear more closely related to each other on the basis of gene size and number of exons, it has been proposed from analysis of exon lengths that *CSN1S2* is most closely related to *CSN2* (Groenen *et al.*, 1993).

CSN3 is not evolutionarily related to the 'calcium-sensitive' casein genes, but is physically linked, and functionally important for stabilizing Ca-sensitive caseins in the micelle. The gene structure of *CSN3* is conserved in all species studied (Rijnkels, 2002). The bovine gene consists of five exons separated by very large introns. The first three exons are quite small, 65 bp or less, with the majority of the protein-coding sequence contained within exon 4 (Alexander *et al.*, 1988). Inter-species comparisons demonstrate *CSN3* possesses the highest degree of conservation of all the casein genes, which may be related to its essential function of stabilizing the casein micelle (Alexander *et al.*, 1988). The level of CSN3 in milk influences micelle size, and as such, the properties of milk during digestion as well as in several dairy applications (Swaisgood, 1993). Furthermore, the degree of glycosylation of the GMP influences the physiological properties of milk and the casein micelle (Fiat and Jolles, 1989).

The major whey proteins and their genes

The whey proteins comprise a diverse group of globular polypeptides and, in contrast to caseins, many of their structures have been determined (Wong *et al.*, 1996). The casein and whey proteins differ not only in their physical structure and properties, but also in their genomic organization and postulated evolutionary origins (Bonsing and Mackinlay, 1987; Bawden *et al.*, 1994). Unlike caseins, the whey protein genes are dispersed throughout the genome and a number of pseudogenes have also been characterized (Soulier *et al.*, 1989; Vilotte *et al.*, 1993; Passey and Mackinlay, 1995).

α-Lactalbumin

The whey protein α-lactalbumin is a calcium metalloprotein that in combination with β-1, 4-galactosyltransferase forms the lactose synthase complex situated in the *trans*-Golgi membrane of mammary epithelial cells (MECs) (Bell *et al.*, 1976). The formation of this complex is necessary for synthesis of lactose, the major carbohydrate in milk. As α-lactalbumin is not membrane bound, in contrast to β-1, 4-galactosyltransferase, it is secreted in bovine milk at a concentration of approximately 1.2 mg ml^{-1} (Brew and Hill, 1975). More recently, α-lactalbumin has been implicated in mammary gland involution at termination of lactation, but while secreted in milk is an apoptotic inducer of mammary epithelial cells (Sharp *et al.*, 2008). Interestingly, the α-lactalbumin gene in fur seals, which engage in long foraging bouts for up to 21 days (depending on the species) away from the pup, has evolved into a pseudogene and does not produce α-lactalbumin protein (Sharp *et al.*, 2005, 2008). The genomic sequence of the bovine α-lactalbumin gene (*LALBA*) comprises a transcriptional unit approximately 2 kb in length containing four exons (Vilotte *et al.*, 1987). Comparison with published α-lactalbumin and c-type lysozyme sequences shows that the exon organization of these genes has been highly conserved (Hall *et al.*, 1987). Though functionally dissimilar, the high degree of homology between the amino acid sequences of α-lactalbumin and the c-type lysozymes

suggests they derived from a common ancestral gene (Brew *et al.*, 1970). The tertiary structures of baboon, human and buffalo LALBA have been determined using X-ray crystallography and are almost superimposable with reported lysozyme structures (Acharya *et al.*, 1989, 1991; Calderone *et al.*, 1996). The similarity of these sequences has been dramatically demonstrated by the transfer of lysozyme activity to goat α-lactalbumin following exchange of exon 2 with the same exon from hen lysozyme (Kumagai *et al.*, 1992).

Two pseudogenes for α-lactalbumin have been isolated and sequenced from the bovine genome (Soulier *et al.*, 1989; Vilotte *et al.*, 1993). Both demonstrate approximately 80% homology to the authentic gene and stretch from intron 2 to the 3'-untranslated region of exon 4. The 5' ends of the pseudogenes are also similar to each other yet bear no homology to the native sequence (Fig. 16.3). The α-lactalbumin pseudogenes are present at the same chromosomal location as the authentic gene on chromosome 5 (5q21), supporting the suggestion that they arose via tandem duplication (Lemay *et al.*, 2009).

β-Lactoglobulin

The major whey protein of ruminants is β-lactoglobulin (LGB) and it is present in bovine milk at a concentration of 3.1 mg ml^{-1} (Dalgleish, 1993). The structure of bovine LGB has been determined by X-ray crystallography (Papiz *et al.*, 1986), which revealed it as a member of the lipocalin superfamily. The ability of LGBs from a number of species to bind retinol *in vitro* and its structural homology to retinol-binding protein implicated LGB in the transport of retinol to the suckling infant (Fugate and Song, 1980; Godovac-Zimmermann *et al.*, 1985). However, despite being isolated over 60 years ago the biological role of LGB is still largely unknown (Perez and Calvo, 1995). LGB is absent in milk of primates such as humans, and studies have shown that bovine LGB can cause allergic reactions in humans (Host *et al.*, 1988; Ball *et al.*, 1994; Dunstan *et al.*, 2005). This has promoted the production of a transgenic cow, using microRNA technology, which produces LGB-free milk during lactation (Jabed *et al.*, 2012).

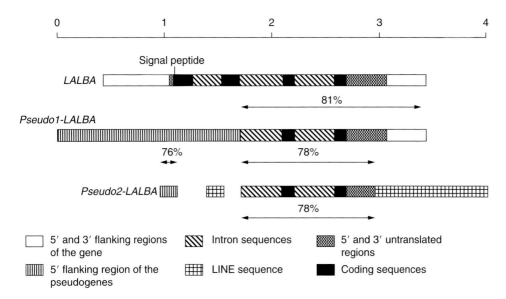

Fig. 16.3. Bovine α-lactalbumin and pseudogenes. Schematic representation of the bovine α-lactalbumin gene (*LALBA*) and comparison with two pseudogenes (*p1LALBA* and *p2LALBA*). Percentage similarity between regions is represented by the arrows (the arrow below *p2LALBA* corresponds to the comparison between *p2LALBA* and *LALBA*). (Reproduced from Vilotte *et al.*, 1993, used by permission of Academic Press, Orlando, Florida.)

The bovine *LGB* gene consists of a 4724-bp transcriptional unit containing seven exons (Alexander *et al.*, 1989). It has been mapped to the short arm of chromosome 3 (3p28) in sheep, and to the homologous chromosome 11 (11q28) in both cattle and goat (Hayes and Petit, 1993). A *LGB*-like pseudogene (ψ-*LGB*) was discovered 14 kb upstream from the authentic bovine *LGB* gene (Passey and Mackinlay, 1995). The pseudogene is of similar length, contains the same number of exons and is in the same orientation as the authentic *LGB* gene. Most of the introns are highly divergent, while exon similarities range from 60 to 92.5%. It is proposed that both genes originated from a common ancestor via gene duplication. Exon 5 of ψ-*LGB* contains an in-frame stop codon but no ψ-*LGB* product has been detected in the lactating mammary glands of cattle. The milk of several species, including the horse and pig, contain major and minor LGB fractions designated LGB I and LGB II, respectively, and the milk of the dog, seal and cat also include a third LGB, designated LGB III (Halliday, 1993). Amino acid analysis confirms these proteins are encoded by separate genes

(Halliday, 1993; Halliday *et al.*, 1991; Conti *et al.*, 1984). The inferred translation product for ψ-*LGB* shows greater similarity to the *LGB* II sequences from the horse and cat than the published bovine sequence. Due to the highly conserved exon sequences, the authors suggest that ψ-*LGB* may have been expressed until relatively recently.

Lactoferrin

Lactoferrin is an 80-kDa iron-binding glycoprotein belonging to the transferrin gene family and is part of the host defence system with a wide range of functions, acting as an antimicrobial, immunomodulatory and antioxidant agent (Oram and Reiter, 1968; Arnold *et al.*, 1977; Yamauchi *et al.*, 2006). The tertiary structure of human lactoferrin has been determined and consists of two globular lobes, each lobe containing a single iron-binding site, which are connected by a hinge region (Anderson *et al.*, 1989). Both lobes possess significant internal homology and are thought to have arisen via internal duplication of an ancestral gene (Park *et al.*, 1985). Part of lactoferrin's antimicrobial activity is due to its highly cationic N-terminal regions. These regions confer bactericidal action by interacting with the negatively charged part of bacterial membranes, which is lipopolysaccharide (LPS) in gram-negative bacteria and lipoteichoic acid in gram-positive bacteria (Yen *et al.*, 2009). Lactoferrin can also compete with LPS for binding of CD14, a part of toll-like receptor (TLR) 4, thereby preventing LPS from activating a pro-inflammatory cascade, which can lead to tissue damage (Actor *et al.*, 2009). Lactoferrin's ability to bind iron not only promotes growth of beneficial low iron requiring bacteria like *Lactobacillus* and *Bifidobacteria* (Yen *et al.*, 2009), but sequestering iron also lowers cellular oxidative stress, thus reducing pro-inflammatory cytokines (Actor *et al.*, 2009). Finally, lactoferrin has targeted control of some cellular processes and can act as a transcription factor and regulate granulopoiesis and DNA synthesis in some cells types (Kanyshkova *et al.*, 2001). The bovine lactoferrin gene is the largest of the characterized milk protein genes containing 17 exons within a 34.5-kb transcriptional unit (Seyfert *et al.*, 1994).

Multi-level Regulation of Milk Protein Genes Expression in the Mammary Gland: Endocrine, Autocrine and Paracrine Regulation

Understanding factors that regulate milk protein synthesis is important for making high quality protein available to the global population. The bovine mammary epithelial cells transcribe four casein genes plus α-lactalbumin and β-lactoglobulin to produce the suite of major proteins secreted into milk. Mammary gland function is controlled by a number of different mechanisms, all of which act concurrently to regulate the synthesis and secretion of milk.

Extensive studies have been conducted on transcriptional regulation of milk protein gene expression and stabilization of transcription products under the influence of lactogenic hormones and various nutritional and environmental factors (Geursen and Grigor, 1987; Yang *et al.*, 2005; Zhou *et al.*, 2008; Johnson *et al.*, 2010). Endocrine factors (hormones and growth factors) stimulate growth of the mammary gland, which together with paracrine influences (e.g. the extracellular matrix; ECM), is essential for the expression and stabilization of mRNA of the milk protein genes (Topper and Freeman, 1980; Streuli *et al.*, 1995a; Sheehy *et al.*, 2004; Riley *et al.*, 2010). Autocrine factors (regulatory molecules which are produced by the mammary gland and subsequently 'feed-back' to stimulate or inhibit mammary function) have also been shown to modulate the effects of endocrine-stimulated milk protein synthesis and secretion (Wilde *et al.*, 1995; Hernandez *et al.*, 2008).

Genome-wide association studies have been undertaken for CSN1S1, CSN1S2, CSN2, CSN3, LALBA and LBG contents in milk using single nucleotide polymorphism (SNP) technologies (Schopen *et al.*, 2011). A total of 1713 Dutch Holstein Friesian cows were genotyped for 50,228 SNPs and associations with milk protein content and composition were detected on 20 bovine autosomes. The main genomic regions associated with milk protein composition or protein percentage were found on chromosomes 5, 6, 11 and 14. Further analysis indicated that three regions had major effects on milk protein composition, in addition to several regions with smaller effects involved in the regulation of milk protein composition. This suggests that milk protein composition is regulated by a few genes with large effects on all milk proteins and several genes with smaller effects on specific milk proteins. The genomic selection of dairy cattle will be discussed further in Chapter 19.

Endocrine regulation

Early studies (Hartmann *et al.*, 1973; Hartmann, 1973) showed that lactogenesis in dairy cattle proceeds in two stages: stage one occurs prior to parturition and the secretion is high protein colostrum. The decline of progesterone at parturition signals stage two: lactogenesis and the copious secretion of milk. This role of progesterone in lactogenesis has been confirmed by *in vitro* studies showing that progesterone inhibits the synthesis and secretion of milk proteins in mammary explants from late pregnant and lactating cows (Gertler *et al.*, 1982; Goodman *et al.*, 1983; Shamay *et al.*, 1987) and enhances the process of autophagy (Motyl *et al.*, 2007). The hormonal regulation of milk protein gene expression varies among mammals, but generally requires prolactin, adrenal glucocorticoids (usually cortisol) and insulin (Neville *et al.*, 2002). The use of the mammary explant culture system, with tissue obtained by biopsy from late pregnant cows shows that casein and whey protein genes are expressed at elevated levels at least 30 days prior to calving, but when this tissue (in phase one lactogenesis) is cultured for 4 days in media containing insulin and cortisol the level of mRNA for the *CSN3*, *CSN2* and *LGB* genes declined to almost undetectable levels (Sheehy *et al.*, 2004). The subsequent inclusion of prolactin in the culture media does not induce milk protein gene expression. However, when explants are prepared from mammary tissue biopsied from cows more than 30 days before calving it is possible to demonstrate induction of these milk protein genes in the presence of insulin, cortisol and prolactin. These results are in alignment with a number of studies that have used mammary gland explants from pregnant and lactating cows (Rabot *et al.*, 2007), primary cell cultures of bovine MECs using mammospheres (Riley *et al.*, 2010) and bovine cell lines (German and Barash, 2002) to demonstrate casein and whey protein genes are induced in response to insulin, cortisol and prolactin.

It has been demonstrated that insulin is essential for casein gene expression in mammary explants from mouse (Bolander *et al.*, 1981) and rat (Kulski *et al.*, 1983). However, the role of insulin in milk protein synthesis in the bovine mammary gland remains equivocal. Early mammary culture experiments indicated that induction of milk protein gene expression required the complement of prolactin, hydrocortisone and insulin (Andersen and Larson, 1970; Djiane *et al.*, 1975; Servely *et al.*, 1982; Choi *et al.*, 1988). Menzies *et al.* (2009) showed insulin was required for maximal induction of milk protein genes in combination with cortisol and prolactin. In contrast, past studies by Sheehy *et al.* (2000) suggested milk protein genes could be expressed in cultured mammary explants from pregnant cows

in the absence of insulin (Sheehy *et al.*, 2000). These results were possibly due to priming the mammary gland with insulin since insulin has been shown to have a priming role in rodent mammary gene expression (Calvert and Clegg, 1996). The mechanism of insulin action has been addressed by assessment of global changes in gene expression in bovine mammary explants in response to insulin (Menzies *et al.*, 2009). Here insulin was shown to induce up-regulation of genes involved in protein synthesis and included the milk protein transcription factor, *ELF5*, translation factors, the folate metabolism genes, *FOLR1* and *MTHFR*, as well as several genes encoding enzymes involved in catabolism of essential amino acids and biosynthesis of non-essential amino acids. These data show that insulin is not only essential for milk protein gene expression, but stimulates milk protein synthesis at multiple levels within bovine mammary epithelial cells.

Local control of milk composition and milk production

The rate of milk secretion in dairy cattle is regulated by the frequency and completeness with which milk is removed from the mammary glands. More frequent milking applied bilaterally to two bovine mammary glands increases milk production in those glands (Wilde *et al.*, 1987). Increased milking frequency during early lactation stimulates an increase in milk production that partially persisted through late lactation, indicating long-term effects on mammary function (Wall and McFadden, 2008). Increased milking frequency of dairy cows is commonly used to increase milk yield and production efficiency. Relative to cows milked twice daily, cows milked three times daily generally produce 15 to 20% more milk, and milk production can be increased an additional 7% by milking four times daily instead of three times (Erdman and Varner, 1995; Stelwagen, 2001; Stockdale, 2006; Wall and McFadden, 2008).

Autocrine modulation of endocrine control may partially explain the developmental responses of the tissue to sustained alterations in milking frequency. In cattle, as in goats

(Wilde *et al.*, 1987), frequency of milking regulates the degree of mammary cell differentiation (Hillerton *et al.*, 1990). It is conceivable that autocrine signalling of MECs may result in changes in milk composition, in addition to milk secretion. Studies using tammar wallaby (*Macropus eugenii*) have suggested that macropodidea marsupials may regulate milk protein gene expression by an autocrine mechanism (Nicholas *et al.*, 1997). The marsupial newborn is immature and the mother has the capacity to alter milk composition significantly during lactation, presumably to meet the nutritional requirements of the developing young. Furthermore, macropodidea marsupials practise asynchronous concurrent lactation, whereby the mother provides milk which differs in all the major components from adjacent mammary glands for two young of different ages. This phenomenon suggests local regulation of mammary function, in addition to endocrine stimuli, is likely to be important for controlling milk composition. Recent studies have shown that the down-regulation of the expression of two protease inhibitor genes in the tammar wallaby is temporally matched to meet the needs of the developing pouch young (Nicholas *et al.*, 1997). It is interesting that the activity of bovine trypsin inhibitor is similarly down-regulated at parturition in cattle and correlates with the onset of the sucking stimulus. However, the potential for autocrine regulation of this protein, and the gene which codes for it, remains to be studied.

The local mechanisms regulating the mammary response to increased milking frequency are poorly understood, although several have been proposed. Experiments in both ruminants and rodents have indicated that the frequency of milk removal influences mammary cell number and activity (Wall and McFadden, 2008). An autocrine mechanism is thought to match the supply of milk to the demands of either the nutritional requirements of suckled offspring or milking timetable in dairy animals. Bovine milk constituents have been screened for the presence of a chemical inhibitor of milk secretion in a mammary tissue explant bioassay, and results indicate that biological activity resides in a small, acidic protein, which has been termed feedback inhibitor of lactation (FIL) (Wilde *et al.*, 1995, 1997).

FIL was reported as a component of the whey fraction of milk, both synthesized and secreted by mammary epithelial cells. It is proposed that FIL is the major autocrine regulator of milk secretion (Peaker *et al.*, 1998). FIL acts in a reversible, concentration-dependent manner to decrease milk accumulation by blocking constitutive secretion in the mammary epithelial cells (Peaker and Wilde, 1987). Later studies provided some evidence that FIL inhibits milk production by interfering with the casein secretory pathway (Rennison *et al.*, 1993; Burgoyne and Wilde, 1994).

Recently, serotonin (Hernandez *et al.*, 2008) and a *CSN2* peptide (Shamay *et al.*, 2002; Silanikove *et al.*, 2009) have been identified as factors that may act on the synthesis and/or secretion of particular milk components.

Fur seals appear to finely regulate the inhibition of milk secretion and avoid mammary gland apoptosis (Sharp *et al.*, 2005). During lactation these animals go through cycles of suckling their young on land and foraging for food at sea for up to 28 days at a time. During foraging, milk secretion is reduced to 19% production on land but resumes to 100% production when they return to shore (Arnould and Boyd, 1995). This decline in milk production is regulated at transcriptional level (Sharp *et al.*, 2006a). It has been suggested that fur seal lactation has evolved to override the influence of local negative feedback mechanisms to accommodate their foraging cycles and continue to rear their offspring.

Engorgement arising from milk accumulation is thought to initiate physiochemical signalling caused by stress between the interaction of the extracellular matrix (ECM) and alveoli epithelial cells, activating independent signalling cascades that act on the apoptotic programme during the first phase of involution (Boudreau *et al.*, 1995; Clark and Brugge, 1995). In this scenario local regulation of milk production is not a chemical stimulus, but the result of intra-alveolar pressure. It is postulated that if the basement membrane (BM) becomes stretched, the molecular interactions with adhesion receptors are altered leading to reduced ligand-binding interacting sites (Banes *et al.*, 1995). The direct attachment of epithelial cells to the ECM occurs through basally located integrins (Alford and Taylor-Papadimitriou, 1996;

Weaver et al., 1997), and studies during transition from lactation to involution in mice have shown that levels of ligand-bound β1 integrin are significantly decreased (McMahon et al., 2004). The affinity modulation of integrin activity and, therefore, a potential inability to respond to survival signals from the basement membrane may contribute to induction of apoptosis at the onset of involution. A candidate mechanism for avoiding alveoli collapse and cell death is up-regulation of ECM components thus avoiding degradation of ECM, preventing the transduction of apoptotic signals (Blatchford et al., 1999).

Regulation by the extracellular matrix

Hormonal cues alone are not sufficient to elicit the expression of milk protein genes within the mammary gland – interactions with the ECM are also required (Lin and Bissell, 1993; Roskelley et al., 1995). The role of ECM in the mammary gland during lactation is well documented and includes modulation of epithelial cell growth (Wang et al., 2011) and differentiation (Streuli, 2009), regulation of specific milk protein expression in various eutherian species (Bissell et al., 1982; Li et al., 1987; Wicha et al., 1982), determination of stromal content (Bissell and Aggeler, 1987) and modulation of mammary myo-epithelial cell function (Deugnier et al., 1999).

Cells of the mammary gland are in intimate contact with the ECM, which provides not only a biochemical context, but also a mechanical context. Cell-mediated contraction allows cells to sense the stiffness of their microenvironment and respond with appropriate mechanosignalling events that act to regulate gene expression. This ECM-mediated signal transduction across the cell membrane, and finally to the nucleus, is facilitated by the integrin family of transmembrane proteins. The first step in this pathway requires the binding of laminin-1 to β_1-integrin, which is then thought to interact with the cytoskeleton (Streuli et al., 1995b). Mammary cells cultured on Engelbreth–Holm–Swarm tumour biomatrix, which is rich in laminin, are able to form spherical alveolar-like structures termed 'mammospheres'. These cells are polarized and secrete both WAP and β-casein preferentially into the lumen of these structures (Li et al., 1987; Barcellos-Hoff et al., 1989). An in vitro bovine mammosphere model grown on Matrigel has been characterized by gene expression profiling (Riley et al., 2010). In this system the initiation of alveoli-like structures was driven by the ECM in the presence of insulin and a glucocorticoid hormone (cortisol), while prolactin was necessary for the induction of milk protein genes. In contrast, when these cells were grown in similar conditions on plastic as a substratum, the cells formed a cobbled-stoned monolayer and failed to induce milk protein genes in the presence of lactogenic hormones. Thus higher order mammary differentiation and regulation by the ECM is required for the expression and secretion of milk protein genes. However, this system does not meet the requirements for the synthesis of α-lactalbumin (Schmidhauser et al., 1995; Riley et al., 2010).

The presence of ECM in culture induces MECs to adopt a spherical morphology, which is absolutely required for prolactin-induced CSN2 gene expression (Roskelley et al., 1994), but the synthesis of lactoferrin is dependent on the presence of insulin and cell rounding alone (Roskelley et al., 1994; Close et al., 1997). Lactoferrin gene expression is repressed at both the transcriptional and post-transcriptional level in flattened cells. The induction of cell rounding by the addition of cytochalasin D, which disrupts the cytoskeleton, was able to relieve this repression of expression (Close et al., 1997). The mechanism by which changes in the cytoskeleton, which is correlated with changes in cell shape, regulate gene expression remains elusive. One theory proposes that changes in the organization of the cytoskeleton could alter the three-dimensional organization of the nuclear matrix to influence gene expression (Getzenberg, 1994). At post-transcriptional level, alterations in the cytoskeleton may affect the stability of mRNA transcripts within the mammary gland (Blum et al., 1989). More than 90% of CSN2 mRNA isolated from mouse MECs is associated with the cytoskeleton. Roskelley et al. (1995) propose that cell rounding represents the lowest level in the hierarchy of ECM-mediated signalling, while higher levels of sophistication approaching the *in vivo* environment are required for *CSN2*, *WAP* and *LALBA* expression, respectively.

Interestingly, it has been proposed that the ECM is species-specific (Sharp *et al.*, 2006b) and provides specialist signals for formation of mammosphere structures and milk protein gene regulation. In the tammar wallaby, it has been shown that ECM extracted from different phases of the lactation cycle can up-regulate expression of specific milk protein genes (Wanyonyi *et al.*, 2013b). Using levels of gene expression for phase specific markers; tELP, tWAP and tLLP-B, representing phases 2A, 2B and 3, respectively, mammary epithelial cells (MECs) extracted from P2B acquired a P3 phenotype when cultured on P3 ECM. Similarly P2A cells acquired the P2B phenotype when cultured on P2B ECM. Mouse studies using ECM extracted from pregnant, lactating or involuting mammary glands showed an influence on gene expression (Schedin *et al.*, 2004; O'Brien *et al.*, 2010).

Intramammary pressure has also been investigated as a potential regulator of milk secretion. Local signals, causing stress between the interaction of the ECM and alveoli epithelial cells are thought to initiate new and independent cascades that activate the apoptotic programme and cause a down-regulation of milk secretion (Boudreau *et al.*, 1995; Clark and Brugge, 1995). During intramammary engorgement, which causes stretching of the basement membrane, it is predicted that molecular interactions with basally located adhesion receptors are altered, leading to reduced ligand-binding interacting sites and a reduction in β1 integrin binding (Banes *et al.*, 1995; McMahon *et al.*, 2004). The affinity modulation of integrin activity and, therefore, a potential inability to respond to survival signals from the basement membrane may contribute to the induction of apoptosis and down-regulation of milk secretion at the onset of involution.

Transcriptional Regulation of Milk Protein Production

Mammary gland-specific expression of milk protein genes involves tissue-unique combinatorial interactions of ubiquitous transcription factors. The production of milk during lactation requires the transcriptional activation of milk protein genes facilitated by binding of nuclear factors to their target sequences, usually located upstream of the coding region. Once bound, these factors interact with RNA polymerase in either a negative or positive manner and thus regulate transcription (Struhl, 1989). The result is a shift in the overall pattern of gene expression, involving quantitative and qualitative alterations. Using the regulation of casein gene expression by peptide and steroid hormones as a model of transcriptional regulation has been extremely valuable for understanding mechanisms by which signalling pathways converge on gene expression (Eisenstein and Rosen, 1988). Much of the information available regarding the nuclear factors involved in the regulation of milk protein gene expression comes from the investigation of the regulation of milk protein gene expression of β-casein and WAP (Rosen *et al.*, 1999). Comparison of different milk protein genes shows that *CSN1* and *CSN2* have highly conserved regulatory motifs, while *CSN3* does not, even though the expression pattern of this gene is similar to the other caseins.

A number of the nuclear factors involved in the regulation of milk gene expression are themselves regulated by hormonal cues and thus act as the link between the hormonal status of the animal and milk production. Many factors demonstrate altered expression profiles and binding activities within the virgin animal and throughout pregnancy, lactation and involution. The occupancy of various nuclear protein-binding sites alters with the stage of lactation as shown by the 'footprint' left by these various proteins within the promoter. The β-casein promoter (−230 bp from the transcription start site) contains binding sites for Stat5 (signal transducer and activator of transcription), CCAAT/enhancer-binding protein β (C/EBPβ), the transcriptional repressor Yin and Yang (YY1) and multiple glucocorticoid response element (GRE) half-sites (Rosen *et al.*, 1999). An evolutionarily conserved distal enhancer with multiple binding sites for Stat5, C/EBPβ, and other factors is located between −6.0 and −1.4 kb from the transcription start site (Winklehner-Jennewein *et al.*, 1998; Rijnkels, 2002).

Prolactin, insulin and glucocorticoid hormones act synergistically to induce β-casein gene expression in mammary epithelial cells (Menzies et al., 2009). The number of transcription factors implicated in the regulation of milk gene expression has risen dramatically with corresponding appreciation of the complexity of interactions involved (Rosen et al., 1996, 1998).

Prolactin-mediated activation of transcription

Early studies using promoter deletion analysis located the hormone response elements necessary for the transcription of the rat and murine Csn2 within the region 221 bp and 258 bp upstream of the transcription initiation site, respectively (Altiok and Groner, 1993; Kanai et al., 1993). Footprinting analysis also revealed that as many as seven different proteins bind to the −258 to +7 portion of the murine promoter (Kanai et al., 1993). One of these proteins, Stat5, is critical in mediating the response of CSN2 to prolactin (Schmitt-Ney et al., 1991).

The prolactin-induced transcriptional regulation of CSN2 proceeds with prolactin binding to the extracellular-binding domain of the long form of the prolactin receptor (PRL-R) located in the cell membrane causing its dimerization (Doppler, 1994). This complex activates the JAK/STAT signal transduction cascade. JAK2 is a cytoplasmic protein tyrosine kinase, which is associated with the intracellular domain of the PRL-R. The enzymatic activity of this kinase is activated when two JAK2 molecules are brought in immediate proximity through receptor dimerization. The activated JAK2 tyrosine phosphorylates PRL-R, resulting in phosphorylated tyrosine residues that act as docking sites for the SH2 domains in cytosolic STAT5. This latent transcription factor is subsequently tyrosine phosphorylated, which promotes dimerization of STAT5 and its subsequent translocation to the nucleus (Duncan et al., 1997; Ihle, 1996; Berchtold et al., 1997). In the nucleus it binds to DNA target sequences known as GAS elements to modulate transcription of target genes

(Gouilleux et al., 1994; Welte et al., 1994) (see Fig. 16.4).

Binding sites for STAT5 dimers have been located between nucleotides −80 to −100 and −130 to −150 of the rat Csn2 gene to sites that correspond to the imperfect palindromic 9-bp consensus sequence 5′-TTCNNNGAA-3′ (Schmitt-Ney et al., 1991; Wakao et al., 1994). Mutations introduced within the proximal Stat5-recognition element abolished the hormonal induction of the Csn2 promoter, confirming its essential role in the hormonal responsiveness of this gene (Schmitt-Ney et al., 1991). This transcription factor has been detected in the mammary gland of lactating cattle and is able to bind to regions within bovine LALBA (Kuys et al., 1996) and CSN3 promoters (Adachi et al., 1996), in addition to bovine CSN1S1 and CSN2 promoters (Wakao et al., 1992). Consensus binding sites for STAT5 are absent from the bovine lactoferrin promoter (Seyfert and Kuhn, 1994), which is in agreement with the observation that the lactoferrin gene can be transcribed in murine MEC culture in the absence of prolactin (Close et al., 1997).

STAT5 DNA-binding activity has been detected in mammary tissue obtained from lactating cattle and is absent in non-lactating, pregnant and involuting cows and liver extracts (Kuys et al., 1996) which agrees with the findings of Schmitt-Ney et al. (1992) in mice. During lactation frequent suckling appears necessary to maintain STAT5 activity in both mice (Schmitt-Ney et al., 1992) and cattle (Kuys et al., 1996). A positive correlation between STAT5 activity and protein concentration in the milk of cattle has also been established (Yang et al., 2000).

Distinct functional domains have been identified within STAT molecules. An amino acid terminal domain mediates cooperatively between STAT molecules, a central domain confers DNA binding specificity of the dimer, an SH-2 domain mediates dimerization through phosphotyrosine recognition and a transactivation domain resides in the carboxyl terminus (Moriggl et al., 1996).

Two STAT5 genes have been identified, STAT5A and STAT5B. These genes are more than 90% identical and probably arose by gene duplication (Liu et al., 1995). Both genes

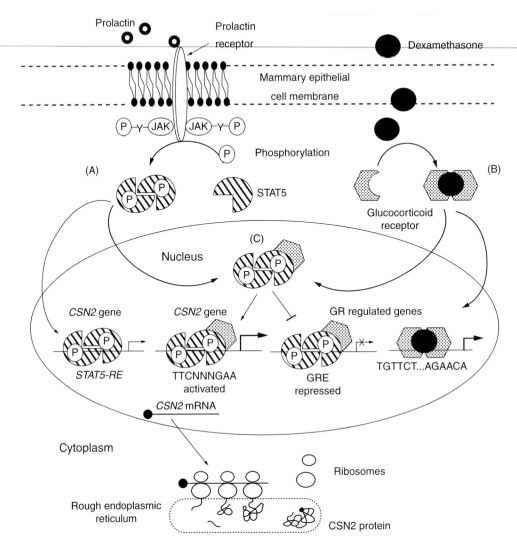

Fig. 16.4. Model of the activation of STAT5 transcriptional regulation. The lactogenic hormones prolactin and glucocorticoid utilize signal transduction pathways that are initiated by ligand receptor interactions. Prolactin binds to the extracellular domain of the PRLR, causing its dimerization and the activation of JAK2, a tyrosine kinase, non-covalently associated with the intracellular domain of the receptor. Phosphorylation of the intracellular domain is thought to cause the recruitment of STAT5 through its SH2 domain to the receptor, phosphorylation of STAT5 on tyrosine 694, dimerization and nuclear localization. Specific DNA binding to a Stat5 response element TTCNNNGAA results in the induction of gene transcription. Lipophilic glucocorticoids molecules can enter the cell and interact with the GR in the cytoplasm. The ligand binding to the receptor causes an allosteric change and dissociation from masking heat shock proteins. The GR activates target genes containing glucocorticoid response element AGAACANNNTGTTCT in their promoters. In the presence of both hormones, STAT5 and the GR form a molecular complex that binds preferentially to the STAT5RE and enhances the induction of target genes compared to the induction observed with prolactin alone. The enhancement requires GR activation and the amino-terminal domain that comprises the AF1 function. The specific DNA-binding function of the GR is not necessary for functional interaction and the GR/STAT5 complex preferentially binds to STAT5 DNA binding sites. The sequestration of the GR is accompanied by a diminished response of glucocorticoid-regulated genes.

encode isoforms that arise by alternative splic-ing. STAT5A and STAT5B differ only in their carboxyl terminus (Liu et al., 1995). Of the two STAT5 isotypes, STAT5A is more impor-tant for PRL-dependent mammary gland devel-opment and lactation, whereas STAT5B is more involved in glucocorticoid hormone sig-nalling (Liu et al., 1997; Udy et al., 1997). The use of a genetically modified mouse with repressed STAT5A expression in the mam-mary gland during lactation showed that STAT5A expression negatively affects the transcription of WAP and LALBA, but not CSN2 (Reichenstein et al., 2011).

The carboxy-truncated isoforms of STAT5 are perhaps the most interesting. These natu-rally occurring carboxy-truncated STAT5 iso-forms lack a functional transactivation domain and act as dominant negative inhibitors of STAT5-dependent transcription. They remain tyrosine phosphorylated and bound to GAS sites for longer periods of time than full-length STAT5 isoforms following prolactin treatment, suggesting that the carboxy-terminal sequences may affect the interaction with a tyrosine phos-phatase (Moriggl et al., 1996, 1997). These variants are able to efficiently recruit nuclear co-repressors to gene promoters and actively participate in the formation of silent chromatin.

Site-directed mutagenesis of a negative control region within the rat CSN2 promoter, which is immediately adjacent to the proximal STAT5-binding site, led to an increased consti-tutive transcription of a CAT reporter construct in culture (Schmitt-Ney et al., 1991; Altiok and Groner, 1993). These data suggested that the regulation of β-casein gene expression included a relief of repression mechanism. Further investigation of one of the proteins that is bound within this region revealed it to be identical to the nuclear protein YY1 (Meier and Groner, 1994; Raught et al., 1994). YY1, a silencing mediator of retinoid and thyroid receptor, and histone deacetylase 3 (HDAC3) have been suggested to play a role in hormone regulation of CSN2 gene expression. YY1 interacts constitutively with the promoter in the absence of hormone, and its dissociation induced by PRL and glucocorticoids is required for activation of CSN2. YY1 interacts constitu-tively with the promoter in the absence of hor-mone. However its dissociation induced by

PRL and glucocorticoids is required for activa-tion of CSN2 as revealed by bandshift assays revealing that STAT5 binds to this negative control region with greater affinity than YY1. This suggests that once STAT5 is activated via JAK2 kinase in response to increased prolactin levels, it is able to displace YY1 from the pro-moter thus relieving its repressive effects and activating transcription (Meier and Groner, 1994; Raught et al., 1994). Murine YY1 is also able to recognize its corresponding bind-ing site within the bovine CSN2 promoter sug-gesting that a similar hormone-mediated relief of repression mechanism may be operating for this gene (Meier and Groner, 1994).

The glucocorticoid receptor

Glucocorticoids alone have little to no ability to induce expression of CSN2. However, they synergize with prolactin through positive coop-erative interactions between the glucocorticoid receptor (GR) and STAT5A (Buser et al., 2007). The GR is a specific transcriptional mediator of glucocorticoid hormone signals that acts directly on DNA to influence tran-scription of downstream target genes. This action involves the uptake of glucocorticoids which are then complexed with the latent form of GR residing in the cytoplasm. This binding induces a conformational change in GR and dissociates heat shock proteins that are associ-ated with the inactive form. Upon glucocorti-coid binding, the now active GR is translocated to the nucleus where it binds to glucocorticoid-response elements (GRE) possessing the pseudo-palindromic consensus sequence GGTACAnnn TGTTCT, and functions as a transcriptional activator (Beato, 1989). GR also can interact with GRE half-sites, a DNA interaction that can be stabilized by other trans acting factors (e.g. STAT5) binding to adjacent sequences.

Studies of CSN2 expression initially showed that the kinetics of induction by prolactin was more rapid than that for glucocorticoids, sug-gesting that these hormones acted via different signal transduction pathways (Doppler et al., 1990). It is now well established that prolac-tin and glucocorticoid hormones cooperate in regulation of milk protein gene transcription

(Wyszomierski and Rosen, 2001), identifying a novel role for the GR. In this regard, GR acts as a transcriptional coactivator for STAT5 and enhances STAT5-dependent transcription of the *CSN2* promoter. This action of GR requires its activation via ligand binding and is dependent on the formation of a complex with activated STAT5 and recruitment to the *CSN2* promoter, but is independent of a GR element (GRE). The formation of the STAT5-GR complex diverts GR from GREs present in other promoters such as MMTV and LTR (Stocklin *et al.*, 1996; Stoecklin *et al.*, 1997). This synergism was first established by co-transfection experiments with *STAT5* and *GR* in addition to deletion and point mutations of the transcription factors, and more recently the essential nature of STAT5 for PRL/glucocorticoid induction of *CSN2* expression was shown by small hairpin RNA knockdown of *STAT5* in primary mammary epithelial cell cultures (Vafaizadeh *et al.*, 2010).

Activated GR is also able to alter the chromatin structure within the promoter and facilitate access of other transcription factors to their recognition sequences (Tsai and O'Malley, 1994), which appears to be the case for nuclear factor 1 (NF1) within the rat WAP promoter (Li and Rosen, 1994).

Classical GREs have not been detected within the promoters of milk protein genes, however, multiple half-sites containing the TGT(T/C)CT motif are present (Yu-Lee *et al.*, 1986; Welte *et al.*, 1993). The functional significance of these half-sites has been demonstrated by their ability to specifically bind purified GR from rat liver at many locations within both the rat *Csn2* and *Wap* promoters (Welte *et al.*, 1993).

Early evidence of this interaction was suggested by mutational inactivation of the proximal Stat5 site within the rat *Csn2* promoter, which interrupted prolactin response and that of glucocorticoids (Schmitt-Ney *et al.*, 1992). Two GR half-sites overlap with the proximal Stat5 element consistent with a role for GR as a co-activator for Stat5.

In addition to cooperative interactions between GR and Stat5A at the promoter, C/EBPβ potentiates Stat5A-mediated transactivation of *CSN2* in a manner dependent on a functional GR. It has been proposed that GR relieves an inhibitory conformation of C/EBPβ within its N-terminal transactivation domain (Wyszomierski and Rosen, 2001). Recent chromatin immunoprecipitation (ChIP) assays demonstrated a correlation between the dynamics of YY1 and HDAC3 dissociation from the *CSN2* promoter upon treatment of cells with PRL plus glucocorticoid, suggesting that YY1 and HDAC3 form a functional complex (Kabotyanski *et al.*, 2009). YY1 has been reported to interact constitutively with STAT5A, and its dissociation has been suggested to also be required for cytokine-induced activation of STAT5 (Nakajima *et al.*, 2001). Thus, hormone activation of *CSN2* transcription involves release of inhibitory factors as well as recruitment of transcriptional activators. Thus, STAT5A, GR and C/EBPβ cooperate to mediate maximal expression of *CSN2* and are thought to act as a unit for recruitment of coactivators such as p300, with histone acetyl transferase activity required for gene activation through acetylation of histones and chromatin remodelling.

The progesterone receptor – a negative regulator of milk protein gene expression

Progesterone plays multiple roles in the mammary gland during pregnancy. Copious milk secretion is repressed until parturition by the action of progesterone. Studies in rodents show a dramatic decline in circulating progesterone at parturition leads to increased milk protein production and tight junction closure (Loizzi, 1985). Progesterone receptor (PR) expression in the mammary gland declines progressively during pregnancy, suggesting that reduced PR at parturition also contributes to the onset of lactation (Ismail *et al.*, 2002). Mice that were ovariectimized during pregnancy were shown to exhibit transient lactogenesis and increased *CSN2* expression. This effect was specifically prevented by progesterone administration at the time of ovariectomy, an effect not observed with other hormones (Kuhn, 1969; Deis and Delouis, 1983). The mechanism of the suppressive effects of progesterone on milk protein gene expression have been examined using cell culture systems reconstituted to express

PR, the PRL-R/STAT5 signalling pathway and GR (Buser et al., 2007). This model enabled evaluation of PR, GR and STAT5 interaction at the CSN2 promoter, and showed that PR, in a progestin-agonist-dependent manner, inhibited PRL- and glucocorticoid-induced transcriptional activation of CSN2 by interfering with the nuclear PRLR/JAK2/STAT5A signalling pathway. Recent studies have used HC-11 cells and chromatin immunoprecipitation assays to examine the influence of progesterone and PR on the synergistic actions of the various transcription factors and coregulatory proteins that control CSN2 expression. PR was shown to be dependent on progesterone for recruitment to the promoter, which in turn inhibited the assembly of an active transcription complex, and produced repressive histone modifications (Buser et al., 2011). These studies suggest a novel mechanism of steroid receptor-mediated transcriptional repression of milk protein genes.

Nuclear factor 1 (NF1)

The NF1 family of transcription factors has been implicated in the tissue-specific expression of a number of genes (Paonessa et al., 1988). Mutational inactivation of two NF1 sites within the distal promoter of rat WAP abolished its mammary-specific expression in transgenic mice (Li and Rosen, 1995). Mammary-specific members of the NF1 family (NF1-A4, NF1-S2 and NF1-X1) have been identified in the mammary gland during lactation, and elicit the tissue-specific expression of milk protein genes (Mukhopadhyay et al., 2001). Differential expression of NF1 isoforms has also been observed within the mammary gland of cattle, where an increase in the binding activity of NF1 complexes can be correlated with their altered subunit composition during lactation.

NF1 in conjunction with another nuclear factor termed mammary cell-activating factor (MAF), were demonstrated to mediate the mammary-specific expression of mouse mammary tumour virus (Mink et al., 1992). The introduction of mutations in the MAF site between −120 and −100 within the mouse WAP promoter was also found to alter the hormone-independent mammary cell-specific

transcription of reporter constructs in cell culture (Welte et al., 1994). Further characterization of the factors that bound within this region showed that they were related to the extensive ETS family of transcription factors. Some ETS proteins are expressed during specific stages of development and are restricted to certain tissues and are thus candidates for the regulation of tissue-specific gene expression (Wasylyk et al., 1993). A number of ETS-related proteins are expressed within MECs during lactation, for example ELF5 can trans-activate the WAP promoter (Thomas et al., 2000) and is important for alveolar cell fate (Oakes et al., 2008; Lee et al., 2013).

The extracellular matrix and transcriptional regulation

Promoter deletion analysis and transfection experiments performed in murine MEC culture discovered the presence of an enhancer region between −1517 and 1677 bp upstream of bovine CSN2, designated CSN2 element-1 (BCE-1), which is able to confer both prolactin and ECM responsiveness to an inactive CSN2 promoter (Schmidhauser et al., 1990, 1992). This regulation was absent when constructs were introduced into Chinese hamster ovary cells and Mabin-Darby Canine Kidney cells, suggesting that the enhancer may be mammary-specific (Lelievre et al., 1996). Elements similar to BCE-1 may not be restricted to CSN2 as an alignment with the bovine LGB revealed a sequence possessing 53% similarity within the region −741 to −619. A 69-bp region approximately 1.5 kb upstream of bovine LALBA, which contains 75% sequence similarity with the 3′ end of BCE-1 has also been identified (Bleck and Bremel, 1993).

Sequence analysis has revealed the presence of a putative STAT5-binding site, a half GRE and two C/EBP-binding sites within BCE-1 (Raught et al., 1995). How does the ECM regulate expression at the transcriptional level? Investigations have revealed that both STAT5 and SARP DNA-binding activity, crucial for the expression of CSN2, are positively regulated by the ECM (Streuli et al., 1995a; Edwards et al., 1996; Altiok and Groner, 1998).

The DNA-binding activity of STAT5 isolated from MEC culture was found to be present in cells cultured in the presence of basement membrane or laminin-1, but missing from those cultured on plastic or collagen (Streuli et al., 1995a). Further studies revealed that the ECM is able to regulate *CSN2* expression by regulating the activity of JAK2 kinase (Edwards et al., 1996). Thus both the ECM and prolactin signalling pathways converge at this junction, which is crucial in the phosphorylation cascade required to activate STAT5. The importance of the ECM can be seen when cells are grown on an appropriate ECM to produce functional mammospheres. In this setting, prolactin is able to induce milk protein expression through activation of STAT5, but only in the presence of laminin-111 (Xu et al., 2009). In addition, expression of *CSN2* cannot be induced in proliferating cultures of HC11 cells, but requires confluent cultures that seem to be able to deposit their own basement membrane (Ball et al., 1988; Chammas et al., 1994).

ECM extracted from the tammar wallaby mammary gland at different phases of the lactation cycle is responsible for the transcriptional regulation of *MaeuCath1a*, an antimicrobial splice variant of the cathelicidin gene (Wanyonyi et al., 2013a). Tammar mammary epithelial cells (WallMECs), grown on ECM extracted from different phases of the wallaby lactation cycle, express the *MaeuCath1a* transcript in a lactation phase-dependent manner. Luciferase reporter-based assays and *in silico* analysis of deletion fragments of the 2245-bp sequence upstream of the cathelicidin translation start site identified ECM-dependent positive regulatory activity in the −709 to −15 bp region and repressor activity in the −919 to −710 bp region. Electrophoretic Gel Mobility Shift Assays using nuclear extract from ECM-treated WallMECs showed differential band shift in the −839 to −710 bp region, suggesting that ECM proteins directly interact with the DNA.

MicroRNA Regulation

A microRNA (miRNA) is a small non-coding RNA molecule (22 nucleotides) involved in gene silencing via negative regulation (transcript degradation and sequestering, translational suppression) of target genes (Chen and Rajewsky, 2007). In animals these miRNAs bind their target mRNA through complementary sequences that confer target specificity (Lewis et al., 2003, 2005). By affecting gene regulation, miRNAs are likely to be involved in most biological processes. In 2010, the first publication on miRNA in breast milk appeared (Kosaka et al., 2010). The study focused on miRNA species that were previously implicated in the down-regulation of immune-related proteins and showed that these miRNAs were relatively resistant to low acidic pH levels similar to those found in the stomach.

Studies have shown that miRNA molecules are found in raw and processed bovine milk (Chen, X. et al., 2010; Hata et al., 2010; Izumi et al., 2012). Mammary miRNAs differ in abundance and type during different stages of the lactation cycle, and also in non-lactating glands, and colostrum shows a greater abundance and variety of exosomal miRNAs compared to mature milk (Gu et al., 2012; Izumi et al., 2012).

Studies of various species, including the cow, strongly support that miRNAs are remarkably resilience to degradation (Gu et al., 2012; Izumi et al., 2012; Zhou et al., 2012). The resistance to degradation may be due to the location of miRNAs, which have been shown to be packaged within exosomes (Hata et al., 2010). Exosomes are small vesicles that form inside cells and are secreted into the fluid surrounding these cells. They contain not only miRNA, but also other biomolecules, including proteins. There is convincing evidence exosomes can fuse into the cell membrane of target cells and release their contents, including miRNA. When exosomes are put together with cells in the laboratory, they have been shown to modify cellular responses of their target cells, including the release of immune modulatory molecules (Admyre et al., 2007; Hata et al., 2010). Studies have shown the presence of plant-derived miRNAs in the bloodstream of adults who had eaten rice (Zhang et al., 2012). A specific rice miRNA (MIR168a) was shown to slow the production of Low Density Lipoprotein Receptor-related Protein Associated Protein (LDLRAP) in the mouse liver.

This demonstrated for the first time that a plant food component potentially regulates mammalian cells. It is, therefore, tempting to speculate that milk exosomal miRNAs are also transferred from the mother's milk to the infant via the digestive tract, and that they play a critical role in the development of the infant immune system (Fig. 16.5).

Different stages of milk storage and processing show considerable variation in the levels of most miRNA molecules in milk. As such miRNAs are now being examined as biomarkers for raw milk quality. Seven miRNA molecules have been identified as relatively stable in raw milk, and these may be useful as indicators of milk processing effects or spoilage (Chen, X. et al., 2010). Although miRNA is relatively resistant to milk processing, it seems to undergo a considerable loss in concentration during the process. Studies of infant formula show miRNA molecules are found at one-tenth the levels of those in unprocessed milk (Izumi et al., 2012).

There are still many questions to be answered, but the high conservation of miRNAs between species suggests it is likely that bovine-derived miRNA will have similar effects to human miRNA. miRNAs in milk may represent another area in the research of designer milk.

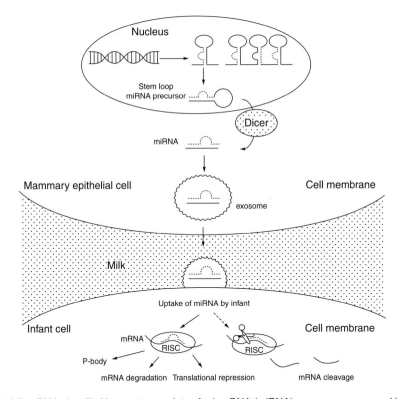

Fig. 16.5. MicroRNAs in milk. Nascent transcripts of microRNA (miRNA) genes are processed into a stem-loop precursor, which is further processed by Dicer into a mature miRNA duplex. The mature miRNA is then packaged into exosomes and secreted into milk and ingested by the young. The exosome resists the acidic pH of the stomach and is taken up by the gut cell whereby the miRNA is released. One strand of the miRNA duplex gets incorporated into the effector complex RISC (RNA-induced silencing complex), which recognizes specific targets through imperfect base-pairing and induces post-transcriptional gene silencing. Several mechanisms have been proposed for this mode of regulation: miRNAs can induce the repression of translation initiation, mark target mRNAs for degradation by deadenylation or sequester targets into cytoplasmic P-body structures, which are involved in mRNA degradation.

The Lactome

In recent years in-depth study of molecular regulation of lactation, particularly in livestock species, has begun (Loor and Cohick, 2009). Studies have focused on large-scale mRNA expression (Bionaz and Loor, 2011) combined with the accumulated knowledge of regulatory mechanisms of milk protein synthesis. This has opened new frontiers for intervention to improve milk protein synthesis.

Study of transcriptomics has been carried out via DNA microarray technology or Next Generation Sequencing (NGS). NGS is replacing microarrays as the tool of choice for functional genomics studies (Bionaz et al., 2012). It allows the analysis of the expression of the entire transcriptome in tissues or cells.

Studies pertaining to the regulation of other minor milk proteins are limited but have gained momentum. For example, lactoferrin expression is known to be induced during mastitis (Moyes et al., 2009), and lactoferrin expression increases dramatically after treatment of mammary epithelial cells with bacteria LPS or double stranded RNA through the PKC, NF-κB and MAPK pathways (Li et al., 2009). Transcript abundance of *LTF* has been shown to be under the control of a specific miRNA (miR-214) (Liao et al., 2010), and butyrophilin expression appears to be strongly regulated by AKT1 (LaRocca et al., 2011). AKT is a serine-threonine protein kinase with a role in multiple pathways, including insulin signalling. The AKT proteins are essential for milk synthesis, as shown by the lack of production of all main milk components in AKT1- and AKT2-deficient mice (Chen, C. et al., 2010). Each of the three known AKT isoforms plays a different role in the mammary gland, with AKT1 being the most important for lactogenesis (Maroulakou et al., 2008).

Transgenic Studies Utilizing Bovine Milk Protein Gene Promoters

Expression studies

One of the advantages to studying transgene expression in animal models is the opportunity to identify genomic regions required for *in vivo* expression that may influence tissue-specific and developmental regulation (Soulier et al., 1992; Burdon et al., 1994; Li and Rosen, 1995). Many of the transgenic experiments performed using milk protein genes have been reviewed (Bawden et al., 1994; Wall et al., 1997; Sabikhi, 2007; Samiec and Skrzyszowska, 2011). Here we will give a brief overview of the transgenic studies performed with bovine mammary-specific DNA promoters, many of which are summarized in Table 16.1. However, it should be kept in mind that many studies utilize chimeric constructs, containing only portions of the bovine promoter, making direct comparisons difficult.

Most bovine milk protein promoters are able to confer mammary-specific expression on native or heterologous protein coding sequences within transgenic organisms (Takahashi and Ueda, 2001; Yen et al., 2008; Cheng et al., 2012). The elements responsible for tissue-specific and developmental regulation appear to be localized less than 1 kb from the transcription start site, while those regions required to direct high-level expression (1 mg ml^{-1} or greater) tend to be further upstream. For example, 750 bp of upstream region is sufficient to drive the tissue-specific expression of bovine *LALBA* (at a maximum of 0.45 mg ml^{-1}) in the milk of transgenic mice (Vilotte et al., 1989). Promoter deletion studies identified the elements required for the tissue-specific expression of this transgene to be located between the region −477 to −220 bp of this promoter (Soulier et al., 1992).

Whey promoters tend to require less upstream region to direct high-level expression in transgenic mice than do the casein genes (Lipnik et al., 2005; Naruse et al., 2006). High-level expression of *LALBA* was achieved with 2 kb of upstream region (Bleck and Bremel, 1994) yet more than twice this amount is insufficient for the expression of *CSN3* (Rijnkels et al., 1995). For this reason the LALBA promoter has been utilized to express bovine casein genes in transgenic mice (Bleck et al., 1995; Choi et al., 1996).

High-level expression has been achieved for a bovine *CSN2* transgene utilizing 6.6 kb of upstream sequence (Brophy et al., 2003), yet very poor results have been obtained for

Table 16.1. Transgenic studies utilizing bovine milk protein promoters in various mammalian species.

Transgene construct	Flanking sequence[b]	Expression per ml milk	Reference
Mice			
α_{S1}-casein-h urokinase	21; 2 kb	1–2 mg	Meade *et al.*, 1990
α_{S1}-casein-ht-PA (cDNA)	1.6 kb; SV40	50 µg	Riego *et al.*, 1993
α_{S1}-casein-h lactoferrin (cDNA)	6.2; 8 kb	0.1–36 µg	Platenburg *et al.*, 1994
α_{S1}-casein (minigene)	1.35; 1.5 kb	<0.1% endog.[c]	Clarke *et al.*, 1994
α_{S1}-casein-h lysozyme (cDNA)	20; 2 kb	to 0.78 mg	Maga *et al.*, 1995
β-casein	4.3; 1.3 kb	<1 µg–3 mg	Bawden *et al.*, 1994
β-casein	16; 8 kb	<0.1–20 mg	Rijnkels *et al.*, 1995
α_{S2}-casein	8; 1.5 kb	<0.1% endog.[c]	Rijnkels *et al.*, 1995
κ-casein	5; 19 kb	<0.03% endog.[c]	Rijnkels *et al.*, 1995
κ-casein-bGH	4.4 kb; bGH	0.1–7.5 ng	Popov *et al.*, 1995
α-lactalbumin	750; 336 bp	2.5 µg–0.45 mg	Vilotte *et al.*, 1989
α-lactalbumin	2; 2.7 kb	100 ng–1.5 mg	Bleck and Bremel, 1994
α-lactalbumin-oTP (cDNA)	740; 450 bp	1 µg	Stinnakre *et al.*, 1991
α-lactalbumin-b β-casein	2 kb; b β-casein	1–12 mg	Bleck *et al.*, 1995
α-lactalbumin-b β-casein[a]	2 kb; b β-casein[a]	2–3 mg	Choi *et al.*, 1996
β-lactoglobulin-hEPO	2.8; 1.9 kb	0.4 µg–0.3 mg	Korhonen *et al.*, 1997
Rats			
α_{S1}-casein-hGH	671 bp; hGH	87 µg–6.5 mg	Ninomiya *et al.*, 1994
β-casein-hGH	1.7 kb; hGH	1 µg–10.9 mg	Ninomiya *et al.*, 1994
α-lactalbumin	0.8; 0.5 kb	0.2 µg–2.4 mg	Hochi *et al.*, 1992
α-lactalbumin-hGH	738 bp; hGH	1 µg–4.4 mg	Ninomiya *et al.*, 1994
Rabbits			
α_{S1}-casein-ht-PA (cDNA)	1.6 kb; SV40	8–50 ng	Riego *et al.*, 1993
α_{S1}-casein-hIGF-1 (cDNA)	2.9; 3.5 kb	0.1–1 mg	Brem *et al.*, 1994
β-lactoglobulin-hEPO	2.8; 1.9 kb	0.35–0.54 mg	Korhonen *et al.*, 1997

[a]Modified protein coding sequence.
[b]Flanking sequence around the designated transcriptional unit (size of 5′ flanking, 3′ flanking sequence or 3′ gene sequence utilized).
[c]As percentage of endogenous bovine expression level.
Abbreviations: h, human; b, bovine; t-PA, tissue plasminogen activator; GH, growth hormone; oTP, ovine trophoblast interferon; EPO, erythropoietin; IGF-1, insulin-like growth factor-1. Adapted and updated from Bawden *et al.* (1994), with permission.

CSN1S2 and *CSN3* (Popov *et al.*, 1995; ijnkels *et al.*, 1995). As these genes are present in cattle within a cluster within in the order *CSN1S1-CSN2-CSN1S2-CSN3*, it has been proposed that major regulatory elements required for the coordinated high-level expression of these genes during lactation may reside upstream of the cluster (Rijnkels *et al.*, 1995) as is the case for the human β-globin locus (Grosveld *et al.*, 1987). This may explain why all transgenics for *CSN1S2* and *CSN3* to date exhibit low expression levels, but can be overcome by the use of *CSN2/CSN3* fusion constructs under the control of the *CSN2* promoter (Brophy *et al.*, 2003).

The alteration of milk composition via transgenesis

Targeting transgene expression to the mammary gland provides an opportunity to modify the composition of all the major milk constituents and subsequently improve the nutritional and manufacturing quality of milk (Sabikhi, 2007). Introduction of additional casein genes to the bovine genome increases the total casein content of milk, thereby increasing its manufacturing value for the production of cheese (Brophy *et al.*, 2003). Expression of less favourable genes, such as those implicated in allergic responses (e.g. LGB) could be down-regulated.

Protein engineering has been performed on milk protein genes such as bovine *CSN3* (Oh and Richardson, 1991) and *CSN2* (Simons *et al.*, 1993; Choi *et al.*, 1996) to improve dairy processing characteristics. Studies were first initiated utilizing transgenic mice as a model system and performed with milk protein genes that had been specifically designed to determine the effects of altered milk protein expression levels on milk production. These include experiments designed to alter lactose levels (Stinnakre *et al.*, 1994; Stacey *et al.*, 1995; L'Huillier *et al.*, 1996), the introduction of antibacterial proteins (Platenburg *et al.*, 1994; Maga *et al.*, 1994) and the alteration of casein expression levels (Kumar *et al.*, 1994; Gutierrez-Adan *et al.*, 1996).

The fast development of transgenic technology has led to the creation of a broad range of transgenic cloned animals for milk production. Three lines of transgenic cloned cattle that specifically express human α-lactalbumin, lactoferrin or lysozyme in milk were created as a result of the integration of a specific transgene into the genome and were cloned by subsequent SCNT (Wang *et al.*, 2008; Yang *et al.*, 2008; Yang *et al.*, 2011). The major changes observed between milk from these animals and milk from conventionally bred animals were the high levels of human milk proteins. In addition, targeted microRNA technologies were recently used to genetically modify a cow in order to knock down LGB expression (Formigli *et al.*, 2000; Jabed *et al.*, 2012). Thus, the milk of these transgenic animals has very different properties from that of conventionally bred cows and provides a unique model for evaluating the effects of exogenous transgenes or gene 'knockdowns' on the profile of the endogenous milk proteins.

α-Lactalbumin

It would be advantageous to reduce lactose levels in bovine milk because much of the world's population is lactose intolerant (Mercier, 1986). Homozygous null mutants for murine α-lactalbumin have been generated via gene targeting in embryonic stem cells by two different laboratories (Stinnakre *et al.*, 1994; Stacey *et al.*, 1995). As predicted, both α-lactalbumin, and thus lactose, were absent from the milk

of these mice. In addition, the milk contained approximately 60% more fat and an 88% increase in protein with respect to wild type milk (Stacey *et al.*, 1995). Unfortunately, the milk obtained from these mice was so viscous that very little could be ejected from the teat (Stinnakre *et al.*, 1994; Stacey *et al.*, 1995). These results confirm the importance of lactose in the regulation of milk volume within the mammary gland.

A ribozyme-mediated approach to the down-regulation of α-lactalbumin has been pursued (L'Huillier *et al.*, 1996). Mice expressing a hammerhead ribozyme designed to interact with the 3' UTR of the bovine α-lactalbumin mRNA were crossed with transgenic mice that expressed bovine α-lactalbumin at a high level. Double transgenic mice exhibited a 78–50% reduction in bovine α-lactalbumin mRNA expression, while the expression of endogenous α-lactalbumin was unaffected.

Lactoferrin and lysozyme

Udder disease caused by bacterial infection not only causes discomfort in the animal, but also leads to loss of milk production and quality (Miles *et al.*, 1992). Milk proteins such as lysozyme and lactoferrin have been demonstrated to possess antimicrobial activity (Flemming, 1922; Arnold *et al.*, 1977) yet are expressed at low levels in bovine milk compared with that of other mammals (Chandan *et al.*, 1968; Masson and Heremans, 1971). Both human lysozyme and lactoferrin have been expressed in the milk of transgenic mice (Maga *et al.*, 1994; Platenburg *et al.*, 1994; Zhang *et al.*, 2001) and in transgenic cows (van Berkel *et al.*, 2002). Milk samples from transgenic mice secreting human lysozyme have been demonstrated to be bacteriostatic against a mastitis-causing strain of *Staphylococcus aureus* in addition to certain cold-spoilage organisms (Maga *et al.*, 1998). Thus, the high-level expression of anti-bacterial proteins in bovine milk could confer mastitis resistance to these cows. Lactoferrin has also been shown to aid gastrointestinal (GI) maturation of human infants and beneficial health effects have been observed when supplementing human and animal diet with lactoferrin. Transgenic milk from cows expressing human lactoferrin at similar levels to those seen in human milk have been fed to young pigs that were used as a model for the GI tract in children (Guilloteau et al., 2010; Cooper et al., 2012). These pigs showed beneficial changes in circulating leukocyte populations, an indicator of decreased systemic inflammation, and overall favourable changes in GI villi architecture; indicating that the consumption of bovine milk containing human lactoferrin has the potential to induce positive changes in the GI tract.

Casein

The primary role of κ-casein in milk is to stabilize the structure of the casein micelle (Waugh and von Hippel, 1956). The chymosin-mediated cleavage of κ-casein leads to destabilization of the micelle and the precipitation of the caseins important in cheesemaking (Mercier et al., 1973). It has been demonstrated that the total amount of κ-casein present in micelles was proportional to the ratio of surface area to volume (McGann et al., 1980). Therefore, it has been proposed that enhanced secretion of κ-casein in milk could reduce micelle size, thereby increasing its surface area, resulting in reduced rennet clotting time (Pearse et al., 1986). Transgenic mice expressing bovine κ-casein have been generated utilizing the caprine β-casein promoter (Gutierrez et al., 1996), however, whereas this milk possessed micelles of a reduced diameter, there was no difference in rennet clotting time compared to milk from non-transgenic mice, though gel strength did increase.

Enhanced milk composition has been achieved by increasing the casein concentration in milk. This was first performed using female fibroblasts, where additional copies of the β- and κ-casein genes were introduced. Independent donor lines then underwent nuclear transfer to produce 11 transgenic calves. The analysis of hormonally induced milk showed substantial expression and secretion of the transgene-derived caseins into milk, with a 8–20% increase in CSN2, a twofold increase in CSN3 levels and a markedly altered CSN3 to total casein ratio (Brophy et al., 2003). This result demonstrated that it is feasible to substantially alter a major component of milk in high-producing dairy cows by a transgenic approach and thus to improve the functional properties of dairy milk.

β-Lactoglobulin

Initial work to generate an LGB-free cow was performed in a mouse model in which microRNA was designed to reduce expression of the LGB gene resulting in reduced protein production of 96% (Jabed et al., 2012). This strategy was then used in a bovine model to produce the first LGB-free milk. Milk from this cow contained no detectable levels of LGB protein; all other major milk proteins were greatly increased, in particular the caseins (α- and β-casein more than twofold and κ-casein approximately fourfold). The LGB-free milk is of particular importance for human consumption as LGB typically causes many allergic reactions to people sensitized to cow's milk. This work validates targeted microRNA expression as an effective strategy to alter milk composition, however the use of such products remain controversial due to current laws surrounding consumption of foods from genetically modified animals.

Conclusion

Understanding the regulation of milk protein gene expression has application in the dairy industry. Traditionally, the genetic improvement of livestock has been achieved by applying the principles of genetics and animal breeding. Advances in gene and quantitative trait mapping coupled with traditional animal breeding approaches, and newer techniques for marker-assisted selection, will be effective in providing superior genetics of livestock that could allow increases in milk yield and increased milk quality. By far the largest economic impact of cloning, or at least the most immediate impact, will be the ability of dairy producers to increase milk yields, since this has an immediate economic return to producers, even more so than enhancing cows for disease resistance and other quality traits. The advancement of modern biotechnology provides new avenues for genetic improvement in production animals and opportunities to produce so-called 'designer milk' for human consumption of 'better for you' products with value-added properties. In the future these 'new milks' may contain therapeutic proteins.

References

Acharya, K.R., Stuart, D.I., Walker, N.P., Lewis, M. and Phillips, D.C. (1989) Refined structure of baboon alpha-lactalbumin at 1.7 Å resolution. Comparison with C-type lysozyme. *Journal of Molecular Biology* 208, 99–127.

Acharya, K.R., Ren, J.S., Stuart, D.I., Phillips, D.C. and Fenna, R.E. (1991) Crystal structure of human alpha-lactalbumin at 1.7 A resolution. *Journal of Molecular Biology* 221, 571–581.

Actor, J.K., Hwang, S.A. and Kruzel, M.L. (2009) Lactoferrin as a natural immune modulator. *Current Pharmaceutical Design* 15, 1956–1973.

Adachi, T., Ahn, J.Y., Yamamoto, K., Aoki, N., Nakamura, R. and Matsuda, T. (1996) Characterization of the bovine kappa-casein gene promoter. *Bioscience, Biotechnology and Biochemistry* 60, 1937–1940.

Admyre, C., Johansson, S.M., Qazi, K.R., Filen, J.J., Lahesmaa, R., Norman, M., Neve, E.P., Scheynius, A. and Gabrielsson, S. (2007) Exosomes with immune modulatory features are present in human breast milk. *Journal of Immunology* 179, 1969–1978.

Alexander, L.J., Stewart, A.F., Mackinlay, A.G., Kapelinskaya, T.V., Tkach, T.M. and Gorodetsky, S.I. (1988) Isolation and characterization of the bovine kappa-casein gene. *European Journal of Biochemistry* 178, 395–401.

Alexander, L.J., Hayes, G., Pearse, M.J., Beattie, C.W., Stewart, A.F., Willis, I.M. and Mackinlay, A.G. (1989) Complete sequence of the bovine beta-lactoglobulin cDNA. *Nucleic Acids Research* 17, 6739.

Alford, D. and Taylor-Papadimitriou, J. (1996) Cell adhesion molecules in the normal and cancerous mammary gland. *Journal of Mammary Gland Biology and Neoplasia* 1, 207–218.

Altiok, S. and Groner, B. (1993) Interaction of two sequence-specific single-stranded DNA-binding proteins with an essential region of the beta-casein gene promoter is regulated by lactogenic hormones. *Molecular and Cellular Biology* 13, 7303–7310.

Altiok, S. and Groner, B. (1998) Regulation of gene expression in mammary epithelial cells by cellular confluence and sequence-specific DNA binding factors. *Biochemical Society Symposium* 63, 115–131.

Andersen, C.R. and Larson, B.L. (1970) Comparative maintenance of function in dispersed cell and organ cultures of bovine mammary tissue. *Experimental Cell Research* 61, 24–30.

Anderson, B.F., Baker, H.M., Norris, G.E., Rice, D.W. and Baker, E.N. (1989) Structure of human lactoferrin: crystallographic structure analysis and refinement at 2.8 A resolution. *Journal of Molecular Biology* 209, 711–734.

Arnold, R.R., Cole, M.F. and McGhee, J.R. (1977) A bactericidal effect for human lactoferrin. *Science* 197, 263–265.

Arnould, J.P.Y. and Boyd, I.L. (1995) Temporal patterns of milk production in Antarctic fur seals (*Arctocephalus gazella*). *Journal of Zoology* 237, 1–12.

Ball, G., Shelton, M.J., Walsh, B.J., Hill, D.J., Hosking, C.S. and Howden, M.E. (1994) A major continuous allergenic epitope of bovine beta-lactoglobulin recognized by human IgE binding. *Clinical and Experimental Allergy* 24, 758–764.

Ball, R.K., Friis, R.R., Schoenenberger, C.A., Doppler, W. and Groner, B. (1988) Prolactin regulation of beta-casein gene expression and of a cytosolic 120-kd protein in a cloned mouse mammary epithelial cell line. *EMBO Journal* 7, 2089–2095.

Banes, A.J., Tsuzaki, M., Yamamoto, J., Fischer, T., Brigman, B., Brown, T. and Miller, L. (1995) Mechanoreception at the cellular level: the detection, interpretation, and diversity of responses to mechanical signals. *Biochemical and Cellular Biology* 73, 349–365.

Barcellos-Hoff, M.H., Aggeler, J., Ram, T.G. and Bissell, M.J. (1989) Functional differentiation and alveolar morphogenesis of primary mammary cultures on reconstituted basement membrane. *Development* 105, 223–235.

Bawden, W.S., Passey, R.J. and Mackinlay, A.G. (1994) The genes encoding the major milk-specific proteins and their use in transgenic studies and protein engineering. *Biotechnology and Genetic Engineering Review* 12, 89–137.

Beato, M. (1989) Gene regulation by steroid hormones. *Cell* 56, 335–344.

Bedo, S., Nikodemusz, E., Percsich, K. and Bardos, L. (1995) Variations in the milk yield and milk composition of dairy cows during lactation. *Acta Veterinaria Hungarica* 43, 163–171.

Bell, J.E., Beyer, T.A. and Hill, R.L. (1976) The kinetic mechansim of bovine milk galactosyltransferase. The role of alpha-lactalbumin. *Journal of Biological Chemistry* 251, 3003–3013.

Berchtold, S., Moriggl, R., Gouilleux, F., Silvennoinen, O., Beisenherz, C., Pfitzner, E., Wissler, M., Stocklin, E. and Groner, B. (1997) Cytokine receptor-independent, constitutively active variants of STAT5. *Journal of Biological Chemistry* 272, 30237–30243.

Bionaz, M. and Loor, J.J. (2011) Gene networks driving bovine mammary protein synthesis during the lactation cycle. *Bioinformatics and Biology Insights* 5, 83–98.

Bionaz, M., Hurley, W.L. and Loor, J.J. (2012) Milk protein synthesis in the lactating mammary gland: insights from transcriptomics analyses. In: Hurley, W.L. (ed.) *Milk Protein*. InTechOpen, Rijeka, Croatia, pp. 285–324.

Bissell, M.J. and Aggeler, J. (1987) Dynamic reciprocity: how do extracellular matrix and hormones direct gene expression? *Progress in Clinical and Biological Research* 249, 251–262.

Bissell, M.J., Hall, H.G. and Parry, G. (1982) How does the extracellular matrix direct gene expression? *Journal of Theoretical Biology* 99, 31–68.

Blackburn, D.E., Hobbs, A.A. and Rosen, J.M. (1982) Rat beta casein cDNA: sequence analysis and evolutionary comparisons. *Nucleic Acids Research* 10, 2295–2307.

Blanc, B. and Isliker, H. (1961) [Isolation and characterization of the red siderophilic protein from maternal milk: lactotransferrin]. *Bulletin de la Société de Chimie Biologique* 43, 929–943.

Blatchford, D.R., Quarrie, L.H., Tonner, E., Mccarthy, C., Flint, D.J. and Wilde, C.J. (1999) Influence of microenvironment on mammary epithelial cell survival in primary culture. *Journal of Cell Physiology* 181, 304–311.

Bleck, G.T. and Bremel, R.D. (1993) Sequence and single-base polymorphisms of the bovine alpha-lactalbumin 5'-flanking region. *Gene* 126, 213–218.

Bleck, G.T. and Bremel, R.D. (1994) Variation in expression of a bovine alpha-lactalbumin transgene in milk of transgenic mice. *Journal of Dairy Science* 77, 1897–1904.

Bleck, G.T., Jiménez-Flores, R. and Bremel, R.D. (1995) Abnormal properties of milk from transgenic mice expressing bovine β-casein under control of the bovine α-lactalbumin 5′ flanking region. *International Dairy Journal* 5, 619–632.

Blum, J.L., Zeigler, M.E. and Wicha, M.S. (1989) Regulation of mammary differentiation by the extracellular matrix. *Environmental Health Perspectives* 80, 71–83.

Bolander, F.F., Jr, Nicholas, K.R., Van Wyk, J.J. and Topper, Y.J. (1981) Insulin is essential for accumulation of casein mRNA in mouse mammary epithelial cells. *Proceedings of the National Academy of Sciences USA* 78, 5682–5684.

Bonsing, J. and Mackinlay, A.G. (1987) Recent studies on nucleotide sequences encoding the caseins. *Journal of Dairy Research* 54, 447–461.

Bonsing, J., Ring, J.M., Stewart, A.F. and Mackinlay, A.G. (1988) Complete nucleotide sequence of the bovine beta-casein gene. *Australian Journal of Biological Science* 41, 527–537.

Boudreau, N., Sympson, C.J., Werb, Z. and Bissell, M.J. (1995) Suppression of ICE and apoptosis in mammary epithelial cells by extracellular matrix. *Science* 267, 891–893.

Brew, K. and Hill, R.L. (1975) Lactose biosynthesis. *Reviews of Physiology, Biochemistry and Pharmacology* 72, 105–158.

Brew, K., Castellino, F.J., Vanaman, T.C. and Hill, R.L. (1970) The complete amino acid sequence of bovine alpha-lactalbumin. *Journal of Biological Chemistry* 245, 4570–4582.

Brophy, B., Smolenski, G., Wheeler, T., Wells, D., L'huillier, P. and Laible, G. (2003) Cloned transgenic cattle produce milk with higher levels of beta-casein and kappa-casein. *Nature Biotechnology* 21, 157–162.

Bulter, L.J. and McGarry Wolf, M. (2010) Economic analysis of the impact of cloning on improving dairy herd composition. *AgBioForum* 13, 194–207.

Burgoyne, R.D. and Wilde, C.J. (1994) Control of secretory function in mammary epithelial cells. *Cell Signalling* 6, 607–616.

Buser, A.C., Gass-Handel, E.K., Wyszomierski, S.L., Doppler, W., Leonhardt, S.A., Schaack, J., Rosen, J.M., Watkin, H., Anderson, S.M. and Edwards, D.P. (2007) Progesterone receptor repression of prolactin/ signal transducer and activator of transcription 5-mediated transcription of the beta-casein gene in mammary epithelial cells. *Molecular Endocrinology* 21, 106–125.

Buser, A.C., Obr, A.E., Kabotyanski, E.B., Grimm, S.L., Rosen, J.M. and Edwards, D.P. (2011) Progesterone receptor directly inhibits beta-casein gene transcription in mammary epithelial cells through promoting promoter and enhancer repressive chromatin modifications. *Molecular Endocrinology* 25, 955–968.

Calderone, V., Giuffrida, M.G., Viterbo, D., Napolitano, L., Fortunato, D., Conti, A. and Acharya, K.R. (1996) Amino acid sequence and crystal structure of buffalo alpha-lactalbumin. *FEBS Letters* 394, 91–95.

Calvert, D.T. and Clegg, R.A. (1996) Responses of glucose metabolism to insulin in perfused mammary tissue of lactating rats: influence of dietary history and recent insulin experience. *Experimental Physiology* 81, 131–140.

Chammas, R., Taverna, D., Cella, N., Santos, C. and Hynes, N.E. (1994) Laminin and tenascin assembly and expression regulate HC11 mouse mammary cell differentiation. *Journal of Cell Science* 107 (Pt 4), 1031–1040.

Chandan, R.C., Parry, R.M. and Shahani, K.M. (1968) Lysozyme, lipase, and ribonuclease in milk of various species1,2. *Journal of Dairy Science* 51, 606–607.

Chen, C.C., Boxer, R.B., Stairs, D.B., Portocarrero, C.P., Horton, R.H., Alvarez, J.V., Birnbaum, M.J. and Chodosh, L.A. (2010) Akt is required for Stat5 activation and mammary differentiation. *Breast Cancer Research* 12, R72.

Chen, K. and Rajewsky, N. (2007) The evolution of gene regulation by transcription factors and microRNAs. *Nature Reviews Genetics* 8, 93–103.

Chen, X., Gao, C., Li, H., Huang, L., Sun, Q., Dong, Y., Tian, C., Gao, S., Dong, H., Guan, D. *et al.* (2010) Identification and characterization of microRNAs in raw milk during different periods of lactation, commercial fluid, and powdered milk products. *Cell Research* 20, 1128–1137.

Cheng, Y., An, L.Y., Yuan, Y.G., Wang, Y., Du, F.L., Yu, B.L., Zhang, Z.H., Huang, Y.Z. and Yang, T.J. (2012) Hybrid expression cassettes consisting of a milk protein promoter and a cytomegalovirus enhancer significantly increase mammary-specific expression of human lactoferrin in transgenic mice. *Molecular Reproduction and Development* 79, 573–585.

Choi, B.-K., Bleck, G.T., Wheeler, M.B. and Jiménez-Flores, R. (1996) Genetic modification of bovine β-casein and its expression in the milk of transgenic mice. *Journal of Agricultural and Food Chemistry* 44, 953–960.

Choi, Y.J., Keller, W.L., Berg, I.E., Park, C.S. and Mackinlay, A.G. (1988) Casein gene expression in bovine mammary gland. *Journal of Dairy Science* 71, 2898–2903.

Clark, E.A. and Brugge, J.S. (1995) Integrins and signal transduction pathways: the road taken. *Science* 268, 233–239.

Close, M.J., Howlett, A.R., Roskelley, C.D., Desprez, P.Y., Bailey, N., Rowning, B., Teng, C.T., Stampfer, M. R. and Yaswen, P. (1997) Lactoferrin expression in mammary epithelial cells is mediated by changes in cell shape and actin cytoskeleton. *Journal of Cell Science* 110 (Pt 22), 2861–2871.

Conti, A., Godovac-Zimmermann, J., Liberatori, J. and Braunitzer, G. (1984) The primary structure of monomeric beta-lactoglobulin I from horse colostrum (*Equus caballus*, Perissodactyla). *Hoppe-Seyler's Zeitschrift fur Physiologische Chemie* 365, 1393–1401.

Cooper, C.A., Nelson, K.M., Maga, E.A. and Murray, J.D. (2013) Consumption of transgenic cows' milk containing human lactoferrin results in beneficial changes in the gastrointestinal tract and systemic health of young pigs. *Transgenic Research*, 22, 571–578.

Dalgleish, D.G. (1993) Bovine milk protein properties and the manufacturing quality of milk. *Livestock Production Science* 35, 75–93.

De Kruif, C.G., Huppertz, T., Urban, V.S. and Petukhov, A.V. (2012) Casein micelles and their internal structure. *Advances in Colloid and Interface Science* 171–172, 36–52.

Deis, R.P. and Delouis, C. (1983) Lactogenesis induced by ovariectomy in pregnant rats and its regulation by oestrogen and progesterone. *Journal of Steroid Biochemistry* 18, 687–690.

Deugnier, M.A., Faraldo, M.M., Rousselle, P., Thiery, J.P. and Glukhova, M.A. (1999) Cell-extracellular matrix interactions and EGF are important regulators of the basal mammary epithelial cell phenotype. *Journal of Cell Science* 112 (Pt 7), 1035–1044.

Djiane, J., Delouis, C. and Denamur, R. (1975) Lactogenesis in organ cultures of heifer mammary tissue. *Journal of Endocrinology* 65, 453–454.

Doppler, W. (1994) Regulation of gene expression by prolactin. *Reviews of Physiology, Biochemistry and Pharmacology* 124, 93–130.

Doppler, W., Hock, W., Hofer, P., Groner, B. and Ball, R.K. (1990) Prolactin and glucocorticoid hormones control transcription of the beta-casein gene by kinetically distinct mechanisms. *Molecular Endocrinology* 4, 912–919.

Duff, G.C. and Galyean, M.L. (2007) Board-invited review: recent advances in management of highly stressed, newly received feedlot cattle. *Journal of Animal Science* 85, 823–840.

Duncan, S.A., Zhong, Z., Wen, Z. and Darnell, J.E., Jr (1997) STAT signaling is active during early mammalian development. *Developmental Dynamics* 208, 190–198.

Dunstan, J.A., Hale, J., Breckler, L., Lehmann, H., Weston, S., Richmond, P. and Prescott, S.L. (2005) Atopic dermatitis in young children is associated with impaired interleukin-10 and interferon-gamma responses to allergens, vaccines and colonizing skin and gut bacteria. *Clinical and Experimental Allergy* 35, 1309–1317.

Edwards, G.M., Wilford, F.H. and Streuli, C.H. (1996) Extracellular matrix controls the prolactin signalling pathway in mammary epithelial cells. *Biochemical Society Transactions* 24, 345S.

Eisenstein, R.S. and Rosen, J.M. (1988) Both cell substratum regulation and hormonal regulation of milk protein gene expression are exerted primarily at the posttranscriptional level. *Molecular and Cellular Biology* 8, 3183–3190.

Erdman, R.A. and Varner, M. (1995) Fixed yield responses to increased milking frequency. *Journal of Dairy Science* 78, 1199–1203.

Farrell, H.M., Jr, Malin, E.L., Brown, E.M. and Mora-Gutierrez, A. (2009) Review of the chemistry of alphaS2-casein and the generation of a homologous molecular model to explain its properties. *Journal of Dairy Science* 92, 1338–1353.

Fiat, A.M. and Jolles, P. (1989) Caseins of various origins and biologically active casein peptides and oligo-saccharides: structural and physiological aspects. *Molecular and Cellular Biology* 87, 5–30.

Flemming, A. (1922) On a remarkable bacteriolytic element found in tissues and secretions. *Proceedings of the Royal Society of London B* 93, 306–317.

Formigli, L., Papucci, L., Tani, A., Schiavone, N., Tempestini, A., Orlandini, G.E., Capaccioli, S. and Orlandini, S.Z. (2000) Aponecrosis: morphological and biochemical exploration of a syncretic process of cell death sharing apoptosis and necrosis. *Journal of Cell Physiology* 182, 41–49.

Fugate, R.D. and Song, P.S. (1980) Spectroscopic characterization of beta-lactoglobulin-retinol complex. *Biochimica et Biophysica Acta* 625, 28–42.

German, T. and Barash, I. (2002) Characterization of an epithelial cell line from bovine mammary gland. *In Vitro Cellular and Developmental Biology – Animal* 38, 282–292.

Gertler, A., Weil, A. and Cohen, N. (1982) Hormonal control of casein synthesis in organ culture of the bovine lactating mammary gland. *Journal of Dairy Research* 49, 387–398.

Getzenberg, R.H. (1994) Nuclear matrix and the regulation of gene expression: Tissue specificity. *Journal of Cellular Biochemistry* 55, 22–31.

Geursen, A. and Grigor, M.R. (1987) Nutritional regulation of milk protein messenger RNA concentrations in mammary acini isolated from lactating rats. *Biochemistry International* 15, 873–879.

Godovac-Zimmermann, J., Conti, A., Liberatori, J. and Braunitzer, G. (1985) The amino-acid sequence of beta-lactoglobulin II from horse colostrum (*Equus caballus*, Perissodactyla): beta-lactoglobulins are retinol-binding proteins. *Biological Chemistry Hoppe Seyler* 366, 601–608.

Goodman, G.T., Akers, R.M., Friderici, K.H. and Tucker, H.A. (1983) Hormonal regulation of alpha-lactalbumin secretion from bovine mammary tissue cultured *in vitro*. *Endocrinology* 112, 1324–1330.

Gouilleux, F., Wakao, H., Mundt, M. and Groner, B. (1994) Prolactin induces phosphorylation of Tyr694 of Stat5 (MGF), a prerequisite for DNA binding and induction of transcription. *EMBO Journal* 13, 4361–4369.

Groenen, M.A., Dijkhof, R.J., Verstege, A.J. and Van Der Poel, J.J. (1993) The complete sequence of the gene encoding bovine alpha s2-casein. *Gene* 123, 187–193.

Grosveld, F., Van Assendelft, G.B., Greaves, D.R. and Kollias, G. (1987) Position-independent, high-level expression of the human beta-globin gene in transgenic mice. *Cell* 51, 975–985.

Gu, Y., Li, M., Wang, T., Liang, Y., Zhong, Z., Wang, X., Zhou, Q., Chen, L., Lang, Q., He, Z., Chen, X., Gong, J., Gao, X., Li, X. and Lv, X. (2012) Lactation-related microRNA expression profiles of porcine breast milk exosomes. *PLoS One* 7, e43691.

Guilloteau, P., Zabielski, R., Hammon, H.M. and Metges, C.C. (2010) Nutritional programming of gastrointestinal tract development. Is the pig a good model for man? *Nutrition Research Reviews* 23, 4–22.

Gutierrez, A., Meade, H.M., Ditullio, P., Pollock, D., Harvey, M., Jimenez-Flores, R., Anderson, G.B., Murray, J.D. and Medrano, J.F. (1996) Expression of a bovine kappa-CN cDNA in the mammary gland of transgenic mice utilizing a genomic milk protein gene as an expression cassette. *Transgenic Research* 5, 271–279.

Gutierrez-Adan, A., Maga, E.A., Meade, H., Shoemaker, C.F., Medrano, J.F., Anderson, G.B. and Murray, J.D. (1996) Alterations of the physical characteristics of milk from transgenic mice producing bovine kappa-casein. *Journal of Dairy Science* 79, 791–799.

Hall, L., Emery, D.C., Davies, M.S., Parker, D. and Craig, R.K. (1987) Organization and sequence of the human alpha-lactalbumin gene. *Biochemistry Journal* 242, 735–742.

Halliday, J.A., Bell, K. and Shaw, D.C. (1991) The complete amino acid sequence of feline beta-lactoglobulin II and a partial revision of the equine beta-lactoglobulin II sequence. *Biochimica et Biophysica Acta* 1077, 25–30.

Halliday, J.A., Bell, K., McAndrew, K. and Shaw, D.C. (1993) Feline beta-lactoglobulins I, II and III, and canine beta-lactoglobulins I and II: amino acid sequences provide evidence for the existence of more than one gene for beta-lactoglobulin in the cat and dog. *Protein Sequence Data Analysis* 5, 201–205.

Hartmann, P.E. (1973) Changes in the composition and yield of the mammary secretion of cows during the initiation of lactation. *Journal of Endocrinology* 59, 231–247.

Hartmann, P.E., Trevethan, P. and Shelton, J.N. (1973) Progesterone and oestrogen and the initiation of lactation in ewes. *Journal of Endocrinology* 59, 249–259.

Hata, T., Murakami, K., Nakatani, H., Yamamoto, Y., Matsuda, T. and Aoki, N. (2010) Isolation of bovine milk-derived microvesicles carrying mRNAs and microRNAs. *Biochemistry and Biophysics Research Community* 396, 528–533.

Hayes, H.C. and Petit, E.J. (1993) Mapping of the beta-lactoglobulin gene and of an immunoglobulin M heavy chain-like sequence to homoeologous cattle, sheep, and goat chromosomes. *Mammalian Genome* 4, 207–210.

Hennighausen, L.G. and Sippel, A.E. (1982) Mouse whey acidic protein is a novel member of the family of 'four-disulfide core' proteins. *Nucleic Acids Research* 10, 2677–2684.

Hernandez, L.L., Stiening, C.M., Wheelock, J.B., Baumgard, L.H., Parkhurst, A.M. and Collier, R.J. (2008) Evaluation of serotonin as a feedback inhibitor of lactation in the bovine. *Journal of Dairy Science* 91, 1834–1844.

Hillerton, J.E., Knight, C.H., Turvey, A., Wheatley, S.D. and Wilde, C.J. (1990) Milk yield and mammary function in dairy cows milked four times daily. *Journal of Dairy Research* 57, 285–294.

Hobbs, A.A. and Rosen, J.M. (1982) Sequence of rat alpha- and gamma-casein mRNAs: evolutionary comparison of the calcium-dependent rat casein multigene family. *Nucleic Acids Research* 10, 8079–8098.

Horne, D.S. (2006) Casein micelle structure: models and muddles. *Current Opinion in Colloid and Interface Science* 11, 6.

Host, A., Husby, S. and Osterballe, O. (1988) A prospective study of cow's milk allergy in exclusively breast-fed infants. Incidence, pathogenetic role of early inadvertent exposure to cow's milk formula, and characterization of bovine milk protein in human milk. *Acta Paediatrica Scandinavia* 77, 663–670.

Ihle, J.N. (1996) Signaling by the cytokine receptor superfamily in normal and transformed hematopoietic cells. *Advances in Cancer Research* 68, 23–65.

Ismail, P.M., Li, J., Demayo, F.J., O'Malley, B.W. and Lydon, J.P. (2002) A novel LacZ reporter mouse reveals complex regulation of the progesterone receptor promoter during mammary gland development. *Molecular Endocrinology* 16, 2475–2489.

Izumi, H., Kosaka, N., Shimizu, T., Sekine, K., Ochiya, T. and Takase, M. (2012) Bovine milk contains microRNA and messenger RNA that are stable under degradative conditions. *Journal of Dairy Science* 95, 4831–4841.

Jabed, A., Wagner, S., McCracken, J., Wells, D.N. and Laible, G. (2012) Targeted microRNA expression in dairy cattle directs production of beta-lactoglobulin-free, high-casein milk. *Proceedings of the National Academy of Sciences USA* 109, 16811–16816.

Jenness, R. (ed.) (1974) *The Composition of Milk.* Academic Press, New York.

Jenness, R. (1979) The composition of human milk. *Seminars in Perinatology* 3, 225–239.

Johnson, T.L., Fujimoto, B.A., Jimenez-Flores, R. and Peterson, D.G. (2010) Growth hormone alters lipid composition and increases the abundance of casein and lactalbumin mRNA in the MAC-T cell line. *Journal of Dairy Research* 77, 199–204.

Jolles, P. (1975) Structural aspects of the milk clotting process. Comparative features with the blood clotting process. *Molecular and Cellular Biology* 7, 73–85.

Kabotyanski, E.B., Rijnkels, M., Freeman-Zadrowski, C., Buser, A.C., Edwards, D.P. and Rosen, J.M. (2009) Lactogenic hormonal induction of long distance interactions between beta-casein gene regulatory elements. *Journal of Biological Chemistry* 284, 22815–22824.

Kanai, A., Nonomura, N., Yoshimura, M. and Oka, T. (1993) DNA-binding proteins and their cis-acting sites controlling hormonal induction of a mouse beta-casein: CAT fusion protein in mammary epithelial cells. *Gene* 126, 195–201.

Kanyshkova, T.G., Buneva, V.N. and Nevinsky, G.A. (2001) Lactoferrin and its biological functions. *Biochemistry (Moscow)* 66, 1–7.

Kishore, A., Mukesh, M., Sobti, R.C., Mishra, B.P. and Sodhi, M. (2013) Variations in the regulatory region of alpha S1-casein milk protein gene among tropically adapted Indian native (*Bos indicus*) cattle. *ISRN Biotechnology* 2013, 10.

Koczan, D., Hobom, G. and Seyfert, H.M. (1991) Genomic organization of the bovine alpha-S1 casein gene. *Nucleic Acids Research* 19, 5591–5596.

Kosaka, N., Izumi, H., Sekine, K. and Ochiya, T. (2010) microRNA as a new immune-regulatory agent in breast milk. *Silence* 1, 7.

Kuhn, N.J. (1969) Progesterone withdrawal as the lactogenic trigger in the rat. *Journal of Endocrinology* 44, 39–54.

Kulski, J.K., Nicholas, K.R., Topper, Y.J. and Qasba, P. (1983) Essentiality of insulin and prolactin for accumulation of rat casein mRNAs. *Biochemistry and Biophysics Research Community* 116, 994–999.

Kumagai, I., Takeda, S. and Miura, K. (1992) Functional conversion of the homologous proteins alpha-lactalbumin and lysozyme by exon exchange. *Proceedings of the National Academy of Sciences USA* 89, 5887–5891.

Kumar, S., Clarke, A.R., Hooper, M.L., Horne, D.S., Law, A.J., Leaver, J., Springbett, A., Stevenson, E. and Simons, J.P. (1994) Milk composition and lactation of beta-casein-deficient mice. *Proceedings of the National Academy of Sciences USA* 91, 6138–6142.

Kumosinski, T.F., Brown, E.M. and Farrell, H.M., Jr (1991) Three-dimensional molecular modeling of bovine caseins: alpha s1-casein. *Journal of Dairy Science* 74, 2889–2895.

Kumosinski, T.F., Brown, E.M. and Farrell, H.M., Jr (1993a) Three-dimensional molecular modeling of bovine caseins: a refined, energy-minimized kappa-casein structure. *Journal of Dairy Science* 76, 2507–2520.

Kumosinski, T.F., Brown, E.M. and Farrell, H.M., Jr (1993b) Three-dimensional molecular modeling of bovine caseins: an energy-minimized beta-casein structure. *Journal of Dairy Science* 76, 931–945.

Kuys, Y.M., Snell, R.G. and Wheeler, T.T. (1996) Binding of nuclear proteins to the bovine α-lactalbumin gene promoter. *Proceedings of the New Zealand Society of Animal Production* 56, 68–70.

Larocca, J., Pietruska, J. and Hixon, M. (2011) Akt1 is essential for postnatal mammary gland development, function, and the expression of Btn1a1. *PLoS One* 6, e24432.

Lee, H.J., Gallego-Ortega, D., Ledger, A., Schramek, D., Joshi, P., Szwarc, M.M., Cho, C., Lydon, J.P., Khokha, R., Penninger, J.M. and Ormandy, C.J. (2013) Progesterone drives mammary secretory differentiation via RankL-mediated induction of Elf5 in luminal progenitor cells. *Development* 140, 1397–1401.

Lefevre, C.M., Sharp, J.A. and Nicholas, K.R. (2009) Characterisation of monotreme caseins reveals lineage-specific expansion of an ancestral casein locus in mammals. *Reproduction, Fertility and Development* 21, 1015–1027.

Lelievre, S., Weaver, V.M. and Bissell, M.J. (1996) Extracellular matrix signaling from the cellular membrane skeleton to the nuclear skeleton: a model of gene regulation. *Recent Progress in Hormone Research* 51, 417–432.

Lemay, D.G., Lynn, D.J., Martin, W.F., Neville, M.C., Casey, T.M., Rincon, G., Kriventseva, E.V., Barris, W.C., Hinrichs, A.S., Molenaar, A.J. *et al.* (2009) The bovine lactation genome: insights into the evolution of mammalian milk. *Genome Biology* 10, R43.

Lewis, B.P., Shih, I.H., Jones-Rhoades, M.W., Bartel, D.P. and Burge, C.B. (2003) Prediction of mammalian microRNA targets. *Cell* 115, 787–798.

Lewis, B.P., Burge, C.B. and Bartel, D.P. (2005) Conserved seed pairing, often flanked by adenosines, indicates that thousands of human genes are microRNA targets. *Cell* 120, 15–20.

L'huillier, P.J., Soulier, S., Stinnakre, M.G., Lepourry, L., Davis, S.R., Mercier, J.C. and Vilotte, J.L. (1996) Efficient and specific ribozyme-mediated reduction of bovine alpha-lactalbumin expression in double transgenic mice. *Proceedings of the National Academy of Sciences USA* 93, 6698–6703.

Li, M.L., Aggeler, J., Farson, D.A., Hatier, C., Hassell, J. and Bissell, M.J. (1987) Influence of a reconstituted basement membrane and its components on casein gene expression and secretion in mouse mammary epithelial cells. *Proceedings of the National Academy of Sciences USA* 84, 136–140.

Li, S. and Rosen, J.M. (1994) Glucocorticoid regulation of rat whey acidic protein gene expression involves hormone-induced alterations of chromatin structure in the distal promoter region. *Molecular Endocrinology* 8, 1328–1335.

Li, S. and Rosen, J.M. (1995) Nuclear factor I and mammary gland factor (STAT5) play a critical role in regulating rat whey acidic protein gene expression in transgenic mice. *Molecular and Cellular Biology* 15, 2063–2070.

Li, Y., Limmon, G.V., Imani, F. and Teng, C. (2009) Induction of lactoferrin gene expression by innate immune stimuli in mouse mammary epithelial HC-11 cells. *Biochimie* 91, 58–67.

Liao, Y., Du, X. and Lonnerdal, B. (2010) miR-214 regulates lactoferrin expression and pro-apoptotic function in mammary epithelial cells. *Journal of Nutrition* 140, 1552–1556.

Lin, C.Q. and Bissell, M.J. (1993) Multi-faceted regulation of cell differentiation by extracellular matrix. *FASEB Journal* 7, 737–743.

Lipnik, K., Petznek, H., Renner-Muller, I., Egerbacher, M., Url, A., Salmons, B., Gunzburg, W.H. and Hohenadl, C. (2005) A 470 bp WAP-promoter fragment confers lactation independent, progesterone regulated mammary-specific gene expression in transgenic mice. *Transgenic Research* 14, 145–158.

Liu, X., Robinson, G.W., Gouilleux, F., Groner, B. and Hennighausen, L. (1995) Cloning and expression of Stat5 and an additional homologue (Stat5b) involved in prolactin signal transduction in mouse mammary tissue. *Proceedings of the National Academy of Sciences USA* 92, 8831–8835.

Liu, X., Robinson, G.W., Wagner, K.U., Garrett, L., Wynshaw-Boris, A. and Hennighausen, L. (1997) Stat5a is mandatory for adult mammary gland development and lactogenesis. *Genes and Development* 11, 179–186.

Loizzi, R.F. (1985) Progesterone withdrawal stimulates mammary gland tubulin polymerization in pregnant rats. *Endocrinology* 116, 2543–2547.

Loor, J.J. and Cohick, W.S. (2009) ASAS centennial paper: lactation biology for the twenty-first century. *Journal of Animal Science* 87, 813–824.

McGann, T.C., Donnelly, W.J., Kearney, R.D. and Buchheim, W. (1980) Composition and size distribution of bovine casein micelles. *Biochimica et Biophysica Acta* 630, 261–270.

McMahon, C.D., Farr, V.C., Singh, K., Wheeler, T.T. and Davis, S.R. (2004) Decreased expression of beta1-integrin and focal adhesion kinase in epithelial cells may initiate involution of mammary glands. *Journal of Cell Physiology* 200, 318–325.

Maga, E.A., Anderson, G.B., Huang, M.C. and Murray, J.D. (1994) Expression of human lysozyme mRNA in the mammary gland of transgenic mice. *Transgenic Research* 3, 36–42.

Maga, E.A., Anderson, G.B., Cullor, J.S., Smith, W. and Murray, J.D. (1998) Antimicrobial properties of human lysozyme transgenic mouse milk. *Journal of Food Protection* 61, 52–56.

Maroulakou, I.G., Oemler, W., Naber, S.P., Klebba, I., Kuperwasser, C. and Tsichlis, P.N. (2008) Distinct roles of the three Akt isoforms in lactogenic differentiation and involution. *Journal of Cell Physiology* 217, 468–477.

Masson, P.L. and Heremans, J.F. (1971) Lactoferrin in milk from different species. *Comparative Biochemistry and Physiology Part B* 39, 119–129.

Meier, V.S. and Groner, B. (1994) The nuclear factor YY1 participates in repression of the beta-casein gene promoter in mammary epithelial cells and is counteracted by mammary gland factor during lactogenic hormone induction. *Molecular and Cellular Biology* 14, 128–137.

Menzies, K.K., Lefevre, C., Macmillan, K.L. and Nicholas, K.R. (2009) Insulin regulates milk protein synthesis at multiple levels in the bovine mammary gland. *Functional and Integrative Genomics* 9, 197–217.

Mercier, J.C. (1986) Genetic engineering applied to milk producing animals: some expectations. In: Smith, C., King, J.W.B. and McKay, J.C. (eds) *Exploiting New Technologies in Animal Breeding.* Oxford University Press, Oxford, pp. 122–131.

Mercier, J.C. and Vilotte, J.L. (1993) Structure and function of milk protein genes. *Journal of Dairy Science* 76, 3079–3098.

Mercier, J.C., Brignon, G. and Ribadeau-Dumas, B. (1973) [Primary structure of bovine kappa B casein. Complete sequence]. *European Journal of Biochemistry* 35, 222–235.

Miles, H., Lesser, W. and Sears, P. (1992) The economic implications of bioengineered mastitis control. *Journal of Dairy Science* 75, 596–605.

Mink, S., Hartig, E., Jennewein, P., Doppler, W. and Cato, A.C. (1992) A mammary cell-specific enhancer in mouse mammary tumor virus DNA is composed of multiple regulatory elements including binding sites for CTF/NFI and a novel transcription factor, mammary cell-activating factor. *Molecular and Cellular Biology* 12, 4906–4918.

Moriggl, R., Gouilleux-Gruart, V., Jahne, R., Berchtold, S., Gartmann, C., Liu, X., Hennighausen, L., Sotiropoulos, A., Groner, B. and Gouilleux, F. (1996) Deletion of the carboxyl-terminal transactivation domain of MGF-Stat5 results in sustained DNA binding and a dominant negative phenotype. *Molecular and Cellular Biology* 16, 5691–5700.

Moriggl, R., Berchtold, S., Friedrich, K., Standke, G.J., Kammer, W., Heim, M., Wissler, M., Stocklin, E., Gouilleux, F. and Groner, B. (1997) Comparison of the transactivation domains of Stat5 and Stat6 in lymphoid cells and mammary epithelial cells. *Molecular and Cellular Biology* 17, 3663–3678.

Motyl, T., Gajewska, M., Zarzynska, J., Sobolewska, A. and Gajkowska, B. (2007) Regulation of autophagy in bovine mammary epithelial cells. *Autophagy* 3, 484–486.

Moyes, K.M., Drackley, J.K., Morin, D.E., Bionaz, M., Rodriguez-Zas, S.L., Everts, R.E., Lewin, H.A. and Loor, J.J. (2009) Gene network and pathway analysis of bovine mammary tissue challenged with

Streptococcus uberis reveals induction of cell proliferation and inhibition of PPARgamma signaling as potential mechanism for the negative relationships between immune response and lipid metabolism. *BMC Genomics* 10, 542.

Mukhopadhyay, S.S., Wyszomierski, S.L., Gronostajski, R.M. and Rosen, J.M. (2001) Differential interactions of specific nuclear factor I isoforms with the glucocorticoid receptor and STAT5 in the cooperative regulation of WAP gene transcription. *Molecular and Cellular Biology* 21, 6859–6869.

Murphy, J.J. and O'Mara, F. (1993) Nutritional manipulation of milk protein concentration and its impact on the dairy industry. *Livestock Production Science* 35, 117–134.

Nakajima, H., Brindle, P.K., Handa, M. and Ihle, J.N. (2001) Functional interaction of STAT5 and nuclear receptor co-repressor SMRT: implications in negative regulation of STAT5-dependent transcription. *EMBO Journal* 20, 6836–6844.

Naruse, K., Yoo, S.K., Kim, S.M., Choi, Y.J., Lee, H.M. and Jin, D.I. (2006) Analysis of tissue-specific expression of human type II collagen cDNA driven by different sizes of the upstream region of the beta-casein promoter. *Bioscience, Biotechnology and Biochemistry* 70, 93–98.

Neville, M.C., McFadden, T.B. and Forsyth, I. (2002) Hormonal regulation of mammary differentiation and milk secretion. *Journal of Mammary Gland Biology and Neoplasia* 7, 49–66.

Nicholas, K., Simpson, K., Wilson, M., Trott, J. and Shaw, D. (1997) The tammar wallaby: a model to study putative autocrine-induced changes in milk composition. *Journal of Mammary Gland Biology and Neoplasia* 2, 299–310.

Oakes, S.R., Naylor, M.J., Asselin-Labat, M.L., Blazek, K.D., Gardiner-Garden, M., Hilton, H.N., Kazlauskas, M., Pritchard, M.A., Chodosh, L.A., Pfeffer, P.L., Lindeman, G.J., Visvader, J.E. and Ormandy, C.J. (2008) The Ets transcription factor Elf5 specifies mammary alveolar cell fate. *Genes and Development* 22, 581–586.

O'Brien, J., Fornetti, J. and Schedin, P. (2010) Isolation of mammary-specific extracellular matrix to assess acute cell-ECM interactions in 3D culture. *Journal of Mammary Gland Biology and Neoplasia* 15, 353–364.

Oftedal, O.T. (2012) The evolution of milk secretion and its ancient origins. *Animal* 6, 355–368.

Oh, S. and Richardson, T. (1991) Genetic engineering of bovine .kappa.-casein to improve its nutritional quality. *Journal of Agricultural and Food Chemistry* 39, 422–427.

Oram, J.D. and Reiter, B. (1968) Inhibition of bacteria by lactoferrin and other iron-chelating agents. *Biochimica et Biophysica Acta* 170, 351–365.

Palmer, A.H. (1934) The preparation of a crystalline globulin from the albumin fraction of cow's milk. *Journal of Biological Chemistry* 104, 359–372.

Paonessa, G., Gounari, F., Frank, R. and Cortese, R. (1988) Purification of a NF1-like DNA-binding protein from rat liver and cloning of the corresponding cDNA. *EMBO Journal* 7, 3115–3123.

Papiz, M.Z., Sawyer, L., Eliopoulos, E.E., North, A.C., Findlay, J.B., Sivaprasadarao, R., Jones, T.A., Newcomer, M.E. and Kraulis, P.J. (1986) The structure of beta-lactoglobulin and its similarity to plasma retinol-binding protein. *Nature* 324, 383–385.

Park, I., Schaeffer, E., Sidoli, A., Baralle, F.E., Cohen, G.N. and Zakin, M.M. (1985) Organization of the human transferrin gene: direct evidence that it originated by gene duplication. *Proceedings of the National Academy of Sciences USA* 82, 3149–3153.

Passey, R.J. and MacKinlay, A.G. (1995) Characterisation of a second, apparently inactive, copy of the bovine beta-lactoglobulin gene. *European Journal of Biochemistry* 233, 736–743.

Peaker, M. and Wilde, C.J. (1987) Milk secretion – autocrine control. *News in Physiological Sciences* 2, 124–126.

Peaker, M., Wilde, C.J. and Knight, C.H. (1998) Local control of the mammary gland. *Biochemical Society Symposium* 63, 71–79.

Pearse, M.J., Linklater, P.M., Hall, R.J. and MacKinlay, A.G. (1986) Effect of casein micelle composition and casein dephosphorylation on coagulation and syneresis. *Journal of Dairy Research* 53, 381–390.

Perez, M.D. and Calvo, M. (1995) Interaction of beta-lactoglobulin with retinol and fatty acids and its role as a possible biological function for this protein: a review. *Journal of Dairy Science* 78, 978–988.

Platenburg, G.J., Kootwijk, E.P., Kooiman, P.M., Woloshuk, S.L., Nuijens, J.H., Krimpenfort, P.J., Pieper, F.R., De Boer, H.A. and Strijker, R. (1994) Expression of human lactoferrin in milk of transgenic mice. *Transgenic Research* 3, 99–108.

Popov, L.S., Korobko, I.V., Andreeva, L.E., Dvorianchikov, G.A., Kaledin, A.S. and Gorodetskii, S.I. (1995) [The functional value of the 5'-region of the bovine kappa-casein gene]. *Doklady Akademii Nauk* 340, 111–113.

Rabot, A., Wellnitz, O., Meyer, H.H. and Bruckmaier, R.M. (2007) Use and relevance of a bovine mammary gland explant model to study infection responses in bovine mammary tissue. *Journal of Dairy Research* 74, 93–99.

Raught, B., Khursheed, B., Kazansky, A. and Rosen, J. (1994) YY I represses beta-casein gene expression by preventing the formation of a lactation-associated complex. *Molecular and Cellular Biology* 14, 1752–1763.

Raught, B., Liao, W.S. and Rosen, J.M. (1995) Developmentally and hormonally regulated CCAAT/enhancer-binding protein isoforms influence beta-casein gene expression. *Molecular Endocrinology* 9, 1223–1232.

Reichenstein, M., Rauner, G. and Barash, I. (2011) Conditional repression of STAT5 expression during lactation reveals its exclusive roles in mammary gland morphology, milk-protein gene expression, and neonate growth. *Molecular Reproduction and Development* 78, 585–596.

Rennison, M.E., Kerr, M., Addey, C.V., Handel, S.E., Turner, M.D., Wilde, C.J. and Burgoyne, R.D. (1993) Inhibition of constitutive protein secretion from lactating mouse mammary epithelial cells by FIL (feedback inhibitor of lactation), a secreted milk protein. *Journal of Cell Science* 106 (Pt 2), 641–648.

Rijnkels, M. (2002) Multispecies comparison of the casein gene loci and evolution of casein gene family. *Journal of Mammary Gland Biology and Neoplasia* 7, 327–345.

Rijnkels, M., Kooiman, P.M., Krimpenfort, P.J., De Boer, H.A. and Pieper, F.R. (1995) Expression analysis of the individual bovine beta-, alpha s2- and kappa-casein genes in transgenic mice. *Biochemistry Journal* 311 (Pt 3), 929–937.

Rijnkels, M., Kooiman, P.M., De Boer, H.A. and Pieper, F.R. (1997) Organization of the bovine casein gene locus. *Mammalian Genome* 8, 148–152.

Riley, L.G., Gardiner-Garden, M., Thomson, P.C., Wynn, P.C., Williamson, P., Raadsma, H.W. and Sheehy, P.A. (2010) The influence of extracellular matrix and prolactin on global gene expression profiles of primary bovine mammary epithelial cells *in vitro*. *Animal Genetics* 41, 55–63.

Rosen, J.M., Li, S., Raught, B. and Hadsell, D. (1996) The mammary gland as a bioreactor: factors regulating the efficient expression of milk protein-based transgenes. *American Journal of Clinical Nutrition* 63, 627S–632S.

Rosen, J.M., Zahnow, C., Kazansky, A. and Raught, B. (1998) Composite response elements mediate hormonal and developmental regulation of milk protein gene expression. *Biochemical Society Symposium* 63, 101–113.

Rosen, J.M., Wyszomierski, S.L. and Hadsell, D. (1999) Regulation of milk protein gene expression. *Annual Review of Nutrition* 19, 407–436.

Roskelley, C.D., Desprez, P.Y. and Bissell, M.J. (1994) Extracellular matrix-dependent tissue-specific gene expression in mammary epithelial cells requires both physical and biochemical signal transduction. *Proceedings of the National Academy of Sciences USA* 91, 12378–12382.

Roskelley, C.D., Srebrow, A. and Bissell, M.J. (1995) A hierarchy of ECM-mediated signalling regulates tissue-specific gene expression. *Current Opinions in Cell Biology* 7, 736–747.

Rullo, R., Di Luccia, A., Chianese, L. and Pieragostini, E. (2010) Hot topic: Gene duplication at the α-lactalbumin locus: finding the evidence in water buffalo (*Bubalus bubalus* L.) *Journal of Dairy Science* 93, 2161–2167.

Sabikhi, L. (2007) Designer milk. *Advances in Food and Nutrition Research* 53, 161–198.

Samiec, M. and Skrzyszowska, M. (2011) Transgenic mammalian species, generated by somatic cell cloning, in biomedicine, biopharmaceutical industry and human nutrition/dietetics–recent achievements. *Polish Journal of Veterinary Science* 14, 317–328.

Schedin, P., Mitrenga, T., McDaniel, S. and Kaeck, M. (2004) Mammary ECM composition and function are altered by reproductive state. *Molecular Carcinogenesis* 41, 207–220.

Schmidhauser, C., Bissell, M.J., Myers, C.A. and Casperson, G.F. (1990) Extracellular matrix and hormones transcriptionally regulate bovine beta-casein 5′ sequences in stably transfected mouse mammary cells. *Proceedings of the National Academy of Sciences USA* 87, 9118–9122.

Schmidhauser, C., Casperson, G.F., Myers, C.A., Sanzo, K.T., Bolten, S. and Bissell, M.J. (1992) A novel transcriptional enhancer is involved in the prolactin- and extracellular matrix-dependent regulation of beta-casein gene expression. *Molecular Biology of the Cell* 3, 699–709.

Schmidhauser, C., Myers, C.A., Mossi, R., Casperson, G.F., Sanzo, K.T., Bolten, S. and Bissell, M.J. (eds) (1995) *Extracellular Matrix Dependent Gene Regulation in Mammary Epithelial Cells*. Plenum Press, New York.

Schmitt-Ney, M., Doppler, W., Ball, R.K. and Groner, B. (1991) Beta-casein gene promoter activity is regulated by the hormone-mediated relief of transcriptional repression and a mammary-gland-specific nuclear factor. *Molecular and Cellular Biology* 11, 3745–3755.

Schmitt-Ney, M., Happ, B., Ball, R.K. and Groner, B. (1992) Developmental and environmental regulation of a mammary gland-specific nuclear factor essential for transcription of the gene encoding beta-casein. *Proceedings of the National Academy of Sciences USA* 89, 3130–3134.

Schopen, G.C., Visker, M.H., Koks, P.D., Mullaart, E., Van Arendonk, J.A. and Bovenhuis, H. (2011) Whole-genome association study for milk protein composition in dairy cattle. *Journal of Dairy Science* 94, 3148–3158.

Servely, J.L., Teyssot, B., Houdebine, L.M., Delouis, C., Djiane, J. and Kelly, P.A. (1982) Induction of beta-casein mRNA accumulation by the putative prolactin second messenger added to the culture medium of cultured mammary epithelial cells. *FEBS Letters* 148, 242–246.

Seyfert, H.M. and Kuhn, C. (1994) Characterization of a first bovine lactoferrin gene variant, based on an EcoRI polymorphism. *Animal Genetics* 25, 54.

Seyfert, H.M., Tuckoricz, A., Interthal, H., Koczan, D. and Hobom, G. (1994) Structure of the bovine lactoferrin-encoding gene and its promoter. *Gene* 143, 265–269.

Shamay, A., Zeelon, E., Ghez, Z., Cohen, N., Mackinlay, A.G. and Gertler, A. (1987) Inhibition of casein and fat synthesis and alpha-lactalbumin secretion by progesterone in explants from bovine lactating mammary glands. *Journal of Endocrinology* 113, 81–88.

Shamay, A., Shapiro, F., Mabjeesh, S.J. and Silanikove, N. (2002) Casein-derived phosphopeptides disrupt tight junction integrity, and precipitously dry up milk secretion in goats. *Life Sciences* 70, 2707–2719.

Sharp, J.A., Cane, K., Arnould, J.P. and Nicholas, K.R. (2005) The lactation cycle of the fur seal. *Journal of Dairy Research* 72 Spec No, 81–89.

Sharp, J.A., Cane, K.N., Lefevre, C., Arnould, J.P. and Nicholas, K.R. (2006a) Fur seal adaptations to lactation: insights into mammary gland function. *Current Topics in Developmental Biology* 72, 275–308.

Sharp, J.A., Cane, K.N., Mailer, S.L., Oosthuizen, W.H., Arnould, J.P.Y. and Nicholas, K.R. (2006b) Species-specific cell-matrix interactions are essential for differentiation of alveoli like structures and milk gene expression in primary mammary cells of the Cape fur seal (*Arctocephalus pusillus pusillus*) *Matrix Biology* 25, 430–442.

Sharp, J.A., Lefevre, C. and Nicholas, K.R. (2008) Lack of functional alpha-lactalbumin prevents involution in Cape fur seals and identifies the protein as an apoptotic milk factor in mammary gland involution. *BMC Biology* 6, 48.

Sheehy, P.A., Nicholas, K.R. and Wynn, P.C. (2000) An investigation of the role of insulin in bovine milk protein gene expression in mammary explant culture. *Asian-Australian Journal of Animal Science*, 272–275.

Sheehy, P.A., Della-Vedova, J.J., Nicholas, K.R. and Wynn, P.C. (2004) Hormone-dependent milk protein gene expression in bovine mammary explants from biopsies at different stages of pregnancy. *Journal of Dairy Research* 71, 135–140.

Silanikove, N., Shapiro, F. and Shinder, D. (2009) Acute heat stress brings down milk secretion in dairy cows by up-regulating the activity of the milk-borne negative feedback regulatory system. *BMC Physiology* 9, 13.

Simons, G., Van Den Heuvel, W., Reynen, T., Frijters, A., Rutten, G., Slangen, C.J., Groenen, M., De Vos, W.M. and Siezen, R.J. (1993) Overproduction of bovine beta-casein in *Escherichia coli* and engineering of its main chymosin cleavage site. *Protein Engineering* 6, 763–770.

Skowronski, J., Plucienniczak, A., Bednarek, A. and Jaworski, J. (1984) Bovine 1.709 satellite. Recombination hotspots and dispersed repeated sequences. *Journal of Molecular Biology* 177, 399–416.

Soulier, S., Mercier, J.C., Vilotte, J.L., Anderson, J., Clark, A.J. and Provot, C. (1989) The bovine and ovine genomes contain multiple sequences homologous to the alpha-lactalbumin-encoding gene. *Gene* 83, 331–338.

Soulier, S., Vilotte, J.L., Stinnakre, M.G. and Mercier, J.C. (1992) Expression analysis of ruminant alpha-lactalbumin in transgenic mice: developmental regulation and general location of important cis-regulatory elements. *FEBS Letters* 297, 13–18.

Stacey, A., Schnieke, A., Kerr, M., Scott, A., McKee, C., Cottingham, I., Binas, B., Wilde, C. and Colman, A. (1995) Lactation is disrupted by alpha-lactalbumin deficiency and can be restored by human alpha-lactalbumin gene replacement in mice. *Proceedings of the National Academy of Sciences USA* 92, 2835–2839.

Stelwagen, K. (2001) Effect of milking frequency on mammary functioning and shape of the lactation curve. *Journal of Dairy Science* 84, E204–E211.

Stewart, A.F., Willis, I.M. and MacKinlay, A.G. (1984) Nucleotide sequences of bovine alpha S1- and kappa-casein cDNAs. *Nucleic Acids Research* 12, 3895–3907.

Stewart, A.F., Bonsing, J., Beattie, C.W., Shah, F., Willis, I.M. and MacKinlay, A.G. (1987) Complete nucleotide sequences of bovine alpha S2- and beta-casein cDNAs: comparisons with related sequences in other species. *Molecular Biology and Evolution* 4, 231–241.

Stinnakre, M.G., Vilotte, J.L., Soulier, S. and Mercier, J.C. (1994) Creation and phenotypic analysis of alpha-lactalbumin-deficient mice. *Proceedings of the National Academy of Sciences USA* 91, 6544–6548.

Stockdale, C.R. (2006) Influence of milking frequency on the productivity of dairy cows. *Australian Journal of Experimental Agriculture* 46, 965–974.

Stocklin, E., Wissler, M., Gouilleux, F. and Groner, B. (1996) Functional interactions between Stat5 and the glucocorticoid receptor. *Nature* 383, 726–728.

Stoecklin, E., Wissler, M., Moriggl, R. and Groner, B. (1997) Specific DNA binding of Stat5, but not of glucocorticoid receptor, is required for their functional cooperation in the regulation of gene transcription. *Molecular and Cellular Biology* 17, 6708–6716.

Streuli, C.H. (2009) Integrins and cell-fate determination. *Journal of Cell Science* 122, 171–177.

Streuli, C.H., Edwards, G.M., Delcommenne, M., Whitelaw, C.B., Burdon, T.G., Schindler, C. and Watson, C.J. (1995a) Stat5 as a target for regulation by extracellular matrix. *Journal of Biological Chemistry* 270, 21639–21644.

Streuli, C.H., Schmidhauser, C., Bailey, N., Yurchenco, P., Skubitz, A.P., Roskelley, C. and Bissell, M.J. (1995b) Laminin mediates tissue-specific gene expression in mammary epithelia. *Journal of Cell Biology* 129, 591–603.

Struhl, K. (1989) Molecular mechanisms of transcriptional regulation in yeast. *Annual Review of Biochemistry* 58, 1051–1077.

Swaisgood, H.E. (1993) Review and update of casein chemistry. *Journal of Dairy Science* 76, 3054–3061.

Takahashi, R. and Ueda, M. (2001) The milk protein promoter is a useful tool for developing a rat with tolerance to a human protein. *Transgenic Research* 10, 571–575.

Thomas, R.S., Ng, A.N., Zhou, J., Tymms, M.J., Doppler, W. and Kola, I. (2000) The Elf group of Ets-related transcription factors. ELF3 and ELF5. *Advances in Experimental Medicine and Biology* 480, 123–128.

Topper, Y.J. and Freeman, C.S. (1980) Multiple hormone interactions in the developmental biology of the mammary gland. *Physiology Review* 60, 1049–1106.

Tsai, M.J. and O'Malley, B.W. (1994) Molecular mechanisms of action of steroid/thyroid receptor superfamily members. *Annual Review of Biochemistry* 63, 451–486.

Udy, G.B., Towers, R.P., Snell, R.G., Wilkins, R.J., Park, S.H., Ram, P.A., Waxman, D.J. and Davey, H.W. (1997) Requirement of STAT5b for sexual dimorphism of body growth rates and liver gene expression. *Proceedings of the National Academy of Sciences USA* 94, 7239–7244.

Vafaizadeh, V., Klemmt, P., Brendel, C., Weber, K., Doebele, C., Britt, K., Grez, M., Fehse, B., Desrivieres, S. and Groner, B. (2010) Mammary epithelial reconstitution with gene-modified stem cells assigns roles to Stat5 in luminal alveolar cell fate decisions, differentiation, involution, and mammary tumor formation. *Stem Cells* 28, 928–938.

Van Berkel, P.H., Welling, M.M., Geerts, M., Van Veen, H.A., Ravensbergen, B., Salaheddine, M., Pauwels, E.K., Pieper, F., Nuijens, J.H. and Nibbering, P.H. (2002) Large scale production of recombinant human lactoferrin in the milk of transgenic cows. *Nature Biotechnology* 20, 484–487.

Vilotte, J.L., Soulier, S., Mercier, J.C., Gaye, P., Hue-Delahaie, D. and Furet, J.P. (1987) Complete nucleotide sequence of bovine alpha-lactalbumin gene: comparison with its rat counterpart. *Biochimie* 69, 609–620.

Vilotte, J.L., Soulier, S., Stinnakre, M.G., Massoud, M. and Mercier, J.C. (1989) Efficient tissue-specific expression of bovine alpha-lactalbumin in transgenic mice. *European Journal of Biochemistry* 186, 43–48.

Vilotte, J.L., Soulier, S. and Mercier, J.C. (1993) Complete sequence of a bovine alpha-lactalbumin pseudogene: the region homologous to the gene is flanked by two directly repeated LINE sequences. *Genomics* 16, 529–532.

Wakao, H., Schmitt-Ney, M. and Groner, B. (1992) Mammary gland-specific nuclear factor is present in lactating rodent and bovine mammary tissue and composed of a single polypeptide of 89 kDa. *Journal of Biological Chemistry* 267, 16365–16370.

Wakao, H., Gouilleux, F. and Groner, B. (1994) Mammary gland factor (MGF) is a novel member of the cytokine regulated transcription factor gene family and confers the prolactin response. *EMBO Journal* 13, 2182–2191.

Wall, E.H. and McFadden, T.B. (2008) Use it or lose it: enhancing milk production efficiency by frequent milking of dairy cows. *Journal of Animal Science* 86, 27–36.

Wall, R.J., Kerr, D.E. and Bondioli, K.R. (1997) Transgenic dairy cattle: genetic engineering on a large scale. *Journal of Dairy Science* 80, 2213–2224.

Wang, J., Yang, P., Tang, B., Sun, X., Zhang, R., Guo, C., Gong, G., Liu, Y., Li, R., Zhang, L., Dai, Y. and Li, N. (2008) Expression and characterization of bioactive recombinant human alpha-lactalbumin in the milk of transgenic cloned cows. *Journal of Dairy Science* 91, 4466–4476.

Wang, P., Ballestrem, C. and Streuli, C.H. (2011) The C terminus of talin links integrins to cell cycle progression. *Journal of Cell Biology* 195, 499–513.

Wanyonyi, S.S., Lefevre, C.A.S.J. and Nicholas, K.R. (2013a) The extracellular matrix regulates MaeuCath1a gene expression. *Developmental and Comparative Immunology* 40, 289–299.

Wanyonyi, S.S., Lefevre, C.A.S.J. and Nicholas, K.R. (2013b) The extracellular matrix locally regulates asynchronous concurrent lactation in tammar wallaby (*Macropus eugenii*). *Matrix Biology* 32, 342–351.

Wasylyk, B., Hahn, S.L. and Giovane, A. (1993) The Ets family of transcription factors. *European Journal of Biochemistry* 211, 7–18.

Watanabe, Y., Tsukada, T., Notake, M., Nakanishi, S. and Numa, S. (1982) Structural analysis of repetitive DNA sequences in the bovine corticotropin-beta-lipotropin precursor gene region. *Nucleic Acids Research* 10, 1459–1469.

Waugh, D.F. (ed.) (1971) *Formation and Structure of Casein Micelles*. Academic Press, New York.

Waugh, D.F. and Von Hippel, P.H. (1956) κ-Casein and the stabilization of casein micelles. *Journal of the American Chemical Society* 78, 4576–4582.

Weaver, V.M., Petersen, O.W., Wang, F., Larabell, C.A., Briand, P., Damsky, C. and Bissell, M.J. (1997) Reversion of the malignant phenotype of human breast cells in three-dimensional culture and *in vivo* by integrin blocking antibodies. *Journal of Cell Biology* 137, 231–245.

Welte, T., Philipp, S., Cairns, C., Gustafsson, J.A. and Doppler, W. (1993) Glucocorticoid receptor binding sites in the promoter region of milk protein genes. *Journal of Steroid Biochemistry and Molecular Biology* 47, 75–81.

Welte, T., Garimorth, K., Philipp, S., Jennewein, P., Huck, C., Cato, A.C. and Doppler, W. (1994) Involvement of Ets-related proteins in hormone-independent mammary cell-specific gene expression. *European Journal of Biochemistry* 223, 997–1006.

Wicha, M.S., Lowrie, G., Kohn, E., Bagavandoss, P. and Mahn, T. (1982) Extracellular matrix promotes mammary epithelial growth and differentiation *in vitro*. *Proceedings of the National Academy of Sciences USA* 79, 3213–3217.

Wilde, C.J., Calvert, D.T., Daly, A. and Peaker, M. (1987) The effect of goat milk fractions on synthesis of milk constituents by rabbit mammary explants and on milk yield *in vivo*. Evidence for autocrine control of milk secretion. *Biochemistry Journal* 242, 285–288.

Wilde, C.J., Addey, C.V., Boddy, L.M. and Peaker, M. (1995) Autocrine regulation of milk secretion by a protein in milk. *Biochemistry Journal* 305 (Pt 1), 51–58.

Wilde, C.J., Addey, C.V., Li, P. and Fernig, D.G. (1997) Programmed cell death in bovine mammary tissue during lactation and involution. *Experimental Physiology* 82, 943–953.

Williams, S.G. and Lovell, S.C. (2009) The effect of sequence evolution on protein structural divergence. *Molecular Biology and Evolution* 26, 1055–1065.

Winklehner-Jennewein, P., Geymayer, S., Lechner, J., Welte, T., Hansson, L., Geley, S. and Doppler, W. (1998) A distal enhancer region in the human beta-casein gene mediates the response to prolactin and glucocorticoid hormones. *Gene* 217, 127–139.

Wong, D.W., Camirand, W.M. and Pavlath, A.E. (1996) Structures and functionalities of milk proteins. *Critical Reviews in Food Science and Nutrition* 36, 807–844.

Wyszomierski, S.L. and Rosen, J.M. (2001) Cooperative effects of STAT5 (signal transducer and activator of transcription 5) and C/EBPbeta (CCAAT/enhancer-binding protein-beta) on beta-casein gene transcription are mediated by the glucocorticoid receptor. *Molecular Endocrinology* 15, 228–240.

Xu, R., Nelson, C.M., Muschler, J.L., Veiseh, M., Vonderhaar, B.K. and Bissell, M.J. (2009) Sustained activation of STAT5 is essential for chromatin remodeling and maintenance of mammary-specific function. *Journal of Cell Biology* 184, 57–66.

Yamauchi, K., Wakabayashi, H., Shin, K. and Takase, M. (2006) Bovine lactoferrin: benefits and mechanism of action against infections. *Biochemical and Cellular Biology* 84, 291–296.

Yang, B., Wang, J., Tang, B., Liu, Y., Guo, C., Yang, P., Yu, T., Li, R., Zhao, J., Zhang, L., Dai, Y. and Li, N. (2011) Characterization of bioactive recombinant human lysozyme expressed in milk of cloned transgenic cattle. *PLoS One* 6, e17593.

Yang, J., Kennelly, J.J. and Baracos, V.E. (2000) Physiological levels of Stat5 DNA binding activity and protein in bovine mammary gland. *Journal of Animal Science* 78, 3126–3134.

Yang, J., Zhao, B., Baracos, V.E. and Kennelly, J.J. (2005) Effects of bovine somatotropin on beta-casein mRNA levels in mammary tissue of lactating cows. *Journal of Dairy Science* 88, 2806–2812.

Yang, P., Wang, J., Gong, G., Sun, X., Zhang, R., Du, Z., Liu, Y., Li, R., Ding, F., Tang, B., Dai, Y. and Li, N. (2008) Cattle mammary bioreactor generated by a novel procedure of transgenic cloning for large-scale production of functional human lactoferrin. *PLoS One* 3, e3453.

Yen, C.C., Lin, C.Y., Chong, K.Y., Tsai, T.C., Shen, C.J., Lin, M.F., Su, C.Y., Chen, H.L. and Chen, C.M. (2009) Lactoferrin as a natural regimen for selective decontamination of the digestive tract: recombinant porcine lactoferrin expressed in the milk of transgenic mice protects neonates from pathogenic challenge in the gastrointestinal tract. *Journal of Infectious Diseases* 199, 590–598.

Yen, C.H., Yang, C.K., Chen, I.C., Lin, Y.S., Lin, C.S., Chu, S. and Tu, C.F. (2008) Expression of recombinant Hirudin in transgenic mice milk driven by the goat beta-casein promoter. *Biotechnology Journal* 3, 1067–1077.

Yu-Lee, L.Y., Richter-Mann, L., Couch, C.H., Stewart, A.F., MacKinlay, A.G. and Rosen, J.M. (1986) Evolution of the casein multigene family: conserved sequences in the 5' flanking and exon regions. *Nucleic Acids Research* 14, 1883–1902.

Zhang, L., Hou, D., Chen, X., Li, D., Zhu, L., Zhang, Y., Li, J., Bian, Z., Liang, X., Cai, X. *et al.* (2012) Exogenous plant MIR168a specifically targets mammalian LDLRAP1: evidence of cross-kingdom regulation by microRNA. *Cell Research* 22, 107–126.

Zhang, P., Sawicki, V., Lewis, A., Hanson, L., Nuijens, J.H. and Neville, M.C. (2001) Human lactoferrin in the milk of transgenic mice increases intestinal growth in ten-day-old suckling neonates. *Advances in Experimental Medicine and Biology* 501, 107–113.

Zhou, Q., Li, M., Wang, X., Li, Q., Wang, T., Zhu, Q., Zhou, X., Gao, X. and Li, X. (2012) Immune-related microRNAs are abundant in breast milk exosomes. *International Journal of Biological Science* 8, 118–123.

Zhou, Y., Akers, R.M. and Jiang, H. (2008) Growth hormone can induce expression of four major milk protein genes in transfected MAC-T cells. *Journal of Dairy Science* 91, 100–108.

17 Genetics of Fatty Acid Composition in Bovine Milk and Beef

R.A. Nafikov,[1] H. Soyeurt[2] and D.C. Beitz[1]

[1]Iowa State University, Ames, Iowa, USA;
[2]University of Liège, Gembloux, Belgium

Introduction

The main reasons for modifying fatty acid composition in beef and milk are to make those foods healthier for human consumption and to control their processing properties. The potentially adverse effects of dietary saturated fatty acids (SFAs) on plasma cholesterol concentrations and incidence of cardiovascular diseases in humans have been known about for some time (Noakes et al., 1996; Kris-Etherton and Yu, 1997; Krauss et al., 2000; Mensink et al., 2003; Mozaffarian et al., 2010; Hooper et al., 2011). Specifically, high dietary palmitic (16:0) and myristic (14:0) acids predominantly found in dairy products, beef tallow and lard are associated with undesirable plasma cholesterol profiles in humans (Kris-Etherton and Yu, 1997). There are, however, some uncertainties

about the deleterious effects of SFAs on human health (Skeaff and Miller, 2009; Siri-Tarino et al., 2010; Huth and Park, 2012). Nevertheless, foods with low concentrations of SFAs and high concentrations of monounsaturated (MUFAs) and polyunsaturated (PUFAs) fatty acids are preferred by consumers. Decreasing the ratio of n-6:n-3 fatty acids in milk and beef is also desirable for human health because it promotes biosynthesis of eicosanoids, which possess anti-inflammatory properties (Wall et al., 2010). The predominant trans fatty acid in milk and beef is vaccenic acid (18:1 trans-11), which can be converted by stearoyl-CoA desaturase (SCD) to conjugated linoleic acid (CLA; 18:2 cis-9, trans-11), and both fatty acids are assumed to have several potential health benefits in humans (Willett et al., 1993, Pietinen et al., 1997). Besides the ability to modify the healthfulness of dairy foods and beef, changes in fatty acid composition influencing melting point or firmness of adipose tissue, colour, flavour, tenderness, and shelf life of beef (Wood *et al.*, 2004) and spreadability of butter (Kaylegian and Lindsay, 1992; Bobe *et al.*, 2003; Couvreur *et al.*, 2006; Hurtaud *et al.*, 2010) are of great significance as well. All together, this underlines the importance of having the ability to control fatty acid composition in bovine milk and beef.

Fatty Acid Composition of Bovine Milk and Beef

In cattle, adipose tissue depots can be classified by location into intramuscular, intermuscular and subcutaneous, and these depots can make up 1.7, 8.4 and 2.2% of carcass weight, respectively (Aldai *et al.*, 2007). The SFA and MUFA concentrations are usually high in subcutaneous and intermuscular adipose tissues and low in intramuscular adipose tissue (Aldai *et al.*, 2007; Fig. 17.1). In contrast, the highest concentration of PUFAs is in intramuscular adipose tissue, where they are mainly present in the phospholipid fraction (Duckett *et al.*, 1993). The major fatty acid in beef is oleic acid (18:1 *cis*-9), followed by palmitic (16:0) and stearic (18:0) acids (Fig. 17.1). Linoleic acid (18:2 *cis*-9, *cis*-12) is the major PUFA in beef, primarily present in the phospholipid fraction of intramuscular adipose tissue. Another major PUFA in beef adipose tissues is α-linolenic acid (18:3 *cis*-9, *cis*-12, *cis*-15). The most abundant *trans* fatty acids in beef are vaccenic acid (18:1 *trans*-11) and CLA (18:2 *cis*-9, *trans*-11), present in high concentrations in subcutaneous and intermuscular adipose tissues (Aldai *et al.*, 2007).

Lipids in bovine milk constitute 3–5% (Jensen *et al.*, 1990) and are emulsified in the form of globular structures 2–4 μm in diameter surrounded by a protective phospholipid

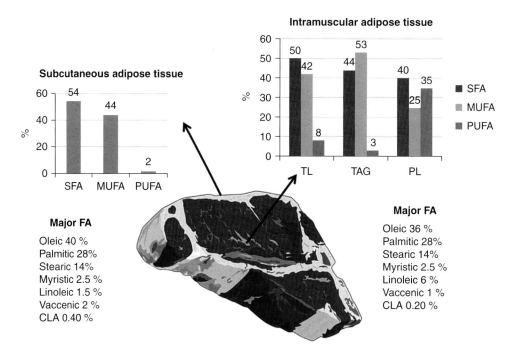

Fig. 17.1. Beef fatty acid composition. CLA, conjugated linoleic acid (18:2 *cis*-9, *trans*-11); FAs, fatty acids; MUFAs, monounsaturated fatty acids; PLs, phospholipids; PUFAs, polyunsaturated fatty acids: SFAs, saturated fatty acids; TAGs, triacylglycerols; TL, total lipids. (From Eichhorn *et al.*, 1986; Zembayashi *et al.*, 1995; De Smet *et al.*, 2004; Indurain *et al.*, 2006; Aldai *et al.*, 2007; Zhang *et al.*, 2008; Nogi *et al.*, 2011; Hoehne *et al.*, 2012.)

membrane derived from the secretory mammary epithelial cells. The triacylglycerol (TAG) fraction makes up about 98% of milk total lipids, with phospholipids accounting for only 0.5–1.0 % of milk total lipids. Fatty acid composition of milk is very diverse and might include up to 400 different fatty acids (Jensen, 2002). Milk total lipids contain about 70% SFAs, 25% MUFAs and 5% PUFAs (Grummer, 1991; Fig. 17.2). The major fatty acid in milk is palmitic acid (16:0; 25–30%) followed by oleic (18:1 *cis*-9; 20–25%), stearic (18:0; about 10%) and myristic (14:0; 8–10%) acids. Bovine milk SFAs also contain substantial amounts of fatty acids with 4–10 carbon chains (Jensen *et al.*, 1990). The major PUFA in milk is linoleic acid (18:2 *cis*-9, *cis*-12) comprising about 2% of milk total lipids. Concentrations of vaccenic acid (18:1 *trans*-11) and CLA (18:2 *cis*-9, *trans*-11) in milk are about 3% and between 0.4 and 1.1%, respectively (Fig. 17.2).

Environmental Factors Influencing Fatty Acid Composition

The environmental factors known to affect beef and milk fatty acid composition should be taken into account during statistical analyses of fatty acid composition data and also used as management tools, in addition to genetics, to influence fatty acid composition. Diet is one of the major

environmental factors influencing fatty acid composition in both beef and milk. A conversion of dietary unsaturated fatty acids (UFAs) into beef and milk UFAs in ruminants is hampered by ruminal biohydrogenation (Doreau and Chilliard, 1997), which makes genetic approaches to changing fatty acid composition more appealing. Nevertheless, diet can affect the microbial population of the rumen and as a consequence alter fatty acid composition in beef and milk. Grass-based feeding of German Simmental and German Holstein bulls, for example, significantly increased concentrations of n-3 fatty acids, vaccenic acid (18:1 *trans*-11), and CLA (18:2 *cis*-9, *trans*-11) and decreased n-6:n-3 fatty acid ratio compared with cattle fed concentrate-based diets (Nuernberg *et al.*, 2005). Feeding high grain diets to dairy cattle depresses milk fat percentage and decreases concentrations of palmitic acid (16:0) and short- and medium-chain fatty acids, while increasing the concentration of 18-carbon UFAs (Bauman and Griinari, 2003). Increased production of CLA (18:2 *trans*-10, *cis*-12) was proved to be the cause of milk fat depression.

The duration of feeding a high-concentrate finishing diet on fatty acid composition of the longissimus muscle was studied in Angus and Hereford crossbred yearling steers, which, after growing on a grass-based diet, were fed the finishing diet for 196 days then slaughtered in 28-day intervals to determine fatty acid composition of beef (Duckett *et al.*, 1993). The percentage of intramuscular adipose tissue had doubled within the 84- to 112-day period, primarily because of the rise in the intramuscular TAG concentration caused by an increase in oleic acid (18:1 *cis*-9) concentration. There were no changes in the concentration of intramuscular phospholipids but PUFA concentrations decreased within the phospholipid fraction during the finishing period (Duckett *et al.*, 1993). As a result of being on a high-concentrate finishing diet, steers had significantly higher MUFA and lower PUFA concentrations in the longissimus muscle during the finishing period. Finishing British and continental crossbred steers for 90 days on a concentrate-based diet significantly increased backfat thickness and marbling score (Camfield *et al.*, 1997). A significant decrease in the concentrations of stearic (18:0)

SFA = 70 %
MUFA = 25 %
PUFA = 5 %

2–5% Fat

Milk FA
Palmitic 25–30%
Oleic 20–25%
Stearic 10%
Myristic 8–10%
Lauric 3%
Linoleic 2%
Vaccenic 3%
CLA 0.4–1.1%

Fig. 17.2. Milk fatty acid composition. CLA, conjugated linoleic acid (18:2 *cis*-9, *trans*-11); FAs, fatty acids; MUFAs, monounsaturated fatty acids; PUFAs, polyunsaturated fatty acids; SFAs, saturated fatty acids. (From Jensen, 2002; Couvreur *et al.*, 2006; Mansson, 2008; Stoop *et al.*, 2008; Nafikov *et al.*, 2013a.)

and linoleic (18:2 *cis*-9, *cis*-12) acids and an increase in the concentration of oleic acid (18:1 *cis*-9) were observed in intramuscular adipose tissue of the longissimus muscle as time on finishing diet progressed.

Other environmental factors influencing beef fatty acid composition are gender and beef cut. After adjusting fatty acid composition data of TAG fractions from the longissimus thoracis muscle and subcutaneous adipose tissue for differences in fatness of a carcass, heifers had significantly higher concentrations of oleic acid (18:1 *cis*-9), MUFAs and UFAs and lower concentrations of SFAs, myristic (14:0) and palmitic (16:0) acids compared with steers (Zembayashi *et al.*, 1995). Comparisons between eight different beef cuts with respect to fatty acid composition revealed that brisket had the lowest concentrations of palmitic (16:0) and stearic (18:0) acids and one of the lowest concentrations of myristic (14:0) acid (Turk and Smith, 2009). At the same time brisket had significantly higher concentrations of different MUFAs including oleic acid (18:1 *cis*-9) and a higher MUFA:SFA ratio.

Another important environmental factor influencing milk fatty acid composition is stage of lactation. In ruminants not in negative energy balance, *de novo* lipogenesis contributes only 50% of fatty acids in milk (Neville and Picciano, 1997). The rest of the fatty acids taken up from blood by the mammary gland are derived from non-esterified fatty acids (NEFAs) and very low density lipoproteins of intestinal origin (Barber *et al.*, 1997). The blood NEFA comes mainly from lipolysis in adipose tissue, which accounts for less than 10% of fatty acids used for milk lipid biosynthesis unless cows are in negative energy balance when lipolysis in adipose tissue provides the majority of fatty acids for milk lipid biosynthesis (Pullen *et al.*, 1989; Bauman and Griinari, 2000). Other fatty acids are derived from the diet. Stage of lactation affects milk fatty acid composition because it coincides with changes in energy balance that determine whether cows use predominantly adipose tissue or both adipose tissue and *de novo* lipogenesis in the mammary gland to provide fatty acids for milk TAG biosynthesis. In early lactation, cows usually are in negative energy balance, relying more extensively on lipolysis in adipose tissue to provide fatty acids for milk lipid biosynthesis (Palmquist *et al.*,

1993). This situation results in the concentration of short- and medium-chain fatty acids (C_6 to C_{14}) in milk being low at the beginning of the lactation because of the inhibitory effects of mobilized long-chain fatty acids on *de novo* lipogenesis in mammary tissue. With the progression of lactation, the concentration of short- and medium-chain fatty acids in milk gradually increases. Because of the predominant role of lipolysis in adipose tissue in supplying fatty acids for milk TAG biosynthesis in earlier lactation, concentration of oleic acid (18:1 *cis*-9) in milk is usually higher during this time period. There were no obvious seasonal effects on milk fatty acid composition but the trend was to have lower concentrations of short- and medium-chain fatty acids during warm months of the year probably because of the higher dietary fat intake at that time (Palmquist *et al.*, 1993).

Fatty Acid Heritabilities and Genetic Correlations

Heritabilities of fatty acids in different adipose tissue depots of beef cattle, described in detail in Chapter 22, were low to moderate (Malau-Aduli *et al.*, 2000; Pitchford *et al.*, 2002; Nogi *et al.*, 2011) indicating feasability of using selection to influence fatty acid composition in beef. More studies, however, are needed to obtain more precise estimates of fatty acid heritabilities in beef. Estimates of fatty acid heritabilities in milk were moderate to high (Karijord *et al.*, 1982; Soyeurt *et al.*, 2007, 2008; Stoop *et al.*, 2008; Nafikov *et al.*, 2013a) underlining potential for selection to influence milk fatty acid composition (Table 17.1). High heritability values were observed for SFAs, MUFAs and PUFAs and individual medium- and long-chain SFAs such as capric (10:0), lauric (12:0), myristic (14:0), palmitic (16:0) and stearic (18:0) acids, with palmitic acid (16:0) being the most heritable. Odd-chain SFAs and individual MUFAs, except for oleic acid (18:1 *cis*-9), elongation and disaturation indices had very low heritability values (Nafikov *et al.*, 2013a). Among the UFAs only oleic (18:1 *cis*-9) and linoleic (18:2 *cis*-9, *cis*-12) acids had high heritability values. Milk fatty acid heritabilities were also dependent on stage

Table 17.1. Heritability estimates for milk fatty acids and milk fat percentage. (Adapted from Arnould and Soyeurt, 2009.)

Trait	Karijord[a]	Soyeurt[b]		Soyeurt[c]	Stoop[d]	Nafikov[e]
	g/100g fat	g/100g milk	g/100g fat	g/100g fat	wt (%)	wt (%)
N	7,000	40,007	40,007	52,950	1,918	3,881
Milk fat (%)	0.09	0.32	0.32	0.33	0.47	0.23
4:0					0.35	0.08
6:0	0.11				0.39	0.07
8:0	0.13				0.48	0.05
10:0	0.16				0.54	0.19
12:0	0.17	0.29	0.09		0.35	0.17
14:0	0.07	0.31	0.19	0.15	0.49	0.18
14:1 *cis*-9	0.26			0.20		0.04
16:0	0.15	0.38	0.20	0.15	0.31	0.34
16:1 *cis*-9	0.12			0.22		0.07
18:0	0.15	0.30	0.28	0.16	0.19	0.19
18:1 *cis*-9	0.06	0.05	0.15	0.17	0.18	0.21
18:2 *cis*-9, *cis*-12	0.11	0.20	0.15		0.13	0.21
CLA					0.21	0.05
18:3 *cis*-9, *cis*-12, *cis*-15	0.09				0.09	
SFA		0.36	0.14			0.28
MUFA		0.15	0.24	0.17		0.26
PUFA						0.30

[a]Karijord *et al.*, 1982; [b]Soyeurt *et al.*, 2007; [c]Soyeurt *et al.*, 2008; [d]Stoop *et al.*, 2008; [e]Nafikov *et al.*, 2013a; CLA, conjugated linoleic acid (18:2 *cis*-9, *trans*-11); MUFAs, monounsaturated fatty acids; PUFAs, polyunsaturated fatty acids; SFAs, saturated fatty acids.

of lactation, with higher heritability values observed towards the end of lactation (Bastin *et al.*, 2011; Nafikov *et al.*, 2013a; Fig. 17.3). Such differences in milk fatty acid heritability values might be explained by a smaller contribution of *de novo* lipogenesis and higher contribution of lipolysis from adipose tissue towards milk TAG biosynthesis at the beginning of lactation.

Genetic correlations between beef fatty acids and other meat quality traits are described in detail in Chapter 22 and will not be discussed here. The genetic correlation between palmitic acid (16:0) and the percentage of milk fat was high and positive ($r = 0.65$–0.74; Stoop *et al.*, 2008; Mele *et al.*, 2009), implying that selection for high milk fat percentage will result in increased palmitic acid (16:0) concentration in milk. The concentration of MUFA with 18 carbon atoms, however, had a negative genetic correlation with the percentage of milk fat ($r = -0.55$ to -0.74; Stoop *et al.*, 2008; Mele *et al.*, 2009). Other fatty acids that had a positive genetic correlation with

milk fat percentage were lauric (12:0) and stearic (18:0) acids ($r = 0.55$ and 0.84, respectively; Soyeurt *et al.*, 2007).

Breed Effects on Fatty Acid Composition in Bovine Milk and Beef

Breed effects on fatty acid composition in beef are predetermined by marbling score or percentage of total lipids in muscle (Wood *et al.*, 2008), which, in turn, depends on breed specific SCD activity in adipose tissues (Smith *et al.*, 2006). Breeds such as Japanese Black, Wagyu and Korean Hanwoo known for high marbling score of their muscle tissue have higher concentrations of oleic acid (18:1 *cis*-9) and subsequently higher MUFA concentration in their beef compared with other breeds (Smith *et al.*, 2009). Higher PUFA and linoleic acid (18:2 *cis*-9, *cis*-12) concentrations and lower oleic acid (18:1 *cis*-9) concentration in intramuscular adipose tissue are present in breeds such as Belgian Blue and Piedmontese

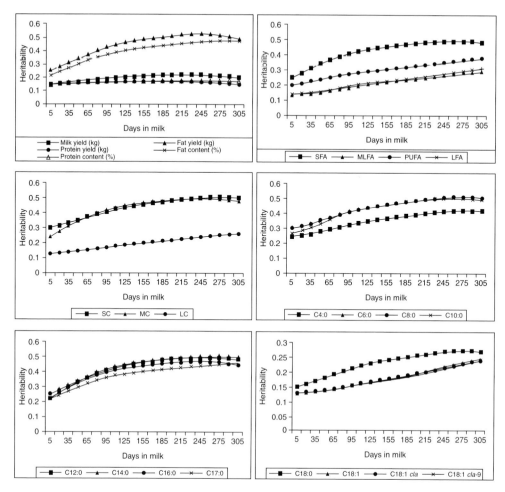

Fig. 17.3. Daily heritabilities of milk, fat and protein yields, fat and protein contents, and groups and individual fatty acid contents: saturated (SFA), monounsaturated (MUFA), unsaturated (UFA), short-chain (SC), medium-chain (MC), long-chain (LC) fatty acids, C4:0, C6:0, C8:0, C10:0, C12:0, C14:0, C16:0, C17:0, C18:0, C18:1, C18:1 *cis*, and C18:1 *cis*-9. (From Bastin *et al.*, 2011.)

known for their double-muscling phenotype (Wood *et al.*, 2008). The observed differences in beef fatty acid composition in these breeds are caused by a double-muscling-dependent increase in muscle phospholipid concentrations and relative leanness of muscle tissue from those animals. Crossbreeding is one approach to changing fatty acid composition in beef. Thus, MUFA concentration in the longissimus dorsi muscle and subcutaneous adipose tissue was higher in Jersey- and Wagyu-sired crossbred cattle compared with Angus-, Hereford-, South Devon-, Limousin- and Belgian Blue-sired crossbreds (Pitchford *et al.*, 2002).

Brahman-sired cattle shown to have higher UFA concentration in subcutaneous adipose tissue and 2.5°C lower melting point of their beef total lipids compared with *Bos taurus*-sired crossbreds (Perry *et al.*, 1998) are another example of crossbreeding changing fatty acid composition in beef.

Breed effects on milk fatty acid composition are mainly predetermined by the percentage of milk fat that in turn could be a function of SCD activity in those animals (Soyeurt *et al.*, 2008). Jersey and Guernsey cows known for producing milk with one of the highest fat percentages (5.13 and 4.87%; Gaunt, 1980) had significantly higher concentrations of capric (10:0), lauric (12:0) and palmitic (16:0) acids and lower concentrations of palmitoleic (16:1 cis-9) and oleic (18:1 cis-9) acids compared with Holstein cows (Stull and Brown, 1964). Other studies reported similar trends with higher concentrations of short- and medium-chain SFAs and palmitic (16:0) and stearic (18:0) acids and lower concentrations of palmitoleic (16:1 cis-9) and oleic (18:1 cis-9) acids in milk from Jersey cows compared with Holsteins (Beaulieu and Palmquist, 1995; DePeters et al., 1995; White et al., 2001; Maurice-Van Eijndhoven et al., 2013). Holstein cows, however, had higher CLA (18:2 cis-9, trans-11) concentration in their milk compared with Jersey cows (White et al., 2001; Kelsey et al., 2003). Breeds such as Brown Swiss with a milk fat percentage of 4.16% (Gaunt, 1980), intermediate in value between milk fat percentages of Holstein and Jersey cows, have milk fatty acid concentrations that are intermediate to values observed for Holsteins and Jerseys (DePeters et al., 1995). A detailed comparison between breed effects on milk fatty acid composition is presented in Table 17.2 as percentage deviations from milk fatty acid concentrations in Holstein cows (Soyeurt and Gengler, 2008).

Genetic Polymorphisms and Fatty Acid Composition

Most early association studies on beef and milk fatty acid composition used a candidate gene approach and focused on small areas of the bovine genome because tools for genome-wide association studies in cattle were not available. Selection of genome areas was based either on previously discovered quantitative trait loci (QTLs) or knowledge about the biology of lipid biosynthesis in adipose tissues and the mammary gland.

Stearoyl-CoA desaturase

The SCD located on BTA26 catalyses the desaturation of long-chain fatty acids at Δ9 carbon atom to their corresponding MUFAs (Ntambi,

1999) and is one of the most extensively studied genes with respect to beef fatty acid composition both in subcutaneous and intramuscular adipose tissues (Taniguchi et al., 2004; Ohsaki et al., 2009; Barton et al., 2010; Matsuhashi et al., 2011; Narukami et al., 2011). The SCD:g.878C>T polymorphism discovered as a non-synonymous SNP with Ala to Val amino acid substitution at 293 residue of SCD polypeptide chain was determined to be significantly associated with MUFA concentration in the trapezius muscle of Japanese Black steers (Taniguchi et al., 2004). The MUFA concentration, calculated based on only the concentrations of myristoleic (14:1 cis-9), palmitoleic (16:1 cis-9) and oleic (18:1 cis-9) acids, was 1.7 wt % higher in the muscle tissue of cattle with the C allele as compared with the T allele. The higher MUFA concentration in beef from cattle with the C allele of SCD:g.878C>T was also confirmed by the 2.2°C lower melting point of beef total lipids in those animals. Subsequent studies in different populations of Japanese Black cattle (Ohsaki et al., 2009; Matsuhashi et al., 2011) confirmed significant associations of C allele of SCD:g.878C>T with higher MUFA concentration, which was achieved because of significant associations of the C allele with higher myristoleic (14:1 cis-9) and oleic (18:1 cis-9) acid concentrations in intramuscular and subcutaneous adipose tissues. Associations of the C allele of SCD:g.878C>T with higher MUFA concentrations in beef might be explained by higher activity of SCD enzyme in those animals leading to increased desaturation of myristic (14:0) and stearic (18:0) acids, whose concentrations in beef were decreased in cattle with the C allele. Similar results with respect to the SCD:g.878C>T polymorphism were observed in intramuscular adipose tissue of Holstein steers (Narukami et al., 2011), but pairwise comparisons between genotypes did not reveal any significant differences pointing to weaker effects of the association in dairy cattle adipose tissue as compared with beef cattle.

After discovery of significant associations of SCD:g.878C>T polymorphism with beef fatty acid composition, there were a few studies describing the effects of the same SNP on milk fatty acid composition in dairy cattle, but results were inconsistent (Mele et al., 2007; Moioli et al., 2007; Schennink et al., 2008).

Table 17.2. Breed differences in milk fatty acid composition as compared with Holstein cows. (Adapted from Soyeurt and Gengler, 2008.)

Breed differences in milk fatty acid concentrations compared to Holsteins (%)

	Guernsey	Jersey			Brown Swiss		Montbeliarde	Normande
	Stull	Stull	Beaulieu	DePeters	DePeters	Kelsey	Lawless	Lawless
N	25	10	8	23	29	106	29	27
4:0	+20.73	+8.54	−2.43	−4.90	−1.47	+12.36	−5.50	−2.75
6:0	+13.16	+15.79	+16.67	+3.32	+2.21	+7.32	−2.54	+0.85
8:0	+14.29	+34.10	+38.46	+7.55	+5.03	+13.13	+1.02	+5.10
10:0	+12.5	+70.83	+43.33	+13.59	+4.08	+14.22	+6.98	+9.30
10:1							−16.67	0.00
12:0	+7.59	+36.90	+42.86	+16.90	+6.34	+14.41	+6.46	+10.77
14:0	+5.64	+9.26	+8.62	+2.36	+2.14	+4.66	+2.61	+1.87
14:1 *cis*-9	−11.31	−4.76				−1.64	−28.09	−10.11
15:0	−6.80	−2.04				−6.76		
16:0	+7.20	+5.63	−6.79	−1.24	−1.70	+0.96	−11.49	−8.15
16:1 *cis*-9	−7.14	−16.67		−9.55	−1.51	−13.08		
18:0	+4.64	+1.12	+12.50	+6.61	−6.83	−3.42	+10.89	+14.93
18:1 *cis*-9	−11.15	−12.92	−12.72	−9.51	+3.91	−1.96	+5.37	+1.37
18:2 *cis*-9, *cis*-12	−4.92	−4.64	0.00	+1.58	−4.74	−5.80	+5.94	+3.96
CLA	−19.79	−32.29	−16.67	+15.50	−6.98	−6.82	+13.07	−5.11
18:3						−2.56	+1.22	−6.10

Stull and Brown, 1964; Beaulieu and Palmquist, 1995; DePeters *et al.*, 1995; Lawless *et al.*, 1999; Kelsey *et al.*, 2003.
CLA, conjugated linoleic acid (18:2 *cis*-9, *trans*-11).

Holstein cows with the C allele of SCD:g.878C>T had significantly higher concentrations of caproleic (10:1 cis-4) and myristoleic (14:1 cis-9) acids (Moioli $et\ al.$, 2007) and higher concentrations of MUFA, myristoleic (14:1 cis-9) and oleic (18:1 cis-9) acids (Mele $et\ al.$, 2007). In a different study, the C allele of SCD:g.878C>T was significantly associated with lower concentrations of capric (10:0), lauric (12:0), myristic (14:0) and palmitoleic (16:1 cis-9) acids and of CLA (18:2 cis-9, $trans$-11), and higher concentrations of caproleic (10:1 cis-4), lauroleic (12:1 cis-4), myristoleic (14:1 cis-9), stearic (18:0) and vaccenic (18:1 $trans$-11) acids in milk (Schennink $et\ al.$, 2008). Inconsistencies with associations between SCD:g.878C>T and milk fatty acid composition might be explained by the fact that MUFAs in milk are not only synthesized by SCD in the mammary gland but are also of dietary origin and from TAG hydrolysis in adipose tissues.

Diacylglycerol O-acyltransferase 1

The diacylglycerol O-acyltransferase 1 ($DGAT1$) gene is located on BTA14 and encodes an enzyme catalysing the rate-limiting step in TAG biosynthesis by adding acyl-CoA to the sn-3 position on the glycerol backbone (Yen $et\ al.$, 2008). Following the discovery of the $DGAT1$ gene location (Cases $et\ al.$, 2001), it was demonstrated that knock-out mice for $DGAT1$ had impaired lactation (Howell $et\ al.$, 1997), prompting researchers to look at the effects of mutations in $DGAT1$ on milk fat synthesis. A non-synonymous substitution of Lys by Ala at 232 amino acid of DGAT1 was identified as being responsible for the effects of that enzyme on milk fat yield, with the lysine-encoding dinucleotide AA allele being associated with higher milk fat yield (Grisart $et\ al.$, 2002; Winter $et\ al.$, 2002). This mutation(s) involves two neighbouring nucleotides. A functional conformation of $DGAT1$:g.10433-10434AA>GC mutation was made later by showing that the AA allele was characterized by higher V_{max} for the DGAT1-catalysed reaction compared with the GC allele, which explained why animals with the AA allele had higher milk fat yield (Grisart $et\ al.$, 2004). The GC allele of $DGAT1$:g.10433-10434AA>GC

was shown to be significantly associated with lower concentrations of SFA and palmitic acid (16:0), but higher concentrations of 18-carbon UFA, CLA (18:2 cis-9, $trans$-11) and myristic acid (14:0; Schennink $et\ al.$, 2007). Genetic correlations were significantly high and negative between milk fat percentage and concentrations of 18-carbon UFA ($r = -0.72$), CLA (18:2 cis-9, $trans$-11; $r = -0.58$) and myristic acid (14:0; $r = -0.43$), and high and positive between milk fat percentage and palmitic acid (16:0; $r = 0.65$) concentration. Such coefficients indicate that selection against high palmitic acid (16:0) concentration in milk will inevitably lead to low milk fat percentage, which might be undesirable for the dairy industry from an economic perspective. A similar situation will occur during selection for high 18-carbon UFA concentrations in milk. Fatty acid desaturation indices defined for fatty acids with the same number of carbon atoms as the ratios of MUFA concentrations to the sum of MUFA and SFA concentrations were also studied for milk fatty acids with respect to $DGAT1$:g.10433-10434AA>GC polymorphism (Schennink $et\ al.$, 2008). The GC allele of $DGAT1$:g.10433-10434AA>GC was significantly associated with smaller desaturation indices for capric (10:0), lauric (12:0), myristic (14:0) and palmitic (16:0) acids, and higher desaturation indices for stearic acid (18:0) and CLA (18:2 cis-9, $trans$-11), and with higher total desaturation index. Until now, associations between $DGAT1$:g.10433-10434AA>GC polymorphism and milk fatty acid composition are the most consistent and largest in terms of size effects on the traits.

Fatty acid synthase

One of the first QTL studies for beef fatty acid composition in subcutaneous adipose tissue was conducted in crossbred steers and heifers either ¾-Jersey or ¾-Limousin (Morris $et\ al.$, 2007). Using linkage analysis based on microsatellite markers, a QTL with significant effects on myristic acid (14:0) concentration in subcutaneous adipose tissue was located on BTA19. The fatty acid synthase ($FASN$) located on BTA19 was identified as a candidate gene for the QTL. Significant associations between A alleles of *FASN*:g.15531C>A and *FASN*:g.15603G>A and decreased concentration of myristic acid (14:0) in subcutaneous adipose tissue were determined (Morris *et al.*, 2007). In a different study the earlier discovered non-synonymous *FASN*:g.17924A>G (Morris *et al.*, 2007) and synonymous *FASN*:g.18663T>C, both located in the thioesterase domain of *FASN*, were tested for associations with beef fatty acid composition (Zhang *et al.*, 2008). The G allele of *FASN*:g.17924A>G was significantly associated with higher concentrations of oleic acid (18:1 *cis*-9) and MUFA, and lower concentrations of myristic (14:0) and palmitic (16:0) acids and SFA in TAG and total lipid fractions of the longissimus dorsi muscle. Associations of SFA, MUFA, and myristic (14:0) and oleic (18:1 *cis*-9) acid concentrations in TAG and total lipid fractions with *FASN*:g.18663T>C were also observed. Both *FASN*:g.17924A>G and *FASN*:g.18663T>C were significantly associated with oleic acid (18:1 *cis*-9) and MUFA concentrations, and *FASN*:g.18663T>C was additionally associated with PUFA concentrations in the phospholipid fraction of intramuscular total lipids. The observed associations between *FASN*:g.17924A>G located in the thioesterase domain of *FASN* and beef fatty acid composition were attributed to a possible effect of this mutation on a chain length of *de novo* synthesized fatty acids (Zhang *et al.*, 2008). Premature release of elongating fatty acid carbon chains by the thioesterase domain of *FASN* might lead to production of myristic acid (14:0) instead of palmitic acid (16:0), the final product of the FASN-catalysed reaction.

Additional non-synonymous *FASN*:g. 16024A>G and *FASN*:g.16039T>C mutations, significantly associated with beef fatty acid composition, were identified in exon 34 of *FASN* (Abe *et al.*, 2009). The *FASN*: g.16024A>G and *FASN*:g.16039T>C were responsible for amino acid changes from Thr to Ala and from Trp to Arg. These two physically close mutations were in linkage disequilibrium producing two haplotypes, AT and GC. Haplotype AT of *FASN* was significantly associated with lower concentrations of myristic (14:0), myristoleic (14:1 *cis*-9), palmitic (16:0) and palmitoleic (16:1 *cis*-9) acids, and higher concentrations of stearic (18:0) and oleic (18:1

cis-9) acids and higher MUFA:SFA ratio in the longissimus dorsi muscle as compared with haplotype GC. There were also significant associations of *FASN* haplotypes with the same fatty acids in subcutaneous and intermuscular adipose tissue depots. Genotyping Japanese Black cattle for *FASN*:g.17924A>G revealed a predominant presence of the G allele in those animals. Since the G allele of *FASN*:g.17924A>G was previously shown (Zhang *et al.*, 2008) to be associated with higher concentrations of MUFA and oleic acid (18:1 *cis*-9) in the longissimus dorsi muscle of Angus cattle, it might explain higher than average intramuscular concentrations of these fatty acids in Japanese Black cattle compared with other breeds (Zembayashi *et al.*, 1995). It also shows that *FASN*:g.17924A>G can be used together with *FASN*:g.16024A>G and *FASN*:g.16039T>C to increase intramuscular concentrations of MUFA and oleic acid (18:1 *cis*-9) in beef. A comparison of *FASN* haplotype frequencies between Japanese Black, Holstein, Angus and Hereford breeds revealed that mainly the GC haplotype was present in Angus, Hereford and Holstein cattle (Abe *et al.*, 2009). Crossbreeding between breeds containing predominantly GC haplotype of *FASN* with Japanese Black cattle might increase AT haplotype frequencies in crossbred animals, leading to beef with higher concentrations of oleic acid (18:1 *cis*-9) and MUFA. In general, polymorphisms in *FASN* can be used to control concentrations of myristic (14:0), palmitic (16:0) and oleic (18:1 *cis*-9) acids and MUFA in beef.

Polymorphisms in *FASN* were the focus of recent association studies on milk fatty acid composition (Roy *et al.*, 2006; Morris *et al.*, 2007; Schennink *et al.*, 2009a). Earlier studies showed the presence of QTLs associated with milk fat percentage and yield on BTA19 (Khatkar *et al.*, 2004), and *FASN*:g.763G>C and *FASN*:g.16009A>G located in exons 1 and 34, respectively, were associated with milk fat yield in dairy cattle (Roy *et al.*, 2006). One of the first association studies for milk fatty acid composition (Morris *et al.*, 2007) determined significant associations for A alleles of *FASN*:g.15531C>A and *FASN*:g.15603G>A with increased concentration of myristic acid (14:0). Another study determined significant associations between A and G alleles of earlier discovered *FASN*:g.16009A>G (Roy *et al.*, 2006) and *FASN*:g.17924A>G (Morris *et al.*, 2007), and decreased concentration of myristic (14:0) acid in milk (Schennink *et al.*, 2009a). In dairy cattle genetic polymorphisms in *FASN* were predominantly associated with myristic acid (14:0) concentration in milk.

Sterol regulatory element binding factor 1

Because of its known ability to regulate lipogenesis and expression of *SCD* (Brown and Goldstein, 1997; Horton, 2002), the sterol regulatory element binding protein 1 (SREBP1) transcription factor, a product of sterol regulatory element binding factor 1 (*SREBF1*) gene, became the focus of some association studies on beef and milk fatty acid composition. The 84-bp indel in intron 5 of *SREBF1* was determined to be significantly associated with MUFA concentration in the trapezius muscle of Japanese Black steers (Hoashi *et al.*, 2007). The absence of the 84-bp indel ≥ was associated with 1.3 wt % higher MUFA concentration that consisted of myristoleic (14:1 *cis*-9), palmitoleic (16:1 *cis*-9) and oleic (18:1 *cis*-9) acids and a 1.6°C lower melting point of intramuscular adipose tissue. In a subsequent study with Japanese Black steers and heifers from two different populations of cattle, there were no significant associations determined between the 84-bp indel in intron 5 of *SREBF1* and fatty acid composition in perirenal and intramuscular adipose tissues (Ohsaki *et al.*, 2009). This controversy confirmed the low reliability of using the 84-bp indel in intron 5 of *SREBF1* for the selection of animals with desirable beef fatty acid composition in different populations of cattle (Mannen, 2011). Additional studies testing associations between *SREBF1* polymorphisms and beef fatty acid composition did not reveal any significant associations in populations of Japanese Black and Brangus cattle (Matsuhashi *et al.*, 2011; Baeza *et al.*, 2013). There were, however, significant associations between *SREBF1*:g.13495T>C and concentrations of a number of SFAs and UFAs determined in the longissimus muscle of Hanwoo cattle (Lee

et al., 2013). Polymorphisms in *SREBF1* might not be useful as a tool to control fatty acid composition in beef.

Genotyping of Holstein cows for polymorphisms in *SREBF1* was used to reconstruct intragenetic haplotypes to determine significant associations for milk fatty acid composition, while evaluating the data from both the entire 10-month lactation and from the first 3 and the last 7 months of lactation (Nafikov *et al.*, 2013a). Significant associations between *SREBF1* haplotypes and concentrations of lauric (12:0) and myristic (14:0) acids in milk were determined for the last 7 months of lactation, and haplotype H1, which can be distinguished by T allele of *SREBF1*:g.13495T>C tag SNP was significantly associated with lower concentrations of lauric (12:0) and myristic (14:0) acids compared with other haplotypes. Selection of animals producing milk with low concentrations of lauric (12:0) and myristic (14:0) acids using *SREBF1*:g.13495T>C tag SNP, however, might be a disadvantage to dairy producers because of significant associations of the same SNP with lower milk production in dairy cattle (Nafikov *et al.*, 2013a). Significant associations between *SREBF1* haplotypes and myristic acid (14:0) concentration in milk were also determined for the entire 10-month lactation period. In a different study, the only significant association was determined between *SREBF1*:g.13495T>C and milk fat percentage in a herd with milk fat depression (Rincon *et al.*, 2012). Nevertheless, polymorphisms in *SREBF1* might be used to control concentrations of lauric (12:0) and myristic (14:0) acids in milk and the use of haplotypes instead of individual SNPs might be a more reliable approach to determine significant associations for milk and beef fatty acid compositions.

Fatty acid transport genes

The fatty acid binding protein 4 (*FABP4*) gene, responsible for transporting fatty acids inside a cell (Furuhashi and Hotamisligil, 2008), recently received some attention because of previously reported associations with marbling score in beef cattle (Michal *et al.*, 2006;

Barendse *et al.*, 2009; Lee *et al.*, 2010), although effects were not always consistent (Pannier *et al.*, 2010). Significant associations between a non-synonymous *FABP4*:g.220A>G and concentrations of palmitoleic acid (16:1 *cis*-9) and C16 desaturation index in the longissimus dorsi muscle of Japanese Black steers and heifers were detected, with the A allele being associated with higher values of the above traits (Hoashi *et al.*, 2008). In a study with Holstein steers, however, only association between *FABP4*:g.220A>G polymorphism and palmitic acid (16:0) concentration in intramuscular adipose tissue from the diaphragm area was detected (Narukami *et al.*, 2011).

In dairy cattle *FABP3*, *FABP4* and solute carrier family 27, isoform A6 (*SLC27A6*) fatty acid transport protein genes were tested for associations with milk fatty acid composition (Nafikov *et al.*, 2013b). After testing *SLC27A6* haplotypes, haplotype H3, distinguished by an intronic *SLC27A6*:g.15740A>C tag SNP, was significantly associated with lower SFA and lauric acid (12:0) concentrations and lower SFA:UFA ratio, but higher MUFA concentration in milk during the first 3 months of lactation. Haplotype H3 of *FABP4*, distinguished by a non-synonymous *FABP4*:g.3691G>A tag SNP, was significantly associated with lower concentrations of lauric (12:0) and myristic (14:0) acids and SFA and lower SFA:UFA ratio, but higher MUFA concentration in milk during the first 3 months of lactation (Nafikov *et al.*, 2013b). Polymorphisms in *FABP4* and *SLC27A6* were associated with milk fatty acid composition primarily during the first 3 months of lactation when the contribution of lipolysis in adipose tissue towards milk TAG biosynthesis is the greatest (McNamara, 1989), and high expression of those genes is necessary to facilitate efficient fatty acid uptake by mammary epithelial cells from blood and transport those fatty acids inside the cells. Despite high abundance of *FABP3* mRNA in the bovine mammary gland (Bionaz and Loor, 2008a, 2008b), there were no significant associations between *FABP3* haplotypes and milk fatty acid composition confirming the primary role of *FABP3* in channelling fatty acids for β-oxidation (Furuhashi and Hotamisligil, 2008).

Genome-wide Association Studies and Milk Fatty Acid Composition

The Illumina BovineSNP50 BeadChip array was used in a number of genome-wide association studies conducted in beef and dairy cattle to discover QTLs significantly associated with beef and milk fatty acid composition (Uemoto *et al.*, 2010; Bouwman *et al.*, 2011; Ishii *et al.*, 2013). One of the first genome-wide association studies in dairy cattle, however, was conducted by simply genotyping animals for a predetermined set of SNPs (Schennink *et al.*, 2009b; Stoop *et al.*, 2009). A single QTL for oleic acid (18:1 *cis*-9) concentration in the trapezius muscle of Japanese Black cattle was discovered between 49 and 55 Mbp on BTA19 (Uemoto *et al.*, 2010). The closest and only lipogenic gene located close to the discovered QTL was *FASN*. In a different study a QTL for intramuscular concentrations of myristic (14:0), myristoleic (14:1 *cis*-9), palmitoleic (16:1 *cis*-9) and oleic (18:1 *cis*-9) acids in Japanese Black cattle was discovered to be on BTA19 between 49 and 52 Mbp (Ishii *et al.*, 2013). Despite close proximity of the QTL to *FASN*, it was speculated, however, that other genes rather than *FASN* might be candidate genes for the QTL. Other genomic areas contained QTL for myristoleic acid (14:1 *cis*-9) concentration on BTA26 where *SCD* was identified as a candidate gene for the QTL, for oleic acid (18:1 *cis*-9) concentration on BTA23, and for palmitic acid (16:0) concentration on BTA25. However, there were no candidate genes suggested for the last two QTLs (Ishii *et al.*, 2013).

One of the first genome-wide association studies conducted in dairy cattle (Schennink *et al.*, 2009b; Stoop *et al.*, 2009) determined a number of significant QTLs for milk fatty acid composition that were located in different areas of the genome and among which QTLs on BTA14, BTA19 and BTA26 were the most significant. Candidate genes such as *DGAT1* for a QTL on BTA14 and *SCD* for a QTL on BTA26 were identified with no obvious candidate gene for the QTL on BTA19. In a different study by Bouwman *et al.* (2011) there were 54 genomic regions located on 29 chromosomes, which were significantly associated with concentrations of one or more milk fatty acids. The most significant QTLs were located on BTA14, BTA19 and BTA26, and previously discussed genes such as *DGAT1*, *SREBF1*, *FASN* and *SCD* were determined to be some of the candidate genes for those QTLs (Bouwman *et al.*, 2011).

Conclusion

The ability to control fatty acid composition in bovine milk and beef would allow the production of foods that are healthier to consumers. Changing fatty acid composition by dietary means is not very practical because of rumenal biohydrogenation of UFAs to SFAs. Moderate to high heritabilities of milk and beef fatty acids, however, are a good indicator for potential manipulation of fatty acid composition through selection. Polymorphisms in *SCD* and *DGAT1* might be considered as those with the most profound effects on beef and milk fatty acid composition, respectively. Polymorphisms in *FASN* have strong associations with fatty acid composition in both beef and milk. Results of genome-wide association studies were not always consistent with candidate gene studies especially with regard to beef fatty acid composition. Such discrepancies are most likely because of inadequate numbers of animals and markers, partly due to the difficulties in collecting fatty acid composition phenotypes on sufficiently large cohorts of animals. More studies are needed to discover polymorphisms associated with milk and beef fatty acid compositions and to validate those discoveries in different populations of cattle.

References

Abe, T., Saburi, J., Hasebe, H., Nakagawa, T., Misumi, S., Nade, T., Nakajima, H., Shoji, N., Kobayashi, M. and Kobayashi, E. (2009) Novel mutations of the FASN gene and their effect on fatty acid composition in Japanese Black beef. *Biochemical Genetics* 47, 397–411.

Aldai, N., Najera, A.I., Dugan, M.E., Celaya, R. and Osoro, K. (2007) Characterisation of intramuscular, intermuscular and subcutaneous adipose tissues in yearling bulls of different genetic groups. *Meat Science* 76, 682–691.

Arnould, V.M. and Soyeurt, H. (2009) Genetic variability of milk fatty acids. *Journal of Applied Genetics* 50, 29–39.

Baeza, M.C., Corva, P.M., Soria, L.A., Pavan, E., Rincon, G. and Medrano, J.F. (2013) Genetic variants in a lipid regulatory pathway as potential tools for improving the nutritional quality of grass-fed beef. *Animal Genetics* 44, 121–129.

Barber, M.C., Clegg, R.A., Travers, M.T. and Vernon, R.G. (1997) Lipid metabolism in the lactating mammary gland. *Biochimica Biophysica Acta* 1347, 101–126.

Barendse, W., Bunch, R.J., Thomas, M.B. and Harrison, B.E. (2009) A splice site single nucleotide polymorphism of the fatty acid binding protein 4 gene appears to be associated with intramuscular fat deposition in longissimus muscle in Australian cattle. *Animal Genetics* 40, 770–773.

Barton, L., Kott, T., Bures, D., Rehak, D., Zahradkova, R. and Kottova, B. (2010) The polymorphisms of stearoyl-CoA desaturase (SCD1) and sterol regulatory element binding protein-1 (SREBP-1) genes and their association with the fatty acid profile of muscle and subcutaneous fat in Fleckvieh bulls. *Meat Science* 85, 15–20.

Bastin, C., Gengler, N. and Soyeurt, H. (2011) Phenotypic and genetic variability of production traits and milk fatty acid contents across days in milk for Walloon Holstein first-parity cows. *Journal of Dairy Science* 94, 4152–4163.

Bauman, D.E. and Griinari, J.M. (2000) Regulation and nutritional manipulation of milk fat. Low-fat milk syndrome. *Advances in Experimental Medicine and Biology* 480, 209–216.

Bauman, D.E. and Griinari, J.M. (2003) Nutritional regulation of milk fat synthesis. *Annual Reviews of Nutrition* 23, 203–227.

Beaulieu, A. and Palmquist, D. (1995) Differential effects of high fat diets on fatty acid composition in milk of Jersey and Holstein cows. *Journal of Dairy Science* 78, 1336–1344.

Bionaz, M. and Loor, J.J. (2008a) ACSL1, AGPAT6, FABP3, LPIN1, and SLC27A6 are the most abundant isoforms in bovine mammary tissue and their expression is affected by stage of lactation. *Journal of Nutrition* 138, 1019–1024.

Bionaz, M. and Loor, J.J. (2008b) Gene networks driving bovine milk fat synthesis during the lactation cycle. *BMC Genomics* 9, 366.

Bobe, G., Hammond, E.G., Freeman, A.E., Lindberg, G.L. and Beitz, D.C. (2003) Texture of butter from cows with different milk fatty acid compositions. *Journal of Dairy Science* 86, 3122–3127.

Bouwman, A.C., Bovenhuis, H., Visker, M.H. and van Arendonk, J.A. (2011) Genome-wide association of milk fatty acids in Dutch dairy cattle. *BMC Genetics* 12, 43.

Brown, M.S. and Goldstein, J.L. (1997) The SREBP pathway: regulation of cholesterol metabolism by proteolysis of a membrane-bound transcription factor. *Cell* 89, 331–340.

Camfield, P., Brown, A., Lewis, P., Rakes, L. and Johnson, Z. (1997) Effects of frame size and time-on-feed on carcass characteristics, sensory attributes, and fatty acid profiles of steers. *Journal of Animal Science* 75, 1837–1844.

Cases, S., Stone, S.J., Zhou, P., Yen, E., Tow, B., Lardizabal, K.D., Voelker, T. and Farese, R.V., Jr (2001) Cloning of DGAT2, a second mammalian diacylglycerol acyltransferase, and related family members. *Journal of Biological Chemistry* 276, 38870–38876.

Couvreur, S., Hurtaud, C., Lopez, C., Delaby, L. and Peyraud, J.L. (2006) The linear relationship between the proportion of fresh grass in the cow diet, milk fatty acid composition, and butter properties. *Journal of Dairy Science* 89, 1956–1969.

De Smet, S., Raes, K. and Demeyer, D. (2004) Meat fatty acid composition as affected by fatness and genetic factors: a review. *Animal Research* 53, 81–98.

DePeters, E., Medrano, J. and Reed, B. (1995) Fatty acid composition of milk fat from three breeds of dairy cattle. *Canadian Journal of Animal Science* 75, 267–269.

Doreau, M. and Chilliard, Y. (1997) Digestion and metabolism of dietary fat in farm animals. *British Journal of Nutrition* 78 Suppl 1, S15–S35.

Duckett, S., Wagner, D., Yates, L., Dolezal, H. and May, S. (1993) Effects of time on feed on beef nutrient composition. *Journal of Animal Science* 71, 2079–2088.

Eichhorn, J., Coleman, L., Wakayama, E., Blomquist, G., Bailey, C. and Jenkins, T. (1986) Effects of breed type and restricted versus ad libitum feeding on fatty acid composition and cholesterol content of muscle and adipose tissue from mature bovine females. *Journal of Animal Science* 63, 781.

Furuhashi, M. and Hotamisligil, G.S. (2008) Fatty acid-binding proteins: role in metabolic diseases and potential as drug targets. *Nature Reviews Drug Discovery* 7, 489–503.

Gaunt, S. (1980) Genetic variation in the yields and contents of milk constituents. *Bulletin, International Dairy Federation*, 73–82.

Grisart, B., Coppieters, W., Farnir, F., Karim, L., Ford, C., Berzi, P., Cambisano, N., Mni, M., Reid, S., Simon, P. *et al.* (2002) Positional candidate cloning of a QTL in dairy cattle: identification of a missense mutation in the bovine DGAT1 gene with major effect on milk yield and composition. *Genome Research* 12, 222–231.

Grisart, B., Farnir, F., Karim, L., Cambisano, N., Kim, J.J., Kvasz, A., Mni, M., Simon, P., Frere, J.M., Coppieters, W. and Georges, M. (2004) Genetic and functional confirmation of the causality of the DGAT1 K232A quantitative trait nucleotide in affecting milk yield and composition. *Proceedings of the National Academy of Sciences USA* 101, 2398–2403.

Grummer, R.R. (1991) Effect of feed on the composition of milk fat. *Journal of Dairy Science* 74, 3244–3257.

Hoashi, S., Ashida, N., Ohsaki, H., Utsugi, T., Sasazaki, S., Taniguchi, M., Oyama, K., Mukai, F. and Mannen, H. (2007) Genotype of bovine sterol regulatory element binding protein-1 (SREBP-1) is associated with fatty acid composition in Japanese Black cattle. *Mammalian Genome* 18, 880–886.

Hoashi, S., Hinenoya, T., Tanaka, A., Ohsaki, H., Sasazaki, S., Taniguchi, M., Oyama, K., Mukai, F. and Mannen, H. (2008) Association between fatty acid compositions and genotypes of FABP4 and LXR-alpha in Japanese Black cattle. *BMC Genetics* 9, 84.

Hoehne, A., Nuernberg, G., Kuehn, C. and Nuernberg, K. (2012) Relationships between intramuscular fat content, selected carcass traits, and fatty acid profile in bulls using a F2-population. *Meat Science* 90, 629–635.

Hooper, L., Summerbell, C.D., Thompson, R., Sills, D., Roberts, F.G., Moore, H. and Davey, Smith, G. (2011) Reduced or modified dietary fat for preventing cardiovascular disease. *Cochrane Database of Systematic Reviews*, CD002137.

Horton, J.D., Goldstein, J.L. and Brown, M.S. (2002) SREBPs: activators of the complete program of cholesterol and fatty acid synthesis in the liver. *Journal of Clinical Investigation* 109, 1125–1131.

Howell, W.H., McNamara, D.J., Tosca, M.A., Smith, B.T. and Gaines, J.A. (1997) Plasma lipid and lipoprotein responses to dietary fat and cholesterol: a meta-analysis. *American Journal of Clinical Nutrition* 65, 1747–1764.

Hurtaud, C., Faucon, F., Couvreur, S. and Peyraud, J.L. (2010) Linear relationship between increasing amounts of extruded linseed in dairy cow diet and milk fatty acid composition and butter properties. *Journal of Dairy Science* 93, 1429–1443.

Huth, P.J. and Park, K.M. (2012) Influence of dairy product and milk fat consumption on cardiovascular disease risk: a review of the evidence. *Advances in Nutrition* 3, 266–285.

Indurain, G., Beriain, M.J., Goni, M.V., Arana, A. and Purroy, A. (2006) Composition and estimation of intramuscular and subcutaneous fatty acid composition in Spanish young bulls. *Meat Science* 73, 326–334.

Ishii, A., Yamaji, K., Uemoto, Y., Sasago, N., Kobayashi, E., Kobayashi, N., Matsuhashi, T., Maruyama, S., Matsumoto, H., Sasazaki, S. and Mannen, H. (2013) Genome-wide association study for fatty acid composition in Japanese Black cattle. *Animal Science Journal* 84, 675–682.

Jensen, R.G. (2002) The composition of bovine milk lipids: January 1995 to December 2000. *Journal of Dairy Science* 85, 295–350.

Jensen, R.G., Ferris, A.M., Lammi-Keefe, C.J. and Henderson, R.A. (1990) Lipids of bovine and human milks: a comparison. *Journal of Dairy Science* 73, 223–240.

Karijord, Ø., Standal, N. and Syrstad, O. (1982) Sources of variation in composition of milk fat. *Zeitschrift für Tierzüchtung und Züchtungsbiologie* 99, 81–93.

Kaylegian, K.E. and Lindsay, R.C. (1992) Performance of selected milk fat fractions in cold-spreadable butter. *Journal of Dairy Science* 75, 3307–3317.

Kelsey, J., Corl, B., Collier, R. and Bauman, D. (2003) The effect of breed, parity, and stage of lactation on conjugated linoleic acid (CLA) in milk fat from dairy cows. *Journal of Dairy Science* 86, 2588–2597.

Khatkar, M.S., Thomson, P.C., Tammen, I. and Raadsma, H.W. (2004) Quantitative trait loci mapping in dairy cattle: review and meta-analysis. *Genetics Selection Evolution* 36, 163–190.

Krauss, R.M., Eckel, R.H., Howard, B., Appel, L.J., Daniels, S.R., Deckelbaum, R.J., Erdman, J.W., Jr, Kris-Etherton, P., Goldberg, I.J., Kotchen, T.A. *et al.* (2000) AHA Dietary Guidelines: revision 2000: A statement for healthcare professionals from the Nutrition Committee of the American Heart Association. *Circulation* 102, 2284–2299.

Kris-Etherton, P.M. and Yu, S. (1997) Individual fatty acid effects on plasma lipids and lipoproteins: human studies. *American Journal of Clinical Nutrition* 65, 1628S–1644S.

Lawless, F., Stanton, C., L'Escop, P., Devery, R., Dillon, P. and Murphy, J. (1999) Influence of breed on bovine milk cis-9, trans-11-conjugated linoleic acid content. *Livestock Production Science* 62, 43–49.

Lee, S.H., van der Werf, J.H., Lee, S.H., Park, E.W., Oh, S.J., Gibson, J.P. and Thompson, J.M. (2010) Genetic polymorphisms of the bovine fatty acid binding protein 4 gene are significantly associated with marbling and carcass weight in Hanwoo (Korean cattle). *Animal Genetics* 41, 442–444.

Lee, Y., Oh, D., Lee, J., La, B. and Yeo, J. (2013) Novel single nucleotide polymorphisms of bovine SREBP1 gene is association with fatty acid composition and marbling score in commercial Korean cattle (Hanwoo). *Molecular Biology Reports* 40, 247–254.

McNamara, J.P. (1989) Regulation of bovine adipose tissue metabolism during lactation. 5. Relationships of lipid synthesis and lipolysis with energy intake and utilization. *Journal of Dairy Science* 72, 407–418.

Malau-Aduli, A., Edriss, M., Siebert, B.D., Bottema, C.D.K. and Pitchford, W. (2000) Breed differences and genetic parameters for melting point, marbling score and fatty acid composition of lot-fed cattle. *Journal of Animal Physiology and Animal Nutrition* 83, 95–105.

Mannen, H. (2011) Identification and utilization of genes associated with beef qualities. *Animal Science Journal* 82, 1–7.

Mansson, H.L. (2008) Fatty acids in bovine milk fat. *Food and Nutrition Research* 52, doi: 10.3402/fnr. v52i0.1821.

Matsuhashi, T., Maruyama, S., Uemoto, Y., Kobayashi, N., Mannen, H., Abe, T., Sakaguchi, S. and Kobayashi, E. (2011) Effects of bovine fatty acid synthase, stearoyl-coenzyme A desaturase, sterol regulatory element-binding protein 1, and growth hormone gene polymorphisms on fatty acid composition and carcass traits in Japanese Black cattle. *Journal of Animal Science* 89, 12–22.

Maurice-Van Eijndhoven, M.H., Bovenhuis, H., Soyeurt, H. and Calus, M.P. (2013) Differences in milk fat composition predicted by mid-infrared spectrometry among dairy cattle breeds in the Netherlands. *Journal of Dairy Science* 96, 2570–2582.

Mele, M., Conte, G., Castiglioni, B., Chessa, S., Macciotta, N.P., Serra, A., Buccioni, A., Pagnacco, G. and Secchiari, P. (2007) Stearoyl-coenzyme A desaturase gene polymorphism and milk fatty acid composition in Italian Holsteins. *Journal of Dairy Science* 90, 4458–4465.

Mele, M., Dal Zotto, R., Cassandro, M., Conte, G., Serra, A., Buccioni, A., Bittante, G. and Secchiari, P. (2009) Genetic parameters for conjugated linoleic acid, selected milk fatty acids, and milk fatty acid unsaturation of Italian Holstein-Friesian cows. *Journal of Dairy Science* 92, 392–400.

Mensink, R.P., Zock, P.L., Kester, A.D. and Katan, M.B. (2003) Effects of dietary fatty acids and carbohydrates on the ratio of serum total to HDL cholesterol and on serum lipids and apolipoproteins: a meta-analysis of 60 controlled trials. *American Journal of Clinical Nutrition* 77, 1146–1155.

Michal, J.J., Zhang, Z.W., Gaskins, C.T. and Jiang, Z. (2006) The bovine fatty acid binding protein 4 gene is significantly associated with marbling and subcutaneous fat depth in Wagyu x Limousin F2 crosses. *Animal Genetics* 37, 400–402.

Moioli, B., Contarini, G., Avalli, A., Catillo, G., Orru, L., De Matteis, G., Masoero, G. and Napolitano, F. (2007) Short communication: effect of stearoyl-coenzyme A desaturase polymorphism on fatty acid composition of milk. *Journal of Dairy Science* 90, 3553–3558.

Morris, C.A., Cullen, N.G., Glass, B.C., Hyndman, D.L., Manley, T.R., Hickey, S.M., McEwan, J.C., Pitchford, W.S., Bottema, C.D. and Lee, M.A. (2007) Fatty acid synthase effects on bovine adipose fat and milk fat. *Mammalian Genome* 18, 64–74.

Mozaffarian, D., Micha, R. and Wallace, S. (2010) Effects on coronary heart disease of increasing polyunsaturated fat in place of saturated fat: a systematic review and meta-analysis of randomized controlled trials. *PLoS Medicine* 7, e1000252.

Nafikov, R.A., Schoonmaker, J.P., Korn, K.T., Noack, K., Garrick, D.J., Koehler, K.J., Minick-Bormann, J., Reecy, J.M., Spurlock, D.E. and Beitz, D.C. (2013a) Sterol regulatory element binding transcription factor 1 (SREBF1) polymorphism and milk fatty acid composition. *Journal of Dairy Science* 96, 2605–2616.

Nafikov, R.A., Schoonmaker, J.P., Korn, K.T., Noack, K., Garrick, D.J., Koehler, K.J., Minick-Bormann, J., Reecy, J.M., Spurlock, D.E. and Beitz, D.C. (2013b) Association of polymorphisms in solute carrier family 27, isoform A6 (SLC27A6) and fatty acid-binding protein-3 and fatty acid-binding protein-4 (FABP3 and FABP4) with fatty acid composition of bovine milk. *Journal of Dairy Science* 96, 6007–6021.

Narukami, T., Sasazaki, S., Oyama, K., Nogi, T., Taniguchi, M. and Mannen, H. (2011) Effect of DNA polymorphisms related to fatty acid composition in adipose tissue of Holstein cattle. *Animal Science Journal* 82, 406–411.

Neville, M.C. and Picciano, M.F. (1997) Regulation of milk lipid secretion and composition. *Annual Review of Nutrition* 17, 159–183.

Noakes, M., Nestel, P.J. and Clifton, P.M. (1996) Modifying the fatty acid profile of dairy products through feedlot technology lowers plasma cholesterol of humans consuming the products. *American Journal of Clinical Nutrition* 63, 42–46.

Nogi, T., Honda, T., Mukai, F., Okagaki, T. and Oyama, K. (2011) Heritabilities and genetic correlations of fatty acid compositions in longissimus muscle lipid with carcass traits in Japanese Black cattle. *Journal of Animal Science* 89, 615–621.

Ntambi, J.M. (1999) Regulation of stearoyl-CoA desaturase by polyunsaturated fatty acids and cholesterol. *Journal of Lipid Research* 40, 1549–1558.

Nuernberg, K., Dannenberger, D., Nuernberg, G., Ender, K., Voigt, J., Scollan, N.D., Wood, J.D., Nute, G.R. and Richardson, R.I. (2005) Effect of a grass-based and a concentrate feeding system on meat quality characteristics and fatty acid composition of longissimus muscle in different cattle breeds. *Livestock Production Science* 94, 137–147.

Ohsaki, H., Tanaka, A., Hoashi, S., Sasazaki, S., Oyama, K., Taniguchi, M., Mukai, F. and Mannen, H. (2009) Effect of SCD and SREBP genotypes on fatty acid composition in adipose tissue of Japanese Black cattle herds. *Animal Science Journal* 80, 225–232.

Palmquist, D.L., Beaulieu, A.D. and Barbano, D.M. (1993) Feed and animal factors influencing milk fat composition. *Journal of Dairy Science* 76, 1753–1771.

Pannier, L., Mullen, A.M., Hamill, R.M., Stapleton, P.C. and Sweeney, T. (2010) Association analysis of single nucleotide polymorphisms in DGAT1, TG and FABP4 genes and intramuscular fat in crossbred *Bos taurus* cattle. *Meat Science* 85, 515–518.

Perry, D., Nicholls, P. and Thompson, J. (1998) The effect of sire breed on the melting point and fatty acid composition of subcutaneous fat in steers. *Journal of Animal Science* 76, 87–95.

Pietinen, P., Ascherio, A., Korhonen, P., Hartman, A.M., Willett, W.C., Albanes, D. and Virtamo, J. (1997) Intake of fatty acids and risk of coronary heart disease in a cohort of Finnish men. The Alpha-Tocopherol, Beta-Carotene Cancer Prevention Study. *American Journal of Epidemiology* 145, 876–887.

Pitchford, W.S., Deland, M., Siebert, B., Malau-Aduliand, A. and Bottema, C.D.K. (2002) Genetic variation in fatness and fatty acid composition of crossbred cattle. *Journal of Animal Science* 80, 2825–2832.

Pullen, D.L., Palmquist, D.L. and Emery, R.S. (1989) Effect on days of lactation and methionine hydroxy analog on incorporation of plasma fatty acids into plasma triglycerides. *Journal of Dairy Science* 72, 49–58.

Rincon, G., Islas-Trejo, A., Castillo, A.R., Bauman, D.E., German, B.J. and Medrano, J.F. (2011) Polymorphisms in genes in the SREBP1 signalling pathway and SCD are associated with milk fatty acid composition in Holstein cattle. *Journal of Dairy Research*, 1–10.

Roy, R., Ordovas, L., Zaragoza, P., Romero, A., Moreno, C., Altarriba, J. and Rodellar, C. (2006) Association of polymorphisms in the bovine FASN gene with milk-fat content. *Animal Genetics* 37, 215–218.

Schennink, A., Stoop, W.M., Visker, M.H., Heck, J.M., Bovenhuis, H., van der Poel, J.J., van Valenberg, H.J. and van Arendonk, J.A. (2007) DGAT1 underlies large genetic variation in milk-fat composition of dairy cows. *Animal Genetics* 38, 467–473.

Schennink, A., Heck, J.M., Bovenhuis, H., Visker, M.H., van Valenberg, H.J. and van Arendonk J.A. (2008) Milk fatty acid unsaturation: genetic parameters and effects of stearoyl-CoA desaturase (SCD1) and acyl CoA: diacylglycerol acyltransferase 1 (DGAT1). *Journal of Dairy Science* 91, 2135–2143.

Schennink, A., Bovenhuis, H., Leon-Kloosterziel, K.M., van Arendonk, J.A. and Visker, M.H. (2009a) Effect of polymorphisms in the FASN, OLR1, PPARGC1A, PRL and STAT5A genes on bovine milk-fat composition. *Animal Genetics* 40, 909–916.

Schennink, A., Stoop, W.M., Visker, M.H., van der Poel, J.J., Bovenhuis, H. and van Arendonk, J.A. (2009b) Short communication: genome-wide scan for bovine milk-fat composition. II. Quantitative trait loci for long-chain fatty acids. *Journal of Dairy Science* 92, 4676–4682.

Siri-Tarino, P.W., Sun, Q., Hu, F.B. and Krauss, R.M. (2010) Meta-analysis of prospective cohort studies evaluating the association of saturated fat with cardiovascular disease. *American Journal of Clinical Nutrition* 91, 535–546.

Skeaff, C.M. and Miller, J. (2009) Dietary fat and coronary heart disease: summary of evidence from prospective cohort and randomised controlled trials. *Annals of Nutrition and Metabolism* 55, 173–201.

Smith, S.B., Lunt, D.K., Chung, K.Y., Choi, C.B., Tume, R.K. and Zembayashi, M. (2006) Adiposity, fatty acid composition, and delta-9 desaturase activity during growth in beef cattle. *Animal Science Journal* 77, 478–486.

Smith, S.B., Gill, C.A., Lunt, D.K. and Brooks, M.A. (2009) Regulation of fat and fatty acid composition in beef cattle. *Asian-Australasian Journal of Animal Sciences* 22, 1225–1233.

Soyeurt, H. and Gengler, N. (2008) Genetic variability of fatty acids in bovine milk. *Biotechnologie Agronomie Société et Environnement* 12, 203–210.

Soyeurt, H., Gillon, A., Vanderick, S., Mayeres, P., Bertozzi, C. and Gengler, N. (2007) Estimation of heritability and genetic correlations for the major fatty acids in bovine milk. *Journal of Dairy Science* 90, 4435–4442.

Soyeurt, H., Dehareng, F., Mayeres, P., Bertozzi, C. and Gengler, N. (2008) Variation of Delta 9-desaturase activity in dairy cattle. *Journal of Dairy Science* 91, 3211–3224.

Stoop, W.M., van Arendonk, J.A., Heck, J.M., van Valenberg, H.J. and Bovenhuis, H. (2008) Genetic parameters for major milk fatty acids and milk production traits of Dutch Holstein-Friesians. *Journal of Dairy Science* 91, 385–394.

Stoop, W.M., Schennink, A., Visker, M.H., Mullaart, E., van Arendonk, J.A. and Bovenhuis, H. (2009) Genome-wide scan for bovine milk-fat composition. I. Quantitative trait loci for short- and medium-chain fatty acids. *Journal of Dairy Science* 92, 4664–4675.

Stull, J. and Brown, W. (1964) Fatty acid composition of milk. II. Some differences in common dairy breeds. *Journal of Dairy Science* 47, 1412.

Taniguchi, M., Utsugi, T., Oyama, K., Mannen, H., Kobayashi, M., Tanabe, Y., Ogino, A. and Tsuji, S. (2004) Genotype of stearoyl-coA desaturase is associated with fatty acid composition in Japanese Black cattle. *Mammalian Genome* 15, 142–148.

Turk, S.N. and Smith, S.B. (2009) Carcass fatty acid mapping. *Meat Science* 81, 658–663.

Uemoto, Y., Abe, T., Tameoka, N., Hasebe, H., Inoue, K., Nakajima, H., Shoji, N., Kobayashi, M. and Kobayashi, E. (2010) Whole-genome association study for fatty acid composition of oleic acid in Japanese Black cattle. *Animal Genetics* 42, 141–148.

Wall, R., Ross, R.P., Fitzgerald, G.F. and Stanton, C. (2010) Fatty acids from fish: the anti-inflammatory potential of long-chain omega-3 fatty acids. *Nutrion Reviews* 68, 280–289.

White, S., Bertrand, J., Wade, M., Washburn, S., Green, Jr J. and Jenkins, T. (2001) Comparison of fatty acid content of milk from Jersey and Holstein cows consuming pasture or a total mixed ration. *Journal of Dairy Science* 84, 2295–2301.

Willett, W.C., Stampfer, M.J., Manson, J.E., Colditz, G.A., Speizer, F.E., Rosner, B.A., Sampson, L.A. and Hennekens, C.H. (1993) Intake of trans fatty acids and risk of coronary heart disease among women. *Lancet* 341, 581–585.

Winter, A., Kramer, W., Werner, F.A., Kollers, S., Kata, S., Durstewitz, G., Buitkamp, J., Womack, J.E., Thaller, G. and Fries, R. (2002) Association of a lysine-232/alanine polymorphism in a bovine gene encoding acyl-CoA:diacylglycerol acyltransferase (DGAT1) with variation at a quantitative trait locus for milk fat content. *Proceedings of the National Academy of Sciences USA* 99, 9300–9305.

Wood, J., Richardson, R., Nute, G., Fisher, A., Campo, M., Kasapidou, E., Sheard, P. and Enser, M. (2004) Effects of fatty acids on meat quality: a review. *Meat Science* 66, 21–32.

Wood, J.D., Enser, M., Fisher, A.V., Nute, G.R., Sheard, P.R., Richardson, R.I., Hughes, S.I. and Whittington, F.M. (2008) Fat deposition, fatty acid composition and meat quality: a review. *Meat Science* 78, 343–358.

Yen, C.L., Stone, S.J., Koliwad, S., Harris, C. and Farese, R.V., Jr (2008) Thematic review series: glycerolipids. DGAT enzymes and triacylglycerol biosynthesis. *Journal of Lipid Research* 49, 2283–2301.

Zembayashi, M., Nishimura, K., Lunt, D. and Smith, S. (1995) Effect of breed type and sex on the fatty acid composition of subcutaneous and intramuscular lipids of finishing steers and heifers. *Journal of Animal Science* 73, 3325–3332.

Zhang, S., Knight, T.J., Reecy, J.M. and Beitz, D.C. (2008) DNA polymorphisms in bovine fatty acid synthase are associated with beef fatty acid composition. *Animal Genetics* 39, 62–70.

18 Genetic Improvement of Beef Cattle

B.P. Kinghorn,[1] R.G. Banks[1] and G. Simm[2]

[1]University of New England, Armidale, New South Wales, Australia;
[2]Scottish Agricultural College, Edinburgh, UK

Introduction

Systems that produce beef are quite heterogeneous, involving different breed types, widely differing environments, a range of market demands and considerable production derived from dairy industries. This contrasts with the pig and poultry industries where there are generally few decision makers, well-directed pursuit of goals and much less variation in production systems, the environment and, in most cases, the market.

In major temperate beef-producing countries such as the USA, Canada, parts of South America, New Zealand and parts of Australia, beef production is based on extensively grazed or ranched cows, mainly of pure British beef breeds like the Hereford, Angus and Shorthorn, or crosses of these breeds. In some of these countries, like the USA, Canada and parts of Australia, this extensive pre-weaning regime is usually followed by a more intensive finishing period in feedlots. The extensive nature of many production systems, and the widespread use of crossbred animals in the commercial sector of most beef industries means that performance recording and genetic improvement are usually concentrated in a relatively small sector of the population.

Bos indicus and Sanga beef breeds have been widely used in tropical areas in developing countries. Their use in tropical and subtropical regions of developed countries such as Australia has increased markedly in the last few decades, but more emphasis is now being placed on their crosses with *Bos taurus* breeds in an attempt to increase productivity and product quality.

In most European countries over 50% of beef production is from pure dairy or dual-purpose breeds, either from cull cows, from male calves or from surplus female calves not required as dairy herd replacements. Traditionally, beef breeding objectives and criteria were usually considered in dairy and dual-purpose breeds. However, there is often little or no emphasis on beef traits today, in dairy breeds. In addition to this direct contribution from dairy herds, there is an indirect contribution to beef production through crossing of dairy cows to beef bulls. This produces beef × dairy calves for slaughter and, in some countries, beef × dairy suckler cows (i.e. cows kept for rearing beef calves). However, some European countries, including the UK, have dedicated beef industries, and breeds such as Limousin and Charolais that are used as lean-beef terminal sires.

The aim of this chapter is to discuss the breeding objectives, genetic resources and methods of genetic improvement used in beef cattle breeding.

Breeding Objectives

From the brief introduction above it is apparent that there are two broad categories of beef production in many countries: (i) beef production from specialized beef herds; and (ii) beef production from dairy and dual-purpose herds. Within the specialized beef sector, there is further differentiation into terminal sire and maternal breeds, crosses or lines. Terminal sire breeds are also used in dairy and dual-purpose herds. Each of these categories of use requires a distinct set of beef breeding objectives, or at least different priorities, and these are discussed below. However, some more general issues of formulation of breeding objectives are described first.

Formulation of breeding objectives

Breeding objectives stipulate the animal characteristics to be improved and the desired direction for genetic change. They should be constructed in a manner that allows them to play an appropriate role, together with parameters such as heritability and correlations, as part of a genetic evaluation system, in order to facilitate ranking of animals on genetic merit and implementation of effective breeding programme design. To this end, breeding objectives are generally expressed as economic weightings that describe the economic impact of a unit change in each trait of commercial importance. These economic weightings can be used directly to help evaluate different breeds and crosses, or, more commonly, they can be used in conjunction with genetic parameters and knowledge of population structure to rank animals on an index of genetic merit in monetary units.

The breeding objective traits are not necessarily the same as the selection criterion traits that are measured and used to make selection decisions. For example, lean percentage may be a breeding objective (McNeill *et al.*, 2012), and ultrasonically measured backfat thickness (Veneroni-Gouveia *et al.*, 2012) a selection criterion. Knowing the genetic relationship between these two traits permits selection index methods to target the former using data on the latter.

There are two approaches to calculating these weightings: the economically rational approach, and the desired gains approach.

The economically rational approach

The classic approach to calculating economic weightings is economically rational – it takes no account of genetic parameters. This makes sense in that the value of making a unit change in a given trait should not be influenced by how difficult it is to generate this change. These difficulties can be handled appropriately at the genetic evaluation phase. In this setting, breeding objectives should reflect the costs and returns involved in a *production* system, and should not consider costs and gains generated in a *breeding* programme.

HELP FROM BIOLOGICAL MODELLING. The economically rational approach assumes that we know the genetic parameters (heritabilities, genetic correlations and phenotypic correlations) for all traits that are measured and/or of economic importance. However, this is often not the case in beef production systems, where it is extremely difficult to measure many of the traits of true importance, such as mature size, shape of the growth and feeding curves, and the patterns of tissue deposition. Such traits are often ignored when developing breeding objectives and yet their direct or indirect effect on profit can be large. In particular, the effect of mature size on production efficiency is such that selecting for efficiency as measured between fixed ages, or fixed weights can be quite misleading (Kinghorn, 1985).

The economically rational approach generally assumes that the biological interactions among traits are linear in nature. However, this is often not the case in meat production systems, where relationships can be complex, such as the effect of fatness on maternal ability and juvenile survival in heterogeneous environments. It is possible that relationships are neutral at the current levels of trait expression, but that with genetic change in selected traits thresholds are passed and/or relationships develop.

Biological modelling of production systems can be used to predict such changes. This modelling usually involves a mixture of mechanistic and empirical features (Ball et al., 1998; Doeschl-Wilson et al., 2007; Kinghorn, 2012). Mechanistic features give powers of extrapolation beyond what we get through use of empirically derived parameters such as heritabilities and linear correlations. However, biological modelling cannot be used to reliably separate predictions of genetic relationships and phenotypic relationships, and this casts doubt on its power to help set breeding objectives. In practice, it seems that biological modelling can play a quality control role, to predict any deleterious effects of breeding objectives

set through use of an economically rational approach.

UNITS OF EXPRESSION. All economic weightings in a breeding objective should have the same basis for units of expression, such as 'dollars per head'. Choice of this basis can have an important influence on the consequences of using the breeding objective. A simple basis for unit of expression, such as 'dollars per head' can be used for situations in which all traits are *directly* related to economic costs or returns, and thus exclude reproductive traits, whose effect is at least partly manifested through progeny. A less simple basis is 'dollars per breeding cow per year', which accommodates both production and reproduction traits. In all cases, each trait should use the same basis. Delays in returns due to expression in progeny can be accommodated by considering the pattern of flow of genes through the population, and discounting future returns to give current values (McClintock and Cunningham, 1974).

Economic weights calculated on a 'dollars per head' or 'dollars per breeding cow per year' basis suffer a potentially important drawback. They relate to dollars per livestock unit, rather than dollars per resource unit, such as 'dollars per hectare'. As an example, consider two breeds of beef cattle, one small and one large, with properties as shown in Table 18.1. The large breed would be targeted by a breeding objective based on 'dollars per head'. However, a breeding objective based on a measure of financial efficiency, such as 'dollars per hectare', would target the small breed. A breeding objective based on dollars per resource unit will usually be more appropriate, as long as proper account is made of any fixed costs per head.

Economic values can be calculated from several different perspectives, e.g. with the aim of maximizing the profitability of an enterprise for an individual producer, or with the aim of

Table 18.1. Assumed properties of a small breed and a large breed to illustrate discordance between profit per head and financial efficiency.

Breed	Value of weight at slaughter (US$)	Value of food consumed (US$)	Profit per head (US$)	Financial efficiency
Small	1000	500	500	2:1
Large	1800	1000	800	1.8:1

improving the efficiency of a national live-stock industry. Amer (1994) and Weller (1994) discuss these different approaches and the attempts to unify them. In the former category, increasingly sophisticated models have been proposed for deriving economic values, including enterprise models which re-optimize management following genetic improvement (e.g. Amer *et al.*, 1996, 1998).

Wolfová *et al.* (2007a, b) give a comprehensive approach to calculating economic weightings for beef cattle, with software available at: http://www.vuzv.cz/index.php?p=ecoweight&site=GenetikaSlechteni_en.

The desired gains approach

An alternative approach to developing breeding objectives, the 'desired gains' approach, involves declaration of the relative magnitudes of genetic gain desired in the traits of importance. The breeding objective calculations still result in relative weights, but these are now influenced by genetic parameters, with generally greater weightings for traits that are more difficult to change. A simple subset of this approach is the restricted index, in which the objective is set up to give a predicted zero genetic change in one or more nominated traits. Examples are restrictions for no change in backfat or birthweight.

Brascamp (1984) describes methods that can be used for both restriction and desired gains. He also shows how to use a mixture of the economically rational and the desired gains approaches, with some traits constrained to pre-chosen levels of response, and others influenced just by production economics. In all cases, relative weights are calculated, which is useful for demonstrating the 'effective economic weights' that nominated desired gains or restrictions imply.

Breeding objectives for beef production systems

Beef breeding objectives in dairy and dual-purpose breeds

At first sight it seems efficient to breed for both milk and meat production from the same type of animal. However, most of the evidence suggests that there is an unfavourable genetic correlation between milk production and growth or carcass characteristics (e.g. Pirchner, 1986).

Some breeds or strains, like the Simmental strains in several continental European countries, have achieved fairly high productivity in both milk and beef traits, as a result of many generations of selection. Even for these strains, it is difficult to compete nationally and internationally with both specialized milk and specialized beef breeds. As a result there is a general trend towards milk production from more specialized dairy cattle breeds and strains. In some countries, there is still an attempt to limit the expected deterioration in beef merit by performance testing dairy bulls for growth and conformation, and pre-selecting bulls on these traits prior to progeny testing for milk production. In other countries, the deterioration in beef merit of the specialized dairy strains is compensated for, at least partially, by crossing those females not required to breed replacement dairy heifers to specialized beef breeds. So, in temperate dairying countries with large-scale specialized industries, breeding objectives in dairy breeds have little or no emphasis on beef traits. Even in dual-purpose breeds, the emphasis on beef traits is likely to be secondary to that on milk traits (Simm, 1998).

Terminal sires for use in dairy herds and specialized beef herds

Terminal sire beef breeds (i.e. those specially selected to sire the slaughter generation of animals) are used in dairy herds for two main purposes. The first is to mate to dairy heifers to reduce the risk of calving difficulties, compared to that following matings to a dairy sire. The second is to mate to mature dairy cows that are not required to breed replacement dairy heifers.

Difficult calvings are costly, both directly and because they delay rebreeding, depress milk production, and compromise both cow and calf survival and welfare. Hence, dairy heifers have often been mated to bulls from one of the easier-calving beef breeds, such as the Hereford, Angus and Limousin. However, mating dairy heifers to a beef bull is becoming less common as more dairy producers recognize that their heifers are often the highest genetic merit animals in the herd, and hence valuable as dams of replacements. Also, the wider availability of calving ease evaluations in dairy breeds means that it is easier to select a dairy sire suitable for mating to heifers.

As the incidence of calving difficulties is lower in mature cows than heifers, there is more scope to select beef bulls for other attributes to maximize returns from calf sales. Many beef cross calves born on dairy farms are sold at a young age. So increasing calf weight and conformation (muscularity or shape) are important breeding objectives for dairy farmers choosing a beef breed, or individual beef sire – though increasing weight and conformation tends to conflict with the aim of reducing calving difficulties.

The performance of beef cross calves in later life is of little direct concern to most dairy farmers, although, in theory, sire breeds or individual sires with high genetic merit for later performance ought to result in higher rewards in the market place. These market signals work reasonably well at the level of sire breed. There is less widespread discrimination among sires within a breed, although in some countries AI companies, beef breed societies or recording agencies have schemes to identify and promote beef sires for use in dairy herds that combine acceptable calving ease with good growth and carcass characteristics.

In many of the specialized beef production systems in temperate countries there is widespread use of crossbreeding. Often this is to achieve complementary use of breeds. Usually small or medium sized breeds or crosses are used as dam lines, and larger breeds are used as terminal sires. Larger breeds are valuable as terminal sires as they usually have faster growth rates, and produce leaner carcasses at a given weight than smaller breeds. Although ease of calving is still important when terminal sire breeds are used in specialized beef breeding herds, their main role is to improve the growth and carcass characteristics of their crossbred offspring.

The definition of carcass merit depends to some extent on whether commercial animals are sold at live auctions, or directly to abattoirs, but it usually encompasses some measure of weight, fatness and conformation. (Breed and sex may also modify the price.) In theory, good communication between sectors of the industry should mean that breeding objectives are similar whether animals are marketed dead or alive. However, in practice they often differ.

In many North American and East Asian markets, a premium is paid for high marbling – that is, high levels of visible intramuscular fat in the eye muscle. Particularly in North America, this premium for marbling is based on its value as an indicator of good eating quality. Recently, interest in marbling in several exporting countries has been fuelled by its importance in the lucrative Japanese beef market.

Meat eating quality is becoming an increasingly important issue with consumers and the meat industry in richer countries (Melucci et al., 2012). The post-slaughter treatment of carcasses, especially chilling rate, ageing and method of hanging, are known to have important effects on eating quality (Dikeman, 1990; Cuthbertson, 1994). However, there is less information on pre-slaughter effects on beef eating quality, such as breed, breeding value within breed or production system. The information that is available suggests that there are breed differences in indirect measures of meat quality, especially marbling, colour and fibre type. There are differences in tenderness between breed types: double-muscled breeds generally have tender meat, followed by other Bos taurus breeds, with Bos indicus breeds ranking lowest. There are less consistent differences in tenderness between the non-double-muscled Bos taurus breeds, or between any of the breed types, in juiciness and flavour. Despite this, there are consistent reports of substantial within-breed genetic variation in both indirect and direct measures of eating quality (Kemp, 1994). This indicates that there is scope for improvement through within-breed selection, though in the absence of good live animal predictors of eating quality this is difficult to achieve without progeny testing. Molecular markers of eating quality may allow more efficient selection programmes.

Breeding replacement females for specialized beef herds

The main breeding objectives for cows in specialized beef herds, in addition to adequate growth and carcass merit, are good fertility, ease of calving, good maternal ability (which includes adequate milk production and good mothering ability) and low or intermediate mature size, to reduce cow maintenance requirements. These individual objectives are sometimes aggregated into measures like *weight of calf weaned per cow per annum*, or *weight of calf weaned per kg cow mature weight per annum*.

The ability of animals to withstand extreme climates and to tolerate low quality feed and periods of feed shortage is also important in some areas, and there is often concern about possible genotype × environment interactions for these 'adaptation' traits. These traits are often difficult to define, and the most practical route for within-breed improvement is often simply to record and select based on performance in the harsh environment concerned (Simm *et al.*, 1996). The emphasis on each of these traits will vary depending on the production system and breed or crossbred type of cow used. In some cases the traits of importance will be best improved by selection, in others they will be best improved by crossbreeding. For instance, the fertility of crossbred cows is usually high as a result of heterosis, and so is of somewhat less concern in selection within the component breeds (Simm, 1998).

Breed Resources and Crossbreeding

Evaluating breed resources

Genetic evaluation of breed resources is relatively simple wherever good estimates of mean performance are available for the environment and production systems of interest. This is because the effects involved can be measured with high accuracy from the large amounts of data available, and can be treated as fixed effects. These effects constitute an inventory of genetic resources, and the economic value of each breed genotype can be estimated by simply multiplying predicted performance for each trait by its corresponding economic weight, and summing across traits. In contrast, when we come to evaluating the genetic merit of individual animals, or individual haplotypes across the genome, fewer data are available per estimate,

and the random nature of breeding values makes the process much more challenging, especially for traits that are difficult to measure, such as feed conversion efficiency and disease resistance.

Breed evaluations can be extended to evaluation of different crossbreeding systems, with breeding objectives being calculated according to the specific role of each component breed or cross. For example, the breeding objective for a terminal sire breed would involve little or no pressure on female fertility traits, as these will only be important within that breed, which will constitute only a small part of the total system.

An overview of crossbreeding theory

The value of crossbreeding

The key reasons for crossbreeding are listed here:

*   *The averaging of breed effects.* For example, to get an animal of intermediate size to fit a particular pasture cycle or market demand. This may involve either regular systems of crossing, or the creation of composite breeds (e.g. Cundiff *et al.*, 1986).
*   *Direct heterosis.* Crossbred individuals often exhibit heterosis. Heterosis is measured as the extra performance of the crossbreds over the weighted average of their parent breeds. The percentage increase in performance ranges from about 0–10% for growth traits and 5–25% for fertility traits (e.g. Gregory *et al.*, 1991). The effect of heterosis on the total production system can be even more than this, as effects accumulate over traits (e.g. Cundiff *et al.*, 1986).
*   *Maternal heterosis.* Crossbred cows can exhibit considerable heterosis in their ability to raise fast growing, viable offspring.
*   *Sire–dam complementation.* A good crossbreeding system aims to use breeding cows that are of small or intermediate mature size (but not so small for dystocia to be a problem) as well as fertile. When a large terminal sire breed is used the proportion of feed directed to growing animals is increased, and the production system

benefits accordingly. This is especially important in lowly fecund species such as beef cattle, where maternal costs are proportionally much higher than in, for example, pigs, poultry and fish.

- *Possibly cheap source of breeding animals.* This is evident in some crossing systems, for example, in the British and Irish beef industries where many suckler cows have come from matings between beef bulls and dairy cows.

The genetic basis of heterosis

We need to know the genetic basis of heterosis in order to predict the value of untested crossbred genotypes. There are two genetic mechanisms postulated as causing heterosis effects:

- *Dominance.* Where the individual's parents come from two different breeds the individual will carry a wider range of alleles, sampled from two breeds rather than just one. It is thought that this better equips the individual to perform well, especially under a varying or stressful environment. We would thus expect dominance to be a positive effect, and there is much evidence to support this.
- *Epistasis.* When we cross breeds, alleles have to interact or 'cooperate' with alleles at other loci, which they are 'not used to'. The crossbred animal may thus be out of harmony with itself, and we predict that epistasis, if important, is a negative effect.

The dominance model of heterosis is widely assumed and used, and so this model will be taken here. It should be borne in mind that epistatic loss could cause errors in prediction based on the dominance model alone.

Breed dominance is greatest when all loci consist of two alleles derived from different breeds as in a first cross (F_1). Breed heterozygosity is in fact proportional to actual alleleic heterozygosity, and this is why we can use breed heterozygosity to make predictions of expression of heterosis. Other crosses show a proportion of this F_1 heterosis equal to the proportion of loci that are heterozygous with respect to breed of origin. This can be seen in the column D_d (Dominance for the direct subtrait) in Table 18.2.

Table 18.2 shows how to predict the merit of untested crossbred genotypes given estimates of crossbreeding effects. These are additive (A) or 'purebreed' effects for each of the three example breeds, and dominance (D) or heterotic effects, here assumed equal for each pair of breeds. Subscripts denote the direct subtrait (d) and the maternal subtrait (m) – both of these being of some importance for weaning weight in cattle. A least squares analysis of the form $\hat{\beta} = (X'X)^{-1} X'Y$ can be used to estimate the crossbreeding effects (in vector $\hat{\beta}$) from Merit (Y), where X is the matrix formed by the body of Table 18.2. The section 'Evaluating breed resources' outlines the simple approach that can be used to consider all breeding objective traits to help predict economic merit of different crosses.

Choice of crossing system

Gregory and Cundiff (1980) reported maternal and individual dominance effects between *Bos taurus* breeds at 14.8% and 8.5%, respectively, for weight of calf weaned per cow exposed. This indicates the importance of crossbred cows in the production system, even though maternal dominance is generally reduced at older ages, for example at slaughter age. Gregory and Cundiff (1980) used these figures to estimate the genetic merit of a wide range of crossing systems in beef cattle.

The best crossing system to use depends to a large extent on the value of the breeds available, as well as the amount of heterosis expressed in crossbred animals. This is illustrated in Table 18.3 by describing the conditions under which each crossbred genotype is worthy of choice.

Of course, care should be taken to consider factors other than the predicted genetic merit of candidate crosses for the traits of importance. The key factor here is the cost of maintaining structured crossing systems, where separate breeding units are required to give ongoing supply of purebred and/or crossbred parents. These costs often outweigh the genetic benefits of more structured crosses, especially in low fecundity species such as cattle, where the parental breeding units must be relatively large to supply the final cross.

Table 18.2. Example prediction of merit of weaning weight from estimated crossbreeding parameters.

Effects:	Mean	A_{d1}	A_{d2}	A_{d3}	A_{m1}	A_{m2}	A_{m3}	D_d	D_m	
Values (kg):	280	+20	0	−20	−6	−1	+7	20	10	MERIT
Breed 1	1	1	0	0	1	0	0	0	0	294.0
Breed 2	1	0	1	0	0	1	0	0	0	279.0
Breed 3	1	0	0	1	0	0	1	0	0	267.0
Best F₁ (1 × 2)	1	.50	.50	0	0	1	0	1	0	309.0
Best 3 Breed-X (1 × 23)	1	.50	.25	.25	0	.50	.50	1	1	318.0
Best Backcross (1 × 12)	1	.75	.25	0	.50	.50	0	.50	1	311.5
Balanced (1,2)	1	.50	.50	0	.50	.50	0	.50	.50	301.5
Synthetics (1,2,3)	1	.33	.33	.33	.33	.33	.33	.67	.67	300.0
Optimum (1,2)	1	.63	.37	0	.63	.37	0	.47	.47	302.4
Synthetics (1,2,3)	1	.57	.31	.12	.57	.31	.12	.56	.56	303.0
Rotations (1,2)	1	.50	.50	0	.50	.50	0	.67	.67	306.5
(1,2,3)	1	.33	.33	.33	.33	.33	.33	.86	.86	305.7

Multiply the coefficients shown in the body of the table by the values of the corresponding effects (see text). Adding the products gives the prediction of weaning weight, MERIT in the last column.
Synthetics at equilibrium; rotations at equilibrium and averaged over years.

Table 18.3. General recommendations on use of crossbreeding.

Purebreed	When no cross is better. The Holstein Friesian is a good example
F₁ cross	When direct heterosis is important
Three breed cross	When both direct and maternal heterosis are important
Four breed cross	When paternal heterosis is important as well
Backcross	When only two good parental breeds are available and/or when direct heterosis is not important
Rotational cross	When females are too expensive to either buy in or to produce in the same enterprise
Open or closed synthetic	When both males and females are too expensive. A few initial well-judged importations establish the synthetic, and it can then either be closed (which helps to establish a breed 'type'), or left open to occasional well-judged importations

Breeds and crosses used in beef production

Clearly the predominance of the Holstein Friesian in the dairy industry means that they are major contributors to beef output both directly through surplus calves and cull cows and, in some countries, indirectly through their contribution to the genetic makeup of suckler cows. However, the increasing specialization for milk production in black and white strains means that their predominance is often seen as a disadvantage in beef production. Because of the economic incentive towards specialization for milk production in most temperate countries, the biggest opportunity to improve beef output from dairy breeds is through crossing surplus females to specialized beef breeds.

Of the specialized beef breeds in Europe, the French breeds, particularly the Charolais and Limousin, and to a lesser extent the British breeds, particularly the Hereford and Angus, have been most common (Simm, 1998). The popularity of the French breeds is probably due to their high growth rates or high lean meat yield, while the popularity of the British breeds is probably due to their relatively low incidence of calving difficulties (Liboriussen, 1982; Thiessen *et al.*, 1984; Cundiff *et al.*, 1986; Gregory *et al.*, 1991; Amer *et al.*, 1992). Also, the traditional

British breeds, especially the Angus, have had something of a renaissance, because of perceived benefits in eating quality. This is now being backed up by implementation of increasingly progressive breeding programmes, and widespread use of Angus semen.

The increased use of the specialized French breeds as terminal sires in Europe, often at the expense of the traditional British breeds, is mirrored in many other temperate beef-producing countries. However, the British breeds remain important in breeding herds, either as purebreds or as components of crossbred maternal lines, in many of these beef producing countries (e.g. the USA, Canada, Australia, New Zealand, Argentina).

Although statistics on numbers of animals are useful, several less numerous breeds have a disproportionate influence through the use of AI, especially in dairy herds. For example, in the UK there are relatively small numbers of purebred Belgian Blue cattle (referred to as British Blues, derived from Belgian Blues imported in the 1980s), but this breed is apparently responsible for 45% of the UK's dairy beef inseminations (http://www.britishbluecattle.org/the_breed/index.html). The growth in importance of this breed is due to its ability to leave high conformation crossbred calves, with acceptable levels of calving ease, when mated to dairy cows.

In several major beef-producing countries (e.g. the USA, Australia) there has been strong interest in the use of composite breeds, especially as maternal lines. The use of these animals is efficient when rotational crossing is impractical or when several breeds have important contributions to overall merit.

Bos indicus and Sanga breeds have been increasingly used in crossing systems in tropical beef production regions (Mourão et al., 2008). There is a general trend to keep the proportional contribution from these breeds low in order to avoid deleterious effect on meat quality. However, research on factors affecting meat quality suggests that more variation in meat quality is caused by management and processing factors than by the proportion of *Bos indicus* (Hearnshaw et al., 1998).

The use of genomics to target increases in heterosis

An extension to the concept of genomic selection can be used to target improvement in the total genetic merit of crossbreds, including increases in heterosis. This was illustrated by Kinghorn et al. (2010) using a modelling framework similar to that of Ibáñez-Escriche et al. (2009). The approach is to split the genotype of each crossbred animal into the separate components (individual alleles) contributed by the sperm and by the egg. Selection in the sire breed is then based on genomic estimated breeding values (GEBVs) calculated using the crossbred phenotypes and the genome-wide sperm alleles, and likewise for the dam breed, using a standard additive model for calculating GEBVs. By doing this, alleles that are common in one breed, but rare in the other, will generally attract more positive weightings in that breed compared to the other. For simple illustration consider that no additive effect prevails at a locus, but that dominance does: if one allele is at frequency say 0.7 in the sire breed, and 0.4 in the dam breed, then that allele will be selected for in the sire breed, increasing the proportion of heterozygous progeny exhibiting dominance, and the alternative allele will be selected in the dam breed. This promotes heterozygosity and therefore heterosis.

Kinghorn et al. (2011) showed that positive response in heterosis does not require overdominance, and that the impact is greater for more quantitative trait loci (QTLs) affecting the trait of interest. Zeng et al. (2013) fitted dominance deviations explicitly when calculating GEBVs, and achieved a much longer-lasting high rate of response from a single genomic calibration. This approach could prove useful if epistasis is sufficiently unimportant, and if not, if recalibrations are maintained across breeds.

Given the positive technical performance of this approach, application in beef would require a long-term commitment to maintenance of specialized sire and dam breeds or lines, with combined marketing of the crossing system that the overall breeding programme is targeting. This in turn would require large-scale planning, and possibly corporatization of significant portions of the beef industries.

Selection within Breeds

Evaluating individuals

What causes an exceptional animal to be so much better than its contemporaries? There are two basic reasons:

1. The alleles it has inherited are more favourable, and/or they are present in more favourable combinations, making the animal genetically superior.
2. It has probably experienced a better 'environment', through good management or good luck, and possible impact of the animals in its herd, particularly its mother.

In seeking genetic change we are not really interested in how much 'environmental advantage' an animal has had because that source of superiority cannot be transmitted to the next generation, except where maternal genetic effects or epigenetic effects are involved. Moreover, in selection programmes we are generally not interested in the combination of alleles, as, in general, these combinations cannot be transmitted to the next generation (in the case of intra-locus *dominance*), or are only weakly transmitted (in the case of inter-locus *epistasis*).

So, in general, we want to be able to choose the animals with alleles that will have the most beneficial effect on progeny, and we do this by selecting animals on the basis of their estimated breeding values. Breeding value (denoted by *A* signifying additivity of effect) is a description of the value of an animal's alleles to its progeny. In general, we do not know which alleles an animal carries, so we can never fully *know* what an animal's breeding value is. However, we can estimate it from a wide range of information sources involving phenotypic measures and genetic markers.

The simplest estimate of an animal's breeding value is that based on just its phenotypic superiority (*P*, phenotype as a deviation from the contemporary mean):

$$\hat{A} = \frac{V_A}{V_P} P = h^2 P \qquad (18.1)$$

Where $\hat{A}$ is estimated breeding value (EBV) and ^ denotes 'estimate', and $h^2 = \frac{V_A}{V_P}$ is heritability of the trait concerned. In conceptual terms,

the phenotypic superiority of the animal, *P*, is regressed or shrunk according to the proportion of phenotypic variation in the trait concerned, which is due to effects that cannot be transmitted between generations. Selection on phenotype gives a percentage response that depends on:

- Selection intensity – the smaller the proportion retained for breeding the higher the response.
- Generation interval – the younger the average age of parents, the faster the rate of response.
- Heritability – the higher the heritability the higher the response.
- Coefficient of variation (CV) – the higher the CV the higher the response.

The last two factors generally differ between traits. Table 18.4 gives estimates of these for a number of traits in beef cattle.

Table 18.4. Coefficient of variation (CV, phenotypic standard deviation divided by mean) and heritability estimates for a range of traits in beef cattle. (Condensed from Simm, 1998.)

Trait	CV (%)	Heritability (%)
Age at first calving	5.7	6
Conception rate[a]	61.8	17
Perinatal mortality as trait of cow[a]	674.1	10
Scrotal circumference	8.0	48
Birth weight – direct	12.3	31
Birth weight – maternal	12.6	14
Weaning weight – direct	12.3	24
Weaning weight – maternal	13.6	13
Post-weaning gain	13.7	31
Mature cow weight	12.1	50
Gross food conversion ratio	11.0	32
Backfat depth at constant age	24.5	44
Dressing percentage	3.2	39
Marbling score, constant age	34.1	38
Eye muscle area, constant age	10.1	42
Tenderness	18.2	29

[a]These traits are binomially distributed with high mean, making CV figures less meaningful.

Use of information from relatives – BLUP

In selecting animals to act as parents, we are interested in choosing those with the most favourable alleles. An animal's own performance gives an indication of the value of its alleles to its progeny. However, some of this animal's alleles are also carried by each of its relatives, and so the performance of an animal's relatives can be used to give a more accurate assessment of the alleles it carries.

Thus progressive breeding programmes make use of information from all known relatives. This is of most value when heritability is low – when an animal's own performance is a poor indicator of breeding value. As heritability increases, there is a diminishing proportional value of information from relatives, until at a heritability of unity an animal's own performance is a perfect indicator of its breeding value, with no room for improvement due to relatives' information.

The method of choice for predicting breeding values, which is an extension of selection index methods, is known as best linear unbiased prediction (BLUP). Kennedy (1981) and Van Vleck et al. (1989) give digestible descriptions of BLUP techniques. This section will not review these, but the following describes the key properties of BLUP EBVs.

- *EBVs are generally additive.* For example, if a bull has an EBV of $\hat{A} = + 20.0$ kg and a cow has $\hat{A} = + 10.0$ kg for 400-day weight, then the prediction is that progeny will have a 400-day weight superiority of $(20.0 + 10.0) / 2 = 15.0$ kg. This is actually a prediction of progeny genetic value, but as progeny dominance deviation and environmental deviation are unknown and thus have 'expectations' of zero, it is also a prediction of progeny breeding value and phenotype. Note also that the proportion of parental superiority in EBV that is transmitted to progeny is unity, after accounting for halving due to meiosis. Thus the heritability of EBVs is unity, as they have been pre-regressed.
- *BLUP makes full use of information from all relatives.* It does this by use of the numerator relationship matrix, which describes the predicted number of alleles per locus shared by descent between each pair of animals. BLUP does not have to give separate attention to sib testing, progeny testing, own performance, etc. Use of information from all relatives (even those long dead) is simultaneously handled. This gives greater flexibility, more accurate EBVs and more selection response.

- *BLUP predicts breeding values and accounts for fixed environmental effects simultaneously* (management group, herd, season, year, etc.). This means that animals can be compared across groups, giving wider scope for selection. For example, comparing across age groups means that older animals have to prove their competitiveness at every round of selection. This property of BLUP usually accounts for most of its advantage over less powerful methods.
- *BLUP gives genetic trends.* The ability to compare the EBVs of animals born and measured in different years means that year mean EBVs can be calculated and genetic trends reported.
- *BLUP can cater for non-random mating.* Bulls can be compared via their progeny even if some were allocated better cows. This can only be done where the cows were allocated on the basis of their recorded performance, such that BLUP can account for their EBVs when evaluating the bulls concerned.
- *BLUP can account for selection bias.* For example, consider ranking bulls on the weaning weights of their daughters at their first two calvings. The worse bulls, who had worse daughters, will have benefited more from culling of daughters on first weaning performance. However, BLUP accounts for this, given that the information used to make selection decisions (first weaning results in this case) is included in the data set.

Outputs from a BLUP analysis includes EBVs (or $\hat{A}$s) for the each of the traits fitted – which can include both measured criterion traits and breeding objective traits, even if there is missing information on the latter. The breeder only needs to weight EBVs for the objective traits by their economic weights to provide a selection index that s/he can select on: Index = $a_1\hat{A}_1 + a_2\hat{A}_2 + a_3\hat{A}_3 + ...$. The selection index is itself an EBV – an estimate of breeding value for economic merit.

Some traits are mediated through the maternal environment. For example, weaning weight is not only influenced genetically by the genes in the calf, but also by the genes in its mother, mediated through the maternal environment (e.g. milk supply). Thus the numerator relationship matrix for maternal effects on weaning weight is determined by relationships among the *dams* of the calves measured. This means that a single set of observations on weaning weight can give rise to both direct EBVs and maternal EBVs. If a breeder is selecting a terminal sire, s/he should ignore the maternal EBV, as this source of genetic merit will never be expressed. However, in order to maximize the weaning weight of the selected bull's grand-progeny via daughters, selection should be based on $\frac{1}{2}$EBV$_{maternal}$ + $\frac{1}{4}$EBV$_{growth}$. This is actually a prediction of the performance of these grandprogeny, and the coefficients result from the fact that the grandprogeny benefit on average from $\frac{1}{4}$ of the bull's genes for direct effects, and $\frac{1}{2}$ of the bull's genes, in their mothers, for maternal effects.

Use of genomic information

The development and early stages of implementation of genomic technologies in beef cattle breeding are covered in Chapter 19, but it is useful to highlight the likely key features and challenges for beef genetic evaluation.

As methods of reading DNA information have progressed, it has become clear that for the overwhelming majority of traits, the infinitesimal model or a close approximation holds. This has actually facilitated integration of genomic information into the BLUP methods mentioned above, using genomic information to help quantify animal relationships. However, a consequence of this polygenic profile is that, far from being able to detect and utilize a small number of QTLs that together explain a large proportion of genetic variance, methods must involve use of large numbers of single nucleotide polymorphisms (SNPs) (or other units of genomic information) and calibrate variation in

that information against considerable amounts of phenotypic information. Goddard and Hayes (2009) provide predictions of numbers of phenotyped and genotyped animals required for specified levels of accuracy of prediction of breeding value.

Extending this point, to date it has not proved possible to generate genomic predictors that maintain accuracy when used outside the population in which the calibration was carried out – so that genomic predictions developed in (say) Angus cattle in the USA, will not achieve the same level of accuracy when used in Angus cattle in Australia, and even less so when used in a different breed.

The underlying cause of this is that in general the accuracy of genomic predictions is proportional to the degree of genomic relationship or similarity between the candidate animals and those in the population measured and genotyped for the trait(s). As this declines, so does the accuracy of prediction, and the decline can be either through less related current animals (such as the examples above of different populations of the same breed, or different breeds), or as animals change in relationship over time within the breed. This last point means that the phenotyping effort needs to be maintained to include current generation animals.

The great attraction of genomic predictions in all species is to allow prediction of genetic merit in young animals, and particularly for traits that are otherwise difficult to measure (referred to as hard-to-measure traits). Examples relevant to beef cattle include feed intake and measures of eating quality. Genomic predictions have been developed for traits such as these for a limited set of circumstances – those where appropriate volumes of phenotypic data have been collected.

These features of genomic prediction bring to very sharp focus a fundamental challenge in animal breeding, and one which has been particularly acute for beef cattle breeding – the collection of phenotypes. As early steps into the genomic era are made in beef cattle breeding, this challenge is starting to force breeders to rethink their approaches to providing genetic evaluations. Whereas to date, individual breeders have contributed phenotypic data and received EPDs or EBVs for those animals and those traits, it is now becoming possible to simply genotype animals and draw on the pool of phenotypes for that breed to obtain EBVs or expected progeny differences (EPDs) based solely on genotype. This immediately creates potential for free-riding – those breeders who take only genotypes are dependent on and benefiting from the efforts of those who phenotype. A range of approaches to dealing with this situation exist, and will need to be explored to find solutions most appropriate to each breed and country.

Predictions of the impact of using genomic information are rather positive (e.g. Van Eenennaam *et al.*, 2011; Bolormaa *et al.*, 2013), and a generally positive outlook prevails (e.g. Hayes *et al.*, 2013). However, realized impact has yet to be well documented.

EBVs across breeds

There is an increasing interest in genetic evaluations using information from crossbred animals, and genetic evaluations on crossbred animals. Pollak and Quaas (1998) give the technical basis of this and a description of example cases. As a simple concept, analysis can be done to estimate all breed and heterosis effects, and to simultaneously fit breeding values in a BLUP analysis. This leads to the prediction of progeny merit from any mating pair, based on the breed constitution of the progeny and the EBVs (free of breed and heterosis effects) of the parents.

However, without a wide range of breeds and crosses, it is very difficult to get a reliable splitting of breed direct and maternal effects. Moreover, the genetic correlation between breeding values over different breeds of mate may be significantly less than unity – such that, for example, the EBV ranking of a group of Angus bulls might depend on what breed of cow they are to be mated to.

One problem with implementation is the general need to rank breeds and crosses on the breeding objective traits. There is much room for argument over the publishing and use of such values. This is one reason that genetic evaluations across breeds may take place more readily behind the closed doors of large breeding corporations.

Even if this contribution to genetic variance proves useful, there will still be a need for phenotyping of animals from each breed of interest for the traits of interest, and this phenotyping will need to include 'head-to-head' comparisons. A useful outcome would be that the total numbers required for phenotyping to underpin across-breed prediction be smaller than the sum of those required for separate individual breeds.

The larger challenge for developing and maintaining such across-breed predictions may be identifying whose interests are served by the resulting predictions, and hence who will need to make the continuing investment in phenotyping. It is not at all obvious that it is in the interest of any individual breed (or its members) to participate in across-breed evaluation, and so it may therefore be necessary for other sectors of the industry to fund such phenotyping.

Systems of testing

Most beef cattle genetic improvement programmes are based on performance testing or progeny testing. Both of these depend on performance recording. Essentially this involves recording the identity, pedigree, birth date, sex, performance and genotype of individual animals, plus any major management groupings or treatments likely to influence performance.

Performance testing

Since many of the traits of interest in beef cattle can be recorded in both sexes and prior to sexual maturity, there is a fairly long history of performance recording and performance testing in beef breeding. This dates from the 1940s and 1950s in the USA, and to slightly later in many other countries. Performance testing has usually the responsibility of breed associations (e.g. in the USA), government departments or agencies receiving some government support (e.g. in many European countries) or private agencies, either alone or in partnership with each other.

Compared to the situation in dairy cattle breeding, a relatively low proportion of beef cattle are performance recorded. This is partly because of the greater distinction between commercial and breeding herds than in the dairy industry – especially in countries where crossbreeding is widespread. For example, performance recorded animals comprise less than 2% of the total beef cattle population in the USA (Middleton and Gibb, 1991), Australia and the UK. However, even within the purebred beef sector, there is usually a much lower proportion of recording than in the dairy industry.

Most performance testing schemes involve recording the pre-weaning performance of all animals on-farm. In some countries post-weaning performance continues to be measured on-farm. In others, central performance testing is used. Central testing of beef cattle has been quite widely used worldwide since the 1950s, especially in the USA, Canada and Europe. It involves submitting some animals, especially higher-performing bulls, from the breeders' own farms to a central station, where they are compared with bulls from other herds in a uniform environment. Despite the potential benefits of this, the correlations between the performance of bulls in central stations and the subsequent performance of their progeny is often lower than expected. This is often attributed to large pre-test environmental effects.

The concept of performance testing in beef cattle is now being taken over by moves to more fully exploit genomic information. Initial moves to form nucleus herds that are both phenotyped and genotyped, to provide resources for genomic evaluations, have been funded by government and industry bodies.

Progeny testing

In many countries there is a deliberate strategy of first performance testing, then progeny testing bulls, with selection at each stage. As with performance testing, progeny testing schemes either operate on-farm or at central testing stations.

Sequential testing is particularly common in the specialized beef breeds in France. Large numbers of purebred animals are performance recorded on farm for weights at birth, 120 and 210 days, and for muscular and skeletal development at weaning (Ménissier, 1988; Bonnett *et al.*, 1994). The best males from on-farm recording are brought to central testing stations after weaning, and tested further from 8–14 months of age. About 35 of the best of these bulls go on to be progeny tested to assess their daughters' maternal ability, in central progeny test stations

Progeny testing causes an increase in generation interval with potentially negative effects on overall selection response. Appropriate breeding programme design is thus needed to balance effects on increased selection accuracy and increased generation interval. In some cases, the high accuracies generated by progeny testing are themselves of commercial value in the seedstock marketplace, and this should also be taken into account.

Cooperative breeding schemes

During the 1970s and 1980s, particularly in Australasia, a range of approaches to cooperative breeding schemes evolved, partly to enable across-herd genetic evaluation, and to a lesser extent to make better use of elite genetic material.

Rather than evolving into sustained breeding structures or businesses, such schemes have proved to be a step towards essentially whole-breed genetic evaluation, by enhancing across-herd genetic linkage, particular via AI.

Schemes making use of elite young sires in 'referencing' – multiple herds using the same bull or bulls – have been implemented, ranging in numbers of participating herds and in duration, in a number of countries through the 1990s and 2000s, but in no case have they become a constant or widespread element of breeding infrastructure. This is despite the fact that in essentially all cases, they demonstrate that young bulls with superior estimated genetic merit 'prove' that merit when progeny tested, and that genetic progress is enhanced by widespread use of such sires.

There is a real possibility that this phase of genetic evaluation and improvement, which is simply summarized as using and enhancing genetic linkage to enable across-herd evaluation, may now evolve rapidly into something broadly along the lines of the intended structure and operation of cooperative schemes, driven by the need for phenotypes to underpin genomic prediction. As noted above, implementation of genomic prediction makes explicit the need for large volumes of phenotypic records, and hence the investment needed to achieve this. This in turn highlights the value of optimizing phenotyping (and genotyping), and this in turn increases the focus on broader optimization of breeding programmes at the across-herd, or whole-breed scale.

Early initiatives that address this challenge include the Information Nucleus concept pioneered in Australia (Banks, 2011), in which genetically diverse and elite young bulls are progeny tested, including for hard-to-measure traits, building and maintaining a reference population to underpin genomic predictions. As breeds start to learn from these initiatives, a likely evolution is towards planned or coordinated investment and selection across herds, aiming to maximize cost-effectiveness of phenotyping.

Traits recorded

Generally on-farm performance recording schemes around the world have concentrated on measuring live weights at regular intervals (or growth rates between these), together with visual scores of muscularity and measurements or scores of height or skeletal development. The development of mobile, reasonably accurate ultrasonic scanners in the 1970s and 1980s allowed measurements of fat and muscle depths or areas to be included in some on-farm recording schemes. Typically these measurements are taken on or over the eye muscle at one of the last ribs, or in the loin region of animals at about a year or 400 days of age. At least in theory, one of the benefits of central testing is that it permits more frequent and more comprehensive measurements to be made. For example, it is rarely practical to measure feed intake of individual animals on farms, but it is fairly common in central performance test stations. See Crowley et al. (2011) for an example study. Similarly, progeny testing allows actual carcass measurements to be obtained.

Terminal sire characteristics have generally dominated beef breeding schemes in Europe. With the exception of some breeding schemes in France, few maternal characteristics like fertility have been recorded. As a result, what little objective selection there has been for maternal characteristics has been on traits like calving ease, birth weight and 200-day weight, which are of importance in both terminal sire and maternal lines. However, until recently, methods of separating direct and maternal genetic influences on these traits have not been in widespread use. Maternal traits have received more attention in North America, Australia and New Zealand, where specialized beef herds account for a far higher proportion of beef output. Genetic evaluations for scrotal size (which is an indicator of both male and female fertility, and age at puberty) and female fertility (measured as days from the start of the mating period to calving) have been introduced recently for some breeds in Australia and New Zealand. Evaluations for scrotal size and mature cow weight have been introduced for some breeds in the USA.

Many of the traits concerned with reproduction have fairly low heritabilities. However, many are economically important, and there is substantial variation in them, so there is both the incentive and scope for genetic improvement.

Direct heritabilities of growth traits tend to be moderately high, while maternal heritabilities tend to be slightly lower. The heritabilities of carcass traits tend to be even higher than those for growth traits. However, carcass traits have to be assessed either indirectly on live candidates for selection (e.g. by ultrasonic measurements), or directly on progeny or other relatives of the candidates for selection, so they are not as easy to improve as it seems at first sight. For more details of genetic parameters see Koots et al. (1994a,b).

Evaluations across herds, breeds and countries

To be able to compare BLUP EBVs fairly across contemporary groups and years, genetic links are needed between groups and years. Hickey et al. (2008) and Phocas and Laloë (2004) evaluate the performance of methods used to generate such EBVs across groups.

In dairy herds, strong links occur automatically because of the very widespread use of AI. In some countries there is little use of AI in specialized beef breeds, and this has limited the introduction of national across-herd genetic evaluations.

However AI use is higher in other countries. For example, between 20 and 50% of births in pedigree herds of the major beef breeds in Britain are the result of AI. Also, the recent introduction of foreign breeds to a country, or the popularity of imported strains within a breed, tend to increase the use of AI. In such cases there will often be strong enough genetic links between herds and years to make reliable comparisons of EBVs across herds and years.

A major technical limitation to performing evaluations across breeds is that animals of different breeds are rarely kept as contemporaries under similar management and feeding systems. However, as indicated above, across-breed evaluations are becoming feasible using information from crossbred animals, or from designed breed comparisons, together with estimates of genetic trends in each of the purebred populations since the breed comparison was made (Amer *et al.*, 1992; Benyshek *et al.*, 1994).

Compared to the situation in dairy cattle, there has been less effort to date in developing international conversions of EBVs or EPDs for beef cattle, or performing international genetic evaluations. However, there is growing interest in this area. For example, international conversions have been produced for some beef breeds in use in Canada and the USA. Also, across-country evaluations are being investigated or performed routinely for several breeds in the USA and Canada, France and Luxembourg, and Australia and New Zealand (Benyshek *et al.*, 1994; Graser *et al.*, 1995; Journaux *et al.*, 1996).

These across-herd, -breed and -country genetic evaluations are starting to have an important impact. They give credible objective comparison between seedstock sources, which in other industries has led to altered buying patterns and a shakeout in the seedstock sector.

Indices of overall economic merit

As noted previously, the selection index provides a means to maximize response in the breeding objective. Briefly, the selection index apportions selection emphasis in the most appropriate way, based on the relative economic importance of traits in the breeding objective, and on the strength of genetic associations between measured traits and breeding objective traits. The emphasis in beef cattle breeding in North America has been on using sophisticated methods to produce individual trait EBVs. However, this has generally not been progressed to indices of overall merit, constraining the impact of this technology (Garrick *et al.*, 2009). In contrast, in Europe, while less sophisticated methods of evaluation were used until recently, selection indexes have been quite widely used in both specialized beef breeds and in dairy and dual-purpose breeds.

Much of the emphasis in Europe has been on producing indexes for terminal sire characteristics. For example, a terminal sire index was introduced in Britain in the mid-1980s, and used in most breeds until 1997. The selection objective of this index was to maximize the margin between saleable meat yield and feed costs, taking into account the costs of difficult calvings (Allen and Steane, 1985). Index scores were calculated from the animal's own records of calving difficulty score, 200- and 400-day weight and a visual muscling score. If they were recorded, additional measurements of birth weight, feed intake and ultrasonically measured fat thickness were included, to increase the accuracy of the index.

Indices currently used are generally more closely linked to market returns (i.e. using associations with carcass weight, fat class and conformation class rather than with saleable meat yield). For example, separate indexes are used for calving performance and for growth and carcass performance of British terminal sires. The calving value index ranks animals on genetic merit for calving ease, based mainly on records of birth weight, calving ease and gestation length, while the beef value index ranks them on genetic merit for growth and carcass traits, based mainly on records for weights, fat depth, muscle depth and muscle score. These two indexes can be added together to rank animals on overall merit for calving ease and production together. The contributions which the calving value and beef value make to overall merit vary depending on the importance of calving ease, and on variation in the component traits, in the breed concerned. However, typically, calving value accounts for about 16% of the variation in overall merit (Amer *et al.*, 1998).

Indexes combining BLUP EBVs for reproduction, growth and carcass traits have been developed in Australia. An important feature of these indexes is that the economic values applied can be tailored or customized to individual breeders' requirements. This is achieved via a computer software package that uses data on returns and costs of beef production for individual producers or production systems (Barwick et al., 1994).

Evidence of genetic improvement and its value

Estimates of genetic change achievable

In theory, changes of at least 1% of the mean per annum are possible following selection for weight or growth traits in beef cattle. However, in practice, rates of change are often lower than this. For example, a review of several beef cattle selection experiments showed that average changes of 0.6% and 0.8% per annum were achieved with selection for weaning and yearling weight, respectively (Mrode, 1988).

The increased uptake of across-herd BLUP genetic evaluations over the last decade has permitted more widespread estimation of genetic trends in industry breeding schemes. For example, Crump et al. (1997) show estimated genetic trends in birth weight, 200- and 400-day growth since 1980, for the most numerous performance-recorded beef breeds in Britain. The changes in 200- and 400-day weights ranged from 0.15 to 0.5% of the breed mean per annum for the different breeds.

Trends similar to or lower than these have been reported in several breeds in Canada and Australia (Graser et al., 1984; de Rose and Wilton, 1988). Slightly higher trends in weaning weight have been reported in the US Angus and Hereford breeds (Benyshek et al., 1994). This may be explained partly by the earlier availability of BLUP methods in the US beef industry. It is probably also partly due to the higher herd and population sizes for these breeds in the US. Similar trends in weaning weight (from about 0.2 to 1.1 kg per annum), and positive trends in muscularity, have been reported for the major French breeds between 1991 and 1995 (Journaux et al., 1996).

More recent studies have found evidence of somewhat higher rates of progress (Johnston, 2007; Amer et al., 2012), and allowed some comparison between breeds and countries. Where information is available, it suggests that the Angus breed is making faster progress than others in countries where it is numerically important, and there is evidence that Angus populations are increasing as a result (D. Garrick, personal communication, 20 July 2012; D.L. Johnston, personal communication, 7 August 2012).

Amer et al. (2012) discuss some of the possible reasons for variation between countries, including differences in production systems and markets, as well as considering briefly the various approaches to delivering genetic evaluation, including the role and use of indexes.

In most of these studies of industry trends, the rates of change achieved are well below those theoretically possible, and below those actually achieved in selection experiments. The apparently low rate of change is partly explained by the fact that selection has not been solely for weight traits, and with many decision makers in the beef breeding business, there is a range of objectives targeted. However, it is also partly due to the relatively low use of objective methods of selection, and the fact that, in at least some of the countries mentioned, only within-herd comparisons could be made for some or most of the period concerned.

The economic value of genetic improvement

There have been relatively few studies of the value of genetic improvement in beef cattle, although these do show favourable estimates of cost–benefit (Barlow and Cunningham, 1984). A study of the costs and benefits of implementation of across-herd BLUP and index selection in the terminal sire sector of the British beef industry showed that estimated discounted returns exceeded the costs of implementation, including research, within a few years of introduction. Estimated annual discounted returns were expected to reach about £18 million per annum, and exceed annual costs of implementation by a factor of 30:1, about 20 years after introduction of these technologies (Simm et al., 1998).

More recently, the Australian industry has formally evaluated investment in R&D for beef genetic improvement in general, and into reference populations in particular. Farquharson *et al.* (2003) and Banks (2011) evaluated overall return on investment, and that into reference populations, and drew the following conclusions:

- Over a 40-year period, the net present value (NPV) of genetic improvement via within-breed selection, taking account of R&D costs, was AUS$944m (to the Australian economy), compared to AUS$520m estimated as the NPV resulting from cross-breeding and breed change, and AUS$31bn from the introduction of tropically adapted *Bos indicus* genotypes.
- NPV of investment in reference populations for the main breeds is estimated at AUS$211m over 25 years, assuming a slightly faster doubling of the current rate of genetic progress.

These specific results are consistent with conclusions from a broad range of analyses of investment into genetic improvement across many species:

- Even at modest rates of genetic improvement, return on investment is very favourable over the medium to long term.
- Even relatively small improvements in the rate of genetic progress result in substantial increase in return on investment.

Amer *et al.* (2012) found some suggestion that rates of progress are rising, and there are signs in most beef breeding countries that the necessity of generating sufficient data to underpin genomic prediction is focusing more attention on the investment required and the necessity to accelerate genetic progress in order to achieve reasonable returns. This could turn out to be a very important indirect result of genomic technologies – that by highlighting the investment in data required, they stimulate much greater focus on maximizing rate of genetic progress to deliver appropriate return on that investment.

Reproductive Methods

MOET and IVF schemes

The potential value of multiple ovulation and embryo transfer (MOET) in accelerating response to selection was first reported for beef cattle by

Land and Hill (1975). They estimated that responses to selection for growth rate could be doubled by the use of MOET, albeit with higher rates of inbreeding. As in dairy cattle, these original estimates of the benefits of MOET are believed to be on the high side. However, 30% extra progress is likely possible, compared to a conventional scheme of similar size and with the same rate of inbreeding (Villanueva *et al.*, 1995).

While MOET has been used widely in beef cattle as a means of importing and exporting genetic material, and to multiply newly imported breeds or valuable individuals more rapidly than possible with natural reproduction, it has not been used widely in structured breed improvement programmes to date.

One of the earliest intended uses of *in vitro*-produced embryos was to improve the beef merit of calves from dairy or suckler cows, by creating a supply of beef embryos. Initially, the main source of eggs was the ovaries of slaughtered beef heifers. Eggs were collected from beef heifers with a high proportion of continental beef breeds in their genetic makeup, and embryos produced from these by maturing them and then fertilizing them with semen from high merit proven bulls, using *in vitro* fertilization (IVF). These embryos were then marketed for transfer into beef suckler cows or dairy cows. Transfers were made either singly, or to create twins either by transferring an *in vitro*-produced embryo into cows already carrying a natural embryo, or by transferring two *in vitro*-produced embryos. Despite a ready supply of ovaries from slaughtered heifers, early techniques produced few transferable embryos per ovary. Also, some *in vitro* culture techniques are implicated in the birth of very large calves, generally with associated calving difficulties (Kruip and den Daas, 1997).

Techniques have been developed to allow the recovery of unfertilized eggs directly from the ovaries of live cows (see Chapter 13 for a review of these and related techniques). These techniques involve collection of eggs through an ultrasonically guided needle inserted into the ovary, usually via the vagina (Kruip, 1994). This type of recovery is called *in vivo* aspiration of oocytes, or ovum pick up (OPU). It has several potential advantages compared to recovery of eggs from slaughtered cows, or to conventional embryo recovery techniques. In particular purebred animals of high genetic merit can be used as donors, so the technique is of potential benefit in genetic improvement and not just in dissemination. Moreover, eggs can be collected from donors on a weekly basis, allowing tens or potentially hundreds of embryos to be produced from the same donor. The resulting *in vitro* fertilization allows for cross-classified mating of males and females, which gives a useful boost in selection accuracy under juvenile breeding schemes, in which young animals are selected before measurement, on the basis of their parents' EBVs (Kinghorn *et al.*, 1991).

Sexing and cloning

Sexing of semen or embryos has long been a dream for animal scientists. There has been a long history of effort in this area, and practical implementation is underway, especially for sexed semen in the dairy industry. Used appropriately, sexing of either semen or embryos can be of good value for improving production efficiency and in the dissemination of genetic improvement. However, it is generally of little value for increasing the rate of genetic improvement. For example, increasing the proportion of females to help balance selection intensities between the sexes only adds a few per cent to predicted selection response (Kinghorn, 2000a).

Use of affordable sexed semen could be the key factor in development of beef cattle breeding operations specializing in production of first cross females. The technology is arguably of most benefit in this situation, because the low fecundity of cattle means that the relative size of such parental units must be larger than for the pig and poultry industries, and this is one reason why they are not widespread today.

Cloning is an extreme form of reproductive boosting. Clonal propagation has long been used in plant breeding – it exploits the genes in the best individuals, but it also exploits the favourable way in which these genes work with each other in these individuals. These favourable partnerships can be broken down when they are mixed with other genes in the normal breeding cycle. Thus, clones are somewhat static – they are good at providing high productivity, but not so good at creating future

generations of better performing individuals. The latter requires genetic variation – variation from which the new elite can be chosen. This variation is generally not lost with other forms of reproductive boosting such as oocyte pickup, described above. These can lead to some of the direct benefits of clones, through widespread use of elite individuals, but with maintenance of genetic diversity, which can lead to further gains in the following generations.

Cloning could conceivably be used to produce many animals of the same genotype in order to improve the accuracy of evaluation, or to allow evaluation of traits normally measured post-slaughter on some members of the cloned group. One factor to consider here is that clone testing can give accurate estimates of an individual's genetic value (value of alleles to self), but the limit in accuracy of estimating breeding value (value of alleles to progeny) from clone testing is $\sqrt{V_A / V_G}$, where V_A is variance due to breeding values and V_G is variance due to genetic values. Moreover, if cloning is considered only in the context of closed breeding schemes, with fixed numbers of animals tested, then the expected benefits generally diminish or disappear, as keeping more identical animals means that fewer different families can be kept, and so selection intensities will be reduced (Villanueva and Simm, 1994).

There has been speculation on farming cohorts of clones in order to reduce variation in product. However, this will only be of value for quite highly heritable traits, and the effect is unlikely to be dramatic. Kinghorn (2000b) shows that for typical parameter values, the range in trait expression can only be reduced by just over a quarter through using clones rather than unrelated animals. Compounding these issues is the continuing high cost of cloning. This makes it unlikely that we will have cloned animals in the food chain in the foreseeable future, and that any role for cloning in the breeding system will be limited to the parental or grandparental generation.

In many countries there is public concern over the application of new technologies in animal production. Most people accept the use of animals for a range of purposes including food production, providing that the animals are treated humanely. However, it is often difficult to decide whether or not a particular treatment is humane. For discussion of these issues with respect to reproductive technologies, see MAFF (1995).

Conclusions

Genetic improvement in beef cattle has progressed more slowly than in the intensive industries, largely due to the high number of decision-makers and greater variations in environment and market. Genomic information has given optimism that such handicaps can be reduced by providing accurate information on a wide range of traits covering a range of objectives and environments. Indeed there are likely to be increasing benefits from this direction, but the task is more challenging than for dairy, where only a few breeds are involved and widespread use of sufficient sires leads to highly accurate evaluations.

Increased ability to manage genetic gains and improved signals in supply chains could help attract coordinated investment in breeding and downstream operations. This will bring a more directed and professional approach to genetic change. For example, in some appropriate industry sectors this could lead to development and use of specific sire and dam lines, with a seedstock marketplace for first cross females, as in the poultry and pig industries.

References

Allen, D.M. and Steane, D.E. (1985) Beef selection indices. *British Cattle Breeders Club Digest No. 40*, 63–70.

Amer, P.R. (1994) Economic theory and breeding objectives. *Proceedings of the 5th World Congress on Genetics Applied to Livestock Production* 18, 197–204.

Amer, P.R., Kemp, R.A. and Smith, C. (1992) Genetic differences among the predominant beef cattle breeds in Canada: an analysis of published results. *Canadian Journal of Animal Science* 72, 759–771.

Amer, P.R., Lowman, B.G. and Simm, G. (1996) Economic values for reproduction traits in beef suckler herds based on a calving distribution model. *Livestock Production Science* 46, 85–96.

Amer, P.R., Emmans, G.C. and Simm, G. (1997) Economic values for carcase traits in UK commercial beef cattle. *Livestock Production Science* 51, 267–281.

Amer, P.R., Crump, R.E. and Simm, G. (1998) A terminal sire selection index for UK beef cattle. *Animal Science* 67, 445–454.

Amer, P.R., Banks, R.G. and Garrick, D.J. (2012) Costs and effectiveness of various national cattle breeding structures. Proceedings of the 38th International Committee for Animal Recording Conference, 28 May–1 June 2012, Cork. Available at: http://www.icar.org/Cork_2012/index.htm (accessed 21 March 2014).

Ball, A.J., Thompson, J.M. and Kinghorn, B.P. (1998) Breeding objectives for meat animals: use of biological modelling. *Australian Society of Animal Production* 22, 94–97.

Banks, R.G. (1997) Genetics of lamb and meat production. In: Ruvinsky, A. and Piper, L. (eds) *Genetics of the Sheep*. CAB International, Wallingford, UK, pp. 505–522.

Banks, R.G (2011) Progress in implementation of a beef information nucleus portfolio in the Australian beef industry. *Proceedings of the Australian Association of Animal Breeding and Genetics* 19, 399–402.

Barlow, R. and Cunningham, E.P. (1984) Benefit–cost analyses of breed improvement programmes for beef and sheep in Ireland. *Proceedings of the Second World Congress on Sheep and Beef Cattle Breeding*, Vol II, paper P31.

Barwick, S.A., Henzell, A.L. and Graser, H.-U. (1994) Developments in the construction and use of selection indexes for genetic evaluation of beef cattle in Australia. *Proceedings of the 5th World Congress on Genetics Applied to Livestock Production* 18, 227–230.

Benyshek, L.L., Herring, W.O. and Bertrand, J.K. (1994) Genetic evaluation across breeds and countries: prospects and implications. *Proceedings of the 5th World Congress on Genetics Applied to Livestock Production* 17, 153–160.

Bolormaa, S., Pryce, J.E., Kemper, K., Savin, K., Hayes, B.J., Barendse, W., Zhang, Y., Reich, C.M., Mason, B.A., Bunch, R.J. *et al.* (2013) Accuracy of prediction of genomic breeding values for residual feed intake and carcass and meat quality traits in *Bos taurus*, *Bos indicus*, and composite beef cattle. *Journal of Animal Science* 91, 3088–3104.

Bonnett, J.N., Journaux, L., Mocquot, J.C. and Rehben, E. (1994) Breeding cattle for the next millennium. *British Cattle Breeders Club Digest No. 49*, 10–18.

Brascamp, E.W. (1984) Selection indices with constraints. *Animal Breeding Abstracts* 52, 645–654.

Crowley, J.J., Evans, R.D., McHugh, N., Pabiou, T., Kenny, D.A., McGee, M., Crews, D.H., Jr and Berry, D.P. (2011) Genetic associations between feed efficiency measured in a performance test station and performance of growing cattle in commercial beef herds. *Journal of Animal Science* 89, 3382–3393.

Crump, R.E., Simm, G., Nicholson, D., Findlay, R.H., Bryan, J.G.E. and Thompson, R. (1997) Results of multivariate individual animal model genetic evaluations of British pedigree beef cattle. *Animal Science* 65, 199–207.

Cundiff, L.V., Gregory, K.E. and Koch, R.M. (1982) Effects of heterosis in Hereford, Angus and Shorthorn rotational crosses. Beef Research Program Progress Report No. 1. (ARM-NC-21) Roman L. Hruska US Meat Animal Research Center, Clay Center, Nebraska, pp. 3–5.

Cundiff, L.V., Gregory, K.E., Koch, R.M. and Dickerson, G.E. (1986) Genetic diversity among cattle breeds and its use to increase beef production efficiency in a temperate environment. *Proceedings of the 3rd World Congress on Genetics Applied to Livestock Production* IX, 271–282.

Cuthbertson, A. (1994) Enhancing beef eating quality. *British Cattle Breeders Club Digest No. 49*, 33–37.

de Rose, F.P. and Wilton, J.W. (1988) Estimation of genetic trends for Canadian station-tested beef bulls. *Canadian Journal of Animal Science* 68, 49–56.

Dikeman, M.E. (1990) Genetic effects on the quality of meat from cattle. *Proceedings of the 4th World Congress on Genetics Applied to Livestock Production* XV, 521–530.

Doeschl-Wilson, A.B., Knap, P.W., Kinghorn, B.P. and van der Steen, H. (2007) Using mechanistic animal growth models to estimate genetic parameters of biological traits. *Animal* 1, 489–499.

Farquharson, R.J., Griffith, G.R., Barwick, S.A., Banks, R.G. and Holmes, W.E. (2003) Estimating the returns from past investment into beef cattle genetic technologies in Australia, Economic Research Report No. 15, NSW Agriculture, Armidale. Available at: http://www.dpi.nsw.gov.au/__data/assets/pdf_file/0009/146592/err-15-Estimating-returns-from-past-investment-into-beef-cattle-genetic-technologies-in-Australia.pdf (accessed 21 March 2014).

Garrick, D.J. and Golden, B.L. (2009) Producing and using genetic evaluations in the United States beef industry of today. *Journal of Animal Science* 87 (14 Suppl), E11–18.

Goddard, M.E. and Hayes, B.J. (2009) Mapping genes for complex traits in domestic animals and their use in breeding programmes. *Nature Reviews Genetics* 10, 381–391.

Graser, H.-U., Hammond, K. and McClintock, A.E. (1984) Genetic trends in Australian Simmental. *Proceedings of the Australian Association of Animal Breeding and Genetics*, 4, 86–87.

Graser, H.-U., Goddard, M.E. and Allen, J. (1995) Better genetic technology for the beef industry. *Proceedings of the Australian Association of Animal Breeding and Genetics* 11, 56–64.

Gregory, K.E. and Cundiff, L.V. (1980) Crossbreeding in beef cattle: evaluation of systems. *Journal of Animal Science* 51, 1224–1242.

Gregory, K.E., Cundiff, L.V. and Koch, R.M. (1991) Breed effects and heterosis in advanced generations of composite populations for preweaning traits of beef cattle. *Journal of Animal Science* 69, 947–960. (Also, see other papers by the same authors in Volumes 69 and 70.)

Hayes, B.J., Lewin, H.A. and Goddard, M.E. (2013) The future of livestock breeding: genomic selection for efficiency, reduced emissions intensity, and adaptation. *Trends in Genetics* 29, 206–214.

Hearnshaw, H., Gursansky, B.G., Gogel, B., Thompson, J.M., Fell, L.R., Stephenson, P.D., Arthur, P.F., Egan, A.F., Hoffman, W.D. and Perry, D. (1998) Meat quality in cattle of varying Brahman content: the effect of post-slaughter processing, growth rate and animal behaviour on tenderness. International Congress of Meat Science and Technology, Spain, pp. 1048–1049.

Hickey, J.M., Keane, M.G., Kenny, D.A., Cromie, A.R., Mulder, H.A. and Veerkamp, R.F. (2008) Estimation of accuracy and bias in genetic evaluations with genetic groups using sampling. *Journal of Animal Science* 86, 1047–1056.

Ibáñez-Escriche, N, Fernando, R.L., Toosi, A. and Dekkers, J.C.M. (2009) Genomic selection of purebreds for crossbred performance. *Genetics Selection Evolution* 41, 12–21.

Johnston, D.L. (2007) Genetic trends in Australian beef cattle – making real progress. *Proceedings of the Australian Association of Animal Breeding and Genetics* 17, 8–15.

Journaux, L., Rehben, E., Laloë, D. and Ménissier, F. (1996) Main results of the genetic evaluation IBOVAL96 for the beef cattle sires. Edition 96/1. Institut de l'Elevage, Département Génétique

Identification et Contrôle de Performances, Paris, and Institut National de Recherche Agronomique, Station de Génétique Quantitative et Appliquée, Jouy-en-Josas.

Kemp, R.A. (1994) Genetics of meat quality in cattle. *Proceedings of the 5th World Congress on Genetics Applied to Livestock Production* 19, 439–445.

Kennedy, B.W. (1981) Variance component estimations and prediction of breeding values. *Canadian Journal of Genetics and Cytology* 23, 565–578.

Kinghorn, B.P. (1985) Modelled relationships between animal size and production efficiency. *Journal of Animal Breeding and Genetics* 102, 241–255.

Kinghorn, B.P. (2000a) Reproductive technology and designs to exploit it. In: Kinghorn, B.P., Van der Werf, J.H.J. and Ryan, M. (eds) *Animal Breeding – Use of New Technologies*. Post Graduate Foundation in Veterinarian Science of the University of Sydney, Sydney.

Kinghorn, B.P. (2000b) Animal production and breeding systems to exploit cloning technology. In: Kinghorn, B.P., Van der Werf, J.H.J. and Ryan, M. (eds) *Animal Breeding – Use of New Technologies*. Post Graduate Foundation in Veterinarian Science of the University of Sydney, Sydney.

Kinghorn, B.P. (2012) The use of genomics in the management of livestock. *Animal Production Science* 52, 78–91.

Kinghorn, B.P., Smith, C. and Dekkers, J.C.M. (1991) Potential genetic gains with gamete harvesting and *in vitro* fertilization in dairy cattle. *Journal of Dairy Science* 74, 611–622.

Kinghorn, B.P., Hickey, J.M. and van der Werf, J.H.J. (2010) Reciprocal recurrent genomic selection for total genetic merit in crossbred individuals. *Proceedings of the 9th World Congress on Genetics Applied to Livestock Production.* Paper 0036, 8 pp.

Kinghorn, B.P., Hickey, J.M. and van der Werf, J.H.J. (2011) Long-range phasing and use of crossbred data in genomic selection. Paper 1. 7th European Symposium on Poultry Genetics, 5–7 October 2011, Peebles Hydro, near Edinburgh.

Koots, K.R., Gibson, J.P., Smith, C. and Wilton, J.W. (1994a) Analyses of published genetic parameter estimates for beef production traits. 1. Heritability. *Animal Breeding Abstracts* 62, 309–338.

Koots, K.R., Gibson, J.P. and Wilton, J.W. (1994b) Analyses of published genetic parameter estimates for beef production traits. 2. Phenotypic and genetic correlations. *Animal Breeding Abstracts* 62, 825–853.

Kruip, T.A.M. (1994) Oocyte retrieval and embryo production *in vitro* for cattle breeding. *Proceedings of the 5th World Congress on Genetics Applied to Livestock Production* 20, 172–179.

Kruip, T.A.M. and den Daas, J.H.G. (1997) *In vitro* produced and cloned embryos: effects on pregnancy, parturition and offspring. *Theriogenology* 47, 43–52.

Land, R.B. and Hill, W.G. (1975) The possible use of superovulation and embryo transfer in cattle to increase response to selection. *Animal Production* 21, 1–12.

Liboriussen, T. (1982) Comparison of paternal strains used in crossing and their interest for increasing production in dairy herds. *Proceedings of the 2nd World Congress on Genetics Applied to Livestock Production* V, 469–481.

McClintock, A.E. and Cunningham, E.P. (1974) Selection in dual purpose cattle populations: defining the breeding objective. *Animal Production* 18, 237–248.

McNeill, S.H., Harris, K.B., Field, T.G. and Van Elswyk, M.E. (2012) The evolution of lean beef: identifying lean beef in today's US marketplace. *Meat Science* 90, 1–8.

Melucci, L.M., Panarace, M., Feula, P., Villarreal, E.L., Grigioni, G., Carduza, F., Soria, L.A., Mezzadra, C.A., Arceo, M.E., Papaleo Mazzucco, J. *et al.* (2012) Genetic and management factors affecting beef quality in grazing Hereford steers. *Meat Science* 92, 768–774.

Ménissier, F. (1988) La sélection des races bovines à viande spécialisées en France. *Proceedings of the 3rd World Congress on Sheep and Beef Cattle Breeding* 2, 215–236.

Middleton, B.K. and Gibb, J.B. (1991) An overview of beef cattle improvement programs in the United States. *Journal of Animal Science* 69, 3861–3871.

Ministry of Agriculture, Fisheries and Food (MAFF) (1995) *Report of the Committee to Consider the Ethical Implications of Emerging Technologies in the Breeding of Farm Animals.* HMSO, London.

Mourão, G.B., Ferraz, J.B., Eler, J.P., Bueno, R.S., Balieiro, J.C., Mattos, E.C. and Figueiredo, L.G. (2008) Non-additive genetic effects on weights and performance of a Brazilian Bos taurus x Bos indicus beef composite. *Genetics and Molecular Research* 7, 1156–1163.

Mrode, R.A. (1988) Selection experiments in beef cattle. Part 2. A review of responses and correlated responses. *Animal Breeding Abstracts* 56, 155–167.

Phocas, F. and Laloë, D. (2004) Should genetic groups be fitted in BLUP evaluation? Practical answer for the French AI beef sire evaluation. *Genetics Selection Evolution* 36, 325–345.

Piper, L.R. and Bindon, B.M. (1982) Genetic segregation for fecundity in Booroola merino sheep. *Proceedings of the 1st World Congress on Sheep and Beef Cattle Breeding* 1, 395–400.

Pirchner, F. (1986) Evaluation of industry breeding programs for dairy cattle milk and meat production. *Proceedings of the 3rd World Congress on Genetics Applied to Livestock Production* IX, 153–164.

Pollak, E.J. and Quaas, R.L. (1998) Multibreed genetic evaluations of beef cattle. *Proceedings of the 6th World Congress on Genetics Applied to Livestock Production* 23, 81–88.

Roughsedge, T., Thompson, R., Villanueva, B. and Simm, G. (2001) Synthesis of direct and maternal genetic components of economically important traits from beef breed-cross evaluations. *Journal of Animal Science* 79, 2307–2319.

Simm, G. (1998) Genetic improvement of cattle and sheep. *Farming Press*. Old Pond Publishing, Ipswich.

Simm, G., Conington, J., Bishop, S.C., Dwyer, C.M. and Pattinson, S. (1996) Genetic selection for extensive conditions. *Applied Animal Behaviour Science* 49, 47–59.

Simm, G., Amer, P.R. and Pryce, J.E. (1998) Returns from genetic improvement of sheep and beef cattle in Britain. In: *SAC Animal Sciences Research Report 1997*. Scottish Agricultural College, Edinburgh, pp. 12–16.

Thiessen, R.B., Hnizdo, E., Maxwell, D.A.G., Gibson, D. and Taylor, St. C.S. (1984) Multibreed comparisons of British cattle. Variation in body weight, growth rate and food intake. *Animal Production* 38, 323–340. (Also, see other papers by the same authors in later volumes of *Animal Production*.)

Van Eenennaam, A.L., van der Werf, J.H., Goddard, M.E. (2011) The value of using DNA markers for beef bull selection in the seedstock sector. *Journal of Animal Science* 89, 307–320.

Van Vleck, L.D., Pollack, E.J. and Oltenacu, E.A.B. (1989) *Genetics for the Animal Sciences*. W.H. Freeman and Co., New York.

Veneroni-Gouveia, G., Meirelles, S.L., Grossi, D.A., Santiago, A.C., Sonstegard, T.S., Yamagishi, M.E., Matukumalli, L.K., Coutinho, L.L., Alencar, M.M., Oliveira, H.N. and Regitano, L.C. (2012) Whole-genome analysis for backfat thickness in a tropically adapted, composite cattle breed from Brazil. *Animal Genetics* 43, 518–524.

Villanueva, B. and Simm, G. (1994) The use and value of embryo manipulation techniques in animal breeding. *Proceedings of the 5th World Congress on Genetics Applied to Livestock Production* 20, 200–207.

Villanueva, B., Simm, G. and Woolliams, J.A. (1995) Genetic progress and inbreeding for alternative nucleus breeding schemes for beef cattle. *Animal Science* 61, 231–239.

Weller, J.I. (1994) *Economic Aspects of Animal Breeding*. Chapman and Hall, London.

Williams, J.L., Aguilar, I., Rekaya, R. and Bertrand, J.K. (2010) Estimation of breed and heterosis effects for growth and carcass traits in cattle using published crossbreeding studies. *Journal of Animal Science* 88, 460–466.

Wolfová, M., Wolf, J., Pribyl, J., Zahradková, R. and Kica, J. (2005a) Breeding objectives for beef cattle used in different production systems. 1. Model development. *Livestock Production Science* 95, 201–215.

Wolfová, M., Wolf, J., Zahradková, R. and Pribyl, J., Dano, J., Krupa, E. and Kica, J. (2005b) Breeding objectives for beef cattle used in different production systems. 2. Model application to production systems with the Charolais breed. *Livestock Production Science* 95, 217–230.

Zeng, J., Toosi, A., Fernando, R.L., Dekkers, J.C. and Garrick, D.J. (2013) Genomic selection of purebred animals for crossbred performance in the presence of dominant gene action. *Genetics Selection Evolution* 45, 11.

19 Genomic Prediction and Genome-wide Association Studies in Beef and Dairy Cattle

Dorian J. Garrick and Rohan Fernando

Iowa State University, Ames, Iowa, USA

Introduction

Genetic improvement is the straightforward result of selecting genetically superior animals to be the parents of the next generation. In practice this involves many challenges, beginning with identification of the breeding objective that defines the list of traits that determine superiority and the relative emphasis of the traits in the list. Subsequent challenges include determining the nature and frequency of trait measurement, and the timely evaluation of selection candidates for all the traits in the breeding objective. The accepted scientific approach to determine genetic superiority or inferiority of an animal is by estimating its breeding value (BV), and this estimate is referred to as the estimated BV (EBV). However, cattle industries in different countries are not unanimous in how this information is communicated. In North and South America it is more common to report one-half of the EBV, which is variously known as expected progeny difference (EPD) in beef cattle, or predicted transmitting ability (PTA) in dairy cattle.

The conventional animal breeding definition of BV is the genetic potential of an individual judged by the average performance of its offspring.

> If an individual is mated to a number of individuals taken at random from the population, then its breeding value is twice the mean deviation of the progeny from the population mean. The deviation has to be doubled because the parent in question provides only half the genes in the progeny, the other half coming at random from the population.
>
> (Falconer and MacKay, 1996, p. 114)

In contrast, there is also another definition of BV based not on performance but on quantitative genetics, as an extension of the concept of BV applied to a single locus. That definition states that 'the breeding value of an individual is equal to the sum of the average effects of the genes it carries, the summation being made over the pairs of alleles at each locus and over all loci' (Falconer and Mackay, 1976, p.115). That definition clearly has a whole genome framework, as it involves summation over all loci influencing a trait, but the definition is not helpful without understanding the 'average effects of the genes'.

In a random mating population, 'the average effect of a particular gene (allele) is the mean deviation from the population mean of individuals which received that gene from one parent, the gene received from the other parent having come at random from the population'. It is this definition of BV that explains the underlying philosophy regarding genomic prediction and genome-wide association studies (GWASs).

Genomic prediction involves the use of historical data to estimate the substitution effects (the difference between the average effects of one allele and the other) at all genotyped loci and use of these estimates to predict the merit of new animals based on knowledge of the alleles they inherited. GWASs are aimed at using historical data to identify the genes and corresponding alleles that contribute to variation in a particular trait. That information has direct relevance to genomic prediction, but also is of interest from a biological viewpoint. The goal of genomic prediction is simply to use molecular information in the calculation of EBVs, in order to improve the accuracy of predictions, particularly from young animals. This would be expected to translate into faster rates of genetic progress as a result of selection.

Conventional Pedigree-based Prediction

Genetic evaluation is based on a model that describes the factors that influence performance. The model includes three components: the model equation and the first and second moments (means and variances) of variables. The model equation is little more than an expansion of the concept that 'phenotype = genotype + environment'. In matrix notation it is:

$$y = Xb + Zu + e \tag{19.1}$$

Where for example y is a vector of phenotypic observations, b is a vector of fixed effects such as herd-year, or age of dam, u is a vector of breeding values, X and Z are known incidence matrices respectively relating fixed and random effects to phenotypes and e is a vector of random residual effects.

Commonly used expectations are $E[u] = 0$, $E[e] = 0$, and as a consequence $E[y] = Xb$. The second moments are $var[u] = G$, $var[e] = R$, cov$[u,e] = 0$, so that $var[y] = V = ZGZ' + R$. This is known as a mixed model because it includes fixed effects other than the mean, and random effects other than the residual.

Generalized least squares and best linear unbiased prediction

Best linear unbiased estimates (BLUEs) of estimable functions of fixed effects can be obtained using generalized least squares (GLS) by solving the equations:

$$[X'V^{-1}X]b = X'V^{-1}y \qquad (19.2)$$

Then a solution $\hat{b}$ to these equations can be used to adjust the phenotypes to obtain best linear unbiased predictions (BLUPs) of random effects using:

$$\hat{u} = GZ'V^{-1}[y - X\hat{b}] \qquad (19.3)$$

In practice, getting evaluations using Eqns 19.2 and 19.3 was not practical for large V because of the computational effort required to invert a matrix of order equal to the number of observations. Such equations can be set up and solved using modern computers for systems of up to about 200,000 observations, but the limit was more like 10,000 or much less just 20 years ago. Many beef and dairy evaluations comprise millions of animals with phenotypes.

One of the most remarkable inventions that revolutionized cattle improvement was made by Henderson (1948) and became known as 'the mixed model equations (MMEs)'. These involved forming the GLS equations as if the random effects were fixed and the only random effect was the residual (such that $var[y] = V = R$), but then adding the inverse of the variance-covariance matrix for the random effects, (i.e. G^{-1}), to the partition of the GLS equations corresponding to the random effect. That is:

$$\begin{bmatrix} X'R^{-1}X & X'R^{-1}Z \\ Z'R^{-1}X & Z'R^{-1}Z + G^{-1} \end{bmatrix} \begin{bmatrix} b \\ u \end{bmatrix} = \begin{bmatrix} X'R^{-1}y \\ Z'R^{-1}y \end{bmatrix} \quad (19.4)$$

The properties of the solutions to these equations were not known until later when it was discovered (Henderson et al., 1959) that the estimates of the fixed effects were the same as from GLS in Eqn 19.2 and the predictions of the random effects were the same as in Eqn 19.3. This discovery opened the door for routine use

of mixed models in cattle evaluation, because it required R^{-1} and G^{-1} rather than V^{-1}. This simplifies computing because R is a diagonal matrix in single trait evaluations and block-diagonal matrix in multiple trait evaluation, and is therefore straightforward to invert. In circumstances where sires were being evaluated, the order of G can be the number of sires, which is much fewer than the number of observations, enabling G^{-1} to be obtained by brute force inversion, or further simplified by ignoring some relationships. All that changed when Dr Henderson made another breakthrough in 1976.

Identity by descent and the numerator relationship matrix

The quantitative-genetic basis for constructing $var[u] = G$ was based on the idea that covariance between relatives depended upon identity by descent (IBD). In a single trait setting $G = A\sigma_g^2$ where element a_{ij} represents the additive or numerator relationship between the individuals in the ith row and jth column. The additive relationship is defined as twice the fraction of genes identical by descent, and gives rise to the natural coefficients that non-inbred individuals have a relationship with themselves of unity, that unrelated parents produce offspring with an additive relationship of one-half with their parents, and that offspring of non-inbred unrelated parents share a full-sib relationship of one-half or a half-sib relationship of one-quarter. The matrix A can be formed from a pedigree using path coefficients (Wright, 1934), or using the tabular method (Cruden, 1949; Emik and Terrill, 1949), but is not trivial to compute for large datasets, and tends to become very dense in most livestock populations.

Dr Henderson noticed that A^{-1} was sparse, had non-zero elements only in locations related to parentage and mate combinations and in non-inbred circumstances involved coefficients of 0.5, −1 and 2. He identified an algorithm (Henderson, 1976) that could directly construct A^{-1} from simple rules applied to the pedigree, which was further enlightened in a companion publication by his colleague (Quaas, 1976), and extended to inbred pedigrees. This remarkable algorithm, in conjunction with the MMEs, allowed cattle evaluations to be extended to entire populations for both single and multiple trait evaluations. The approach spread throughout the world and became the basis for BV prediction of livestock up until the genomic era.

Prediction of breeding values and associated reliabilities

Inspection of the second equation in Eqn 19.4 provides insight as to the manner in which MMEs generate sensible linear combinations of information in order to predict BVs. It shows that the prediction of an individual with no phenotypic information and no offspring is simply the average of its parents. It shows that if such an individual then has its own phenotype recorded, its prediction becomes a weighted function of its parent average and the deviation of its own performance from its contemporaries, adjusted for the merit of its contemporaries. Subsequently, should such an animal produce offspring and later descendants, its prediction can be written as a weighted function of three sources of information, its parent average BV, its own performance deviation and the average BV of its offspring, adjusted for the merit of its mates (VanRaden and Wiggans, 1991).

In the same way that there is no consensus on the descriptor for genetic merit (BV, EBV, EPD, PTA), there is also a lack of consensus on how the precision of EBVs is reported. One option is the prediction error variance (PEV), which can be obtained from that submatrix of the inverse of the MME that corresponds to the random effects (Henderson, 1975, 1984). However, the units of PEV are trait specific, making PEV difficult to interpret. The PEV can be transformed to obtain the so-called reliability of the prediction, defined as the square of the correlation between true and predicted BV. That expression tends to be used in dairy industries throughout the world. In beef cattle and some other livestock industries, it is more common to report the accuracy, which is the correlation between true and predicted BV, or square root of reliability. In the US beef industry, the Beef Improvement Federation (BIF, www.beefimprovement.org) accuracy is used which is a more conservative measure of accuracy. In practice, MMEs are solved iteratively (Berger et al., 1989), using Gauss-Seidel iteration

or more commonly pre-conditioned conjugate gradient, without forming the inverse of the coefficient matrix required to obtain PEV to use in deriving accuracy or reliability. The elements of the diagonal of the inverse are therefore approximated using various means (e.g. Harris and Johnson, 1998), and the more useful prediction error covariances required for deriving the reliability of differences between selection candidates are never obtained or reported.

Limitations of conventional prediction

The rate of genetic gain resulting from selection can be predicted using what Lush referred to as 'the breeders equation', or its straightforward extension to four pathways of selection more relevant to dairy cattle (Rendel and Robertson, 1950) and some beef cattle improvement programmes. In its simplest representation of annual improvement in units of trait measurement, using Δ to represent 'change' and G to denote 'genetic merit' (rather than $var[u]$ as following Eqn 19.1):

$$\Delta G = \frac{ir\sigma_g}{L} \qquad (19.5)$$

Where i represents the intensity of selection determined by the proportion of available candidates selected as parents, r denotes the correlation between true and predicted merit (i.e. accuracy of prediction), L denotes the length of time in years for a generation of parents to be replaced by their offspring and σ_g reflects the amount of variation in the population expressed in genetic standard deviations.

Inspection of Eqn 19.5 demonstrates a number of opportunities to increase the annual rate of genetic gain. First, the intensity of selection could be increased by screening a greater number of potential selection candidates, or by choosing fewer to use as parents. Second, the accuracy of selection could be increased by measuring more phenotypic information on the candidates or their offspring. Third, the generation interval could be decreased by reducing the average age of the parents when offspring are born. In a few cases it might also be possible to manipulate the genetic variance in the population. Unfortunately, the first three options tend to be antagonistic, and also tend to increase the cost of genetic improvement. Increasing the accuracy of selection typically means delaying decisions until phenotypic measurements can be obtained on the individual, or as in the case of dairy bulls, delaying selection until phenotypes can be measured on daughters. In the pre-genomic era, most breeding schemes were manipulated to try and optimize these factors. Although it was possible to achieve highly accurate prediction of BVs, this was expensive both in terms of financial costs and in terms of extended generation intervals.

Corresponding to the fact that the BLUP BV of a young animal with only parent information is the average of its parental BVs, it follows that all full sibs of a particular pair of parents will have identical EBVs. In reality, their true BVs could vary widely, but the only way to identify those half of the offspring that were better than parent average was through investment in additional phenotypic or offspring measurement. The reliability of the offspring evaluated from parent average can be calculated as one-quarter of the sum of the parental reliabilities. This means that even when the BVs of the sire and dam are known perfectly ($r^2 = 1$), the reliability of the offspring is one-half. This makes sense if one considers an equation for describing the BV of an offspring:

$$u_{offspring} = 0.5u_{sire} + 0.5u_{dam} + \phi_{offspring} \quad (19.6)$$

Where $\phi_{offspring}$ represents the Mendelian sampling or deviation of the offspring BV from parent average. Assuming no covariance between either parental BV and the chance Mendelian sampling term:

$$
\begin{aligned}
var(u_{offspring}) = {} & 0.25var(u_{sire}) + \\
& 0.25var(u_{dam}) + \\
& 0.5cov(u_{sire}, u_{dam}) + \\
& var(\phi_{offspring}) \quad (19.7)
\end{aligned}
$$

If sires and dams are mated at random with no covariance between their BVs, it is apparent that $var(\phi_{offspring})$ must equal half the genetic variance in order for the genetic variance of the population to remain constant from one generation to the next. That is, choice of sires dictates one-quarter of the genetic variance, choice of dam dictates one-quarter of the genetic variance and chance Mendelian sampling accounts for the remainder. According to Mendelian inheritance, perfect knowledge of sire and dam cannot provide information on Mendelian sampling, so reliability of an offspring is at most one-half, when both sire and dam BVs are known perfectly.

The aim of genomic prediction is to use molecular information so that the additive relationships between individuals do not need to be based on the expected fraction of genes identical by descent, but can be quantified based on actual sharing of marker genotypes. This will allow resemblance of an individual to all other animals in the population to inform the prediction of its merit, and will raise the ceiling of reliability above 0.5. This does not mean that all genomic predictions will be more reliable than conventional EBVs, as many sires achieve high reliabilities without genomic information – rather it means that the reliability of their predictions at young ages will exceed that possible at the same age using conventional methods. Genomic prediction might increase the annual rate of genetic gain (Eqn 19.5) by a variety of different mechanisms. This increase might result from an increase in intensity of selection, an increase in the reliability, or a reduction in generation interval. Increasing reliability may not be possible when conventional breeding programmes have used progeny testing to obtain high reliability predictions, as in dairy cattle. In that case, genetic gain might be increased with a reduction in reliability provided the generation interval is reduced by a sufficient amount (Schaeffer et al., 2006).

Prediction using Genomic Information

Ideal model

The MMEs represent a versatile approach for fitting a huge variety of mixed models (Henderson, 1984). The conventional MME used in genetic evaluation is based on a variance–covariance matrix determined by the pedigree, but any other appropriate variance–covariance matrix could be used. If the polygenic infinitesimal model involving a very large number of genes, each with a very small effect, was the appropriate model for determining BVs, and all the infinite numbers of genes segregated independently, the numerator relationship matrix would be the appropriate description of the variance–covariance among relatives, even in the presence of molecular information. The additive relationship among full sibs resulting from non-inbred unrelated parents would always be exactly one-half, while the relationship between half sibs or between grandparent and grand offspring would also be exactly one-quarter, etc.

However, genes do not all segregate independently. They are physically located on one pair of 30 pairs of chromosomes in cattle. There is an average of about one crossover event on each chromosome at meiosis, with a small proportion of chromosomes being inherited intact, and a few exhibiting two or even three crossover events. This means that with respect to a particular chromosome, a pair of half-sibs could share an entire chromosome, or not share anything. Averaged over 30 chromosome pairs the fraction shared identical by descent would average the additive relationship based on expected fraction IBD, but some individuals would share more, and others would share less.

Not all traits are infinitesimal – some are monogenic, while others may be oligogenic or involve perhaps only hundreds of 30,000 or so genes. In that case, it is the similarity at just that fraction of the genome where these segregating causative variants exist that will determine the true similarity among relatives.

If we knew the locations of every segregating variant that influenced performance of a particular trait, and knew the causal variant themselves, and their contribution to variation in the trait, we could derive the true genomic relationship matrix for every trait, and use that knowledge in genomic prediction. We may not be able to use the form of MME in Eqn 19.4, as the genomic relationship matrix will not be full rank if there are more animals than causal variants, but an alternative form not requiring existence of G^{-1} could be used.

In practice, even though we may know the physical location of most genes, and know their variants that exist in the population, we have relatively little knowledge of the variants that influence any particular complex trait. Accordingly, we need to use some kind of analytical method to approximate the true genomic relationship matrix for any particular trait.

Practical models based on mixed model equations (SNP effects or genomic relationships)

Predicting BVs using molecular information requires knowledge of substitution effects.

These can be derived from historical data by regressing BVs, deregressed EBVs or phenotypes, on the number of copies of one of the alleles at a locus. Illumina single nucleotide polymorphism (SNP) genotypes can be labelled as an 'A' or 'B' allele, so regression could be done on the number of copies of the B allele. Single-marker regression can produce spurious results when the historical data, known as the training or discovery population, contain some kind of population structure. It is more common in animal breeding circumstances to undertake multiple regression, fitting many if not all markers simultaneously. It is not possible to fit more markers in the model than there are individuals with observations if the markers are treated as fixed effects. Accordingly, it is more common to fit markers as random effects.

If the causative loci were known, an obvious model might be to fit Eqn 19.1 using the MME in Eqn 19.4, with the incidence matrix Z being replaced by a matrix Q having each column representing a covariate for the number of copies of the B allele at that quantitative trait locus (QTL), and u being replaced by a vector q of unknown allele substitution effects (i.e. the difference between the average effects of allele B and allele A, Falconer and Mackay, 1996). The model equation is:

$$y = Xb + Qq + e \tag{19.8}$$

In this model, the BV is defined as Qq, which is simply a matrix version of summing the substitution effects. In order to form MME for this model, $var[q]$ and $var[e]$ need to be known. In a single trait setting, we typically assume homogeneous residual variances, so $var[e] = R = I\sigma_e^2$, and the left- and right-hand sides of the equations can be multiplied by the scalar σ_e^2, whether known or not. It remains to define $var[q]$. It is not unreasonable to assume zero covariance between average effects at different loci, resulting in a diagonal $var[q]$ matrix. The simplest model for the QTL effects would be to also assume homogeneous variance, so $var[q] = I\sigma_q^2$, leading to the following MME:

$$\begin{bmatrix} X'X & X'Q \\ Q'X & Q'Q + I\lambda \end{bmatrix} \begin{bmatrix} b \\ q \end{bmatrix} = \begin{bmatrix} X'y \\ Q'y \end{bmatrix} \tag{19.9}$$

Where $\lambda = \sigma_e^2/\sigma_q^2$, showing that only the variance ratio is required to fit this model and not the individual residual and QTL effect variances. These equations in Eqn 19.9 are exactly like ordinary least squares (OLS) equations, except that a scalar λ has been added to the diagonal elements corresponding to the QTL effects. Such equations are also known as ridge regression, and have been found in other regression problems to improve predictive ability in circumstances with highly correlated covariates. However, in those circumstances, the ridge parameter is often found by trial and error, whereas in our model it is determined by the ratio of variance components.

In their landmark paper considering the analysis of whole genome markers Meuwissen *et al.* (2001) solved the MMEs in Eqn 19.9 as if the SNP markers (or flanking marker haplotypes) on the genotyping panel represented QTLs. The authors referred to this method as 'BLUP', and assumed the variance ratio was known. Their model equation was:

$$y = Xb + Ms + e \qquad (19.10)$$

Where M represents the covariates or allele dosage for markers, and s represents the allele substitution effects for the fitted SNPs. The MMEs are:

$$\begin{bmatrix} X'X & X'M \\ M'X & M'M + I\lambda \end{bmatrix} \begin{bmatrix} b \\ s \end{bmatrix} = \begin{bmatrix} X'y \\ M'y \end{bmatrix} \qquad (19.11)$$

Genomic predictions of animals, known as direct genomic values (DGVs), were then computed as $\hat{g} = M\hat{s}$. The rationale for this model was that, provided the markers were sufficiently dense across the whole genome, every QTL should be in linkage disequilibrium (LD) with some markers, and the marker genotypes would therefore act as surrogates for the QTL genotypes. In fact, it has generally been accepted that the predictive ability of this and related methods that follow are strongly influenced by the extent of LD. Goddard and Hayes (2009) developed formulae to predict the accuracy of prediction as a function of LD according to effective population size, heritability and size of the reference population. However, as we will show later, LD between each QTL and any one marker locus is not required to achieve good predictive ability with this method.

The model represented by Eqn 19.10 with MMEs in Eqn 19.11 bears little resemblance to Eqn 19.1 and its MMEs in Eqn 19.4. Instead of estimating effects for every animal, it estimates effects for every locus. The left-hand side or coefficient matrix was sparse in Eqn 19.4 but will typically be dense in Eqns 19.9 or 19.11, with order more or less depending upon the number of animals relative to the number of genotyped loci. However, a trivial modification to Eqn 19.10, pre-multiplying the incidence matrix M by an identity matrix and substituting $g = Ms$, gives:

$$y = Xb + I(Ms) + e = Xb + Ig + e \qquad (19.12)$$

Which is an equivalent model (Henderson, 1984; Strandén and Garrick, 2009), and provided there are more markers than animals and $M'M$ is full rank, has MMEs given by:

$$\begin{bmatrix} X'X & X'I \\ I'X & I + (MM')^{-1}\lambda \end{bmatrix} \begin{bmatrix} b \\ g \end{bmatrix} = \begin{bmatrix} X'y \\ I'y \end{bmatrix} \qquad (19.13)$$

The equations in Eqn 19.13 are often referred to as genomic BLUP or 'GBLUP', although in practice the variance ratio which has a different value from the variance ratio in Eqn 19.11 is never known, and an estimate from prior pedigree-based analyses, or by fitting Eqn 19.13 in a restricted maximum likelihood (REML) context is used. An appealing feature of this model is that existing evaluation software need only be modified to replace A^{-1} with $(MM')^{-1}$ to provide genomic predictions. However, this approach will require brute-force inversion of the genomic relationship matrix MM', and it is not a sparse matrix. In some cases, genotype covariates that form columns of M are not allele counts of 0, 1 or 2, but are centred to have mean 0 (e.g. VanRaden, 2008). In that case, $M'1 = 0$ indicating M does not have full row rank, and therefore MM' will not be full rank and Eqn 19.13 is not appropriate. Some researchers (e.g. VanRaden, 2008) use the inverse of a weighted mean of A and MM' in Eqn 19.13.

Bayesian regression methods

Inspection of the model in Eqn 19.10 and its MMEs in Eqn 19.11, and consideration of the nature of the markers used in M raise some options for more sophisticated models. The features included in M are determined by content used to populate marker panels, and not the association of any particular marker with phenotype. Accordingly, many markers might have zero true association with any particular trait. The estimated effects of those loci might be small, but non-zero. The ridge parameter λ in Eqn 19.11 serves to 'shrink' the estimate of each effect towards zero, the extent of shrinkage determined by the magnitude of λ in relation to the magnitude of the corresponding diagonal element of $M'M$. One option would be to vary λ for each marker (i.e. replace it with λ_i for marker i), using larger values for poorly associated markers and smaller values for highly associated markers. Meuwissen et al. (2001) invented such a model in this context, which they referred to as BayesA. They proposed a Bayesian rather than a conventional mixed model analysis because the latter would require the value λ_i for each marker to have been known prior to the analysis. A Bayesian analysis allows λ_i to be treated as an unobservable ratio whose numerator and denominator are assumed to a priori follow scaled inverse chi-square distributions. The idea in Bayesian analyses is that as the data set increases in size, more information as to the posterior distribution of a parameter should come from the data, and the effect of the prior should be washed out.

Another concept for diminishing the impact of markers that are not truly associated with phenotype is to fit a mixture model. In this context, Meuwissen et al. (2001) assumed that any particular marker effect might have zero effect, or alternatively, a non-zero substitution effect. A marker will have zero effect when it's not a causal mutation and when it is not in LD with a causal mutation. A frequentist statistician might do a significance test, and based on the test statistic either accept the hypothesis that the marker has zero effect, or reject it and fit the marker in the model. In contrast a Bayesian statistician assumes a prior assumption that reflects the probability that a marker has zero or non-zero effect (Meuwissen et al., 2001, referred to this parameter as π). The Bayesian will make inferences using the posterior probability that the marker has non-zero

effect. Meuwissen et al. (2001) superimposed this mixture model on BayesA and referred to it as BayesB.

An obvious extension of Eqn 19.11 would be to fit all markers in the model, but to estimate the variance ratio λ using a Bayesian approach. Kizilkaya et al. (2010) developed such a model, which is now referred to as BayesC0. This is like 'BLUP' except that the variance ratio is not assumed to be known. It also seems reasonable to extend that model by further superimposing a mixture model with known mixture parameter π. Kizilkaya et al. (2010) developed that model known as BayesC. In practice, prior knowledge of likely values for the mixture parameter π may not be available, and Habier et al. (2011) demonstrated that π could be estimated from the data, and refer to that method as BayesCπ.

In the Bayesian models the posterior distributions of various parameters are used for inference, and these distributions are constructed using Markov chain Monte Carlo (MCMC) techniques. Basically, each iteration of a Markov chain generates a sample of a parameter, and many such samples are used for inference. Most commonly, inference is based on the posterior mean of the sample values. Precision of the estimates can be quantified by considering the variance of the corresponding posterior distribution, and confidence intervals can also be obtained from the posterior itself. This differs from frequentist use of the MMEs, where the solutions to the equations are the posterior means, and the variance is obtained from elements of the inverse of the coefficient matrix, which are typically approximated from the coefficient matrix itself.

Validation of Predictive Ability

In practice, there are many more animals with phenotypic data than with genotypic data. One approach is to analyse the phenotypic data only on animals with genotypic data. This is not useful in dairy cattle where genotyped bulls do not have lactation records, but have daughters with lactation information. It makes more sense to use all the information on relatives in the genomic analyses, and therefore seems sensible to use EBVs as training data. However, BLUP estimates of EBVs using Eqn 19.1 are shrunk to varying degrees (i.e. $var[\hat{u}] < var[u]$) when different animals have differing number of offspring. One solution to this problem is to use deregressed EBVs (DEBVs), which involves standardizing the genetic variance of the deregressed data by dividing the EBV by its reliability, and then using different weights to account for the heterogeneity of the residual variance (Garrick et al., 2009).

Predictive ability cannot be determined directly from training analyses, but must be determined in separate validation analyses. In dairy cattle populations, it is common for virtually all recent ancestral sires to be included in the training analysis, and validation is typically undertaken in the most recent generation (VanRaden et al., 2009). Since AI is widely used, the most recent generation therefore represents direct descendants of the training population. This makes sense since the direct descendants are the population for inference using genomic predictions. Research has shown that predictions in close relatives will be better than predictions in distant relatives (Habier et al., 2010b). In contrast to dairy cattle, AI is much less widely used in beef cattle, and many animals are sired by natural mating. Beef cattle breeders with little or no use of AI may be interested in genomic predictions of their animals, whereas validation in direct descendants of the training population may overstate predictive ability. Accordingly, it has been more common in beef cattle to report predictive ability in validation animals that are distantly related to the training individuals.

One approach to validation in real beef cattle data is to partition the training population into groups, and undertake training analyses in all but one group, then use the group left out for validation. Groups can be formed at random, but it is more common to cluster animals into groups on the basis of relationships, so close relatives are in the same group, such that validation animals will not have close relatives in training. The training analyses that leave out one group can be repeated with each group being left out of training once, so every animal is used in validation (Garrick, 2011). A single value for validation accuracy can be reported by estimating the genetic correlation between the phenotypes or DEBVs and the genomic predictions or DGV (Saatchi et al., 2011). In dairy cattle when the validation individuals are all reliably evaluated progeny tested sires, the raw correlation between DGV and progeny test EBVs may be used. In simulated data, the correlation between true and predicted performance in a validation population can be obtained directly.

Within-breed predictions

The various genomic prediction methods (BLUP, GBLUP, BayesA, BayesB, BayesC0, BayesC, BayesCπ, etc.) have been applied to the analysis of beef (Saatchi et al., 2013) and dairy cattle data (Habier et al., 2011). In simulated data with small numbers of QTLs, mixture models such as BayesB and BayesC tend to outperform models that fit all markers like GBLUP, BayesA and BayesC0 (Daetwyler et al., 2010). In real data analysis of complex traits, there tends to be less difference between the methods, and GBLUP or BayesC tend to perform as well or almost as well as other methods (Moser et al., 2009; Saatchi et al., 2013). Nevertheless relative performance of different methods can vary from one trait to another, and traits influenced by one or a few major genes tend to perform better with BayesB (Hayes et al., 2010).

Results of within-breed genomic predictions for a range of routinely recorded traits are in Table 19.1 for US beef cattle and Table 19.2 for US dairy cattle. The columns in both tables are sorted by descending size of the training population, which demonstrates the tendency for accuracy to improve with larger training sets. Predictive ability can vary widely from one trait to another, likely reflecting differences in genomic architecture, as well as the large sampling variance that can exist, particularly in smaller validation populations. Typical findings for both beef cattle (data not shown) and dairy cattle (Table 19.2) are that GEBVs and DGV are more accurate than conventional parent average EBVs. That is, selection using GEBVs or DGV would result in greater genetic progress than using parent average EBVs.

Table 19.1. Genetic correlations between direct genomic values (DGV) and deregressed estimated breeding values (DEBV) from k-fold cross-validation.

Trait	Angus[a]	Simmental[b]	Limousin[b]	Brangus[c]	Red Angus[c]	Hereford[d]
Number genotyped	3570	2703	2239	1362	1274	1081
Birth weight	0.56	0.65	0.58	0.82	0.66	0.45
Weaning weight direct	0.33	0.52	0.58	0.66	0.55	0.32
Weaning weight maternal	0.32	0.34	0.46	0.51	0.54	0.26
Yearling weight	0.36	0.45	0.76	0.70	0.57	0.32
Calving ease direct	0.49	0.45	0.52	–	0.59	0.40
Calving ease maternal	0.42	0.32	0.51	–	0.37	0.18
Fat thickness	0.60	0.29	–	0.54	0.85	0.45
Marbling	0.69	0.63	0.65	–	0.77	0.33
Rib eye muscle area	0.60	0.59	0.63	0.79	0.71	0.42
Scrotal circumference	0.49	–	0.45	0.39	–	0.26

Saatchi *et al.*, 2011; [b]Saatchi *et al.*, 2012; [c]Saatchi, unpublished; [d]Saatchi *et al.*, 2013.

Table 19.2. Correlations derived from the square root of the coefficient of determination for regression of June 2010 daughter deviations on predictions including genomic and conventional information (GEBVs) or conventional parent average (PA) transmitting ability from training on young bulls without daughter information as at August 2006. (From Wiggans *et al.*, 2012.)

Trait	Holstein		Jersey		Brown Swiss	
	DGV	PA	DGV	PA	DGV	PA
Milk yield	0.64	0.44	0.70	0.62	0.49	0.23
Fat yield	0.66	0.42	0.62	0.55	0.46	0.27
Protein yield	0.63	0.45	0.64	0.58	0.47	0.25
Productive life	0.56	0.40	0.44	0.33	0.47	0.31
Somatic cell score	0.56	0.40	0.43	0.32	0.48	0.35
Daughter pregnancy rate	0.54	0.47	0.48	0.31	0.31	0.18

Across-breed predictions

The greatest benefit from genomic prediction would occur if training could be undertaken in one subpopulation comprising many economically important phenotypes, and the resulting prediction equation applied to unrelated animals such as those in distant populations or different breeds. Provided genotype–environment interactions are not important, and the same QTLs are segregating in different breeds, with the same markers predictive of alternate genotypes for those QTLs, useful predictive ability in distant relatives should be practical. This has been investigated using simulated and real data.

Simulation of realistic genotypes in different breeds is problematic, as most breeds have gone through various bottlenecks, and their genomes have been subject to both natural and artificial selection, and these circumstances are hard to realistically specify in terms of simulation parameters. Kizilkaya *et al.* (2010) avoided this problem by using actual 50K marker genotypes in a purebred and admixed population of about 1000 animals each. Some markers were randomly chosen to represent QTLs, then phenotypes were simulated by randomly sampling substitution effects for each QTL, and adding a residual effect so that the heritability was 0.5. Training was undertaken in one or other of the populations, with validation in the other, and vice versa. The results of five replicates of those analyses are in Table 19.3 for the simulation using only 50 QTLs. These results show that accurate (i.e. accounting for >50% genetic variance) across-breed pre-dictions can be achieved provided the marker panel includes the actual QTLs (first three rows of results in Table 19.3). However, accuracies dropped considerably when training using a small 50 marker panel consisting of 1 marker for each QTL, each of the 50 markers chosen to have the highest LD with a corresponding QTL in the training data, or a larger panel con-sisting of all 50K markers except those chosen to represent QTLs. In that case, less than 20% genetic variance could be explained from the 50K marker panel. When ≥100 QTLs were responsible for variation in the trait, predic-tions were even poorer, accounting for <10% genetic variance.

Table 19.3. Accuracy of beef cattle across-breed prediction from five replicates of multibreed or purebred training using simulated phenotypes based on 50 QTLs for a trait with heritability 0.50 in a population of about 1000 animals validated in another population of about 1000 animals.

	Accuracy	
Marker panel	Multibreed	Purebred
50 actual QTLs	0.95	0.96
50 actual QTLs and 50 highest LD markers	0.93	0.94
All 50K markers (including 50 QTLs)	0.77	0.84
Only 50 highest LD markers	0.57	0.49
All 50K markers except 50 QTLs	0.39	0.42

genetic variance. These results suggest that the 50K panel does not contain enough markers to guarantee that LD will be retained across breeds.

Across-breed predictions in beef cattle for weaning and yearling weights are in Table 19.4 based on training in 2713 Angus, 897 Hereford or 1670 Limousin, validated using whole-herd data (Kachman et al., 2013). Validation data included 962 Angus, 599 Limousin and less than 200 animals in each of the other breeds. Training data were edited to exclude animals in the pedigrees represented for the whole-herd data. It is apparent that predictions derived in one breed have little utility in other breeds. Standard errors of predictive ability tend to be large except in Angus and Limousin, which had large validation populations. The most accurate predictions are likely to be those developed in the same breed.

Across-breed predictions in dairy cattle from training in Holstein, Jersey or Brown Swiss then applying the predictions obtained within breed to either of the two other breeds were reported by Olson et al. (2012). Predictive ability was reported in terms of the coefficients of determination for predicting post-2005 performance from both parent average (PA) and DGV based on records avail-able prior to that time. The increase in coeffi-cient of determination (adjusted for degrees of freedom) for using PA and DGV compared to using PA alone was typically 0.00 or 0.01. Correlations between DGV using marker effects from different breeds were <0.30 for

Table 19.4. Estimated genetic correlations and standard errors for within-breed training on the breeds in the columns evaluated in independent data from sampled herds of seven breeds[a]. (From Kachman et al., 2013.)

	Weaning weight DGV			Yearling weight DGV		
Breed	Angus	Hereford	Limousin	Angus	Hereford	Limousin
Angus	**0.36 ± 0.07**	0.13 ± 0.08	−0.06 ± 0.08	**0.51 ± 0.07**	0.26 ± 0.08	−0.11 ± 0.09
Red Angus	0.16 ± 0.16	0.09 ± 0.16	0.20 ± 0.16	0.08 ± 0.18	−0.11 ± 0.17	0.15 ± 0.19
Charolais	−0.17 ± 0.19	−0.13 ± 0.20	0.35 ± 0.19	0.09 ± 0.18	−0.29 ± 0.18	0.61 ± 0.12
Gelbvieh	0.12 ± 0.14	0.33 ± 0.13	−0.12 ± 0.14	0.10 ± 0.16	0.27 ± 0.16	0.16 ± 0.17
Hereford	0.05 ± 0.21	**0.42 ± 0.18**	0.27 ± 0.21	0.05 ± 0.22	**0.06 ± 0.22**	0.24 ± 0.21
Limousin	0.02 ± 0.09	0.23 ± 0.09	**0.40 ± 0.08**	0.06 ± 0.09	0.17 ± 0.09	**0.28 ± 0.08**
Simmental	−0.14 ± 0.13	0.14 ± 0.14	−0.02 ± 0.14	−0.11 ± 0.17	0.07 ± 0.18	−0.38 ± 0.16

[a]Genetic correlations and their standard errors are bold when the within-breed DGV is evaluated in the breed it was trained.

all traits tested, and only 0.07 for milk when DGVs for Jersey validation animals were predicted from Jersey or Holstein marker effects. Consistent with the beef cattle results, predictions in one breed have little or no utility in other breeds.

Training using various methods (BayesA, BayesB, BayesC) on US deregressed PTA for milk yield on 8513 Holstein bulls resulted in correlations of 0.14–0.19 when validated in Brown Swiss and 0.18–0.24 when validated on Jersey (Garrick, unpublished). Using the same populations but pooling data from two breeds to predict the third using BayesC ($\pi = 0.99$) resulted in correlations of 0.25 in Holstein, 0.20 in Jersey and 0.08 in Brown Swiss. Applying similar approaches to New Zealand data gave better results, with training on milk yield in 2007 Holstein Friesian bulls giving correlations in 1287 Jersey cows of 0.47, down from 0.69 when predicting 5718 Holstein Friesian cows. Training on 1255 Jersey bulls resulted in correlations of 0.45 in Holstein Friesian cows, down from 0.56 in Jersey cows (D.J. Garrick, unpublished). Repeating the New Zealand analyses with 700K rather than 50K data slightly increased within-breed predictions from 0.69 to 0.70 in Holstein Friesians and 0.47 to 0.59 in Jerseys, but actually decreased across-breed predictions for both breeds (D.J. Garrick, unpublished). The New Zealand data are slightly unusual in that AI sires do not need to be purebred, they can be produced from bull dams that include 1/8 another breed, and this might explain the better across-breed predictions in that context.

Across-breed prediction involving training in one breed to predict performance in another breed does not seem practical for complex traits based on current marker panels and current analytical methods. This is unfortunate, as it limits options for developing genomic predictions for expensive or hard-to-measure traits such as disease resistance, greenhouse gas emissions or feed intake.

Prediction using admixed breeds

Options to increase the size of the training data include pooling with animals of the same breed from other countries, as has been widely exploited in dairy cattle improvement, ultimately leading to formation of several mutually exclusive syndicates. This has not been achieved in beef cattle, except to a minor extent between the USA and Canada. Dairy cattle live in more standard conditions than beef cattle. The role of AI in dairy is higher, including import/export of semen from different countries. Beef cattle in different countries like the USA, UK, Australia, New Zealand, Argentina and so on are relatively more isolated than their dairy cattle counterparts. This may lead to significant genotype–environment interactions. Thus the same measurement in different countries may not represent the same trait.

Another option is to admix training populations from different breeds. In beef cattle, composite and admixed breeds are commonly used in many countries. Admixture reduces long range LD (Toosi et al., 2010), reducing the likelihood of spurious associations in training data from markers on different chromosomes from QTLs. However, admixture also reduces short range LD (Toosi et al., 2010), which might be argued to improve or erode detection of strongly associated markers. Using simulated data, Toosi et al. (2010) demonstrated that accurate genomic prediction could be achieved using admixed data, whereas Ibáñẽz-Escriche et al. (2009) showed that breed specific effects of SNP alleles need not be modelled even when combining breeds that did not have common origin.

Analysis of admixed beef cattle data (Table 19.5) from Kachman et al. (2013) shows that admixture typically did not improve or erode predictive ability compared to within-breed analyses. Unpublished data using larger training sets (Table 19.6) shows that there can be modest increases in accuracy when pooling genotypes from similar breeds, e.g. mean accuracy in Red Angus increased from 0.64 to 0.69 when adding (black) Angus data. Results are less consistent when pooling genotypes from more disparate breeds such as Limousin, which segregates a myostatin allele with major effect, had an average accuracy of 0.54, which reduced to 0.49 when Angus/Red Angus data was admixed. Similarly, the average Brangus (composite of Bos indicus and Bos taurus) correlation reduced from 0.63 to 0.61 when

Angus and Red Angus data were included in training. The admixed training data in Table 19.6 do not pool EBVs from different breed associations, but uses multibreed deregressed EBVs from outcross matings within the same evaluation of the corresponding breed association.

Similar results showing slight increases or slight decreases in predictive ability have been reported in dairy cattle. Table 19.7 presents results from within breed compared to admixed breed training data for 5331 Holstein, 1361 Jersey and 506 Brown Swiss animals (Olson et al., 2012). Holstein and Jersey breeds showed small decreases or no changes in coefficients of determination relating parent average and genomic information to subsequent performance, but increases were seen for

Brown Swiss, the smallest data set with only 506 genotyped training animals.

Understanding Predictive Ability

Sources of information

Genomic regression methods and their equivalent models such as GBLUP were widely considered to have their predictive ability determined by the extent of LD between the markers and QTLs (Meuwissen et al., 2001), and the amount of training data (Goddard and Hayes, 2009). It was soon recognized that additive relationships between the training and validation data (Habier

Table 19.5. Estimated genetic correlations and standard errors for across-breed DGV trained using admixed Angus, Hereford, Red Angus, Limousin and Simmental data validated using independent data from sampled herds of seven breeds whose pedigree ancestors were excluded from admixed training. (From Kachman et al., 2013.)

| Breed | Admixed | | Within breed | |
	Weaning weight	Yearling weight	Weaning weight	Yearling weight
Angus	**0.36 ± 0.07**[a]	**0.45 ± 0.08**	0.36 ± 0.07	0.51 ± 0.07
Red Angus	**0.33 ± 0.14**	**−0.03 ± 0.17**		
Charolais	0.06 ± 0.19	−0.02 ± 0.18		
Gelbvieh	0.47 ± 0.12	0.21 ± 0.16		
Hereford	**0.46 ± 0.17**	**0.05 ± 0.23**	0.42 ± 0.18	0.06 ± 0.22
Limousin	**0.34 ± 0.08**	**0.39 ± 0.08**	0.40 ± 0.08	0.28 ± 0.08
Simmental	**0.11 ± 0.14**	**0.09 ± 0.17**		

[a]Estimated genetic correlations and standard errors are in bold when evaluated in breeds that were part of the admixed training.

Table 19.6. Genetic correlations from validation in purebred Red Angus (RAN), Limousin (LIM) and Brangus (BRG) from training in purebred or admixed populations augmenting the purebred data with Angus (+AAN), Red Angus (+RAN) and Simmental (+SIM) genotyped sires. (From: M. Saatchi, unpublished.)

Validation breed	RAN	RAN	LIM	LIM	BRG	BRG
Training breed(s)	RAN	+AAN+SIM	LIM	+AAN	BRG	+AAN+RAN
Birth weight	0.66	0.75	0.58	0.60	0.82	0.83
Carcass weight	0.62	0.75	0.55	0.55	0.90	0.84
Fat thickness	0.85	0.90	0.54	0.45	0.53	0.52
Marbling	0.77	0.85	0.75	0.58		
Rib eye muscle area	0.71	0.75	0.68	0.57	0.79	0.79
Weaning weight direct	0.55	0.67	0.57	0.56	0.66	0.65
Weaning weight maternal	0.54	0.51	0.46	0.43	0.51	0.44
Yearling weight	0.57	0.69	0.43	0.42	0.70	0.69
Average	0.64	0.69	0.54	0.49	0.63	0.61

Table 19.7. Coefficients of determination for predicting post 2004 performance from conventional parent average EBV and either within or admixed breed training data obtained through 2004. (From: Olson *et al.*, 2012.)

Trait	Holstein		Jersey		Brown Swiss	
	Within	Admixed	Within	Admixed	Within	Admixed
Milk yield	0.50	0.46	0.55	0.52	0.11	0.15
Fat yield	0.48	0.45	0.48	0.44	0.09	0.13
Protein yield	0.50	0.47	0.49	0.47	0.10	0.13
Productive life	0.43	0.42	0.31	0.29	0.42	0.43
Somatic cell score	0.45	0.45	0.35	0.35	0.32	0.41
Daughter pregnancy rate	0.46	0.46	0.38	0.38	0.42	0.48

et al., 2010b) had marked impact on predictive ability, and it was shown that predictive accuracy equal to conventional parent average EBVs can be obtained even when markers and QTLs are on different chromosomes and therefore in linkage equilibrium (Habier *et al.*, 2007), demonstrating that LD is not necessary for genomic prediction. That is, a sufficient number of any independent markers can capture pedigree relationships between individuals across training and validation populations. Habier *et al.* (2013) have demonstrated this using simple models where GBLUP exploited both pedigree relationships (like conventional evaluations using the relationship matrix) and cosegregation, which is not exploited using the additive relationship matrix that assumes average rather than realized fractions of alleles identical by descent. Cosegregation is created by linkage, even in linkage equilibrium, i.e the complete absence of LD, and leads to joint inheritance of nearby alleles within families. Further, Habier *et al.* (2013) showed that information from relationships and information from cosegregation can decline with increases in the size of the training set. This means that accuracy can increase as the sizes of existing families in training increase, but can decrease with increases in the number of families. These results help explain some observations from field data comparing different studies, but not for comparisons of traits in the same study.

Further consideration is required to elucidate those differences between traits that have been trained and validated in the same populations because the same LD among markers, degree of relationships and cosegregation would exist for all traits. The answer involves

the fact that different QTL locations, numbers of QTL, their sizes and distributions likely vary among traits. Furthermore, these QTL alleles are inherited jointly with alleles at neighbouring markers, in chromosome segments known as haplotypes.

Haplotypes and their distribution

A biallelic locus has two alleles and two such loci can produce four combinations of alleles known as haplotypes. In general, k loci can produce 2^k different haplotypes, so there are more than 1 million haplotypes possible for 20 loci. In a training population of 1000 individuals, it would only be possible to observe at most 2000 of those haplotypes. When loci are closely linked so that recombination between them seldom occurs, only a relatively small number of different haplotypes will be observed. For a sufficiently dense set of markers, the actual number of different haplotypes will depend upon the length of the haplotype and the effective population size. The effective population size varies between cattle breeds, most notably being larger in *B. indicus* than *B. taurus* breeds. Small effective population sizes as are common in many cattle breeds will result in few haplotypes being present, even in large data sets with millions of animals.

Genotyping does not expose the haplotypes. In order to determine haplotypes, the genotypes need to be phased, for example using Beagle (Browning and Browning, 2007). Inspection of the haplotype distribution in any 1 Mb window in 941 US Hereford cattle based on 50K genotypes, which represent about 20 markers per Mb, indicates that there are on average about 20 common haplotypes; many rare haplotypes are observed only once. The number of common haplotypes (>0.5%, those observed ≥10 times) in any particular 1 Mb window varies up to about 40, but their distribution is skewed, typically dominated by a small number of the most common ones. In theory, for m haplotypes, there can be $m*(m+1)/2$ pairwise combinations of haplotypes, known as diplotypes. Diplotypes are to haplotypes like genotypes are to alleles; a specific combination of haplotypes. Corresponding to 20 haplotypes we could observe 210 diplotypes, but since many haplotypes are quite rare, diplotypes involving two rare haplotypes are seldom observed. Thus the number of observed diplotypes is fewer than the possible number, and is more like twice the number of haplotypes. So based on the 50K panel, a typical 1 Mb region contains about 20 markers and at least for US Hereford cattle, it segregates about 20 repeatedly observed haplotypes and has about 40 diplotypes. Next we consider the implications of fitting a marker-based model as in Eqn 19.10, when not all allelic combinations are segregating in the population.

Marker-based models

For simplicity of explanation, suppose a region contained just four haplotypes, labelled h_1, h_2, h_3 and h_4. Suppose our training population animals had been genotyped in that region for six marker loci generating observations for m_1–m_6, as in Table 19.8. Assume h_1 contains a causative mutation (QTL allele) that increases performance 7.5 units, whereas h_2–h_4

contains the wild type allele that reduces performance 2.5 units, resulting in a QTL substitution effect of 10.0. Suppose this QTL accounts for 1% phenotypic variance and data are observed on $n = 1000$ randomly chosen training animals, with the 10 diplotypes (h_1h_1, h_1h_2, h_1h_3, h_1h_4, h_2h_2, h_2h_3, h_2h_4, h_3h_3, h_3h_4, h_4h_4) occurring at their expected Hardy-Weinberg ratios. We now use the MMEs in Eqn 19.11 with simulated data to estimate effects for the six marker loci, using a small value for λ = 0.01, to provide little shrinkage of the OLS estimates. That analysis does not require any knowledge of the four underlying haplotypes, it uses only the diploid genotypes to form marker covariates.

The estimates of the six marker effects were –2.1, –1.9, –4.0, –2.0, –4.1 and –4.0. Some researchers try and interpret marker effects with large absolute values as evidence of QTLs, but in this example, none of these markers is a QTL. Even with a lot of training data some effects like m_3, m_5 and m_6 in this example, appear quite large. Since we know the marker alleles that form the haplotypes, we can estimate the four haplotype effects as linear combinations of the marker effects. The relevant linear combinations are those in columns for m_1–m_6 in the rows of Table 19.8 corresponding to each haplotype. The estimates of the four haplotype effects are 7.6, –2.5, –2.4 and –2.7, very close to their real values of 7.5, –2.5, –2.5 and –2.5, despite the fact that the QTL genotype itself was not one of the markers.

Models including the QTL as one of the markers

We can repeat the same analysis with the addition of the QTL genotype used as a

Table 19.8. Haplotype frequency and allele values for 14 marker loci and 1 QTL (Q = 7.5, q = –2.5) at a hypothetical locus represented by four haplotypes.

	Frequency	m_1	m_2	m_3	m_4	m_5	m_6	QTL	m_7	m_8	m_9	m_{10}	m_{11}	m_{12}	m_{13}	m_{14}
h_1	0.4	0	0	0	1	1	1	Q	0	0	0	1	1	1	1	0
h_2	0.1	0	1	1	0	0	1	q	0	0	1	0	1	1	0	1
h_3	0.3	1	0	1	0	1	0	q	0	1	0	0	1	0	1	1
h_4	0.2	1	1	0	1	0	0	q	1	0	0	0	0	1	1	1

marker (like m_{10}) along with m_1–m_6. In that case, the seven marker effects become –1.3, –1.2, –2.5, –1.3, –2.6 –2.5 and 3.8, with the causal genotype having the largest substitution effect (3.8), but still well short of the true QTL substitution effect (10.0), whose effect has been spread among the other six linked markers simultaneously fitted in the model. Even with the QTL genotype or a marker in perfect LD with the QTL on the marker panel, its effect may not be apparent from the estimates of marker effects. Using these seven marker effects results in identical estimates of the four haplotype effects to those obtained previously using six marker effects, without the QTL genotype, indicating that the causal genotype did not add any new information to the analysis. As explained below, this is not surprising, given knowledge of linear algebra.

The four covariates that uniquely identify the haplotypes (i.e. m_7–m_{10}) in the population of n genotyped animals are orthogonal vectors that span only a four-dimensional subspace of an n dimensional space. Any vector in four-dimensional space can be written as a trivial linear function of the orthogonal vectors. Similarly, any set of four linearly independent covariates from the same subspace will form a basis for that subspace. Such sets of linearly independent vectors do not have to be orthogonal, that is they can be correlated, such as most marker covariates. So once you have any four linearly independent covariates in four-dimensional space, any new covariate in that subspace does not bring any new information, it is simply a linear function of those already present. Thus the covariate representing the QTL genotype was implicitly present in the analysis as a linear function of the six markers already fitted. There are many possible linear functions of the six markers that could represent the QTL genotype, two of those being $h_4 + 0.5h_3 - 0.5h_1 - 0.5h_2$ and $0.5h_3 + 0.5h_4 - 0.5h_1$. The dataset itself is composed of covariates for the ten diplotypes, rather than covariates for the four haplotypes, but since the ten diplotypes are simply linear functions of the four haplotypes, an additive model fitted to this data only requires four linearly independent covariates to be fully parameterized.

Equivalent models with fewer markers

Since only four linearly independent covariates are required for four haplotypes, fitting six marker covariates was already over-parameterizing this analysis. Accordingly, any one or two of the six marker covariates can be deleted without having any effect on the genomic prediction of the haplotype effects, and therefore without having any effect on the DGV of animals in training data. Such deletions of markers will also not have any effect on predictions on validation individuals, provided the validation population includes only the same four haplotypes present in the training data.

All k haplotypes, even rare ones, can be estimated from any set of k linearly independent markers observed on the animals that carry the k haplotypes. Some of the k haplotypes might be very rare. However, it is not required that any of the marker alleles be rare. This is apparent in Table 19.8, where h_2 is a rare haplotype with frequency 0.1, but all of markers m_1–m_6 have higher allele frequency. Thus linear combinations of high minor allele frequency (MAF) markers can predict QTL alleles with low MAF. Accordingly, it is not required that marker panels be populated with markers of low MAF. Marker panels with high MAF will be more informative for imputation and genomic selection, provided the marker density allows all haplotypes, whether common or rare, to be uniquely identified as linear combinations of the markers.

Multi marker-QTL LD versus single marker-QTL LD

Multi marker-QTL LD will always be perfect, provided the number of linearly independent markers in a genomic window is equal to the number of haplotypes in that window. Multi marker-QTL LD can be perfect, even without high single marker-QTL LD. The fact that many different sets of at least four of the 14 marker haplotypes in Table 19.8 all give rise to different marker effects, but identical genomic predictions using BLUP, or its equivalent GBLUP model, shows that LD between individual markers in the panel and QTL (single marker-QTL LD) does not in itself have any effect on GBLUP predictions. Even when the causal mutation or a marker in perfect LD (like m_{10}) is added to a panel, the predictions will be unchanged provided the markers previously there represent a linearly independent set of the same size as the number of haplotypes. Thus, in GBLUP it is not LD in the conventional sense, but multi marker-QTL LD that is critical to its predictive ability.

Models with inadequate markers

Suppose the marker panel had not included all of m_1–m_6 in Table 19.8. If it contained less than four markers, then linear combinations of those markers would not span the four-dimensional subspace represented by the four haplotypes. That is, multi marker-QTL LD may not be perfect. In that case, animals with the same marker genotypes may not all share the same pair of haplotypes. If two animals have the same marker genotypes and different haplotypes, they could also have different QTL genotypes. The predictive ability would therefore be compromised unless the haplotypes that appear the same have the same QTL alleles. Mutations are most likely to occur in the most common haplotypes, and this could result in two haplotypes that appear identical at the genotyped markers but contain different QTL alleles. If there are too few markers within the haplotypes that contain QTLs, the effects of the QTLs will be picked up by more distant markers, perhaps even on other chromosomes. The effects of distant markers will provide spurious predictions when applied to distant relatives, but may still retain some of their predictive ability in close relatives, such as immediate offspring.

As mentioned previously, the 50K panel has about 20 markers per 1 Mb, and there are on average about 20 common recognizable haplotypes per 1 Mb, but the genome is quite heterogeneous, with some regions having many more haplotypes and others having many fewer. This suggests that the current 50K panel is likely to have more markers than required in some regions, and too few markers in others. The impact of such panel inadequacy will depend upon whether the regions with too few markers coincide with regions with large effect QTLs.

The 700K panel should do a better job in regions with too few markers, but will at the same time put many more unnecessary markers in other regions. The example presented based on data in Table 19.8 used a small λ relative to the sums of squares of the covariates that form the diagonals of the MMEs. The performance of 50K compared to 700K predictions depends upon the value of λ, as that influences the extent of shrinkage. Generally speaking, λ will be larger when more markers are fitted, and this will increase shrinkage.

Provided there is sufficient marker density on the genotyping panel and the training population is sufficiently large, BLUP/GBLUP should, through its estimates of the marker effects, obtain good estimates of the effects of all the haplotypes present in the training data. However, such predictions of marker effects will have no utility when used to estimate the effects of haplotypes that were not present in the training data. Inspection of haplotypes in QTL regions show that many are breed-specific, and this alone would explain poor predictive ability when training is in one breed and validation in another. However, provided the training data contains all the target breeds, all the haplotype effects could be observed and estimated in the training data, the reliability of the estimates depending upon the frequency of each haplotype and the extent that haplotypes at different loci are mutually confounded.

Models with haplotype covariates

The problem of too many markers in some regions and too few markers in others relative to the number of recognizable haplotypes can be partially resolved by fitting haplotype covariates rather than marker covariates. In large datasets this may not improve the prediction of haplotypes in regions that have adequate marker density. It may reduce the computational effort, by reducing the number of redundant markers being fitted (de Roos et al., 2011).

The BLUP analysis outlined using data in Table 19.8 explicitly estimated marker effects, which were then used to obtain implicit estimates of the four haplotype effects, even though the analysis used marker genotypes, and not phased haplotypes. For appropriately chosen variance ratios, the analysis was equivalent to one which explicitly fitted the haplotypes. In practice, there will typically be many more than four haplotypes to fit. However, it is possible that many QTLs will be biallelic, in which case information is being lost by fitting the four haplotypes when some of them have identical QTL alleles. A more appropriate analysis might be to fit a mixture model at each genomic window, that mixture model comprising an effect for each QTL allele, rather than an effect for each marker allele. Any particular marker or haplotype would in the biallelic case belong to the distribution corresponding to the Q allele, or the distribution corresponding to the q allele. Such models are currently being developed, and would have computational advantages compared to currently used marker-based models.

Blending Genomic and Pedigree Information

Only a small fraction of typical beef and dairy cattle populations are currently being genotyped. This fraction may increase in future, as more cost-effective options are obtained by lower-density genotyping and imputation, and new methods of low-density genotyping become available. Low-density genotyping may also become cheaper simply through economies of scale. In the meantime, to fully use all available data, information from genotyped and non-genotyped animals needs to be combined. There are several methods currently available.

Correlated traits

Conventional genetic evaluations have exploited genetic correlations between hard-to-measure and easy-to-measure traits to improve the accuracy of prediction. In beef cattle, ultrasound measures on live animals have been used to improve predictions for carcass traits, and in dairy cattle somatic cell scores from milk have been used to improve predictions for resistance to mastitis. Further, many multiple trait software packages have been developed that facilitate the inclusion of correlated traits in an MME framework.

The DGV can be considered to be a 'trait' and its genetic correlation with observed performance for the same attribute can be estimated, as in MacNeil *et al.* (2010) and Tables 19.1, 19.4, 19.5 and 19.6. One approach to combine DGVs obtained in genomic analyses with conventional phenotypic information is therefore to use the DGV as a correlated observation. This approach has been implemented by the American Angus Association, allowing DGVs from either or both of two competing companies (Zoetis and Igenity) to be used in the analyses, given knowledge of the correlation between each DGV and the trait, as well as the correlation between the two sources of DGVs.

There are however a number of issues with this approach. First, the usual multi-trait MMEs are parameterized in terms of residual variance–covariance matrices. However, if the genotypes are accurate, heritability of DGVs will be unity and there will be no residual variance. Second, the approach assumes that all DGVs are equally informative. This is not true, as the reliability of DGVs depends upon heterozygocity and the relatedness of the validation animals to those in the training population. Third, the DGV is a shrunk estimate (e.g. from BLUP analyses) and will therefore exhibit varying degrees of shrinkage according to information content. It would be better to use deregressed DGV as correlated information. Fourth, MMEs including correlated DGVs on only a small fraction of the animals in the pedigree can exhibit convergence problems during iteration. Fifth, the approach applied to a large set of multi-trait MMEs as used in beef cattle by Breedplan, will greatly increase the number of equations to set up and solve simultaneously. Nevertheless, the method does have some immediate appeal in terms of simplicity of implementation.

Selection index

Another approach to combine two or more sources of information is best linear prediction, also known as selection index (Hazel, 1943).

That approach requires knowledge of the phenotypic and genetic variance–covariance matrices for the sources of information used in the prediction. The method can be applied to any sources of information provided the relevant variance–covariance matrices are used. It can be used to combine DGVs and EBVs. It is used by the USDA in their dairy evaluation (VanRaden *et al.*, 2009), and by the American Hereford Association in the beef industry.

An advantage of this selection index approach is that it is computationally straightforward once the variance–covariance matrices are known. This is particularly true if the DGVs and EBVs are treated as being independent, in which case the selection index simplifies to weighting the two sources of information as a function of their respective reliabilities.

External EPDs

In developing multiple trait evaluations for composite beef cattle, it is apparent that in the absence of strong genotype–environment interactions, that the information from external purebred analyses should improve accuracy when purebred sires are repeatedly introduced to the composite population. This is the case in American Simmental Association analyses, where for example Angus and Red Angus sires are frequently used. One option to improve accuracy would be to include all the Angus and Red Angus phenotypic data in the analyses, but this is not an option if Breed Associations are not prepared to share their raw data. Golden *et al.* (1994) developed a method to form genetic groups for external animals. Quaas and Zhang (2001) developed a method implemented in the American Simmental Association that includes external EPDs as prior information in a Bayesian context, with the strength of the prior being dictated by the accuracies of the external EPDs. The same approach can be used to introduce DGVs as external information. An advantage of this approach is that it allows each DGV to have its own accuracy. However, the method increasingly faces computational issues, as it requires matrix inversion of the order of the number of animals with external DGVs/EPDs.

Single-step GBLUP

A preferred method to combine data is to include it all in a unified analysis. Conventional genetic evaluations have never been single-step analyses, they have first involved occasional estimation of variance components (heritabilities, genetic and residual variances and correlations), followed by routine analyses assuming the estimated variance components represent true values. Single-step GBLUP refers to a model for such two-step routine analysis, assuming variance parameters are already known. It is a straightforward application of the 'usual' MMEs in Eqn 19.4 except for the manner in which $var[u] = G$ is defined. In the single-step GBLUP literature (Misztal *et al.*, 2009; Aguilar *et al.*, 2010), the variance–covariance matrix is redefined as $var[u] = H$. Suppose the animals in u are reordered so that non-genotyped animals (subscript 1) come before genotyped animals (subscript 2). In a conventional analysis

$$var[u] = var \begin{bmatrix} u_1 \\ u_2 \end{bmatrix} = \begin{bmatrix} A_{11} & A_{12} \\ A_{21} & A_{22} \end{bmatrix} \sigma_g^2,$$

$$\text{with inverse} \begin{bmatrix} A^{11} & A^{12} \\ A^{21} & A^{22} \end{bmatrix} \sigma_g^{-2}.$$

Misztal *et al.* (2009) proposed that for genotyped animals, the variance–covariance be represented by a genomic relationship matrix G_{22}, in place of A_{22}. Legarra *et al.* (2009) further suggested that the covariances between genotyped and non-genotyped animals be modified, and defined such a representation for H.

Aguilar *et al.* (2010) showed that the inverse of H, had a convenient representation. That is,

$$H^{-1} = \left\{ A^{-1} + \begin{bmatrix} 0 & 0 \\ 0 & G^{-1} - A_{22}^{-1} \end{bmatrix} \right\} \sigma_g^{-2}.$$

This has some computational appeal because it simply involves the usual A^{-1}, which can be constructed from rules (Henderson, 1976; Quaas, 1976), except that the submatrix corresponding to genotyped animals is augmented by the addition of the difference between the brute force inverse of the genomic relationship matrix and the brute force inverse of the relationship matrix for the genotyped animals. Provided that relatively few (<200,000) animals are genotyped, this computation is feasible. In practice, there are a number of issues with this approach. First, G is singular if there are any two animals with identical genotypes, or there are more genotyped animals than markers, and it can also be made singular in the process of centring it to adjust its calculation for the allele frequencies of the markers. Second, the concepts of the foundation population or base are not necessarily identical. In the relationship matrix, the founders are assumed to be unrelated and non-inbred, and these form the base population. In the genomic relationship matrix, the base population is defined by the allele frequencies used in its centring, but the allele frequencies in the typically non-genotyped founders are not known, so the frequencies in the genotyped population are often used. This creates a number of problems, which are exacerbated when the concept is applied in a multibreed setting with different allele frequencies in different breeds (Harris and Johnson, 2010).

The approach has appeal from the viewpoint of implementation, as it basically involves enhancing existing routines for deriving the inverse variance–covariance matrix with an extra component to augment that inverse with a matrix of differences. In practice, a number of other often ad hoc modifications have been applied to try and improve predictive ability. The method has the difficulty that it involves the brute force inversion and manipulation of some dense matrices and will not therefore scale with projected increases in genotyping. In fact, with 200,000 genotyped animals, the US dairy industry is already close to the technical limit for this method. Nevertheless, it has proven to deliver some modest increases in predictive ability (Tsuruta et al., 2013).

There are a number of problems with single-step GBLUP. First, it is not possible to obtain accurate reliabilities, which require elements of the inverse of the coefficient matrix of the MMEs. Reliabilities can be approximated, but not as easily as is the case with conventional MMEs based on the inverse numerator relationship matrix. Second, the marker effects are not explicit with this method, so extra activity backsolving marker effects from animals effects (Strandén and Garrick, 2009) is required to use the results of the analysis for GWASs. In this sense, single-step GBLUP is a method that hides marker and haplotype effects in a black box. Finally, and more importantly, GBLUP uses all markers and allocates them equal variance as in

BayesC0, when it has been shown that better results can often be obtained by fitting variable selection or mixture models (e.g. BayesC, BayesCπ) or by allocating markers different variances (e.g. BayesA), or doing both simultaneously (e.g. BayesB). What is really required is an extension of single-step GBLUP to allow for these alternative models, remove the computational limitations, and provide direct prediction of marker or haplotype effects. Such a model can be derived in a Bayesian framework.

Single-step Bayesian regression

Let us consider the original model in Eqn 19.1, but as for single-step GBLUP with the animals in u reordered and partitioned into non-genotyped animals u_1 and genotyped animals u_2. The usual model equation would then be expanded by partitioning other relevant matrices to:

$$y = \begin{bmatrix} y_1 \\ y_2 \end{bmatrix} = Xb + \begin{bmatrix} Z_1 & 0 \\ 0 & Z_2 \end{bmatrix} \begin{bmatrix} u_1 \\ u_2 \end{bmatrix} + e \quad (19.14)$$

Now recognizing the genetic merit of the genotyped animals can be expressed as the sum of their substitution effects, i.e. $u_2 = M_2 S$, and following Legarra et al. (2009), the genetic merit of the non-genotyped animals can be partitioned into a component representing the regression of non-genotyped animals on the genotyped animals, plus an orthogonal residual genetic effect, which we denote using ϵ:

$$\begin{aligned} u_1 &= A_{12}A_{22}^{-1}u_2 + (u_1 - A_{12}A_{22}^{-1}u_2) \\ &= A_{12}A_{22}^{-1}u_2 + \epsilon \end{aligned} \quad (19.15)$$

Substituting this in Eqn 19.14 gives:

$$\begin{aligned} \begin{bmatrix} y_1 \\ y_2 \end{bmatrix} = Xb &+ \begin{bmatrix} Z_1 & 0 \\ 0 & Z_2 \end{bmatrix} \begin{bmatrix} A_{12}A_{22}^{-1}M_2 s \\ M_2 s \end{bmatrix} \\ &+ \begin{bmatrix} Z_1 & 0 \\ 0 & 0 \end{bmatrix} \begin{bmatrix} \epsilon \\ 0 \end{bmatrix} + e \end{aligned} \quad (19.16)$$

And simplifies to:

$$\begin{bmatrix} y_1 \\ y_2 \end{bmatrix} = Xb + \begin{bmatrix} Z_1 A_{12}A_{22}^{-1}M_2 \\ Z_2 M_2 \end{bmatrix} s + \begin{bmatrix} Z_1 \\ 0 \end{bmatrix} \epsilon + e \quad (19.17)$$

Note that $A_{12}A_{22}^{-1}M_2$ is the regression of the non-genotyped animals on the marker genotypes of their genotyped relatives, which we will label M_1 and this can be thought of as being like an imputed genotype. That matrix can be efficiently computed taking advantage of the sparsity of the inverse relationship submatrices, and can be done by marker locus (i.e. column) in parallel. Defining $W' = [W'_1 \ W'_2] = [M'_1 Z'_1 \ M'_2 Z'_2]$, the corresponding MMEs (Fernando et al., 2014) are:

$$\begin{bmatrix} X'X & X'W & X'_1 Z_1 \\ W'X & W'W + \Lambda & W'_1 Z_1 \\ Z'_1 X_1 & Z'_1 W_1 & Z'_1 Z_1 + A^{11}\lambda \end{bmatrix} \begin{bmatrix} b \\ s \\ \epsilon \end{bmatrix} = \begin{bmatrix} X'y \\ W'y \\ Z'_1 y_1 \end{bmatrix}$$

(19.18)

Where Λ is a diagonal matrix containing the variance ratios for each marker. Those diagonal elements will all be the same in single-step BayesC, but can all be different as in single-step BayesA. Furthermore, this model can be extended to a mixture model for the marker effects, so the equations are also relevant to single-step BayesB, or BayesCπ etc. The MMEs in Eqn 19.18 fit the usual fixed effects, marker effects and a residual effect only for each non-genotyped animal. These equations do not increase in order as the number of genotyped animals increases. These equations do not require the brute-force inversion of any matrices, using only the sparse A^{11} which can be created directly from the pedigree. Ignoring the third row and column, these equations are generalizations of Eqn 19.11, except that the coefficient matrix involves the sums of squares and cross-products of the actual genotypes and the 'imputed' genotypes. Ignoring the second row and column, these equations are just like those used in conventional analyses, except that only the submatrix of the entire inverse relationship matrix corresponding to non-genotyped animals is required.

In the special case where λ is 'known' and diagonal elements of Λ are all $\lambda 2\overline{pq}k$, and $2\overline{pq}$ is the mean of $2pq$ for all k markers, these equations are equivalent to single-step GBLUP. However, Eqns 19.14 can be fitted in a Bayesian context, where λ and diagonal elements of Λ are treated as unknowns with appropriate prior distributions, thus making this representation a true single-step approach.

After obtaining the posterior means for $\hat{b}, \hat{s}$ and $\hat{\epsilon}$, for example by solving Eqn 19.18,

genomic predictions for genotyped animals are $\hat{u}_2 = M_2 \hat{s}$, whereas for non-genotyped animals they are $\hat{u}_1 = M_1 \hat{s} + Z_1 \hat{\epsilon}$. The posterior distributions of the effects in Eqn 19.18 can alternatively be obtained by MCMC methods, and these can be used to obtain the posterior means for prediction, but also provide much richer information for inference. For example, the posterior distributions allow the prediction error variances and therefore accuracies and reliabilities to be obtained without approximations, and without having to invert the coefficient matrix in Eqn 19.18.

Extension of Eqn 19.18 to multi-trait applications should be possible, and is the focus of current research, as is comparison of alternative MCMC approaches for most efficiently implementing Eqn 19.18 in both beef and dairy cattle national evaluations.

Many analyses can be biased by selection, but mixed model approaches can be unbiased provided the information used in the selection is appropriately included in the analyses. Genomic analyses are not exempt from selection bias. In a conventional progeny test, the average EBV of progeny tested sires is expected to be their parent average EBV. Sires are expected to be equally likely to increase or decrease in EBV as a result of additional information being obtained during the progeny test. This will not be the case in genomic prediction scenarios if animals are prescreened based on genomic information, and only those with above average DGVs are used as parents, unless the genotypes on the failed candidates are included in the analysis. There is already evidence that such biases are occurring in dairy evaluations (Patry and Ducrocq, 2011b) although procedures are being developed to account for this (Patry and Ducrocq, 2011a). Using all the phenotypic, pedigree and genomic data in Eqn 19.18 should avoid genomic preselection bias.

Genome-wide Association Studies

Major gene effects

Conventional GWAS analyses in the context of human data have typically involved fitting every marker individually, while simultaneously fitting various other factors to account for population structure. Estimated effects with experiment-wise significance have then been the subject of publication (Hindorff *et al.*, 2013) and further validation. Genomic prediction of disease resistance, for example, then uses only the validated markers. This contrasts sharply with the approach to genomic prediction in animals, whereby following Meuwissen *et al.* (2001) all markers are typically used in prediction, regardless of statistical significance. Experience with polygenic traits suggests that discarding some of the poorest markers has no impact on predictive ability, while discarding many of the poorest markers can result in erosion of predictive ability. Even in variable selection models such as BayesB, BayesC and BayesCπ, where π denotes the probability a marker has zero effect, all markers are used. When $\pi = 0.99$, we would expect a fraction 0.01 of markers to be included in the model in any particular iteration of a Markov chain, but different iterations will include different sets of markers, so every marker will usually have a non-zero posterior mean substitution effect and will therefore contribute to genomic prediction. One approach to GWASs in livestock is to carry out analyses in the same manner as are used in human studies (e.g. Snelling *et al.*, 2011). However, another approach is to use Bayesian genomic prediction analyses as introduced in this chapter and covered in formal detail by Fernando and Garrick (2013), and practical detail by Garrick and Fernando (2013).

Individual marker effects from multiple marker analyses are sometimes used as indicators of QTLs. However, these are not good indicators (Gianola, 2013). Even if a marker represents the causal mutation, its effect will be diminished by simultaneous fitting of other markers, as described earlier. The more markers that are fitted, the smaller the effects of any particular marker. Further, the genome is heterogeneous and marker density and/or LD could vary greatly in the vicinity of different QTLs. Ranking the markers on the size of their effects may not therefore reflect the relative sizes of the QTLs. A QTL with small effect but in high LD with one marker in a region with low LD among markers may result in a larger marker effect than a QTL with large effect in a region with high LD among markers. A much better indication of QTL size can be obtained

from estimating the haplotype effects as linear combinations of the marker effects, but this requires that the haplotypes are known. In the absence of knowledge as to haplotypes, prediction of genomic fragments or windows (e.g. 1 Mb) can be used to provide evidence of QTLs. The prediction of a genomic window is obtained in the same manner as whole genome predictions, namely by summing up the B allele minus A allele substitution effects multiplied by the number of copies of the B allele, except that summation is only over the loci in the window rather than the whole genome. Computing the variance of breeding values for each genomic window provides quantifiable indication of QTL size. This approach was used to characterize QTLs expressed as five marker windows in Brangus cattle for growth and yearling ultrasound measures (Peters *et al.*, 2012). Significance levels were obtained by bootstrapping.

In Bayesian analyses calculation of the variance of breeding values for every genomic window can be repeated in many MCMC samples of marker effects rather than just once using the posterior mean of the marker effects. Such repeated samples of window variance are samples from their posterior distributions and provide rich inference as they can be used to construct confidence intervals, significance levels or for hypothesis testing. This approach was used to characterize in 1 Mb windows QTLs for heifer reproductive performance (Peters *et al.*, 2013). Unpublished studies based on observed beef cattle marker genotypes and simulated QTLs and phenotypes have shown that the posterior distribution of window variance can be used to obtain the posterior probability of association of a genomic region with a nearby QTL, however the QTL may sometimes be 1–2 Mb up- or downstream of windows explaining the highest genetic variance.

Incorporation of Sequence Information

Genomic prediction is expected to be more accurate when based on causal variants than when relying on linked markers (Kizilkaya *et al.*, 2010; Meuwissen and Goddard, 2010).

Many researchers are next generation rese-quencing individual animals, and aligning the resultant sequence to the reference genome, with the hope this will ensure all causal variants are available for analysis. Some researchers are collaborating in order to exchange sequence on other animals (e.g. www.1000bullgenomes. org). Eventually, enough animals may be sequenced to allow the direct use of sequence variants in genomic training, but in the short to medium term, a sufficiently large training pop-ulation will only be possible by imputation of sequence from marker panels. Meuwissen and Goddard (2010) showed for 3 or 30 simulated QTLs, that improvements in accuracy were obtained using a variable selection method (i.e. BayesB), but were not apparent using GBLUP. The inability of GBLUP to benefit from the causal mutation was demonstrated by example previously in this chapter using the data in Table 19.8.

Variable selection models have the oppor-tunity to identify the causal mutation if they can markedly reduce the number of variables that need to be fitted in the model. Consider a genomic region with three haplotypes present in the data and containing a biallelic QTL. Under additive gene action, the causal muta-tion could perfectly identify the three QTL genotypes (qq, Qq and QQ) using one covari-ate (e.g. with values 0, 1 and 2). Any set of three linearly independent marker covariates could also perfectly identify the three QTL genotypes, but this would require fitting of more parameters than would be the case for the causal mutation. Provided the data have sufficient statistical support for the model with fewer parameters, and no other marker is con-founded with the actual causal variant, the causal mutation could be identified. This will be more easily achieved if there are many haplo-types in one region and only a single causal mutation.

Computing using whole genome sequence information will provide some challenges in vari-able selection and other approaches. Using sin-gle precision storage, the genotype matrix for 10,000 animals with 1 million variants will require about 40 Gb. Closer to 1 Tb will be required for 10,000 animals with 20+ million variants. Bayesian regression methods such as BayesB, as typically implemented, require at least one multiplication of the genotype matrix with a vector, for each iteration. In that case, run times might take 100 times longer than those using current density SNP panels. Alternative algorithms are available that after forming MMEs are invariant to the number of animals geno-typed (Fernando et al., 2014), but these are not well suited for analyses with >100k markers.

Another approach to using sequence data would be targeted imputation only in QTL regions (e.g. Littlejohn et al., 2014). That approach is somewhat problematic for subse-quent BayesB as it results in regional variation in marker density, and compromises the assumption of every marker having zero effect with the same probability π. A region- or density-specific π seems more appropriate, and such methods are under development.

Rather than fitting all the sequence vari-ants in variable selection models, one could use just the haplotypes that are present in the data. This reduces the number of features to con-sider in the model, especially if effective popu-lation size is small. Identifying haplotypes has some challenges, including the definition of haplotype boundaries, particularly in the pres-ence of genotyping errors. Sequence variants can be screened for concordance with esti-mated haplotype effects rather than being required directly in the analysis. A similar approach was used to identify causal mutations for DGAT1 (Grisart et al., 2002) and PLAG1 (Karim et al., 2011), but those analyses first focused on genomic sequence within the QTL regions of just those sires that could be shown to be segregating the QTLs.

Another alternative is to fit QTL effects directly in the model (Perez-Enciso, 2003; Habier et al., 2010a). Such models are chal-lenging because the QTL genotypes are not observed, but are inferred based on various approaches that take advantage of pedigree information, linkage information, and/or link-age disequilibrium. The advantage is that you don't have to fit the effect of every marker.

Once identified, causal variants could be used in place of markers, with the ultimate goal that the features used in prediction would migrate from evenly spaced genome-wide markers as used at present, to a smaller number of causal variants or markers in perfect LD with causal variants. This is equivalent to migrating a marker-based genomic relationship matrix to a QTL-based relationship matrix.

Sequencing of individual sires will also facilitate identification of loss of function mutations, which in homozygous form might compromise embryo development, growth, production or reproduction. It has been shown that carrier frequencies for such mutations can reach disturbing levels (e.g. >7%, Charlier et al., 2012), and such effects cannot always be found from whole-genome analysis of training data, but can be detected when statistical testing is carried out based on a particular genotype.

Immediate and Future Challenges

Genomic prediction is an advancing but immature technology. Some challenges for genomic prediction are markedly different in beef cattle compared to dairy cattle. Genetic improvement of dairy cattle has a history based on progeny testing that has necessitated the industry to consolidate its improvement efforts around the activities of AI companies. The dairy industry also has a strong focus on multiple trait economic indexes, since the value of milk typically depends upon the values of individual components such as protein, fat and somatic cell concentrations. Prior to genomic prediction, those companies have had to be well resourced in terms of capital because progeny testing is expensive, and had to be well organized because of the time delays between selection of young bulls for progeny testing and their ranking based on progeny tested daughters. Accordingly, investment in genomic training populations was rapidly adopted given the possibilities of faster genetic progress and savings on progeny test costs (Schaeffer, 2006). Further, the dairy industry is dominated by the Holstein breed and wide use of AI so that relatively few sires are used on a global basis, and training populations can include virtually all sires and maternal grandsires. Finally, the target population for genomic prediction are the immediate offspring of the training population.

The beef cattle industry is characterized by a wider diversity of environments than are present for dairy production, and accordingly no one breed has dominated the global beef industry. Many beef breeding cows are managed in extensive conditions where natural mating is the only practical approach to obtaining pregnancies. Even in seedstock herds, natural mating is widely used. Selection has focused on independent culling for multiple traits such as birth weight, calving ease, sale weight, mature size, carcass traits, with little use of economic indexes except for marketing. Beef bulls have been sourced by AI companies through private treaty from mostly small seedstock operations, rather than through direct investment in improvement programmes. Accordingly, the value proposition has made it a slow process to develop training populations for genomic analyses. Finally, the target population for prediction will include many animals that are only distantly related to those in training.

National and international evaluation in the dairy industry has been undertaken by skilled individuals in organizations with critical mass for development of new procedures. These organizations have rapidly acquired genotypes on large training populations, and some are now challenged by having too many animals (>100,000 genotyped) to continue using their current analytical methods. In contrast, beef cattle evaluations are often done in small organizations that lack the funding or expertise to develop new approaches to evaluation. Many beef cattle breed associations have a rich history of pedigree recording, and the relative contribution of genotyped animals is very small, and has failed to motivate approaches to adopt genomic prediction unless it involves simple modifications such as post analysis blending. Genotyping for parentage and genetic defects is widely used, and the advent of reasonably priced low density marker panels is leading to rapid growth in the fraction of animals being genotyped.

Current genomic prediction is based on imputation of some animals from low- to high-density markers, and construction of marker-based genomic relationships assuming additive models. Future challenges will be to migrate the features used in prediction towards causal mutations obtained from sequence, and to reduce the number of anonymous features, so that QTL-based relationships are used. Identification of QTLs also facilitates research into dominance at individual loci, and epistatic interactions between the QTLs. This work will also open up new opportunities for mate selection (Kinghorn, 2011) to simultaneously manage multiple QTLs, loss of function alleles and polygenic effects for multiple traits.

Conclusions

Genomic prediction promises more accurate assessment of breeding merit in young animals than can be obtained using pedigree-based parent-average methods. Genomic prediction uses knowledge of genome-wide markers to infer genetic covariances between selection candidates and their relatives who have recorded phenotypes. Understanding the intricacies of genomic prediction requires knowledge of linear models and quantitative genetics. It is anticipated that genomic prediction using causal variants rather than anonymous markers will improve the accuracy of predicting breeding merit in unrelated animals including animals in other breeds. The first step to identifying causal variants is discovering QTLs, which can be achieved in genome-wide association studies that utilize the same statistical approaches as genomic prediction. There remain many challenges associated with fine-mapping and moving from associated regions to causal mutations, and these include issues such as sequence analysis, genotype imputation and other aspects of bioinformatics. Genomic prediction is an immature, but maturing technology, and is likely to rapidly evolve over the next decade, leading to considerable change in the nature and structure of cattle improvement programmes.

References

Aguilar, I., Misztal, I., Johnson, D.L., Legarra, A., Tsuruta, S. and Lawlor, T.J. (2010) Hot topic: a unified approach to utilize phenotypic, full pedigree, and genomic information for genetic evaluation of Holstein final score. *Journal of Dairy Science* 93, 743–752.

Berger, P.J., Luecke, G.R. and Hoekstra, J.A. (1989) Iterative algorithms for solving mixed model equations. *Journal of Dairy Science* 72, 514–522.

Browning, S.R. and Browning, B.L. (2007) Rapid and accurate haplotype phasing and missing data inference for whole genome association studies using localized haplotype clustering. *American Journal of Human Genetics* 81, 1084–1097.

Charlier, C., Agerholm, J.S., Coppieters, W., Karlskov-Mortensen, P., Li, W., de Jong, G., Fasquelle, C., Karim, L., Cirera, S., Cambisano, N. *et al.* (2012) A deletion in the bovine *FANCI* gene compromises fertility by causing fetal death and brachyspina. *PLoS ONE* 7, e43085.

Cruden, D. (1949) The computation of inbreeding coefficients for closed populations. *Journal of Heredity* 40, 248–251.

Daetwyler, H.D., Pong-Wong, R., Villanueva, B. and Woolliams, J.A. (2010) The impact of genetic architecture on genome-wide evaluation methods. *Genetics* 185, 1021–1031.

de Roos, A.P.W., Schrooten, C. and Druet, T. (2011) Genomic breeding values estimation using genetic markers, inferred ancestral haplotypes, and the genomic relationship matrix. *Journal of Dairy Science* 94, 4708–4714.

Emik, L.O. and Terrill, C.E. (1949) Systematic procedures for calculating inbreeding coefficients. *Journal of Heredity* 40, 51–55.

Falconer, D.S. and Mackay, T.F.C. (1996) *Introduction to Quantitative Genetics.* Fourth Edition. Prentice Hall, Pearson Education Ltd, London.

Fernando, R.L. and Garrick, D.J. (2013) Bayesian methods applied to GWAS. In: Gondro, C., van der Werf J.H.J. and Hayes, B. (eds) *Genome-Wide Association Studies and Genomic Prediction.* Springer Verlag, Berlin, pp. 237–274.

Fernando, R.L., Dekkers, J.C.M. and Garrick, D.J. (2014) A Bayesian method to combine large numbers of genotyped and non-genotyped animals for whole genome analyses. *Genetics Selection Evolution* 46, 50.

Garrick, D.J. (2011) The nature, scope and impact of genomic prediction in beef cattle. Invited Review. *Genetics Selection Evolution* 43, 17.

Garrick, D.J. and Fernando, R.L. (2013) Implementing a QTL detection study (GWAS) using genomic pre-diction methodology. In: Gondro, C., van der Werf, J.H.J. and Hayes, B. (eds) *Genome-Wide Association Studies and Genomic Prediction.* Springer Verlag, Berlin, pp. 275–298.

Garrick, D.J., Taylor, J.F. and Fernando, R.L. (2009) Deregressing estimated breeding values and weighting information for genomic regression analyses. *Genetics Selection Evolution* 41, 55.

Gianola, D. (2013) Priors in whole-genome regression: the Bayesian alphabet returns. *Genetics* 194, 573–596.

Goddard, M.E. and Hayes, B.J. (2009) Mapping genes for complex traits in domestic animals and their use in breeding programmes. *Nature Reviews Genetics* 10, 381–391.

Golden, B.L., Bourdon, R.M. and Snelling, W.M. (1994) Additive genetic groups for animals evaluated in more than one breed association national cattle evaluation. *Journal of Animal Science* 10, 2559–2567.

Grisart, B., Coppieters, W., Farnir, F., Karim, L., Ford, C., Berzi, P., Cambisano, N., Mni, M., Reid, S. and Simon, P. (2002) Positional candidate cloning of a QTL in dairy cattle: identification of a missense mutation in the bovine DGAT1 gene with major effect on milk yield and composition. *Genome Research* 12, 222–231.

Habier, D., Fernando, R.L. and Dekkers, J.C.M. (2007) The impact of genetic relationship information on genome-assisted breeding values. *Genetics* 177, 2389–2397.

Habier, D., Totir, L.R. and Fernando, R.L. (2010a) A two-stage approximation for analysis of mixture genetic models in large pedigrees. *Genetics* 185, 655–670.

Habier, D., Tetens, J. Seefried, F.-R., Lichtner, P. and Thaller. G. (2010b) The impact of genetic relationship information on genomic breeding values in German Holstein cattle. *Genetics Selection Evolution* 42, 5.

Habier, D., Fernando, R.L. Kizilkaya, K. and Garrick, D.J. (2011) Extension of the Bayesian alphabet for genomic selection. *BMC Bioinformatics* 12, 186.

Habier, D., Fernando, R.L. and Garrick, D.J. (2013) Genomic-BLUP decoded: a look into the black box. *Genetics* 194, 597–607.

Harris, B.L. and Johnson, D.L. (1998) Approximate reliability of genetic evaluations under an animal model. *Journal of Dairy Science* 81, 2723–2728.

Harris, B.L. and Johnson, D.L. (2010) Genomic predictions for New Zealand dairy bulls and integration with national genetic evaluation. *Journal of Dairy Science* 93, 1243–1252.

Hayes, B.J., Pryce, J. Chamberlain, A.J. Bowman, P.J. and Goddard, M.E. (2010) Genetic architecture of complex traits and accuracy of genomic prediction: coat colour, milk-fat percentage, and type in Holstein cattle as contrasting model traits. *PLoS Genetics* 6, e1001139.

Hazel, L.N. (1943) The genetic basis for constructing selection indexes. *Genetics* 28, 476–490.

Henderson, C.R. (1948) Estimation of general, specific, and maternal abilities. PhD thesis, Iowa State University, Ames, Iowa.

Henderson, C.R. (1975) Best linear unbiased estimation and prediction under a selection model. *Biometrics* 31, 423–447.

Henderson, C.R. (1976) A simple way for calculating the inverse of a numerator relationship matrix used in prediction of breeding values. *Biometrics* 32, 69–83.

Henderson, C.R. (1984) *Applications of Linear Models in Animal Breeding.* University of Guelph, Guelph, Ontario.

Henderson, C.R., Kempthorne, O., Searle, S.R. and Von Krosigk, C.N. (1959) Estimation of environmental and genetic trends from records subject to culling. *Biometrics* 13, 192–218.

Hindorff, L.A., MacArthur, J. (European Bioinformatics Institute), Morales, J. (European Bioinformatics Institute), Junkins, H.A., Hall, P.N., Klemm, A.K. and Manolio, T.A. (2013) A catalog of published genome-wide association studies. Available at: www.genome.gov/gwastudies (accessed 18 June 2013).

Ibáñez-Escriche, N., Fernando, R.L., Toosi, A. and Dekkers, J.C.M. (2009) Genomic selection of purebreds for crossbred performance. *Genetics Selection Evolution* 41, 12.

Kachman, S.D., Spangler, M.L., Bennett, G.L., Hanford, K.J., Kuehn, L.A., Pollak, E.J., Snelling, W.M., Thallman, R.M., Saatchi, M., Garrick, D.J. *et al.* (2013) Comparison of molecular breeding values based on within- and across-breed training in beef cattle. *Genetics Selection Evolution* 45, 30.

Karim, L., Takeda, H., Lin, L., Druet, T., Arias, J.A.C., Baurain, D., Cambisano, N., Davis, S.R., Farnir, F., Grisart, B. *et al.* (2011) Variants modulating the expression of a chromosome domain encompassing PLAG1 influence bovine stature. *Nature Genetics* 43, 405–413.

Kinghorn, B.P. (2011) An algorithm for efficient constrained mate selection. *Genetics Selection Evolution* 43, 4.

Kizilkaya, K., Fernando, R.L. and Garrick, D.J. (2010) Genomic prediction of simulated multi-breed and purebred performance using observed 50k SNP genotypes. *Journal of Animal Science* 88, 544–551.

Legarra, A., Aguilar, I. and Misztal, I. (2009) A relationship matrix including full pedigree and genomic information. *Journal of Dairy Science* 92, 4656–4663.

Littlejohn, M.D., Tiplady, K., Lopdell, T., Law, T.A., Scott, A., Harland, C., Sherlock, R., Henty, K., Obolonkin, V.,
Lehnert, K., MacGibbon, A., Spelman, R.J., Davis, S.R. and Snell, R.G. (2014) Expression variants of
the lipogenic AGPAT6 gene affect diverse milk composition phenotypes in *Bos taurus*. *PLoS ONE*
9, e85757.

MacNeil, M.D., Nkrumah, J.D., Woodward, B.W. and Northcutt, S.L. (2010) Genetic evaluation of Angus
cattle for carcass marbling using ultrasound and genomic indicators. *Journal of Animal Science* 88,
517–522.

Meuwissen, T.H.E. and Goddard, M. (2010) Accurate prediction of genetic values for complex traits by
whole-genome resequencing. *Genetics* 185, 623–631.

Meuwissen, T.H.E., Hayes, B.J. and Goddard, M.E. (2001) Prediction of total genetic value using genome-
wide dense marker maps. *Genetics* 157, 1819–1829.

Misztal, I., Legarra, A. and Aguilar, I. (2009) Computing procedures for genetic evaluation including phenotypic, full pedigree, and genomic information. *Journal of Dairy Science* 92, 4648–4655.

Moser, G., Tier, B., Crump, R.E., Khatkar, M.S. and Raadsma, H.W. (2009) A comparison of five methods
to predict genomic breeding values of dairy bulls from genome-wide SNP markers. *Genetics Selection
Evolution* 41, 56.

Olson, K.M., VanRaden, P.M. and Tooker, M.E. (2012) Multibreed genomic evaluations using purebred
Holsteins, Jersey and Brown Swiss. *Journal of Dairy Science* 95, 5378–5383.

Patry, C. and Ducrocq, V. (2011a) Accounting for genomic pre-selection in national BLUP evaluations in
dairy cattle. *Genetics Selection Evolution*, 43, 30.

Patry, C. and Ducrocq, V. (2011b) Evidence of biases in genetic evaluations due to genomic preselection in
dairy cattle. *Journal of Dairy Science* 94, 1011–1020.

Perez-Enciso, M. (2003) Fine mapping of complex trait genes combining pedigree and linkage disequilibrium information: a Bayesian unified framework. *Genetics* 163, 1497–1510.

Peters, S.O., Kizilkaya, K., Garrick, D.J., Fernando, R.L., Reecy, J.M., Weaber, R.L., Silver, G.A. and
Thomas, M.G. (2012) Heritability and Bayesian genome-wide association study of first service conception and pregnancy in Brangus heifers. *Journal of Animal Science* 90, 3398–3409.

Peters, S.O., Kizilkaya, K., Garrick, D.J., Fernando, R.L., Reecy, J.M., Weaber, R.L., Silver, G.A. and
Thomas, M.G. (2013) Bayesian genome-wide association analyses of growth and yearling ultrasound
measures of carcass traits in Brangus heifers. *Journal of Animal Science* 91, 605–612.

Quaas, R.L. (1976) Computing the diagonal elements and inverse of a large numerator relationship matrix.
Biometrics 32, 949–953.

Quaas, R.L. and Zhang, Z.W. (2001) Incorporating external information in multiple breed genetic evaluation. *Journal of Animal Science* 79 (Suppl. 1), 342 (Abstract).

Rendel, J.M. and Robertson, A. (1950) Estimation of genetic gain in milk yield by selection in a closed herd
of dairy cattle. *Journal of Genetics* 50, 1–8.

Saatchi, M., McClure, M.C., McKay, S.D., Rolf, M.M., Kim, J., Decker, J.E., Taxis, T.M., Chapple, R.H.,
Ramey, H.R. Northcutt, S.L. *et al.* (2011) Accuracy of genomic breeding values in American Angus
beef cattle using K-means clustering for cross-validation. *Genetics Selection Evolution* 43, 40.

Saatchi, M., Schnabel, R.D., Rolf, M.M., Taylor, J.F. and Garrick, D.J. (2012) Accuracy of direct genomic
breeding values for nationally evaluated traits in US Limousin and Simmental beef cattle. *Genetics
Selection Evolution* 44, 38.

Saatchi, M., Ward, J. and Garrick, D.J. (2013) Accuracies of direct genomic breeding values in Hereford beef
cattle using national or international training populations. *Journal of Animal Science*, 91, 1538–1551.

Schaeffer, L.R. (2006) Strategy for applying genome-wide selection in dairy cattle. *Journal of Animal
Breeding and Genetics* 123, 218–223.

Snelling, W.M., Allan, M.F., Keele, J.W., Kuehn, L.A., Thallman, R.M., Bennett, G.L., Ferrell, C.L., Jenkins, T.G.,
Freetly, H.C., Nielsen, M.K. and Rolfe, K.M. (2011) Partial-genome evaluation of postweaning feed
intake and efficiency of crossbred beef cattle. *Journal of Animal Science* 89, 1731–1741.

Strandén, I. and Garrick, D.J. (2009) Derivation of equivalent computing algorithms for genomic predictions
and reliabilities of animal merit. *Journal of Dairy Science* 92, 2971–2975.

Toosi, A., Fernando R.L. and Dekkers, J.C.M. (2010) Genomic selection in admixed and crossbred populations. *Journal of Animal Science* 88, 32–46.

Tsuruta, S., Misztal, I. and Lawlor, T.J. (2013) Genomic evaluations of final score for US Holsteins benefit
from the inclusion of genotypes on cows. *Journal of Dairy Science* 96, 1–4.

VanRaden, P.M. (2008) Efficient methods to compute genomic predictions. *Journal of Dairy Science* 91, 4414–4423.

VanRaden, P.M. and Wiggans, G.R. (1991) Derivation, calculation, and use of national animal model information. *Journal of Dairy Science* 74, 2737–2746.

VanRaden, P.M., Van Tassell, C.P., Wiggans, G.R., Sonstegard, T.S., Schnabel, R.D., Taylor J.F. and Schenkel, F.S. (2009) Invited review: reliability of genomic predictions for North American Holstein bulls. *Journal of Dairy Science* 92, 16–24.

Wiggans, G.R., VanRaden, P.M. and Cooper, T.A. (2011) The genomic evaluation system in the United States: past, present, future. *Journal of Dairy Science* 94, 3202–3211.

Wright, S. (1934) The methods of path coefficients. *Annals of Mathematical Statistics* 5, 161–215.

20 Genetics of Feed Intake and Efficiency

D.P. Berry,[1] E. Kennedy[1] and J.J. Crowley[2]

[1]*Teagasc, Moorepark, Fermoy, Ireland;*
[2]*University of Alberta, Edmonton, Alberta, Canada*

Introduction

The growing demand for food, coupled with ever-increasing competition by non-food industries for land, as well as societal pressures for cognisance of environmental footprint, imply that without compromising the environment, more food will need to be produced from a decreasing land base available to agriculture. Although many sources of efficiency gains exist across the whole food chain, improved efficiency of animal production has, no doubt, a major role in achieving these goals. Here we review the scientific literature of feed intake and efficiency in dairy and beef cattle, identify potential gains and pitfalls from selection on either trait, discuss alternative options to achieve these gains, and identify future gaps in knowledge. This chapter draws on many of the opinions and results from a meta-analysis of the scientific literature presented by Berry and Crowley (2013).

Trait Definitions

Feed intake measurement

Depending on diet and system of feeding, the optimum methods to measure feed intake vary. To date, the main technologies used to measure individual feed intake in feedlot-type systems are Calan gates (American Calan Inc., Northwood, New Hampshire, USA), Griffith Elder (Griffith Elder & Co Ltd, Bury St Edmunds, UK), Insentec (Insentec B.V., Marknesse, The Netherlands) and Growsafe (GrowSafe Systems Ltd, Airdrie, Canada) systems.

© CAB International 2015. *The Genetics of Cattle,*
2nd Edn (eds D.J. Garrick and A. Ruvinsky)

More recent feeding systems allow multiple animals per feed station and use radio frequency identification tags to track when an animal is eating. Coupled with weigh cells on the feed bunk this allows feed intake to be measured in real time facilitating the generation of additional phenotypes on the animal, like eating behaviour.

While the aforementioned systems are well suited to feedlots, alternative approaches must be used to measure feed intake in grazing production systems. Numerous methods exist to estimate individual animal feed intake at pasture such as calculating faecal output (total collection or use of markers such as n-alkanes), diet digestibility, weighing animals or recording grazing behaviour.

Hydrocarbons of plant cuticular wax (predominantly odd-chain n-alkanes) together with orally dosed even-chain n-alkanes have been used successfully as markers for estimating intake (Mayes et al., 1986; Dove and Mayes, 1991). The use of n-alkanes to estimate intake of herbage, as a sole feed (Dillon, 1989; Stakelum and Dillon, 1990), and herbage intake when supplemented with a known quantity of a concentrate supplement (Stakelum and Dillon, 1990) has been adequately validated. Briefly, when using the n-alkane technique all animals are dosed twice daily for 12 consecutive days with a paper filter (Carl Roth, GmbH and Co. KG, Karlesruhe, Germany) containing 500 mg of dotriacontane (C_{32}). From day 7 of dosing, faecal grab samples are collected from each animal twice daily for the remaining 6 days. The fecal grab samples are then bulked (12 g of each collected sample) and dried for 48 h in a 40°C oven in preparation for chemical analysis. In conjunction with faecal collection, the grass offered to the animals is also sampled on days 6 to 11 (inclusive). Grass contains C_{33}, a natural alkane. The ratio of herbage C_{33} (tritriacontane) to dosed C_{32} is used to estimate dry matter intake.

Feed efficiency traits

Feed efficiency traits can be broadly categorized as ratio traits or residual traits and the traits defined differ between growing and lactating/mature animals.

Growing animals

Feed conversion ratio (FCR), defined as the ratio of average daily feed intake to average daily gain, is one of the most commonly used definitions of feed efficiency. Animals with lower FCR are deemed to be more efficient. Other ratio traits include partial efficiency of growth (PEG), relative growth rate (RGR) and Kleiber ratio (KR).

Residual feed intake is defined as the difference (i.e. residual) between measured and predicted feed intake (Fig. 20.1) and is increasing in popularity as a measure of feed efficiency (Berry, 2009). Predicted feed intake is based on performance and various energy sinks of the individual. Traditionally the energy sinks in the multiple regression model for RFI were metabolic live weight and average daily gain (Koch et al., 1963; Crowley et al., 2010), but more recently it has been recommended to include some measure of body composition (Baker et al., 2006; Basarab et al., 2011) to account for differential energy demands of protein and fat deposition and turnover. Including body composition in the regression model is also important to minimize the impact of selection solely for RFI on mature size or subsequent female reproduction due to the relationship between puberty and fat deposition.

Coefficients in the regression model (i.e. the energy cost per incremental change in the energy sink) are usually estimated using least squares ensuring independence of the residuals (i.e. RFI) from the regressor variables (i.e. the energy sinks). Standard feed tables (e.g. National Research Council, 2001) or other information sources (Arthur et al., 2001) may alternatively be used to predict energy demand of each energy sink but if this approach is undertaken, independence between RFI and the energy sinks is not guaranteed.

Koch et al. (1963) also proposed a residual trait termed residual body weight gain (RG). Residual gain is defined as the residuals from multiple regression of ADG on feed intake and other energy sinks. Therefore RG is the difference between the actual growth rate of an individual and the predicted gain

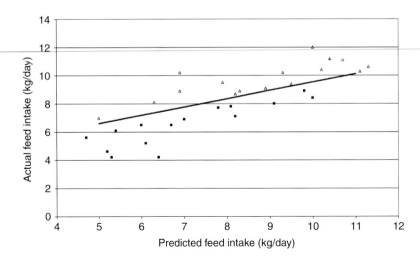

Fig. 20.1. Graphical illustration of residual feed intake (RFI), which represents the residuals (i.e. difference) of actual feed intake and predicted feed intake from the energy sinks; open triangles represent positive RFI animals, while closed squares represent negative RFI animals.

based on the feed intake and likely maintenance requirements of the individual. In contrast to RFI where negative values are deemed to be more efficient animals, more positive RG values (i.e. animals growing faster than predicted) are deemed to be more efficient.

Berry and Crowley (2012) proposed an amalgamation of RFI and RG, which they termed residual intake and gain (RIG). Their justification for deriving such a trait was the zero correlation between RFI and ADG such that slow growing animals could have favourable RFI, whereas industry typically prefers fast growing animals. Due to its construction RIG is independent of metabolic live weight.

Berry and Crowley (2013) outlined some of the issues that need to be considered when modelling RFI, RG or RIG. Considerations included correct representation of metabolic live-weight and testing for non-linear associations among the dependent and independent variables. Moreover, if including, for example, ultrasound fat depth as a measure of body composition, the interaction between the ultrasound measure and body weight should be considered in the model (Savietto *et al.*, 2014). For example, two animals of different weight but the same ultrasound fat depth will have a different mass of body fat; similarly animals with the same ultrasound fat levels but growing

at a different rate will be depositing different quantities of fat. The same is true if activity is included in the multiple regression model; it requires more energy to move a heavier animal than a light animal.

Lactating animals

Relative to growing animals, there has been less research on individual animal feed efficiency in lactating animals. This is most likely due to complications of lactation profiles of energy intake, demand and cyclic patterns in body tissue mobilization to energy kinetics, which complicates the definition of feed efficiency in lactating animals. Because body tissue mobilization contributes energy to the animal, cognisance of this contribution must be considered when defining feed efficiency in lactating animals (Berry, 2009). Not properly accounting for the contribution of changes in body weight or body condition score to the animal energy availability results in a trait that is mathematically equivalent to energy balance (Veerkamp, 2002; Savietto *et al.*, 2014). Energy balance is a term commonly used in lactating dairy cattle (Berry *et al.*, 2006) and known to be correlated with fertility and health (Beam and Butler, 1999; Collard *et al.*, 2000). Feed efficiency in cows must consider the entire inter-calving period, if not the entire life-time of the animal (including the period as a growing heifer).

Several ratio-type measures to reflect feed efficiency in cows have been proposed. Such ratio traits include milk or milk solids yields per unit intake – commonly referred to as feed conversion efficiency (Nieuwhof et al., 1992; Coleman et al., 2010; Prendiville et al., 2011; Vallimont et al., 2011), milk production per kilogramme of body weight (Coleman et al., 2010; Prendiville et al., 2011) and intake per kilogramme of body weight (Coleman et al., 2010; Prendiville et al., 2011). Although not a feed efficiency trait per se, the weight of calf weaned relative to weight of cow is a com-monly used measure of efficiency in beef and sheep production systems.

Using a similar mathematical principle to RFI defined in growing animals, RFI has also been defined in lactating dairy cows by regress-ing feed intake on energy sinks and additional energy sources (i.e. body tissue mobilization). Coleman et al. (2010) defined RFI in lactating Holstein Friesian dairy cows as the residuals from regressing daily dry matter intake on daily milk yield, fat yield, protein yield, lactose yield, metabolic live-weight, body condition score and change in body weight. Coleman et al. (2010) used piece-wise regression on body weight to account for the differential in energy required or generated to gain 1 kg or lose 1 kg of body weight (O'Mara, 2000). Vallimont et al. (2011) fitted body condition score change rather than body weight change and interac-tions between body weight and body condition score change and days in milk. Veerkamp et al. (1995) in one definition of RFI also included interactions between body condition score and live-weight, and body condition score and live-weight change. Those interac-tions are important since body condition score (BCS) only reflects subcutaneous fat, while energy demand or energy released is depend-ent on the entire mass of fat and protein so therefore also dependent on body size/weight.

Analogous to RG in growing animals, Coleman et al. (2010) proposed residual solids production (RSP) as residuals from regressing milk solids production on the remaining energy sinks or sources plus dry matter intake. Residual solids production may be defined as the actual milk solids produced relative to expected solids production based on the individual animal's feed intake and other energy sinks (e.g. main-tenance, growth) or energy sources (e.g. body tissue mobilization). In contrast to RFI, positive RSP values are indicative of more feed-efficient animals.

Advantages and disadvantages of the different measures of feed efficiency

The main advantage of ratio traits such as FCR and feed conversion efficiency is their ease of calculation and interpretation, and ability to compare feed efficiency statistics across popu-lations. None the less, comparison of such sta-tistics across populations should be undertaken with caution given the possible differences in systematic environmental effects (e.g. age of animals in the different populations, breed dif-ferences). Disadvantages of ratio traits include: (i) an increase in the error variance as a pro-portion of the total variance; (ii) strong cor-relations between the ratio trait and its component traits; and (iii) no distinction is made between the energy used for separate functions. Accordingly, expected responses to selection on ratio traits are difficult to predict (Gunsett, 1984) because a desirable response can occur in either the numerator or the denominator and their relative selection pres-sures are unknown. A disproportionate selec-tion pressure will be exerted on the trait in the ratio with the greater genetic variance (Sutherland, 1965); for FCR this is ADG since it generally has a greater coefficient of genetic variation. Another disadvantage of ratio traits especially in lactating animals is that contribu-tions of body tissue catabolism or anabolism to energy kinetics are not accounted for, espe-cially if undertaken over a relatively short time period. McCarthy et al. (2007) and Roche et al. (2006) in Irish and New Zealand dairy cattle, respectively, reported greater BCS loss in Holstein Friesian dairy cattle originating from North American ancestry and of greater genetic merit for milk production. Using feed conversion efficiency in early lactation as a measure of feed efficiency would clearly identify such a strain of animal as the most efficient. However, greater loss of BCS in early lactation has implications for both cow fertility (Roche *et al.*, 2007) and health (Berry *et al.*, 2007), resulting in a less efficient production system.

An often cited advantage of residual-type feed efficiency traits is their independence from the independent variables in the regression model and therefore the perceived lack of an association with mature cow weight. First, genetic independence with the regressor traits only exists if the regression is undertaken at the genetic level. If using genetic regression, genetic independence to the regressor trait itself (i.e. live-weight in growing animals) is guaranteed but not mature weight; this is especially true if some measure of fat is not included in the multiple regression model. The variation in the residual trait provides an excellent indication of the extent of variation in feed efficiency that can be altered without impacting performance.

One of the main disadvantages of the residual traits is the greater difficulty in their interpretation and explanation. Wulfhorst *et al.* (2010) in their social assessment of producers' perception of RFI in the US, concluded that 'the RFI concept is complex and not readily understood when first encountered, even for trained scientists'.

RFI and RSP in lactating animals, although measuring feed efficiency per se, do not accurately reflect production efficiency. This is because the models used to calculate both residual traits do not account for the partitioning of energy into the individual components, some of which are more economically important (e.g. milk fat and protein yield) than others (metabolic live-weight).

The choice of which measure of feed efficiency is best will depend on its proposed use. If comparing herds or even phenotypic differences among animals, the ratio traits (in particular FCR or feed conversion efficiency) may be most appropriate. If however attempting to understand and explain differences in feed efficiency among animals, the residual traits may be most appropriate since the prediction process should be minimally influenced by the energy sinks (e.g. growth rate differences between animals). The decision on which trait

may be most appropriate for breeding programmes will be discussed later.

Genetics of Feed Intake and Efficiency

Heritability estimates and genetic correlations between measures

Heritability estimates for feed intake and efficiency from the literature on growing animals are summarized in Table 20.1. Pooled heritability estimates from up to 45 different studies or populations of growing animals varied from 0.23 (FCR) to 0.40 (feed intake). Moreover, considerable variation in heritability estimates existed across populations. This is not unexpected given the diversity in breeds and feeding systems contributing to the meta-analysis. Pooled genetic correlation estimates from the scientific literature between the different measures of feed intake and efficiency are summarized in Table 20.2. Many of the efficiency traits were strongly genetically correlated with each other; for example, the absolute value of the genetic correlations between FCR, RFI, RG and RIG varied from 0.46 to 0.89. Residual feed intake was not correlated with RGR (−0.01) and was weakly correlated with KR (−0.19). Feed conversion ratio was moderately correlated with feed intake, which is not unexpected given the part–whole relationships between them (i.e. feed intake is included in the calculation of FCR). Variation however existed in the extent of the genetic correlations between RFI and both ADG and live-weight. One of the cited advantages of residual traits like RFI is its independence to the regressor variables. However, derivation of RFI using phenotypic regression does not necessarily imply genetic independence (Kennedy *et al.*, 1993) unless the heritability estimates of the feed intake and the production traits are identical and the residual covariance between the traits is equal to the genetic covariance. In general, as the genetic correlation between the regressor traits and feed intake becomes more positive, the genetic correlation between RFI and the regressor traits also becomes more positive (i.e. unfavourable) unless the residual

Table 20.1. Number of studies (N), pooled heritability, minimum and maximum heritability estimates for average daily gain (ADG), weight (WT), feed intake (FI), residual feed intake (RFI), feed conversion ratio (FCR), residual gain (RG), Kleiber ratio (KR), relative growth rate (RGR) and residual intake and gain (RIG) from a review of the literature in growing animals.

	ADG	WT	DMI	RFI	FCR	RG	KR	RGR	RIG
N	35	25	37	36	34	2	5	4	1
Pooled (standard error)	0.31 (0.014)	0.39 (0.010)	0.40 (0.012)	0.33 (0.013)	0.23 (0.013)	0.28 (0.030)	0.35 (0.030)	0.26 (0.041)	0.36 (0.06)
Min	0.06	0.30	0.06	0.07	0.06	0.28	0.21	0.14	0.36
Max	0.65	0.88	0.70	0.62	0.46	0.62	0.52	0.33	0.36

Table 20.2. Number of studies/populations (N), pooled heritability, minimum and maximum heritability estimates for weight (WT), and feed intake (FI), residual feed intake (RFI) and feed conversion ratio (FCR) from a review of the literature in mature animals.

	WT	FI	RFI	FCR
N	10	7	11	7
Pooled heritability (standard error)	0.63 (0.008)	0.06 (0.008)	0.04 (0.008)	0.06 (0.010)
Min heritability	0.20	0.02	0.00	0.05
Max heritability	0.72	0.28	0.38	0.32

correlation between feed intake and the regressor traits is also strongly positive. Negative genetic correlations between RFI and the regressor traits are evident when the genetic correlation between feed intake and the regressor traits is weak and the residual correlation is strongly positive.

Pooled heritability estimates from the scientific literature for feed intake and efficiency traits in mature animals are in Table 20.3; heritability estimates in cows were considerably lower than those reported in growing animals. The lower heritability estimates for mature animals is likely attributable to increased contribution of random noise to the residual variance attributable to potential errors in the collection of the data (e.g. estimated grass feed intake at pasture as well as the influence of gut fill on live-weight measures) and an incomplete or inappropriate statistical models (i.e. not properly accounting for body tissue mobilization).

Genetic correlations between feed intake and efficiency with other performance measures

Male and female reproduction

Few studies have estimated genetic correlations between feed efficiency and either male or female fertility. Koots *et al.* (1994) in their review of the literature in growing beef cattle reported no genetic correlation (0.04) between FCR and scrotal circumference. Although associated with relatively large standard errors (due primarily to a relatively small dataset size), Crowley *et al.* (2011a) documented unfavourable genetic correlations between three different measures of feed efficiency (FCR, RFI and RG) measured in performance tested bulls and age at first calving in related beef cows (−0.55

± 0.14, −0.29 ± 0.14 and 0.36 ± 0.15, respectively); no measure of body fat was included in the derivation of RFI. In the same study, the genetic correlations of these traits with calving interval (0.07, 0.01 and −0.01) and calving to first service interval (0.21, −0.03 and −0.15) were not different from zero. Crowley *et al.* (2011a) also documented genetic correlations ranging from −0.15 to 0.31 (standard errors ranged from 0.17 to 0.22) between FCR and survival to next lactation, up to lactation 5; genetic correlations between RFI and survival were near zero (Crowley *et al.*, 2011a). Arthur *et al.* (2005), following a divergent selection experiment, documented a 5-day later calving date in low RFI cows compared to high RFI cows, and similarly Basarab *et al.* (2007) reported that cows producing low RFI progeny calved 5 to 6 days later than cows producing medium to high RFI progeny. These results suggest possible genetic antagonisms between feed efficiency and reproductive performance.

Carcass traits

Although differences existed among studies, Berry and Crowley (2013) concluded from their review of seven studies in beef cattle that both RFI and FCR in growing animals were negatively correlated with carcass conformation (pooled genetic correlation of −0.47 and −0.30 for FCR and RFI, respectively). This is substantiated by genetic correlations reported between both FCR and RFI with muscularity, scored in live animals, which showed that genetically superior FCR or RFI animals had, on average, superior genetic merit for muscularity (Bouquet *et al.*, 2010; Crowley *et al.*, 2011b). Although these genetic correlations can be interpreted to imply that selection for FCR or

Table 20.3. Pooled genetic correlations (below the diagonal) with associated standard errors and pooled phenotypic correlations (above the diagonal) from the scientific literature between measures of feed intake and efficiency.

Trait	FI	ADG	WT	FCR	RFI	RG	RGR	KR	RIG
FI		0.42	0.60	0.23	0.66	0.00	0.05	0.19	−0.34
ADG	0.78 (0.02)		0.45	−0.52	0.00	0.7	0.74	0.85	0.41
WT	0.75 (0.02)	0.68 (0.03)		−0.01	−0.01	0.00	−0.35	0.08	0.00
FCR	0.39 (0.04)	−0.62 (0.04)	−0.03 (0.04)		0.39	−0.71	−0.35	−0.74	−0.66
RFI	0.72 (0.02)	0.02 (0.05)	−0.01 (0.04)	0.75 (0.02)		−0.40	−0.01	−0.01	−0.85
RG	−0.03 (0.13)	0.82 (0.05)	0.07 (0.12)	−0.89 (0.03)	−0.46 (0.11)		0.65	0.72	0.85
RGR	−0.18 (0.06)	0.77 (0.06)	−0.54 (0.09)	−0.66 (0.07)	−0.01 (0.07)	0.61 (0.08)		0.86	0.36
KR	−0.04 (0.03)	0.80 (0.03)	−0.09 (0.08)	−0.78 (0.06)	−0.19 (0.07)	0.76 (0.06)	0.97 (0.01)		0.41
RIG	−0.35 (0.10)	0.47 (0.10)	0.11 (0.10)	−0.80 (0.05)	−0.87 (0.03)	0.83 (0.04)	0.26 (0.12)	0.37 (0.11)	

RFI alone will improve muscularity and conformation, they can also be interpreted to imply that selection for improved muscularity and conformation, as undertaken in many beef breeding programmes, is indirectly selecting for improved feed efficiency.

There was a lack of consistency in the literatures of beef cattle on the genetic correlation between RFI and animal fat (Berry and Crowley, 2013). Across most studies reviewed by Berry and Crowley (2013), FCR was negatively genetically correlated with body fat, assessed either based on ultrasound measurements on live animals or subcutaneous fat level of carcasses. Genetic correlations between RFI and body fat reported in the literature of growing beef and dairy animals (i.e. bulls, heifers, steers) from multiple breeds (Berry and Crowley, 2013) were more variable with a pooled genetic correlation from 12 studies of 0.20 (range of −0.79 to 0.48) with ultrasound body fat and a pooled correlation from 7 studies of 0.06 (range of −0.37 to 0.33) with carcass fat. The large range in documented genetic correlations can be attributable to the different breeds and animal types (i.e. steers, bulls, heifers) used in the analyses as well as the timing of measurement. Barwick *et al.* (2009) reported a genetic correlation between RFI with scanned rib fat depth when entering a feedlot of −0.23 and a genetic correlation of 0.40 with scanned rib depth when exiting the feedlot; the latter correlation was 0.16 when estimated in Brahman steers and 0.60 when estimated in a tropical composite. Furthermore, large standard errors were associated with some presented correlations; when the correlation between RFI and ultrasound fat of −0.79 reported by Mujibi *et al.* (2010) with a standard error of 1.15 was discarded the correlations between RFI and ultrasound fat varied from −0.24 to 0.48. Veerkamp *et al.* (1995) reported positive genetic correlations (0.26 to 0.36) between RFI in lactating animals and subjectively scored body condition score.

There is a paucity of information on the genetic correlation between feed efficiency and meat quality. Although solely based on phenotypic analysis of 54 steers obtained from a single seedstock producer (with no apparent selection criteria imposed), Baker *et al.* (2006) reported no difference in a range of meat quality and palatability traits between animals divergent in RFI with the exception of cooking loss (weight change pre- and post-cooking) and the yellowness (i.e. muscle reflectance colour b*; Prache and Theriez, 1999) colour of the steak. For cooking loss, there was no difference between the high (19.1%) and low (21.2%) RFI animals; however, the mid-RFI animals had the lowest cooking loss (17.0%; $P = 0.005$). High RFI animals had steaks with a greater yellow colour ($P = 0.02$) indicating a greater fat content; this is consistent with other studies showing more feed efficient animals have less body fat (discussed previously).

Although the studies available are few, there is conflicting evidence on the association between RFI and calpastatin activity in the meat of animals differing in RFI. There is clear evidence of an association between increased levels of calpastatin in beef and reduced meat tenderness (Chapter 22; Shackelford *et al.*, 1991; Wulf *et al.*, 1996). However, greater calpastatin levels may be associated with lower protein turnover, which is one of the hypothesized possible contributors to differences in RFI among animals (Richardson and Herd, 2004), with lower RFI animals having lower protein turnover. Baker *et al.* (2006) reported no difference in calpastatin activity between Angus steers differing in RFI, while McDonagh *et al.* (2001) reported 13% greater calpastatin in muscle tissue from low RFI animals. One contributing factor to the apparent discrepancies may be that McDonagh *et al.* (2001) measured calpastatin activity immediately post-slaughter, while Baker *et al.* (2006) measured calpastatin activity 24 h post-slaughter, which may influence calpastatin (Koohmaraie, 1992). Also, the relative contribution of protein turnover to differences in RFI may differ between both populations.

Animal size

Few studies have estimated genetic correlations between feed efficiency in growing animals and mature size. There was a concern that aggressive selection for FCR alone would result in larger mature animals thereby increasing maintenance requirement, and this was one of the main reasons for the interest in RFI. As previously discussed, however, genetic independence between the RFI and the independent traits in the multiple regression model only exists if RFI is defined using genetic regression. Furthermore, live-weight measured during performance testing may not be genetically the same trait as mature live-weight; McHugh et al. (2011) reported a weak genetic correlation (0.16 ± 0.05) between live-weight of commercial beef animals measured at weaning and as cows. Moreover, if some measure of body composition is not included in the multiple regression model to derive RFI, animals depositing less fat as a proportion of their ADG may be deemed more efficient and if compared to animals of similar age could result in the selection for older maturing (and therefore probably larger) animals. When comparing multiple breeds of growing beef animals on the same scale without including fat in the multiple regression model for RFI, Crowley et al. (2010) reported, on average, superior RFI in the later maturing Continental breeds compared to the British breeds.

Environmental load

Berry and Crowley (2013) failed to identify any study that estimated genetic correlations between feed efficiency and environmental load in cattle. None the less, a favourable association between feed efficiency and methane production is expected given that methane production represents a source of energy loss (Johnson and Johnson, 1995), and therefore inefficiency. In direct contrast however, the likely improved digestive ability of more efficient animals could result in greater methane emissions per unit feed intake. Berry and Crowley (2013) stressed the importance of an appropriate phenotype when evaluating the true impact on environmental load. For example, superior RFI animals, on average, eat less and this will likely result in less daily methane emissions. However, if superior RFI animals have improved digestibility it may result in greater methane emissions per unit dry matter intake.

Metagenomic analysis of the rumen microbial populations can provide an inexhaustible insight into the prokaryotic ecosystem within the rumen and elucidate the biological rational for any differences in methane output or other rumen characteristics for individuals divergent for feed efficiency. Metagenomic approaches are now more feasible with the development of second and third generation sequencing reducing the cost of sequencing (Mardis, 2008). Differences in the bacterial species in the rumen of high and low RFI animals have already been identified (Luo Guan et al., 2008; Zhou et al., 2009). It is none the less important that as many contributors to differences in RFI are measured (e.g. body composition, activity) as well as very accurate assessment and modelling of RFI are undertaken so that a truer reflection of the true differences in feed efficiency among animals are available on which to undertake subsequent detailed analyses.

Milk yield

Very little is known on the genetic associations between feed intake and efficiency in growing animals and milk yield of their relatives as lactating cows. Crowley et al. (2011a) failed to identify any genetic correlation between RFI and maternal weaning weight, a proxy for milk yield in beef cows, but documented a negative genetic correlation −0.61 ± 0.25 between FCR and maternal weaning weight. Arthur et al. (2005) reported no difference in milk yield, on average, 60 days post-calving in 122 Angus cows selected for 1.5 generations divergent for post-weaning RFI. Strong positive genetic and phenotypic correlations in dairy cattle have been reported between milk yield and feed efficiency defined as the ratio of milk production to feed intake (Vallimont et al., 2011); this however is expected given the statistical part–whole relationship between the two traits. Phenotypic correlations between milk production, live-weight and RFI assessed in lactating dairy cattle, however, are near zero (Veerkamp et al., 1995), but this is because traits reflecting milk production were included in the multiple regression model defining RFI.

Animal health

There is a large gap in knowledge on the genetic associations between feed efficiency and animal health in both beef and dairy cattle. Although many of the genetic correlations had large associated standard errors, Waasmuth *et al.* (2000) reported negative genetic correlations existed between FCR in bulls and mastitis (−0.79 to −0.15) in Danish Friesian and Jersey cows as well as with ketosis in Danish Jersey cows (−0.37). Research on the genetic relationships between feed efficiency and animal health needs immediate prioritization.

Breeding Programmes for Feed Intake and Efficiency

Past breeding programmes

Feed intake is explicitly included in few beef breeding objectives and is not directly included in any national dairy cow breeding objective. Despite this, however, significant (indirect) efficiency gains have been achieved in both beef and dairy cattle populations. Selection for increased growth rate in beef cattle will improve FCR because of the genetic correlation that exists between both traits. The strength of the genetic correlation between ADG and FCR (or any ratio trait) is dependent on the relative differences in the genetic variation of ADG and feed intake (as part of the FCR ratio) as well as

the genetic correlation between ADG and feed intake (Sutherland, 1965). The expected genetic correlation between ADG and FCR is illustrated in Fig. 20.2 where the ratio of variation in ADG to feed intake differs but also the genetic correlation between ADG and feed intake changes. When the genetic variation in ADG is greater than that of feed intake (which is usually the case) then the genetic correlation between ADG and FCR will always be strong. Such a phenomenon has resulted in improved FCR in breeding programmes that selected for increased growth rate and has been especially observed in pig and poultry breeding programmes (Rauw *et al.*, 1998).

Past dairy cow breeding programmes that selected for increased milk production have also indirectly selected for improved feed conversion efficiency. Milk output is positively genetically correlated (0.69; Veerkamp and Brotherstone, 1997) with feed intake (i.e. the denominator of the feed efficiency variable) and therefore genetic gain in the efficiency of milk production is expected to have improved with selection for milk production. Feed conversion efficiency does not however take cognisance of reproductive performance (and in particular body tissue mobilization) and thus overall system efficiency.

None the less, genetic selection for production traits is unlikely to influence genetic gain in residual feed intake because of its independence

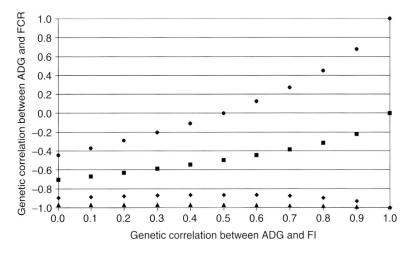

Fig. 20.2. Expected genetic correlation between average daily gain (ADG) and feed conversion efficiency (FCR) as the correlation between ADG and feed intake (FI) varies from 0 to 0.1. The ratio of the genetic variation of ADG relative to feed intake was 4 (▲), 2 (♦),1(■) and 0.5 (•).

to these energy sinks. In growing animals though, the reported genetic correlations between animal muscularity/carcass conformation and RFI indicate that breeding programmes that selected for improved animal conformation will, on average, have also improved genetic merit for RFI.

Breeding programmes including feed intake and efficiency

The importance of feed intake and efficiency in breeding programmes is undisputed. How to optimally include such measures in a breeding goal however is not clear; should feed itself (as well as other energy sinks) or feed efficiency be included in a breeding goal? Including either feed intake (plus the energy sinks) or RFI in a breeding goal or selection index is mathematically equivalent. Much of the discussions therefore are on what is likely to be best understood and accepted by stakeholders.

The advantages of including feed intake itself in a breeding goal include: (i) the trait itself is well understood by the end user; (ii) it is probably less susceptible to genotype by environment (G × E) interactions compared to feed efficiency traits like RFI, which are an index and can exhibit re-ranking across environments even if no re-ranking exists in the component traits (Namkoong, 1985); (iii) the understanding of the economic value is relatively easy and development of customized indexes by individual producers is relatively simple since all they have to do is alter the economic value on the feed intake goal trait. The disadvantages of using feed intake in a breeding goal, all of which are related include: (i) that it is correlated with performance and therefore animals with negative estimated breeding values (EBVs) for feed intake will, on average, also have reduced performance; (ii) independent culling levels on EBVs for feed intake can therefore have unfavourable effects on genetic gain for other traits; (iii) it is difficult to identify feed efficient animals; and (iv) a perception that inclusion of feed intake in a breeding goal will select for reduced performance.

The advantages of including a feed efficiency trait like RFI directly in the breeding goal instead of feed intake include: (i) RFI may be independent of (many of) the other goal traits if derived using genetic regression, although the independence may not hold among a group of animals (e.g. elite animals) other than the population in which the parameters were derived (i.e. performance tested animals); and (ii) the economic value for RFI is the same as that for feed intake. The main disadvantages of including RFI in a breeding goal include: (i) a lack of understanding of the RFI term by the stakeholder as discussed by Wulfhorst et al. (2010) and the implications this may have on its use in the field; (ii) it is likely to be more susceptible to G×E as discussed previously; (iii) development of customized indexes where the value of feed differs (e.g. grazing versus confinement production systems) is not trivial since some of the economic value for feed intake is on the different energy sinks; and (iv) the difficulty of properly accounting for body tissue mobilization in lactating animals. Of course a combination of including, for example, feed intake in the breeding goal and including RFI as a separate trait for identification of efficient animals may also be used. This is not dissimilar to what is currently undertaken in for example dairy cattle where an overall breeding goal value is published for each animal but also the individual predicted transmitting ability (PTA) for (many of) the individual component traits of the breeding goal.

Irrespective of whether feed intake or feed efficiency is included in a breeding goal, accurate measures of genetic merit for feed intake are required. This can be achieved from phenotypic data or genomic information, or both using a selection index approach.

The genetic gain in reducing dry matter intake that could potentially be achieved can be approximated using relatively simple calculations based on four selection pathways with differing selection intensities (Table 20.4). The accuracy of selection of sires to produce sires and sires to produce dams was altered from 0.50 to 0.90 reflecting more information available to identify genetically different animals. Here we used the genetic standard deviation of RFI as a proxy for a trait representing feed intake independent of the breeding goal; potential genetic correlations between RFI and reproduction and health traits if included in the breeding goal were ignored. A genetic standard deviation of RFI in growing cattle and lactating cows was assumed to be 0.058 kg

Table 20.4. Genetic gain achievable (in standard deviation units per generation) for alternative scenarios of accuracy of selection.

Selection pathway	Generation interval	Percentage selected	Selection intensity	Scenario 1		Scenario 2	
				Accuracy of selection	Intensity × accuracy	Accuracy of selection	Intensity × accuracy
Sires of sires	8.15	1%	2.66	0.5	1.33	0.9	2.39
Dams of sires	3.94	20%	1.40	0.2	0.28	0.2	0.28
Sires of dams	7.63	1%	2.66	0.5	1.33	0.4	1.06
Dams of dams	4.03	90%	0.20	0.2	0.04	0.9	0.18
Σ	23.75				2.98		3.91

DM/day (~11-month-old cattle; Berry and Crowley, 2012) and 10.87 MJ/day (Veerkamp *et al.*, 1995; assumed to equate to 1 kg DM/day), respectively. Assuming an accuracy of selection of 0.50 or 0.90, after 10 years' selection, beef animals would be expected to eat, on average, 0.64 to 1.09 kg DM per day less with no apparent effect on performance. Assuming the genetic standard deviation at 11 months of age is the average across a growing animal's lifetime up to 24 months at slaughter (assumes the animal eats only solid food for 18 months for simplicity of illustration), this equates to a reduction in lifetime feed intake after 10 years of between 349 and 598 kg DM per animal. Assuming an accuracy of selection of 0.50 or 0.90, after 10 years' selection, dairy cows would be expected to eat, on average, 1.25 to 2.15 kg DM less per day with no apparent effect on performance. Assuming a cow produces for on average 3.5 lactations, this equates to 2068 to 3545 kg DM over the lifetime of cow (including the period of 24 months as a replacement heifer based on the data generated above for a growing animal but now for 24 months). Hence, considerable gains in reducing feed without impacting performance could be achieved, although monitoring of the breeding programme should be undertaken to ensure no deterioration in traits not routinely measured (e.g. immuno-competence).

Genotype by environment interactions for feed intake and efficiency

Few studies have attempted to quantify the extent of GxE in feed efficiency in growing (Durunna *et al.*, 2011a,b, 2012) and lactating (Coleman *et al.*, 2010; Prendiville *et al.*, 2011) animals. Recent research from Canada evaluated the change in animal ranking for RFI, FCR and KR in beef steers (Durunna *et al.*, 2011a) and heifers (Durunna *et al.*, 2012) across time and/or diet. Irrespective of whether the animals were steers (Durunna *et al.*, 2011a) or heifers (Durunna *et al.*, 2012), re-ranking among the same animals tended to occur between two test periods, close together in the lifetime of the animals, yet based on the same diet. The heifers (*n* = 190) were fed a high roughage diet (Durunna *et al.*, 2012) and the steers (*n* = 159; steers that remained on the same diet in both test periods) were fed a higher energy density diet (Durunna *et al.*, 2011a). In both studies the animals were stratified into groups based on calculated RFI within the first test period as <0.5 standard deviations (SD) from the mean RFI, ±0.5 SD from the mean RFI and >0.5 SD from the mean RFI. In both studies, approximately 50% of animals changed RFI group from period 1 to period 2 indicating re-ranking. Re-ranking also occurred for FCR and KR. This re-ranking was despite the very short period in time between the two test periods and the same diet being fed in both periods. Durunna *et al.* (2011a) also observed re-ranking of steers for feed efficiency when the diets between both test periods changed. Although the number of animals differed between the treatments imposed, the proportion of animals re-ranking from extremely good RFI to extremely poor RFI, or vice versa, did not differ considerably between whether the diets differed between the test period or remained the same. The possible

existence of G × E interactions for feed efficiency has important implications especially
for production systems that rely heavily on
grazed grass (e.g. Ireland).

None the less, although re-ranking of animals may occur between some environments
or different stages of maturity, the necessity to
genetically evaluate animals based on performance in differing environments may not be
justifiable; this is especially true for a trait like
feed intake or efficiency, which requires considerable resources for measurement. The feasibility of a separate breeding programme is a
function of, amongst others, the genetic correlation between environments for feed efficiency as well as the intensity of selection. This
can be illustrated using a simple example of
two environments: (i) a high-input system; and
(ii) a low input system; similarly younger and
older maturity stages can be used. For the purposes of simplicity the generation interval,
genetic variance and accuracy of selection is
assumed to be identical in both environments.
The genetic correlation between feed efficiency
in both environments is assumed to be either
0.7, 0.8, 0.9 or 1.0. It is assumed that the top
10, 20 and 30 elite sires are chosen from
100 candidate sires evaluated in the high-input
environment (i.e. selection intensity of 1.76,
1.40 and 1.16, respectively). Table 20.5 summarizes the number of sires requiring evaluation in the low-input environment to achieve
similar genetic progress as obtained from using
sires evaluated in the high-input environment.
A smaller testing programme in the low-input

environment is needed to achieve similar
genetic gains from using sires selected in the
high-input environment as the correlations
between environments are reduced and selection intensity is increased. The question then
arises as to the ability of the low-input environment to generate the progeny group sizes
required to achieve a high accuracy of estimating a sire's breeding value as may be achievable in the high-input environment. This can be
easily modelled using the approach described.
Smaller progeny group sizes will reduce genetic
gain in the low input environment. Consideration
must also be given to the monetary cost of
implementing a separate breeding programme
in the low-input environment. Alternatively
stronger genetic correlations may exist between
another environment (i.e. another country) and
the low-input environment implying that more
rapid genetic gain can be achieved when using
germplasm from the alternative environment.

Genomics of Feed Intake and Efficiency

Several genome-wide association studies have
been undertaken for feed intake or feed efficiency, although most are from populations of
limited size most likely due to the large resources
required to phenotype large populations for
feed intake. Moreover, because of the lack of
primarily phenotypic data, validation of identified putative quantitative trait loci (QTLs) in an
independent population are generally not possible since maximizing the statistical power in
the detection dataset requires as many individuals as possible. Barendse et al. (2007) however
did undertake a validation of detected putative
QTLs for feed efficiency. Not validating (i.e.
re-testing detected single nucleotide polymorphism (SNP) associations) in independent populations is likely to increase the number
of false positive associations (i.e. Type I errors)
and over-predict the proportion of variation in
the phenotypes explained by the genetic markers. For example, only one segregating SNP
was common in the association studies of
Sherman et al. (2008b) and Barendse et al.
(2007) in beef cattle from Canada and
Australia, respectively; only 78 SNPs residing

Table 20.5. Number of young test sires that
must be tested in an environment to realize the
same genetic gain achieved by selecting among
100 young test sires in a different environment
with different selection intensities and genetic
correlations between environments.

Correlation	No. test sires		
	10	20	30
0.6	28	42	54
0.7	37	54	64
0.8	51	61	70
0.9	71	74	83
1.0	100	100	100

in previously reported QTL regions for RFI were evaluated in the study of Sherman *et al.* (2008b), while Barendse *et al.* (2007) used 8786 SNP genotypes. This SNP was associated (*P* = 0.0032) with RFI in the Australian population (Barendse *et al.*, 2007) but was not associated with RFI in the Canadian population (Sherman *et al.*, 2008b). This does not however imply that it was a false positive in tagging a functional mutational in the Australian population, but could simply reflect different linkage phases in the different populations but also the different contribution of underlying biological traits to differences in RFI in both populations (i.e. G × E interactions).

Using the same Canadian population of 464 steers used by Sherman *et al.* (2008b), but with a larger number of SNPs tested (i.e. 2633 SNPs) four SNPs significantly associated with RFI were also associated with RFI in the Australian population. Rolf *et al.* (2010) using a population of 698 US Angus steers documented significant associations with RFI for seven SNPs that were also reported to be associated with RFI in an Australian (Barendse *et al.*, 2007) and a Canadian (Nkrumah *et al.*, 2007) population; these SNPs were located on BTA3, BTA5, BTA6, BTA12, BTA15, BTA17 and BTA21.

Overestimation of genetic marker effects can be observed when the calculated proportion of phenotypic variance in a trait explained by the genetic markers from a model assuming additive allelic effects with no epistasis (Sherman *et al.*, 2010) is considerably greater than the narrow sense heritability estimated using the same animals (Nkrumah *et al.*, 2007); this should not be the case if the estimate of heritability is correct. Validation in an independent, unrelated population should provide a better estimate of the proportion of variation in feed efficiency explained by a set of genetic markers.

Interrogation of the quantitative trait loci database (http://www.animalgenome.org/cgi-bin/QTLdb/index; release 17 April 2012) revealed approximately 60 putative QTL regions for RFI and 38 putative QTL regions for FCR in cattle. These numbers were based on assuming possible overlapping QTL regions from different studies were tagging the same QTL and therefore our numbers reported here

should only be treated as an estimate; furthermore not all available genome-wide association studies (e.g. Barendse *et al.* 2007; Pryce *et al.*, 2012) were represented in the database. Putative QTLs in QTLdb for either RFI or FCR have been identified on all chromosomes except BTA27; no QTL associated with FCR was identified on BTA14, BTA19, BTA23 or BTA29. Similarly no QTL was identified for feed intake on BTA27. Barendse *et al.* (2007) in their genome-wide association analysis for feed efficiency in beef animals reported that the SNPs they identified as being most strongly associated with RFI were represented in microRNA motifs, mRNA sequences or promoter sequences suggesting a likely role of non-genic variants in explaining phenotypic variation in feed efficiency. Despite the reported association between mitochondrial activity and RFI (Kolath *et al.*, 2006a) no association between variations in the mitochondrial genome and RFI have been identified (Kolath *et al.*, 2006b).

To our knowledge there is no genomic study for feed efficiency in lactating animals. Veerkamp *et al.* (2012) using feed intake data and genotype information (37,590 SNPs following editing) on lactating primiparous Holstein Friesian dairy cows collated from research herds in four European countries identified many putative QTLs for feed intake. Olfactory genes and genes involved in the sensory smell process were overrepresented in the regions flanking significant SNP associations. Veerkamp *et al.* (2012) suggested indoleamine 2,3-dioxygenase 2 (IDO2) as a potential positional candidate gene because of the role it plays in tryptophan metabolism. Tryptophan is an essential amino acid and has been linked to feed intake in both humans (Wolfe *et al.*, 1997) and cattle (Choung and Chamberlain, 1992).

Candidate gene studies (Sherman *et al.*, 2008a; Magee *et al.*, 2010) have struggled to conclusively identify associations between mutations in different genes and feed efficiency. Sherman *et al.* (2008a) using up to 464 beef animals did not report any statistically strong associations between their 24 SNPs in 11 candidate genes and either FCR or RFI. However, the animals used were genetically very diverse and therefore the linkage disequilibrium between loci was expected to be low, hence genotyped markers would need to be close to the functional mutation to detect strong signals. Magee *et al.* (2010) reported a relatively strong signal of a putative quantitative trait nucleotide in the Zinc finger imprinted 2 (ZIM2) gene on BTA18 with RFI in a small dataset of 97 pure-bred Limousin cattle; pedigree structure was accounted for through the use of a relationship matrix at least four generations deep, where available. Seven SNPs in the Zinc finger imprinted 2 (ZIM2) gene were associated with RFI (although two of the SNPs were in complete linkage disequilibrium).

Future Research Priorities

Accurate quantification of individual animal feed intake, ideally at a low cost on a large population of animals is required irrespective of the definition of feed efficiency or whether or not the inclusion of feed intake in a breeding goal is preferred over the inclusion of feed efficiency per se. Although the developments in available genomic tools and technologies as well as the associated statistical and mathematical algorithms are advancing at a rapid pace, direct information on the genome of an animal is unlikely to explain all of the present genetic variation, at least in the short to medium term. Therefore the necessity to generate accurate phenotypes on a large population of animals remains. Several methods to predict feed intake are currently under research, some of which show promise. McParland *et al.* (2011, 2012) proposed the use of the spectra generated from mid-infrared (MIR) spectroscopy analysis of individual milk samples as predictors of energy intake (as well as energy balance) in lactating dairy cows. McParland *et al.* (2011, 2012) reported correlations between actual and MIR predicted energy intake in Holstein Friesian dairy cows across two contrasting production systems of up to 0.75 when assessed using external validation. Similar accuracy of predicting daily dry matter intake was observed from near-infrared spectroscopy analysis of faecal samples in growing Angus bulls (Huntington *et al.*, 2011). Moreover, methods such as infrared thermography (Montanholi *et al.*, 2009), although at an early stage of

research, may have some potential in explaining at least some of the variation in RFI among animals. The correlations between the average temperature of the feet of cattle estimated from a thermal image and RFI was 0.36 to 0.43; it has to be remembered that a likely substantial proportion of the variation in 'RFI' is due to measurement error (Robinson, 2005) and therefore unity correlations are not expected. Also, the utility of group feed intakes in genetic evaluations for feed efficiency (i.e. FCR) is under investigation (Tedeschi *et al.*, 2006; Cooper *et al.* 2010). Feeding behaviour is, by its very nature, associated with feed intake; animals cannot eat without being present at the feed bunker. Berry and Crowley (2013) in their meta-analysis of the literature documented that 12 to 14% of the phenotypic variance in dry matter intake could be explained by daily feeding duration and feeding frequency. Automatic measurement of feeding behaviour is relatively simple by the use of sensor technology with a sensor on the animal and a receiver at the feed bunk.

Feed efficiency traits represented as residuals from a regression model of usually energy intake on the various energy sinks (i.e. RFI) is increasing in popularity. If all life functions that require energy were individually known and measured precisely as well as the individual animal energy cost associated with each energy sink, then the residual effect of all individuals should be zero. This, however, is not the case because: (i) not all energy sink terms are known and even if they were it would not be feasible to measure in all animals; (ii) the individual-specific energy cost of each energy sink is not precisely known; and (iii) there are likely to be errors in the measurement of the individual traits. Understanding however the components of residual-type feed efficiency traits can be useful in reducing the variance of the residual and therefore alter selection from RFI itself (with all its associated unknowns) to selection on the individual components of feed intake. Moreover, most research to date on feed efficiency has focused on animals in the linear phase of growth; considerable research and discussions need to be undertaken on the most appropriate definition of feed efficiency (if such exists) for lactating cows. Accurate knowledge of the genetic correlations between growing and lactating animals has implications for breeding programmes, especially in dairy cattle, since selection of animals on feed efficiency during the growing phase may impact the selection intensity on the breeding goal usually derived for lactating animals. This will possibly reduce overall genetic gain and thus profitability, which is contrary to the purpose of selection on feed efficiency in the first place.

Conclusions

Irrespective of definition, there is unequivocal evidence of the existence of exploitable genetic variation in feed efficiency in both growing and lactating animals. Past breeding programmes in dairy and beef cattle have indirectly improved efficiency as measured through FCR (growing animals) or feed conversion efficiency (lactating animals). There is not yet a consensus on how best to directly incorporate feed efficiency in national breeding goals and maybe alternative approaches may be more suitable for different systems. There is no doubt that feed efficiency will play an important role in feeding the growing human global population from a constrained land area and that breeding for improved feed efficiency will be a critical and fruitful component of this global strategy.

References

Arthur, P.F., Renand, G. and Krauss, D. (2001) Genetic and phenotypic relationships among different measures of growth and feed efficiency in young Charolais bulls. *Livestock Production Science* 68, 131–139.

Arthur, P.F., Herd, R.M., Wilkins, J.F. and Archer, J.A. (2005) Maternal productivity of Angus cows divergently selected for post-weaning residual feed intake. *Australian Journal of Experimental Agriculture* 45, 985–993.

Baker, S.D., Szasz, J.I., Klein, T.A., Kuber, P.S., Hunt, C.W., Glaze, J.B., Falk, D., Richard, R., Miller, J.C., Battaglia, R.A. and Hill, R.A. (2006) Residual feed intake of purebred Angus steers: effect on meat quality and palatability. *Journal of Animal Science* 84, 938–945.

Barendse, W., Reverter, A., Bunch, R.J., Harrison, B.E., Barris, W. and Thomas, M.B. (2007) A validated whole-genome association study of efficient food conversion in cattle. *Genetics* 176, 1893–1905.

Barwick, S.A., Wolcott, M.L., Johnston, D.J., Burrow, H.M. and Sullivan, M.T. (2009) Genetics of steer daily and residual feed intake in two tropical beef genotypes, and relationships among intake, body composition, growth and other post-weaning measures. *Animal Production Science* 49, 351–366.

Basarab, J.A., McCartney D., Okine, E. and Baron, V.S. (2007) Relationships between progeny residual feed intake and dam productivity traits. *Canadian Journal of Animal Science* 87, 489–502.

Basarab, J.A., Colazo, M.G., Ambrose, D.J., Novak, S., McCartney, D. and Baron, V.S. (2011) Residual feed intake adjusted for backfat thickness and feed frequency is independent of fertility in beef heifers. *Canadian Journal of Animal Science* 91, 573–584.

Beam, S.W. and Butler, W.R. (1999) Effects of energy balance on follicular development and first ovulation in postpartum dairy cows. *Journal of Reproduction and Fertility Supplement* 54, 411–424.

Berry, D.P. (2009) Improving feed efficiency in cattle with residual feed intake. In: Garnsworthy, P.C. and Wiseman, J. (eds) *Recent Advances in Animal Nutrition*. Nottingham University Press, Nottingham, pp. 67–99.

Berry, D.P. and Crowley, J.J. (2012) Residual intake and gain; a new measure of efficiency in growing cattle. *Journal of Animal Science* 90, 109–115.

Berry, D.P. and Crowley, J.J. (2013) Genetics of feed efficiency in dairy and beef cattle. *Journal of Animal Science* 91, 1594–1613.

Berry, D.P., Veerkamp, R.F. and Dillon, P.G. (2006) Phenotypic profiles for body weight, body condition score, energy intake, and energy balance across different parities and concentrate feeding levels. *Livestock Science* 104, 1–12.

Berry, D.P., Lee, J.M., Macdonald, K.A., Stafford, K., Matthews, L. and Roche, J.R. (2007) Associations among body condition score, body weight, somatic cell count, and clinical mastitis in seasonally calving dairy cattle. *Journal of Dairy Science* 90, 637–648.

Bouquet, A., Fouilloux, M.-N., Renand, G. and Phocas, F. (2010) Genetic parameters for growth, muscularity, feed efficiency and carcass traits of young beef bulls. *Livestock Science* 129, 38–48.

Choung, J.J. and Chamberlain, D.G. (1992) Protein nutrition of dairy-cows receiving grass-silage diets – effects on silage intake and milk-production of postruminal supplements of casein or soya-protein isolate and the effects of intravenous infusions of a mixture of methionine, phenylalanine and tryptophan. *Journal of the Science of Food and Agriculture* 58, 307–314.

Coleman, J., Berry, D.P., Pierce, K.M., Brennan, A. and Horan, B. (2010) Dry matter intake and feed efficiency profiles of 3 genotypes of Holstein-Friesian within pasture-based systems of milk production. *Journal of Dairy Science* 93, 4318–4331.

Collard, B.L., Dekkers, J.C.M., Petitclerc, D. and Schaeffer, L.R. (2000) Relationships between energy balance and health traits of dairy cattle in early lactation. *Journal of Dairy Science* 83, 2683–2690.

Cooper, A.J., Ferrell, C.L., Cundiff, L.V. and van Vleck, L.D. (2010a) Prediction of genetic values for feed intake from individual body weight gain and total feed intake of the pen. *Journal of Animal Science* 88, 1967–1972.

Crowley, J.J., McGee, M., Kenny, D.A., Crews, D.H. Jr, Evans, R.D. and Berry, D.P. (2010) Phenotypic and genetic parameters for different measures of feed efficiency in different breeds of Irish performance tested beef bulls. *Journal of Animal Science* 88, 885–894.

Crowley, J.J., Evans, R.D., McHugh, N., Kenny, D.A., McGee, M., Crews, D.H. Jr and Berry, D.P. (2011a) Genetic relationships between feed efficiency in growing males and beef cow performance. *Journal of Animal Science* 89, 3372–3381.

Crowley, J.J., Evans, R.D., McHugh, N., Pabiou, T., Kenny, D.A., McGee, M., Crews, D.H. Jr and Berry, D.P. (2011b) Genetic associations between feed efficiency measured in a performance-test station and performance of growing cattle in commercial beef herds. *Journal of Animal Science* 89, 3382–3393.

Dillon, P. (1989) The use of dosed and herbage n-alkanes to estimate herbage intake with dairy cows. MSc thesis, University College, Cork.

Dove, H. and Mayes, R.W. (1991). The use of plant wax alkanes as marker substances in studies of the nutrition of herbivores: a review. *Australian Journal of Agricultural Research* 42, 913–952.

Durunna, O.N., Mujibi, F.D.N., Goonewardene, L., Okine, E.K., Basarab, J.A., Wang, Z. and Moore, S.S. (2011a) Feed efficiency differences and reranking in beef steers fed grower and finisher diets. *Journal of Animal Science* 89, 158–167.

Durunna, O.N., Plastow, G., Mujibi, F.D.N., Grant, J., Mah, J., Basarab, J.A., Okine, E.K., Moore, S.S. and Wang, Z. (2011b) Genetic parameters and genotype × environment interaction for feed efficiency traits in steers fed grower and finisher diets. *Journal of Animal Science* 89, 3394–3400.

Durunna, O.N., Colazo, M.G., Ambrose, D.J., McCartney, D., Baron, V.S. and Basarab, J.A. (2012) Evidence of residual feed intake reranking in crossbred replacement heifers. *Journal of Animal Science* 90, 734–741.

Gunsett, F.C. (1984) Linear index selection to improve traits defined as ratios. *Journal of Animal Science* 59, 1185–1193.

Huntington, G.B., Leonard, E.S. and Burns, J.C. (2011) Use of near-infrared reflectance spectroscopy to predict intake and digestibility in bulls and steers. *Journal of Animal Science* 89, 1163–1166.

Johnson, K.A. and Johnson, D.E. (1995) Methane emissions from cattle. *Journal of Animal Science* 73, 2483–2492.

Kennedy, B.W., van der Werf, J.H.J. and Meuwissen, T.H.E. (1993) Genetic and statistical properties of residual feed intake. *Journal of Animal Science* 71, 3239–3250.

Koch, R.M., Swiger, L.A., Chambers, D. and Gregory, K.E. (1963) Efficiency of feed use in beef cattle. *Journal of Animal Science* 22, 486–494.

Kolath, W.H., Kerley, M.S., Golden, J.W. and Keisler, D.H. (2006a) The relationship between mitochondrial function and residual feed intake in Angus steers. *Journal of Animal Science* 84, 861–865.

Kolath, W.H., Kerley, M.S., Golden, J.W., Shahid, S.A. and Johnson, G.S. (2006b) The relationships among mitochondrial uncoupling protein 2 and 3 expression, mitochondrial deoxyribonucleic acid single nucleotide polymorphisms, and residual feed intake in Angus steers. *Journal of Animal Science* 84, 1761–1766.

Koohmaraie, M. (1992) The role of calcium-dependent proteases (calpains) in post mortem proteolysis and meat tenderness. *Biochimie (Paris)* 74, 239–245.

Koots, K.R., Gibson, J.P. and Wilton, J.W. (1994) Analyses of published genetic parameter estimates for beef production traits. 2. Phenotypic and genetic correlations. *Animal Breeding Abstracts.* 62, 825–853.

Luo Guan, L., Nkrumah, J.D., Basarab, J.A. and Moore, S.S. (2008) Linkage of microbial ecology to pheno-type: correlation of rumen microbial ecology to cattle's feed efficiency. *FEMS Microbiology Letters* 288, 85–91.

Magee, D.A., Berkowicz, E.W., Sikora, K.M., Berry, D.P., Park, S.D.E., Kelly, A.K., Sweeney, T., Kenny, D.A., Evans, R.D., Wickham, B.W., Spillane, C. and MacHugh, D.E. (2010) A catalogue of validated single nucleotide polymorphisms in bovine orthologs of mammalian imprinted genes and associations with beef production traits. *Animal* 4, 1958–1970.

Mardis, E.R. (2008) Next-generation DNA sequencing methods. *Annual Reviews. Genomics Human Genetics* 9, 387–402.

Mayes, R.W., Lamb, C.S. and Colgrove, P.M. (1986) The use of dosed and herbage n-alkanes as markers for the determination of herbage intake. *Journal of Agricultural Science, Cambridge* 107, 161–170.

McCarthy, S., Berry, D.P., Dillon, P.G., Rath, M. and Horan, B. (2007) Effect of strain of Holstein-Friesian and feed system on calving performance, blood parameter and overall survival. *Livestock Science* 111, 218–229.

McDonagh, M.B., Herd, R.M., Richardson, E.C., Oddy, V.H., Archer, J.A. and Arthur, P.F. (2001) Meat qual-ity and the calpain system of feedlot steers after a single generation of divergent selection for residual feed intake. *Australian Journal of Experimental Agriculture* 41, 1013–1021.

McHugh, N., Evans, R.D., Amer, P.R., Fahey, A.G. and Berry, D.P. (2011) Genetic parameters for cattle price and body weight from routinely collected data at livestock auctions and commercial farms. *Journal of Animal Science* 89, 29–39.

McParland, S., Banos, G., Wall, E., Coffey, M.P., Soyeurt, H., Veerkamp, R.F. and Berry, D.P. (2011) The use of mid-infrared spectrometry to predict body energy status of Holstein cows. *Journal of Dairy Science* 94, 3651–3661.

McParland, S., Banos, G., McCarthy, B., Lewis, E., Coffey, M.P., O'Neill, B., O'Donovan, M., Wall, E. and Berry, D.P. (2012) Validation of mid-infrared spectrometry in milk for predicting body energy status in Holstein-Friesian cows. *Journal of Dairy Science* 95, 7225–7235.

Montanholi, Y.R., Swanson, K.C., Schenkel, F.S., McBride, B.W., Caldwell, T.R., and Miller, S.P. (2009) On the determination of residual feed intake and associations of infrared thermography with efficiency and ultrasound traits in beef bulls. *Livestock Production Science* 125, 22–30.

Mujibi, F.D.N., Moore, S.S., Nkrumah, D.J., Wang, Z., and Basarab, J.A. (2010) Season of testing and its effect on feed intake and efficiency in growing beef cattle. *Journal of Animal Science* 88, 3789–3799.

Namkoong, G. (1985) The influence of composite traits on genotype by environment relations. *Theoretical and Applied Genetics* 70, 315–317.

National Research Council (2001) *Nutrient Requirements of Dairy Cattle*, 7th rev. edn. National Academy of Sciences, Washington, DC.

Nieuwhof, G.J., van Arendonk, J.A.M., Vos, H. and Korver, S. (1992) Genetic relationships between feed intake, efficiency and production traits in growing bulls, growing heifers and lactating heifers. *Livestock Production Science* 32, 189–202.

Nkrumah, J.D., Sherman, E.L., Li, C., Marques, E., Crews, D.H. Jr, Bartusiak, R., Murdoch, B., Wang, Z., Basarab, J.A. and Moore, S.S. (2007) Primary genome scan to identify putative quantitative trait loci for feedlot growth rate, feed intake, and feed efficiency of beef cattle. *Journal of Animal Science* 85, 3170–3181.

O'Mara, F. (2000) *A Net Energy System for Cattle and Sheep*. Version 1.2. National University of Ireland, Dublin.

Prache, S. and Theriez, M. (1999) Traceability of lamb production systems: carotenoids in plasma and adipose tissue. *Animal Science* 69, 29–36.

Prendiville, R., Pierce, K.M., Delaby, L. and Buckley, F. (2011) Animal performance and production efficien-cies of Holstein-Friesian, Jersey and Jersey x Holstein-Friesian cows throughout lactation. *Livestock Science* 138, 25–33.

Pryce, J.E., Arias, J., Bowman, P.J., Davis, S.R., Macdonald, K.A., Waghorn, G.C., Wales, W.J., Williams, Y.J., Spelman, R.J. and Hayes, B.J. (2012) Accuracy of genomic predictions of residual feed intake and 250-day body weight in growing heifers using 625,000 single nucleotide polymorphism markers. *Journal of Dairy Science* 95, 2108–2119.

Rauw, W.M., Kanis, E., Noordhuizen-Stassen, E.N. and Grommers, F.J. (1998). Undesirable side effects of selection for high production efficiency in farm animals: a review. *Livestock Production Science* 56, 15–33.

Richardson, E.C. and Herd, R.M. (2004) Biological basis for variation in residual feed intake in beef cattle. 2. Synthesis of results following divergent selection. *Australian Journal of Experimental Agriculture* 44, 431–440.

Robinson, D.L. (2005) Accounting for bias in regression coefficients with example from feed efficiency. *Livestock Production Science* 95, 155–166.

Roche, J.R., Berry, D.P. and Kolver, E.S. (2006) Holstein-Friesian strain and feed effects on milk production, body weight, and body condition score profiles in grazing dairy cows. *Journal of Dairy Science* 89, 3532–3543.

Roche, J.R., Berry, D.P., Lee, J.M., MacDonald, K.A. and Boston, R.C. (2007) Describing the body condition score change between successive calvings: a novel strategy generalizable to diverse cohorts. *Journal of Dairy Science* 90, 4378–4396.

Rolf, M.M., Taylor, J.F., Schnabel, R.D., McKay, S.D., McClure, M.C., Northcutt, S.L., Kerley, M.S. and Weaber, R.L. (2010) Impact of reduced marker set estimation of genomic relationship matrices on genomic selection for feed efficiency in Angus cattle. *BMC Genetics* 11, 24.

Rolfe, K.M., Snelling, W.M., Nielsen, M.K., Freetly, H.C., Ferrell, C.L. and Jenkins, T.G. (2011) Genetic and phenotypic parameter estimates for feed intake and other traits in growing beef cattle, and opportunities for selection. *Journal of Animal Science* 89, 3452–3459.

Savietto, D., Berry, D.P. and Friggens, N.C. (2014) Towards an improved estimation of the biological components of residual feed intake. *Journal of Animal Science* 92, 467–476.

Shackelford, S.D., Koohmaraie, M., Whipple, G., Wheeler, T.L., Miller, M.F., Crouse, J.D. and Reagan, J.O. (1991) Predictors of beef tenderness: development and verification. *Journal of Food Science* 56, 1130–1140.

Sherman, E.L., Nkrumah, J.D., Murdoch, B.M., Li, C., Wang, Z., Fu, A. and Moore, S.S. (2008a) Polymorphisms and haplotypes in the bovine neuropeptide Y, growth hormone receptor, ghrelin, insulin-like growth factor 2, and uncoupling proteins 2 and 3 genes and their associations with measures of growth, performance, feed efficiency, and carcass merit in beef cattle. *Journal of Animal Science* 86, 1–16.

Sherman, E.L., Nkrumah, J.D., Murdoch, B.M. and Moore, S.S. (2008b) Identification of polymorphisms influencing feed intake and efficiency in beef cattle. *Animal Genetics* 39, 225–231.

Sherman, E.L., Nkrumah, J.D. and Moore, S.S. (2010) Whole genome single nucleotide polymorphism associations with feed intake and feed efficiency in beef cattle. *Journal of Animal Science* 88, 16–22.

Stakelum, G. and Dillon, P. (1990) Influence of sward structure and digestibility on the intake and performance of lactating and growing cattle. In: Mayne, C.S. (ed.) *Management Issues for the Grassland Farmer in the 1990's*. Proceedings of the British Grassland Society Winter Meeting held at Malvern, Worcestershire, 26–27 November, pp. 30–44.

Sutherland, T.M. (1965) The correlation between feed efficiency and rate of gain, a ratio and its denominator. *Biometrics* 21, 739–749.

Tedeschi, L.O., Fox, D.G., Baker, M.J. and Kirschten, D.P. (2006) Identifying differences in feed efficiency among group-fed cattle. *Journal of Animal Science* 84, 767–776.

Vallimont, J.E., Dechow, C.D., Daubert, J.M., Dekleva, M.W., Blum, J.W., Barlieb, C.M., Liu, W., Varga, G.A., Heinrichs, A.J. and Baumrucker, C.R. (2011) Short communication: heritability of gross feed efficiency and associations with yield, intake, residual intake, body weight, and body condition score in 11 commercial Pennsylvania tie stalls. *Journal of Dairy Science* 94, 2108–2113.

Veerkamp, R.F. (2002) Feed intake and energy balance in lactating animals. 7th World Congress on Genetics Applied to Livestock Production, Session 10, 19–23 August, Montpellier, France, pp. 0–8.

Veerkamp, R.F. and Brotherstone, S. (1997) Genetic correlations between linear type traits, food intake, live-weight and condition score in Holstein Friesian dairy cattle. *Animal Science* 64, 385–392.

Veerkamp, R.F., Emmans, G.C., Cromie, A.R. and Simm, G. (1995) Variance components for residual feed intake in dairy cows. *Livestock Production Science* 41, 111–120.

Veerkamp, R.F., Coffey, M.P., Berry, D.P., de Haas, Y., Strandberg, E., Bovenhuis, H., Calus, M.P.L. and Wall, E. (2012) Genome-wide associations for feed utilisation complex in primiparous Holstein-Friesian dairy cows from experimental research herds in four European countries. *Animal* 6, 1738–1749.

Waasmuth, R., Boelling, D., Madsen, P., Jensen, J. and Anderson, B.B. (2000) Genetic parameters of disease incidence, fertility and milk yield of first parity cows and the relation to feed intake of growing bulls. *Acta Agriculture Scandinavia Section A. Animal Science* 50, 93–102.

Wolfe, B.E., Metzger, E.D. and Stollar, C. (1997) The effects of dieting on plasma tryptophan concentration and food intake in healthy women. *Physiology and Behaviour* 61, 537–541.

Wulf, D.M., Tatum, J.D., Green, R.D., Morgan, J.B., Golden, B.L. and Smith, G.C. (1996) genetic influences on beef longissimus palatability in Charolais- and Limousin-sired steers and heifers. *Journal of Animal Science* 74, 2394–2405.

Wulfhorst, J.D., Ahola, J.K., Kane, S.L., Keenan, L.D. and Hill, R.A. (2010). Factors affecting beef cattle producer perspectives on feed efficiency. *Journal of Animal Science* 88, 3749–3758.

Zhou, M., Hernandez-Sanabria, E. and Luo Guan, L. (2009) Assessment of the microbial ecology of ruminal methanogens in cattle with different feed efficiencies. *Applied and Environmental Microbiology* 75, 6524–6533.

21 Genetics of Growth and Body Composition

Michael L. Thonney

Cornell University, Ithaca, New York, USA

Introduction

Growth and body composition are inextricably linked. As in most mammals, cattle growth follows a sigmoid curve that is defined primarily by adult size. Body composition among breeds, strains and genders of cattle is similar at any point along the growth curve. Cattle breeders have exploited phenotypic variation to change growth since Bakewell first wrote about selective breeding in the eighteenth century (Orel and Wood, 1998; Cobb, 2006). Most selection for growth changed the growth curve by decreasing or increasing adult size. Selection rarely changed the shape of the growth curve or the underlying biological relationships between growth and body composition. Heritabilities and genetic correlations of carcass traits are given in Chapter 22. The objective of this chapter is to review the relationships of traits commonly used for selection to the underlying biological associations among growth, adult size and body composition.

Growth Traits and Growth Curves

A multibreed experiment examined growth and feed intake of 25 British breeds from birth to 72 weeks of age (Thiessen *et al.*, 1984). Weights and individual feed intakes from 292 heifers (5 to 16 per breed) fed *ad libitum* were obtained every 2 weeks from weaning at 12 weeks. The resulting growth curves were slightly sigmoid-shaped and very uniform for each breed. Except for a few cases of crossing over of mid-range breed weights, breeds that were heavier at 12 weeks were heavier at all ages and vice versa. Observed variation in growth and feed intake in these animals can be partitioned into components representing the variation among breed means, which is genetic, and the variation within breeds, which results from both genetic and environmental causes. Proportion of total variation of body weight accounted for by breeds was 0.44, 0.47, 0.59, 0.68, 0.70 and 0.71 at 12, 24, 36, 48, 60 and 72 weeks of age, respectively. Proportion of total variation of feed intake accounted for by breeds was lower at 0.25, 0.47, 0.53, 0.45

and 0.48 during the intervals from 12 to 24, 24 to 36, 36 to 48, 48 to 60 and 60 to 72 weeks of age, respectively. Results from this multi-breed experiment suggest that genetic variation in projected adult weight accounts for much of the variation in weight and feed intake during growth.

Cattle growth is measured most frequently by a few weights per animal along the growth curve. Traditionally, as indicated in Fig. 21.1, these include birth weight and – using linear age adjustments – weaning weight at 205 days of age, and yearling weight at 365, 452 or 550 days of age (Crews et al., 2010). The growth curves in Fig. 21.1 were taken from the literature based upon the post-natal equation from Brody (1945):

$$W_t = A(1-be^{-kt}) \qquad (21.1)$$

Where W_t is the weight at a given time in days (t), A is the asymptotic or adult weight, b is an estimate of the time-scale parameter, e is the base of the natural logarithm and k is the logarithmic rate of change of weight per day. The Brody equation does not model the early inflection point and is progressively less reliable for weights less than 30% of adult weight (Taylor, 1980b). In most datasets, insufficient young weights have been recorded to capture the

early inflection point because it is relatively close to birth weight in cattle.

The curves of the four lightest groups of cows in Fig. 21.1 are depictions of the equations of DeNise and Brinks (1985) for Hereford and Red Angus cows and from the equations of Beltran et al. (1992) for Angus cattle. Six growth curves from more recent, heavier cows depict equations from Freetly et al. (2011). The purpose of Fig. 21.1 is to demonstrate that cattle growth traits are all snapshots of the underlying growth curve. Thus, weights at different ages are necessarily highly correlated. The first derivative of the Brody equation is:

$$ADG = Abke^{-kt} \qquad (21.2)$$

This provides the expected value of the instantaneous rate of gain at a given number of days of age. The growth curve legend in Fig. 21.1 is ordered within study by estimated adult size, A. The shapes of the growth curves are similar. Growth rates at a given age (Table 21.1) generally are related to adult size. Notable exceptions in Fig. 21.1 are the growth curves for the Tuli and Boran cows, which had faster earlier growth than expected for their predicted adult sizes, indicating some potential to alter growth rate independent of adult size.

Fig. 21.1. Post-natal growth curves based upon Eq 21.1 of Brody (1945) with parameters from the literature for cows (Beltran et al., 1992; DeNise and Brinks, 1985; Freetly et al., 2011) or from Taylor's (1980b) standardized growth equation at adult weights of 500 or 600 kg. The equations are applicable at greater than 30% of adult weight (Taylor, 1965).

Table 21.1. Brody (1945) and Taylor (1980b) growth curve equation parameter estimates and growth rates at Beef Improvement Federation (Crews *et al.*, 2010) ages.

Reference	Cow group	Equation parameter estimates			Age (days)			
		A	b	k	205	365	452	550
					------ ADG, g/day ------			
Freetly *et al.*, 2011	Hereford-sired	598	0.918	0.00214	758	538	447	362
	Angus-sired	575	0.923	0.00233	767	529	432	344
	Brahman-sired	576	0.906	0.00203	699	505	423	347
	Belgian Blue-sired	557	0.905	0.00221	709	497	410	330
	Tuli-sired	524	0.908	0.00207	645	463	387	316
	Boran-sired	515	0.899	0.00216	642	454	377	305
DeNise and Brinks, 1985	Linecross	522	0.940	0.00185	622	462	394	328
	Inbred	495	0.943	0.00177	574	433	371	312
Beltran *et al.*, 1992	Line A Angus	457	0.910	0.00169	497	379	327	277
	Line K Angus	429	0.910	0.00185	494	368	313	261
Taylor, 1980b	Standardized	500			623	462	393	327
	Standardized	600			734	552	473	397

Expected growth rates in Table 21.1 are slower than typically observed growth rates that approach 2 kg/day in crossbred feedlot cattle out of terminal sires with heavy adult weights. Cattle growth weights are influenced by many environmental factors, including digestible feed intake, stress due to inclement weather, weighing conditions and physical exertion. After accounting for these and other environmental factors in genetic evaluations, weights derive from the underlying genetic growth curves of individual cattle. Therefore, selection to increase weight or growth rate at a given age is associated with heavier adult weight with consequent increased breeding herd feed requirements and the prospect of reduced fertility (Owens *et al.*, 1993; Arango and Van Vleck, 2002).

Scientists have explored the possibility of modifying the dependence of the growth curve on adult weight (Eisen, 1976; Fitzhugh, 1976). Fitzhugh (1976) indicated that changing the shape of cattle growth curves could:

- 'Resolve genetic antagonisms between fast, efficient, early growth rate of animals for meat with smaller size and lower maintenance cost of adult cows (Cartwright, 1970; Dickerson *et al.*, 1974; Gregory, 1965).
- Improve efficiency by increasing the rate of maturation (Taylor and Young, 1966; Blaxter, 1968) and, thus, lower age to first breeding.
- Reduce calving difficulty (Monteiro, 1969).'

Fitzhugh (1976) reported that from 10 to 92% of genetic variation in maturing rate (k) was independent of adult weight (A). Eisen (1976) thoroughly reviewed the literature on changing the growth curves of mice and rats, where far shorter generation intervals allow for more rapid change. He concluded that 'Selection for increased preweaning or early postweaning growth while holding mature size constant or decreasing it presents formidable problems.' Recently, Boligon *et al.* (2012) modelled growth of Nelore cattle using a random regression model. They concluded that: 'The likelihood to modify the growth curve in Nelore cattle in order to obtain animals with fast growth at young ages and moderate-to-low mature cow weight is restricted.' The major obvious alternative to selection for increasing growth rate of feeder cattle is to maintain moderate-sized cows and breed a portion of the older cows to larger terminal sires, perhaps using a rotational-terminal breeding system (Cundiff and Gregory, 1977).

Bull selection based upon expected progeny differences for heavier weaning weight and lighter yearling weight might favourably change the shape of the cattle growth curve. Genetic trends indicate that cattle breeders have not given this priority except to maintain birth weight while weights recorded at later ages have increased. The example for Angus cattle in Fig. 21.2 shows that weaning and yearling weight expected progeny differences (EPDs) increased linearly, while – after about 1992 – birth weight EPDs levelled off and even declined slightly. Surprisingly, while yearling weight EPDs increased at twice the rate of weaning weight EPDs, yearling height EPDs followed the same trend as birth weight EPDs. Given that bone growth is early maturing compared to weight (Taylor, 1985), this may indicate that in recent years Angus breeders have increased early rate of growth while holding adult size constant.

Another approach to account for adult size when selecting for faster growth is to assume an average breed adult weight and use weights from young ages to estimate time taken to mature (Taylor and Fitzhugh, 1971). Selecting cattle with shorter than expected times to mature could result in fast early growth without increasing adult size. This approach to account for adult size differences ultimately led Taylor (1980a, 1985) to propose the following genetic size-scaling rules:

- 'Convert all age and time variables to metabolic age, $\ominus$, by dividing days since conception minus 3.5 days (when a fertilized egg enters the uterus) and scaling by $A^{0.27}$, where A is adult body weight.
- Convert all cumulated variables such as weight and feed intake to proportions of A. The proportion of weight to A is degree of maturity, μ.'

In a companion paper, Taylor (1980b) applied these genetic size scaling rules to convert Brody's post-natal growth equation to include A as the only parameter:

$$W_t = A(1-e^{-0.01A^{-0.27}t+0.50})$$ (21.3)

Where W_t is the weight at a given time in days (t) since 3.5 days after conception, A is the asymptotic or adult weight and e is the base of the natural logarithm. Growth curves for A of 500 kg (283 days gestation) and A of 600 kg (290 days gestation) are in Fig. 21.1. The curve for 500 kg is remarkably similar to that derived from cows of moderate adult size (DeNise and Brinks, 1985; Beltran et al., 1992). The birth weight predicted by the curve

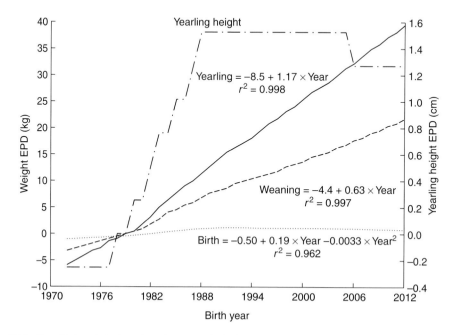

Fig. 21.2. Genetic trends in American Angus cattle for birth weight, weaning weight and yearling weight, and height EPDs by birth year along with linear or, if significant, quadratic equations. (From: the American Angus Association, 2013.)

for 600 kg begins at a lighter than expected amount; therefore the curve is steeper than the curve from the larger cattle of Freetly *et al.* (2011). Assuming that Taylor's (1980b) equation represents expected growth for most mammalian species, the growth curves of the larger cows (Freetly *et al.*, 2011) can be assumed to: (i) deviate in a positive genetic fashion for faster earlier growth than expected based upon their adult sizes; and (ii) possibly reflect high intake of digestible feed in early growth.

The rates of gain in Table 21.1 for Taylor's (1980b) growth equation were computed from the first derivative of Eq 21.3:

$$ADG = 0.01A^{0.73}e^{0.5-0.01t/A^{0.27}} \qquad (21.4)$$

For adult weight of 500 kg, the growth rates from Taylor's (1980b) growth equation were similar to those of DeNise and Brinks (1985), which had adult weights near 500 kg. At an A of 600 kg, the growth rates from Taylor's (1980b) growth equation were slightly slower at lighter weights and slightly faster at heavier weights than those of the 598-kg Hereford cows of Freetly *et al.* (2011) shown in Table 21.1, reflecting differences in growth curve shape. The cattle of Freetly *et al.* (2011) were several decades more recent than those from DeNise and Brinks (1985). Although their adult weights were heavier, their early growth was faster than expected compared to the generalized growth equation of Taylor (1980b). This may indicate that selection for early growth in recent decades has not increased adult weight as much as anticipated based upon genetically standardized growth, or it could reflect more digestible feed intake during early growth. Cattle breeders should be cognisant of the relationship of various growth and body composition (see below) traits to adult weight. One way to do this would be to fit the five Beef Improvement Federation-adjusted (Crews *et al.*, 2010) weights (Fig. 21.1) to Eqn 21.1 or to use an estimate of adult weight – perhaps based upon the adult weights of parents – in Eqn 21.3 to calculate growth curves for individual cattle. Those that deviate positively from standardized curves for early growth compared to their expected adult weights could be selected as replacements (Taylor, 1980b; Arango and Van Vleck, 2002).

The fitting of random regression models (Meyer, 1999) is a more recent alternative to jointly analysing growth curves of individual animals. Using this approach with B-spline functions to model growth of Nelore cattle (Boligon *et al.*, 2012) resulted in heritabilities from 0.23 for birth weight to about 0.38 for adult weight. Although the direct genetic correlations decreased as the time between weights increased, they were higher than 0.5 for most ages, indicating that selecting for heavy weights at young ages would increase adult weights (Boligon *et al.*, 2012). A subsequent paper examined genetic trends in Nelore cattle (Boligon *et al.*, 2013). While genetic trends for weaning index scores (heavily weighted for gain from birth to weaning) and for yearling index scores (heavily weighted for gain from birth to weaning and for post-weaning gain) were 0.26 and 0.27 genetic standard deviation units per year, mature cow weight increased only 0.01 genetic standard deviation units (0.35 kg) per year. These results imply that it is possible to select for heavier weights at young ages without dramatically increasing adult weight.

Body Composition

Body composition is directly related to stage of growth (immature weight, W, divided by mature weight, A; given the symbol μ) and therefore adult size because animals at earlier μ are leaner than animals at later μ (Berg and Butterfield, 1968). Chemical composition is the proportion of water, protein, lipid and ash in the empty (gut contents removed) body. The chemical composition of steers from Haecker (1920) is plotted in Fig. 21.3. Water and protein (the primary chemicals in muscle) and ash (primarily in bone) increase with increasing cattle weight until they plateau to adult amounts, while lipid (the primary chemical in fat tissue) is gained at an increasing rate. Chemical composition is important when defining nutritional needs for cattle and to describe nutritional value of meat from cattle, but cattle breeders are primarily interested in physical composition. The physical components of cattle carcasses, however, reflect the underlying chemical composition shown in Fig. 21.3.

Physical composition is the proportion of carcass and non-carcass components. Non-carcass components include partially digested

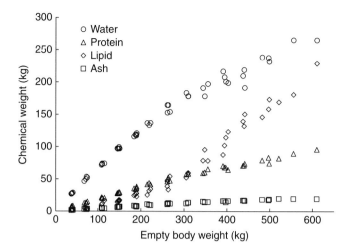

Fig. 21.3. Chemical components in the empty body (gut contents removed) of 49 beef steers. (From the data of Haecker, 1920.)

feed in the stomach and intestines, visceral fat, organs, hide, head, feet and legs. More angular dairy breeds have a higher proportion of non-carcass components, especially visceral fat (as reviewed by Comerford *et al.* (1992)) in the live weight (Nour *et al.*, 1983), but few differences were reported among beef breeds (Gregory *et al.*, 1994). Carcasses are divided into muscle, fat and bone.

Muscle, fat and bone

Proportions of muscle, fat and bone are the primary determinants of carcass value. The key genetic determinant of response to selection to change carcass value is the extent of genetic variation in proportions of these tissues at economical carcass weights. A large proportion of this genetic variation is associated with degree of maturity, μ. A recent report concluded that even the effect of prenatal maternal under-nutrition on carcass characteristics were minimal when adjusted for weight and sire breed (Robinson *et al.*, 2013), which constitute μ.

Butterfield (1983) described maturity coefficients for muscle, fat and bone of Merino sheep, which he then presented in a figure showing tissue growth as a proportion of mature tissue plotted against μ (Butterfield,

1988). Taylor and Murray (1987) called this the genetically standardized allometry square. This concept is presented in Fig. 21.4 to demonstrate the proportions of muscle, fat and bone as carcass weight increases. As cattle gain carcass weight, muscle and bone weights increase at decreasing rates (Fig. 21.4). Therefore – because muscle, fat, and bone must sum to carcass weight – fat in the carcass must increase at an increasing rate. A genotype with a larger adult size has more muscle and slightly more corresponding bone to gain. Therefore, at a given carcass weight, a genotype with a larger adult size will be leaner than a smaller genotype (Fig. 21.4). The data in Fig. 21.5 show that Friesian cattle, with larger potential adult size than the Hereford cattle of the time, had more muscle and bone and less fat at a given carcass weight.

Sex differences are related to μ. Bulls grow to be larger than heifers. Unless steers are allowed to get very old, they grow to weights intermediate of bulls and heifers. As shown in Fig. 21.6, muscle, fat and bone in bull, steer and heifer carcasses grow as expected based upon the genetically-standardized allometry square scaled for weight in Fig. 21.4. Individual data from the dissection of the 9th through 11th rib section by Nour *et al.* (1981) for traditional small Angus compared with Holstein steers are shown in Fig. 21.7. Those data are consistent with the effect of

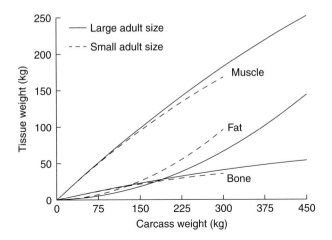

Fig. 21.4. Transformation of the allometry square (Taylor and Murray, 1987) to demonstrate growth of cattle carcasses to mature proportions of 56, 32 and 12% (Berg and Butterfield, 1976) using maturity coefficients (q) of 1.25, 0.073 and 1.40 (Butterfield *et al.*, 1983) for muscle, fat and bone, respectively. Solid lines represent a large adult-sized animal with a 450-kg carcass. Dashed lines represent a small adult-sized animal with a 300-kg carcass. Note that muscle, fat and bone weights must add to carcass weight.

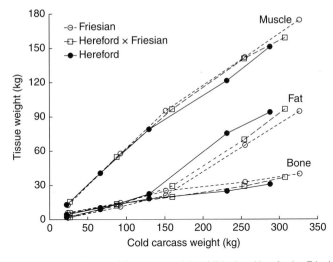

Fig. 21.5. Growth of tissues relative to cold carcass weight of Friesian, Hereford × Friesian and Hereford steers. (From the data of Gooch, 1966.)

adult size demonstrated in Fig. 21.4. Carcass cutout data often are reported (Gregory *et al.*, 1995; Pabiou *et al.*, 2009). Without anatomical dissection, such data are difficult to interpret biologically because they are fraught with differences in extent of fat trimmed and bone removed. Ultrasonic (Nalaila *et al.*, 2012) or video imaging (Pabiou *et al.*, 2011) data are even further removed from describing the underlying biology. Such data may, however, provide estimates of genetic covariation that are required for genetic prediction of carcass traits adjusted for the influence of adult size.

The growth patterns of carcass muscle, fat and bone in relation to μ have many consequences. Carcasses from less mature cattle yield a higher proportion of lean cuts. Consuming meat from such animals is important in obese human populations where lower energy diets are desirable. In human populations where highly digestible, nutrient-rich food is scarce, consumption of meat from more mature cattle may be more desirable. This is particularly so when the cattle can be fed material that is unfit for human consumption, such as native pasture. Thus, the objectives of cattle breeding and the importance of adult size depend upon the human population for which the cattle are being selected.

Beef cattle breeders tend to focus on increasing muscling because lean meat is the primary end product. Muscling is a nebulous term. Until the middle of the 20th century, muscling was a visual perception of meatier cattle. Meatier cattle were those that were small and blocky, representing selection for

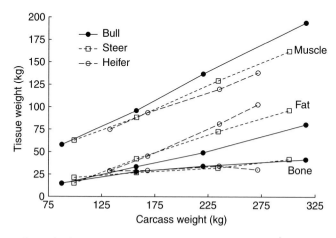

Fig. 21.6. Patterns of muscle, fat and bone growth of heifers, steers and bulls. (From the data of Berg *et al.*, 1979.)

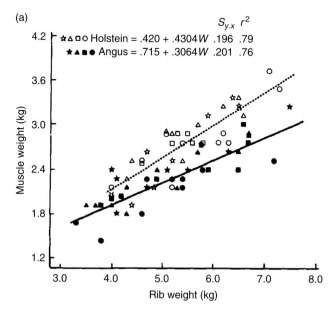

Fig. 21.7. Muscle (a), fat (b) and bone (c) in the 9th through 11th rib section of small-type Angus and Holstein cattle. (From: Nour *et al.*, 1981.)

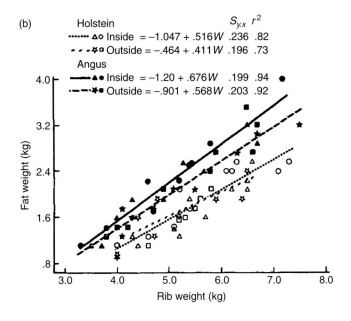

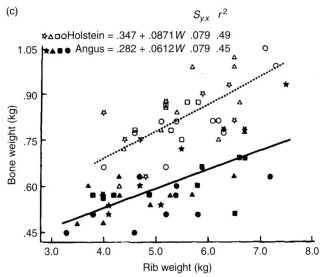

Fig. 21.7. Continued.

fatness and not muscle. The physical work that most people did at least until the early 20th century used the extra calories provided by consumption of the fat. This can be inferred from Haecker's (1920) monumental investigation of the composition of cattle from birth to 700 kg. Although Haecker (1920) carefully recorded weights of gut contents, inedible offal, edible offal, bone, cartilage and tendons, blood and skin, the entire boneless carcass was described as flesh with no division into muscle and fat. By the late 1960s nutritionists and medical practitioners had started to caution the more sedentary population about high caloric intake. This, combined with exportation of larger European continental breeds to the

Americas, New Zealand and Australia, brought about rapid change in beef cattle selection. Emphasis was placed upon faster growth and leaner carcasses, both accompanied by concomitant selection for larger adult size (Figs 21.3–21.7).

Body composition, and therefore – μ – drives efficiency of gain. About 75 kJ of metabolizable energy is required to gain a gram of either pure protein or pure lipid (Thonney et al., 1991a). Muscle tissue contains about 70% water, while fat tissue contains only about 20% water. The remainder is mainly protein and lipid. Therefore, a unit of fat gain requires 2.67 times as much metabolizable energy as a unit of muscle gain. At lighter weights, before much fat has accumulated, there is not much difference in composition. This was reflected by the lower proportion of variation in weight and feed intake accounted for by the 25 breeds of Theisen et al. (1984) at young ages compared with older ages, indicating that environmental effects were relatively more important at light weights. At comparable weights, cattle with larger potential adult sizes are less mature (earlier μ) than cattle with smaller potential adult sizes, so weight they gain contains more muscle and less fat (Fig. 21.4). At earlier μ, cattle may have slightly higher maintenance requirements, but they consume at least as much feed as cattle at later μ of the same weight (Thonney et al., 1981; Taylor et al., 1986), so the cost of maintenance and gain is diluted by more water (muscle) gain.

Muscling

In 1965 a regression equation to predict lean meat yield based upon carcasses measured in the 1950s (Murphey et al., 1960) was implemented in USDA grading standards. Although it has since been shown that the equation results in biased predictions across breed types representing different adult sizes and shapes (Crouse et al., 1975; Thonney, 2003; Lawrence et al., 2012), the equation is still used a half-century later. The equation includes cross-sectional area of the longissimus dorsi muscle (rib eye area) as a positive factor and

three negative factors: carcass weight; estimated kidney, pelvic and heart fat percentage; and subcutaneous fat thickness.

Crouse et al. (1975) analysed data from 786 steers sired by Angus, Hereford, Charolais, Limousin, Simmental, South Devon and Jersey bulls. Rib eye area had a phenotypic correlation of only 0.18 with actual percentage of the carcass that was retail cuts trimmed to 8 mm of fat. The data shown in Fig. 21.8 are from a study that compared muscle, fat and bone in the 9th through 11th rib section representing the same anatomical proportion of Holstein and traditional Angus carcasses (Nour et al., 1981). At the same rib eye area, Holstein rib sections contained more muscle than Angus rib sections, indicating that the Holstein rib sections (and the carcasses) were longer. This demonstrates the fallacy of using a two-dimensional measurement to represent a three-dimensional quantity. Rib eye area could be used as an indicator of muscle shape, but it is unlikely that muscle shape adds value to meat cuts (Thonney et al., 1991b) so rib eye area should not be used to predict muscle yield.

Based upon the relationships in Fig. 21.4, carcass weight is a logical predictor of muscle yield for genotypes with similar adult weights. As is obvious from the relationships in Fig. 21.4, however, carcass weight is a biased predictor of muscle yield across genotypes with different adult weights. Carcass weight could be used to predict muscle yield within genotypes of similar adult weights, but if selection is from among cattle with different potential adult weights, it should not be used.

Kidney, pelvic and heart fat percentage should be a good negative predictor of muscle yield, but this is only subjectively estimated in most carcasses. In addition, these fat depots should be removed during carcass processing and not included in carcass weight (Savell et al., 1989). This subjective estimate should not be used as a predictor of muscle yield.

Subcutaneous fat thickness directly represents most of what is removed from the carcass to obtain muscle yield. It is unbiased and measured easily. Of the four factors used in the USA, fat thickness is the best indicator of the proportion of the trimmed carcass that will be muscle. Professor Rex Butterfield (personal

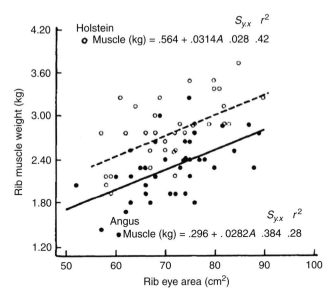

Fig. 21.8. Relationship of dissectible rib muscle weight in the 9th through 11th rib section to rib eye area. (From: Nour *et al.*, 1981.)

communication, 2004) offered the following observation about evaluation of beef carcasses for yield:

> Don Charles was the senior vet officer in charge of Cannon Hill Meat Works in Brisbane on the top floor of which was the original CSIRO Meat Research Lab. where I did most of my dissection. He eased my way through the problems of the meat works and spent lots of time chatting to me while I dissected. He eventually left the meat inspection service and became a staff member of the Vet School. I was reminded of him by a sentence in your notes re the importance of fat as an index of carcass composition. Way back in 1963 at a conference in Melbourne he was asked what information he needed for a commercially useful description of a carcass??? His answer was: 'Carcass weight and fat thickness. Fullstop.'

Nevertheless, the USDA yield grade prediction itself, as well as two of the underlying factors (rib eye area and fat thickness), have been adopted by most breed associations as traits indicating carcass value. Ultrasound methods developed in the 1960s (Stouffer *et al.*, 1961) have been widely used to predict these traits along with intramuscular fat (marbling score) in live animals. Instrument grading of carcasses has enabled large amounts of data to be

transmitted to genetic databases (Gray *et al.*, 2012). Despite wide use of USDA grading factors for selection, muscling should be defined by more precise quantitative methods.

One quantitative definition of muscling is proportion of the carcass that is muscle. Because fat increases as μ increases, muscle yield is mostly defined by adult size for the range of acceptable carcass weights (Fig. 21.4). An example is a comparison, at three fat trim levels, of the yield of lean meat from carcasses of English, European continental, *Bos indicus* and Holstein steers (Knapp *et al.*, 1989). When external fat was trimmed to leave 25.4 mm, Holstein carcasses had lower yields of major cuts because the other breed types were credited with the fat up to 25.4 mm that the Holstein carcasses did not have. When fat was trimmed to 6.4 mm, however, Holstein yields were numerically higher than English breed yields and there were no significant breed differences. Because the potential adult size of Holstein steers was likely to have been greater than that of English breed cattle, this demonstrates the effect of μ on fat shown in Fig. 21.4. Therefore, genetic differences in muscle yield should be identified that are independent of μ.

In addition to genetic variation in yield of carcass muscle that is independent of μ, another quantitative definition of muscling is the proportion of dissectible muscle that is in the most valuable cuts of the carcass. Many aspects of this topic were examined in detail by Berg and Butterfield (1976). An example is shown in Table 21.2 for Hereford, Shorthorn, and Hybrid genotypes selected for beef and Brown Swiss, Holstein, and Jersey breeds selected for milk. There was little variation in muscle distribution among breeds differently shaped and from widely disparate selection strategies.

Muscle weights as percentages of dissected muscles from Holstein and Hereford cattle are shown in Table 21.3 from the data of Garcia-de-Siles et al. (1977b). There were few differences in muscle distribution. Kauffman et al. (1973) compared lipid-free muscle weights of the pelvic limbs of 12 'muscular' (beef breed) with those from 12 'non-muscular' (dairy breed) cattle. There were no differences between cattle types in proportions of total pelvic muscle represented by biceps femoris,

Table 21.2. Distribution of muscle weight in breed groups of bulls and steers. (Modified from Table 5.7 of Berg and Butterfield, 1976; eCommons@cornell: http://hdl.handle.net/1813/1008.)

Item	Hereford	Shorthorn crossbred	Hybrid and other crosses	Holstein	Jersey
Bulls					
Number of animals	13	12	22	8	8
Days of age	461	361	430	386	407
Live weight (kg)	465	386	489	415	294
Expensive muscles (%)	53.2	53.3	53.2	53.7	53.2
Hind quarter (%)	46.7	47.8	47.6	48.7	47.6
Fore quarter (%)	53.3	52.2	52.4	51.3	52.4

	Hereford	Shorthorn crossbred	Hybrid and other crosses	Brown Swiss crossbred	Holstein
Steers					
Number of animals	11	22	32	14	6
Days of age	402	383	434	404	480
Live weight (kg)	373	376	461	456	466
Expensive muscles (%)	54.3	54.2	54.4	54.7	54.4
Hind quarter (%)	50.2	49.9	49.5	49.7	49.5
Fore quarter (%)	49.8	50.1	50.5	50.3	50.5

Table 21.3. Individual muscle weight as a percentage of dissected muscle weight in medium- and heavy-weight Holstein and Hereford carcasses. (Adapted from Garcia-de-Siles et al., 1977b.)

	Holstein		Hereford		SD	Breed P-value
Carcass weight group (kg):	227	275	237	280	12.3	ns
Muscle	- - - - - - - % of dissected muscle - - - - - -					
Adductor	5.4	5.3	5.2	4.7	0.23	ns
Quadriceps femoris	17.9	17.4	16.6	16.1	2.16	ns
Biceps femoris	19.5	18.6	18.8	19.1	1.41	ns
Semitendinosus	7.8	8.0	7.3	7.3	0.61	<0.05
Semimembranosus	15.8	16.1	16.0	16.0	1.40	ns
Longissimus	18.9	20.3	21.1	20.4	1.79	ns
Triceps brachii laterale	3.1	3.4	3.5	5.6	1.05	<0.01
Triceps brachii longum	11.6	10.9	11.5	10.8	1.19	ns

gastrocnemius, quadriceps femoris, semimembranosus and adductor or the remainder of the muscles.

These small or non-existent differences in muscle distribution among widely disparate breeds cast doubt about whether there is enough measurable genetic variation for selection to increase the proportion of muscle in high value parts of the carcass. Currently, carcass traits defined by factors in the USDA yield grade or by standardized cutting procedures may be viewed as sufficient to define traits of economic importance. But it is not clear that they will always represent the underlying value of cattle for human food. For example, selection based upon ultrasound and direct measurements of rib eye area as well as production of larger carcasses has increased average rib eyes to the point that steaks may be too large for the average consumer (Rutherford, 2013). Although dissection studies are expensive, it is surprising that none have been done more recently to measure genetic variation in muscle distribution within and among breeds.

A higher proportion of muscle is found in the carcass – particularly in the more valuable cuts – in cattle with mutations in the bovine myostatin gene, which cause muscular hyperplasia or 'double-muscling' (Grobet et al., 1997; McPherron and Lee, 1997; Marchitelli et al., 2003). The meat is more tender than meat

from cattle without myostatin mutations (Arthur, 1995). Myostatin mutations have been selected in Belgian, French and Italian breeds. Keele and Fahrenkrug (2001) developed mating systems to reduce the cost of calving difficulty (Arthur, 1995) and pre-weaning calf mortality (Casas et al., 2004) in the production of double-muscled cattle. However, outside of Europe, cattle with the mutation do not bring a premium sufficiently high to pay for such production losses.

Muscle:bone ratio is often used as an indicator of carcass muscling (Berg and Butterfield, 1966; Purchas, 2002), but muscle:bone ratios can be dramatically misinformative of carcass value. Breed percentages of carcass muscle plotted against muscle:bone ratios demonstrate an inconsistent relationship (Fig. 21.9). Berg and Butterfield (1976) reported in their Table 1.2 that, although Friesian and Charolais × Friesian steers both had about 63.7% muscle in the carcass, the muscle:bone ratio was 4.0 in Friesians and 4.3 in Charolais × Friesians. In Table 2.5 of Berg and Butterfield (1976), percentages of muscle increased (as expected based upon μ) from heifers (54.9), to steers (63.6) to bulls (67.8). But muscle:bone ratio values were 4.9, 4.8 and 5.3, respectively. The relationship between the proportion of the carcass that is muscle and muscle:bone ratio is inconsistent because muscle:bone ratios do not account for fat differences.

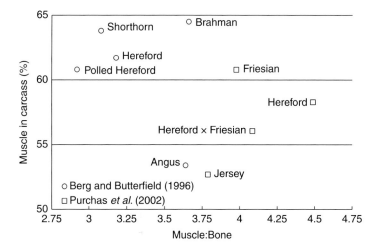

Fig. 21.9. Relationship of dissectible carcass muscle to muscle:bone ratio. (From the data of Berg and Butterfield, 1966 and Purchas et al., 2002.)

Intramuscular lipid

Intramuscular lipid, as visualized in the cut surface of the longissimus dorsi muscle between the 12th and 13th ribs, is known as 'marbling'. As described in Chapter 22, molecular genomic methods are being developed to identify genes that improve the healthfulness of fatty acid profiles in intramuscular lipid (Buchanan et al., 2013). Currently, the amount of marbling is regarded as an important trait in the beef industry. Along with physiological age or 'maturity', marbling is a primary factor used to classify cattle carcasses into USDA quality grades. Quality grades are higher for carcasses with more marbling and younger physiological maturity. Maturity is based principally upon ossification of the vertebrae in the split carcass. European cattle carcasses are graded on 'muscling' (primarily based upon carcass conformation) and degree of fatness (EEC, 1991; Dunbia Supplier Group, 2013). Instead of grade categories, each Australian beef carcass is tagged with the traceback ID, weight, sex, tropical breed content, hanging method, hormonal growth promotant use, ossification, marbling, rib fat, pH and temperature, and meat colour (Meat and Livestock Australia, 2007).

Cattle of large adult size have the advantage of being less mature and leaner than cattle of small adult size at acceptable carcass weights (Fig. 21.4). Marbling is directly related to total carcass fat (Johnson et al., 1972) so the amount is directly related to μ (Fig. 21.4). Therefore, cattle with larger potential adult size would be expected to have lower marbling scores at acceptable carcass weights. Some breed differences in marbling are independent of adult size. Angus are known to exhibit higher marbling than other beef breeds, but Jersey cattle had higher marbling scores – partly reflecting their small adult size – and Holstein cattle had higher marbling scores than expected for their adult size (Marshall, 1994). The heritability of marbling is high (0.68), but selection for marbling is unlikely to improve tenderness (Dikeman et al., 2005). The genetic correlation with fat trim varied from 0.82 to 1.00 (Cundiff et al., 1971), which means that there is a conflict between marbling and muscle yield.

How much marbling is necessary for beef with good palatability attributes? Marbling explained less than 7 percentage units of the variation in tenderness from data used to create the original photographic standards for

Table 21.4. Tenderness, juiciness and flavour scores and shear force values for marbling scores of steers and heifers. (Adapted from Garcia-de-Siles et al., 1977a.)

				Marbling score groups			
Item[a]	Traces	Slight	Small	Modest	Moderate	Slightly abundant	SD
Steers							
Number	22	49	163	140	58		
Tenderness	5.4	5.5	5.6	5.6	5.9		1.18
Juiciness	5.6	5.4	5.7	5.8	6.0		0.87
Flavour	5.9	5.9	6.0	6.0	6.1		0.64
Shear force (kg)	8.3	8.2	7.9	7.8	7.5		2.33
Heifers							
Number		34	88	148	108	25	
Tenderness[b]		6.0	5.7	5.7	6.1	5.9	1.16
Juiciness		5.8	5.8	5.9	6.1	6.1	0.84
Flavour		6.0	6.0	6.1	6.3	6.2	0.70
Shear force (kg)		8.0	8.6	8.5	8.6	8.0	2.34

[a]Tenderness, juiciness, and flavour assigned scores of 1 (very unacceptable) to 9 (very acceptable) by a trained taste panel.
[b]Difference ($P < 0.01$) between modest and moderate, neither of which were significantly different from other marbling score groups.

USDA marbling scores (Wellington and Stouffer, 1959). A USDA-trained taste panel evaluation of 496 steer carcasses from a variety of breeds (Campion et al., 1975) showed that marbling explained from 6 to 9 percentage units of variation in tenderness, from 3 to 4 percentage units of variation in flavour, from 9 to 10 percentage units of variation in juiciness and from 6 to 9 percentage units of variation in overall acceptability. Across a wide weight range for traditional Angus and Holstein cattle, marbling explained 3.6, 2.0, 1.2 and 1.0% of the variation in trained taste panel assessments of flavour, juiciness, tenderness and the number of chews before 2-cm cubes were ready to swallow (Armbruster et al., 1983). The data from Garcia-de-Siles et al. (1977a) in Table 21.4 show that the effect of marbling on sensory attributes or shear force values was inconsistent or non-existent. Smith et al. (1987) collected trained taste panel sensory ratings for steaks from 1005 carcasses across the entire range of USDA quality grades. Prime (high marbling), Choice, Good and Standard (very little marbling) grades accounted for 30 to 38% of the variation in flavour, tenderness, juiciness and overall palatability for loin steaks. But only 3.9 to 14.3% of the variation in these palatability attributes in top round steaks could be accounted for by USDA quality grade and none of the variation was explained for bottom round or eye of round.

Neither trained taste panel evaluations nor mechanical shear force values were highly correlated with in-home consumer ratings of steaks in top choice (modest and moderate marbling), low choice (small marbling), high select (slight+ marbling) and low select (slight− marbling) USDA quality grade categories (Lorenzen et al., 2003). Neely et al. (1998) summarized consumer rating data from the same study and found significant USDA quality grade (marbling) × carcass cut and USDA quality grade × city interactions for overall like, tenderness, flavour and flavour intensity of steaks. The effect of USDA quality grade within cities was small and inconsistent. Consumers in 540 households across three cities evaluated top loin steaks for overall desirability ratings using a 9-point scale (Savell et al., 1987). Regressions on marbling scores

ranging from traces to slightly abundant were highly significant ($P < 0.001$), but the effect was minor. Mean overall desirability ratings for traces were 6.8 in San Francisco, 6.9 in Kansas City and 6.7 in Philadelphia, while they were 7.3 for slightly abundant marbling in all three cities. Neely et al. (1998) concluded that: 'Findings from this study and that of Smith et al. (1987) suggest that USDA quality grade may be limited in the sorting of products for the marketplace derived from the longissimus muscle, and that it has less effect on the remaining major muscles of the beef carcass.' Yet, beef packers continue to discount carcasses with marbling below small (low choice quality grade) so that breeders continue to select for higher marbling or predictions of higher marbling from ultrasound measurements and genotyping (Elzo et al., 2013).

Genetic Markers

The generation interval is dramatically increased by the wait for weights at older ages needed for selection to favourably alter the shape of the growth curve. In addition, the cost of obtaining precise progeny data to effectively change body composition is high and also increases the generation interval. Selection using DNA markers may help to overcome some of these limitations to make more rapid, meaningful genetic changes in cattle growth.

Illinois researchers carried out one of the first studies to report quantitative trait loci (QTLs) that affect cattle growth and composition (Beever et al., 1990). Of the six markers they examined in a paternal half-sib family of Angus cattle, two markers significantly affected weaning weight, yearling weight, fat thickness and rib eye area. Taylor et al. (1998) found a QTL on chromosome 19 of Angus × Brahman cattle that affected fat thickness and concentration of lipid in the longissimus muscle and another QTL that influenced the proportions of saturated to unsaturated fatty acids in subcutaneous fat. A half-sib family from a Brahman × Hereford sire was examined for QTLs for growth and composition (Casas et al., 2003). Significant QTLs for birth weight were found on chromosomes

5 and 21, for rib eye area on chromosomes 5 and 6, for dissectible retail yield on chromosome 9 and for marbling score on chromosome 23.

The effects of QTLs on growth was studied using 385 heifers and 398 steers from Jersey and Limousin back-cross cattle finished in feedlots in Australia or on pasture in New Zealand (Morris et al., 2009, 2010). There were 18 traits common to both countries that had significant QTLs on a genome wise level with eight of the QTLs significant in both countries. Significant QTLs for growth traits were mostly on chromosome 14. Most of the significant carcass weight and organ weight QTLs were on chromosome 2, which contains the myostatin gene. One copy of the F941 myostatin allele, which had been found to increase carcass yield (Sellick et al., 2007), was found in the six first-cross sires used to create the backcross cattle.

Commercial Angus from 1769 registered bulls and 1622 steers in 38 half-sib families were screened for QTLs for post-natal growth and carcass traits (McClure et al., 2010). Significant QTLs were found for birth weight, fat thickness, marbling, weaning weight, rib eye area, yearling height, yearling weight, mature height and mature weight. A single QTL explained, at most, 3% of the genetic variation for a trait. But, an average of 58% of the genetic variation for each trait was explained by multiple QTLs across many of the 29 autosomes. This suggested that marker-assisted selection could be used effectively for growth and body composition traits.

Parent-of-origin (POE) QTL effects (imprinting) on growth and carcass traits were identified by Imumorin et al. (2011) in Angus × Brahman cattle. A cluster of POE QTLs on chromosome 2 were found for weight-related traits. Fat-related POE QTLs were on chromosomes that had few known human or mouse imprinted gene homologues.

The 'C' allele of a single nucleotide polymorphism (SNP) in the promoter region of the oxidized low-density lipoprotein receptor 1 (OLR1) gene was associated with decreased residual feed intake (more feed efficient) and larger rib eye area in 567 Angus, but not in 423 Charolais or 456 Angus × Charolais steers (Vinsky et al., 2013). The authors suggested that the effect of the 'C' allele was related to body composition. Angus cattle had smaller potential mature size than the Charolais and Angus × Charolais cattle. Therefore, the Angus cattle were at later μ so they could gain more fat (Fig. 21.4). Because the OLR1 gene is involved in lipid metabolism, the authors suggested that the effect of the 'C' allele in Angus was to reduce fat synthesis, which caused improved feed efficiency and reduced residual feed intake. Steers with Charolais ancestry were at earlier μ and, therefore, not able to deposit enough fat for the 'C' allele to have any effect. This proposition is similar to the observation by Nour et al. (1981) that small-type Angus steers accumulated more fat when housed inside in a warm environment compared with those housed outside in a cold environment because more metabolizable nutrients were needed for maintenance in the colder environment. For Holsteins, the environment had no effect (Fig. 21.7) because they were at earlier μ where additional fat could not be accumulated in the warm environment.

A strong QTL signal on chromosome 14 for height at 18 months and weight after 24 months of age was identified using 864 F2 Holstein Friesian × Jersey cows and 3570 outbred Holstein Friesian cows (Karim et al., 2011). This QTL explained 7% and 2% of the genetic variance for live weight in Holstein Friesian and Jersey cattle, respectively. The QTL was biallelic, with QQ cows being 20 kg heavier, Qq cows being average and qq cows being 24 kg lighter after 24 months of age. In the parental population, q had a frequency of 95% in Jersey and Q had a frequency of 78% in Holstein Friesian cows. The QTL was localized to a 780-kb interval encompassing the Pleomorphic adenoma gene 1 (PLAG1 gene), suggesting that PLAG1 is an important regulator of stature in cattle. An SNP in intron 2 of the PLAG1 gene was genotyped in 921 New Zealand Holstein heifers with individual intakes of alfalfa cubes over a 50-day period after they were 6 to 8 months old (Littlejohn et al., 2012). The SNP genotype significantly affected birth weight as well as the peripubertal weights measured during the 50-day feeding study. However, feed intake, feed efficiency and residual feed intake were not associated with the SNP. Growth rate expressed as a ratio to mid-weight of the feeding study was not associated with the SNP in the *PLAG1* gene. The authors suggested that the effect of *PLAG1* alleles on growth rate, therefore, were through an effect on adult size, perhaps starting with fetal development. Based upon their results and those of Hensen *et al.* (2004) for *PLAG1* knockout mice, the authors concluded that *PLAG1* markers could be used to select faster-growing, larger beef cattle or, conversely, dairy cattle with lighter adult weights to minimize maintenance costs. Although the evidence for dominant allelic effect of *PLAG1* was not strong (Littlejohn *et al.*, 2012), presumably selecting the *PLAG1* allele for faster growth in beef cattle would be most useful in a terminal sires to keep from increasing the adult size and the associated cost of maintaining beef cows.

Conclusions

The primary underlying biological factor that regulates growth and body composition of cattle is adult size. The consequence of selection to improve most growth and carcass traits can be increased cow size with associated higher breeding herd feed costs and possible lower fertility. Some recent growth curves based upon data for cattle selected on growth indexes, however, and some genetic trend data show that cow sizes may not increase as much as might be expected. Terminal sire crossbreeding systems can be used to increase growth rate, feed efficiency and leanness at acceptable carcass weights, while avoiding increases in cow size.

The proportions of muscle, fat and bone in carcasses are directly related to stage of maturity, μ, defined by genetic size-scaling as weight divided by adult weight. This means that, across the range of acceptable weights, carcasses from cattle with larger potential adult sizes will contain less fat and more muscle than carcasses from cattle with smaller potential adult sizes. Because the water content of muscle tissue is about three times that of fat tissue, cattle with larger potential adult sizes generally will produce the same carcass weight more efficiently and faster than genetically smaller cattle.

Carcass traits traditionally used for selection should be scrutinized carefully. Three of the four factors in the USDA yield grade equation are questionable predictors, with fat thickness being the best predictor of yield of lean meat from the carcass. Muscle:bone ratio inconsistently predicts muscle yield because it ignores carcass fat. Marbling has explained little of the variation in sensory attributes of beef in large multi-city consumer rating studies and in trained taste panel evaluations.

Given sufficient phenotypic data, powerful molecular genomics methodology should help to find and exploit genetic exceptions to the underlying principles of growth and body composition. These advances will facilitate selection for traits with meaningful biological importance that enhance the value of cattle for agricultural production.

Acknowledgements

Thanks to Natasha Pettifor and Raluca Mateescu for editing, Chris Posbergh for entry of data for figures and St C.S. (Clair) and Jackie Taylor for their inspiration and review of this chapter.

References

American Angus Association (2013) Angus genetic trend by birth year. American Angus Association, St Joseph, Missouri.

Arango, J.A. and Van Vleck, L.D. (2002) Size of beef cows: early ideas, new developments. *Genetics and Molecular Research* 1, 51–63.

Armbruster, G., Nour, A.Y.M., Thonney, M.L. and Stouffer, J.R. (1983) Changes in cooking losses and sensory attributes of Angus and Holstein beef with increasing carcass weight, marbling score or longissimus ether extract. *Journal of Food Science* 48, 835–840.

Arthur, P.F. (1995) Double muscling in cattle: a review. *Australian Journal of Agricultural Research* 46, 1493–1515.

Beever, J.E., George, P.D., Fernando, R.L., Stormont, C.J. and Lewin, H.A. (1990) Associations between genetic markers and growth and carcass traits in a paternal half-sib family of Angus cattle. *Journal of Animal Science* 68, 337–344.

Beltran, J.J., Butts, W.T., Olson, T.A. and Koger, M. (1992) Growth patterns of two lines of Angus cattle selected using predicted growth parameters. *Journal of Animal Science* 70, 734–741.

Berg, R.T. and Butterfield, R.M. (1966) Muscle: bone ratio and fat percentage as measures of beef carcass composition. *Animal Science* 8, 1–11.

Berg, R.T. and Butterfield, R.M. (1968) Growth patterns of bovine muscle, fat and bone. *Journal of Animal Science* 27, 611–619.

Berg, R.T. and Butterfield, R.M. (1976) *New Concepts of Cattle Growth*. John Wiley and Sons, New York.

Berg, R.T., Jones, D.M., Price, M.A., Fukuhara, R., Butterfield, R.M. and Hardin, R.T. (1979) Patterns of carcass fat deposition in heifers, steers and bulls. *Canadian Journal of Animal Science* 59, 359–366.

Blaxter, K.L. (1968) The effect of dietary energy supply on growth. In: Lodge, G.A. and Lamming, G.E. (eds) *Growth and Development of Mammals*. Pelnum Press, New York.

Boligon, A.A., Mercadante, M.E.Z., Lobo, R.B., Baldi, F. and Albuquerque, L.G. (2012) Random regression analyses using B-spline functions to model growth of Nellore cattle. *Animal* 6, 212–220.

Boligon, A.A., Carvalheiro, R. and Albuquerque, L.G. (2013) Evaluation of mature cow weight: genetic correlations with traits used in selection indices, correlated responses, and genetic trends in Nelore cattle. *Journal of Animal Science* 91, 20–28.

Brody, S. (1945) *Bioenergetics and Growth*. Reinhold, New York.

Buchanan, J.W., Garmyn, A.J., Hilton, G.G., VanOverbeke, D.L., Duan, Q., Beitz, D.C. and Mateescu, R.G. (2013) Comparison of gene expression and fatty acid profiles in concentrate and forage finished beef. *Journal of Animal Science* 91, 1–9.

Butterfield, R.M. (1988) *New Concepts of Sheep Growth*. University of Sydney, Epping.

Butterfield, R.M., Griffiths, D.A., Thompson, J.M., Zamora, J. and James, A.M. (1983) Changes in body composition relative to weight and maturity in large and small strains of Australian Merino rams 1. Muscle, bone and fat. *Animal Production* 36, 29–37.

Campion, D.R., Crouse, J.D. and Dikeman, M.E. (1975) Predictive value of USDA beef quality grade factors for cooked meat palatability. *Journal of Food Science* 40, 1225–1228.

Cartwright, T.C. (1970) Selection criteria for beef cattle for the future. *Journal of Animal Science* 30, 706–711.

Casas, E., Shackelford, S.D., Keele, J.W., Koohmaraie, M., Smith, T.P.L. and Stone, R.T. (2003) Detection of quantitative trait loci for growth and carcass composition in cattle. *Journal of Animal Science* 81, 2976–2983.

Casas, E., Bennett, G.L., Smith, T.P.L. and Cundiff, L.V. (2004) Association of myostatin on early calf mortality, growth, and carcass composition traits in crossbred cattle. *Journal of Animal Science* 82, 2913–2918.

Cobb, M. (2006) Heredity before genetics: a history. *Nature Review Genetics* 7, 953–958.

Comerford, J.W., House, R.B., Harpster, H.W., Henning, W.R. and Cooper, J.B. (1992) Effects of forage and protein source on feedlot performance and carcass traits of Holstein and crossbred beef steers. *Journal of Animal Science* 70, 1022–1031.

Crews, D.H., Dikeman, M.E., Northcutt, S.L., Garrick, D.J., Marston, T.T., MacNeil, M.D., Olson, L.W., Paschall, J.C., Rouse, G.H., Weaber, R.L. *et al.* (2010) Animal evaluation. In: Cundiff, L.V., Van Vliet, G. and Hohenboken, W.D. (eds) *Guidelines for Uniform Beef Improvement Programs*. Beef Improvement Federation, Raleigh, North Carolina, pp. 16–65.

Crouse, J.D., Dikeman, M.E., Koch, R.M. and Murphey, C.E. (1975) Evaluation of traits in the U.S.D.A. yield grade equation for predicting beef carcass cutability in breed groups differing in growth and fattening characteristics. *Journal of Animal Science* 41, 548–553.

Cundiff, L.V. and Gregory, K.E. (1977) *Beef Cattle Breeding*. US Department of Agriculture, Agricultural Research Service, Washington, DC.

Cundiff, L.V., Gregory, K.E., Koch, R.M. and Dickerson, G.E. (1971) Genetic relationships among growth and carcass traits of beef cattle. *Journal of Animal Science* 33, 550–555.

DeNise, R.S.K. and Brinks, J.S. (1985) Genetic and environmental aspects of the growth curve parameters in beef cows. *Journal of Animal Science* 61, 1431–1440.

Dickerson, G.E., Künzi, N., Cundiff, L.V., Koch, R.M., Arthaud, V.H. and Gregory, K.E. (1974) selection criteria for efficient beef production. *Journal of Animal Science* 39, 659–673.

Dikeman, M.E., Pollak, E.J., Zhang, Z., Moser, D.W., Gill, C.A. and Dressler, E.A. (2005) Phenotypic ranges and relationships among carcass and meat palatability traits for fourteen cattle breeds, and heritabilities and expected progeny differences for Warner-Bratzler shear force in three beef cattle breeds. *Journal of Animal Science* 83, 2461–2467.

Dunbia Supplier Group (2013) Dunbia beef grading. Available at: http://www.dunbiafarmers.com/Beef-Grading.aspx (accessed 27 March 2014).

EEC (1991) European Economic Community classification scale for the carcases of adult bovine animals. Available at: http://eur-lex.europa.eu/LexUriServ/LexUriServ.do?uri=CONSLEG:1981R1208:19910429:EN:PDF (accessed 21 June 2013).

Eisen, E.J. (1976) Results of growth curve analyses in mice and rats. *Journal of Animal Science* 42, 1008–1023.

Elzo, M.A., Martinez, C.A., Lamb, G.C., Johnson, D.D., Thomas, M.G., Misztal, I., Rae, D.O., Wasdin, J.G. and Driver, J.D. (2013) Genomic-polygenic evaluation for ultrasound and weight traits in Angus–Brahman multibreed cattle with the Illumina3k chip. *Livestock Science* 153, 39–49.

Fitzhugh, H.A. (1976) Analysis of growth curves and strategies for altering their shape. *Journal of Animal Science* 42, 1036–1051.

Freetly, H.C., Kuehn, L.A. and Cundiff, L.V. (2011) Growth curves of crossbred cows sired by Hereford, Angus, Belgian Blue, Brahman, Boran, and Tuli bulls, and the fraction of mature body weight and height at puberty. *Journal of Animal Science* 89, 2373–2379.

Garcia-de-Siles, J.L., Ziegler, J.H. and Wilson, L.L. (1977a) Effects of marbling and conformation scores on quality and quantity characteristics of steer and heifer carcasses. *Journal of Animal Science* 44, 36–46.

Garcia-de-Siles, J.L., Ziegler, J.H., Wilson, L.L. and Sink, J.D. (1977b) Growth, carcass and muscle characters of Hereford and Holstein steers. *Journal of Animal Science* 44, 973–984.

Gooch, R.E.S. (1966) *A Comparison of the Growth of Different Types of Cattle for Beef Production.* Royal Smithfield Club, London.

Gray, G.D., Moore, M.C., Hale, D.S., Kerth, C.R., Griffin, D.B., Savell, J.W., Raines, C.R., Lawrence, T.E., Belk, K.E., Woerner, D.R. et al. (2012) National Beef Quality Audit – 2011: survey of instrument grading assessments of beef carcass characteristics. *Journal of Animal Science* 90, 5152–5158.

Gregory, Keith E. (1965) Symposium on performance testing in beef cattle: evaluating postweaning performance in beef cattle. *Journal of Animal Science* 24, 248–254.

Gregory, K.E., Cundiff, L.V., Koch, R.M., Dikeman, M.E. and Koohmaraie, M. (1994) Breed effects and retained heterosis for growth, carcass, and meat traits in advanced generations of composite populations of beef cattle. *Journal of Animal Science* 72, 833–850.

Gregory, K.E., Cundiff, L.V. and Koch, R.M. (1995) Genetic and phenotypic (co)variances for growth and carcass traits of purebred and composite populations of beef cattle. *Journal of Animal Science* 73, 1920–1926.

Grobet, L., Martin, L.J., Poncelet, D., Pirottin, D., Brouwers, B., Riquet, J., Schoeberlein, A., Dunner, S., Ménissier, F., Massabanda, J., Fries, R., Hanset, R. and Georges, M. (1997) A deletion in the bovine myostatin gene causes the double-muscled phenotype in cattle. *Nature Genetics* 17, 71–74.

Haecker, T.L. (1920) Investigations in beef production. I. The composition of steers at various stages of growth and fattening. *Minnesota Agricultural Experiment Station Bulletin* 193.

Hensen, K., Braem, C., Declercq, J., Van Dyck, F., Dewerchin, M., Fiette, L., Denef, C. and Van de Ven, Wim J.M. (2004) Targeted disruption of the murine *Plag1* proto-oncogene causes growth retardation and reduced fertility. *Development, Growth and Differentiation* 46, 459–470.

Imumorin, I.G., Kim, E.H., Lee, Y.M., De Koning, D.J., Van Arendonk, J., De Donato, M., Taylor, J.F. and Kim, J.J. (2011) Genome scan for parent-of-origin QTL effects on bovine growth and carcass traits. *Frontiers in Genetics* 2, 1–13.

Johnson, H.R., Butterfield, R.M. and Pryor, W.J. (1972) Studies of fat distribution in the bovine carcass. I. The partition of fatty tissues between depots. *Australian Journal of Agricultural Research* 23, 381–388.

Karim, L., Takeda, H., Lin, L., Druet, T., Arias, J.A., Baurain, D., Cambisano, N., Davis, S.R., Farnir, F., Grisart, B. et al. (2011) Variants modulating the expression of a chromosome domain encompassing PLAG1 influence bovine stature. *Nature Genetics* 43, 405–413.

Kauffman, R.G., Grummer, R.H., Smith, R.E., Long, R.A. and Shook, G. (1973) Does live-animal and carcass shape influence gross composition? *Journal of Animal Science* 37, 1112–1119.

Keele, J.W. and Fahrenkrug, S.C. (2001) Optimum mating systems for the myostatin locus in cattle. *Journal of Animal Science* 79, 2016–2022.

Knapp, R.H., Terry, C.A., Savell, J.W., Cross, H.R., Mies, W.L. and Edwards, J.W. (1989) Characterization of cattle types to meet specific beef targets. *Journal of Animal Science* 67, 2294–2308.

Lawrence, T.E., Elam, N.A., Miller, M.F., Brooks, J.C., Hilton, G.G., VanOverbeke, D.L., McKeith, F.K., Killefer, J., Montgomery, T.H., Allen, D.M. *et al.* (2012) Predicting red meat yields in carcasses from beef-type and calf-fed Holstein steers using the United States Department of Agriculture calculated yield grade. *Journal of Animal Science* 88, 2139–2143.

Littlejohn, M., Grala, T., Sanders, K., Walker, C., Waghorn, G., Macdonald, K., Coppieters, W., Georges, M., Spelman, R., Hillerton, E. *et al.* (2012) Genetic variation in PLAG1 associates with early life body weight and peripubertal weight and growth in *Bos taurus*. *Animal Genetics* 43, 591–594.

Lorenzen, C.L., Miller, R.K., Taylor, J.F., Neely, T.R., Tatum, J.D., Wise, J.W., Buyck, M.J., Reagan, J.O. and Savell, J.W. (2003) Beef customer satisfaction: trained sensory panel ratings and Warner-Bratzler shear force values. *Journal of Animal Science* 81, 143–149.

McClure, M.C., Morsci, N.S., Schnabel, R.D., Kim, J.W., Yao, P., Rolf, M.M., McKay, S.D., Gregg, S.J., Chapple, R.H., Northcutt, S.L. and Taylor, J.F. (2010) A genome scan for quantitative trait loci influencing carcass, post-natal growth and reproductive traits in commercial Angus cattle. *Animal Genetics* 41, 597–607.

McPherron, A.C. and Lee, S.J. (1997) Double muscling in cattle due to mutations in the myostatin gene. *Proceedings of the National Academy of Sciences USA* 94, 12457–12461.

Marchitelli, C., Savarese, M.C., Crisà, A., Nardone, A., Marsan, P.A. and Valentini, A. (2003) Double muscling in Marchigiana beef breed is caused by a stop codon in the third exon of myostatin gene. *Mammalian Genome* 14, 392–395.

Marshall, D.M. (1994) Breed differences and genetic parameters for body composition traits in beef cattle. *Journal of Animal Science* 72, 2745–2755.

Meat and Livestock Australia (2007) *MSA Standards Manual for Beef Grading.* Meat Standards Australia, Sydney.

Meyer, K. (1999) Estimates of genetic and phenotypic covariance functions for postweaning growth and mature weight of beef cows. *Journal of Animal Breeding and Genetics* 116, 181–205.

Monteiro, L.S. (1969) The relative size of calf and dam and the frequency of calving difficulties. *Animal Production* 11, 293–306.

Morris, C.A., Pitchford, W.S., Cullen, N.G., Esmailizadeh, A.K., Hickey, S.M., Hyndman, D., Dodds, K.G., Afolayan, R.A., Crawford, A.M. and Bottema, C.D.K. (2009) Quantitative trait loci for live animal and carcass composition traits in Jersey and Limousin back-cross cattle finished on pasture or feedlot. *Animal Genetics* 40, 648–654.

Morris, C.A., Bottema, C.D.K., Cullen, N.G., Hickey, S.M., Esmailizadeh, A.K., Siebert, B.D. and Pitchford, W.S. (2010) Quantitative trait loci for organ weights and adipose fat composition in Jersey and Limousin back-cross cattle finished on pasture or feedlot. *Animal Genetics* 41, 589–596.

Murphey, C.E., Hallett, D.K., Tyler, W.E. and Pierce, J.C., Jr (1960) Estimating yields of retail cuts from beef carcasses. *Journal of Animal Science* 19, 1240.

Nalaila, S.M., Stothard, P., Moore, S.S., Li, C. and Wang, Z. (2012) Whole-genome QTL scan for ultrasound and carcass merit traits in beef cattle using Bayesian shrinkage method. *Journal of Animal Breeding and Genetics* 129, 107–119.

Neely, T.R., Lorenzen, C.L., Miller, R.K., Tatum, J.D., Wise, J.W., Taylor, J.F., Buyck, M.J., Reagan, J.O. and Savell, J.W. (1998) Beef customer satisfaction: role of cut, USDA quality grade, and city on in-home consumer ratings. *Journal of Animal Science* 76, 1027–1033.

Nour, A.Y.M., Thonney, M.L., Stouffer, J.R. and White Jr, W.R.C. (1981) Muscle, fat and bone in serially slaughtered large dairy or small beef cattle fed corn or corn silage diets in one of two locations. *Journal of Animal Science* 52, 512–521.

Nour, A.Y.M., Thonney, M.L., Stouffer, J.R. and White Jr, W.R.C. (1983) Changes in carcass weight and characteristics with increasing weight of large and small cattle. *Journal of Animal Science* 57, 1154–1165.

Orel, Y. and Wood, R.J. (1998) Empirical genetic laws published in Brno before Mendel was born. *Journal of Heredity* 89, 79–82.

Owens, F.N., Dubeski, P. and Hanson, C.F. (1993) Factors that alter the growth and development of ruminants. *Journal of Animal Science* 71, 3138–3150.

Pabiou, T., Fikse, W.F., Näsholm, A., Cromie, A.R., Drennan, M.J., Keane, M.G. and Berry, D.P. (2009) Genetic parameters for carcass cut weight in Irish beef cattle. *Journal of Animal Science* 87, 3865–3876.

Pabiou, T., Fikse, W.F., Cromie, A.R., Keane, M.G., Näsholm, A. and Berry, D.P. (2011) Use of digital images to predict carcass cut yields in cattle. *Livestock Science* 137, 130–140.

Purchas, R.W. (2002) Relationships between beef carcass shape and muscle to bone ratio. *Meat Science* 61, 329–337.

Robinson, D.L., Cafe, L.M. and Greenwood, P.L. (2013) Meat science and muscle biology symposium: developmental programming in cattle: consequences for growth, efficiency, carcass, muscle, and beef quality characteristics. *Journal of Animal Science* 91, 1428–1442.

Rutherford, B. (2013) Bigger cattle; smaller steaks. *Beef*. Available at: http://beefmagazine.com/retail/bigger-cattle-smaller-steaks (accessed 9 June 2013).

Savell, J.W., Branson, R.E., Cross, H.R., Stiffler, D.M., Wise, J.W., Griffin, D.B. and Smith, G.C. (1987) National consumer retail beef study: palatability evaluations of beef loin steaks that differed in marbling. *Journal of Food Science* 52, 517–519.

Savell, J.W., Knapp, R.H., Miller, M.F., Recio, H.A. and Cross, H.R. (1989) Removing excess subcutaneous and internal fat from beef carcasses before chilling. *Journal of Animal Science* 67, 881–886.

Sellick, G.S., Pitchford, W.S., Morris, C.A., Cullen, N.G., Crawford, A.M., Raadsma, H.W. and Bottema, C.D.K. (2007) Effect of myostatin F94L on carcass yield in cattle. *Animal Genetics* 38, 440–446.

Smith, G.C., Savell, J.W., Cross, H.R., Carpenter, Z.L., Murphey, C.E., Davis, G.W., Abraham, H.C., Parrish, F.C. and Berry, B.W. (1987) Relationship of USDA quality grades to palatability of cooked beef. *Journal of Food Quality* 10, 269–286.

Stouffer, J.R., Wallentine, M.V., Wellington, G.H. and Diekmann, A. (1961) Development and application of ultrasonic methods for measuring fat thickness and rib-eye area in cattle and hogs. *Journal of Animal Science* 20, 759–767.

Taylor, J.F., Coutinho, L.L., Herring, K.L., Gallagher, D.S., Brenneman, R.A., Burney, N., Sanders, J.O., Turner, J.W., Smith, S.B., Miller, R.K., Savell, J.W. and Davis, S.K. (1998) Candidate gene analysis of GH1 for effects on growth and carcass composition of cattle. *Animal Genetics* 29, 194–201.

Taylor, St C.S. (1965) A relation between mature weight and time taken to mature in mammals. *Animal Production* 7, 203–220.

Taylor, St C.S. (1980a) Genetic size-scaling rules in animal growth. *Animal Production* 30, 161–165.

Taylor, St C.S. (1980b) Genetically standardized growth equations. *Animal Production* 30, 167–175.

Taylor, St C.S. (1985) Use of genetic size-scaling in evaluation of animal growth. *Journal of Animal Science* 61 (Suppl. 2), 118–143.

Taylor, St C.S. and Fitzhugh, H.A. (1971) Genetic relationships between mature weight and time taken to mature within a breed. *Journal of Animal Science* 33, 726–731.

Taylor, S.C.S. and Murray, J.I. (1987) *Butler Memorial Lecture: Genetic Aspects of Mammalian Growth and Survival in Relation to Body Size*. University of Queensland, Brisbane.

Taylor, St C.S. and Young, G.B. (1966) Variation in growth and efficiency in twin cattle with live weight and food intake controlled. *Journal of Agricultural Science* 66, 67–85.

Taylor, St C.S., Moore, A.J. and Thiessen, R.B. (1986) Voluntary food intake in relation to body weight among British breeds of cattle. *Animal Production* 42, 11–18.

Thiessen, R.B., Hnizdo, E., Maxwell, D.A.G., Gibson, D. and Taylor, St C.S. (1984) Multibreed comparisons of British cattle. Variation in body weight, growth rate and food intake. *Animal Production* 38, 323–340.

Thonney, M.L. (2003) Letter to Editor: Consideration of fat thickness in models to predict beef carcass cutability. *Journal of Animal Science* 81, 2103–2104.

Thonney, M.L., Heide, E.K., Duhaime, D.J., Nour, A.Y.M. and Oltenacu, P.A. (1981) Growth and feed efficiency of cattle of different mature sizes. *Journal of Animal Science* 53, 354–362.

Thonney, M.L., Arnold, A.M., Ross, D.A., Schaaf, S.L. and Rounsaville, T.R. (1991a) Energetic efficiency of rats fed low or high protein diets and grown at controlled rates from 80 to 205 grams. *Journal of Nutrition* 121, 1397–1406.

Thonney, M.L., Perry, T.C., Armbruster, G., Beermann, D.H. and Fox, D.G. (1991b) Comparison of steaks from Holstein and Simmental X Angus steers. *Journal of Animal Science* 69, 4866–4870.

Vinsky, M., Islam, K., Chen, L. and Li, C. (2013) Short Communication: Association analyses of a single nucleotide polymorphism in the promoter of OLR1 with growth, feed efficiency, fat deposition, and carcass merit traits in hybrid, Angus and Charolais beef cattle. *Canadian Journal of Animal Science* 93, 193–197.

Wellington, G.H. and Stouffer, J.R. (1959) Beef marbling. Its estimation and influence on tenderness and juiciness. *Cornell University Agricultural Experiment Station Bulletin* 941, 1–30.

22 Genetics of Meat Quality

Raluca G. Mateescu

University of Florida, Gainesville, Florida, USA

Introduction

Meat quality is determined by a multitude of component traits, including body composition at harvest, physico-chemical attributes, as well as visual and sensory traits. Understanding what consumers want and how these demands are changing is critical for the survival of any industry in a competitive market. Consumers are increasingly becoming more quality and health conscious, a trend that is reflected in the growing public interest in the nutritional and health-promoting value of the diet. This provides an opportunity for the beef industry to respond to consumers through increased emphasis on producing beef of high quality, nutritional value and healthfulness; and selection is an obvious option to generate a permanent and cumulative improvement of these traits. Classic approaches based on pedigree and performance records, successfully implemented for growth and other easily measured traits with adequate heritability and a recording system in place, may not be practical for improving meat quality. Visual and sensory traits, nutritional value or health-fulness traits are difficult and costly to measure post-harvest on a sufficient number of close relatives of selection candidates. Reliable genomic prediction leveraging recent developments in genome resequencing, genotype imputation and high-throughput genotyping offers the only real promise to address this goal.

The aim of this chapter is to summarize current knowledge of quantitative and molecular genetic aspects of meat quality and to discuss implications of genetic improvement of meat quality for the cattle industry.

Meat Quality Traits

Meat quality can be described and categorized in various ways and has evolved over time to include new traits of interest for farmers, processors or consumers. In this chapter, meat quality will be defined broadly to include four main categories: carcass composition, fat composition and qualities, meat physico-chemical qualities and meat sensory qualities. The large number of traits used to describe meat quality, the variation in quality between muscles within the same carcass and different ways meat quality can be changed post mortem, make it difficult to define a superior animal for meat quality. An added complication is the variation in consumer preference. Consumers of beef in developed countries demand products of high and consistent quality. In recent years increased awareness regarding diet, health, environment and other socio-cultural values have led to rising interest among consumers in new quality attributes such as product safety, healthfulness, environmental impact and animal welfare. In order for the beef industry to be able to respond to these demands, a good understanding of all attributes that contribute to meat quality is necessary, along with the ability to predict, control and manipulate the quality of beef to address these emerging requirements (Hocquette et al., 2012).

Carcass composition

From a North American perspective, carcass composition can be described by hot carcass weight; dressing percentage; backfat thickness; ribeye area; kidney, pelvic, and heart fat percentage; marbling score, yield grade, predicted and actual retail product weight and percentage; fat weight and percentage; bone weight and percentage. Other systems, such as those used in Europe, New Zealand or Australia, are based on somewhat different criteria.

Hot carcass weight is the unchilled weight of the animal after slaughter and removal of hide, head, intestinal tract and internal organs. Hot carcass weight ranges from 60 to 64% of the live weight for the majority of feedlot fed cattle, and is used to determine dressing percentage and yield grade. Hot carcass weight divided by live weight (expressed as a percentage) is referred to as dressing percentage and is a major factor in determining total revenue when the animals are sold on a live weight basis. Backfat thickness is a measure of the thickness of external fat on the carcass and is taken over the longissimus dorsi between the 12th and 13th ribs at a point three-fourths the length of the ribeye from the split chine bone. Ribeye or longissimus dorsi area is the measure of the total area of the loin or longissimus dorsi between the 12th and 13th ribs. Percentage kidney, pelvic and heart fat is the percentage of total fat in the regions of the kidneys, pelvis, and heart found in the body cavity. Although it could be calculated as the change in carcass weight after removal of fat from these areas, it is usually estimated visually and ranges from 1 to 4% in most cattle. Yield grade is an estimate of the high value portion of the carcass (boneless retail cuts from the round, loin, rib and chuck) and is calculated based on hot carcass weight, backfat thickness, ribeye area and percentage kidney, pelvic and heart fat. Yield grade is expressed as a numeric score from 1 to 5, representing the greatest and lowest percentage of boneless, closely trimmed, retail cuts (greater cutability), respectively. Marbling score evaluates the amount of intramuscular fat at the cut surface of the ribeye on the 12th rib surface. There are nine marbling scores (1 = practically devoid, 9 = abundant) and each is divided into 100 subunits and assigned as subscripts to the scores ranging from 00 to 99, representing the least and greatest amount of marbling within the score. Retail product weight and percentage is a useful measure of the saleable portion of carcass beef and it is most often predicted or estimated based on backfat thickness, ribeye area, kidney and pelvic fat, hot carcass weight and marbling score.

Fat composition and qualities

Intramuscular fat is a major factor determining sensory attributes like flavour, juiciness and texture of meat as well as its nutritional value. Fat composition and qualities could be described by fatty acid composition, the fat melting point, and by several indices including the atherogenic, desaturation and elongation index.

The evaluation of beef fat composition is primarily based on determining its fatty acid composition. Most frequently, this involves derivatizing the fatty acids in the various lipid classes to their fatty acid methyl esters, which can then be analysed by gas chromatography, high performance liquid chromatography or a combination of both. Beef intramuscular fat consists on average of 45–48% saturated fatty acids (SFAs), 35–45% monounsaturated fatty acids (MUFAs) and up to 5% polyunsaturated

fatty acids (PUFAs) (Scollan *et al.*, 2006). The predominant SFAs are myristic acid (C14:0), palmitic acid (C16:0) and stearic acid (C18:0), the latter representing 30% of total SFAs. Oleic acid (C18:1n-9) is the most prominent MUFA, with the remainder of the MUFAs occurring mainly as cis- and trans- isomers of 18:1, while linoleic (C18:2) and α-linolenic acids (C18:3) are the main PUFAs (Fig. 22.1). In general, a diet with a ratio of PUFAs to SFAs above about 0.45 and a ratio of omega-6 to omega-3 below about 4.0 is recommended to prevent various 'lifestyle diseases' such as coronary heart disease and cancers (Williams, 2000; Simopoulos, 2004). The omega-6 to omega-3 ratio for beef is beneficially low, reflecting the considerable amounts of beneficial omega-3 PUFAs (particularly C18:3n-3) and the long chain PUFAs, eicosapentaenoic acid (EPA; 20:5n-3) and docosahexaenoic acid (DHA; 22:6n-3) it contains. The atherogenic index can be viewed as a measure of healthfulness of lipid composition and is calculated as a ratio of the 4×C14:0 and C16:0 to the total MUFA and PUFA (Ulbricht and Southgate, 1991). The desaturation index is a product-to-precursor ratio used to indirectly

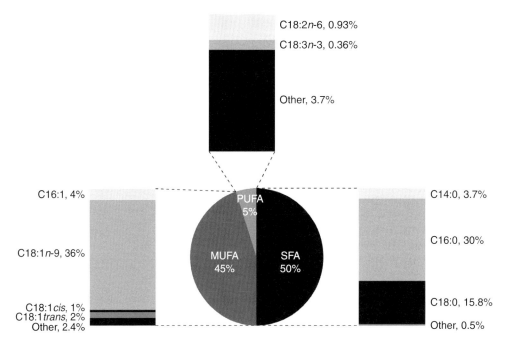

Fig. 22.1. Percentages of predominant fatty acids contributing to the main saturated (SFA), monounsaturated (MUFA) and polyunsaturated (PUFA) fatty acids classes in beef intramuscular fat.

calculate the activity of enzymes responsible for desaturation of fatty acids (C16:1/C16:0 and C18:1/C18:0), while the elongation index represents the conversion of fatty acids from 16 to 18 carbon atoms.

Beef is a highly nutritious food and an excellent source of good protein and micronutrients (vitamins A, B6, B12, D, E, iron, zinc, selenium, essential omega-3 fatty acids, etc.) (Biesalski, 2005; Scollan et al., 2006), therefore playing a critical role in a healthy diet. However, current dietary recommendations based on the belief that red meat, as a source of saturated fat, contributes significantly to cardiovascular disease, may detract from recognition of the key positive attributes of beef and its importance for the human diet. In a recent review, McNeill et al. (2012) indicate that, although taste is the most common reason Americans consume beef, 41% of consumers decreased consumption of beef between 2002 and 2008, and 53% of US adults in 2000 perceived red meat as the 'least healthy' protein source among red meat, chicken/poultry, fish/seafood or pork, supporting the hypothesis that public perception of beef with respect to health is an important determinant of consumption.

Consistent evidence from clinical trials suggests that lean beef, in the context of a well-balanced diet, is equally effective in the reduction of low-density lipoprotein cholesterol as is lean white meat (McNeill et al., 2012). A systematic review of red meat studies provides supportive evidence that fresh red meat from both grass-fed and grain-finished animals is associated with reduced low-density lipoprotein cholesterol in both healthy and mildly hypercholesterolaemic individuals when included as part of a diet low in saturated fat ($\leq$10%) (Li et al., 2005). In addition, a recent meta-analysis of prospective cohort studies failed to identify an association between saturated fat and increased risk of heart disease (pooled relative risk estimate 1.07 between intake quartiles) (Siri-Tarino et al., 2010).

Meat physico-chemical qualities

Physical and chemical properties are objective measures of meat quality and include shear force, intramuscular lipid percentage, ultimate pH, myofibrillar fragmentation, lean colour reflectance, and mineral and micronutrient concentration.

Beef tenderness has been the most important determinant of the consumer eating experience and is determined objectively by the Warner-Bratzler shear force test, which measures the force required to shear a cooked steak after post mortem ageing. It has been estimated to range from 0.86 to 2.99 kg within different breeds of cattle (Dikeman et al., 2005). This range represents an improvement (lower shear force indicating more tender meat) over previous estimates of 3.22 to 7.39 kg (Ramsbottom and Strandine, 1948) and 2.2 to 5.0 kg (Smith et al., 1978). Given that no specific programme to improve tenderness has been in place, the improvement is likely a secondary consequence of increases in intramuscular fat/marbling score during this period. Intramuscular lipid percentage is an objective measurement that quantifies the total fat content usually within the ribeye muscle by chemical extraction of lipid. There is a strong relationship between percentage of intramuscular fat, marbling score and USDA quality grade (Table 22.1).

Generally, meat ultimate pH in the range of 5.4–5.6 has the most desirable properties, while an elevated ultimate pH results in lower overall quality characterized by a darker colour (ultimate pH above 6.2), shorter shelf life due to increased growth of spoilage bacteria at higher pH, and weaker beef flavour (Lawrie, 1966). Myofibrillar fragmentation represents the extent of fragmentation of myofibrils caused by homogenization, and has been shown to be highly correlated with other measures of meat tenderness such as shear force and sensory panel tenderness. Higher myofibrillar fragmentation could be achieved with longer ageing and results in more tender meat.

Meat colour, often used by consumers to select or reject products, revolves around myoglobin, the primary red pigment in meat. Meat colour is objectively measured using colorimeters or spectrophotometers measuring reflected light and reporting $L^*a^*b^*$ values, where the a^* axis extends from green (−60) to red (+60), the b^* axis from blue (−60) to yellow (+60), and the L^* axis measures the brightness (100 is white and 0 is black) according to the

Table 22.1. Relationship between percentage of intramuscular fat, marbling score and USDA quality grades.

Percentage of intramuscular fat	USDA quality grade	Degree of marbling/marbling score
<2.30	Standard	Traces/3.0–3.9
2.30–3.00	Select–	Slight/4.0–4.9
3.10–3.99	Select+	
4.00–5.79	Choice–	Small/5.0–5.9
5.80–7.69	Choice0	Modest/6.0–6.9
7.70–9.89	Choice+	Moderate/7.0–7.9
9.90–12.10	Prime–	Slightly abundant/8.0–8.9
>12.10	Prime0	Moderately abundant/9.0–9.9

procedures of the Commission Internationale d'Eclairage (CIE, 1976).

Meat chemical composition comprises 56–72% water, 15–22% protein, 5–34% fat and 3.5% soluble non-protein substances including carbohydrates, organic salt, dissolved nitrogen substances, minerals and vitamins (Lawrie, 1966). Mineral and micronutrient concentration has received increased attention given that iron deficiency was identified in 2012 by the World Health Organization as the most common and widespread nutritional disorder in the world. In adolescents, non-heme iron and zinc absorption from a beef meal is significantly greater than that from a soy protein meal (Etcheverry et al., 2006).

Meat sensory qualities

Sensory factors that influence consumer preference and purchase can be evaluated through taste test panels assessing tenderness, juiciness, connective tissue amount, flavour, off flavour and overall acceptability. Although previous research suggests that degree of marbling contributes to meat tenderness (Smith et al., 1985; Platter et al., 2003), still little is known about the exact role of intramuscular fat in cooked meat tenderness. There are close relationships between marbling, juiciness and tenderness, whereby meat samples that readily release fat and maintain juiciness are perceived as tender. Juiciness is directly related to differences in pH, water-holding capacity, fatness and firmness (Lawrie, 1966). Beef flavour results from a combination of sensation of odour, taste, texture, temperature and pH (Lawrie, 1966).

Beef tenderness has been recognized as the major determinant of the consumers' eating experience. The 2011 National Beef Quality Audit assessed the status and progress being made towards quality and consistency of USA cattle, carcasses and beef products (Igo et al., 2013). The only quality category identified by packers, food service buyers and retailers as one for which to pay a premium was 'eating satisfaction'.

Sensory attributes are in general evaluated by a multi-member panel on steak samples cooked to a medium degree of doneness. Samples are evaluated in duplicate on standard ballots from the American Meat Science Association (AMSA, 1995). The tenderness, juiciness and amount of connective tissue are evaluated on an eight-point scale (1 = extremely tough, extremely dry or abundant and 8 = extremely tender, extremely juicy or none). The scale used for beef flavour and off-flavour intensity is 1 = not detectable, 2 = slightly detectable and 3 = strong.

Heritability of Meat Quality Traits

Carcass composition traits have been the most studied meat quality traits due to their ease of measurement and their direct economic importance to cattle breeders. Knowledge of heritability estimates for these traits provide insight into likely response to direct selection. Response to selection determines the benefits of investing in selection schemes that increase harvest premiums and reduce discounts. As the consumers' focus on meat quality has increased, more studies reporting heritability estimates for the physico-chemical and sensory qualities of meat have become available. More recently, nutritional and healthfulness qualities of beef in different populations and management conditions are being investigated, but few studies have been published to date.

Carcass composition

Heritability estimates for carcass traits adjusted to different endpoints are available for numerous breeds and crossbreeds. Utrera and Van Vleck (2004) exhaustively reviewed heritability estimates for carcass traits of cattle and examined the effect of different finish endpoints (slaughter age, slaughter weight and backfat thickness) on these estimates. These estimates are in Table 22.2, along with averages from studies reported after this review. Carcass composition traits are moderately to highly heritable, and their heritabilities are not very sensitive to finish endpoints. Rumph et al. (2007) compared heritability estimates adjusted to different endpoints and showed little re-ranking of sires for most carcass composition traits when using alternative endpoints except for percentage retail cuts and longissimus dorsi muscle area. Further investigation is needed to determine whether a particular endpoint results in a more predictive estimated progeny difference (EPD) than the traditional age at harvest endpoint.

Fat composition and qualities

Consumer interest in nutritional and healthfulness aspects of their diet has stimulated efforts to understand the role beef plays in a healthful modern diet and determine opportunities to improve nutritional and healthfulness values of beef. Relatively few studies report heritabilities for nutritional and healthfulness traits, those being summarized in Table 22.3. Heritability estimates for different fatty acids range from low to high, suggesting that opportunities exist for changing lipid composition by selection in order to enhance healthfulness value of beef.

Intramuscular fat content and fatty acid composition are key determinants of the nutritional

and health value of beef (Wood et al., 2008). Fatty acid composition is a complex system determined by tens of individual fatty acids concentrations (i.e. oleic acid, palmitic acid, stearic acid), which could be grouped in several major classes (i.e. saturated, mono- and polyunsaturated fatty acids) or used in calculating several indices (i.e. atherogenic index, desaturation index, elongation index). Complex relationships also exist between individual traits at every level of the system (i.e. oleic acid is produced by the desaturation of stearic acid, therefore the level of oleic acid is dependent on the availability of stearic acid, as well as the activity of stearoyl-CoA desaturase) with the concentration of any particular fatty acid being therefore the collective result of processes that add to its concentration, less processes that convert it to other fatty acids.

A consensus is emerging regarding impact of dietary fat on human health with trans-fats and some SFAs (particularly C12:0 and C14:0) having a negative effect, while MUFAs and PUFAs have a positive effect. Acknowledging the complexity of the system, a challenge for the beef industry is to explore opportunities for changing fatty acid composition to improve healthfulness of beef using genetics as well as other means and to develop marketing strategies emphasizing the key positive attributes of red meat.

Meat physico-chemical qualities

Heritability estimates for physico-chemical traits and for subjective evaluations of sensory traits are in Tables 22.4 and 22.5, respectively. Intramuscular fat content measured by chemical analysis as amount of lipids in the longissimus dorsi muscle is highly heritable (0.54), similar to subjectively measured marbling score (0.29–0.45, Table 22.2). Different measures of tenderness, including Warner-Bratzler shear force (a physical measure of tenderness), calpastatin activity and myofibrillar fragmentation (a biochemical measure of tenderness), are moderately heritable. Processing conditions at harvest (e.g. electrical stimulation) and post mortem treatments (e.g. ageing), influence heritability of beef tenderness. No apparent trends have been reported for effect of ageing time on

Table 22.2. Heritability (h^2) estimates of cattle carcass composition traits.

Trait[a]	2004 review[b]		Post 2004[c]	
	Mean h^2	n[d]	h^2	n[d]
Hot carcass weight (A)	0.42	36	0.46	9
Hot carcass weight (W)	0.37	8		
Hot carcass weight (F)	0.35	12	0.34, 0.41, 0.57	3
Dressing % (A)	0.28	18		
Dressing % (W)	0.38	11		
Dressing % (F)	0.36	3		
Backfat thickness (A)	0.39	34	0.25	7
Backfat thickness (W)	0.33	23	0.26	6
Backfat thickness (F)	0.29	6		
Longissimus muscle area (A)	0.41	36	0.35	8
Longissimus muscle area (W)	0.37	19	0.27	1
Longissimus muscle area (F)	0.41	11	0.26, 0.50	2
Kidney, pelvic and heart fat % (A)	0.48	8	0.13	4
Kidney, pelvic and heart fat % (W)	0.19	2	0.23	1
Kidney, pelvic and heart fat % (F)	0.34	4	0.26	1
Marbling score (A)	0.45	29	0.52	6
Marbling score (W)	0.29	15	0.21, 0.27, 0.4, 0.41	4
Marbling score (F)	0.30	12	0.27, 0.35, 0.37	3
Yield grade (A)	0.60	4	0.3, 0.38, 0.46	3
Yield grade (W)			0.24	1
Yield grade (F)	0.74	2	0.30	1
Predicted retail product % (A)	0.28	8	0.30	1
Predicted retail product % (W)	0.41	6	0.30	1
Predicted retail product % (F)	0.48	3	0.24	1
Retail product weight (A)	0.51	11	0.42	1
Retail product weight (W)	0.42	1	0.50	1
Retail product weight (F)	0.50	1	0.32	1
Fat weight (A)	0.52	7	0.27, 0.28	2
Fat weight (W)	0.37	1	0.35	1
Fat weight (F)	0.50	1	0.29	1
Bone weight (A)	0.51	6	0.43, 0.75	2
Bone weight (W)	0.39	1	0.33	1
Bone weight (F)			0.35	1
Actual retail product % (A)	0.54	9	0.23, 0.42, 0.66	3
Actual retail product % (W)	0.50	8	0.21, 0.44	2
Actual retail product % (F)			0.32, 0.41	2
Fat % (A)	0.51	7	0.31	1
Fat % (W)	0.51	2	0.35	1
Fat % (F)			0.26	1
Bone % (A)	0.45	7	0.28	1
Bone % (W)	0.35	1	0.33	1
Bone % (F)			0.35	1

[a]Letter in parentheses indicates that the trait was measured, or adjusted to, constant age (A), weight (W) or backfat thickness (F).
[b]Source: review of Utrera and Van Vleck (2004).
[c]Sources: Dikeman et al., 2005; Rios-Utrera et al., 2005; Rumph et al., 2007; Smith et al., 2007; Crews et al., 2008; Pabiou et al., 2009; Gill et al., 2010b; Inoue et al., 2011; Nogi et al., 2011; Oyama, 2011; Zuin et al., 2012.
[d]Number of estimates included in the mean. When fewer than five estimates were available, all estimates are presented.

Table 22.3. Heritability (h^2) estimates of fatty acid concentrations in beef cattle (only most predominant fatty acids are shown).

Fatty acid[a]	Breed					
	Crossbred[b]	Japanese Black[c]	Japanese Black[d]	Piedmontese[e]	Multiple[f]	Angus[g]
C14:0	0.18	0.82	0.70	0.19	0.55	0.57
C14:1		0.86	0.60		0.51	0.50
C16:0	0.21	0.65	0.65	0.17	0.43	0.51
C16:1		0.66	0.76	0.09	0.38	0.49
C18:0	0.14	0.71	0.59	0.12	0.44	0.52
C18:1	0.17	0.73	0.78	0.20		
C18:1c9					0.56	0.55
C18:1c11					0.21	0.11
C18:2		0.34	0.58	0.15	0.06	0.22
C18:3n3			0.00			0.14
CLAc9t11						0.11
SFA	0.27		0.66	0.15	0.54	0.57
MUFA		0.66	0.68	0.2	0.53	0.49
PUFA			0.47	0.19	0.05	0.19
MUFA/SFA		0.75	0.63			
PUFA/SFA			0.47			0.21
Desaturation index (%)	0.18		0.67			
Elongation index (%)	0.16	0.67	0.80			

[a]g/100 g fatty acids.
[b]Pitchford *et al.*, 2002.
[c]Inoue *et al.*, 2011.
[d]Nogi *et al.*, 2011.
[e]Cecchinato *et al.*, 2012.
[f]Kelly *et al.*, 2013.
[g]J. Reecy, 2013, personal communication.

heritability of shear force. Two detailed studies on heritabilities of shear force at 1, 4, 7, 14, 21 and 35 days report variable estimates of 0.2, 0.7, 0.47, 0.27, 0.36 and 0.19 (O'Connor *et al.*, 1997), and 0.12, 0.28, 0.12, 0.29, 0.32 and 0.14 (Wulf *et al.*, 1996). Electrical stimulation and the level of voltage applied have been shown to affect heritability estimates (Johnston *et al.*, 2001), and electrical stimulation was more beneficial in reducing shear force for *Bos indicus* or crossbred *B. indicus* than for *Bos taurus* beef (Ferguson *et al.*, 2000).

Beef is a significant source of essential micronutrients and several bioactive compounds. Red meat is one of the best human dietary sources of readily absorbed iron and zinc and a contributor of several other essential minerals and trace elements (Zanovec *et al.*, 2010; O'Neil *et al.*, 2011). High heritability (0.54) was reported for iron content (Mateescu

et al., 2012a); whereas, sodium, magnesium, phosphorus, potassium and zinc had low to moderate heritabilities (0.009–0.12). Other bioactive compounds, or health-promoting active ingredients, present in beef include carnitine, creatine, creatinine, carnosine and anserine (Muller *et al.*, 2002; Wyss *et al.*, 2007; Hipkiss, 2009). Almost no genetic variation was observed in carnitine or creatinine (0.015 and 0.002, respectively), while creatine, carnosine and anserine had moderate heritabilities with estimates of 0.43, 0.38 and 0.53, respectively (Mateescu *et al.*, 2012b).

Meat sensory qualities

It is important to point out that, once cooked, consumer satisfaction is primarily determined by meat tenderness, juiciness and flavour

Table 22.4. Heritability (h^2) estimates of physico-chemical quality traits of beef.

Trait	2004 review[a] Mean h^2	Post 2004[b] Mean h^2	n[c]	Range[d]
Intramuscular lipid %	0.54	0.47	7	0.34–0.77
Cooked lipid %		0.48	2	0.17, 0.79
Shear force	0.25	0.28	7	0.09–0.42
Calpastatin activity	0.43	0.26	2	0.07, 0.45
Myofibrillar fragmentation	0.39			
Lean colour reflectance				
L*a*b* lightness	0.29	0.17	4	0.16, 0.17, 0.17, 0.18
L*a*b* redness	0.17	0.13	2	0.13, 0.13
L*a*b* yellowness	0.11			
Ultimate pH	0.15	0.02	1	0.02–0.02
Warner-Bratzler shear force				
7-day shear force		0.22	2	0.14, 0.29
14-day shear force		0.17	2	0.14, 0.20
21-day shear force		0.06	1	0.06
35-day shear force		0.46	2	0.28, 0.63

[a]Source: review of Utrera and Van Vleck, 2004.
[b]Sources: Reverter *et al.*, 2000; Newman *et al.*, 2002; Pitchford *et al.*, 2002; Johnston *et al.*, 2003b; Reverter *et al.*, 2003; Riley *et al.*, 2003; Dikeman *et al.*, 2005; Rumph *et al.*, 2007; Smith *et al.*, 2007; Gill *et al.*, 2010b; Inoue *et al.*, 2011; Zuin *et al.*, 2012.
[c]Number of estimates included in the mean.
[d]When fewer than 5 estimates were available, all estimates are presented.

(Glitsch, 2000). Beef consumers rate tenderness as the most important palatability trait (Huffman *et al.*, 1996; Robinson *et al.*, 2001; Watson *et al.*, 2008). Moderate to low heritability estimates are consistently reported for subjective sensory traits, among which panel tenderness score generally has the highest heritability (Table 22.5). Sensory and tenderness measures are affected by several factors such as carcass processing method (i.e. electrical stimulation, tenderstretching or ageing) or the specific muscle analysed, and this is reflected in the relatively wide ranges for reported estimates.

Genetic Correlations with Meat Traits

There is no well-defined goal for national beef cattle improvement at the present time. Garrick and Golden (2009) suggested that the goal should be to produce beef that is nutritious, healthful, and desirable. A comprehensive analysis of genetic correlations between meat quality traits, as well as between meat quality and

Table 22.5. Heritability (h^2) estimates[a] of meat sensory quality.

Trait	Mean h^2	n[b]	Range
Tenderness	0.25	9	0.06–0.46
Juiciness	0.22	9	0.00–0.46
Connective tissue amount	0.12	1	0.12
Flavour intensity	0.10	8	0.00–0.32
Flavour desirability	0.32	1	0.32
Off flavour	0.01	2	0.00–0.01
Overall acceptability	0.27	6	0.10–0.46

[a]Sources: O'Connor *et al.*, 1997; Pitchford *et al.*, 2002; Johnston *et al.*, 2003a; Riley *et al.*, 2003; Dikeman *et al.*, 2005; Rumph *et al.*, 2007; Gill *et al.*, 2010b; Wheeler *et al.*, 2010.
[b]Number of estimates included in the mean.

other economically relevant traits are important in determining the role for meat quality traits in the development of such a goal. These genetic correlations are summarized in this section and are described on the basis of sign (positive or negative, whereby the increase in one trait is associated with an increase or decrease, respectively, in the other trait), strength (weak correlation if less than 0.4, moderate between 0.4 and 0.7, and strong if higher than 0.7) and effect (favourable or unfavourable, whereby the direction of variation of the two traits coincides or not with our goals of improvement).

Carcass composition

The relationships of carcass weight, cutability (percentage of boneless, closely trimmed, retail cuts) and marbling are probably the most economically relevant from a selection standpoint, as, at least in North America, they directly influence carcass value. Genetic correlation estimates of marbling score with fat thickness and lean yield reviewed by Marshall (1999) ranged from moderately unfavourable to slightly favourable (0.44 to –0.12 for fat thickness and marbling, respectively; –0.60 to 0.12 for lean yield and marbling, respectively). These correlations suggest that increased marbling might be achieved without increasing external fat, but backfat thickness alone may not be a good predictor of marbling ability.

Meat sensory and physico-chemical qualities

Relatively few studies since Marshall (1999) have reported genetic correlations between sensory traits and other important carcass or meat quality traits. Recent studies on meat palatability traits in different cattle breeds (Riley et al., 2003; Dikeman et al., 2005; Gill et al., 2010b) reported only heritability and phenotypic or residual correlations, due to inadequate sample or population size for reliable estimation of genetic correlations.

Two reports by Reverter et al. (2003) and Wheeler et al. (2010) on genetic correlations between intramuscular fat and tenderness in temperate breeds reported similar moderate and favourable estimates: 0.40 and 0.61 for intramuscular fat and consumer panel tenderness, and –0.52 and –0.38 for intramuscular fat and tenderness evaluated by Warner-Bratzler shear force, respectively. In contrast, reported genetic correlations between tenderness measured by

Warner-Bratzler shear force and intramuscular fat measured either as percentage of lipid or marbling score in Brahman cattle are suggestive of a different fat–tenderness relationship. Riley et al. (2003) reported moderate and positive genetic correlations (i.e. more fat, higher shear force) of 0.45, 0.39 and 0.41 between percentage of intramuscular lipids and shear force at 7, 14 and 21 days, respectively, for Brahman cattle. This unique fat–tenderness relationship in Brahman cattle seems to be supported by the residual correlations between Warner-Bratzler shear force and marbling estimated by Dikeman et al. (2005), which were slightly negative (–0.21) when progeny of 14 breeds were included, and slightly positive (0.15) when only Bos indicus progeny were included.

Favourable and moderate genetic correlations between tenderness (measured as shear force or panel tenderness) and muscle colour (L^* darkness–lightness and a^* green–red) were reported (Johnston et al., 2003b) for both tropical and temperate breeds: –0.4 and –0.64 for shear force with L^* and a^*, respectively; 0.54 and 0.22 for tenderness score with L^* and a^*, respectively. This suggests presence of a genetic component for the previously reported phenotypic or residual correlations estimates between shear force and L^* colour (Dubeski et al., 1997; Wulf et al., 1997; Wulf and Page, 2000).

The relationship between meat concentrations of important minerals, micronutrients and other bioactive compounds and palatability traits has been recently investigated in Angus cattle. Only a few significant genetic correlations were identified between minerals and all were positive and favourable (Mateescu et al., 2012a). Strong genetic correlations were found between magnesium and phosphorus (0.88), magnesium and potassium (0.68) and phosphorus and potassium (0.69). A moderate genetic correlation was identified between iron and zinc (0.49). As indicated earlier, beef is a major source of easily absorbable iron and zinc in human diet. High heritability for iron and its favourable genetic correlation with zinc and other minerals suggest that genetic improvement of iron content in beef is feasible and should have substantial positive effect on human health without negative consequences on palatability traits. Other compounds reported to potentially have protective effects on human health, including carnitine, creatine, creatinine, carnosine and anserine were also investigated. Several significant genetic correlations were identified between these compounds but no significant associations with beef palatability traits in longissimus dorsi of beef cattle were detected (Mateescu et al., 2012b).

Fat composition and qualities

There are few published estimates of genetic parameters, especially genetic correlations, for fat qualities and fatty acid composition traits in beef cattle. Correlations between fat traits and carcass weight were in general weak, the largest being with fat melting point (0.34) (Pitchford et al., 2002), an important factor affecting beef quality. Fat melting point was negatively genetically correlated with level of MUFAs (−0.42, indicating a favourable correlation: low melting point and high MUFA content) and desaturation index (−0.46, favourable correlation as well, given that both a low melting point and a high desaturation index are desirable), but not correlated with elongation index (−0.05). Genetic correlations between specific fatty acids and carcass weight, fat depth, intramuscular fat and fat colour (Pitchford et al., 2002) were generally weak (−0.25 to 0.28), with the exception of correlation between intramuscular fat and palmitic acid (C16:0, 0.43 ± 0.28) and oleic acid (C18:1, −0.48 ± 0.31). Strong genetic correlations were found between specific fatty acids and melting point and other fat index traits (MUFA, desaturation index and elongation index). Genetic variation in melting point was most closely associated with variation in palmitoleic acid (C16:1, −0.74) and stearic acid (C18:0, 0.54). In Angus cattle, myristic acid (C14:0), which is the SFA in beef with the most negative effect on low-density lipoprotein/high-density lipoprotein ratio, was shown to have favourable (decreased myristic acid content) and weak genetic correlations to carcass weight (−0.23), 12–13th rib subcutaneous fat thickness (0.27) and Warner-Bratzler shear force (0.31), but unfavourable genetic correlation with marbling score (0.31) (Tait et al., 2008). In Japanese

Black cattle, genetic correlations between myristic acid (C14:0) and carcass weight, long-issimus muscle area, rib thickness, subcutaneous fat thickness and marbling score were essentially zero, ranging from −0.07 to 0.05 (Nogi et al., 2011). Oleic acid (C18:1), the most prominent MUFA in beef, suggested to be positively related to fat melting point and beef flavour (Wood et al., 2004), was found to have a strong favourable genetic correlation with marbling score (0.83, Tait et al., 2008; 0.19, Nogi et al., 2011) but a weak to moderate antagonistic genetic relationship with hot carcass weight (−0.14), 12–13th rib subcutaneous fat thickness (0.18) and Warner-Bratzler shear force (0.12) (Tait et al., 2008).

Improving the nutritional and health values of beef offers opportunities to make a positive contribution towards reducing the risk of some chronic diseases through a healthier diet. Beef is a major dietary source of conjugated linoleic acid (CLA), which possesses a range of health-promoting biological properties (reviewed in Hennessy et al., 2011). Beef also contains small amounts of the long chain C20/22 polyunsaturated, omega-3 fatty acids: EPA (eicosapentaenoic acid) and DHA (docosahexaenoic acid), and also a beneficially low omega-6 to omega-3 ratio. Initial studies suggest fatty acid composition is heritable and sufficient genetic variation exists for improvement by selection. However, more detailed studies on the genetic correlation among fatty acids, as well as genetic correlations with carcass and palatability traits are needed in order to predict correlated changes.

Major Genes and Gene Markers Associated with Meat Quality Traits

Over the last decade, research on the genes determining meat quality traits has transitioned from candidate gene approaches for traits with extensive phenotypic and pedigree information to large-scale approaches of genome analysis including nucleotide sequence, gene analysis and expression, proteomics, metabolomics and biochemical modelling. A number of genome-wide association studies have detected quantitative trait loci for meat quality traits in cattle.

The comprehensive animal quantitative trait locus (QTL) database (Hu *et al.*, 2013), developed to facilitate integration and mining of available genomic information, currently contains 904 curated QTLs for 96 individual meat quality traits and 459 curated QTLs for 42 carcass characteristics traits (AnimalQTLdb, http://www.animalgenome.org/QTLdb). A number of genes and gene markers have been identified, and their effects on different meat quality traits have been evaluated in different breeds and these are discussed in this section.

Double-muscling

Documented more than 200 years ago in Durham cattle and described by Kaiser in 1888 (Kaiser, 1888), double-muscling syndrome was proved to be caused by a single gene effect when Charlier *et al.* (1995) mapped the genetic defect to the centromeric end of BTA2. A dramatic doubling in weight of individual muscles has been observed when this gene, which was initially called growth/differentiation factor-8 (GDF-8), was disrupted by gene targeting in mice. The gene was renamed myostatin because of its function as a negative regulator of skeletal muscle development (McPherron *et al.*, 1997).

Analysis of myostatin gene in the Belgian Blue double-muscled cattle identified an 11-base-pair deletion leading to a non-functional myostatin protein (Grobet *et al.*, 1997) and the bovine myostatin gene was mapped to BTA2 (Smith *et al.*, 1997). This frame-shift 11-base-pair mutation (*p.D273RfsX13*) exists in the Belgian Blue, Blonde d'Aquitaine, Limousin and South Devon breeds, occurs in the third exon of the myostatin gene, and results in a premature STOP codon leading to the truncation of most of the bioactive carboxyterminal domain (Fig. 22.2). Subsequently, other loss-of-function mutations have been identified in double-muscled cattle (Grobet *et al.*, 1998; Karim *et al.*, 2000; see Table 22.6) and include four mutations due to a premature STOP codon: three single nucleotide substitutions (*p.Q204X* in Charolais and Limousin, *p.E226X* in Maine-Anjou, *p.E291X* in Marchigiana) and one insertion–deletion, where a 7-base-pair deletion is replaced with an unrelated stretch of 10 base pairs (*p.F140X*=c.419- 426delTTAAA TTTinsAAGCATACAA, in Maine-Anjou) (Grobet *et al.*, 1998; Marchitelli *et al.*, 2003). Another mutation in Piedmontese cattle (*p. C313Y*) leading to substitution of a carboxyterminal cysteine with a tyrosine, affects the intramolecular disulphite bridge controlling the stability of the bioactive domain.

Other mutations that do not lead to non-functional myostatin have been identified, and the most notable one is *p.F94L* – a missense mutation with no demonstrated effect on myostatin function, and predominant in Limousin cattle, a breed known for its pronounced (but not extreme) muscularity. The *p.F94L* mutation has been shown to be associated with lower fat thickness, fat yield, marbling scores and carcass fat (Wiener *et al.*, 2002; Casas *et al.*, 2004; Esmailizadeh *et al.*, 2008).

Myostatin is secreted as a 52-kDa precursor protein, which is proteolytically processed into a mature 12-kDa polypeptide and a 40-kDa amino-terminal inhibitory propeptide (latency-associated peptide). This latency-associated

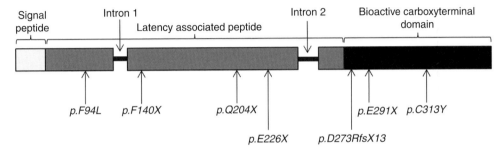

Fig. 22.2. Schematic representation of the myostatin gene and the approximate location of seven most important mutations.

Table 22.6. Summary of major mutations in the myostatin gene.

Mutation name	Type of mutation	Change at gene level	Change at protein level	Breed
p.F94L	Silent	C → A transversion at position 282	Conservative phenylalanine-to-leucine substitution in the N-terminal latency-associated peptide	Limousin
p.F140X	Loss-of-function	Insertion–deletion (7 base pair deletion replaced with an unrelated stretch of 10 base pairs) at position 419	Truncated protein due to a premature STOP codon in the N-terminal latency-associated peptide	Maine-Anjou
p.Q204X	Loss-of-function	C → T transition at position 610	Truncated protein due to a premature STOP codon in the N-terminal latency-associated peptide	Charolais, Limousin
p.E226X	Loss-of-function	G → T transition at position 676	Truncated protein due to a premature STOP codon in the N-terminal latency-associated peptide	Maine-Anjou
p.D273RfsX13	Loss-of-function	Deletion of 11 base pairs at position 821	Truncated protein due to a premature STOP codon in the bioactive carboxyterminal peptide	Belgian Blue, Blonde d'Aquitaine, Limousin, South Devon
p.E291X	Loss-of-function	G → T transversion at position 874	Truncated protein due to a premature STOP codon in the bioactive carboxyterminal peptide	Marchigiana
p.C313Y	Change of structure and function of the protein	G → A transition at position 938	Substitution of a cysteine involved in intramolecular disulphite bridge in the bioactive carboxyterminal peptide	Piedmontese

peptide has the ability to bind to the mature myostatin polypeptide and inhibit its signalling activity (Li *et al.*, 2010). The primary role of myostatin as a negative regulator of myogenesis is accomplished upon binding of the myostatin to activin type II receptors and altering the transcription of several target genes. The end result is a negative regulation of the G1 to S progression in the cell cycle of myoblasts, negative regulation of myoblast differentiation and negative regulation of satellite cell activation (Langley *et al.*, 2002; McCroskery *et al.*, 2003). The inhibition of myostatin activity during fetal development due to the loss-of-function mutations relieves these negative regulations and promotes myofibre formation (Joulia-Ekaza and Cabello, 2007). This leads to hyperplasia (increased myofibril number), to a certain extent hypertrophy (increased myofibre area), and increased proportion of fast-twitch glycolytic fibres (Bellinge *et al.*, 2005; Hennebry *et al.*, 2009).

In double-muscled cattle, traits other than muscling are affected. While some muscles exhibit hypertrophy, a 3 to 10% decrease in bone weight has been observed and an overall reduction in skeletal size (Bellinge et al., 2005), with the net result being a higher dressing percentage and higher cutability (proportion of saleable lean meat) in double-muscled individuals (Arthur et al., 1989; Fiems et al., 1995). The proportion of connective tissue or collagen is reduced by approximately 20–30% (Boccard, 1982; Uytterhaegen et al., 1994) and the collagen has a lower proportion of stable cross-links, which is overall associated with an increased tenderness (Wheeler et al., 2001). However, other studies report contradictory results regarding the tenderness of double-muscled cattle (Casas et al., 1998; Esmailizadeh et al., 2008; Rourke et al., 2009; Wiener et al., 2009). There is also a decrease in the subcutaneous fat and the intramuscular fat, which could be reduced by almost 50%. In addition to lower fat content, fatty acid composition of intramuscular fat is also different in double-muscled animals, which have a higher polyunsaturated to saturated fatty acids ratio due to increased concentration of polyunsaturated and decreased concentration of saturated fatty acids (Raes et al., 2001; Aldai et al., 2007; Wiener et al., 2009). The extent that myostatin causes the change in intramuscular fatty acid composition is difficult to determine because lower levels of intramuscular fat are normally associated with a higher unsaturated to saturated fatty acids ratio.

Birth weight of double-muscled calves is higher than normal calves due in part to a longer gestation period (Hanset, 1991) and this leads to higher incidence of dystocia, or calving difficulties. In addition, the cow pelvis has been shown to have a higher level of muscling and the pelvic opening of the double-muscled dams was 6–10% lower, contributing to higher mortality rates in unassisted births (Wiener et al., 2002). These conditions have resulted in most of the births in homozygote animals requiring caesarean section, which has no negative effects on the health of the calf, but it is associated with a significant reduction in subsequent pregnancy rate and longer calving interval (Arthur, 1995; Kolkman et al., 2007). This is probably the main reason why double-muscling is not systematically selected for and, mostly in some European beef breeds, it is tolerated at intermediate frequencies.

Calpain system

Tenderness is considered one of the most important factors in determining consumer satisfaction and eating quality, and the post-slaughter ageing process plays a major role in the quality achieved in the final meat product. This tenderization process includes proteolytic degradation of proteins in the meat. The calpain–calpastatin enzyme complex regulates the rates of degradation of various structural proteins in the muscle (Koohmaraie et al., 2002). Three well-characterized members of the calpain system are the μ-calpain and m-calpain – two Ca^{2+}-dependent cysteine proteases; and calpastatin, a protein that specifically inhibits the activity of μ- and m-calpain (Goll et al., 2003). The calpain-mediated proteolysis of myofibril proteins has been identified as responsible for breakdown of myofibril protein, which is associated with improved meat tenderness during post mortem storage of carcasses (Ouali and Talmant, 1990; Goll et al., 1992; Huff-Lonergan et al., 1996). Although both μ- and m-calpain target and breakdown the same myofibrillar proteins, they are activated at different times post mortem. μ-calpain is activated early post mortem when proteolysis of key microfibrils occurs (Taylor et al., 1995), but it is less stable than m-calpain, which persists longer. The μ- and m-calpain are encoded by calpain 1 (CAPN1) and calpain 2 (CAPN2) genes, respectively, and both genes are activated in vitro by micromolar and millimolar calcium concentrations (Goll et al., 2003). The calcium binding causes changes in the calpain molecule resulting in activation but also allowing calpastatin to interact and inhibit the enzyme. High levels of calpastatin reduce the proteolysis activity of calpain and have been correlated with decreased meat tenderness (Koohmaraie et al., 1995; Pringle et al., 1997). Shackelford et al. (1994) estimated that 40% of the variation in tenderness after conditioning could be explained by differences in calpastatin activity in the muscle 24 h after slaughter.

Several markers for beef tenderness have been developed for genes in the calpain system (Table 22.7) and some have been commercialized through genetic tests like GeneSTAR Tenderness and Igenity Tender-GENE marker panels. It is believed that none of these polymorphisms represent the functional mutation given that only two combinations of the identified polymorphisms have been associated with improved tenderness while the others failed to have consistent effects.

Van Eenennaam (2007) conducted a study on *Bos taurus* and *Bos indicus* cattle in which associations of the calpastatin and calpain markers with tenderness were validated. In that study, a 1-kg difference in Warner-Bratzler shear force, representing approximately 0.66 phenotypic standard deviations, was observed between the most and least tender genotypes. Johnston *et al.* (2010) analysed the effect of *CAST T1*, *CAPN 316*, *CAPN 4751* and *CALP 3* across a range of beef cattle breeds including temperate breeds (Angus, Hereford, Murray Grey and Shorthorn) and tropically adapted breeds (Brahman, Santa Gertrudis and Belmont Red),

across production systems. Out of the four markers investigated, *CAST T1* and *CAPN 316* had significant and consistent effects on shear force across all breeds and populations studied (ranging from –0.13 to –0.19 kg for *CAST T1* and from –0.15 to –0.20 kg for *CAPN 316*), even though the size of the marker effect was different across muscles and reduced upon tenderstretch hanging. Marker *CAPN 4751* had a significant effect for shear force in tropically adapted breeds (–0.12 kg) but was not significant in temperate breeds, while *CAPN3* was only significant for shear force in two tropical breeds (0.13 kg in Santa Gertrudis and –0.11 kg in Brahman) each having a different favourable allele. The estimated combined effect of all four markers on the phenotypic variation of shear force in M. longissimus thoracis et lumborum ranged from 1.5 to 4.7%.

In addition to effects on tenderness, calpain system gene markers have been investigated for effects on production traits in Brahman cattle (Cafe *et al.*, 2010). No effects on intake or feed efficiency were identified, however, there was a small but significant effect of *CAST T1*

Table 22.7. Genetic markers for beef tenderness in genes in the calpain system.

Gene	BTA	SNP name	SNP location	SNP position/nucleotide substitution	Amino acid substitution	Trait
Micro-Calpain	29	CAPN1-530[a]	Exon 14	G/A	Valine/ isoleucine	Shear force[a]
		CAPN1-316[a]	Exon 9	CAPN1:c.947C > G	Glycine/ alanine	Shear force[a,b,c,d,e] Tenderness score[d]
		CAPN1-4751[f]	Intron 17	CAPN1:g.6545C > T		Shear force[c,f] Tenderness score[g,h] Flavour[g] Overall liking[h]
Calpain 3	10	CAPN3[l]	Intron	CAPN3:c.1538+225G>T		Tenderness score[h] Shear force[i] Juiciness[h] Flavour[h]
Calpastatin	7	CAST-T1[m]	3' UTR	CAST:c.2832A > G		Tenderness score[g,h] Shear force[b,c,g,i] Overall liking[h] Juiciness[g] Flavour[g]
		UoG- CAST[j]	Intron 5	CAST g.282C>G		Shear force[c,j]
		CAST-155[k]		CAST:c.155C>T	Proline/ leucine	Shear force[k]

[a]Page *et al.*, 2002; [b]Morris *et al.*, 2006; [c]Van Eenennaam *et al.*, 2007; [d]Gill *et al.*, 2009; [e]Melucci *et al.*, 2012; [f]White *et al.*, 2005; [g]Casas *et al.*, 2006; [h]Robinson *et al.*, 2012; [i]Cafe *et al.*, 2010; [j]Schenkel *et al.*, 2006; [k]Barendse *et al.*, 2007; [l]Barendse *et al.*, 2008; [m]Barendse, 2002.

or *CAPN3* on rib fat, body weight and flight speed in some populations. Few or no effects of the calpain system gene markers were found on steer finishing and meat quality traits and heifer puberty traits in Brahman cattle (Wolcott and Johnston, 2009). The only significant effects were on some fat depth and marbling traits in steers.

Thyroglobulin

Thyroglobulin, the precursor to thyroid hormones, plays an important role in fat metabolism by regulating adipocyte development. The thyroglobulin gene (*TG*) is in a QTL region for backfat thickness on BTA14 (Casas *et al.*, 2000; Moore *et al.*, 2003) and genetic variation in the 5′ promoter region has been associated with marbling score and fat deposition traits in beef cattle (Barendse *et al.*, 2004). The SNP upstream of the promoter resulting in a C/T transition and estimated to account for 6.5% of the residual variance for marbling phenotype is the basis of the commercially available GeneSTAR Marbling™ (Barendse *et al.*, 2004), whereby cattle with TT or CT genotypes appear to have higher marbling scores than cattle with the CC genotype. Although a number of studies reported contradictory results regarding the effect of this SNP (Barendse *et al.*, 2004; Rincker *et al.*, 2006; Casas *et al.*, 2007), a meta-analysis of 11 independent studies provided support for the association of the TG marker with marbling score (Wood *et al.*, 2006). The frequency of the favourable T allele has been reported to be highest in the Wagyu breed, followed by other *Bos taurus* and then *Bos indicus* (Van Eenennaam *et al.*, 2007). Four additional SNPs in the 3′ flanking region of the *TG* gene were significantly associated with marbling score (Gan *et al.*, 2008). An epistatic additive association has been identified between the TG and casein gene markers (Bennett *et al.*, 2013). It has been suggested that differences among populations and inconsistencies in association studies could be the result of underlying complex associations and epistatic interactions. The TG gene markers may simply be in linkage disequilibrium with the causal gene. The region of BTA14, where the thyroglobulin is located, is rich in genes involved in lipid metabolism and adipocyte differentiation.

Other candidate genes

Leptin, one of the most important adipose-derived hormones, plays an important role in regulation of energy intake and expenditure, appetite and body composition (Houseknecht *et al.*, 1998; Baile *et al.*, 2000). Serum leptin concentration has been associated with adipose depots and carcass characteristics in beef cattle such as marbling, 12th rib back fat thickness, kidney pelvic heart fat and quality grade (Minton *et al.*, 1998; Geary *et al.*, 2003). The leptin gene (*LEP*) is located on BTA4 (Stone *et al.*, 1996) and has three exons, with the first one not being transcribed into the 167-amino acid protein. Significant associations have been identified between leptin genotypes and carcass fat (Buchanan *et al.*, 2002; Schenkel *et al.*, 2005; Lusk, 2007), body weight (Lusk, 2007), feed intake (Lagonigro *et al.*, 2003), meat quality (Gill *et al.*, 2009) and growth rate (Nkrumah *et al.*, 2005) in beef cattle, as well as fertility and milk production traits in dairy cattle (Clempson *et al.*, 2011). The C/T SNP resulting in an arginine to cysteine amino acid substitution has a significant association with carcass characteristics. Homozygous CC animals have the lowest backfat thickness and are expected to require an average of 7 more days on feed to reach 12-mm backfat than TT animals (Woronuk *et al.*, 2012). However, other studies failed to identify an association between that leptin gene SNP and cattle marbling and fatness traits (Barendse *et al.*, 2005) or backfat thickness and total lipids (Fortes *et al.*, 2009; Pannier *et al.*, 2009).

The *DGAT1* (*diacylglycerol acyltransferase 1*) gene maps to BTA14, near the thyroglobulin gene and CSSM66, a marker for marbling score (Barendse *et al.*, 1997). The *DGAT1* gene encodes acyl CoA:diacylglycerol acyltransferase, a key enzyme in triacylglycerol synthesis (Cases *et al.*, 1998). A dinucleotide substitution (AA/GC) in exon 8 of *DGAT1* resulting in a non-synonymous lysine (K) to alanine (A) substitution is suspected to have direct impact on the enzyme activity (Grisart *et al.*, 2002). This polymorphism associates with milk fat content and other milk characteristics in Holstein Friesian cattle (Grisart et al., 2002). Association with intramuscular fat deposition and marbling in beef cattle has been hypothesized (Thaller et al., 2003) based on a 1.62% difference in intramuscular fat content between the AA and other (AG + GG) genotypes. However, the small number of animals in that study (5 AA, 23 AG + GG) did not allow for a strong conclusion about the effect of this polymorphism. An association with sirloin weight and fat depth around the sirloin has been identified in Aberdeen Angus-sired cattle (Gill et al., 2009) where animals with the AA genotype had heavier sirloins (1.17 kg difference between the AA and GG genotypes) and more fat surrounding the sirloin (4.58 mm difference between the AA and GG genotypes). This effect was not confirmed in other B. taurus cattle breeds (Pannier et al., 2009) and no associations with carcass fat traits have been identified in Brahman cattle (Casas et al., 2005) or other B. indicus-influenced cattle (Fortes et al., 2009).

An association between polymorphisms in the growth hormone gene on BTA19 and fat-related traits have been observed in several studies. These associations include intramuscular and rump fat distribution in feedlot cattle (Barendse et al., 2006) and carcass traits, marbling score and fatty acid composition in Japanese Black cattle (Tatsuda et al., 2008; Ardiyanti et al., 2009). In these studies, the C allele was associated with lower marbling and had an average allele substitution effect of −0.22 phenotypic standard deviations, higher rump fat (average effect of 0.11 standard deviations), higher marbling scores (substitution effect of 17% of the phenotypic standard deviation), higher C18:1 and MUFA percentages, and lower SFA percentages. Although the associations with carcass weight and fat-related traits were not validated in a population of Aberdeen Angus-cross cattle, polymorphisms in the growth hormone gene had a significant effect on eye muscle area length as a percentage of sirloin muscle length, eye muscle length and eye muscle area (Gill et al., 2010a). A polymorphism in intron 3 of the growth hormone gene was found to be related to meat quality traits in Piedmontese cattle where a gene substitution effect of 2.39% for cooking loss and 0.65 kg for 11-day tenderness has been calculated

(Di Stasio et al., 2003). In addition to the growth hormone gene, an A to T substitution in exon 8 of the growth hormone receptor gene that results in a non-synonymous substitution of phenylalanine with a tyrosine residue (Blott et al., 2003), has been associated with drip loss (Di Stasio et al., 2005), marbling score (Hale et al., 2000) and meat odour judged by taste panel members. The same study failed to identify association with any carcass quality traits (Gill et al., 2009).

Recent studies on fat composition and quality investigated a number of candidate genes known to play key roles in lipid metabolism and fatty acids synthesis and regulation in adipose tissue. These genes are: adipose fatty acid binding protein (FABP4), stearoyl-coA desaturase (SCD), fatty acid synthase (FASN) and sterol regulatory element-binding protein 1 (SREBP1).

FABP4 is expressed in adipose tissue and has an important role in lipid metabolism and homeostasis in adipocytes. The FABP4 gene is located on BTA14 in a QTL region for meat production traits fat thickness, yield grade, and marbling (Casas et al., 2003). Two SNPs in this gene were found to have a significant effect on carcass weight, marbling score and subcutaneous fat depth (Michal et al., 2006; Cho et al., 2008; Barendse et al., 2009; Lee et al., 2010). The A allele of g.3473T>A SNP increased carcass weight by 5.01 ± 2.2 kg and explained 0.8% of phenotypic variance, the G allele of g.3631A>G SNP increased marbling score by 0.21 ± 0.07, explaining 1.3% of phenotypic variance, and the G allele at g.2502C>G locus had an effect of 0.3% of variation in intramuscular fat. However, a number of studies failed to identify associations between markers in FABP1 gene and intramuscular fat levels and marbling scores (Pannier et al., 2009; Tizioto et al., 2012).

The enzyme that catalyses the desaturation of saturated to unsaturated fatty acids is encoded by the SCD gene, reported to play a key role in conversion of stearic (C18:0) and palmitic acid (C16:0) to oleic acid (C18:1) (Kim and Ntambi, 1999). In Wagyu cattle, a T/C substitution at position 878 (exon 5) resulting in a valine to alanine substitution at position 293 in the SCD protein has been associated with MUFA content and the melting point of intramuscular fat (Taniguchi *et al.*, 2004). Melting point is an important indicator of meat quality, as it is associated with favourable beef flavour and tenderness (Melton *et al.*, 1982; Smith *et al.*, 2006). The substitution effect for the *SCD* gene in this study was 0.81% on MUFA percentage. Analysis of 3 other SNP in the 3′ untranslated region of the *SCD* gene revealed a positive association with beef marbling score (ranging in a difference of 0.56 to 0.87 marbling scores between the two homozygotes), amount of MUFAs (ranging from 0.67 to 1.36% difference between the two homozygotes) and conjugated linoleic acid content (ranging from 6.31 to 11.76 mg/100 g dry tissue difference between the two homozygotes), but negative association with amount of SFA, which ranged from −1.01 to −1.87% difference between homozygous genotypes (Jiang *et al.*, 2008).

The expression of the *SCD* gene is known to be regulated by sterol regulatory element binding protein-1 (*SREBP*) (Bene *et al.*, 2001). A *SREBP1* polymorphism has been related to carcass weight (Ohsaki *et al.*, 2009), but no significant effects on meat yield traits were identified in Japanese Black cattle (Matsuhashi *et al.*, 2011). An 84-bp insertion/deletion in intron 5 of the *SREBP1* gene has been suggested to affect MUFA content through regulation of *SCD* gene expression (Hoashi *et al.*, 2007), with favourable *SREBP1* and *SCD* genotypes showing 1.3% and 2.1% higher MUFA content, respectively, and 1.6°C and 2.5°C lower melting point, respectively.

Fatty acid synthase (*FASN*) is a multifunctional enzyme catalysing *de novo* synthesis of long-chain SFA in cells (Smith, 1994). The *FASN* gene is on BTA19 in a QTL region for fatty acid composition of adipose tissue in beef cattle and milk fat in dairy cattle (Morris *et al.*, 2007). Polymorphisms in the *FASN* gene have been associated with fatty acid composition of backfat, intermuscular and intramuscular fat in Japanese Black and Limousin cattle (Abe *et al.*, 2008; Matsuhashi *et al.*, 2011). Amino acid substitutions in the thioesterase domain of the *FASN* gene were associated with fatty acid composition in Angus cattle (Zhang *et al.*, 2008), while five SNPs were associated with increased MUFA,

marbling score and decreased SFA in Korean cattle (Oh *et al.*, 2012).

Conclusions

Consumer preferences are becoming more complex. New traits of interest include nutritional and health value, high and consistent quality of beef products as well as concern regarding the environmental impact and animal welfare associated with beef production systems. There seems to be a growing discrepancy between demands of consumers and industry response, which tend to focus the improvement effort on traits directly associated with cost of production or revenue (economically relevant traits). As expressed by Garrick and Golden (2009), the goal of the industry should be to produce beef that is nutritious, healthful and desirable and be produced in a manner that is respectful of the resources used in its production, including the environment and welfare of the animals. If this goal is adopted, then meat quality, nutritional value and healthfulness should be important components. There is currently no attempt in the beef industry to improve quality, nutritional or healthfulness traits, primarily because these traits are difficult and expensive to measure on large numbers of seedstock animals, there is no phenotypic measure accepted across the supply chain, and correspondingly no market incentives. Recent advances in biotechnology such as whole-genome sequencing, high-density genotyping panels and next generation sequencing provide new opportunities for progress toward the goal of identifying the genes and gene networks controlling the variation in meat quality traits. As shown in this chapter, genetic factors influence mineral content, fatty acid composition, as well as other meat quality traits, so it stands to reason that these traits can be improved via marker assisted or genomic selection.

Producers have difficulty seeing how they can benefit financially from improvement in these quality traits. It is clear that consumers' perception of quality, nutritional value and healthfulness has a great effect on their preferences and, therefore, demand for the product.

To address these issues a strong economic case for improving these traits can be made by quantifying the economic consequences associated with changes in consumers' demand for beef and relating the changes with quality traits. In addition, practical measures of these traits would need to be developed and associated with product specification across the supply chain, from sale of animals at weaning and onwards, including at point of consumer sale.

References

Abe, T., Saburi, J., Hasebe, H., Nakagawa, T., Kawamura, T., Saito, K., Nade, T., Misumi, S., Okumura, T., Kuchida, K. *et al.* (2008) Bovine quantitative trait loci analysis for growth, carcass, and meat quality traits in an F2 population from a cross between Japanese Black and Limousin. *Journal of Animal Science* 86, 2821–2832.

Aldai, N., Najera, A.I., Dugan, M.E., Celaya, R. and Osoro, K. (2007) Characterisation of intramuscular, intermuscular and subcutaneous adipose tissues in yearling bulls of different genetic groups. *Meat science* 76, 682–691.

AMSA (1995) Research guidelines for cookery, sensory evaluation, and instrumental tenderness measurements of fresh meat. National Live Stock and Meat Board. Chicago, IL.

Ardiyanti, A., Oki, Y., Suda, Y., Suzuki, K., Chikuni, K., Obara, Y. and Katoh, K. (2009) Effects of GH gene polymorphism and sex on carcass traits and fatty acid compositions in Japanese Black cattle, *Animal Science Journal* 80, 62–69.

Arthur, P. (1995) Double muscling in cattle: a review. *Australian Journal of Agricultural Research* 46, 1493–1515.

Arthur, P.F., Makarechian, M., Price, M.A. and Berg, R.T. (1989) Heterosis, maternal and direct effects in double-muscled and normal cattle: I. Reproduction and growth traits. *Journal of Animal Science* 67, 902–910.

Baile, C.A., Della-Fera, M.A. and Martin, R.J. (2000) Regulation of metabolism and body fat mass by leptin. *Annual Review of Nutrition* 20, 105–127.

Barendse, W. (2002) DNA markers for meat tenderness. International patent application PCT/AU02/00122., edited by I. p. p. W. A1.

Barendse, W., Vaiman, D., Kemp, S.J., Sugimoto, Y., Armitage, S.M., Williams, J.L., Sun, H.S., Eggen, A., Agaba, M., Aleyasin, S.A. *et al.* (1997) A medium-density genetic linkage map of the bovine genome. *Mammalian Genome* 8, 21–28.

Barendse, W., Bunch, R., Thomas, M., Armitage, S., Baud, S. and Donaldson, N. (2004) The TG5 thyroglobulin gene test for a marbling quantitative trait loci evaluated in feedlot cattle. *Australian Journal of Experimental Agriculture* 44, 669–674.

Barendse, W., Bunch, R.J. and Harrison, B.E. (2005) The leptin C73T missense mutation is not associated with marbling and fatness traits in a large gene mapping experiment in Australian cattle. *Animal Genetics* 36, 86–88.

Barendse, W., Bunch, R.J., Harrison, B.E. and Thomas, M.B. (2006) The growth hormone 1 GH1:c.457C>G mutation is associated with intramuscular and rump fat distribution in a large sample of Australian feedlot cattle. *Animal Genetics* 37, 211–214.

Barendse, W., Harrison, B.E., Hawken, R.J., Ferguson, D.M., Thompson, J.M., Thomas, M.B. and Bunch, R.J. (2007) Epistasis between calpain 1 and its inhibitor calpastatin within breeds of cattle. *Genetics* 176, 2601–2610.

Barendse, W., Harrison, B.E., Bunch, R.J. and Thomas, M.B. (2008) Variation at the Calpain 3 gene is associated with meat tenderness in zebu and composite breeds of cattle. *Biomedical Genetics* 9, 41.

Barendse, W., Bunch, R.J., Thomas, M.B. and Harrison, B.E. (2009) A splice site single nucleotide polymorphism of the fatty acid binding protein 4 gene appears to be associated with intramuscular fat deposition in longissimus muscle in Australian cattle. *Animal Genetics* 40, 770–773.

Bellinge, R.H., Liberles, D.A., Iaschi, S.P., O'Brien, A. and Tay, G.K. (2005) Myostatin and its implications on animal breeding: a review. *Animal Genetics* 36, 1–6.

Bene, H., Lasky, D. and Ntambi, J.M. (2001) Cloning and characterization of the human stearoyl-CoA desaturase gene promoter: transcriptional activation by sterol regulatory element binding protein and repression by polyunsaturated fatty acids and cholesterol. *Biochemical and Biophysical Research Communications* 284, 1194–1198.

Bennett, G.L., Shackelford, S.D., Wheeler, T.L., King, D.A., Casas, E. and Smith, T.P. (2013) Selection for genetic markers in beef cattle reveals complex associations of thyroglobulin and casein1-s1 with carcass and meat traits. *Journal of Animal Science* 91, 565–571.

Biesalski, H.K. (2005) Meat as a component of a healthy diet – are there any risks or benefits if meat is avoided in the diet? *Meat Science* 70, 509–524.

Blott, S., Kim, J.J., Moisio, S., Schmidt-Kuntzel, A., Cornet, A., Berzi, P., Cambisano, N., Ford, C., Grisart, B., Johnson, D. *et al.* (2003) Molecular dissection of a quantitative trait locus: a phenylalanine-to-tyrosine substitution in the transmembrane domain of the bovine growth hormone receptor is associated with a major effect on milk yield and composition. *Genetics* 163, 253–266.

Boccard, R. (1982) Relationship between muscle hypertrophy and the composition of skeletal muscles. In: King, J.W.B. and Ménissier, F. (eds) *Muscle Hypertrophy of Genetic Origin and its use to Improve Beef Production.* Springer, Amsterdam, pp. 148–161.

Buchanan, F.C., Fitzsimmons, C.J., Van Kessel, A.G., Thue, T.D., Winkelman-Sim, D.C. and Schmutz, S.M. (2002) Association of a missense mutation in the bovine leptin gene with carcass fat content and leptin mRNA levels. *Genetics, Selection, Evolution* 34, 105–116.

Cafe, L.M., McIntyre, B.L., Robinson, D.L., Geesink, G.H., Barendse, W. and Greenwood, P.L. (2010) Production and processing studies on calpain-system gene markers for tenderness in Brahman cattle: 1. Growth, efficiency, temperament, and carcass characteristics. *Journal of Animal Science* 88, 3047–3058.

Casas, E., Keele, J.W., Shackelford, S.D., Koohmaraie, M., Sonstegard, T.S., Smith, T.P., Kappes, S.M. and Stone, R.T. (1998) Association of the muscle hypertrophy locus with carcass traits in beef cattle. *Journal of Animal Science* 76, 468–473.

Casas, E., Shackelford, S.D., Keele, J.W., Stone, R.T., Kappes, S.M. and Koohmaraie, M. (2000) Quantitative trait loci affecting growth and carcass composition of cattle segregating alternate forms of myostatin. *Journal of Animal Science* 78, 560–569.

Casas, E., Shackelford, S.D., Keele, J.W., Koohmaraie, M., Smith, T.P. and Stone, R.T. (2003) Detection of quantitative trait loci for growth and carcass composition in cattle. *Journal of Animal Science* 81, 2976–2983.

Casas, E., Bennett, G.L., Smith, T.P. and Cundiff, L.V. (2004) Association of myostatin on early calf mortality, growth, and carcass composition traits in crossbred cattle, *Journal of Animal Science* 82, 2913–2918.

Casas, E., White, S.N., Riley, D.G., Smith, T.P., Brenneman, R.A., Olson, T.A., Johnson, D.D., Coleman, S.W., Bennett, G.L. and Chase, C.C. Jr (2005) Assessment of single nucleotide polymorphisms in genes residing on chromosomes 14 and 29 for association with carcass composition traits in *Bos indicus* cattle. *Journal of Animal Science* 83, 13–19.

Casas, E., White, S.N., Wheeler, T.L., Shackelford, S.D., Koohmaraie, M., Riley, D.G., Chase, C.C. Jr, Johnson, D.D. and Smith, T.P. (2006) Effects of calpastatin and micro-calpain markers in beef cattle on tenderness traits. *Journal of Animal Science* 84, 520–525.

Casas, E., White, S.N., Shackelford, S.D., Wheeler, T.L., Koohmaraie, M., Bennett, G.L. and Smith, T.P. (2007) Assessing the association of single nucleotide polymorphisms at the thyroglobulin gene with carcass traits in beef cattle. *Journal of Animal Science* 85, 2807–2814.

Cases, S., Smith, S.J., Zheng, Y.W., Myers, H.M., Lear, S.R., Sande, E., Novak, S., Collins, C., Welch, C.B., Lusis, A.J. *et al.* (1998) Identification of a gene encoding an acyl CoA:diacylglycerol acyltransferase, a key enzyme in triacylglycerol synthesis. *Proceedings of the National Academy of Sciences USA* 95, 13018–13023.

Cecchinato, A., De Marchi, M., Penasa, M., Casellas, J., Schiavon, S. and Bittante, G. (2012) Genetic analysis of beef fatty acid composition predicted by near-infrared spectroscopy. *Journal of Animal Science* 90, 429–438.

Charlier, C., Coppieters, W., Farnir, F., Grobet, L., Leroy, P.L., Michaux, C., Mni, M., Schwers, A., Vanmanshoven, P., Hanset, R. *et al.* (1995) The mh gene causing double-muscling in cattle maps to bovine Chromosome 2. *Mammalian Genome* 6, 788–792.

Cho, S., Park, T.S., Yoon, D.H., Cheong, H.S., Namgoong, S., Park, B.L., Lee, H.W., Han, C.S., Kim, E.M., Cheong, I.C. *et al.* (2008) Identification of genetic polymorphisms in FABP3 and FABP4 and putative association with back fat thickness in Korean native cattle. *BMB Reports* 41, 29–34.

Clempson, A.M., Pollott, G.E., Brickell, J.S., Bourne, N.E., Munce, N. and Wathes, D.C. (2011) Evidence that leptin genotype is associated with fertility, growth, and milk production in Holstein cows. *Journal of Dairy Science* 94, 3618–3628.

Crews, D.H., Jr, Enns, R.M., Rumph, J.M. and Pollak, E.J. (2008) Genetic evaluation of retail product percentage in Simmental cattle. *Journal of Animal Breeding and Genetics* 125, 13–19.

Di Stasio, L., Brugiapaglia, A., Destefanis, G., Albera, A. and Sartore, S. (2003) GH1 as candidate gene for variability of meat production traits in Piedmontese cattle. *Journal of Animal Breeding and Genetics* 120, 358–361.

Di Stasio, L., Destefanis, G., Brugiapaglia, A., Albera, A. and Rolando, A. (2005) Polymorphism of the GHR gene in cattle and relationships with meat production and quality. *Animal Genetics* 36, 138–140.

Dikeman, M.E., Pollak, E.J., Zhang, Z., Moser, D.W., Gill, C.A. and Dressler, E.A. (2005) Phenotypic ranges and relationships among carcass and meat palatability traits for fourteen cattle breeds, and heritabilities and expected progeny differences for Warner-Bratzler shear force in three beef cattle breeds. *Journal of Animal Science* 83, 2461–2467.

Dubeski, P.L., Aalhus, J.L., Jones, S.D.M., Robertson, W.M. and Dyck, R.S. (1997) Meat quality of heifers fattened to heavy weights to enhance marbling. *Canadian Journal of Animal Science* 77, 635–643.

Esmailizadeh, A.K., Bottema, C.D., Sellick, G.S., Verbyla, A.P., Morris, C.A., Cullen, N.G. and Pitchford, W.S. (2008) Effects of the myostatin F94L substitution on beef traits. *Journal of Animal Science* 86, 1038–1046.

Etcheverry, P., Hawthorne, K.M., Liang, L.K., Abrams, S.A. and Griffin, I.J. (2006) Effect of beef and soy proteins on the absorption of non-heme iron and inorganic zinc in children. *Journal of the American College of Nutrition* 25, 34–40.

Ferguson, D.M., Jiang, S.T., Hearnshaw, H., Rymill, S.R. and Thompson, J.M. (2000) Effect of electrical stimulation on protease activity and tenderness of M. longissimus from cattle with different proportions of *Bos indicus* content. *Meat Science* 55, 265–272.

Fiems, L.O., Hoof, J.V., Uytterhaegen, L., Boucque, C.V. and Demeyer, D. (1995) Comparative quality of meat from double-muscled and normal beef cattle. In: Ouali, A. Demeyer, D. and Smulders, F. (eds) *Expression of Tissue Proteinases and Regulation of Protein Degradation as Related to Meat Quality*, ECCEAMST, Utrecht, pp. 381–391.

Fortes, M.R., Curi, R.A., Chardulo, L.A., Silveira, A.C., Assumpcao, M.E., Visintin, J.A. and de Oliveira, H.N. (2009) Bovine gene polymorphisms related to fat deposition and meat tenderness. *Genetics and Molecular Biology* 32, 75–82.

Gan, Q.F., Zhang, L.P., Li, J.Y., Hou, G.Y., Li, H.D., Gao, X., Ren, H.Y., Chen, J.B. and Xu, S.Z. (2008) Association analysis of thyroglobulin gene variants with carcass and meat quality traits in beef cattle. *Journal of Applied Genetics* 49, 251–255.

Garrick, D.J. and Golden, B.L. (2009) Producing and using genetic evaluations in the United States beef industry of today. *Journal of Animal Science* 87 (14 Suppl), E11–E18.

Geary, T.W., McFadin, E.L., MacNeil, M.D., Grings, E.E., Short, R.E., Funston, R.N. and Keisler, D.H. (2003) Leptin as a predictor of carcass composition in beef cattle. *Journal of Animal Science* 81, 1–8.

Gill, J.L., Bishop, S.C., McCorquodale, C., Williams, J.L. and Wiener, P. (2009) Association of selected SNP with carcass and taste panel assessed meat quality traits in a commercial population of Aberdeen Angus-sired beef cattle. *Genetics, Selection, Evolution* 41, 36.

Gill, J.L., Bishop, S.C., McCorquodale, C., Williams, J.L. and Wiener, P. (2010a) Associations between single nucleotide polymorphisms in multiple candidate genes and carcass and meat quality traits in a commercial Angus-cross population. *Meat Science* 86, 985–993.

Gill, J.L., Matika, O., Williams, J.L., Worton, H., Wiener, P. and Bishop, S.C. (2010b) Consistency statistics and genetic parameters for taste panel assessed meat quality traits and their relationship with carcass quality traits in a commercial population of Angus-sired beef cattle. *Animal* 4, 1–8.

Glitsch, K. (2000) Consumer perceptions of fresh meat quality: cross-national comparison. *British Food Journal* 102, 177–194.

Goll, D.E., Thompson, V.F., Taylor, R.G. and Zalewska, T. (1992) Is calpain activity regulated by membranes and autolysis or by calcium and calpastatin? *Bioessays* 1–4, 549–556.

Goll, D.E., Thompson, V.F., Li, H., Wei, W. and Cong, J. (2003) The calpain system. *Physiological Reviews* 1990, 731–801.

Grisart, B., Coppieters, W., Farnir, F., Karim, L., Ford, C., Berzi, P., Cambisano, N., Mni, M., Reid, S., Simon, P. *et al.* (2002) Positional candidate cloning of a QTL in dairy cattle: identification of a missense mutation in the bovine DGAT1 gene with major effect on milk yield and composition. *Genome Research* 1–2, 222–231.

Grobet, L., Martin, L.J., Poncelet, D., Pirottin, D., Brouwers, B., Riquet, J., Schoeberlein, A., Dunner, S., Menissier, F., Massabanda, J., Fries, R., Hanset, R. and Georges, M. (1997) A deletion in the bovine myostatin gene causes the double-muscled phenotype in cattle. *Nature Genetics* 17, 71–74.

Grobet, L., Poncelet, D., Royo, L.J., Brouwers, B., Pirottin, D., Michaux, C., Menissier, F., Zanotti, M., Dunner, S. and Georges, M. (1998) Molecular definition of an allelic series of mutations disrupting the myostatin function and causing double-muscling in cattle. *Mammalian Genome* 9, 210–213.

Hale, C.S., Herring, W.O., Shibuya, H., Lucy, M.C., Lubahn, D.B., Keisler, D.H. and Johnson, G.S. (2000) Decreased growth in Angus steers with a short TG-microsatellite allele in the P1 promoter of the growth hormone receptor gene. *Journal of Animal Science* 78, 2099–2104.

Hanset, R. (1991) The major gene of muscular hypertrophy in the Belgian Blue cattle breed. In: Axford, R.F.E. and Owen, J.B. (eds) *Breeding for Disease Resistance in Farm Animals*. CAB International, Wallingford, UK, pp. 467–478.

Hennebry, A., Berry, C., Siriett, V., O'Callaghan, P., Chau, L., Watson, T., Sharma, M. and Kambadur, R. (2009) Myostatin regulates fiber-type composition of skeletal muscle by regulating MEF2 and MyoD gene expression. *American Journal of Physiology – Cell Physiology* 296, C525–C534.

Hennessy, A.A., Ross, R.P., Devery, R. and Stanton, C. (2011) The health promoting properties of the conjugated isomers of alpha-linolenic acid. *Lipids* 46, 105–119.

Hipkiss, A.R. (2009) Carnosine and its possible roles in nutrition and health. *Advances in Food and Nutrition Research* 57 87–154.

Hoashi, S., Ashida, N., Ohsaki, H., Utsugi, T., Sasazaki, S., Taniguchi, M., Oyama, K., Mukai, F. and Mannen, H. (2007) Genotype of bovine sterol regulatory element binding protein-1 (SREBP-1) is associated with fatty acid composition in Japanese Black cattle. *Mammalian Genome* 1–8, 880–886.

Hocquette, J.F., Botreau, R., Picard, B., Jacquet, A., Pethick, D.W. and Scollan, N.D. (2012) Opportunities for predicting and manipulating beef quality. *Meat Science* 92, 197–209.

Houseknecht, K.L., Baile, C.A., Matteri, R.L. and Spurlock, M.E. (1998) The biology of leptin: a review. *Journal of Animal Science* 76, 1405–1420.

Hu, Z.L., Park, C.A., Wu, X.L. and Reecy, J.M. (2013) Animal QTLdb: an improved database tool for livestock animal QTL/association data dissemination in the post-genome era. *Nucleic Acids Research* 41 (Database Issue), D871–D879.

Huff-Lonergan, E., Mitsuhashi, T., Beekman, D.D., Parrish, F.C., Olson, D.G. and Robson, R.M. (1996) Proteolysis of specific muscle structural proteins by mu-calpain at low pH and temperature is similar to degradation in postmortem bovine muscle. *Journal of Animal Science* 74, 993–1008.

Huffman, K.L., Miller, M.F., Hoover, L.C., Wu, C.K., Brittin, H.C. and Ramsey, C.B. (1996) Effect of beef tenderness on consumer satisfaction with steaks consumed in the home and restaurant. *Journal of Animal Science* 74, 91–97.

Igo, J.L., Vanoverbeke, D.L., Woerner, D.R., Tatum, J.D., Pendell, D.L., Vedral, L.L., Mafi, G.G., Moore, M.C., McKeith, R.O., Gray, G.D., Griffin, D.B., Hale, D.S., Savell, J.W. and Belk, K.E. (2013) Phase I of The National Beef Quality Audit – 2011: quantifying willingness-to-pay, best-worst scaling, and current status of quality characteristics in different beef industry marketing sectors. *Journal of Animal Science* 91, 1907–1919.

Inoue, K., Kobayashi, M., Shoji, N. and Kato, K. (2011) Genetic parameters for fatty acid composition and feed efficiency traits in Japanese Black cattle. *Animal* 5, 987–994.

Jiang, Z., Michal, J.J.,Tobey, D.J., Daniels, T.F., Rule, D.C. and Macneil, M.D. (2008) Significant associations of stearoyl-CoA desaturase (SCD1) gene with fat deposition and composition in skeletal muscle. *International Journal of Biological Sciences* 4, 345–351.

Johnston, D.J., and Graser, H.U. (2010) Estimated gene frequencies of GeneSTAR markers and their size of effects on meat tenderness, marbling, and feed efficiency in temperate and tropical beef cattle breeds across a range of production systems. *Journal of Animal Science* 88, 1917–1935.

Johnston, D.J., Reverter, A., Robinson, D.L. and Ferguson, D.M. (2001) Sources of variation in mechanical shear force measures of tenderness in beef from tropically adapted genotypes, effects of data editing and their implications for genetic parameter estimation. *Australian Journal of Experimental Agriculture* 41, 991–996.

Johnston, D.J., Reverter, A., Ferguson, D.M., Thompson, J.M. and Burrow, H.M. (2003a) Genetic and phenotypic characterisation of animal, carcass, and meat quality traits from temperate and tropically adapted beef breeds. 3. Meat quality traits. *Australian Journal of Agricultural Research* 54, 135–147.

Johnston, D.J., Reverter, A., Burrow, H.M., Oddy, V.H. and Robinson, D.L. (2003b) Genetic and phenotypic characterisation of animal, carcass, and meat quality traits from temperate and tropically adapted beef breeds. 1. Animal measures. *Australian Journal of Agricultural Research* 54, 107–118.

Joulia-Ekaza, D. and Cabello, G. (2007) The myostatin gene: physiology and pharmacological relevance. *Current Opinions in Pharmacology* 7, 310–315.

Kaiser, D. (1888) Uber die sogenannten doppellendigen Rinder. In: Thiel, H. (ed.) *Senerhebung Landwirtschaftliche Jahrbuch*. Verlag von Paul Parey, Berlin, pp. 387–403.

Karim, L., Coppieters, W., Grobet, L., Georges, M. and Valentini, A. (2000) Convenient genotyping of six myostatin mutations causing double-muscling in cattle using a multiplex oligonucleotide ligation assay. *Animal Genetics* 31, 396–399.

Kelly, M.J., Tume, R.K., Newman, S. and Thompson, J.M. (2013) Genetic variation in fatty acid composition of subcutaneous fat in cattle. *Animal Production Science* 53, 129–133.

Kim, Y.C. and Ntambi, J.M. (1999) Regulation of stearoyl-CoA desaturase genes: role in cellular metabolism and preadipocyte differentiation. *Biochemical and Biophysical Research Communications* 266, 1–4.

Kolkman, I., De Vliegher, S, Hoflack, G., Van Aert, M., Laureyns, J., Lips, D., De Kruif, A. and Opsomer, G. (2007) Protocol of the caesarean section as performed in daily bovine practice in Belgium. *Reproduction in Domestic Animals* 42, 583–589.

Koohmaraie, M., Shackelford, S.D., Wheeler, T.L., Lonergan, S.M. and Doumit, M.E. (1995) A muscle hypertrophy condition in lamb (callipyge): characterization of effects on muscle growth and meat quality traits. *Journal of Animal Science* 73, 3596–3607.

Koohmaraie, M., Kent, M.P., Shackelford, S.D., Veiseth, E. and Wheeler, T.L. (2002) Meat tenderness and muscle growth: is there any relationship? *Meat Science* 62, 345–352.

Lagonigro, R., Wiener, P., Pilla, F., Woolliams, J.A. and Williams, J.L. (2003) A new mutation in the coding region of the bovine leptin gene associated with feed intake. *Animal Genetics* 34, 371–374.

Langley, B., Thomas, M., Bishop, A., Sharma, M., Gilmour, S. and Kambadur, R. (2002) Myostatin inhibits myoblast differentiation by down-regulating MyoD expression. *Journal of Biological Chemistry* 277, 49831–49840.

Lawrie, R.A. (1966) The eating quality of meat. In: Lawrie, R.A. (ed.) *Meat Science*. Pergamon Press, London.

Lee, S.H., van der Werf, J.H., Lee, S.H., Park, E.W., Oh, S.J., Gibson, J.P. and Thompson, J.M. (2010) Genetic polymorphisms of the bovine fatty acid binding protein 4 gene are significantly associated with marbling and carcass weight in Hanwoo (Korean Cattle). *Animal Genetics* 41, 442–444.

Li, D., Siriamornpun, S., Wahlqvist, M.L., Mann, N.J. and Sinclair, A.J. (2005) Lean meat and heart health. *Asia Pacific Journal of Clinical Nutrition* 14, 113–119.

Li, Z., Zhao, B., Kim, Y.S., Hu, C.Y. and Yang, J. (2010) Administration of a mutated myostatin propeptide to neonatal mice significantly enhances skeletal muscle growth. *Molecular Reproduction and Development* 77, 76–82.

Lusk, J.L. (2007) Association of single nucleotide polymorphisms in the leptin gene with body weight and backfat growth curve parameters for beef cattle. *Journal of Animal Science* 85, 1865–1872.

McCroskery, S., Thomas, M., Maxwell, L., Sharma, M. and Kambadur, R. (2003) Myostatin negatively regulates satellite cell activation and self-renewal. *Journal of Cell Biology* 162, 1135–1147.

McNeill, S.H., Harris, K.B., Field, T.G. and Van Elswyk, M.E. (2012) The evolution of lean beef: identifying lean beef in today's U.S. marketplace. *Meat Science* 90, 1–8.

McPherron, A.C., Lawler, A.M. and Lee, S.J. (1997) Regulation of skeletal muscle mass in mice by a new TGF-beta superfamily member. *Nature* 387, 83–90.

Marchitelli, C., Savarese, M.C., Crisa, A., Nardone, A., Marsan, P.A. and Valentini, A. (2003) Double muscling in Marchigiana beef breed is caused by a stop codon in the third exon of myostatin gene. *Mammalian Genome* 14, 392–395.

Marshall, D.M. (1999) Genetics of meat quality. In: Fries, R. and Ruvinsky, A. (eds) *The Genetics of Cattle*. CAB International, Wallingford, pp. 605–636.

Mateescu, R.G., Garmyn, A.J., Tait, R.G. Jr, Duan, Q., Liu, Q., Mayes, M.S., Garrick, D.J., Van Eenennaam, A.L., VanOverbeke, D.L., Hilton, G.G., Beitz, D.C. and Reecy, J.M. (2012a) Genetic parameters for concentrations of minerals in longissimus muscle and their associations with palatability traits in Angus cattle. *Journal of Animal Science* 91, 1067–1075.

Mateescu, R.G., Garmyn, A.J., O'Neil, M.A., Tait, R.G. Jr, Abuzaid, A., Mayes, M.S., Garrick, D.J., Van Eenennaam, A.L., VanOverbeke, D.L., Hilton, G.G., Beitz, D.C. and Reecy, J.M. (2012b) Genetic parameters for carnitine, creatine, creatinine, carnosine, and anserine concentration in longissimus muscle and their association with palatability traits in Angus cattle. *Journal of Animal Science* 90, 4248–4255.

Matsuhashi, T., Maruyama, S., Uemoto, Y., Kobayashi, N., Mannen, H., Abe, T., Sakaguchi, S. and Kobayashi, E. (2011) Effects of bovine fatty acid synthase, stearoyl-coenzyme A desaturase, sterol regulatory element-binding protein 1, and growth hormone gene polymorphisms on fatty acid composition and carcass traits in Japanese Black cattle. *Journal of Animal Science* 89, 12–22.

Melton, S.L., Amiri, M., Davis, G.W. and Backus, W.R. (1982) Flavor and chemical characteristics of ground-beef from grass-finished, forage-grain-finished and grain-finished steers. *Journal of Animal Science* 55, 77–87.

Melucci, L.M., Panarace, M., Feula, P., Villarreal, E.L., Grigioni, G., Carduza, F., Soria, L.A., Mezzadra, C.A., Arceo, M.E., Papaleo Mazzucco, J. *et al.* (2012) Genetic and management factors affecting beef quality in grazing Hereford steers. *Meat Science* 92, 768–774.

Michal, J.J., Zhang, Z.W., Gaskins, C.T. and Jiang, Z. (2006) The bovine fatty acid binding protein 4 gene is significantly associated with marbling and subcutaneous fat depth in Wagyu x Limousin F2 crosses. *Animal Genetics* 37, 400–402.

Minton, J.E., Bindel, J.S., Titgemeyer, E.C., Grieger, D.M. and Hill, C.M. (1998) Serum leptin is associated with carcass traits in finishing cattle. *Journal of Animal Science* 76 (Suppl.), 231.

Moore, S.S., Li, C., Basarab, J., Snelling, W.M., Kneeland, J., Murdoch, B., Hansen, C. and Benkel, B. (2003) Fine mapping of quantitative trait loci and assessment of positional candidate genes for backfat on bovine chromosome 14 in a commercial line of *Bos taurus. Journal of Animal Science* 81, 1919–1925.

Morris, C.A., Cullen, N.G., Hickey, S.M., Dobbie, P.M., Veenvliet, B.A., Manley, T.R., Pitchford, W.S., Kruk, Z.A., Bottema, C.D. and Wilson, T. (2006) Genotypic effects of calpain 1 and calpastatin on the tenderness of cooked M. longissimus dorsi steaks from Jersey x Limousin, Angus and Hereford-cross cattle. *Animal Genetics* 37, 411–414.

Morris, C.A., Cullen, N.G., Glass, B.C., Hyndman, D.L., Manley, T.R., Hickey, S.M., McEwan, J.C., Pitchford, W.S., Bottema, C.D. and Lee, M.A. (2007) Fatty acid synthase effects on bovine adipose fat and milk fat. *Mammalian Genome* 1–8, 64–74.

Muller, D.M., Seim, H., Kiess, W., Loster, H. and Richter, T. (2002) Effects of oral L-carnitine supplementation on *in vivo* long-chain fatty acid oxidation in healthy adults. *Metabolism: Clinical and Experimental* 51, 1389–1391.

Newman, S., Reverter, A. and Johnston, D. (2002) Purebred-crossbred performance and genetic evaluation of postweaning growth and carcass traits in *Bos indicus* x *Bos taurus* crosses in Australia. *Journal of Animal Science* 80, 1801–1808.

Nkrumah, J.D., Li, C., Yu, J., Hansen, C., Keisler, D.H. and Moore, S.S. (2005) Polymorphisms in the bovine leptin promoter associated with serum leptin concentration, growth, feed intake, feeding behavior, and measures of carcass merit. *Journal of Animal Science* 83, 20–28.

Nogi, T., Honda, T., Mukai, F., Okagaki, T. and Oyama, K. (2011) Heritabilities and genetic correlations of fatty acid compositions in longissimus muscle lipid with carcass traits in Japanese Black cattle. *Journal of Animal Science* 89, 615–621.

O'Connor, S.F., Tatum, J.D., Wulf, D.M., Green, R.D. and Smith, G.C. (1997) Genetic effects on beef tenderness in *Bos indicus* composite and *Bos taurus* cattle. *Journal of Animal Science* 75, 1822–1830.

Oh, D., Lee, Y., La, B., Yeo, J., Chung, E., Kim, Y. and Lee, C. (2012) Fatty acid composition of beef is associated with exonic nucleotide variants of the gene encoding FASN. *Molecular Biology Reports* 39, 4083–4090.

Ohsaki, H., Tanaka, A., Hoashi, S., Sasazaki, S., Oyama, K., Taniguchi, M., Mukai, F. and Mannen, H. (2009) Effect of SCD and SREBP genotypes on fatty acid composition in adipose tissue of Japanese Black cattle herds. *Animal Science Journal* 80, 225–232.

O'Neil, C.E., Zanovec, M., Keast, D.R., Fulgoni III, V.L. and Nicklas, T.A. (2011) Nutrient contribution of total and lean beef in diets of US children and adolescents: National Health and Nutrition Examination Survey 1999–2004. *Meat Science* 87, 250–256.

Ouali, A. and Talmant, A. (1990) Calpains and calpastatin distribution in bovine, porcine and ovine skeletal muscles. *Meat Science* 28, 331–348.

Oyama, K. (2011) Genetic variability of Wagyu cattle estimated by statistical approaches. *Animal Science Journal* 82, 367–373.

Pabiou, T., Fikse, W.F., Nasholm, A., Cromie, A.R., Drennan, M.J., Keane, M.G. and Berry, D.P. (2009) Genetic parameters for carcass cut weight in Irish beef cattle. *Journal of Animal Science* 87, 3865–3876.

Page, B.T., Casas, E., Heaton, M.P., Cullen, N.G., Hyndman, D.L., Morris, C.A., Crawford, A.M., Wheeler, T.L., Koohmaraie, M., Keele, J.W. and Smith, T.P. (2002) Evaluation of single-nucleotide polymorphisms in CAPN1 for association with meat tenderness in cattle. *Journal of Animal Science* 80, 3077–3085.

Pannier, L., Sweeney, T., Hamill, R.M., Ipek, F., Stapleton, P.C. and Mullen, A.M. (2009) Lack of an association between single nucleotide polymorphisms in the bovine leptin gene and intramuscular fat in *Bos taurus* cattle. *Meat Science* 81, 731–737.

Pitchford, W.S., Deland, M.P., Siebert, B.D., Malau-Aduli, A.E. and Bottema, C.D. (2002) Genetic variation in fatness and fatty acid composition of crossbred cattle. *Journal of Animal Science* 80, 2825–2832.

Platter, W.J., Tatum, J.D., Belk, K.E., Chapman, P.L., Scanga, J.A. and Smith, G.C. (2003) Relationships of consumer sensory ratings, marbling score, and shear force value to consumer acceptance of beef strip loin steaks. *Journal of Animal Science* 81, 2741–2750.

Pringle, T.D., Williams, S.E., Lamb, B.S., Johnson, D.D. and West, R.L. (1997) Carcass characteristics, the Calpain proteinase system, and aged tenderness of Angus and Brahman crossbred steers. *Journal of Animal Science* 75, 2955–2961.

Raes, K., De Smet, S. and Demeyer, D. (2001) Effect of double-muscling in Belgian Blue young bulls on the intramuscular fatty acid composition with emphasis on conjugated linoleic acid and polyunsaturated fatty acids. *Animal Science* 2, 253–260.

Ramsbottom, J.M. and Strandine, E.J. (1948) Comparative tenderness and identification of muscles in wholesale beef cuts. *Food Res* 13, 315–330.

Reverter, A., Johnston, D.J., Graser, H.U., Wolcott, M.L. and Upton, W.H. (2000) Genetic analyses of live-animal ultrasound and abattoir carcass traits in Australian Angus and Hereford cattle. *Journal of Animal Science* 78, 1786–1795.

Reverter, A., Johnston, D.J., Ferguson, D.M., Perry, D., Goddard, M.E., Burrow, H.M., Oddy, V.H., Thompson, J.M. and Bindon, B.M. (2003) Genetic and phenotypic characterisation of animal, carcass, and meat quality traits from temperate and tropically adapted beef breeds. 4. Correlations among animal, carcass, and meat quality traits. *Australian Journal of Agricultural Research* 54, 149–158.

Riley, D.G., Chase, C.C. Jr, Hammond, A.C., West, R.L., Johnson, D.D., Olson, T.A. and Coleman, S.W. (2003) Estimated genetic parameters for palatability traits of steaks from Brahman cattle. *Journal of Animal Science* 81, 54–60.

Rincker, C.B., Pyatt, N.A., Berger, L.L. and Faulkner, D.B. (2006) Relationship among GeneSTAR marbling marker, intramuscular fat deposition, and expected progeny differences in early weaned Simmental steers. *Journal of Animal Science* 84, 686–693.

Rios-Utrera, A., Cundiff, L.V., Gregory, K.E., Koch, R.M., Dikeman, M.E., Koohmaraie, M. and Van Vleck, L.D. (2005) Genetic analysis of carcass traits of steers adjusted to age, weight, or fat thickness slaughter endpoints. *Journal of Animal Science* 83, 764–776.

Robinson, D.L., Ferguson, D.M., Oddy, V.H., Perry, D. and Thompson, J. (2001) Genetic and environmental influences on beef tenderness. *Australian Journal of Experimental Agriculture* 41, 997–1003.

Robinson, D.L., Cafe, L.M., McIntyre, B.L., Geesink, G.H., Barendse, W., Pethick, D.W., Thompson, J.M., Polkinghorne, R. and Greenwood, P.L. (2012) Production and processing studies on calpain-system gene markers for beef tenderness: consumer assessments of eating quality. *Journal of Animal Science* 90, 2850–2860.

Rourke, B.A.O., Dennis, J.A., Healy, P.J., McKiernan, W.A., Greenwood, P.L., Cafe, L.M., Perry, D., Walker, K.H., Marsh, I., Parnell, P.F. and Arthur, P.F. (2009) Quantitative analysis of performance, carcass and meat quality traits in cattle from two Australian beef herds in which a null myostatin allele is segregating. *Animal Production Science* 49, 2–97.

Rumph, J.M., Shafer, W.R., Crews, D.H. Jr, Enns, R.M., Lipsey, R.J., Quaas, R.L. and Pollak, E.J. (2007) Genetic evaluation of beef carcass data using different endpoint adjustments. *Journal of Animal Science* 85, 1120–1125.

Schenkel, F.S., Miller, S.P., Ye, X., Moore, S.S., Nkrumah, J.D., Li, C., Yu, J., Mandell, I.B., Wilton, J.W. and Williams, J.L. (2005) Association of single nucleotide polymorphisms in the leptin gene with carcass and meat quality traits of beef cattle. *Journal of Animal Science* 83, 2009–2020.

Schenkel, F.S., Miller, S.P., Jiang, Z., Mandell, I.B., Ye, X., Li, H. and Wilton, J.W. (2006) Association of a single nucleotide polymorphism in the calpastatin gene with carcass and meat quality traits of beef cattle. *Journal of Animal Science* 84, 291–299.

Scollan, N., Hocquette, J.-F., Nuernberg, K., Dannenberger, D., Richardson, I. and Moloney, A. (2006) Innovations in beef production systems that enhance the nutritional and health value of beef lipids and their relationship with meat quality. *Meat Science* 74, 17–33.

Shackelford, S.D., Koohmaraie, M., Cundiff, L.V., Gregory, K.E., Rohrer, G.A. and Savell, J.W. (1994) Heritabilities and phenotypic and genetic correlations for bovine postrigor calpastatin activity, intramuscular fat content, Warner-Bratzler shear force, retail product yield, and growth rate. *Journal of Animal Science* 72, 857–863.

Simopoulos, A.P. (2004) Omega-6/omega-3 essential fatty acid ratio and chronic diseases. *Food Reviews International* 20, 77–90.

Siri-Tarino, P.W., Sun, Q., Hu, F.B. and Krauss, R.M. (2010) Meta-analysis of prospective cohort studies evaluating the association of saturated fat with cardiovascular disease. *American Journal of Clinical Nutrition* 91, 535–546.

Smith, G.C., Culp, G.R. and Carpenter, Z.L. (1978) Postmortem aging of beef carcasses. *Journal of Food Science* 43, 823–826.

Smith, G.C., Carpenter, Z.L., Cross, H.R., Murphey, C.E., Abraham, H.C., Savell, J.W., Davis, G.W., Berry, B.W. and Parrish, F.C. (1985) Relationship of USDA marbling groups to palatability of cooked beef. *Journal of Food Quality* 7, 289–308.

Smith, S. (1994) The animal fatty acid synthase: one gene, one polypeptide, seven enzymes. *FASEB Journal* 8, 1248–1259.

Smith, S.B., Lunt, D.K., Chung, K.Y., Choi, C.B., Tume, R.K. and Zembayashi, M. (2006) Adiposity, fatty acid composition, and delta-9 desaturase activity during growth in beef cattle. *Animal Science Journal* 77, 478–486.

Smith, T., Domingue, J.D., Paschal, J.C., Franke, D.E., Bidner, T.D. and Whipple, G. (2007) Genetic parameters for growth and carcass traits of Brahman steers. *Journal of Animal Science* 85, 1377–1384.

Smith, T.P., Lopez-Corrales, N.L., Kappes, S.M. and Sonstegard, T.S. (1997) Myostatin maps to the interval containing the bovine mh locus. *Mammalian Genome* 8, 742–744.

Stone, R.T., Kappes, S.M. and Beattie, C.W. (1996) The bovine homolog of the obese gene maps to chromosome 4. *Mammalian Genome* 7, 399–400.

Tait, R.J., Zhang, S., Knight, T.J., Strohbehn, D.R., Beitz, D.C. and Reecy, J.M. (2008) Genetic correlations of fatty acid concentrations with carcass traits in Angus-sired beef cattle. *Animal Industry Report* AS 654, ASL R2285.

Taniguchi, M., Utsugi, T., Oyama, K., Mannen, H., Kobayashi, M., Tanabe, Y., Ogino, A. and Tsuji, S. (2004) Genotype of stearoyl-CoA desaturase is associated with fatty acid composition in Japanese Black cattle. *Mammalian Genome* 15, 142–148.

Tatsuda, K., Oka, A., Iwamoto, E., Kuroda, Y., Takeshita, H., Kataoka, H. and Kouno, S. (2008) Relationship of the bovine growth hormone gene to carcass traits in Japanese Black cattle. *Journal of Animal Breeding and Genetics* 125, 45–49.

Taylor, R.G., Geesink, G.H., Thompson, V.F., Koohmaraie, M. and Goll, D.E. (1995) Is Z-disk degradation responsible for postmortem tenderization? *Journal of Animal Science* 73, 1351–1367.

Thaller, G., Kuhn, C., Winter, A., Ewald, G., Bellmann, O., Wegner, J., Zuhlke, H. and Fries, R. (2003) DGAT1, a new positional and functional candidate gene for intramuscular fat deposition in cattle. *Animal Genetics* 34, 354–357.

Tizioto, P.C., Meirelles, S.L., Tulio, R.R., Rosa, A.N., Alencar, M.M., Medeiros, S.R., Siqueira, F., Feijo, G.L., Silva, L.O., Torres R.A. Jr and Regitano, L.C. (2012) Candidate genes for production traits in Nelore beef cattle. *Genetics and Molecular Research* 11, 4138–4144.

Ulbricht, T.L. and Southgate, D.A. (1991) Coronary heart disease: seven dietary factors. *The Lancet* 338, 985–992.

Utrera, A.R. and Van Vleck, L.D. (2004) Heritability estimates for carcass traits of cattle: a review. *Genetics and Molecular Research* 3, 380–394.

Uytterhaegen, L., Claeys, E., Demeyer, D., Lippens, M., Fiems, L.O., Boucque, C.Y., Van de Voorde, G. and Bastiaens, A. (1994) Effects of double-muscling on carcass quality, beef tenderness and myofibrillar protein degradation in Belgian Blue White bulls. *Meat Science* 38, 255–267.

Van Eenennaam, A.L., Li, J., Thallman, R.M., Quaas, R.L., Dikeman, M.E., Gill, C.A., Franke, D.E. and Thomas, M.G. (2007) Validation of commercial DNA tests for quantitative beef quality traits. *Journal of Animal Science* 85, 891–900.

Watson, R., Gee, A., Polkinghorne, R. and Porter, M. (2008) Consumer assessment of eating quality – development of protocols for Meat Standards Australia (MSA) testing. *Australian Journal of Experimental Agriculture* 48, 1–360.

Wheeler, T.L., Shackelford, S.D., Casas, E., Cundiff, L.V. and Koohmaraie, M. (2001) The effects of Piedmontese inheritance and myostatin genotype on the palatability of longissimus thoracis, gluteus medius, semimembranosus, and biceps femoris. *Journal of Animal Science* 79, 3069–3074.

Wheeler, T.L., Cundiff, L.V., Shackelford, S.D. and Koohmaraie, M. (2010) Characterization of biological types of cattle (Cycle VIII): carcass, yield, and longissimus palatability traits. *Journal of Animal Science* 88, 3070–3083.

White, S.N., Casas, E., Wheeler, T.L., Shackelford, S.D., Koohmaraie, M., Riley, D.G., Chase, C.C. Jr, Johnson, D.D., Keele, J.W. and Smith, T.P. (2005) A new single nucleotide polymorphism in CAPN1 extends the current tenderness marker test to include cattle of *Bos indicus*, *Bos taurus*, and crossbred descent. *Journal of Animal Science* 83, 2001–2008.

Wiener, P., Smith, J.A., Lewis, A.M., Woolliams, J.A. and Williams, J.L. (2002) Muscle-related traits in cattle: the role of the myostatin gene in the South Devon breed. *Genetics, Selection, Evolution* 34, 221–232.

Wiener, P., Woolliams, J.A., Frank-Lawale, A., Ryan, M., Richardson, R.I., Nute, G.R., Wood, J.D., Homer, D. and Williams, J.L. (2009) The effects of a mutation in the myostatin gene on meat and carcass quality. *Meat Science* 83, 127–134.

Williams, C.M. (2000) Dietary fatty acids and human health. *Annales de Zootechnie* 49, 165–180.

Wolcott, M.L. and Johnston, D.J. (2009) The impact of genetic markers for tenderness on steer carcass and feedlot exit and heifer puberty traits in Brahman cattle. *Proceedings of the Association for Advancement of Animal Breeding and Genetics* 18, 159–162.

Wood, I.A., Moser, G., Burrell, D.L., Mengersen, K.L. and Hetzel, D.J. (2006) A meta-analytic assessment of a thyroglobulin marker for marbling in beef cattle. *Genetics, Selection, Evolution* 38, 479–494.

Wood, J.D., Richardson, R.I., Nute, G.R., Fisher, A.V., Campo, M.M., Kasapidou, E., Sheard, P.R. and Enser, M. (2004) Effects of fatty acids on meat quality: a review. *Meat Science* 66, 21–32.

Wood, J.D., Enser, M., Fisher, A.V., Nute, G.R., Sheard, P.R., Richardson, R.I., Hughes, S.I. and Whittington, F.M. (2008) Fat deposition, fatty acid composition and meat quality: a review. *Meat Science* 78, 343–358.

Woronuk, G.N., Marquess, F.L., James, S.T., Palmer, J., Berryere, T., Deobald, H., Howie, S. and Kononoff, P.J. (2012) Association of leptin genotypes with beef cattle characteristics. *Animal Genetics* 43, 608–610.

Wulf, D.M. and Page, J.K. (2000) Using measurements of muscle color, pH, and electrical impedance to augment the current USDA beef quality grading standards and improve the accuracy and precision of sorting carcasses into palatability groups. *Journal of Animal Science* 78, 2595–2607.

Wulf, D.M., Tatum, J.D., Green, R.D., Morgan, J.B., Golden, B.L. and Smith, G.C. (1996) Genetic influences on beef longissimus palatability in Charolais- and Limousin-sired steers and heifers. *Journal of Animal Science* 74, 2394–2405.

Wulf, D.M., O'Connor, S.F., Tatum, J.D. and Smith, G.C. (1997) Using objective measures of muscle color to predict beef longissimus tenderness. *Journal of Animal Science* 75, 684–692.

Wyss, M., Braissant, O., Pischel, I., Salomons, G.S., Schulze, A., Stockler, S. and Wallimann, T. (2007) Creatine and creatine kinase in health and disease – a bright future ahead? *Sub-Cellular Biochemistry* 46, 309–334.

Zanovec, M., O'Neil, C.E., Keast, D.R., Fulgoni, V.L. 3rd and Nicklas, T.A. (2010) Lean beef contributes significant amounts of key nutrients to the diets of US adults: National Health and Nutrition Examination Survey 1999–2004. *Nutrition Research* 30, 375–381.

Zhang, S., Knight, T.J., Reecy, J.M. and Beitz, D.C. (2008) DNA polymorphisms in bovine fatty acid synthase are associated with beef fatty acid composition. *Animal Genetics* 39, 62–70.

Zuin, R.G., Buzanskas, M.E., Caetano, S.L., Venturini, G.C., Guidolin, D.G., Grossi, D.A., Chud, T.C., Paz, C.C., Lobo, R.B. and Munari, D.P. (2012) Genetic analysis on growth and carcass traits in Nelore cattle. *Meat Science* 91, 352–357.

23 Genetic Aspects of Cattle Adaptation in the Tropics

H.M. Burrow

CSIRO Animal, Food and Health Sciences, Armidale, New South Wales, Australia

Introduction

Improving the quality of life of many people living in tropical regions throughout the world is a very high global priority. For many of those people, this requires significant improvements to the productivity of their cattle herds, which comprise a major part of their household income (Payne and Hodges, 1997). Morgan and Tallard (2007) indicate that globally, there are ~1.5 billion head of cattle, with more than 65% located in the tropics and subtropics. Cattle make a major contribution to milk, meat and hide supply and additionally, in some regions, they are a major source of draught power and manure for use as fuel and maintaining soil

fertility. Improving productivity in tropical cattle herds therefore needs to consider not only the use of existing and new genetic technologies, but also the application of those technologies to the specific socioeconomic systems and the cultural values of the people who own the cattle and who will therefore benefit from the genetic improvement strategy (Payne and Hodges, 1997).

Newman and Coffey (1999) provide considerable discussion on the differing definitions of 'tropical adaptation'. For the purposes of this chapter though, tropical adaptation is simply defined as an animal's ability to survive, grow and reproduce in the presence of endemic stressors of tropical environments. The economic implications for production systems due to lack of adaptation include production losses, mortalities, treatment costs where treatment is feasible and potential loss of markets (Burrow et al., 2001; Prayaga et al., 2006).

Cattle grazed on pasture in tropical and subtropical environments encounter numerous stressors including ectoparasites (cattle ticks; horn flies, buffalo flies, screw-worm and tsetse flies; other biting insects), endoparasites (gastrointestinal helminths or worms), seasonally poor nutrition, high heat and humidity and diseases that are often transmitted by parasites. Often the impact of each individual stressor on production and animal welfare is multiplicative rather than additive, particularly when animals are already undergoing physiological stress such as lactation (e.g. Turner and Short, 1972; Turner, 1982; Frisch and Vercoe, 1982, 1984; Frisch and O'Neill, 1998b). Under the extensive production systems common in the tropics, it is generally not possible to control the stressors through management strategies alone. Even if intervention strategies were feasible, the treatments themselves often cause their own problems. For example, chemical treatments to control parasites generate concern about residues in beef products. In addition, the parasites acquire resistance to the chemical treatments, creating additional parasite control problems. In intensive feedlot systems and live cattle exports across (sub) tropical regions, high heat and humidity, even in the absence of other stressors, can become critically important for both production and animal welfare reasons. In such cases, the management interventions that are essential for poorly adapted cattle may be possible, but are difficult and/or expensive to implement. The best way to reduce the impacts of these stressors to improve productivity and animal welfare is therefore to breed cattle that are well adapted to the stressors, thereby negating the need for management interventions (Burrow, 2012).

This chapter examines the role of adaptation of beef cattle to the stressors of tropical environments and highlights the genetic basis of those traits and their relationships with productive attributes. It also examines the role of different genetic approaches (crossbreeding, within-breed selection and use of DNA-based technologies) in genetically improving the adaptation of cattle grazed in the tropics and sub-tropics. In doing so, though, the discussion is focused primarily on the improvement of tropically adapted cattle (i.e. Bos indicus and tropically adapted taurine breeds) grazed in the tropics and subtropics. Use of tropically adapted breeds provides more cost-effective opportunities for beef and dairy cattle producers in the tropics and the subtropics than the British or European breeds, at least based on currently available technologies. Because of the existence of significant genotype × environment (G × E) interactions in poorly adapted Bos taurus breeds across temperate and tropical environments and the lack of importance of such interactions in tropically adapted breeds (Burrow, 2012), no attempt is made to examine the use of British or European breeds as purebreds in the tropics or to review the genetic improvement of resistance of any cattle breeds to traits important for milk and meat production in temperate environments (e.g. high mountain/altitude disease; mastitis, etc.). This means that issues such as resistance to eye cancers and improving resistance to parasites in British and European breeds grazed on pasture or improving resistance of those breeds to heat stress during feedlot finishing are not examined in this chapter.

Measuring Adaptation of Cattle to Tropical Environments

Most scientific reports relating to adaptation of livestock refer to animals developing 'resistance' to environmental stressors. Gibson and Bishop (2005) summarize the earlier discussions that differentiate 'resistance' and 'tolerance' of animals to environmental stressors, suggesting an individual host may be infected by a parasite (including viruses, bacteria and protozoa – pathogens or microparasites; and helminths, flies and ticks – macroparasites) but suffer little or no harm. They refer to that situation as 'tolerance'. By contrast, they define 'resistance' as the ability of the individual host to resist infection or control the parasite lifecycle. Whilst recognizing and accepting such differentiation, this chapter continues to uses the term 'resistance' interchangeably with 'tolerance' to align with common usage in the scientific literature.

Table 23.1 provides examples of some common environmental stressors experienced by

Table 23.1. Examples of stressors experienced by cattle grazed at pasture in tropical and subtropical environments and methods of recording resistance of animals to those stressors.

Stressor	Measurement(s)	Impact(s)
Ticks	Single or repeated *counts* of number of ticks on one side of each animal following artificial or field infestation (Wharton and Utech, 1970) or single/repeated *scores* on a 0–5 (low–high)-point scale of the number of engorging ticks >4.5 mm on one side of the animal following artificial or field infestation (Prayaga *et al.*, 2009)	Anaemia impacting on productive and reproductive performance (Lehmann, 1993) Depressed appetite and reduced dietary nitrogen utilization, reducing growth rates, milk production and reproductive performance (Seebeck *et al.*, 1971; O'Kelly and Kennedy, 1981; O'Kelly *et al.*, 1988) Tick saliva transmits diseases that suppress immune function and result in cattle mortality and morbidity (Lehmann, 1993) Hide and udder damage that reduces/negates sale value of hides and reduces milking ability Cattle welfare impacts
Buffalo or horn flies	Repeated *counts* of number of flies on one side of each animal, with each animal being recorded by two observers on each occasion (Bean *et al.*, 1987) or fly lesion *score* recorded on a 1–5-point scale on one side of the animal when flies are prevalent (1 = no visible lesions and 5 = multiple lesions >35 cm² in size on 4+ sites; Prayaga *et al.*, 2009)	Reduced feed intake during periods of heavy infestation, resulting in short-term impacts on live weight gain (Bean *et al.*, 1987; Pruett *et al.*, 2003) and milk production (Drummond, 1987; Steelman *et al.*, 1991; Anziani *et al.*, 2000) Transmission of diseases that cause cattle morbidity and occasionally indirect impacts such as accidental deaths through diseases that cause blindness in animals Hide damage that reduces/negates sale value of hides Major cattle welfare impacts with visibly irritated animals during periods of fly infestation
Gastro-intestinal worms	Single or repeated counts of number of worm eggs per gram of faeces (Roberts and O'Sullivan, 1950)	Anaemia impacting on productive and reproductive performance (O'Kelly *et al.*, 1988; Mirkena *et al.*, 2010) Depressed appetite and reduced dietary nitrogen utilization, reducing growth rates, milk production and reproductive performance (Bremner, 1961; Vercoe and Springell, 1969) Cattle welfare impacts
High ambient temperatures	Traditionally heat stress has been recorded using repeated measures of rectal temperature of animals under conditions of heat stress (being in full sun for a period of at least 3 h when ambient temperatures are >30°C; Turner, 1982, 1984). Increasingly in developed countries though, body temperature is now more often recorded using remote monitoring through electronic data loggers implanted or attached to the animals	Depressed appetite and increased protein catabolism impacting on live weight gain and milk production (Vercoe, 1969) Failure of homeostasis, resulting in increased respiration rate, reduced sweating capacity and increased metabolic heat production (Blackshaw and Blackshaw, 1994) Cattle welfare impacts
Coat thickness	Coat score of animals recorded during summer months on a scale of 1 = extremely short, sleek coat to 7 = very woolly coat; scores are further ranked as + or −, giving a continuous 21-point scale applied to the numeric 1–7 score (Turner and Schleger, 1960)	See impact of high ambient temperatures above

Continued

Table 23.1. Continued.

Stressor	Measurement(s)	Impact(s)
Coat colour	Subjective score on a scale of 1 = light to 6 = dark as an attribute of heat tolerance resulting from greater absorption of solar radiation by darker-coloured animals leading to greater rate of environmental heat gain at the skin (Finch *et al.*, 1984)	See impact of high ambient temperatures above
Endemic diseases transmitted by parasites (e.g. trypanosomiasis)	Resistance of cattle to endemic diseases is particularly important in low-input livestock production systems in the developing world (Bishop *et al.*, 2002). Measurement of disease resistance under pastoral conditions is often very difficult. The simple presence/absence of disease can be subjectively assessed by a skilled recorder when animals are observed during routine handling procedures. However infrequent observation of animals means that incidence of disease generally goes unrecorded except in the case of specific diseases such as trypanosomiasis, where the key indicators of trypanotolerance (e.g. the ability of cattle to control parasitaemia and to resist the development of anaemia, measured by packed red cell volume) can be recorded under research conditions (Trail *et al.*, 1991a,b; Baker and Rege, 1994)	See sections relating to specific ecto-parasites which transmit such diseases
Nutritional deficiency	Except for the wet tropics, most tropical and subtropical areas experience periods of seasonally poor nutrition (i.e. a deficiency in quality and/or quantity of feed). Amongst the adaptive traits, this deficiency is the easiest to quantify, being adequately reflected in live weights and period live weight gains	Nutritional inadequacy impacting on live weight gain and milk production (Frisch and Vercoe, 1977, 1984) Can result in cattle deaths particularly when other physiological stressors (e.g. lactation, parasites) are also present Cattle welfare impacts
Temperament	Single or repeated flight time/speed or 'exit velocity', which is the time taken for an animal to cover a short, fixed distance after leaving a weighing crush (Burrow *et al.*, 1988). Several subjective measurements of temperament also exist as described in the review of Burrow (1997)	Reduced feed intake under intensive management systems resulting in reduced live weight gain (Burrow and Dillon, 1997) and possibly milk production Poor conception rates in artificial insemination (AI) programmes due to failure of nervous cattle to demonstrate oestrus in presence of an observer (Burrow *et al.*, 1988) Reduced glycogen in nervous animals pre-slaughter, resulting in tougher beef (Kadel *et al.*, 2006) Cattle welfare impacts

dairy and beef cattle grazed in tropical and subtropical environments, the impacts of those stressors and the methods used to measure an animal's resistance to them. As is evident from the table, most of these adaptive traits are very difficult to measure during routine animal husbandry procedures due both to the intermittent nature of the stressors and/or the difficulty of measurement, suggesting that genetic improvement of adaptive traits is also likely to be very difficult.

No attempt has been made to ensure Table 23.1 is complete, though the major stressors of tropical environments are included. There are numerous additional stressors experienced by cattle in tropical environments, with many of them being specific to local regions across the globe. As well, there are numerous additional ways of measuring an animal's resistance to them. For example, characteristics of cattle adapted to tropical heat and humidity include lower fasting metabolism, more and larger sweat glands, larger skin surfaces (e.g. navel, dewlap), small amounts of subcutaneous fat and a smooth, short hair coat (Oyenuga and Nestel, 1984). In addition to these factors, several early studies (Frisch and Vercoe, 1969, 1977, 1984; Seebeck et al., 1971; Turner, 1984) show that differences in growth rates of animals experiencing high temperatures or parasites are expressed through aspects of feed utilization such as depression in feed intake, digestion or metabolism. Measurement of one or more of these factors in addition to those identified in Table 23.1 could be considered when examining animal differences in adaptive traits, though most are impractical to measure under field conditions.

It is also recognized that aspects of cattle behaviour such as temperament are not specific to cattle reared in the tropics. However, temperament is included as an adaptive trait in this chapter because animals reared in extensive environments are generally handled less frequently than those in temperate environments. Hence, inherent differences in temperament between animals are often exaggerated due to the lack of routine handling, meaning temperament becomes an important trait for management reasons under the extensive pastoral systems commonly found in the tropics and subtropics (Burrow, 2012). Poor temperament affects the profitability of beef enterprises by increasing production costs as well as increasing the risk of injury to animals and their handlers. In addition, with the increasing emphasis placed by consumers on ethical production systems, it is essential to consider the animal welfare implications of handling animals with poor temperaments.

Breed Differences in Adaptive Traits

Breed groupings and comparative performance of the breeds

Cattle evolved into two distinct geographic groupings around 610–850,000 year b.p. (MacHugh et al., 1997). B. taurus breeds are adapted mostly to temperate environments in Europe and the Near East and include British and European breeds most suited for milk (e.g. Holstein Friesian, Jersey) or beef (e.g. Angus, Hereford, Charolais) production. Zebu or B. indicus breeds evolved in more tropical environments in southern Asia (MacHugh et al., 1997) and include breeds that have evolved for specialist milk (e.g. Sahiwal, Red Sindhi) and beef (e.g. Brahman, Nelore) production. A third distinct grouping evolved more recently in tropical environments and these are now commonly referred to as tropically adapted taurine breeds. These are true B. taurus (Frisch et al., 1997; Hanotte et al., 2002; Gibbs et al., 2009) that retain some of the productive attributes of B. taurus, but they are better adapted to tropical environments than European B. taurus. They include the southern African Sanga breeds (e.g. Africander, N'guni, Tuli), West African humpless breeds (e.g. N'dama) and Criollo breeds of Latin America and the Caribbean (e.g. Romosinuano). Historical reports describing these breeds suggested they were admixtures of B. indicus and B. taurus. However based on recently available molecular genetic tools, it is now accepted these breeds are true B. taurus.

Numerous historical and more recent studies indicate that large differences exist between cattle breeds in resistance to a wide range of tropical environmental stressors and factors that impact on animal performance.

These studies were used to develop a summary table shown herein as Table 23.2 and include:

- Ticks (Hewetson, 1972; Seifert, 1971a; Frisch 1981a, 1987; Frisch and O'Neill, 1998b; Wambura et al., 1998; Frisch et al., 2000; Berman, 2011);
- Gastrointestinal helminths (Seifert, 1971b; Turner and Short, 1972; Frisch 1981a, 1987; Frisch and O'Neill, 1998b; Frisch et al., 2000; Gasbarre and Miller, 2000);
- Endemic diseases such as eye cancers and trypanosomiasis (Frisch, 1975, 1987; Nishimura and Frisch, 1977; Trail et al., 1991a,b; Baker and Rege, 1994; D'Ieteren et al., 2000);
- High temperatures and humidity (Turner, 1972, 1982, 1984; Finch et al., 1984; Finch, 1985, 1986, 1987; Hammond et al., 1996, 1998; Frisch and O'Neill, 1998b; McManus et al., 2009; Berman, 2011);

- Poor quality feed (Ashton, 1962; Frisch, 1973, 1987; Hunter and Siebert, 1985; Berman, 2011),
- Appetite and fasting metabolic rate (Frisch and Vercoe, 1969, 1977, 1978, 1984; Frisch, 1987; Berman, 2011).

These studies show that in temperate environments there are substantial differences in growth, milking ability, reproduction and product quality between different cattle breeds. However in cattle grazed on pasture in tropical environments, the differences in performance are generally masked by the effects of environmental stressors on those productive attributes. This led to recommendations that, for most purposes in the tropics, comparisons of performance should be made across general breed types or groupings (B. taurus – British and European; B. indicus; and tropically adapted taurine) rather than across specific breeds

Table 23.2. Comparative rankings of different breed types for productive traits in temperate and tropical environments and for adaptation to selected stressors of tropical environments. (From: Burrow et al., 2001; the higher the number, the higher the value for the trait.)

	Bos taurus		Tropical B. taurus	Bos indicus		F₁ Brahman ×
Breed type	British	European[e]	Sanga	Indian	African	British
Temperate[a]						
Growth	4	5	3	3	2	4
Fertility	5	4	4	3	4	5
Tropical[a]						
Growth	2	2	3	4	2	4
Fertility	2	2	5	3	4	5
Mature size	4	5	3	4	3	4
Meat quality[b]	5	4	5	3	4	4
Resistance to environmental stressors						
Cattle ticks[c]	1	1	4	5	5	4
Worms[d]	3	3	3	5	4	4
Eye disease	2	3	3	5	4	4
Heat	2	2	5	5	5	5
Drought	2	1	5	5	5	4

[a] A temperate environment is assumed to be one free of environmental stressors, while tropical environment rankings apply where all environmental stressors are operating. Hence, while a score of, for example, 5 for fertility in a tropical environment indicates that breed type would have the highest fertility in that environment, the actual level of fertility may be less than the actual level of fertility for breeds reared in a temperate area, due to the effect of environmental stressors that reduce reproductive performance.

[b] Principally meat tenderness.

[c] *Rhipicephalus (boophilus) microplus.*

[d] Specifically *Oesophagostomum, Haemonchus, Trichostrongylus* and *Cooperia* spp.

[e] Data from purebred European breeds are not available in tropical environments and responses are predicted from the CSIRO Rockhampton crossbreeding data.

(Burrow *et al.*, 2001). These results also allowed development of comparative rankings of the different breed types for different productive and adaptive attributes in both temperate and tropical environments, reproduced here as Table 23.2, and provide further support for the primary focus on tropically adapted cattle in this chapter. From this table, it is clear that any breeding programme designed for cattle grazed at pasture in tropical environments must consider the impacts of both productive and adaptive attributes, even though adaptive traits (and some productive traits) are generally very difficult and/or expensive to measure (Burrow, 2012).

Impact of heterosis on adaptive traits

Considerable historical evidence exists that on-farm productivity in temperate and subtropical regions can be substantially improved in dairy and beef cattle by exploiting the heterosis that occurs in the progeny of crosses between *B. taurus* (British and European) breeds (e.g. Dickerson, 1969; Cundiff *et al.*, 1974a,b) and *B. taurus* and *B. indicus* crosses (e.g. Koger, 1963; Dickerson, 1969; Madalena, 1981, 1993; Cunningham and Syrstad, 1987). Despite this opportunity, the use of purebred Angus and Holstein Friesian cattle currently dominates beef and dairy production systems in temperate, developed regions of the world, primarily because of the strong genetic gains made in production traits through within-breed selection specifically in those two breeds over recent decades, relative to the much slower genetic gains in other breeds that might offer complementarity in crossbreeding systems.

The systematic contribution of heterosis to improving the productivity of herds in tropical environments remains low (Frisch and O'Neill, 1998a), with substantial scope to increase the productivity of both tropical dairy and beef cattle herds by effectively utilizing the well-documented benefits of heterosis for productive traits such as growth and reproduction. However, there remains a paucity of information relating to the magnitude of heterosis on adaptive traits due to the lack of designed experiments that allow valid estimation of heterosis from crossbred and parental breeds managed as contemporaries and also measured for resistance traits.

A single study known to examine this issue was undertaken in beef cattle in northern Australia using *B. indicus* (Brahman), *B. taurus* (Hereford × Shorthorn cross) and first cross (F$_1$) and subsequent generations of crosses between those breeds (Frisch and O'Neill, 1998b). By comparing least squares means of parental and crossbred populations, those authors reported that heterosis was consistently significant for live weights and live weight gains and for tick counts, but not for worm egg counts. Subsequently Prayaga (2003) analysed the Frisch and O'Neill (1988b) populations and estimated the genetic effects (direct and maternal additive and dominance effects) as partial regression coefficients using data pooled across breed types (rather than comparing the individual breeds as reported by Frisch and O'Neill (1988b).

Prayaga (2003) found that with the exception of flight time, all adaptive traits (tick and worm egg count, rectal temperature, coat score) benefited from crossing, demonstrated by significant and favourable direct dominance effects in taurine × indicine and taurine × taurine crosses. As shown in Fig. 23.1, estimates of heterosis were significant and favourable for adaptive traits in Zebu and British crosses. Heterosis percentages in F$_1$ genotypes ranged from –40% to 7% for tick count, –20% to –9% for worm egg count, –0.32% to 0.04% for rectal temperatures, –11.6% to –1.1% for coat scores and –6.6% to 0.3% for flight time (Prayaga, 2003). Negative heterosis was desirable for the resistance traits, and positive heterosis was desirable for flight time and demonstrates the superiority of the tropically adapted breeds for resistance traits and their poorer temperaments.

Direct and maternal additive and dominance effects provide critical evidence of the direction and magnitude of breed contributions to the traits of interest. The variability amongst these effects for the different adaptive traits in the wide range of British, European, tropically adapted taurine and *B. indicus* breeds in this study can be used to specifically design optimal crossbreeding systems for cattle grazed in the tropics (Prayaga, 2003). However, some caution must also be observed as, to date, this is the only study known to have examined these genetic effects for adaptive and temperament traits in beef cattle. Further research is therefore needed to enhance knowledge in different tropical

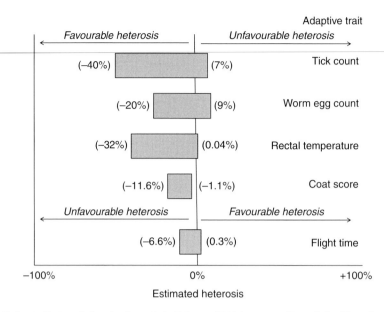

Fig. 23.1. Estimated heterosis for adaptive traits in Zebu and British crosses. (Data derived from Prayaga, 2003.)

environments and for different breeds and production systems. Regardless, these results strongly emphasize the importance of including adaptive traits in well-designed breeding programmes for tropical environments.

To maximize productivity of beef herds in tropical regions, Frisch and O'Neill (1988a) suggest that cattle producers need access to a range of breed types that are unrelated (to generate heterosis), tropically adapted (to cope with the environmental stressors) and that have desirable production characteristics (to satisfy markets and production systems). Because of the lack of highly resistant breeds available to complement the *B. indicus* breeds, those authors recommend that efforts continue to better characterize tropically adapted taurine breeds for adaptive traits, with the aim of identifying highly productive breeds that also have high resistance to all the stressors of tropical environments.

The only study known to examine heterosis for an adaptive trait in tropically adapted dairy cattle was reported by Madalena *et al.* (2012). Their Brazilian study compared contemporaneous females of six Holstein Friesian × Guzerá (*B. taurus* × *B. indicus*) crosses that were specifically hybridized to estimate the effects of heterosis in the resultant crosses, which ranged from 25% to ≥31/32 Holstein

Friesian. Crossbred cows were milked twice a day on 67 high- and low-input farms. The study showed important favourable heterosis effects for milk solids yield, fertility, mortality, herd life, weight, weight/height ratio and tick resistance.

The authors concluded that heterosis was an extremely important effect for overall farm profitability, particularly in lower-input farms. Table 23.3 shows that profit per day of herd-life was higher in F_1 cows than for other crossbreeding strategies. In low-management farms, rotational crossbreeding achieved 59% of the F_1 profit, *inter se* mating of 5/8 Holstein-Guzerá crosses achieved 30% of the F_1 profit and ≥31/32 Holstein Friesian were not economic (−21% of the F_1 profit). In better-managed farms a strategy of repeating the Holstein Friesian sires for two generations followed by one generation of Zebu sires attained 75% of the F_1 profit, the same as the ≥31/32 Holstein Friesian, whilst the 5/8 Holstein-Guzerá crosses showed a negative profit of −18%. Madalena *et al.* (2012) therefore concluded that new breeds developed from crossbred foundations would suffer from loss of heterosis. They suggested a strategy of continuous replacement with F_1 *B. taurus* × *B. indicus* heifers may be very profitable depending on price and availability of such animals.

Table 23.3. Profit per day of herd-life under alternative strategies of crossbreeding of Holstein (H) × Zebu (Z) (profit per day = (income − expense)/current herd life). (From: Madalena *et al.*, 1990, 2012.)

	Management level	
	High	Low
F_1 profit, equivalent kg milk/day	1.8	4.6
Crossbreeding strategy	Percentage of F_1 profit	
Continuous replacement with F_1	100	100
H-H-Z rotation	75	48
H-Z rotation	41	59
Upgrading to H	75	−21
New breed (5/8H: 3/8Z)	−18	30

Exploiting breed differences and heterosis for adaptive traits to improve performance of cattle in the tropics

As with any difficult- or expensive-to-measure trait, it is likely that cattle breeders will be better able to exploit the large differences between breeds for adaptive traits than to identify and select genetically superior animals within a breed. It also appears that significant benefits are likely to be achieved by systematically crossing complementary breeds to exploit heterosis not only for growth and reproduction, but also to directly improve resistance to the numerous stressors of tropical environments. Based on reviews of the scientific literature, Burrow (2006, 2012) suggested a number of 'rules of thumb' to allow cattle producers in the tropics to optimize their crossbreeding systems. Those 'rules of thumb' include:

- Depending on the severity of the environment and the level of stressor challenge, cattle in the tropics should comprise somewhere between 25% and 75% of 'adapted genes' for optimal production; only exceptionally stressful environments require 100% 'adapted genes'.
- 'Adapted genes' can be derived from *B. indicus* and their crosses, as well as the tropically adapted taurine breeds, providing opportunities to capture benefits from heterosis and to maximize productivity without reducing resistance to environmental stressors below levels acceptable for the particular environment in which production occurs.

- For most tropical environments, optimal levels of productivity and adaptation will be achieved using a combination of multiple breed types (e.g. *B. indicus*, tropically adapted taurine, British, European).
- In harsh tropical environments, some *B. indicus* content is required to ensure sufficient adaptation to ticks and worms. The harsher and wetter the environment, the greater the need for *B. indicus* content.
- Even in the harshest tropical environment, it is possible to combine *B. indicus* and adapted taurine breeds to optimize heterosis and maximize productivity. If 60–100% 'adapted genes' are required, a purebred tropically adapted taurine breed (e.g. Sanga or Criollo) could be combined with a *B. indicus* or *B. indicus*-derived breed to provide high levels of adaptation and capture the advantages of heterosis. If lower levels of adaptation (e.g. 25–60%) are required, an adapted taurine composite (e.g. Bonsmara, Senepol) could be combined with a *B. indicus* or *B. indicus*-derived breed to achieve desired levels of production and adaptation.

Within-breed Selection to Improve Adaptive Traits

Heritabilities of adaptive traits

To genetically improve traits through breeding, the traits being selected must be under direct or indirect genetic control. Direct genetic control is assessed by estimating the heritability of traits.

Indirect genetic control relies on the heritability of the traits as well as the favourable or unfavourable associations (genetic correlations or co-heritabilities) between different traits. This section examines the heritabilities of adaptive traits to determine whether they can be improved directly, while the next section examines opportunities to improve adaptation of cattle to tropical environments using indirect selection.

To date, there are relatively few reports of the heritability of the resistance of cattle to the stressors of tropical environments, probably due to the difficulties of measuring the very large numbers of animals required for such studies. Hence, most estimates currently available are derived from beef herds in northeastern Australia. Table 23.4 summarizes available estimates of heritabilities of resistance to ticks, worms, buffalo flies, heat stress, seasonally poor nutrition and temperament in beef and dairy cattle grazed on pasture in tropical environments. In addition, Stear et al. (2001) reviewed the livestock production literature and reported that, across species, the heritability of traits associated with resistance to many important diseases and parasites was often high, with considerable variation existing among animals. However, their review did not differentiate between livestock reared in temperate and tropical environments or under extensive and intensive production systems. They reported significant heritabilities for resistance to nematode species, tick infestations and buffalo fly numbers as well as resistance to trypanosomes and also indicated the heritability of immune response traits could be substantial in pigs and dairy cattle.

These estimates of heritability indicate that most adaptive traits examined herein are moderately to highly heritable, suggesting there is ample opportunity to directly improve these traits through within-breed selection, even though they may be difficult to include in genetic evaluation systems because of their difficulty of measurement.

Genetic correlations between adaptive and productive traits

Not unexpectedly, there are even fewer estimates in the scientific literature of genetic correlations amongst various adaptive traits and amongst adaptive and productive traits than there are estimates of heritabilities for the adaptive traits. Knowledge of these genetic correlations is essential for the effective design of breeding programmes for both beef and dairy cattle in the tropics.

Burrow (2001) reported that in Tropical Composite beef cattle grazed on pasture in the tropics, genetic correlations between tick and worm counts (r_g = 0.30) and tick count and rectal temperature (r_g = 0.22) were both favourable, although the genetic correlation between worm count and rectal temperature was close to zero. Correlations between tick and worm counts and growth, male and female fertility and flight times were all close to zero, indicating that tick and worm resistance were largely independent traits, except for correlations between the different resistance traits. Genetic correlations between rectal temperatures and most weights and period weight gains were favourable (r_g = −0.20 to −0.49), though genetic correlations between rectal temperatures and birth weight, mature cow weight and dry season gain were lower (r_g = −0.08 to −0.12). Low to moderate favourable genetic correlations existed between rectal temperature and pregnancy rate (r_g = −0.16) and days to calving (r_g = 0.16), although phenotypic correlations were close to zero.

Analyses of genetic trends in those same Tropical Composite populations showed that selection for high growth in a more resistant line (AXBX-UPWT, comprising 25% each of the Africander, Brahman, Hereford and Shorthorn breeds) and a more susceptible line (AX comprising 50% Africander and 25% of each of the Hereford and Shorthorn breeds) increased period weight gains and the additive and maternal genetic components of weights between birth and 18 months of age, but did not always increase mature cow weights (Fig. 23.2; Burrow and Prayaga, 2004). Calf birth and mature cow weights were restricted due to environmental effects. Resistance to heat stress improved through selection for high growth rate (Fig. 23.3), whilst resistance to ticks improved in a more tick-susceptible line but did not change in a more resistant line (Fig. 23.4a). Resistance to worms consistently decreased (worm counts increased) in the better adapted line as a result of selection, but did not change

Table 23.4. Estimates of heritabilities of a range of adaptive traits. (From: Prayaga *et al.*, 2006 and subsequent reports.)

Reference	Location	Breed[a]	Measure[a]	h² (se)[a]
Resistance to ticks				
Wharton *et al.*, 1970	Australia	Shorthorn	Count	0.39
Madalena *et al.*, 1985	Brazil	Holstein Friesian	Count	0.20
Mackinnon *et al.*, 1991	Australia	AX, AXBX	Count	0.34 (0.05)
Burrow, 2001	Australia	AX, AXBX	Count	0.44
Henshall *et al.*, 2001	Australia	AX, AXBX	Count	0.42
Henshall, 2004	Australia	HS	Count	0.44
Prayaga and Henshall, 2005	Australia	Multibreed crosses	Count	0.13 (0.03)
Bueno *et al.*, 2006	Brazil	Tropical Composite	Score	0.06
Prayaga *et al.*, 2009	Australia	Brahman, Tropical Composite	Score	0.09–0.15
Budeli *et al.*, 2009	South Africa	Bonsmara	Count	0.05–0.17
Resistance to worms				
Mackinnon *et al.*, 1991	Australia	AX, AXBX	FEC	0.28 (0.03)
Burrow, 2001	Australia	AX, AXBX	FEC	0.35
Henshall *et al.*, 2001	Australia	AX, AXBX	FEC	0.57
Henshall, 2004	Australia	HS	FEC	0.41
Prayaga and Henshall, 2005	Australia	Multibreed crosses	FEC	0.24 (0.03)
Prayaga *et al.*, 2009	Australia	Brahman, Tropical Composite	FEC	0.07–0.40
Resistance to buffalo flies				
Mackinnon *et al.*, 1991	Australia	AX, AXBX	Count	0.06
Burrow, 2001	Australia	AX, AXBX	Count	0.36
Prayaga *et al.*, 2009	Australia	Brahman, Tropical Composite	Lesion score	0.04–0.19
Resistance to heat stress (rectal temperature)				
Turner, 1982	Australia	HS, BX	°C	0.25 (0.12)
Turner, 1984	Australia	HS, BX	°C	0.33
Mackinnon *et al.*, 1991	Australia	AX, AXBX	°C	0.19 (0.02)
Burrow, 2001	Australia	AX, AXBX	°C	0.17
Prayaga and Henshall, 2005	Australia	Multibreed crosses	°C	0.12 (0.03)
Prayaga *et al.*, 2009	Australia	Brahman, Tropical Composite	°C	0.21–0.22
Resistance to heat stress (coat score and colour)				
Da Silva *et al.*, 1988	Brazil	Jersey	Thickness	0.23 (0.12)
Da Silva *et al.*, 1988	Brazil	Jersey	Length	0.08 (0.08)
Prayaga and Henshall, 2005	Australia	Multibreed crosses	Score	0.26 (0.03)
Bueno *et al.*, 2006	Brazil	Tropical Composite	Score	0.46–0.50
Prayaga *et al.*, 2009	Australia	Brahman, Tropical Composite	Score	0.62–0.64
Prayaga *et al.*, 2009	Australia	Brahman, Tropical Composite	Colour	0.61–0.84
Resistance to seasonally poor nutrition (dry season weight gain)				
Mackinnon *et al.*, 1991	Australia	AX, AXBX	ADG	0.34 (0.07)
Burrow, 2001	Australia	AX, AXBX	ADG	0.17
Barwick *et al.*, 2009	Australia	Brahman, Tropical Composite	ADG	0.14–0.18
Resistance to disease				
Janssen-Tapken *et al.*, 2010	East Africa	N'dama and Boran	Tryps	0.30
Janssen-Tapken *et al.*, 2010	East Africa	N'dama and Boran	PCV	0.40
Janssen-Tapken *et al.*, 2010	East Africa	N'dama and Boran	Paras	0.20

Continued

Table 23.4. Continued.

Reference	Location	Breed[a]	Measure[a]	h[2] (se)[a]
Temperament				
Burrow *et al.*, 1988	Australia	AX, AXBX	FT	0.21–0.54
Fordyce *et al.*, 1996	Australia	Brahman crosses	Score	0.08–0.14
Fordyce *et al.*, 1996	Australia	Brahman crosses	FD	0.32–0.70
Burrow, 1997, review	Worldwide	Various	Unre-strained	0.18–0.70
Burrow, 1997, review	Worldwide	Various	Restrained	0.00–0.67
Burrow and Corbet, 1999	Australia	Brahman, SG, BR, crossbreds	FT	0.35
Burrow and Corbet, 1999	Australia	Brahman, SG, BR, crossbreds	Visual FT	0.08
Burrow and Corbet, 1999	Australia	Brahman, SG, BR, crossbreds	Crush score	0.30
Burrow and Corbet, 1999	Australia	AX, AXBX	FT	0.48–0.50
Burrow, 2001	Australia	AX, AXBX	Flight time	0.44
Figueiredo *et al.*, 2005	Brazil	Nelore	Score	0.17
Kadel *et al.*, 2006	Australia	Brahman, SG, BR	FT	0.30–0.34
Kadel *et al.*, 2006	Australia	Brahman, SG, BR	Visual FT	0.21
Kadel *et al.*, 2006	Australia	Brahman, SG, BR	Crush score	0.15–0.19
Prayaga *et al.*, 2009	Australia	Brahman, Tropical Composite	FT	0.17–0.31

[a]HS, F_{5+} Hereford × Shorthorn (*Bos taurus*) cross; BX, F_{5+} Brahman (*Bos indicus*) × HS; AX, F_{5+} Africander (Sanga, a tropically adapted taurine breed) × HS; AXBX, F_{3+} AX × BX cross; SG, Santa Gertrudis (nominally 3/8 Brahman, 5/8 Shorthorn stabilized breed); BR, Belmont Red (registered AX and AXBX animals); Bonsmara, comparable to AX and BR – Corbet *et al.*, 2006a,b; FEC, faecal egg count; ADG, average daily gain; Tryps, trypanotolerance; PCV, minimum packed cell volume after disease challenge; Paras, mean of natural logarithm of parasitaemia count; FT, flight time; FD, flight distance; unrestrained measures of temperament include FD and FT; restrained measures of temperament include crush score; h[2], heritability; se, standard error.

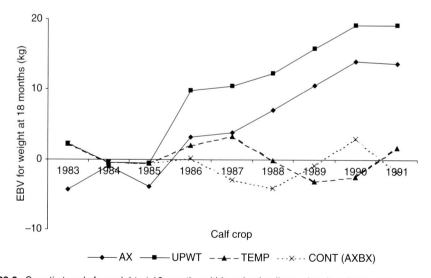

Fig. 23.2. Genetic trends for weight at 18 months within selection lines, showing direct response per year to selection for high EBV for weight at 600 days within AX and UPWT and correlated responses to selection within CONT (AXBX) and TEMP. (From: Burrow and Prayaga, 2004.)

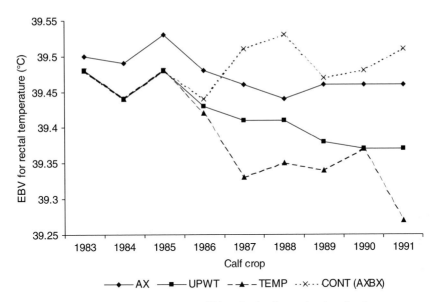

Fig. 23.3. Genetic trends for rectal temperature within selection lines, showing direct response per year to selection for low EBV for rectal temperature within TEMP and correlated responses to selection within AX, UPWT and CONT. (From: Burrow and Prayaga, 2004.)

in the more susceptible line (Fig. 23.4b). However worm numbers over the experimental period were low and may not have biologically impacted on growth rates. There were no general changes in male or female fertility traits, carcass and meat quality attributes or feed intake and feed conversion ratio (kilograms of feed/kilograms gain) due to selection for high growth (Burrow et al., 1991; Burrow and Prayaga, 2004).

In those same populations, selection for low estimated breeding values (EBVs) for rectal temperature reduced rectal temperatures in selected animals (AXBX-TEMP; Fig. 23.3) relative to the control line (AXBX-CONT). Small correlated improvements in resistance to ticks and worms also occurred in response to selection for low rectal temperature (Figs 23.4a and 23.4b). As well, changes occurred in the pattern of fat distribution in the carcass, with animals selected for low rectal temperature having carcasses with higher marbling than control animals. There were no significant changes to other attributes through selection for low EBV for rectal temperature (Burrow and Prayaga, 2004). The authors believed that selection for low rectal temperature in this line

may have reflected selection for factors other than resistance to heat stress per se (e.g. temperament, differences in mobility of animals).

In different tropically adapted crossbreed populations also grazed on pasture in northern Australia, Prayaga and Henshall (2005) reported that genetic correlations between tick or worm counts and growth traits were close to zero. However, the correlations between growth traits and heat tolerance traits (rectal temperatures and coat scores) were moderately negative implying that as the ability of an animal to handle heat stress improves, growth also increases at the genetic level. Genetic correlations between tick or worm counts and rectal temperatures were moderately favourable (Prayaga and Henshall, 2005). A significant negative genetic correlation between rectal temperature and flight time suggests that cattle with high heat tolerance also have desirable temperaments, similar to the findings reported by Burrow and Prayaga (2004).

In independent Brahman and Tropical Composite cattle populations, Prayaga et al. (2009) reported that phenotypic correlations between a range of adaptive and productive traits were low and, in general, genetic correlations

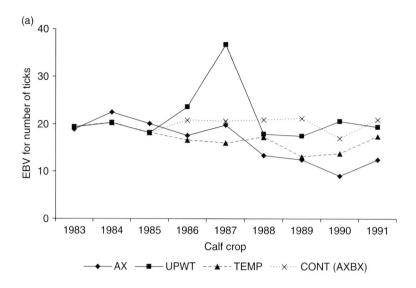

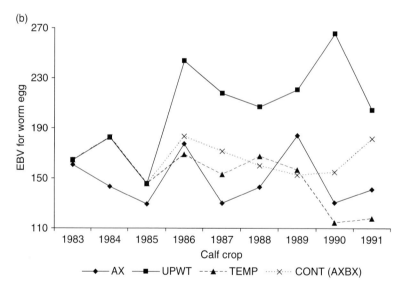

Fig. 23.4. Genetic trends for (a) tick counts and (b) worm eggs per gram of faeces within selection lines, showing correlated responses per year to selection within UPWT, TEMP, AX and CONT. (From: Burrow and Prayaga, 2004.)

were not significant, implying the traits were largely independent. Genetic correlations between worm egg counts and weight traits ($r_g = 0.29$ to 0.44) indicated there was no deleterious effect of worms on the growth of animals at a genetic level. This was particularly true in the Tropical Composites. Negative genetic correlations were found between coat and body condition score ($r_g = -0.33$ to -0.48) indicating that animals of both genotypes with sleek coats had a significant genetic advantage in the tropics. A strong, favourable genetic correlation between coat score and the age at the first-observed corpus luteum ($r_g = 0.73$) in Brahmans indicated that Brahman heifers with sleeker coats matured sexually at an earlier age than those with longer and thicker coats (Prayaga et al., 2009). Further, sleeker coats were associated with lower weights and reduced fat cover at puberty in Brahmans (Prayaga et al., 2009). Subsequent analyses of the extended dataset showed that coat score is also a useful genetic indicator of pregnancy and weaning rates and days to calving following first joining as well as lactation anoestrus interval and lifetime annual calving rate in a multi-trait genetic evaluation in Brahmans (Wolcott et al., 2014). Wolcott et al. (2014) concluded that selecting Brahmans for a sleeker coat would have only favourable impacts on female reproductive performance for first and second joinings as well as for lifetime reproductive performance.

In feedlot-finished Brahman and Tropical Composite steers, scanned fat measures at rump and rib sites were strongly genetically correlated with rectal temperatures in half-sib heifers of both breeds grazed on pasture in the tropics (r_g = 0.50 to 0.58), indicating that genetically fatter animals had genetically lower resistance to heat stress (Prayaga et al., 2009). In Brahmans, a strong negative genetic correlation (r_g = –0.97) was also observed between steer retail beef yield and rectal temperatures in heifers, indicating a favourable genetic association between increased yield and improved heat resistance (Prayaga et al., 2009).

The latest results from these same Brahman populations indicate that opportunities exist to use early measures of tropical adaptation as genetic indicators of female reproduction, with buffalo fly lesions and coat scores being significantly genetically related to lifetime annual calving rate (r_g = –0.76 and –0.51, respectively; Wolcott et al., 2014). The genetic correlations between tropical adaptation traits and female reproduction were weaker in Tropical Composites than those observed for Brahmans and presented less opportunity to exploit these measurements as genetic indicators of lifetime cow reproductive performance (Wolcott et al., 2014).

Based on these several different studies from northern Australia, it can be concluded that selection to improve resistance to any one stressor of tropical environments will improve resistance to other stressors. This is particularly true for resistance to ticks, worms and heat stress, where genetic correlations have been consistently moderately positive, suggesting the same or closely linked genes affect all three traits.

The same is not true of correlations between adaptive and productive traits. Except for heat stress measured by rectal temperatures under conditions of high ambient temperatures, resistance to most environmental stressors appears to be largely independent of productive traits such as growth, reproduction and product quality, albeit the conclusions are drawn from a small number of Australian studies. Genetic correlations between resistance to heat stress and growth and reproduction traits (Turner, 1982, 1984; Burrow, 2001; Burrow and Prayaga, 2004) are generally significantly negative (favourable), emphasizing there are many genes in common between genes controlling growth and reproduction in the tropics and rectal temperatures when ambient temperatures are high.

Hence from these studies, it can be concluded that most economically important productive and adaptive traits are at least moderately heritable, indicating they will respond to selection. In addition, there are no major strongly antagonistic correlations between the traits that would preclude simultaneous genetic improvement in all the traits in tropical beef breeding objectives (Burrow, 2012). Although there are no data available on the correlations between productive and adaptive traits in dairy cattle grazed in tropical environments, there is no reason to suspect that this conclusion would not also apply to tropically adapted breeds of dairy cattle.

Role of Molecular Information in Improving Adaptive Traits

Prior to availability of the bovine genome sequence (Gibbs et al., 2009; Elsik et al., 2009), livestock genomics research groups globally were aiming to identify individual genetic markers associated with economically important productive or adaptive traits. Their intent was to develop a small number of diagnostic tests (e.g. five to ten genetic markers per trait) that would collectively account for a large proportion (e.g. >50%) of the genetic variation for the trait of interest (e.g. Cunningham, 1999; Morris, 2000). Gibson and Bishop (2005) suggested that adaptive traits were ideal candidate traits for a genomic approach due to the major difficulty and expense of collecting the essential phenotypes needed for conventional phenotype-based selection. Those authors predicted that use of genomic information based on single nucleotide polymorphisms (SNPs) or quantitative trait loci (QTLs) would be most beneficial for traits of low heritability or traits that are difficult, expensive or impossible to record. They also suggested that use of genetic markers could be particularly beneficial in the low- to medium-input systems of the developing world, where disease resistance and adaptation of livestock are critically important for the sustainable livelihoods of poor farmers. Whilst their predictions about usage remain valid, it has since become clear that very few SNPs will account for a significant proportion of the genetic variation for economically important traits, as was hypothesized at that time.

Since 2005, genome-wide association studies (GWASs) have been routinely used to study the genetics of complex traits based on phenotypic measurements of the traits of interest and SNPs from genome-wide dense genetic markers. Those studies map SNPs associated with the phenotype and are used to identify QTLs associated with the traits of interest or to develop equations that predict an individual's breeding value from SNPs. In livestock, the latter approach is known as genomic selection (Meuwissen et al., 2001) and it has proven to be successful in spite of the failure to identify individual QTLs or the causal mutations for the traits of interest. Subsequently Villumsen et al. (2009) suggested that SNPs will be used to build genomic selection prediction models based on either single SNP associations or the effects of haplotypes of closely linked SNPs, with haplotypes having an optimal length based on the distance between the SNP and the extent of linkage disequilibrium (LD) in the population.

Early results from GWASs showed that very few QTLs were being detected in livestock studies, and those QTLs that were detected accounted for much smaller amounts of genetic variation than had been expected. Subsequently in a study of height in humans, Yang et al.

(2010) demonstrated that hundreds or thousands of SNPs were required to collectively account for the significant amounts of genetic variation in economically important traits necessary to achieve genetic improvement, thereby solving the problem that had become known as 'the missing heritability'.

In dairy cattle, where most production globally is based on a single breed, the use of 50,000 SNPs is sufficient to account for up to 80% of genetic variation in milk production traits in Holstein Friesian cattle grazed in temperate environments (Hayes et al., 2008, 2009). In beef cattle though, there is a need to predict performance across multiple breeds in a range of environments including the tropics and subtropics. In those populations, a minimum of at least 150,000 SNPs evenly spread across the entire genome is required to successfully predict performance (Hayes et al., 2007). In the near future though, due to the rapid decrease in the cost of sequencing and the ability to impute full genome sequence data from lower density SNP panels (Goddard and Hayes, 2009; Meuwissen et al., 2010a), the need to consider minimum numbers of SNPs will no longer be an important consideration, as it is highly likely that use of full genomic sequence data will rapidly replace SNP data to significantly increase the accuracy of genomic selection in both beef and dairy cattle (Meuwissen and Goddard, 2010a,b).

Until now genomic selection has not made any use of biological information about the specific genes that affect the trait being predicted, or the sites within those genes that might impact on the function of the gene. However, in research to discover individual causal mutations of large effect, such biological information is routinely used. It is highly likely that the same type of biological information will be useful in future in identifying causal polymorphisms of smaller effect, thereby explaining more of the variation in complex traits. By way of example, GWASs for height in humans found significant associations with SNPs in or near genes at which major mutations cause dwarfism and a number of the significant SNPs were found to be in LD with sites known to affect the expression of a nearby gene on the same chromosome (Lango Allen et al., 2010).

A related study of human height showed that causal polymorphisms have different properties to SNPs because only a proportion of the genetic variance is tracked by the SNPs, possibly because they have a lower average minor allele frequency (Yang *et al.*, 2010). This failure of SNPs to track the causal polymorphisms may also occur in cattle when data from many breeds are combined. For mutations of large effect, this approach has already been successful in humans (Ng *et al.*, 2009). Identifying the specific sites affecting a trait of interest and including the biological information relating to causal mutations in genomic prediction equations should increase the accuracy of genomic selection in future.

Hence research needs to continue to identify areas of the genome responsible for the genetic control of economically important traits with the aim of identifying causal mutations. Research reported to be underway to identify causal mutations related specifically to adaptive traits includes the following studies (note though, the reported QTLs may not be independent of other reported QTLs and hence the sizes of effect of individual QTLs may also not be independent).

Resistance to trypanosomiasis

Hanotte *et al.* (2003) found that putative QTLs for resistance to trypanosomiasis in tropically adapted African cattle were mapped to 18 autosomes at a false discovery rate of <0.20. The results were consistent with a single QTL on 17 chromosomes and two QTLs on BTA16. Individual QTL results ranged from ~6% to 20% of the phenotypic variance of the trait.

Resistance to Bovine Respiratory Disease

Neibergs *et al.* (2013) used four different statistical methods to analyse data from Holstein dairy calves. Their preliminary results suggested that genomic best linear unbiased prediction (GBLUP) SNP effects explained around 20% of the variation in Bovine Respiratory Disease incidence. All analytical approaches identified concordant single SNP associations on BTA3, BTA15 and BTA23. Twelve additional chromosomes provided evidence for association with two or more approaches. When chromosomal regions rather than single SNPs were compared, 29 regions on 13 chromosomes were associated with BRD, including those identified in the single SNP association comparison. Twelve regions were identified by all analyses and 17 by two analyses.

Resistance to heat stress

Mariasegaram *et al.* (2007) reported evidence that the slick hair gene, which endows its carriers with a slick hair coat to improve their heat tolerance was located on BTA20 in Senepol cattle.

Resistance to gastrointestinal nematodes (worms)

Gasbarre *et al.* (2004) reported that QTLs for faecal egg count and immune response to infection with *Ostertagia ostertagi* in Angus cattle were located at BTA3, BTA5 and BTA6, with the gene encoding interferon gamma being one of the candidate genes located on BTA5.

Resistance to ticks

Several studies are underway in different populations of beef and dairy cattle to identify genes associated with resistance to the cattle tick (*Riphicephalus (Boophilus) microplus*). Gasparin *et al.* (2007) undertook a study in F2 *B. taurus* × *B. indicus* animals in Brazil and reported different genomic regions controlling tick resistance on BTA5, BTA7 and BTA14. The significance of these QTLs was dependent on the season in which the ticks were counted, suggesting the QTLs may depend on environmental factors (Fig. 23.5).

Porto Neto *et al.* (2010a) identified SNPs associated with tick resistance on BTA3 and confirmed the association of the intronic SNP rs29019303 and its gene (*ELTD1*) with tick burden in taurine dairy and tropically adapted beef cattle in northern Australia. They found that rs29019303 was significantly ($P < 0.05$)

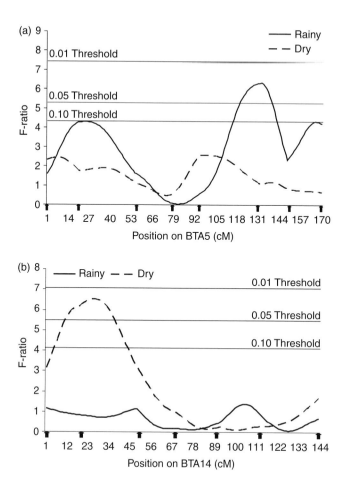

Fig. 23.5. Mapping of quantitative trait loci controlling tick (*Riphicephalus (Boophilus) microplus*) resistance on bovine chromosomes 5 (a) and 14 (b). (From: Gasparin *et al.*, 2007.)

associated with tick burden in both cattle types with the same favourable allele. A second SNP in the same genomic region was also significantly associated with tick burden in each cattle type. Associations using haplotypes were stronger than for single markers, including a haplotype of nine tag SNPs that was highly significantly ($P = 0.0008$) associated with tick counts in the dairy animals.

In a further study, Porto Neto *et al.* (2010b) identified a QTL on BTA10 linked to tick burden. To confirm the association, genotypes of 17 SNPs from BTA10 were tested in beef (Brahman and Tropical Composite) and dairy (Holstein Friesian) cattle. Three of the genotypes were obtained by sequencing part of the *Integrin alpha 11* (*ITGA11*) gene. In total, seven SNPs

were significantly ($P < 0.05$) associated with tick burden in any of the samples. One SNP, ss161109814, was significantly ($P < 0.05$) associated with tick burden in both the taurine and the Brahman sample, but the favourable allele was different. Haplotypes for three and for ten SNPs were more significantly ($P < 0.001$) associated with tick burden than SNPs analysed individually. The analyses confirmed the location of a QTL affecting tick burden on BTA10 and positioned it close to the *ITGA11* gene. The presence of a significant association in such widely divergent animals suggests that further SNP discovery in this region to detect causal mutations would be warranted.

To test the correlations between tick burden and milk production in the dairy populations reported by Porto Neto et al. (2010a,b), Turner et al. (2012) found that polymorphisms of four genes associated with milk yield (ABCG2, DGAT1, GHR and PRLR) were not significantly related to tick burden, suggesting that selection on markers for tick burden or milk component yield may have no undesirable effect on the other trait.

In a final study, Porto Neto et al. (2011) found that gene expression studies strongly indicated that both immune and non-immune mechanisms were associated with tick resistance in cattle, but no single causal variant was identified. In their studies, most of the genetic markers associated with tick burden explained a relatively small proportion of the variance, typical of DNA markers for quantitative traits. Hence the authors suggested that panels of multiple markers for tick resistance rather than a single marker would most likely be developed in future, possibly involving specific panels for zebu and taurine breeds.

Adaptive Traits in Breeding Programmes

Options to genetically improve economically important traits in beef and dairy cattle include within-breed selection, systematic crossbreeding and/or the use of composite populations. Genomic information is used to improve the accuracy of selection within those options, rather than being used as an alternative method of genetic improvement.

Within-breed selection is a very effective tool over the medium to long term to improve traits that are moderately to highly heritable (e.g. growth, carcass attributes, resistance to ticks, worms and heat stress). However, although the benefits of selection are cumulative across generations, the process is slow. Hence the best way to optimize use of within-breed selection in the first instance is to combine the breeds and breed types that are most appropriate for the target markets and production systems. In tropical environments, this means using tropically adapted breeds, unless the poorly adapted cattle reliably produce niche products that provide lucrative returns to adequately compensate for their higher costs of production.

Systematic crossbreeding systems can be planned to optimize the use of different breeds and breed types to make rapid genetic progress. There are numerous combinations of crossbreeding systems that could be used, and as with individual breeds, none is perfect for all markets and production systems and none is suitable for all herds. In designing a crossbreeding system, cattle breeders need to consider herd size, facilities available to ensure different groups of crossbred cattle remain segregated, the amount and skill of labour available and the breeds that best suit the markets, available feed and other resources. The most essential requirement of an effective crossbreeding system (or formation of a composite breed) is that the system must be properly designed.

Crossbreeding allows the best features of different breeds to be combined through complementarity (Cundiff et al., 1974a,b), which combines direct and maternal breed and heterosis effects to optimize performance. It also helps to match genetic potential for growth rate, mature size, reproduction and maternal ability and carcass and meat quality attributes with climatic environment, feed resources and market requirements. Crossbreeding should be used in conjunction with selection in the parental breeds to achieve long-term genetic improvement. However major difficulties are often encountered with systematic crossbreeding programmes in tropical environments primarily due to the difficulties of managing the different generational groups under extensive pastoral systems and maintaining sufficient levels of adaptation in some generations, unless both parental breeds are themselves highly adapted to the stressors of tropical environments.

To overcome these difficulties, composite populations offer the option of maintaining significant levels of heterosis (though less than rotational crossing) but with the ease of management of within-breed selection. In any breeding programme, genetic progress will be maximized through simultaneous selection of parental breeds for the traits of interest. However, breeders must be cognisant of the genetic antagonisms that exist both within and between breeds, to ensure that unintended correlated responses do not retard genetic progress. Regardless of what option is used to genetically improve beef and dairy cattle, accurate recording of all the traits of interest is essential.

Franklin (1986) suggests two broad options exist for within-breed improvement (including composites) of tropically adapted livestock, namely:

1. The breeding programme may simply concentrate on maximizing production and allow natural selection to maintain adaptation. Franklin (1986) suggests this method will be successful if the environment in which the animals are selected is the same environment in which the animals will be used. However results reported by Frisch (1981b), Hetzel *et al.* (1990) and Burrow (2012) provide different examples showing that if poorly adapted cattle are selected for use in a tropical environment, then the consequences of such selection may not be desirable, at least during the decades over which the population acquires general resistance to the stressors of the tropical environment.

2. The breeding programme may simultaneously select for adaptation and production using a multiple trait selection procedure. This approach allows improvement of both adaptation and production, but requires considerably more information including indicators of the economics of production, genetic parameters and a genetic evaluation scheme that allows accurate identification of selection candidates. In particular, it is highly likely that, due to the differing performance of different cattle breeds in various environments, the economic weightings and genetic parameters for use in selection indexes will need to be specifically developed for each breed type, environment and production system (Barwick *et al.*, 2009).

In their review of the literature, Baker and Rege (1994) argue that a programme aimed at improving both adaptive and performance traits (i.e. option 2 above) is to be preferred to one aimed at improving performance only. Evidence from the literature reviewed in previous sections of this chapter shows there is sufficient variation in adaptive traits in beef cattle and potentially also in dairy cattle, and

that the heritabilities of these traits are high enough to suggest that significant genetic progress can be achieved by within-breed selection. However, as suggested by Gasbarre and Miller (2000), a critical consideration is whether the environmental stressors exert sufficient economic pressure to warrant such selection. This consideration would need to be considered for each specific breeding programme based on the breed(s), market(s) and environmental stressor(s) operating in the production system. However if adaptive traits are to be included in the breeding objective, a further consideration is how best to incorporate the traits in the breeding objective.

Morris (2000) suggests that, in practice, the incentives to apply genetic selection principles to adaptive traits in breeding programmes derive from the following:

- a realization that the breeding objective is often incomplete without the trait(s);
- the increasing cost or decreasing availability of effective management strategies and drug therapies;
- ethical concerns about continuing to treat animals with drugs, although with perhaps equal ethical concerns to minimize the suffering experienced by diseased animals; and
- increasing consumer preferences for residue-free animal products.

Perhaps though the greatest difficulty to be addressed in improving adaptive traits by breeding is measuring the traits under commercial production environments in the tropics. As suggested by Crawford *et al.* (2000) at the extreme, measurement of resistance is a binary measurement (i.e. affected/not affected) with that classification only crudely reflecting an underlying distribution of susceptibility. This type of measurement does not provide a satisfactory level of discrimination for use in breeding programmes. Additional complications arise because responses are often affected by the level of challenge at the time of measurement, the health status of the host and previous exposure to the parasite or disease. Ideally, the resistance measurement should be on a continuous scale and highly correlated with actual field resistance (Crawford *et al.*, 2000).

Assuming these difficulties can be overcome, further consideration needs to be given to the method of incorporating the resistance traits into the breeding objective. To date the most practical approach (primarily in research herds) has been to use independent culling levels, mainly due to the lack of accurate genetic parameters that would allow formal inclusion of the traits in a comprehensive breeding objective. However, as Newman and Coffey (1999) suggest, optimization in the form of a multiple-trait selection objective is perhaps the only logical method of balancing productivity with environmental challenges.

Breeding objectives are generally expressed as economic weightings describing the economic impact of a unit change in each trait of commercial importance. Objective traits are often not the same as the selection criteria, which are the traits being measured and used to make selection decisions. Knowing the genetic correlations between the objective and selection traits allows selection index methods to target the objective traits using the selection traits. Kinghorn (1999) suggests there are two approaches to calculating the economic weightings: the economically rational approach and the desired gains approach. The desired gains approach involves declaration of the relative magnitudes of genetic gain desired in the traits of importance (Kinghorn, 1999). The breeding objective calculations still result in relative economic weights as in the economically rational approach, but under a desired gains approach, they are now influenced by genetic parameters with generally greater economic weightings for the traits that are more difficult to change. Once relative economic weights have been calculated, they can be used to drive breeding programmes in the appropriate direction (Kinghorn, 1999). However, inclusion of adaptive/resistance traits in formal breeding objectives (regardless of method of calculating economic weightings) is still not a practical reality, though at least in northern Australian beef herds, the genetic correlations amongst some of the objective and selection traits are now beginning to be understood. New research is required to confirm the value of these preliminary Australian results for use in other tropical areas worldwide.

Regardless of the approaches to including adaptive traits in formal breeding programmes, the greatest limitation to genetic improvement of these traits for the foreseeable future is likely to be the lack of accurate phenotypes on which selection can be made.

Implications

Although there are relatively few studies on which to base firm recommendations across differing breed types and a diverse range of tropical environments and production systems, it appears from the studies cited herein, that in breeds that are well adapted to their production environment, there are very few antagonistic correlations that would preclude simultaneous genetic improvement of both productive and adaptive traits through selection to maximize herd profitability. The major constraint to genetic improvement under commercial production systems in tropical environments is the difficulty and expense of accurately identifying appropriate fixed effects (e.g. contemporary groups) and measuring the full range of economically important productive and adaptive traits required to achieve a balanced breeding objective.

Genomic selection, which is currently based on SNP panels but in the relatively near future likely on sequence data and possibly knowledge of a limited number of causal mutations, offers new opportunities for commercial breeders to better breed and manage their cattle. However, achieving the levels of accuracy required by commercial breeders depends on the availability of very large cattle resource populations that have been accurately recorded for all the economically important traits, including adaptive/resistance traits. The major limitation for the next 5–10 years at least is likely to be the lack of large numbers of accurate phenotypes and accurate knowledge of the factors that affect them. This need has been recognized in several countries, with specifically designed, industry-relevant programmes now being developed to generate these phenotypes, and collaborative efforts to pool data across countries are underway to ensure new growth opportunities for the beef and dairy cattle industries can be captured.

References

Anzianl, O.S., Zimmermann, G., Gugllelmone, A.A., Forchleri, M. and Volpogni, M.M. (2000) Evaluation of insecticide ear tags containing ethion for control of pyrethroid resistant *Haematobia irritans* (L.) on dairy cattle. *Veterinary Parasitology* 91, 147–151.

Ashton, G.C. (1962) Comparative nitrogen digestibility in Brahman, Brahman x Shorthorn, Africander x Hereford and Hereford steers. *Journal of Agricultural Science (Cambridge)* 58, 333–342.

Baker, R.L. and Rege, J.E.O. (1994) Genetic resistance to diseases and other stresses in improvement of ruminant livestock on the tropics. *Proceedings of the 5th World Congress on Genetics Applied to Livestock Production* 20, 405–412.

Barwick, S.A., Johnston, D.J., Burrow, H.M., Holroyd, R.G., Fordyce, G., Wolcott, M.L., Sim, W.D. and Sullivan, M.T. (2009) Genetics of heifer performance in 'wet' and 'dry' seasons and their relationships with steer performance in two tropical beef genotypes. *Animal Production Science* 49, 367–382.

Bean, K.G., Seifert, G.W., Macqueen, A. and Doube, B.M. (1987) Effect of insecticide treatment for control of buffalo fly on weight gains of steers in coastal central Queensland. *Australian Journal of Experimental Agriculture* 27, 329–334.

Berman, A. (2011) Are adaptations present to support dairy cattle productivity in warm climates? *Journal of Dairy Science* 94, 2147–2158.

Bishop, S.C., De Jong, M. and Gray, D. (2002) Opportunities for incorporating genetic elements into the management of farm animal diseases: policy issues. In: Food and Agricultural Organization Study Paper No. 18. Commission on Genetic Resources for Food and Agriculture, Rome, 36 pp.

Blackshaw, J.K. and Blackshaw, A.W. (1994) Heat stress in cattle and the effect of shade on production and behaviour: a review. *Australian Journal of Experimental Agriculture* 34, 285–295.

Bremner, K.C. (1961) A study of pathogenic factors in experimental bovine oesophagostomosis. 1. An assessment of the importance of anorexia. *Australian Journal of Agricultural Research* 12, 498–512.

Budeli, M.A., Nephawe, K.A., Norris, D., Selapa, N.W., Bergh, L. and Maiwashe, A. (2009) Genetic parameter estimates for tick resistance in Bonsmara cattle. *South African Journal of Animal Science* 39, 321–327.

Bueno, R.S., Ferraz, J.B.S., Eler, J.P., Torres, R.A., Mourão, G.B., Baliero, J.C.C., Mattos, E.C. and Pedrosa, V.B. (2006) Genetic parameters of growth and adaptive traits in a composite beef cattle population (*Bos taurus x Bos indicus*). *Proceedings of the 8th World Congress on Genetics Applied to Livestock Production*. CD ROM Communication No. 16-576-1173.

Burrow, H.M. (1997) Measurements of temperament and their relationships with performance traits of beef cattle. *Animal Breeding Abstracts* 65, 477–495.

Burrow, H.M. (2001) Variances and covariances between productive and adaptive traits and temperament in a composite breed of tropical beef cattle. *Livestock Production Science* 70, 213–233.

Burrow, H.M. (2006) Utilization of diverse breed resources for tropical beef production. *Proceedings of the 8th World Congress on Genetics Applied to Livestock Production*. CD ROM Communication No. 32-01, ISBN 85-60088-01-6, 8 pp.

Burrow, H.M. (2012) Importance of adaptation and genotype x environment interactions in tropical beef breeding systems. *Animal* 6, 729–740.

Burrow, H.M. and Corbet, N.J. (1999) Genetic and environmental factors affecting temperament of zebu and zebu derived beef cattle grazed at pasture in the tropics. *Australian Journal of Agricultural Research* 51, 155–162.

Burrow, H.M. and Dillon, R.D. (1997) Relationships between temperament and growth in a feedlot and commercial carcass traits of *Bos indicus* crossbreds. *Australian Journal of Experimental Agriculture* 37, 407–411.

Burrow, H.M. and Prayaga, K.C. (2004) Correlated responses in productive and adaptive traits and temperament following selection for growth and heat resistance in tropical beef cattle. *Livestock Production Science* 86, 143–161.

Burrow, H.M., Seifert, G.W. and Corbet, N.J. (1988) A new technique for measuring temperament in cattle. *Proceedings of the Australian Society of Animal Production* 17, 154–157.

Burrow, H.M., Gulbransen, B., Johnson, S.K., Davis, G.P., Shorthose, W.R. and Elliott, R.F. (1991) Consequences of selection for growth and heat resistance on growth, feed conversion efficiency, commercial carcass traits and meat quality of zebu crossbred cattle. *Australian Journal of Agricultural Research* 42, 1373–1383.

Burrow, H.M., Moore, S.S., Johnston, D.J., Barendse, W. and Bindon, B.M. (2001) Quantitative and molecular genetic influences on properties of beef. *Australian Journal of Experimental Agriculture* 41, 893–919.

Corbet, N.J., Shepherd, R.K., Burrow, H.M., Prayaga, K.C., van der Westhuizen, J. and Strydom, P.E. (2006a) Evaluation of Bonsmara and Belmont Red cattle breeds in South Africa. 1. Productive performance. *Australian Journal of Experimental Agriculture* 46, 199–212.

Corbet, N.J., Shepherd, R.K., Burrow, H.M., Prayaga, K.C., van der Westhuizen, J. and Bosman, D.J. (2006b) Evaluation of Bonsmara and Belmont Red cattle breeds in South Africa. 2. Genetic parameters for growth and fertility. *Australian Journal of Experimental Agriculture* 46, 213–224.

Crawford, A.M., Dodds, K.G. and McEwan, J.C. (2000) DNA markers, genetic maps and the identification of QTL: general principles. In: Axford, R.F.E., Bishop, S.C., Nicholas, F.W. and Owen, J.B. (eds) *Breeding for Disease Resistance in Farm Animals*. CAB International, Wallingford, pp. 3–26.

Cundiff, L.V., Gregory, K.E. and Koch, R.M. (1974a) Effects of heterosis on reproduction in Hereford, Angus and Shorthorn cattle. *Journal of Animal Science* 38, 711–727.

Cundiff, L.V., Gregory, K.E., Schwulst, F.J. and Koch, R.M. (1974b) Effects of heterosis on maternal performance and milk production in Hereford, Angus and Shorthorn cattle. *Journal of Animal Science* 38, 728–745.

Cunningham, E.P. (1999) The application of biotechnologies to enhance animal production in different farming systems. *Livestock Production Science* 58, 1–24.

Cunningham, E.P. and Syrstad, O. (1987) Crossbreeding *Bos indicus* and *Bos taurus* for milk production in the tropics. Food and Agriculture Organization of the United Nations Animal Production and Health Paper No. 68. FAO, Rome.

Da Silva, R.G., Arantes-Neto, J.G. and Holtz-Filho, S.V. (1988) Genetic aspects of the variation of the sweating rate and coat characteristics of Jersey cattle. *Revue Brasilia de Genetica* 11, 335–347.

Dickerson, G. (1969) Experimental approaches in utilizing breed resources. *Animal Breeding Abstracts* 37, 191–202.

D'Ieteren, G.D.M., Authié, E., Wissocq, N. and Murray, M. (2000) Exploitation of resistance to trypanosomes and trypanosomiasis. In: Axford, R.F.E., Bishop, S.C., Nicholas, F.W. and Owen, J.B. (eds) *Breeding for Disease Resistance in Farm Animals*, 2nd Edition. CAB international, Wallingford, UK, pp. 195–216.

Drummond, R.O. (1987) Economic aspects of ectoparasites in North America. *Proceedings of the XXIII World Veterinary Congress*, pp. 9–24.

Elsik, C.G., Tellam, R.L., Worley, K.C. and the Bovine Genome Sequencing Analysis Consortium (2009) The genome sequence of Taurine cattle: a window to ruminant biology and evolution. *Science* 324, 522–528.

Figueiredo, L.G.G., Eler, J.P., Mourão, G.B., Ferraz, J.B.S., de Balieiro, J.C. de C. and de Mattos, E.C. (2005) Análise genética do temperamento em uma população da raça Nelore. *Livestock Research for Rural Development* 17, 1–7.

Finch, V.A. (1985) Body temperature in beef cattle: its control and relevance to production in the tropics. *Journal of Animal Science* 62, 531–542.

Finch, V.A. (1986) Comparison of non-evaporative heat transfer in different cattle breeds. *Australian Journal of Agricultural Research* 36, 497–508.

Finch, V.A., Bennett, I.L. and Holmes, C.R. (1984) Coat colour in cattle: effect on thermal balance, behaviour and growth, and relationship with coat type. *Journal of Agricultural Science* 102, 141–147.

Fordyce, G., Howitt, C.J., Holroyd, R.G., O'Rourke, P.K. and Entwistle, K.W. (1996) The performance of Brahman-Shorthorn and Sahiwal-Shorthorn beef cattle in the dry tropics of northern Queensland. 5. Scrotal circumference, temperament, ectoparasite resistance and the genetics of growth and other traits in bulls. *Australian Journal of Experimental Agriculture* 36, 9–17.

Franklin, I.R. (1986) Breeding ruminants for the tropics. *Proceedings of the 3rd World Congress on Genetics Applied to Livestock Production* 11, 451–461.

Frisch, J.E. (1973) Comparative drought resistance of *Bos indicus* and *Bos taurus* crossbred herd in Central Queensland. 2. Relative mortality rates, calf birth weights and weight changes of breeding cows. *Australian Journal of Experimental Agriculture and Animal Husbandry* 13, 117–126.

Frisch, J.E. (1975) The relative incidence and effect of bovine infectious keratoconjunctivitis in *Bos indicus* and *Bos taurus* cattle. *Animal Production* 21, 265–274.

Frisch, J.E. (1981a) Factors affecting resistance to ecto- and endoparasites of cattle in tropical areas and the implications for selection. In: *Isotopes and Radiation in Parasitology IV*. International Atomic Energy Agency, Vienna.

Frisch, J.E. (1981b) Changes occurring in cattle as a consequence of selection for growth rate in a stressful environment. *Journal of Agricultural Science (Cambridge)* 96, 23–38.

Frisch, J.F. (1987) Physiological reasons for heterosis in growth of *Bos indicus* x *Bos taurus*. *Journal Agricultural Science (Cambridge)* 109, 213–230.

Frisch, J.E. and O'Neill, C.J. (1998a) Comparative evaluation of beef cattle breeds of African, European and Indian origins. 1. Live weights and heterosis at birth, weaning and 18 months. *Animal Science* 67, 27–38.

Frisch, J.E. and O'Neill, C.J. (1998b) Comparative evaluation of beef cattle breeds of African, European and Indian origins. 2. Resistance to cattle ticks and gastrointestinal nematodes. *Animal Science* 67, 39–48.

Frisch, J.E. and Vercoe J.E. (1969) Liveweight gain, food intake and eating rate in Brahman, Africander and Shorthorn x Hereford cattle. *Australian Journal of Agricultural Research* 20, 1189–1195.

Frisch, J.E. and Vercoe J.E. (1977) Food intake, eating rate, weight gains, metabolic rate and efficiency of feed utilization in *Bos taurus* and *Bos indicus* crossbred cattle. *Animal Production* 25, 343–358.

Frisch, J.E. and Vercoe J.E. (1978) Utilizing breed differences in growth of (beef) cattle in the tropics. *World Animal Review* 25, 8–12.

Frisch, J.E. and Vercoe, J.E. (1982) Consideration of adaptive and productive components of productivity in breeding beef cattle for tropical Australia. *Proceedings of the Second World Congress on Genetics Applied to Livestock Production* 6, 307–321.

Frisch, J.E. and Vercoe, J.E. (1984) An analysis of growth of different cattle genotypes reared in different environments. *Journal of Agricultural Science (Cambridge)* 103, 137–153.

Frisch, J.E., Drinkwater, R., Harrison, B. and Johnson, S. (1997) Classification of the southern African Sanga and East African shorthorned zebu. *Animal Genetics* 28, 77–83.

Frisch, J.E., O'Neill, C.J. and Kelly, M.J. (2000) Using genetics to control cattle parasites – the Rockhampton experience. *International Journal for Parasitology* 30, 253–264.

Gasbarre, L.C. and Miller, J.E. (2000) Genetics of helminth resistance. In: Axford, R.F.E., Bishop, S.C., Nicholas, F.W. and Owen, J.B. (eds) *Breeding for Disease Resistance in Farm Animals*. Second Edition. CAB International, Wallingford, UK, pp. 129–152.

Gasbarre, L.C., Sonstegard, T., Van Tassell, C.P. and Araujo, R. (2004) Using host genomics to control nematode infections. *Veterinary Parasitology* 125, 155–161.

Gasparin, G., Miyata, M., Coutinho, L.L., Martinez, M.L., Teodoro, R.L., Furlong, J., Machado, M.A., Silva, M.V.G.B., Sonstegard, T.S. and Regitano, L.C.A. (2007) Mapping of quantitative trait loci controlling tick [*Riphicephalus (Boophilus) microplus*] resistance on chromosomes 5, 7 and 14. *Animal Genetics* 38, 453–459.

Gibbs, R.A., Taylor, J.F., Van Tassell, C.P., Barendse, W., Eversole, K.A., Gill, C.A., Green, R.S., Hamernik, D.L., Kappes, S.M., Lien, S. and the Bovine HapMap Consortium (2009) Genome-wide survey of SNP variation uncovers the genetic structure of cattle breeds. *Science* 324, 528–532.

Gibson, J.P. and Bishop, S.C. (2005) Use of molecular markers to enhance resistance of livestock to disease: a global approach. *Revue Scientifique et Technique de l'Office International des Epizooties* 24, 343–353.

Goddard, M.E. and Hayes, B.J. (2009) Mapping genes for complex traits in domestic animals and their use in breeding programmes. *Nature Reviews Genetics* 10, 381–391.

Hammond, A.C., Olson, T.A., Chase, C.C. Jr, Bowers, E.J., Randel, R.D., Murphy, C.N., Vogt, D.W. and Tewolde, A. (1996) Heat tolerance in two tropically adapted *Bos taurus* breeds, Senepol and Romosinuano, compared with Brahman, Angus and Hereford cattle in Florida. *Journal of Animal Science* 74, 295–303.

Hammond, A.C., Chase, C.C. Jr, Bowers, E.J., Olson, T.A. and Randel, R.D. (1998) Heat tolerance in Tuli-, Senepol- and Brahman-sired F1 Angus heifers in Florida. *Journal of Animal Science* 76, 1568–1577.

Hanotte, O., Bradley, D.G., Ochieng, J.W., Verjee, Y., Hill, E.W. and Rege, J.E.O. (2002) African pastoralism: genetic imprints of origins and migrations. *Science* 296, 336–339.

Hanotte, O., Ronin, Y., Agaba, M., Nilsson, P., Gelhaus, A., Horstmann, R., Sugimoto, Y., Kemp, S., Gibson, J., Korol, A., Soller, M. and Teale, A. (2003) Mapping of quantitative trait loci controlling trypanotolerance in a cross of tolerant West African N'dama and susceptible East African Boran cattle. *Proceedings of the National Academy of Sciences USA* 100, 7443–7448.

Hayes, B.J., De Roos, A.P.W. and Goddard, M.E. (2007) Predicting genomic breeding values within and between populations. *Proceedings of the Association for the Advancement of Animal Breeding and Genetics* 17, 296–303.

Hayes, B.J., Bowman, P.J., Chamberlain, A.J. and Goddard, M.E. (2008) Invited Review: Genomic selection in dairy cattle: progress and challenges. *Journal of Dairy Science* 92 433–443.

Hayes, B.J., Bowman, P.J., Chamberlain, A.C., Verbyla, K. and Goddard, M.E. (2009) Accuracy of genomic breeding values in multi-breed dairy cattle populations. *Genetics Selection Evolution* 41, 51.

Henshall, J.M. (2004) A genetic analysis of parasite resistance traits in a tropically adapted line of *Bos taurus*. *Australian Journal of Agricultural Research* 55, 1109–1116.

Henshall, J.M., Burrow, H.M. and Tier, B. (2001) Segregation analysis for major genes affecting adaptive traits in cattle grazed in the tropics. *Proceedings of the Association for the Advancement of Animal Breeding and Genetics* 14, 167–170.

Hetzel, D.J.S., Quaas, R.L., Seifert, G.W., Bean, K.G., Aspden, W.J. and Mackinnon, M.J. (1990) Evidence of natural selection in a herd of Hereford-Shorthorn cattle in the tropics. *Proceedings of the Australian Association of Animal Breeding and Genetics* 8, 451–454.

Hewetson, R.W. (1972) The inheritance of resistance by cattle to cattle tick. *Australian Veterinary Journal* 48, 299–303.

Hunter, R.A. and Siebert, B.D. (1985) Utilization of low-quality roughage by *Bos taurus* and *Bos indicus* cattle. *British Journal of Nutrition* 53, 637–648.

Janssen-Tapken, U., Dekkers, J.C.M. and Kadarmideen, H.N. (2010) Genetic analyses of cattle breeding schemes including genetic markers for trypanotolerance. *Proceedings of the 9th World Congress on Genetics Applied to Livestock Production*, Leipzig, Germany. Available at: http://www.kongressband. de/wcgalp2010/assets/pdf/0193.pdf (accessed 3 April 2014).

Kadel, M.J., Johnston, D.J., Burrow, H.M., Graser, H.-U. and Ferguson, D.M. (2006) Genetics of flight time and other measures of temperament and their value as selection criteria for improving meat quality traits in tropically adapted breeds of beef cattle. *Australian Journal of Agricultural Research* 57, 1029–1035.

Kinghorn, B.P. (1999) Description and targeting of breeding objectives. In: Kinghorn, B., van der Werf, J. and Ryan, M. (eds) *Animal Breeding – Use of New Technologies*. Post-graduate Foundation in Veterinarian Science of the University of Sydney, Sydney, pp. 266–276.

Koger, M. (1963). Breeding for the American tropics. In: Cunha, T.J., Koger, M. and Warnick, A.C. (eds) *Crossbreeding Beef Cattle*. University of Florida Press, Gainesville, FL, p. 41.

Lango Allen, H., Estrada, K., Lettre, G., Berndt, S.I., Weedon, M.N., Rivadeneira, F., Willer, C.J., Jackson, A.U., Vedantam, S., Raychauduri, S. *et al.* (2010) Hundreds of variants clustered in genomic loci and biological pathways affect human height. *Nature* 467, 832–838.

Lehmann, T. (1993) Ectoparasites: direct impact on host fitness. *Parasitology Today* 9, 8–13.

Mackinnon, M.J., Meyer, K. and Hetzel, D.J.S. (1991) Genetic variation and covariation for growth, parasite resistance and heat tolerance in tropical cattle. *Livestock Production Science* 27, 105–122.

MacHugh, D.E., Shriver, M.D., Loftus, R.T., Cunningham, P. and Bradley, D.G. (1997) Microsatellite DNA variation and the evolution, domestication and phylogeography of taurine and zebu cattle (*Bos taurus* and *Bos indicus*). *Genetics* 146, 1071–1086.

McManus, C., Prescott, E., Paludo, G.R., Bianchini, E., Louvandini, H. and Mariante, A.S. (2009) Heat tolerance in naturalized Brazilian cattle breeds. *Livestock Science* 120, 256–264.

Madalena, F.E. (1981) Crossbreeding strategies for dairy cattle in Brazil. *World Animal Review* 38, 23–30.

Madalena, F.E. (1993) A simple scheme to utilize heterosis in tropical dairy cattle. *World Animal Review* 74–75, 17–25.

Madalena, F.E., Teodoro, R.L., Lemos, A.M. and Oliveira, G.P. (1985) Causes of variation of filed burdens of cattle tick (*Boophilus microplus*). *Revue Brasilia Genetics* 8, 361–375.

Madalena, F.E., Teodoro, R.L., Lemos, A.M., Monteiro, J.B.N. and Barbosa, R.T. (1990) Evaluation of strategies for crossbreeding of dairy cattle in Brazil. *Journal of Dairy Science* 73, 1887–1901.

Madalena, F.E., Peixoto, M.G.C.D. and Gibson, J. (2012) Dairy cattle genetics and its applications in Brazil. *Livestock Research for Rural Development* 24, Article 97, available at: http://www.lrrd.org/lrrd24/6/ made24097.htm (accessed 3 April 2014).

Mariasegaram, M., Chase, C.C. Jr, Chaparro, J.X., Olson, T.A., Brenneman, R.A. and Niedz, R.P. (2007) The slick hair coat locus maps to chromosome 20 in Senepol-derived cattle. *Animal Genetics* 38, 54–59.

Meuwissen, T.H.E. and Goddard, M.E. (2010a) The use of family relationships and linkage disequilibrium to impute phase and missing genotypes in up to whole genome sequence density genotypic data. *Genetics* 185, 1441–1449.

Meuwissen, T.H.E. and Goddard, M.E. (2010b) Accurate prediction of genetic values for complex traits by whole genome resequencing. *Genetics* 185, 623–631.

Meuwissen, T.H.E., Hayes, B.J. and Goddard, M.E. (2001) Prediction of total genetic value using genome-wide dense marker maps. *Genetics* 157, 1819–1829.

Mirkena, T., Duguma, G., Haile, A., Tibbo, M., Okeyo, A.M., Wurzinger, M. and Sölkner, J. (2010) Genetics of adaptation in domestic farm animals: a review. *Livestock Science* 132, 1–12.

Morgan, N. and Tallard, G. (2007) Cattle and beef international commodity profile. Background paper for the competitive commercial agriculture in Sub-Saharan Africa study. Food and Agriculture Organization of the United Nations, Rome.

Morris, C.A. (2000) Genetics of susceptibility in cattle and sheep. In: Axford, R.F.E., Bishop, S.C., Nicholas, F.W. and Owen, J.B. (eds) *Breeding for Disease Resistance in Farm Animals*, 2nd Edition. CAB International, Wallingford, UK, pp. 343–355.

Neibergs, H.L., Seabury, C.M., Taylor, J.F., Wang, Z., Scraggs, E., Schnabel, R.D., Decker, J., Wojtowicz, A., Davis, J.H., Lehenbauer, T.W., Van Eenennaam, A.L., Aly, S.S., Blanchard, P.C., Crossley, B.M. and the Bovine Respiratory Disease Consortium (2013) Identification of loci associated with Bovine Respiratory Disease in Holstein calves. *Proceedings of the Plant and Animal Genomics Conference XXI*, Paper 7995, available at: https://pag.confex.com/pag/xxi/webprogram/Paper7995.html (accessed 3 April 2014).

Newman, S. and Coffey, S.G. (1999) Genetic aspects of cattle adaptation in the tropics. In: Fries, R. and Ruvinsky, A. (eds) *The Genetics of Cattle*. CABI, Wallingford, pp. 637–656.

Ng, S.B., Buckingham, K.J., Lee, C., Bigham, A.W., Tabor, H.K., Dent, K.M., Huff, C.D., Shannon, P.T., Wang, J.E., Nickerson, D.A., Shendure, J. and Bamshad, M.J. (2009) Exome sequencing identifies the cause of a Mendelian disorder. *Nature Genetics* 42, 30–35.

Nishimura, H. and Frisch, J.E. (1977) Eye cancer and circumocular pigmentation in Bostaurus, Bosindicus and crossbred cattle. *Australian Journal of Experimental Agriculture and Animal Husbandry* 17, 709–711.

O'Kelly, J.C. and Kennedy, P.M. (1981) Metabolic changes in cattle due to the specific effect on the tick, *Boophilus microplus*. *British Journal of Nutrition* 45, 557–566.

O'Kelly, J.C., Post, T.B. and Bryan, R.P. (1988) The influence of parasitic infestations on metabolism, puberty and first mating performance of heifers grazing in a tropical area. *Animal Reproduction Science* 16, 177–189.

Oyenuga, V.A. and Nestel, B. (1984) Development of animal production systems in humid Africa. In: Nestel, B. (ed.) *World Animal Science, Development of Animal Production Systems*. Elsevier, London, pp. 189–199.

Payne, W.J.A. and Hodges, J. (1997) *Tropical Cattle: Origins, Breeds and Breeding Policies*. Blackwell Science, Oxford.

Porto Neto, L.R., Bunch, R.J., Harrison, B.E. and Barendse, W. (2010a) DNA variation in the gene eltd1 is associated with tick burden in cattle. *Animal Genetics* 42, 50–55.

Porto Neto, L., Bunch, R., Harrison, B.E., Prayaga, K. and Barendse, W. (2010b) Haplotypes that include the integrin alpha 11 gene associated with tick burden in cattle. *BMC Genetics* 11, 55.

Porto Neto, L.R., Jonsson, N.N., D'Occhio, M.J. and Barendse, W. (2011) Molecular genetic approaches for identifying the basis of variation in resistance to tick infestation in cattle. *Veterinary Parasitology* 180, 165–172.

Prayaga, K.C. (2003) Evaluation of beef cattle genotypes and estimation of direct and maternal genetic effects in a tropical environment. 2. Adaptive and temperament traits. *Australian Journal of Agricultural Research* 54, 1027–1038.

Prayaga, K.C. and Henshall, J.M. (2005) Adaptability in tropical beef cattle: genetic parameters of growth, adaptive and temperament traits in a crossbred population. *Australian Journal of Experimental Agriculture* 45, 971–983.

Prayaga, K.C., Barendse, W. and Burrow, H.M. (2006) Genetics of tropical adaptation in northern Australian cattle. *Proceedings of the 8th World Congress on Genetics Applied to Livestock Production*. CD Rom Communication No. 16-01, ISBN 85-60088-01-6.

Prayaga, K.C., Corbet, N.J., Johnston, D.J., Wolcott, M.L., Fordyce, G. and Burrow, H.M. (2009) Genetics of adaptive traits in heifers and their relationship to growth, pubertal and carcass traits in two tropical beef cattle genotypes. *Animal Production Science* 49, 413–425.

Pruett, J.H., Steelman, C.D., Miller, J.A., Pound, J.M. and George, J.E. (2003) Distribution of horn flies on individual cows as a percentage of the total horn fly population. *Veterinary Parasitology* 116, 251–258.

Roberts, F.H.S. and O'Sullivan, P.J. (1950) Methods for egg counts and larval cultures for strongyles infesting the gastrointestinal tract of cattle. *Australian Journal of Agricultural Research* 1, 99–102.

Seebeck, R.M., Springell, P.H. and O'Kelly J.C. (1971) Alterations in host metabolism by the specific and anorectic effects of the cattle tick (*Boophilus microplus*) I. Food intake and body weight growth. *Australian Journal of Biological Sciences* 24, 373–380.

Seifert, G.W. (1971a) Variations between and within breeds of cattle in resistance to field infestations of the cattle tick (*Boophilus microplus*). *Australian Journal of Agricultural Research* 22, 159–168.

Seifert, G.W. (1971b) Ecto- and endoparasitic effects on the growth rates of Zebu crossbred and British cattle in the field. *Australian Journal of Agricultural Research* 22, 839–850.

Stear, M.J., Bishop, S.C., Mallard, B.A. and Raadsma, H. (2001) The sustainability, feasibility and desirability of breeding livestock for disease resistance. *Research in Veterinary Science* 71, 1–7.

Steelman, C.D., Brown, A.H., Gbur, E.E. and Tolley, G. (1991) Interactive response of the horn fly (*Diptera: Muscidae*) and selected breeds of beef cattle. *Journal of Economic Entomology* 84, 1275–1282.

Trail, J.C.M., D'Ieteren, G.D.M., Feron, A., Kakiese, O., Mulungo, M. and Pelo, M. (1991a) Effect of trypanosome infection, control of parasitaemia and control of anaemia development on productivity of N'dama cattle. *Acta Tropica* 48, 37–45.

Trail, J.C.M., D'Ieteren, G.D.M., Maille, J.C. and Yangari, G. (1991b) Genetic aspects of control of anaemia development in trypanotolerant N'dama cattle. *Acta Tropica* 48, 285–291.

Turner, H.G. (1972) Selection of beef cattle for tropical Australia. *Australian Veterinary Journal* 48, 162–166.

Turner, H.G. (1982) Genetic variation of rectal temperature in cows and its relationship to fertility. *Animal Production* 35, 401–412.

Turner, H.G. (1984) Variation in rectal temperature in a tropical environment and its relation to growth rate. *Animal Production* 38, 417–427.

Turner, H.G. and Schleger, A.V. (1960) The significance of coat type in cattle. *Australian Journal of Agricultural Research* 11, 645–663.

Turner, H.G. and Short, A.J. (1972) Effects of field infestations of gastrointestinal helminths and of the cattle tick (*Boophilus microplus*) on growth of three breeds of cattle. *Australian Journal of Agricultural Research* 33, 177–193.

Turner, L.B., Harrison, B.E., Bunch, R.J., Porto Neto, L.R., Li, Y. and Barendse, W. (2012) A genome-wide association study of tick burden and milk composition in cattle. *Animal Production Science* 50, 235–245.

Vercoe, J.E. (1969) The effect of increased rectal temperature on nitrogen metabolism in Brahman cross and Shorthorn x Hereford steers fed on lucerne chaff. *Australian Journal of Agricultural Research* 20, 607–612.

Vercoe, J.E. and Springell, P.H. (1969) Effect of subclinical helminthosis on nitrogen metabolism in beef cattle. *Journal of Agricultural Science (Cambridge)* 63, 203–209.

Villumsen, T.M., Janss, L. and Lund, M.S. (2008) The importance of haplotype length and heritability using genomic selection in dairy cattle. *Journal of Animal Breeding and Genetics* 12, 3–13.

Wambura, P.N., Gwakisa, P.S., Silayo, R.S. and Rugaimumamu, E.A. (1998) Breed-associated resistance to tick infestation in *Bos indicus* and their crosses with *Bos taurus*. *Veterinary Parasitology* 77, 63–70.

Wharton, R.H. and Utech, K.B.W. (1970) The relation between engorgement and dropping of *Boophilus microplus* (Canestrini) (Ixodidae) to the assessment of tick numbers on cattle. *Journal of the Australian Entomological Society* 9, 171–182.

Wharton, R.H., Utech, K.B.W. and Turner, H.G. (1970) Resistance to the cattle tick *Boophilus microplus* in a herd of Australian Illawarra Shorthorn cattle: its assessment and heritability. *Australian Journal of Agricultural Research* 21, 163–181.

Wolcott, M.L., Johnston, D.J. and Barwick, S.A. (2014) Genetic relationships of female reproduction with growth, body composition, maternal weaning weight and tropical adaptation in two tropical beef genotypes. *Animal Production Science* 54, 60–73.

Yang, J., Beben, B., McEvoy, B.P., Gordon, S., Henders, A.K., Nyholt, D.R., Madden, P.A., Heath, A.C., Martin, N.G., Montgomery, G.W., Goddard, M.E. and Visscher, P.M. (2010) Common SNPs explain a large proportion of the heritability for human height. *Nature Genetics* 42, 565–569.

24 Standard Genetic Nomenclature

Zhi-Liang Hu,[1] James M. Reecy,[1] Fiona McCarthy[2] and
Carissa A. Park[1]
[1]Iowa State University, Ames, Iowa, USA;
[2]University of Arizona, Tucson, Arizona, USA

Introduction

Genetics includes the study of genotypes and phenotypes, the mechanisms of genetic control between them, and information transfer between generations. Genetic terms describe processes, genes and traits with which genetic phenomena are examined and described. While the genetic terminologies are extensively discussed in this book and elsewhere, the standardization of their names has been an ongoing process. Therefore, this chapter will only concentrate on discussions about the issues involved in the standardization of gene and trait terminologies. Readers may wish to refer to online resources (see Table 24.1 for URLs) for lists of the glossaries currently in use.

A standardized genetic nomenclature is vital for unambiguous concept description, efficient genetic data management and effective communications not only among scientists, but also among those who are involved in cattle production and genetic improvement. This issue has become even more critical in the post-genomics era due to rapid accumulation of large quantities of genetic and phenotypic data, and the requirement for data management and computational analysis, which increases the need for precise definition and interpretation of gene and trait terms.

For example, the Myostatin (*MSTN*) gene is known as Growth and Differentiation Factor 8 (*GDF8* or *GDF-8*) in some literature and is also referred to as the 'muscle hypertrophy' or 'double-muscling' locus in cattle. While the interchangeable use of all these names in the literature can cause confusion, it gets more complicated when one considers paralogous gene duplications across species, which led Rodgers *et al.* (2007) to propose MSTN-1 and MSTN-2. Unfortunately, this naming scheme does not follow the Human Genome Organization (HUGO) Gene Nomenclature Committee (HGNC) guidelines, which indicate that these paralogues should be named *MSTN1* and *MSTN2*, respectively.

In terms of traits, an example that would benefit from consistent nomenclature is the longissimus dorsi muscle area, which is also referred to as the loin eye area (LEA), loin muscle area (LMA), meat area (MLD), ribeye area (REA), etc. Each of these is known to certain researchers as their default name for the trait. Complexity is further increased by variation in anatomic locations, physiological stages and methods used to measure a given trait. This may seem manageable at first, but once one starts to compare data across different laboratories, publications or species, it quickly becomes very confusing.

The 'standard genetic nomenclature' recommendations made by the Committee on Genetic Nomenclature of Sheep and Goats (COGNOSAG) in the 1980s and 1990s initially covered sheep and goats and were later extended to cattle (Broad *et al.*, 1999). Dolling (1999) summarized these efforts and abstracted guidelines for practical use. In 2009, an international meeting to discuss coordination of gene names across vertebrate species was held in Cambridge, UK (Bruford, 2010). While we may hesitate to dictate how genetic terms are defined, adopting a standardized genetic nomenclature system enables researchers to more easily manage and compare their data, both within and across species. The emergence of the use of ontologies in biological research has contributed a new way to effectively organize biological data and facilitate analysis of large datasets. Adopting standardized nomenclature will further enable researchers to unambiguously organize and manage their data. When genomic information must be transferred across species to perpetuate genetic discoveries, the role of a standardized genetic nomenclature becomes even more important.

The goal of this chapter is to clearly state guidelines for nomenclature, with the hope that they will facilitate comparison of results between experiments and, most importantly, prevent confusion.

Locus and Gene Names and Symbols

Locus name and symbol

The following guidelines for cattle gene nomenclature are adapted and abbreviated from the HUGO Gene Nomenclature Committee (HGNC; see Table 24.1 for URL).

A gene is defined as 'a DNA segment that contributes to phenotype/function. In the absence of demonstrated function a gene may be characterized by sequence, transcription or homology.' A locus is not synonymous with a gene. It is defined as 'a point in the genome, identified by a marker, which can be mapped by some means. A locus could be an anonymous non-coding DNA segment or a cytogenetic feature.' A single gene may have numerous loci within it (each may be defined by different markers).

A gene name should be short and specific, and convey the character or function of the gene. Gene names should be written using American spelling and contain only Latin letters or a combination of Latin letters and Arabic numerals.

A gene symbol should start with the same letter as the gene name. The gene symbol should consist of upper-case Latin letters and possibly Arabic numerals. Gene symbols must be unique.

A locus name should be in capitalized Latin letters or a combination of Latin letters and Arabic numerals.

A locus symbol should consist of as few Latin letters as possible or a combination of Latin letters and Arabic numerals. The characters of a symbol should always be capital Latin characters and should begin with the initial letter of the name of the locus. If the locus name is two or more words, then the initial letters of each word should be used in the locus symbol.

Gene and locus names and symbols should be printed in italics whenever possible; otherwise they should be underlined.

When assigning cattle gene nomenclature, the gene name and symbol should be assigned based on existing HGNC nomenclature when 1:1 human:bovine orthology is well established. Recognized members of gene families should be named following existing naming schemes. Initial efforts to provide information about genes predicted during the cattle genome sequencing project resulted in the assignment of standardized names for 5757 cattle genes based on human gene nomenclature (Bovine Genome Sequencing and Analysis Consortium, 2009).

There are two categories of novel cattle genes: (i) novel genes predicted by bioinformatic gene prediction programs; and (ii) novel genes that have been studied prior to the completion of the cattle genome. In addition, it is anticipated that, in the future, additional novel genes will be identified by RNA-sequencing experiments. In cases where no strict 1:1 human orthologue exists that has been assigned nomenclature, the NCBI LOC# or Ensembl ID should be used as a temporary gene symbol for predicted genes with no known function. In order to assign a symbol/name to novel genes, they will need to be manually curated and assigned a unique symbol/name following these guidelines.

Allele name and symbol

These guidelines for allele nomenclature are adapted from Dolling (1999) and mouse genome nomenclature guidelines (see Table 24.1 for URL), consistent with HGNC guidelines.

Alleles do not have to be named, but should be assigned symbols. An allele symbol should always be written following the locus symbol. It can consist of Latin letters or a combination of Latin letters and Arabic numerals. An allele name should be as brief as possible, and should convey the variation associated with the allele. If a new allele is similar to one that has already been named, it should be named according to the breed, geographic location or population of origin. If new alleles are to be named for a recognized locus, they should conform to nomenclature established for that locus. The first letter of the allele name should be lower case.

The allele name and symbol may be identical for a locus detected by biochemical, serological or nucleotide methods. The HGNC guideline recommends that 'allele designation should be written on the same line as gene symbol separated by an asterisk e.g. *PGM1*1*, the allele is printed as **1*'. The wild-type allele can be denoted with a + (e.g. *MSTN+*). Neither + nor – symbols should be used in alleles detected by biochemical, serological or nucleotide methods. Null alleles should be designated by the number zero. A single nucleotide polymorphism (SNP) allele should be designated based on its dbSNP_id, followed by a hyphen and the specific nucleotide (e.g. *MSTNrs1234567-T*). If the SNP occurs outside of an identified gene, the SNP locus can be designated using the dbSNP_id as the locus symbol, followed by a hyphen and the nucleotide allelic variants as in *rs1234567-T*.

The allele name and symbol should be printed in italics whenever possible; otherwise they should be underlined.

Genotype terminology

The genotype of an individual should be shown by printing the relevant locus and allele symbols for the two homologous chromosomes concerned, separated by a slash, e.g. *MSTNrs1234567-T/rs1234567-C*. Unlinked loci should be separated by a semicolon, e.g. *CD11RsaI-2400/2200*; *ESRPvuII-5700/4200*. Linked loci should be separated by a space or dash and listed in linkage order (e.g. *POU1F1A/G–STCHC/G–PRSS7A/T*), or in alphabetical order if the linkage order is not known. For X-linked loci, the hemizygous case should have a /Y following the locus and allele symbol, e.g. *AR-Eco57I-1094/*Y. Likewise, Y-linked loci should be designated by /X following the locus and allele symbol.

Gene annotations and the gene ontology (GO)

Advances in genomic technologies require that researchers be able to functionally analyse large, high-throughput datasets to gain insight into the complex systems they are studying. By using the same nomenclature and procedures to describe gene function, gene components can be consistently linked to function in a way that facilitates effective computational analysis and promotes comparative genomics. In 1998, the GO Consortium was formed to standardize functional annotation in the form of gene ontologies that can be used across all eukaryotes (Gene Ontology Consortium, 2000). This effort not only provided a standard method for functional annotation but also promoted data sharing and enabled modelling of functional genomics datasets. The GO consists of three separate ontologies: Biological Process, Cellular Component, and Molecular Function. Genes or gene products are associated with GO terms that represent gene attributes.

A GO term is defined with a term name, a unique identifier and a definition (preferably indicating which of the three sub-ontologies it belongs to, information about its relationships to other GO terms and cited sources). GO terms may also have synonyms, database cross-references and comments to provide more detailed information. A unique GO identifier consists of the prefix 'GO' followed by a colon and six to eight numerical digits, e.g. GO:0000016. It serves as a key to reference GO terms in a GO database. An example of a GO term is shown in Fig. 24.1.

Standard GO annotations are maintained by the GO Consortium (see Table 24.1 for URL), which provides updates of quality-checked data for public access. The GO annotations are used by secondary source databases like Entrez Gene (see Table 24.1 for URL; Sayers et al., 2012) and UniProt (UniProt Consortium, 2010), genome browsers like Ensembl (see Table 24.1 for URL; Flicek, 2013), and analysis tools like DAVID (see Table 24.1 for URL; Huang, 2009), among other publicly accessible resources and tools. A growing number of model organism and livestock animal species (including bovine) databases and working groups contribute annotation sets to the GO repository (McCarthy, 2007; Reese, 2010).

GO annotations are created by capturing the gene product information (database, database accession, name and symbol, type of gene product and species taxon), its associated GO term, GO sub-ontology and evidence for the assertion with references. The current practice for bovine GO annotation is to provide names and symbols based upon a combination of NCBI Entrez Gene and UniProtKB names. In instances where there is no suitable gene symbol, database accessions are used. Continued efforts are made to improve the accuracy of the bovine GO annotations by transferring GO annotations from better annotated proteins in human and mouse based on Ensembl orthology. As of September 2012, GO annotation for bovine (McCarthy, 2007) comprises 306,746 annotation entries for 41,637 gene products; 86.7% of these annotations

```
        id:   GO:0000016
      name:   lactase activity
 namespace:   molecular_function
       def:   "Catalysis of the reaction: lactose + H2O = D-glucose + D-
              galactose." [EC:3.2.1.108]
   synonym:   "lactase-phlorizin hydrolase activity" BROAD [EC:3.2.1.108]
   synonym:   "lactose galactohydrolase activity" EXACT [EC:3.2.1.108]
      xref:   EC:3.2.1.108
      xref:   MetaCyc:LACTASE-RXN
      xref:   Reactome:20536
      is_a:   GO:0004553 ! hydrolase activity, hydrolyzing O-glycosyl
              compounds
```

Fig. 24.1. An example of a GO term. (For further information, see Table 24.1 for GO website URL.)

are computationally derived (AgBase: see Table 24.1 for URL).

To contribute annotations to the GO, or for a complete list of bovine GO data, users are encouraged to contact either the GO Consortium or AgBase at their respective websites.

Trait and Phenotype Terminology

Cattle traits are conventionally named based on performance (e.g. body weight), physiological parameters (e.g. blood cholesterol level), anatomic locations/dissections (e.g. loin muscle area), physical-chemical properties (e.g. milk protein content), livelihood soundness (e.g. immune capacity) and exterior appearance (e.g. coat colour), etc. As such, there is a good chance a trait will be named differently by different people, even within a species community. Furthermore, traits have been studied across many species, which adds additional complexity to their naming. The study of traits may also involve the study of underlying genes and markers, environments and management protocols that contribute to the manifestation of a trait. Therefore, it is obvious that factors that contribute to the naming of a trait are multi-dimensional. As the amount of trait information associated with a gene or chromosomal region is growing exponentially, we cannot overemphasize the need for a standard nomenclature to be used by researchers to communicate as consistently and unambiguously as possible, with the aid of bioinformatics tools.

Traits

Cattle trait terms can be found ubiquitously throughout journal articles, farm reports and daily communications among scientists and cattle industry personnel. A trait term can be created by anyone, and each person may have a slightly different definition for any given term. As such, hundreds of thousands of terms can be found in the literature with various naming conventions used. Previously, there was no central repository where the uniqueness of a trait term could be maintained and checked, until two relatively recent database development

efforts emerged: the Online Mendelian Inheritance in Animals (OMIA) database and the Animal QTL database (QTLdb).

OMIA (see Table 24.1 for URL) was initiated in 1978. To date, it contains >400 cattle trait variations and/or abnormalities from cattle genetic research publications (Nicholas, Chapter 5). The Animal QTLdb (see Table 24.1 for URL) has a collection of 470 cattle traits, including measurement method variations (Hu *et al.*, 2013), of which 407 traits have at least one QTL. Curators at both OMIA and Animal QTLdb made efforts to make each database entry unique in terms of the names and their representations. Expanded from the QTLdb development, an Animal Trait Ontology (ATO) project at Iowa State University (see Table 24.1 for URL) has been launched to standardize traits for livestock species including cattle. Its initial purpose was to help with organization and management of trait information through the use of a controlled vocabulary to facilitate comparison of QTL results and standardize trait data annotation and retrieval (Hu *et al.*, 2005, 2007). It was soon introduced to the community (Hughes *et al.*, 2008).

Super-traits

Compared to standard gene nomenclature, trait name standardization is far more complex, not only because the same trait can be named differently (e.g. 'loin eye area' versus 'ribeye area'), but also because many factors contribute to how a trait is defined under various circumstances. For example, Fig. 24.2 shows a list of 10 'backfat thickness' variations, each of which is defined by their specific measurement methods, measuring time and specific anatomic locations, which may contribute to trait comparison difficulties and increase the potential for confusion.

One attempt to simplify the comparisons was by introduction of the concept of 'trait types' or 'super-traits'. Hu *et al.* (2005) described trait type as a general physical or chemical property of, or the processes that lead to, or types of measurements that result in, an observation (phenotype). The 'trait types'

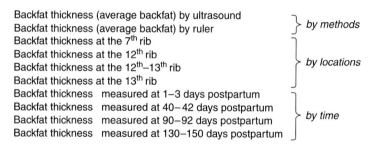

Backfat thickness (average backfat) by ultrasound
Backfat thickness (average backfat) by ruler   } by methods
Backfat thickness at the 7th rib
Backfat thickness at the 12th rib
Backfat thickness at the 12th–13th rib   } by locations
Backfat thickness at the 13th rib
Backfat thickness   measured at 1–3 days postpartum
Backfat thickness   measured at 40–42 days postpartum
Backfat thickness   measured at 90–92 days postpartum } by time
Backfat thickness   measured at 130–150 days postpartum

Fig. 24.2. An example of the trait name variations by different modifiers such as measurement methods, time and sampling locations. This variation can easily add difficulties for accurate and unambiguous trait comparisons.

or 'super-traits' were initially used to serve as a general concept for a trait, regardless of possible variations in trait names based on measurement times, locations or methods. As the ATO project progressed, the factors in the methods of trait measurements, such as point in time or time span, anatomic locations, instruments, etc., were classified as 'trait modifiers', because they do not constitute a component of a trait, but only affect the way a trait is described. Therefore, the 'super-trait' may only be employed to categorize variations in how a trait is defined or named. For example, 'rib eye area', 'rib-eye area', 'rib muscle area', 'longissimus dorsi muscle area', 'longissimus muscle area', 'loin eye area', 'loin muscle area', etc. can be unified as 'longissimus dorsi muscle area (LMA)'. 'Backfat', 'backfat depth', 'backfat thickness', 'backfat above muscle dorsi', 'backfat intercept', 'backfat linear', etc. may all simply be referred to as 'subcutaneous fat thickness'.

Trait hierarchy and ontology

In order to compare QTL across experiments, the Cattle QTLdb uses a trait hierarchy (Fig. 24.3) to provide a framework for organizing the traits and easily locating them (Hu *et al.*, 2013). This approach simplifies the procedures by which traits are defined, linked and compared. Subsequently, a computer program could be implemented to automatically process the database searches, so that when a user queries for a trait by keywords, the database can gather and retrieve related trait names and

their associated QTL, put them together and present them to the user in real time.

However, people of different disciplines may see the need for a different trait hierarchy, which may better capture the subtleties required in their field. For example, for body weight gained over a period of time (e.g. average daily gain, ADG), a farmer considers it a production trait, a nutritionist may see it as an indicator for feed conversion efficiency and a veterinarian may find it a health status parameter. Similarly, blood cholesterol levels may be used to predict meat quality by beef producers, and may also be used as a parameter to predict coronary heart disease by those who use cattle as an animal model for human heart disease research. Therefore, a simple hierarchy may be helpful to reduce the complexity in some cases, although may not be adequate in all cases. In addition, due to the existence of multiple overlapping hierarchies for cattle traits, the management of such data may introduce one more dimension of complexity to the ontology structure.

Ontologies are controlled vocabularies used to describe objects and relationships between them in a formal manner. In an ontology, the Directed Acyclic Graph (DAG), a mathematical graphic modelling method, is used to solve data management problems with complex hierarchical structures. For example, the trait 'marbling' may belong to the 'meat quality', 'adipose trait' or 'muscular system physiology' hierarchies. Computer tools have been developed and are freely available to manage such ontology data with DAG structures. The two most popular tools that are likely to be useful to the cattle genetics community

Fig. 24.3. A simple cattle trait class hierarchy used in the Animal QTLdb for users to browse for traits of interest. (See Table 24.1 for URL.)

are AmiGO and OBO-Edit (Gene Ontology Tools, see Table 24.1 for URL). AmiGO is an ontology browser adapted to the ATO database, which allows users to share and view trait data stored in ATO with any web browser on the internet. OBO-Edit is a java-based ontology data editor that can be used by anyone to edit ontology term definitions and relationships, and to export data in Open Biological/Biomedical Ontologies (OBO) format to share data.

Current status of research

The ATO has been a successful project since its development from the QTLdb several years ago. Recently, the developers of ATO have begun working with Mouse Genome Informatics, the Rat Genome Database, European Animal Disease Genomics Network of Excellence (EADGENE) and the French National Institute for Agricultural Research (INRA) to incorporate the Mammalian Phenotype Ontology (MPO) and the ATO into a unified Vertebrate Trait (VT) Ontology (Park *et al.*, 2013; see Table 24.1 for URL). To reach a proper granularity level of the trait ontology, Product Trait (PT) Ontology (see Table 24.1 for URL) and Clinical Measurement Ontology (CMO; Shimoyama *et al.*, 2012; see Table 24.1 for URL) were introduced. By reuse of existing ontologies and integration of production-specific livestock traits, researchers at INRA have also launched an Animal Trait Ontology for Livestock (ATOL) site, containing over 1000 traits including those of cattle (Golik *et al.*, 2012).

Current efforts have been aimed at enhancing the ability to standardize trait nomenclature within and across species. For example, a disease such as mastitis in dairy cattle may have been considered a 'trait' in classical animal genetic studies. In fact, in terms of concept specifications, it is not a characteristic cattle trait observable in the general population, but rather an abnormal manifestation in some cattle (in fact, resistance to mastitis is a trait). In addition, a trait name may have variations because it is 'modified' by measurement time or method (Fig. 24.2), but the names actually represent the same trait. The separation of diseases from traits reflects the efforts toward a well-defined and standardized trait nomenclature. Standardization of the trait nomenclature will undoubtedly help the cattle genomics community make meaningful trait comparisons, as well as facilitate the transfer of genomics information from some well-studied species. The challenge of using ontologies to standardize and manage trait nomenclature is not only a technical issue, but a community issue, in the sense that it has to be commonly recognized, mutually agreed upon, and widely shared.

Trait and phenotype nomenclature

Until an international committee issues rules for trait and phenotype nomenclature, a good practice with wide acceptability is to follow the 'norm' in published materials. Listed in Table 24.1 are some of the best trait reference resources available to date (see table footnote for details). Since this has been an active research area in recent years, it is highly recommended that users check multiple databases for the best and most up-to-date information.

Phenotype is the actual manifestation of observable traits. A phenotype is a trait observed in an individual. It usually consists of a trait with characteristic features (e.g. twinning), variations that can be described (e.g. black spots on the body) or qualities that can be measured (e.g. birth weight of 30 kg). Since there are so many variations as to how a phenotype can be 'observed' (often such observation is made indirectly with instruments or through tests) and obtained, a technical guide for recording each trait might be ideal. Often a description of comments for a phenotype record may be necessary to correctly understand and use the data. For example, when blood samples are taken, the number of hours the animal is fasted might be an important co-factor for the measurement of blood cholesterol concentration.

When a phenotype is a reflection of a certain genotype, the phenotype symbol should be the same as the genotype symbol. The difference is that the characters should not be underlined or in italics, and they should be written with a space between locus characters and allele characters, instead of an asterisk. Square brackets [ ] may also be used.

In classical genetics, phenotypes were sometimes used to denote Mendelian genotypes. This was done using an abbreviation of the trait, post-fixed with a plus (+) or minus (–) sign to represent 'presence' or 'absence' of certain trait features. For example, halothane-negative was denoted as 'Hal–', and halothane-positive as 'Hal+'. A phenotype denotation can also be used to represent genetic haplotypes, such that 'K88ab+, ac+, ad–' are written together as an entire denotation. Likewise, numbers or letters may be used to denote alleles when polymorphisms are observed, for example, ApoB1/2, ApoB2/3, etc. (Note the difference from recording genotypes, where italics or asterisks are required.)

Future Prospects

The Gene Ontology and Mammalian Phenotype Ontology are already playing a role in robust annotation of mammalian genes and phenotypes in the context of mutations, quantitative trait loci, etc. (Smith et al., 2005). Undoubtedly, a standardized cattle genetic nomenclature will more effectively facilitate efficient cattle genome annotation and transfer of knowledge from information-rich species such as humans and mouse, and make it possible for new bioinformatics tools to easily streamline data management and genetic analysis. Meanwhile, it is noteworthy to mention that the term 'phene' for 'trait' is being used more frequently in the scientific literature in recent years. It is interesting that in terms of etymology lineage, 'phene' is to 'phenotype' and 'phenome' as 'gene' is to 'genotype' and 'genome' (Wikipedia, 2012), where 'phene' is an equivalent term for 'trait'. However, Dr Frank Nicholas from the University of Sydney has used the term 'phene' in OMIA in a slightly different but more concise context, namely 'phene is to gene as phenotype is to genotype', where 'phene' refers to a set of phenotypes that correspond to a set of genotypes determined by a gene. This is practically very useful in light of the future structured genetic terminology standardization in the genomics era.

Several genome databases, such as ArkDB, Animal QTLdb, Bovine Genome Database, Ensembl and NCBI GeneDB, have played a role in the usage of commonly accepted gene/trait notations. Undoubtedly, existing and new genome databases and tools will further develop and evolve. As such, a standardized genetic nomenclature in cattle will definitely become crucial for information sharing and comparisons between different research groups, across experiments and even across species. Recently the *Animal Genetics* journal has updated its Author Guidelines insisting that proper gene nomenclature be followed: 'All gene names and symbols should be italicized throughout the text, table and figures'; 'Locus

Table 24.1. Internet URL addresses for the web resources used in this chapter and cattle trait glossary information.

Data source	URL
AgBASE	http://www.agbase.msstate.edu/cgi-bin/information/Cow.pl
Animal QTLdb	http://www.animalgenome.org/QTLdb
Animal Trait Ontology project	http://www.animalgenome.org/bioinfo/projects/ATO/
ATOL	http://www.atol-ontology.com
Cattle trait hierarchy	http://www.animalgenome.org/QTLdb/export/cattle_traits
CMO project	http://bioportal.bioontology.org/ontologies/CMO (BioPortal)
	http://www.animalgenome.org/bioinfo/projects/cmo
DAVID	http://david.abcc.ncifcrf.gov
Ensembl	http://www.ensembl.org
Entrez Gene	http://www.ncbi.nlm.nih.gov/gene
Gene Ontology Tools	http://neurolex.org/wiki/Category:Resource:Gene_Ontology_Tools
Genetic glossaries	http://www.animalgenome.org/genetics_glossaries
GO Consortium	http://www.geneontology.org
GO structure	http://www.geneontology.org/GO.ontology.structure.shtml
HGNC guidelines	http://www.genenames.org/guidelines.html
Mouse genome nomenclature guidelines	http://www.informatics.jax.org/mgihome/nomen/gene.shtml
OMIA	http://omia.angis.org.au/
PT project	http://www.animalgenome.org/bioinfo/projects/pt
UniProt	http://www.uniprot.org
VT project	http://bioportal.bioontology.org/ontologies/VT (BioPortal)
	http://www.animalgenome.org/bioinfo/projects/vt

VT, Vertebrate Trait Ontology is a controlled vocabulary for the description of traits (measurable or observable characteristics) pertaining to the morphology, physiology or development of vertebrate organisms. CMO, Clinical Measurement Ontology is designed to be used to standardize morphological and physiological measurement records generated from clinical and model organism research and health programmes. PT, Product Trait Ontology is a controlled vocabulary for the description of traits (measurable or observable characteristics) pertaining to products produced by or obtained from the body of an agricultural animal or bird maintained for use and profit. QTLdb, Animal QTLdb is a database to house all QTL data for all livestock species. OMIA, Online Mendelian Inheritance in Animals is a comprehensive collection of phenotypic information on heritable animal traits and genes in a comparative context, relating traits to genes where possible. ATOL, Animal Trait Ontology for Livestock is aimed at defining livestock traits, with a focus on the main types of animal production in line with societal priorities.

symbols used in Animal Genetics publications must be confirmed with HGNC' and 'non-human gene names should be checked against NCBI's Entrez Gene database'. This is a good move towards educating the community on the proper use of standardized genetic nomenclatures. Active development and use of a standardized genetic nomenclature will surely help to improve data quality and reusability, and facilitate data comparisons between experiments, laboratories, even species.

Acknowledgements

The authors wish to thank Dr Frank Nicholas from the University of Sydney for useful discussions, inputs and kind review of the draft.

References

Bovine Genome Sequencing and Analysis Consortium *et al.* (2009) The genome sequence of taurine cattle: a window to ruminant biology and evolution. *Science* 324, 522–528.

Broad, T.E., Dolling, C.H.S., Lauvergne, J.J. and Millar, P. (1999) Revised COGNOSAG guidelines for gene nomenclature in ruminants 1998. *Genetics, Selection, Evolution* 31, 263–268.

Bruford, E.A. (2010) Highlights of the 'Gene Nomenclature Across Species' meeting. *Human Genomics* 4, 213–217.

Dolling, C.H.S. (1999) Standardized genetic nomenclature for cattle. In: Fries, R. and Ruvinsky, A. (eds) *The Genetics of Cattle*. CAB International, Wallingford, UK, pp. 657–666.

Flicek, P., Ahmed, I., Amode, M.R., Barrell, D., Beal, K., Brent, S., Carvalho-Silva, D., Clapham, P., Coates, G., Fairley, S. *et al.* (2013) Ensembl 2013. *Nucleic Acids Research* 41 (Database issue), D48–D55.

Gene Ontology Consortium (2000) Gene Ontology: tool for the unification of biology. *Nature Genetics* 25, 25–29.

Golik, W., Dameron, O., Bugeon, J., Fatet, A., Hue, I., Hurtaud, C., Reichstadt, M., Meunier-Salaün, M.C., Vernet, J., Joret, L. *et al.* (2012) ATOL: the multi-species livestock trait ontology. 6th International Conference on Metadata and Semantic Research (MTSR'12), Cadiz, Spain, 28–30 November.

Hu, Z.-L., Dracheva, S., Jang, W.-H., Maglott, D., Bastiaansen, J., Rothschild, M.F. and Reecy, J.M. (2005) A QTL resource and comparison tool for cattle: PigQTLDB. *Mammalian Genome* 16, 792–800.

Hu, Z.-L., Fritz, E.R. and Reecy, J.M. (2007) AnimalQTLdb: a livestock QTL database tool set for positional QTL information mining and beyond. *Nucleic Acids Research* 35 (Database issue), D604–D609.

Hu, Z.-L., Park, C.A., Wu, X.-L. and Reecy, J.M. (2013) Animal QTLdb: an improved database tool for live-stock animal QTL/association data dissemination in the post-genome era. *Nucleic Acids Research* 41, D871–D879.

Huang, D.W., Sherman, B.T. and Lempicki, R.A. (2009) Systematic and integrative analysis of large gene lists using DAVID bioinformatics resources. *Nature Protocols* 4, 44–57.

Hughes, L.M., Bao, J., Hu, Z.-L., Honavar, V.G. and Reecy, J.M. (2008) Animal Trait Ontology (ATO): The importance and usefulness of a unified trait vocabulary for animal species. *Journal of Animal Science* 86, 1485–1491.

McCarthy, F.M., Bridges, S.M., Wang, N., Magee, G.B., Williams, W.P., Luthe, D.S. and Burgess, S.C. (2007) AgBase: a unified resource for functional analysis in agriculture. *Nucleic Acids Research* 35 (Database issue), D599–D603.

Park, C.A., Bello, S.M., Smith, C.L., Hu, Z.-L., Munzenmaier, D.H., Nigam, R., Smith, J.R., Shimoyama, M., Eppig, J.T. and Reecy, J.M. (2013) The Vertebrate Trait Ontology: a controlled vocabulary for the annotation of trait data across species. *Journal of Biomedical Semantics* 4, 13.

Reese, J.T., Childers, C.P., Sundaram, J.P., Dickens, C.M., Childs, K.L., Vile, D.C. and Elsik, C.G. (2010) Bovine Genome Database: supporting community annotation and analysis of the *Bos taurus* genome. *BMC Genomics* 11, 645.

Rodgers, B.D., Roalson, E.H., Weber, G.M., Roberts, S.B. and Goetz, F.W. (2007) A proposed nomenclature consensus for the myostatin gene family. *American Journal of Physiology – Endocrinology and Metabolism* 292, E371–E372.

Sayers, E.W., Barrett, T., Benson, D.A., Bolton, E., Bryant, S.H., Canese, K., Chetvernin, V., Church, D.M., Dicuccio, M., Federhen, S. *et al.* (2012) Database resources of the National Center for Biotechnology Information. *Nucleic Acids Research* 40 (Database issue), D13–D25.

Shimoyama, M., Nigam, R., McIntosh, L.S., Nagarajan, R., Rice, T., Rao, D.C. and Dwinell, M.R. (2012) Three ontologies to define phenotype measurement data. *Frontiers in Genetics* 3, 87.

Smith, C.L., Goldsmith, C.A. and Epcattle, J.T. (2005) The Mammalian Phenotype Ontology as a tool for annotating, analyzing and comparing phenotypic information. *Genome Biology* 6, R7.

UniProt Consortium (2010) The Universal Protein Resource (UniProt) in 2010. *Nucleic Acids Research* 38 (Database issue), D142–D148.

Wikipedia (2012) Phene. Available at:http://en.wikipedia.org/wiki/Phene (accessed 30 March 2013).

Index

Note: Page numbers in **bold** refer to figures; page numbers in *italic* refer to tables.